TRANSACTIONS

OF THE

INTERNATIONAL

ASTRONOMICAL UNION

VOLUME XIXA

REPORTS

INTERNATIONAL COUNCIL OF SCIENTIFIC UNIONS
INTERNATIONAL ASTRONOMICAL UNION
UNION ASTRONOMIQUE INTERNATIONALE

TRANSACTIONS
OF THE
INTERNATIONAL ASTRONOMICAL UNION
VOLUME XIXA

REPORTS
ON
ASTRONOMY

Edited by

RICHARD M. WEST
General Secretary of the Union

D. REIDEL PUBLISHING COMPANY
A MEMBER OF THE KLUWER ACADEMIC PUBLISHERS GROUP
DORDRECHT / BOSTON / LANCASTER

Library of Congress Cataloging in Publication Data

Main entry under title:

Reports on astronomy.

(Transactions of the International Astronomical Union; v. 19A)
English and French.
"Contains the reports by presidents of commissions, covering the period July 1981–June 1984"–Pref.
1. Astronomy. 2. International Council of Scientific Unions. 3. International Astronomical Union. I. West, Richard M., 1941– II. International Astronomical Union. III. Series.
QB1.I6 vol. 19A [QB43.2] 560 s [520] 85-8210
ISBN 90-277-2039-8

Published on behalf of
the International Astronomical Union
by
D. Reidel Publishing Company, P. O. Box 17, 3300 AA Dordrecht, Holland

Sold and distributed in the U.S.A. and Canada
by Kluwer Academic Publishers,
190 Old Derby Street, Hingham, MA 02043, U.S.A.

In all other countries, sold and distributed
by Kluwer Academic Publishers Group,
P. O. Box 322, 3300 AH Dordrecht, Holland

Printed in The Netherlands

P R E F A C E

Volume XIX A of IAU Transactions contains the reports by Presidents of Commissions, covering the period July 1981 - June 1984. The topics of IAU Commissions deal with all of contemporary astronomy and the present volume therefore constitutes an exhaustive and unique record of astronomical research during this interval. It is particularly useful for astronomers and other scientists who want to gain an overview of a certain field, not necessarily near their own research area.

Each Commission President was requested, by early 1984, to begin preparations for the compilation of his/her Commission report. All members of Commissions were asked to supply details about their individual research programmes. The ensuing, very substantial task of concatenating was undertaken in late 1984 by the Presidents, often supported by a team of authors, all recognized authorities in their fields. In order to preserve some measure of uniformity - which is in any case very difficult with so many authors - editorial guidelines were sent to Presidents, including the number of pages allotted to each Commission.

I am most thankful to all involved, Presidents and Members of Commissions, for having put so much work into the preparation of these reports and for having complied so well with the guidelines. It is due to their great efforts that it is possible to send this volume to the printer only a few weeks after the originally indicated deadline. It will now become available well before the XIXth General Assembly in Delhi, India, on 19-28 November 1985 and shall serve as a valuable basis for the Commission meetings there.

I believe that all readers will concur about the extraordinary importance of these reports. They contain a wealth of details, not only about past and ongoing projects, but, what is perhaps even more important, also about future trends in the various disciplines. The extensive bibliographical lists form a major part, although it is a pity that the definitive recommendations for journal abbreviations etc., to be incorporated in the new IAU Style Book, were not yet available. The lack of uniformity in this respect should not, however, distract from the other virtues of this volume.

It is a pleasure to acknowledge the friendly and expert collaboration of the IAU Publisher, D. Reidel Publishing Co., Dordrecht, during the various phases of the editorial work.

Richard M. West
General Secretary

C O N T E N T S

REPORTS OF COMMISSIONS

4. EPHEMERIDES (ÉPHÉMÉRIDES)

PRESIDENT: T Lederle

VICE-PRESIDENT: B L Morando

ORGANIZING COMMITTEE: V K Abalakin, S Aoki, R L Duncombe, J H Lieske, A Orte, P K Seidelmann, A M Sinzi, G A Wilkins, B D Yallop

I. Introduction

The period of this Report includes 1984 January 1, the date which was probably the most drastic caesura in the history of astronomical almanacs. It seemed, therefore, appropriate to concentrate here to the general aspects rather than to describe the works going on at the particular almanac offices. It is, however, hoped that the past years with their developments and changes will be followed by a period of consolidation and continuity. This would be also of great benefit for the users of the almanacs who still need some time for getting accustomed to so many innovations.

II. International and National Ephemerides

1. GENERAL CHANGES

According to the Resolutions adopted at the IAU General Assemblies 1976, 1979, and 1982 many changes had to be introduced from 1984 January 1 onwards, namely:

(a) the IAU (1976) System of Astronomical Constants (Trans. IAU XVIB, 58);

(b) the 1980 IAU Theory of Nutation (Trans. IAU XVIIIA, 8);

(c) the new definitions of time and time scales (Trans. IAU XVIB, 58; XVIIB, 69; XVIIIB, 72);

(d) the Cartographic Coordinates and Rotational Elements of the Planets and Satellites (33.091.004);

(e) the new fundamental reference frame (Trans. IAU XVIB, 58).

Each ephemeris and data set as given in an almanac is affected by one or more of these changes. It should be remarked that slight modifications of the adopted values for the astronomical constants may be introduced either indirectly by the use of certain fundamental ephemerides or for other reasons which are mentioned in the explanation of the almanac.

2. THE FUNDAMENTAL SYSTEM

A particular remark is necessary with regard to item (e); in the Resolution adopted in 1976, it was recommended that the new fundamental reference frame should be defined by the positions and proper motions of the FK5 catalogue. This catalogue which is currently in preparation at the Astronomisches Rechen-Institut Heidelberg, is not yet available at the time when the almanacs for 1984 and some subsequent years are prepared. Only the correction of the FK4 equinox and its fictitious motion was available (Trans. IAU XVIIIA, 3) and could be applied whenever necessary. Nevertheless the FK5 can be regarded de jure as the fundamental reference system from 1984 January 1 onwards. Corrections for the transition from the

FK4 to the FK5 system will be given in both analytical and tabulated form as soon as the FK5 catalogue is available, for those years from 1984 onwards in which the ephemerides in the almanacs had still to be based on the FK4 system.

In particular, this will also be true for the volumes "Apparent Places of Fundamental Stars". They are still based on the FK4 (except of the equinox correction); but for each year from 1984 until the first volume based on the FK5, for each star the reduction from FK4 to FK5 will be given containing both the systematic and the individual corrections.

3. BASIS OF THE EPHEMERIDES

Intentionally one has not adopted a certain theory or method as the basis for the ephemerides of the bodies of the solar system. Of course, there is a common agreement that the results should "best fit" the observations. There are in principle two different ways to approach that goal:

(1) The new fundamental ephemerides are derived by numerical integration and represent a most accurate, consistent set that can be calculated based on currently available observational data. They were cooperatively prepared by the Jet Propulsion Laboratory, Pasadena, and the U. S. Naval Observatory, Washington. The data as given in most of the almanacs are derived from these fundamental ephemerides. Ref.: E M Standish, P K Seidelmann (31.047.021); E M Standish (31.091.048); P K Seidelmann, R L Duncombe (31.091.030); X X Newhall et al. (34.046.003).

(2) The other method, which is used by the Bureau des Longitudes, Paris, consists in the development of new theories for the motions of the Sun, Moon, and planets as well as for the Galilean Satellites. The data as given in the Connaissance des Temps is based on these theories. Ref.: J-F Arlot (31.099.138); P Bretagnon (32.042.038); J-L Simon (33.042.026); G Francou et al. (34.046.007); M Chapront-Touzé, J Chapront (34.094.002); J Chapront, D T Vu (Astron. Astrophys. 141, 131).

The question which of these both methods is better and preferable, appears to be idle because each has its merits: The first may furnish more accurate results for the present period of time; the second permits to improve our knowledge on the long-term development and changes of the orbits. In any case, because both methods are based on practically the same observational material, the comparison of the results would also facilitate to reveal any defects or neglected effects.

4. ACCURACY, PRESENTATION, AND EXPLANATION

Improvement of the measurements required also a corresponding increase of accuracy in computing the ephemerides as given in the almanacs. This demand could be met partly by more powerful calculating means available; but also new computing techniques had to be developed and additional small effects which could be neglected in the past, were included.

Moreover, in some almanacs the presentation of the ephemerides and of other data has been more or less drastically modified. The main reason was to accommodate the form to the modern possibilities which the users had at their disposition, in particular the assistance of small computers. In many cases the most practicable realization of that aim was to tabulate interpolation coefficients in suitable intervals of time (e. g. The Astronomical Almanac, Connaissance des Temps, and, particularly, the Almanac for Computers). On the other hand, for the needs of occasional users the classical method of tabulation of the required data in a fixed interval (eventually with differences) offers some advantage (e. g. Astronomical Ephemeris of the USSR, Apparent Places of Fundamental Stars).

The introduction of all these changes of different kind required corresponding modifications of the explanation or even its re-writing for the 1984 editions of the almanacs. It seems, therefore, not necessary to enumerate in detail the changes which had been made in each almanac. Two examples only should be mentioned as exemplary:

(a) The Astronomical Almanac contains quite practice-orientated explanations given with each section of tabulation and accompanied by very useful examples.

(b) Both The Astronomical Almanac and the Japanese Ephemeris have added to the 1984 and 1985 edition respectively a Supplement which gives the basis of the new Ephemeris, i. e. a theoretical description of all fundamentals of the almanac including the necessary formulae and numerical values.

III. Recent Developments

The Commission should not restrict its activity to the almanacs and their production. There are now two quite distinct groups which do no longer use the ephemerides which are given in the almanacs:

(1) Formerly these ephemerides had been used not only for preparing the observations and for facilitating to find the objects to be observed on the sky, but also for comparing the ephemeris values with the observations, with the goal to improve the theory or at least some basic elements or constants, by analyzing the O-C's. For that use, the ephemerides had to be at least of the same precision as the observations. This utilization of the almanacs is, however, widely superseded. Most of the institutions which do such fundamental work have now large computers available; they are used not only for the reduction of the observations, but also for calculating the data directly from the theory rather than to interpolate them in an almanac. For some ephemerides one could therefore reduce the precision such that it is sufficient for prediction purposes.

(2) The other quite different group consists of those who have a small computer at their disposal. They would be able and willing to calculate by themselves the data which they formerly took from the almanacs. They are therefore interested to find formulae and practical prescriptions for this, which are not offered in the classical astronomical litterature. I do not think that this Commission should ignore that development (sometimes perhaps regarding it even as somewhat as a competition to the printed almanacs). The Commission should rather be interested that publications should be available which present the material in accordance with the adopted methods and constants. Good examples are T C van Flandern and F K Pulkkinen (26.021.032), Y Kubo, T Fukushima (Report Hydrogr. Res. Nos 15 and 16), and J Meeus (33.003.071).

IV. Long-term Ephemerides

Almanacs usually appear 1 - 3 years in advance. But in some cases one is interested to have ephemerides which are extended for a longer interval of time into the future. Without claiming completeness, some of them may be enumerated:

(1) Planetary and Lunar Coordinates for the years 1984 - 2000, by H. M. Nautical Almanac Office, Herstmonceux, (34.046.009) is the continuation of the volume for the years 1980 - 1984. The publication contains coordinates of the Sun, Moon, and planets to low precision for J2000.0 and some other predictions.

(2) Compact Data for Navigation and Astronomy for 1986 - 1990, to be published by H. M. Stationery Office, London.

(3) Another interesting publication which also covers the time up to 2000, giving also a star catalogue, but no positions of the Moon is: Star and Planet Catalogue (Equinox 2000.0) by S Nakano and A Otawara (34.002.082).

(4) Advanced predictions of eclipses for the next few years are, as usually,

given in U.S. Nav. Obs. Circulars; they are currently prepared by A D Fiala and M R Lukac. Eclipse predictions for the years 1981 - 2000 are published also as Supplements to the Japanese Ephemeris 1980 to 1983.

(5) Longer intervals are covered by the Canon of Solar Eclipses -2003 to +2526 (Wien, 1983) and the Canon of Lunar Eclipses -2002 to +2526 (Wien, 1983), both by J Meeus and H Mucke, as well as the Canon of Lunar Eclipses from 1000 B.C. to A.D. 3000 by B-l Liu (34.002.068).

V. Reduction to apparent place

Due to the changes mentioned above, there is an urgent need for precise instructions for the reduction of mean positions (old and new system) to apparent positions (new system) and vice versa; special problems arise in the case of stars where reductions include those from one equinox to another and also from one epoch to another (i. e. taking account of proper motions). Several publications appeared on this subject, e. g. by

P Bretagnon, J Chapront (30.043.004);
G H Kaplan (31.043.001);
E M Standish (32.043.003);
S Aoki et al. (34.041.033);
F Chollet (37.041.007);
W Melbourne et al. (37.044.055);
T Lederle, H Schwan (37.046.018).

Unfortunately there are some, even small differences between the various proposed methods. The most divergent point is the question whether the transition from the old to the new system (particularly concerning the numerical values for precession) has to become effective on 1984 January 1 or at another epoch. Those who are not so familiar with the problems may be highly disturbed by all the nuances. It seems to be not necessary that the Commission expresses its opinion in favour of one method by a formal recommendation; but it would be highly desirable that the experts agree to one unified procedure (even by some compromise) - and that this procedure be communicated by a common publication of these experts.

VI. Formulae relating time scales

Resolution 3 of Commissions 4, 19 and 31 on the expression of UT1 in terms of GMST as adopted in 1979 (Trans. IAU XVIIB, 69) has been superseded by the Resolution 3 of Commissions 4, 19 and 31 on the expression for GMST at 0^hUT1 as adopted in 1982 (Trans. IAU XVIIIB, 72). S Aoki has emphasized that in consequence of that modification one has also to change the figures as given in the Note to the Resolution (Trans. IAU XVIIB, 70). Although the expressions with the corrected numerical values have been already published by S Aoki et al. (31.044.003), it appears appropriate to give here the new version of the Note according to the text proposed by S Aoki:

"Note: The followings are frequently used quantities which are also affected by the Resolution:

a) The interval of mean sidereal time in a mean solar day becomes

$$24^h + \frac{8\ 640\ 184^s.812\ 866 + 0^s.186\ 208T_u - 1^s.86\cdot 10^{-5}T_u^2}{36525}$$

$$= 86\ 636^s.555\ 367\ 908\ 72 + 0^s.000\ 005\ 098\ 10T_u - 5^s.1\cdot 10^{-10}T_u^2,$$

b) The ratio of a sidereal day of 86400 mean sidereal seconds to this interval becomes

$$\frac{\text{mean sidereal day}}{\text{mean solar day}} =$$

$$0.997\ 269\ 566\ 329\ 0840 - 5.868\ 40\cdot 10^{-11}T_u + 5.86\cdot 10^{-15}T_u^2,$$

c) The ratio of the mean solar day to the mean solar day becomes

$$1.002\ 737\ 909\ 350\ 7954 + 5.900\ 58\cdot 10^{-11}T_u - 5.89\cdot 10^{-15}T_u^2,$$

d) Disregarding the secular variations, the equivalent measures of the lengths of the day at J2000.0 are

1 mean sidereal day = $23^h56^m04^s.090\ 530\ 8329$

= $86\ 164^s.090\ 530\ 8329$ of mean solar time,

and 1 mean solar day = $24^h03^m56^s.555\ 367\ 9087$

= $86\ 636^s.555\ 367\ 9087$ of mean sidereal time."

VII. IAU/IAG/COSPAR Working Group on Cartographic Coordinates and Rotational Elements of the Planets and Satellites

The 1982 Report of the IAU Working Group presented at the XVIIIth General Assembly has been published by M E Davies et al. (33.091.004). Tables give the recommended values for the directions of the north poles of rotation and the prime meridians of the planets and satellites referred to both the B1950.0 and J2000.0 standard coordinate systems. References for mapping these bodies are described. An appendix discusses the principal changes to the tables since 1979.

The International Association of Geodesy (IAG) suggested the establishment of a joint IAU/IAG working group to define reference systems for the planets and satellites. Since this falls into the activity of this Working Group, the IAU Executive Committee and the IAG agreed with the Joint Working Group which has been enlarged by members nominated by the IAG. COSPAR would endorse the activities of the Joint Working Group. The new name will therefore be as indicated above in the title.

T LEDERLE

President of the Commission

COMMISSION 5. DOCUMENTATION AND ASTRONOMICAL DATA
DOCUMENTATION ET DONNEES ASTRONOMIQUES

PRESIDENT: W.D.Heintz.
VICE-PRESIDENT: G.A.Wilkins.
ORGANIZING COMMITTEE: O.B.Dlushnevskaya, B.Hauck, C.Jaschek, P.Lantos, S.Mitton, L.Schmadel, F.Spite, C.E.Worley.

The report period 1982-84 was characterized by an again increased volume of material processed in data and abstracting centers, and by a growing clientele particularly of online services. The Working Groups of Commission 5 seek continued consultation with research object commissions so that the advanced documentation technology be efficiently employed toward specific demands of subject areas as to indexing, tagging, comprehensive, selective and inter-discliplinary retrievals. The guideline library for these purposes has been augmented by the First Dictionary of the Nomenclature of Celestial Objects by A.Fernandez, M.-C. Lortet, and F.Spite (Astr.Astrophys.Suppl. 52, no.4, 1983) and by the Guide to the Presentation of Astronomical Data by G.A.Wilkins (CODATA Bull.46, 1982); a new draft of the IAU Style Manual is before the IAU EC.

Following the 1981 Strasbourg conference on data retrieval (IAU Colloquium 64, published 1982: Reidel), some topics of cataloguing have been discussed at the IAU Symposium 111 on Calibration of Fundamental Stellar Quantities in Como (Italy), May 24-29, 1984 (chaired by A.H.Batten and L.Pasinetti). Astronomical catalogs were the specific subject of a colloquium arranged by the Astronomical Council of the USSR Academy of Sciences (USSR Data Center, O.Dlushnevskaya) and the Abastumani Observatory, September 10-16, 1984, near Tbilissi. The Commission is arranging for an open meeting on celestial object designations (chair: C.Jaschek) during the XIX GA. Among the topics proposed for discussion are particularly the nomenclature of specific objects in other galaxies, designation practices of objects observed from satellites, establishments of lists of acronyms, and procedures of error correction.

The Working Groups present the following Progress Reports:

Astronomical Data (B.Hauck, IAU Representative in CODATA):

The most visible activity of the WG was the arrangement of a course on data handling in astronomy and astrophysics, which was held in Trieste, July 9-13, 1984, and co-sponsored mainly by CNR (Italy), CODATA, and the Osservatorio Astronomico di Trieste. There were over 50 participants, and the chief topics were: Data management for data banks; astronomical catalogs; information transfer systems and standards; extraction methods for observational data; data handling in other sciences. The proceedings will be published in the Mem.Soc.Astr.Ital.

Two meetings in Strasbourg relating to astronomical data were titled "Statistical Methods in Astronomy" and "The Future of Non-stellar Astronomical Data". The proceedings of the Astronet conference (Rome 1982) have appeared in Mem.Soc.Astr. Ital. vol.53.

To characterize the expanding activities of the existing datacenters: About 450 astronomical catalogs are now available at the CDS Strasbourg. The Catalogue of Stellar Identifications (CSI) now contains about 550,000 objects, and has commenced inclusion of non-stellar objects, among them 50,000 galaxies. The USSR datacenter has at present 268 catalogs, including 13 prepared in the USSR, and four new ones dispatched in 1983. The collaboration between several observatories permitted extensive activity in relation to data in the USSR, and many scientific papers are being published in that field. Nauchnye Informatsii no.5 was devoted to the activities of the datacenter as a special issue. The compendium "Informatics in astronomy and Geodesy" (in Russian, ed. E.Mustel) was published by Nauka (Moscow 1982).

Among a number of new catalogues and atlases, wide interest attaches to the new edition of the Bright Star Catalog (Hoffleit and Jaschek; Yale 1982) with data on 9096 stars brighter than V = 6.50, with a 1983 Supplement. The Yale parallax catalog (to be published) contains - according to the entries completed in 1984 - trigonometric parallaxes for 7435 stars. Bibliographies of photographic catalogs of star positions, and of atlases and charts are found in Bull.Astr.O.Beograd no. 29 (1982). The new edition of the General Catalogue of Variable Stars is being completed in the USSR. Two astronomical satellites (the Space Telescope and Hipparcos) are at the origin of very large catalog projects. A massive guide-star list of some 10^7 objects is needed for the first mission, and comprehensive information on some 10^5 program stars for the second. The International Ultraviolet Explorer obtained many spectra, and a recently-published IUE Low-dispersion Spectra Reference Atlas contains 229 stars, the aim being to establish and to illustrate spectral reference sequences in the UV range.

Bibliography is an important complement of catalog data; in particular, the Bibliographical Index of the Strasbourg CDS (BSI) collects all citations of stars or of galaxies.

CODATA activities: The 8th International CODATA Conference was held in Jachranka (Poland), October 4-7, 1982. The Proceedings ("Data for Science and Technology", ed. Ph.Glaeser) were published by North-Holland Co. The 13th General Assembly took place just after this conference. The 9th International Conference, followed by the 14th GA, was held in Jerusalem, June 24-28, 1984. This assembly adopted a new policy concerning task groups, and some of the groups were discontinued, or modified in scope. The next conference will be prepared for July 14-17, 1986 at the Univ. of Ottawa (Canada), the subject being Computer Handling and Dissemination of Data. IAU members interested in this topic may contact the WG Chairman. CODATA Bulletins 45 through 55 were published during the report period. Apart from the "Guide to the Presentation of Astronomical Data" (G.A.Wilkins, no. 46, cited earlier), the following issues may be mentioned:

- No.48: CODATA Directory of Data Sources for Science and Technology, chap.7: Nuclear and elementary particle physics (June 1982).
- No.49: Directory chap.8: Molecular spectroscopy (July 1982).
- No.52: Directory chap.10: Geodesy (Jan 1984).
- No.54: 9th Conference Jerusalem 1984), Scientific Program and Abstracts (March 1984)

In addition, CODATA has sponsored the publication of "Database Management in Science and Technology", eds. J.R.Rumble and V.E.Hampel (North Holland Co.), which may be useful to astronomers. It is designed as an introduction for scientists and engineers to the use of computers to store, manipulate, and distribute collections of, numerical data.

Standardisation of Data Exchange (Special Task Force; P.Grosbol):

The Task Force on the Flexible Image Transport System (FITS) was created by Commission 5 during the XVIII GA in 1982, where Resolution C1 recommended the FITS tape format for the interchange of astronomical image data and other digital arrays between observatories. In cooperation of the North American AAS group and the European Software Coordination group, the two main functions of the Task Force are: (1) to channel comments and suggestions on the use of FITS for data interchange, and (2) to explore the possibilities for extending the FITS tape format to include table and catalog data. The current results, including presentations at the Course in Trieste (1984), are as follows:

(1) Close contacts between the groups have been established; the problems, comments and suggestions raised are collected, and discussed at intervals. The main items that came up in the groups and in a Panel Discussion at Trieste were: (a) a need for a booklet on how to use FITS; (b) a list of commonly used non-standard FITS keywords and their meaning would also be helpful; (c) concern on the 'small' physical block size on FITS tapes has been expressed by a number of major institutions.

(2) After extensive discussion a proposal for Generalized FITS Extensions (with applications to tables and catalogs) was published in the Astronomical Image Processing Circular no.10, and presented at Trieste. The proposed FITS extension format has been implemented in prototype versions at NRAO and ESO. Table data exchange between these institutions proceeded without problems. The extension is being examined by datacenters for the purpose of catalog exchange. Sample programs for encoding and decoding have been prepared in addition to a test tape which contains the Nilson Uppsala General Catalogue of Galaxies.

Designations (C.Jaschek):

The "First Dictionary of Nomenclature ..." (Fernandez, Lortet, Spite 1983, cited above) contains about 1200 acronyms and 900 references, and is expected to become the standard reference for catalog abbreviations. Preparation of the first supplement is well under way, as is similar work on objects in M31 and in the Magellanic Clouds.

Circulars were addressed to the editors of astronomical periodicals to call their attention to the designation problem. All contacted magazines answered favorably and requested additional desk copies of the "Dictionary" which were provided. It is hoped that the actions promised to be taken will yield visible results in the future. Input from other commissions is desired. Designation practices were also the subject of a paper by J.Mead and a discussion by Jaschek at Trieste. The WG had an exchange of letters with regard to the designations of objects observed by EXOSAT.

Classification Systems and Information Retrieval (P.Lantos):

The IAU Vocabulary, on which a report was presented to the CODATA 1984 conference, is awaiting approval by the WG. The translation into French is planned. Contacts to journal editors are intended in order to explore a feasible way of disseminating the Vocabulary.

Abstracting Guidelines (L.Schmadel):

The Task Force is still constituting itself, but preliminary material has been circulated. A progress report will be completed in time for discussion at the XIX GA.

The Astronomy and Astrophysics Abstracts (AAA) are approaching a volume of 20,000 papers per year. Compilation has been transferred to a modified ITT 3030 system which is very satisfactory and has permitted further reduction of the publication time of the volumes. As a result, the 3rd Five-year Index (vol.35/36) has already appeared, and vol.38 is in progress. A tape service is in preparation. The code ASTHMA, under which the latest volumes of AAA should be available by mid-1984, will be a standard tape (9-track unlabeled, ASCII, 1600 bpi, block size 80, record length 4000) of high redundancy and simple structure, so that it can be operated on computer systems of low capacity. Marketing is still being negotiated.

VINITI (All-Union Institute for Scientific and Technical Information, Dept. Astronomy and Geodesy, Moscou; I.S.Shcherbina-Samoilova) :

The Department continued publication of Abstract Journals (Referativnyj Zhurnal) in three series: 51.Astronomy, 52.Geodesy, and 62.Space Research. The total number of abstracts amounted to 22,000 annually.

Itogy Nauky (state-of-the-art reviews) have the following new volumes:
(Astronomy Series)
17. Radio Exploration of Moon and Terrestrial Planets, part II (1981).
18. Problems of Extragalactic Astronomy (1981).
19. Radiation Receivers for Ground-based Astronomy (1982).
20. Celestial Mechanics.
21. Stellar Flares. Supernovae. X Ray Sources. The Sun (1982).
22. Peculiar Stars, Strong Magnetic Fields, and Gamma Astronomy (1983).
23. Problems of Modern Astronomy (1983).
24. Astrophysics and Cosmic Physics (1983).
25. Solar Physics (1984).
(Space Research Series)
17. Physics and Evolution of Stars (1981).
18. Magnetism of Planets (1982).
19. Key Problems of Planet Exploration in the Solar System (1983).
20. Molecules in Astrophysics (1983).
21. Materials Science in Space (1984).

The descriptor vocabulary "Space research" has been prepared and is used for storage and machine readable output of Abstract Journal 62. The corresponding work for Astronomy is nearing completion.

Astronomy and space research reviews are contained in a newly established database for reviews.

NASA/GSFC Astronomical Data Center (Greenbelt MD, USA; W.Warren):

Request activity at the ADC, NASA Goddard Space Flight Center, continues to increase. During the report period, 1300 requests were completed by the ADC, with 900 machine readable catalogs disseminated. New or corrected/modified machine readable catalogs received numbered 141, while detailed descriptive comments were produced for 64 catalogs. In brief description, significant work completed at the ADC includes these items:

Machine readable versions of the entire Cordoba (613959 stars) and Cape Photographic (454877 stars) Durchmusterungen were completed. Zones -01° to $+19^\circ$ of the Bonner Durchmusterung (111323 stars) were also finished.

New versions of the Smithsonian Astrophys.Obs. Star Catalog (SAOC) and the SAO-HD-GC-DM Cross Index were completed following five years of part time work (Roman and Warren, ADC Bull. 1,212). A new machine version of the Henry Draper Catalogue has also been prepared (Roman and Warren, Bull. AAS 16, 502, 1984).

The NASA version of the retrieval system of the Bibliographical Star Index (Carpuat et al., Bull.Inf.CDS 26, p.83, 1984), available on-line to the scientific community, was considerably improved and updated with new BSI data (Hill, Mead, and Nagy, ADC Bull. 1,220, 1983; Hill, Brotzman, and Mead, Bull.AAS 16, 503,1984).

A Bibliographical Index of Objects Observed by IUE 1978-82, containing 523 papers, has been prepared (Mead, Kondo, and Boggess, Bull.AAS 16,502). A new version of the Catalog of Infrared Observations has been completed in both machine and printed forms, and published as NASA Reference Publ. 1118 (May 1984).

ICSTI (International Council for Scientific and Technical Information, the former ICSU Abstracting Board) reconstituted itself at the GA in Philadelphia in June 1984, the IAU being represented on its Executive Committee by W.Heintz. The organisation is resuming its wide-field and inter-disciplinary studies (Aggregate lists of serials; thesauri and translations; downloaded machine readable databases; copyrights) and intensifying the cooperation with CODATA.

Astronomy has a problem, in the opinion of this author, with some bibliographical databases (other than those prepared by astronomical services) which are less than helpful when used for searches. They may be incompletely compiled or downloaded (particularly on noncommercial literature), confusingly classified, or poorly cross-referenced. Diversity of publishing policies in the primary literature, especially with respect to vocabularies, and disagreements on classification schemes do not help to alleviate the problem. A fact accepted with resignation appears to be the "unmanageable" volume of primary publications. Hardware and techniques for the requisite secondary processing are in place, but the conventions toward optimal use of these facilities are incomplete.

CDS Strasbourg and the USSR Data Center in Moscou report in the Bulletins d' information of CDS. Attention is called to other publications in these Bulletins, notably the Holding Aggregates (no.25, p.87, 1983) with supplments on microfiches, and the paper by D.Egret (no.24, p.109) on the organisation, contents and use of the CDS database. Literature relevant to documentation is referenced in chapter 002 of the A.A.Abstracts.

W.D.HEINTZ
President of the Commission.

6. ASTRONOMICAL TELEGRAMS (TÉLÉGRAMMES ASTRONOMIQUES)
(Committee of the Executive Committee)

PRESIDENT: M. P. Candy
VICE-PRESIDENT: A. Mrkos
DIRECTOR OF THE BUREAU: B.G. Marsden, Smithsonian Astrophysical Observatory, 60 Garden Street, Cambridge, MA 02138, USA (TWX 710-320-6842 ASTROGRAM CAM)
ASSOCIATE DIRECTOR OF THE BRUEAU: vacant
ASSISTANT DIRECTORS OF THE BUREAU: C. M. Bardwell, D. W. E. Green

I. INTRODUCTION

The bulk of the Commission's work has, as in previous years, been carried out most efficiently by the I.A.U. Bureau, especially by its Director, Dr. B.G. Marsden.

As predicted in the report by our previous President, rapid transmission of information by electronic means is now with us; details can be found in the Report of the Central Bureau below. With the ever-increasing cost of telegrams (and their occasional loss or misdirection) more astronomers will be attracted to the new methods.

The Commission is particularly grateful to the Smithsonian Astrophysical Observatory for its continued support.

M.P. Candy
PRESIDENT OF THE COMMISSION

II. REPORT OF THE CENTRAL BUREAU FOR ASTRONOMICAL TELEGRAMS

During the triennium the Central Bureau for Astronomical Telegrams issued "telegram books" and *Circulars* on the following occasions:

	Telegrams	*Circulars*
1981 (from 24 Oct.)	4	9 (Nos. 3642-3655)
1982	54	64 (Nos. 3656-3758)
1983	60	70 (Nos. 3759-3904)
1984 (to 31 Oct.)	42	53 (Nos. 3905-4002)

The exceptional activity during 1983 was due to the record 23 announcements of discoveries and recoveries of comets. The announcement of comet IRAS-Araki-Alcock (1983d) one week before it made the closest approach of any comet to the earth since 1770 caused the Bureau to have its busiest time ever. The Bureau made quick checks on some 50 tentative reports from IRAS of unusual moving objects, one of which was the 'Geminid parent' 1983 TB. Other highlights during the triennium were the recovery of P/Halley and the visual discovery by the Australian amateur R.O. Evans of several supernovae.

Increases in subscription rates, in June 1982 to 55c and 33c, and a year later to 60c and 36c, for 'regular' and 'special' accounts, respectively, as well as a deterioration in the speed with which the *Circulars* are printed and delivered, caused the number of subscribers to drop to around 730 during the latter part of 1983. There was subsequently a significant improvement in printing and delivery, however, and the number of subscribers increased to around 780 by October 1984.

Beginning in October 1982, the copy for the *Circulars* was prepared by computer. Beginning in January 1984, the *Circulars* have been available directly in computer form to subscribers to the Bureau's computer service. For an additional charge of 30c per day the computer service allows subscribers to 'dial in' to read complete *Circulars* as soon as they are issued, to leave and receive messages, and to obtain orbital elements and ephemerides. Since telephone charges are necessarily involved also, the vast majority of the 30 computer-service subscribers are in North America, but there is also a subscriber in each of West Germany and Japan. Computer dissemination of information is clearly and rapidly becoming the way of the future, and with its introduction the Bureau discontinued the use of the generally unsatisfactory and already rather old-fashioned system of automatic recording of telephoned messages.

The Bureau's telegram service continues to operate, however, and although the cost is greater and much less information is provided than in the computer service, it has the advantage of being 'passive', whereas a computer-service subscriber may either just miss a particularly urgent message or find no change since his last call; furthermore, while telegrams may be issued at weekends, *Circulars* are not. Subscription rates for telexed and mailgram messages in North America were increased to \$4.00 and \$3.00, respectively, in January 1984, while overseas telexes and cablegrams and telephoned domestic telegrams continue to cost a minimum of \$7.00. From early 1983, World Data Center A for Solar-Terrestrial Physics, Boulder, Colorado, was no longer able to relay telegrams for the Bureau, so messages are now telexed directly to its counterpart in Moscow for distribution in the USSR. Telexes are also sent to the counterpart in Sydney, and following the retirement of F.E. Cook, the further distribution of messages in Australia and New Zealand has been handled by P. Davies; this generosity of the Ionospheric Prediction Service has resulted in the anomaly that these subscribers are paying only \$2.00 per message.

B.G. MARSDEN
Director of the Bureau

7 - MECANIQUE CELESTE (CELESTIAL MECHANICS)

PRECIDENT: J. Kovalevsky
VICE-PRESIDENT: V.A. Brumberg
ORGANIZING COMMITTEE: E.P. Aksenov, J. Chapront, S. Ferraz-Mello, J. Hadjidemetriou, J. Henrard, W.H. Jefferys, Y. Kozai, H. Scholl, P.K. Seidelmann, A. Sinclair.

A - INTRODUCTION

Celestial Mechanics has undergone, in the past three years, a considerable development that has been marked by a significant increase of published papers and by the recent organization of several specialized colloquia or symposia. The reasons of this increase of interest may be found among the following circumstances :

- The discovery of new dynamical situations in the Solar system: ellipticity of ringlets, shepherd satellites.
- The increase of interest in the problem of the dynamical evolution of the solar system or of its sub-systems.
- The present necessity to include relativistic effects in the theory of motion of celestial bodies.
- New promising results in the analysis of the properties of dynamical systems.

Several books or monographs in Celestial Mechanics were published during these three years:

- Akim, E.L., Bazhinov, I.K., Pavlov, V.P. and Pochukaev, V.N., (1984), "The gravitational field of the Moon and the motion of its artificials satellites", in russian, Mashinostroenie, Moscow.
- Bordovitzina, T.V. (1984), "Modern numerical methods in the problems of Celestial Mechanics", in russian, Nauka, Moscow.
- Dragomir, V.C., Ghitan, D.N., Mihailescu, M.S. and Rotoru, M.G. (1984), "The theory of Earth's shape", Elsevier, Amsterdam.
- Duboshin, G.N. (1983), "Celestial Mechanics: Methods of the theory of motion of artificial celestial bodies", in russian, Nauka, Moscow.
- Emelyanov, N.V. (1983), "Methods for the construction of algorithms and programs in problems of Celestial Mechanics", in russian, Nauka, Moscow.
- Ferraz-Mello, S. (1983), "Dynamics of the Galilean satellites", russian translation, Mir, Moscow.
- Kane, T.R., Likins, P.W. and Levinson, D.A. (1982), "Spacecraft dynamics", Mc. Graw Hill, New York.
- Kovalevsky, J. (1984), "Introduction to Celestial Mechanics", chinese translation.
- de Luca, N. (1982), "Mecânica Celeste", in portuguese, Curitiba.
- Nazarenko, A.I. and Skrebushevsky, B.S. (1981), "Evolution and stability of satellite systems", in russian, Mashinostroenie, Moscow.
- Roy, A.E. (1982), "Orbital motion", 2nd edition, Heyden and Sons, Philadelphia; russian translation of the first edition (1981), Mir, Moscow.
- Szebehely, V. (1982), "Theory of orbits: the restricted problem of three bodies", russian translation, Nauka, Moscow.
- Timoshkova, E.I. and Kholshevnikov, K.V. (1982), "Asymptotic methods in Celestial Mechanics", in russian, Itogui Nauki i techniki, ser.astron., vol.20, VINITI, Moscow.

The great diversity of research under way in the various fields of Celestial Mechanics does not permit to draw a complete picture of its development in the few pages that have been allocated to the commission report. So, following the example of the 1982 report by Y. Kozai, and in harmony with the wishes of the organizing committee, no attempt was made to give a quasi-exhaustive bibliography. We chose to critically describe in more details the achievements in some specific domains of Celestial Mechanics, leaving aside large fields of interest in which quite important contributions have been made. Hopefully, they will be included in the next report as a part of a six year development description. In the present report, we put the emphasis on two domains: the dynamics and the evolution of the solar system and the structure of the solutions of dynamical systems in Celestial Mechanics and particularly in the 3 or n-body problem. In doing so, no mention is made of the achievements in relativistic Celestial Mechanics (this will be done in the incomming IAU symposium N°114), resonance theory, systems involving finite bodies (including artificial satellite motion, structure or rotation of celestial bodies), methods for solving the equations (asymptotic methods, various developments, Hamiltonian mechanics, perturbations theory, etc...), variable masses, etc...

In particularly thank J. Chapront, J. Hadjidemetriou and C. Marchal who have accepted to review the achievements in their respective fields and have greatly contributed to this report, and also all the colleagues who sent their reports.

N.B. In the course of the report, A and A. Abstracts numbers are used except for the most usual titles for which an abreviated designation followed by the volume number (Colloquium or Symposium number for IAU Coll. or Symp.) and the page number is used. In addition, results presented in three recent non IAU colloquia are indicated as follows :

BER : "The Big Bang and Georges Lemaître"
A. Berger ed. (Louvain, Oct. 1983), Reidel Publ.Co.

SZB : "The stability of the Solar System"
V. Szebehely ed. (Cortina d'Ampezzo, Aug. 1984), in press.

FME : Resonances in the motion of planets, satellites and asteroids"
S. Ferrza-Mello and W. Sessin eds. (Sào Paulo, Nov. 1984), in press.

B - PLANETARY AND LUNAR THEORIES

(Main contributor : J. Chapront)

The last three years have been characterized, in these fields, by the introduction in the "Connaissance des Temps" of the planetary and lunar theories recently constructed in the Bureau des Longitudes and a series of improvements and accuracy analyses of these theories.

I. SECULAR VARIATION PLANETARY THEORIES

These theories are characterized by the presence of time in the coefficients of the periodic terms. They are valid for an interval of time of the order of one thousand years. The planetary theories constructed at Bureau des Longitudes have resulted in the solutions TOP82 for giant planets (Simon, AA 120.197) and VSOP82 for telluric and giant planets (Bretagnon, AA 114.278). Since the 1984 edition of the Connaissance des Temps, the solutions TOP82 and VSOP82 replace the theories by Le Verrier and Gaillot in the computation of the ephemerides of the Sun and major planets. The solution VSOP82 was compared to planetary observations (transits of Mercury and Venus, occultations of Venus and Mars) over the interval 1717-1973 by Krassinky et al (submitted to AA). The relativistic drifts that have been introduced for the perihelion of Mercury have been verified to an accuracy of a second of arc. The ephemerides of Mars constructed from an earlier solution (VSOP80) have been compared to 1622 observations performed between 1935 and 1976 by Niimi (33.042.090). A set of constants of integration was determined for the Earth and Mars. A particular attention was directed towards the modelling of the phase effect.

Presently, the improvement of these new theories is undertaken either in trying to improve the precision or in extending the interval of validity. The effects of lunar perturbations on the barycenter of the Earth-Moon system and on all major planets have been recomputed with a better precision. Perturbations due to Ceres, Pallas and Vesta on the major planets have been determined up to the second order of masses (Bretagnon, CM in press). On an interval of 1000 years around J2000, the accuracies range between 0."001 and 0."1 according to the planet. The interval of validity of the theories of Jupiter and Saturn has been extended to 6000 years around J2000 using methods that have been developed during the construction of TOP82 (Simon and Bretagnon, AA 138.169). The accuracy of the theory of Jupiter is better than 8" in the interval - 4000 to +8000. For Saturn, it is better than 23". The interval of validity of the theories of the Sun, Mercury, Venus and Mars has also been extended to 6000 years using, in particular, secular developments deduced from the general theory of Laskar (AA in press). For the Sun and these planets, the accuracy ranges between 1" and 5" over all the interval -4000 to +8000 years.

J. Chapront (CM in press) has proposed a new method of solving the problem of the motion of the Neptune-Pluto. He renounced the successive approximation method using Fourier series developments in mean longitudes which diverges. If one accepts to limit the validity of the solution to a thousand years - and this eliminates the 3.2 resonance problem -, the solution may be represented in terms of mixed series called Fourier-Chebyshev series (Chapront, CM 28.415).

II. GENERAL PLANETARY THEORIES

These theories provide long periodic variations of orbital elements of planets and their validity covers a time interval of the order of one million years. They have been developed in different directions. Bretagnon (CM in press) has refined his 1974 previous solution, introducing relativistic and lunar perturbations. Duriez (25.42.100) has constructed an autonomous system limited to the four giant planets. Laskar has continued this work (AA in press) and has included all the eight planets. He performed a numerical integration of the autonomous system over one million years instead of looking for a solution in quasi-periodic functions of time that have a poor convergence. The relativistic and lunar perturbations were included. The accuracy obtained allows a development of the solution in powers of time around the origin which has permitted to obtain the secular variations of the elements that were used to extend VSOP82 over 6000 years.

An important effort was initiated by Kamel in order to compute litteral expressions for a general theory of the motion of giant planets. It was prepared by the development of powers of the inverse distance between two planets (Kamel MP 26.339) and by the elimination of the critical terms in Jupiter-Saturn and Uranus-Neptune theories (Kamel MP 24.137, 27.407, 28.221 and Kamel and Bakry MP 24.261). A theory to the second order of masses of the Jupiter-Saturn system was computed using Poincaré variables and the Von-Zeipel method up to the fourth order in the eccentricities and inclinations (Kamel and Bakry MP 27.417 and 28.113).

III. NUMERICAL THEORIES AND METHODS

Most of the international planetary ephemerides are now based upon the numerical integration on the solar system performed in JPL (see Standish CM 26.181) and called DE200/LE200. In USSR, work is progressing in improving high accuracy numerical theories of the motion of inner planets and the Moon (Krassinsky and Sveshnikov CM 26.171; Krassinsky et al. AA in press; Sveshnikov 32.041.001, Trudy ITA 19). They are based upon soviet radar observations (1961-80), Washington meridian observations of the Sun (1961-70), passages of Mercury on the Sun (1717-1913), occultations of stars and inner planets by the Moon (1713-1980), laser observations of the Moon in McDonald (1970-1973). Among the results already obtained a number of parameters have been improved such as the motion of the equinox, the tidal acceleration of the mean motion of the Moon, Universal time scale during the XVII-XIXth centuries, the relativistic motion of the perihelion of Mercury, the secular variation of the solar radius, etc... During this work, rare but very valuable modern photoelectric observations of occultations of stars by planets were used. A simple such observation has an efficiency that is equivalent to a long series of classical astrometric observations. Planning of such observations in the future, might be very useful for improving the theories in Celestial Mechanics. In parallel, in the Institute for Theoretical Astronomy of Leningrad, a general information center on the motion of bodies of the solar system is being set up.

Kinoshita and Nakai (CM in press) have integrated the motion of the outer planets for five million years. The libration of Pluto's perihelion about 90° with a period of 4 million of years was assessed. It avoids close approaches to Neptune. This integration was used by Milani (CM in press) who traced the time variations of c^2h (angular momentum squared times energy) for various sub-systems and discovered a secular resonance locking between the perihelion of Uranus and the aphelion of Jupiter with a 1.100.000 years period. This locking is controlled by a complex 4 planets interaction.

Much work was done in the domaine of improving the method of numerical integration of the equations of Celestial Mechanics. They are developed by Bordovitsina in 30.021.22 and in a monograph (see Section A). Many experiments are described in this book. Most of the participants to these experiments showed their preference to the method of Everhart. The representation of the motion of planets or of the Moon by series of finite functions such as Chebyshev polynomials was studied in several places. The Bureau des Longitudes algorithm has been improved and a single formula may represent one year of the motion of the Earth with the desired accuracy. Batrakov and Fursenko (31.042.32) have developed a theory of Chebyshev slipnes to represent such motions; Schwartz and Walker have continued their studies of the applications of recurrence methods (CM 27.191) and Dvorak and Hanslmeier used Lie series method for numerical integration (AA 132.203 and 34.042.122), and found that this method is very advantageous.

IV. LUNAR THEORY

The semi-analytical theory of the motion of the Moon ELP2000-82 constructed by Chapront-Touzé and Chapront (CM 26.082 and AA 124.50) was adjusted to the JPL.LE 200 numerical integration. The angular and radius vector residuals do not exceed respectively 0".01 and 15 meters in a century. This solution was introduced in Connaissance des Temps starting 1984 in replacement of the Brown-Eckert theory. Recently, improvements were brought by the computation of perturbations due to the shape of the Moon and the effects induced by the libration (Chapront-Touzé AA 119.256). It is shown that, if third or higher order harmonics of the potential are taken into account, physical libration has to be introduced, otherwise secular drift appears. Kubo (Report of Hydrographic Res., Japan, 18) also computed perturbations due to the shape of the Moon and compared his results with ELP and Henrard's series. They agree better with EPL. Relativistic perturbations were evaluated by Lestrade and Chapront-Touzé (AA 116.75) and by Brumberg and Ivanova (Trudy ITA, 19). Both solutions agree. Comparisons of ELP with occultations of stars by the Moon have been made by Krassinsky et al (AA in press) and by Sôma (submitted to CM). The first authors discuss the value of the secular acceleration which is in good agreement with lunar laser data. The second author discusses the values of the lunar and solar parameters. The tidal term obtained is not inconsistent with the modern occultation observations and the motions of the perigee and the nodes coincide with the theoretical values.

Analytical theories of the libration of the Moon have been recently improved. Eckhard (29.094.023 and IAU 63.193) has developed a theory that includes planetary perturbations, Earth's flattening and the effect of the Earth on the fourth order harmonics of the lunar potential. Moons (CM 26.131 and 33.094.23) has developed a more general theory, retaining more physical parameters in a litteral form and her expressions are more precise. However, it is not yet completed and some perturbations are still missing. The difference between the common parts of both theories is of the order of 2 meters on the surface of the Moon and are mainly due to two resonant terms of the perturbation in parallax.

C - SMALL BODIES IN THE SOLAR SYSTEM

(main contributor : J. Kovalevsky)

The detection of many new satellites in the Jupiter and Saturn systems and the analysis of the detailed structure of the ring systems have given rise to a number of theories to explain the new dynamical features so discovered. In parallel many efforts have been devoted to the understanding of some features of the solar system from the dynamical evolution point of view.

I. MOTION OF NATURAL SATELLITES

The dynamical characteristics of natural satellites and their consequences on the theory of their motion have been described by Kozai (CM 23.265). Many new orbits of satellites have since been completed. However, the results will not be reviewed here and we shall consider here only general problems or methods.

Galilean satellites. The work on the theory of the motion continues in several directions. Arlot (AA 107.305) has obtained new integration constants for the Sampson-Lieske theory. The possibility to extend a general planetary theory to a resonant case and particularly to the Galilean satellites was shown by Duriez (CM 26.231) and the results agree in a first approximation with known results. After the elimination of all periodic terms to the third order, Brown (CM 23.203) obtained and solved the equations for the secular effects on the elements. New values of the semi-major axes were obtained by Vu (CM 26.265), while a complete second order theory, based on Sagnier approach has been constructed by Thuillot and Vu (34.099.112). In particular, the influence of dynamical parameters on the secular terms was computed by Thuillot (CM in press).

Saturn satellites. The dynamics of the newly discovered satellites have been subject to numerous studies in the frame of the discussion of horse-shoe librating orbits (Dermott and Murray, IC 48.12; Yoder et al. IC 53.431). Harrington and Seidelmann (IC 47.97) showed numerically that the system is stable and that the satellites never approach one to another by more than 6° during the 3000 days libration of S3 around S1. The dynamics of the co-orbiting satellites was studied by Spiris and Waldfogel (SZB in press) who showed that the problem reduces to the Hill's problem with appropriate boundary conditions at infinity. They predicted that S1 and D3 exchange orbits at close encounters, while S26 and S27 do not. I. Stellmacher (CM 28.381) applied her algorithm for the construction of periodic orbits in the Mimas-Tethys case, the satellites being considered as two oscillators coupled by gravitational interaction. Bec-Borsenberger derived a litteral theory of the motion of Phoebe (CM 26.271).

II. RING DYNAMICS

The fine structure of Saturn rings found by Voyager spacrafts as well as the ellipticity of some Saturn and Uranus rings has brought to Celestial Mechanics a number of new problems which have now found satisfactory answers (Aknes 31.100.058 and SZB in press; Goldreich and Tremaine 32.091.025). The dynamics of elliptical rings for which a non zero inclination was found, have been discussed by Borderies, Goldreich and Tremaine (AJ 88.226). They show that by self-gravity, the ring may maintain by external forces (planetary oblatenes or satellite action). The remarkable satellite-ring interaction known as "shepherd satellites" by Goldreich and Tremaine in 1979 has been analysed and the role of dissipation was clarified by

Greenberg (IC 53.207). Borderies et al. discussed globally the dynamics of elliptical rings taking into account external forces due to the oblateness of the planet and the sphepherd satellites, self-gravity and viscous forces due to interparticle collisions (AJ 88.1560). The picture is as follows: the mean eccentricity grows under the action of corotation resonances with the sphepherd satellites and is damped by viscosity so that an equilibrium is established. The uniform precession is induced by self-gravity.

The general ring picture has been also refined by Borderies et al. (IC 55.124) who computed the stress tensor due to particle collisions and explained sharp edges of rings and the decay of density waves. Other tentative explanations exist also for the radial diffusion or density and bending waves (Borderies, CM in press) and fine ring structure enhanced by viscosity (Michel 31.100.43). Salo (37.100.63) has applied the Hameen-Anttila's theory of bimodal gravitating systems to obtain steady state solutions for the optical thikness of Saturn's rings. The elliptic properties of particles determine the behaviour of gaps while the existence of ringlets would depend on the size and density of the particles.

The accretion of centimeter sized particles into large aggregates has been proved to be possible in Saturn's rings on a time scale of weeks by Davis, Weidenschilling et al. (37.100.44). These aggregates are later disrupted by tidal stresses, so that the mass of the ring system is processed through a population of large "dynamic ephemeral bodies" which are continously forming and desintegrating. Seidelmann, Harrington and Szebehely (IC 58.169) have studied analytically and numerically the dynamical behaviour of the E ring which extends from the orbit of Mimas to the orbit of Rhea and consequently interacts with Tethys, Dione and their Trojan companions. These detailed studies do not superseed the classical interpretation of ring density distribution by resonance with satellites. For instance, Wiesel (IC 51.149), including also Saturn oblateness, has explained in detail the structure of Cassini division. Other investigation in this direction have been made by Lissaper and Cuzzi (AJ 87.1051) who tabulated all the resonance locations and strengths from the most recent and reliable values of the masses and orbital parameters. These theories have also their equivalent in the studies of the structure of Uranus rings (Aksnes, IAU 75 in press).

III. TROJAN PLANETS AND ASSOCIATED PROBLEMS

Garfinkel (31.098.038; CM 30.373) is constructing a general analytical theory of the motion of Trojan planets. Formal long periodic orbits have been obtained and their periods expressed in terms of a mass parameter and of the normalised Jacobian constant. The motion of perihelion of Trojan orbits was investigated by Erdi (CM 24. 377; IAU 47.165) and Erdi and Varadi (34.098.059). Several studies of horseshoe periodic orbits have been made in connection with this problem (Dermott and Murray, IC 48.1 and 12; Taylor, AA 103.288).

IV. EVOLUTIONARY PROBLEMS

Under the word evolution, are meant the dynamical processes that tend to modify the fundamental structure of planetary or satellite systems: secular drifts in semi-major axis, trapping into resonance, escapes from, or capture into a system, etc... Many investigations have been made in the recent years in an attempt to understand the present structure of the solar system (an introduction to the problem is found in Burns, 32.091.053).

<u>Asteroidal belt</u>. Several different models have been proposed to explain the Kirkwood gaps in the distribution of Minor Planets as a result of gravitational or cosmological processes. Hadjidemetriou (CM 27.305 and 31.042.051) and the same author with Ichtiaroglou (IAU 74.141 and AA 131.20) have based their study on the behaviour of resonant solutions of the non-averaged plane restricted problem. Unstable regions

have been identified in the phase space. Wisdom (AJ 87.577) starts from the same model for the 3.1 resonance, then he models high frequency terms as Dirac impulsions and constructs a mapping of the phase space on itself. This permits him to study the behaviour of an asteroid over much longer times. He finds that the eccentricity oscillates and its maximum value increases so that the planet reaches the orbit of Mars and leaves the resonant region. In a second paper (IC 56.51) he traces the chaotic region in the elliptic restricted region: it outer boundary coincides with the boundary of the 3/1 Kirkwood gap in the actual distribution of asteroids. The extension to other gaps is not straght forward because of the specific role played by Mars. Other attempts have been to explain the gaps in introducing non conservative effects. Torbett and Smoluchowski (IC 44.722, AA 110.43 and 127.345) and Gonczi et al. (IC 51.639) have investigated the effect of resonance sweeping due to a proto-solar accretion disk or Pointing-Robertson effect, by numerical integration. Henrard and Lemaître (IC 55.482 and BER 217) and Lemaître (IAU 74.189 and CM in press) developed an analytical model that they applied to the gravitational effects of the proto-solar nebula. The adiabatic invariant theory and the introduction of an area index (the algebraic area enclosed by the trajectory) are used to monitor the evolution of the orbit in time. Comparing the statistical results of the evolution of a number of asteroids initially resonant with the actual distribution has shown a satisfactory agreement.

Satellites. Several papers were devoted to the time evolution of satellite systems. Sinclair (IAU 74.19) has reexamined the tidal hypothesis for Saturn resonant satellites. For Mimas-Tethys, this assumption is consistent with their anomalously high inclinations. It is not sufficient for the Enceladus-Dione system, where one should in addition consider tidal dissipation within Enceladus. Henrard has developed the theory of capture in a general one degree of freedom Hamiltonian system and connected it to the adiabatic invariant theory (CM 27.3 and 31.042.054). He applied it to the orbital evolution of the Galilean satellites (IC 53.55), while the same theory was applied by Borderies and Goldreich to capture probabilities for the j+1:j and j+2:j orbit-orbit resonance problem (CM 32.127).

For the non-resonant evolution of satellites, Mignard has completed his study of the past secular evolution of the Earth-Moon system (MP 24.189) and of the Martian satellites (MN 194.365). In both cases, the dissipation in the satellite is a critical parameter in distinguishing various evolutionaly paths. He shows that, in any case, Deimos has not undergone a significant tidal evolution, while Phobos may have evolved from a high eccentricity orbit. Kovalevsky (SZB and FME in press) has shown that planetary terms of the lunar theory may undergo a passage through resonance while the tidal friction makes the lunar semi-major axis to evolve. If its coefficient is sufficiently large such a term may become resonant and stop the evolution of the semi-major axis during several libration periods, then an escape from the resonant region occurs. For a given term the same situation may repeat a great number of time while the Earth's eccentricity undergoes long periodic variations. A model of capture into and escape from resonance is being studied by Sidlichovsky.

Comets. The evolution of cometary orbits at close approaches with planets has been studied by several authors. Alekseev and Osipov (Ergod.Th. and Dyn.Sys.2,263) have estimated rigorously the variation of parameters during a close approach to Jupiter. Nakamura (IC 45.529 and IAU 74.97) has estimated, by numerical methods, the final orbital distribution of extinct comets. The depletion of the Oort cloud by a dynamical process has been studied by several authors. Fernandez (IC 42.406 and 47.470) has studied numerically the perturbations of cometary orbits by stars and outer planets. Weissman (33.102.034 and IAU 61.637) developed a model of stellar perturbation in terms of a velocity impulse and obtained rates of depletion and of formation of comets entering the planetary system. Rémy (IAU 83 in press) has removed two assumptions from the Weissman model: the original eccentricities are assumed to hase a wide distribution between 0 and 1 and not only close approaches are considered. She estimated analytically the main depletion of long periodic comets.

D - PERIODIC ORBITS AND STABILITY

(Main contributor : J.D. Hadjidemetriou)

Research in periodic orbits and stability has been influenced by recent developments in non-linear dynamics and this interaction has been proven very fruitful. Integrability, chaotic motion, infinite bifurcation, Lyapounov characteristic exponents have been successfully used in the study of old problems of Celestial Mechanics. Resonance phenomena and their relation to the stability of planetary systems were widely studied in connection with a variety of problems (see also the section on the small bodies in the solar system). The study of dynamical systems with three degrees of freedom attracted the attention of many investigators. Finally, the stability of planetary or stellar systems was also widely studied by a variety of methods and definitions of stability. The method of work varies from abstract mathematical proofs to numerical computations.

I. METHODS OF NON-LINEAR DYNAMICS

Bifurcation theory. Bifurcations of families of periodic orbits in a dynamical non integrable system are now studied in connection with the understanding of the onset of chaotic motion. This work started after a paper by Feigenbaum (J.Stat.Phys.19,25) who found that dissipative systems have infinite period doubling bifurcations and that this leads to chaotic motion. This result was extended to Hamiltonian systems, which mostly interest Celestial Mechanics, and also to area preserving mappings. It was found that, in this latter case, there exists a universal ratio of the successive intervals between the (infinite) bifurcations, equal to 8.72 (Benettin et al., lett.Nuovo Cim., 28,1 and Bountis, Physica 3D,577 for area preserving mappings). Such a sequence of infinite period doubloing patchwork bifurcations has been found by Contopoulos and Pinotsis (AA 133.49) for two simple families of periodic orbits of the restricted circular three body problem, and the same universal ratio was found. Successive bifurcations in Hamiltonian systems have been also studied by Contopoulos in letters to Nuovo Cimento (30.498; 37.149 and 38.257). Infinite period doubling bifurcations have also been studied by Heggie (CM 29.207) for a family of periodic orbits of a dynamical system with two degrees of freedom, representating two coupled oscillators. These bifurcations were of different nature than those mentionned above, and it was proved that no quantitative universality exists in this case. The bifurcations occur when the stability changes along a family of periodic orbits, for a system with two degrees of freefom. In a system with three degrees of freedom, the periodic orbits along a family may loose stability through complex instability. Contopoulos (let.Nuovo.Cim.38.257) showed that, at this point, when the bifurcation sequence terminates, a complex instability appears. Heggie (CM in press) studied the bifurcations at this transition point to complex instability and showed that what bifurcates is not a new family of periodic orbit, but a family of invariant tori. This was detected in the general planar three body problem. Contopoulos also showed that in a systems with three degrees of freedom, the bifurcation sequence terminates at an inverse bifurcation.

Integrability. The problem of integrability is important in the study of chaotic and ordered motion. Necessary conditions for the existence of algebraic first integrals in a Hamiltonian system with two degrees of freedom have been studied by Yoshida (CM 31. 363 and 381). He showed that the existence of an algebraic first integral controls the Kovalevskaya exponent characterizing a singularity of the solution and that the appearance of irrational or imaginary exponents proves the non-existence of a sufficient number of first integrals. In this way, the three body problem and the Hénon-Heiles system are proved to be non-integrable. The integrability of a Hamiltonian system with three degrees of freedom at a resonance 1:2:ω with ω:1,2,3,4 was studied by Van der Aa (CM 31.163) by making use of the Birkhoff normal form. He found that, in general, no quadratic or cubic integral exists, but asymptotic integrability occurs for special values of the parameter. A new integrable system was found by Mignard and Hénon (CM 33.239) in their study of the motions of a particle around a planet taking into account the solar radiation pressure.

Chaotic motion. Generation of chaotic motion and its relation to be interaction of resonances is one of the problems of non-linear dynamics. Yokoyama (CM 33.99) studied a non-linear Hamiltonian system where ordered motion reappears as the energy increases and this was explained by the criterion of the interaction of resonances. Yi-Sui Sun (CM 30.7 and 30.111) studied the appearance of ordered or chaotic motion in a three dimensional area conserving mapping. Chaotic motion in the restricted three body problem was found by Hadjidemetriou and Ichtiaroglou (IAU 74.141 and AA 131.20) for the circular case by considering a mapping on a surface of section. Milani and Nobili (AA in press) found that the chaotic regions are more important and escape orbits exist in the elliptic case.

Another method to detect chaotic motion is by the use of the Lyapounov characteristic exponents which allow a precise quantitative definition of the stochasticity of the orbit. Gonczi and Froeschlé (CM 25.271) applied this method to the restricted three-dimensional three body problem, while Huang and Innanen (33.42.001 and 37) discussed the stability of the general planar three body problem, arguing that there exist local additional integrals when the initial elements are situated inside the stable region.

II. DYNAMICAL SYSTEMS WITH THREE OR MORE DEGREES OF FREEDOM

The general behaviour of dynamical systems with three or more degrees of freedom presents properties that do not appear in systems with two degrees of freedom. However, due to the complexity of the problem, papers deal with non-linear systems which are simpler than those which appear in actual problems of Celestial Mechanics, but which show typical features present in the systems with three degrees of freedom in general. For instance, systems of three coupled oscillators near various resonances were studied by Martinet et al (CM 25.93) and Magnenet (CM 28.319). Other types of dynamical systems with three degrees of freedom were studied by Contopoulos and his colleagues (32.151.087 and CM in press).

Very few things have been done for systems with more than three degrees of freedom. Stellmacher (CM 32.23 and 247) studied analytically the existence and the stability of families of periodic orbits in a Hamiltonian system with N degrees of freedom, applying her results to 3 degrees of freedom. Hadjidemetriou and Michalodimitrakis (AA 93.204) have studied the general planar 4 body problem. Several families of periodic orbits and their stability were computed for the actual masses of Jupiter and three Galilean satellites: one periodic orbit, at resonance 1:2:4 is very close to the actual motion and is found to be stable. Grigorelis is presently computing families of periodic orbits of the general 4 body problem with equal masses.

III. PERIODIC ORBITS

In the last three years, new existence proofs for particular types of periodic orbits were given and generalizations of existing proofs were made. Much numerical computations of periodic orbits were also made recently, though the computations of families of periodic orbits of the general three body problem was not as popular as some years ago. In addition, analytic expansions using computerized litteral algebra were obtained. A review on periodic orbits by J. Hadjidemetriou has been prepared (FME in press).

Proofs of existence. Meyer (J.Diff.eq. 39.2 and CM 23.69) proved the existence of periodic orbits of the general n-body problem, starting from a periodic solution of a restricted (n-1)-body problem, generalizing older results on the three body problem. Message (CM 28.107) proved the existence of periodic orbits of the third sort in the general problem of three bodies. Kammeyer (CM 30.329) proved the existence of symmetric periodic orbits in the rectilinear three body problem with the middle mass larger than the others. Belburno (CM 25.195 and 397) proved the existence of periodic orbits in the restricted three-dimensional 3-body problem by continuation of collision orbits.

Analytical methods. Several investigators developed methods to construct analytically a periodic orbit. Presler and Broucke (Comp.M.Apl. 7.451 and 473) and Davoust (CM 31.241 and 293) used the Linstedt method. Wiesel (CM 23.231) gave a formal solution for the motion near a periodic solution, extending Floquet's theory. Stellmacher (CM 28.351) gave an algorithm to construct a nearly circular periodic orbit for the motion of a satellite around an oblate planet. Ding and Tong (37.042.100) generalized the Poincaré theorem of analytic continuation of periodic orbits to the case of multiple parameters and showed that the importance of the oblateness of the primary cannot be neglected in the restricted three body problem.

Numerical methods. Many families of periodic orbits in the general three body problem were computed by Broucke et al (CM 24.63), Markellos (CM 25.3) and Delibaltas (CM 29.191) or in the restricted 3 dimensional problem by Robin (CM 23.97) and Taylor (CM 29.51 and 75). Brown's conjecture concerning the termination of the long period family around L_4 was disproved numirically by Henrard (CM 31.115) and analytically by Garfinkel (SZB in press). Other studies around Lagrangian points were made by Doubochine (CM 33.21) and Gomez and Noguera (CM in press). In addition, periodic orbits of particular interest to space flight were computed by several investigators.

IV. STABILITY

Various concepts and definitions of stability have been proposed, ranging from the stability in the neighborhood of a periodic orbit or an equilibrium position to a global view. Both empirical and rigorous approaches were used. A review of the stability concepts was given by Szebehely (SZB in press).

Stability near a periodic solution. Yoshida (CM 32.73) showed that, in certain cases, the characteristic exponents can be found, and thus the linear stability can be established. Sknol (CM 33.159) studied the motion around an equilibrium point in an autonomous Hamiltonian system with two degrees of freedom near the main resonances. The regions of non-linear stability around the equilibrium points L_4 and L_5 in the restricted three body problem were studied by Szebehely and McKenzie (CM 23.131). The relation between resonance and instability in a planetary system with two or more planets has been studied by Hadjidemetriou (CM 27.305) and the mechanism by which the instability is generated was found.

Global stability. The classical Hill stability appears in the circular restricted three body problem for large values of the Jacobi constant. The Hill type stability is an extension of this kind of stability to the general three body problem. If the product of the energy by the square of the angular momentum C is below some negative function of the masses, the possible domain of evolution in the phase space is divided into three disconnected parts and the motion remains for ever in one of them. There exists then a binary that the third body can neither approach nor disrupt (however, this kind of stability does not prevent the escape of the third mass, but motions of exchange type cannot exist). Many studies have been done in the last ten years on this problem. After the last of a series of papers by Szebehely (CM 22.7) more precise results have been obtained by Markellos and Roy (CM 23.269) who showed that Hill's stability in the circular case is valid for direct satellites but underestimates the actual possible maximum distances from the planet by at least a factor 2. Valsecchi et al (CM 32.217) extended this numerically to elliptic orbits and foundsignificant differences with the circular case. C. Huang extented the notion to non spherical bodies (33.042.070), while Marchal and Bozis (CM 26.311) have generalized the criterion to zero and positive energy. They found the forbidden zones and also that stable systems with negative energy exist. They form a hierarchical configuration (close binary and a distant body) that is not disrupted. It appears thenthat most of the triple stellar systems are Hill stable. On the contrary, Ding and Huang (31.042.120), studying the general three-body, showed that most natural satellites are Hill unstable. Nugeyre and Bouvier (CM 25.51) have also studied hierarchical structures in stellar systems and showed that the choice of suitable coordinates results in simpler formalism for stability.

Much work has recently been done by Roy and his collaborators on the stability of hierarchical systems. Roy (31.042.053, IAU 74.277 and SZB in press), Walker and Roy (CM 24.195, 29.117 and 29.267), Walker (CM 20.149 and 215) and Milani and Nobili (CM 31.213 and 241; BER 219) have applied the empirical parameter method to the problem of the stability of N body hierarchical dynamical systems and are using it to approach the stability of the solar system. Szebehely and Whipple (CM 32.137 and BER 195) studied the stability of the restricted problem of n massive primaries and ν small bodies, by making use of the integrals of motion. A similar method is also used by Szebehely to study the stability of binary asteroids.

V. THE INVERSE PROBLEM

This problem arises when a set of orbits is given and one wants to determine a force field or the potential. The problem was proposed by Szebehely in 1974 who reduced it to be solution of partial differential equations that he established. Several people have worked recently on various aspects of this problem using Cartesian coordinates with two or three degrees of freedom in an inertial or a rotating frame or using generalized coordinates. Szebehely and Broucke (CM 24.23) studied the determination of the potential field from a given family of periodic orbits in a rotating frame. Bozis, in a series of papers (CM 28.367, 29.329, 31.43 and 129, AA 134.360) and Xanthopoulos and Bozis (AA 122.251, IAU 74.253) studied the problem of determining the potential (or force field if the system is not conservative) from a given mono-parametric or two parametric family of orbits in the plane or in three dimensions. Erdi (CM 28.209) generalized Szebehely's equations to three dimensional orbits. Varadi and Erdi (CM 39.395) considered a two parametric family in three dimensions and the corresponding system of partial differential equations for the determination of the potential was studied by methods of differential geometry. Melis and Piras (CM 32.87) studied the determination of the potential generating a given family of orbits in the N dimensional configuration space of a holonomic system. Puel (CM 32.319) showed that Szebehely's equation of the inverse problem is equivalent to a multiple variational problem deduced from Maupertuis' principle. Similarly, Gonzalez-Gascon et al. (CM 33.85) studied the connection of this equation with the equations of Dainelly and Whittaker on similar problems.

E - QUALITATIVE ANALYSIS IN CELESTIAL MECHANICS

(Main contributor : C. Marchal)

In Celestial Mechanics, as in many other domains, the quantitative and the qualitative analyses are complementary. The quantitative analysis gives excellent informations of the future or the past of some particular solutions of interest but is usually unable to give general informations and its accuracy generally decreases and becomes null for large times. On the contrary, the qualitative analysis gives partial, but rigorously demonstrated properties, that are valid for very long periods of time, and even often for infinite time.

The qualitative analysis especially deals with the problems of integrals of motion, symmetries, periodic orbits, final evolutions, structure of a set of solutions, analysis and regularization of singularities, escaping motions, bounded motions, oscillotary motions, asymptotic motions, etc... It has been initiated by

Poincaré, Sundman, Chazy, Khilmi, Sitnikov, Merman, Alekseev, etc... The Kolmogorov-Arnold-Moser theorem can be considered as the result of a qualitative analysis.

The fantastic progress of computers has led to many improvements in the quantitative analysis of the n-body problem. These improvements have disclosed the extreme complexity of the set of different solutions and have given new orientations and a new impetus to the qualitative analysis, especially in the domains of final evolutions and tests of escape. Let us consider the different questions successively.

I. FINAL EVOLUTIONS IN THE N-BODY PROBLEM

The existence of five main types of final evolution are now well established (see Alekseev 30.042.079, Marchal and Saari, J. of diff.eq. 20, p 150).

1. The triple or multiple collision at some time t_0. When t goes to t_0, all n bodies go to a definite final position at a bounded distance, and the bodies going to the collision approach a "central configuration" (Marchal 31.042.056). This type of final evolution is exceptional and has a measure zero in the phase space. Several studies of particular cases of multiple collisions have recently been published. Eschbach (CM 27.157) has analysed the n-tuple collisions in the n-body problem and discussed various cases in function of the energy h of the system. Lacomba and Simó have constructed the linear and rectangular degenerate cases of the four body problem leading to quadruple collisions (CM 28.49).

2. The infinite expansion in a bounded interval of time. This type of final evolution requires $n \geqq 4$ and very strong oscillations. An example was given by Mather and McGehee in 1974. It is conjectured that this type of expansion is infinitely rare.

3. The super-hyperbolic expansion. It is analogous to the previous type of motion and is problably also infinitely rare.

4. The hyperbolic expansion. It is very usual type and corresponds to most of the phase space. The limit of the ratio of the largest mutual distance R to the time t is bounded and non-zero. Masses may cluster in sub-systems in which the energy and the angular momentum have bounded limits.

5. The parabolic or sub-parabolic expansion. All mutual distances remain of the order of $t^{2/3}$ at most and the energy integral h in the axes of the center of masses is negative or zero. If h=0, the expansion is parabolic and non-infinitesimal masses approach a central configuration. If h<0, many possibilities exist: parabolic expansion, sub-parabolic expansion, bounded motions, formation of subsystem (clusters), periodic, asymptotic, oscillating, quasi-periodic, chaotic motions, etc... It is known that this type corresponds to a set of positive measure, but a complete analysis of these cases still remains to be done. Some partial studies were recently done by Lacomba and Simó (CM 28.37) who assigned a boundary (total collision) manifold to each energy surface. In the case h<0, there exists an infinity manifold components which reflect the complexity of the situation encountered when the total energy is negative.

II. REGULARIZATION OF THE SINGULARITIES OF THE N-BODY PROBLEM

The n-body problem has two types of singularities: the collision of two or several bodies and the infinite expansion in a bounded interval of time (see case I.2 above). The regularization of singularities answers to the question: "How to extend naturally a solution after a singularity?". It has a theoretical interest and is also useful for numerical integrations of nearly singular solutions. Two types of regularization exist (McGehee, in Dynamical Systems, theory and Applications, J. Moser ed., p 550, 1975), the analytical (or Siegel's) and the topological or by continuity (Easton's regularization). They are independent and may be different as shown by McGehee in 1975. It is well known that the collision of two bodies is always regularizable in both senses. The infinite expansion is an essential singularity and is never regularizable. It has been found that the possible cases of regularization (Marchal 31.042.056 and Richa, Thèse de 3e cycle, Univ. de Paris 6, "Cas de régularisation des singularités du problème des n corps") are very rare: they may appear only in the analysis of rectilinear motions. In addition, all but perhaps one masses are infinitesimal masses and the latter have particular mass ratio (Simó CM 21.25). However, in other cases of triple collisions, regularized coordinates on plane collision manifold may be found (Waldogel, CM 28.69) and permit numerical studies of the collision.

III. FINAL EVOLUTIONS IN THE THREE BODY PROBLEMS

Ten final evolutions of three non-infinitesimal masses have been identified:

1. The triple collision requiring a zero angular momentum.

2.3.4. The hyperbolic, the hyperbolic-parabolic types that require a positive energy h and the hyperbolic elliptic type for which h may be zero or negative. They present all an hyperbolic expansion.

5. The tri-parabolic expansion with $h=0$.

6. The parabolic elliptic expansion with $h<0$.

7. The bounded type: all distances and velocities remain finite and $h<0$.

8. The Khilmi and Sitnikov oscillating type: $h<0$ and the velocities remain bounded while the largest mutual distance is such that lim sup $R=+\infty$ and lim inf $R=<+\infty$.

9. The second oscillating type: $h<0$. The mutual distances remain bounded, but the largest velocity is not bounded. An infinite number of arbitrarily close approaches occurs.

10. The third oscillating type: it is a combination of types 8 and 9.

Several sub-types cal also be identified.

Presently, it is known that cases 1,3,5 and 6 are infinitely rare. It is conjectured that this is true for cases 8 and 10. The others correspond to sets of positive measure in the phase space. An important consequence of the positive measure of type 9 which presents an infinite number of arbitrarily close passages is that many more collisions can happen in this indirect way than in the direct mode (Marchal 25.151.093). This type of motion could be found in new born triple stellar systems, but not in old triple systems: the latter would already have undergone a collision.

For any non-zero values of the mass ratios, all combinations of original (decreasing time) evolution and final (increasing time) evolution are possible, provided that they correspond to the same sign of energy integral and that some evolutions beginning and/or ending by the type 1 be excluded. The cases corresponding to sets of positive measure in phase space are any combination of types 2 and 4 (with or without exchange) and also the evolutions 7→7, 9→9 and perhaps 8→8, 10→10. New criteria for hyperbolic and hyperbolic-elliptic motions were proposed by Merman (31.042.029) and considerable successive progresses have appeared recently in the domain of the tests of escape (Zare, 1981 (CM 24.345), Laskar, 1982 (unpublished), Laskar and Marchal, 1984 (CM 32.15), Marchal, Yoshida and Sun-Yi-Sui, 1984 (CM 33.193). It is to be noted that the smallest mass is the most unstable and the most subject to be ejected and the present tests seen to approach very near to true limit of escape. However, no general test of bounded motion is known and seems to be difficult to obtain: because of the conservation of measure in phase space, their limit would be a set of solutions. In any case, it appears, from the above mentionned results, that the bounded solutions are much more rare than it was generally expected.

IV. STRUCTURE OF A SET OF SOLUTIONS IN THE THREE BODY PROBLEM - CONJECTURES

The numerous studies made on the general structure of a set of solutions have led to distingich three types or orbits:

A. Periodic and quasi-periodic orbits,
B. Semi-ergodic or chaotic orbits,
C. Open orbits leading to an escape or a collision (evolution 1 to 6 of section III).

There are also some exceptional orbits such as asymptotic orbits. The structure of the phase space is usually presented through some simple area preserving mappings (see also the section on periodic orbits and stability) where similar behaviour exists (Easton 31.021.020). The ergodic throrem being extended to open systems, it can be shown that almost all orbits remain in the same class during the evolution for increasing and decreasing time (Marchal, J. of Diff.Eq. 23, p 387). However, only the orbits of class A and B have the usual strong stability properties. It seems that, as in most time independent Hamiltonian problems, orbits of class A are the backbone of the set of solutions: the chaotic orbits fill densely the bounded holes left by the periodic and quasi-periodic orbits, while the open orbits are found in the unbounded holes. However, it seems that bounded holes of chaotic motions are exceptional and correspond to exceptional cases like the restricted problem. Generally, holes appear to be unbounded and almost all apparently chaotic orbits end by an escape. This "Arnold diffusion conjecture" would mean that the escape orbits of type C are dense everywhere in phase space. This conjecture is to be put together with the still not demonstrated Poincaré conjecture that the periodic orbits are dense in the set of bounded orbits and also the conjecture on the structure of solutions (for non infinitesimal masses, apart from a set of measure zero in phase space, all bounded and oscillating orbits belong to the Arnold torus and are either periodic or quasi-periodic: other orbits have least one escape for both $t \to + \infty$). These three conjectures are in the center of the progress in the analysis of the three body problem. The last conjecture is not true in the restricted three body problem: this can be interpreted as the situation in which one of the masses is infinitesimal may not be representative of the complete problem. Recently, a weakened version of Poincaré's conjecture was proved by Gomez and Llibre (CM 24.325).

J. Kovalevsky
President of the Commission.

8. POSITIONAL ASTRONOMY (ASTRONOMIE DE POSITION)

PRESIDENT: J.A. Hughes VICE PRESIDENT: Y. Requieme

ORGANIZING COMMITTEE: G. Billaud, G. Carrasco, W. Fricke, V.S. Gubanov, E. Høg, B.L. Klock, J.A. Lopez, Luo Ding-jiang, M. Miyamoto, L.V. Morrison, I. Nikoloff, G. Teleki, Y.S. Yatskiv

INTRODUCTION

The president thanks those members contributing material to this report. The volume of the material necessitated some editing, but no substantive omissions occurred. Whenever available, AAA numbers are used in lieu of complete titles of publications to help conserve space.

IAU Symposium No. 109 Astrometric Techniques, was held in Gainesville, Florida in January 1984. Although the Proceedings of that meeting are not now available (January 1985), commission members and other interested parties are urged to secure access to that volume when it appears since so many facets of the commission's work are addressed therein.

With regard to the coming 1985 General Assembly, a joint meeting of Commissions 8 and 24 and a Joint Discussion on Reference Frames involving Commissions 7, 8, 19, 24, 31, 33 and 40 will be scheduled.

The following material is arranged by country in alphabetical order followed by special reports of various groups and committees.

REPORTS FROM OBSERVATORIES AND COMMISSION MEMBERS

ARGENTINA - The Meridian Circle Department of the Felix Aguilar Observatory, San Juan published, 1) The Second San Juan Meridian Circle (SJMC) Catalogue of positions of 617 FKSZ (+30° to -90°) with std. dev., $\varepsilon(\alpha)\cos\delta = \pm^{s}.021$, $\varepsilon(\delta) = \pm 0''.39$; 2) San Juan 72 of positions of 7190 SRS (-40° to -90°) with std. dev. $\varepsilon(\alpha)\cos\delta = \pm 0^{s}.024$, $\varepsilon(\delta) = \pm 0''.46$; 3) The Fourth SJMC Catalogue of positions of 364 FK4 (-30° to -90°) with std. dev., $\varepsilon(\alpha)\cos\delta = \pm 0^{s}.023$, $\varepsilon(\delta) = \pm 0''.33$; 4) The Fifth SJMC of positions of 689 FK4 (+30° to -90°) with std. dev. $\varepsilon(\alpha)\cos\delta = \pm 0.020$, $\varepsilon(\delta) = \pm 0''.31$. The SJMC continues to observe FK4 and has also begun the observation of 54 radio stars. (CARESTIA)

Using both actual and simulated observation of major and minor planets, Branham has concluded that an L_1 solution is competitive with classic least squares. He recommends the initial use of L_1 together with criteria for the treatment of discordant observations, followed by solutions based upon either normal equations or orthogonal transformations. Simulated observations of the Sun, Moon and planets Mercury through Neptune for a period of ten years gauged the precision of equator and equinox determinations. Theoretically, the Moon should contribute strongly, but in practice this is not so. The Sun, Mercury and Venus give the best results, while a fundamental program using minor planets alone is possible. This follows from analysis of results from El Leoncito. (BRANHAM)

CHILE - At the Cerro Calan National Observatory 651 FKSZ and 1217 FK4 stars (+40° to -90°) are being observed by the Repsold Meridian Circle. Observations are made in four zones with 40 FKSZ and 82 FK4 stars (-70° to -90°) observed at both culminations. The intent is to obtain six observations of each star and complete the program in June 1985. A quasi-absolute method will be used to determine RA and DEC. The compilation of RA observations is in progress. Two additional programs involve 367 double stars (-25° to -47°), and 31 radio sources recommended by Commission 24. The latter include 9 FK4 and 2 FK4 Supp. Mars, Jupiter, Saturn, Uranus and Neptune plus minor planets 1-4 have been observed since 1983. In all, a total of 16 233 observations were made by three observers in these programs during the reporting periods.

A grant was received from the Comision Nacional de Investigacion Cientifica y Technologica to improve RA data acquisition; and a grant from the Departmento de Investigacion y Bibliotecas Universidad de Chile to design and construct a machine using a photodiode array to measure divided circle films and thence to install a similar device on the telescope to measure the circle directly. Publications include: Santiago 67 Catalogue 7610 stars (-25° to -47°), Carrasco and Loyola (33.002.064), and Meridian Circle Observations of FK4 Radio Stars, Carrasco, Costa and Loyola (33.041.012). (CARRASCO)

The program of the Danjon Astrolabe of Santiago is a joint project between the University of Chile and the European Southern Observatory (ESO) and has been in progress for the last three years. Observations for a Second Astrolabe Catalogue of 525 FK4 stars were completed during 1984. Four galactic radio sources: FK4 Nos. 306, 309, 616, and 9 Sgr of the list of IAU Commission 24 WG on identification of Radio/Optical Astrometric Sources, have been observed regularly since 1980. Some results were presented at Symposium 109. The Uranus observations campaign has been in progress during 1982-1984; Mars, Jupiter and Saturn have also been observed. The results of observations of Uranus obtained during the period 1980-1982 were published in A. and A. Suppl. Ser. 56, p. 281, 1984. (NOEL)

CHINA, NANJING - At the Beijing Observatory, a catalogue of absolute RA's of stars observed with the photoelectric transit instrument during 1981-1983, and reduced by the Gubanov chain method, is in press. The second photoelectric astrolabe catalogue of 905 stars (1976-1981) was compiled (34.002.070). Observations of more than 500 stars were obtained (during 1981-1984) with the photoelectric astrolabe Mark II. Reductions were started at the beginning of 1984. Observations of radio stars have continued since 1980. Other papers relating to fundamental astrometry are: 30.031.517; 31.041.041; 31.002.059.

In the Nanjing Astronomical Instrument Factory, development of the photoelectric astrolabe, Mark III continues. With the cooperation of Copenhagen University Observatory and Shaanxi Observatory, the construction of Høg's GMC is being done in Denmark and China.

At Shaanxi Observatory, a contribution to the celestial reference frame was presented (32.043.008). Mean positions of 1800 stars for the beginning of the years from 1051 BC to 2050 AD have been computed and published (Publ. of Shaanxi Observatory 6, 1983, No. 2,55). A photoelectric astrolabe catalogue of 820 stars has been published (34.041.041). Sixteen radio stars have been observed since 1981.

At Shanghai Observatory, observations of Mars, Jupiter and Saturn with the Danjon astrolabe in 1964-78 were published (33.041.041; 33.041.043). A synthetic catalogue of stars observed with the Danjon astrolabe, and proper motions of individual stars were presented (33.041.042). Five optical radio positions have been determined (33.041.044). A photoelectric astrolabe catalogue of 537 stars (1979-81), and observations of 15 radio stars with astrolabes have been published in the Annals of Shanghai Observatory, No. 4, 1982. For the study of the coordinate frame, photographic observation of radio sources and observations of solar system objects with astrolabes are under consideration.

At Tianjin Latitude Station, a declination catalogue of 99 zenith stars observed with the zenith telescope ZTL-180 has been presented (34.002.071). At Wuchang Time Observatory corrections for FK4 star positions were derived from observations by the Danjon astrolabe in 1965-1975 (Acta Geodaetica et Geophisica, No. 3, 1981, 27), and RA corrections for 464 stars were obtained with the photoelectric transit instrument in 1976-1979 (34.041.046).

At Yunnan Observatory, a method for the absolute determination of the azimuth of transit instruments at low-latitudes was presented (34.032.048). Five papers appeared in the Publ. of Yunnan Observatory, 1982, No. 2. A photoelectric astrolabe catalogue of 293 stars and a paper on astrolabe catalogues were in the same publication.

A catalogue of 1579 stars from photoelectric astrolabes in China (GCPA) (34.041.044; 34.002.069) was compiled from astrolabe catalogues from Beijing, Shaanxi, Shanghai, Yunnan and Wuchang. A detailed summary of GCPA was given at the meeting of Commission 8 during the General Assembly XVIII IAU, 1982. A machine readable form of GCPA was sent to Heidelberg.

Papers presented at the IAU Symposium No. 109 included: Astrometry in China; A new method of determining absolute azimuth and latitude; The consideration of a new type of meridian circle; Application of vacuum chambers in astrometric instruments; Design of a large transit circle with reflecting optics; Determination of the equatorial plane of astrolabe catalogues by observations in the prime vertical; Linkage of radio and optical coordinate reference frames; On the implementation of absolute meridian observations at low- latitude stations; Photoelectric astrolabes and astrolabe star catalogues; The 1.56 meter astrometric telescope of Shanghai Observatory and the observing program.

DENMARK AND GREAT BRITAIN - The Carlsberg Automatic Transit Circle carried out two test programmes of observation from August 1981 to March 1983 at Brorfelde. The instrument has a photoelectric moving-slit micrometer, automatic setting and a circle reading system under computer control. Two catalogues of star positions and magnitudes resulting from these tests were published and discussed at IAU Symposium 109. The internal mean errors of a single observation are $0\overset{s}{.}014$ in right ascension and $0\overset{''}{.}22$ in declination.

Following this test period, the telescope was shipped to Spanish Observatorio del Roque de los Muchachos del Instituto de Astrofisica de Canarias situated on the island of La Palma at an altitude of 2400m. By December 1983 the telescope was assembled. Its building is a light-weight structure made of panels of wood and compressed glasswool with a steel framework to give it rigidity. A run-off section to the west gives a 2m opening on the meridian. This structure provides a stable thermal environment for the telescope. The telescope was linked to the two minicomputers in February 1984. It was aligned and the pivot and circle division errors measured during March and April. Regular observations began in May 1984, with almost 18 000 observations made by August. The CATC is operated jointly by Copenhagen University Observatory, Royal Greenwich Observatory and the Instituto y Observatorio de Marina, San Fernando.

The programme comprises the following, with a limiting declination of -56° (in descending order of priority): major planets, 23 minor planets, FK4, IRS, faint reference stars in fields of some 250 radio sources, radio stars, NPZT + Stromlo, Gliese catalogue, AGK3, and SAO. The usual scan time of a star is about 15 s. This is increased fourfold for the faint reference stars in benchmark fields. The limiting magnitude is near 13.0 in Johnson V. The average internal rms error within 50° of the zenith is close to $0\overset{''}{.}20$ in right ascension and declination. It is $0\overset{m}{.}05$ in magnitude.

The Third Greenwich Catalogue of Stars, Sun, Planets and Moon for 1950.0 was published in 1983 as R. Gr. Obs. Bulletin No. 187. It contains results of observations made at Greenwich with the Airy Transit Circle from 1942 to 1954.

The First, Second and Third Herstmonceux Catalogues of Stars for 1950.0 were published in 1983 as R. Gr. Obs. Bulletin No. 189. These catalogues contain the results of observations of stars made at the Royal Greenwich Observatory (Herstmonceux) with the Cooke Transit Circle from 1957 to 1980. This instrument was taken out of operation in 1982. (MORRISON)

References for Denmark and Great Britain

Clausen, J.V., L. Helmer, L.V. Morrison, C.A. Murray, A. Orte, and L. Quijano - 1982, The Carlsberg Automatic Transit Circle on La Palma, The Observatory, 102, 9-20.

Helmer, L., C. Fabricius, O. H. Einicke and C. Thoburn - 1983, Meridian Observations made with the Carlsberg Automatic Meridian Circle at Brofeld (Copenhagen University Observatory) 1981-1982, Astron. Astrophys. Suppl. Ser. 53, 223-245.

Helmer, L., C. Fabricius, O.H. Einicke, C. Thoburn and L.V. Morrison - 1984, Meridian observations made with the Carlsberg Automatic Meridian Circle at Brorfeld (Copenhagen University Observatory) 1982-1983, Astron. Astrophys. Suppl. Ser. 55, 87-102.

Tucker, R.H., M.E. Buontempo, P. Gibbs and R.H.D. Swifte - 1983, Third Greenwich Catalogue of Stars, Sun, Planets and Moon for 1950.0, R. Greenwich Obs. Bull. 187.

Tucker, R.H., M.E. Buontempo, P. Gibbs and R.H.D. Swifte - 1983, First, Second and Third Herstmonceux Catalogues of Stars for 1950.0, R. Greenwich Obs. Bull. 189.

FRANCE - The automation of the Bordeaux photoelectric meridian circle was completed in 1982 with accurate automatic setting. Two determinations of the division errors by the Benevides-Boczko method were carried out in December 1982 and in March 1983: the std. dev. between the two sets of corrections is about $\pm0''.015$. The reobservation of the NPZT programme has been completed (11 600 positions, m.s.e. = $\pm0''.11$ in α and $\pm0''.16$ in δ). This new catalogue and the individual FK4 star deviations are available on magnetic tape and have been sent to the A.R.I. at Heidelberg for inclusion in the FK5 catalogue. Preliminary lists of faint HIPPARCOS stars with no astrometric history have been regularly observed. However, the observation rate (which is usually 15 000 positions/year) is drastically reduced by the very bad positions given in these lists of astrophysical stars. The automatic meridian circle has also been used for observations of Mars, Uranus and Neptune, Galilean satellites and the Saturnian satellites, Titan and Iapetus. Furthermore, 700 positions of 55 minor planets from the HIPPARCOS program have been obtained.

During the next three years, the instrument will be mainly devoted to the preparation of the HIPPARCOS Input catalogue (about 7000 northern faint stars). (REQUIEME)

Observations of the sun continues at CERGA with the solar astrolabe (31.041.29; 28.032.015). The analysis of the data involves not only the coordinates (equinox correction), but also the diameter (fluctuations of $0''.5\pm0''.5$ in a period of 800-900 days); (31.041.035; 34.041.009). Preliminary observations of solar positions in June-July, 1984 gave a dispersion of order $0''.4$ for one observation. An improved mounting is being installed.

Analysis of observations of Mars (29.097.014; 30.097.036; 31.097.114) caused Chollet (31.097.001) to propose replacing the empirical correction for "phase effect" by one deduced from a model of the distribution of luminous intensity on the planet's surface (in press, Astron. and Astrophys.).

For these observations, and for stars, the HIPPARCOS catalog should permit reduction to a system free of internal error, and for which the residual rotation may be estimated from solar observations (31.043.019). Reduction to other reference systems, and finally to the FK4 (or FK5) will be facilitated by bright radio stars (28.041.078). Finally, possible other uses of astrolabes have been the object of a synthesis study. (CHOLLET)

During the last three years the CERGA astrolabe team has been mainly devoted to perfecting a strictly impersonal astrolabe. Observations of 188 groups of 40 stars each were made from July 1983 through June 1984. The results will soon be published. Comparison with B.I.H. results indicate a dispersion oscillating between 1.9 to 5.6 ms. (BILLAUD)

FEDERAL REPUBLIC OF GERMANY - At Heidelberg about 250 catalogues of observation thought to be suitable for contributing to the construction of the FK5 were placed on magnetic tape. For the improvement of the system of the FK4 about 40 catalogues with absolute or fundamental observations turned out to be qualified. For the determination of the equinox of the FK5 including the elimination of the motion of the equinox in the FK4, Fricke (31.041.002) has employed observations of the Sun and of planets from 1900 to 1977 and lunar occultations including the results of 42 authorities. For differential corrections to the FK4 about 80 catalogues were analyzed. The determination of the differences Cat-FK4 have been computed in all cases by other methods developed by Bien et al. (25.041.067). For the construction of a system of positions and proper motions Schwan (34.041.052) has compiled a pilot program which has been tested on the systematic differences between the FK4 and the Washington Six-inch Transit Circle catalogues from 1910 to 1975. The systematic differences between these catalogues and the FK4 were given by Schwan. Furthermore, the differences between Greenwich and the FK4 were determined for the catalogues from 1915 to 1981 by Bien (34.041.047), the differences Cape minus FK4 from 1918 to 1974 were determined by Jahreib (34.041.048), and the differences

Brorfelde minus FK4 from 1964 to 1976 by Roser and Fricke (31.041.004). The results of the analysis of other catalogues of observation will be published later. It is expected that the construction of the system of the FK5 can be completed in 1985 and the determination of the individual corrections to FK4 positions and proper motions in 1986. By that time the extension of the FK5 to fainter stars should be completed also. (FRICKE)

Additional work at Heidelberg included the compilation and examination of radio and optical astrometric data for radio stars in order to help determine the suitability of the latter for relating the HIPPARCOS and radio reference systems (31.041.038). Within the framework of HIPPARCOS data reduction the concept of the determination of astrometric parameters was described (34.036.229), (4.036.231). A data base containing meridian circle observations of nearly 50 000 stars was established by Hering and Walter with the objective of extending the FK4 to fainter stars (32.041.017). (WALTER)

At Hannover, Pilowski continued his study in astrometry. Due to high internal precision and considering systematic effects, Zenith Cameras offer excellent possibilities for fundamental optical observations. See Astron. Station, Universitat Hannover, Nos. 13 and 15, 1981 and 1983 and the Festschrift 100 Jahre Geodasie in Hannover, 1981. (PILOWSKI)

Observations of fundamental stars and major and minor planets have continued at the Munich Observatory using the vertical circle. Positions of multiple stars between +70° and +90° were measured to derive positions with sufficient accuracy for HIPPARCOS. (SCHMEIDLER)

At Hamburg the determination of precise optical positions of extragalactic sources and radio stars continues. Also continued is the work with the MERIT-PZT catalog +52°.5. The CPC2 catalog of the Cape Zone -40° to -52°, is finished and has been given to CDS-Strassbourg (Nicholson et al. MNRAS 208, 911; 1984). Extensive work has commenced for the HIPPARCOS Input Catalog, especially for the the extragalactic reference link (Subgroup 2130), and special objects, e.g., star clusters. (DE VEGT)

ITALY - Observations of FK4 and catalog stars were made with the Danjon astrolabe in the period 1982-84. Research continued to deduce the diurnal and anomalous refraction using observations of the solar disk. Publications by Poma, Proverbio and Mancuso appear in Sun and Planetary Systems, Reidel, 1982 pp. 477-82, and by Proverbio and Uras in Nota Tecnica, SAIL, 1984 (in press). PZT reference star lists were developed for observers on the international parallel. See, Proverbio, Pub., SAIL (in press). (PROVERBIO)

JAPAN - At the Tokyo Observatory, Prof. H. Yasuda retired as the director of the meridian astronomy division in April 1982. His successor is Prof. M. Miyamoto.

The Tokyo automated photoelectric meridian circle (TPMC), manufactured by Carl Zeiss Oberkochen, was installed at Mitaka in September 1982. The instrument was then investigated. Over 100 FK4 stars were observed in excess of ten times each in ideal weather, and preliminary results have shown average internal accuracies for a single observation of $\pm0^s.008$ in RA and $\pm0''.13$ in DEC. These values coincide closely with the limiting values expected from image motion (Cf. Høg, 1968). Corrections for the graduated circle of the TPMC have been determined four times, twice within two weeks (end of May, beginning of June, 1984). Results show that the circle corrections are determined within an accuracy of $\pm0''.01$.

In order to detect and correct for the so-called anomalous (local) refraction, which occasionally reaches about 0''.1, and is caused by an inhomogeneous temperature distribution in the pavilion, more than 40 thermistors have been installed around the TPMC. They are located on lattice points separated by 3m, and have an accuracy of about $\pm0.05^\circ$C. A correction for anomalous refraction is applied to each 2 minute observation made with the TPMC. Routine TPMC observations of stars to the 13th magnitude, the sun and planets will commence in 1985. Since 1982 the Gautier MC has not been active. (MIYAMOTO)

At the Mizusawa International Latitude Observatory (MILO) Fujii and Murakami (1982) derived position corrections for the MILO PZT star list using data from the new PZT. Sato et al.

(1984) compiled a PZT star catalog for the Mizusawa, Washington and ILS PZT chain to be used as a standard in recalculating past observations and for initial observations by the ILS instruments. Positions of stars observed by PZT's were corrected internally and reduced to the MC3 and MC3T systems referenced to the FK4 system.

Yokoyama (1984) presented a method of reducing optical astrometric observations of earth orientation parameters at individual stations and at international centers in accordance with the FK5 and the IAU (1976) System of Astronomical Constants. This method was presented at the 2nd MERIT workshop and was approved as the MERIT standard. Yokoyama and Tanikawa produced a FORTRAN program for computing apparent places based upon the new constants and sent it to all countries where optical astrometric instruments operate.

Aoki et al. (1984) presented a method of converting from the old to the new reference frame in order to establish the positions of radio sources. Sato et al. (1984) showed that internal corrections of star positions used for earth orientation parameters are related to the adopted nutation and they proposed the re-reduction of past observations in the new system.

References for Japan

Aoki, S., M. Soma, K. Niimi, M. Fujishita, Y. Takahashi - 1984, Proc. IAU Symposium No. 109 (in press), Astrometric Techniques, The Conversion from the B1950 FK4-based Position to the J200 Position of Celestial Object.

Fujii, S., and G. Murakama - 1982, Poc. Int. Latit. Obs. Mizusawa, 21, 1, On the Declination Error of the Mizusawa PZT Star Catalogue.

Sato, K., K. Yokoyama, K. Tanikawa and Y. Goto - 1984, Proc. IAU Symposium 109 (in press), Astrometric Techniques, The Observations of the Earth Rotation and Stellar System.

Sato K., H. Kitago, Y. Goto and D.D. McCarthy - 1984, Publ. Int. Latit. Obs. Mizusawa, 18 (in press), A Compiled Catalog of PZT Stars at Mizusawa and Washington.

Yokoyama, K., - 1984, Publ. Int. Latit. Obs. Mizsawa 17, 47, Guidelines for the Reduction of Optical Astrometry Observations during the Main Campaign in Project MERIT.

Fukaya, R., and M. Yoshizawa, A numerical Integration of the Astronomical Refraction, Publications of the Astronomical Society of Japan (in press).

Yoshizawa, M., and M. Miyamoto, The Design of Modern Meridian Circles for the Observation of Faint Objects, Proceedings, IAU Sym. 109 (in press).

Miyamoto, M., and M. Yoshizawa, A New Method for Zenith Distance Determination in Meridian Observations, Proecedings IAU Sym. 109 (in press).

Yasuda, H., et al., Meridian Observations of Major and Minor Planets (1977-79), Tokyo Astro. Bull., 2nd Series, No. 270, 1983.

Hara, H., et al., Meridian Observations of the Moon, 1963-77, Tokyo Astron. Bull., 2nd Series, No. 268, 1983.

Miyamoto, M., and H. Ishii, A Scheme for Determining Division Corrections for a Photoelectric Meridian Circle, Publ. Astron. Soc. Japan, 34, 1982.

Miyamoto, M., and C. Kühne, An accurate derivation of the division corrections in a photoelectric meridian circle, Astron. Astrophys. Suppl. Ser., 50, 1982.

Yoshizawa, M., and H. Yasuda, PMC 190: A New Photoelectric Meridian Circle of Tokyo Astronomical Observatory, Parts I and II, Ann. Tokyo Astron. Obs., 2nd Series, XVIII, No. 4 (1982).

Yasuda, H., et al., Northern PZT Stars Catalog (NPZT 74), Ann. Tokyo Astron. Obs., 2nd Series, XVIII, No. 4.

ROMANIA - The meridian team of the Bucharest Observatory completed the reduction of the observations for the Bucharest Stellar Catalogue of NPZT Stars for the zones $+41^{\circ}$ to $+63^{\circ}$, and for $+39^{\circ}$. The material was prepared for publication and sent to interested astronomical observatories. Observations of reference stars for the Prague PZT were completed in May 1983. The reduction of these observations will be completed at the end of 1984. (TOMA)

SPAIN - Observations for the fifth catalog of the San Fernando Astrolabe (CASFS), and of 14 radio sources from the Commission 24 Working Group list continued through 10 May 1983. The astrolabe will be converted to a full pupil instrument. Difficulties with the micrometer drive have hindered observations, but repairs are in progress.

The San Fernando Meridian Circle observed transits of the planets; Mars through Neptune in addition to stars. Observations of 38 radio sources from the first list of the C24WG were completed and a provisional catalog of positions in the FK4 system was formed and will be published. A campaign to observe 55 radio sources began in September 1982.

Cooperation with the CUO-RGO project for the CATC on La Palma island continues. Two observers from San Fernando have been assigned to the CATC. Pivot errors and their effect on instrumental constants were determined. The cooperation includes the calculation of astronomical refraction. A prepared selection of faint stars in 103 southern radio source fields is to be observed by the CATC. Star positions have been measured on transparencies of POSS and SERC plates and on plates from the Gautier Astrograph.

The "Carte du Ciel" astrograph is observing 20 minor planets of the ITA program and, beginning in 1983, 55 minor planets for the HIPPARCOS project. (QUIJANO, SANCHEZ)

USA - At the Univ. of Florida, Eichhorn published an outline of the theory of systems of standards (1982; 32.021.006). He defines a "system" as a set of estimates, and discusses the theoretical difficulties in stating that a particular set of estimates is on a particular system. With co-authors Googe, Lukac and Murphy, a catalog of 20 457 stars was published (1983; 33.041.004) with a std. error about 0".15 in each coordinate, in the zone -48° to -54°. (EICHHORN)

Observations for the Six-inch Transit Circle W6(50) Washington catalog were completed at the U.S. Naval Observatory (USNO). The W5(50) of this series was published (Hughes and Scott) in 1982. Other published catalogs include: Precise Positions in the FK4 System for 120 Radio Source Reference Stars (Dick and Holdenried, AJ, 87, 1982, 1374-78) and, Precise Positions in the FK4 System for Ten Optical Counterparts of Extragalactic Radio Sources (Harrington, et al., AJ, 88, 1983, 1376). Smith made the selection of the IRS supplemental list. His work on the SRS with Jackson is covered in the appropriate committee report. Corbin selected possible candidates for the faintward extension of the FK5, and worked on the proper motions of the SRS. He presented a paper, The Determination of Fundamental Proper Motions at Sym. 111. The discussion included absolute, quasi-absolute and differential TC observations and the compilation of fundamental positions and proper motions from these data.

The Seven-inch Transit Circle was shipped to New Zealand in early 1984 and the Black Birch Astrometric Observatory (BBAO) was thus established. Observations should commence in early 1985; the eight-inch twin astrograph will be sent to BBAO in late 1985.

Proposals were submitted to ESA regarding observations of IRS, IRS Supp., NPZT and FFC stars by HIPPARCOS. Hughes and Smith collaborate with INCA.

A CCD array scanner is being constructed to read divided circles. It is anticipated that such a device will be used on all USNO TC's. The image dissector micrometer has performed well in its initial tests, but full evaluation awaits observations made at BBAO. The transit times of solar system objects will be determined in TAI as well as in LST. Precise siderostats and other components are being constructed in the USNO instrument shop for use on a 20m baseline, wide-band, optical interferometer to be installed at the Mt. Wilson Observatory. This work is in collaboration with the Smithsonian Astrophysical Observatory, Massachusetts Institute of Technology and the Naval Research Laboratory. A dual frequency laser interferometer has been set up at USNO to monitor the motion of settled concrete piers and its effect upon baseline stability. Stone published a detailed study of color effects in transit circle observations (Astron. Astrophys. 138, 2, 1984). The use of "two color" observations with the image dissector micrometer was discussed.

At University of California, Berkeley, Townes and his group are making good progress with the portable long baseline infrared (11 microns) interferometer. (HUGHES)

USSR - Present and future astrometry, especially inertial and other coordinate systems were considered by Yatskiv (34.041.010; 34.041.049). Polozhentsev and Yatskiv proposed ways of weighting catalogues for proper motions of faint fundamentals (FKSZ). Centennial proper motions were derived with errors, S(μ) = ±0^s.162 and S(μ') = ±0".323. (YATSKIV)

The Sun and major planets were observed with the Wanschaff vertical circle at the Main Astronomical Observatory of the Academy of Sciences (MAOAS Ukr. SSR). Observation of an absolute catalogue of 2000 fundamental stars, bright and faint continues. Differential catalogues of 35 bright radio sources, 100 Mikhailov equatorial stars, and 254 reference stars for extragalactic radio sources were derived. Observations at zenith distances ≤86° checked the refraction tables. (KHARIN, MINYAILO, LASORENICO, NENAKHOVA)

A discussion of 40 000 observations of the Sun and major planets made by various instruments in 1960-80 to estimate the various precisions was completed. (KHARIN, MINYAILO, VORONKEVICH)

Current and future determinations of the orientation of coordinate systems, and the periodic errors of fundamental catalogues were discussed using optical observations. Recommendations and estimates of the correction to the equinox including its motion were given. Instruments for observing high-orbit satellites with errors of 0".3 to 0".5 have been designed. (DUMA)

Various southern catalogues were completed at MAOAS, Pulkovo, including: Absolute right ascensions of 2256 Blacklund-Hough stars (+32° to -90°) observed in 1928-42 at Melbourne Observatory (MO) [1]; absolute right ascensions of 144 latitude stars of the Kimura list observed in 1933-35 at MO [2]; absolute right ascensions of 35 stars from the Kopff list (+32° to -90°) observed in 1928-40 at MO [3]; absolute right ascensions of 1960 bright (FK4) and faint southern stars observed in 1969-73 at Cerro Calan Observatory (CCO) [4]; declinations of 5512 SRS, 829 BS, 513 DS, 280 FK4 stars (+47° to -90°) observed in 1963-68 at CCO [5,6].

Catalogues of bright and faint fundamental stars include: a compiled catalogue of fundamental faint stars PFKSZ-2 (-20° to +90°) [7] (MAO AS Ukr. SSR, Goloseevo); right ascensions of 567 FKSZ stars observed in 1954-55 in Pulkovo [8] the MAO AS USSR; right ascensions of 586 FKSZ stars (-20° to +90°) observed in 1974-76 [9] at Nikolaev; right ascensions of 433 FK4 stars observed in 1971-73 and of 376 FK4 stars observed in 1976-78 (-20° to +90°) [10,11] at the Astronomical Institute of the Uzbek Academy of Sciences (AI AS Uzb. SSR).

Other catalogues: right ascensions of 1244 PZT stars (+25° to +56°) observed in 1968-70 [12] and declinations of 507 stars of Pulkovo latitude programs (starting from 1904) of the Pulkovo zenith telescope ZTF-135 [13]. From the Sternberg Astronomical Institute: compiled right ascensions of 187 circumpolar stars using over 100 catalogues published in 1790-1975; using Moscow observations, absolute right ascensions of 187 circumpolar stars observed in 1972-76 [14,15]; declinations of 200 bright stars (+70° to +90°) and 51 Cepheids (-7° to +70°) observed in 1974-77 [16]; declinations of 529 double stars and high luminocity stars observed in 1981-82 [17]; declinations of 3695 stars of the latitude programs for zenith telescopes and PZT's observed in 1965-75 [18]; declinations of 426 zenith zone stars observed in 1970-77 [19]. A catalogue of 9580 zodiacal stars within 15° of the ecliptic observed in 1969-72 at Nikolaev [20] and positions of 5976 SRS and 727 BS stars (0° to -20°) also at Nikolaev [21] were prepared at the Nikolaev Department of the MAO AS USSR. From the V.P. Engelhardt Astronomical Observatory (Kazan): a compilation of positions of 10946 reference KCZ stars (-5° to +90°) based on the improvement of the AGKSR catalogue using faint star observations at observatories in Bucharest, Kiev, Moscow and Kazan [22,23]; a compiled catalogue of declinations and proper motions of 75 stars of the latitude programs of the AOE and the Poltava gravimetric observatory of the Ukr. SSR AS [24,25]. MAO AS Ukr. SSR prepared declinations of 254 stars in the neighborhood of 72 extragalactic radio sources observed in 1979-82 in Goloseevo [26,27]. Positions of red variable stars observed in Odessa [28] were prepared at Odessa University; declinations of 1407 circumpolar stars and differential declinations of 75 circumpolar FK4 stars [29,31] were prepared at Kharkov University.

Observations of the Sun and major planets were carried out at Nikolaev and lately at the AI AS Uzb. SSR. Reductions and analysis were undertaken at the MAO AS Ukr. SSR [32-39]. Studies of the orientation and zero points of fundamental systems were done at the MAO AS Ukr. SSR [40,42,43] and at the Institute of Theoretical Astronomy of the USSR Academy of Sciences (Leningrad) [41,44-46]. Studies of new observational program methods and reduction procedures were concentrated at the astronomical institutions of Moscow and Leningrad. Work by astrometrists from Pulkovo [46-49], Moscow [50,51] and others [52,53] are included.

Work on automating instruments and the observational process [54-62] done at the Sternberg Institute, MAO AS USSR and its Nikolaev Department, involve new observational programs as well as instrumentation, e.g., the design and construction of automatic meridian instruments, and moving instruments to sites with good observing conditions. In 1981, for the first time in the USSR, a stationary meridian instrument was installed at a high altitude [65,66]. This followed studies developing criteria for the choice of a site for an astrometric instrument [63,65]. The first observational results obtained with the Repsold meridian circle of the Sternberg Institute High Altitude Middle Asia Observatory have been published [67]. (SHAMAEV)

References for USSR

Kharin, A. - 1981, An analysis of 9-year sequences of planetary and solar observations with the Wanschaff vertical circle.

Kharin, A. - 1981, Organizing problems of positional observations of major planets by means of astrometry. In the book, "Opredelenie coordinat nebesnych tel." Riga, Latvinian University, pp. 113-121.

Kharin, A. - 1983, On the refraction anomalies from meridian observations. In the book "Refraccia opticheskich voln v atmosfere." Tomsk, pp. 174-184.

Minyailo, N. - 1981, Catalogue of declinations of 100 equatorial stars. Astrometrija i Astrofisika, Kiev, 43, pp. 84-88.

Lasorenko, P. - 1982, Choice of 315 stars in 87 areas with extragalactic radio sources for meridian observations in the FK4 system. Astrometria i Astrofisika, Kiev, 46, pp. 73-79.

Duma, D. - 1982, A problem of the determination of the coordinate system orientation and of periodic errors of the star catalogues from optical observations. Astron. Zh., tom 59, vyp. 5, p. 1012-1019. In Russian.

Duma, D. - 1984, Determination of high-precision coordinates of satellites by the photographic method, Astron., Zh, tom 61, vyp. 1 p. 184-190. In Russian.

Note for numbered references ([1] to [67]) for USSR

Since the latter portion of the information from the USSR was unfortunately received by the President after the commission material had been delivered to the General Secretary on 31 January 1985, it proved impossible to accommodate the numbered references in the space allotted to the commission, although re-editing and revision did finally make it possible to include the additional text. The reference list is available upon request by post from the President, or finally, at the XIX General Assembly.

YUGOSLAVIA - Finishing the observation of 308 bright stars (+70° to +90°) with the Belgrade Vertical Circle, Bozhichkovich and Trajkovska started observations of the Sun, planets and FK4 stars in 1983. Teleki continued his refraction research. Atanachkovich began investigations of the effect of atmospheric turbulence on astrometric observations. (TELEKI)

Sadzakov and Dacic finished the observations of 1750 Double Stars and 523 FK4 stars, with the Belgrade Meridian Circle. Positions will be published in Publ. Obs. Astron. Belgrade (POAB). They also finished observations of 315 stars in 87 planes with extragalactic radio sources. The positions will be published in POAB. Sadzakov and Dacic regularly observe the Sun, Mercury, Venus and Mars with the Meridian Circle. Results are published in Bull. OAB. (SADZAKOV)

The vacuum meridian marks were fully used for the first time for observations (completed early in 1983) of 308 bright polar stars (+65° to +90°) made by the Belgrade Large Transit Inst. for absolute right ascensions. Reductions will be finished by the end of the year in Pulkovo. The std. dev. of a reading of a vacuum meridian mark is $0^s.005$. (PAKVOR)

References for Yugoslavia

Kolaczek and Teleki discussed reference systems (29.045.024). Sevarlic, Teleki and Knezevic published bibliographies on photographic catalogues, stellar atlases, etc. (34.002.061, 34.002.062). Teleki (31.082.078), and Teleki and Saastamoinen (31.082.074), reported on problems with refraction and the former published material on historical astronomical activity (31.004.090) and developments in Yugoslavia (33.013.095). Teleki and Atanackovich wrote on refraction (33.082.074); while Mijatov, Bozhichkovich, and Trajkovska published investigations of the Belgrade Verticle Circle. Sadzakov published several papers concerning catalogues of NPZT stars (29.041.044), (31.041.026, (32.041.016), (33.041.009), (33.041.017), (33.041.045), (33.041.052), (27.041.031). Additional reports by the same author on solar and planetary observations appear in (31.041.036), (32.041.012), (33.041.053), (34.041.034), while work involving catalogue comparisons (e.g., AGK3, latitude star observations, etc.) and proper motions is found in (32.041.024), (27.111.023). Other publications include (33.013.009) and (32.014.018).

SPECIAL REPORTS

HIPPARCOS-TYCHO PROJECT

In 1984 the project entered phase C/D, i.e., the final design stage and hardware construction with launch scheduled for 1988 for a 2.5 yr. mission. The HIPPARCOS and TYCHO catalogues are expected in 1993. Expected accuracies are detailed in earlier reports. Details of observing lists are given in the following. Four Consortia are responsible for producing the catalogues: INCA will form the Input Catalogue (IC) based on over 200 proposals from the international scientific community and on priorities assigned by the ESA Selection Committee; FAST, and NDAC, will analyze data from the main mission, working independently, but comparing data and results during the development and execution of the analysis to ensure producing a single, unique HIPPARCOS catalogue; TDAC will produce the TYCHO catalogue. (HØG)

The **INCA** (INput CAtalogue) Consortium involves institutes in Belgium, France, Western Germany, Netherlands, Spain, Switzerland, the UK and USA. Task leaders are: Turon (Team leader, Paris-Meudon), Gomez (Paris-Meudon, proposal management and successive versions of the IC), Fricke (Heidelberg, astrometric data), Requieme (Bordeaux, astrometric observations), Egret (Strasbourg, photometric data), Grenon (Geneve, photometric observations), Creze (Besancon, mission simulation). Working groups coordinate activities related to double stars (Dommanget, Bruxelles), variable stars (Mennessier, Montpellier), minor planets (Bec-Borsenberger, Paris), and the link to an extragalactic reference frame (Argue, Cambridge). Reports are in the Bulletin d'Information du Centre de Donnees Stellaires (Strasbourg).

Due to constraints resulting from the basic principles of the satellite (two fields of view, continuous scanning, etc.) stars must be carefully selected in advance, evenly distributed over the sky, respect a given magnitude repartition (most brighter than 11), and their positions and magnitudes must be known with a precision of $\pm1''.5$ and $\pm0^m.5$ or better. Moreover, as many stars as possible rated "high-priority" stars by the ESA Selection Committee should be included in the IC. In order to optimize the IC with respect to the satellite's observing possibilities, and the scientific return of the mission, it will be improved by successive iterations after numerical simulations of the mission. Details are given in IAU Symposium No. 109 and the Codata Int. Course on Data Handling in Astron. and Astrophys. (Trieste, July 1984, Crifo et al.) The IC will contain two basic parts: a "survey" including about 60 000 stars complete to about 1.5 stars/square degree (limiting magnitude ranging from V=7.7 in the galactic plane to V=8.7 towards the poles), and a selection of about 40 000 fainter stars chosen for their specific astronomical and/or astrophysical interest.

About 20 000 stars need astrometric measurements to obtain precise positions; new northern observations will come from automatic Meridian Circles (mainly Bordeaux); southern measurements will be made on Schmidt Sky Survey plates (mainly ESO Survey). Cooperation with the Space Telescope Astrometry Team and with the VLBI JPL observers will select the objects best suited for linking the HIPPARCOS system to an extragalactic reference frame. (TURON)

The **FAST** (Fundamental Astronomy by Space Techniques) Consortium for the reduction of HIPPARCOS data includes 17 teams from 5 countries. It started in 1981 under Kovalevsky. The work is divided into 14 tasks. The main reduction chain involves 6 teams: 1) Satellite/attitude determination using the star-mapper data (Donati, Torino); 2) Computation of image positions on the grid from observed photon counts (Donati, Torino); 3) Determination of the abcissae on a fixed great circle of all stars observed during 6 to 12 consecutive hours (Van Daalen, Delft); 4) Synthesis of all observations and determination of the reference system (Galligani, Bologna); 5) Computation of the positions, proper motions and parallaxes (Walter, Heidelberg); and 6) Construction of the software system controlling the reduction (J-L. Pieplu, Toulouse). For all these tasks, studies have been conducted and mock-up programs have been written and are being verified. The methods are described in the report of the FAST Thinkshop held in Asiago in May 1983 (P.L. Bernacca, ed., published by the Padova University).

In addition to the main tasks, a simulation of the mission has been programmed (Froeschle, Grasse). Other tasks underway include: 1) Reduction of double star data (Delaney, Bari); 2) Reduction of minor planet data (Morando, Paris); 3) Reduction of photometric data (Wesselius, Groningen); 4) Modelling of the optics and calibration (J-Y. Le Gall, Marseille); 5) First look examination of data (Schrijver, Utrecht); 6) Connection with FK5 system (Roser, Heidelberg); 7) Connection with the VLBI extragalactic system of reference (Preston, Pasadena). (KOVALEVSKY)

The **NDAC** (Northern Data Analysis Consortium) includes 15 members from institutes in Denmark, the UK and Sweden. Task leaders are: Høg (chairman, Copenhagen); Cruise (Holmbury St. Mary, London); Lindegren (Lund); Murray and van Leeuwen (Herstmonceux), Lund (Lyngby, Copenhagen).

Tasks are divided into major blocks. The British reduce the raw data into a compact form, describing the observations relative to the smooth scanning motion of the instrument which is still inaccurately known. The observations are then synthesized by the Danish/Swedish group, using a method akin to the photographic plate overlap technique, yielding an accurate instrumental motion and astrometric parameters. The tasks of algorithm development, coding and testing are similarly divided. Work to date includes computer simulation of some processes, a description of reduction tasks and data interfaces, and coding of basic routines, e.g., for data management, estimation methods, and various transformations. The astrometric effects of general relativity will be given particular attention. Numerical experiments show that six non-isotropic metric coefficients according to a theory by S. Cowling can be solved for, in addition to the isotropic light-deflection, without weakening the astrometric solution. (HØG)

The **TYCHO** Data Analysis Consortium (TDAC) involves institutes in Denmark, France, Holland, Italy, Sweden, Western Germany, UK, and USA.

Working from a TYCHO Input Catalogue to contain the two million brightest stars in the sky, star mapper signals will be extracted and analyzed in order to derive magnitudes and positions for stars brighter than B=11 (about 500 000 stars). The Input Catalogue will be a subset of the Guide Star Catalogue, at the Space Telescope Science Institute, supplemented with stars from the Centre de Donnees Stellaires at Strasbourg. Using coarse attitude information, the photon records from the star mapper slits will be used for a point-source deconvolution in small maps of 12" x 12" centered on each star position of the Input Catalogue. The deconvolution includes components of double or multiple systems down to about B=11 and V=10.3, and will be fairly complete for separations >2". The two million stars will correspond to a data extraction from 22 square degrees, (0.05% of the sky). To increase the coverage, data extraction from the original full data set using a signal-to-noise criterion is conceived but will not be pursued with full priority. The photometry will be calibrated by available ground-based photometric standard stars. The reference frame will be that of HIPPARCOS. (GREWING)

STUDY GROUP ON HORIZONTAL MERIDIAN CIRCLES

Chairman: Høg; Members: van Herk, Leiden; Klock, USA; Li Zhigang, Shaanxi; Morrison, RGO; Osorio, Oporto; Pinigin/Timashkova, Pulkovo; Xu Tong-qi, Shanghai; and Hu Ning-sheng, Nanjing.

Pinigin reports that the Pulkovo HMC, after being automated, is expected to determine coordinates of fundamental stars plus faint stars in the neighborhood of radio sources. A horizontal transit circle for observing in the prime vertical has been described by Pinigin and Shornikov in Astrometria i Astrofizika, Vol. 49, 1983, p. 75-82. An axial meridian circle is being constructed at Kiev, Main Astronomical Observatory. Yatskiv reports that the telescope tube is placed horizontally in the prime vertical and revolves in bearings located at each end of the tube. In front of the 18 cm objective a 45° flat mirror is mounted in rigid connection with the tube. Osorio reports from Oporto that a new automatic device to collect data from the micrometer is being implemented. A new study of the stability of the mirror is underway. Høg reports that the Glass Meridian Circle (GMC) is being manufactured in China in cooperation with astronomical institutes in China (Shaanxi and Nanjing) and Denmark. The instrument will be tested in Brorfelde before it is put into operation at Shaanxi.

Internal refraction was detected during the development of the GMC and has been confirmed in the flexure measurements at the USNO by Høg and Miller. A tangential ventilation system to remove the effect will be installed on the CAMC on La Palma. (HØG)

SRS PROJECT

The SRS Committee is charged with the task of furthering cooperation between the various observatories participating in observations and reductions. The Committee consists of the following members: Fomin, Pulkovo; Fricke, Heidelberg (Chairman); Hughes, Washington; Lederle, Heidelberg; Polozhentsev, Pulkovo; Smith, Washington. The SRS Committee will meet at the Pulkovo Observatory in the summer of 1985 for a discussion of the results of the SRS project. The discussions and conclusions will be reported to Commission 8 at the XIX General Assembly.

The SRS project has reached its final phase, the construction of the SRS catalogue. The Pulkovo and U.S. Naval Observatories have completed the exchange of data such that both were enabled to complete the analysis of the observations. The U. S. Naval Observatory will present a catalogue of observed positions on the system of the FK4. The Pulkovo Observatory has followed a plan to present the SRS catalogue on an improved FK4 system where the improvements shall result from absolute observations made within the SRS project. The SRS catalogue will be transformed to the system of the FK5 as soon as the transformation is known. (FRICKE)

WORKING GROUP ON ASTRONOMICAL REFRACTION

The Group consists of: Teleki (Chairman), Brunner, Currie, Garfinkel, Hu Ningsheng, Huang Kun-Yi, Hughes, Kolchinskij, Mavridis, Nefed'eva, Saastamoinen, Sergienko, Spoelstra, Sugawa and Tengstrom. The proceedings of the Workshop, "Three-dimensional Refraction," were published by Reidel in "Sun and Planetary System," 1982, pp. 453-530. Teleki (Bull. Obs. Astron. Belgrade, 134, 1984, pp. 39-77) reported on investigations published during 1979-82. This report was presented in part to Commission 8 in Patras in 1982. Shamaev published (Itogi Nauki i Tehniki, Astronomiya, Tom 23, Ed. VINITI, Moscow, 1983, pp. 61-101) a review paper on refraction investigations (bibliography of 264 papers).

The Group has begun organizing a Workshop on the Determination of Refraction in Optical and Radio Astrometry, to be held at the Pulkovo Observatory, Leningrad, in May 1985. The Scientific Organizing Committee: Hughes, Sergienko and Teleki (chairman). Two main topics are suggested: 1) Refractional standards (terminology, tables, atmosphere models, algorithms, etc.); and 2) Determination of refraction in the real atmosphere. The new Pulkovo Refraction Tables (Fifth Edition) will be finished in 1984 and will be discussed as part of topic No. 1.

J. A. HUGHES
PRESIDENT OF THE COMMISSION

9. INSTRUMENTS AND TECHNIQUES (INSTRUMENTS ET TECHNIQUES)

PRESIDENT: W.C. Livingston
VICE-PRESIDENT: C.M. Humphries
ORGANIZING COMMITTEE: J.C.Bhattacharyya, J. Davis, J.L. Heudier, I.M. Kopylov, G. Lelievre, J.P. Maillard, A.B. Meinel, N.N. Mikhel'son, E.H. Richardson, J. Ring, J. Rösch, N. Steshenko, R.G. Tull, M.F. Walker, W.L. Wilcock, and G. Wlerick

I. MEETINGS, SYMPOSIA, AND GENERAL COMMENTS

Between the Patras and New Delhi General Assemblies no meetings were directly sponsored by Commission 9 because the discipline was amply covered by the following:

- "Eighth Symposium on Photoelectronic Image Devices," Imperial College (London) 5-7 September 1983 (B. L. Morgan, Ed., Academic Press, London, 1984).

- "Advanced Technology Optical Telescopes II" Imperial College, 5-7 September 1983.

- "Instrumentation in Astronomy V" 7-9 September 1983 (A. Boksenberg and D. Crawford, Eds., Proc. SPIE 445).

- IAU Symposium No. 109: "Astrometric Techniques," 9-12 January 1984, Gainsville (Florida).

- "Astronomical Photography 1984," Edinburgh, 4-6 April 1984 (E. Sim and K. Ishida, Eds., Number 14 of Occasional Reports of the Royal Observatory, Edinburgh).

- "IAU Colloquium No. 79: "Very Large Telescopes, Their Instrumentation and Programs," Garching bei München, 9-12 April 1984 (M. H. Ulrich and K. Kjär, Eds.)

Following the New Delhi General Assembly a large turnout is expected at:

- IAU Symposium No. 118: "Instrumentation and Research Programs for Small Telescopes," Christchurch, 2-6 December 1985.

Our report period, 1981-84, covers many advances in astronomical instruments. Outstanding was the success of the Infrared Astronomical Satellite, a 57-cm aperture telescope cooled to 2-5 K permitting high resolution mapping of the entire sky to a wavelength as long as 100μm. Perhaps most significant to the ground-based astronomer has been the pervasive application of CCD detectors. Now even small instruments can tackle problems that formerly were the reserve of the great telescopes. Of course the efficiency of large aperture instruments has likewise soared, particularly through the techniques of multi-object spectroscopy. No longer are galaxies observed one at a time but rather 50 to 100 are simultaneously recorded via, for example, fiber optic feeds. A system planned for the Special Astrophysical Observatory 6-m telescope will accommodate up to 500 objects.

Several very large telescopes in the 7m+ class are being planned. Some are even financed! Particularly noteworthy toward such endeavors is the large mirror casting facility being developed by R. Angel and colleagues at the University of Arizona. Seldom before have astronomers proven so self-reliant.

Spatial interferometry and solar/stellar oscillations are two subjects that have blossomed during the past three years. The promise of this new generation of large aperture telescopes is especially favorable for speckle, both in the visible and IR. The possibility of probing to the very centers of the sun and stars by observing a periodic surface heaving a few cm s^{-1} in amplitude was unthinkable a decade ago. We are fortunate to have received and included comprehensive reports on interferometry by John Davis and oscillation techniques by Jack Harvey.

We can recognize a new trend in instrument design whereby the relatively inexpensive "microprocessor" becomes central to instrument control, and assuming the role formerly occupied by large computers. The latter may still be essential to data gathering, but not at the operational level.

One dream of the telescope makers is proving very difficult to realize: the restoration of images (as distorted by atmospheric seeing) through the use of active optics. Several talented people are working on this problem but so far the atmosphere has not cooperated. Perhaps high resolution imagery will prove ultimately to be the domain of space telescopes.

II. FUTURE VERY LARGE TELESCOPES
J. Beckers

Many countries are planning large ground-based telescopes for visible and infrared astronomy. This renaissance in large ground-based telescopes is at least partly due to the development of new technologies related to telescope control, telescope buildings, and mirror making. Innovations in these areas, as well as the rapid increase in the use of computer control of telescopes and instruments, have increased the capabilities of telescopes while often reducing the cost.

The European Southern Observatory will build a 3.50-m New Technology Telescope (NTT) in Chile which uses a relatively thin (1:15) low expansion glass or, possibly, a metal mirror blank. It also has plans for a Very Large Telescope (VLT), consisting of an interferometric array of four 8-m aperture telescopes. Great Britain plans to dedicate shortly its 4.2-m Herschel Telescope in the Canary Islands. The Royal Greenwich Observatory is developing a plan for a multiple mirror type telescope using six 7.5-8-m mirrors and plans for an interferometric array of large telescopes are being studied at the Royal Edinburgh Observatory. Iraq will acquire a 3.5-m telescope, identical to the German Calar Alto Telescope in Spain. The Tokyo Astronomical Observatory in Mitaka, Japan, has submitted a proposal to the Japanese government for a 7-8-m telescope to be built on Mauna Kea. In the United States the University of California and the California Institute of Technology are well along the way with the planning and funding of a 10-m aperture segmented mirror telescope (TMT) to be placed on Mauna Kea. The University of Texas is proposing a 300-in. telescope to be located at or near the McDonald Observatory. A consortium of universities (Princeton, Chicago, Washington State and New Mexico State) will shortly start construction of a 3.5-m aperture telescope in the Sacramento Mountains of New Mexico. The National Optical Astronomy Observatories is proposing a Southern Optimized IR Telescope to be located at Cerro Tololo, Chile. It is also now working on the final design and proposal for the National New Technology Telescope (NNTT), which will be of the multiple mirror kind having four 7.5-m diameter mirrors. Sites on Mauna Kea and Mt. Graham (Arizona) are being tested for its location. In the USSR a 5-m class telescope is being considered by the Leningrad University to be located in Central Asia. The Crimea Astrophysical Observatory is developing plans for a 25-m aperture segmented mirror telescope to be built at an, as yet, undefined location.

III. SOLAR TELESCOPES

O. Engvold

At the Canary Islands, three interesting new solar telescopes are nearing completion.

A 60-cm Vacuum Tower Telescope of the Kiepenheuer-Institute, Freiburg, is expected to be operational at the Izaña site in the summer of 1986. The feed system is a classical 80-cm coelostat on top of a 38-m high, double walled concrete tower. An auxiliary telescope system uses a fraction of the main aperture and forms a full disk image of the Sun for tracking. Piezoelectric-driven tilting of the second coelostat mirror is used to correct tracking errors and stabilize the main solar image. This telescope will be equipped with a multi-channel slit-jaw camera unit. There is also a large, grating spectrograph to be outfitted with a French Multi-channel Subtracting Double Pass (MSDP) arrangement. Italian solar astronomers will provide a narrow band, wavelength stabilized, universal filter.

A 40-cm Gregory-Coude vacuum tower telescope of Göttingen University, will be operational at Izaña by June 1985. This telescope was earlier at the German station at Locarno, and it has been refurbished for the installation at the new site.

Finally, the solar telescope of the Swedish station at La Palma is being reconstructed and is expected also to be operational in summer 1985. An alt-azimuth turret at the top of a 16-m high tower will feed an f/45 50-cm achromatic objective lens which at the same time serves as entrance window for the vacuum system.

In addition there is the large French telescope "Telescope Heliographique pour L'Etude du Magnetisme et des Instabilites Solares" (THEMIS) which has been designed for accurate polarization measurements of the Sun. It will have a 90-cm Ritchey-Chretien telescope feeding a solar magnetograph. THEMIS is planned for the Izaña site.

An open solar telescope with emphasis on stability is being developed in Utrecht. Its construction could be finished by the end of 1985, and the telescope might possibly be erected at La Palma.

A 35-cm aperture solar telescope-magnetograph is being developed by the Nanjing Astronomical Instrument Factory for the Beijing Observatory. This telescope has a refractive imaging system with a doublet objective that gives a 4x6 $arcmin^2$ field-of-view. The polarization measuring system is made up of a KD*P modulator and specially designed narrow band birefringent filters.

The design study of the Large European Solar Telescope (LEST) has been completed. A 2.4-m Gregorian-type telescope with an additional ellipsoidal mirror provides a 20-arcmin diameter diffraction limited field-of-view. Two flat mirrors reflect the light out and along the elevation and azimuth axes of the telescope mounting so that the final image is fixed (except in rotation about its center) with respect to its auxiliary instruments. These will be located near the bottom of a 25-m high double concrete tower. A thin entrance window seals the telescope light path which will be filled with helium gas. The aim is a telescope planned for high spatial resolution and low instrumental polarization measurements. The detailed design of the telescope will be complete in 1-1/2 to 2 years. By the end of 1984 the following countries have joined the project: FRG, Israel, Italy, Norway, Sweden, Switzerland, and Australia.

At the Baikal Astrophysical Observatory in the Soviet Union, the construction of a large solar vacuum telescope is nearing completion. This telescope has a 76-cm doublet objective 760 with a focal length of 40 m that is fed by a siderostat 100 cm in diameter. The siderostat is placed atop a 20 m tower on the steep slope of a mountain. A vacuum tank of 40-m length comes down to the foundation of the tower along the polar axis. This tank includes the objective lens in the upper part and a diagonal flat mirror that directs the light beam horizontally onto the spectrograph slit. The vacuum tank is closed from above and below by optical windows 100 cm in diameter. Outside the lower window the solar image is fed to a photoelectric guiding and scanning carriage, while the diagonal mirror directs a minor portion of the solar image onto the slit. The telescope optics and optical windows are now installed, and the tank can be evacuated as low as 10 mm. Investigation of the optical and thermal properties of the entrance window are underway. White-light pictures at the Newtonian focus are being taken. Construction of a spectrograph and other auxiliary instruments is also well along.

At the Sayan Solar Observatory a modernization of the horizontal solar telescope is completed. A new coelostat mounting with mirrors 80 cm in diameter are installed. The 80-cm primary of the telescope has a focal length of 20 m.

The first attempt to achieve high spatial resolution of a solar image for extended periods (more than 80 hours) is to be made from a NASA satellite scheduled for launch in July 1985 as Spacelab II. A Lockheed Research Laboratory instrument called a Solar Optical Universal Polarimeter (SOUP) and having a 30-cm aperture, is intended to map solar magnetic fields in visible light and to study their evolution to 1/2 arc-sec. While this resolution has been obtained occasionally on the ground, the significant aspect is that resolution is expected to be constant during the entire flight. Complementing this is the Naval Research Laboratory High Resolution Telescope Spectrometer (HRTS) which maps the transition region redward from Ly α at a similar spatial resolution.

IV. OTHER NEW INSTRUMENTATION

Limitation of space does not permit a full listing of all the new and planned instrumentation that has come to the attention of Commission No. 9. We therefore mention only a sampling.

- In India, at Kavalur, a 2.3-m reflector is nearing completion by the Indian Institute Astrophysics. This telescope was fathered by Vainu Bappu, former President of the IAU.

- In Argentina a 2.15-m Boller and Chivens Reflector, purchased in 1970, is now being installed by Compeljo Astromico "El Leonitio."

- In the Peoples Republic of China, a 1.56-m Cervit reflector will begin operation next year in a suburb of Shanghai.

- On the southwest ridge of Kitt Peak a 2.4 m reflector of the McGraw-Hill Observatory expects first light in June 1985.

- At Siding Spring, Australia, a new 2.3-m "Multiple-mirror-telescope/ advanced technology telescope," is now operational. Innovations include a rotating domeless building similar to Arizona's MMT.

- A multi-slit spectrograph is being developed for the 6-m Zehelchuk, Special Astrophysical Obs., that allows the simultaneous observation of 500 objects.

- A 24 element segmented active mirror with phase correction is being tested by Robert Smithson for seeing correction at the Sac Peak Vacuum Tower.

- The National Optical Astronomy Observatory continues to support a Gratings Laboratory built around the MIT "C" engine. Mounted on a vibration isolated slab and temperature controlled to ~ .003 C, a variety of unique gratings have been produced. Three attempts have been made at a 300x400-mm, 632 groove-mm^{-1} ruling for solar work. The machine improvements that followed from these attempts make the venture still promising. When success is realized replicas of the sub-master will be available to astronomical community.

- A new focal reducer (f/15 to f/8) has been developed by G. Courtes at the Observatoire de Haute Provence in order to obtain an 18 arc-min field with the 80-mm Lallemand-Wlerick electronic camera. This focal reducer is equipped with two spectrographic modes, 1500A mm^{-1} and 40A mm^{-1} multislit and slitness combinations.

V. PHOTOGRAPHIC PROBLEMS
Report by the Chairman of the Working Group, E. Sims

The principal activity of the Working Group in the period under review was the meeting held in Edinburgh, Scotland, from April 4 to 6, 1984. About 40 papers and posters were presented to 65 participants from 16 countries. Topics ranged from the nature and behaviour of emulsions through their use at the telescopes and in the laboratory, to the use of photographic and machine based techniques for rapid extraction of information from original photographs.

Highlights of the meeting included a discussion on the possibility that a new technique of emulsion manufacture which can control the morphology of the light-sensitive grains ("T-grains") may be applicable to spectroscopic emulsions. If this proves to be possible, it may lead to a new technology emulsion with a significant improvement in speed/grain characteristics over the present IIIa-J and IIIa-F emulsions. The problem of the occurrence of microspots on processed IIIa-J plates was discussed at length, but it is clear that we do not yet understand the cause of this problem. Several papers described specially designed photographic laboratories and some of the techniques used in them to make information from the original plates more widely or more readily available, such as unsharp masking and image enhancement. Descriptions and reports were given of some new, and not so new, fast measuring machines (including three PDS machines and COSMOS) and their ability to produce high quality photometry, astrometry and spectrophotometry from measurements of large photographic plates. There was general agreement that a good, atlas-quality copy of an original plate can be as reliable and useful as the original plate for any of these applications. During the period of the meeting a conducted tour of the Royal Observatory, Edinburgh included visits to the Plate Library, housing all the UK 1.2 metre Schmidt photographs, the COSMOS measuring machine, and the extensive Photographic Laboratories.

Also of interest to astronomical photographers was IAU Colloquium number 78, "Astronomy with Schmidt-type Telescopes" (Asiago 1983). At that meeting the following Resolution was passed:

Participants in the IAU Colloquium number 78: "Astronomy with Schmidt-type Telescopes," in concordance with the IAU Commission 9 Working Group on Photographic Problems <u>acknowledging</u> the impressive advances during the past decade resulting from the introduction of new emulsions of increased sensitivities, and over a wide spectral range, and <u>recalling</u> the continued interest in our science shown by several manufacturers of photographic materials, in particular the

Eastman Kodak Company, and noting the continuing and growing need for photographic emulsions in support of both ground-based and space astronomy, especially with wide-field, high resolution sky survey telescopes for which it is unlikely in the near future that panoramic electronic detectors will have the detector size and resolution available on photographic materials, urge all manufacturers to initiate and continue research with the aim of providing new emulsions with improved characteristics - for example quantum efficiency, sensitivity and spectral range."

In response to this Resolution, Mr. L. J. Thomas, Senior Vice-President of Eastman Kodak Company and Director of Kodak Research Laboratories, replied to the General Secretary (25 October, 1983) with the following encouraging letter:

"Eastman Kodak Company is proud of its long association with the astronomical community. Improved plates have continued to be developed such as the introduction in the 1970's of KODAK spectroscopic plates Type IIIa-J and Type IIIa-F, based on the percepts of Marchant and Millikan. Also, Kodak's research effort as reflected by the papers of James, Lewis, Babcock et al. on the reduction of low intensity reciprocity failure in model systems, the $AgNO_3$ hypersensitization of KODAK Spectroscopic Plate, Type IV-N, by Jenkins and Farnell; the continuing participation by a member of our scientific staff in both the AAS and IAU photographic working groups; and the corporate support of the new Palomar Sky Survey in concert with the Alfred Sloan Foundation and the National Geographic Society are important roles for Kodak.

I believe you are aware of the recent introduction in the color negative and medical X-ray markets of films which utilize a new emulsion technology employing silver halide crystals of specific morphology that we refer to as "T-grains". Our research laboratories intend to explore the possible application of this new emulsion technology to astronomical photography. Although it is not prudent to anticipate in specific terms the results of research not yet complete, we share your hope that the work will lead to advances useful in photographic astronomy."

In the Soviet Union a new high speed infrared film with a maximum sensitivity at 10600A is available. Sample tests at the Crimean Observatory by Scherbakova indicate a sizable gain over Kodak IZ. The film does not need to be hypersensitized.

VI. HIGH ANGULAR RESOLUTION INTERFEROMETRY
Report by the Chairman of the Working Group, J. Davis

a) Introduction

The following report has been based on replies received to a circular letter sent to all groups and workers known to be active in the field of high angular resolution interferometry. The assistance of F. Roddier in preparing the section on speckle interferometry and image reconstruction is gratefully acknowledged.

The interest in high angular resolution interferometry and its potential astronomical applications has increased significantly with several meetings including sessions on interferometry in their programs. These include IAU Colloquium No. 62, "Current Techniques in Double and Multiple Star Research," IAU Symposia Nos. 109 and 111, "Astrometric Techniques," and "Calibration of Fundamental Stellar Quantities," and a combined IAU-URSI symposium on "Indirect Imaging" [1]. The Optical Society of America included a symposium on Michelson spatial interferometry in its 1982 annual meeting and there have been a number of workshops held on various aspects of the subject.

b) Speckle Interferometry and Image Reconstruction

Major reviews of speckle interferometry and image reconstruction have been published by Bates [2] and Dainty [3]. As noted by Dainty, observational speckle interferometry is now over a decade old and some 80 papers primarily concerned with astronomical results have been published. Examples of its applications include the study of binary stars where it has produced a wealth of data. It has also been used in studies of Pluto and its moon Charon, to determine the dimensions of asteroids and to study solar granulation. In extragalactic studies one of the components of the "triple" quasar PG 1115+08 has been resolved as a binary and new limits on fine structure of the nucleus of the Seyfert galaxy NGC 1068 has been obtained in the visual and in the IR.

Extension of speckle interferometry to the IR has led to the detection of unseen companions to nearby stars. A method of exponential filtering [4] has been used in the processing of the speckle data to overcome the 180° ambiguity in the relative orientation of the components of binary system [5]. IR speckle interferometry has also been used to measure the size and shape of dust shells surrounding either very young stars (possibly protostars) or evolved stars (cool giants and supergiants). The prospects for the discovery and study of protoplanets and protostars as well as for the understanding of the mechanism responsible for the mass loss of evolved giant stars appear to be excellent, especially with the new generation of giant telescopes now being planned.

Photon counting combined with high speed digital signal processing is now a standard technique in speckle experiments. However, the technique is still limited by the performance of available detectors and new detector systems are being developed. At Georgia State University an intensified CCD camera [6] is now used for a wide range of optical speckle interferometry programs. P. Nisenson, C. Papaliolios and R. Stachnik report that major hardware efforts at the Harvard-Smithsonian Center for Astrophysics include the development of computer-controlled CCD, intensified CCD and PAPA [7] photon counting speckle cameras. The PAPA (Precision Analog Photon Address) Detector is a 512x512 element sensor capable of generating accurate photon x-y (and t, if required) coordinates at megahertz rates. At Steward Observatory a real-time photoelectron event-detecting video system [8] has been developed based on an intensified Plumbicon camera and a Grinnell video digitizer. Improvements to the detector and data processing systems have been discussed by the Steward Observatory group [9].

Rotation shearing interferometry has been proposed as an alternative to speckle interferometry and has been successfully used to observe α Ori [10]. Beckers has described a new technique called differential speckle interferometry [11,12] which has been shown by R. Petrov, F. Roddier & C. Aime to yield a higher signal-to-noise ratio and higher angular resolution than conventional speckle interferometry. The importance of simultaneous observations at several wavelengths is now widely recognized and a method for obtaining objective prism spectra with diffraction limited resolution (speckle spectroscopy) has been described by Weigelt [13].

Image reconstruction from interferometric data has continued to be an area of considerable theoretical and practical activity. As a result the possibilities of phase retrieval from amplitude data alone are now better understood [1,14,15]. Phases have been successfully retrieved from speckle data using the Knox-Thompson algorithm [16] and the signal-to-noise ratio of the procedure has been discussed [17,18]. Phases have also been retrieved using the speckle masking algorithm [19] and B. Wirnitzer has shown that speckle masking has the same limiting magnitude as speckle interferometry. An image of α Ori has been reconstructed in the continuum using rotation shearing interferometry by C. &

F. Roddier and in Hα using differential speckle interferometry by J. Beckers & K. Hege. A well attended workshop on image reconstruction was held in Tucson in April 1983.

Speckle methods, especially roll deconvolution (G. Weigelt), and rotation shearing interferometry (F. Roddier & J. Breckenridge) are now being considered for improving the performance of large space telescopes. Adaptive optics are also under consideration as a means of improving the results of passive image reconstruction methods.

c) Visual Long Baseline Interferometry

The 11-m baseline prototype modern Michelson interferometer under development at the University of Sydney [1,20] was nearing completion in June 1984 with the installation of component parts essentially complete and final optical alignment in progress. The prototype features coelostats at each end of an 11-m baseline to direct starlight into the instrument, path compensation, active wavefront tilt correction, and rapid signal sampling to overcome the problem of phase fluctuations. This prototype is the first stage in the development of a major instrument with baselines up to ~1 km. A search for a suitable site and the design of the major instrument are under way.

The CERGA prototype long baseline interferometer (I2T) [21,22], which employs two 26-cm aperture telescopes, has been used with baselines in the range 6 to 40 m to measure the angular sizes of 10 stars with accuracies in the range 7 to 14% [23,24]. The more recent determination of diameters of equivalent uniform disks are based on the visual detection of very low contrast fringes for the largest possible baseline. These measurements are the first step in a program aimed at measuring the angular sizes of 50 stars of spectral types F-M with baselines up to 67 m. Labeyrie has continued his program to develop a large aperture synthesis array [22,25] and the second 1.5-m aperture telescope and central interferometer have been constructed. Alignment of a complete two aperture interferometer is in progress.

Shao reports that data have been obtained with the Mark II astrometric interferometer indicating that fringe visibility measurements are repeatable at the ±2% level and that ±0.5% is possible if corrections for seeing changes are made. Construction of the Mark III astrometric interferometer has commenced. This is a joint effort involving the Smithsonian Astrophysical Observatory, MIT, the U.S. Naval Research Laboratories and the U.S. Naval Observatory. The instrument will have three baselines: 20 m N-S, 15 m NE-SW, and 15 m NW-SE.

d) Infrared Long Baseline Interferometry

The CERGA I2T has been used as a Michelson type interferometer at 2.2 μm with baselines in the range 6 to 41 m to measure the angular sizes of giant stars [26] with accuracies in the range 2 to 14%. Future plans include extending the range of IR wavelengths from 1 to 5 μm with increased sensitivity at 2.2 μm.

An IR interferometer to be used in conjunction with the 1.5-m aperture telescopes at CERGA is being developed at the Observatoire de Lyon for use in the wavelength range 2 to 10 μm.

C. H. Townes reports that plans for a large two-telescope IR interferometer will come to fruition in 1984-1985. Based on preliminary work carried out with a 5.5-m baseline prototype, the instrument, now under construction, involves high-precision movable telescopes of 1.65-m aperture and will use heterodyne detection in the 10 μm atmospheric window. The instrument will be capable of both high angular resolution and precision astrometry.

The Multiple Mirror Telescope has been used in a number of observations and experiments designed to demonstrate its potential for high angular resolution research as an optical-IR interferometer and phased array [27,28].

The possibilities of long baseline interferometry with a coherent array of large ground-based telescopes have been discussed in a "Versatile Array" concept [29] and in connection with the ESO VLT Project [30,31]. D. Enard reports that the possibility of coherent coupling in the IR is to be a prime requirement in the development of the linear array concept for the VLT.

e) Interferometry From Space

The study of possible designs for high angular resolution interferometers for operation above the atmosphere has been pursued both in the U.S. and in Europe [32].

R. Stachnik reports on two interferometers being studied in the U.S. One is the Space Platform Interferometer (SPI) which is designed as a low cost, Explorer-class experiment which would put a Michelson spatial interferometer having a baseline of 2 to 50 m into space on board one of a number of space platforms under development. The second system is called the Spacecraft Array for Michelson Spatial Interferometry (SAMSI) and comprises two 1-m telescopes and a central station, for fringe measurement, allowing baselines up to 10 km to be scanned as the three spacecraft orbit the earth. SPI would provide a testbed for much of the technology required for more ambitious space interferometers and would operate from the UV to near-IR. A magnitude limit of 15 is predicted for 25 cm apertures. Fringe measurement, including phase determination, at a number of position angles would permit experiments in two dimensional aperture synthesis for some larger stars. SAMSI is estimated to have a limiting magnitude of 19 and it appears feasible to move the 1-m telescopes in spirals as large as 500 m in diameter to allow synthesis of very large two-dimensional apertures.

In Europe a group sponsored by ESA is also studying an array of three spacecraft which has been named TRIO [32]. TRIO shares many common features with SAMSI but differs primarily in its approach to the problem of propulsion. It is proposed to use solar sails and radiation pressure to rotate and position TRIO's spacecraft in order to avoid possible problems of vibrational noise associated with accessories such as gyroscopes and inertia wheels.

Cooperative studies between the U.S. and European groups are in progress. A workshop on "High Angular Resolution Optical Interferometry From Space," was held during the 164th AAS Meeting in June 1984 and a colloquium on "Kilometric Optical Arrays In Space," was to be held in October 1984 with support from ESA.

VII. PHOTO-ELECTRONIC IMAGE DEVICES

Report Prepared by B. Fort, W. Livingston, D. Monet, V. V. Prokof'eva, A. G. Scherbakov, G. Wlerick and B. Ye

a) Solid-State Arrays

CCD systems have proliferated to telescopes everywhere. Suddenly small telescopes have become much more powerful research tools producing results that formerly was obtainable only at the major installations. No doubt this topic will be expanded on at the forthcoming IAU Symposium No. 118.

At the Toulouse Observatory B. Fort and J. P. Picat have developed and applied a Cine-CCD camera for comet occulation and a new CCD camera for spectroscopy between 5000-11000Å. These cameras use the commercial Thomson

TX31133 buried-channel chip (576x384, 23-μm pixels, readout noise ~ 20 e r.m.s.). See Fort et al. [33]. Besides having reasonable photometric properties, the Thompson chip looms important for general use because it is readily available and has a history of low blemishes.

In the Soviet Union a new CCD array (A-1033) has been evaluated by Anoshkin et al. [34]. The A-1033 has a size of 1024x1024, 15-μm pixels, with a full well capacity of 10^6e. They report the readout noise is about 20 e at -196°C. Another array of 288x232 has been used by Abramenko, et al. [35] for stellar photometry.

A new CCD system utilizing the RCA 512x320 array has been used for the early detection of Halley's Comet by the PRC 1-m telescope at Yunnan Observatory.

D. Monet, at the U.S. Naval Observatory in Flagstaff, reports on the applicability of CCD images for the measurements of stellar trigonometric parallaxes. Monet and Dahn [36] find that such images are comparable to the best photographs measured on interferometric measuring machines (~ 5 milli-arcsec per exposure). Work continues to improve on this astrometric accuracy.

Experiences with the linear Reticon RL 1024B as installed at the Sayan Observatory magnetograph has been given by Markov [37, 38]. Likewise, Ye reports the RL 1024 is being tested on the Cass spectrograph of Yunnan's 1-m telescope.

K. Dowdney [39] has prepared an engineering evaluation of the Texas Instruments 800x800 3-phase CCD. This report is recommended for its thoroughness, describing in detail the equipment, procedures used, and the results. The device proved to have excellent linearity (.02% @ 8000e) low read noise (15e) and good vertical and horizontal charge transfer efficiency. An area of about 300 columns near the center of the chip is defect-free and the remaining irregularities should flat-field out very well. This is the CCD being distributed by the U.S. NSF Astronomy Grants Program.

The University of Arizona is constructing a unique CCD/transit instrument, for example, to detect supernovae by sampling 20 square degrees of sky per night to a magnitude limit of 22. A fixed 1.8-m primary feeds two RCA 512x320 CCDs operated in the time-delay and integration mode. That is, the CCD is locked out at the same rate that the image drifts across, so that residual flat-field errors are minimized [40].

A ripple of excitement ran through the instrument community with the announcement that Tektronix was planning to market a 2048x2048 CCD. According to the specifications the chip would have 32 μm square pixels, a full well capacity of 10^6e, a read-out noise of 2e, etc. It would be a thinned device with 2, 3, or 4 phase clocking. Whether or not this particular device materializes, it does signal the approaching obsolescence of our data handling techniques. To read out the n(10^6) element arrays of the future, and then to extract the desired information in a timely fashion will demand innovative methods. Simply scaling up our present techniques, i.e., ordering more mag tapes, will become totally inadequate.

b) Electronography

Lelievre et al. [41] continue to apply the large field electronographic camera to extra-galactic problems at the C.F.H. 3.6-m telescope. Details of the M87 jet have been uniquely recorded.

At the Vilnius Observatory (USSR) Drazdis et al. [42] have developed a Kron-type camera. They report that with a 35-cm aperture telescope a 14.5 mag object through a V filter gave images in 3 min.

P. P. Petrov [43] continues to apply the EIC-1 for spectroscopy at the Nasmith focus of the Crimean 2.6-m reflector. At a dispersion of 45A/mm, 4300-5400A, with a widening of 0.3 mm, 2 arc-sec seeing, 1 arc-sec slit (spectral res. ~ 2A), film R-type (fine grain, low sensitivity; DQE with EIC-1 about 5%), he records FG Sge B = 10.8 mag in 20 min.

c) Image Photon-Counting

C. Papaliolios has built a 2-D system which has been applied successfully for speckle imaging. The camera has a 512x512 pixel field which can operate at a photon detection rate as high as $2(10^5)\ s^{-1}$.

The PAPA (Precision Analog Photon Address) camera uses a high-gain image intensifier for photon detection. The position of a detected photon in the image plane is determined by a set of 19 photomultipliers, each looking at an image of the output face of the intensifier through a coded mask. The camera functions properly only if a single photon is detected. The rate at which this process can be repeated is determined primarily by the phosphor decay time of the intensifier, which is measured to be about 200 nsec for the P47 phosphor used.

Two photon counting spectrometers are operational on the Soviet 6-m reflector (Markelov *et al.* [44]; Belaga *et al.* [45]; Kopylov *et al.* [46]). In one case there is a 1000 pixel format having a dynamic range of 30-8000 photons per channel per hr. The second system employs two linear arrays of 500 pixels each; one for the object and one for the sky.

d) Television-Type Systems

Efforts continue at Alma-Ata and the Crimea to employ the image orthicon, isocon, and the antiisocon for photometry and astrometry. These programs are reviewed by Galinsky [47], Abramenko *et al.* [48], and Alexandrin *et al.*[49].

VIII. SOLAR/STELLAR SEISMOLOGY INSTRUMENTS

J. Harvey

Investigation of the interior structure and dynamics of the sun and other stars by observation and analysis of global oscillations is a rapidly developing area of astrophysics [50,51,52]. Large amplitude, pulsating variable stars have long been observed but exhibit only one or a few different modes of oscillation from which the average properties of the star can typically be determined [53]. The goal of the new research is to study stars which oscillate in a large number of different regions of the star's interior. Analysis of the different oscillations allows the interior properties of the star to be determined with resolution in the radial, and potentially the latitudinal, directions.

New instrumental and observational techniques are required because the photospheric amplitudes of the oscillation modes are small (e.g., for the sun, ~ 10 cm s^{-1} in velocity and ~ 10^{-6} in intensity). Three types of measurements are currently made: Doppler shifts, intensity fluctuations and changes in the intensity profile near the solar limb. While general purpose telescopes have been used, a number of specialized telescopes have been developed and more are planned. Long time series observations are required to resolve the frequency structure of the oscillation modes. Observation interruptions caused by the day-night cycle have been overcome by using multiple instruments around the world, observing from the geographic south pole, and by observing from spacecraft.

Measurements of Doppler shift offer the best ratio of oscillation amplitude to background solar noise and are probably the preferred oscillation diagnostic in a wide range of stars. Accordingly, most instrumental effort is devoted to precise Doppler shift measurements. Since the oscillation amplitudes are of the order of cm s^{-1} and periods are > 100s, the emphasis is on stability and sensitivity. Measurements have been made with grating spectrographs using telluric or laboratory lines for reference [54,55]. With a resolved solar image, differential measurements are made by comparing Doppler shifts from different parts of the disk to overcome the intrinsic instability of existing grating spectrographs [56,57,58,59]. It is a frequent practice to employ more than one spectrum line to enhance the typically poor efficiency of grating spectrographs [55,60]. A number of interesting results have come from grating spectrograph observations but the difficulties of achieving two-dimensional imaging and intrinsic instability make these instruments nearly obsolete for solar work. Reflecting starlight from a stable Fabry-Perot etalon before it passes into a grating spectrograph has been proposed as a way of inserting reference absorption lines to enable precise Doppler shift measurements [61]. A ingenious technique that uses a large part of the spectrum of a star formed by a grating spectrograph which is then referenced to a laboratory stabilized laser has been proposed [62]. There is a plan to pursue this idea at the National Optical Astronomy Observatories.

Three other approaches have been used for precise Doppler measurements. Light is selectively scattered by hot alkaline metal vapors at wavelengths of the resonance spectrum lines. This principle has been used to obtain Doppler shift measurements with very good stability. The method is relatively inefficient and is restricted to solar observations with little or no angular resolution [63,64]. A variation of the idea, which depends on anomalous dispersion of light transmitted through the vapor, permits imaging and is also relatively efficient [65] A long series of solar observations with a transmission resonance device was obtained in 1984 at Mt. Wilson Observatory [66]. A similar spectrometer, designed to sense both of the sodium D lines, was used to measure apparent oscillations of α Cen A [67].

Another relatively old spectrometer is the Fabry-Perot etalon. The demands for etalon stability are great but one was recently successfully used for differential solar measurements [68].

A number of investigators have selected the Michelson interferometer operated at a nearly fixed path length difference as the basis for precise Doppler shift measurements [69,70,71]. It is possible to achieve good stability with a temperature-compensated design employing selected glasses [72]. An absolute wavelength reference can be provided by a laboratory light source such as a stabilized laser. The devices are efficient and compact. Both solar and stellar observations have been attempted.

Other Doppler-sensitive spectrometers have been tried such as Lyot filters and heterodyne systems but most current activity is centered on grating spectrographs for stellar observations, resonance scattering and transmission devices and specialized Michelson interferometers for solar and stellar work.

The relative merits of measuring oscillations by Doppler shift compared with intensity have been discussed [73,74]. Considering available equipment, intensity measurements offer advantages for all but the brightest stars and most solar observations. Measurements of the total solar irradiance from the Solar Maximum Mission satellite greatly improved our knowledge of solar oscillations [75]. Three-color measurements of solar oscillations were obtained from a balloon in the stratosphere [76]. Solar oscillations were apparently detected with what must be the smallest aperture telescope ever built, 10 μm [77,78]. Additional intensity measurements of the sun, now with some angular resolution, were made at the Crimean Astrophysical Observatory in 1983 using a 16x16 integral detector array [79]. Some spectral features show an order of magnitude larger intensity fluctuations than the continuum. Thus the CN band at 3883 Å and the

CaII H and K lines have been used to observe solar oscillations with angular resolution [80,81,82].

The prospects of directly observing weak intensity oscillations of stars from the ground are not encouraging because of noise from scintillation and extinction fluctuations [83,84,85]. An attempt to detect solar-like oscillations on other stars using conventional photometry was not successful [86]. Proposals for reducing the effects of atmospheric noise by using reference stars and different wavelengths of the star have been advanced [87,88]. This second idea has already been used to detect apparent oscillations of ε Eri [89] and is the basis of instruments under development by the Center for Astrophysics [90] and the National Optical Astronomy Observatories.

Measurements of changes in the limb darkening function of the sun caused by global oscillations have been undertaken by three groups. The SCLERA group in Tucson have added several detectors to their system to enable many limb positions to be measured simultaneously [91]. A group from the Sacramento Peak Observatory, observing from the south pole measured the entire solar limb and obtained several remarkably long unbroken observation sequences [92,93]. A group observing at the Pic du Midi continue to develop their limb instrument [94,95].

In the near future, the problem of obtaining long, unbroken time sequences of oscillation observations is being addressed by solar and stellar observations from the south pole [96] and by networks of instruments deployed at different longitudes at low latitude. Two networks for observing solar oscillations with little or no angular resolution are under development [97,98]. Both networks will employ resonance scattering spectrometers. Another network, with good angular resolution, has been proposed by the National Optical Astronomy Observatories [99]. A more modest, two-station approach is under development by the California Institute of Technology [100].

Space is an obvious place to achieve continuous observing and also freedom from noise caused by the earth's atmosphere. A number of proposals have been made for both solar and stellar oscillation observations from space. A joint French-Soviet experiment has been proposed to observe solar oscillations with low angular resolution during the cruise phase of the Soviet Phobos mission to Mars [100]. The Solar and Heliospheric Observatory proposed to the European Space Agency includes both imaging and non-imaging solar oscillation experiments [102]. A working group recently completed a study of the advantages of solar oscillation observations from space for the National Aeronautics and Space Administration [103]. Stellar obscillation observations from space have been discussed by several authors [104,105,106,107]. Probably the most widely-discussed proposal is the Probing Stellar Interiors via Variability and Activity (PSIVA) project which was a response to ESA's interest in new mission concepts [108].

REFERENCES

1. "Indirect Imaging," 1984, Proceedings of an International Symposium, (ed. J. A. Roberts), Cambridge University Press.
2. Bates, R. H. T., 1982, "Astronomical Speckle Imaging," Physics Reports **90**, pp. 203-297.
3. Dainty, J. C., 1984, "Stellar Speckle Interferometry," Topics in Applied Physics **9**, pp. 254-320.
4. Walker, J. G., 1982, Appl. Opt. **21**, 3132.
5. Lippincott, S. L. et al., 1983, P.A.S.P. **95**, 271.
6. McAlister, H. A. et al., 1982, Proc. SPIE **331**, 113.
7. Papaliolios, C., and Mertz, L., 1982, Proc. SPIE **331**, 360.
8. Macklin, R. H. et al., 1982, Proc. SPIE **359**.

9. Hege, E. K. et al., 1983, Proc. SPIE **445**, 469.
10. Roddier, C., and Roddier, F., 1983, Ap. J. **270**, L23.
11. Beckers, J. M., 1982, Optica Acta, **29**, 361.
12. Beckers, J. M. et al., 1983, Proc. SPIE **445**, 462.
13. Weigelt, G., 1981, "Scientific Importance of High Angular Resolution at Infrared and Optical Wavelengths," ESO, p. 95.
14. Fienup, J. R., 1982, Appl. Opt. **21**, 2758.
15. Bates, R. H. T., and Fright, W. R., 1983, J.O.S.A. **73**, 358.
16. Stachnik, R. V. et al., 1983, Ap. J. **271**, L37.
17. Nisenson, P., and Papaliolios, C., 1983, Opt. Commun. **47**, 91.
18. Weigelt, G., and Wirnitzer, B., 1983, Opt. Lett. **8**, 389.
19. Deron, R., and Fontanella, J. C., 1984, J. Opt. **15**, 15.
20. Davis, J., 1979, New Zealand J. Sci. **22**, 451.
21. Labeyrie, A., 1975, Ap. J. **196**, L71.
22. Labeyrie, A., 1982, Sky & Telescope, **63**, 334.
23. Bonneau, D. et al., 1981, Astron. Astrophys. **103**, 28.
24. Faucherre, M. et al., 1983, Astron. Astrophys. **120**, 263.
25. Labeyrie, A., 1978, Ann Rev. Astron. Astrophys. **16**, 77.
26. Di Benedetto, G. P., & Conti, G., 1983, Ap. J. **268**, 309.
27. McCarthy, D. W. et al., 1982, Proc. SPIE **332**, 57.
28. Beckers, J. M., et al, 1983, Proc. SPIE **440**.
29. Woolf, N. J. et al., 1983, Proc. SPIE **444**, 78.
30. Weigelt, G., 1983, "ESO's Very Large Telescope," ESO Conference and Workshop Proceedings, No. 17, p. 121.
31. Dugue, M. et al., 1983, "ESO's Very Large Telescope," ESO Conference and Workshop Proceedings No. 17, p. 141.
32. Stachnik, R. & Labeyrie, A., 1984, Sky & Telescope **67**, 205.
33. Fort, B., Picat, J. P., Cailloux, M., Mauron, N., Dreux, M., and Fauconnier, T., 1984, Astron. Astrophys **135**, 356.
34. Anoshkin, V. A., Petrov, G. G., and Savvin, A. V., 1984, Leningrad, p. 65.
35. Abramenko, A. N., Balyagin, A. V., Vereschagin, S. V., Pavlenko, E. P., and Prokof'eva, V. V., 1984, Leningrad, L. p. 77.
36. Monet, D. G., and Dahn, C. C., 1983, Astron. J. **88**, 1489.
37. Markov, V. S., 1982, "Research in Geomagnetism, Astronomy, and Physics of the Sun," **60**, 79.
38. Markov, V. S., 1984, Leningrad, p. 69.
39. Dowdney, K. C., 1984, "AURA Engineering Technical Report No. 75," available from NOAO, P.O. Box 26732, Tucson, Arizona 85726, USA.
40. McGraw, J. T., Stockman, H. S., Angel, J. R. P., Epps, H., and Williams, J. T., 1984, Steward Observatory Preprint No. 382.
41. Lelievre, G., Nieto, J.-L., Wlerick, G., Servan, B., Renard, L., and Horville, D., 1983, Comp. Rendu **296**, 1779.
42. Drazdis, R., Blazhavichus, Ch., and Yukonis, I., 1984, Leningrad, p. 24.
43. Petrov, P. P. 1983, Izv. Crimean Astrophys. Obs. **67**, 173.
44. Markelov, S. V., Nebelitsky, V. B., Somov, N. N., Somova, T. A., Spividonova, O. I., and Fomenko, A. F., 1984, Leningrad, p. 9.
45. Belaga, Yu. Yu., Kasperovich, A. N., Popov, Yu. A., Smov, N. N., and Fomenko, A. F., 1983, Izv. Crimean Astrophy. Obs. **67**, 176.
46. Kopylov, I. M., Somov, N. N., and Somova, T. A., 1984, Leningrad, p. 5.
47. Galinsky, N. D., 1983, Astron. Astrophys. **50**, 81.
48. Abramenko, A. N., Agapov, E. S., Anisimov, V. F., Galinsky, N. D., Prokof'eva, V. V., and Sinenok S. M., 1984, "Television Astronomy," (ed. B. V. Nikonov), Nauka, Moscow, p. 272.
49. Alexandrin, Yu. S., Kachmin, V. A., Sinenok, S. M., and V. N. Yakushin, 1984, Leningrad, p. 45.
50. Brown, T. M., Mihalas, B. W., and Rhodes, E. J., Jr., 1985 (In Press), Physics of the Sun (eds. P. A. Sturrock, T. E. Holzer, D. Mihalas, and R. K. Ulrich), D. Reidel.
51. Deubner, F.-L., and Gough, D., 1984, Ann. Rev. Astron. Astrophys. **22**, 593.

52. Scherrer, P. H., 1982, Pulsations in Classical and Cataclysmic Variable Stars, (eds. J. P. Cox and C. J. Hansen), JILA, 83.
53. Christensen-Dalsgaard, J., Meudon, 11.
54. Linsky, J., Personal Communication.
55. Smith, M. A., 1983, Astrophys. J. **265**, 325.
56. Kotov, V. A., Severny, A. B., and Tsap, T. T., 1982, Izv. Krymsk. Astrofiz. Obs. **66**, 3.
57. Scherrer, P. H., Wilcox, J. M., Christensen-Dalsgaard, J., and Gough, D. O., 1983, Solar Phys. **82**, 75.
58. Rhodes, E. J., Jr., Ulrich, R. K., Harvey, J. W., and Duvall, T. L., Jr., 1980, Solar Instrumentation: What's Next? (ed. R. B. Dunn), Sacramento Peak Observatory, 37.
59. Duvall, T. L., Jr., and Harvey, J. W., 1983, Nature **302**, 24.
60. Harvey, J. W., and Duvall, T. L., Jr., Snowmass.
61. Cochran, W. D., Smith, H. J., and Smith, W. H., 1982, Proc. SPIE **331**, 315.
62. Connes, P., 1984 (Preprint), "Absolute Astronomical Accelerometry."
63. Grec, G., Fossat, E., and Vernin, J., 1976, Astron. Astrophys. **50**, 221.
64. Brookes, J. R., Isaak, G. R., and van der Raay, H. B., 1978, Mon. Not. Roy. Astron. Soc. **185**, 1.
65. Cacciani, A., and Rhodes, E. J., Jr., Snowmass.
66. Rhodes, E. J., Jr., 1984, Personal Communication.
67. Fossat, E., Grec, G., Gelly, B., and Decanini, Y., 1984, The Messenger, No. 36, 20.
68. Appourchaux, T., 1984, Thesis, Universite de Paris.
69. Brown, T. M., Snowmass.
70. Forrest, A. K., 1984, Proc. SPIE **445**, 543.
71. Shepherd, G. G. et al., 1984, Appl. Opt. (Submitted).
72. Title, A. M., and Ramsey, H., 1980, Appl. Opt. **19**, 2046.
73. Fossat, E., 1984, Mem. Soc. Astron. Italiana **55**, 47.
74. Fossat, E., Meudon, 77.
75. Woodard, M., 1984, Thesis, University of California.
76. Fröhlich, C., 1984, Mem. Soc. Astron. Italiana, **55**, 237.
77. Winkler, C., Meudon, 125.
78. Schmidt-Kaler, T., and Winkler, C., 1984, Astron. Astrophys. **136**, 299.
79. Kotov, V., 1984, Personal Communication.
80. Kneer, F., Newkirk, G., Jr., and von Uexküll, M., 1982, Astron. Astrophys. **113**, 129.
81. Dame, L., Gouttebrouze, P., and Malherbe, J.-M., 1984, Astron. Astrophys. **130**, 331.
82. Pomerantz, M. A., Harvey, J. W., and Duvall, T. L., Jr., 1982, Antarctic J. U. S. **17**(5), 232.
83. Kurtz, D. W., 1984 Preprint, "The Ground Based Photometric Limitations to the Search for Light Variations due to 5-minute Solar Type Oscillations in Other Stars."
84. Heintze, J. R. W., de Jager, C., and Van der Veen, W., Meudon, 157.
85. Schmidt-Kaler, T., Meudon, 169.
86. Deubner, F.-L., and Isserstedt, J., 1983, Astron. Astrophys. **126**, 216.
87. Le Contel, J. M., and Valtier, J. C., Meudon, 213.
88. Noyes, R. W., Meudon, 113.
89. Noyes, R. W., Baliunas, S. L., Belserene, E., Duncan, D., Horne, J., and Widrow, L., 1984, Astrophys. J. **285**, L23.
90. Nissenson, P. and Noyes, R. W., 1984, Personal Communication.
91. Bos, R. J., and Hill, H. A., 1983, Solar Phys. **82**, 89.
92. Stebbins, R. T., and Wilson, C., 1983, Solar Phys. **82**, 43.
93. Stebbins, R., and Mann, R., 1983, Antarctic, J. U. S. **18**(5), 268.
94. Rösch, J., and Yerle, R., 1983, Solar Phys. **82**, 139.
95. Yerle, R., 1984, Mem. Soc. Astron. Italiana **55**, 123.
96, Anonymous, 1984, Appl. Opt. **23**, 3938.
97. Claverie, A., Isaak, G. R., McLeod, C. P., van der Raay, H. B., Palle, P. L., and Roca Cortes, T., 1984, Mem. Soc. Astron. Italiana **55**, 63.

98. Fossat, E., 1984, Personal Communication.
99. Global Oscillation Network Group, 1984, Project Proposal, National Optical Astronomy Observatories.
100. Zirin, H., 1984, Personal Communication.
101. Severny, A. B., and Kotov, V. A., 1985 Adv. Space Res. (In Press).
102. Anonymous, 1983, SOHO Assessment Study, European Space Agency, SCI(83)3.
103. Probing the Depths of a Star: The Study of Solar Oscillations from Space, 1984, (eds. R. W. Noyes, and E. J. Rhodes, Jr.), Jet Propulsion Laboratory.
104. Lemaire, P., Meudon, 177.
105. Schmidt-Kaler, T., Meudon, 191.
106. Hudson, H., Meudon, 197.
107. Praderie, F., and Mangeney, A., Meudon, 379.
108. Praderie, F., Mangeney, A., and Lemaire, P., 1985, Adv. Space Res. (In Press).

Note: Leningrad ≡ "New Techniques in Astronomy," Leningrad, Nauka, 1984.

Meudon ≡ "Proceedings of the Workshop on Space Research Prospects in Stellar Activity and Variability," (eds. A. Mangeney and F. Praderie) Observatoire de Paris-Meudon, 1984.

Snowmass ≡ "Solar Seismology from Space," (ed. R. K. Ulrich), Jet Propulsion Laboratory, In Press.

COMMISSION 10. SOLAR ACTIVITY (ACTIVITÉ SOLAIRE)

President: E. Tandberg-Hanssen Secretary: V. Gaizauskas
Vice President: M. Pick
Organizing Committee: A. Bhatnagar, V. Bumba, E.A. Gurtuvenko, M. Machado, D.B. Melrose, E.R. Priest, K. Tanaka, S.H. Ye, and H. Zirin

In preparing the present report, which covers the period July 1, 1981, to June 30, 1984, close collaboration has taken place between Commissions 10 and 12, the two solar commissions, in order to avoid duplications and to insure that pertinent subjects are treated. The reader is referred to the report of Commission 12 for further solar topics. It is a pleasure to acknowledge the excellent work of the reviewers who wrote the different sections of this report, and all the members of the commission who provided information on research to be included.

I. Ground-Based Optical Solar Instrumentation
(P. Mein)

A. NEW TELESCOPES AND CORONAGRAPHS

A new 50-cm horizontal telescope, built by C. Zeiss Jena and similar to the Ondrejov instruments, has been installed by the Slovak Academy of Sciences at Stara Lesna Observatory. The solar tower telescope of Nanjing University, China, is now in operation, and the telescope of Purple Mountain and its control system were renewed. In the framework of the Joint Organization for Solar Observations (JOSO), several telescopes are set up in the Canary Island Observatory. The Swedish instrument is operating at the Roque de Los Muchachos, and a 45-cm open telescope from the Utrecht Institute should be erected also at La Palma. The 45-cm Gregory-Coudé from the Göttingen University and the 60-cm Vacuum Tower Telescope from the Kiepenheuer Institute will be ready at Izana in 1985 (JOSO report 1983). On the same site, the 90-cm French polarization-free telescope THEMIS is planned before 1990. As a final goal of JOSO, a Large European Solar Telescope (LEST) is presently promoted by the LEST Foundation. The aperture should be 240 cm. Active and adaptive optics should be included in this new generation instrument. Attempts for defining a Large Multipurpose Solar Telescope were also made at Pulkovo Observatory; seeing conditions are analyzed in this context at the Kislovodsk Station. A one-dimensional aperture telescope is presently under construction in Nice. It is especially suited for brightness and velocity speckle-interferometry measurements (Aime et al. 1983). The stability of telescope drives has been investigated by Hammerschlag with a special view to wind effects on open telescopes (Adv. Tech. Opt. Tel., SPIE 1982). Improvements were undertaken at the tower of the Sacramento Peak Observatory (SPO). Tests have been made to use helium instead of vacuum inside the telescope. The window could be replaced by a thinner one. In order to compensate image motions due to seeing effects, a correlation tracker was realized with a 32 x 32 pixel Reticon array; the rate of FFT calculations is around 1 kHz (von der Luke 1983). Electronic detectors are used more and more in solar astronomy. The multiple diode array of SPO includes eight units; a flat field source calibrates the interference fringes. Two new coronagraphs are operated by the High Altitude Observatory (HAO) at Mauna Loa, Hawaii. One of them (23 cm) allows inner corona observations with linear diode array detectors; the other (12.5 cm) is devoted to Hα limb activity (Fisher et al. 1981). A new 20-cm emission-line coronagraph is also in use at SAO; the level of scattered light is remarkably small (SPIE 1984). Electron density in coronal active regions can be measured at the Pic-du-Midi by the K-coronameter (Noens & Leroy 1981). The Huancayo Observatory (Peru) is setting up a 12-cm coronagraph at the

Cosmos Station; because of financial difficulties, Fabry-Perot observations of λ5303 will be carried out in 1985 during only 2 months.

B. SPECTROMETERS AND FILTERS

Most of the new telescopes are equipped with spectrographs. A long Cerny-Turner Spectrograph (10 m) is associated with the Slovak instrument. A multiple spectrograph is fed by the Nanjing telescope, and a wide wavelength range can be covered at the Purple Mountain Observatory (Acta Ast. Sinica 16, 1975; 24, 1983). The spectrograph installed on the 50-cm refractor of the turrett dome at the Pic-du-Midi can be used in the mode of a multichannel subtractive double pass, which allows high spatial resolution in two-dimensional spectroscopy. The Meudon spectroheliograph is presently renewed. At the Canary Island Observatory, the Gregory-Coudé and the VTT will have 9.3 and 15 m spectrographs respectively. The latter will be equipped with a MSDP device built in Meudon (JOSO 1983). The SPO universal birefringent filter (UBF) has been improved. Absolute calibration is provided by laser source, and the filter can be tuned to a selected wavelength in less than 1 s. The Arcetri Observatory is developing a UBF in view to the German VTT set up in the Canary Islands (JOSO 1982). The filter is tried now at the Pic-du-Midi, and the wavelength control should be digitized in the near future. The universal birefringent filter FPSS of Meudon is designed for velocity and magnetic field observations (Huntsville workshop 1984). A high resolution (2 to 5 cm) computer-controlled universal filter is in progress at Culgoora. It will be mounted on a 30-cm telescope.

C. POLARIMETERS AND MAGNETOGRAPHS

The vector magnetograph of the Okayama Observatory, Japan, has been in operation since 1982; the polarization of the 65-cm Coudé telescope is compensated (Huntsville Workshop, 1984). The Mount Wilson Magnetograph has been renewed (Howard 1983), and a wavelength calibration device is included (Bruning & Howard 1983). At Big Bear Observatory, the videomagnetograph was also improved by a new calibration system using solar rotation (Patterson 1983). The large coronagraph (53 cm) of the Crimean Observatory is now equipped with a linear polarization analyzer. A vector magentograph using KD*P crystals as modulators is now developed at Beijing Observatory. For special study of flares, linear polarization analysis is also available at Meudon Observatory on the refractor devoted to Hα observations.

D. CHROMOSPHERE AND SOLAR ACTIVITY OBSERVATIONS

Many instruments observe the solar activity daily throughout the world. Some of them are well known. Others must be mentioned here, even though they are not recent. At Purple Mountain Observatory, a 20-cm refractor is devoted to Hα patrol, with the possibility of local enlargement. Relative numbers of sunspots are given by the 8-cm instrument of Huancayo. The Manila Observatory (Philippines) makes daily white-light observations and Hα-K spectroheliograms. Two 6-cm refractors are used at Oelsnitz Observatory (RDA) for the study of the frequency of spots versus latitude. A double telescope (20 and 15 cm) provides white-light and Hα observations at the Stara Lesna Observatory. An eclipse-spectrograph built by Beijing and Nanjing Observatories was operating in June 1983 for chromosphere and corona observations. Some improvements in processing or acquisition of data can also be mentioned. A sunspot area meter using a TV camera is in use at Debrecen. Spectral observations of spicules are optimized at the Abastumani Observatory, Georgian SSR, by the use of automatization and special devices for line position measurements.

E. GLOBAL OSCILLATIONS, LARGE-SCALE MOTIONS, AND SOLAR DYNAMO

Helioseismology can be supported by networks of small telescopes measuring velocity oscillations without any interruption. Following the South Pole experiments, a project called IRIS and using resonance cell detectors is presently being developed by the Nice group. A similar project is planned by the

Global Oscillation Network group. Further data can be obtained with instruments observing the solar disk with various spatial resolutions. This is the case of the filter built by Cacciani et al. (1981). For high degree oscillations, we must mention the Mount Wilson system (Rhodes et al. 1983) using a CID camera. A second version of the Fourier tachometer developed by SPO and HAO is now operating; with the 100 x 100 Reticon array, the velocity noise is less than 5 ms^{-1} for each pixel (a few cm s^{-1} for the full-disk average velocity). The spatial distribution of velocity fields can be provided by the two-dimensional solar spectrometer of Birmingham, with a view to oscillations as well as large-scale motions. The resonance-cell instrument developed in Bordeaux (Robillot et al. 1984) is also suited for differential rotation measurements. Solar diameter fluctuations complement velocities for helioseismology studies. The Pic-du-Midi heliometer is designed for high accuracy measurements. The solar diameter monitor (Brown et al. 1982) is mainly devoted to long-term effects (1 arc sec/century), as is the Astrolabe of CERGA Observatory, France (Laclare 1983). Observations of global brightness variations of the photosphere are achieved at the Crimean Observatory where a 16 x 16 detector matrix is used for center-limb measurements. For stellar comparisons, the solar cycle can be investigated by spectrometers observing the Sun as a whole. In this context, the Selective Solar Irradiance Spectrometer (Oranje 1982) realizes spectrograms of Ca II lines from the entire solar disk, as well as from smaller areas.

References

Acta Astron. Sinica: 1975, Vol. 16, 225.

Acta Astron. Sinica: 1983, Vol. 24, 189.

Aime, C., Demarcq, J., Martin, F., and Ricort, G.: 1983, Optical Engr. 22(2), 224-226.

Brown, T.M., Elmore, D.F., Lacey, L., and Hull, H.: 1982, Appl. Optics 21(19), 3588-3596.

Bruning, D.H. and Howard, R.: 1983, BAAS 15, 701.

Cacciani, A., Croce, V., Fortini, T., and Torelli, M.: 1981, Solar Phys. 74, 543-546.

Fisher, R.R., Lee, R.H., MacQueen, R.M., and Poland, A.I.: 1981, Appl. Optics 20(6), 1094-1101.

Hammerschlag, R.H.; 1982, SPIE 332, Adv. Tech. Opt. Tel.

Howard, R., Byden, J.E., Bruning, D.H., Clark, M.K., Crist, H.W., and Labonte, R.J.: 1983, Solar Phys. 87, 195-203.

Joint Organization of Solar Observations: 1982, p. 67, Inst. Theoret. Astrophys., Oslo.

Joint Organization of Solar Observations: 1983, Inst. Theoret. Astrophys., Oslo.

Laclare, F.: 1983, Astron. Astrophys. 125, 200-203.

Noens, J.-C. and Leroy, J.-L.: 1981, Solar Phys. 73, 81-87.

Oranje, B.J.: 1982, Astron. Astrophys. 109, 32-36.

Patterson, A.P.: 1983, BAAS 15, 718.

Rhodes, E.J., Jr., Howard, R.F., Ulrich, R.K., and Smith, E.J.: 1983, Solar Phys. 82, 245-258.

Robillot, J.M., Bocchia, R., Fossat, E., and Grec, G.: 1984, Astron. Astrophys. 137, 43-50.

SPIE: 1984, Vol. 388, 395.

von der Luke, O.: 1983, Astron. Astrophys. 119, 85-94.

Workshop on Measurements of Solar Vector Magnetic Fields, Huntsville: 1984, M.J. Hagyard (ed.), NASA CP (in press).

II. Solar Maximum Year

(D.M. Rust)

A. ACTIVITIES

Work begun during the Solar Maximum Year (SMY) (1980-1981) was the focus of several sympsoia and workshops starting in March 1981 with the International SMY Workshop in Simferopol, Crimea (Obridko 1981). Results of the Annecy SMY Workshop (Annecy, France, October 1981) were presented in May 1982 with later work presented at the first SMY symposium (Švestka et al. 1982) in Ottawa, Canada. A special symposium for the Study of Travelling Interplanetary Phenomena was held in Maynooth, Ireland, in August 1982 (Shea et al. 1984). Another SMY workshop was held in November 1983 in Kunming, China (Chen 1984), and the second SMY symposium was held in June 1984 in Graz, Austria (Simon 1985). The scientific progress of the SMY is detailed in the above-mentioned publications and in the 1982 U.S.-Japan Seminar "Recent Advances in the Understanding of Solar Flares" (Kane et al. 1983). Highlights of the work of the SMY's Flare Build-Up Study, Study of Energy Release in Flares, and Study of Travelling Interplanetary Phenomena are presented below.

B. RESEARCH

1. Flare Build-Up

a. The Magnetic Field Situation. An important development has been the increasing utilization of measurements of the transverse component of the photospheric magnetic fields beneath flares. Although the interpretation of the polarization measurements from which the magnetic fields are inferred is still beset with uncertainties (Stenflo 1984), maps of the transverse field outside sunspots may give useful clues to the magnetic field configuration in flares. When measured transverse fields differ significantly from potential fields, one says that the degree of shear is large. Hagyard et al. (1984) found that several flares occurred at persistent sites of high photospheric field shear. Krall et al. (1982) used the proper motions of sunspots in the same series of observations to infer how much energy is in sheared magnetic fields. Their results on the best-studied (Strong et al. 1985) active region of the SMY (AR 2372, April 6-14, 1980) agree with earlier studies in implying that the transverse motions of sunspots cause storage of $\sim 10^{32}$ ergs day^{-1} in the magnetic fields. The flares in AR 2372 were homologous (Martres 1985) in that the morphology of each flare differed only slightly from the preceding and following flares. Most of the flares in the region originated at a growing bipolar feature that moved laterally, thus distorting the photospheric neutral line in the sunspot region. This familiar scenario was studied analytically by Low (1982), who showed that it leads to the creation of magnetic free energy, severe shearing of low-lying loops, and the emergence and disappearance of flux. Low showed how to find the magnetic free energy from vector magnetograph observations and potential field calculations. The theoretical work by Low and others and the renewed interest in vector magnetographs have not produced any breakthroughs yet. But the models are better, e.g., three-dimensional, non-constant-α force free fields versus the old two-dimensional constant-α approximation, and terms are being more clearly defined. The next IAU report may include news of a convincing, quantitative demonstration of magnetic free energy accumulation and release.

b. Preflare Emission. The sites of magnetic free energy accumulation might be detectable by increased emission or turbulence, and many new reports of preflare brightening and filament activation were received (Simon 1985, Simon et al. 1984). Lang & Willson (1985) used the VLA at 20 cm to study preburst heating. Their results are consistent with other studies in showing preflare temperatures T_e 2 x 10^7 K and preflare densities n_e 10^{10} cm^{-3}. Models of chromospheric heating by electron beams are hard to reconcile with such a high preflare density.

2. Flare Energy Release

a. Impulsive Phase. The improved time resolution of instruments placed in regular use for the first time during the SMY has yielded important new data on the fundamental processes of the impulsive phase. Kaufmann et al. (1984) achieved a time resolution of ~1 ms at 22 and 44 GHz and reported on a spike burst that had quasiperiodic structure at a frequency of 16-20 s^{-1}. The frequency is too high to have been detected with the SMM Hard X-Ray Burst Spectrometer, which also recorded the burst. However, the SMM record did show fine structure at the 0.3 s level, suggestive of unresolved higher frequencies. The authors believe that the ultra-fast time structures resulted from discrete emission events from individual flare kernels. If this interpretation is correct, the kernels emit at extremely high radio brightness temperatures. A maser mechanism was suggested by Melrose & Dulk (1984) to explain high radio burst brightness temperatures and the apparent cross-field spread of flare excitation. Their theory is further discussed below. Other microwave observations of ultra-fast structures were made by Takakura et al. (1983), who found that the millimeter structures correlated very well with hard x-ray structures recorded at the Hinotori satellite, and by Zhao (1982), Fu et al. (1984), and Li et al. (1984). Ultra-fast fine structures may be absent for several hundred milliseconds and then reappear in clusters of spike bursts. It is not clear, then, whether the bursts reflect a fundamental phenomenon, such as emission from elementary flare kernels, or a secondary phenomenon, such as Type III bursts. The high brightness temperatures (10^{10}-10^{15} K) inferred for the microwave bursts mean that the radio emission must result from a non-thermal population of electrons. Melrose & Dulk (1982) suggested that even the mild anisotropies that result from precipitation from flare loops of electrons with small pitch angles can lead to maser radiation. Maser radiation at the first harmonic of the local gyrofrequency would not escape the immediate vicinity of the flaring loop. Melrose and Dulk suggest that it is by absorption of this intense microwave radiation that loops adjacent to a primary energy release site are heated. The bright microwave bursts would, in their view, be second harmonic gyroradiation. This and other interpretations of the ultra-fast microwave emission were reviewed by Holman (1982). The most important result about the impulsive phase, described in the previous IAU review period, was the detection of hard x-ray emission at the feet of flare loops and of microwave emission at the apex of the loops (Rust 1982). Now, with many more observations in hand, the picture we had of an electron beam accelerated at the top and precipitated at the feet has become cloudy. Kundu's (1984) latest review of VLA and hard x-ray flare images includes nearly every permutation of microwave and x-ray sources that five-color figures allow. It shows that the idealizations used earlier to interpret the first results cannot be relied upon. Current interpretation of the SMM results is reviewed by Machado in this report.

b. Thermal Conduction and Electron Beams. The physical processes that transport energy from the flaring corona to the chromosphere were the subjects of many SMY studies. Henoux et al. (1983) found linear polarization in chromospheric lines (Hα and SI 1437 Å) during the flare gradual phase and found that it was almost certainly not caused by direct excitation by 10-100 keV beamed electrons. The polarization is best understood as a result of anisotropic excitation by thermal electrons carrying heat from the corona through the transition zone and into the chromosphere. The conductive heat flux can be deduced from the observed degree of linear polarization. Henoux et al. concluded that the conductive flux was of the same order of magnitude as the chromospheric radiation that finally carries the added heat away. Rust et al. (1985) found fast-moving fronts in sequences of soft x-ray images from the Utrecht-Birmingham x-ray imager on SMM, and interpreted them as thermal conduction fronts. The observed velocities were 900-1700 km s^{-1}, and in one case for which simultaneous Hα data were available, they concluded the chromospheric radiation losses could not be accounted for by the energy in an electron beam. The Hα emission was sustained by thermal conduction after the first 30 s of the flare. Kämpfer & Magun (1983) also investigated simultaneous Hα and coronal emissions. A solid-state, two-dimensional array detector recorded Hα brightness variations with 1.4 s time resolution while a radiospectrograph recorded 8.4-19.6 GHz microwaves with 100 ms resolution. The result was that, although one Hα kernel brightened simultaneously with the radio source, other kernels experienced delays of up to 30 s. Evidently, beams were effective in transporting energy, probably by thick-target collisions, in one kernel, while most energy was carried to the chromosphere by slower-moving conductive processes. Another approach to differentiating between electron beam and thermal processes in flares was taken by Brown et al. (1983). They compared x-ray flux ratios predicted for five flares seen by two widely-separated spacecraft, ISEE-3 and Pioneer-Venus Orbiter, in order to determine the height structure of the sources of hard x rays. In each of the five cases, one spacecraft saw radiation from the entire flare, while the other saw radiation from only the upper portion. Brown et al. found that the energy dependence of the occultation ratio was not consistent with that predicted by the thick-target model. In three events, non-collisional losses must have been important; and, in two events, bremsstrahlung emission other than that from a beam must have been important. The effect of electron beams and of thermal conduction on the chromosphere was considered in a number of papers from the San Diego group (Ricchiazzi & Canfield 1983, Canfield et al. 1982, Canfield et al. 1984a). Using numerical models of static flare chromospheres and theoretical Hα line profiles to interpret observations obtained at Sacramento Peak, they were able to distinguish between the effects of intense electron beam heating and thermal conduction. In two small events Canfield et al. (1984b) found that the thick-target input power implied by the Hα observations was indistinguishable from that inferred from the corresponding hard x-ray bursts. Other studies of hard x-ray bursts and chromospheric phenomena, including chromospheric evaporation, were published by Antonucci & Dennis (1983), Antonucci (1982), Somov et al. (1982), and many others. At the present time, the relative importance of thermal and non-thermal phenomena in the impulsive phase of flares is still not clear. It may be that beams play an energetically minor role in chromospheric heating (Rust 1985), albeit an interesting and eye-catching role. A better understanding of the preflare corona seems to be required, because the models are strongly dependent on assumptions about preflare density and the magnetic field configuration.

c. White-light Flares. An area of outstanding progress was in observations of white-light flares. Before the SMY, only one or two flares had been observed spectroscopically, but an intensive program at the Sacramento Peak Observatory and at the Big Bear Observatory yielded many broadband filter observations as well as some superb high-resolution spectra (Neidig & Wiborg 1984). The Paschen jump near 8500Å was seen for the first time and it indicates that the Paschen continuum may be a substantial contributor to the visible flare continuum. A complete understanding of white-light flare emission is not available yet, since the blue excess (Neidig 1983) is not understood. Also, some white-light flare features are closely associated with hard x-ray spikes, while others are found in the main phase ribbons (Boyer et al. 1985). White-light flares had been thought to be caused by protons, but the SMM gamma-ray observations of the July 1, 1980, flare (Ryan et al. 1983) showed conclusively that the gamma-ray emission due to protons was not correlated with the optical continuum emission (Zirin & Neidig 1981). Models to explain white-light flare emissions were published by Livshits et al. (1981) and Dame & Cram (1983).

d. Main Phase and Coronal Transients. Understanding of the main phase of flares has reached a very high level, with the model of Kopp & Pneuman (1976) undergoing continual refinements and observational tests. It now appears that coronal magnetic field lines do open outward with the eruption of an Hα filament and a coronal transient (Low et al. 1982), and it is the reconnection and contraction of the associated magnetic field back into a system of loops that drives the main phase of flares. The process has most recently been modeled by Forbes & Priest (1982), Kopp & Poletto (1984), and Pneuman (1982). For a discussion of the coronal transient and its passage into interplanetary space, the reader is referred to Howard (1985) and Shea et al. (1984).

References

Antonucci, E.: 1982, Mem. Soc. Astr. Ital. 53, 495.

Antonucci, E. and Dennis, B.R.: 1983, Solar Phys. 86, 67.

Boyer, R., Machado, M.E., Rust, D.M., and Sotirovski, P.: 1985, Solar Phys. (submitted).

Brown, J.C., Carlaw, V.A., Cromwell, D., and Kane, S.R.: 1983, Solar Phys. 88, 281.

Canfield, R.C., Gunkler, T.A., Hudson, H.S., Acton, L.W., Leibacher, J.W., and Kiplinger, A.L.: 1982, Adv. Space Res. 2, 145.

Canfield, R.C., Gunkler, T.A., and Kiplinger, A.L.: 1984a, Adv. Space Res. (in press).

Canfield, R.C., Gunkler, T.A., and Ricchiazzi, P.J.: 1984b, Astrophys. J. (in press).

Chen, M.B. (ed.): 1984, Proc. Int. Workshop on Solar Phys. and Interplanet. Travelling Phenom., Kunming, China, Acad. Sinica (in press).

Dame, L. and Cram, L.: 1983, Solar Phys. 87, 329.

Forbes, T.G. and Priest, E.R.: 1982, Solar Phys. 81, 303.

Fu, Q. et al.: 1984, Proc. Int. Workshop on Solar Phys. and Interplanet. Travelling Phenom., Kunming, China, Acad. Sinica (in press).

Hagyard, M.J., Smith, J.B., Jr., Teuber, D., and West, E.A.: 1984, Solar Phys. 91, 115.

Henoux, J.-C., Chambe, G., Heristchi, D., Semel, M., Woodgate, B., Shine, D., and Beckers, J.: 1983, Solar Phys. 86, 115.

Holman, G.D.: 1982, Adv. Space Res. 2, 181.

Howard, R.: 1985, Reports on Astronomy 1985 (this volume).

Kämpfer N. and Magun, A.: 1983, Astrophys. J. 274, 910.

Kane, S.R., Uchida, Y., Tanaka, K., and Hudson, H.S. (eds.): 1983, "Recent Advances in the Understanding of Solar Flares," Solar Phys. 86(1/2).

Kaufmann, P., Correia, E., Costa, J.E.R., Dennis, B.R., Hurford, G.J., and Brown, J.C.: 1984, Solar Phys. 91, 359.

Kopp, R.A. and Pneuman, G.W.: 1976, Solar Phys. 50, 85.

Kopp, R.A. and Poletto, G.: 1984, Solar Phys. 93, 351.

Krall, K.R., Smith, J.B., Jr., Hagyard, M.J., West, E.A., and Cummings, N.P.: 1982, Solar Phys. 79, 59.

Kundu, M.R.: 1984, Adv. Space Res. (in press).

Lang, K.R. and Willson, R.F.: 1985, Adv. Space Res. (in press).

Li, C. et al.: 1984, Proc. Int. Workshop on Solar Phys. and Interplanet. Travelling Phenom., Kunming, China, Acad. Sinica (in press).

Livshits, M.A., Badalyan, O.G., Kosovichev, A.G., and Katsova, M.M.: 1981, Solar Phys. 73, 269.

Low, B.C.: 1982, Solar Phys. 77, 43.

Low, B.C., Munro, R.H., and Fisher, R.R.: 1982, Astrophys. J. 254, 335.

Martres, M.J.: 1985, Adv. Space Res. (in press).

Melrose, D.B. and Dulk, G.A.: 1982, Astrophys. J. 259, 844.

Melrose, D.B. and Dulk, G.A.: 1984, Astrophys. J. 282, 308.

Neidig, D.F.: 1983, Solar Phys. 85, 285.

Neidig, D.F. and Wiborg, P.H., Jr.: 1984, Solar Phys. 92, 217.

Obridko, V.N. (ed.): 1981, Solar Maximum Year, Proc. of Int. Workshop, IZMIRAN, Acad. Sciences, USSR, Moscow.

Pneuman, G.W.: 1982, Solar Phys. 78, 229.

Ricchiazzi, P.J. and Canfield, R.C.: 1983, Astrophys. J. 272, 739.

Rust, D.M.: 1982, in P.A. Wyman (ed.) Reports on Astronomy, D. Reidel, Dordrecht, p. 72.

Rust, D.M.: 1985, Adv. Space Res. (in press).
Rust, D.M., Simnett, G.M., and Smith, D.F.: 1985, Astrophys. J. (in press).
Ryan, J.M., Chupp, E.L., Forrest, D.J., Matz, S.M., Rieger, E., Reppin, C., Kanbach, G., and Share, G.H.: 1983, Astrophys. J. Letters 272, L61.
Shea, M.A., Smart, D.F., and McKenna-Lawlor, S.M.P. (eds.): 1984, STIP Symposium on Solar/Interplanetary Intervals, Book Crafters, Chelsea, Michigan.
Simon, P. (ed.): 1985, "Solar Maximum Analysis," in Adv. Space Res. (in press).
Simon, G., Mein, N., Mein, P., and Gesztelyi, L.: 1984, Solar Phys. 93, 325.
Somov, B.V., Sermulina, B.J., and Spektor, A.R.: 1982, Mem. Soc. Astr. Ital. 53, 495.
Stenflo, J.: 1984, in M.J. Hagyard (ed.) Workshop on Measurements of Solar Vector Magnetic Fields, Huntsville, NASA CP (in press).
Strong, K.T., Smith, J.B., Jr., McCabe, M., Machado, M.E., and Simnett, G.M.: 1985, Adv. Space Res. (in press).
Švestka, Z., Rust, D.M., and Dryer, M. (eds.): 1982, "Solar Maximum Year," in Adv. Space Res. 2(11).
Takakura, T., Kaufmann, P., Costa, J.E.R., Deganokar, S.S., Ohki, K., and Nitta, N.: 1983, Nature 302, 317.
Zhao, R-Y.: 1982, Adv. Space Res. 2, 177.
Zirin, H. and Neidig, D.F.: 1981, Astrophys. J. Letters 248, L45.

III. Solar Maximum Mission Results

(M.E. Machado)

The ongoing research carried out by the solar community has been reported in the proceedings of several recent symposia, seminars and workshops, as well as in scientific journals (Kane et al. 1983, Švestka et al. 1982a, Shea et al. 1984, Kundu & Woodgate 1984, Simon 1984). We summarize here some of the novel results with reference to flare research as far as SMM data analysis is concerned. Understanding of impulsive phase phenomena was one of the primary goals of the SMM. The early reports from the analysis of the first ever obtained high-resolution images in the $\leqslant 30$ keV energy range stressed the fact that some flares showed hard x-ray (HXR) bright sources at the feet of coronal loops (Hoyng et al. 1981a,b, Machado et al. 1982, Duijveman et al. 1982), the so-called HXR "footpoints," favoring the thick-target beam mechanism for the production of HXRs, and indicating acceleration efficiencies $\geqslant 20\%$ during the early impulsive phase. This phenomenon was shown to be accompanied by soft x-ray (SXR) line broadening, indicative of strong turbulence, and the immediate appearance of blue shifted spectral lines, which shows that plasma heated to $\geqslant 10^7$ K rises from the footpoints of loops with velocities to 300 km s^{-1} (Antonucci et al. 1982, Antonucci et al. 1984a). This result provides a strong indication of the chromospheric evaporation phenomenon, which has been confirmed in analyses of combined SXR and Hα observations (Acton et al. 1982, Gunkler et al. 1984).

High temporal resolution HXR observations ($E \leqslant 300$ keV) have revealed the existence of millisecond time variations (FWHM ≈ 45 ms)during the impulsive bursts. These place new constraints on the physical nature of the HXR source. Particularly, in a non-thermal scenario the observed time scales are upper limits to the temporal scale of the acceleration process itself, while within the context of thermal HXR models, it follows that they lose their energetic advantage over the non-thermal hypotheses (Kiplinger et al. 1983). Along the same lines, the combined analysis of HXR and UV impulsive bursts, particularly those observed in the OV line at 1371 Å, formed at $T \simeq 2.5 \times 10^5$ K, has shown a simultaneity in the burst structures to within $\leqslant 1$ s, limited by the UV temporal resolution (Woodgate et al. 1983). There is also very good correlation between the appearance of HXR and UV spikes, and the density of the UV burst region as determined from the Si IV/O IV line ratio (Cheng et al. 1981) yields a range $5 \times 10^{12} \leqslant n \leqslant 10^{13}$ cm^{-3}, confirming that UV burst emission originates below the height of the preflare transition region. An important aspect stemming from this comparison arises from the fact that hydrodynamic simulations have established that the response of the transition region should be different in conductively heated and particle heated models. Only the latter predicts a strong temporal correlation of UV intensity with energy input, as measured by the HXR flux (Emslie & Nagai 1984). It is thus obvious that the HXR/UV correlation seems to favor the beam model, a fact supported by the spatial coincidence of HXR and UV footpoints in cases where common data sets have been analyzed (Machado et al. 1982, Duijveman et al. 1982, Poland et al. 1982).

Important information concerning the acceleration of ions has been provided by the gamma-ray observations. Simultaneous peaking of <100 keV and >10 MeV photons has been observed, accompanied by direct evidence of ion acceleration revealed by line emission (<10 MeV), meaning that all energetic particles were injected into and interacted with the target medium at the same time (≤ 1 s), or they were produced (accelerated) simultaneously (Forrest & Chupp 1983, Chupp 1983, 1984), in contradiction with widely accepted concepts before the SMM launch. High-energy neutrons were also observed for the first time by the SMM, indicating the rapid acceleration ($\ll 60$ s) of GeV protons during the impulsive phase. The limb flare observations require densities $\leq 5 \times 10^{15}$ cm^{-3} in the interaction region, so that the flux is not greatly attenuated. On the other hand, prompt gamma-ray lines have to be produced in a region with density $>10^{11}$ cm^{-3}. The SMM events with significant emission above 10 MeV show a concentration towards the limb (Chupp et al. 1982), a fact that provides evidence for downward beaming in agreement with previous statements. Although the evidence for particle beaming is strong, the imaging observations have also indicated that a substantial fraction of the HXR output may be thermal in nature (Rieger et al. 1983). The presence of hot ($\geq 4 \times 10^7$ K) sources embedded within the flaring loops or at the boundary between field structures has been detected (Hoyng et al. 1981b, Machado et al. 1982, Duijveman 1983, Machado 1983, Machado & Lerner 1984). In the case of a two ribbon flare observations, these have been interpreted as the heating byproduct of the reconnection of field lines blown open during the filament eruption. The same x-ray imaging observations have also provided some observational evidence for thermal wave fronts, whose flux may comprise a large fraction of the released flare energy (Rust 1984, Rust et al. 1984, Švestka & Poletto 1984).

Two other aspects of SXR line spectroscopy during impulsive and gradual flare phases have been considered. The spectral analysis leads to densities of the order of 10^{12} cm^{-3} in the 10^7 K emitting region (Phillips et al. 1982, Wolfson et al. 1983). At the same time it has been determined that, due to the high densities of flaring loops, ionization balance is very close to steady state over almost all phases of flares for which statistically good data are available (Antonucci et al. 1984b). Finally, the behavior of the Fe Kα emission near 1.94 Å has been shown to be consistent with photoionization by x-rays with energy >7.11 keV (fluorescence). In a limited number of cases, the Kα light curve showed impulsive behavior coincident with HXRs, which could be indicative of electron impact ionization, but these results are rare (Culhane et al. 1981, Parmar et al. 1984) perhaps because they are masked by the overwhelming effect of fluorescence. With respect to the origin of all these energy release by-products, the SMM observations have pointed out that a very large fraction of the x-ray flares are composed of two or more loop structures which intersect at a common point (Machado et al. 1983, Rust & Somov 1984, Acton et al. 1983). Only in few cases (Tandberg-Hanssen et al. 1984), flaring may occur spontaneously within a simple nonpotential arch. This fact has thus stressed the importance of flux emergence and evolution in the production of flares through magnetic reconnection. Magnetic shear (Hagyard et al. 1984) is also well correlated, and defines which of the possible flaring structures may possess the strongest high-energy phenomena. The association of Type III/V radio events, whose components may appear at widely different locations with strong HXR bursts, also suggests that accelerated particles have immediate access to various field structures. Thus, the acceleration/injection site may comprise a rather large volume of interacting field lines (Pick & Raoult 1984, and Chapter 2 of Kundu & Woodgate 1984). In the high corona, broad expansions of the SXR emitting regions have been detected right at the maximum of the flares, a newly discovered phenomenon which has been called "coronal explosion" (de Jager & Boelee 1984, de Jager et al. 1984). Another discovery was that of the gigantic post flare arches which appear after strong two ribbon flares (Švestka et al. 1982b,c). These seem to be related to Type I noise storms observed for many years in the radio records, and sometimes undergo in situ variations in SXR high in the corona, without an obvious chromospheric counterpart (Švestka et al. 1983). They demonstrate the importance of large-scale magnetic fields in relation to flaring activity. Regions of high temperature plasma ($\geq 10^7$ K) in non-flaring active regions have also been detected in the imaging data (Schadee et al. 1983), while spectra show line widths broader than the Doppler width corresponding to the local electron temperature (Acton et al. 1981).

Finally, the SMM data have provided new information on coronal mass ejections (CMEs) (Wagner 1982, 1984), which are a clearly important manifestation of activity, with mass and energy content which may exceed other flaring manifestations (Wagner et al. 1981). The CMEs may be related either to the flare explosions or eruptive prominences and, surprisingly, their frequency did not change by more than 20% as compared with the Skylab (1973-74) period (Wagner 1984, Hundhausen et al. 1984). This seems to be related to a similar phenomenon in eruptive prominences, and its explanation is still controversial. The observations have also shown that the shape of these transients favors in many cases a three-dimensional bubble geometry over the commonly accepted planar loop. Comparison of CME observations with radio data (Stewart et al. 1982, Gary et al. 1984), as well as extrapolated estimates of starting times, casts strong doubts on the physical association of this phenomenon with flare blast shocks. The observations have also revealed (House et al. 1981) frequent Hα emission from insulated cool remnants of eruptive prominences seen in the outer corona. There is a type of CMEs, those most commonly observed, which appears to be unassociated with flares and eruptive prominences (Wagner 1984). These are of great interest in themselves, not only because they represent a distinct physical class, but also due to the fact that they may be indicators of the evolution of large-scale fields related to dynamo processes internal to the Sun (Wagner 1984).

References

Acton, L.W., Culhane, J.L., Gabriel, A.H., Wolfson, C.J., Rapley, C.G., Phillips, K.J.H., Antonucci, E., Bentley, R.D., Hayes, R.W., Joki, E.G., Jordan, C., Kayat, M.A., Kent, B., Leibacher, J.W., Nobles, R.A., Parmar, A.N., Strong, K.T., and Veck, N.J.: 1981, Astrophys. J. Letters 244, L137.

Acton, L.W. Canfield, R.C., Gunkler, T.A., Hudson, H.S., Kiplinger, A.L., and Leibacher, J.W.: 1982, Astrophys. J. 263, 409.

Acton, L.W., Brown, W.A., Bruner, M.E.C., Haisch, B.M., and Strong, K.T.: 1983, Solar Phys. 86, 79.

Antonucci, E., Gabriel, A.H., Acton, L.W., Culhane, J.L., Doyle, J.G., Leibacher, J.W., Machado, M.E., Orwig, L.E., and Rapley, C.G.: 1982, Solar Phys. 78, 107-123.

Antonucci, E., Gabriel, A.H., and Dennis, B.R.: 1984a, Astrophys. J. (in press).

Antonucci, E. Gabriel, A.H., Doyle, J.G., Dubau, J., Faucher, P., Jordan, C., and Veck, N.: 1984b, Astron. Astrophys. 133, 239.

Cheng, C.-C., Tandberg-Hanssen, E., Bruner, E.C., Orwig, L., Frost, K.J., Woodgate, B.E., and Shine, R.A.: 1981, Astrophys. J. Letters 248, L39.

Chupp, E.L.: 1983, Solar Phys. 86, 384.

Chupp, E.L.: 1984, Ann. Rev. Astron. Astrophys. 22,

Chupp, E.L., Forrest, D.J., Ryan, J.M., Heslin, J., Reppin, C., Pinkau, K.,

Kanbach, G., Rieger, E., and Share, G.H.: 1982, Astrophys. J. Letters 263, L95.
Culhane, J.L., Gabriel, A.H., Acton, L.W., Rapley, C.G., Phillips, K.J.H., Wolfson, C.J., Antonucci, E., Bentley, R.D., Catura, R.C., Jordan, C., Kayat, M.A., Kent, B.J., Leibacher, J.W., Parmar, A.N., Sherman, J.C., Springer, L.A., Strong, K.T., and Veck, N.J.: 1981, Astrophys. J. Letters 244, L141.
de Jager, C. and Boelee, A.: 1984, Solar Phys. 92, 227.
de Jager, C., Boelee, A., and Rust, D.M.: 1984, Solar Phys. 92, 245.
Duijveman, A.: 1983, Solar Phys. 84, 189.
Duijveman, A., Hoyng, P., and Machado, M.E.: 1982, Solar Phys. 81, 137.
Emslie, A.G. and Nagai, F.: 1984, Astrophys. J. (in press).
Forrest, D.J. and Chupp, E.L.: 1983, Nature 305, 291.
Gary, D.E. Dulk, G.A., House, L., Illing, R., Sawyer, C., Wagner, W.J., McLean, D.J., and Hildner, E.: 1984, Astron. Astrophys. 134, 222.
Gunkler, T.A., Canfield, R.C., Acton, L.W., and Kiplinger, A.L.: 1984, Astrophys. J. (in press).
Hagyard, M.J., Moore, R.L., and Emslie, A.G.: 1984, Adv. Space Res. (in press).
House, L.L., Wagner, W.J., Hildner, E., Sawyer, C.B., and Schmidt, H.U.: 1981, Astrophys. J. Letters 244, L117.
Hoyng, P., Machado, M.E., Duijveman, A., de Jager, C., Fryer, R., Boelee, A., and 18 co-authors: 1981a, Astrophys. J. Letters 244, L153.
Hoyng, P., Duijveman, A., Machado, M.E., Rust, D.M., Švestka, Z., Boelee, A., de Jager, C., Frost, K.J., Lafleur, H., Simnett, G.M., van Beek, H.F., and Woodgate, B.E.: 1981b, Astrophys. J. Letters 246, L155.
Hundhausen, A.J., Sawyer, C.B., House, L.L., Illing, R.M.E., and Wagner, W.J.: 1984, J. Geophys. Res. 89, 2639.
Kane, S.R., Uchida, Y., Tanaka, K., and Hudson, H.S. (eds.): 1983, Solar Phys. 86(1/2).
Kiplinger, A.L., Dennis, B.R., Emslie, A.G., Frost, K.J., and Orwig, L.E.: 1983, Astrophys. J. Letters 265, L99.
Kundu, M.R. and Woodgate, B.E. (eds.): 1984, Solar Flares, A Monograph from the SMM Workshop (in press).
Machado, M.E.: 1983, Solar Phys. 89, 133.
Machado, M.E. and Lerner, G.: 1984, Adv. Space Res. (in press).
Machado, M.E., Duijveman, A., and Dennis, B.R.: 1982, Solar Phys. 79, 85.
Machado, M.E., Somov, B.V., Rovira, M.G., and de Jager, C.: 1983, Solar Phys. 85, 157.
Parmar, A.N., Wolfson, C.J., Culhane, J.L., Phillips, K.J.H., Acton, L.W., Dennis, B.R., and Rapley, C.G.: 1984, Astrophys. J. 279, 866-874.
Phillips, K.J.H., Leibacher, J.W., Wolfson, C.J., Parkinson, J.H., Fawcett, B.C., Kent, B.J., Mason, H.E., Acton, L.W., Culhane, J.L., and Gabriel, A.H.: 1982, Astrophys. J. 256, 774.
Pick, M. and Raoult, A.: 1984, Adv. Space Res. (in press).
Poland, A.I., Machado, M.E., Wolfson, C.J., Frost, K.J., Woodgate, B.E., Shine, R.A., Kenny, P.J., Cheng, C.-C., Tandberg-Hanssen, E.A., Bruner, E.C., and Henze, W.: 1982, Solar Phys. 78, 201-213.
Rieger, E. Reppin, C., Kanbach, G., Forrest, D.J., Chupp, E.L., and Share, G.H.: 1983, 18th Internat. Cosmic Ray Conf. Vol. 4, p. 83 (Tata Inst. Fundamental Res.: Bombay).
Rust, D.M.: 1984, Adv. Space Res. (in press).
Rust, D.M. and Somov, B.V.: 1984, Solar Phys. 93, 73.
Rust, D.M., Simnett, G.M., and Smith, D.F.: 1984, Astrophys. J. (submitted).
Schadee, A., de Jager, C., and Švestka, Z.: 1983, Solar Phys. 89, 287.
Shea, M.A., Smart, D.F., and McKenna-Lawlor, S.M.P. (eds.): 1984, STIP Symposium on Solar/Interplanetary Intervals, Book Crafters, Chelsea, Michigan.
Simon, P. (ed.): 1984, Proceedings of the SMA Symposium, Adv. Space Res. (in press).
Stewart, R.T., Dulk, G.A., Sheridan, K.V., House, L.L., Wagner, W.J., Sawyer, C., and Illing, I.R.: 1982, Astron. Astrophys. 116, 217-223.

Švestka, Z. and Poletto, G.: 1984, Adv. Space Res. (in press).

Švestka, Z., Rust, D.M., and Dryer, M. (eds.): 1982a, Adv. Space Res. 2(11).

Švestka, Z., Stewart, R.T., Hoyng, P., Van Tend, W., Acton, L.W., Gabriel, A.H., Rapley, C.G., and 8 co-authors: 1982b, Solar Phys. 75, 305.

Švestka, Z., Dennis, B.R., Pick, M., Raoult, A., Rapley, C.G., Stewart, R.T., and Woodgate, B.E.: 1982c, Solar Phys. 80, 143.

Švestka, Z., Schrijver, J., Somov, B., Dennis, B.R., Woodgate, B.E., Furst, E., Hirth, W., Klein, L., and Raoult, A.: 1983, Solar Phys. 85, 313-338.

Tandberg-Hanssen, E., Kaufmann, P., Reichmann, E.J., Teuber, D.L., Moore, R.L., Orwig, L.E., and Zirin, H.: 1984, Solar Phys. 90, 41-62.

Wagner, W.J.: 1982, Adv. Space Res. 2(11), 203.

Wagner, W.J.: 1984, Ann. Rev. Astron. Astrophys. 22, 267.

Wagner, W.J., Hildner, E., House, L.L., Sawyer, C.B., Sheridan, K.Y., and Dulk, G.A.: 1981, Astrophys. J. Letters 244, L123.

Wolfson, C.J., Doyle, J.G., Leibacher, J.W., and Phillips, K.J.H.: 1983, Astrophys. J. 269, 319.

Woodgate, B.E., Shine, R.A., Poland, A.I., and Orwig, L.E.: 1983, Astrophys. J. 265, 530.

IV. Results from Hinotori and P78-1

(K. Tanaka)

Hinotori observed 720 flares through its operation February 1981-October 1981. General discussions of the results were given in two symposia: the Hinotori Symposium (ISAS 1982) and the U.S.-Japan Seminar (de Jager & Švestka 1983). The hard x-ray imaging made at the effective energy 20-35 keV showed a wide variety of morphology. Many flares (22 out of 30 events) showed single source structure, either compact (12) or extended (10) in the spatial resolution of 15 arc sec (Takakura et al. 1983a, Ohki et al. 1983, Takakura 1984). Evidences are given in some limb events that the main source is located in the high corona (1-4 x 10^4 km) (Takakura et al. 1983b). The extended single source could be the whole coronal loops which may include footpoints, but the maximum brightness is near the loop top. In some events (8 out of 30), weak subsource(s) which sould be identified with the footpoint(s) appear intermittently (Tsuneta et al. 1983). Takakura et al. (1984) found that the extended single source becomes compact and slightly shifts to higher altitudes in later phases of the impulsive burst (5 out of 10 events). Tanaka (1984), Takakura (1984), and Tanaka & Zirin (1984) argued that the hard x-ray morphology of the impulsive burst is consistent with the non-thermal electron beam model in high density corona. Sakurai (1983) investigated magnetic field structures of the hard x-ray sources based on the potential field calculations. Timing between the hard x-ray and microwave in the impulsive bursts was examined by Takakura et al. (1983c), who found correlated subsecond time structures and also by Takakura et al. (1983d), who found a long (5-10 s) delay of the peaks at 17 GHz and $E > 300$ keV to the peak at $E < 100$ keV. Kurokawa (1983) showed detailed coincidence between the hard x-ray spikes and Hα brightenings.

The flares observed from Hinotori were classified into three types (Types A, B, and C) according to combined characteristics of the morphology, spectrum, and time profile (Tanaka 1984, Tanaka et al. 1983). With contrast to the impulsive burst (Type B), the gradual hard burst (Type C) shows an extended duration without impulsive spike component, and a spectral hardening is observed with a large delay (30 s-1 min) of the gamma-ray peak to the hard x-ray peak (Yoshimori et al. 1983). Tsuneta et al. (1984a) showed that its hard x-ray image is an extended, stationary source located high in the corona. Microwave interferometric (one-dimensional) observations by Kawabata et al. (1983) showed the microwave structure co-spatial with the hard x-ray source suggesting that the high-energy electrons in the MeV range are trapped due to magnetic mirroring. Kai et al. (1985) found a larger microwave to hard x-ray flux ratio for this type compared with the impulsive burst. Nitta et al. (1983) studied dynamic spectra of the hard x-ray burst and found a characteristic change from the exponential to power law spectra in the initial phase of the impulsive burst, but the spectrum in the gradual hard burst remains power law all the time. The hot thermal burst (Type A) is characterized by a gradual enhancement in the low-energy (17-40 keV) hard x-ray flux of thermal origin from the middle phase of flare (Tsuneta et al. 1984b). Tanaka et al. (1982a) found strong emission of Fe XXVI lines whose spectrum and intensity show an electron temperature of 30-40 million K and a large emission measure of the order of 10^{49} cm^{-3}, and showed that this component is responsible for the low-energy, hard x-ray excess. This hot thermal component shows different temporal behavior from the previously known hot plasma which is diagnosed from the Fe XXV spectrum to have electron temperature in the range of 13-25 million K, and can be found in about half (7 out of 13) of large flares (Tanaka & Akita, 1984). Independently Lin et al. (1981) discovered this new thermal component from the balloon observation of very high-resolution, hard x-ray continuum spectrum. Tsuneta (1984) proposed a mechanism to explain the formation of this hot thermal component by direct energy conversion from the magnetic field due to the joule heating in the high density region in the corona, while Tanaka & Zirin (1984) attributed the origin to the collisional heating by low-energy electron beams.

A wealth of data have been obtained for the high-resolution, soft x-ray line spectra from P78-1 (Doschek et al. 1979, McKenzie et al. 1980), SMM (Culhane et al. 1981), and Hinotori (Tanaka et al. 1982a). Doschek et al. (1980) and Feldman et al. (1980) derived general spectral properties of X and M class flares, respectively. The electron temperature for Fe XXV (Ca XIX) is constant near 22 (16) million K or increases somewhat during the rise phase for the X class flare and is 13-20 million K in the M class flare. Blue shifted (400 km/s) components and large non-thermal broadenings (130 km/s) were discovered during the rise phase. The results from SMM and Hinotori confirmed these, and Hinotori's Fe XXVI spectra revealed an existence of an abundant higher temperature (30-40 million K) component (Tanaka et al. 1982a, Tanaka & Akita 1984). From line ratios of the density-sensitive line (O VII), Doschek et al. (1981) found a rapid increase in the electron density to 10^{12} cm^{-3} for the 3 million K plasma before the peak flux in the Fe XXV and Ca XIX lines. Using the same line McKenzie & Landecker (1982a) found the densities of non-flare active regions to be in the range of 3-13 x 10^{9} cm^{-3}. Feldman et al (1982a) studies impulsive soft x-ray flares with the rise and decay times less than 1 min and concluded that the 20 million K plasma cools by radiation in the high-density ($>10^{12}$ cm^{-3}) corona. From these results Feldman et al. (1982b) discussed observational constraints for a soft x-ray flare model. On the other hand Tanaka et al. (1983) found that the soft x-ray source thermal energy agrees with the electron energy deposition above 30-50 keV, the threshold depending on the preflare coronal density, and Tanaka et al. (1982b) showed a casual relationship between soft and hard x-ray flux increases for the elementary flare bursts. Tanaka & Zirin (1984) studied the temporal behavior of the blue shifts in Fe XXV lines and showed that mass of the stationary hot cloud emitting Fe XXV line is explained by the mass input from the rising plasma. They argue that the chromposheric evaporation caused by the electron (50 keV) bombardment provides the coronal hot cloud. Akita et al. (1983) derived the upper limit for the mean polarization of Fe XXV line in six flares, which is 4% and seems to contradict the high polarization degree obtained from the Intercosmos measurement (Krutov et al. 1981). In a very impulsive flare observed by Hinotori a high (10%) polarization was reported for a short period (30 s) by Tanaka et al. (1982c). An unusually intense Fe Kα emission well correlated with an impulsive hard x-ray burst was reported by Tanaka et al. (1984), who attributed it to the fluorescence from the non-thermal (power law) flux extending to 7 keV observed from the continuum spectra. Watanabe et al. (1983) studied thermal evolution of many small flares from the high-resolution, soft x-ray (1-25 keV) continuum spectra which resolve line emissions. McKenzie & Landecker (1982b) identified many lines from ionized Ca and Cr ions in the 15-23 A region from the spectra obtained from P78-1.

References

Akita, K., Tanaka, K., and Watanabe, T.: 1983, Solar Phys. 86(1/2), 101-105.
Culhane, J.L., Gabriel, A.H., Acton, L.W., Rapley, C.G., Phillips, K.J.H., Wolfson, C.J., Antonucci, E., Bentley, R.D., Catura, R.C., Jordan, C., Kayat, M.A., Kent, B.J., Leibacher, J.W., Parmar, A.N., Sherman, J.C., Springer, L.A., Strong, K.T., and Veck, N.J.: 1981, Astrophys. J. Letters 244, L141.
de Jager, C. and Švestka, Z. (eds.): 1983, "Recent Advances in the Understanding of Solar Flares," Solar Phys. 86(1/2).
Doschek, G.A., Kreplin, R.W., and Feldman, U.: 1979, Astrophys. J. Letters 233, L157.
Doschek, G.A., Feldman, U., Kreplin, R.W., and Cohen, L.: 1980, Astrophys. J. 239, 725.
Doschek, G.A., Feldman, U., Landecker, P.B., and McKenzie, D.L.: 1981, Astrophys. J. 249, 372.
Feldman, U., Doschek, G.A., Kreplin, R.W., and Mariska, J.T.: 1980, Astrophys. J. 241, 1175.
Feldman, U., Doschek, G.A., and Kreplin, R.W.: 1982a, Astrophys. J. 260, 885.
Feldman, U., Cheng, C.-C., and Doschek, G.A.: 1982b, Astrophys. J. 255, 320.
ISAS: 1982, Hinotori Symposium on Solar Flares, Institute of Space and Aeronautical Science, Tokyo, Japan.
Kai, K. et al.: 1985, Publ. Astron. Soc. Japan (in press).
Kawabata, K., Ogawa, H., and Suzuki, I.: 1983, Solar Phys. 86(1/2), 247-252.
Krutov, V.V., Korneev, V.V., Mandelstam, S.L., Shlyaptseva, A.S., Urnov, A.M., Vinogradov, A.V., and Zhitnik, I.A.: 1981, Lebedev Physical Institute Preprint No. 133.
Kurokawa, H.: 1983, Solar Phys. 86(1/2), 195-196.
Lin, R.P. Schwartz, R.A., Pelling, R.M., and Hurley, K.C.: 1981, Astrophys. J. Letters 251, L109.
McKenzie, D.L., Landecker, P.B., Broussard, R.M., Rugge, H.R., Young, R.M., Feldman, U., and Doschek, G.A.: 1980, Astrophys. J. 241, 409.
McKenzie, D.L. and Landecker, P.B.: 1982a, Astrophys. J. 259, 372.
McKenzie, D.L. and Landecker, P.B.: 1982b, Astrophys. J., 254, 309.
Nitta, N., Takakura, T., Ohki, K., and Yashimori, M.: 1983, Solar Phys. 86(1/2), 241-246.
Ohki, K., Takakura, T., Tsuneta, S., and Nitta, N.: 1983, Solar Phys. 86(1/2), 301-312.
Sakurai, T.: 1983, Solar Phys. 86(1/2), 339-344.
Takakura, T.: 1984, Int. Workshop on Solar Phys. and Interplanet. Travelling Phenom., Kunming, China, Acad. Sinica (in press).
Takakura, T., Ohki, K., Tsuneta, S., and Nitta, N.: 1983a, Solar Phys. 86(1/2), 323-331.
Takakura, T., Tsuneta, S., Ohki, K., Nitta, N., Makishima, K., Murakami, T., Ogawara, Y., Oda, M., and Miyamoto, S.: 1983b, Astrophys. J. Letters 270, L83.
Takakura, T., Kaufmann, P., Costa, J.E.R., Deganokar, S.S., Ohki, K., and Nitta, N.: 1983c, Nature 302, 317.
Takakura, T., Degaonkar, S.S., Ohki, K., Kosugi, T., and Enome, S.: 1983d, Solar Phys. 89, 379-390.
Takakura, T., Ohki, K., Nitta, N., and Wang, J.L.: 1984, Astrophys. J. Letters 281, L51.
Tanaka, K.: 1984, Activities in Red Dwarf Stars, Proc. IAU Colloquium 71, Catania, Italy, p. 307.
Tanaka, K. and Akita, A.: 1984, Astrophys. J. (submitted).
Tanaka, K. and Zirin, H.: 1984, Astrophys. J. (submitted).
Tanaka, K., Watanabe, T., Nishi, K., and Akita, K.: 1982a, Astrophys. J. Letters

254, L59.
Tanaka, K., Nitta, N., and Watanabe, T.: 1982b, in Hinotori Symp. on Solar Flares (ISAS: Tokyo, Japan), pp. 20-23.
Tanaka, K., Akita, K., Watanabe, T., and Nishi, K.: 1982c, in Hinotori Symp. on Solar Flares (ISAS: Tokyo, Japan), pp. 43-57.
Tanaka, K., Nitta, N., Akita, K., and Watanabe, T.: 1983, Solar Phys. 86(1/2), 91-100.
Tanaka, K., Watanabe, T., and Nitta, N.: 1984, Astrophys. J. 282, 793.
Tsuneta, S.: 1984, Astrophys. J. (in press).
Tsuneta, S., Takakura, T., Nitta, N., Ohki, K., Makishima, K., Murakami, T., Oda, M., and Ogawara, Y.: 1983, Solar Phys. 86(1/2), 313-321.
Tsuneta, S., Takakura, T., Nitta, N., Ohki, K., Tanaka, K., Makishima, K., Murakami, T., Oda, M., Ogawara, Y., and Kondo, I.: 1984a, Astrophys. J. 280, 887-891.
Tsuneta, S., Nitta, N., Ohki, K., Takakura, T., Tanaka, K., Makishima, K., Murakami, T., Oda, M., and Ogawara, Y.: 1984b, Astrophys. J. 284, 827-832.
Watanabe, T., Tanaka, K., Akita, K., and Nitta, N.: 1983, Solar Phys. 86(1/2), 107-113.
Yoshimori, M., Okudaira, K., Hirasima, Y., and Kondo, I.: 1983, 18th Internat. Cosmic Ray Conf. Vol. 4, pp. 85-88 (Tata Inst. Fundamental Res.: Bombay).

V. Sunspots

(V. Bumba)

Several proceedings of scientific meetings on sunspots appeared during the 1981-1984 period [The Physics of Sunsots, Cram and Thomas (eds.) 1981; see also reports of regional meetings, e.g., Third European Solar Meeting, Oxford 1981; Nordic Astronomy Meeting, O. Hauge (ed.), Oslo 1983; 11th Regional Consultation on Solar Physics, L. Dezsö and B. Kalman (eds.), Debrecen 1983]. New interest in sunspots was aroused through observations of EUV sunspot spectra from space and was also inspired by the growing number of observations of starspots and other stellar activities [IAU Symposium No. 102, Solar and Stellar Magnetic Fields: Origin and Coronal Effects, J.O. Stenflo (ed.) 1983; Colloquium IAU No. 71 Activity in Red Dwarf Stars, Catania 1982]. Other reasons for the increased interest in sunspots and their energetics were prompted by the correlation between sunspot occurrence and the variations of the solar constant (Hudson et al. 1982) and by the use of sunspot positions for determining solar differential rotation and its change with latitude, depth, and time (Howard et al. 1984, Godoli & Mazzucconi 1982, Balthasar et al. 1984, Tuominen & Kyröläinen 1982, Adam 1983, Koch 1984).

A. EVOLUTION OF SUNSPOTS IN THE FRAME OF ACTIVE REGION'S MAGNETIC FIELD

Current studies define with more precision than before the modes of sunspot development from individual nuclei, sunspot rotation, and the course of penumbral formation, strongly dependent on field topology, etc. (Bumba & Suda 1983a, 1984a,b, Liggett & Zirin 1983). The studies describe the complicated changes in umbral and penumbral forms due to the action of colliding magnetic entities of the same or opposite polarities (Bumba et al. 1982). The structure of the magnetic field and its evolution around sunspots visualized in photospheric and chromospheric fine structures has been studied during the normal development of sunspots (Gopasyuk & Kartashova 1981), in cases of more complicated magnetic situations (Karlicky & Suda 1983), during the appearance of "δ spots" (Patty 1981, Tang 1983), and during still more complex magnetic field topologies (Lin & Wang 1982, Yang & Chang 1983), as well as in association with flare occurrence (Ikhsanov 1982). The investigation of the behavior of sunspot magnetic flux during the time of sunspot disappearance brought a new enigma concerning the manner in which sunspot magnetic fields vanish. Some observers (Wallenhorst & Howard 1982, Wallenhorst & Topka 1982) do not see any evidence for spreading or diffusion of the field into the supergranular network, although others see small spots and pores as the result of secondary effects caused by the formation (Parfinenko 1983) or the decay (Garcia de la Rosa 1981, 1983) of larger active regions.

B. HIGH RESOLUTION AND SPECTRAL OBSERVATIONS OF THE PHOTOSPHERE, CHROMOSPHERE, AND TRANSITION REGION IN UMBRAE AND PENUMBRAE

The study of bright umbral dots continued. Adjabshirzadeh & Koutchmy (1982) demonstrated that the dots do not show the foreshortening effect typically observed in granular structures near the solar limb. The same authors (Adjabshirzadeh & Koutchmy 1983) constructed from their spectrophotometrical measurements a preliminary model of a two-component umbra. Parfinenko (1981a) measured the mean contrast of bright points; Soltan (1981) demonstrated a morphological relation between umbral dots and bright penumbral filaments. Lawrence (1983), using a statistical analysis, showed that umbral dots mapped by different observers have different distributions, probably due to the different resolution of the instruments used. Moore (1981a) proposed that umbral dots and penumbral grains are essentially the same phenomenon. Abdusamatov (1980) concluded that bright umbral points cover about 10% of the total umbral area and that their corrected mean brightness is smaller than that of the undisturbed photosphere. From new observations of umbral limb-darkening Albregtsen et al. (1984) found a significant decrease in the umbra-photosphere intensity ratio toward the limb. Wittmann (1983) derived corrected relative intensities of umbral dots, penumbral filaments, pores, and photospheric granulation as a function of wavelength. Van Ballegooijen (1984) constructed models of the temperature stratification in the deep layers of the umbra from a high-resolution infrared spot spectrum. Only a few papers appeared dealing with molecular spectra were studied (Glenar et al. 1983, Wöhl et al. 1983, Sinha 1983, Sinha & Joshi 1983, Lee et al. 1982a,b, Yun & Kim 1983). Abdusamatov (1984) demonstrated with the use of photometric studies the absence of the exterior as well as of the interior bright rings in studied sunspots. Two main types of sunspot light-bridges have been confirmed by Parfinenko (1981b), and links of Hα-flare emission above sunspots with these light-bridges have been shown (Bumba et al. 1981). With the help of the speckle image reconstruction procedures Stachnik et al. (1983) demonstrated that the finest structures in sunspots appear to be embedded in penumbral filaments. The mean width and distribution of these filaments were estimated during the solar eclipse of June 1973 (Bonet et al. 1982). A good summary of properties of penumbral fine structures was published by Muller (1981). Rapid variations of contrast with time of penumbral bright features were found by Parfinenko (1982). Moore (1981b) concluded that some of the dark filaments located at the outer edge of a well observed penumbra may lie several hundred kilometers above the base of the quiet photosphere. This concept has been theoretically (Cram et al. 1981) and observationally (Lites 1984) developed but needs further observational confirmation, because many observed details of sunspot structure are not in good mutual agreement (Bumba & Suda 1983b).

Using the repeated intensity and velocity images of a large isolated sunspot in both a chromospheric and a photospheric line, Lites et al. (1982) showed a relationship between the propagating umbral disturbances and the running penumbral waves: the oscillations in the sunspot are very regular and the phase relationship between the velocity and intensity of the chromospheric oscillations is radically different from that of the quiet Sun. Ye & Jin (1981a,b) found that spectral lines formed in the deep layers of a sunspot may be notably influenced by Alfvén waves, while this influence is weak on lines originating in its upper layers. The widths and intensities of Ca II H and K emission lines across a sunspot were estimated (Yun et al. 1982, Yun & Beebe 1982), and Gurman (1984) published observations of the Mg II H line in sunspot umbrae. The dependence of umbral flashes on the magnetic field structure of sunspots and other parameters was established (Turova et al. 1983, Turova 1984, Teplitskaya & Turova 1983). Slow and fast variations in the brightness oscillations of a sunspot umbra in the center of the Hα line were shown (Loskutnikov 1984). From observations of brightness oscillation in a sunspot umbra in both the Hα and Ca II K lines, a model was constructed which explains the oscillations by resonant transmission of slow-mode magnetoacoustic waves semi-trapped in a chromospheric cavity above the umbra (Zhugzhda & Makarov 1982, Zhugzhda et al. 1984). Many papers trying to explain the umbral oscillations and photospheric and chromospheric penumbral waves were published, e.g., Cally & Adam 1983, Zhugzhda et al. 1983, Thomas & Scheuer 1982, and Cally 1983). Spectra of several strong lines obtained in sunspots (Beebe et al. 1982) allowed discussions of models of the chromosphere above a sunspot umbra or penumbra (Yun et al. 1984). The observations of intensities of Lα and Lβ above sunspots yielded estimates of the pressure in the upper chromosphere and transition region above the umbra, the values of which are somewhat lower than the average quiet coronal pressure (Lites & Skumanich 1981). The UV observations from the SMM satellite gave the physical and dynamic characteristics of the transition region above sunspots: low intensities, mass flow, electron densities (Kingston et al. 1982), presence of both subsonic and supersonic flow, reduced electron temperature, and energy budget (Nicolas et al. 1982). Oscillations with periods of about 2 min of line-of-sight velocity and intensity of transition-region lines were observed (Gurman et al. 1982, Henze et al. 1984).

C. OBSERVATIONS OF MAGNETIC FIELDS

Huseynov (1983) studied the Zeeman splitting of Fe II and Fe I lines in several sunspots. Fluctuations of up to 1000 Gauss between the individual bright and dark penumbral regions were found by Stellmacher & Weihr (1981). Brants & Zwaan (1982), measuring the magnetic field strengths in small umbrae and pores using the Ti I line formed in umbrae, found values between 1900-2600 Gauss in their darkest parts. Using different temperature-sensitive lines Huseynov & Russkih (1983) studied the distribution of field strength across the umbrae. The degree of polarization for π-components of two photospheric lines in sunspot umbrae was estimated by Vyalshin & Kulagin (1981). The circular polarization of sunspots with high sensitivity ($\sim 3 \times 10^{-6}$) in a wide spectral range (3700-4500 Å) was measured by Kemp & Henson (1983); in the same way Faraday rotation during measurements of the orientation of sunspot transverse magnetic field was investigated in detail (West & Hagyard 1983). The influence of instrumental polarization on such measurements was discussed by Kawakami (1983). Coordinated ground-based and SMM satellite observations of sunspot magnetic fields were carried out by (Hagyard et al. 1983), and (Henze et al. 1982). Koval & Stepanjan (1983), estimated that there are several modes of magentic field distribution with height above sunspots. Obashev et al. (1982) estimated the angle of inclination of the sunspot symmetry axis to the solar surface from obsevations of the Wilson depression. Giovanelli & Jones (1982) studying the magnetic canopies above sunspots, demonstrated that these canopies extend from sunspots at a height of a few hundred kilometers above the outer edge of the sunspot penumbrae beyond the sunspots, rising slowly, almost horizontally, into the height of several hundred kilometers interconnecting various spots.

D. OBSERVATIONS OF MOTIONS IN AND ABOVE SUNSPOTS

Tönjes and Wöhl (1982) confirmed that penumbral bright grains lasting 1.5 to 3 hours move toward sunspot umbrae with velocities of 80 to 330 m s^{-1}. The motions in different depths of the umbral photosphere were studied by Adam (1983) and Gopasyuk (1982). From space observations of spectral lines originating in the transition region above sunspots Athay et al. (1982) demonstrated the steady flow patterns in the reversed Evershed sense with a substantial vertical component. Nicolas (1981) detected systematic flows in the EUV spectrum that extend to about 10^4 km outside the penumbra. Moore (1981c) presented his model of dynamic phenomena in sunspots, for which Thomas (1981) attempted to find a theoretical explanation within the framework of magnetohydrodynamic theory. Spruit (1981a) reviewed the role of convective processes in the small-scale umbral and penumbral phenomena. Several papers were devoted to the study of spot motions in sunspot groups (e.g., Doval 1983). Two types of such motions were discussed by Dezsö & Kovacs (1983). Quasiperiodic oscillations of relative distances of small spots and pores in a group were found by Selenzneva (1981). Bumba & Suda (1983c) pointed out the low efficiency of magnetic forces during the formation, motion, and collision of sunspots and the importance of hydrodynamic forces during these processes. The irregular rotation of several spots was studied by Kučera (1982) and Antalova (1983).

E. RADIO ASTRONOMICAL OBSERVATIONS OF SUNSPOTS

Measurements estimating the physical conditions above sunspots, especially the magnetic field intensity and structure with the aid of radio astronomical methods, have been made by Abramov-Maksimov & Gelfreikh (1983), and Bogod et al. (1982) demonstrated that the magnetic field intensity above sunspots at the bottom of the corona either coincided with the photospheric value or was smaller by about 30%. Akhmedov et al. (1982a) studied the S-component of the solar radio emission and showed that the height of its sources above umbrae did not generally exceed 2000 km. They also estimated the magnetic field gradient which is not less than 0.25 Gauss km^{-1}. Akhmedov et al. (1982b, 1983) compared radio observations with model calculations. Comparing the radio data with UV and soft x-ray data Shibasaki et al. (1983) identified the radio emission mechanisms and their height and magnetic field strength above sunspots. The absence of polarized emission above the sunspot umbra and the circularly polarized emission above the penumbra were detected at 6 cm by Lang & Willson (1982). Additional radio emission above bipolar sunspot group was observed by Akhmedov & Gelfreikh (1981).

F. THEORETICAL MODELS OF SUNSPOTS

Yoshimura (1983) presented a new concept of sunspot cooling based on the size and depth of magnetic field tubes. Albregtsen & Maltby (1981) drew attention to the possibility of distinguishing between different sunspot theories by observing the umbra/photosphere intensity ratio as a function of spot size, and Elboth & Engvold (1983) estimated the contrast of this ratio for a large sunspot with high spectral resolution. Avrett & Kurucz (1983) used the 3860-3880 Å band for a detailed study of the sunspot/photosphere intensity ratio and suggested that in sunspots it is formed at the temperature minimum with a brightness temperature of about 3350 K. Schatten (1981) proposed as a sunspot cooling mechanism the gravitational draining in which the magnetic field of pores provides a conduit wherein cooled gases can return more easily to their heat source at the bottom of the convective zone. Staude (1981) derived horizontally averaged working models for both the umbra of a large stable spot and its mean undisturbed environment for reference. Fowler et al. (1983) studied the intensity distribution around sunspots for comparison with the isotherms predicted by convective blocking models of heat flow. The magnetic field played a dominant role in many newly proposed sunspot models. Solovev (1982) examined the force-free magnetic flux rope with the uniform twisting in which the cross-section radius of the magnetic surface varies along the symmetry axis of the system. Huseynov (1982) looked for sunspot models and conditions where equilibrium between the total pressure of a sunspot and the pressure of the surrounding photosphere is reached. Spruit (1981b) reviewed the current theories concerning the pressure equilibrium and thermal balance of sunspots, their hydromagnetic stability, energy transport mechanisms, and wave propagations. Solovev (1983) discussed the problem of static confinement in magnetic flux tubes of variable diameter.

A new magnetohydrostatic theory of sunspots, taking into account the magnetic field lines which return to the photosphere, was discussed by Osherovich (1982). Flaa et al. (1982) and Osherovich (1984) applied this model to a typical sunspot; Osherovich & Flaa (1983) included the azimuthal component of the magnetic field into the model, which only slightly influenced the sunspot structure. Another extension of this model into a "double return flux model" was presented by Osherovich & Lawrence (1983). In a series of papers Solovev (1981, 1982b, 1983) and Mordvinov (1983) discussed the physical behavior of a boundary layer between the sunspot magnetic flux tube and the photosphere under different assumptions. A theoretical analysis of torsional oscillations of sunspots in a bipolar group observed by Gopasyuk (1981) was presented by Solovev (1984). From a derived simple relation between the period of oscillations and the effective depth of the sunspots, a depth of a few thousand kilometers was estimated. Thomas et al. (1982) interpreted the observations of sunspot umbra 5-min oscillations splitting into several individual modes of different periods as the response of the sunspot to forcing by the 5-min p-modes in the surrounding quiet atmsophere. Thomas (1984) interpreted the fundamental 3-min umbral oscillations as due to a photospheric resonance. His theory of umbral oscillations was supported by calculating the resonant modes in a much more detailed model of the umbral atmsophere (Thomas & Scheuer 1982). Bel & Leroy (1981) showed that the cooling of sunspots by ascending Alfvén waves is not possible. Krivodubskij (1983), studying the efficiency of different mechanisms of sunspot magnetic field dissipation, found that in the deep layers of the sunspot umbra the Alfvén oscillations are prevalent, while in the upper layers ohmic dissipation seems to be more effective. Singh et al. (1983), investigating the dynamical properties of convective overstability and Kelvin-Helmholtz instability in a vertical magnetic field with a downdraft, demonstrated that both agents produce upward/downward propagating Alfvén waves depending on the magnitude of the kinematic viscosity coefficient and the thermometric conductivity. The electric current density in a sunspot was calculated from the computed distribution of the magnetic field vector compared with its actual distribution in a real spot by Denisenko et al. (1982). Similar computations of the distribution of vertical electric current density were compared with UV emissions by deLoach et al. (1984). The electrical conductivity, thermal conductivity, and viscosity of models both of the quiet photosphere and of the sunspot umbra were calculated by Kovitya & Cram (1983).

References

Abdusamatov, K.I.: 1980, Soln. Dann. 11, 99.

Abdusamatov, K.I.: 1984, Soln. Dann. 6, 80.

Abramov-Maksimov and Gelfreikh, G.B.: 1983, Pisma Astron. Zh. 9, 244.

Adam, M.G.: 1983, M.N.R.A.S. 203, 9-18.

Adjabshirzadeh, A. and Koutchmy, S.: 1982, Solar Phys. 75, 71-74.

Adjabshirzadeh, A. and Koutchmy, S.: 1983, Astron. Astrophys. 122, 1-8.

Akhmedov, Sh. B. and Gelfreikh, G.B.: 1981, Soln. Dann. 6, 107.

Akhmedov, Sh. B., Gelfreikh, G.B., Bogod, V.M., and Korzhavin, A.N.: 1982a, Solar Phys. 79, 41-58.

Akhmedov, Sh. B. et al.: 1982b, Soln. Dann. 10, 72.

Akhmedov, Sh. B., Gelfreikh, G.B., Fürstenberg, F., Hilderbrant, J., and Krüger, A.: 1983, Solar Phys. 88, 103-108.

Albregtsen, F. and Maltby, P.: 1981, Solar Phys. 74, 147-151.

Albregtsen, F., Joras, P.B., and Maltby, P.: 1984, Solar Phys. 90, 17-30.

Antalova, A.: 1983, Bull. Astron. Inst. Czech. 33, 345.

Athay, R.G., Gurman, J.B., Henze, W., and Shine, R.A.: 1982, Astrophys. J. 261, 684.

Avrett, E.H. and Kurucz, R.L.: 1983, Rep. Inst. Theoret. Astrophys., Blindern, Olso 59, 43-58.

Balthasar, H., Lustig, G., and Wöhl, H.: 1984, Solar Phys. 91, 55-59.

Beebe, H.A., Baggett, W.E., and Yun, H.S.: 1982, Solar Phys. 79, 31-39.

Bel, N. and Leroy, B.: 1981, Astron. Astrophys. 104, 203-206.

Bogod, V.M., Vyalshin, G.F., Gelfriekh, G.B., and Petrova, N.W.: 1982, Soln. Dann. 1, 104.

Bonet, J.A., Ponz, J.D., and Vazquez, M.: 1982, Solar Phys. 77, 69-75.

Brants, J.J. and Zwaan, C.: 1982, Solar Phys. 80, 251-258
Bumba, V. and Suda, J.: 1983a, Bull. Astron. Inst. Czech. 34, 349.
Bumba, V. and Suda, J.: 1983b, Bull. Astron. Inst. Czech. 34, 155.
Bumba, V. and Suda, J.: 1983c, Bull. Astron. Inst. Czech. 34, 29
Bumba, V. and Suda, J.: 1984a, Bull. Astron. Inst. Czech. 35, 28.
Bumba, V. and Suda, J.: 1984b, Bull. Astron. Inst. Czech. 35, 224.
Bumba, V. et al.: 1981, Bull. Astron. Inst. Czech. 32, 286.
Bumba, V. et al.: 1982, Bull. Astron. Inst. Czech. 33, 160.
Cally, P.S.: 1983, Solar Phys. 88, 77-102.
Cally, P.S. and Adam, J.A.: 1983, Solar Phys. 85, 97-111.
Cram, L.E. and Thomas, J.H. (eds.): 1981, The Physics of Sunspots, Sacramento Peak Observatory, Sunspot, New Mexico.
Cram, L.E., Nye, A.H., and Thomas, J.H.: 1981, in Phys. of Sunspots, pp. 384-388.
deLoach, A.C., Hagyard, M.J., Rabin, D., Moore, R.L., Smith, J.B., Jr., West, E.A., and Tandberg-Hanssen, E.: 1984, Solar Phys. 91, 235-242.
Denisenko, V.V., Kotov, V.A., Romanov, V.A., and Sokolov, V.S.: 1982, Solar Phys. 81, 217-229.
Dezsö, L. and Kalman, B. (eds.): 1983, Proc. 11th Regional Consultation on Solar Physics, Publ. Debrecen Obs. 5.
Dezsö, L. and Kovacs, A.: 1983, Publ. Debrecen Heliophys. Obs. 5, 317-322.
Doval, P.: 1983, Soln. Dann. 12, 81.
Elboth, D. and Engvold, O.: 1983, Rep. Inst. Theoret. Astrophys., Blindern, Oslo 59, 15.
Flaa, T., Osherovich, V.A., and Skumanich, A.: 1982, Astrophys. J. 261, 700.
Fowler, L.A., Foukal, P., and Duvall, T., Jr.: 1983, Solar Phys. 84, 33-44.
Garcia de la Rosa, J.I.: 1981, Solar Phys. 74, 117-123.
Garcia de la Rosa, J.I.: 1983, Solar Phys. 89, 51-62.
Giovanelli, R.G. and Jones, H.P.: 1982, Solar Phys. 79, 267-278.
Glenar, D.A., Deming, D., Jennings, D.E., Kostiuk, T., and Mumma, M.J.: 1983, Astrophys. J. 269, 309-318.
Godoli, G. and Mazzucconi, F.: 1982, Astron. Astrophys. 116, 188-189.
Gopasyuk, S.I.: 1981, Izv. Krymsk. Astrofiz. Obs. 64, 108.
Gopasyuk, S.I.: 1982, Izv. Krymsk. Astrofiz. Obs. 66, 77.
Gopasyuk, S.I. and Kartashova: 1981, Izv. Krymsk. Astrofiz. Obs. 63, 25.
Gurman, J.B.: 1984, Solar Phys. 90, 13-15.
Gurman, J.B., Leibacher, J.W., Shine, R.A., Woodgate, B.E., and Henze, W.: 1982, Astrophys. J. 253, 939-948.
Hagyard, M.J., Teuber, D., West, E.A., Tandberg-Hanssen, E., Henze, W., Jr., Beckers, J.M., Bruner, M., Hyder, C.L., and Woodgate, B.E.: 1983, Solar Phys. 84, 13-31.
Hauge, O. (ed.): 1983, Rep. Inst. Theoret. Astrophys., Blindern, Oslo 59.
Henze, W., Jr., Tandberg-Hanssen, E., Hagyard, M.J., Woodgate, B.E., Shine, R.A., Beckers, J.M., Bruner, M., Gurman, J.B., Hyder, C.L., and West, E.A.: 1982, Solar Phys. 81, 231-244.
Henze, W., Tandberg-Hanssen, E., Reichmann, E.J., and Athay, R.G.: 1984, Solar Phys. 91, 33-44.
Howard, R., Gilman, P.A., and Filman, P.I.: 1984, Astrophys. J. 283, 373-384.
Hudson, H.S., Silva, S., and Woodward, M.: 1982, Solar Phys. 76, 211-219.
Huseynov: 1982, Izv. Krymsk. Astrofiz. Obs. 66, 83.
Huseynov: 1983, Izv. Krymsk. Astrofiz. Obs. 68, 36.
Huseynov and Russkih: 1983, Izv. Krymsk. Astrofiz. Obs. 68, 31.
Ikhsanov, V.N.: 1982, Soln. Dann. 11, 88.
Karlicky, M. and Suda, J.: 1983, Bull. Astron. Inst. Czech. 34, 282-285.
Kawakami, H.: 1983, Publ. Astron. Soc. Japan 35, 459-489.
Kemp, J.C. and Henson, G.D.: 1983, Astrophys. J. Letters 266, L69.
Kingston, A.E., Doyle, J.G., Dufton, P.L., and Gurman, J.B.: 1982, Solar Phys. 81, 47-58.
Koch, A.: 1984, Solar Phys. 93, 53-72.
Koval, I.K. and Stepanjan, N.N.: 1983, Izv. Krymsk. Astrofiz. Obs. 68, 3.

Kovitya, P. and Cram, L.: 1983, Solar Phys. 84, 45-48.
Krivodubskij: 1983, Soln. Dann. 11, 51.
Kučera: 1982, Bull. Astron. Inst. Czech. 33, 345.
Lang, K.R. and Willson, R.F.: 1982, Astrophys. J. Letters 255, L111-L117.
Lawrence, J.K.: 1983, Solar Phys. 87, 1-5.
Lee, S.-W., et al.: 1982a, J. Korean Astron. Soc. 14, 19.
Lee, S.-W., et al.: 1982b, J. Korean Astron. Soc. 14, 79.
Liggett, M. and Zirin, H.: 1983, Solar Phys. 84, 3-11.
Lin, R.P. and Wang, J.-l.: 1982, Sci. Sinica Series A 25, 89.
Lites, B.W.: 1984, Solar Phys. 90, 1-12.
Lites, B.W. and Skumanich, A.: 1981, in Phys. of Sunspots, pp. 152-166.
Lites, B.W., White, O.R., and Packman, D.: 1982, Astrophys. J. 253, 386-392.
Loskutnikov: 1984, Soln. Dann. 6, 91.
Moore, R.L.: 1981a, Space Sci. Rev. 28, 387-421.
Moore, R.L.: 1981b, Astrophys. J. 249, 390-393.
Moore, R.L.: 1981c, in Phys. of Sunspots, pp. 259-311.
Mordvinov: 1983, Soln. Dann. 5, 76.
Muller, R.: 1981, in Phys. of Sunspots, pp. 340-342.
Nicolas, K.R.: 1981, in Phys. of Sunspots, pp. 336-339.
Nicolas, K.R., Kjeldseth-Moe, O., Bartoe, J.-D., and Brueckner, G.E.: 1982, Solar Phys. 81, 253-280.
Obashev, S.O., Gainullina, R.K., Minasyants, T.M., and Minasyants, G.S.: 1982, Solar Phys. 78, 59-66.
Osherovich, V.A.: 1982, Solar Phys. 77, 63-68.
Osherovich, V.: 1984, Solar Phys. 90, 31-35.
Osherovich, V. and Flaa, T.: 1983, Solar Phys. 88, 109-115.
Osherovich, V.A. and Lawrence, J.K.: 1983, Solar Phys. 88, 117-135.
Parfinenko, L.D.: 1981a, Soln. Dann. 2, 96.
Parfinenko, L.D.: 1981b, Soln. Dann. 12, 79.
Parfinenko, L.D.: 1982, Soln. Dann. 5, 86.
Parfinenko, L.D.: 1983, Soln. Dann. 3, 102.
Patty, S.R.: 1981, in Phys. of Sunspots, pp. 64-74.
Schatten, K.H: 1981, Astrophys. J. Letters 247, L139-L142.
Selezneva, L.M.: 1981, Soln. Dann. 9, 94.
Shibasaki, K., Chiuderi-Drago, F., Melozzi, M., Slottje, C., and Antonucci, E.: 1983, Solar Phys. 89, 307-321.
Singh, P.D. et al: 1983, Acta Astron. Sinica 89, 207.
Sinha, R.P.: 1983, Bull. Astron. Soc. India 10, 223.
Sinha, R.P. and Joshi, M.N.; 1983, Bull. Astron. Soc. India 10, 329.
Solovev, A.A.: 1981, Soln. Dann. 12, 71.
Solovev, A.A.: 1982a, Astron. Zh. 59, 380.
Solovev, A.A.: 1982b, Soln. Dann. 1, 86.
Solovev, A.A.: 1983, Soln. Dann. 3, 65.
Solovev, A.A.: 1984, Soln. Dann. 1, 73.
Soltan, D.: 1981, Astron. Astrophys. 107, 211.
Spruit, H.C.: 1981a, in Phys. of Sunspots, pp. 359-368.
Spruit, H.C.: 1981b, Space Science Rev. 28, 435-448.
Stachnik, R.V., Nisenson, P., and Noyes, R.W.: 1983, Astrophys. J. Letters 271, L37.
Staude, J.: 1981, Astron. Astrophys. 100, 284-290.
Stellmacher, G. and Wiehr, E.: 1981, Astron. Astrophys. 103, 211-215.
Stenflo, J.O. (ed.): 1983, IAU Symp. No. 102, D. Reidel Publ. Co., Dordrecht.
Tang, F.: 1983, Solar Phys. 89, 43-50.
Teplitskaya, R.B. and Turova, I.P.: 1983, Soln. Dann. 8, 102.
Thomas, J.H.: 1981, in Phys. of Sunspots, pp. 345-358.
Thomas, J.H. and Scheuer, M.A.: 1982, Solar Phys. 79, 19-29.
Thomas, J.H., Cram, L.E., and Nye, A.H.: 1982, Nature 297, 485.
Thomas, J.H.: 1984, Astron. Astrophys. 135, 188.
Tönjes, K. and Wöhl, H.: 1982, Solar Phys. 75, 63-69.

Tuominen, J. and Kyröläinen, J.: 1982, Solar Phys. 79, 161-172.
Turova, I.P.: 1984, Solar Phys. 91, 51-54.
Turova, I.P., Teplitskaja, R.B., and Kuklin, G.V.: 1983, Solar Phys. 87, 7-22.
Van Ballegooijen, A.A.: 1984, Solar Phys. 91, 195-217.
Vyalshin, G.F. and Kulagin, E.S.: 1981, Soln. Dann. 9, 97.
Wallenhorst, S.G. and Howard, R.: 1982, Solar Phys. 76, 203-209.
Wallenhorst, S.G. and Topka, K.P.: 1982, Solar Phys. 81, 33-46.
West, E.A. and Hagyard, M.J.: 1983, Solar Phys. 88, 51-64.
Wöhl, H., Engvold, O., and Brault, J.W.: 1983, Rep. Inst. Theoret. Astrophys., Blindern, Oslo 56.
Wittmann, A.: 1983, Mitt. Astron. Ges. 60, 280.
Yang, H.-S., Chang, H.-M., and Harvey, J.W.: 1983, Solar Phys. 84, 139-151.
Ye, S.-H. and Jin, W.-J.: 1981a, Acta Astron. Sinica 22, 365.
Ye, S.-H. and Jin, W.-J.: 1981b, Acta Astron. Sinica 23, 276.
Yoshimura, H.: 1983, Solar Phys. 87, 251-259.
Yun, H.S. and Beebe, H.A.: 1982, Solar Phys. 78, 347-349.
Yun, H.S. and Kim, Y.H.: 1983, J. Korean Astron. Soc. 16, 19.
Yun, H.S., Beebe, H.A., and Baggett, W.E.: 1982, Proc. Southwest Reg. Conf., Astron. Astrophys. 8, 31.
Yun, H.S., Beebe, H.A., and Baggett, W.E.: 1984, Solar Phys. 92, 145-151.
Zhugzhda, Y.D. and Makarov, V.I.: 1982, Solar Phys. 81, 245-251.
Zhugzhda, Y.D., Locans, V., and Staude, J.: 1983, Solar Phys. 82, 369-378.
Zhugzhda, Y.D., Staude, J., and Locans, V.: 1984, Solar Phys. 91, 219-234.

VI. Solar Radiophysics

(R.T. Stewart)

A. INTRODUCTION

The past 3 years since the last solar maximum have witnessed an unprecedented number (>200) of published scientific papers on many aspects of solar radiophysics. These contributions are the result of an intense research effort mounted during the first Solar Maximum Mission of 1980 and continued until the present. Excellent x-ray, EUV, and visible light observations of the disturbed corona and transition region have been obtained from the SMM, Hinotori, P78-1, and ISEE-3 spacecraft. ISEE-3 also has provided very low-frequency radio observations of solar bursts in the interplanetary medium. Ground-based radio support for space experiments has been provided by many observatories throughout the world. In particular, many collaborative studies using x-ray and radio observations of solar flares have been reported. The outstanding radio instrument during this period has been the VLA, operating at 2, 6, and 20 cm with a time resolution of 10 s and both modes of circular polarization. The two-dimensional spatial resolution of the radio images is a few seconds of arc, almost as good as the best resolution obtained so far at any wavelength in the solar spectrum. To complement the rather poor time resolution of the VLA one-dimensional arrays such as the WSRT at 6 cm wavelength and the Nobeyama interferometer at 17 GHz have been used successfully. In addition, a number of very-high-time resolution radiometers have been built at different locations. At meter wavelengths two-dimensional arrays at Clark Lake, Culgoora, and Nancay and a one-dimensional array at Nobeyama have been employed. The little known region of the solar spectrum at decimeter wavelengths is being investigated by the Zürich spectrograph. It is pleasing to see Chinese participation in solar radiophysics.

Looking ahead there remains only one part of the solar spectrum, from 20 to 1 MHz and corresponding to coronal heights from 3 to 30 $R_{\odot}$, to be explored. This spectral gap hinders the merging of coronal and interplanetary radiophysics. Another obvious need in the future is for a high time and spatial resolution radioheliograph operating at microwave frequencies, i.e., near 17 GHz. Such an instrument combined with high spatial hard x-ray observations could settle some important questions about particle acceleration and propagation in solar flares, because both the radio and x-ray observations would be viewing emission from the same electron population. For a review of current ideas on particle acceleration in solar flares see report no. VIII by E.R. Priest. For a review on spacecraft observations see report no. IV by K. Tanaka. A review on advances in solar radiophysics has been published recently (Kundu 1982), and a new book on meter-wavelength solar radio astronomy is being published soon (McLean & Labrum 1985).

B. SOLAR FLARE EMISSION

1. Pre-Flare Phase

Observations of active regions showing changes in microwave intensity and circular polarization prior to flare activity have been reported previously. VLA observations show changes in the polarization structure of the flare region just prior to the impulsive phase. These changes may be due to emerging magnetic flux and reconnection (Kundu 1983). Hard x-ray spikes have been correlated with decimeter Type III bursts during this phase (Benz et al. 1983) and show evidence for a downwards motion of the acceleration site as proposed earlier to explain the shift in starting frequencies of Type III bursts (Kane & Raoult 1981). The Type III bursts may be generated by secondary acceleration of electrons outside a flaring loop (Sprangle & Vlahos 1983). Millisecond spike bursts at microwaves occur predominantly during the pre-flare phase (Zhao & Jin 1982). The high degree of circular polarization and brightness temperature ($\sim 10^{15}$ K) can be explained by masering at the second harmonic of the gyrofrequency (Melrose & Dulk 1982, Sharma et al. 1982) from a loss cone distribution of energetic electrons in a magnetic loop. The conditions for enhanced spike emission require the plasma temperature to be $<10^7$ K, which may explain why this type of emission seems to be confined to the early stages of the flare event (Holman 1983). Significant plasma heating may occur because of the absorption of electron-cyclotron maser emission at the second harmonic level (Melrose & Dulk 1984). Other models, including plasma emission from runaway electrons, have been proposed (Kuijpers et al. 1981). The size of the source region (and burst brightness temperature) has yet to be confirmed. VLBI measure-ments (Tapping et al. 1983) indicate that the radio flare region can be very small.

2. Impulsive Phase

Theoretical models suggest that the main release of flare energy and particle acceleration occurs at the top of a magnetic loop. Microwave spatial observations, particularly with the VLA, have tended to confirm this picture, in that 2- and 6-cm wavelength impulsive sources often appear near the magnetic neutral line between Hα kernels (Lang et al. 1981). Sometimes bipolar sources are observed, suggesting that the emission comes from the sides of the loop rather than from the top (Kundu et al. 1982). The impulsive burst at microwaves is thought to be due to gyrosynchrotron radiation from trapped electrons in a magnetic loop. Mirroring and other collisions will produce a non-thermal distribution and a general hardening of the electron spectrum even if the initial accelerated distribution is thermal (Holman et al. 1982). Hard x-ray bursts will also be produced by collisions of high-energy electrons in dense regions of the loop. Some evidence from joint hard x-ray and microwave spatial observations exists for streaming of electrons in a loop (Hoyng et al. 1981). Correlation studies between microwave and hard x-ray bursts have been reported by many authors. Some studies (Kane et al. 1983, Kai et al. 1984) argue for a common population of electrons producing the x-ray and microwave bursts at slightly different times because of propagation effects. However, the radio emissivity is critically dependent on the magnetic field strength and spectral index, which makes the close correlation found between hard x-ray and microwave fluxes seen even more remarkable (Kai et al. 1984). Short time delays of $\leqslant 0.5$ s between hard x-ray peaks and decimeter Type III bursts have also been reported (Kane et al. 1982, Dennis et al. 1984). Such short delays argue for an acceleration region at a considerable height, ~3 x 10^9 cm, above the chromosphere (Dennis et al. 1984). A strong case of weak turbulence effects in the generation of Type III bursts has been determined (Grognard 1982), based on ISEE-3 measurements of the electron distribution function for the electron streat at 1 AU (Lin et al. 1981). A detailed review of Type III burst theory and observations has also been published (Goldman 1983). The controversy between fundamental and second harmonic plasma frequency emission at low frequencies continues.

3. Main Flare Phase

With the large coverage of coronal mass outflows by white-light coronagraphs aboard SMM and P78 it has been possible to study the relationship between the Type II burst shock wave and the leading edge of the coronal transient. In several events there is evidence that the Type II radio emission occurs behind the transient leading edge rather than in front of it, where a piston-driven shock might be expected to form. This result has been interpreted as evidence that the Type II disturbance is a blast wave ejected from the flare, independent of the coronal transient, but producing radio emission upon its intersection with the outflowing material (Wagner & MacQueen 1983, Gary et al. 1984). Statistical studies have shown that 60-70% of Type II bursts are accompanied by coronal transients (Sheeley et al. 1984, Kahler et al. 1984) so that at least 30-40% may be blast waves. Type II bursts appear to be more closely associated with filament channels than with flare regions (Stewart 1984a), suggesting that coronal streamers may be the origin of fast mass transients and Type II bursts. At frequencies below 1 MHz very few Type II bursts have been detected, and these usually have low starting frequencies (Robinson et al. 1984) and are accompanied by interplanetary distrubances such as the magnetic clouds (McKenna-Lawlor & Richter 1983) associated with the homologous Type II bursts of June 27-29, 1980 (Stewart 1984), or with interplanetary mass outflows detected by radio scintillation effects (Watanabe & Kahinuma 1984). The coronal source region of the moving Type IV radio bursts has been observed (Stewart et al. 1982, Gergely et al. 1984) and the density enhancement estimated from white-light measurements to be 30 to 60 times greater than the quiet corona. Such large source densities are consistent with the radio emission arising from Langmuir wave conversion at the second harmonic plasma frequency, although gyrosynchrotron emission cannot be ruled out as an alternative mechanism (Stewart et al. 1982). Other arguments for plasma emission have been advanced (Duncan 1981). A model for electron acceleration in a coronal loop transient suggests that lower hybrid waves are excited at the shock front which then accelerates the tail of the electron distribution within the loop. The resulting plasma emission may be sufficient to explain moving and stationary Type IV bursts (Vlahos et al. 1982). Other evidence for shock acceleration of particles in the corona are the low-frequency SA events observed by ISEE-3 (Cane et al. 1981) at the time of Type II bursts and an electron event associated with the behind-the-limb flare which produces a Type II burst and a meter-wavelength flare continuum (Hudson et al. 1982). Shock-drift acceleration of particles has been suggested as a possible mechanism to explain the herringbone and split-band fine structure in Type II bursts (Holman & Pesses 1983). The close temporal association between gradual hard x-ray and meter continuum bursts during large flares has been advanced as evidence that the x-ray and radio sources are linked by a large magnetic arch and that a continuous common acceleration process takes place. The details of this process are not specified (Klein et al. 1983).

4. Post-Flare Phase

During the decay phase of the solar flare, expanding magnetic loop systems have been observed in Hα, soft x-rays and more recently in microwaves (Velusamy &

Kundu 1981, Nakjima 1982). The radio source region is hotter than the Hα loops and possibly comparable in temperature with the x-ray source, where T $\approx 10^6$ K. The degree of circular polarization is small (Nakajima 1982), indicating thermal emission. For 20-cm wavelength emission the magentic induction is estimated to be B 100 G (Velusamy & Kundu 1981). The radio source initially moves outwards with a speed of ~10 km s^{-1}, comparable with the underlying colder (T $\approx 10^4$ K) Hα loop systems. The post-burst increase may last ~4 h at 17 GHz, thus requiring a continuous heat input to the source region. An exciting discovery by SMM was the gigantic soft x-ray loop systems of arches extending out to heights ~1.5 x 10^5 km above the flare site and visible 6-30 h after a major flare (Švestka et al. 1982a,b). The arch system was stable and associated with an overlying stationary noise storm source region extending out to at least 1.5 x 10^6 km (~2 $R_\odot$). The casue of the noise storm electron acceleration is not certain but it may be due to the emergence of magnetic flux into the corona from the photosphere or due to slow magnetic reconnection following a coronal transient (McLean 1981). Several theories for noise storms have been proposed recently which involve the generation of microturbulence by emerging magnetic flux. The Type I bursts are thought to arise either from the coalescence of ion-acoustic waves and plasma waves (Benz & Wentzel 1981) or from the mode coupling between lower and upper hybrid waves (Spicer et al. 1981). Evidence for a connection between noise storms and emerging flux is the close association beetween noise storms and coronal transients (Lantos et al. 1981, Kerdraon et al. 1983, Duncan 1983), Hα surges (Garczynska et al. 1982), and rapidly growing EUV loop systems (Brueckner 1983). Electron acceleration during the late phase of the flare might occur as a result of magnetic reconnection above post-flare loops. This could explain the apparent temporal association between secondary peaks in microwave outbursts and decameter continuum events (Cliver 1983).

5. Active Regions

The radio emission from active regions, commonly referred to as the slowly varying component, has been studied in considerable detail at microwaves (Alissandrakis & Kundu 1982, Lang & Willson 1982, Dulk & Gary 1983, McConnell & Kundu 1983, Felli et al. 1981, Shibashi et al. 1983, Urpo et al. 1982). The emission seems to be a mixture of bremsstrahlung and gyrosynchrotron emission from thermal distributions of electrons (Seehafer et al. 1983, Pallavicini et al. 1981), with the strongest emission occurring as a broken ring around the sunspot penumbra (Alissandrakis & Kundu 1982, Lang & Willson 1982). Thermal cyclotron lines may also be emitted (Willson 1983). The decrease in microwave emission above the center of the sunspot appears to be due to the presence of a cooler region (Strong et al. 1984). Bright radio sources with brightness temperatures ~3 x 10^6 K appear to be associated with arcades of corodnal loops above magnetic neutral lines (Schmahl et al. 1982). However, some of these bright microwave sources overlie regions of weak magnetic field, which is difficult to explain by gyrosynchrotron emission from thermal electrons unless the loops carry current or unless there is a quasi-continuous low level of particle acceleration from active regions (Webb et al. 1983). It has been suggested that non-thermal electrons might be accelerated in bursts and trapped in magnetic fields of ~100 G for ~1 h to produce the bright microwave sources by gyrosynchrotron emission (Chiuderi & Melozzi 1984). The extension of active region structures outwards into the interplanetary medium has been studied using very low-frequency observations of Type III storms (Bougeret et al. 1984). These storms are closely associated with long-lived Type I noise storms at meter wavelengths and with the slowly varying component at microwaves (Bougeret et al. 1984, Bratova et al. 1983). At meter and decameter wavelengths the Type III storm bursts are polarized in the same sense as the Type I noise storms, but the degree of circular polarization varies considerably, suggesting that the emission may sometimes occur at the fundamental and sometimes at the second harmonic of the plasma frequency (Dulk et al. 1984). Noise storms at metre wavelengths appear to form magnetic sector structures similar to those in interplanetary space (Stewart 1984).

6. Quiet Sun

There has been little work published on this subject over the past 3 years (Kundu et al. 1983, Sastry et al. 1983). Microwave observations of the supergranulation network have been reported (Gelfreikh 1982); the radio brightness temperature is ~10^4 K and the number of sources an order of magnitude less than the number of x-ray bright points.

References

Alissandrakis, C.E. and Kundu, M.R.: 1982, Astrophys. J. Letters 253, L49.

Benz, A.O. and Wentzel, D.G.: 1981, Astron. Astrophys. 94, 100.

Benz, A.O., Barrow, C.H., Dennis, B.R., Pick, M., Raoult, A., and Simnett, G.: 1983, Solar Phys. 83, 267-283.

Bougeret, J.-L., Fainberg, J., and Stone, R.G.: 1984, Astron. Astrophys. 136, 865.

Bratova, T.A., Gelfreikh, G.B., and Gnezdilov, A.A.: 1983, Soviet Astron. Letters 9, 260.

Brueckner, G.E.: 1983, Solar Phys. 85, 243.

Cane, H.V., Stone, R.G., Fainberg, J., Stewart, R.T., Steinberg, J.L., and Hoang, S.: 1981, Geophys. Res. Letters 8, 1285.

Chiuderi D. and Melozzi, M.: 1984, Astron. Astrophys. 131, 103.

Cliver, E.W.: 1983, Solar Phys. 84, 347.

Dennis, B.R., Benz, A.L., Ranieri, M., and Simnett, G.M.: 1984, Solar Phys. 90, 383-399.

Dulk, G.A. and Gary, D.E.: 1983, Astron. Astrophys. 124, 103.

Dulk, G.A., Suzuki, S., and Sheridan, K.V.: 1984, Astron. Astrophys. 130, 39.

Duncan, R.A.: 1981, Solar Phys. 73, 191.

Duncan, R.A.: 1983, Solar Phys. 89, 63.

Felli, M., Lang, K.R., and Willson, R.F.: 1981, Astrophys. J. 247, 325.

Garczynska, I.N., Rompolt, B., Benz, A.O., Slottje, C., Tlamicha, A., and Zanelli, C.: 1982, Solar Phys. 77, 277-283.

Gary, D.E., Dulk, G.A., House, L., Illing, R., Sawyer, C., Wagner, W.J., McLean, D.J., and Hildner, E.: 1984, Astron. Astrophys. 134, 222.

Gelfreikh, G.B.: 1982, in "Sun and Planetary System" (eds. W. Fricke and G. Teleki), p. 109.

Gergely, T.E., Kundu, M.R., Erskine, F.T., III, Sawyer, C., Wagner, W.J., Illing, R., House, L.L., McCabe, M.K., Stewart, R.T., Nelson, G.J., Koomen, M.J., Michels, D., Howard, R., and Sheeley, N.: 1984, Solar Phys. 90, 161-176.

Goldman, M.V.: 1983, Solar Phys. 89, 403.

Grognard, R.J.-M.: 1982, Solar Phys. 81, 173.

Holman, G.: 1983, Adv. Space Res., 2(11), 181.

Holman, G.D. and Pesses, M.E.: 1983, Astrophys. J. 267, 837.

Holman, G., Kundu, M.R., and Papadopoulos, K.: 1982, Astrophys. J. 257, 354.

Hoyng, P., Duijveman, A., Machado, M.E., Rust, D.M., Švestka, Z., Boelee, A., de Jager, C., Frost, K.J., Lafleur, H., Simnett, G.M., van Beek, H.F., and Woodgate, B.E.: 1981, Astrophys. J. Letters 246, L155.

Hudson, H., Lin, R.P., and Stewart, R.T.: 1982, Solar Phys. 75, 245.

Kahler, S., Sheeley, N.R., Jr., Howard, R.A., Koomen, M.J., and Michels, D.J.: 1984, Solar Phys. 93, 133-141.

Kai, K., Kosugi, T., and Nitta, N.: 1984, Publ. Astron. Soc. Japan (in press).

Kane, S.R. and Raoult, A.: 1981, Astrophys. J. Letters 273, L95.

Kane, S.R., Benz, A.O., and Treumann, R.A.: 1982, Astrophys. J. 263, 423.

Kane, S.R., Kai, K., Kosugi, T., Enome, S., Landecker, P.G., and McKenzie, D.L.: 1983, Astrophys. J. 271, 376.

Kerdraon, A., Pick, M., Trottet, G., Sawyer, C., Illing, R., Wagner, W., and House, L.: 1983, Astrophys. J. Letters 265, L19.

Klein, L., Anderson, K., Pick, M., Trottet, G., Vilmer, N., and Kane, S.: 1983, Solar Phys. 84, 295-310.
Kuijpers, J., Van der Post, P., and Slottje, C.: 1981, Astron. Astrophys., 103, 331.
Kundu, M.R.: 1982, Rep. Prog. Phys. 45, 1435.
Kundu, M.R.: 1983, Solar Phys. 86, 205.
Kundu, M.R., Schmahl, E.J., Velusamy, T., and Vlahos, L: 1982, Astron. Astrophys. 108, 188.
Kundu, M.R., Erickson, E.C., Gergely, T.E., Mahoney, M.J., and Turner, P.J.: 1983, Solar Phys. 83, 385-389.
Lang, K. and Willson, R.F.: 1982, Astrophys. J. Letters 255, L111.
Lang, K.R., Willson, R.F., and Felli, M.: 1981, Astrophys. J. 247, 338.
Lantos, P., Kerdraon, A., Rapley, C.G., and Bentley, R.D.: 1981, Astron. Astrophys. 101, 33.
Lin, R.P., Potter, D.W., Gurnett, D.A., and Scarf, F.L.: 1981, Astrophys. J. 251, 364-373.
McConnell, D. and Kundu, M.R.: 1983, Astrophys. J. 269, 698.
McKenna-Lawlor, S.M.P. and Richter, A.K.: 1983, Adv. Space Res. 2(11), 239.
McLean, D.J.: 1981, Proc. Astron. Soc. Austl. 4, 132.
McLean, D.J. and Labrum, N.R. (eds.): 1985, Cambridge Univ. Press (in press).
Melrose, D.B. and Dulk, G.A.: 1982, Astrophys. J. 259, 844.
Melrose, D.B. and Dulk, G.A.: 1984, Astrophys. J. 282, 308.
Nakajima, H.: 1982, Publ. Astron. Soc. Japan 34, 351.
Pallavicini, R., Sakurai, T., and Vaiana, G.S.: 1981, Astron. Astrophys. 98, 316.
Robinson, R.D., Stewart, R.T., and Cane, H.V.: 1984, Solar Phys. 91, 159.
Sastry, Ch.V., Shevgaonkar, R.K., and Ramanuja, M.N.: 1983, Solar Phys. 87, 391.
Schmahl, E.J., Kundu, M.R., Strong, K.T., Bentley, R.D., Smith, J.B., Jr., and Krall, K.R.: 1982, Solar Phys. 80, 233-249.
Seehafer, J. et al.: 1983, Publ. Debrecen Heliophysical Obs. 5, 431.
Sharma, R.R., Vlahos, L., and Popadopoulos, K.: 1982, Astron. Astrophys. 112, 377.
Sheeley, N.R., Jr., Stewart, R.T., Robinson, R.D., Howard, R.A., Koomen, M.J., and Michels, D.J.: 1984, Astrophys. J. 279, 839-847.
Shibashi, K., Chiuderi-Drago, F., Melozzi, M., Slottje, C., and Antonucci, E.: 1983, Solar Phys. 89, 307-321.
Spicer, D.S., Benz, A.O., and Huba, J.D.: 1981, Astron. Astrophys. 105, 221.
Sprangle, P. and Vlahos, L.: 1983, Astrophys. J. Letters 273, L95.
Stewart, R.T.: 1984a, Solar Phys. 92, 343.
Stewart, R.T.: 1984b, Solar Phys. 94, 379.
Stewart, R.T.: 1984c, Proc. Third Asian-Pacific Meeting of IAU, Kyoto, October 1984 (in press).
Stewart, R.T., Dulk, G.A., Sheridan, K.V., House, L.L., Wagner, W.J., Sawyer, C., and Illing, I.R.: 1982, Astron. Astrophys. 116, 217-223.
Strong, K.T., Alissandrakis, C.E., and Kundu, M.R.: 1984, Astrophys. J. 277, 865.
Švestka, Z., Dennis, B.R., Pick, N., Raoult, A., Rapley, C.G., Stewart, R.T. and Woodgate, B.E.: 1982, Solar Phys. 80, 143-159.
Švestka, Z., Stewart, R.T., Hoyng, P., van Tend, W., Acton, L.W., Gabriel, A.H., Rapley, C.G., Boelee, A., Bruner, E.C., de Jager, C., Lafleur, H., Nelson, G., Simnett, G.M., van Beek, H.F., and Wagner, W.J.: 1982, Solar Phys. 75, 305-329.
Tapping, K.F., Kuijpers, J., Kaastra, J.S., van Nieuwkoop, G.D., and Slottje, C.: 1983, Astron. Astrophys. 122, 177-180.
Urpo, S. et al.: 1982, Physica Solariterr., Potsdam 19, 5.
Velusamy, T. and Kundu, M.R.: 1981, Astrophys. J. Letters 243, L103.
Vlahos, L., Gergely, T.E., and Papadopoulos, K.: 1982, Astrophys. J. 258, 812.
Wagner, W.J. and MacQueen, R.M.: 1983, Astron. Astrophys. 120, 136.
Watanabe, T. and Kakinuma, T.: 1984, COSPAR Proceedings in Adv. Space Res. (in press).
Webb, D.F., Davis, J.M., Kundu, M.R., and Velusamy, T.: 1983, Solar Phys. 85,

267-283.
Willson, R.F.: 1983, Solar Phys. 89, 103.
Zhao, R. and Jin, A.: 1982, Sci. Sin. A25, 422.

VII. Active Regions: Structure and Evolution
(V. Gaizauskas)

The corona above active regions is now recognized as an assemblage of magnetically confined loops of plasma. This advance in understanding the active upper atmosphere is documented in the monograph resulting from the Third Skylab Workshop (Solar Active Regions, 1981, ed., F.Q. Orrall). International collaborative programs during the Solar Maximum Year (SMY) have further stimulated the study of active regions with emphasis on the search for the underlying causes of solar flares. Scores of analyses of individual regions, combining space- and ground-based observations, have been published. We have as a result an improved picture of interactions between active regions: from creation of shear in the magnetic topology to inter-region connections via the corona. A revived interest in the phenomena of recurrent active regions and sunspot decay has highlighted a basic problem for solar magnetism: the removal of magnetic flux from the solar surface. The interpretation of temporary dips in the solar irradiance caused by active regions continues to generate lively debate.

A. SMALL-SCALE MAGNETIC FIELDS

Recent developments in measuring the fine-scale magnetic structure, with stress on the problems of interpreting those measurements, are reviewed by Stenflo (1984a,b,c). A major advance is the simultaneous recording of fully resolved, circularly polarized spectra of hundreds of spectral lines with the Fourier Transform Spectrometer at the McMath Telescope (Stenflo et al. 1984). Plasma diagnostics can thus be derived over a wide range of excitation conditions. In comparing plage and network areas with this technique, Solanki & Stenflo (1984) confirm that their magnetic field strengths are approximately equal; they find similar velocity structures in plage and network, but network flux tubes are hotter in their lower layers as compared to plages. If there is a chaotic field between the intermittent strong field elements, Stenflo (1982) can put a lower limit on it of 10 Gauss from a first attempt to measure the depolarization of scattered radiation on the fringe of active regions by the Hanle effect. The speed and sensitivity gained by integrating videomagnetograms make it possible to follow interactions between fine-scale fragments of solar magnetic fields (Martin 1984). When like polarities collide, they merge without obvious change in net magnetic flux; when fragments of opposite polarities collide, there is a gradual loss of flux in both fragments until the smaller one disappears. It is not clear whether the loss of flux occurs through reconnection or submergence of flux loops. Wilson & Simon (1983) found large and rapid fluctuations in small unipolar magnetic features with no observable changes in the fragments of strong fields in the opposite polarity. Daras-Papamargaritas & Koutchmy (1983) have measured the magnetic flux in a "rosette" ($\approx 10^{20}$ Mx) and estimate a flux of 5×10^{18} Mx per Hα fibril. The anti-correlation between coronal bright points and sunspot number has been confirmed (Davis 1983); the lack of direct correspondence between bright points and ephemeral regions (Tang et al. 1983) remains enigmatic. A solar cycle dependence is indicated for the density of photospheric network elements (Müller & Roudier 1984) and of chromospheric granules (Fang et al. 1984).

B. GROWTH AND DISAPPEARANCE

Observations at high spatial and spectral resolution (Zwaan et al. 1984, Brants 1985) reveal that a new pore strengthens by adding new flux at the edge facing the center of its growing region; the rule that an emerging flux region grows outward from its center extends right to its birth. Garcia de la Rosa

(1983, 1984) finds intrinsic differences between small and large ($\approx 5 \times 10^{22}$ Mx) regions which he attributes to different depths and modes of birth: as loosely packed flux tubes submerged among the supergranules for the small ones; as tightly wound flux ropes at the bottom of the convection zone for the large ones; Parker (1984a) estimated the depth of origin quantitatively by assuming that regions disappear by being pulled back into the Sun and that surface fields are controlled at unique subsurface anchor points. Evidence which can be interpreted as submerging flux (Rabin & Moore 1984) keeps mounting, although conclusive evidence has never been reported, such as reversal of the normal spreading of a bipolar pair. In his review of extensive observations at Ondrejov on the motions and changes in shapes of evolving sunspots, Bumba (1983) cites patterns of development which he cannot reconcile with the concept of emerging flux. Akasofu (1984) is developing a model in which hydrodynamic forces act on the large-scale background magnetic fields at the surface to create a photospheric dynamo. The term "naked sunspot" was coined by Ligget & Zirin (1983) to describe aged, large sunspots without accompanying plage. The plage-associated fields are presumed to disappear first through local reconnections in the large unipolar areas which such spots prefer. Because these spots disappear without trace, the remnant dispersed fields of active regions are likely left by dissolving plages, not by sunspots. Active regions do not form at random over the Sun. Ligget & Zirin (1984) measured a rate of flux emergence 27 times higher within active regions than in quiet background areas at the same latitudes. Active regions cluster in space and time in distinct entities called "complexes of activity"; yet flux does not accumulate in a complex (Gaizauskas et al. 1983). Flux disappears locally at a rate which balances the rate of emergence in the same complex. Each complex rotates around the Sun with its own period. At times complexes are spaced regularly around the active belts of latitude in bands of alternating polarity. Parker (1984a) shows that the gaps between complexes should block reconnection and prevent the escape of any substantial amounts of flux by this process. Loss of flux from the solar surface remains a more subtle problem than imagined; its resolution has profound consequences for concepts of the solar dynamo.

C. CORONAL INTERCONNECTIONS

Coronal loops track active region evolution; new loops connect with older sets as new active regions appear (Sheeley 1981). The transient brightenings observed in these interconnecting loops are interpreted by Spicer & Švestka (1983) as due to either excitation of the fast tearing mode in young, newly born loops or possibly to anomalous Joule heating in the old loop connections. Hot plasma ($\approx 10^7$ K) above active regions has been found even in the absence of flares (Schadee et al. 1983). The character of faint x-ray (≈ 3.5 keV) emission depends upon the stage of the region's development. Interconnection of widely spaced, non-flaring active regions has also been observed (Farnik & van Beek 1984). Large-scale C IV velocity patterns have been found in close correspondence with photospheric magnetic fields in active regions: over sunspots it is in the reverse Evershed sense with a substantial vertical component (Athay et al. 1982); outside of sunspots it is nearly horizontal with a preference for downflow (≈ 10 km/s) in both legs of flat loops (Athay et al. 1983). Systems of C IV loops last many hours over active regions, often rising into the corona. A geometrical technique for reconstructing the true shapes of solar loops observed on the disk is described in a series of papers: Loughhead et al. (1983a), Chen & Loughhead (1983), Wang et al. (1984), Loughhead et al. (1984). During a subflare observed at 6 cm and in Hα (Kundu et al. 1983), simultaneous brightenings at both wavelengths were observed $\approx 10^5$ km away from the primary site of energy release. Using co-temporal microwave (1.8 cm, spatially resolved) and hard x-ray (HXRBS) observations, Nakajima et al. (1985) found transient events with secondary microwave bursts 10^5-10^6 km away from the primary sites of energy release. The distant bursts could be produced by electron beams with energies 10-100 keV channeled along a connecting loop; two of these events had "sympathetic" flares triggered at remote sites. Simultaneous eruptions are reported in many studies:

between adjacent active regions (Golovko et al. 1981, Gaizauskas 1983, Loughhead et al. 1983b, Machado et al. 1983); between widely spaced regions (Ogir 1981); and even as coronal transients on opposite limbs (Wagner & Wagner 1984). Stresses built up by the normal evolution of active regions are sensed by the global network of coronal loops; their release is complicated by this interdependence. If sites of energy storage do exist, they need not coincide with the sites of initial energy release.

D. FLARE-ASSOCIATED EVOLUTION

The source of flare energy is believed to be the free energy stored in magnetic fields stressed by evolutionary trends in active regions, i.e., by emergence of additional flux and by proper motions of sunspots (Priest 1984a,b). The elucidation of these trends is the goal of the Flare Build-Up Study. Many accounts illustrate increased flare activity in locations of sheared magnetic and velocity fields: Krall et al. (1982), Nagy (1983), Kalman & Nagy (1983), Kalman (1984), Dezsö et al. (1984), Zirin (1983, 1984). An MHD model of sheared loops by Wu et al. (1984) shows a growth rate roughly proportional to the shearing speed and a buildup of magnetic energy near the neutral lines. Calculations of magnetic shears (Hagyard et al. 1984) and of current densities (Gopasryuk et al. 1983) for observed active regions show that flares occur where continued magnetic evolution of the region push these parameters to critical values. In their study of the development of shear, Athay et al. (1985a) found very little flare activity despite sustained strong shear along a magnetic neutral line, except for minor disruptions associated with emerging flux. The strongly flare productive δ-regions (Knoska & Krivsky 1983) often form by collision of neighboring, growing bipolar regions (Tang 1983, Gesztelyi & Kondas 1983, Gesztelyi 1984). Athay et al. (1985b) followed a growing δ-region from a state of weak to a state of strong magnetic shear in localized zones. Although large flares occurred in some strongly sheared zones, none occurred in other zones of strong shear in the same region. Magnetic shear alone is not sufficient to produce large flares. The role of magnetic flux is ambivalent. Emergence of a bipole is accompanied by lateral spreading which can lead to complex interactions with pre-existing fields either nearby or at remote locations. Although emerging flux is closely associated with small flares inside an EFR (Martin 1983) and is often invoked as a trigger for larger flares (Moore et al. 1984, Simon et al. 1985), there are also clear examples of flares with strong filament eruptions and no emerging flux (Gaizauskas 1984).

E. RADIO EMISSION

The rapid progress toward an understanding of the emission processes in coronal structures above active regions has been greatly stimulated by the application of supersynthesis arrays to solar observations (see review by Kundu 1982). The bipolar character of magnetic fields above active regions is the dominant factor in shaping and insulating structures in the multi-thermal coronal plasma (Dulk & Gary 1983, Lang & Willson 1983). The model-dependent identification of emission processes becomes more problematic at the shorter wavelengths which originate at lower levels near regions of intense fields. A new thermal model incorporating a force-free field extrapolation has been developed by Staude et al. (1983) and applied to observations by Urpo et al. (1982), Seehafer et al. (1983), Kaverin et al. (1983), Akhmedov et al. (1983), and Hildebrandt et al. (1984). A new model by Alissandrakis & Kundu (1984) has been applied to the center-to-limb variation of a pair of large active regions. A non-thermal model has been proposed by Chiuderi-Drago & Melozzi (1984) to account for high temperature radio sources uncorrelated with sunspots or x-ray sources. Bandiera (1982) has developed a diagnostic for deducing magnetic fields above bipolar regions which needs no assumptions about the emission mechanism. The structures of radio sources and their associations with x-ray and optical structures are so complex (Webb et al. 1983) that little can be offered as yet by way of simplification. Other multi-wavelength observations confirm the conclusion of Webb et al. that the presence of a strong magnetic field is not a sufficient condition for bright microwave emission; they further show that the 20-cm emission is the microwave thermal counterpart of soft x-ray emission from large loops straddling a bipolar region, while the 6-cm emission is attributed to gyro-resonant emission, usually from the legs of loops associated with sunspots, but also from arcades of loops in locations of strong transverse magnetic fields (Lang et al. 1983, McConnell & Kundu 1983, 1984, Shibasaki et al. 1983, Strong et al. 1984, Kundu & Alissandrakis 1984). The bewildering profusion of radio sources with very different structures suggests that clarifying knowledge of the long-term evolution of microwave emission from a single active region observed at high spatial resolution is needed. A three-dimensional structure for a large coronal loop over an active region has been presented by Shevgaonkar & Kundu (1984). Schmahl et al. (1984) have mapped a region with an anomalously high spectral index which they explain in terms of emission in cyclotron lines. Brueckner (1983) found that all Type I noise storms observed during the Skylab period were caused by changes in coronal magnetic field structure; all disk changes were correlated with emerging flux.

F. VARIABILITY IN SOLAR IRRADIANCE

The blocking by active regions of heat transported from the convection zone continues to stir debate. The background and future directions of this topic are summarized in reviews by Newkirk (1983) and Willson (1984) and in the proceedings of the Caltech Workshop on Solar Irradiance Variations on Active Region Time Scales. Opinion is sharply divided between two alternatives: either the blocked enrgy is stored in the convection zone for long periods of time (Foukal et al. 1983) or it is re-routed and quickly re-radiated by faculae (Oster et al. 1982, Chapman et al. 1984). The former alternative leads to a solar luminosity modulated weakly at the 11-year period of the solar cycle, while the latter accounts only for daily changes in solar irradiance. Tests to resolve this issue by appealing to archival records are as yet inconclusive (Eddy 1984). Much of the dispute arises from the poor quality of information about faculae, particularly about their role in the evolution of active regions. Attempts to solve the controversy will remain speculative until a prolonged data base is available from irradiance monitors with modern detectors (Hudson et al. 1984). Fowler et al. (1983) find marginally significant bright rings of non-facular origin around sunspots, much weaker than predicted from conventional mixing-length theory; significantly lower values of the eddy thermal conductivity are indicated. An upper limit of 1.5 K rms can be placed on any thermal shadow over the area occupied by an active region 1 day before it appears (Foukal 1984). The growth and breakup of plages measured during the solar cycle seem insufficient to account for the cycle-dependent increase of Ca K emission which has been modeled by Skumanich et al. (1984). The effect on solar UV irradiance by the evolution of active regions has been determined by Donnelly et al. (1982); UV variability cannot be neglected in determining the facular contribution to the total irradiance (Donnelly et al. 1984).

References

Akasofu, S.-I.: 1984, Planet. Space Sci. 32(10), 1257-1261.

Akhmedov, Sh.B, Gelfreikh, G.B., Fürstenberg, F., Hildebrandt, J., and Krüger, A.: 1983, Solar Phys. 88, 103-108.

Alissandrakis, C.E. and Kundu, M.R.: 1984, Astron. Astrophys. 139, 271-284.

Athay, R.G., Gurman, J.B., Henze, W., and Shine, R. A.: 1982, Astrophys. J. 261, 684.

Athay, R.G., Gurman, J.B., and Henze, W.: 1983, Astrophys. J. 269, 706-714.

Athay, R.G. et al.: 1985a, Astrophys. J. (in press).

Athay, R.G. et al.: 1985b, Astrophys. J. (in press).

Bandiera, R.: 1982, Astron. Astrophys. 112, 52-60.

Brants, J.J.: 1985, Solar Phys. 95, 15-36.

Brueckner, G.E.: 1983, Solar Phys. 85, 243-265.
Bumba, V.: 1983, in Proc. 11th Regional Consultation on Solar Physics, eds., L. Deszö and B. Kalman, Publ. Debrecen Obs. 5, 47.
Chapman, G.A., Herzog, A.D., Lawrence, J.K., and Shelton, J.C.: 1984, Astrophys. J. Letters 282, L99.
Chen, C.-I. and Loughhead, R.E.: 1983, Proc. Astron. Soc. Aust. 5(2), 204-208.
Chiuderi-Drago, F. and Melozzi, M.: 1984, Astron. Astrophys. 131, 103-110.
Daras-Papamargaritas, H. and Koutchmy, S.: 1983, Astron. Astrophys. 125, 280-286.
Davis, J.M.: 1983, Solar Phys. 88, 337-342.
Dezsö, L. et al.: 1984, in Solar Maximum Analysis Symp. 3, XXV COSPAR, Graz, Adv. Space Res. (in press).
Donnelly, R.F., Heath, D.F., and Lean, J.L.: 1982, J. Geophys. Res. 87(A12), 10,318-10,324.
Donnelly, R.F., Heath, D.F., Lean, J.L., and Rottman, G.J.: 1984, in Solar Irradiance Variatons on Active Region Time Scales, NASA CP-2310, eds., B.J. LaBonte, G.A. Chapman, H.S. Hudson, and R.C. Willson (NASA: Washington, D.C.), pp. 233-242.
Dulk, G.A. and Gary, D.E.: 1983, Astron. Astrophys. 124, 103-107.
Eddy, J.A.: 1984, in Solar Irradiance Variatons on Active Region Time Scales, NASA CP-2310, eds., B.J. LaBonte, G.A. Chapman, H.S. Hudson, and R.C. Willson (NASA: Washington, D.C.), pp. 213-229.
Fang, C., Mouradian, Z., Banos, G., Dumont, S., and Pecker, J.C.: 1984, Solar Phys. 91, 61-70.
Farnik, F. and van Beek, H.F.: 1984, in Solar Maximum Analysis Symp. 3, XXV COSPAR, Graz, Adv. Space Res. (in press).
Foukal, P.: 1984, in Solar Irradiance Variatons on Active Region Time Scales, NASA CP-2310, eds., B.J. LaBonte, G.A. Chapman, H.S. Hudson, and R.C. Willson (NASA: Washington, D.C.), pp. 97-117.
Foukal, P., Fowler, L.A., and Livshits, M.: 1983, Astrophys. J. 267, 863-871.
Fowler, L.A., Foukal, P., and Duvall, T., Jr.: 1983, Solar Phys. 84, 33-44.
Gaizauskas, V.: 1983, Adv. Space Res. 2, 17.
Gaizauskas, V.: 1984, in Proc. Internat. Workshop on Solar Physics and Interplanetary Travelling Phenomena, Kunming, China (in press).
Gaizauskas, V., Harvey, K.L., Harvey, J.W., and Zwaan, C.: 1983, Astrophys. J. 265, 1056-1065.
Garcia de la Rosa, J.I.: 1983, Solar Phys. 89, 51-62.
Garcia de la Rosa, J.I.: 1984, Solar Phys. 92, 161-172.
Gesztelyi, L.: 1984, in Solar Maximum Analysis Symp. 3, XXV COSPAR, Graz, Adv. Space Res. (in press).
Gesztelyi, L. and Kondas, L.: 1983, in Proc. 11th Regional Consultation on Solar Physics, eds., L. Deszö and B. Kalman, Publ. Debrecen Obs. 5, 133.
Golovko et al.: 1981, in Proc. Internat. Workshop on the Solar Maximum Year, ed., V.N. Obridko, Simferopol, p. 197.
Gopasryuk, S.I. et al.: 1983, in Proc. 11th Regional Consultation on Solar Physics, eds., L. Deszö and B. Kalman, Publ. Debrecen Obs. 5, 249.
Hagyard, M.J., Smith, J.B., Jr., Teuber, D., and West, E.A.: 1984, Solar Phys. 91, 115-126.
Hildebrandt, J., Seehafer, N., and Krüger, A.: 1984, Astron. Astrophys. 134, 185-188.
Hudson, H.S., Chapman, G.A., and LaBonte, B.J.: 1984, in Solar Irradiance Variatons on Active Region Time Scales, NASA CP-2310, eds., B.J. LaBonte, G.A. Chapman, H.S. Hudson, and R.C. Willson (NASA: Washington, D.C.), pp. 311-313.
Kalman, B.: 1984, in Solar Maximum Analysis Symp. 3, XXV COSPAR, Graz, Adv. Space Res. (in press).
Kalman, B. and Nagy, I.: 1983, in Proc. 11th Regional Consultation on Solar Physics, eds., L. Deszö and B. Kalman, Publ. Debrecen Obs. 5, 207.
Kaverin et al.: 1983, in Proc. 11th Regional Consultation on Solar Physics, eds., L. Deszö and B. Kalman, Publ. Debrecen Obs. 5, 631.

Knoska, A. and Krivsky, L.: 1983, in Proc. 11th Regional Consultation on Solar Physics, eds., L. Deszö and B. Kalman, Publ. Debrecen Obs. 5, 557-566.
Krall, K.R., Smith, J.B., Jr., Hagyard, M.J., West, E.A., and Cummings, N.P.: 1982, Solar Phys. 79, 59-75.
Kundu, M.R.: 1982, Rep. Prog. Phys. 45, 1435-1541.
Kundu, M.R. and Alissandrakis, C.E.: 1984, Solar Phys. 94, 249.
Kundu, M.R., Rust, D.M., and Bobrowsky, M.: 1983, Astrophys. J. 265, 1084-1089.
Lang, K.R. and Willson, R.F.: 1983, Astron. Astrophys. 127, 135-139.
Lang, K.R., Willson, R.F., and Gaizauskas, V.: 1983, Astrophys. J. 267, 455-464.
Liggett, M. and Zirin, H.: 1983, Solar Phys. 84, 3-11.
Liggett, M. and Zirin, H.: 1984, Solar Phys. 91, 259-267.
Loughhead, R.E., Wang, J.-L., and Blows, G.: 1983a, Astrophys. J. 274, 883-899.
Loughhead, R.E., Wang, J.-L, and Duncan, R.A.: 1983b, Solar Phys. 83(2), 257-266.
Loughhead, R.E., Chen, C.-L., and Wang, J.-L.: 1984, Solar Phys. 92, 53-65.
Machado, M.E., Somov, B.V., Rovira, M.G., and de Jager, C.: 1983, Solar Phys. 85, 157-184.
Martin, S.: 1983, Adv. Space Res. 2, 39.
Martin, S.: 1984, Big Bear Solar Observatory Preprint #288.
McConnell, D. and Kundu, M.R.: 1983, Astrophys. J. 269, 698-705.
McConnell, D. and Kundu, M.R.: 1984, Astrophys. J. 279, 421-426.
Moore, R.L., Hurford, G.J., Jones, H.P., and Kane, S.R.: 1984, Astrophys. J. 276, 379-390.
Müller, R. and Roudier, Th.: 1984, Solar Phys. 94, 33-47.
Nagy, I.: 1983, in Proc. 11th Regional Consultation on Solar Physics, eds., L. Deszö and B. Kalman, Publ. Debrecen Obs. 5, 107.
Nakajima, H. et al.: 1985, Astrophys. J. (in press).
Newkirk, G., Jr.: 1983, Ann. Rev. Astron. Astrophys. 21, 429-467.
Ogir, M.R.: 1981, in Proc. Internat. Workshop on the Solar Maximum Year, ed., V.N. Obridko, Simferopol, p. 197.
Oster, L., Schatten, K.H., and Sofio, S.: 1982, Astrophys. J. 256, 768-773.
Parker, E.N.: 1984a, Astrophys. J. 280, 423-427.
Parker, E.N.: 1984b, Astrophys. J. 281, 839-845.
Priest, E.R.: 1984a, in Proc. Internat. Workshop on Solar Physics and Interplanetary Travelling Phenomena, Kunming, China (in press).
Priest, E.R.: 1984b, in Solar Maximum Analysis Symp. 3, XXV COSPAR, Graz, Adv. Space Res. (in press).
Rabin, D. and Moore, R.: 1984, Astrophys. J. 285, 359-367.
Schadee, A., de Jager, C., and Švestka, Z.: 1983, Solar Phys. 89, 287-305.
Schmahl, E.J., Shevgaonkar, R.K., Kundu, M.R., and McConnell, D.: 1984, Solar Phys. 93, 305-315.
Seehafer, J. et al.: 1983, in Proc. 11th Regional Consultation on Solar Physics, eds., L. Deszö and B. Kalman, Publ. Debrecen Obs. 5, 431.
Sheeley, N.R., Jr.: 1981, in Solar Active Regions--A Monograph from Skylab Solar Workshop III, ed., F.Q. Orrall, Colorado Associated Univ. Press, Boulder, Colorado, pp. 17-42.
Shevgaonkar, R.K. and Kundu, M.R.: 1984, Astrophys. J. 283, 413-420.
Shibasaki, K., Chiuderi-Drago, F., Melozzi, M., Slottje, C., and Antonucci, E.: 1983, Solar Phys. 89, 307-321.
Simon, G., Mein, N., Mein, P., and Gesztelyi, L.: 1984, Solar Phys. 93, 325-336.
Skumanich, A., Lean, J.L., White, O.R., and Livingston, W.C.: 1984, Astrophys. J. 282, 776-783.
Solanki, S.K. and Stenflo, J.O.: 1984, Astron. Astrophys. 140, 185-198.
Solar Active Regions--A Monograph from Skylab Solar Workshop III: 1981, F.Q. Orrall (ed.), Colorado Associated Univ. Press, Boulder, Colorado.
Solar Irradiance Variatons on Active Region Time Scales: 1984, NASA CP-2310, eds., B.J. LaBonte, G.A. Chapman, H.S. Hudson, and R.C. Willson (NASA: Washington, D.C.).
Spicer, D.S. and Švestka, Z.: 1983, Solar Phys. 87, 271-278.
Staude, J., Fürstenberg, F., Hildebrandt, J., Krüger, A., Jakimiec, J., Obridko,

V.N., Siarkowski, M., Sylwester, B., and Sylwester, J.: 1983, Acta Astronomica 33, 431-460.
Stenflo, J.O.: 1982, Solar Phys. 80, 209-226.
Stenflo, J.O.: 1984a, in Proc. Internat. Workshop on Solar Physics and Interplanetary Travelling Phenomena, Kunming, China (in press).
Stenflo, J.O.: 1984b, in Solar Maximum Analysis Symp. 3, XXV COSPAR, Graz, Adv. Space Res. (in press).
Stenflo, J.O.: 1984c, in M.J. Hagyard (ed.), MSFC Workshop on Measurements of Solar Vector Magnetic Fields, NASA CP (in press).
Stenflo, J.O., Harvey, J.W., Brault, J.W., and Solanki, S.: 1984, Astron. Astrophys. 131, 333-346.
Strong, K.T., Alissandrakis, G.E., and Kundu, M.R.: 1984, Astrophys. J. 277, 865-873.
Tang, F.: 1983, Solar Phys. 89, 43-50.
Tang, F. et al.: 1983, Adv. Space Res. 2, 65.
Urpo, S. & 5 others:1982, Phys. Solariterr., Potsdam 19, 5.
Wagner, W.J. and Wagner, J.J.: 1984, Astron. Astrophys. 133, 288-292.
Wang, S.-G. et al.: 1984, in Proc. Internat. Workshop on Solar Physics and Interplanetary Travelling Phenomena, Kunming, China (in press).
Webb, D.F., Davis, J.M., Kundu, M.R., and Velusamy, T.: 1983, Solar Phys. 85, 267-283.
Willson, R.C.: 1984, Space Science Rev. 38, 203-242.
Wilson, P.R. and Simon, G.W.: 1983, Astrophys. J. 273, 805-821.
Wu, S.T., Hu, Y.Q., Krall, K.R., Hagyard, M.J., and Smith, J.B., Jr.: 1984, Solar Phys. 90, 117-131.
Zirin, H.: 1983, Astrophys. J. 274, 900-909.
Zirin, H.: 1984, Astrophys. J. 281, 884-885.
Zwaan, C. et al.: 1985, Solar Phys. 95, 3-14.

VIII. Theory of Flares

(E.R. Priest)

Magnetohydrodynamic (MHD) theory for the initiation and development of solar flares has developed considerably over the past 3 years and represents one of the liveliest areas of solar physics (Hood & Priest, 1981a, Priest 1983a,b, Schindler 1982, Van Hoven 1982, Syrovatskii et al. 1983). This has been stimulated by a thorough analysis of the Skylab observations and also by the startling new observations from the Solar Maximum Mission (SMM). In addition, the realization that flares appear to form two basic types, namely, small simple-loop flares and large two-ribbon flares, has focussed the imagination of theorists (e.g., Priest 1981, 1982), even though reality may be somewhat more complex. In the former type, a single-loop structure brightens up and decays without moving; whereas in the latter, an active region filament erupts and then two ribbons of chromospheric emission form and separate, with an arcade of hot and cool loops joining them.

The basic theory for hydrodynamic flow in a rigid loop and for magnetic field reconnection has been studied in depth, as summarized below. Major theoretical problems have been to try and understand how the magnetic field can become unstable and so initiate a flare in the two basic geometries, namely, a loop and an arcade. Also, the creation of post-flare loops by magnetic reconnection as the magnetic field closes back down in the main phase of a two-ribbon event has been modeled, and the roles of emerging flux are being clarified.

All these are at present active topics and one expects much theoretical progress over the next few years. In particular, the coupling of loop flow to the magnetic field should be studied, and the details of the new fast reconnection regions and of the nonlinear development of tearing should be worked out. Coupling the MHD to the various mechanisms for particle acceleration, such as turbulence and shock waves, is ready to be attempted, and the details of the flux emergence need to be studied. Also, a study of the nonlinear development of the MHD instabilities thought to initiate a flare has hardly begun.

A. LOOP HYDRODYNAMICS

Many authors have been tackling the hydrodynamic problem of the thermal response of plasma in a rigid magnetic tube to a sudden deposition of heat due to, for example, particle beams or magnetic reconnection. The fluid equations of mass continuity, momentum, and energy for one-dimensional flow in a loop of given shape are solved numerically for a variety of heat sources and loop shapes and lengths, representing an extension of the earlier pioneering work on chromospheric evaporation by Antiochos and Sturrock (1982) and Somov et al. (1977). After an additional heat source is suddenly switched on, pairs of conduction fronts and shock waves propagate down both legs of the loop away from the site of the source. When they reach the loop footpoints, the chromosphere is heated and expands upward (i.e., evaporates) to fill the loop with hot dense plasma. After some time, this plasma cools by radiation and conduction and then drains back down. Superimposed on this process, one finds oscillations with a period equal to the time it takes to propagate along the loop. [For reviews, see Craig (1981), Somov & Spektor (1982), and Pallavicini et al. (1983).] Recent work in numerical simulation can be classified into two main groups. One group involves the hydrodynamic response to the injection of an electron beam (Somov et al. 1981, Nagai & Emslie 1984). The other group comprises conduction-heated models (Wu et al. 1981, Smith & Harmony 1982, Pallavicini & Peres 1983, Nagai et al. 1983, Cheng et al. 1983, Serio 1982). Peres et al. (1982) have modeled the optically thick footpoints and find that chromospheric heating is much more efficient at evaporating material than coronal heating. With a Lagrangian formalism, Craig et al. (1982) show how a summit heat pulse raises the summit temperature and later (by evaporation) the summit density, after which both decline as the plasma drains back down. Antiochos & Sturrock (1982) suggest that supersonic downflows in the late phase could be driven by footpoint cooling. MacNeice et al. (1984) and Pallavicini et al. (1983) have predicted the spectroscopic consequences of heating and evaporation (see also Doschek et al. 1982, 1983). In a series of papers, McClymont & Canfield (1983a,b) and collaborators (Canfield et al. 1983, An et al. 1983, Ricchiazzi & Canfield 1983) have investigated flare-loop radiative hydrodynamics with the aim of examining the role of various energy-transport mechanisms in a flaring loop.

B. MAGNETIC RECONNECTION

Our basic understanding of reconnection has been changed recently by some numerical experiments, which have linked the two previous strands of theory, namely, tearing mode instability and fast Petschek-Sonnerup reconnection, and have presented us with new surprises (e.g., Vasyluinas 1983, Heyvaerts 1983, Priest 1985).

1. Tearing Mode Instability

The theory for this reconnection instability in a current sheet or a sheared magnetic field has been recently extended. The effect of photospheric line tying may be to stabilize a loop (Mok & Van Hoven 1982) or an arcade (Migliuolo & Cargill 1983). An important development is the discovery of a radiative tearing mode, which is typically a hundred times faster than normal tearing for coronal parameters (Van Hoven et al. 1983, Tachi et al. 1983, Van Hoven & Steinolfson 1983, Steinolfson & Van Hoven, 1982, 1984). The temperature dependence of the magnetic diffusivity couples the induction equation to an energy equation containing joule heating, thermal conduction, and optically thin radiative losses. This modifies considerably the normal tearing and radiative modes. Van Hoven (1981) has discussed the energetics of tearing. The tearing mode may develop nonlinearly along several pathways, depending on the geometry and the parameter regime. At large values of the Lundquist number, Steinolfson & Van Hoven (1983a) find Rutherford saturation when the wavelength is twice the shear length, but a much faster growth at larger wavelengths. In a flux tube, the coupling of modes on different surfaces can be important. Also, if several islands form, they may rapidly combine by the coalescence instability (Bhattacharjee et al. 1983, Biskamp 1982). When the boundary conditions are free enough Forbes & Priest (1982a, 1983a) have shown that it is possible for tearing to develop nonlinearly into the fast steady state of Petschek-Sonnerup reconnection. Steinolfson & Van Hoven (1983b) have followed the nonlinear development of tearing at Lundquist numbers of 10°. They find a considerably slowing of the growth, with 80% of the magnetic energy going into thermal energy and a maximum electric field of only one-thousandth of the Dreicer field. Also, at long wavelengths, secondary vortices can create a new magnetic island at the X points.

2. Fast Reconnection

The second main theme of reconnection theory has been the fast nonlinear state of Petschek-Sonnerup reconnection, which may either develop from linear tearing or be driven by pushing separate flux systems together. The current sheet bifurcates into two pairs of slow shock waves, which stand in the flow and convert inflowing magnetic energy into the heat and kinetic energy of hot fast jets. A rigorous compressible treatment of the Petschek extreme has now been completed (Soward & Priest 1982). In general the external boundary conditions at the sources of inflowing plasma produce a hybrid Petschek-Sonnerup regime. However, when reconnection develops locally from the tearing mode, the nonlinear state tends to be Petschek-like (Forbes & Priest 1982a), whereas when the reconnection is driven from outside, the nonlinear state can be closer to the pure Sonnerup regime (Forbes 1984). The steady Petschek-Sonnerup mode is possible when the inflow speed (v) of plasma at large distances is less than a maximum speed (v_{max}), whose value depends on the magnetic Reynolds number and external boundary conditions. For pure Petschek reconnection, it is typically 0.01 v_A, but for pure Sonnerup reconnection, it is roughly the Alfvén speed (v_A) at large distances. A new discovery is that fast-mode shocks may be present in the outflowing hot jets (Forbes & Priest 1983a). These have the effect of degrading the kinetic energy to heat and may be very efficient at accelerating particles. Recent numerical experiments (Biskamp 1982, Forbes & Priest 1982a, 1983a,b) have revealed two new regimes of fast unsteady reconnection when the Petschek-Sonnerup mechanism breaks down. The flux pile-up regime occurs when the inflow of plasma is faster than v_{max}, such as when reconnection is driven by an ideal instability. The impulsive bursty regime occurs when the central current sheet goes unstable to secondary tearing and produces a more rapid energy release in a series of bursts as the islands coalesce.

C. THE FLARE INSTABILITY

1. Kink Instability of Loop

The preflare magnetic configuration has been modeled by a single loop, and its magnetic stability has been investigated, both for simple-loop and two-ribbon flares. Usually, the loop curvature has been neglected, but recently the important stabilizing effect of photospheric line tying has been included. It keeps the loop stable until the amount of twist in the loop exceeds a critical value, typically between 2π and 4π, depending on the particular equilibrium, the plasma beta, and the details of the line tying. The most complete analysis has been presented by Hood & Priest (1981b) and Einaudi & Van Hoven (1983), building on earlier work (e.g., Van Hoven et al. 1981, Van Hoven 1981, Illing et al. 1981, Song & Cao 1983a,b). Recently, An (1983) investigated the radiative modes in a loop and found them unstable for small twist.

2. Eruptive Instability of Arcade

For two-ribbon flares the preflare magnetic configuration has recently been modeled more accurately by a force-free coronal arcade, and its stability has been tested (including photospheric line tying) by Birn & Schindler (1981), Schindler et al. (1983), Ray & Van Hoven (1982), Hood (1983a), and Migliuolo & Cargill (1983). The original analysis of Hood & Priest (1980) found that a simple arcade with its magnetic axis below the photosphere tends to be unstable. However, arcades with their magnetic axis above the photosphere are configurations within which an active region (or plage) filament may well form. Such filaments invariable erupt just before a two-ribbon event, slowly at first and then much more rapidly at flare onset. This type of coronal arcade has been found by Hood & Priest (1981b) to become unstable when either the height of the magetic axis (and therefore the filament) or the amount of shear become too great. This suggests that the eruption of the arcade may be caused by a spontaneous eruptive instability when the filament height or magnetic shear become too great. Several equilibrium models for magnetostatic arcades with a balance between the Lorentz force, a pressure gradient, and gravity have been set up by Heyvearts et al. (1982), Zweibel & Hundhausen (1982), and Melville et al. (1983). Using the energy method, Schindler et al. (1983) have obtained a sufficient condition for stability of such fields. When there is no axial field, the conditions also becomes necessary, but Migliuolo et al. (1984) have shown that interchange modes with very short wavelengths along the arcade are unstable, which may be important for generating small-scale structure. In addition, the arcade becomes unstable when the plasma pressure gradient is too large. Zweibel (1982) has derived a sufficient condition for stability and for one particular configuration has shown a tendancy for instability when a magnetic island is present. Hood (1983b) has extended her analysis and investigated the instability threshold in more detail by including the effect of magnetic tension, which tends to make the field stable for small plasma pressure. Flares are often accompanied by a coronal transient, for which a theory has been developed by Low (1983) based on a simple self-similar model. Numerical simulations have been presented by Steinolfson (1982), and Wu et al. (1983).

D. CREATION OF FLARE LOOPS

During the preflare phase of a two-ribbon flare, the filament embedded in a magnetic arcade rises slowly, possibly due to the above eruptive instability. The onset of the flare itself coincides with the start of the much more rapid eruption of the filament, which probably occurs because the magnetic field lines of the stretched-out arcade start to reconnect below the filament (Priest 1981). During the main phase, the reconnection continues and creates hot post-flare loops with Hα ribbons at their footpoints as the field closes down. The source of the immense mass of plasma that is subsequently observed to be falling down along cool post-flare loops is an upflow of plasma from the chromosphere along the open field lines before they reconnect. The cause of the upflow may well be evaporation driven by thermal conduction or by fast particles that are accelerated at the shocks associated with the reconnection process (Forbes & Priest 1983b). In the original Kopp-Pneuman model, the plasma was heated by gas dynamic shocks to only a few million degrees. Instead, Cargill & Priest (1982, 1983) have suggested that the shocks should be slow magnetoacoustic in character, which can release extra energy from the magnetic field and can heat the upflowing plasma to the temperature of 20 million degrees or more observed in hot loops. Forbes has conducted a numerical experiment on the line-tied reconnection thought to be occurring below the erupting filament (Forbes & Priest 1982a, 1983a,b). Open field lines that are stretched out go unstable to tearing, which develops nonlinearly into a quasi-steady Petschek regime with a fast shock in the downflowing jet. The field lines close down, with the X type neutral point rising and a plasmoid being ejected upward. In the subsequent development, the sheet thins and the Petschek mode goes unstable, being transformed into the impulsive bursty regime. Secondary tearing and coalescence creates and annihilates neutral point pairs repetitively, with fast energy release in an impulsive manner. Forbes has also set up a simple kinematic model which describes the upward motion of the neutral line and the underlying hot loops in the main phase of a two-ribbon flare (Forbes & Priest 1982b).

E. ROLE OF EMERGING FLUX AND SATELLITE SUNSPOTS

Small regions of emerging flux or satellite sunspots are often observed before flares, signifying the interaction of separate magnetic flux systems by either vertical or horizontal motions and the creation of current sheets at the interface between the two flux systems. The first role of such flux evolution is to create small flares; it has been estimated by solving the energy balance within the sheet (Milne & Priest 1981). The second role is to build up magnetic energy in the coronal reservoir by increasing the complexity and shear in the active region. The third role is to trigger large flares by initiating energy release in the more extensive overlying field (e.g., Hood & Priest 1981c). In particular, it may push up the filament or tear away some of the overlying field lines, or initiate large-scale reconnection (Priest 1984). Forbes has conducted a numerical experiment on flux emergence and finds that the whole process can be much more complicated and dynamic than previously thought (Forbes 1984). The flux emerges at a high speed and reconnects at first in a flux pile-up regime. Later, the flux pinches off and forms a plasmoid. In the future, it would be interesting to see whether experiments at magnetic Reynolds numbers higher than 2000 would give a more turbulent reconnection state. There has been a continuation of the earlier work on the behavior of the current sheet at the interface between new and old flux. Forbes et al. (1982) considered the evolution of the sheet in response to an enhanced turbulent resistivity. Sakai & Washimi (1981, 1982) suggested that fast magnetoacoustic waves generated by the emerging flux may drive tearing with a faster growth than normal. The state of the high temperature turbulent current sheet appropriate to the main phase has been studied by Somov and co-workers (Somov 1981, 1983, Somov & Titov 1983, Somov et al. 1982). They include an anomalous conductivity and heat flux as well as the effect of transverse and longitudinal field components, and demonstrate that the sheet liberates enough power for all phases of the flare. An improved model for the transverse structure of a current sheet has been established by Milne & Priest (1981) who deduce the way that the critical sheet height for flare triggering depends on magnetic field strength and emergence speed.

References

An, C.-H.: 1983, Astrophys. J. 264(1), 302-308.

An, C.-H., Canfield, R.C., Fisher, G.H., and McClymont, A.N.: 1983, Astrophys. J. 267(1), 421-432.

Antiochos, S.K. and Sturrock, P.A.: 1982, Astrophys. J. 254, 343-348.

Bhattacharjee, A., Brunel, F., and Tajima, T.: 1983, Phys. Fluids 26(11), 3332-3337.

Birn, J. and Schindler, K.: 1981, in E.R. Priest (ed.), Solar Flare Magnetohydrodynamics, Gordan and Breach, London, Chapter 6.

Biskamp, D.: 1982, Phys. Letters 87A(7), 357-360.

Canfield, R.C., Fisher, G.H., and McClymont, A.N.: 1983, Astrophys. J. 265(1), 507-518.

Cargill, P.J. and Priest, E.R.: 1982, Solar Phys. 76, 357-375.

Cargill, P.J. and Priest, E.R.: 1983, Astrophys. J. 266(1), 383-389.

Cheng, C.-C., Oran, E.S., Doschek, G.A., Boris, J.P., and Mariska, J.T.: 1983, Astrophys. J. 265(2), 1090-1102.

Craig, I.J.D.: 1981, in E.R. Priest (ed.), Solar Flare Magnetohydrodynamics, Gordan and Breach, pp. 217-336.

Craig, I.J.D., Robb, T.D., and Rollo, M.D.: 1982, Solar Phys. 76, 331-355.

Doschek, G.A., Boris, J.P., Cheng, C.-C., Mariska, J.T., and Oran, E.S.: 1982, Astrophys. J. 258, 373-383.
Doschek, G.A., Cheng, C.-C., Oran, E.S., Boris, J.P., and Mariska, J.T.: 1983, Astrophys. J. 265(2), 1103-1119.
Einaudi, G. and Van Hoven, G.: 1983, Solar Phys. 88(1/2), 163-177.
Forbes, T.G.: 1984, in Proc. Solar Maximum Analysis Symp., Graz (in press).
Forbes, T.G. and Priest, E.R.: 1982a, Solar Phys. 81, 303-324.
Forbes, T.G. and Priest, E.R.: 1982b, Planet. Space Sci. 30, 1183-1197.
Forbes, T.G. and Priest, E.R.: 1983a, Solar Phys. 84(1/2), 169-188.
Forbes, T.G. and Priest, E.R.: 1983b, Solar Phys. 88(1/2), 211-218.
Forbes, T.G., Priest, E.R., and Hood, A.W.: 1982, J. Plasma Phys. 27(1), 157-176.
Heyvaerts, J.: 1983, Adv. Space Res. 2(9), 21-26.
Heyvaerts, J., Lasry, J.M., Schatzman, M., and Witomsky, P.: 1982, Astron. Astrophys. 111, 104-112.
Hood, A.W.: 1983a, Solar Phys. 87(2), 279-299.
Hood, A.W.: 1983b, Solar Phys. 89(2), 235-242.
Hood, A.W. and Priest, E.R.: 1980, Solar Phys. 66, 113-134.
Hood, A.W. and Priest, E.R.: 1981a, in R.M. Bonnet and A.K. Dupree (eds.), NATO Advanced Study Institutes Series C, Vol. 68, D. Reidel Publ. Co., Dordrecht, pp. 509-531.
Hood, A.W. and Priest, E.R.: 1981b, Geophys. Astrophys. Fluid Dyn. 17, 297-318.
Hood, A.W. and Priest, E.R.: 1981c, Solar Phys. 73, 289-311.
Illing, R.M.E., House, L.L., Wagner, W.J., and Sawyer, C.: 1981, Bull. AAS, 13, 862.
Low, B.C.: 1983, in J.O. Stenflo (ed.), IAU Symp. No. 102, D. Reidel Publ. Co., Dordrecht, pp. 467-471.
MacNeice, P., McWhirter, R.W.P., Spicer, D.S., and Burgess, A.: 1984, Solar Phys 90(2), 357-382.
McClymont, A.N. and Canfield, R.C.: 1983a, Astrophys. J. 265(1), 483-496.
McClymont, A.N. and Canfield, R.C.: 1983b, Astrophys. J. 265(1), 497-506.
Melville, J.P., Hood, A.W., and Priest, E.R.: 1983, Solar Phys. 87(2), 301-307.
Migliuolo, S. and Cargill, P.G.: 1983, Astrophys. J. 271(2), 820-831.
Migliuolo, S., Cargill, P.G., and Hood, A.W.: 1984, Astrophys. J. 281, 413-418.
Milne, A.M. and Priest, E.R.: 1981, Solar Phys. 73, 157-181.
Mok, Y. and Van Hoven, G.: 1982, Phys. Fluids 25, 636.
Nagai, F. and Emslie, A.G.: 1984, Astrophys. J. 279, 896-908.
Nagai, F., Wu, S.-T., and Tandberg-Hanssen, E.: 1983, Solar Phys. 84(1/2), 271-283.
Pallavicini, R. and Peres, G.: 1983, Solar Phys. 86(1/2), 147-157.
Pallavicini, R., Peres, G., Serio, S., Vaiana, G., Acton, L., Leibacher, J., and Rosner, R.: 1983, Astrophys. J. 270(1), 270-287.
Peres, G., Rosner, R., Serio, S., and Vaiana, G.S.: 1982, Astrophys. J. 252, 791-799.
Priest, E.R. (ed.): 1981, Solar Flare Magnetohydrodynamics, Gordan and Breach, London.
Priest, E.R.: 1982, Fundam. Cosmic Phys. 7(4), 363-445.
Priest, E.R.: 1983a, in Activity in Red Dwarf Stars, IAU Colloq. 71, pp. 545-558.
Priest, E.R.: 1983b, Solar Phys. 86(1/2), 33-45.
Priest, E.R.: 1984, in Proc. Solar Maximum Analysis Symp., Graz (in press).
Priest, E.R.: 1985, Rep. Prog. Phys. (in press).
Ray, A. and Van Hoven, G.: 1982, Solar Phys. 79, 353-364.
Ricchiazzi, P.J. and Canfield, R.C.: 1983, Astrophys. J. 272(2), 739-755.
Sakai, J.-I. and Washimi, H.: 1981, Report IPPJ-540, Nagoya University, Japan.
Sakai, J.-I. and Washimi, H.: 1982, Astrophys. J. 258, 823-824.
Schindler, K.: 1982, Phys. Scripta T2(1), 163-168.
Schindler, K., Birn, J., and Janicke, L.: 1983, Solar Phys. 87(1), 103-133.
Serio, S.: 1982, Mem. Soc. Astron. Ital. 53(4), 865-870.
Smith, D.F. and Harmony, D.W.: 1982, Astrophys. J. 252, 800-809.
Somov, B.V.: 1981, Fiz. Inst. AN SSSR. Preprint 15.

Somov, B.V.: 1983, Zemlya Vselennaya 6, 13-17.
Somov, B.V. and Spektor, A.R.: 1982, Space Sci. Rev. 32, 27-41.
Somov, B.V. and Titov, V.S.: 1983, Pisma Astron Zh. 9(1), 48-52 (in Russian).
Somov, B.V., Spektor, A.R., and Syrovatskii, S.I.: 1977, Izv, AN SSSR. Ser. Fiz 41, 273-287.
Somov, B.V., Syrovatskii, S.I., and Spektor, A.R.: 1981, Solar Phys. 73, 145-155.
Somov, B.V., Stepanov, V.E., Stepanyan, N.N., and Tomozov, V.M.: 1982, Issledovaniya po geomagnetizmu aeronomii i fizike Solntsa, Moskva 60, 112-131 (in Russian).
Song, M.-t. and Cao, T.-j.: 1983a, Acta Astron. Sin. 24(1), 1-9.
Song, M.-t. and Cao, T.-j.: 1983b, Chin. Astron. Astrophys. 7(3), 159-164
Soward, A.M. and Priest, E.R.: 1982, J. Plasma Phys. 28(2), 335-367.
Steinolfson, R.S.: 1982, Astron. Astrophys. 115, 39-49.
Steinolfson, R.S. and Van Hoven, G.: 1982, Phys. Fluids 26, 2590-2602.
Steinolfson, R.S. and Van Hoven, G.: 1983a, Phys. Fluids 26(1), 117-123.
Steinolfson, R.S. and Van Hoven, G.: 1983b, Bull. AAS 15(2), 710.
Steinolfson, R.S. and Van Hoven, G.: 1984, Astrophys. J. 276, 391.
Syrovatskii, S.I., Bulanov, S.V., and Dogel', V.A.: 1983, Astrophys. Space Phys. Rev. 2, 385.
Tachi, T., Steinolfson, R.S., and Van Hoven, G.: 1983, Phys. Fluids 26(10), 2976-2982.
Van Hoven, G.: 1981, in E.R. Priest (ed.), Solar Flare Magnetohydrodynamics, Gordon and Breach, London, Chapter 4.
Van Hoven, G.: 1982, Mem. Soc. Astron. Ital. 53(2), 441-460.
Van Hoven, G. and Steinolfson, R.S.: 1983, Bull. AAS 15(2), 700.
Van Hoven, G., Ma, S.S., and Einaudi, G.: 1981, Astron. Astrophys. 97, 232-234.
Van Hoven, G., Steinolfson, R.S., and Tachi, T.: 1983, Astrophys. J. 268(2), 860-864.
Vasyliunas, V.M.: 1983, in R.L. Carovillano and J.M. Forbes (eds.), Astrophysics and Space Science Library, Vol. 104, D. Reidel Publ. Co., Dordrecht, pp. 255-260.
Wu, S.-T., Kan, L.C., Nakagawa, Y., and Tandberg-Hanssen, E.: 1981, Solar Phys. 70, 137-148.
Wu, S.-T., Wang, S., Dryer, M., Poland, A.I., Sime, D.G., Wolfson, C.J., Orwig, L.E., and Maxwell, A.: 1983, Solar Phys. 85(2), 351-373.
Zweibel, E.G.: 1982, Astrophys. J. Letters 258, L53-L56.
Zweibel, E.G. and Hundhausen, A.J.: 1982, Solar Phys. 76, 261-299.

E. Tandberg-Hanssen
President of Commission

COMMISSION 12: RADIATION AND STRUCTURE OF THE SOLAR ATMOSPHERE (RADIATION ET STRUCTURE DE L'ATMOSPHERE SOLAIRE)

PRESIDENT: R. W. Noyes
VICE-PRESIDENT: M. Kuperus
ORGANIZING COMMITTEE: Y. Uchida (Past President), Chen Biao, F. Deubner, E. Fossat, J. Harvey, V.A. Kotov, J. Leibacher, K.R. Sivaraman, J.O. Stenflo, P. Wilson

I. INTRODUCTION

(R.W. Noyes)

The scope of Commission 12 has broadened somewhat in recent years, to include not only the structure of the solar atmosphere, but that of the solar interior as well. The scientific purview of this commission, and of the present report, are complementary to those of Commission 10 (solar activity). Rather than attempting to review all progress in solar structure studies over the past triennium, this report deals with six topics of great current interest, in which there is a great deal of current work.

Section II, on solar oscillations, emphasizes observational aspects. Reference is made here to related reviews of theoretical aspects of stellar oscillations in the reports of Commissions 35 and 27.

The President of Commission 12 wishes to take this opportunity to thank the authors of the remaining sections of this report for their conscientious and effective work in preparing these reviews. In addition, he wishes to thank the Organizing Committee of the Commission for their support during the past three years.

II. SOLAR OSCILLATIONS

(T.R. Duvall, Jr.)

The study of solar oscillations, now becoming known as helioseismology, is rapidly expanding our knowledge of the interior structure of the Sun. Questions addressed include the Sun's internal thermal structure, rotation, initial helium abundance, giant cell convection, and gravitational quadrupole moment. Recent progress has been reviewed by Deubner and Gough (1984) and Brown *et al.* (1984), and a comprehensive bibliography was published (GONG, 1984). Several conference proceedings are available (Gough, 1983; Gabriel and Noels, 1984; Belvedere and Paterno, 1984; Ulrich, 1984) .

At least two restoring forces, pressure and gravity, are important for global oscillation modes. The *p*-modes have pressure as the dominant restoring force while the *g*-modes have gravity as the restoring force. The *f*-mode, or fundamental, is essentially a surface gravity wave. An oscillation mode is described by an eigenfunction of, say, vertical velocity which is a product of a radial function, a spherical harmonic $Y_{lm}(\theta,\phi)$ and a harmonic function of time. The radial function is, in general, an oscillatory function of radius in a restricted range of radius, exponentially decaying away from this region. This leads to the concept of a resonant cavity in radius in which the wave energy is trapped. Waves can propagate in the interior of the cavity and are reflected at the boundaries. For *p*-modes the upper boundary of the cavity is just below the visible surface and the reflection is caused by the large density gradient. The inner reflection is caused by the refraction of obliquely propagating waves away from the vertical by the increase of sound speed with depth. The inner reflection radius is mode-dependent which gives us the basic depth diagnostic capability.

1. *Thermal Structure*

One way to investigate the thermal structure of the Sun is to compare mode frequencies computed for a solar model to those observed. By trial and error, the uncertain input parameters to the model are varied to obtain a better agreement with the observed frequencies. In addition, some models can be safely excluded because of the large discrepancies between computed and observed frequencies. This procedure has been generally successful in obtaining mode frequencies accurate to better than 1% (Ulrich and Rhodes, 1983; Shibahashi *et al.*, 1983). The general result is that a relatively standard model fits the data best. In particular, a convection zone depth of 0.3 R_o, a helium abundance of $Y = 0.25$ and a heavy element abundance of $Z = 0.02$ are preferred.

Many of the observed mode frequencies have relative accuracies of 0.1% or better, but none of the models predict frequencies within the observational uncertainties. This has led to suggestions that the physics in the models is incomplete (Gough, 1984). Christensen-Dalsgaard and Gough (1984) have shown the differences between the computed (model 1 of Christensen-Dalsgaard, 1982) and observed frequencies are similar functions of frequency at constant degree. For the different degrees ($l = 1 - 200$) the positions of the bottom of the cavities range over a large fraction of the solar radius. The only area that these modes have in common is the outer few percent by radius and so it was concluded that the dominant error in the models is in this region. A second effect seen was a rather abrupt change in the frequency errors between degrees 20 and 40. This suggests another error in the model between the bottoms of the $l = 20$ and $l = 40$ cavities, or roughly $r/R_o = 0.6 - 0.7$.

Another approach to the investigation of the thermal structure is to attempt to obtain the internal structure directly from the observed frequencies. This procedure, known as inverse theory, has been successfully employed in terrestrial seismology and should be applicable to the solar problem. The initial attempts to invert the observed frequencies utilize asymptotic relations for the frequencies (Gough, 1984; Christensen-Dalsgaard *et al.*, 1984) that are not as accurate as the full eigenmode calculations. The initial conclusions are that the dominant errors in the models are very near the surface.

2. *Rotation*

The rotation of the solar interior has intrigued investigators for nearly two decades since the suggestion of Dicke (1974) that a rapidly rotating solar interior would influence the relativistic interpretation of the precession of planetary orbits through an enhanced solar gravitational quadrupole moment. In addition, differential rotation is thought to drive the eleven-year activity cycle and so radial and latitudinal profiles of rotation are needed to understand this phenomenon. The oscillation modes are sensitive to the interior rotation in a well-defined way. For *p*-modes a mode frequency is shifted by an amount proportional to the product of the azimuthal order, m, and the integral of a kernel function multiplied by the rotation frequency, over the interior of the star. Since different modes sample different parts of the star, we can in principle obtain the interior rotation from the measurements of a sufficient number of frequency shifts.

The initial attempts to measure subsurface rotation utilized sectoral modes ($|m| = l$) of high degree l (Deubner *et al.*, 1979). These are the natural modes to start with, as the frequency shifts are large (and thus easily measurable) due to the large value of $|m|$. The sectoral modes of high l are confined to the equatorial zone and so information is obtained on the depth dependence of the equatorial rotation. Modes of high degree are confined to a rather shallow layer near the surface and so the depth range covered is small. The original depth variation seen by Deubner *et al.* (1979) was not confirmed by subsequent observations (Rhodes *et al.*, 1983) although Hill *et al.* (1983) have argued that giant cell convection may cause time variations in the observed rotational splitting.

At low degrees, the observational problems are magnified although the information content in the modes is much higher. An $l = 1$ *p*-mode with a period near 5 minutes is confined in a cavity that ranges over 95% of the Sun's radius. An $l = 1$ *g*-mode is confined mostly to the deep interior and is actually a much better probe of that region. The frequency shifts of the low-degree modes are, however, much smaller because of the reduced value of $|m|$. In addition, most of the effort at low degrees have been on observations with little or no spatial resolution and hence the prograde and retrograde modes are not clearly separated as they are in the high-degree case. The initial rotational splitting observations for low-degree *p*-modes in the five-minute band (Claverie *et al.*, 1981) have not been confirmed by the later observations of Woodard (1984). And, in fact, Woodard has made a powerful argument that the $l = 1$ and 2 *p*-mode splitting cannot be resolved without spatial resolution because of the finite lifetime of the modes with periods near five minutes.

The advantage of spatial resolution has been demonstrated by the sectoral mode splittings of low to intermediate degree ($l = 1 - 100$) observed by Duvall and Harvey (1984). These data have been used to derive a radial rotation profile by Duvall *et al.* (1984). The results show that over most of the radius ($r/R_o > .4$) the rotation rate is close to the surface rate with a slight decline with radius. The gravitational quadrupole moment, J_2, calculated from this rotation profile makes a negligible contribution to the precession of planetary orbits.

3. *g-modes*

Measurements of *g*-modes would be an important step in helioseismology. The *g*-mode frequencies are determined mainly by conditions in the deep interior (where internal gravity waves can propagate) in contrast to the *p*-modes, whose frequencies are very sensitive to the structure near the surface. Several observational problems are associated with the detection of *g*-modes. Theoretical calculations tell us that the evanescent region in the convection zone will reduce the eigenfunctions of high-degree *g*-modes to an unobservable level at the surface and so we would probably only be able to observe the lowest few values of spherical harmonic degree l.

The frequencies of *g*-modes are low ($\nu < 0.3$ mHz) and for moderate resolution $\Delta\nu \approx 0.1$ mHz the temporal power spectrum of these low $- l$ modes should be crowded. The solar background (active regions, supergranulation, granulation) may contribute significant power at the frequencies in question. In contrast, the lifetimes of the *g*-modes could be considerably longer than any of the data strings currently being analyzed and so progress could be made with longer data sequences. Several groups claim to have detected *g*-modes (Delache and Scherrer, 1983; van der Raay *et al.*, 1984; Fröhlich and Delache, 1984; Kotov *et al.*, 1984; Hill *et al.*, 1982). The different observations are not in agreement (where there is overlap) on the frequencies of modes present and the mode identifications. For a given spherical harmonic degree, the *g*-modes are asymptotically (high n) equally spaced in period as a function of radial order n (Berthomieu *et al.*, 1978). There is also not general agreement on the fundamental period spacing.

4. *Conclusions*

Much has been learned in the study of global solar oscillations and much remains to be done. Interesting physical problems, such as how the latitudinal differential rotation varies with depth, have not yet begun to be attacked. Observationally there are several impediments to further progress, including the day-night cycle, the need for two-dimensional imaging and atmospheric seeing. Atmospheric seeing is a problem only for the high-degree modes which require high spatial resolution to observe. The day-night cycle introduces spurious peaks into the spectra from a single mid-latitude observing site. This difficulty can be adequately overcome with a network of mid-latitude sites judiciously placed around the globe. The Birmingham group has been operating a two-station network with considerable success during the last several summers. The Birmingham and Nice groups are proposing to build extensive networks of stations for the study of oscillations in unimaged sunlight. A network of imaging stations that would observe the low and intermediate degree modes is being proposed by the Global Oscillation Network Group (GONG). The problem of seeing as well could be solved by going to space, and both the European (ESA, 1983) and US (Noyes and Rhodes, 1984) space agencies are studying the possibilities. Major advances in helioseismology should come from the implementation of any or all of the network or space proposals.

References

Belvedere, G. and Paterno, L., eds., 1984, *Mem. Soc. Astron. Ital.* **55**.

Berthomieu, G., Gonczi, G., Graff, P.H., Povost, J. and Rocca, A. 1978, *Astron. Astrophys.*, **70**, 597.

Brown, T.M., Mihalas, B.W. and Rhodes, E.J., Jr., 1984, "Waves and Oscillations," in *Physics of the Sun*, eds. P.A. Sturrock, T.E. Holzer, D. Mihalas and R.K. Ulrich, Reidel, in press.

Christensen-Dalsgaard, J. 1982, *M.N.R.A.S.* **199**, 735.

Christensen-Dalsgaard, J. and Gough, D.O. 1984, in *Proc. Conf. on Solar Seismology from Space*, ed. R.K. Ulrich, JPL, in press.

Christensen-Dalsgaard, J., Duvall, T.L., Jr., Gough, D.O. and Harvey, J.W. 1984, *Nature*, submitted.

Claverie, A., Isaak, G.., McLeod, C.P., van der Raay, H.B. and Roca Cortes, T. 1981, *Nature*, **293**, 443.

Delache, P. and Scherrer, P.H. 1983, *Nature*, **306**, 651.

Deubner, F.-L., Gough, D.O., 1984, *Ann. Rev. Astron. Astrophys.*, **22**, 593.

Deubner, F.-L., Ulrich, R.K., Rhodes, E.J., Jr. 1979, *Astron. and Astrophys.*, **72**, 177.

Dicke, R.H. 1974, *Science*, **184**, 419.

Duvall, T.L., Jr., Dziembowski, W.A., Goode, P.R., Gough, D.O., Harvey, J.W., Leibacher, J.W. 1984, *Nature*, **310i**, 22.

Duvall, T.L., Jr., Harvey, J.W. 1984, *Nature*, **310**, 19.

ESA, 1983, Solar and Heliospheric Observatory Assessment Study

Fröhlich, C. and Delache, P. 1984, in *Oscillations as a Probe of the Sun's Interior*, eds. G.

Belvedere, L. Paterno, *Mem. Soc. Astron. Ital*, **55**.
Gabriel, M. and Noels, A. 1984, eds., *Theoretical Problems in Stellar Stability and Oscillations*, Obs. de Liege, Liege, in press.
GONG (Global Oscillation Network Group), 1984, report No. 2, National Solar Observatory, Tucson, AZ, USA.
Gough, D.O. 1983, ed., *Problems of Solar and Stellar Oscillations*, IAU Colloq. No. 66, *Solar Phys.*, **82**.
Gough, D.O. 1984, to appear in *Advances in Space Research*, Proc. 25th COSPAR plenary mtg., Graz, ed. H.S. Hudson.
Hill, H.A., Bos, R.J. and Goode, P.R. 1982, *Phys. Rev. Lett.*, **49**, 1794.
Hill, F., Toomre, J., November, L.J. 1983, *Solar Phys.*, **82**, 411.
Kotov, V.A., Severny, A.B. and Tsap, T.T. 1984, *Mem. Soc. Astron. Ital*, **55**, 117.
Noyes, R.W. and Rhodes, E.J., Jr. 1984, *Probing the Depths of a Star: The Study of Solar Oscillations from Space*, Report of the NASA Science Working Group on the Study of Solar Oscillations from Space, JPL, NASA.
Rhodes, E.J., Jr., Harvey, J.W., Duvall, T.L., Jr. 1983, *Solar Phys.*, **82**, 111.
Shibahashi, H., Noels, A. and Gabriel, M. 1983, *Astron. Astrophys.*, **123**, 283.
Ulrich, R.K. and Rhodes, E.J., Jr. 1983, *Ap. J.*, **265**, 551.
van der Raay, H.R., Claverie, A., Isaak, G.R., McLeod, J., Woodard, M. 1984, *Nature*, **309**, 530.

III. SOLAR ROTATION

(Robert F. Howard)

1. *Spectroscopic Studies*

The study of solar rotation using the Doppler effect in solar spectrum lines continues to be an active and fruitful area of research. Duvall (1982) has confirmed a slower rotation rate for the photosphere than for sunspots, which bears on an earlier controversy regarding this point between the Stanford Solar Observatory and some other observatories. The issue is not settled, however.

The analysis of Doppler data has been discussed by Kubicela and Karabin (1983), who have proposed a new vector formulation for the reduction of solar disk data, including the effects of the Earth's orbit on the velocity signal. Snodgrass *et al.* (1983) have discovered an error in the calibration of the Mt. Wilson velocity signal which resulted from an error in the published wavelengths of the solar spectrum lines that have been used in calibrating the observations. This error, which was present in the Stanford Solar Observatory data as well, resulted in an over-estimate of the rotation velocity in all earlier published results from these observatories of 0.55%.

A comprehensive summary of the Mt. Wilson Doppler rotation results starting in 1967 was published by Howard *et al.* (1983). This work gives the rotation rate and latitude dependence over the time interval in one-rotation averages. Other characteristics of the large-scale velocity fields are also listed in this paper.

Balthasar (1983) has analyzed the depth dependence of the solar rotation rate using 63 Fraunhofer lines observed with the Fourier transform spectrometer at the National Solar Observatory in the visible region of the spectrum. He finds, in agreement with other observers in earlier years, that the rotation rate is slightly higher at higher elevations in the solar atmosphere. This result still remains a puzzle, with no theoretical explanation.

The rotation rate of a faintly discernible pattern of polar velocity field — large-scale cellular pattern centered on the pole — ith a rotation period of 30 days was found in Doppler observations by Cram, Durney, and Guenther (1983). This is the first evidence of any such pattern. This interesting result, which would have profound implications in the study of interior structure and dynamics, deserve verification and further study.

Variations with time of the Doppler velocity signal were the topic of a study by Kuveler and Wöhl (1983). These authors detected a decrease of nearly 2% in the equatorial rotation rate of the Sun between 1981 and 1982, in agreement with unpublished results from other observatories. The variations seen on a daily basis and the absolute value of the rotation are not in such good agreement between the various sites. Evidently instrumental effects affect the daily determinations of rotation rate significantly at most or all observatories. To what extent long-term averages of rotation rate are affected by instrumental effects is not yet known. The measurement of the rotation rate of the Sun by spectroscopic techniques is still a very uncertain process, and systematic errors are evidently not yet totally negligible.

Perez-Garde *et al.* (1981), using the λ6301.5 line of Fe I, found a large-scale velocity pattern during a few days in September 1978 between +48° and −30° with a period of about 45° in longitude. They also found an equatorward meridional flow with an amplitude of about 20 m/sec. This meridional motion is opposite in direction to that found previously at other observatories. This is another indication that measurements of small wavelength shifts in the solar spectrum may be plagued by errors of an unknown nature. These authors found an equatorial rotational velocity of 2.881 rad/sec.

Snider (1983), using an atomic beam resonance scattering technique for the λ7699 line of potassium, found an equatorial rotation rate in the interval 1979 to 1982 of 13.8 − 0.2°/day sidereal. There were no significant variations in this rather low rate. These observations were made at two sites — Oberlin and Mt. Wilson.

2. *Rotation from Tracers.*

Considerable interest in recent years has centered on the rotation of sunspots. The extensive Greenwich data set has been much utilized in several studies of sunspot motions. These data have several important advantages: They are quite homogeneous, they cover long periods of time, and there are very few missing days. A conspicuous disadvantage of the Greenwich data set is that it contains information on sunspot groups onl — t least in recent times. Group areas and positions are tabulated in the Greenwich data, but individual spot characteristics are not available, except for the case of single-spot groups, which are not typical of either groups or individual spots. Often investigators have not stressed this distinction.

Among the studies using the Greenwich data is that of Arevalo *et al.* (1982). In this analysis, it was found that the rotation rate of groups was higher in the interval 1874 to 1902 than in the interval 1940 to 1968. Furthermore, meridional flows as seen in spot group motions were northward in cycle 12, while in cycle 13 they were southward.

The positional correlation of Greenwich spot groups and high-speed solar wind streams has been studied by Balthasar and Schössler (1982). In general, an anticorrelation between these quantities is observed, strongest at the post-maximum phase of the cycle. Consecutive pairs of cycles show conspicuous similarities in this respect, which led the authors to conclude that there is a tendency for activity to occur in preferred longitudes. It is known that high-speed solar wind streams tend to originate at coronal holes, so it may be that this result is to some extent a matter of definition. Coronal holes tend to avoid active regions (and therefore sunspots) so one would not expect to find a positive correlation between the two features.

Two groups (Godoli and Mazzucconi 1983; Tuominen, Tuominen, and Kyröläinen 1983), using the Greenwich data, find the same rotational latitude shear seen in the velocity maps showing torsional oscillations. These results are barely above the noise level, and further confirmation would be desirable. This is a very important issue, because the nature of the torsional oscillation is still undefined. If these small-amplitude motions are shared by the strong photospheric magnetic fields, then this gives us more information on the depth of the oscillations.

Other investigations using the Greenwich data in this interval include that of White (1982) on the velocity dispersions of single spots (groups). He found that the velocity dispersion does not vary with latitude, is twice as large in longitude as in latitude, and is smaller for smaller spots. Kopecky (1982) has listed the high-latitude spot groups $\geq$ 40) for the interval 1874 to 1976 from the Greenwich data. Tang (1981) has studied high latitude spots in the interval 1978 to 1979 for rotation rate and finds a slightly slower rotation rate at the high latitudes than predicted from the Newton and Nunn (1951) formula. A similar result was found by Landman and Takushi (1981) using data from 1966 to 1968. They found high latitude rotation in agreement with the Newton and Nunn result from the years 1934 to 1944.

The 17th century records of Scheiner and Hevelius have been examined (Yallop, *et al.* 1982) for rotation information. These authors find no significant difference in rotation rate from modern times. A similar result was found by Abarbanell and Wöhl (1981).

The Mt. Wilson white light photographs from 1921 to 1982 have been measured for positions and areas of all sunspots (Howard, Gilman, and Gilman 1984). Individual sunspot positions have been matched with data from the following day to obtain returns of the same sunspots. In order to do this a computer algorithm had to be devised to match the pattern of sunspots on consecutive days. Thus, for the first time, rotation rates and latitude drifts of individual sunspots could be derived on a massive scale over a long time interval. One of the first results to come from this study was that the largest spots rotate more slowly by several percent than the smallest spots. This may represent a different depth at which the magnetic fields of large and small spots are linked. Using the same data set, Gilman and Howard (1984) discovered a systematic variation of the rotation rate of sunspots with phase in the solar activity cycle. A slight rise in the rotation rate is detected at solar minimum, and a smaller and less well-defined increase is seen near solar maximum. In the interval 1967 to 1982, for which there are Doppler velocity data for solar rotation from the Mt. Wilson magnetograph, the Doppler-determined rotation rate of the Sun shows a similar twin-peaked behavior. An increased rotation rate at minimum was found by Rovithis (1982) using Greenwich data for the 18th cycle (1945-54), who also found a slight difference between the rotation rates of the north and south hemispheres, with the north being slightly faster. Arevalo *et al.* (1982), from an examination of Greenwich spot group data over several early sunspot cycles, show a plot with a twin-peaked cyclic variation of spot rotation similar to that found in the Mt. Wilson data.

Another long set of non-Greenwich data is that of Kanzelhohe. Lustig (1982, 1983) used data from 1947 to 1979 to study the rotation rates of individual sunspots. He found that group types H and J, which are larger than group types C and D, also have slower rotation rates. Using individual spot data, Lustig determined that the differential rotation gradient declined slightly during the interval. Also, he found that the equatorial rotation rate is slightly faster during minimum than during the maximum years.

Another study of individual spots is that of Koch, Wöhl, and Schröter (1981). They determined that the rotation rate of any one spot was remarkably constant in time but that different spots differed in their rotation rates significantly.

Ternullo, Zappala, and Zuccarello (1981), using spot groups observed at Catania, found that young spots rotate faster than old spots. It is possible that this is the same effect as the size-rotation correlation discussed above. The older spots tend to be larger and rotate slower than the smaller spots, which don't live as long. It is not easy to determine which is the important factor in this case. Judging from the results of Koch *et al.* (1981) described above, it seems most likely that individual spots do not change their rates as they age. Whether or not they change rotation rate as they change size is an open question.

Magnetic Field Rotation

A special type of tracer is the weak magnetic field pattern seen in solar magnetogram observations. Snodgrass (1983) has examined the Mt. Wilson daily magnetic data covering 1967-1982 and used a cross-correlation technique in narrow latitude zones to determine day-to-day differences in longitude corresponding to the rotation rate of the patterns. The result is a rate remarkably constant with time and agreeing well with the Newton and Nunn (1951) value near the equator. At higher latitudes, where sunspots are scarce, the magnetic pattern results agreed well with the Mt. Wilson Doppler results. The errors in these determinations varied from 0.1% at low latitudes to 1.1% at very high latitudes. The cycle-related variations seen in the spot and Doppler results are not seen in the magnetic field data.

4. *Rotating Distortions*

Dicke (1982) has suggested that earlier observations of the figure of the solar limb, which showed a cyclic variation with a period of close to 12 days, result from the rotation of the solar core distorted by a strong magnetic field. Dicke calculates that such a field would experience a torsional oscillation within a period of years. Libbrecht (1984) has made more recent observations of the shape of the solar limb with an improved version of the instrument first used by Dicke and his collaborators. Libbrecht's observations were made at a much better site than the original set, and they show no more than the distortion expected by a reasonable extrapolation of the surface rotation; moreover they show no significant rotating distortion. Solar oscillation data (reviewed in Sec. II above) presently give conflicting results about whether there is rapid rotation in the deep solar interior.

5. *Torsional Oscillations*

Some further work was carried out in this interval on torsional oscillations of the Sun. LaBonte and Howard (1982) used the latitude zone Doppler velocity data to search for the torsional oscillations. This method does not depend on the latitude series solution for the rotation, so it is not subject to possible influence from the choice of terms in the latitude expansion. Using the latitude-strip data only, the torsional oscillations may be seen, although against a rather noisy background. Another feature that shows up in this plot (Figure 8 of the LaBonte and Howard paper) is a one-cycle-per-hemisphere torsional wave which is characterized by a faster rotation at high latitudes near solar minimum and a faster rotation at low latitudes near solar maximum. The amplitude is roughly that of the traveling torsional wave, i.e. $\sim 5 \mathrm{m\ s}^{-1}$. The high latitude portion of this pattern was found earlier by Livingston and Duvall (1979).

There is evidence for a 1/2 cycle–per–hemisphere oscillation in a comparison of equatorial rotation rates determined from north and south data separately (Howard *et al.* 1983). The amplitude is comparable to those of the two modes mentioned above.

The cyclic variation of the rotation rate of the Sun determined from sunspot and Doppler data discussed above is essentially another mode of torsional oscillation. In this case, the amplitude is about 200 ms^{-1}, which is much larger than the other modes.

References

Abarbanell, C. and Wöhl, H. 1981, *Solar Phys.*, **70**, 197.
Arevalo, M.J. *et al.* 1982, *Astron. and Astrophys.*, **111**, 266.
Balthasar, H. and Schössler, M. 1982, *Solar Phys.*, **87**, 23.
Balthasar, H. 1983, *Solar Phys.*, **84**, 371.
Cram, L.E., Durney, B.R. and Guenther, D.B. 1983, *Ap. J.*, **267**, 442.
Dicke, R.H. 1982, *Solar Phys.*, **78**, 3.
Duvall, T.L., Jr. 1982, *Solar Phys.*, **76**, 137.
Gilman, P.A. and Howard, R. 1984, *Ap. J.*, **283**, 285.
Godoli, G. and Mazzucconi, F. 1983, *Solar Phys.*, **83**, 339.
Howard, R. *et al.* 1983, *Solar Phys.*, 87, 195.
Howard, R., Gilman, P.A. and Gilman, P.I. 1984, *Ap. J.*, **283**, 373.
Koch, A., Wöhl, H. and Schröter, E.H. 1981, *Solar Phys.*, **71**, 295.
Kopecky, M. 1982, *Bull. Astron. Inst. Czech.*, **33**, 285.
Kubicela, A. and Karabin, M. 1983, *Solar Phys.*, **84**, 389.
Kuveler, G. and Wöhl, H. 1983, *Astron. and Astrophys.*, **123**, 29.
LaBonte, B.J. and Howard, R. 1982, *Solar Phys.*, **75**, 161.
Landman, D.A. and Takushi, J.T. 1981, *Solar Phys.*, **73**, 379.
Libbrecht, K. 1984, Ph.D. Thesis: Princeton University.
Livingston, W. and Duvall, T.L., Jr. 1979, *Solar Phys.*, **61**, 219.
Lustig, G. 1982, *Astron. and Astrophys.*, **106**, 151.
Lustig, G. 1983, *Astron. and Astrophys.*, **125**, 355.
Newton, H.W. and Nunn, M.L. 1951, *M.N.R.A.S.*, **111**, 413.
Perez-Garde, M. *et al.* 1981, *Astron. and Astrophys.*, **93**, 67.
Rovithis, P. 1982, *Astrophysics and Space Science*, **88**, 5.
Snider, J.L. 1983, *solar Phys.*, **84**, 377.
Snodgrass, H.B. 1983, *Ap. J.*, in press.
Ternullo, M., Zappala, R.A., and Zuccarello, F. 1981, *Solar Phys.*, **74**, 111.
Tuominen, J., Tuominen, I., and Kyröloinen, J. 1983, *M.N.R.A.S.*, **205**, 69.
Yallop, B.D., Hohenkerk, C., Murdin, L. and Clark, D.H. 1982, *Quart. J.R.A.S.*, **23**, 213.

IV. SOLAR GRANULATION

(R. Muller)

During the period 1981-1984, major developments have been made in the analysis of the shape and shifts of photospheric lines used as a diagnostic of the convection in the outer layers of the Sun. The first three-dimensional numerical simulation of the solar granulation has been developed, and it is able to reproduce with some success most of the observed morphological and spectroscopic properties of granulation. It is becoming clear that the properties of the solar granulation are variable over the solar cycle. Many questions remain to be solved; further progress will rely heavily on expected improvements of the spatial resolution of the observations and the availability of larger computers.

1. *Morphological structure of the solar granulation.*

Wittmann (1981), analysing high resolution Spektro-Stratoskop photographs (Mehltretter, 1978), confirms several accepted values like the fractional area of granules (47.1%), the corrected contrast (22%), and the asymmetry of intensity histograms. However, he finds a somewhat larger granular size than usually accepted, and a shorter lifetime (5.8 min).

It appears that granules are more evenly spaced than random points (Lawrence, 1983) and form bright chains, with contrast exceeding the mean contrast of granulation by several percent and of longer lifetime (Perfinenko, 1981). Confirming and extending the last result, Oda (1984) finds that the bright and exploding granules tend to form chains, defining a cellular pattern comparable in size to that of the mesogranulation. Isophotal contour maps of a single granule are presented by Bray and Loughhead (1984).

Recent measurements of the granulation contrast are in reasonable agreement around 5500A, but are very uncertain in the blue and the red, so that they do not allow comparison with models (Bray, 1982).

2. *Statistical analysis of brightness fluctuations.*

New evaluations of rms brightness fluctuations (ΔI_{rms}), corrected for instrumental and atmospheric blurring fall near the lower limit of the range 10 – 20% from previous evaluations (Schmidt *et al.* 1981, solar eclipse observation at Izana; Durrant *et al.*, 1983, Spektro-Stratoskop pictures). As these new values are not safer than the previous ones, the true ΔI_{rms} of the solar granulation remains dubious. It is confirmed that the change of rms fluctuation from the centre of the disk to $\mu = 0.3$ is small. Fitting the observed spread function of Deubner and Mattig (1975) by the sum of two Lorentzians, rather than by two Gaussians, Nordlung (1984a) re-evaluated the ΔI_{rms} of the solar granulation which is thus increased from 12.7% (Deubner and Mattig, 1975) to about 20% at 6070 Angstrom. This value is in good agreement with the ΔI rms of the Nordlung's tri-dimensional numerical model (Nordlung, 1982) of 20 – 25% at 6000 Angstrom.

Bi-dimensional power spectra obtained by Schmidt *et al.* (1981) and Durrant *et al.* (1983) are in good agreement with previously published single peak power spectra, while the bi-dimensional power spectrum, derived from one dimensional scans by Wittmann (1981), exhibits a multi-component peak; similar peak splitting was already reported in the past by a few authors but its existence as a property of the solar granulation remains doubtful.

The correct formulation of statistical analysis in the general case of a tilted image is derived by Wiesmeier and Durrant (1981) both for uni- and bi-dimensional observations; restoration by the point spread function is included in the formulation. V. der Lühe (1981) demonstrated that Optical Fourier Transformation, if carefully handled, may be a reliable alternative to the usual numerical procedures for deriving the power spectrum of the solar granulation. Ricort *et al.* (1982) performed estimations of Fried's parameter r_o, simultaneously using the same telescope, from the observed solar granulation contrast and from the variance of angle-of- arrival fluctuations; r_o obtained by both methods are well correlated.

Fluctuations of the limb position, due to brightness fluctuations of the granular pattern, do not exceed a few thousandths of a second of arc (Lites, 1982).

3. *Empirical models of temperature and velocity.*

Temperature models. Durrant *et al.* (1981) have shown that if the "geometrical smoothing" toward the limb is taken into account, recent empirical temperature models of the solar granulation, in which the temperature fluctuation is rapidly vanishing in the lower photosphere, are able to reproduce the centre-to-limb variation of the continuum intensity, as well as the intensity fluctuations in spectral lines, observed with the Spektro-Stratoskop. Karpinsky (1981) reports on measurements of the horizontal gradient of brightness of granules of 1.5 to 2.0 K/km, the peak values reaching 2.7 K/km; the values can be as high as 5.0 to 10.0 K/km when corrected for distortions.

Velocity models. The steepness of the decrease with height in the photosphere of the rms vertical velocity field associated with the solar granulation is a much debated question (see, for example, the flat gradient model of Durrant *et al.*, 1979 and the steep gradient model of Keil, 1980). It seems that the discrepancy is a result of the different procedures used to separate the convective and oscillatory components. The results of Bässgen and Deubner (1982) who analyse a high resolution time series of spectra obtained at Sacramento Peak Observatory, and use the resulting $k - \omega$ diagram to separate convective and oscillating powers, favor the flat gradient model. The coherence of the vertical convective velocities is maintained at least up to 300 km in the photosphere (Durrant and Nesis, 1982; Pravdjuk, 1982), confirming previous results. The penetration of convective cells to such high levels also favors a flat gradient model. It is found by Durrant and Nesis that the velocity coherence of the structures of sizes in the range 2 – 6 arcsec is maintained high up into the photosphere while it is lost more rapidly for smaller structures.

It is confirmed that the coherence between continuum intensities and vertical velocities is restricted to the lower photosphere (Durrant and Nesis, 1982; Nesis *et al.*, 1983). The horizontal velocity patterns are not coherent with the vertical velocity, or with the brightness pattern, indicating lack of horizontal heat transport. It is not possible to conclude how the horizontal velocity and its variation with the height in the atmosphere is related to the structure of granules (Nesis *et al.*, 1983).

An analysis of the Doppler shift fluctuations in the profile of the strong Na I D_1 line, instead of the more usual analysis of fluctuations in the core of lines of different strengths, failed to provide new information about the structure of the velocity field in the photosphere (Edmonds and Jin-Chung Hsu, 1983).

4. *Shift and asymmetry of photospheric line profiles.*

Spatially unresolved photospheric line profiles are asymmetric (C-shape of the line bisector) and blueshifted; these properties are signatures of the presence of convection and are used to get information about the vertical and horizontal structure of the photosphere; but this is not a trivial problem.

Line shift and asymmetry as tests for convection models. Kaisig and Durrant (1982) investigate in detail the information content of the shifts of mean line profiles with the aid of a perturbation analysis and a two-stream model. They demonstrate that the characteristic C-shape of mean solar line profiles is due not to the larger velocity amplitude of the downflow as proposed by Dravins *et al.* (1981), but to a complex opacity effect. The analysis of shifts of photospheric lines does not allow separation of the depth dependence of velocity, $\delta v(h)$, and temperature, $\delta T(h)$, fluctuations, but it is useful as tests of convective models (Kaisig and Durrant, 1982, Kaisig and Schröter, 1983; Kostick, 1983).

Methods based on the analysis of the third central moment M_3 of the line profiles (Marmolino and Severino, 1981) and on a linear analysis of the bisector (Buonaura and Caccin, 1982) have been developed to retrieve $\delta v(h)$. The main limitation of both methods is that they are applicable only to data with infinite resolution. Pierce (1984) shows that comparing an observed Fraunhofer line to a fitted gaussian yields more information on shape and asymmetry of the solar line than the simple bisector method.

Line asymmetry and blue-shift-parameters dependence. Balthasar (1984) finds that lines with cores formed in higher layers show larger asymmetries. Analysis of lines of ionized elements (Fe II) confirms the weak increase of the blue-shift with increasing excitation potential found for neutral elements (Dravins and Larsson, 1983). These authors also report that lines formed in shorter wavelength regions are more blue-shifted than lines in the red. In their investigation, Dravins and Larsson used the Kitt Peak Wavelength Table by Pierce and Breckinridge (1973), while Balthasar used observations obtained with the Fourier-Transform-Spectrometer (FTS) at the Mac Math Telescope of the Kitt Peak National Observatory (KPNO).

Some authors find that line asymmetry changes with time (Roca-Cortes *et al.*, 1983, but the line KI 7699 investigated is formed high in the photosphere and is more sensitive to oscillations than to convection), while others do not (Cavallini *et al.*, 1982).

Accurate measurements of the shape of Fe I 6301.5, performed with a Fabry-Perot spectrometer, agree with previous measurements (Cavallini *et al.*, 1982). The shape of the KI 7699 line profile is also confirmed by new measurements (Roca-Cortes *et al.*, 1983.

Limb effect. The blue-shift progressively decreases as one moves away from the disk centre and may even become a supergravitational redshift at the extreme limb: this is known as the limb effect. Balthasar (1984) confirms the existence of supergravitational red-shifts at the limb for many photospheric lines; in addition he finds that the red-shifts for lines from ionized elements are larger than those from neutral elements. Brandt and Schröter (1982), analysing spectra with iodine reference lines obtained at Locarno Observatory, confirm earlier findings that the limb effect is a combination of shifts of the line centre and the centre-to-limb variation of the shape; at the limb the asymmetry disappears. The relative blue-shift of many lines at $\cos\Theta = 0.8$ compared to the disk center is confirmed by Brandt and Shröter (1982) and by Balthasar (1984).

Latitude variation. It appears that there is a significant difference between the limb-effect curves along the polar and equatorial diameters for $\cos \Theta \leq 0.4$ and a rather strong indication of a latitude dependence of the line bisector C-shape (Brandt and Schröter, 1982). These authors come to the conclusion that this latitude dependence may account for the so-called "ears" observed by Howard *et al.* (1980).

Line asymmetry and blue-shift in plages. The C-shape of line bisector is flatter in plages than in the quiet Sun (Livingston, 1982; Kaisig and Schröter, 1983; Brandt and Schröter, 1984; Cavallini *et al.*, 1984). There is presently disagreement about the relative *shift* of the line bisector in plage regions compared to the quiet Sun: Livingston (1982), analysing spectra obtained with the FTS at KPNO and Cavallini *et al.* (1984), analysing the Fabry-Perot measurements made at Arcetri Observatory, find a red-shift; Kaisig and Schröter (1983) find a blue-shift. These authors interpret the blue-shift as a penetration of convective flux into higher photospheric levels in active regions, while the former interpret the measured red-shift as a retardation or an inhibition of convection.

In the quiet network, at supergranular boundaries, the wings of three lines ($g = 0$) are red-shifted by between 75 – 200 m s^{-1} relative to the cell profiles. The cores are relatively unshifted (Miller *et al.*, 1984). Such trends can result if granular convection is suppressed near the network flux tubes, so that there is little downflow in the vicinity of the flux tubes.

Accuracy of absolute measurements. The accuracy of Doppler-shift measurements is limited by uncertainties of the position of reference lines. Causes of reference line shifts are analysed, and solutions allowing an improvement of measurement accuracy are given by Balthasar *et al.* (1982) for telluric reference lines and by Koch and Wöhl (1984) for iodine reference lines.

5. *Theoretical models of the solar granulation*

We will restrict ourselves to those theoretical models of convection which are closely related to the solar granulation.

Nordlung (1982) develops a full three-dimensional numerical simulation of the solar granulation. He solves the hydrodynamic equations, including the radiative transfer equations, in the anelastic approximation; turbulent viscous terms are included in the equation of motion; the importance of spectral lines on the energy balance in the upper photosphere is stressed.

The results of these simulations show the main observed characteristics of the granulation (Dravins *et al.*, 1981; Nordlung, 1984 b): granules surrounded by intergranular lanes; downward motion in the lanes of larger amplitude than the upward motion in the granules; granules increasing their horizontal dimension in time; larger granules breaking apart into smaller ones, which in turn grow and merge. Synthetic spectral lines calculated both for the solar disc center and for the Sun as a star are in good agreement with the widths, strengths and shapes of observed spectral lines. The spectacular fit of the simulations with observations makes Nordlung's tri-dimensional model very appealing, although it is inherent to the model not to be able to predict a turbulent cascade toward the smaller scales. In this model granules are driven by the buoyancy force, horizontal motions are driven by pressure gradients, and the centres of granules are cooled while they expand and brake. The model does not contain any arbitrary parameters: the typical amplitudes of velocity and temperature fluctuations depend only on effective temperature, surface gravity, and the chemical abundance.

Under a different approach (Narasimha and Antia, 1982), convective modes in the solar envelope are investigated in the frame of the mixing length theory. The structure of the convection zone calculated according to this theory is found to be consistent with the transport of convective flux by a linear superposition of statistically independent unstable convective modes, provided the effects of turbulent conductivity and viscosity are taken into account. The resultant vertical velocity is in reasonable agreement with observed granular velocity. In a subsequent paper Antia *et al.* (1983) investigated the stability of linear convective modes in the solar envelope model by incorporating in a very crude manner the effects of turbulent pressure and eddy transport coefficients calculated in the mixing length approximation. It is demonstrated that for a reasonable choice of parameters there occur two peaks in the plot of growth rate versus horizontal wavenumber, which correspond to the observed features (horizontal size and lifetime) associated with the granulation and supergranulation.

6. *Variation of the properties of solar granulation over the activity cycle.*

A set of converging results of different origin have recently been published, indicating that the structure of the solar granulation varies over the solar cycle.

The mean size of granules, either measured by the distance between the centres of granules, or by the number of granules per unit surface area, decreases with increasing activity (Macris and Rösch, 1983; Macris *et al.*, 1984, Muller and Roudier, 1984). The granule-intergranular lane intensity ratio at 5200 Å varies with time, being minimum (1.10) around the minimum of activity, and maximum (1.30) around activity maximum. All these measurements were made on high resolution photographs of the solar granulation performed with the 50 cm refractor at Pic du Midi Observatory.

The equivalent width of photospheric lines, measured with the 13.5m spectrometer at Kitt Peak National Observatory, decreases by a factor ranging from 0 to 2.3% from 1976 to 1980 (Livingston and Holweger, 1982). The authors come to the conclusion that the weakenings are due to global variations of surface properties, namely a reduction of the temperature gradient through the low photosphere. Bisectors of strong iron lines in the full disk Fraunhofer spectrum, measured with the Fourier Transform Spectrometer at Kitt Peak National Observatory, are observed to diminish in curvature as the activity cycle proceeds from minimum (1976) to maximum (1979-1982). The implication is reduced convection on a global scale in response to an increase of total magnetic flux (Livingston *et al.* 1982; Livingston, 1983, 1984).

References

Alissandrakis, C.E., Macris, C.J., and Zacharidis, G. 1982, *Solar Phys.* **76**, 129.
Antia, H.M., Chitre, S.M., and Narasimha, D. 1983, *M.N.R.A.S.* **187**, 23.
Balthasar, H. 1984, *Solar Phys.*, **93**, 219.
Balthasar, H., Thiele, U., and Wöhl, H. 1982, *Astr. Ap.* **114**, 357.
Bässgen, M., and Deubner, F.L. 1982, *Astr. Ap.* **111**, L1.
Brandt, P.N., and Schröter, E.H. 1982, *Solar Phys.* **79**, 3.
Brandt, P.N., and Schröter, E.H. 1984, in Keil, 1984.
Bray, R.J., 1982, *Solar Phys.* **77**, 299.
Bray, R.J., and Loughhead, R.E., 1984, *Astr. Ap.* **133**, 409.
Buonaura, B. and Caccin, B., 1982, *Astr. Ap.* **111**, 113.
Cavallini, F., Ceppatelli, G., and Righini, A. 1982, *Astr. Ap.* **109**, 233.
Cavallini F., Ceppatelli, G., and Righini, A. 1984, in Keil, 1984.
Deubner, F.L. and Mattig, W. 1975, *Astr. Ap.*, **45**, 167.
Dravins, D. and Larsson B. 1984, in Keil, 1984.
Dravins, D., Lindegren, L., Nordlung, A. 1981, *Astr. Ap.*, **96**, 345.
Durrant, C.J., Kneer, F. and Maluck, G. 1981, *Astr. Ap.*, **104**, 211.
Durrant, C.J., Mattig, W., and Nesis, A. 1979, *Solar Phys.* **61**, 251.
Durrant, C.J., Mattig, W., Nesis, A., and Schmidt, W. 1983, *Astr. Ap.*, **123**, 319.
Durrant, C.J. and Nesis, A. 1982, *Astr. Ap.* **111**, 272.
Edmonds, F.N., and Jin-Chung Hsu 1983, *Solar Phys.* **83**, 217.
Howard, R., Boyden, J.E., and LaBonte, B.J. 1980, *Solar Phys.* **66**, 167.
Kaisig, M. and Durrant, C.J., 1982, *Astr. Ap.* **116**, 332.
Kaisig, M. and Schröter, E.H. 1983, *Astr. Ap.* **117**, 305.
Karpinsky, V.N. 1981, *Soln. Dannye.* **10**, 88.
Keil, S. 1980, *Ap. J.* **207**, 1024.
Keil, S. 1984 (ed), *Small-Scale Dynamical Processes in the Solar and Stellar Atmospheres* (Sacramento Peak Observatory: Sunspot, NM), in press.
Koch, A., and Wohl, H. 1984, *Astr. Ap.* **134**, 134.
Kostick, R. 1983, *Publication of the Academy of Science of Ukraine*, **83**, 63.
Lawrence, J.K. 1983, *Solar Phys.* **87**, 1.
Lites, B.W. 1983, *Solar Phys.* **85**, 193.
Livingston, W. 1983, in *Solar and Stellar Magnetic Fields*, J.O. Stenflo Ed.
Livingston, W. 1984, in Keil, 1984.
Livingston, W., and Holweger, H. 1982, *Ap. J.* **252**, 375.
Macris, C.J., and Rösch, J. 1983, *C.R. Acad. Sc. Paris*, **296**, 265.
Macris, C.J., and Rösch, J., and Roudier, T. 1984, in Keil, 1984.
Marmolino, C. and Severino, G. 1981, *Astr. Ap.* **100**, 191.
Mehltretter, J.P. 1978, *Astr. Ap.* **62**, 311.
Miller, P., Foukal, P., and Keil, S. 1984, *Solar Phys.* **92**, 33.
Muller, R. and Roudier, T. 1984 in *Hydromagnetics of the Sun*, Proc. 4th European Mtg. on Solar Physics, Noordwijk, in press.

Narasimha, D., and Antia, H.M. 1982, *Ap. J.* **262**, 358.
Nesis, A., Durrant, C.J. and Mattig, W., 1984, in Keil, 1984.
Nordlung, A. 1982, *Astr. Ap.* **107**, 1.
Nordlung, A. 1984a, in Keil, 1984.
Nordlung, A. 1984b, in Keil, 1984.
Oda, N. 1984, *Solar Phys.* **93**, 243.
Perfinenko, L.D., 1981, *Soln. Dannye*, **10**, 101.
Pierce, A.K. 1984, *Solar Phys.* **90**, 195.
Pierce, A.K. and Breckinridge, J.B., 1973, *Kitt Peak Nat. Obs. Contrib.* 559.
Pravdjuk, L.M. 1982, *Soln. Dannye*, **2**, 102.
Ricort, G., Borgnino, J., and Aime C. 1982, *Solar Phys.* **75**, 377.
Schmidt, W., Knölker, M., and Schröter, E.H. 1981, *Solar Phys.* **73**, 217.
Von der Lühe, O. 1981, *Astr. Ap.* **101**, 277.
Wiesmeier, A., and Durrant, C.J. 1981, *Astr. Ap.* **104**, 207.
Wittman, A. 1981, *Astr. Ap.* **99**, 90.

V. DYNAMICS OF THE CHROMOSPHERE AND TRANSITION REGION

(R. Grant Athay)

One of the more interesting aspects of the chromosphere-corona transition region is its tendency to exhibit large Doppler shifts. Both the non-thermal velocity component of line widths and the velocity displacement of line positions tend to maximize at temperatures near 10^5 K. The increase in velocity amplitudes with increasing temperatures below 10^5 K is readily understood in terms of the increasing sound speed and decreasing densities associated with the outwardly increasing temperature. Why the observed velocity amplitudes should decrease at still higher temperatures is not at all clear, however, and it seems very likely that this phenomenon is indicative of fundamental differences in the dynamics of the upper transition region and corona from those in the lower transition region and chromosphere.

The possibility remains, of course, that the apparent velocity decrease at high temperatures is only partially a solar effect. At temperatures above 10^5 K, increases in temperature, in general, are associated with increasing amounts of radiating material as a result of the decreasing temperature gradients. The resulting increase in path length over which a given spectral line forms tends to blend together regions of differing Doppler shift. As a result, the lines are broadened in preference to overall wavelength displacements. The apparent decrease in Doppler shifts very likely is due partially to this effect. However, since no marked increase in the non-thermal component of line broadening has been observed for temperatures above 10^5 K, it is evident that the steady increase in velocity amplitude up to 10^5 K does not continue at the same rate into the 10^6 K regime of the corona. Thus, at least part of the effect appears to be of solar origin.

A further unusual property of the temperature regime below about 10^5 K is the tendency for the solar plasma to exhibit large scale systematic flows as well as both periodic and highly transient localized flows. This review concentrates on the observational aspects of these flows as reported from 1981 to mid-1984.

Vertical Flow.

A number of observers have continued to report systematic downflows observed in spectral lines formed in the lower transition region and chromosphere. Gebbie *et al.* (1981) report average downflow velocities for the quiet sun ranging from 1.4 km s^{-1} in C II to 4.2 km s^{-1} in C IV, which is consistent with momentum conservation. Roussel-Dupre and Shine (1982) find mean redshifts of 12 km s^{-1} in C IV and Si IV at disk center, and Dere (1982a) finds an average red shift of 5.4 km s^{-1} for the quiet sun in C IV.

In active regions, Feldman, Cohen and Doschek (1982) find mean red shifts ranging from 4 to 17 km s^{-1} in ions formed at different temperatures with the maximum occurring between 5×10^4 and 10^5 K and decreasing to 2 km s^{-1} in lines of Si II, S II and C II formed in the upper chromosphere, and Brueckner (1981) finds downflows of 10-60 km s^{-1} in C IV.

Over sunspot umbrae the situation is less clear. Several authors (Brueckner 1981, Nicholas *et al.* 1982, Dere 1982b and Athay et al. 1982) have reported downflows in C IV generally exceeding those in the quiet sun. However, in subsequent studies Athay, Gurman and Henze (1983) found several spots with upflow and Gurman and Athay (1983) found from a study of 8 sunspots a small net average upflow. Also, Mein *et al.* (1982) found strong upflow in C IV in three sunspots;

Kingston *et al.* (1982) found only a weak net downflow in two sunspots; and Henze *et al.* (1984) found both upflow and downflow in different portions of a single spot. It seems clear that some sunspots have downflow, but that perhaps an equal number have upflow at transition region temperatures. Within a given spot the flow direction may well be time dependent.

Evidence that the downflows in the quiet sun are associated at least partially with the network seems quite conclusive. Positive correlations between red shift amplitude and intensity of C IV have been found by Gebbie *et al.* (1981), and Athay *et al.* (1983). Similar correlations have been found in active regions by Mein *et al.* (1982), Simon *et al.* (1982), and Athay *et al.* (1982, 1983). On the other hand, Dere (1982a) found no correlation between velocity and intensity and Athay *et al.* (1982, 1983) found that the correlation in C IV was often restricted to intermediate intensities. Furthermore, the latter authors noted that whereas the network features tended to have a red shift the reverse correlation was much weaker. Much of the red shifted areas are near to or below average brightness. Thus, it appears that the red shifted areas include most of the network but, in addition, include much of the supergranule cell interior. Also, the very brightest features in active regions often show blue shifts (Athay et al. 1983). This is consistent with high resolution observations reported by Brueckner (1981) and Dere (1982a, 1982b) that show many bright blue shifted features in C IV.

Although the preponderant downflow at transition region temperatures seems well established in both the quiet and active sun, the form of the downflow is unclear. In both low resolution and high resolution data C IV lines show a variety of profile types ranging from asymmetric profiles composed of a a strong unshifted component and weaker Doppler shifted components in the red wing (Dere 1982), to symmetric profiles with notable Doppler shift (Athay *et al.* 1983). It seems probable, therefore, that the Doppler shifted areas have scale sizes ranging from sub-arcsecond to many arcseconds.

An attempt by Rottman, Orrall and Klimchuk (1981) to measure the radial flow velocity in a coronal hole in OV showed a blue shift of 3 km s^{-1} in the coronal hole relative to the average of the observed area outside the coronal hole. Since the average sun is very probably redshifted in OV, as in other lines formed at similar temperatures, it is unclear whether, in fact, the observed coronal hole had a net outflow in OV or just a reduced redshift.

The high spatial resolution HRTS data from the U.S. Naval Research Laboratory reveal a complex array of moving features including small blue shifted jets of relatively short lifetime. Brueckner (1981), Dere (1982a, 1982b) and Brueckner and Bartoe (1983) identify "chromospheric jets" as a class of objects with blue shifts at 10-20 km s^{-1} and "coronal jets" as a second class with blue shifts up to 400 km s^{-1}. Little is known about the morphology and the associations of these jets with other solar features due to the limited spatial sampling in the HRTS data. The discovery of these interesting high speed jets together with the unanswered questions about the structure of the downflows underscores the need to pursue high resolution studies of the highly dynamic transition region.

Horizontal Flow.

As is clear on general physical grounds and supported for some time by observations, flows with dominant horizontal components play an important role in the dynamics of the solar atmosphere. The difficulty of observing such flows has notably restricted what is known concerning them. Again, however, recent transition region studies have provided interesting new results.

Nicholas and Kjeldseth-Moe (1981) and Athay *et al.* (1982, 1983) report flow patterns observed in C IV around large sunspots indicative of a reverse Evershed effect. On the sun-center sides of the spots, the gas is red shifted, and on the limbward side it is blue shifted. The red shifted area is usually larger in size than the blue shifted area and the red shift amplitude usually exceeds the blue shift amplitude, which suggests that the flow follows a pronounced Wilson depression.

At transition region temperatures, active regions show horizontal flow patterns that are coherent over large spatial domains and that are closely associated with the overall magnetic field structure of the active region (Athay *et al.* 1982, Athay, Gurman and Henze 1983). Near the magnetic neutral line separating strong field regions of opposite polarity the flow is usually divergent with opposite sign either side of the neutral line. However, near the neutral lines separating the weaker field areas of opposite polarity bordering the strong field areas the flow is often convergent towards the neutral line. In many cases the local flow around individual spots appears to be a coherent part of the large scale pattern, but in other cases the flow appears to be peculiar to the spot itself.

Oscillations.

Periodic oscillations in a sunspot umbra were reported by Zhugzhda and Makarov (1982). Brightness fluctuations observed in the Hα and K lines show series of wave trains lasting for 1.5 to 2 hours. Individual wave trains start at periods near 200 s and decrease to about 150 s as the train dies out. The wave trains are somewhat less pronounced in Hα than in the K line, but otherwise they are very similar in the two lines. The authors interpret the pulses of wave trains and the changing wave period in terms of an alternate compression and expansion of the chromospheric resonant cavity.

Umbral oscillations in C IV have been reported by Gurman *et al.* (1982) and Henze *et al.* (1984). Of the eight sunspots studied by Gurman et al., all showed velocity and brightness oscillations. Periods ranged from 130 to 170 s and velocity amplitudes from 0.8 to 3.5 km s^{-1}. For four of the spots, maximum velocity and maximum blue shift were in phase, consistent with adiabatic acoustic waves. In the one spot studied by Henze *et al.* different umbral pixels showed oscillatory periods from 110 to 200 s in both brightness and velocity. The pixel with the best defined period showed peak intensity leading peak blue shift approximately 45°.

Successful attempts to construct quiet sun $k - \omega$ diagrams in the chromospheric components of the K line have been reported by Kneer and von Uexküll (1983) and by Dame, Gouttebroze and Malherbe (1984). Both sets of observations show modal structure in the approximate wave number ranges 0.25 to 2.5 Mm^{-1} and at frequencies beginning near 15 mHz ($P \approx 420$ s). Modal structure is resolved up to frequencies of approximately 30 mHz ($P \approx 210$ s) in the observations of Kneer and von Uexküll and to approximately 45 mHz (140 s) in those of Dame, Gouttebroze and Malherbe. The latter authors also find faint but inconclusive evidence for longer period waves in the gravity wave regime.

References

Athay, R.G. *et al.* 1982, *Ap. J.*, **261**, 684.
Athay, R.G., Gurman, J.B. and Henze, W. 1983, *Ap. J.*, **269**, 706.
Athay, R.G. *et al.* 1983, *Ap. J.*, **265**, 519.
Brueckner, G.E. 1981, *Solar Active Regions*, ed. F.Q. Orrall, Colo. Univ. Press: Boulder.
Dame, L., Gouttebroze, P. and J.-M. Malherbe 1984, *Astron. and Astrophys.*, **130**, 331.
Dere, K.P. 1982a, *NASA Conference Publ. 2280*, 33, ed. M. Neugebauer, NASA: Wash.D.C.
Dere, K.P. 1982b, *Solar Phys.*, **77**, 77.
Feldman, U., Cohen, L. and Doschek, G.A. 1982, *Ap. J.*, **255**, 325.
Gebbie, K.B. *et al.* 1981, *Ap. J.*, **251**, L115.
Gurman, J.B. *et al.* 1982, *Ap.J.*, **253**, 939.
Gurman, J.B. and Athay, R.G. 1983, *Ap. J.*, **273**, 374.
Henze, W. *et al.* 1984, *Solar Phys.*, **91**, 33.
Kingston, A.E. *et al.* 1982, *Solar Phys.*, **81**, 47.
Kneer, F. and von Uexküll, M. 1983, *Astron. and Astrophys.*, **119**, 124.
Mein, P. *et al.* 1982, *Astron. and Astrophys.*, **111**, 136.
Nicholas, K.R. and Kjeldseth-Moe, O. 1981, *The Physics of Sunspots*, ed. L.E. Cram and J.H. Thomas, Sacramento Peak Obs: Sunspot.
Nicholas, K.R. *et al.* 1982, *Solar Phys.*, **81**, 253.
Rottman, G.J., Orrall, F.Q. and Klimchuk, J.A. 1981, *Ap. J.*, **247**, L135.
Roussel-Dupre, D. and Shine, R.A. 1982, *Solar Phys.* **77**, 329.
Simon, G. *et al.* 1982, *Astron. and Astrophys.*, **115**, 367.
Zhugzhda, Y.D. and Makarov, V.I. 1982, *Solar Phys.*, **81**, 245.

VI. THE PHYSICS OF CORONAL FLUX TUBES

(B. Roberts)

Recent texts dealing with background aspects of this topic include Priest (1982), Noyes (1982) and Giovanelli (1984). Here we concentrate on the equilibrium structure of coronal loops, their oscillations and instabilities, and their heating.

Equilibrium Structure of Loops

The structuring of the coronal atmosphere by magnetic fields, with density, pressure and temperature differences giving rise to the appearance in X-ray and EUV-lines of individual loops, has drawn attention to the need to explore magnetic equilibria and their stability properties. Loops are undoubtedly not static, exhibiting both flows and variations in intensity on a variety of timescales. Nonetheless, it is important to understand the simplest static behavior that is possible under coronal conditions, and to inquire as to the limitations imposed by stability.

Static configurations give rise to simple scaling relations (e.g. Rosner, Tucker, and Vaiana 1978; Chiuderi, Einaudi, and Torricelli-Ciamponi 1981; Pallavicini *et al.* 1981; Withbroe 1981). A number of extensions are possible, such as the inclusion of gravity (Wragg and Priest 1981; Serio *et al.* 1981). In uniform (zero gravity) pressure loops, simple power-law scalings pertain (e.g. $T \approx (pL)^{1/3}$). Gravity is important in long loops, with lengths exceeding the coronal scale-height. In loops longer than about 10^8 cm, the hydrostatic fall-off in height of density and pressure modifies the simple power laws by the inclusion of an exponential dependence of temperature upon the loop length (Wragg and Priest 1981).

Scaling relations originally offered the hope of providing detailed information on the physical mechanisms at work in heating the coronal atmosphere. However, it has transpired that such relations are somewhat insensitive to direct testing against the observations and, moreover, the temperature structure along a loop is poorly determined. Golub (1982, 1983) has reviewed the matter and emphasized the possible use of stellar data, including relations which depend upon magnetic field strength. For stellar sources, Galeev *et al.* (1981) argue that quiescent emission levels arise principally from loops of length comparable with the atmosphere's pressure scale-height; shorter, newly emerged, loops tend to give rise to fluctuations in emission. The use of emission measure as a diagnostic tool, rather than direct use of temperature, length, etc., has been emphasized by Torricelli-Ciamponi, Einaudi, and Chiuderi (1982).

Magnetic equilibria are of obvious general interest, and so have been investigated for a variety of circumstances, including force-free (Sakurai 1981; Low 1982; Low and Hundhausen 1983; Melville, Hood, and Priest 1983; Browning and Priest 1983) and gravitational (Melville, Hood, and Priest 1984; Browning and Priest 1984a) equilibria.

1. *Instabilities*

Loops and arcades of loops are subject to a number of instabilities, including both hydromagnetic and thermal. Ideal magnetic instabilities have been reviewed by Van Hoven (1981), Hood and Priest (1981), Priest (1981; 1983a, b; 1984), and Einaudi (1984). Regarding a loop as a twisted magnetic cylinder, and including the important stabilizing influence of line-tying in the dense photosphere (e.g. Raadu 1972; Hood and Priest 1981; An 1982, 1984; Spicer 1982), allows a rather full analysis of the possible ideal magnetohydrodynamic instabilities (Raadu 1972; Hood and Priest 1979, 1981; Van Hoven, Ma, and Einaudi 1981; Priest 1981, 1983; Einaudi and Van Hoven 1983). Similarly, arcades of loops have been investigated in some detail (Hood and Priest 1980; Ray and Van Hoven 1982; Hood 1983; Schindler, Birn, and Janicke 1983; Migliuolo and Cargill 1983; Migliuolo, Cargill, and Hood 1984) as also the question of a general magnetic atmosphere (Birn and Schindler 1981; Heyvaerts *et al.* 1982; Zweibel and Hundhausen 1982; Priest 1983a, b).

The aim of instability studies is to understand how, on the one hand, loops and arcades may be stable on timescales of perhaps a day and, on the other hand, may cause or contribute to the onset of flares or other dynamical events. The main difficulty in these analyses is in a realistic modeling of the equilibria, the choice of boundary conditions at the footpoints of loops, and the appropriate treatment of the transition zone. Uncertainties on all these issues remain.

Analysis of magnetic instabilities is generally based upon two standard approaches: either the method of normal modes or the energy principle, the latter usually being simpler. There are several obvious similarities with analyses of laboratory plasmas but the coronal case provides a distinct set of problems in its own right, so it is not possible simply to transpose results from the laboratory to the coronal plasma (see reviews by Spicer (1982), Priest (1983a, b), and Einaudi (1984)).

A loop may be free from hydromagnetic instabilities and yet liable to thermal instabilities (or nonequilibrium) if the optically-thin radiation is insufficient to balance with heating. Some analytical (Wragg and Priest 1982; Martens and Kuperus 1982) and numerical (McClymont and Canfield

1982) studies are available (see the review by Craig (1981)) but, as with hydromagnetic instabilities, it is presently unclear how to treat properly the footpoints of loops and how sensitive results are to such a treatment of boundary conditions.

2. *Oscillations in the Corona*

A. Observations

Oscillations are reported in a variety of coronal structures and possess a variety of periodicities, from seconds to tens of minutes. Oscillations have been detected in hard X-ray wavelengths (Orwig, Frost, and Dennis 1981; Dennis, Frost, and Orwig 1981; Kiplinger *et al.* 1983a, b), simultaneously in hard X-ray and microwave (Takakura *et al.* 1983; Kane *et al.* 1983), in EUV emission (Antonucci, Gabriel, and Patchett 1984), and in the green coronal line (Koutchmy, Zhugzhda, and Locans 1983; Pasachoff and Landman 1984). Radio wavelengths have revealed coronal oscillations for over a decade; recent reports include a moving Type IV event (Trottet *et al.* 1981). Prominences are also seen to oscillate (Strauss, Kaufman, and Opher 1980; Malville and Schindler 1981; Malherbe, Schmieder, and Mein 1981; Bashkirtsev, Kopanov, and Mashnich 1983; Bashkirtsev and Mashnich 1984), with periods typically in tens of minutes.

B. Theoretical Aspects

Magnetically Structured Atmosphere. The behavior of magnetoacoustic waves in a magnetically structured atmosphere has been investigated by Roberts (1981), Spruit (1981, 1982), Edwin and Roberts (1982, 1983), Rae and Roberts (1982, 1983a, b), Gordon and Hollweg (1983), Roberts, Edwin, and Benz (1983, 1984), following on from earlier studies (e.g. Wentzel 1979; Wilson 1980). The general conclusion is that structured media support both body waves and surface waves, the forms and speeds of which depend upon the particular values of sound and Alfven speeds. Both slab and cylinder geometries are amenable to investigation, with applications to uniform pressure coronal loops as well as chromospheric fibrils and photospheric flux tubes. In addition to the sound and Alfven speeds, two other velocities turn out to be important in structured media: a tube or cusp speed, which is both subsonic and subAlfvenic; and a mean Alfven speed, which is associated with both a single magnetic interface and with asymmetrical (kink) modes in tubes. Reviews are available in Spruit (1981), Roberts (1982, 1984) and Spruit and Roberts (1983).

The inhomogeneous nature of the corona means that regions of low Alfven speed may act as wave guides. In particular, dense loops may act as ducts for fast magnetoacoustic waves. Roberts, Edwin, and Benz (1983, 1984) have worked out the temporal signature of an impulsively generated fast wave in a magnetic slab or cylinder. The signature of the wave exhibits short timescales, roughly the loop diameter divided by the internal Alfven speed. The signature exhibits three phases: first a periodic phase, then a quasi-periodic phase of higher frequency and stronger amplitude and, finally, a rapid decay phase (with the frequency of the oscillation maintained). The explanation for this distinctive signature lies in the dispersive nature of the mode in a duct. Periods in the event are typically on the order of a second (assuming loop diameters of $\sim 10^3$ km and Alfven speed $\sim 10^3$ km s^{-1}). Such signatures offer the possibility of use as a diagnostic of *in situ* coronal conditions (such as the magnetic field strength). Dissipative effects have yet to be incorporated in the calculations of the dynamical response, but the analysis by Gordon and Hollweg (1983) suggests that these modes are most likely to persist and propagate for large distances in regions of strong (above 10 gauss) field strengths.

Ducted waves may be either kink or sausage modes, accordingly as the tube's axis is displaced or undisturbed in the oscillation (e.g. Roberts 1981). The above described signature is specifically for the sausage mode but higher harmonics than the first of the kink wave will behave similarly. The exception is the principal kink mode which is only weakly dispersive and so may propagate as a single pulse. Both sausage and kink modes are analogous to other, non-magnetic, systems; the sausage mode to Pekens waves in oceanography and the kink modes to Love waves in seismology (Edwin and Roberts 1982, 1983). Additionally, there are close analogies with fibre optics (Roberts, Edwin, and Benz 1983, 1984). Loops may also act as ducts for sound waves but these are virtually non-dispersive.

The oscillation of a loop as a whole is also of general interest. The period is given by the length of the loop and the mean Alfven speed; for a dense loop of length L the period is $2L/v_A$, where v_A is the Alfven speed within the loop (Spruit 1982; Edwin and Roberts 1983; Roberts, Edwin, and Benz 1984). Thus periods of the order of minutes may arise.

Stratified Atmospheres. The above discussion concentrates on the role of magnetic structuring, ignoring gravity. Other aspects of wave behavior have been investigated by ignoring structuring and instead retaining the effect of gravity. Leroy (1981) has examined the propagation of Alfven waves in a stratified medium and concluded that Alfven waves observed in interplanetary space cannot have been generated in the convection zone. Bel and Leroy (1981) have argued that sunspots are not the seat of, nor are cooled by, a large flux of outwardly propagating Alfven waves. (This is at variance with the suggestion of Mullen and Owens (1984) that observed interplanetary Alfven waves may have arisen from sunspots.) Schwartz, Cally, and Bel (1984) have examined Alfven waves in an oblique field, showing that a two-layer model atmosphere represents remarkably well the behavior in multi-layered media. They go on to analyze the behavior of a spreading magnetic field (see also Cally 1983), demonstrating an order-of-magnitude increase in coronal flux when field strengths are high ($\sim$ 3000 gauss).

Leroy and Schwartz (1982), Schwartz and Leroy (1982), and Zhugzhda and Dzhalilov (1984) take up the question of the propagation of magnetoacoustic waves, examining the non-WKB limit. Leroy and Schwartz conclude that such modes make only a small contribution to coronal energetics. Schwartz and Bel (1984a,b) examine propagation in an oblique field, concluding that Alfven waves are the most viable wave option for coronal heating. When the oblique field is almost horizontal, Schwartz and Bel (1984b) find that there is little evidence of the singular critical level behavior associated with the horizontal field case. This raises the question of how physically relevant is the usual treatment of the horizontal case. General reviews of magnetic waves in stratified media have been given by Adam (1981), Thomas (1983), and Campos (1983a, b).

2. *Coronal Heating*

With the recognition that acoustic waves are not responsible for the heating of the corona (Athay and White 1978; Mein and Mein 1980) has come a general effort to understand the ways in which magnetic fields may prove an essential ingredient. There has been good progress, with a number of promising effects explored in enough detail that they can no longer be described as "hand-waving," the most popular means of heating the corona! Nonetheless, a clear picture of how the corona is in fact heated has yet to emerge and so a description of the various contenders must suffice for the present. Recent reviews are: Kuperus, Ionson, and Spicer (1981), Chiuderi (1983), Hollweg (1983), Priest (1983a, b), and Heyvaerts (1984).

Two processes have been explored for coronal heating: the dissipation of hydromagnetic waves, and the continuous reconnection of stressed magnetic fields.

A. Hydromagnetic Waves

The equations of magnetohydrodynamics admit an Alfven wave and two compressive (fast and slow) waves. The properties of these waves in a uniform medium are well known but in a non-uniform medium (such as in loops or flux tubes) require re-appraisal (see, for example, Roberts 1981, 1984). Strong inhomogeneities render the propagation of characteristics more complex; indeed, the behavior of mhd waves in a nonuniform media is not fully understood (Goedbloed 1984). The simplest mode to investigate is the Alfven wave. Heyvaerts and Priest (1983) have demonstrated the phase-mixing of an Alfven wave propagating in an inhomogeneous medium. Phase-mixing leads to the reduction of the effective length-scale of the wave and consequently to dissipation under the enhanced effectiveness of viscosity and diffusivity (see also Nocera, Leroy, and Priest 1983). Damping of propagating waves by phase-mixing is most efficient for short-period (10 s) waves in a weak (1 gauss) magnetic field; for longer periods or stronger fields the mechanism is not efficient enough to heat the corona (e.g., Priest 1983a, b).

In loops standing waves are important and these are also damped by phase-mixing, but now in time rather than in space. Estimates (e.g. Priest 1983a) give a time of typically 20 wave periods for damping under coronal conditions. Dissipation may be further enhanced by local Kelvin-Helmholtz instabilities generating turbulence (Heyvaerts and Priest 1983; Browning and Priest 1983) and a turbulent cascade (Hollweg 1983).

In addition to the Alfven wave a non-uniform medium supports compressive modes. The presence of inhomogeneities again allows the build-up of sharp velocity gradients and thus enhanced dissipation (Rae and Roberts 1981, 1982), making magnetoacoustic waves a candidate for coronal heating. The complexity of the behavior of magnetoacoustic waves in a nonuniform medium makes the complementary approaches of analytical (e.g., Sakurai and Granik 1984) and numerical (e.g., Steinolfson 1984) analyses here invaluable.

A somewhat different approach to the coronal heating problem has been taken by Ionson (1982, 1983, 1984), who has argued that the linear mhd equations may be cast in the form of an equivalent LRC circuit, with the circuit's coefficients being determined by the plasma's geometry and physical state; photospheric conditions provide a driver. The use of equivalent circuits has also been discussed by Spicer (1981, 1982). Because of the general complexity of the problem, the application of equivalent circuit concepts provides a valuable means of investigating the physics of coronal heating. Though it remains presently unclear how representative such circuits are of the actual coronal physics, and how much of that physics is averaged-out in the derivation of the circuit equations, quite separate analyses by Heyvaerts and Priest (1983, 1984) would seem to provide strong support for the circuit approach.

Waves may also heat coronal loops by damping in a resonant cavity (Hollweg 1981, 1983, 1984; Zhugzhda and Locans 1982). Hollweg (1984) has considered a three-layer model of a coronal loop, the two end layers representing the photosphere and the middle layer the corona. Resonant transmission peaks arise if the middle-layer cavity is driven externally by incident waves from the photosphere. The analysis permits a derivation of the heating rate which agrees with that obtained by Ionson (1982) from a circuit approach. However, estimates of qualities are much smaller than those in the circuit approach of Ionson (1982), though an allowance for leakages is possible (Ionson 1984). Hollweg and Sterling (1983) have compared the predictions of the resonant loop model with available coronal loop data, finding agreement.

B. DC Currents

Rapid motions in the photosphere may give rise to waves which propagate into the corona and dissipate their energy in the various ways outlined in (A). On the other hand, slow photospheric motions may build up stresses in the corona, the resulting field perhaps being force-free. Such DC current systems represent a source of energy for heating the corona if the stressed field can be dissipated or reconnected at an adequate rate (as compared with the rate for build-up of the field stresses). Direct dissipation of magnetic energy by joule heating is inadequate to meet the coronal power needs unless scales as small as centimeters (or meters, if microturbulence is admitted) are envisaged (Heyvaerts and Priest 1984). Heating may also result from the resistive tearing modes (Galeev *et al.* 1981; Chiuderi 1981; Kuperus, Ionson, and Spicer 1981; Sturrock and Uchida 1981; Parker 1984).

Heyvaerts and Priest (1984) have recently considered the application of Taylor's hypothesis (familiar in laboratory plasma physics; Taylor 1974) to solar or stellar conditions (see also Norman and Heyvaerts 1983). The application of the hypothesis implies that at any given time the coronal magnetic field is a linear force-free one, determined from the evolution of magnetic helicity (see also Berger 1983). Heyvaerts and Priest demonstrate that in this theory heating results from photospheric motions which build-up field stresses at just that rate which is comparable to the rate at which the field is relaxed by reconnection; in this theory, photospheric motions that are either too fast or too slow give negligible heating. The efficiency of heating is limited by the geometrical ratio of the scale of the driving velocity cells to the scale of the magnetic field, and also by the ratio of the reconnection (dissipation) timescale to the transit time of the driving velocity across the base of the magnetic structure (Parker 1983a, b; Heyvaerts and Priest 1984). Thus, for loops with very small cross-sections in the photospheric base the efficiency of this process is strongly reduced.

In general terms, the view that slow small-scale shufflings of magnetic fields in the photosphere results in a significant heating in the corona has been advocated by Parker (1982, 1983a, b), without resource to Taylor's hypothesis. Parker demonstrates that wrapping and winding of magnetic field at the structure's footpoints inevitably leads to reconnection, there being no static equilibrium available to the field (Parker 1972). This, Parker concludes, is the basic cause of the X-ray corona. Parker (1984) points out that the nonequilibrium of the magnetic field is closely akin to the phenomenon of magnetohydrodynamic turbulence (Matthaeus and Montgomery 1981).

Both Parker (1982) and Heyvaerts and Priest (1984) build their scenarios of coronal heating on the transference and subsequent dissipation of magnetic stresses in the corona, the stresses being induced by motions in the photosphere. Consequently, the heating rate is essentially the same in both calculations, though the geometrical scaling factor (between the velocity cells and the scale of the magnetic field) is squared in Heyvaerts and Priest (1984), presumably as a consequence of the restrictions brought about by the Taylor hypothesis, and linear in Parker (1982). Since this factor is small, such a difference may be important.

The notion that the manipulated photospheric field is inevitably subject to nonequilibrium (Parker 1972, 1981, 1982; Tsinganos 1982) has been taken up by others, who point out that small motions of the footpoints may in fact result in fields which possess large variations in boundary layers near the footpoints (Sakurai and Levine 1981), and that small perturbations which change the field's topology are in fact excluded in Parker's discussion (Rosner and Knobloch 1982; see also Priest 1984).

Finally, it should be noted that DC heating can also be treated from a circuit viewpoint (Ionson 1982, 1984), though two or more circuits may be needed for an adequate representation; see the discussion in Heyvaerts and Priest (1984).

References

Adam, J.A. 1981, *Astrophys. and Space Sci.*, **78**, 293.
An, C.-H. 1982, *Solar Phys.*, **75**, 19.
An, C.-H. 1984, *Ap. J.*, **281**, 419.
Antonucci, E., Gabriel, A.H., and Patchett, B.E. 1984, *Solar Phys.*, **93**, 85.
Athay, R.G., and White, O.R. 1978, *Ap.J.*, **226**, 1135.
Bashkirtsev, V.S., Kobanov, N.I., and Mashnich, G.P. 1983, *Solar Phys.*, **82**, 443.
Bashkirtsev, V.S., and Mashnich, G.P. 1984, *Solar Phys.*, **91**, 83.
Bel, N., and Leroy, B. 1981, *Astr. Ap.*, **104**, 203.
Berger, M. 1984, *Bull. Am. Astr. Soc.*, **16**, 528.
Birn, J., and Schindler, K. 1981, in *Solar Flare Magnetohydrodynamics*, ed. E.R. Priest (New York: Gordon and Breach), p. 337.
Browning, P.K., and Priest, E.R. 1984a, *Astr. Ap.*, **131**, 283.
Browning, P.K., and Priest, E.R. 1984b, *Solar Phys.*, **92**, 173.
Cally, P.S. 1983, *Solar Phys.*, **88**, 77.
Campos, L.M.B.C. 1983a, *J. Phys. A: Math. Gen.*, **16**, 417.
Campos, L.M.B.C. 1983b, *Solar Phys.*, **82**, 355.
Chiuderi, C. 1981, in *Solar Phenomena in Stars and Stellar Systems*, eds. R.M. Bonnet and A.K. Dupree, Dordrecht: Reidel), p. 269.
Chiuderi, C. 1983, in *Solar and Stellar Magnetic Fields: Origins and Coronal Effects*, ed. J.O. Stenflo (Dordrecht: Reidel), p. 375.
Chiuderi, G., Einaudi, G., and Torricelli-Ciamponi, G. 1981, *Astr. Ap.*, **97**, 27.
Craig, I.J.D. 1981, in *Solar Flare Magnetohydrodynamics*, ed. E.R. Priest (New York: Gordon and Breach), p. 277.
Dennis, B.R., Frost, K.J., and Orwig, L.E. 1981, *Ap. J.*, **244**, L167.
Edwin, P.M., and Roberts, B. 1982, *Solar Phys.*, **76**, 239.
Edwin, P.M., and Roberts, B. 1983, *Solar Phys.*, **88**, 179.
Einaudi, G. 1984, in *The Hydromagnetics of the Sun*, ed. T. D. Guyenne (Paris: ESA), in press.
Einaudi, G., and Van Hoven, G. 1983, *Solar Phys.*, **88**, 163.
Galeev, A.A., Rosner, R., Serio, S., and Vaiana, G.S. 1981, *Ap. J.*, **243**, 301.
Giovanelli, R.G. 1984, *Secrets of the Sun*, (Cambridge: Cambridge Univ. Press).
Goedbloed, J.P. 1984, *Physica*, **12D**, 107.
Golub, L. 1982, in *Coll Stars, Stellar Systems, and the Sun*, eds. M.S. Giampapa and L. Golub (Smithsonian Special Report No. 392), p. 39.
Golub, L. 1983, in *Solar and Stellar Magnetic Fields: Origins and Coronal Effects*, ed. J.O. Stenflo (Dordrecht: Reidel), p. 345.
Heyvaerts, J. 1984, in *The Hydromagnetics of the Sun*, ed. T.D. Guyenne (Paris: ESA), in press.
Heyvaerts, J., Lasry, J.M., Schatzman, M., and Witomsky, P. 1982, *Astr. Ap.*, **111**, 104.
Heyvaerts, J., and Priest, E.R. 1983, *Astr. Ap.*, **117**, 220.
Heyvaerts, J., and Priest, E.R. 1984, *Astr. Ap.*, **137**, 63.
Hollweg, J.V. 1981, *Solar Phys.*, **70**, 25.
Hollweg, J.V. 1983, in *Solar Wind Five*, ed. M. Neugebauer (Washington, NAA), p. 5.
Hollweg, J.V. 1984, *Ap. J.*, **277**, 392.
Hollweg, J.V., and Sterling, A.C. 1983, *Bull. Am. Astr. Soc.*, **15**, 994.
Hood, A.W. 1983, *Solar Phys.*, **87**, 279; **89**, 235.
Hood, A.W., and Priest, E.R.1979, *Astr. Ap.*, **77**, 233.
Hood, A.W., and Priest, E.R. 1980, *Solar Phys.*, **66**, 113.
Hood, A.W., and Priest, E.R. 1981, in *Solar Phenomena in Stars and Stellar Systems*, eds. R.M. Bonnet and A.K. Dupree (Dordrecht: Reidel), p. 509.

Ionson, J.A. 1984, *Ap. J.*, **254**, 318.
Ionson, J.A. 1983, in *solar and Stellar Magnetic Fields: Origins and Coronal Effects*, ed. J.O. Stenflo (Dordrecht: Reidel), p. 391.
Ionson, J.A. 1984, *Ap. J.*, **276**, 357.
Kane, S.R., Kai, K., Kosugi, T., Enome, S., Landecker, P., and McKenzie, D. 1983, *Ap. J.*, **271**, 376.
Kiplinger, A.L., Dennis, B.R., Emslie, A.G., Frost, K.J., and Orwig, L.E. 1983a, *Ap. J.*, **265**, L99.
Kiplinger, A.L., Dennis, B.R., Frost, K.J., and Orwig, L.E. 1983b, *Ap. J.*, **273**, 783.
Koutchmy, S., Zhugzhda, Y.D., and Locans, V. 1983, *Astr. Ap.*, **120**, 185.
Kuperus, M., Ionson, J.A., and Spicer, D.S. 1981, *Ann. Rev. Astr. Ap.*, **19**, 7.
Leroy, B. 1981, *Astr. Ap.*, **97**, 245.
Leroy, B., and Schwartz, S.J. 1982, *Astr. Ap.*, **112**, 84.
Low, B.C. 1982, *Rev. Geophys. Sp. Phys.*, **20**, 145.
Low, B.C., and Hundhausen, A. 1983, *Phys. Fluids*, **26**, 2731.
Malherbe, J.M., Schmieder, B.M., and Mein, P. 1981, *Astr. Ap.*, **102**, 124.
Malville, J.M.K., and Schindler, M. 1981, *Solar Phys.*, **70**, 115.
Martens, P.C.H., and Kuperus, M. 1982, *Astr. Ap.*, **113**, 324.
Matthaeus, W.H., and Montgomery, D. 1981, *J. Plasma Phys.*, **25**, 11.
McClymont, A.N., and Canfield, R.C. 1983, *Ap. J.*, **265**, 497.
Mein, N., and Mein P. 1980, *Astr. Ap.*, **84**, 96.
Melville, J.P., Hood, A.W., and Priest, E.R. 1983, *Solar Phys.*, **87**, 301.
Melville, J.P., Hood, A.W., and Priest, E.R. 1984, *Solar Phys.*, **92**, 15.
Migliuolo, S., and Cargill, P.J. 1983, *Ap. J.*, **271**, 820.
Migliuolo, S., Cargill, P.J., and Hood, A.W. 1984, *Ap. J.*, **281**, 413.
Mullan, D.J., and Owens, A.J. 1984, *Ap. J.*, **280**, 346.
Nocera, L., Leroy, B., and Priest, E.R. 1984, *Astr. Ap.*, **133**, 387.
Norman, C., and Heyvaerts, J. 1983, *Astr. Ap.*, **124**, L1.
Noyes, R.W. 1982, *The Sun, Our Star* (Cambridge: Harvard Univ. Press).
Orwig, L.E., Frost, K.J., and Dennis, B.R. 1981, *Ap. J.*, **244**, L163.
Pallavicini, R., Peres, G., Serio, S., Vaiana, G.S., Golub, L., and Rosner, R. 1981, *Ap. J.*, **247**, 692.
Parker, E.N. 1972, *Ap. J.*, **174**, 499.
Parker, E.N. 1981, *Ap. J.*, **244**, 631, 644.
Parker, E.N. 1982, *Geophys. Ap. Fluid Dyn.*, **22**, 195.
Parker, E.N. 1983a, *Ap. J.*, **264**, 635, 642.
Parker, E.N. 1983b, *Geophys. Ap. Fluid Dyn.*, **24**, 79.
Parker, E.N. 1984, in *Magnetic Reconnection in Space and Laboratory*, ed. E. Hones (Washington: Amer. Geophys. Union), p. 32.
Pasachoff, J.M., and Landman, D.A. 1984, *Solar Phys.*, **90**, 325.
Priest, E.R. 1981, in *Solar Active Regions*, ed. F. Orrall (Boulder: Colorado Assoc. Univ. Press), p. 213.
Priest, E.R. 1982a, *Solar Magnetohydrodynamics* (Dordrecht: Reidel).
Priest, E.R. 1982b, *Fundamentals of Cosmic Physics*, **7**, 363.
Priest, E.R. 1983a, *Plasma Phys.*, **25**, 161.
Priest, E.R. 1983b, *Solar Phys.*, **86**, 33.
Priest, E.R. 1984, *Reports Progress Physics*, in press.
Raadu, M.A. 1972, *Solar Phys.*, **22**, 425.
Rae, I.C., and Roberts, B. 1981, *Geophys. Ap. Fluid Dyn.*, **18**, 197.
Rae, I.C., and Roberts, B. 1982, *Mon. Not. Roy. Astr. Soc.*, **201**, 1171.
Rae, I.C., and Roberts, B. 1983a, *Solar Phys.*, **84**, 99.
Rae, I.C., and Roberts, B. 1983b, *Phys. Fluids*, **26**, 269.
Ray, A., and Van Hoven, G. 1982, *Solar Phys.*, **79**, 353.
Roberts, B. 1981a, in *Physics of Sunspots*, eds. L.E. Cram and J.H. Thomas (Sunspot: Sacramento Peak Observatory), p. 369.
Roberts, B. 1981b, *Solar Phys.*, **69**, 27, 39.
Roberts, B. 1984, in *Hydromagnetics of the Sun*, ed. T.D. Guyenne (Paris: ESA), in press.
Roberts, B., Edwin, P.M., and Benz, A.O. 1983, *Nature*, **305**, f1688.
Roberts, B., Edwin, P.M., and Benz, A.O. 1984, *Ap. J.*, **279**, 857.
Rosner, R., and Knobloch, E. 1982, *Ap. J.*, **262**, 349.
Rosner, R., Tucker, W.H., and Vaiana, G.S. 1978, *Ap. J.*, **220**, 643.

Sakurai, T. 1981, *Solar Phys.*, **769**, 343.
Sakurai, T., and Granik, A. 1981, *Ap. J.*, **277**, 404.
Sakurai, T., and Levine, R.H. 1981, *Ap. J.*, **248**, 817.
Schindler, K., Birn, J., and Janicke, L. 1983, *Solar Phys.*, **87**, 103.
Schwartz, S.J., and Bel, N. 1984a, *Astr. Ap.*, **137**, 128.
Schwartz, S.J., and Bel, N. 1984b, *Solar Phys.*, **92**, 133.
Schwartrz, S.J., Cally, P.S., and Bel, N. 1984, *Solar Phys.*, **92**, 81.
Schwartz, S.J., and Leroy, B. 1982, *Astr. Ap.*, **112**, 93.
Serio, S., Peres, G., Vaiana, G.S., Golub, L., and Rosner, R. 1981, *Ap. J.*, **243**, 288.
Spicer, D.S. 1981, *Solar Phys.*, **71**, 115.
Spicer, D.S. 1982, *Space Sci. Rev.*, **31**, 351.
Spruit, H.C. 1981, in *The Sun as a Star*, ed. S. Jordan (Washington: NASA SP-450), p. 385.
Spruit, H.C. 1982, *Solar Phys.*, **75**, 3.
Spruit, H.C., and Roberts, B. 1983, *Nature*, **304**, 401.
Steinolfson, R.S. 1984, *Phys. Fluids*, **27**, 781.
Strauss, F.M., Kaufmann, P., and Opher, R. 1980, *solar Phys.*, **67**, 83.
Sturrock, P.A., and Uchida, Y. 1981, *Ap. J.*, **246**, 331.
Takakura, T., Kaufmann, P., Costa, J.E.R., Degaonkar, S.S., Ohki, K., and Nitta, N. 1983, *Nature*, **302**, 317.
Taylor, J.B. 1974, *Phys. Rev. Letters*. **33**, 1139.
Thomas, J.H. 1983, *Ann. Rev. Fluid Mech.*, **15**, 321.
Torricelli-Ciamponi, G., Einaudi, G., and Chiuderi, C. 1982, *Astr. Ap.*, **105**, L1.
Trottet, G., Kerdraon, A., Benz, A.O., and Treymann, R. 1981, *Astr. Ap.*, **93**, 129.
Tsinganos, K.C. 1982, *Ap. J.*, **259**, 832.
Van Hoven, G. 1981, in *Solar Flare Magnetohydrodynamics*, ed. E.R. Priest (New York: Gordon and Breach), p. 217.
Van Hoven, G., Ma, S., and Einaudi, G. 1981, *Astr. Ap.*, **99**, 232.
Wentzel, D.G. 1979, *Astr. Ap.*, **76**, 20.
Wilson, P.R. 1980, *Astr. Ap.*, **87**, 121.
Withbroe, G. 1981, in *Solar Active Regions*, ed. F. Orrall (Boulder: Colorado Assoc. Univ. Press).
Wragg, M.A., and Priest, E.R. 1981, *Solar Phys.*, **70**, 293.
Wragg, M.A., and Priest, E.R. 1982, *Astr. Ap.*, **113**, 269.
Zhugzhda, Y.D., and Dzhalilov, N.S. 1984, *Astr. Ap.*, **132**, 45, 52.
Zhugzhda, Y.D., and Locans, V. 1982, *Solar Phys.*, **76**, 77.
Zweibel, E.G., and Hundhausen, A. 1982, *Solar Phys.*, **76**, 261.

VII. THE SUN AS A STAR

(K. R. Sivaraman)

An important topic of current interest both in solar and stellar physics is the possibility of identifying solar phenomena in stars through an understanding of these on a global scale on the sun. We concentrate here on the recent research concerning the variability in solar irradiance from disc integrated observations (sun as a star) in the white light as well as in monochromatic wavelengths.

1. *Luminosity variation in white light*

The principal scientific motivation for monitoring the solar constant is to detect any variability that could explain the climate changes over the earth. For an excellent review see Newkirk (1983). Measurements in 1980 by the precision radiometers on NIMBUS-7 and SMM have revealed temporal dips of 0.1 to 0.3% in the measured solar irradiance which had eluded previous detection. An empirical model ascribing 90% of the short time irradiance fluctuations to sunspot blocking reproduces best the ACRIM data between 1980-1982. The solar irradiance reconstructed from the archived sunspot areas from 1874 through 1981 using this model imply that the average solar constant during an active year can be as much as 0.1% lower than during a quiet year (Hoyt and Eddy, 1983). Photospheric heat flow can also be modulated by large laminar convective cells causing temperature perturbations between up and down flowing regions on scales of $\sim 1 \times 10^5$ km and life times of months (Gilman, 1978). Also the subsurface magnetic concentrations might cause 'thermal shadows' on the photosphere (Spruit, 1977; Foukal *et al.* 1983). But the continuum maps outside spots and faculae show that temperature inhomogeneities on these scales are below $2-3^\circ$ K, which is far below the spot signal (Foukal and Fowler, 1984). The heat blocked by sunspots is apparently rapidly redistributed for storage throughout the convection zone over time scales at least comparable to the spot life times. The efficient storage and slow release explains the spot induced dips in the radiometry data. (Foukal *et al.* 1983).

2. *Luminosity variations in monochromatic radiation*

2.1. *Photospheric variability*

Changes in luminosity are intimately related to the possible slow global changes in photospheric temperature which can be detected by monitoring the equivalent widths of temperature sensitive spectral lines originating in the photosphere. During the ascending phase of solar activity (1976-1980) the full disc equivalent widths of CI 5380A and four Fe I lines representing the low and high photosphere showed a steady decrease, but remained constant since 1980 (Livingston and Holweger, 1982; Holweger *et al.* 1983). This calls for more data to decide the presence of any 11-year cycle modulation. Similar behaviour of all the diagnostic lines can be reconciled only by a slight flattening of the photospheric temperature gradient as the carbon line has a temperature sensitivity opposite to that of iron lines. Since the temporal behaviour of these lines has no correlation with short lived features like active regions, it seems to reflect true global changes accompanying solar activity.

Another diagnostic is the spectral line asymmetry which is a measure of the granular convection. The C-shaped line bisector is more blue shifted relative to the core over quiet regions than in magnetic regions. The full disc Fe I 5250.6 mean bisectors from the Kitt Peak 13.5m spectrograph show a mid-C shift towards the core by $\sim$ 2mA between 1976 and 1981. Better observations with the FTS of ten similar Fe I lines confirm this trend between 1980-1982, suggesting a global decrease in the net upward velocity of convection in the photosphere with increase in solar activity (Livingston, 1982). Pic du Midi observations showing that the size of granules is reduced at that time support this (Macris and Rösch, 1983). Far infrared radiometry of the full disc indicates a warming of the temperature minimum accompanying solar activity (Müller *et al.* level in the atmosphere contradicts this finding (Cook *et al.* 1980).

2.2. *Chromospheric variability*

K-emission. The first attempt to demonstrate the variability in the Ca II K line emission from the 'Sun as a star' was by Bumba and Topolova (1967). The continuing disc integrated K emission measurements commencing in 1969 at Kodaikanal provide data on the variability of the K-line parameters over a full cycle (Bappu and Sivaraman, 1971; Sivaraman *et al.* 1984). Other published results, although for a limited period are those of White and Livingston (1978) and Keil and Worden (1984). The 1A emission index for the years 1975-1981 suggest that the excess emission in the ascending phase of the solar cycle arises from plages (White and Livingston, 1978). The quiet sun is represented well by the summation of the contributions from the network and the cell interior forming one component. The addition of two or more components, the plages and the active network (remnants of decayed plages) gives good agreement with full disc emission except near the solar maximum (Skumanich *et al.* 1984).

He I 10830A. Full disc He I 10830A equivalent width measurements since 1974 show it was lowest at 24 mA in mid 1975, a year earlier than sunspot minimum and steadily rose afterwards to 83 mA in late 1981, two years after the sunspot minimum. The rotational modulation reveals active longitudes. The apparent rotation period increased from 1977 to 1981 and has been shorter since 1981 which contradicts the equatorial migration of active regions during the course of solar cycle (Harvey, 1983).

Lyman Alpha. The ground based K index serves as a good proxy to space measures of Lyα and the EUV. The Lyα flux computed using a three component model of K with the photometric contrasts at Lyα agrees well with the measured flux in the years 1969-1980 except near the solar maximum. High resolution observations may bring better agreement as, at 1"$\times$1" resolution 33% of the disc would have enhanced Lyα emission at solar maximum (Lean and Skumanich, 1983).

3. *Radius variation*

A change in the solar output would require a change in the solar radius. A day to day variation of 0.4" is detected by the dedicated solar diameter telescope in Boulder (Brown 1982). LaBonte and Howard (1981) from analysis of 5 years magnetograph data report that R is constant to within $\pm$0.1". The main problem is to be able to relate any apparent diameter change to change in luminosity with certainty. To what accuracy must $\Delta R/R$ be measured to produce a change of $\Delta L/L \sim 0.1\%$? Unless this is settled, the results of the radius change measurements cannot be linked to luminosity changes.

4. *Rotation*

A plot of the daily mean plage areas would reveal the 27 day modulation (Bappu and Sivaraman, 1971). Temporal power spectra of the daily plage index for the period 1976-1982 as a whole as well as of the broken data of one year each show only two peaks at 27 day and 13 day periods, (Keil and Worden, 1984). Analysis of the Mt. Wilson disc integrated magnetic flux also shows the 27 day and its harmonics with an accuracy of $\sim 2\%$ (LaBonte, 1982). But none of the attempts have revealed the differential rotation of the Sun from full disc data.

5. *Solar mean magnetic field*

The only measurement of the magnetic field of the sun integrated over the entire disc is the one using the Fe I 5250A line for the years 1968-1981 due to Severny and his co-workers. Year to year values of the yearly means of the mean magnetic field show changes as much as by factors 3 to 4 (Kotov and Severny, 1983).

References

Bappu, M.K.V.; and Sivaraman, K.R.; 1971, *Solar Phys.* **17**, 316.
Bappu, M.K.V.; and Sivaraman, K.R.; 1977, *Mon. Not. R. Astr. Soc.* **78**, 279.
Brown, T.M.; Elmore, D.F.; Lacey, L.; and Hull, H.; 1982, *Appl. Opt.* **21**, 3588.
Bumba, V.; and Topolova, B.R.; 1967, *Solar Phys.* **1**, 216.
Cook, J.W.; Brueckner, G.E.; and Van Hoosier, M.E.; 1980, *Astron. Astrophys.* **92**, L7.
Foukal, P.; Fowler, L.; and Livshits, M.; 1983, *Astrophys. J.* **267**, 863.
Foukal, P.; and Fowler, L.; 1984, *Astrophys. J.* **281**, 442.
Gilman, P.; 1978, *Geophys. Astrophys. Fluid Dyn.* **11**, 157.
Harvey, J.W.; 1983, in *Workshop on 'Solar Variations on Active Region Time Scales'* eds. Chapman, Hudson and LaBonte, Pasadena.
Holweger, H.; Livingston, W.; and Steenbock, W.; 1983, *Nature,* **302**, 125.
Hoyt, D.V.; and Eddy, J.A.; 1983, *Geophys. Res. Lett.* **10**, 509.
Keil, S.L.; and Worden, S.P.; 1984, *Astrophys. J.* **276**, 766.
Kotov, V.A.; and Severny, A.B.; 1983, *Materials of the World Data Centre B,* Moscow.
LaBonte, B.J.; and Howard, R.; 1981, *Science,* **214**, 907.
LaBonte, B.J.; 1982, *Astrophys. J.* **260**, 647.
Lean, J.L.; and Skumanich, A.; 1983, *Geophys. Res.* **88**, 5751.
Livingston, W.; 1982, *Nature,* **297**, 208.
Livingston, W.; and Holweger,H.; 1982, *Astrophys. J.* **252**, 375.
Macris, C.J.; and Rösch, J.; 1983, *C.R.* **296**, 265.
Müller, E.A.; and Kneubühl, F.K.; Rast, J.; and Stettler, P.; 1980, *Astron. Astrophys.* **87**, L3.
Newkirk, Jr. G.; 1983, *Ann Rev. Astron. Astrophys.* **21**, 429.
Sivaraman, K.R.; Jagdev Singh; Bagare, S.P.; and Gupta, S.S.; 1984, *Astrophys. J.* **282**, 776.
Spruit, H.; 1977, *Solar Phys.* **55**, 3.
White, O.R.; and Livingston, W.; 1978, *Astrophys. J.* **226**, 679.

COMMISSION 14: ATOMIC AND MOLECULAR DATA (DONNÉES ATOMIQUES ET MOLÉCULAIRES)

PRESIDENT : A. H. Gabriel
VICE PRESIDENT : R. W. Nicholls
ORGANIZING COMMITTEE: D. R. Johnson, S. L. Mandel'shtam, H. Nussbaumer, J. G. Phillips, S. Sahal, K. Takayanagi, W. Wiese.

Introduction

As the pool of fundamental data available to astronomers continues to increase, the question of how best to promote the necessary cross-discipline interaction becomes increasingly important. Commission 14 has traditionally played an important role in this activity, by publishing triennial reports in the IAU Proceedings, as well as by responding to more specific requests for data. We are fortunate in having the support for these activities of some energetic Working-Groups and Chairmen, whos contributions to the present report are very gratefully acknowledged. With the expansion of available data it is appropriate that these reports take on more and more the form of references to review articles and other more specific data bases. The question of whether the field of activity of the Commission should be enlarged was discussed at Patras and will be reviewed again at the Delhi meeting. One possibility is to include nuclear processes and fundamental particle physics. On the other hand a rationale for limiting the scope of our activities might be the direct application to astronomical observations. Astronomical theorists are usually better placed to access the fundamental data themselves. The interaction between fundamental physics and astronomy will in general take two forms. There is the essential service role of making data available in a usable form. However, we should surely aim to stimulate the other very profitable mode, in which the two disciplines are brought together to form real scientific collaborations, in order to research the problems of astronomy.

WORKING GROUP 1: WAVELENGTH STANDARDS

This report gives some highlights of recent work in the field of wavelength standards; it is by no means to be considered a complete review.

In October, 1983, the Seventeenth General Conference of Weights and Measures ratified a new definition of the International Metre as follows: "The Metre is the length of the path travelled by light in vacuum during a time interval of 1/299792458 of a second". This action at once fixed the speed of light by convention at exactly 299792458 m/s and increased the accuracy of the definition of the Metre to about one part in 10^{13}, as given by the second, defined in terms of the Cs standard radiation. Any future improvement resulting from a redefinition of the second will automatically be reflected in the definition of the Metre.

It follows from the new definition that the wavelengths of electromagnetic radiation are given by 299792458m (exactly) divided by their frequency. The latter can be obtained for certain suitable radiations, by a process of direct frequency comparison techniques, in terms of the Cs standard. Working wavelength standards can be measured by reference to such radiations either by frequency or wavelength comparison methods. The latter are at present advantageous for wavelengths shorter than those in the near infrared, but because of diffraction effects, are progressively less accurate at longer wavelengths, becoming quite impractical in the microwave region of the Cs standard, where comparison by frequency techniques is far more accurate and simpler.

6. Reader, J. and Acquista, N.: 1984, unpublished measurements.
7. Persson, W. and Reader, J.: 1985, J. Opt. Soc. Am. (to be submitted).
8. Epstein, G.L. and Reader, J.: 1982, J. Opt. Soc. Am. 72 pp. 476.

K.M. Baird
Chairman of the Committee

WORKING GROUP 2: ATOMIC TRANSITION PROBABILITIES

The Data Center on Atomic Transition Probabilities at the National Bureau of Standards, Gaithersburg, Md., is continuing its critical compilation work and the updating of the bibliographical data base on atomic transition probabilities. Work is in progress on the revision of the existing NBS critical data compilations (Refs. A-D) for all allowed and forbidden transitions in Fe-group elements. A single volume containing all these data for the Fe-group elements Sc to Ni (Vol. III of the NBS series of atomic transition probabilities) is planned for publication in the near future. However, this compilation work is proceeding slower than expected due to numerous recent additions and revisions in the data base.

In Table 1 the most recent literature references covering the period since the last working group report (August 1981) up to the present time (summer 1984) are presented, ordered according to element and stage of ionization. For brevity, the references are identified there only by the running number of the general reference list, given at the end of this section.

In the general reference list (Refs. 1-427), the recent literature is ordered alphabetically according to authors. Each reference contains some code letter(s), indicating the method(s) applied by the author. Specifically, the code letters are defined as follows:

THEORETICAL METHODS:

Q quantum mechanical (including self-consistent field) calculations.
I - interpolation within isoelectronic sequences, spectral series, or homologous atoms; also, data that are presented in graphical, rather than tabular form.

EXPERIMENTAL METHODS:

E - measurements in emission (arc, furnace, discharge tube, shock tube, etc.).
A - measurements in absorption (King furnace, absorption tube, etc.).
L - lifetime measurements (including Hanle-effect).
H - anomalous dispersion (hook) measurements.
M - miscellaneous experimental methods (for example, Stark effect, astrophysical measurements, etc.).

OTHER:

C - additions or suggested revisions to data in previous articles, comments on particular theoretical or experimental methods, etc.
Cp- data compilations.
R - relative (non-absolute) oscillator strengths have been tabulated.
F - data on forbidden (i.e., other than electric dipole) transitions have been tabulated.

Accurate frequencies and wavelengths, referred to the Cs standard, have been obtained for a number of highly reproducible standards in the visible and infrared by the use of direct frequency comparisons. These values have been approved and published by the International Committee of Weights and Measures(1), and are considered accurate to about one part in 10^{10}.

There are a number of active programs, principally at National Standards Laboratories, that have the aim of transferring the one in 10^{13} accuracy of the Cs standard to the near infrared and visible parts of the spectrum. These involve both the development of improved stabilized lasers (using, for example, cooled trapped ion absorbers) and the development of systems for better direct frequency comparisons of microwave and optical radiations. A notable example of the latter is the recent successful operation of CO_2, 10μm lasers phase locked to the Cs primary standard(2). Another example is the recent measurement of the frequencies of a grid of OsO_4 absorption lines at 10μm(3). Examples of source development are the programs at the Joint Institute of Laboratory Astrophysics and at the National Bureau of Standards in Boulder, U.S.A. These include work on laser-cooled trapped ions for absorption line reference frequencies, and the use of modulation side techniques to avoid low frequency noise in order to obtain extremely high resolution of narrow line profiles and servo control of lasers with respect to their centres(4,5).

Wavelength standards in the VUV and XUV regions are measured by length based comparison with visible standards. An example of recent important work is given in the following excerpt from a letter from W.C. Martin of the U.S. National Bureau of Standards:

> "1. We are making high-accuracy measurements of the spectrum of a platinum hollow-cathode discharge for the region 110-330 nm(6). We expect to determine about 3000 wavelengths for Pt I, II and Ne I, II with uncertainties of ± 0.0002 nm or less for many lines. The wavelengths will be used for on-board calibration of High-Resolution Spectrograph for the Hubble Space Telescope. Our measurements will also allow use of the Pt hollow cathode as a source of calibration wavelengths for other spectroscopy in the above region.
>
> 2. Our recently completed determinations of an expensive system of energy levels for Y VI have given Ritz-type wavelengths standards in the XUV region 20-40 nm(7). Together with wavelengths to be similarly derived from our Y IV analysis(8) and work on Y V in progress, these wavelengths will comprise of a system of yttrium standards covering the range 18-70 nm with uncertainties of ± 0.001 to ± 0.0003 nm."

References

1. Documents Concerning the New Definition of the Metre, Metrologia: 1984, 19 p. 163-177.
2. Whitford, B.G.: 1984, Appl. Phys. B. 35 p. 119-122.
3. B. Dahmani and A. Clairon.: 1983, IEEE Trans. Instr. Meas. Vol. IM 32 no. 1.
4. Proceedings of the Workshop on Spectroscopic Applications of Slow Atomic Beams, held at NBS, Gaitherburg, MD, 14-15 April 1983.: 1983, Ed. by W.D. Phillips, Natl. bur. Stand. (U.S.) Spec. Publ. 653.
5. J. Hough, D. Hils, M.D. Rayman, MA L.-S., L. Hollberg, and J.L. Hall.: Joint Institute for Laboratory Astrophysics, University of Colorado and National Bureau of Standards, Boulder, CO 80309, U.S.A.

Some research groups have communicated their work in progress. At the Center for Astrophysics, Cambridge, Massachusetts, lifetime measurements are underway formetastable levels of C II, C III, Si V, and Al II, utilizing laser-plasma excitation and ion traps to store the metastable ions. Also, transition probabilities for Si II and Co II are being measured with the hook technique. At the University of Lyon, lifetime measurements are underway for He-like, Li-like and Be-like ions of Cr, Fe and Cu, and for Li-like Al. At NBS, Gaithersburg, Maryland, transition probabilities are calculated for the $2s^2\ {}^1S_0 - 2s3p\ {}^{1,3}P_1^o$ transitions of Be-like ions using both non-relativistic and relativistic correlated wavefunctions.

References for Introductory Discussion

A. Smith, M.W., Wiese, W.L.: 1973, J. Phys. Chem. Ref. Data 2, p. 85.

B. Wiese, W.L., Fuhr, J.R.: 1975, J. Phys. Chem. Ref. Data 4, p. 263.

C. Younger S.M., Fuhr, J.R., Martin, G.A., Wiese, W.L.: 1978, J. Phys. Chem. Ref. Data 7, p. 495.

D. Fuhr, J.R., Martin, G.A., Wiese, W.L., Younger, S.M.: 1981, J. Phys. Chem. Ref. Data 10, p. 305.

TABLE 1

Recent literature sources for atomic transition probability data of astrophysical interest

This table covers the period since the publication of our last IAU report (Transact. IAU XVIII A, 116, (1982)) to the present (August 1984). The table is arranged in alphabetical order of elements symbols with further sub-divisions according to stage of ionization (I,II, etc.). Theoretical papers containing data for some transitions along an isoelectronic sequence are separately identified. The numbers are the running numbers of the reference list following this table.

Ag I: 38,195,300,328,373,378,379
Ag II: 328

Al sequence: 215,383
Al I: 124,132,185,226,267,274,307 400,401
Al II: 96
Al III: 92,291,373
Al IV: 79,185
Al VII: 5

Ar I: 62,63,64,65,140,218,236 303,416
Ar II: 58,85,99,100,304,376,416
Ar III: 146
Ar V: 145
Ar VI: 71,124,127,147
Ar VII: 71,126
Ar VIII: 71
Ar IX: 72
Ar XV: 156
Ar XVI: 118,264
Ar XVII: 118
Ar XVIII: 178
Ar XXIII: 156

B sequence: 134,290,313,347
B I: 267,290,341
B II: 266,322
B III: 70,88,264,266,322,324
B IV: 268,341

Ba I: 17,18,21,42,143,241,318
Ba II: 318,373
Ba III: 84,95

Be sequence: 11,12,59,121,128,161 164,166,169,224,313, 361,364
Be$^-$: 20,25
Be I: 20,234
Be II: 33,34,35,89,135,233,306,324 335,369
Be III: 341

C sequence: 5,59,129,144,313
C I: 45,211,289,374,395
C II: 96,173,290
C III: 36,37,96,157,160,239
C IV: 81,118,256,335
C V: 46,118

Ca I: 28,141,200,217,253,287,318, 321,377,398,406,407,421,
Ca II: 280,318,368,373
Ca V: 146
Ca VIII: 127
Ca IX: 126
Ca XI: 329
Ca XVII: 43,112,121,156,353
Ca XVIII: 32,168,269,353,426
Ca XIX: 187,268,382

Cl sequence: 214,383,411
Cl I: 367
Cl II: 146
Cl III: 23
Cl IV: 23,145
Cl V: 23,124,127,147
Cl VI: 23,126
Cl VII: 23,227,291
Cl VIII: 23
Cl XVI: 46

Co sequence: 419
Co I: 77,97,186,190
Co II: 198
Co IX: 391,394

Cr I: 57,212,248,401,427
Cr II: 251
Cr IV: 338
Cr VI: 390,393
Cr XII: 127
Cr XIII: 126
Cr XIX 348
Cr XX: 75
Cr XXI: 75,389
Cr XXII: 389
Cr XXIII: 46,389

References

1. Aashamar, J., Luke, T.M., Talman, J.D.: 1983, Phys. Scr. 27, p. 267. Q
2. Afanaseva, N.V.: 1982, Opt. Spectrosc. (USSR) 52, p. 465. Q
3. Agentoft, M., Andersen, T., Froese Fischer, C., Smentek-Mielczarek, L.: 1983, Phys. Scr. 28, p. 45. Q
4. Aggarwal, K.M., Baluja, K.L.: 1983, J. Phys. B 16, p. 107. Q
5. Aggarwal, K.M.: 1984, Sol. Phys. 90, p. 281. I,Q
6. Aleksandrov, Yu.M., Kozlov, M.G., Makhov, V.N., Fedorchuk, R.V., Yakimenko, M.N.: 1982, Opt. Spectrosc. (USSR) 52, p. 129. A
7. Aleksandrov, Yu.M., Gruzdev, P.F., Kozlov, M.G., Loginov, A.V., Makhov, V.N., Fedorchuk, R.V., Yakimenko, M.N.: 1983, Opt. Spectrosc. (USSR) 54, p. 4. A,Q
8. Alexa, B., Baig, M.A., Connerade, J.P., Garton, W.R.S., Hormes, J., Stavarakas, T.A.: 1983, Nucl. Instrum. Methods 208, p. 841. A
9. Altas, L., Atac, T.: 1982, Kandilli Observatory, Heliophysics Service, Istanbul, Turkey, p. 1-9. Cp
10. Anderson, E.K., Anderson, E.M.: 1981, Opt. Spectrosc. (USSR) 51, p. 221. Q,QF
11. Anderson, E.K., Anderson, E.M.: 1982, Opt. Spectrosc. (USSR) 52, p. 478. Q,QF
12. Anderson, E.M., Anderson, E.K.: 1981, Bull. Acad. Sci. USSR, Phys. Ser. 45, No. 12, p. 16. Q,QF
13. Armour, I.A., Silver, J.D., Träbert, E.: 1981, Phys. B 14, p. 3563. L
14. Arqueros, F., LaRocque, P.E., O'Sullivan, M.S., Stoicheff, B.P.: 1984, Opt. Lett. 9, p. 82. A,FR
15. Artru, M.C., Jamer, C., Petrini, D., Praderie, F.: 1981, Astron. Astrophys., Suppl. Ser. 44, p. 171. Q
16. Aspromallis, G, Nicolaides, C.A., Beck, D.R.: 1983, Phys. Rev. A. 28, p. 1879. Q
17. Aymar, M., Champeau, R.-J., Delsart, C., Keller, J.-C: 1981, J. Phys. B. 14, p.4489. L
18. Aymar, M., Grafström, P., Levinson, C., Lundberg, H., Svanberg, S.: 1982, J. Phys. B 15, p. 877. L
19. Aynacioglu, A.S., Oppen, G.V., Perschmann, W.-D., Szostak, D.: 1981, Z. Phys. A 303, p. 97. L
20. Baba, C.V.K., Betigeri, M.G., Datar, V.M., Kurup, M.B., Sharma, R.P., Singh, P.: 1982, Pramana, 18, p. 295.
21. Bachor, H.-A., Kock, M.: 1981, J. Phys. B. 14, p. 2793. H
22. Bassani, F., Vignale, G.: 1982, Nuovo Cimento Soc. Ital. Fis., D 1D, p. 519. Q,QF 23.Bandinet-Robinet, Y., Dumont, P.D., Garnir, H.P.: 1982, Nucl. Instrum. Methods 202, p. 33. L
24. Baumann, M., Liening, H., Loos, H.: 1981, Z. Naturforsch, Teil A 36, p. 778. L
25. Beck, D.R., Nicolaides, C.A., Aspromallis, G.: 1981, Phys. Rev. A 24, p. 3252. Q
26. Beck, D.R.: 1982, Int. J. Quantum Chem., Symp. No. 16, p. 345.
27. Beck, D.R., Nicolaides, C.A.: 1983, Phys. Rev. A 28, p. 3112. Q
28. Beck, D.R., Nicolaides, C.A.: 1983, J. Phys. B 16, p. L627. QF
29. Bely-Dubau, F., Dubau, J., Faucher, P., Steenman-Clark, L.: 1981, J. Phys. B 14, p. 3313.
30. Bely-Dubau, F., Faucher, P., Steenman-Clark, L., Bitter, M., von Goeler, S., Hill, K.W., Camhy-Val, C., Dubau, J.: 1982, Phys. Rev. A 26, p. 3459. Q
31. Bely-Dubau, F., Dubau, J., Faucher, P., Gabriel, A.H.: 1982, Mon. Not. R. Astron. Soc. 198, p. 239. Q
32. Bely-Dubau, F., Dubau, J., Faucher, P., Gabriel, A.H., Loulergue, M., Steenman-Clark, L., Volonte, S., Antonucci, E., Rapley, C.G.: 1982, Mon. Not. R. Astron. Soc. 201, p. 1155. Q

33. Bentzen, S.M., Andersen, T., Poulsen, O.: 1981, J. Phys. B 14, p. 3435. L
34. Bentzen, S.M., Anderson, T., Poulsen, O.: 1982, J. Phys. B 15, p. L71. L
35. Bentzen, S.M., Anderson, T., Poulsen, O.: 1982, Phys. Rev. A 26, p. 2639. L
36. Berrington, K.A., Burke, P.G., Dufton, P.L., Kingston, A.E.: 1981, At. Data Nucl. Data Tables 26, p. 1. Q
37. Berry, H.G., Brooks, R.L., Cheng, K.T., Harids, J.E., Ray, W.: 1982, Phys. Scr. 25, p. 391. L
38. Bezuglov, N.N., Gorshkov, V.N., Osherovich, A.L., Phekhotkina, G.L.: 1982, Opt. Spectrosc. (USSR) 53, p. 239. L
39. Bhalla, C.P., Tunnell, T.W.: 1981, Z. Phys. A 303, p. 199. QF
40. Bhatia, A.K., Mason, H.E.: 1981, Astron. Astrophys. 103, p. 324. QF
41. Bhatia, A.K.: 1982, J. Appl. Phys. 53, p. 59. Q,QF
41. Bhatia, A.K.: 1982, J. Appl. Phys. 53, p. 59. Q,QF
42. Bhatia, K.: 1981, Z. Phys. A 303, p. 1. L
43. Bhatia, A.K., Mason, H.E.: 1983, Astron. Astrophys., Suppl. Ser. 52, p. 115. Q,QF
44. Biemont, E., Grevesse, N., Hannaford, P., Lowe, R.M.: 1981, Astrophys. J. 248, p. 867. L,E
45. Biemont, E.: 1981, Bull. Cl. Sci., Acad. R. Belg. 67, p. 80. Q
46. Biemont, E.: 1982, Nucl. Instrum. Methods 202, p. 283. Q
47. Biemont, E., Grevesse, N., Kwiatkowski, M., Zimmermann, P.: 1982, Astron. Astrophys. 108, p. 127. L
48. Biemont, E., Bromage, G.E.: 1983, Mon. Not. R. Astron. Soc. 205, p. 1085. Q,F
49. Biemont, E., Grevesse, N., Kwiatkovski, M., Zimmermann, P.: 1984, Astron. Astrophys. 131, p. 364. L
50. Bitter, M., von Goeler, S., Cohen, S., Hill, K.W., Sesnic, S., Tenney, F., Timberlake, J., Safronova, U.I., Vainshtein, L.A., Dubau, J., Loulergue, M., Bely-Dubau, F., and Steenman-Clark, L.: 1984, Phys. Rev. A 29, p. 661. Q
51. Blackwell, D.E., Petford, A.D. Shallis, M.J., Simmons, G.J.: 1982, Mon. Not. R. Astron. Soc. 199, p. 43. A
52. Blackwell, D.E., Petford, A.D., Shallis, M.J., Leggett, S.: 1982, Mon. Not. R. Astron. Soc. 199, p. 21. A
53. Blackwell, D.E., Petford, A.D., Simmons, G.J.: 1982, Mon. Not. R. Astron. Soc. 201, p. 595. A
54. Blackwell, D.E., Menon, S.L.R., Petford, A.D., Shallis, M.J.: 1982, Mon. Not. R. Astron. Soc. 201, p. 611. A 55.Blackwell, D.E., Menon, S.L.R., Petford, A.D.: 1982, Mon. Not. R. Astron. Soc. 201, p. 603. A
56. Blackwell, D.E., Menon, S.L.R., Petford, A.D.: 1983, Mon. Not. R. Astron. Soc. 204, p. 883. A
57. Blackwell, D.E., Menon, S.L.R., Petford, A.D.: 1984, Mon. Not. R. Astron. Soc. 207, p. 533. A
58. Blagoev, K.B.: 1983, J. Phys. B 16, p. 33. L
59. Bodashko, P.G., Zapryagaev, S.A., Safronova, U.I., Senashenko, V.S.: 1981, Opt. Spectros. (USSR) 51, p. 9. Q
60. Bogdanova, I.P., Kazantsev, S.A., Chaika, M.P.: 1983, Opt. Spectrosc. (USSR) 54, p. 362. L
61. Booth, A.J., Blackwell, D.E., Petford, A.D., Shallis, M.J.: 1984, Mon. Not. R. Astron. Soc. 208, p.147. A
62. Borge, M.J.G., Campos, J.: 1982, Phys. Rev. A 25, p. 271. L
63. Borge, M.J.G., Campos, J.: 1983, Phys. Rev. A 27, p. 1910. L,E,Q
64. Borge, M.J.G., Campos, J.: 1983, Physica C 119, p. 359. L,E,ER
65. Borge, M.J.G., Campos, J.: 1983, J. Quant. Spectrosc. Radiat. Transfer 30, p. 433. E,ER
66. Brandenberger, J.R.: 1984, Phys. Rev. A 29, p. 1208. L
67. Brandus, L.: 1983, Rev. Roum. Phys. 28, p. 595. Q
68. Brault, J.W., Litzen, U.: 1983, Phys. Scr. 28, p. 475. QR

69. Brooks, R.L., Hardis, J.E., Berry, H.G., Curtis, L.J., Cheng, K.T., Ray, W.: 1980, Phys. Rev. Lett. 45, p. 1318. L
70. Bruch, R., Chung, K.T., Träbert, E., Heckmann, P.H., Raith, B., Müller, H.R.: 1984, J. Phys. B. 17, p. 333. Q
71. Buchet-Poulizac, M.-C, Buchet, J.-P, Ceyzeriat, P.: 1982, Nucl. Instrum. Methods 202, p. 13. L
72. Buchet-Poulizac, M.-C., Buchet, J.-P.: 1983, Phys. Scr. 27, p. 99. L
73. Bunge, C.F.: 1980, Phys. Rev. Lett. 44, p. 1450. Q
74. Bunge, C.F.: 1980, Phys. Rev. A 22, p. 1. Q
75. Burkhalter, P.G., Charatis, G., Rocket, P.D., Newman, D.: 1984, J. Opt. Soc. Am. B 1, p. 155. Q
76. Can, C., Bhalla, C.P.: 1983, IEEE Trans. Nucl. Sci. 30, p. 1090. Q
77. Cardon, B.L., Smith, P.L., Scalo, J.M., Testerman, L., Whaling, W.: 1982, Astrophys. J 260, p. 395. E,H
78. Ceyzeriat, P.: 1982, J Opt. Soc. Am. 72, p. 110. Q
79. Ceyzeriat, P., Aymar, M.: 1982, J. Opt. Soc. Am. 72, p. 116. Q
80. Chaleard, C., Dubreuil, B., Catherinot, A.: Phys. Rev. A 26, p. 1431. L
81. Chem, M.H., Crasemann, B., Mark, H.: 1983, Phys. Rev. A 27, p. 544. Q,QF
82. Chen, M.H., Crasemann, B., Mark, H.: 1982, Phys. Rev. A 26, p. 1441. Q,QF
83. Chen, M.H., Crasemann, B., Mark, H.: 1981, Phys. Rev. A 24, p. 1852. Q,QF
84. Cheng, K.T., Froese-Fischer, C.: 1983, Phys. Rev. A 28, p. 2811. Q
85. Christiansen, P.G., Dalton, B.J.: 1982, J. Phys. B 15, p. 633. Es
86. Chung, K.T.: 1981, Phys. Rev. A 24, p. 1350. Q
87. Chung, K.T.: 1982, Phys. Rev. A 25, p. 1596. Q
88. Chung, K.T., Bruch, R., Träbert, E., Heckmann, P.H.: 1984, Phys. Scr. 29, p. 108. Q
89. Chung, K.T., Davis, B.F.: 1984, Phys. Rev. A 29, p. 1871. Q
90. Chung, S. Lin, C.C., Fischer, C.F., Lee, E.T.P.: 1982, J. Chem. Phys. 76, p. 498. QF
91. Coetzer, F.J., Kotze, P.B., Van der Westhuizen, P.: 1982, Nucl. Instrum. Methods 202, p. 19. L
92. Coetzer, F.J., Mostert, F.J., Van der Westhuizen, P.: 1983, Spectrochim. Acta, Part B 38, p. 885.
93. Cohen, M.: 1981, Chem. Phys. Lett. 81, p. 513. C
94. Collins, C.B., Curry, S.M., Johnson, B.W., Mirza, M.Y., Chellehmalzadh, M.A., Anderson, J.A., Popescu, D., Popescu, I.: 1976, Phys. Rev. A 14, p. 1662. AFR
95. Connerade, J.P., Mansfield, M.W.D.: 1982, Phys. Rev. Lett. 48, p. 131. Q
96. Cowan, R.D., Hobbs, L.M., York, D.G.: 1982, Astrophys. J 257, p. 373; 1983, Astrophys. J 265, p. 582. Q
97. Cowley, C.R., Corliss, C.H.: 1983, Mon. Not. R. Astron. Soc. 203, p. 651. C
98. Cowley, C.R.: 1983, Mon. Not. R. Astron. Soc. 202, p. 417. C
99. Danzmann, K., Kock, M.: 1982, J. Opt. Soc. Am. 72, p. 1556. ER
100. Danzmann, K., Kock, M.: 1983, J. Quant. Spectrosc. Radiat. Transfer 29, p. 517. E
101. Davis, B.F., Chung, K.T.: 1982, Phys. Rev. A 25, p. 1328. Q
102. Davis, B.F., Chung, K.T.: 1982, Phys. Rev. A 26, p. 2743. Q
103. Dehmelt, G., Georgiadis, A., Gerdtell, L., Sträter, D., Brentano, P., Ahmed-Bitar, R.N.: 1982, Phys. Lett. A 89, p. 193. L
104. Deschepper, Ph., Lebrun, P., Palffy, L., Pellegrin, P.: 1981, Phys. Rev. A 24, p. 1633. L,F
105. Deschepper, P., Lebrun, P., Palffy, L., Pellegrin, P.: 1982, Phys. Rev. A 26, p. 1271. L
106. Dohmann, H.D., Mann, R., Pfeng, E.: 1982, Z. Phys. A 309, p. 101. L
107. Doschek, G.A., Feldman, U., Cowan, R.D.: 1981, Astrophys. J 245, p. 315. Q
108. Dubau, J., Gabriel, A.H., Loulergue, M., Steenman-Clark, L., Volonté, S.: R. Astron. Soc. 195, p. 705. Q

109. Dubreuil, B., Chaleard, C.: 1984, Phys. Rev. A 29, p. 958. L
110. Dufton, P.L., Hibbert, A., Kingston, A.E., Doschek, G.A.: 1982, Astrophys. J 257, p. 338. Q
111. Dufton, P.L., Hibbert, A., Kingston, A.E., Tully, J.A.: 1983, Mon. Not. R. Astron. Soc. 202, p. 145. Q
112. Dufton, P.L., Kingston, A.E., Scott, N.S.: 1983, J. Phys. B 16, p. 3053. Q
113. Dufton, P.L., Hibbert, A., Kingston, A.E., Doschek, G.A.: 1983, Astrophys. J 274, p. 420. Q
114. Duquette, D.W., Salin, S., Lawler, J.E.: 1981, Radiative Lifetimes in W I Using a Novel Atomic-Beam Source, Phys. Rev. A 24, p. 2847. L
115. Duquette, D.W., Lawler, J.E.: 1982, Phys. Rev. A 26 p. 330. L
116. Duquette, D.W., Salih, S., Lawler, J.E.: 1982, Phys. Rev. A 25, p. 3382. L
117. Duquette, D.W., Salih, S., Lawler, J.E.: 1982, Phys. Rev. A 26, p. 2623. L
118. Duston, D., Rogerson, J.E., Davis, J., Blaha, M.: 1983, Phys. Rev. A 28, p. 2968. Q
119. Ekberg, J.O.: 1981, Phys. Scr. 23, p. 7. QF
120. Ekberg, J.O., Engstrom, L., Bashkin, S., Denne, B., Huldt, S., Johannsson, S., Jupen, C., Lintzen, U., Trigueiros, A., Martinson, I.: 1983, Phys. Scr. 27, p. 425. L
121. Ellis, D.G.: 1983, Phys. Rev. A 28, p. 1223. QI
122. Engstrom, L., Denne, B., Ekberg, J.O., Jones, K-W., Jupén, C., Litzén, U., Meng, W.T., Trigueiros, A., Martinson, I: 1981, Phys. Scr. 24, p. 551. L
123. Erdman, P.W., Zipf, E.C.: 1983, J. Geophys. Res., Sect. A 88, p. 7245. ER
124. Farrag, A., Luc-Koenig, E., Sinzelle, J.: 1981, J. Phys. B 14, p. 3325. Q
125. Farrag, A., Luc-Koenig, E., Sinzelle, J.: 1982, At. Data Nucl. Data Tables 27, p. 539. Q
126. Fawcett, B.C.: 1983, At. Data Nucl. Data Tables 28, p. 579. Q
127. Fawcett, B.C.: 1983, At. Data Nucl. Data Tables 28, p. 557. Q
128. Fawcett, B.C.: 1984, At. Data Nucl. Data Tables 30, p. 1. Q
129. Feldman, U., Doschek, G.A., Bhatia, A.K.: J. Appl. Phys. 53, p. 8554. IFR
130. Feldman, U., Doschek, G.A., Bhatia, A.K.: 1981, Astrophys. J. 250, p. 799. Q,QF
131. Feldman, U., Doschek, G.A., Cowan, R.D.: 1981, Mon. Not. R. Astron. Soc. 196, p. 517. Q
132. Fischer, C.F.: 1981, Phys. Scr. 23, p. 38. Q
133. Fischer, C.F.: Godefroid, M.: 1982, Nucl. Instrum. Methods 202, p. 307. Q
134. Fischer, C.F.: 1983, J. Phys. B 16, p. 157. QF
135. Fischer, C.F.: 1982, Phys. Rev. A 26, p. 2627. Q
136. Froese-Fischer, C., Saha, H.P.: 1983, Phys. Rev. A 28, p. 3169. QF
137. Froese-Fischer, C., Saha, H.P.: 1984, J. Phys. B 17, p. 943. Q
138. Fujimoto, T., Goto, C., Fukuda, K.: 1981, Opt. Commun. 40, p. 23. L
139. Fujimoto, T., Goto, C., Fukuda, K.: 1982, Phys. Scr. 26, p. 443. L
140. Fujimoto, T., Goto, C., Uetani, Y., Fukuda, K.: 1983, Phys. Scr. 28,p. 617. L
141. Fukuda, K., Ueda, K.: 1982, J. Phys. Chem. 86, p. 676. AF
142. Gaillard, M.L., Pegg, D.J., Bingham, C.R., Carter, H.K., Mlekodaj, R.L., Cole, J.D.: 1982, Phys. Rev. A 26, p. 1975. L
143. Gallagher, T.F., Gounand, R., Kachru, R., Tran. N.H., Pillet, P.: 1983, Phys. Rev. A 27, p. 2485. L
144. Ganas, P.S.: 1981, Astron. Astrophys., Suppl. Ser. 46, p. 101. Q
145. Ganas, P.S.: 1981, Physica C (Amsterdam) 111, p. 365. Q
146. Ganas, P.S.: 1982, Phys. Lett A 87, p. 394. Q
147. Ganas, P.S.: 1982, Int. J. Theor. Phys. 21, p. 363. Q
148. Ganas, P.S.: 1981, J. Appl. Phys. 52, p. 6482. Q
149. Ganas, P.S.: 1983, Physica C (Amsterdam) 119, p. 337. Q
150. Garstang, R.H.: 1981, Publ. Astron. Soc. Pac 93, p. 641. Q
151. Garton, W.R.S., Connerade, J.P., Baig, M.A., Hormes, J., Alexa, B.: 1983, J. Phys. 16, p. 389. M

152. Gauppe, A., Kuske, P., Andrä, H.J.: 1982, Phys. Rev. A 26, p. 3351. L
153. Gil, T., Heldt, J.: 1983, Z. Phys. A 312, p. 343. QF
154. Glass, R.: 1981, Physica C (Amsterdam) 104, p. 434. Q
155. Glass, R.: 1981, Aust. J. Phys. 34, p. 147. Q
156. Glass, R.: 1981, Z. Phys. A 302, p. 203. Q
157. Glass, R.: 1981, Z. Phys. A 303, p. 27. Q
158. Glass, R.: 1982, Mon. Not. R. Astron. Soc. 199, p. 435. Q
159. Glass, R.: 1982, Z. Phys. A 306, p. 25. Q
160. Glass, R.: 1982, Sol. Phys. 78, p. 29. Q
161. Glass, R.: 1982, Astrophys. Space Sci. 87, p. 41. QF
162. Glass, R.: 1982, Aust. J. Phys. 35, p. 693. Q
163. Glass, R.: 1982, Sol. Phys. 80, p. 321. Q
164. Glass, R.: 1983, Astrophys. Space Sci. 91, p. 417. QF
165. Glass, R.: 1983, Mon. Not. R. Astron. Soc. 204, p. 735. Q
166. Glass, R.: 1983, Astrophys. Space Sci. 92, 307 (1983).
167. Godefroid, M., Froese-Fischer, C.: 1984, J. Phys. B 17, p. 681. QF
168. Goett, S.J., Sampson, D.H., Clark, R.E.H.: 1984, Astrophys. J., Suppl. Ser. 54, p. 115. Q
169. Goett, S.J., Clark, R.E.H., Sampson, D.H.: 1980, At. Data Nucl. Data Tables 25, p. 185. Q
170. Goett, S.J., Sampson, D.H., Clark, R.E.H.: 1983, At, Data Nucl. Data Tables 28, p. 279. Q
171. Goett, S.J., Sampson, D.H.: 1983, At. Data Nucl. Data Tables 29, p. 535. Q
172. Goldsmith, S., Boxman, R.L., 1981, J. Phys. B 14, p. 3031. Q
173. Goly, A., Weniger, S.: 1982, J. Quant. Spectrosc. Radiat. Transfer 28, p. 389. E
174. Gordon, H., Hobby, M.G., Peacock, N.J.: 1980, J. Phys. B 13, p. 1985 and supplement from the British Library Lending Division, ref. SUP 70028). Q
175. Berry, H.G., Brooks, R.L., Hardis, J.E. and Ray, W.J.: 1982, Nucl. Instrum. Methods 202, p. 73.
176. Gough, D.S., Hannaford, P., Lowe, R.M.: 1982, J. Phys. B 15, p. L431. L
177. Gough, D.S., Hannaford, P., Lowe, R.M.: 1983, J. Phys. B 16, p. 785. LE
178. Gould, H., Marrus, R.: 1983, Phys. Rev. A 28, p. 2001. LF
179. Graftstrom, P., Zhan-Kui, J., Jonsson, G., Levinson, C., Lundberg, H., Svanberg, S.: 1983, Phys. Rev. A 27, p. 947. L
180. Green, J.M., Collins, G.J., Webb, C.E.: 1973, J. Phys. B 6, p. 1545. Q
181. Grevesse, N., Biemont, E., Hannaford, P., Lowe, R.M.: 1981, "Upper Main Sequence CP Stars," 23rd Liege Astrophysical Colloquium, pp. 211-222, Universite de Liege. C,E
182. Griffin, D.C., Bottcher, C., Pindzola, M.S.: 1982, Phys. Rev. A 25, p. 1347. Q
183. Gruzdev, P.F., Denisov, V.I.: 1982, Opt. Spectrosc. (USSR) 52, p. 8. CA
184. Gruzdev, P.F., Loginov, A.V.: 1983, Opt. Spectrosc. (USSR) 54, p. 348. CA,Q
185. Gruzdev, P.F.: 1982, Akad. Nauk, SSSR, Otd. Obshch. Fiz. Astron., Nauch. Sov. Spektrosk., "Atomic and Ionic Spectra in X-ray and Ultraviolet Regions," Moscow, pp. 4-35. Q
186. Guern, Y., Lotrian, J.: 1982, J. Phys. B 15, p. 713. E
187. Gurchumeliya, A.D., Tsirekidze, T.A., Tsirekdze, M.A., Khutsishvili, O.G.: 1983, Soobshch. Akad. Nauk Gruz. SSR. 111, p. 285. Q
188. Gurtovenko, E.A., Kostik, R.I.: 1981, Astron. Suppl. Ser. 46, p. 239. M
189. Gurtovenko, E.A., Kostik, R.I.: 1982, Astron. Astrophys. Suppl. Ser. 47, p. 193. M
190. Gurtovenko, E.A., Kostyk, R.I., Orlova, T.V.: 1983, Sov. Astron.-AJ 27, p. 439. M
191. Hannaford, P., Lowe, R.M.: 1982, J. Phys. B 15, p. 65. L
192. Hannaford, P., Lowe, R.M., Grevesse, N., Biemont, E., Whaling, W.: 1982, Astrophys. J. 261, p. 736. LE

193. Hannaford, P., Series, G.W.: 1981, "Laser Spectroscopy V", Eds., A.R.W. Mckellar, T. Oda, and P. Stoicheff, Springer-Verlag, New York, pp. 94-102. L
194. Hannaford, P., Lowe, R.M.: 1983, J. Phys. B 16, p. L43. L
195. Hannaford, P., Lowe, R.M.: 1983, Opt. Eng. 22, p. 532. L
196. Hannaford, P., Lowe, R.M.: 1983, Opt. Eng. 22 p. 532. L
197. Hannaford, P., Lowe, R.M.: 1983, J. Phys. B 16, p. 4539. L
198. Hansen, J.E., Raassen, A.J.J., Uylings, P.H.M.: 1984, Astrophys. J. 277, p. 435. QF
199. Hansen, W.: 1983, J. Phys. B 16, p. 933. L
200. Hansen, W.: 1983, J. Phys. B 16, p. 2309. L
201. Hardis, J.E., Curtis, L.J., Ramanujam, P.S., Livingston, A.E., Brooks, R.L.: 1983, Phys. Rev. A 27, p. 257. L
202. Hartmetz, P., Schmoranzer, H.: 1983, Phys. lett. A 93, p. 405. E
203. Hata, J., Grant, I.P.: 1982, Mon. Not. R. Astron. Soc. 198, p. 1081. Q
204. Hayes, M.A., Nussbaumer, H.: 1983, Astron. Astrophys. 124, p. 279. CF
205. Hibbert, A., Kingston, A.E., Tiwary, S.N.: 1982, J. Phys. B 15, p. L643. Q
206. Hibbert, A.: 1982, Nucl. Instrum. Methods 202, p. 323. Q
207. Hibbert, A., Dufton, P.L., Murrary, M.J., York, D.G.: 1983, Mon. Not. R. Astron. Soc. 205, p. 535. Q
208. Ho, Y.K., Henry, R.J.W.: 1983, Astrophys. J. 264, p. 733. Q
209. Ho, Y.K., Henry, R.J.W.: 1983, Astrophys. J. 267, p. 886. Q
210. Ho, Y.K., Henry, R.J.W.: 1984, Quant. Spectrosc. Radiat. Transfer 31, p. 57. Q
211. Hofsaess, D.: 1982, J. Quant. Spectrosc. Radiat. Transfer 28, p. 131. Q
212. Hollander, Tj., DeLeeuw, A., Ter Horst, E.: 1983, Spectrochim. Acta, Part B 38, p. 691. ER
213. Hovis, F.E., Gelbwachs, J.A.: 1983, J. Chem. Phys. 78, p. 6680. L
214. Huang, K.-N., Kim, Y.-K., Cheng, K.T., Desclaux, J.P.: 1983, At. Data Nucl. Data Tables 28, p. 355. Q,QF
215. Huang, K.-N.: 1984, At. Data Nucl. Data Tables 30, p. 313. Q,QF
216. Husain, D., Schifino, J.: 1982, J. Chem. Soc., Faraday Trans. 2 78, p. 2083. L
217. Husain, D., Schifino, J.: 1983, J. Chem. Soc., Faraday Trans. 2 79, p. 1265. L
218. Inoue, G., Setser, D.W., Sadeghi, N.: 1982, J. Chem. Phys. 76, p. 977. L,ER
219. Irwin, A.W.: 1983, Astron. Astrophys. 117, p. 173. C
220. Jamieson, M.J., Watts, R.S.: 1981, Chem. Phys. Lett. 84, p. 560. Q
221. Jacques, C., Druetta, M., Berry, H.G., Knystautas, E.J.: 1982, Nucl. Instrum. Methods 202, p. 45. L
222. Johansson, S., Litzén, U., Lundin, L. and Mannervik, S.: 1981, Phys. Scr. 24, p. 30. L
223. Johnson, C., Kingston, A.E., Dufton, P.L.: 1984, Mon. Not. R. Astron. Soc. 207, p. 7P. CF
224. Johnson, W.R., Huang, K.-N.: 1982, Phys. Rev. Lett. 48, p. 315. Q
225. Jönsson, G., Lundberg, H., Svanberg, S.: 1983, Phys. Rev. A 27, p. 2930. L
226. Jönsson, G., Lundberg, H.: 1983, Z. Phys. A 313, p. 151. L
227. Jupén, C., Engström, L., Huldt, S., Trigueiros, A., Ekberg, J.O., Litzén, U., Martinson, I.: 1984, Phys. Scr. 29, p. 226. L
228. Kandela, S.A.: 1984, Physica C (Amsterdam) 123, p. 370. E
229. Dandela, S.A., Schmoranzer, H.: 1981, Phys. Lett. A 86, p. 101. L
230. Karim, K.R., Chen, M.H., Crasemann, B.: 1983, Phys. Rev. A 28, p. 3355. Q
231. Kastner, S.O., Behring, W.E., Bhatia, A.K.: 1983, Astrophys. J. Suppl. Serv. 53, p. 129. Q,QF
232. Keenan, F.P., Berrington, K.A., Burke, P.G., Kingston, A.E., Dufton, P.L.: 1984, Mon. Not. R. Astron. Soc. 207, p. 459. I

233. Kelly, H.P.: 1974, Phys. Rev. A 9, p. 1582. Q
234. Kerkhoff, H., Schmidt, M., Zimmerman, P.: 1980, Phys. Lett. A 80, p. 11. L
235. Kernhan, J.A., Simpson, F.R.: 1982, Nucl. Instrum. Methods 202, p. 49. L
236. Keto, J.W., Kuo, C.-Y.: 1981, J. Chem. Phys. 74, p. 6188. L
237. Kingston, A.E., Tayal, S.S.: 1983, J. Phys. B 16, p. 3465. Q
238. Khashan, M.A., Nassif, A.Y.: 1984, Physica C (Amsterdam) 124, p. 114. HR
239. Kholtygin, A.F.: 1980, Astrophys (USSR) 16, p. 77. Q
240. Klimchitskaya, G.L., Safronova, U.I., Labzovskii, L.M.: 1974, Akad. Nauk. SSSR, Ord. Obshch. Fiz. Astron., Preprint #7 (Mosk. Oblast.). Q,QF
241. Klimovskii, I.I., Minaev, P.V., Morozov, A.V.: 1981, Opt. Spectrosc. (USSR) 50, p. 464. AF
242. Knight, R.D.: 1982, Phys. Rev. Lett. 48, p. 792. L
243. Kono, A., Hattori, S.: 1982, J. Opt. Soc. Am. 72, p. 601. L,E,Q
244. Kono, A., Hattori, S.: 1982, J. Quant. Spectrosc. Radiat. Transfer 28, p. 383. L,E
245. Komninos, Y., Nicolaides, C.A.: 1983, Phys. Scr. 28, p. 472. Q
246. Kostenko, V.A., Kasyanenko, S.V., Tolmachev, Yu.A.: 1983, Opt. Spectrosc. (USSR) 54, p. 440. E
247. Kostyk, R.I.: 1982, Astrometrijy Astrofiz. 46, p. 58. M
248. Kostyk, R.I.: 1981, Astrometriya Astrofiz. 45, p. 3. M
249. Kostyk, R.I.: 1982, Sov. Astron.--AJ 26, p. 422. M
250. Kostyk, R.I.: Orlova, T.V.: 1982, Astrometriya Astrofiz. 47, p. 32. M
251. Kostyk, R.I., Orlova, T.V.: 1983, Astrometriya Astrofiz. 49, p. 39. M
252. Kricke, C., Brenn, R.: 1982, Nucl. Instrum. Methods 202, p. 107. QF
253. Krogulec, M., Wojthowiak, J., Heldt, J.: 1981, Acta Phys. Pol. A 60 p. 591. L
254. Kubota, M. Fujishiro, Y., Ishida, R.: 1981, Spectrochim. Acta., Part B, 36, p. 697. C
255. Kupliauskiene, A.V., Kupliauskis, Z.J., Tutlis, V.I.: 1982, Sov. Phys.--Collect. 22, No. 5, p. 11. Q
256. Kupliauskene, A.V., Kuplyauskis, Z.I.: 1983, Opt. Spectrosc. (USSR) 54, p. 26. Q
257. Kwiatkowski, M., Micali, G., Werner, K., Zimmermann, P.: 1981, Phys. Lett. A 85, p. 273. L
258. Kwiatowski, M., Micali, G., Werner, K., Zimmermann, P.: 1981, Astron. Astrophys. 103, p. 108. L
259. Kwiatkowski, M., Micali, G., Werner, K., Schmidt, M., Zimmermann, P.: 1982, Z. Phys. A 304, p. 197. L
260. Kwiatkowski, M., Micali, G., Werner, K., Zimmerman, P.: 1982, J. Phys. B 15, p. 4357. L
261. Kwiatkowski, M., Zimmerman, P., Biemont, E., Grevesse, N.: Astron. Astrophys. 112, p. 337. L
262. Kwong, H.S., Smith, P.L., Parkinson, W.H.: Phys. Rev. A 25, p. 2629. H,A
263. Kwong, H.S., Johnson, B.C., Smith, P.L., Parkinson, W.H.: Phys. Rev. A 27, p. 3040. L
264. LaGattuta, K.J., Hahn, Y.: 1982, Phys. Rev. A 25, p. 411. Q
265. Larsson, M., Mannfors, B., Pendleton, W.R., Jr.: 1983, Phys. Rev. A 28, p. 3371. L
266. Laughlin, C.: 1983, J. Phys. B 16, p. 3329. Q
267. Lindgard, A.: 1982, Phys. Scr. 26, p. 84. Q
268. Lisina, T.G., Safronova, U.I.: 1981, Opt. Spectrosc. (USSR) 50, p. 354. Q
269. Lisina, T.G., Safronova, U.I.: 1982, Akad. Nauk SSSR, Otd. Obsch. Fiz. Astron., Nauch. Sov. Spektrosk., "Atomic and Ionic Spectra in X-ray and Ultraviolet Regions," pp. 134-45. I
270. Livingston, A.E., Hinterlong, S.J.: 1982, Nucl. Instrum. Methods 202, p. 103. L
271. Livingston, A.E., Hinterlong, S.J.: 1980, Phys. Lett. A 80, p. 372.
272. Lokner, V., Vadla, C., Vujnovic, V.: 1983, J. Quant. Spectrosc. Radiat. Transfer 30, p. 187. ER

273. Lombardi, G.G., Cardon, B.L., Kurucz, R.L.: 1981, Astrophys. J 248, p. 1202. HA
274. Lotrian, J., Guern, Y.: 1982, J. Phys. B 15, p. 69. E,ER
275. Lugger, P., Barker, E., York, D.G., Oegerle, W.: 1982, Astrophys. J. 259, p. 67. M
276. Luke, T.M.: 1982, J. Phys. B 15, p. 1217. L
277. Magazzu, A., Pirronello, V., Strazzulla, G.: 1983, Astron. Astrophys. 120, p. 139. Q
278. Magazzu, A., Pirronello, V., Strazzulla, G.: 1983, J. Quant. Spectrosc. Radiat. Transfer 29, p. 375. Q
279. Mannervik, S., Jelenkovic, B., Kisielinski, M.: 1981, Phys. Lett. A 86, p. 10. L
280. Mannervik, S., Cederquist, H.: 1983, Phys. Scr. 27, p. 175. L
281. Mansfield, M.W.D., Connerade, J.P.: 1982, J. Phys. B 15, p. 503. Q
282. Martin, P., Campos, J.: 1983, J. Quant. Spectrosc. Radiat. Transfer 30, p. 131. E,R,L
283. Martinson, I., Denne, B., Ekberg, J.O., Engstrom, L., Suldt, S., Jupen, C., Litzen, U., Mannervik, S., Triqueiros, A.: 1983, Phys. Scr. 27, p. 201. L
284. Martinson, I.: 1983, private communication quoted in Berry, H.G., Phys. Scr. T3, p. 56. E,L
285. Mason, H.E., Bhatia, A.K.: 1983, Astron. Astrophys., Suppl. Ser 52, p. 181. Q
286. Mathur, M.S., Kelly, F.M.: 1982, Can. J. Phys. 60, p. 1237. L
287. Maujean, M., Descoubes, J.-P.: 1967, C.R. Acad. Sci., Ser. B 264, p. 1653. L
288. McEachran, R.P., Cohen, M.: 1982, J. Quant. Spectrosc. Radiat. Transfer 27, p. 119. Q
289. McEachran, R.P., Cohen, M.: 1982, J. Quant. Spectrosc. Radiat. Transfer 27, p. 111. Q
290. McEachran, R.P., Cohen, M.: 1983, J. Phys. B 16, p. 3125. Q
291. Mendoza, C., Zeippen, C.J.: 1982, Mon. Not. R. Astron. Soc. 198, p. 127. QF
292. Mendoza, C., Zeippen, C.J.: 1982, Mon. Not. R. Astron. Soc. 199, p. 1025. QF
293. Mendoza, C.: 1982, J. Phys. B 15, p. 867. Q
294. Mendoza, C., Zeippen, C.J.: 1983, Mon. Not. R. Astron. Soc. 202, p. 981. QF
295. Migdelek, J., 1983, J. Quant. Spectros. Radiat. Transfer 30, p. 169. Q
296. Moity, J.: 1983, Astron. Astrophys., Suppl. Ser. 52, p. 37. E
297. Morrison, M.D., Cunningham, A.J.: 1983, J. Geophys. Res. 88, p. 3233. ER
298. Morrison, M.D., Cunningham, A.J., Christensen, A.B.: 1983, J. Quant. Spectros. Radiat. Transfer 29, p. 137. ER
299. Muradov, V.G., Kudryavtsev, Yu.N., Muradova, O.N.: 1978, Sov. Phys. J. 21 1244. AR
300. Muradov, V.G., Kudryavtsev, Yu.N., Muradova, O.N.: 1982, Opt. Spectrosc. (USSR) 52, p. 252. AR
301. Muradova, O.N., Muradov, V.G.: 1979, J. Appl. Spectrosc. (USSR) 29, p. 1168.
302. Musiol, K., Jones, D.W., Wiese, W.L.: 1983, J. Quant. Spectros. Radiat. Transfer 29, p. 321. ER
303. Nick, K.P., Helbig, V.: 1983, J. Quant. Spectrosc. Radiat. Transfer 29, p. 465. E,C
304. Nicolaides, C.A., Komninos, Y., Beck, D.R.: 1981, Chim. Chronika 10, p. 35. Q
305. Nicolaides, C.A., Komninos, Y., Beck, D.R.: 1982, Phys. Rev. A 27, p. 3044. Q
306. Nicolaides, C.A.: 1983, Chem. Phys. Lett. 101, p. 435. Q
307. Nussbaumer, M., Perrini, M., Storey, P.S.: 1981, Astron. Astrophys. 102, p. 351. Q,QF

308. Nussbaumer, H.: 1973, Astron. Astrophys. 27, p. 303. Q
309. Nussbaumer, H., Storey, P.J.: 1982, Astron. Astrophys. 110, p. 295. QF
310. Nussbaumer, H. Storey, P.J.: 1982, Astron. Astrophys. 115, p. 205. Q,QF
311. Obbarius, H.U., Kock, M.: 1982, J. Phys. B 15, p. 527. ER
312. Oboladze, H.S., Safronova, U.I., Khristenko, S.V.: 1981, Akad. Nauk SSSR, Otd. Obshch. Fiz. Astron., Inst. Spektrosk., Preprint No. 6 (Mosk. Oblast). Q
313. Oke, J.B., Lauer, T.R.: 1979, Astophy. J. 230, p. 360. ER
314. Omidvar, K.: 1982, Phys. Rev. A 26, p. 3053. Es
315. Ortiz, M., Campos, J.: 1980, J. Chem. Phys. 72, p. 5635. L
316. Osherovick, A.L., Piekhotkina, G.L., Obidin, V.R.: 1981, Opt. Spectrosc. (USSR) 50, p. 576. L
317. Osherovich, A.L., Verolainen, Ya. F., Nikolaich, A. Ya., Privalov, V.I., Pulkin, S.A., Tezikov, V.V.: 1977, Prikl. Spektrosk., pp. 33-36 (Akad. Nauk SSSR, Otd. Obsch. Fiz. Astron., Moscow). L
318. Parpia, F.A., Johnson, W.R.: 1982, Phys. Rev. A 26, p. 1142. QF
319. Pascu, M.L., Dang, T.M., Dumbraveanu, G.: 1982, Rev. Roum. Phys. 27, p. 659. AR
320. Pence, W.R., Leone, S.R.: 1981, J. Chem. Phys. 74, p. 5707. L
321. Petrini, D.: 1981, J. Phys. B 14, p. 3839. Q
322. Pinnington, E.H., Ansbacher, W., Meenakshi Raja Rao, P.: 1983, Phys. Lett. A 97, p. 337. L
323. Pipin, J., Woznicki, W.: 1983, Chem. Phys. lett. 95, p. 392. Q
324. Pirronello, V., Strazzulla, G.: 1981, Astron. Astrophys. 93, p. 411. Q
325. Pirronello, V., Strazzulla, G.: 1980, Mem. Soc. Astron. Ital. 51, p. 59. Q
326. Plekhotkin, G.A., Shukhtin, A.M., Verolainen, Ya.F.: 1983, Opt. Spectrosc. (USSR) 54, p. 569. L,H,HR
327. Plekhotkina, G.L.: 1981, Opt. Spectrosc. (USSR) 51, p. 106. L
328. Pokleba, A.K., Safronova, U.I.: 1982, Opt. Spectrosc. (USSR) 53, p. 7. Q,I
329. Pokleba, A.K., Safronova, U.I.: 1982, Akad. Nauk SSSR, Otd. Obshch. Fiz. Astron. Nauchn. Sov. Spektrosk., pp. 84-128. Q,I
330. Pokleba, A.K., Safronova, U.I.: 1981, Akad. Nauk SSSR, Otd. Obshch. Fiz. Astron., Inst. Spektrosk., Preprint No. 11 (Mosk, Oblast). Q,I
331. Polkeba, A.K., Safronova, U.I.: 1984, Int. J. Quantum Chem. 25, p. 69. I
332. Poulsen, O., Anderson, T., Bentzen, S.M., Nielson, U.: 1981, Phys. Rev. A 24, p. 2523. L
333. Poulsen, O., Anderson, T., Bentzen, S.M., Koleva, I.: 1982, Nucl. Instrum. Methods 202, p. 139. L
334. Pradhan, A.K.: 1982, Phys. Rev. A 25, p. 592. Q
335. Pradhan, A.K.: 1983, Phys. Rev. A 28, p. 2113. Q
336. Pradhan, A.K.: 1983, Phys. Rev. A 28, p. 2128. Q
337. Ramonas, A.A., Uspalis, K.K., Yanukonene, O.Yu.: 1979, VINITI (USSR) N1299-77, Deposited Doc. 32 pp. Q
338. Rebolledo, M.A., Sanz, J.J.: 1984, Phys. Rev. A 29, p. 1983. L
339. Ricca, D., Bassani, F.: 1983, Nuovo Cimento D 2, p. 801. Q,QF
340. Roginsky, D.V.T., Klapisch, M., Cohen, M.: 1983, Chem. Phys. Lett. 95, p. 568. Q
341. Rose, S.J.: 1982, Rutherford Appleton Laboratory Report (Chilton, Didcot, Oxon, OX11 0QX) RL-82-114. I,Q
342. Rudolph, J., Helbig, V.: 1982, J. Phys. B 15, p. L1. L
343. Rudolph, J. Helbig, V.: 1982, Phys. Lett. A 89, p. 339. L
344. Rudolph, J., Helbig, V.: 1982, Z. Phys. A 306, p. 93. L
345. Rudolph, J., Helbig, V.: 1982, J. Phys. B 15, p. L599. L
346. Rudzikas, Z.B., Kaniauskas, J.M., Merkelis, G.V., Savicius, E.H.: 1983, J. Phys. B 16, p. 2879. Q
347. Rudzikas, Z.B.: 1984, Int. J. Quantum Chem. 25, p. 47. Q
348. Runge, S., Valance, A.: 1983, Chem. Phys. Lett. 95, p. 564. Q
349. Rutten, R.J., Van der Zalm, E.B.J.: 1994, Astron. Astrophys., Suppl. Ser. 55, p. 143. M

350. Safronova, U.I., Urnov, A.M.: 1980, J. Phys. B 13, p. 869. Q
351. Safronova, U.I., Senashenko, V.S.: 1982, Phys. Scr. 25, p. 37. Q
352. Safronova, U.I., Senashenko, V.S.: 1981, Akad. Nauk SSSR, Otd. Obshch. Fiz. Astron., Inst. Spektrosk., Preprint No. 9 (Mosk. Oblast).
353. Safronova, U.I., Tsirekidze, T.A.: 1981, Akad. Nauk SSSR, Otd. Obshch. Fiz. Astron., Inst. Spektrosk., Preprint No. 18 (Mosk. Oblast).
354. Saha, H.P., Trefftz, E.: 1982, J. Phys. B 15, p. 1089. Q
355. Saha, H.P., Trefftz, E.: 1982, Astron. Astrophys. 116, p. 224. QF
356. Saha, H.P., Trefftz, E.: 1982, Z. Naturforsch., Teil A 37, p. 744. Q
357. Saha, H.P., Trefftz, E.: 1983, Sol. Phys. 87, p. 233. QF
358. Salih, S., Duquette, D.W., Lawler, J.E.: Phys. Rev. A 27, p. 1193. L
359. Sahil, S., Lawler, J.E.: 1983, Phys. Rev. A 28, p. 3653. L
360. Sampson, D.H., Clark, R.E.H., Goett, S.J.: 1981, Phys. Rev. A 24, p. 2979. Q
361. Sampson, D.H., Goett, S.J., Clark, R.E.H.: 1983, At. Data Nucl. Data Tables 28, p. 299. Q
362. Sampson, D.H., Goett, S.J., Clark, R.E.H.: 1983, At. Data Nucl. Data Tables 29, p. 467. Q
363. Sampson, D.H., Goett, S.J., Clark, R.E.H.: 1984, At. Data Nucl. Data Tables 30, p. 125. Q
364. Sasaki, W., Ueda, H., Ohta, T.: 1983, IEEE J. Quantum Electron. 19, p. 218. L
365. Schnehage, S.E., Danzmann, K., Kunnemeyer, R., Kock, M.: 1983, J. Quant. Spectrosc. Radiat. Transfer 29, p. 507. E,HL
366. Schwab, J.J., Anderson, J.G.: 1982, J. Quant. Spectrosc. Radiat. Transfer 27, p. 445. A
367. Scott, P., Kingston, A.E., Hibbert, A.: 1983, J. Phys. B 16, p. 3945. Q
368. Serrao, J.M.P.: 1982, J. Phys. B 15, p. 2009. Q
369. Shapochkin, M.V., Smirnov, Y.M.: 1981, Opt. Spectrosc. (USSR) 51, p. 532. E
370. Shemansky, D.E.: 1980, Astrophys. J. 236, p. 1043. CA
371. Shemansky, D.E., Smith, G.R.: 1981, J. Geophys. Res., Sect. A 86, p. 9179. CA
372. Shevelko, V.P.: 1974, Fiz. Inst. Akad. Nauk, Ordea Lenina im. P.N. Lebedeva, Preprint No. 120. Q
373. Shine, R.A., Lites, B.W., Chipman, E.G.: 1978, Astrophys. J. 224, p. 247. Cp
374. Shull, J.M., van Steenberg, M., Seab, C.C.: 1983, Astrophys. J. 271, p. 408. M
375. Smid, H., Hansen, J.E.: 1983, J. Phys. B 16, p. 3339. Q
376. Smith, G., St. J. Raggett, D.: 1981, J. Phys. B 14, p. 4015. A
377. Soltanolkotabi, M., Gupta, R.: 1983, Phys. Lett. A 96, p. 399. L
378. Soltanolkotabi, M., Gupta, R.: 1984, Physica C (Amsterdam) 123, p. 386. L
379. Spencer, W.P., Vaidyanathan, A.G., Kleppner, D., Ducas, T.W.: 1981, Phys. Rev. A 24, p. 2513. L
380. Spencer, W.P. Vaidyanathan, A.G., Kleppner, D., Ducas, T.W.: 1982, Phys. Rev. A 25, p. 380. L,M
381. Steenman-Clark, L., Faucher, P.: 1984, J. Phys. B 17, p. 73. Q
382. Sugar, J., Kaufaman, V.: 1984, J. Opt. Soc. Am. B 1, p. 218. QF
383. Sureau, A., Guennou, H., Cornille, M.: 1984, J. Phys. B 17, p. 541. Q
384. Szulkin, M., Karwowski, J.: 1981, J. Phys. B 14, p. 4729. Q
385. Tanarro, I., Arqueros, F., Campos, J.: 1982, J. Chem. Phys. 77, p. 1826. L
386. Tanarro, I., Arqueros, F., Campos, J.: 1983, Phys. Rev. A 27, p. 2533. L
387. Taszner, A.: 1981, Appl. Phys. 25, p. 31. L
388. TFR Group, Dubau, J., Loulergue, M.: 1981, J. Phys. B 15, p. 1007. Q
389. Tiwary, S.N.: 1982, Chem. Phys. Lett. 93, p. 47. Q
390. Tiwary, S.N.: 1983, Chem. Phys. Lett. 76, p. 333. Q
391. Tiwary, S.N., Kingston, A.E., Hibbert, A.: 1983, J. Phys. B 16, p. 2457. Q

392. Tiwary, S.N.: 1983, Astrophys. J. 269, p. 803. Q
393. Tiwary, S.N.: 1983, Astrophys. J. 272, p. 781. Q
394. Tozzi, G.P., Huber, M.C.E., Pauls, U.: 1983, Astron. Astrophys. 126, p. 320. ER
395. Träbert, E., Jones, K.W., Johnson, B.M., Gregory, D.C., Kruse, F.H.: 1982, Phys. Lett. A 876, p. 336. L
396. Träbert, E., Schneider, G., Heckman, P.H.: 1983, Phys. Scr. 27, p. 407. E
397. Troyan, V.I.: 1981, Astrometriya Astrofiz. 44, p. 19. MR
398. Tunnel, T.W., Can, C. Bhalla, C.P.: 1982, J. Quant. Spectrosc. Radiat. Transfer 27, p. 405. Q,QR
399. Uchida, H., Kosinski, M.A. Omenetto, N., Winefordner, J.D.: 1983, Spectrochim. Acta. Part B 38, p. 529. L
400. Uchida, H., Kosinski, M.A., Omenetto, N., Winefordner, J.D.: 1984, Spectrochim. Acta, Part B 39, p. 63. L
401. Ueda, K., Imura, H., Karasawa, M., Fukuda, K.: 1981, J. Phys. Soc. Japan 50, p. 3545. H
402. Ueda, K., Karasawa, M., Fukuda, K.: 1981, J. Phys. Soc. Japan 51, p. 2267. H,A
403. Ueda, K., Ashizawa, Y., Fukuda, K.: 1982, J. Phys. Soc. Japan 51, p. 1936. H
404. Ueda, K., Karasawa, M., Fukuda, K.: 1982, J. Phys. Soc. Japan 51, p. 2267. H
405. Ueda, K., Hamaguchi, Y., Fukuda, K.: 1982, J. Phys. Soc. Japan 51, p. 2973. H
406. Ueda, K., Hamaguchi, Y., Fukuda, K.: 1983, J. Phys. Soc. Japan 52, p. 2666. H
407. Vainshtein, L.A., Safronova, U.I.: 1980, At. Data Nucl. Data Tables 25, p. 311. Q
408. Vainshtein, L.A., Safronova, U.I., Shlyaptsev, V.N.: 1982, Fiz. Inst. Akad. Nauk, Ordena Lenina im. P.N. Lebedeva, Preprint No. 27. Q
409. Vainshtein, L.A., Safronova, U.I.: 1976, Fiz. Inst. Akad. Nauk, Ordena Lenina im. P.N. Lebedeva, Preprint No. 146. Q
410. Vajed-Samii, M., MacDonald, K.: 1981, At. Data Nucl. Data Tables 26, p. 467. Q
411. Verheijen, M.J., Beijerinck, H.C.W., Eenshuistra, P.J., Kroon, J.P.C., Verster, N.F.: 1983, Opt. Commun. 45, p. 336. L
412. Verolainen, Y.F., Plekhotkina, G.L., Privalov, V.I.: 1982, Opt. Spectrosc. (USSR) 53, p. 586. L
413. Wang, C.C., Shirinzadah, B.: 1983, Phys. Rev. A 28, p. 1166. A
414. Want, J.-S., Griem, H.R.: 1983, Phys. Rev. A 27, p. 2249. ER
415. Wiese, W.L.: 1983, "The Physics of Ionized Gases", (Ed., Pichler, G., Institute of Physics, Zagreb, Yugoslavia), pp. 435-456. C
416. Wosinski, L.: 1983, Acta Phys. Pol. A 64, p. 471. E
417. Wujec, T., Weniger, S.: J. Quant. Spectros. Radiat. Transfer 28, p. 113. E
418. Wyart, J.-F., Klapisch, M., Schwob, J.-L., Schweitzer, N.: 1982, Phys. Scr. 26, p. 141. QR
419. York, D.G., Spitzer, L., Bohlin, R.C., Hill, J., Jenkins, E.B., Savage, B.D., Snow, T.P.: 1983, Astrophys. J., 266, p. L55. M
420. Zajonc, A.G.: 1982, Phys. A 25, p. 2830. E
421. Zaki Ewiss, M.A., Snoek, C., Dönszelmann, A.: 1983, Astron. Astrophys. 121, p. 327. L
422. Zeippen, C.J.: 1982, Mon. Not. R. Astron. Soc. 198, p. 111. QF
423. Zettl, F., Neger, T., Jäger, H.: J. Phys. B 17, p. 1755. EHR
424. Zhechev, D.Z.: 1982, J. Phys. Lett. (Paris) 43, p. L.67. L
425. Zilitis, V.A.: 1983, Opt. Spectrosc. (USSR) 55, p. 127. Q
426. Zimmerman, P.: 1982, Nucl. Instrum. Methods 202, p. 167. L

W.L. Weise
Chairman of Working Group

WORKING GROUP 3: COLLISION CROSS SECTIONS AND LINE BROADENING

A. LINE BROADENING

The proceedings of the 6th and 7th International Conference on Spectral Line Shapes shown the general state of advances of the subject. Though a great part of the actual content lies outside the present interest, a number of useful results for astrophysical purposes can be found therein and will be quoted hereafter.

1. Stark Broadening of Hydrogen or Hydrogenic Lines

1.1 HYDROGEN LINES AT MODERATE DENSITIES AND TEMPERATURES

Hydrogen line profiles at low densities have been revisited. New calculations have been performed for Hα at typical astrophysical densities (smaller than 10^{14} cm^{-3}), demonstrating the validity of the impact model in the center and the neighbouring wings for proton collisions and electrons as well: this has been done by comparing results obtained with model microfield method with standard impact (semi-classical perturbation method and quantum exact resonance method) theories. Analytical formulae have been given(4,5) and have been shown to be now in agreement(6) with available recent experiments (contrary to previous calculations based on impact electrons and quasistatic ions). On the other hand new measurements of the cores of Hα and Hβ have been published[7,8] which are in agreement with other recent experiments(9).

The first calculations of hydrogen Stark broadening in the presence of a magnetic field have been obtained(10,11) by using unified classical straight path theory: case of Lyα, Lyβ, Hα for $N_e \sim 10^{15}$ cm^{-3}, T = 5 10^3 to 2 10^4 K, B = 2.5 10^3 to 4 10^4 Gauss and two orientations of the field (longitudinal and orthogonal). Since the influence of the strength and orientation of the field appears to be not negligible, the use of those results appears to be highly advisable for analysis of mangetic stellar atmospheres: Ion dynamic effects are shown to be relevant in the near wings whereas the ions can be considered as static in the far wings(12). Extensive tables are provided(13).

Plasma satellites in the presence of non resonant, coherent and polarized radiation(14) and in the presence of non coherent and unpolarized radiation field(15) have been investigated: the effect of radiation shifts the satellites and enhances their intensities. Observational evidence of that effect among chromospheric lines has been indicated(16).

1.2 HIGH TEMPERATURES AND DENSITIES AND HYDROGENIC LINES

A very important effort is actually directed toward very hot and dense plasmas, in view of controlled fusion studies. Inertially confined, laser-produced plasmas are concerned.

This effort should benefit to astrophysicists studying hot and dense stars. Theoreticians focus their attention on strong correlation effects. Due to the exceptionally intense electric and magnetic field of laser compressed plasmas, consequent perturbations of spectra occur (for Z > 10 a few bound states may remain) and the physics of the plasma itself is complicated by strong interactions among the constituant species and by near quantum degeneracy. These high density effects on atomic and ionic plasmas have been discussed(17). The electric micro-field distribution is the object of various theoretical studies. Though quantum corrections to the electron screening of the ions have been shown to be weak (less than 10%) for actually laser produced plasmas(18), quantum microfield distributions studies seem to be relevant(1,19) in view of the future development of higher temperatures and densities: quantum effects of both diffraction and degeneracy appear when the electron gas Fermi temperature approaches the temperature of the system. Concerning classical electric microfield distributions, the case of strongly coupled plasmas deserves attention: a new method based on the expression of the Fourier transform of the microfield distribution gives a good agreement with computer simulations(20). A critical comparison of several approximations has been provided(21), including actual versions of the nearest neighbour, next nearest neighbour approximations together with two approximations based on independent perturber models; comparisons with Monte Carlo simulation techniques have been made. The formal theory of line shifts due to quasistatic ions has been reconsidered, suggesting that a self-consistent treatment of plasma correlations requires a much more complete theory(22). In these plasmas, dynamic terms may be as important as static ones when second and higher order terms are taken into account for the calculations of line shifts in hydrogen(23). By using a systematic instead of a phenomenological justification for long range cutoffs and treatments of initial correlations, it has been confirmed for ionized emitters that the dominant collective effect of the long ranged electron-electron Coulomb interaction is to screen the binary radiator-electron collision(24). The plasma polarization shift has a region of interest, because it is expected to be large at high Z and high densities. In fact recent measurements(25) (Pα of He II at kT = 4 eV and Ne = 1-6 10^{23} m^{-3}) seem to infer that the most probable cause of the shift is the electron impact effect, provided it is properly calculated by including plasma polarization, and that a quasistatic approximation cannot be used; on the other hand line profiles of Pα and Pβ of He II have been measured(26) (Ne = 10^{23} to 5 10^{23} m^{-3} and kT $<$ 11 eV) in a gas liner pinch: the results can be explained by a combination of the electron impact collisions and by the quadrupole quasistatic effect which should be responsible for the obtained asymmetry. Using a pulsed linear discharge, Stark profiles of Balmer α, Balmer β, and Pα line of He II have been measured(27) at kT = 3.8 eV and Ne = 5.5 10^{22} m^{-3}: these results show the importance of ion dynamics effects and also of fine structure effects for Balmer α. Radiator ion motions and low frequency electron produced fields have been included in a standard theory using impact electrons and quasistatic ions(1) and the results fit more favourably the experiments than the MMM method: profiles of Lyα and Lyγ of Ar^{17+} (Ne = 5 10^{23} cm^{-3}, T = 4.6 10^6 K) have been calculated. On the other hand, Lyα and Lyβ of Ne^{9+} and Ar^{17+} profiles have been computed(28),(29) (N_e = 2 10^{23} - 10^{24} cm^{-3}, T = 1.2 10^6 and 1.2 10^7 K) by means of a Baranger-Mozer quasistatic distribution revised and extended to dense and hot binary ionic mixtures: A synthetic spectrum model and fast codes have been developed(30), using standard theories (impact electrons and quasistatic ions): line shapes of the Lyman and helium series have thus been calculated for Aℓ, Na, Si, Cℓ, Ar, Ca, Ti. Typical results have been drawn and the fitting method based on the comparison of the synthetic to the measured spectrum (Te = 750 ± 100 eV, Ne = 2 ± 0.3 10^{21} cm^{-3}) seems to be accurate for determining temperature and densities. A new field of measurements has opened up with the advent of multi-beam lasers capable of compressing symmetrically spherical targets: densities as high as 1.5 10^{24} cm^{-3} and temperatures ~ 1 keV can be attained(31,32): line profiles of Lyα, Lyβ of Ar^{17+} and lines of Ar^{16+} are reported.

2. Stark Broadening of Overlapping Lines of Helium

The basic ideas of the model microfield method (MMM) have been reviewed as well as various theoretical and experimental works devoted to ion dynamics effects of overlapping helium line profiles(33). Improved Stark profiles of the He I 4471 Å have been computed using the MMM at astrophysical densities(34). New experiments(35) concerning He I 4471 and 4922 Å have been made at Ne = 10^{21} to 3 10^{21} m^{-3}, T = 2 10^4 to 3 10^4 K, extending existing measurements towards higher temperatures and densities: these results are rather well described by the MMM; ion dynamic effects considerably modify the line shape of the forbidden components. Simple empirical formulae useful for determination of electron densities of helium plasmas have been proposed from the interpretation of experiments(36): the only parameters which enter the formula leading to N_e are

TABLE 1

Stark broadening and shift of isolated line: reference data
E = experimental, T = theory, SC = semi-classical, SE = semi-empirical, MMM = model microfield method.

Lines	References and method	Range of plasmas parameters Ne cm^{-3}, T_e °K
CI $2p3s-2p4p$ ($^1P^o-^1D$, $^1P^o-^1P$, $^1P^o-^1S$, $^3P^o-^3S$, $^3P^o-^3P$)	(38),E(wall stabilized arc)	7.4-8.3 10^{16},11350,12330
HeI 56 lines (UV,visible,IR)	(39), T (SC)	10^{16}, 2.5 10^3-8 10^4
HeI 36 lines ($4<n_{upper}<10$)	(40), T (SC)	10^{16}, 2.5 10^3-8 10^4
HeI 42 lines ($2<n_{upper}<6$)	(41), T (SC)	10^{16}, 5$10^3$-4$10^4$
HeI 6678,5876,3889 Å	(42), E(laser)+T(MMM)	10^{17}-2 10^{18}, 2 10^4
ArI 4s-4p° array	(2), E(wall-stabilized arc)	10^{17}, 1.3 10^4
ArI 4300 Å	(43), E(wall-stabilized arc)	1.1-11 10^{16},9 10^3-1.5 10^4
XeI 6s-6p', 6s-7p	(44), E(Shock-tube)+T(SC in j-ℓ coupling(45))	2.2-10^{16}, 9.2 10^3-1.25 10^4
Heavy elements (BrI,CdI,GeI, HgI,PbI,RbI,SnI,ZnI)	(46), T (SC)	10^{16}, 5 10^3-4 10^4
Heavy ionic solar lines (TiII,MnII)	(47), T (SC,SE)	
F I	(48), E(wall-stabilized arc) + T (SC)	1.4 10^{17}, 1.4 10^4
MgI 2852 Å, MgII 2795,5,2802.7 Å	(49), E(wall-stabilized arc)	1.1-1.6 10^{17}, 1.3-1.4 10^4
CaII 3933.7, 3968.5 Å	(50), E(wall-stabilized arc)	10^{17}, 1.3 10^4
KI ns-4p (n=7-10), nd-4p (n=5-8)	(51), E(wall-stabilized arc)	2.10 10^{15}, 2.9 10^3

Lines	References and method	Range of plasmas parameters Ne cm^{-3}, T_e ^{o}K
SiIII 3p ^2D-4p $^2P^o$, 4s ^2S-4p $^2P^o$, 3d ^2D -4p $^2F^o$ $^2P^o$-5s ^2S, 4p $^2P^o$-4d ^2D	(52), E(Shock-tube)	0.3-1.4 10^{17}, 1.6-2-2, 10^4
SiII, SiIII	(53), T (SE)	
Xe II 5339,5372,5419,5439 Å (6s-6p, 6p-6d)	(54), E (Shock-tube)	0.1-1.5 10^{17}, 7.8-8.4 10^3
XeII 6s-6p, 5d-6p	(55), (Shock-tube)	0.6-1.8 10^{17}, 1.0-1.2 10^4 K
SII 5320,5453,4525 Å	(56), (wall-stabilized arc)	7 10^{16}, 1.2 10^4
SIII, Cl III, S IV	(57)	
OIII,SIII;OII,ClIII,ArIV;FII, ArIII;OIII,SIII: several lines of homologous structure	(58), T (SE)	
CIV, 12 lines (985-5805 Å)	(26), E(gas liner pinch)	10^{17}-5 10^{17}, kT = 11 eV
CrII 4s ^4D-4p $^4F^o$	(59), E(Shock-tube)	2.55 10^{17}, 1.8 10^4
Be like ions	(60)	
TiII,MnII + others already available	(61), T (SE,SC)	10^{17}, 2.5 10^3-2 10^4

the ratio of the forbidden peak to the allowed peak intensity, the ratio of the minimum intensity between the two lines to the allowed peak, and the forbidden to allowed peak separation Ne = 10^{21} to 10^{22} m^{-3} ± 15%, Te = 10^4 to 1.5 10^4.

3. Stark Broadening of Isolated Lines in Moderately Dense Plasmas

A critical review and a tabulation of selected data concerning experimental Stark widths and shifts has been carried out(37) for non hydrogenic spectral lines of ionized and neutral atoms: it concerns the period 1976-1982. A number of new data of astrophysical interest have been obtained and are collected in Table 1. Systematic trends of Stark broadening parameters as function of the principal quantum number have been studied(2),(39),(40),(52) and can be used for obtaining missing data, by interpolation or extrapolation (when possible). The increase of accuracy of the measurements permits now the determination of rather small features which were not attainable even in the recent past and thus broadening by ions becomes measurable: asymmetry patterns of Stark broadening of several CI lines have been measured (N_e = 7.4 - 8.3 10^{16} cm^{-3}, T = 11350 - 12230 K), and are in qualitative agreement with the quasistatic theory of ion broadening due to the qudratic Stark effect; the same conclusion arises from(44). Measured Stark broadening parameters are in general satisfactory agreement with semi-classical impact calculations of MMM calculations when they exist. The use of j-ℓ coupling instead of L-S coupling for the case of heavy atoms is responsible for great differences(44), especially when quadrupole effects are important. However, a number of experimental results still disagree with theoretical calculations and this remains to be explained; in(26) (case of C IV lines) the disagreement holds especially for lines originating from the first excited levels, for which quantum effects may be important. This has to be compared to the fact that the new experiments performed on the case of Mg II 2s-2p(49) and Ca II 4s-4p(50) agree now with quantum close coupling calculations, whereas they disagree with the semi-classical ones: this had been expected for a long time but had been contradicted by previous measurements.

4. Line Broadening by Foreign Gases and Molecular Line Broadening

In astrophysics, the interest in molecular line broadening is actually increasing, owing to the increasing importance of spectroscopic diagnostics in planetary atmospheres and aeronomy. However, from the quantum chemistry point of view, the subject is very difficult and very few scattered results can be theoretically attained. Furthermore the physicists' actual preoccupations are not generally directed towards astrophysical applications except in certain cases. As a matter of fact the elucidation of detailed molecular dynamics is becoming increasingly fruitful, owing to the rapid advances in molecular beam technology which make possible experiments free from macroscopic constraints such as temperature, pressure and impurity concentrations. There is in fact a greater need for an augmenting theory based on the description of detailed dynamics than for results based on traditional pressure broadening theories and on the statistical effects of collisions. In this way a semi-classical approach to spontaneous emission of molecular collision systems has been developed(63): it involves the derivation of a dynamical theory of fluorescence line shapes which is valid for all frequency regimes and not just in the line wings.

In view of stellar atmospheres diagnostics, the broadening and shift of the sodium D lines perturbed by atomic hydrogen have been measured in a combustion driven shock tube(64); the broadening by Ar and H_2 has also been obtained and agrees with previous measurements made in different experimental conditions; for atomic hydrogen, however, these new values differ significantly from the available theoretical results, casting doubt on the validity of the Na-H interatomic potential used in the calculations. The important effort concerning noble gas broadening is continuing: molecular-structure calculations for alkali-metal atom

+ He systems have been performed(65), leading to pseudo-potentials relevant for ground states and numerous excited states: they are in overall good agreement with recent available experimental data, indicating a large improvement on all previous calculations for these systems. Quantum close coupling impact broadening and shift of the D lines of sodium perturbed by He have been performed: the hyperfine components of the same fine structure lines have the same width and shift(66) as expected; the temperature dependence ($800<T<4500$ K) has also been studied, and is in satisfactory agreement with available experiments(67). The broadening of resonance lines of Cs by noble gases has been measured at low(68) and high temperatures(69); also concerning Rb ns and nd(70), and Sr 5s ns $^1S_0$, 5d and $^1D_2$, $^3D_2$, by He and Xe(71). Detailed analysis of profiles of the calcium 4227 Å line perturbed by Krypton and obtained by laser induced fluorescence show departures from the usual reduced Voigt form(72). Though far from direct astrophysical application, this paper is quoted there because the obtained profiles show an asymmetry of the Anderson-Talman unified theory type, due to the radiation emitted during the collision: these results emphasize the need for a sophisticated description of the profile when accurate measurements are made in the wings or in the adjacent continuum(73),(74).

Concerning line width parameters in molecular broadening, various results have been obtained: the rotational spectrum of NO_2 at 300 K and $10^{-3}<p<0.2$ torr has been carried out(75); similarly for CH_3 CN(76). Pressure broadening in the infrared lines of CO($v = 1 \rightarrow 0$) perturbed by H_e and H_2 at liquid nitrogen and room temperature have been measured(77): these data which are of direct astrophysical interest have been compared to *ab initio* quantum calculations: the accuracy of the CO-H_e data have allowed to choose between the two available potential energy surfaces, but the rather poor sensitivity of the pressure broadening to the anisotropic interaction has to be pointed out.

Inspired by the highly successful observations made aboard Voyager I space craft, and ground based high resolution astronomical measurements, a wide ranging series of high resolution (0.04 cm^{-1}) measurements has been performed in the laboratory at NASA Ames Research Center, using an infrared Fourier transform spectrometer. In view of applications to lines formed in Jupiter, Saturn, Titan and Earth atmospheres in mind, N_2 has been used as the broadening gas. Comparisons of observed and computed spectral transmittances have yielded line strengths, N_2 broadened half-widths, and their variation with temperature (100 K$<T<297$ K); the measured widths are considerably larger than those used in the AF GL compilation; the ν_4 fundamental band of $^{12}CH_4$(78), $^{13}CH_4$(79), $^{12}C_2$ H_2, ^{12}C $^{13}CH_2$ in the 13.7 μm region(80), C_2 H_2 in the 7.53 μm region(81) and CH_3D (ν_2 fundamental)(82) have been carried out. Other molecules are in progress ($^{12}C_2$ H_4, C_2 H_6, $^{12}C_3$ H_8, ^{14}N H_3 and ^{15}N H_3).

5. Collisional Redistribution of Radiation, Transfer of Radiation and Related Topics

Recent developments in high resolution and high sensitivity spectroscopy and atomic excitation by tunable laser sources have increased the interest in collisional redistribution of radiation and related molecular dynamic studies. They can be divided into two main parts: first, the redistribution of weak radiation field is of direct astrophysical interest; the atom-field interaction can be treated by perturbation theory and induced emission can be neglected. Secondly, the redistribution of strong radiation field, which does not seem to be of direct astrophysical interest, occurs when induced emission is important and when the atom field interaction cannot be treated by perturbation theory: the formalism of the dressed atom has then to be used. In the present report we will focus our attention on the redistribution of weak radiation.

Earlier work based on the unified theory of the redistribution of radiation by collisions for isolated lines has been extended to overlapping lines: and explicit expression for the frequency redistribution has been given for Lyα, valid even in the non impact regime(83). Following this work, a consistent formulation for the Hα, Lyα, Lyβ system is ready to be published(84). It must be pointed out that redistribution in the hydrogen line wings requires a very complicated non-Markovian formulation. Experiments(85) have demonstrated the importance of lower state radiative decay as a mechanism for redistribution in the absence of collisions, leading to the conclusion, contrary to Lyα, that complete redistribution may be a good approximation for Hα.

The fundamental of the formulation of the related topics of radiative transfer and statistical equilibrium equations have been examined. The equations for redistribution have been derived in a quantum mechanical impact framework(86) and the equation of transfer has been derived for broad band fields from Maxwell's equations(87). The effects of stimulated emission in the astrophysical context have been clarified(88) and a work is also in progress on clarification of velocity changing effects.

An attempt to formulate, in a general way, the problem of the transfer of line radiation by multilevel atoms allowing for the effects of partial redistribution has been presented(89); the importance of the knowledge of the distributions of photons and particles is emphazised(90), velocity-averaged line profile coefficients have been considered(91) and explicit expressions have been obtained for a three-level atom in terms of the local radiation intensity field. The concept of the redistribution function has been generalized in the impact framework to other resonance two photon processes including resonance Raman scattering, resonance two photon absorption or emission, and inverse Raman scattering(92).

Farther from astrophysical applications one may quote a number of theoretical works of fundamental interest. A model microfield theory redistribution of intense radiation has been developed(93). Basic studies of the fluorescence spectrum and the probe absorption spectrum of an atom in a radiation field and a sea of binary perturbers have been performed(94),(95),(96).

A non adiabatic theory of atomic neutral line broadening is being developed: the theory of atomic scattering in the presence of a radiation field is used to incorporate simultaneously in the close coupled equations the effects of inelastic scattering and dipole coupling due to the field in an original manner(1),(97),(98). These studies (see also(99),(100)) are relevant both for radiation redistribution studies, radiatively-aided collisions and "half-collisions" analysis of atomic line shapes studies(1),(2). A review paper which concerns this field of research has been written(101) (cf. also the invited talks of (1) and (2) relative to this subject) and evinces the increasing importance of this field of research which appears to be rich in potentialities for future astrophysical applications.

References

1. Proceedings of the Sixth International Conference (Boulder 1982): 1983, "Spectral line Shapes, volume 2", Ed. K. Burnett, de Gruyter, Berlin-New York.
2. Proceedings of the Seventh International Conference (Aussois 1983): 1985, "Spectral line Shapes, volume 3", Ed. Rostas, de Gruyter, Berlin-New York.
3. Proceedings of the Fifth International Conference (Berlin 1980): 1981, "Spectral line Shapes", Ed. B. Wende, de Gruyter, Berlin-New York.

49. Goldbach, C., Nollez, G., Plomdeur, P., Zimmermann, J.P.: 1982, Phys. Rev. A 25, p. 2596.
50. Goldbach, C., Nollez, G., Plomdeur, P., Zimmermann, J.P.: 1983, Phys. Rev. A 28, p. 234.
51. Hohimer, J.P.: 1984, Phys. Rev. A 30, p. 1449.
52. Lesage, A., Rathore, B.A., Lakicević, I.S. and Purić, J.: 1983, Phys. Rev. A 28, p. 2264.
53. Dimitrijević, M.S.: 1983, Astron. Astrophys. 127, p. 68.
54. Richou, J., Manola, S., Lebrun, J.L. and Lesage, A.: 1984, Phys. Rev. A 29 p. 3181.
55. Miller, M.H., Lesage, A. and Abadie, D.: 1982, Phys. Rev. A 25, 2064.
56. Mar, S., Czernichowski, A. and Chapelle, J. (reference 3).
57. Dimitrijević, M.S. and Konjević, N.: 1982, J.Q.S.R.T., 27, p. 203.
58. Hey, J.D. and Breger, P.: 1981 (reference 1).
59. Rathore, B.A., Lakicević, I.S., Cuk, M. and Purić, J.: 1984, Phys. Lett. 100A, p. 31.
60. Dimitrijević, M.S.: 1984, Astron. Astrophys. 131, p. 237.
61. Dimitrijević, M.S.: 1982, in the Physics of ionized gases: SPIG 82, (Dubrovnik), ed. G. Pichler (Zagreb, Yugoslavia).
62. Lakicević, I.S. and Purić, J.: 1983, J. Phys. B 16, p. 1525.
63. Kai-Shue Lam and George, T.F.: 1982, J. Chem. Phys. 76, p. 3396.
64. Lemaire, J.L., Chotin, J.L. and Rostas, J.: 1984, J. Phys. B, in press.
65. Pascale, J.: 1982, Phys. Rev. A 26, p. 3709 and 28, p. 632.
66. Roueff, E. and Spielfiedel, A.: 1982, J. Phys. B 15, 2627 and 1983, J. Phys. B 16, p. 506.
67. Spielfiedel, A. and Roueff, E.: 1981, J. Phys. B 14, L795.
68. Siegling, F. and Niemax, K.: 1984, Z. Naturforsch. 39a, p. 447.
69. Siegling, F. and Niemax, K.: 1984, Z. Naturforsch. 39a, p. 455.
70. Weber, K.H. and Niemax, K.: 1982, Z. Phys. A 307, p. 13.
71. Weber, K.H. and Niemax, K.: 1982, Z. Phys. A 309, p. 19.
72. Harris, M., Lewis, E.L., McHugh, D. and Shannon, I.: 1984, J. Phys. B 17, in press.
73. Allard, F. and Kielkopf. J.: 1982, Rev. Mod. Phys. 54, p. 1103.
74. Allard, N.F. and Biraud, Y.G.: 1982, J. Physique 43, p. 501.
75. Moazzen-Ahmadi, M.N. and Roberts, J.A.: 1982, J. Mol. Spectr. 96, p. 336.
76. Messer, J.K., Khoobehi, B. and Roberts, J.A.: 1982, J. Chem. Phys. 76, p. 2914.
77. Picard Bersellini, A., Charneau, R. and Brechignac, P.: 1982, J. Mol Struct. 80, p. 441.
78. Varanasi, P., Giver, L.P. and Valero, F.P.J.: 1983, J.Q.S.R.T., 30, p. 481.
79. Varanasi, P., Giver, L.P. and Valero, F.P.J.: 1983, J.Q.S.R.T., 30, p. 491.
80. Varanasi, P., Giver, L.P. and Valero, F.P.J.: 1983, J.Q.S.R.T., 30, p. 497.
81. Varanasi, P., Giver, L.P. and Valero, F.P.J.: 1983, J.Q.S.R.T., 30, p. 505.
82. Varanasi, P., Giver, L.P. and Valero, F.P.J.: 1983, J.Q.S.R.T., 30, p. 511.
83. Yelnik, J.B., Burnett, K., Cooper, J., Ballagh, R.J. and Voslamber, D.: 1981, Astrophys. J. 248, p. 705.
84. Ballagh, R.J. and Cooper, J.: 1984, J. Phys. B., in press.
85. Cooper, J., Lombardi, G. and Kelleher, D.: 1984, Astrophys. J., in press.
86. Cooper, J., Ballagh, R.J., Burnett, K. and Hummer, D.G.: 1982, Astrophys. J. 260, p. 299.
87. Cooper, J. and Zoller, P.: 1984, Astrophys. J. 277, p. 813.
88. Cooper, J., Hubeny, I. and Oxenius, J.: 1983, Astron. Astrophys. 127, p. 224.
89. Hubeny, I.: 1981, Bull. Astron. Inst. Czechosl. 32, p. 271.
90. Hubeny, I., Oxenius, J. and Simonneau, E.: 1983, J.Q.S.R.T., 29, p. 477.
91. Hubeny, I., Oxenius, J. and Simonneau, E.: 1983, J.Q.S.R.T., 29, p. 495.
92. Hubeny, I.: 1982, J.Q.S.R.T., 27, p. 593.
93. Talin, B. and Klein, L.: 1982, Phys. Rev. A 26, p. 2717.
94. Nienhuis, G.: 1982, Acta Phys. Polonica, A61, p. 235.

4. Stehlé, C., Mazure, A., Nollez, G. and Feautrier, N.: 1983, Astron. Astrophys. 127, p.263.
5. Stehlé, C. and Feautrier, N.: 1984, J. Phys. B 17, p.1477.
6. Stehlé, C. and Feautrier, N.: 1984, Ann. Phys. Fr. 9, p.697.
7. Torres, F., Gigosos, M.A. and Mar, S.: 1984, J.Q.S.R.T., 31, 265.
8. Hernandez, J., Torres, F., Mar, S. and Gigosos, M.A.: 1984, J.Q.S.R.T., in press.
9. Helbig, V. and Nick, K.P.: 1981, J. Phys. B 14, p. 3573.
10. Mathys. G.: 1983, Astron. Astrophys. 125, p. 13.
11. Mathys, G.: 1984, Astron. Astrophys., in press.
12. Mathys, G.: 1984, Astron. Astrophys., in press.
13. Mathys, G.: 1984, Astron. Astrophys. Suppl. Series, in press.
14. Drawin, H.W. and Truong Bach: 1982, J. Phys. B 15, p. 4477.
15. Truong Bach and Drawin, H.W.: 1982, J.Q.S.R.T., 27, p. 627.
16. Öhman, Y.: 1983, Solar Phys. 85, p. 53.
17. Burgess, D.D. and Lee, R.W.: 1982, J. Phys. Fr., Colloque C 2, 43, Suppl. 11, C2 413.
18. Iglesias, C.A. and Hooper Jr., C.F.: 1982, Phys. Rev. A 25, 1632.
19. Boercker, D.B. and Dufty, J.W.: 1984, J.Q.S.R.T., in press.
20. Iglesias, C.A., Lebowitz, J.L. and MacGowan, D.: 1983, Phys. Rev. A 28, p. 1667.
21. Iglesias, C.A., Hooper Jr., C.F. and De Witt, H.E.: 1983, Phys. Rev. A 28, p. 361.
22. Iglesias, C.A.: 1984, Phys. Rev. A 29, p. 1366.
23. Boercker, D.B. and Iglesias, C.A.: 1984, Phys. Rev. A, in press.
24. Iglesias, C.A.: 1983, J.Q.S.R.T., 30, p. 55.
25. Fleurier, C. and Le Gall, P.: 1984, J. Phys. B, in press.
26. Ackermann, U., Finken, K.H. and Musielock, J.: 1984, preprint.
27. Piel, A. and Slupeck, J.: 1984, Z. Naturforsch, 39a, in press.
28. Held, B., Deutsch, C. and Gombert, M.M.: 1983, Phys. Rev. A 28, p. 3134.
29. Held, B., Deutsch, C. and Gombert, M.M.: 1984, Phys. Rev. A 29, pp. 880 and 896.
30. Lee, R.W., Whitten, B.L. and Stout, R.E.: 1984, J.Q.S.R.T., 32, p. 91.
31. Yaakobi, B., Skupsky, S., McCrory, R.L., Hooper, C.F., Deckman, H., Bourke, P. and Soures, J.M.: 1981, Phil. Trans. R. Soc. Lond. A300, p. 623.
32. Yaakobi, B., Villeneuve, D.M., Richardson, M.C., Soures, J.M., Hutchinson, R. and Letzring, S.: 1982, Opt. Comm. 43, p. 343.
33. Mazure, A. and Nollez, G.: 1984, Ann. Phys. Fr., 9, p. 675.
34. Mazure, A. and Nollez, G.: 1982, Astrophys. J., 254, p. 823.
35. Piel, A. and Richter, H., 1983, Z. Naturforsch. 38a, p. 37.
36. Czernichowski, A. and Chapelle, J.: 1984, J.Q.S.R.T., in press.
37. Konjevic, N., Dimitrijevic, M.S., Wiese, W.L.: 1984, J. Phys. Chem. Ref. Data, 13, in press.
38. Jones, D.W. and Wiese, W.L.: 1984, Phys. Rev. A 30, 2602.
39. Dimitrijević, M.S. and Sahal-Bréchot, S.: 1984, J.Q.S.R.T., 31, p. 301.
40. Dimitrijević, M.S. and Sahal-Bréchot, S.: 1984, Astron. Astrophys., 136, p. 289.
41. Bassalo, J.M., Cattani, M. and Walder, V.S.: 1983, J.Q.S.R.T., 28, p. 75.
42. Gauthier, J.C., Geindre, J.P., Goldbach, C., Grandjouan, N., Mazure, A. and Nollez, G.: 1981, J. Phys. B 14, p. 2099.
43. Czernichowski, A. and Chapelle, J.: 1983, Acta Phys. Pol. A63, p. 67.
44. Truong Bach, Richou, J., Lesage, A. and Miller, M.H.: 1981, Phys. Rev. A 24, p. 2550.
45. Truong Bach and Drawin, H.W.: 1981, J.Q.S.R.T., 25, p. 285.
46. Dimitrijević, M.S. and Konjevic, N.: 1983, J.Q.S.R.T., 30, p. 45.
47. Dimitrijević, M.S.: 1982, in Sun and Planetary System, ed. Fricke and Teleki, Astrophys. and Space Sci. (Reidel).
48. Vujnović, V., Vadla, C., Lokner, V., Dimitrijević, M.S.: 1983, Astron. Astrophys. 123, p. 249.

95. Nienhuis, G.: 1981, J. Phys. B 14, p. 3117.
96. Nienhuis, G.: 1982, J. Phys. B 15, p. 535.
97. Julienne, P.S.: 1982, Phys. Rev. A 26, p. 3299.
98. Julienne, P.S. and Mies, F.H.: 1984, Phys. Rev. A 30, p. 831.
99. Hutchinson, M. and George, T.F.: 1982, Molec. Phys., 46, p. 81.
100. Hutchinson, M. and George, T.F.: 1983, J. Phys. Chem. 87, p. 2037.
101. Mies, F.H.: 1981, in "Theoretical chemistry advances and perspectives", Vol. 6B, p. 127, Acad. Press Inc.

S. Sahal-Brechot

B. COLLISIONAL CROSS SECTIONS

Because of the magnitude and diversity of material on electron and heavy-particle collisions, I gave references only to published papers of obvious immediate relevance to astronomical research.

1. Electron Collisions

1.1 ELECTRON IMPACT IONIZATION

A valuable compilation of semi-empirical cross sections and rate coefficients for electron impact ionization of a wide range of ions, expressed in convenient analytical form has been prepared(1). An empirical formula for K-shell ionization cross sections was derived(2). Theoretical estimates ''233 for the ionization of complex ions were given (3,4). Experimental values were obtained for $A\ell^+$(5), He(6), Fe^+(7), Mg^+, Ca^+ and Sr^+(8), Be^+(9), Ne^{3+}, Ar^{3+}, Kr^{3+}, Xe^{3+}(10a), Mg^+, Al^{2+} and Si^{3+}(10b), and B^+, C^{2+} and O^{4+}(11). Ionization rate coefficients were measured for NeVI, NeVII, OVI and Ti IX(12) and for nitrogen ions(13). Distorted wave calculations for the ionization of FeX and the iron ions from FeXV through FeXXVI were reported(14).

2. Electron Impact Excitation

An extensive bibliography of data on electron impact excitation (and ionization) was published(15). Many calculations of excitation cross-sections of varying degrees of sophistication were carried out for positive ions since the last IAU report: Inner shell excitation of helium-like ions(16), lithium-like ions(17a,b) and beryllium-like ions(18) were studied. More detailed calculations were performed for SiIX(19,20), OIV(21), NeV(20,22), CaXV(23), CaXVII(24), SIII(25), SiX(26), SXII(27), OVII(28,29), CV and MgXI(30), SV(31), FeXXV(32), Li-like ions (33), SIII(34), MgVII(35), OII and OIII(36), OIII(37), SII(38), FeX and FeXV-XXVI(14), NeIII(39), NIII(40), FeXX(41), CaXVII(42) and CII(43). Cross sections for excitation to high n states of hydrogenic ions have been obtained using the Born approximation(44). Excitation of hydrogen, helium, lithium and beryllium-like ions has been studied(45) and of sodiuim-like ions in(46).

Electron impact excitation rates for neutral helium are given in(47-50) and for neutral hydrogen in(51,52). Fits to many excitation rate coefficients are given in(53-55). Studies of excitation to auto-ionizing levels and satellite line emission are described in(56-67). Experimental work on electron impact excitation may be found in(64-68).

A valuable compilation of electron impact excitation rate coefficients of astrophysically interesting systems has been assembled by Mendoza(69) and a useful survey of low energy collisions with complex atoms and ions has been presented by Burke and Eissner(70).

1.2 RECOMBINATION

Dielectronic recombination of selected ions has received considerable experimental attention(71-73). Theoretical work has continued also(61,74-82). A tabulation of recombination coefficients by Nussbaumer and Storey(83) is of particular importance to models of astrophysical nebulae. Empirical values of the recombination coefficients of Fe X-Fe XIX have been inferred from an analysis of plasma electron densities(84).

Several measurements of the rate coefficients for the dissociative recombination of molecular ions have been reported(85-90). Theoretical studies of H_2^+(91) and H_3^+(92) recombination were performed with the conclusion that both are unusually slow.

1.3 ELECTRON-MOLECULE COLLISIONS

I list only references to the rotational, vibrational and electronic excitation and dissociative attachment of molecular hydrogen by electron impact(93-100).

4. Heavy-Particle Collisions

4.1 CHARGE TRANSFER AT LOW ENERGIES

Some further calculations on low energy charge transfer cross sections of multicharged ions in collisions with atomic hydrogen and helium(101-116) have been completed. Empirical values for O^{3+} in H were derived from observations of planetary nebulae(117). Estimates of charge transfer between ions of oxygen and sulphur with the neutral parents were made (118). Several measurements were reported(119-126) though only those of Church and Holzscheiter(126) refer to thermal energies.

Major reviews of theoretical charge transfer cross sections for partially and fully-stripped ions in hydrogen have appeared(127,128).

4.2 PROTON IMPACT EXCITATIONS

A review of proton impact cross sections was published(129) and calculations of the excitation cross-sections for fine-structure excitation were carried out(130,131).

4.3 CHEMICAL REACTIONS

Lists of chemical reactions which play a role in interstellar circumstellar and shock region chemistry have been assembled in many papers(132-147). Extensive experimental investigations of relevant ion-molecule reactions have been reported(148-171). Theoretical analyses of the important reaction of C^+ with H_2 have been done (172,173). The possible significance of reactions producing and being driven by hot ions or atoms in cold clouds has been raised(174). The connection between endothermic and association reactions has been explored(175).

Radiative association processes are important in molecular clouds. The theory has been developed in(176-180).

A great variety of neutral reactions that may be important in astronomical environment has been studied experimentally(181). A partial listing is references(181-196). Useful reviews are found in(197) and (198). Theoretical considerations are presented in(199-202) and the dependence of the reaction rate on the rotational populations has been investigated(204). Many studies have been made of the reaction of atomic hydrogen and deuterium with H_2, HD and D_2, and of the influence of rotational and vibrational excitation(205-224).

4.4 MOLECULAR EXCITATION

Rotational, vibrational and fine-structure excitation of astrophysically interesting molecules are discussed in(225-252). Transitions in OH are investigated in(253-258).

Collisions of ions with heteronuclear molecules have been studied particularly by Sakimoto(259) (see also 260-263).

4.5 MOLECULAR DISSOCIATION

The collision-induced dissociation of H_2 and of CO has been investigated theoretically(264-267) and the rate coefficients have been shown to depend on density.

4.6 EXCITATION BY NEUTRAL ATOM IMPACT

The cross section for the excitation of $O(^3P)$ to $O(^1D)$ by hydrogen atom impact has been calculated by Federman and Shipsey(268).

References

1. Bell, K.L., Gilbody, H.B., Hughes, J.G., Kingston, A.E. and Smith, F.J.: 1983, J. Phys. Chem. Ref. Data 12, p. 891.
2. Casnati, E., Tartari, A. and Baraldi, C.: 1982, J. Phys. B 15, p. 155.
3. Burgess, A. and Chidichimo, M.C.: 1983, MNRAS 203, p. 1269; Chidichimo, M.C.: 1983, J. Phys. B 506.
4. Shevelko, V.P., Urnov, A.M., Vainshtein, L.A., and Muller, A.: 1983, 203 p. 45.
5. Montague, R.G., Harrison, M.F.A.: 1983, J. Phys. B 16, 3045.
6. Montague, R.G., Harrison, M.F.A. and Smith, A.C.H.: 1984, J. Phys. B 17, p. 3295.
7. Montague, R.G., Dissens, M.J. and Harrison, M.F.A.: 1984, J. Phys. B 17 p. 2085.
8. Shrivastava, S.K. and Roy, B.M.: 1984, J. Phys. B 17, p. 4935.
9. Falk, F.A. and Dunn, G.H.: 1983, Phys. Rev. A 27, p. 754.

10a Gregory, D.C., Dittner, P.F. and Crandall, D.H.: 1983, Phys. Rev. A 27, p. 724.

10b Crandall, D.H., Phaneuf, R.A., Falk, R.A., Belic, D.S. and Dunn, G.H.: 1982, Phys. Rev. A 25, p. 154.

11. Falk, R.A., Stefanis, G., Camilloni, R., Dunn, G.H., Phaneuf, R.A., Gregory, D.C. and Crandall, D.H.: 1983, Phys. Rev. A 28, p. 91.
12. Datla, U. and Roberts, J.R.: 1983, Phys. Rev. 28, p. 2201.
13. Brown, R., Deuchars, W.M., Kidd, D.E., Summers, H.P. and Wood, L.: 1983, J. Phys. B 16, p. 2053.
14. Mann, J.B.: 1983, Atom. Data Nucl. Data Tables 29, p. 407.
15. Itikawa, Y., Takayanagi, K. and Iwai, T.: 1984, Atom. Data Nucl. Tables 31 p. 215.
16. Sampson, D.H., Goett, S.J. and Clark, R.E.H.: 1983, Atom. Data Nucl. Data Tables 28, p. 299.

17a Goett, S.J. and Sampson, D.H.: 1983, Atom. Data Nucl. Data Tables 29, p. 535.

17b Goett, S.J., Sampson, D.H. and Clark, R.E.H.: 1984, Ap. J. Suppl. 54, p. 115.

18. Sampson, D.H., Goett, S.J. and Clark, R.E.H.: 1984, Atom. Data Nucl. Data Tables 30, p. 125.
19. Aggarwal, K.M. and Baluja, K.L.: 1983, J. Phys. B 16, p. 107.
20. Aggarwal, K.M.: 1984, Ap. J. Suppl. 55 p. 1.
21. Hayes, M.A.: 1983, J. Phys. B 16, p. 285; 1982, MNRAS 199, p. 49.
22. Aggarwal, K.M.: 1983, J. Phys. B 16, p. 2405.
23. Aggarwal, K.M.: 1983, J. Phys. B 16, p. 2231.
24. Dufton, P.L., Kingston, A.E. and Scott, N.S.: 1983, J. Phys. B 16, p. 3053.

25. Mendoza, C.: 1982, J. Phys. B 15, p. 867; Ho, Y.K. and Henry, R.J.W.: 1984 Ap. J. 283, p. 816.
26. Saha, H.P. and Trefftz, E.: 1982, J. Phys. B 15, p. 1089.
27. Saha, H.P. and Trefftz, E.: 1982, Zeits.f. Naturf. 37a, p. 744.
28. Kingston, E.E. and Tayal, S.S.: 1983, J. Phys. B 16, p. 3465.
29. Tayal, S.S. and Kingston, A.E.: 1984, J. Phys. B 17, p. 3465.
30. Tayal, S.S. and Kingston, A.E.: 1984, J. Phys. B 17, L.145.
31. Dufton, P.L. and Kingston, A.E.: 1984, J. Phys. B 17, p. 3321.
32. Pradhan, A.K.: 1983, Phys. Rev. A 28, p. 243.
33. Bhadra, K. and Henry, R.J.W.: 1982, Phys. Rev. A 26, p. 1848.
34. Ho, Y.K. and Henry, R.J.W.: 1984, Ap. J. 284, p. 435.
35. Aggarawal, K.M.: 1984, Ap. J. Suppl. 56, p. 303.
36. Ho, Y.K. and Henry, R.J.W.: 1983, Ap. J. 264, p. 733.
37. Aggarwal, K.M.: 1983, Ap. J. Suppl. 52, p. 387; Aggarwal, K.M., Baluja, K.L. and Tully, J.A.: 1982, MNRAS 201, p. 923.
38. Ho, Y.K. and Henry, R.J.W.: 1982, Ap. J. 267, p. 888.
39. Butler, K. and Mendoza, C.: 1983, MNRAS 208, p. 17.
40. Nussbaumer, H. and Storey, P.J.: 1982, Astron. Ap. 106, p. 271.
41. Mason, H.E. and Bhatia, A.K.: 1983, Astron. Ap. Suppl. 52, p. 181.
42. Mason, H.E. and Bhatia, A.K.: 1983, Astron. Ap. Suppl. 52, p. 115.
43. Hayes, M.A. and Nussbaumer, H.: 1982, Astron. Ap. 139, p. 233.
44. Oza, D.H.: 1984, Ap. J. Suppl. 54, p. 395; Oza, D.H., Callaway, J. and Abu-Salbi, N.: 1982, Phys. Rev. A 25 p. 2812.
45. Peek, J.M. and Mann, J.B.: 1982, Phys. Rev. A 25, p. 154.
46. Henry, R.J.W. and Msezane, A.Z.: Phys. Rev. A 26, p. 2545.
47. Berrington, K.A., Burke, P.C., Fon, W.C. and Kingston, A.E.: 1982 MNRAS 200, p. 347.
48. Berringjton, K.A., Fon, W.C. and Kingston, A.E.: 1982, MNRAS 200, p. 347.
49. Freitas, L.C.G., Berrington, K.A., Burke, P.G., Hibbert, A., Kingston, A.E. and Sinfailam, A.L.: 1984, J. Phys. B 17, L303.
50. Aggarwal, K.M., Kingston, A.E. and McDowell, M.R.C.: 1983, Ap. J. 278, p. 874.
51. Rau, A.P.: 1984, J. Phys. B 17, L75.
52. Badnell, N.R.: 1983, J. Phys. B 16, p. 3263.
53. Drake, S.A.: 1983, MNRAWS 205, p. 15.
54. Aggarwal, K.M.: 1983, MNRAS 202, p. 15; erratum: 1984, MNRAS 205, p. 447.
55. Stepney, S. and Guilbert, P.W.: 1983, MNRAS 204, p. 1269.
56. Clark, R.E.H., Magee, N.H., Mann, J.B. and Merts, A.L.: 1982, Ap. J. 254, p. 412.
57. Trefftz, E.: 1983, J. Phys. B 16, p. 1247.
58. Pradhan, A.K.: 1983, Phys. Rev. A 28, p. 2128.
59. Griffin, D.C., Bottcher, C. and Pindzola, M.S.: 1982, Phys. Rev. A 25, p. 154.
60. Foucher, P., Loulergue, M., Steenman-Clark, L. and Volonté, S.: 1982, Astron. Ap. 118, p. 147.
61. Faucher, P. and Volonté, S.: 1982, Astron. Ap. 135, p. 152.
62. Bely-Dubau, F., Dubau, J., Faucher, P. and Gabriel, A.H.: 1982, MNRAS 98, p. 239.
63. Bely-Dubau, F., Faucher, P., Steenman-Clark, L., Dubau, J., Loulergue, M., Gabriel, A.H., Antonucci, E., Volonté, S. and Rapley, C.G.: 1982 MNRAS 201, p. 1158.
64. Wang, J-S., Datla, R.U. and Griem, H.R.: 1984, Phys. Rev. A 29, p. 1558.
65. Chang, C.C., Greve, P., Kolk, K-H and Denze, H-J.: 1984, Phys. Scripta 29, p. 132.
66. Wang, J-S and Griem, H.R.: 1983, Phys. Rev. A 27, p. 2249.
67. Kolbe, G., Huang, Y.W. and Burgess, D.D.: 1982, J. Phys. B. 15, p. 4283.
68. Johnston, A.R. and Burrow, P.D.: 1983, J. Phys. B 15, p. 613.
69. Mendoza, C.: 1983, IAU Symp. 103 (ed. D.R. Flower) p. 143.
70. Burke, P.G. and Eissner, W.: 1982, in "Atoms in Astrophysics" Ed. P.G. Burke, D.G. Hummer and I.C. Percival, Plenum.

71. Williams, J.F.: 1984, Phys. Rev. A 29, p. 2936.
72. Mitchell, J.B.A., Ng, C.T., Forand, J.L., Levac, D.P., Mitchell, R.E., Sen, A., Miko, B. and McGowan, J.W.: 1984, Phys. Rev. Lett. 50 p. 335.
73. Dittner, P.F., Datz, S., Miller, P.D., Moak, C.D., Stelson, P.H., Bottcher, C., Press, W.D., Alton, G.D., Neskovic, N. and Fox, C.M.: 1984, Phys. Rev. Lett. 51, p. 31.
74. McLaughlin, D.J. and Hahn, Y.: 1983, Phys. Rev. A 28, p. 493.
75. Hahn, Y., Gattuta, K.L. and McLaughlin, D.J.: 1982, Phys. Rev. A 26, p. 1385.
76. Nussbaumer, H. and Storey, P.J.: 1982, Astron. Ap. 126, p. 75.
77. Nussbaumer, H. and Storey, P.J.: 1984, Astron. Ap.
78. McLaughliln, D.J. and Hahn, Y.: 1983, J. Phys. B 16, L739.
79. Seaton, M.J.: 1984, J. Phys. B 17, L531.
80. Hickman, A.P.: 1984, J. Phys. B 17, L101.
81. LaGattuta, K.J.: 1984, Phys. Rev. A 30, p. 3072.
82. Alber, G., Cooper, J. and Rau, A.R.P.: 1984, Phys. Rev. A 30, p. 2845.
83. Nussbaumer, H. and Storey, P.J.: 1984, Astron. Ap. Suppl. 56, p. 293.
84. Isler, R.C., Crune, F.C. and Arnurius, D.E.: 1982, Phys. Rev. A 26, p. 2105.
85. MacDonald, J.A., Biondi, M.A. and Johnsen, R.: 1984, Planet. Spa. Sci. 32, p. 651.
86. Mitchell, J.B.A., Forand, J.L., Ng, C.T., Levac, D.P., Mitchell, R.E., Mul, P.M., Claeys, W., Sen, A. and McGowan, J.W.: 1983, Phys. Rev. Lett. 51, p. 885.
87. Smith, D. and Adams, W.G.: 1984, Ap. J. Lett. 284, L13; Adams, N.G., Smith and Alge, E.: 1984, J. Chem. Phys. 81, p. 1778.
88. Mul, P.M., McGowan, J.W. Defrance, P. and Mitchell, J.B.A.: 1983, J. Phys. B 16, p. 3099.
89. Mitchell, J.B.A., Ng., C.T., Forand, L., Janssen, R. and McGowan, J.W.: 1984, J. Phys. 17, L909.
90. Alge, E., Smith, D. and Adams, N.G.: 1983, J. Phys. B 16, p. 1433.
91. Giusti-Suzor, A., Bardsley, J.W. and Derkits, C.: 1983, Phys. Rev. A 28, p. 682.
92. Michels, H.H. and Hobbs, R.H.: 1984, Ap. J. Lett. 286, L27.
93. Kurawaki, J. and Ogawa, T.: 1983, Chem. Phys. Lett. 98, p. 12.
94. Higo, M., Kamata, S. and Ogawa, T.: 1982, Chem. Phys. 66, p. 243 and 73, p. 99.
95. Morrison, M.A., Feldt, A.N. and Saha, B.C.: 1984, Phys. Rev. A 30, p. 2811.
96. Luna, M.A.P., Gibson, T.L., Taktasuka, K. and McKoy, V.: 1984, Phys. Rev. A 30, p. 1741.
97. Morrison, M.A., Feldt, A.N. and Austin, D.: 1983, Phys. Rev. A 29, p. 2518.
98. Wadehra, J.M.: 1983, Phys. Rev. A 29, p. 106.
99. Mu-Tao, L., Lucchese, R.R. and McKoy, V.: 1982, Phys. Rev. A 26, p. 3240.
100. Holley, T.K., Chung, S., Lin, C.C. and Lee, F.T.P.: 1983, Phys. Rev. A 26, p. 1852.
101. Fritsch, W. and Lin, C.D.: 1984, J. Phys. B 17, p. 3271.
102. Hanssen, J., Gayet, R., Harel, C. and Salin, A.: 1984, J. phys. B 17, L323.
103. Hardie, D.J.W. and Olson, R.E.: 1983, J. Phys. B 16, 1983.
104. Mandal, C.R., Dalta, S. and Mukherjee, S.C.: 1984, Phys. Rev. A 30, p. 1104.
105. Janev, R.K., Belic, D.S. and Bransden, B.H.: 1983, Phys. Rev. A 28, p. 1310.
106. Heil, T.G., Butler, S.E. and Dalgarno, A.: 1983, Phys. Rev. A 27, p. 2365.
107. Feickert, C.A., Blint, R.J., Surratt, G.T. and Watson, W.D.: 1984, Ap. J. 286, p. 371.
108. Gargaud, M., McCarroll, R. and Valiron, P.: 1982, Astron. Ap. 106, p. 197.
109. Janev, R.K.: 1983, Comm. Atom. Molec. Phys. 22, p. 277.
110. Bottcher, C. and Heil, T.G.: 1982, Chem. Phys. Lett. 86, p. 506.
114. Grizdanov, T.P. and Belic, D.S.: 1984, Phys. Scripta 30, p. 194.
115. Butler, S.E., Heil, T.G. and Dalgarno, A.: 1984, J. Chem. Phys. 80, p. 4986.
116. Kimura, M. et al.: 1984, J. Phys. Soc. Japan 53, p. 2224.
117. Shields, G.A., Dalgarno, A. and Sternberg, A.: 1983, Phys. Rev. A 28, p. 2137.

118. Johnson, R.E. and Strobel, D.F.: 1982, J. Geophys. Res. 87, p. 10385.
119. Yoshida, J. and O-ohata, K.: 1983, J. Phys. Soc. Japan 52, p. 417.
120. Suzuki, H., Kajikawa, Y., Toshurna, N., Ryufuku, H. and Watadre, T.: 1984, Phys. Rev. A 29, p. 525.
121. Suzuki, H., Toshimo, N., Watanbe, T. and Ryufuku, H.: 1984, Phys. Rev. A 29, p. 529.
122. Phaneuf, R.A.: 1983, Phys. Rev. A 28, p. 1310.
123. Iwai, T., Kaneko, Y., Kimuira, M., Kobayashi, N., Ohtami, S., Okimo, K., Takagi, S., Tawara, H. and Tsurubuchi, S.: 1982, Phys. Rev. A 26, p. 105.
124. Tsurubuchi, S. et al.: 1982, J. Phys. B 15, L733.
125. Okimo, K. et al.: 1983, Phys. Rev. A 28, p. 127.
126. Church, D.A. and Holzscheiter, A.M.: 1982, Phys. Rev. Lett. 49, p. 643.
127. Gallagher, J.W., Bransden, B.H. and Janev, R.K.: 1983, J. Phys. Chem. Ref. Data 12, p. 873.
128. Janev, R.K., Bransden, B.H. and Gallagher, J.W.: 1983, J. Phys. Chem. Ref. Data 12, p. 829.
129. Dalgarno, A.: 1983 in Atoms in Astrophysics, Ed. P.G. Burke, W.S. Eissner, D.G. Hummer and I.C. Percival (Plenum) p. 103.
130. Heil, T.G., Kirby, K. and Dalgarno, A.: 1982, Phys. Rev. A 27, p. 2826.
131. Heil, T.G., Green, S. and Dalgarno, A.: 1982, Phys. Rev. A 26, p. 3293.
132. Mitchell, G.F.: 1984, Ap. J. 287, p. 665.
133. Thorne, L.R., Anicich, V.G., Prasad, S.S. and Huntress, W.T.: 1984, Ap. J. 280, p. 139.
134. Herbst, E.: 1983, Ap. J. Suppl. 53, p. 41.
135. Graedel, T.E., Langer, W.D. and Frerking, M.A.: 1982, Ap. J. Suppl. 48, p. 321.
136. Mitchell, G.F. and Devau, T.J.: 1983, Ap. J. 266, p. 646.
137. Prasad, S.S. and Huntress, W.T.: 1983, Ap. J. 260, p. 590.
138. Suzuki, H.: 1983, Ap. J. 272, p. 579.
139. Yung, Y.L., Allen, M. and Pinto, J.P.: 1984, Ap. J. Suppl. 55, p. 465.
140. Leung, C.M., Herbst, E. and Huebner, W.F.: 1984, Ap. J. Suppl. 56, p. 231.
141. Mitchell, G.F.: 1984, Ap. J. Suppl. 54, p. 81.
142. Nejaid, L.A.M., Millar, T.J. and Freeman, A.: 1982, Astron. Ap. 134, p. 129.
143. Millar, T.J.: 1982, MNRAS 199, p. 309.
144. Clegg, R.E.S., van ljzendroon, L.J. and Allamondola, L.J.: 1983, MNRAS 203, p. 129.
145. Watt, G.D.: 1983, MNRAS 205, p. 321.
146. Mitchell, G.F.: 1983, MNRAS 205, p. 765.
147. Millar, T.J. and Freeman, A.: 1983, MNRAS 207, 405 and 425.
148. Adams, N.G. and Smith, D.: 1983, in Kinetics and Energetics, Ed. A. Fontijn and M.A.A. Clyne (Academic Press) p. 31.
149. Rowe, B.R., Dupeyrot, G., Marquette, J.B., Smith, D., Adams, N.G. and Ferguson, E.E.: 1984, J. Chem. Phys. 80, p. 241.
150. Miller, T.M., Welterskog, R.E. and Paulson, J.F.: 1984, J. Chem. Phys. 80, p. 4922.
151. O'Keefe, A., Parent, D., Mauclaire, G. and Bowers, M.T.: 1984, J. Chem. Phys. 80, p. 4901.
152. Tanaka, K., Kato, T., Guyon, P-M. and Koyano, I.: 1984, J. Chem. Phys. 80, p. 4441.
153. Smith, D., Adams, N.G. and Alge, E.: 1983, J. Chem. Phys. 78, p. 1261.
154. Barlow, S.E., Dunn, G.H. and Schauer, M.: 1984, Phys. Rev. Lett. 52, p. 1610.
155. Herbst, E., Smith, D. and Adams, N.G.: 1982, Astron. Sp. 138, L13.
156. Kusonoki, I. and Ottinger, Ch.: 1984, Chem. Phys. Lett. 109, p. 554.
157. Adams, N.G. and Smith, D.: 1984, Chem. Phys. Lett. 105, p. 604.
158. Tanaka, K., Kato, T., Guyon, P.M. and Koyano, J.: 1983, J. Chem. Phys. 79, p. 4302.
159. Guyer, D.R., Huwel, L. and Leone, S.R.: 1983, J. Chem. Phys. 79, p. 1259.

160. Chatham, H., Hils, D., Robertson, R. and Gallagher, A.C.: 1983, J. Chem. Phys. 79, p. 1301.
161. Kemper, P.R., Bowers, M.T., Parent, D.C., Mauclaire, G., Devai, R. and Marx, R.: 1983, J. Chem. Phys. 79, p. 1301, 79, p. 160.
162. Haudon, M., Capp, N.W. Wareing, D.P., Jones, J.D.C., Birkinshaw, K., and Twiddy, N.: 1982, J. Chem. Phys. Lett. 88, p. 63.
163. Smith, D., Adams, N.G. and Alge, E.: 1982, Ap. J. 263, p. 123.
164. Smith D., Adams, N.G. and Herbst, E.: 1983, Ap. J. 272, p. 365.
165. Herbst, E., Adams, N.G. and Smith, D.: 1982, Ap. J. 269, p. 329.
166. Federer, W., Villinger, H., Howorka, F., Lindinger, W., Tosi, P., Bassi, D. and Ferguson, E.: 1984, Phys. Rev. Lett. 52, p. 2084.
167. Illies, A.J., Jarrold, M.F. and Bowers, M.T.: 1982, J. Chem. Phys. 77, p. 5847.
168. Bass, L.M. and Bowers, M.T.: 1982, Organ. Mass. Spectr. 17, p. 229.
169. Homa, K., Kato, T., Tanaka, K. and Koyano, J.: 1984, J. Chem. Phys. 81, p. 5666.
170. Turner, T. and Lee, Y.T.: 1984, J. Chem. Phys. 81, p. 5638.
171. Turner, T., Dubuit, O. and Lee, Y.T.: 1984, J. Chem. Phys. 81, p. 3475.
172. Chesnavich, W.J., Akin, V.e. and Webb, D.A.: 1984, Ap. J. 287, p. 676.
173. Erwin, K.M. and Armentrout, P.B.: 1984, J. Chem. Phys. 80, p. 2978.
174. Adams, N.G., Smith, D. and Millar, T.J.: 1984, MNRAS 211, p. 857.
175. Ferguson, E.E., Smith, D. and Adams, N.G.: 1984, J. Chem. Phys. 81, p. 742.
176. Bates, D.R.: 1983, Ap. J. Lett. 267, L121.
177. Bates, D.R.: 1983, Ap. J. 270, p. 511.
178. Graff, M.M., Moseley, J.T. and Roueff, E.: 1983, Ap. J. 269, p. 796.
179. Herbst, E.: 1982, Chem. Phys. 65, p. 185.
180. Freed, K.F., Oka, T. and Suzuki, H.: 1982, Ap. J. 263, p. 718.
181. Black, G., Sharpless, R.L. and Slanger, T.G.: 1982, Chem. Phys. Lett. 90 p. 55.
182. Black, G., Sharpless, R.L. and Slanger, T.G.: 1982, Chem. Phys. Lett, 93 p. 598.
183. Clyne, M.A.A. and Ono, Y.: 1983, Chem. Phys. Lett. 94, p. 597.
184. Ravinshankara, A.R. and Thompson, R.L.: 1983, Chem. Phys. Lett. 99, p. 327.
185. Whyte, A.R. and Phillips, L.F.: 1983, Chem. Phys. Lett. 98, p. 590.
186. Sridharan, U.C. and Kaufman, F.: 1983, Chem. Phys. Lett. 102, p. 45.
187. Black, G., Jusinski, L.E. and Slanger, T.G.: 1983, Chem. Phys. Lett. 102, p. 64.
188. Whyte, A.R. and Phillips, I.F.: 1983, Chem. Phys. Lett. 102, p. 451.
189. Meier, U., Grothier, H.H. and Just, Th.: 1984, Chem. Phys. Lett. 106, p. 97.
190. Wu, K.T.: 1984, Chem. Phys. Lett. 107, p. 405.
191. Kleinermanns, K. and Luntz, A.C.: 1983, J. Chem. Phys. 78, p. 3774.
192. Bermann, M.R. and Lin, M.: 1984, J. Chem. Phys. 81, p. 5743.
193. Persky, A. and Brioda, M.: 1984, J. Chem. Phys. 81, p. 4352.
194. Howard, M.J. and Smith, I.W.M.: 1982, J. Chem. Soc. Faraday Trans. 2, 78, p. 1403.
195. Howard, M.J. and Smith, I.W.M.: 1983, Prog. Reaction Kinetics 12, p. 55.
196. Miller, J.C. and Gordon, R.J.: 1983, J. Chem. Phys. 79, p. 1252.
197. Baulch, D.L., Co, R.A., Crutzen, P.J., Hampson, R.F., Kerr, J.A., Troe, J. and Watson, R.T.: 1982, J. Phys. Chem. Ref. Data 11, p. 327.
198. Cohen, N. and Westberg, R.R.: 1983, J. Phys. Chem. Ref. Data 12, p. 531.
199. Rau, S.N. and Truhlar, D.G.: 1983, J. Chem. Phys. 79, p. 6046.
200. Pamidimukkala, K.M. and Skinner, G.B.: 1982, J. Chem. Phys. 76, p. 311.
201. Lee, K.T., Bowman, J.M., Wagner, A.F. and Schatz, G.C.: 1982, J. Chem. Phys. 76, p. 3583.
202. Clary, D.C.: 1984, Molec. Phys. 53, p. 3.
203. Bowman, J.M.: 1984, J. Chem. Phys. 81, p. 1739.
204. Clary, D.C. Werner, H-J.: 1984, Chem. Phys. Lett. 112, p. 346.
205. Bowman, J.M., Lee, K.T. and Walker, R.B.: 1983, J. Chem. Phys. 79, p. 3742.
206. Sun, J.C., Choi, B.H., Poe, R.T. and Tang, K.T.: 1983, J. Chem. Phys. 78 p. 4523 and 79, p. 5376.

207. Borasin, J., Thomas, R. and Borassin, A.: 1983, J. Chem. Phys. 79, p. 1546.
208. Gerrity, D.P. and Valentine, J.J.: 1983, J. Chem. Phys. 79, p. 502.
209. Margolis, S., Garetz, B.A. and Sathyamurthy, N.: 1983, J. Chem. Phys. 79, p. 2736.
210. Schwartz, S.D. and Miller, W.H.: 1983, J. Chem. Phys. 79, p. 3759.
211. Bowman, J.M. and Lee, K.T.: 1983, Chem. Phys. Lett. 94, p. 363.
212. Schatz, G.C.: 1983, Chem. Phys. Lett. 94, p. 183.
213. Abu Salbi, N., Kouri, D.J., Shuma, Y. and Baer, M.: 1984, Chem. Phys. Lett. 105, p. 472.
214. Rozenshtein, V.B., Gershenzon, Yu. M., Ivanov, A.V., Kuchergavii, S.I.: 1984, Chem. Phys. Lett. 105, p. 423.
215. Blais, N.C. and Truhlar, D.G.: 1983, Chem. k. Phys. Lett. 102, p. 120.
216. Choi, B.H., Poe, R.T. and Tang, K.T.: 1984, J. Chem. Phys. 81, p. 4979.
217. Pollak, E. and Wyatt, R.E.: 1984, J. Chem. Phys. 81, p. 1801.
218. Mayne, H.R.: 1984, J. Chem. Phys. 81, p. 2684.
219. Schatz, G.C., Hubbard, L.L., Dardi, P.S. and Miller, W.H.: 1984, J. Chem. Phys. 81, p. 231.
220. Gerrity, D.P. and Valentine, J.J.: 1984, J. Chem. Phys. 81, p. 1298.
221. Connor, J.N.L. and Southall, W.J.E.: 1984, Chem. Phys. Lett. 108, p. 527.
222. Shatz, G.C.: 1984, Chem. Phys. Lett. 108, p. 532.
223. Zamir, E., Levine, R.D. and Bernstein, R.D.: 1984, Chem. Phys. Lett. 107, p. 217.
224. Glass, G.P. and Chatuverdi, B.K.: 1982, J. Chem. Phys. 77, p. 3478.
225. Flower, D.R. and Kirkpatrick, D.J.: 1982, J. Phys. B. 15, L11.
226. Baker, D.J. and Flower, D.R.: 1984, J. Phys. B. 17, p. 119.
227. Bel Bruno, J., Gelfand, J. and Rabitz, H.: 1983, J. Chem. Phys. 78, p. 3990.
228. Green, S. and Chapman, S.: 1983, Chem. Phys. Lett. 98, p. 467.
229. Storey, J.W., Watson, D.M. and Green, S.: 1982, Ap. J. 259, p. 647.
230. Bieniek, R.J. and Green, S.: 1983, Ap. J. 265, L29.
231. Elitzur, M.: 1983, Ap. J. 266, p. 609.
232. Green, S.: 1983, Ap. J. 277, p. 900.
233. Billing, G.d. and Cacciatore, M.: 1982, Chem. Phys. Lett. 86, p. 20.
234. Price, R.J., Clary, D.C. and Billing, G.D.: 1983, Chem. Phys. Lett. 101, p. 269.
235. Teitelbaum, H.: 1984, Chem. Phys. Lett. 106, p. 69.
236. Bhattacharyaja, S.S. and Dickinson, A.S.: 1982, Astron. Ap. 107, p. 26.
237. Monterio, T.: 1984, MNRAS 210, 1.
238. Schinke, R., Meyer, H., Buck, U. and Diercksen, G.H.F.: 1984, J. Chem. Phys. 80, p. 5518.
239. Burns, M.J. and Coy, S.L.: 1984, J. Chem. Phys. 80, p. 3544.
240. Picard-Bersellini, A., Charneau, R. and Brechignac, P.L.: 1983, J. Chem. Phys. 79, p. 8900.
241. Kohler, W.E. and Schaefer, J.: 1983, J. Chem. Phys. 78, p. 6602.
242. Beard, L.H. and Kouri, D.J.: 1983, J. Chem. Phys. 78, p. 220.
243. Jolicard, G. and Grosjean, A.: 1983, J. Chem. Phys. 78, p. 2327.
244. van Hemert, M.C.: 1983, J. Chem. Phys. 78, p. 2345.
245. Klassen, D.B.M., ter, Meulen, J.J. and Dymanus, A.: 1983, J. Chem. Phys. 78, p. 767.
246. Wight, C.A., Magnotta, F. and Leone, S.R.: 1984, J. Chem. Phys. 81, p. 3951.
247. Billing, G.D. and Poulsen, L.L.: 1984, J. Chem. Phys. 81, p. 3866.
248. Roeging, A., Rabitz, H., Gelfand, J. and Miles, R.B.: 1984, J. Chem. Phys. 81, p. 820.
249. Mehrotra, L.: 1984, Astrophys. Spa. Sci. 100, p. 129.
250. Drolshage, G. and Gianturco, F.A.: 1983, Molec. Phys. 48, p. 673.
251. Andres, J., Buck, U., Meyer, H. and Launay, J.M.: 1982, J. Chem. Phys. 76, p. 1417.
252. Wood, C.F., Flynn, G.W. and Weston, R.E.: 1982, J. Chem. Phys. 77, p. 4776.
253. Andresen, P., Hausler, D. and Lulf, H.W.: 1984, J. Chem. Phys. 81, p. 571.
254. Andresen, P., Hausler, D., Lulf, H.W. and Kegel, W.H.: Astron. Ap. 138, L17.
255. Schinke, R. and Andresen, P.: 1984, J. Chem. Phys. 81, p. 5644.

256. Devangan, D.P. and Flower, D.R.: 1982, MNRAS 199 p. 457.
257. Devangan, D.P. and Flower, D.R.: J. Phys. B. 17 p. 2157.
258. Flower, D.R. and Guilloteau, S.: 1982, Astron. Ap. 114 p. 238.
259. Sakimoto, K.: 1982, Chem. Phys. 68 p. 155; 1982, J. Phys. Soc. Japan 51, 2657; 1983, Chem. Phys. 79 p. 137; 1984, Chem. Phys. 85 p. 273.
260. Takayanagi, K.: 1982, J. Phys. Soc. Japan 51 p. 3337.
261. Su, T. and Chesnovich, W.J.: 1982, J. Chem. Phys. 76 p. 5183.
262. Bates, D.R.: 1982, Proc. Roy. Soc. A 384 p. 289.
263. Bates, D.R.: 1983, Chem. Phys. Lett. 97 p. 19.
264. Roberge, W.G. and Dalgarno, A.: 1982, Ap. J. 255 p. 1176.
265. Blais, N.C. and Truhlar, D.G.: 1982, Ap. J. 258, L79.
266. Blais, N.C. and Truhlar, D.G.: 1983, J. Chem. Phys. 78 p. 2388.
267. Lepp, S. and Shull, M.J.: 1983, Ap. J. 270 p. 578.
268. Federman, S.R. and Shipsey, F.J.: 1983, Ap. J. 269 p. 791.

WORKING GROUP 4: STRUCTURE OF ATOMIC SPECTRA

A. Recent Laboratory Results and Research in Progress

A Bibliography on Atomic Energy Levels and Spectra covering the period July 1979 through December 1983 is scheduled for publication in late 1984(1). The references in Table 1 are thus limited to papers published or known to be in press in 1984; the selection was made from references giving data on energy levels, wavelengths, and line classifications for spectra of the elements Z $\leq$ 30. S. Johansson notes that the spectra under investigation in the Physics Department of Lund University, Sweden, include S III, IV, VII, VIII, IX; Cl VI, VII, VIII, X; Sc V; the (Ne I) isoelectronic spectra Sc XII-Fe XVII; Ti I; Cr I, II, V; Fe II, III; and several high-ionization spectra of Ni, the work on Fe III and the Ni spectra being done in collaboration with researchers in the Zeeman Laboratory, Amsterdam. Term analyses in progress at the Lund Institute of Technology, as reported by W. Persson, include Ne III and K II, III. This group has recently completed analyses for Kr IV, Rb V, Sr V, VI and has papers in preparation for Rb IV, Y VI, and Xe II. Fourier-transform spectrometric observations at the National Solar Observatory (Kitt Peak) by J. Brault and collaborators are yielding more accurate and complete line lists for the first and second spectra of several elements, including Mg, Al, Ti, V, Cr, Fe, and Ni. Ongoing extensions of the term analyses for some of these spectra were mentioned above; similar work is underway for Mg I, II, Al I, II, and V I, II (E. Biémont, University of Liége) and for Fe I (R. C. M. Learner, Imperial College, London). B. C. Fawcett's current research at the Rutherford Appleton Laboratory, Oxfordshire, England, includes work on the oxygen-like spectra P VIII through Fe XIX.

B. Compilations, Isoelectronic-Sequence Observations and Predictions

The NBS Atomic Energy Levels Data Center published energy-levels compilations for all the spectra of Fe(29) and Si(30). Revised and updated energy-levels compilations for the 235 spectra of the ten iron-group elements K through Ni will appear in a single volume in 1985(31). A compilation of levels for the P spectra has also been submitted(32), and C. E. Moore has prepared new tables of levels and multiplets for O IV and O III(33).

R. L. Kelly's revised compilation of atomic lines below 2000 Å is scheduled for publication in 1985(34). These tables supersede a similar 1973 compilation and cover the same elements, H through Kr. Fawcett's review(28) includes tables of observed forbidden lines for ions stripped to the n=2 and n=3 shells. Some of the tables of atomic wavelengths in the CRC Handbook were revised and updated for the latest edition(35); about 45000 lines from the first five spectra of all elements in all wavelength ranges are included.

A number of papers give energy levels and/or wavelengths as observed or predicted along isolectronic sequences. Some of the predictions are sufficiently accurate to be used for line identifications in high-ionization spectra. Peacock et al.(15) have measured wavelengths of lines of Ar, Ti, Cr, Fe and Ni in ionization stages belonging to the Li, O, F, Na, or Mg isoelectronic sequences; for the F-like and Mg-like ions, predicted wavelengths are listed for the elements to Mo (Z=42). B. Edlén has further refined his methods to obtain accurate predictions for the n=2 configurations of the Li, Be, O, and F sequences(36), the B sequence(37), the N sequence(38), and the C sequence(39). These papers include critical compilations of pertinent experimental results, as does Edlén's survey of magnetic-dipole lines in the spectra of ions of the elements Cr to Ag(40). Fawcett's tabulations of calculated quantities for the Be(41), F(42), Mg(43), and Al(44) isoelectronic sequences include predicted wavelengths and, in most cases, experimental wavelengths for some transitions. Recent calculations by Hata and Grant(45) give predicted wavelengths for selected lines of He-like, Li-like, and Be-like ions of the elements Ca to Cu. Additional references for isoelectronic-sequence predictions are given in(1) and in the report of Working Group 2.

TABLE 1

Selected references on energy levels, wavelengths, and line classifications for spectra of elements Z $\leq$ 30. This table is supplementary to a recent bibliography (1) and also does not include references described in Sec. II of this report.

Fe VII (2)	V II (17)
	V XIV (14)
Mg I (3)	
	Cr XV (14)
Al I (3)	Cr XX, XXI (18)
Al IV (4)	Cr XXII, XXIII (19)
Al VIII-X (5)	
Al XI (6)	Fe II (20)
	Fe XV (21)
Si I (7,8)	Fe XVII (14,22)
	Fe XVIII-XX (15)
S I (9)	
S III (10)	Co III (23)
S VII (11)	
	Ni XIX (14,22)
Cl VII (12)	
	Cu XXVI-XXVIII (24)
Ar IX (13)	
	Zn IV (25)
Sc XII (14)	Zn VI (26)
	Zn XIV-XIX (27)
Ti XIII (14)	
Ti XIV-XX (15)	
Ti XXI, XXII (16)	

References

1. Musgrove, A. and Zalubas, R.: 1984, Bibliography on Atomic Energy Levels and Spectra, July 1979 through December 1983, NBS Spec. Publ. 363, Suppl. 3 (U.S. Government Printing Office, Washington, D.C. 20402, U.S.A.).
2. Engström, L.: 1984, Phys. Scr. 29, p. 113.
3. Chang, E.S. and Noyes, R.W.: 1983, Astrophys. J. 275, L.11.
4. Madin, M.I.: 1983, Opt. Spectrosc. (USSR) 55, p. 336.
5. Denne, B. and Hinnov, E.: 1984, Phys. Rev. A 29, p. 3442.
6. Buchet, J.P., Buchet-Poulizac, M.C., Denis, A., Desesquelles, J., Druetta, M., Martin, S., Grandin, J.P., Hennecart, D., Husson, X. Lecler, D.: 1984, J. Phys. (Paris) Lett. 45, L361.
7. Inguscio, M., Evenson, K.M., Beltran-Lopez, V. and Ley-Koo, E.: 1984, Astrophys. J. 278, L127.
8. Chang, E.S.: 1984, J. Phys. B 17, L11.
9. Sarma, V.N. and Joshi, Y.N.: 1984, Physica (Utrecht) 123C, p. 349.
10. Smith, P.L., Magnusson, C.E. and Zetterberg, P.O.: 1984, Astrophys. J. 277, L79.
11. Jupén, C., Litzén, U. and Trigueiros, A.: 1984, Phys. Scr. 29, p. 317.
12. Jupén, C., Fremberg, J. and Fawcett, B.: 1984, Phys. Scr. 30, p. 260.
13. Fawcett, B.C.: 1984, Phys. Scr. (submitted for publication).
14. Jupén, C., Litzén, U.: 1984, Phys. Scr. 30, p. 112.
15. Peacock, N.J., Stamp, M.F. and Silver, J.D.: 1984, Phys. Scr. T8, p. 10.
16. Bitter, M., von Goeler, S., Cohen, S., Hill, K.W., Sesnic, S., Tenney, F., Timberlake, J., Safronova, U.I., Vainshtein, L.A., Dubau, J., Loulergue, M., Bely-Dubau, F. and Steenman-Clark, L.: 1984, Phys. Rev. A 29, p. 661.
17. Iglesias, L., Rico, F.R. and Garcia-Riquelme, O.: 1984, Opt. Pura Apl. 17 p. 93.
18. Burkhalter, P.G., Charatis, G., Rockett, P.D. and Newman, D.: 1984, J. Opt. Soc. Am. B 1, p. 155.
19. Grandin, J.P., Huet, M., Husson, X., Lecler, D., Touvet, D., Buchet, J.P., Buchet-Poulizac, M.C., Denis, A., Desesquelles, J. and Druetta, M.: 1984, J. Phys. (Paris) 45, p. 1423.
20. Johnsson, S. and Jordan, C.: 1984, Mon. Not. R. Astron. Soc. 210, p. 239.
21. Briand, J.P., Tavernier, M., Marrus, R. and Desclaux, J.P.: 1984, Phys. Rev. A 29, p. 3143.
22. Jupén, C.: 1984, Mon. Not. R. Astron. Soc. 208, p.1.
23. Raassen, A.J.J. and Orti Ortin, S.: 1984, Physica (Utrecht) 123C., p. 353.
24. Buchet, J.P., Buchet-Poulizac, M.C., Denis, A., Desesquelles, J., Druetta, M., Grandin, J.P., Husson, X. and Lecler, D.: 1984, Nucl. Instrum. Methods (in press).
25. van Kleef, T.A.M., Joshi, Y.N. and Barakat, M.M.: 1984, Phys. Scr. 29, p. 216.
26. van Kleef, T.A.M., Joshi, Y.N., Barakat, M.M. and Meijer, F.G.: 1984, Physica (Utrecht) 125C, p. 97.
27. Burrell, K.H., Groebner, R.J., Brooks, N.H. and Rottler, L.: 1984, Phys. Rev. A 29, p. 1343.
28. Fawcett, B.C.: 1984, J. Opt. Soc. Am. B 1, p. 195.
29. Corliss, C. and Sugar, J.: 1982, J. Phys. Chem. Ref. Data 11, p. 135.
30. Martin, W.C. and Zalubas, R.: 1983, J. Phys. Chem. Ref. Data 12, p. 323.
31. Sugar, J. and Corliss, C.: 1985, Atomic Energy Levels of the Iron Period Elements: Potassium through Nickel (submitted for publication as a Supplement of J. Phys. Chem. Ref. Data).
32. Martin, W.C., Zalubas, R. and Musgrove, A.: 1985 (submitted to J. Phys. Chem. Ref. Data).
33. Moore, C.E.: 1982, Nat. Stand. Ref. Data Ser., Nat. Bur. Stand. (U.S.) 3, Sec. 10 [O IV]; 1984, ibid., Sec. 11 [O III].
34. Kelly, R.L.: 1985, Atomic and Ionic Lines Below 2000 Angstroms (submitted for publication as a Supplement to J. Phys. Chem. Ref. Data).

35. Reader, J. and Corliss, C., Editors: 1984, in CRC Handbook of Chemistry and Physics, 65th Edition (CRC Press, Boca Raton, FL, U.S.A.).
36. Edlén, B.: 1983, Phys. Scr. 28, p. 51.
37. Edlén, B.: 1983, Phys. Scr. 28, p. 483.
38. Edlén, B.: 1984, Phys. Scr. 30, p. 135.
39. Edlén, B.: 1985, Phys. Scr. (to be published)
40. Edlén, B.: 1984, Phys. Scr. T8, p. 5.
41. Fawcett, B.C.: 1984, At. Data Nucl. Data Tables 30, 1; ibid. 30, p. 423.
42. Fawcett, B.C.: 1984, At. Data Nucl. Data Tables (in press).
43. Fawcett, B.C.: 1983, At. Data Nucl. Data Tables 28, p. 579.
44. Fawcett, B.C.: 1983, At. Data Nucl. Data Tables 28, p. 557.
45. Hata, J. and Grant, I.P.: 1984, Mon. Not. R. Astron. Soc. (in press).

W.C. Martin
Chairman of the Working Group

WORKING GROUP 5: MOLECULAR SPECTROSCOPY

A. Compendia. Bibliographies. Atlases

Research in molecular spectroscopy over much of the electromagnetic spectrum has continued intensively over the past three years. It has been stimulated not only by the imperatives of fundamental research programmes in many laboratories, but also by the impact of molecular lasers on the field, and the needs of atmospheric and environmental programmes. The literature is so prolific that it is impossible even to review briefly here all that is relevant to astrophysical needs. Thus most of this report has been compiled from the contributions from individual workers and Research Centres.

The bi-monthly Berkeley Newsletter(1) compiled from molecular spectroscopic publications in more than 30 journals continues to be a prime bibliographic tool for its more than 500 subscribers.

B. Molecular Data

1. ELECTRONIC BAND ANALYSES AND CONSTANTS OF DIATOMIC MOLECULES

Using laser spectroscopic methods Linton et al.(2) in a collaboration of workers from four laboratories, have continued their definitive studies of the CeO spectrum. The energies of 16 low lying states have been determined and quantum numbers assigned with certainty to 14 of them, and identification and analysis tables are provided for many bands.

Hirota and collaborators at the Institute of Molecular Science at Okazaki have an extensive programme on the microwave and infrared spectra of CH_3(3,4), HO_2(5), SiN(6), PO(7,8), HCCN(9) and FeO(10). Hefferlin(11) continues his theoretical and numerical studies on the systematic properties of constants of diatomic molecules.

2. TRANSITION PROBABILITIES, LIFETIMES, INTENSITIES

Dressler (Zurich) reports the following experimental and theoretical studies on f- values, lifetimes, and transition moments for band systems of NO, N_2 and H_2. Experimental and theoretical studies have been made on transitions between excited $^2\Pi$ states of NO by Gallusser and dressler(12). Similar work has been done for transitions between singlet states of N_2 by Stahel, Leoni and Dressler(13). An extensive study of lifetimes and transitions probabilities for a number of H_2 states and transitions between then are also reported(14-18).

Sharp (Munich)(19) has developed an efficient computational method for numerical evaluation of Morse Franck-Condon factors and r-centroids of molecular band systems.

RESEARCH IN PROGRESS

A. From the Herzberg Institute of Astrophysics, National Research Council of Canada (Reported by A.R.W. McKellar)

Studies of the spectra of molecules of astronomical interest have continued to be the major activity in this laboratory. The diatomic molecules CH^+ (20), NH(21), BH^+(22), Geo(23), SiN(24), ClF(25) have been studied in the visible and ultraviolet regions. Infrared studies include the ions HeH^+(26), NeH^+(27), ArH^+ and KrH^+(28), as well as the neutral diatomics HN(29), SO(30), HD(31,32), SH(33), O_2(34), OH(35) and CH(36).

Significant advances have been made in the study of the triplet ground electronic state of the CH_2 radical, mostly involving the laser magnetic resonance technique. References to this work may be found in a paper(37) giving transition frequencies in the infrared and far infrared regions for astronomical purposes. Other triatomic molecules studied include NH_2(38-40), PH_2(41), NCO(42), CNO(43), HSO(44), HNO(45), FNO(46), and H_2S(47). The closely related ions NCO and $NH_2{}^+$ and their deuterated counterparts, were studied in the infrared(48-51). Interest continued in the spectrum of $H_3{}^+$ and its deuterated isotopes(52-54); molecules which are thought to play a key role in interstellar chemistry.

Following on from the study of the Rydberg spectra of neutral triatomic hydrogen, there has been considerable interest in the analogous spectrum of the ammonium (NH_4 and ND_4) radical(55-58). Studies of formaldehyde in the ultraviolet(59-62) and of thioformaldheyde, H_2CS, in the visible(63-70) have continued. In the infrared region, the CH_3 radical(71,72) and the H_3O^+ ion(73-75) have been studied, and extensive spectra of diacetylene have been analyzed(76). Prediction of the structure and spectra of SiH_3(77) and GeH_2(78) have been made and features due to the hydrogen dimer $(H_2)_2$, have been identified in the far infrared spectra of Jupiter and Saturn(79).

B. From the National Bureau of Standards (Reported by F.J. Lovas)

The Molecular Spectroscopy Division of the National Bureau of Standards carries out experimental studies on molecular spectroscopy in the microwave, infrared and visible regions of the spectrum and develops critical reviews on microwave spectra of interstellar molecules and tables of infrared absorption lines for calibration of diode laser measurements.

During the past three years several new reviews have been published or submitted. A review of the microwave spectra of ethanol (C_2H_5OH), and propionnitrile (C_2H_5CN) has appeared(80), and a review of SO_2(81) is in press. An update to a previously published table of rest frequencies for interstellar molecules is nearly finished(82). Compilations of infrared spectra include the species N_2O(83), OCS(84) and H_2O(85). A comprehensive review of vibrational energy levels of transient molecules was recently completed(86).

New laboratory studies in the microwave region have been published on BH_3NH_3(87), CH_2(88), CD_2(89) and H_2NSH(90). A matrix isolation study of CH_3CO and CH_2CHO was completed(91). Gas phase infrared spectra are reported for AlF and KF(92), benzene(93), ethane(94,95), SnO(96), OCS far infrared(97), CO(98) LiF(99) and CO_2(100). Analyses of electronic spectra of H_3(101) and ND_4(102) were also reported.

C. From the Center for Astrophysics, Harvard College Observatory and Smithsonian Astrophysical Observatory (reported by W.H. Parkinson)

Since the last report we have used the 6.65 m vacuum scanning spectrometer/spectrograph(103) to make the following high resolution cross section measurements on molecules of atmospheric and/or astrophysical significance.

(a) Cross sections of the Schumann-Runge bands of O_2 at 300 K have been measured between 179.3 and 201.5 nm with an instrumental full width at half maximum (FWHM) of 0.0013 nm(104). Oscillator strengths of the (12,0) through (1,0) bands have been obtained by direct integration of the measured cross sections.

(b) Oscillator strengths for some rovibronic lines of the Schumann-Runge bands (13,0) through (16,0) or O_2 have been obtained between 176 and 179 nm, and the prospects for their use for Space Telescope observations of interstellar O_2 have been discussed(105).

(c) The absorption cross section of O_2 has been measured between 193.5 and 204.0 nm at 300 K and FWHM of 0.0013 nm. This spectral region contains discrete Schumann-Runge lines and underlying dissociation continua of the weak Herzberg I transition of O_2 and of the O_2 dimer. Analysis of the pressure dependence of the continuum cross section between the lines leads to Herzberg continuum cross sections smaller than previous laboratory values and in accord with recent *in situ* stratospheric measurements (106,107).

(d) Cross sections of O_3 at 195 K have been measured between 240 and 350 nm at a resolution of 0.003 nm, which is sufficient to resolve all of the fine structure present in the Hartley-Huggins bands(108).

(e) Cross sections of SO_2 at 213 K have been measured at a resolution of 0.002 nm from 172 to 240 nm(109). The results should be useful in elucidating the atmospheric compositions of Venus and Io.

(f) Cross sections of N_2O at 295-299 K have been measured between 170 and 222 nm with a resolution sufficient to yield cross sections that are independent of the instrumental function. Previously unresolved details of the banded structure which is superimposed on the continuous absorption between 174 and 190 nm are observed(110).

We have also compiled a photographic atlas of the Schumann-Runge absorption bands of O_2 between 175 and 205 nm(111).

In the microwave region of the spectrum, measurements on the submillimeter Zeeman spectra of OH(112) and HCO(113) have been completed. Fixed frequency lasers are used as radiation sources and the absorption spectra are scanned using the Zeeman effect. For comparison with astronomical submillimeter spectra, detailed theoretical calculations of the molecular Zeeman effect have been carried out by J.W. Brown and colleagues of the University of Southampton for OH. Similar calculation on HCO are almost complete.

D. From the Centre for Research in Experimental Space Science, York University, Toronto (reported by R.W. Nicholls)

An experimental and theoretical research programme continues for the provision of transition probability and structure constant data for astrophysical and atmospheric molecules and applications.

An analysis of the rotational structure of the ClO violet bands has recently been completed by Morreau(114). Studies have continued on the line by line absorption coefficients of the absorption bands of ClO. The realistic spectrum synthesis facility developed in our laboratory has been applied to a wide range of atmospheric and astrophysical absorption spectra from the mm wave to the vacuum UV regions of the spectrum. Very recent studies include the synthesis of very high pressure oxygen(115) and long paths of CO_2(116). Such work has been applied to the extinction of sunlight in the atmosphere(117).

Studies have continued on the systematic properties of Franck-Condon factor arrays(118,119) and the relationship between simple harmonic oscillator Franck-Condon factors and those calculated for more realistic models(120). Studies have also been made on the geometry of principal and subsidiary Condon loci(121,122).

Analytical studies have been made of the wavelength dependence of Mie scattering coefficients(123). This work has led to an analytical form for the interstellar extinction function from the far IR to 100 nm(124,125).

References

1. Phillips, J.G., Davis, S.P. and Eakin, D.M.: Berkeley Newsletter, Departments of Astronomy and Physics, University of California, Berkeley.
2. Linton, C., Dulick, M., Field, R.W., Carette, P., Leyland, P.C. and Barrow, R.F.: 1983, J. Mol. Spectrosc. 102, p. 441.
3. Yamada, C., Hirota, E. and Kawaguchi, K.: 1981, J. Chem. Phys. 75, p. 5256.
4. Amano, T., Bernath, P.F., Yamada, C., Endo, Y. and Hirota, E.: 1982, J. Chem. Phys. 77, p. 5284.
5. Yamada, C., Endo, Y. and Hirota, E.: 1983, J. Chem. Phys. 78, p. 4379.
6. Saito, S., Endo, Y. and Hirota, E.: 1983, J. Chem. Phys. 78, p. 6447.
7. Kawaguchi, K., Saito, S. and Hirota, E.: 1983, J. Chem. Phys. 79, p. 629.
8. Butler, J.E., Kawaguchi, K. and Hirota, E.: 1983, J. Mol. Spectrosc. 101, p. 161.
9. Saito, S., Endo, Y. and Hirota, E.: 1984, J. Chem. Phys. 80, p. 1427.
10. Endo, Y., Saito, S. and Hirota, E.: 1984, Astrophys. J. 278, L131.
11. Hefferrlin, R.: 1985, J. Quant. Spect. Rad. Transf. (in press).
12. Gallusser, R. and Dressler, K.: 1982, J. Chem. Phys. 76, p. 4311.
13. Stahel, D., Leoni, M. and Dressler, K.: 1983, J. Chem. Phys. 79, p. 2541.
14. Wolniewicz, L. and Dressler, K.: 1982, J. Mol. Spectrosc. 96, p. 195.
15. Glass-Maujean, M., Quadrelli, P., Dressler, K. and Wolniewicz, L.: 1983, Phys. Rev. A 28, p. 2868.
16. Glass-Maujean, M., Quadrelli, P. and Dressler, K.: 1984, At. Data and Nucl. Data Tables 30, p. 273.
17. Glass-Maujean, M., Quandrelli, P. and Dressler, K.: 1984, J. Chem. Phys., 80, p. 4355.
18. Wolniewicz, L. and Dressler, K.: 1984, Can J. Phys (Dec 1984- in press)
19. Sharp, C.M.: 1984, Astron. Astrophys. Sup. Ser. 55, p. 33.
20. Carrington, A. and Ramsay, D.A.: 1982, Phys. Scripta 25, p. 272.
21. Ramsay, D.A. and Sarre, P.J.: 1982, J. Mol. Spectrosc. 93, p. 445.
22. Ramsay, D.A. and Sarre, P.J.: 1982, Faraday Trans. (2) 78, p. 1331.
23. Appelblad, O., Fredin, S., Lagerqvist, A. and Alberti, F.: 1983, Phys. Scripta 28, p. 160.

24. Foster, S.: 1984, J. Mol. Spectrosc. 106, p. 369.
25. Alberti, F., Huber, K.P. and Looi, E.C.: 1983, J. Mol. Spectrosc., 102, p. 289.
26. Bernath, P. and Amana, T.: 1982, Phys. Rev. Lett., 48, p. 20.
27. Wong, M., Bernath, P. and Amano, T.: 1982, J. Chem. Phys., 77, p. 693.
28. Johns, J.W.C.: 1984, J. Mol. Spectosc. 106, p. 369.
29. Bernath, P.F. and Amano, T.: 1982, J. Mol. Spectrosc. 95, p. 359.
30. Wong, M., Amano, T. and Bernath, P.: 1982, J. Chem. Phys. 77, p. 2211.
31. Riche, N.H., Johns, J.W.C. and McKellar, A.R.W.: 1982, J. Mol. Spectrosc. 95, p. 432.
32. Riche, N.H. and McKellar, A.R.W.: 1983, Can. J. Phys. 61, p. 1648.
33. Bernath, P.F., Amano, T. and Wong, M.: 1983, J. Mol. Spectrosc. 98, p. 20.
34. Reid, J., Sinclair, R.L., Robinson, A.M. and A.R.W. McKellar: 1981, Phys. Rev. A 24, p. 1944.
35. Amano, T.: 1984, J. Mol. Spectrosc. 103, p. 436.
36. Lubic, K.G. and Amano, T.: 1984, J. Chem. Phys. 81, p. 1655.
37. Sears, T.J., McKellar, A.R.W., Bunker, P.R., Evenson, K.M. and Brown, J.M.: 1984, Astrophys. J. 276, p. 399.
38. Vervloet, M. and Merienne-Lafore, M.F.: 1982, Can. J. Phys. 60, p. 49.
39. Amano, T., Bernath, P.F. and McKellar, A.R.W.: 1982, J. Mol. Spectrosc. 94, p. 100.
40. Merrienne-Lafore, M.F. and Vervloet, M.: 1984, J. Mol. Spectrosc. 108, p. 160.
41. Birss, F.W., Lessard, G., Thrush, B.A. and Ramsay, D.A.: 1982, J. Mol. Spectrosc. 92, p. 269.
42. Barnes, C.E., Brown, J.M., Fackerell, A.D. and Sears, T.J.: J. Mol. Spectrosc. 92, p. 485.
43. Ramsay, D.A. and Winnewisser, M.: 1983, Chem. Phys. Lett. 96, p. 502.
44. Sears, T.J. and McKellar, A.R.W.: 1983, Mol. Phys., 49, p. 25.
45. John, J.W.C., McKellar, A.R.W. and Weinberger, E.: 1983, Can. J. Phys. 61, p. 1106.
46. Foster, S.C. and Johns, J.W.C.: 1984, J. Mol. Spectrosc. 103, p. 176.
47. Flaud, J.M., Camy-Peyret, C. and Johns, J.W.C.: 1983, Can. J. Phys. 61 p. 1462.
48. Amano, T.: 1983, J. Chem. Phys., 79, p. 3595.
49. Kraemer, W.P. and Bunker, P.R.: 1983, J. Mol. Spectrosc. 101, p. 379.
50. Foster, S.C., McKellar, A.R.W. and Sears, T.J.: 1984, J. Chem. Phys. 81, p. 578.
51. Foster, S.C. and McKellar, A.R.W.: 1984, J. Chem. Phys. 81, p. 3424.
52. Oka, T.: 1981, Phil. Trans. Roy. Soc. 303A, p. 543.
53. Watson, J.K.G.: 1984, J. Mol. Spectrosc. 103, p. 350.
54. Amano, T. and Watson, J.K.G.: 1984, J. Chem. Phys. 81, p. 2869.
55. Watson, J.K.G.: 1984, J. Mol. Spectrosc. 107, p. 124.
56. Alberti, F., Huber, K.P. and Watson, J.K.G.: 1984, J. Mol. Spectrosc. 107, p. 133.
57. Herzberg, G.: 1984, J. Astrophys. Astr. 5, p. 131.
58. Watson, J.K.G.: 1984, J. Mol. Spectrosc. 103, p. 125.
59. Jensen, P. and Bunker, P.R.: 1982, J. Mol. Spectrosc. 94, p. 114.
60. Kerr, C.M.L., Moule, D.C. and Ramsay, D.A.: 1983, Can. J. Phys., 61, p. 6.
61. Clouthier, D.J., Craig, A.M. and Ramsay, D.A.: 1983, Can. J. Phys. 61, p. 1073.
62. Clouthier, D.J., Craig, A.M. and Birss, F.W.: 1984, Can. J. Phys. 62, p. 973.
63. Goddard, J.D. and Clouthier, D.J.: 1982, J. Chem. Phys. 76, p. 5039.
64. Clouthier, D.J., Moule, D.C., Ramsay, D.J. and Birss, F.W.: 1982, Can. J. Phys. 60, p. 1212.
65. Jensen, P. and Bunker, P.R.: 1982, J. Mol. Spectrosc. 95, p. 92.
66. Clouthier, D.J. and Kerr, C.M.L.: 1982, Chem. Phys. 70, p. 55.
67. Clouthier, D.J. and Kerr, C.M.L.: 1983, Chem. Phys. 80, p. 299.

68. Ramsay, D.A.: 1983, Ann. Rev. Phys. Chem. 34, p. 31.
69. Clouthier, D.J. and Ramsay, D.A.: 1983, J. Chem. Phys., 79, p. 5851.
70. Fung, K.H. and Ramsay, D.A.: 1984, J. Phys. Chem. 88, p. 39.
71. Amano, T., Bernath, P.F., Yamada, C. Endo, Y. and Hirohita, E.: 1982, J. Chem. Phys. 77, p. 5284.
72. Spirko, V. and Bunker, P.R.: 1982, J. Mol. Spectrosc. 95, p. 405.
73. Bunker, P.R. and Kraemer, W.P. and Spirko, V.: 1983, J. Mol. Spectrosc. 101, p. 180.
74. Spirko, V. and Bunker, P.R.: 1982, J. Mol. Spectrosc. 95, p. 226.
75. Bunker, P.R., Amano, T. and Spirko, V.: 1984, J. Mol. Spectrosc. 107, p. 208.
76. Guelachvili, G., Craig, A.M. and Ramsay, D.A.: 1984, J. Mol. Spectrosc. 105, p. 156.
77. Bunker, P.R. and Olbrich, G.: 1984, Chem. Phys. Lett. 109, p. 41.
78. Bunker, P.R., Phillips, R.A. and Buenker, R.J.: 1984, Chem. Phys. Letts. 110, p. 351.
79. McKellar, A.R.W.: 1984, Can. J. Phys. 62, p. 760.
80. Lovas, F.J.: 1982, J. Phys. Chem. Ref. Data 11, p. 251.
81. Lovas, F.J.: 1983, J. Phys. Chem. Ref. Data (in press).
82. Lovas, F.J.: 1984, In Preparation for J. Phys. Chem. Ref. Data.
83. Olson, W.B., Maki, A.G. and Lafferty, W.J.: 1981, J. Phys. Chem. Ref. Data 10, p. 1065.
84. Maki, A.G., Wells, J.S., Petersen, F.R., Olson, W.B., Fayt, A. and Sattler, J.P.: 1984, Submitted to J. Chem. Phys. Ref. Data.
85. Pine, A.S., Coulombe, M.J., Camy-Peyret, C. and Flaud, J.M.: 1983, J. Chem. Phys. Ref. Data 12, p. 413.
86. Jocox, M.E.: 1984, Submitted to J. Chem. Phys. Ref. Data.
87. Thorne, L.R., Suenram, R.D. and Lovas, F.J.: 1983, J. Chem. Phys. 78, p. 167.
88. Lovas, F.J., Suenram, R.D. and Evenson, K.M.: 1983, Astrophys. J. Letters, 267, L131.
89. Bunker, P.R., Sears, T.J., McKellar, A.R.W., Evenson, K.M. and Lovas, F.J.: 1983, J. Chem. Phys. 79, p. 1211.
90. Lovas, F.J., Suenram, R.D. and Stevens, W.J.: 1983, J. Mol. Spectrosc. 100, p. 316.
91. Jacox, M.E.: 1982, Chem. Phys. 69, p. 407.
92. Maki, A.G. and Lovas, F.J.: 1982, J. Mol. Spectrosc., 95, p. 80.
93. Pliva, J. and Pine, A.S.: 1982, J. Mol. Spectrosc. 93, p. 209.
94. Pine, A.S. and Lafferty, W.J.: 1982, J. Res. NBS 87, p. 237.
95. Henry, L., Valentin, A., Lafferty, W.J., Hougen, J.T., Devi, V.M., Des, P.P. and Rao, K.N.: 1983, J. Mol. Spectrosc. 100, p. 260.
96. Maki, A.G. and Lovas, F.J.: 1983, J. Mol. Spectrosc. 98, p. 146.
97. Wells, J.S., Petersen, F.R. and Maki, A.G.: 1983, J. Mol. Spectrosc. 98, p. 404.
98. Pollock, C.R., Petersen, F.R., Jennings, D.A., Wells, J.S. and Maki, A.G.: 1983, J. Mol. Spectrosc. 99, p. 357.
99. Maki, A.G.: 1983, J. Mol. Spectrosc. 102, p. 361.
100. Petersen, F.R., Wells, J.S., Siemsen, K.J., Robinson, A.M. and Maki, A.G.: 1984, J. Mol. Spectrosc. 105, p. 324.
101. Herzberg, G., Hougen, J.T. and Watson, J.K.G.: 1982, Can. J. Phys. 60, p. 1261.
102. Herzberg, G. and Hougen, J.T.: 1983, J. Mol. Spectrosc. 97, p. 430.
103. Yoshino, K., Freeman, D.E. and Parkinson, W.H.: 1980, Applied Optics, 19, p. 66.
104. Yoshino, K., Freeman, D.E., Esmond, J.R. and W.H. Parkinson: 1983, Planet. Space. Sci. 31, p. 339.
105. Smith, P.L., Griesinger, H.E., Black, J.H., Yoshino, K. and Freeman, D.E.: 1984, Astrophys. J. 277, p. 569.

106. Cheung, A.S.C., Yoshino, K., Parkinson, W.H. and Freeman, D.E.: 1984, Geophys. Res. Lett 11, p. 580.
107. Cheung, A.S.C., Yoshino, K., Parkinson, W.H. and Freeman, D.E.: 1984, Can. J. Phys. Dec. 1984 (in press).
108. Freeman, D.E., Yoshino, K., Esmond, J.R. and Parkinson, W.H.: 1984, Planet. Space. Sci. 32, p. 239.
109. Freeman, D.E., Yoshino, K., Esmond, J.R. and Parkinson, W.H.: 1984, Planet. Space Sci. (in press)
110. Yoshino, K., Freeman, D.E. and Parkinson, W.H.: 1984, Planet. Space. Sci (in press).
111. Yoshino, K., Freeman, D.E. and Parkinson, W.H.: 1984, J. Phys. Chem. Ref. Data 13, p. 207.
112. Brown, J.M., Kerr, C.M.L., Wayne, F.D., Evenson, K.M. and Radford, H.E.: 1981, J. Mol. Spect. 86, p. 544.
113. Brown, J.R., Evenson, K.M. and Radford, H.E.: 1984, in preparation.
114. Morreau, T.: 1984, PhD Thesis, York University.
115. Cann, M.W.P., Shin, J.B. and Nicholls, R.W.: 1984, Can. J. Phys. (December issue) in press.
116. Cann, M.W.P., Nicholls, R.W., Roney, P.L., Blanchard, A. and Findlay, F.D.: 1985, Applied Optics (Submitted).
117. Blake, A.J., Freeman, D.E., Nicholls, R.W., Ogawa, T. and Simon, P.C.: 1985 "The penetration of ultraviolet solar radiation into the middle atmosphere" Conclusions and Recommendations of MAP Study group 7. Middle Atmosphere Handbook, In press.
118. Nicholls, R.W.: 1982, J. Quant. Spect. Rad. Transf. 28, p. 481.
119. Nicholls, R.W.: 1981, Astrophys. J. Supp. 47, p. 279.
120. Amani, M. and Nicholls, R.W.: 1985, submitted to J. Quant. Spect. Rad. Transf.
121. Nicholls, R.W.: 1983, Chem. Phys. Letts, 101, p. 459.
122. Nicholls, R.W.: 1985, Submitted to Chem. Phys. Letts.
123. Nicholls, R.W.: 1984, Applied Optics, 23, p. 1142.
124. Nicholls, R.W.: 1984, Optics News 10, p.66.
125. Nicholls, R.W.: 1985, Submitted to J. Quant. Spect. Rad. Transfer.

R.W. Nicholls
Chairman of the Working Group

COMMISSION 15: PHYSICAL STUDY OF COMETS, MINOR PLANETS, AND METEORITES (L'ETUDE PHYSIQUE DES COMETES, DES PETITES PLANETES ET DES METEORITES)

PRESIDENT: C.R. Chapman VICE PRESIDENT: L. Kresak

ORGANIZING COMMITTEE: B.D. Donn, F. Dossin, H. Fechtig, M. Festou, D.L. Matson, H.U. Schmidt, Z. Sekanina, L.M. Shulman, J. Wasson, G.W. Wetherill, L. Wilkening

I. Introduction

The last three years have witnessed a growing interest in the physical properties of the small bodies in the solar system. Perhaps the most significant impetus to research on small bodies has been the imminent arrival of Comet Halley in the inner solar system. This famous comet, which was recovered in autumn 1982, has been the object of intense study during the past year as it has approached the sun and developed a tail. Much of the international, ground-based astronomical research on Halley has been coordinated through the International Halley Watch program. Spacecraft from several nations have been successfully launched (or soon will be, we hope) and are on their way to intercept the comet and make close-up observations and *in situ* measurements. The commencement of spacecraft study of small bodies marks a new era in comet/asteroid science and, in coordination with ground-based and Earth-orbital observations, will result in unprecedented new knowledge about the origin of the solar system and about solar system processes.

Although Halley is receiving the most attention, interest is also high in Comet Giacobini-Zinner, the vicinity of which will be probed by a diverted American spacecraft in September 1985. Upcoming spacecraft studies of comets through 1986 are described at the end of the comet section of this report. Asteroid exploration by spacecraft is also anticipated to begin in the near future. The trajectory of the NASA Gailieo Mission to Jupiter has been changed to permit close-encounter observations of the large main-belt asteroid 29 Amphitrite in December 1986; these observations will be conducted on a "best effort" basis only a few months after launch of Galileo. Interest is also high in Europe, the Soviet Union, and the United States in possible spacecraft missions to additional comets and asteroids during the 1990's. If these efforts are pursued, there will be a concomitant ground-based effort. The last three years have also witnessed extremely productive efforts to observe small bodies from Earth orbit. For example, the Infrared Astronomical Satellite discovered a comet in 1983, which made the closest approach to the Earth of any comet in centuries. IRAS made important discoveries and measurements of other comets and also assembled an immense database on infrared brightnesses of thousands of numbered and unnumbered asteroids. The planned launch of the Hubble Space Telescope next year highlights the continuing potential for applying extremely powerful instrumental techniques to the study of comets and asteroids from above the Earth's atmosphere.

A large number of conferences have been held during the past three years on comets, asteroids, and related topics. Among the conferences (many of them co-sponsored by Commission 15) were the Tucson conference on comets (resulting in the major reference book *Comets*, L. Wilkening, ed.); conferences on international cometary exploration, modern observational techniques, comets and the origin of life, and ices in the solar system; two meetings in Uppsala on asteroids, comets, and meteors (the first one published as *Asteroids, Comets, and Meteors*, Lagerkvist and Rickman, eds.); a colloquium on dust; a conference on the galaxy and the solar system; a Chapman Conference on magnetic reconnection; a conference on "Physics and Dynamics of Solar System Minor Bodies" held in Dushanbe, USSR, which attracted 200 representatives from various Soviet institutes; a symposium on the search for extra-terrestrial life; plus a variety of other smaller meetings and workshops. The study of meteorites is an important adjunct aspect of Commission 15 work, inasmuch as meteorites are the products of the fragmentation and decay of asteroids and comets -- the chief meetings for the presentation of international research on meteorites are the annual meetings of the Meteoritical Society, the last two of which were held in Mainz and Albuquerque.

Activities of Commission 15 during the past three years have centered around co-sponsoring various meetings and fostering international communication about research on small bodies. The Commission, either by itself or in association with other Commissions, has worked on developing a file of asteroid lightcurve and rotation data, defining a series of standard comet filters, trying to improve the predictions of stellar occultations by small bodies, and developing an improved approach to predicting the magnitudes of asteroids.

Time and space constraints have not permitted us to develop a truly comprehensive report on the last three years of research on the physical properties of asteroids and comets. The field is very rapidly expanding, and our members have been pre-occupied with doing new research rather than with preparing exhaustive summaries of material available in the standard journals. Therefore, we apologize in advance for the somewhat uneven treatment of topics and for omitting many important publications. We have attempted, however, to highlight many of those areas that have been most intensively studied during recent years. The general format of the report follows that of previous years, beginning with topics in cometary physics and concluding with a summary of some asteroidal topics. We omit the traditional two pages on meteoritics, which is really a separate field larger than asteroid and cometary astronomy combined, and call readers' attention to such journals as _Geochimica et Cosmochimica Acta_, _Meteoritics_, and various letter journals for up-to-date information. The President of the Commission appreciates the efforts of the authors of the various sections below for their help in preparing reports with short advance notice. However, the President has made numerous changes and additions to the sections, so the listed authors should be regarded as major contributors to their section, but not blamed for defects or deficiencies.

II. Comets

PHYSICAL PROPERTIES OF COMETARY NUCLEI AND DUST - Z. Sekanina

Whipple (NATO Workshop on Ices in the Solar System, in press) has reviewed the history of the icy-conglomerate model for comet nuclei, emphasizing the rapid progress during the past decade. He suggests (Highlights in Astronomy 6, 323; BAAS 15, 805; Icarus, in press) that a pair of enormous outbursts several weeks apart, as displayed by P/Holmes and P/Tuttle-Giacobini-Kresak, may be triggered by near-tangent collisions between the components of a double comet. Sekanina (AJ 89, 1573) formulates a precession model for the nucleus of P/Kopff, indicating that its spin axis moves currently at a rate of about 20° per revolution. He also argues (BAAS 16, 642) that a limited but growing body of evidence suggests that comet nuclei are generally far from being spherical.

A number of papers deal with the light variations of P/Halley at large solar distances (in 1982-84) and with implied constraints on the physical properties of the nucleus and on the processes on its surface (Belton BAAS 15, 804; Sicardy et al. A&A 121, L4; West and Pedersen A&A 121, L11 and 138, L9; Le Fevre et al. A&A 138, L1; Lecacheux and Le Febre IAUC 3957; Jewitt and Danielson BAAS 16, 636; Sekanina BAAS 16, 636). A majority view is that the photometry has not yet allowed one to determine the comet's spin rate unequivocally. As this report goes to press (Feb. 1985) there are many reports of light variations for Halley, with several favored periods, but no secure result as yet.

Newburn and Reinhard (Adv. Space Res. 2, No. 12, 151) provide nominal model parameters for P/Halley. The comet's mass and lifetime are estimated by Delsemme and Yeomans (BAAS 16, 636). Also see Hughes (MNRAS 204, 1241) and Ferrin (A&A 135, L7); the latter concludes Halley will last ~40 more apparitions. Larson and Sekanina (AJ 89, 571; BAAS 16, 637) present a new image-processing algorithm to improve visibility of features in the head of P/Halley and suggest relatively high spatial densities of dust in its bright jets. Sekanina and Larson (BAAS 15, 651; AJ 89, 1408) introduce the concept of continuous emission of dust from small active areas on the nucleus to explain the evolution of spiral jets into envelopes seen on processed plates, summarize the results of modeling of these features, and determine the comet's spin vector. Whipple (Center for Astrophys. Preprint 1763) explains narrow dust jets in P/Swift-Tuttle as ejecta from small, less active sources embedded in larger, more active areas on the comet's nucleus.

Froeschle et al. (Asteroids, Comets, Meteors, p. 215), Froeschle and Rickman (Asteroids, Comets, Meteors, p. 225), Weissman and Kieffer (JGR, in press), Horanyi et al. (ApJ 278, 449), and Fanale and Salvail (Icarus, in press) propose new thermal outgassing models for comets. The problems addressed include the formation and purging of an insulating surface mantle, heat and vapor transport, rotation effects (diurnal and seasonal), implications for the nongravitational acceleration of the orbital motion, etc. Klinger (Asteroids, Comets, Meteors, p. 205) gives an overview of current trends in modeling the comet nuclei. Sekanina (Icarus 58, 81) concludes that the process of comet dissipation (disintegration before the eyes of observers) could be triggered by an eruptive dust-emission event weeks before demise. Mendis (NATO Workshop on Ices in the Solar System, in press) discusses the roles of dust mantles and dust halos in the nucleus outgassing.

Delsemme (NATO Workshop on Ices in the Solar System, in press) analyzes observational evidence that favors water ice as the dominant component of both new and old comet nuclei. Klinger (Icarus 55, 169) considers the activity of comets as a function of the phase of water ice (amorphous vs. crystalline). Smoluchowski (JGR 87, A422) draws attention to the roles of porosity and the abundance ratio CO_2/H_2O for heat transport in the nuclei. Cowan and A'Hearn (Icarus 50, 53) propose sudden exposure of CO or CO_2 to explain the outbursts of P/Schwassmann-Wachmann 1.

A large number of papers on Comet IRAS-Araki-Alcock (1983d), relevant to the problems of cometary nucleus, including reports and interpretations of its radar observations, were presented at meetings of the AAS Division for Planetary Sciences (cf. BAAS 15, No. 3 and BAAS 16, No. 3, Pt. 1). An extensive radar study of this comet was published by Goldstein et al. (AJ 89, 1745). Radar detection of P/Grigg-Skjellerup was reported by Kamoun et al. (BAAS 14, 753).

An extensive series of papers on cometary dust has appeared in the past three years. Campins et al. (Nature 301, 405) first reported spectrophotometric detection of water ice in a comet, Comet Bowell. Hanner (ApJ 277, L75) later reported detection of a similar absorption feature near 3 microns in Comet 1983ℓ, interpreting it as scattering by icy grains. Hanner et al. (AJ 89, 162, 170) also provide infrared data on P/Comets Stephan-Oterma, Swift-Gehrels, Kearns-Kwee, Gunn, and Grigg-Skjellerup, and on Comet 1981 XV. Sekanina (Adv. Space Res. 2, No. 12, 121) summarizes information on the properties of dust ejecta from P/Halley based on its 1910 and earlier apparitions. Sekanina and Farrell (AJ 87, 1836) model the striated dust tail of 1957 V and find two kinds of submicron-sized particles. Jewitt (Icarus 60, 373) presents the results of his comprehensive study of Comet Bowell (1980b), largely confirming Sekanina's earlier conclusions (AJ 87, 161). A large number of additional papers on dust in this comet are published as referenced by Jewitt, including the results of the IRAS observations (Walker et al. ApJ 278, L111). Sekanina's (AJ 87, 1059) study of the dust tail of Comet 1979 XI that collided with the Sun shows a possible effect of solar corona on the particle motions.

PHOTOMETRY AND SPECTROPHOTOMETRY BETWEEN 0.12 AND 1 μm - M.C. Festou

While the 1970's saw an increase of the spectral range inside which comet ejecta can be studied (access to near UV and vacuum UV, IR and radio emissions), the first years of the 1980's were characterized by the development of a new family of detectors, enhancing our ability to carry out important quantitative studies of the cometary phenomena. Monochromatic images are being obtained, fainter objects are being observed, and better spectral resolutions are being achieved. This technological breakthrough, together with the still increasing effort to observe comets, triggered by the impending return of Halley's comet, should soon yield a completely new picture of what a comet could be like. And more surprises are to be expected when the Space Telescope becomes available.

The photometric technique is used to monitor the gaseous output of the nucleus, using more or less broadband interference filters. A'Hearn et al. (AJ 86, 1559) found that Comet 1979 X was an exceptionally gassy object, while Millis et al. (AJ 87, 1310) found that Comet Stephan-Oterma was, on the contrary one of the dustiest (and showing a very strong brightening in the continuum near opposition). Numerous comets were observed by various groups in a more or less systematic way. The extensive work by Newburn and Spinrad (AJ 89, 289), to be confirmed by other studies, reveals that C_2 and CN production rates do not vary with the same heliocentric distance power law, and that C_2 radicals could be strongly deficient in the comet beyond 1.8 A.U. Two comets were observed by many astronomers using various techniques: Comet Iras-Araki-Alcock (IAA, 1983d) because of its proximity to the Earth, and Comet Bowell 1980b because of its exceptional activity far from the sun. Visible and IR photometric observations are not well understood, e.g., as shown by A'Hearn and Millis (ApJ 282, L43), Hanner et al. (Icarus, in press), and Jewitt et al (AJ 87, 1854).

The photometric technique can be complemented by polarimetric measurements (Michalsky, Icarus 47, 388) or by photopolarimetric measurements (to better isolate the continuum, Myers and Mordsieck, Icarus 58, 431), although the results are very difficult to interpret.

Spectrophotometry is now used more widely and allows a clean separation of emission features from the continuum as well as studies of the spatial extension of cometary emissions. Most studies deal with the determination of abundances, especially that of $O(^1D)$ atoms observed through their red forbidden lines (Spinrad PASP 94, 1008; Goraya et al., Moon & Planets 26, 3; Fink and Jonhson, AJ 89, 1565). The good detector sensitivity achieved in the near IR allowed the study of the spatial extension of OI and NH_2 emissions (Fink and Jonhson, ibid.) as well as the intensity ratios of various bands of NH_2 (Jonhson et al., Icarus 60, 351) and the CN red bands (Danks and Dennefels, AJ 86, 314; Jonhson et al., ibid.). A'Hearn et al. (AJ, in press) combined photometric and spectrophotometric observations of Comet Bowell in the 0.12-1 μm wavelength interval to study the strange behavior of that object.

Spectrophotometry was used to investigate the spectrum of comets in a few atypical cases: Johnson et al (ApJ 270, 759) recorded for the first time the spectrum of a comet in the 0.9-2.5 μm region at a resolution of 4-6 cm^{-1}, taking advantage of the very high gas prodution rate of Comet West around mid-March 1976 (this led to the first detection of the C_2 Ballik-Ramsay and Philipps bands). Comet IAA came very close to the Earth, thus allowing a detailed study of its inner coma. W. Cochran (Icarus 58, 440) made the first unambiguous observation of the green line of OI; Cosmovici and Ortolani (Nature 310, 122) may have detected new lines attributed to HCO, H_2S^+, DCO, S_2 and NH_4; however, the doscovery of S_2 was made by A'Hearn et al. (ApJ 224, L176) with the IUE long wavelength camera. The strange Comet P/ Schwassmann-Wachmann 1 was the object of careful studies which led to the identification of CO^+ emission lines when the comet was either in a dormant state (Larson, ApJ 238, L47) or undergoing an outburst (Cochran et al., AJ 85, 474). The $^{12}C/^{13}C$

ratio was derived by Lambert and Danks (ApJ 268, 428) using a completely new set of C_2 lines.

Fabry-Perot interferometers are not often used; however, they offer the best way to isolate faint lines such as the Balmer alpha emission of HI atoms, thus providing an alternative way to observing the Lyman alpha line from space. Hydrogen production rates can then be deduced (Scherb, ApJ 243, 644; Shih et al., ApJ 279, 453).

The International Ultraviolet Explorer (IUE) telescope has been used to observe more than 20 comets in a few years. Numerous results were obtained and were reviewed by Festou et al. (2nd European Conference, ESA-SP 179) and Feldman (Science 219, 347). References relative to the exceptional input to cometary science that the IUE made are found in the two above-mentioned papers. For example, Feldman et al. (A&A 131, 394) conclude that IUE observations of the ultraviolet coma of Comet Austin (1982g) demonstrate a composition similar to that of the best previously observed comet by IUE, Bradfield (1979X). The derived water production rate varies with heliocentric distance with an exponent between -3 and -4, similar to the 3.7 found for Bradfield.

Theoretical investigations of comet spectra range from the (tentative) identification of new lines (CN, CN^+, and CS, Krisna-Swanny, A&A, 97, 110, CS^+, Singh, A&A 108, 369), fluorescence calculations (OH, Schleicher and A'Hearn, ApJ 258, 864; OH, Elitzur, ApJ 246, 354; CO, Krisna-Swanny, ApJ 267, 882; C_2, Krisna-Swanny and O'Dell, ApJ 251, 805; CN, Tatum, A&A 135, 183; CN, Festou and Zucconi, A&A, 134, L4, for the Comm. 15 comet filters) to the evaluation of the photodissociation lifetime of OD and OH radicals (Singh et al., Icarus 56, 184; Van Dishoeck and Dalgarno, Icarus 59, 305; Van Dishoeck, J. Chem. Phys. 79, 873). A high resolution atlas for fluorescence calculations was compiled by A'Hearn et al. (TR AP83-044, Univ. of MD) from high resolution spectra of the full solar disc obtained by various authors. Results are provided with a 0.005Å resolution in the 2,400-7,000Å wavelength range.

The occultation technique may become a very useful tool for deriving tidal depths in the continuum, although the technique is still quite difficult to master (Lecacheux et al., Icarus 60, 386; Combes et al., Icarus 56, 229; Larson and A'Hearn Icarus 58, 446), and occultations are difficult to predict (Taylor, A&A 135, 181); so are neutral species column densities (Smith et al., Icarus 47, 441).

Comet P/Halley was discovered by Jewitt and Danielson on Oct. 16, 1982, near magnitude 25. Since then, the comet has been studied using the broadband photometry technique. The recorded variability of the light reflected by P/Halley is, up to now, interpreted in terms of rotation of its nucleus (Felenbok et al., A&A 113, L1; Sicardy et al., ibid.; West and Pedersen, ibid.). As this report goes to press, Halley is showing a faint coma according to several preliminary reports, although it has not yet gotten brighter than 20th magnitude.

To complete this report, we note the additional contributions presented at various conferences (Tucson, 1981; Budapest, 1982; Uppsala, 1983; Nice, 1984). Reviews on the topics discussed here by M. A'Hearn and P. Feldman appear in the book Comets, (L.L. Wilkening, ed.).

COMA CHEMISTRY - W.F. Huebner

Reviews on chemical modeling have been published by Rh. Lüst (Topics in Chemistry 99, 73) who relates spectral observations to coma models, Mendis and Houpis (Rev. Geophys. Space Phys. 20 885) who discuss the effects of solar wind-coma interaction on chemistry, and A'Hearn (Chem. Eng. News, May 28, p. 32) who emphasize the complexity of photolytic reactions with water and also the newer species containing sulfur. A series of articles in which various aspects of comet chemistry are discussed are published in the Proceedings of the International Conference on Cometary Exploration: Cometary Exploration, Vol. I (Gombosi, ed.).

In the last three years the suspicion that coma chemistry cannot be isolated from the physics of the coma has been confirmed. Energy balance in the coma, multi-fluid flow, gas-dust interaction, transition from collision dominated flow to free molecular flow, and interaction with the solar wind, all affect the density and velocity profiles of the various chemical species and therefore the coma chemistry (Houpis and Mendis, Moon & Planets 25, 95; Marconi and Mendis, Moon & Planets 27, 27 and 431; Huebner and Keady, Cometary Exploration, Vol. I, Gombosi, ed., 165; Marconi and Mendis, ApJ 273, 381). Although Beushausen and Jockers (Asteroids, Comets, Meteors, Lagerkvist and Rickman, eds., 317) do not discuss the effect of solar wind-coma interaction on the chemistry directly, the indirect effects on the ionization rate coefficients that they do discuss are important to the chemistry in the region of the contact surface. Similarly, effects of energetic neutral molecules on the coma were reviewed by Ip and Hsie (Cometary Exploration, Vol. I, Gombosi, ed., 265). H.U. Schmidt and Wegmann (Max Planck Inst. Astrophys.) model the chemical effects in the solar wind-coma interaction. Transition from collision-dominated flow to free molecular flow in the coma was modeled by means of escape probabilities by Huebner and Keady (A&A 135, 117); accelerations caused by radiation pressure on neutral species and the multiple collision random walk were calculated by Kitamura et al. (Icarus, in press) using the Monte Carlo procedure. Gas-dust interaction in the coma was discussed by Keller (Cometary Exploration, Vol. I, Gombosi, ed., 119). All these effects change the density and velocity profiles of species in the coma and thus the chemical reaction rates.

Festou (Adv. Space Res., in press) investigated the carbon abundance relative to the hydroxyl radical; his conclusion was the C/OH ratio varies greatly from comet to comet. Important to all chemical model calculations are the comparisons to observations as provided by Cucchiaro and Malaise (A&A 114, 102) for CN, C_2, and NH in Comet Bennett (1970 II), by Delsemme and Combi (ApJ 271, 388) for NH_2, $O(^1D)$, and CN in Comet Kohoutek (1973 XII), and by Weaver et al. (A&A 251, 809) for OH, H, and O in Comet Bradfield (1979 X). Festou and Feldman (A&A 103, 154) predict profiles for the forbidden oxygen lines at 5577Å and 6300Å.

New impetus to coma chemistry modeling came from the detection of several molecular species in Comet IRAS-Araki-Alcock (1983d). Altenhoff et al.(A&A 125, L19) probably detected H_2O and NH_3 in the cm wavelength range in a ratio approximately in agreement with chemical modeling. A'Hearn et al. (ApJ 274, L99) discovered S_2 in their UV spectra. Jackson (Howard U.) has modeled the sulfur compounds in comets.

The chemical composition of the nucleus has recently been investigated by Yamamoto et al. (A&A 122, 171). Related to this is the work of Irvine and Hjalmarson (Cosmochemistry and the Origin of Life, Ponnamperuma, ed., 113) who have investigated the chemistry in the formation of comet nuclei in interstellar clouds.

PLASMA TAILS AND THE SOLAR-WIND INTERACTION - M.B. Niedner, Jr.

The research performed in this field since mid-1981 significantly exceeds that of the previous reporting period, undoubtedly the positive influence of Comets Halley and P/Giacobini-Zinner, soon to be visited by spacecraft. Several excellent reviews emphasizing all or several aspects of the cometary/solar-wind interaction have appeared recently. Photographic observations of the plasma tail and physical interpretations were discussed in detail by Brandt (Comets, L.L. Wilkening, ed., 519) and Jockers (The Need for Coordinated Ground-based Observations of Halley's Comet, 193), whereas a somewhat more plasma physics-oriented (less observational) approach was adopted by Ip and Axford (Comets, L.L. Wilkening, ed., 588), Mendis and Houpis (Rev. Geophys. Space Phys. 20, 885), and Breus (Space Sci. Rev. 32, 361).

Reviews addressing more specific problems were those of Mendis - development and time dependence of cometary ionospheres (Comets, Origin of Life, 71); Niedner - data needs for plasma phenomena (Modern Obs. Tech. for Comets, 21); Niedner - disconnection events (DE's) and magnetic reconnection (Magnetic Reconnection, Geophys. Monograph 30, 79); Russell et al.- Venusian/cometary analogues (Comets, L.L. Wilkening, ed., 561); and Schmidt and Wegmann - plasma flow and magnetic fields (Comets, L.L. Wilkening, ed., 538). Published research papers are summarized below.

A. The Sunward Comet/Solar-wind Interaction

Mendis et al. (ApJ 249, 787) examined the access of solar wind to the nuclear surface at large heliocentric distances, concluding for water-dominated comets at $r > 5$ A.U. that the solar wind impinges directly on it, charging the nucleus to small positive potentials on the sunlit portion and to numerically large negative values on the nightside. Electrostatic levitation and blow-off of small, loose dust particles (created by cosmic ray damage to the nucleus) can occur, the process being partly modulated by the strong dependence of the nightside potential on the solar-wind speed. Houpis and Mendis (Moon & Planets 25, 95) studied the properties of the cometary bow shock and ionosphere for CO/CO_2-dominated comets as a function of heliocentric distance and proposed that Comet Humason, which had a prominent plasma tail and was irregularly active at large-r, could be explained by their model. Strazzulla et al. (A&A 123, 93) conducted laboratory tests of the irradiation of thin films of frozen gas with energetic (keV-MeV) particles, and suggested that erosion of ices by fast ion (solar wind) bombardment could explain cometary activity at large-r.

Gombosi et al. (ApJ 268, 889) examined the effect of charge exchange between fast solar-wind ions and cometary neutrals on the pressure balance budget and location of the standoff distance (contact surface); they concluded for a Halley-like comet at 1 A.U. that enough momentum is removed from the solar wind via charge exchange to place the standoff distance at several thousand km. Wallis (The Need for Coordinated Ground-based Observations of Halley's Comet, 171; Asteroids, Comets, Meteors, 303) discussed several aspects of ion formation and MHD modeling of plasma dynamics. The role of Alfven's critical velocity mechanism as an ionization source was studied by Formisano et al. (Planet. Space Sci. 30, 491), who proposed that a self-sustained beam discharge and Alfven's condition may only be met within 10^4 km of the nucleus. Beard (ApJ 245, 743) proposed the existence of a strong shock as the result of a neutral radical falloff made more rapid than ρ^{-2} by solar radiation pressure; the strong shock then causes rapid ionization via electron impact, thus explaining the rapid formation of ion tail rays. However, advocacy of such a steep falloff of neutrals, and especially the required large radiation pressures, is largely confined to Beard.

Marochnik (Moon & Planets 26, 353) proposed that unlike the case of Venus, where solar-wind magnetic fields penetrate the ionosphere as a result of instabilities along the ionopause, the cometary ionosphere is threaded by magnetic fields essentially at all times due to a continuous diffusion of fields across the ionopause. Ershkovich and Mendis (ApJ 269, 743) challenged this view, concluding that if deep field penetration does occur (and this was uncertain in their study), it is likely to be due to an instability (probably Kelvin-Helmholtz) of the ionopause. The response of a CO cometary atmosphere to the passage of interplanetary shocks and high-speed streams was examined in a numerical simulation by Beushausen and Jockers (Asteroids, Comets, Meteors, 317), who predicted that the increased charge exchange should result in the formation of an observable ion cloud similar to those seen in several comets.

B. Plasma Tails: Observation and Theory

Using photographs taken around the world, Jockers (Icarus 47, 397) studied the formation and kinematics of rays, waves, a large-scale disturbance, and other features in the plasma tail of Comet Kohoutek (1973XII) during the period 1974 January 19-21. His conclusions about the interaction of the comet with the solar wind and about the nature of tail disconnection events (DE's) disagree in part with earlier findings by Niedner and Brandt (ApJ 221, 1014; ApJ 223, 655), especially those concerning DE's, which Niedner and Brandt attribute to sector boundary crossings and magnetic reconnection, and around which they propose much of plasma-tail morphology is organized. Tail structure in the same comet (Kohoutek) was also briefly discussed by Jockers (Asteroids, Comets, Meteors, 327).

Le Borgne (A&A 123, 25) and Niedner et al. (Solar Wind V, NASA CP-2280, 737) pointed to the likely influence of a solar flare-generated interplanetary disturbance on a rapid 10° turning of the plasma tail axis of Comet Bradfield (1979X) on 1980 February 6. Directional discontinuities seen in the tail of Comet Austin (1982g) were interpreted by Fulle and Pansecchi (Icarus 57, 410) as resulting from the interaction of the comet with probable (but unobserved) changes in the azimuthal component of the solar-wind velocity. Jockers and Balazs (Astronomy with Schmidt-type Telescopes, 237) obtained Schmidt camera objective prism spectra of Comet Austin (1982g) and noted the significantly greater strength of H_2O^+ emission at λ6500 compared to the CO^+ radiation at λ4270. Li et al. (Acta Astrophys. Sin. 3, 81) collected Shanghai Observatory photographic material of Comets Morehouse 1980III), Brooks (1911V), and Halley (1910II), and analyzed the data for plasma-tail structure and overall morphology.

The well-known helical structures seen in the outer tail of Comet Ikeya-Seki (1964f) during 1965 October-November were interpreted by Krishan and Sivaraman (Moon & Planets 26, 209) as resulting from plasma instabilities excited in a tail containing twisted magnetic fields. Extending the earlier work of Podgorny et al. (Moon & Planets 23, 323), Dubinin et al. (Astron. Zh. 59, 1006; trans. in Sov. Astron. 26, No. 5, 608) used the motion of a plasma structure to derive the strength and topology of the magnetic field in the tail of Comet Halley (1910II). A major conclusion of both studies was that the field strength is greatest near the tail boundary (~75γ) and weakest near the tail center (~0γ) (cf. also Ioffe, Astron. Zh. 61, 121;trans. in Sov. Astron. 28, 72). Ershkovich (Moon & Planets 25, 521) pointed out several internal inconsistencies in Podgorny et al.'s method which vitiate the conclusion about a strong boundary/weak core magnetic model, and advocated the use of hypersonic pressure balance calculations to determine that the field strength in the distant tail is not more than ~2½ times the local interplanetary value. Podgorny et al. (Moon & Planets 27, 135) responded to Ershkovich's criticism and re-advocated their weak core tail model. Russell et al. (Comets, L.L. Wilkening, ed., 561) pointed out the essential correctness of Ershkovich's values based on Pioneer Venus Orbiter measurements of the Venusian magnetotail and the justifiable assumption that the Venus/solar-wind interaction is in many respects comet-like.

Ershkovich (Comets, Origins of Life, 105; MNRAS 198, 297) interpreted folding tail rays as drift motions in convectional electric fields; from the observed folding rates, he derived values for electrical conductivity in the tail plasma which are orders of magnitude less than classical values. Observations of generation rates as well as other tail ray properties were presented by Miller (BAAS 14, 752, 952; ibid. 15, 960).

Ershkovich et al. (ApJ 262, 396) and Niedner et al. (ApJ 272, 362) constructed a hypersonic pressure balance model of plasma tails which, given different values of the tail gas pressure as inputs, satisfactorily reproduced the observed and variable flaring behavior of Comet Bradfield (1974b). A major conclusion of the second study was that MHD instabilities should act to destroy magnetic flux conservation in the distant plasma tail.

Buti (ApJ 252, L43; Astrophys. Space Sci. 85, 35; ApJ 268, 420) studied the effect of high frequency electrostatic turbulence (in the solar wind) on the Kelvin-Helmholtz (KH) instability in cometary plasma tails. She found that hydromagnetic waves (KH) can either be stabilized or destabilized, depending on the direction of propagation of the electrostatic relative to the hydromagnetic waves. In addition, the stabilizing effect of a large tail magnetic field may explain why most waves are observed at large distances from the cometary head, where the expected tail field is small.

Ray (Planet. Space Sci. 30, 245) used KH theory to calculate growth rates for Comets Kohoutek (1973f), Arend-Roland (1956h), and Morehouse (1908c), and essentially agreed with the earlier results of Ershkovich and Heller (Astrophys. Space Sci. 48, 365) that the KH instability should have operated in these comets. Tupchienko and Yukhimuk (Geophys. J. 4, 135) considered the stability of the plasma tail/solar-wind interface and advocated the importance of the whistler mode.

Niedner (ApJ 48, 1) used the 72 plasma tail disconnection events (DE's) contained in his earlier DE catalog (ApJ 46, 141) to derive solar-wind sector boundary orientations and latitudinal extents far out of the ecliptic plane. The study offers a set of predictions to be tested by the International Solar Polar Mission, and emphasizes the unique value of comets as natural probes of the interplanetary medium.

C. Laboratory and Computer Simulations of the Comet/Solar-Wind Interaction

Although primarily concerned with deriving the magnetic field characteristics of Halley's tail from the observed motion of a plasma formation, Dubinin et al. (Astron. Zh. 59, 1006; trans. in Sov. Astron. 26, No. 5, 608) gave a brief summary of the earlier work of Podgorny and collaborators on laboratory simulation of the comet/solar-wind interaction. These simulations have shown that an induced magnetosphere and magnetotail are formed whose characteristics are similar to those predicted many years ago by Alfven (Tellus 9, 92).

Fedder et al. (Icarus, in press) ran 3-dimensional MHD computer simulations of the comet/solar-wind interaction, deriving large-scale plasma flow and magnetic field characteristics for various sets of input solar-wind conditions. Quiet-time models, characterized by a steady solar wind and IMF, yield magnetic field strengths in the head which are several times smaller than the large ($\gtrsim 100\gamma$) values reported by Schmidt and Wegmann (Comets, L.L. Wilkening, 538), and tail values ($\sim 10\gamma$) which are in agreement with Ershkovich's pressure balance and helical wave calculations. Fedder et al. found that a 90° change in the orientation of the IMF corresponding to a rotational discontinuity produced tail ray-like features similar to those seen in the simulations of Schmidt and Wegmann. A 180° reversal corresponding to a sector boundary produced a plasma depletion in the anti-sunward direction which Fedder et al. interpreted as a tail disconnection event (DE), in basic agreement with Niedner and Brandt's (ApJ 223, 655) sector boundary/magnetic reconnection model of DE's.

D. Plans for Ground-based, Wide-field Observations of Comets Halley and Giacobini-Zinner

The Large-Scale Phenomena (LSP) Discipline of the International Halley Watch (IHW) has organized a ground-based network of ≥ 90 observatories with wide-field imaging capability, in order to assemble as complete a photographic record as possible of the comet's rapidly variable plasma tail (e.g., DE's, helical waves, etc.). The network's period of maximum activity will be 1985 November - 1986 May, during which the comet is expected to be bright and possess a plasma tail. Image data obtained from the network will be permanently stored in various forms (digitized images, film copies, prints) in the IHW Archive. The philosophy and goals of the LSP network have been described in detail by Niedner et al. (The Need for Co-ordinated Ground-based Observations of Halley's Comet, 227) and by Brandt et al. (Astronomy with Schmidt-type Telescopes, 233). The LSP network is administered by J.C. Brandt (Code 680, NASA/GSFC, Greenbelt, MD 20771, USA), M.B. Niedner (Code 684.1, NASA/GSFC), and J. Rahe (Remeis Observatory, Bamberg, FRG).

An extension of the entire IHW's activities to include a shorter interval of coverage of Comet Giacobini-Zinner is presently being planned. The period of coverage will include the 1985 September 11 flyby of the comet by the International Cometary Explorer (ICE).

The history, research record, and overall raison d'etre of the Joint Observatory for Cometary Research (JOCR), a key station in the LSP Network, was discussed by Brandt (Modern Observational Techniques for Comets, 171), and Miller (Modern Observational Techniques for Comets, 169) gave a brief history of the wide-field photography of comets.

PHYSICAL STUDY OF COMETS IN THE USSR (INST. OF ASTROPHYSICS, DUSHANBE) - O.V. Dobrovolsky

The secular brightness decrease Δm of comets was shown to be statistically greater for comets with greater perihelion distances q and vice versa (O.V. Dobrovolsky, S.I. Gerasimenko, Kh.I. Ibadinov, C.R. Ac. Sci. Tadjik., No. 6). This is just the opposite of what D.W. Hughes and P.A. Daniels (Icarus 53, 444) expected but failed to find. However, this finding corresponds well to Shulman's (IAU Symp. No. 45, G.A. Chebotarev et al., 271) theory of evolution of nuclei with mineral crusts, even showing Shulman's bifurcation on the Δm - q diagram, and thus giving the first observational proof of the existence of dusty crusts on cometary nuclei. A comet catalogue has been composed, covering the last three centuries, of comets showing signs of nuclear splitting (O.V. Dobrovolsky, S.I. Gerasimenko, Bull. Inst. Astrophys. Dushanbe, No. 77).

There have been a variety of contributions to studies of cometary comas. The Na emission maximum far off the nucleus of Comet Mrkos (1957V) is explained by the cooling of dust grains in the vicinity of the nucleus by outflowing gas (S. Ibadov, C.R. Ac. Sci. Tadjik., 26, 90; Problems of Cosmic Phys., 18, 91). X-ray emission from dusty comas colliding with interplanetary dust is predicted to reach a measurable intensity at heliocentric distances $r \lesssim 1.0$ A.U. (S. Ibadov, C.R. Ac. Sci. Tadjik., in press). Comet head densities deduced from M. Beyers' visual observations exhibit unusual minima at r's corresponding to the orbital radii of planets (R.S. Osherov and M.Z. Markovich, C.R. Ac. Sci. Tadjik., No. 8). Cometary grains exposed to solar protons have been shown to become sources of negative ions involved in comet chemistry (O.V. Dobrovolsky et al., C.R. Ac. Sci. Tadjik., 24, No. 9 = Cometary Exploration, Vol. II, Gombosi, ed., Budapest).

An Invited Review on "Polarimetry of Comets" was presented by O.V. Dobrovolsky, N.N. Kiselev, and G.P. Chernova at a conference on "Polarimetric Methods in Astrophysics" held in October 1983 in Crimea. The wavelength dependence of polarization in the continuum studied by Kiselev et al. was found to be very flat in accordance with the results of J. Michalsky (Icarus 47, 388). A new result is that the phase angle of the polarization sign inversion turned out not to be at all wavelength dependent.

The high pressure C_2 bands (0,5) at 4680 Å and (0,7) at 5413 and 5435 Å were identified by V.F. Esipov and O. Mamadov in Comet Austin (1982g) on a 17 Sept. 1982 spectrogram; these bands are rare in cometary spectra. Some authors reported the C_2 Swan band vibrational temperature to be ~3250 K, and the continuum to appear bluer than the solar one (spectrophotometric gradient 2.08). The solar wind velocity at heliographic latitudes 33°-43° apparently is correlated with solar activity and with the relative acceleration of tail ions, deduced from Comet 1976 VI plates (Kh.I. Ibadinov, M. Rasulova, E. Pittich, Comets & Meteors, No. 38).

The equilibrium temperatures, sublimation rates, dielectric properties, and formation of refractory matrices were examined in the laboratory using models of crystalline water ice with different admixtures. Sublimation rates correspond well to the observed brightness decrease of comets (O.V. Dobrovolsky et al, C.R. Ac. Sci. Tadjik., 26, No. 1, 25). The majority of organic matrices showed densities as great as 0.04 g/cm^3 and compressive strengths as high as 0.1-0.2 kg/cm^2 (Kh.I. Ibadinov, S. Aliev, A. Rakhmonov, C.R. Ac. Sci. Tadjik., No. 9).

ORBITAL DYNAMICS AND EVOLUTION INCLUDING THE OORT CLOUD - D.K. Yeomans

This section briefly notes some of the important work on the dynamics of comets and asteroids as it pertains to their physics and origin. Works relating only to the dynamics of these objects is covered in the Commission 20 report. The dynamical and physical relationships between comets, large meteors, and meteorites are reviewed by Wetherill and ReVelle (Comets, L.L. Wilkening, ed., 297). They conclude that while many cometary meteoroids have measurable strength, almost all are weaker than any recovered meteorites. A possible connection between exhausted comets and Apollo-type asteroids was strengthened when Whipple noted the virtual coincidence of the orbital elements of 1983TB with the mean orbital elements of 19 Geminid meteors photographed with the super-Schmidt meteor cameras (I.A.U. Circular 3881).

Everhart (Comets, L.L. Wilkening, ed., 659) outlined the dynamical evolution of long and short-period orbits via stellar and planetary perturbations and noted that while the evolution of near parabolic comets with perihelion distances near the orbits of the major planets can account for short period comets with much smaller perihelia, there still remains the questions as to where comets originally formed, and whether the flux of near-parabolic new comets is sufficient to account for the number of observed short-period comets.

The dynamical aspects of cometary origin were investigated by Fernandez (AJ 87, 1318) with the result that a cometary origin in the Uranus-Neptune region produces the greatest current rate of cometary passages through the solar system as compared to cometary origins in the Jupiter region or at Oort cloud distances. Greenberg et al. (Icarus 59, 87) studied the origin of comets in the context of the accretional growth of Neptune and concluded that comets are probably physically unaltered planetesimals, not collisional fragments of larger bodies.

Everhart and Marsden (AJ 88, 135) continued their work on original and future reciprocal semimajor axes for cometary orbits. For 28 new osculating cometary orbits derived during the past few years, no convincing evidence could be found for original hyperbolic orbits. Using Monte Carlo techniques, Weissman (Highlights of Astronomy 6, 363) modeled the dynamical evolution of comets in the Oort cloud under the influence of stellar perturbations. The population of the cloud, depleted over the lifetime of the solar system, is estimated to be 1.0 to 1.5 x 10^{12} comets with a total mass of approximately 1.9 Earth masses. An alternative to the theory of a primordial and thermalized Oort cloud has recently been advanced by Clube and Napier (Highlights of Astronomy 6, 355) who suggest that passages of our solar system through galactic giant molecular clouds periodically removes the solar system's current cloud of comets and replaces it with comets captured from the giant molecular clouds.

Bailey, McBreen, and Ray (MNRAS 209, 881) suggest that the distribution of observed cometary orbits is consistent with a comet cloud that contains a very dense inner core containing most of the mass. This mass is $\leq$300 Earth masses for the more stable, centrally condensed models. Sekanina and Yeomans (AJ 89, 154) conducted a computer search for catalogued Earth-approaching comets and derived an average period between collisions of 33-64 million years. The constant rate of the close approaching comets in the last 300 years suggests a lack of long-period comets intrinsically fainter than absolute magnitude ~11. Fernandez (A&A 135, 129) believes that the small number of low-perihelion distance comets (none with $q < 0.34$ A.U.) is understood as a combination of the improbability of Jupiter encounters placing comets in such orbits plus the rapid physical sublimation of comets in such orbits.

SPACECRAFT AND COMETS: 1985-1986 - J.C. Brandt

The direct exploration of comets will be spectacularly inaugurated in 1985 and 1986 with six missions -- five to Halley's Comet and one to Comet Giacobini-Zinner. In addition, other spacecraft will be used for remote observations. The combined effort should produce a large volume of scientific data and should begin a new era in our understanding of comets.

Table 1: KEY DATE SUMMARY

- Halley's Comet Recovered -- October 16, 1982
- International Cometary Explorer (NASA) "Launched" -- Dec. 22, 1983
- Comet Giacobini-Zinner Recovered -- April 3, 1984
- VEGA's (USSR) Launched -- December 1984
- Ms-T5 (Japan) Launched -- January 1985
- VEGA's Encounter Venus -- June 1985
- Giotto (ESA) Launched -- July 1985
- Planet A (Japan) Launched -- August 1985
- Comet G/Z-ICE Encounter -- September 11, 1985
- SMM & Pioneer Venus Observations Centered on Perihelion, Feb. 9, 1986
- Comet Halley Encounters -- 1986

VEGA-1	March 6
Planet A	March 7
MS-T5	March 8
VEGA-2	March 9
Giotto	March 13

- Astro 1 -- March 7 - 14, 1986

The pace of space-related events is indicated by the "Key Date Summary", as shown in Table 1. The International Cometary Explorer (ICE) was sent from the Earth-Moon system on December 22, 1983, on a trajectory to intercept Comet Giacobini-Zinner (G/Z). Two VEGA spacecraft have been launched by the USSR. They will intercept Halley's Comet after encountering Venus. MS-T5 and Planet A are to be launched by Japan, and they travel directly to Halley's Comet. Giotto is to be launched by the European Space Agency (ESA), and it travels directly to Halley's Comet. The encounters of Halley's Comet all occur during a short period in March 1986 with the comet near its orbital node, as dictated by launch energy considerations.

During the time period centered on perihelion, February 9, 1986, the comet is also near the sun on the sky as seen from Earth. Observations will be possible from Venus using the Pioneer Venus Orbiter and by the repaired Solar Maximum Mission (SMM) from Earth orbit. In addition, NASA plans to launch a Spartan payload to observe the comet during this time. Extensive Earth-orbital observations of Halley's Comet by the Astro I Mission are planned for an approximately one-week period in March of 1986. The intent is to provide coverage overlapping some of the times of encounter by the spacecraft carrying out direct exploration.

The ICE spacecraft will be targeted to pass 10,000 km tailward of the nucleus to explore the solar-wind interaction with Comet Giacobini-Zinner and the formation of its plasma tail. The ICE spacecraft was formerly the Third International Sun-Earth Explorer (ISEE-3) and carries instruments well-suited for probing the comet's plasma/magnetic field environment. In particular, we expect important results to be obtained for the magnetic field, plasma and radio waves, energetic electrons, and molecular ions. ICE has no imaging experiment.

The instrumental capabilities and other data relating to the spacecraft to Halley's Comet are summarized in Table 2. All of these spacecraft will be targeted

Table 2: SPACECRAFT TO HALLEY'S COMET

Characteristic	USSR		ESA	Japan	
	Vega-1	Vega-2	Giotto	MS-T5	Planet-A
Launch date	Dec. 1984	Dec. 1984	July 1985	Jan. 1985	Aug. 1985
Encounter date	3-6-86	3-9-86	3-13-86	3-8-86	3-7-86
Distance to nucleus, km	10,000	3,000	500	$4x10^6$	$2x10^5$
Flyby speed, km/sec	80	77	69	74	75
Sun distance, A.U.	0.79	0.83	0.89	0.81	0.80
Earth distance, A.U.	1.16	1.09	0.98	1.11	1.14
Spacecraft stabilization	3-axis	3-axis	Spin	Spin	Spin
Weight of experiments, kg	125	125	50	14	12
Number of experiments	13	13	10	3	2
Imaging (visible, IR, UV)	4	4	2	--	1
Mass Spectrometers (neutrals, ions, dust)	4	4	4	--	--
Dust detectors	1	1	1	--	--
Plasma instruments	5	5	4	3	1

to pass on the sunward side of the nucleus. In addition to the plasma/magnetic field measurements, the dust and neutral gas environment will be measured. Imaging experiments are very important. A goal of Giotto and the VEGA's is to obtain images of Halley's nucleus. The imaging on Planet A will record the hydrogen cloud surrounding the comet.

III. Asteroids

LIGHTCURVES, ROTATIONS, COLLISIONS, AND PHASE RELATIONS - A.W. Harris & V. Zappalá

In the period 1981-1984, the number of main belt asteroids with rotation periods known to a statistically useful accuracy grew from about 215 to about 315. The number of Earth- and Mars-crossing asteroids with similarly known periods increased from about 15 to about 25. Approximately 60 papers reporting asteroid lightcurves were published in this time. The following researchers have been most active in reporting new lightcurve results:

Italy:	Zappalá; Di Martino (Torino Obs.); Barucci (Inst. Astrofisica, Roma)
Sweden:	Lagerkvist; Carlsson; Rickman (Uppsala)
Austria:	Schober (Graz)
Belgium:	Debehogne (Obs. Belg.); Surdej (ESO)
USSR:	Lupishko (Kharkov, Tadzhik)
China:	Zhou (Purple Mt. Obs.)
USA:	Harris (JPL); Binzel (U. Texas)

In addition, in the last year, several amateur astronomers in the USA, Britain, and Australia have begun producing high quality results with photoelectric photometers attached to telescopes in the 0.2-0.5m size range (cf. Ziegler, BAAS 16, 691; Minor Planet Bull. 11, 21). Such contributions by amateurs have been valuable, and promise to be even more so in the future, for supporting observations of asteroids observed by radar and/or by stellar occultations, since these objects tend to be large and within reach of modest aperture telescopes. In addition to the increasing availability of electronic and computer components at modest cost, the rise of amateur participation is a result of the formation of the International Amateur and Professional Photoelectric Photometry Association (IAPPP).

Among the more recent publications of lightcurves and related information are: Schober, A&A Suppl. 48, 57; Schober and Schroll, A&A 107, 402; Debehogne *et al.*, A&A 108, 197; Piironen, A&A 112, 172; Debehogne *et al.*, A&A Suppl. 50, 277; Pavolvski *et al.*, Publ. Astron. Obs. Sarajevo 1, 225; Lagerkvist and Rickman, Moon & Planets 27, 107; Zhou *et al.*, Acta Astron. Sinica 23, 349; Zappalá *et al.*, Icarus 52, 196; Barucci and Dipaoloantonio, A&A 117, 1; Schober, A&A 117, 362; Schober and Schroll, A&A 120, 106; Harris and Young, Icarus 54, 59; Lagerkvist and Kamel, Moon & Planets 27, 463; Surdej *et al.*, A&A Suppl. 52, 203; Cunningham, JRASC 77, 121; Lupishko *et al.*, Vestn. Khar'kov Univ. 232, 54; Schober, A&A Suppl. 53, 71; Schroll and Schober, A&A Suppl. 53, 77; Carlsson and Lagerkvist, A&A Suppl. 53, 157; Debehogne *et al.*, Icarus 55, 236; Binzel and Tholen, Icarus 55, 495; Hahn, *Asteroids, Comets, Meteors*, 35; Poutanen, *Asteroids, Comets, Meteors*, 45; Lupishko *et al.*, Soviet Astron. Lett. 9, 358; Di Martino and Cacciatori, A&A 130, 206; Harris *et al.*, Icarus 57, 294; Harris *et al.*, Icarus 58, 377; Ziegler and Florence, BAAS 16, 691; Di Martino and Cacciatori, Icarus 60, 75; Ostro *et al.*, Icarus 60, 391.

A systematic photoelectric survey on very small asteroids has been started at the Univ. of Texas (Binzel and Mulholland, Icarus 56, 519); it is important to investigate the spin evolution of small, faint bodies thought to be collisional fragments of catastrophic impacts, for which there are still severe observational biases in the available database. A similar project is in progress in Italy (Barucci and Di Martino, A&A Suppl. 57, 103). Worldwide campaigns were devoted to lightcurve photometry of selected objects (Schober and Kristensen, *Sun and Planetary System*,

305; Birch et al., Icarus 54, 1; Tedesco et al., Icarus 54, 23; Tedesco et al., Icarus 54, 30). The unusual fast rotation of M-type asteroids (Lagerkvist and Rickman, Sun and Planetary System, 287; Zappalá et al., A&A Suppl. 50, 23; Surdej et al., A&A Suppl. 54, 371) and the extremely slow rotation of some small asteroids (Harris and Young, BAAS 14, 724; Schober et al., A&A 115, 257) are confirmed (see below).

A Commission 15 Working Group (A.W. Harris, H.J. Schober, E.P. Tedesco, and V. Zappalá) is preparing a detailed data-file on all the existing observations. A preliminary list of rotational properties is already completed and is at disposal of the scientific community; a general lightcurve catalogue is also in preparation (Lagerkvist et al., Asteroids, Comets, Meteors, 27). The latest published list of known rotations, including references, is by Harris and Young (Icarus 54, 59).

New observations have shown recently ambiguity for some rotational periods previously considered to be of high reliability (Zappalá et al., Icarus 56, 319; Schober, A&A 105, 419), and a careful re-analysis of the available data is planned, because of the potential bias on statistical studies. For example, Zappala et al. (A&A 130, 208) have re-observed 21 Lutetia, originally thought to have a spin period of 6.133h, and have found it to be 8.17h. They attribute the incorrect earlier period to confusion caused by the high phase angle at the time of the marginally redundant earlier observations.

Several papers have been published in recent years reporting statistical analyses of the rotation data available (e.g., Farinella et al., A&A 104, 159; Paolicchi et al., Sun and Planetary System, 291; Golubeva and Shestopalov, Sov. Astron. 27, 583; Lagerkvist, Asteroids, Comets, Meteors, 11; Dermott et al., Icarus 57, 14; Binzel, Icarus 57, 294). The results that appear to be agreed upon are that the largest asteroids rotate with a mean period of ~8 hours with a dispersion about the mean which is not significantly different from that expected of a stochastically produced distribution (i.e., Maxwellian in three dimensions). In the size range of ~50 to ~150 km, it appears that there is a significantly longer mean period, ~10 hours. Dobrovolskis and Burns (Icarus 57, 464) suggest that this might be due to loss of angular momentum in collisions due to escape of ejecta thrown in the direction of rotation. Since preferential escape of forward thrown ejecta requires that the ejecta velocity be of the same order as the escape velocity from the asteroid, the mechanism is only effective in the size range of ~100 km diameter. Other possibilities exist, such as variation in the population index of asteroids in different size ranges, but these have not been explored in detail. Among asteroids smaller than ~50 km, the mean rotation period is again ~8 hours, but the dispersion becomes broader than expected for a Maxwellian distribution, becoming even bi-modal at the smallest sizes. Since essentially all known rotations of asteroids of D <10 km are of Earth- or Mars-crossing objects, it is not clear whether this result applies only to that class of asteroids, or to main-belt asteroids as well. A survey of small main-belt asteroids by Binzel and Mulholland (Icarus 56, 519) suggests that it does, but more data is required to reach a firm conclusion. Among the Earth- and Mars-crossing asteroids, the bi-modal distribution in spin rates may indicate that at least some of these bodies are extinct comets, since comets appear to have a bi-modal split in their rotation rates, including a very long-period group (Whipple, Comets, 227). Among asteroids in the main belt of D < 50 km, there appears to be an excess of slowly rotating asteroids, with periods greater than a day, up to 6 days. These slow spins can be explained as being due to tidal friction between an initially more rapidly spinning asteroid and a massive satellite, with the system eventually becoming synchronous, as in the Pluto/Charon system (Farinella et al., op cit.). Because of the limited power of tidal friction on small, solid bodies, evolution to synchronous is expected only if that state is reached before the satellite recedes beyond ~10-15 radii from the primary. Thus, only periods up to ~6 days can be explained this way. Two asteroids, 288 Glauke (Harris, Sky & Tel. 65, 504) and 1220 Crocus (Binzel, Icarus, in press) have apparent periods much longer than 6 days (~60 and ~30 days, respectively). Both exhibit large amplitudes (>0.5 mag.), and the periods appear firmly established. The only explanation which has been offered for these extremely long periods is that they are periods of precession of the spin axes of rather oblate asteroids induced by the presence of massive satellites orbiting about them. The true rotation rates would be considerably shorter. In the case of 1220 Crocus, Binzel observed a smaller amplitude period of ~8 hours when the long period variation is near its minimum level, suggesting that this is the true rotational variation, visible only when the asteroid is seen close to its equatorial aspect.

The plausibility of binary systems among asteroids has been analyzed from different points of view (Lambert, Proc. Southwest Reg. Conf. 7, 127; Scaltriti et al., Sun and Planetary System, 277; Showalter et al., BAAS 14, 725; Zappalá et al., Icarus 53, 458; Schober, Astrophys. Space Sci. 99, 387; Cellino et al., BAAS late paper, 1984). In particular, models of equilibrium figures with mass ratio different from one were studied (Leone et al., A&A 140, 265). As described above, Harris (BAAS 15, 828) suggested that the very slow rotation of small asteroids can be explained with forced precession due to a satellite: a puzzling oscillation found in the period of Ceres has also been proposed to be due to a precession phenomenon (Zappala et al., BAAS late paper, 1984).

Lightcurve data also yield information on the shapes, pole orientations, and phase relations of asteroids, but to obtain such results, one must impose model constraints on the data. One requirement is to separate the effects of solar phase angle from that of the viewing aspect (latitude on the asteroid of the viewing direction). The group led by M. Fulchignoni in Rome has studied in the laboratory the brightness variations of irregular models under various viewing conditions to attempt to separate these effects. For example, they have studied the effects of changing aspect and phase angle on the amplitude of variation for a wide variety of shapes and variable albedo models (Barucci and Fulchignoni, Asteroids, Comets, Meteors, 101). Harris (Icarus 57, 251) has called attention to the fact that the geocentric line of sight to the asteroid is an unrealistic approximation for the effective viewing direction when the solar phase angle is non-zero. A better approximation is the phase angle bisector, the line which bisects the Earth and sun lines to the asteroid. By referencing the aspect (latitude on the asteroid) to this line, one makes a better separation of the effects of viewing aspect and phase relation as the geometry of the Earth and sun change. Work has progressed on improving the magnitude-aspect method of shape and pole determination (Zappalá, Moon & Planets 24, 319; Zappalá and Scaltriti, Sun and Planetary Systems, 303; Zappala, Moon & Planets 24, 319; Zappalá and Kněžević, Icarus 59, 436) so that it now appears to be comparable to the method of photometric astrometry (Taylor and Tedesco, Icarus 54, 13; Tedesco and Taylor, BAAS 16, 700; Taylor, Icarus, in press) in accuracy. An important work which remains is to combine the two methods, since they are complementary in that the former is better for objects whose poles lie near the ecliptic plane, whereas the latter is better for objects which are more nearly pole upright. Both yield accurate results only if the amplitude of variation at some aspect is fairly large. Efforts are also progressing on understanding the potential of speckle interferometry for the same purposes (Drummond et al., BAAS 14, 725; Drummond et al., Steward Obs. Preprint 523), and future work is planned on comparisons with other techniques (Drummond and Hege, BAAS 16, 647).

The classic work of H.N. Russell in 1906 has been reconsidered in the light of our present knowledge that most of the rotational light variation of asteroids is due to shape rather than albedo. Ostro and Connelly (Icarus 57, 443) have considered the idealized case of geometric scattering from a body viewed equatorially to deduce a recipe for inverting the lightcurve to obtain a convex profile of the mean cross-section of the asteroid. The idealizations made require caution in interpreting results. Most important, the work serves to give a feeling for the best viewing conditions under which observations for shape studies should be made, and to understand the uncertainties of shapes and poles determined under more general conditions but which require model assumptions in their results.

Torino Observatory started an international observing campaign, which should produce a considerable amount of new pole positions in a few years. Several results have already been published on rotation axis determinations (Zhou and Yang, Acta Astron. Sinica 22, 378; Burchi and Milano, Moon & Planets 28, 17; Lupishko and Belsksaya, Asteroids, Comets, Meteors, 55; Magnusson, Asteroids, Comets, Meteors, 77; Barucci, A&A Suppl. 54, 471; McCheyne et al., Icarus 59, 286; Binzel, Icarus 59, 456; Zappalá and Knezević, Icarus 59, 436). Pole coordinates will also be derived from the program on equilibrium shapes of the P.S.I. group in Tucson (Chapman et al., BAAS 14, 728; Chapman et al., BAAS 15, 829; Chapman et al., BAAS 16, 708). When the program is completed in about a year, we also expect to have decisive evidence supporting or negating the existence of triaxial quasi-equilibrium figures expected for "rubble-pile" structures.

A considerable amount of work, both observational and theoretical, has been devoted to the study of phase relations and scattering laws. Most theoretical work to date has been restricted to the case of a spherical body, thus again the laboratory work of Fulchignoni and co-workers (e.g., Barucci et al., BAAS 16, 700) is important for comparing results to real observations of non-spherical objects. Hapke (JGR 86, 3039; JGR 86, 3055; Icarus 59, 41) and Lumme and Bowell (AJ 86, 1694; AJ 86, 1705) have developed theories which attempt to define the disk integrated brightness of a body of uniform albedo as a function of phase angle, accounting for such effects as single particle phase function, surface roughness, porosity, and single scattering albedo. Theoretical studies of phase relations have been carried out in the Soviet Union, as well (e.g., Golubeva and Shestopalov, Astron. Zh. 60, 602; Akimov, Vestn. Kharkov Univ. 204, 3; Shevchenko, Astron. Vestn. 16, 209). In addition, the group led by Lupishko has observed a number of asteroids very extensively to measure the phase relations in detail (e.g., Lupishko et al., Astron. Vestn. 15, 25; Pis'ma Astron. Zh. 7, 437; Astron. Vestn. 16, 101; Vestn. Kar'kov Univ. 232, 54). It is only through such detailed observations covering all observable phase angles that we will be able to test the various theoretical models proposed.

In addition to attempts to define asteroud phase relations theoretically, work is underway to define a function purely empirically for the purpose of prediction of magnitude vs. phase angle (e.g., Lumme et al., Proc. Nord. Astron. Mtg.; BAAS 16, 684). This function is especially required for use by IAU Commission 20 in the publication of asteroid ephemerides. However, it will also be useful for interpretation of thermal infrared data (e.g., IRAS asteroid observations), and can serve as a paradigm for judging individual asteroid phase relations as "normal" or unusual.

It is now widely believed that collisions at high relative velocities (several km/s) have been the main process responsible for the evolution of the physical properties of asteroids since their origin. Special efforts have been made to analyze how these collisions have affected the rotational properties and, more generally, how we can recognize among the present asteroids the outcomes of different types of impacts.

Determining the mass ratio of the probable largest collision for target asteroids of different sizes during the solar system's lifetime, Farinella et al. (Icarus 52, 409) concluded that, if the results of laboratory impact experiments can be scaled up to asteroidal sizes, then all the asteroids have undergone collisional events capable of overcoming the material's solid-state cohesion. Such events do not always lead to complete disruption of the targets, and also self-gravitation can easily cause a reaccumulation of the fragments for targets exceeding a critical size, which appears to be of the order of 100 km. It follows that in the intermediate size-range (100 to 300 km) the formation of gravitationally bound "rubble piles", having low strength, is fairly frequent (cf. Davis et al., Asteroids, T. Gehrels, ed., 528), and the collisional angular momentum transfer can be large enough to produce objects with triaxial equilibrium shapes or to cause fission into binary systems. Asteroids smaller than about 100 km are mostly multi-generation fragments, while for diameters larger than 300 km, the collisional process yields nearly spheroidal objects covered by deep regolith layers. These conclusions are in accord with a comparative analysis of the asteroid shapes with those of collisional fragments produced in laboratory experiments (Catullo et al., A&A 138, 464; Capaccioni et al., Nature 309, 832).

The analysis of the collisional outcomes can be applied to study the origin of dynamical Hirayama families (Zappalá et al., Icarus 59, 261). In particular, we can separate the families according to their mass and relative velocity distributions, identifying "asymmetric" families, whose minor members were asymmetrically ejected on one side of the largest collisional remnant, and "dispersed" families, for which the disruption event was nearly isotropic and the largest asteroid is roughly at the center of the fragment mass distribution. The influence of the parent body's self-gravitation in most cases was crucial (Fujiwara, Icarus 52, 434), and in particular it could amplify the effects of ejection asymmetries. These ideas are supported by the available data on families, even though the current family identification schemes have several problems both from the point of view of celestial mechanics and from that of the clustering techniques (cf. Carusi and Valsecchi, A&A 115, 327). Davis et al. (BAAS 16, 697; Icarus, in press) use the characteristics of the major families, as well as the apparent preservation of the presumed ancient basaltic crust on Vesta, to constrain the collisional evolution of the asteroids. In some contradiction to earlier work and the recent work of Farinella et al. mentioned above, Davis et al. find the collisional evolution to have been quite modest since early epochs. The original abundance of asteroids may have been only a few times greater than that of today, which may also be consistent with the number of intermediate-sized asteroids having high angular momenta (Farinella et al., Icarus 52, 409).

There are clear hints that the results of laboratory experiments cannot be fully extrapolated for asteroids without a better physical understanding of the rupture mechanism. Davis et al. (loc cit) pointed out the difficulty of applying a scaling law passing from centimeter to multi-kilometer-sized targets, and have proposed that there is increasing strength for increasing body size due to hydrostatic self-compression; this can explain the unexpectedly large fragment-to-parent body mass ratios observed in the Hirayama families with respect to outcomes of laboratory experiments. Theoretical (Paolicchi et al., Hvar Obs. Bull. 6, 163) and experimental (Matsui et al., Lun. Planet. Sci. XIII, 475; Waza and Matsui, Lun. Planet. Sci. XIV, 838; Bianchi et al., A&A 139, 1) studies of fragmentation processes have been made and are in progress for a general understanding of the physical and rotational properties of the break-up outcomes. The surprising abundance of objects with very long lightcurve periods remains an open problem, despite the mechanisms of angular momentum drain (Dobrovolskis and Burns, Icarus 57, 464) and tidal evolution within binary systems (Harris, BAAS 15, 828; Binzel, BAAS 16, 698) mentioned above, which could play a role in its solution.

Another important puzzle is presented by the bimodal distribution of rotational periods of the Apollo-Amor asteroids. Moreover, Apollo-Amors seem on the average to be more irregularly-shaped than both laboratory fragments and small main-belt asteroids (cf. Capaccioni et al., Nature 309, 832). Catullo et al. (A&A 138, 464) have compared lightcurve amplitude distributions for several groups of asteroids with equivalent shape parameters for fragments generated by high-velocity impact experiments. They find that main-belt asteroids smaller than 100 km diameter match the experimental distribution well. Large asteroids are lacking in extreme amplitudes, suggesting equilibrium configurations due to self-gravity. But Apollo-Amor asteroids have an excess of unusually elongated objects: the interpretation of this latter fact is unclear. Could we be observing two classes of Apollo-Amors, one of "asteroidal" and one of "cometary" origin?

COMPOSITIONS, PHYSICAL PROPERTIES, RELATIONS TO METEORITES - C.R. Chapman

Detailed infrared spectra of 2 Pallas are interpreted by Larson et al. (Icarus 56, 398) in terms of a mineral assemblage similar to CI and CM meteorites, but subjected to greater aquaeous alteration and with an admixture of another component. An earlier study of 1 Ceres, especially of the 3 micron region, by Lebofsky et al. (Icarus 48, 453), suggests a hydrated clay mineralogy for that body. In addition to water of hydration, Lebofsky et al. claim to see a feature due to water ice on Ceres, which would have profound implications if true. Results of the most recent survey of 3-micron water features, and relationships to carbonaceous chondrite mineralogy, are reported by Feierberg et al. (BAAS 16, 692).

What may be the first silicate emission features observed for asteroids in the 8-13 micron spectral range were detected by Feierberg et al. (Icarus 56, 393) for 19 Fortuna and 21 Lutetia; comparable features were absent for several other asteroids.

A continuing question of great importance is whether the S-type asteroids are mineralogically similar to the undifferentiated ordinary chondrites or, instead, they are similar to differentiated stony-iron meteorites. Feierberg et al. (ApJ, 257, 361) interpret mid-IR spectra as implying ordinary chondrite-like mineralogies. Gaffey et al. (Icarus 60, 83; BAAS 16, 698) have studied 8 Flora in exhaustive detail and find that this particular S-type object shows variations in its spectrum as it rotates that are inconsistent with ordinary chondrite mineralogy. Gaffey strongly feels that a differentiated (stony-iron) mineralogy is required and generalizes this particular result as suggesting that most S-types are differentiated. This may be incompatible with the radar work of Ostro et al. (see below), who find that Flora has one of the lowest radar reflectivities of asteroids observed by radar, which probably means that it is incompatible with the large fraction of surficial metal that Gaffey believes is there. Inasmuch as the continuing 8-color spectrophotometry program (see below) has turned up few objects (except those already in Earth-approaching orbits) with colors more like ordinary chondrites than the S-types, it is becoming less and less clear how we can derive so many meteorites from such uncommon main-belt asteroids, unless we interpret many (or at least some) S-types as being ordinary chondrites. The alternative, namely that comets are the parent bodies of ordinary chondrites, is incompatible with current views of condensation temperatures in the region of comet formation. Also, earlier dynamical difficulties with extracting sufficient material from the inner main belt to account for the abundance of ordinary chondrites are evaporating in the face of new work (see below). It is hoped that a major input to this controversy about S-types can be made by the Galileo spacecraft, which may make near-infrared spectral maps and take high resolution images of the surface of 29 Amphitrite in late 1986. Some additional information may be derived from the more intensive studies of S-type asteroids in the infrared, which have been initiated by Brown (see Bell et al., BAAS 16, 692).

It has long been suspected that certain S-like asteroid spectra (some asteroids with these spectra were originally termed R's) could imply an olivine-rich assemblage, which would be of great interest since olivine-rich meteorites are believed to be derived from the mantles or mantle-core interfaces of strongly-heated and geochemically differentiated parent bodies (cf. Greenberg and Chapman, Icarus 57, 267). Definitive spectra of 246 Asporina and another main-belt asteroid by Cruikshank and Hartmann (Science 223, 281) shows the diagnostic signature of olivine with an absence of pyroxene. These are members of a small group of A-type asteroids that are presumably all olivine-rich and pyroxene-poor. JHK photometry of asteroids (e.g., Veeder et al., AJ 87, 834; AJ 88, 1060) continues to supplement visible and near-IR colorimetry in the taxonomic classification of asteroids, including the A-types (Veeder et al., Icarus 55, 177).

A major contribution to asteroid research has been made by the 8-color spectrophotometry program conducted by Tholen, Tedesco, and Zellner. Although the final publication of data on nearly 600 asteroids is still due this year, preliminary interpretation of these data, and of associated radiometry, by Gradie and Tedesco (Science 216, 1405) has identified -- with much better precision than earlier studies -- a well-ordered sequence of predominant taxonomic types of asteroids with semi-major axis. There is a progression of seven distinct peaks in the distribution of types with distance from the sun, in the order E, R, S, M, F, C, P, and D (from 1.0 to 5.2 A.U.), although the distributions do overlap substantially. An important element of this work has been the clear identification of additional sub-types of asteroids previously thought to be C's or M's (e.g., the P type, which has C-like albedos but M-like colors). Tholen, in his 1984 thesis, has further refined the taxonomy of low-albedo asteroids; he adds "T" as intermediate between C and D, "G" as an unusual C-like type, and "B" as intermediate between C and the bluish F-types. The chief conclusion from the preliminary interpretations so far is that the asteroids truly are likely to preserve, in an ordered way, the primordial variation in solar nebular condensates with solar distance. Perhaps one or more types (e.g., the E's) have been implanted from orbits originally closer to 1 A.U.

Further observations of asteroids by Earth-based radar have been made, with the numbers of observed objects now large enough to permit some tentative correlations with such properties as size and taxonomic type (Ostro *et al.*, BAAS 16, 692). Radar studies of 1685 Toro (Ostro *et al.*, AJ 88, 565) suggest that its mean effective diameter is about 3.3 km, that its surface has a porosity comparable to the moon's, and that it is much rougher than the moon on 10-100 meter scales. A comprehensive report on radar observations of 2100 Ra-Shalom has also been presented (Icarus 60, 391).

There has been increasing success during the triennium in measuring asteroid diameters and shapes from studies of stellar occultations, despite the considerable logistical efforts required for success. For example, Millis *et al.* (AJ 89, 592) deduce a mean diameter for 375 Ursula of 216±10 km from a November 1982 event observed at six sites. Ten chords were measured by Millis *et al* (Icarus 61, 124) for 93 Minerva, yielding a diameter of 170 km, in good agreement with the radiometric value. The occultation of a naked-eye star by 2 Pallas in May 1983 resulted in the observation of more than 100 chords, providing by far the most refined profile of an asteroid to date; the southern cap of Pallas was missed, however, due to cloudy weather throughout the southern United States and northern Mexico.

It remains to be seen how fruitful speckle interferometry will prove to be in addressing the questions about asteroid sizes, shapes, and possible satellites. For example, Drummond *et al.* (Steward Obs. Preprint 525) have withdrawn an earlier suggestion that speckle data showed a satellite for 532 Herculina, and now propose a triaxial shape for that body with a superimposed albedo spot. Drummond and Hege (BAAS 16, 697) have also withdrawn the suggestion that speckle data imply the existence of a satellite of 2 Pallas, which was not detected during the widely observed 1983 stellar occultation by that body. Speckle measurements of elliptical shapes have been reported for 29 Amphitrite and 3 Juno by Baier and Weigelt (A&A 121, 137). An important test of the speckle method has been reported for the well-observed, highly elongated object 433 Eros (Drummond *et al.*, Icarus 61, 132).

The Infrared Astronomical Satellite (IRAS) observed a large number of asteroids during its 300-day mission. These data are currently being cataloged for future scientific analysis. Preliminary reports on asteroid results were presented by Tedesco *et al.* and Aumann and Walker (BAAS 16, 689). Of particular interest was the discovery of zodiacal "dust bands", apparently related to collisions in the asteroid belt (cf. Sykes and Greenberg, BAAS 16, 690) or perhaps the Hirayama families.

Froeschlé and Scholl (Asteroids, Comets, Meteors, 115) reviewed the dynamical evolution of the asteroid belt, including possible depletion mechanisms for the Kirkwood gaps. Gravitational perturbations may be able to remove existing bodies in the gaps but not in times on the order of 10^4 years. Gonczi, Froeschlé, and Froeschlé (Asteroids, Comets, Meteors, 137)studied the motion of small grains, in the early solar system evolving under the gravitational effect of the sun-Jupiter system and the Poynting-Robertson drag effect. Their numerical experiments suggest that, if in a given resonance gap there was no preexisting large body, no grains could remain in the gap to form larger bodies.

A number of programs are directed toward the discovery of unusual objects, especially Earth-approachers. Among the important discoveries of asteroids during the period was the object 1982 DB, which is the most accessible of the near-Earth objects known, requiring a delta-V of only 4.45 km/sec (Helin et al., Icarus 57, 42). Another important discovery (by IRAS) was 1983 TB, which Whipple has proposed to be the probable parent of the Geminid meteors (see also Fox et al., MNRAS 208, 11P, and the discussion in Comets section above). The possibility that 1984 BC may be an extinct comet has been raised by Helin et al. (BAAS 16, 691); it apparently has a low albedo and a reddish, D-like spectrum. A recent summary of the population of Earth-crossing asteroids has been presented by Helin (Minor Planet Bull. 11, 19). There are 60 Earth-crossers now known, although the total population of such bodies to visual magnitude 18 is probably about 1000.

Considerable interest has been generated concerning the small Earth-crossing asteroid 2201 Oljato due to unusual colors apparently observed by McFadden (see McFadden et al., Icarus 59, 25, for a presentation of reflectance spectra for 17 Earth-approaching objects). The possibility that Oljato is a nearly extinct comet has been raised, although no diagnostic evidence of well-known cometary emission features exists. There has also been reported a possible connection between events on Venus and the proximity of this object. These unresolved questions are summarized by McFadden et al. (BAAS 16, 691).

There has been considerable work on the important question of how meteorites may be derived from the asteroids. Levin and Simonenko (cf. Bull. Inst. Theo. Astron. 15, 20) believe that there is good evidence that Apollo-Amor objects represent the last parent bodies of most or even all classes of meteorites, whether comets or main-belt asteroids were the predecessors. Greenberg and Chapman (Icarus 55, 455) argue, however, that the observed distribution of meteorite classes can be derived through well-understood cratering processes on main-belt asteroids; they suggest that Earth-approaching asteroids play a minor role. The greatest advance during the period comes from the work by Wisdom (Meteoritics 18, 422), who suggests that "chaotic" behavior in the evolution of orbits of asteroids near the 3:1 Jupiter commensurability results in Earth-crossing orbits; Wetherill (Meteoritical Society Presidential Address, Albuquerque, 1984, in press) thinks this may provide the long-sought-for mechanism of getting ordinary chondrites from the main belt to the Earth.

Clark R. Chapman
President of the Commission.

16. PHYSICAL STUDY OF PLANETS AND SATELLITES (ETUDE PHYSIQUE DES PLANETES ET DES SATELLITES)

PRESIDENT: V.G.Tejfel'

VICE-PRESIDENT: G.E.Hunt

EXECUTIVE SECRETARY: A.Brahic

ORGANIZING COMMITTEE: R.F.Beebe, J.L.Bertaux, J.L.Elliott, D.Gautier, L.V.Ksanfomaliti, M.Y.Marov, H.Masursky, G.Neukum, T.C.Owen, V.V.Shevchenko, B.A.Smith

I. INTRODUCTION

During past three years from 1982 to 1984 we saw the further progress in the planets and satellites research by the space and ground-based technique, in the analysis and interpretation of the observational data. Inspite of some decrease of the activity in the planetary spacecrafts launches during this period (except of two Soviet missions to Venus) many important scientific results were obtained from the continued reduction and analysis of the measurements which were performed by Mariner 10 (Mercury), Pioneer Venus, Venera 13 and 14, Viking (Mars), Pioneer 10 and 11, and Voyager 1 and 2.

It is impossible to review here even the main results of the planetary studies because the number and volumes of recent publications are too great and the simple listing of them would request of many pages.

The summaries of a number of most important papers were published regularly in the Journal MOON AND PLANETS, which is named now EARTH, MOON, AND PLANETS. Each planetary scientist can find the valuable collection of papers in ICARUS (International Journal of Solar System studies) which is published monthly and consists of four volumes per year.

Most of Soviet planetary papers are published in Russian ASTRONOMICHESKIJ VESTNIK, which is translated into English as SOLAR SYSTEM RESEARCH, and the results of planetary space studies are published in special issues of the Journal KOSMICHESKIE ISSLEDOVANIA (SPACE RESEARCH).

Many interesting and useful books have appeared during past three years. General problems of the planetary physics were considered in "Moon and planets" by W.K.Hartmann (Belmont, California, 1983) and in Russian "Introduction to the Physics of the Planets" by Yu.V.Alexandrov (Kiev, 1982),and "Key problems of the studies of the Solar System Planets" by K.Ya.Kondratiev and N.I.Moskalenko (Moscow, 1983). Best planetary photographs are collected in "The Cambridge Photographic Atlas of the Planets" by G.A.Briggs and F.W.Taylor (Cambridge Univ.Press, 1983). The physics of the planetary interiors have discussed in "Inner structure of the Earth and Planets" by V.N.Zharkov ("Nauka", Moscow, 1983).

II. THE STUDY OF THE TERRESTRIAL PLANETS

Two Soviet automatic space probes Venera I5 and I6 launched in June I983 were placed into elliptic polar orbits around Venus. By means of the board side-view radar system the images of Venus surface were obtained simultaneously with the measurements of the altitude differences by radioaltimeter. Very informative radar maps of the Venus surface relief were constructed from these measurements for the north hemisphere of Venus from the latitude 30°N to pole with a resolution about I-I.5 km. It was the first so detailed study of the Venus surface and many unusual relief formations (the impact and volcanic craters, mountain chains, ring-shaped ridges, lava flows ets.) were discovered. The joint experiment of the USSR and GDR scientists for infrared Fourier spectrometry aboard of Venera has confirmed the presence of significant altitudinal variations of the clouds upper boundary on Venus. The variations of the minor atmospheric components abundances and of the thermal flux have been analysed in detail from the measurements of the spectrum at the wavelengths 6 to 36 µm. The main results of these experiments were published preliminary in special issue of "Pis'ma v Astronomicheskij Zhurnal" (SOVIET AJ LETTERS) <u>IO</u>, No II, I984.

The beautifully illustrated review of the Venus study results from Pioneer Venus was published by R.O.Fimmel, L.Colin, and E.Burgess (NASA, I983) with a supplement about the Soviet studies of Venus. From the data of Venus probes the models of the vertical structure of the mesosphere and lower thermosphere were developed by A.Seiff and D.B.Kirk (ICARUS, 49, 49-70, I982). The comprehensive review of the spectral measurements of scattered solar radiation within the Venus atmosphere from Venera II and I2 descent probes has been published by V.I.Moroz et al. in ICARUS 53, 509-537, I983.

Two symposia "Venus atmosphere" and "Planetology of Venus, Mars, and satellites of outer planets" were held in Graz, Austria in June I984.

The series of papers on the results obtained from Venera I3 and I4 were published in "Pis'ma v Astron.Zhurnal" 8, No 7, I982 and in "Kosmicheskie Issledovania" 2I, No 2 and 3, I983.

Most of modern knowledges about Venus is reviewed in the book "Venus" (Eds. D.M.Hunten, L.Colin, T.M.Donahue,and V.I.Moroz; Univ. Arizona Press, I983).

Three monographs were published in Russian: "Photochemistry of the atmospheres of Mars and Venus" by V.A.Krasnopolskij ("Nauka", Moscow, I982), "Chemistry of the atmosphere and surface of Venus" by V.P.Volkov, and "Cryolithosphere of Mars" by R.O.Kuzmin ("Nauka", Moscow, I983).

Double issue of ICARUS (50, No 2-3, I982) was given up to the papers about the Martian geology and climate changes, presented on the workshop that was held in NASA Ames Research Center in February I98I.

III. THE OUTER PLANETS

There was continued the publication of the results of ultraviolet spectrophotometry of Jupiter, Saturn, Uranus, and Neptune from satellite IUE. Very impressive results were obtained from the observations of the radioemission of major planets at the centimeter wavelengths with high angular resolution (to I.5 arcsec) using VLA (I.de Pater et al., ICARUS 50, 88-I02; 5I, 25-38, I982; Adv.Space Res. 3, 39-4I, I983 ets.)

The book "Physics of the Jovian magnetosphere (Ed. by A.J.Dessler, Cambridge Univ.Press, I983) has presented a collection of the review papers about the different topics of the study of magnetic field, charged particles, plasma waves, and electromagnetic emissions from Pioneer and Voyager.

The detailed analysis of polarimetric and photometric measurements from Pioneer for Jupiter and Saturn was carried out by M.G. Tomasko et al. (ICARUS 58,I-34, and 35-73, I984).

The papers presented at the meeting "Saturn" (Tucson, Arizona, May I982) were collected in two special issues of ICARUS (53,No 2, I983, and 54, No2, I983) and main reviews of different topics of Saturn's physical properties and the nature of Titan's atmosphere were compiled in the chapters of the book "Saturn" (Eds. T.Gehrels and M.Matthews, Univ.Arizona Press, I984).

Many results of the Saturn study from Voyager 2 were described in special issue of SCIENCE 2I5 No 4532, I982 and in the book "Voyages to Saturn" by D.Morrison (NASA, I982).

The reviews of modern state of our knowledges about Uranus which has been discussed at the IAU Colloquium No60 in Bath (England) in April I98I are found in the book "Uranus, and the outer planets" (Ed.G.Hunt, Cambridge Univ.Prss, I982). More recent studies of Uranus and Neptune have been presented in Proceedings of a workshop held in Pasadena, California (Ed. J.T.Bergdtralh, JPL, February I984).

The collected book "The physics of planetary atmospheres" with papers about photometry and spectrophotometry of Jupiter and Saturn was published in Russian (Kiev, I982).

Some problems of the outer planets research have been discussed ai the COSPAR Symposium on the Giant Planets and their Satellites in Ottawa (May I982), on scientific sessions in Patras, Greece (August I982), and during the scientific session at the Meeting of WGPSN in Tbilisi, USSR (April I984).

IV. PLANETARY SATELLITES

Most of problems of the physics, chemistry and geology of the Moon are considered at annual Lunar and Planetary Conferences in Houston, Texas. The Polarimetric Atlas of the Moon has been prepared and published by V.P.Dzhapiashvili and A.N.Korol' (Tbilisi,I982).

The numerical data for the model of the vertical structure of Titan's atmosphere from the Voyager I radio occultation experiment have given in the paper of G.F.Lindal et al. in ICARUS 53, 348-363, 1983.

The presence of the water ice on the surface of Miranda was detected by R.H.Brown and R.N.Clark. From the study of the near-infrared spectrum of Triton D.P.Cruikshank et al. have discovered the absorption band at 2.16 μm which is tentatively identified as the density-induced band of liquid nitrogen. These communications asa well as many other papers were presented at the IAU Colloquium No 77 "Natural Satellites" held at Cornell University in July 1983 and published in special issue of ICARUS 58, No2, 1984.

The problems of the structure and dynamics of the rings around Jupiter, Saturn, and Uranus were a subject of the IAU Colloquium No 75 (Toulouse, France, August-September 1982) and described in the book "Planetary Rings" (Eds. R.Greenberg and A.Brahic, Univ. Arizona Press, 1984).

V. COMMISSION BUSINESS

The members of the Organizing Committee have discussed the proposal on the preparation of the International Reference Book - Review "Physical Characteristics of the Planets". The preliminary project of this book with the invitations to participate as the authors of any parts of book and the chapters editors was distributed between the Commission 16 members.

The ephemeris of the future mutual events in the Jovian satellites system in 1985-1986 were computed for several observatories by K.Aksnes.

V.G.Tejfel'

President of Commission

COMMISSION 19: ROTATION OF THE EARTH (ROTATION DE LA TERRE).

President: Ya. S. Yatskiv
Vice-President: W. J. Klepczynski
Organizing Committee: F. Barlier, H. Enslin, C. Kakuta,
B. Kolaczek, I. I. Mueller, V.A. Naumov, P. Paquet,
B. Proverbio, M. G. Rochester, Ye Shu-hua, K. Yokoyama.

Introduction

During the period, work on the problem of the Earth's rotation has continued to expand and increase its scope. The total number of institutions engaged in the determination of the Earth's rotation parameters (ERP) by different techniques has been increased significantly. The rotation of the Earth is currently measured by classical astrometry, Doppler and laser satellite tracking, laser ranging of the Moon, and radio interferometry. Several long time series of the ERP are available from most of these techniques, in particular, those made during the Main Campaign of the MERIT project. The various series have been intercompared and their stability, in the time frame of years to days, has been estimated for the purposes of establishing a new conventional terrestrial reference system (COTES). On the other hand, the difficulties of maintaining a regular operation for laser ranging to the Moon (LLR) have been recognized. It resulted in the proposal to organize an one-month campaign of observations in 1985 in order to complement the COTES collocation program and to allow additional intercomparisons with other techniques.

The international, electronic exchange of data on Earth rotation through the General Electric Mark III Information System computer network has been expanded during the period of this report. Many Time Service publications are exchanged in the RC28 catalog.

Reference systems for study of the Earth's rotation

Different aspects concerned with the definition and the establishment of reference coordinate systems for the Earth's dynamics, in particular COTES, have been considered by the IAG/IAU Working Group (Chairman I.I Mueller). It has been recommended that the reference frame of the future COTES be defined by an adopted set of spatial coordinates of a global network of observing stations.

C. Boucher and M. Feissel (1984) proposed a new realization of the BIH terrestrial system, including the, scale unit origin and orientation of the axes on the basis of the permanent stations used in the monitoring of the Earth's rotation by space geodesy techniques.

In the development of a celestial radio reference frame there are now over 100 sources whose relative positions are known with an average uncertainty less than 5 millarcseconds (J. L. Fanselow et al, 33.043.003).

In the improvement of a celestial stellar reference frame there are now about 100000 stars, whose relative positions will be observed by the HIPPARCOS Space Astrometry Mission (P.L. Bernacca, 1983).

The orientation of the dynamic reference frame realized by the JPL ephemerides DE 200/LE 200 has been considered by E. M. Standish (32.047.016).

The effects of adopting new precession and equinox correction have been discussed in connection with a new definition of UT (31.044.001, 31/044.003, 31.043.010, 33.044.021).

J. Vondrak (31.045.001, 33.044.035, 33.044.061,34.044.004) derived the direct and indirect influences of the planets on nutation. The total effect is comparable with the smallest terms in the luni-solar nutation series adopted by the IAU. Some authors have considered the definition of the celestial ephemeris pole (32.043.010, 33.044.078).

The work on the PZT/DHI 82 star catalogue with improved star positions and proper motions has been completed at DHI, Hamburg. In addition, the improved catalogue PZT8 was completed at ZIPE, Pottsdam.

Yokoyama (1983, 1984) presented a full description of the method for reducing the optical astrometric observations at each station, as well as at the international centers, in accordance with the adoption of the FK5 system and the IAU (1976) System of Astronomical Constants for Jan. 1, 1984 onwards. This new method was presented to the 2nd MERIT Workshop and was approved as the MERIT standards.

Determination of the Earth's rotation parameters from the classical technique

(Research in Instrumentation)

T. Tsubokawa has designed a new type of fully automatized pendulum astrolabe controlled from a remote minicomputer. It has been constructed and tested at Mizusawa (31.032.011, 33.032.036). The photoelectric astrolabe has been improved at CERGA-Grasse (33.032.009). Different origins of errors of observations by photoelectric astrolabes in the People's Republic of China have been studied (33.041.040, 33.041.050, 33.044.044). The photoelectric astrolabe of Yumman Observatory has proven to be the best in time observation of a dozen Chinese instruments. The relation between temperature and the displacements of the focal plane in the Danjon astrolabe of Wuchang Time Observatory has been derived (34.032.050).

At Irkutsk, a latitude service has been established using the observations from two Danjon astrolabes (34.044.009). Regular observations of latitude and UT variations with the use of the Danjon astrolabe were started in Borowiec in 1982 (M. Lehmann).

Studies on the instrumental errors of the PZT are being continued at Tokyo (34.03.046) and at Kitab (34.032.022). Z. M. Malkin (34.032.062) has proposed an optimum construction of a pavilion for the PZT at Kitab.

Many papers concerned with the instrumental errors of photoelectric transit instruments and also with their automation have been published (34.032.011, 34.032.017, 34.032.041), 34.032.042, 34.032.045, 34.032.105, 34.036.106).

Research on instrumental errors of zenith-telescopes is being continued at Kitab (33.032.019-33.032.21). PZT observations at the DHI, Hamburg will be terminated by the end of 1986.

(Analysis of observational data and determination of the ERP)

N. I. Panchenko and A. A. Slavinskaya (34.044.024) investigated mean latitude variations of some stations equipped by astrolabes. Tidal and nutational effects from observations of the Paris astrolabe have been derived (31.043.012, 32.045.011). P. I. Popova has (34.044.022) determined the fortnightly nutation term from observations of bright zenith stars in Poltava. At Pulkovo, non-polar latitude variations derived from observation with two zenith telescopes (ZLL-180 and ZTF-135) have been studied (32.045.025, 32.045.026). New reduction and analysis of observational data by the ZTL-180 at Kitab have been undertaken (34.044.010-34.044.012).

Different systematic effects (wind influence, parallel curvature ets.) have been investigated at Poltawa (34.044.013, 34,044.023) and at Pulkovo (34.044.015).

N. Capitane and N. Xiao (31.044.002) studied the Z-term of latitude and the w-term of UT, as computed by the BIH, for deriving the principal terms of nutation.

G. P. Pil'nik (32.044.040) estimated the power spectrum of UT determinations from 1968 to 1978 and showed that there were diurnal and semidiurnal variations.

R. M. Rasulov (34.044.011) determined corrections to the ILS pole coordinates resulting from micrometer errors for interval 1938.4-1939.2.

S. Takagi (32.031.503), A.A. Korsun' and R. M. Rasulov (34.044.021), D. Yu. Belotserkovskij and M. B. Kaufman (33.044.028) considered the determination of ERP in an uniform system, based on different sets of data. Re-reduction of observations of the Soviet Time Service from 1955 to 1979 has been done by L. I. Yagudin (33.044.036). J. Hefty (31.044.005) derived the Love number K from UT determinations during the years 1967-1978.

Different aspects of Chinese activity in the field of ERP determination and analysis have been presented at the Shanghai Syposium on Astrometry (34.044.032-34.044.059) and have been published in Publs. Shanghai Astron. Obs. (33.044.078-33.044.094) and in Ann. Shanghai Obs. (33.044.065-33.044.070).

P. Fellget (34.044.005) computed the power spectrum of the observed variations of latitude, and calculated the explicit likelihood distribution.

V. L. Gorshkov (34.036.028, 34.036.059) proposed a modification to the traditional chain method in order to determine a correction for instrumental and refeaction errors.

S.-Y. Zu (33.044.029) and S. Okasaki and T. Sakai (34.044.094) considered the possibility of predicting polar motion and variation in the rate of the Earth's rotation respectively. Investigations of local, nonpolar variations in latitude determined by astrometric methods were carried out at the Space Research Center, Warsaw. Annual and semiannual terms of local latitude variations are very changable especially for some stations. Their amplitudes and phases have periodical variations with the periods of 2,3,4, years and longer. Introducing variable annual and semiannual terms of local variations of latitude instead of the annual mean as determined by the BIH ought to improve astrometric determinations of pole motion. About 20 astrometric stations have more stable local variations of latitude and UTO (B. Kolaczek and G. Teleki, 1982; B. Kolaczek, 1982; B. Kolaczek and B. Napiorkowska, 1983)

Determination of the Earth's rotation parameters from the new techniques

(Laser)

Instrumentation of the SLR stations was discussed at the Laser Ranging Workshop at Herstmonceux in September 1984. Different requirements of station distribution for the determination of ERP are discussed with the aim of minimizing the number of SLR stations 34.045.021) and the number of LLR stations (34.044.028).

D.E. Smith et al (34.044.105) derived the variations in the length-of-day from Lageos data.

The McDonald Observatory LLR data have been analysed for determining UT1 and variations of latitude by W. Jin for the period 1971-1979 (32.044.043) and by J.O. Dickey and J. G. Williams for the period 1972-1982. (34.044.029).

At present there are about 10 SLR stations with an accuracy better than 10 cm. One of them is at Simostato Hydrographic Observatory (33.032.037).

D.C. Christodoulidis and D. E. Smith (34.045.023) discussed the role of SLR through the 1990's.

Two programs for analysing SLR data have been developed in the USSR. One used an analytical approach (Sternberg Astronomical Institute, Moscow) and the other used numerical integration (Main Astronomical Observatory, Kiev.)

A permanent Doppler station (TRANET II) for satellite tracking is in operation since August, 1983 at the Observatorio de Marina de San Fernando, Spain. The station took part in the MERIT campaign and the MEDOC campaign. Also, regular observations of Doppler satellites were made at Borowiec with the use of the Astro-receiver DOG-2, constructed at the Space Research Center.

(R a d i o i n t e r f e r o m e t r y)

The present state and future developments for the VLBI technique were discussed at an International Conference held in Toulouse in 1982 (33.012.020). The Mark III VLBI system allows recording and later processing of up to 112 megabits per second from each radio telescope of an interferometer array. The relative delays of the signals recived at each pair of sites can be determined with uncertainties under 50 psec (33.033.010). This information can be used for determination of the ERP (33.033.019) and nutation motion (32.044.005). E.C. Pavlis and I. I Mueller (34.045.006) considered the effect of Earth's rotation errors in baseline determination.

Different aspects of the VLBI observations (frequency standard effect, method of observation and analysis etc.) and CERI have been considered (33.044.023, 34.036.234, 34.036.244, 34.044.081, M. B. Bondarenko and M. I. Yurkina, 1984).

Intercomparison of different techniques for determining the Earth's rotation parameters

Several papers concerned with the intercomparison of different techniques for determining the ERP have been presented at the IAU Coll. No, 63. "High-precision Earth rotation and Earth-Moon dynamics" (31.012.035). The colloquium was organized by O. Calame, CERGA.

M. Feissel (31.044.059) considered the possiblity of determining ERP's independent of the observation technique.

B. Kolaczek et al (1984) compared the pole position determined by different techniques and detected short periodical terms with periods in the following ranges of days: 70-80; 50-55; 30-35, and 15-25.

Research concerned with the rotation of the Earth

Several review papers concerned with dynamics of the Earth and its internal constitution (34.081.001, 34.081.014 - 34.081.020, 34.081.032, 34.044.103 - 34.044.104) have been published.

V. P. Dolgchev (34.044.089) obtained an explicit expression for the axial rotation of the rigid Earth and analysed secular and periodic components in the rate of the Earth's rotation.

M. Bur (32.044.041, 32.044.042) discussed the solution of the equations describing the rotation dynamics of a deformable Earth. He determined the deformation caused by the Earth's rotation.

J. M. Wahr (32.045.002, 34.044.001) examined the effects of the atmosphere and oceans on wobble excitation and variations in the length of day. He found that the atmosphere and ocean were apparently not the primary excitation source. R.T. Barnes et al (34.044.003) re-examined an angular momentum exchange between the atmosphere and solid Earth. Daily values of the atmospheric angular momentum functions have been calculated for period 1981-1982 and have been comapred with the ERP.

J. Hinderer et al (32.045.006) derived the basic equations describing the dynamical effects of the Earth's fluid core for an ellipsoidal Earth model without axial symmetry. This problem has been also studied by M. J. Sevilla and his collaborators at Instituto de Astronomia y Geodesia, Madrid.

S. M. Molodenskij (1984) considered the theory of tide and nutation for real model of the Earth.

S. K. Runcorn (32.044.006) considered the role of the core in irregular fluctuations of the Earth's rotation and the excitation of the Chandler wobble. The torque is supposed to be of an impulse character.

N. N. Parijskij (32.044.002) pointed out a correlation between irregular fluctuation and changes in gravity. However, new observations did not support this suggestion (Parijskij, 1985).

On the other hand, a comparative analysis of the irregularities in the Earth's rotation and the solar activity have been undertaken (32.044.003, 32.044.012).

C. F. Yoder et al (31.044.004) and J. B. Merriam (31.044.007) considered short period terms in the variation of the Earth's rotation. New determinations of secular and decade fluctuation in the Earth's rotation have been carried out by S. J. Goldstein using old solar tables (31.044.060) using timings of lunar eclipses and occultations. Besides the tidal deceleration of the Earth, the study, which is being continued (31.044.061, 33.012.006), found an acceleration component. Atmospheric excitation of the Earth's rotation has been discussed in details at the meetingof the IUGG/IAG Special Study Group 5.98 (Chairman J. O. Dickey). Abstracts for this meeting were printed in EOS, Transaction, AGU, 65, 16, 186, 1984.

The theoretical and observed estimates of the Chandler component in polar motion is being continued. B. Guinot (32.045.005) derived the amplitude, phase and ellipticity of this component based on data from 1900 to 1980. W.E. Carter (32.045.004, 33.004.005, 33.044.057) suggested that the Chandler component may be frequency modulated as a linear function of the polar motion magnitude.

C. Pan (32.044.026) explored the hypothesis on the multiple-frequency Chandlerian nutation. On the other hand, S. O. Kubo (32.045.016) did not find any positive evidence for the variable period hypothesis. He also discussed the damping factor Q, as derived from the Chandle wobble.

Report of the Bureau International de l'Heure

The BIH is organized in two sections: (a) Time , (b) Rotation of the Earth, respectively under the scientific responsibility of B. Guinot and M. Feissel.

This following report covers the activities on the Rotation of the Earth and related problems. The BIH activities on atomic time and some administrative and operational matters common to both sections are reported to Commission 31.

Current evaluation.

The Earth Rotation Parameters (ERP), i.e., Universal Time and the coordinates of the pole, are evaluated regularly on the basis of independent time series obtained by all the existing techniques of observation: Very Long Base Radio-Interferometry (VLBI),

Connected Element Radio-Interferometry (CERI), optical astrometry, Lunar Laser Ranging (LLR), laser ranging to LAGEOS (SLR), and Doppler tracking of artificial satellites.

The BIH has continued the computation of a solution by optical astrometry, from the observations of 88 stations in 22 countries. The IAU 1976 System and the IAU 1980 Theory of Nutation have been introduced in 1983, in order to allow the combination of the optical astrometry solution for universal time with the VLBI results which stated to be available. It was shown that the introduction of the IAU 1980 Theory of Nutation had a negligible effect on the pole coordinates, but corrected a quasi-annual error of maximum amplitude 1.8 ms on universal time (Capitaine and Feissel, 1983).

The other series have been regularly reported to the BIH by the U.S. Defense Mapping Agency (Doppler), the Smithsonian Astrophysical Observatory (SLR), the University of Texas Center for Space Research (SLR), the U.S. Naval Observatory (CERI), and the U.S. National Geodetic Survey (VLBI). Series covering several years have also been provided by some of these institutes, and by the Centre d'Etudes et de Recherches Geodynamiques et Astronomiques (LLR), The Massachusetts institute of Technology (LLR), the Jet Propulsion Laboratory (LLR and VLBI).

Prior to the combination, the error spectra of the individual time series are determined, under the view-point of both the deterministic and the random errors. Only the optical astrometry and the CERI time series require a seasonal correction (annual and semi-annual terms). All series are brought to the BIH System by the correction of a bias (Feissel, 1982). The data of the individual series are weighted according to an unbiased estimate of their uncertainty, derived from the internal analysis of each of the series.

Over the time interval 1982-1984, the contribution of the data from the space techniques has been drastically increased. For polar motion, SLR has contributed over 50% of the total weight. VLBI has had a continuously increasing contribution, in particular for universal time, where it provides about 90% of the total weight in 1984. The precision of the final combined solution at five-day interval for 1984 is estimated to 0.0005s for universal time and 0.002" for the pole coordinates.

The final combined solution is computed for its publication in the Annual Report, issued in June of the year following the year covered. However, when the accumulation of new or improved results justifies it, a revised version of this combination is performed. This was the revised series obtained for optical astrometry was combined from 1978 onwards with the space geodesy series then available. Figure 1 shows the evolution of weights of the different techniques in the final solution at five-day interval.

Besides this final combined solution, the BIH has continued the implementation and the dissemination of two operational solution, available with a shorter time delay: the weekly Rapid Service and the monthly Circular D. computed with the same algorithm as the final solution, applied to the data available to the BIH in due time (4-10 days for the Rapid Service, 4-8 weeks for Circular D). Table 1 gives the quadratic mean of the differences of these solutions with the standard smoothings of the final solution, for 1983.

Table 1 - Agreement of the operational solutions with the final one (standard smoothing)

Solution \ Parameter	UT1	X	Y
Rapid Service	0.0019s	0.010"	0.008"
Circular D	0.0009s	0.004"	0.003"

COORDINATES OF THE POLE

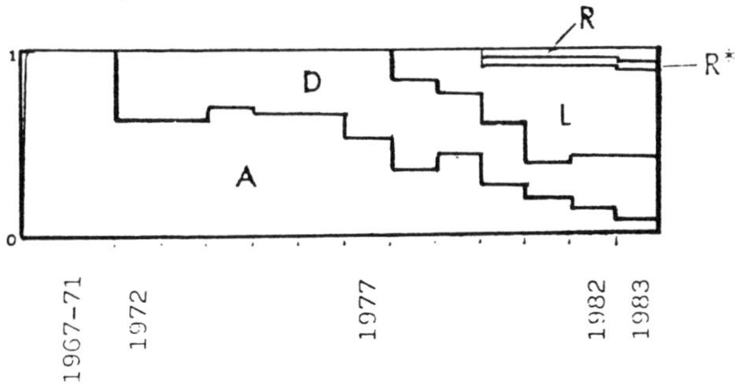

UNIVERSAL TIME

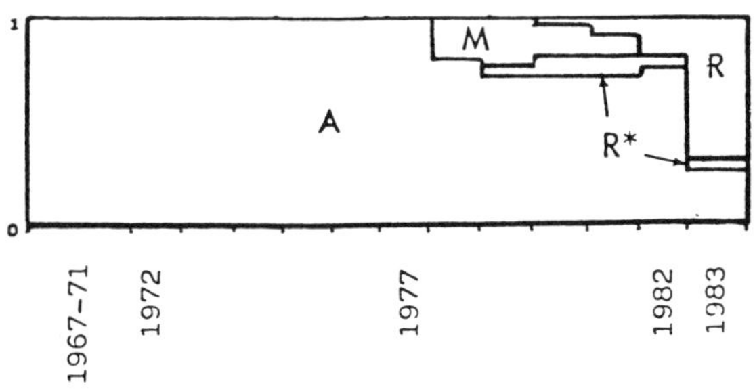

Figure 1. BIH combined solution for the Earth's rotation ; relative weights of the techniques.

A Optical astrometry
L Satellite laser ranging
R VLBI
D Doppler tracking of satellites
M Lunar laser ranging
R* Connected Radio Interferometry

This figure refers to Table 6(6C) of the Annual Report of the BIH.

Analyses.

The stability of the BIH Earth Rotation results since 1962 has been evaluated by internal analysis and by comparison with space geodesy results obtained independently. The BIH series show no detectable drift, the upper limits found being 0,0001 s/y for universal time and 0.0003" for the coordinates of the pole.

The random error spectra of the independent series of the ERP have been investigated, for sampling times ranging from five days to one year (Feisse., 1984). It was shown that the hierarchy of stabilities of the different series is not identical in the short term and in the long term. This confirms the validity of the choice made earlier by the BIH to use different systems of weighting for the long term and the short term stability of their results (Guinot and Feissel, 1968).

A new solution by optical astrometry has been computed for the years 1962-1981, on the basis of revised series of time and latitude sent to the BIH by numerous stations. This work was accomplished by and astronomer of Shanghai Observatory, in stay at the BIH in 1982-1984 (Li, 1984). The complete observational data consisted in 500 000 determinations of time and latitude, organized in 198 time series obtained by 136 stations. The ERP were obtained at five-day interval in a global adjustment, together with the classical auxiliary unknowns z and w with 2600 "group unknowns" to free the solution from catalog or local errors. The solution is published in the Annual Report for 1984.

Experimental determinations of the ERP from SLR observations have been obtained, in cooperation with the Groupe de Recherches de Geodesie Spatiale (GRGS). The results are published in the MERIT reports. The construction of normal points to condense to raw observational data has been studied (Gambis, 1982, 1984a).

Several analyses of the BIH results have been performed;

- The amplitudes of the principal and fortnightly terms of nutation have been obtained with a 0"001 percision using z and w data from 1962.0 to 1981.0. They are in good agreement with the 1980 IAU theory of nutation except the principal term in obliquity which shows a difference of 0"007 (Capitaine and Xiao, 1982).

- A time-varying anomaly of the response of the Earth to the Mr zonal wave has been detected in the UT1 results (Capitaine et al. 1984) which could be of oceanic or atmospheric origin.

- The high correlation of the short term and medium term (30 days to 2 years) variations of the duration of the day with the atmospheric angular momentum has been established (Feissel, Rosen et al., 1982; Nitschelm, 1983).

- It has been shown that short term oscillations of the duration of the day are always present, with total amplitudes of the order of 1 ms and recurrence times between 20 and 60 days (Feissel and Nitschelm, 1984).

Theoretical aspects of the rotation of a deformable Earth with a fluid core have been considered, leading to an explicit formulation of the nutation in terms of the Earth model and of the considered axis (Capitane, 1982).

The use of Guinot's non rotating origin was investigated, in particular with respect to the definition of UT1 (Aoki et al, 1982). It has been pointed out that the definition of the Celestial Ephemeris Pole (newly adopted by IAU) needs a clarification concerning the strictly diurnal term (Capitaine et al, 1984).

A collective program of time and latitude stars was proposed for observation by the ESA satellite HIPPARCOS (Feissel, Aoki et al., 1982). It includes 6082 stars coming from 39 lists provided by 24 instutes. The program was accepted by ESA with some minor restrictions.

In cooperation with the French Institut Geographique National (IGNF), the relationships between the terrestrial frames to which the different series of ERP obtained by space geodesy are referred were studied. As a result of this study, it was proposed that the BIH terrestrial system be redefined on the basis of the coordinates of the sites where space geodesy stations are operated in programs for the determination of the Earth's rotation (Boucher and Feissel, 1984).

MERIT Campaign.

The BIH acted as the coordinator of MERIT project, in particular during the main observing campaign, from 1983 September 1 to 1984 October 31. The different tasks accomplished in this framework were as follows.

- Edition of the final results of the short campaign, (August-October 1980), included in the Report on the short campaign (Wilkins and Feissel (eds), 1982).

- Collection of the precise description and geodetic ties of the sites involved in the measurement of the Earth's rotaion (Feissel and Wilson (eds), of MERIT Sites (DOMES) is kept up-to-date.

- Collection of the Sets of Station Coordinates (SSC) defining the terrestrial frame for each series of ERP.

- Collection of the series of ERP obtained during the campaign by the Operational Centres (see Table 2) and of the series obtained after the end of the campaign by the Analysis Centres.

- Maintainance of a data bank for storing the collected data, dissemination of copies on request.

- Issuing of results from, and information on, the campaign, in the MERIT Monthly Circular, distributed to 180 addresses.

Report of the International Polar Motion Service

During the main campaign in Project MERIT, the IPMS acted as an analysis, as well as an operational center of optical astrometry.

Prior to the commencement of the main campaign of Project MERIT, the IPMS, as an analysis center of optical astrometry, developed a new method for estimating an optimumly smoothed series of Earth orientation parameters (EOP's: polar motion, UT1 and l.o.d.) with one-day interval. This method is based on a firm statistical basis which uses Akaike's information criterion for selecting optimum smoothing parameters (Manabe et al. 1982; Yokoyama 1984a).

Nominal rms errors of the IPMS daily EOP's were estimated as +/-0.003" for polar motion and +/-0.2 ms for both UT1 and l.o.d. Due to the improvement of the analysis technique, precisions of the EOP's derived from optical astrometry were remarkably improved.

This method is still being improved to determine variations of the station coordinates simultaneously with EOP's. The improved method is expected to bring out maximum possibility of optical astrometry as a technique for monitoring Earth rotation.

The IPMS daily EOP's have been published in the succeeding issues of the Monthly Note of the IPMS since 1983 No. 9, in place of the monthly mean EOP's which had so far been published. The new series of the daily EOP's have been reproduced in the Monthly Circular of Project MERIT published by the BIH and also in the Earth Orientation Bulletin published by the USNO.

Table 2 - MERIT Campaign (1983 Sep 1 - 1984 Oct 31)
Results received by the Cooordinating Centre

Earth Rotation	Parameters	Interval	First Transmission	Terrestrial Frame
ERP(NGS) 83 R QL	x, y, UT1	5d	Nov 83 (1)	SSC(NGS) 83 R 01
ERP(NGS) 83 R QL	x, y, UT1	5d	Nov 83 (1)	SSC(NGS) 83 R 01
ERP(NGS) 83 R QL	x, y, UT1	5d	Jul 84 (1)	
ERP(JPL) 82 R 01	, UTO	5-7d	Mar 84	SSC(JPL) 82 R 05
Green Bank Interf.	, UTO	3d	Sep 83 (1)	
Green Bank Interf.	UTO	3d	Apr 84 (1)	
ERP(CERGA)83 M 01	UTO	1-33d	May 84	SSC(CERGA) 83 M 01
ERP(CSR) 81 L 01	x, y, D	5d	Sep 83 (1)	SSC(CSR) 81 L 01
ERP(CSR) 84 L 02	x, y, D	5d	May 84 (1)	
ERP(GRGS) 84 L 01	x, y, D	5d	Jun 84	SSC(GRGS) 84 L 01
ERP(DMA) 77 D 01	x, y,	2d	Sep 83 (1)	
ERP(DMA) 79 D 01	x, y,	2d	Sep 83 (1)	
ERP(DMA) 82 D 01	x, y,	2d	Sep 83 (1)	
ERP(BIH) 83 A 02	x, y, UT1	5d	Sep 83 (1)	SSC(BIH) 83 A 01
ERP(BIH) 84 A 02	x, y, UT1	5d	Feb 84 (1)	SSC(BIH) 84 A 01
ERP(IPMS) 83 A	x, y, UT1, D	1d	Nov 83	

(1) Results in the current combined ERP solution of the BIH.

The Monthly Note of the IPMS also gives daily mean values of latitude and UT1-UTC.

The IMPS daily EOP's from 1962 were computed based on the system of MERIT Standards. These are in preparation for print. Magnetic tape versions of the results were already distributed to several institutions on request.

Intercomparison of the IPMS daily EOP's was made with those estimated by other techniques and centers, as well as by atmospheric data (Yokoyama 1984; Hara and Yokoyama 1984). It turned out that the UT1 and l.o.d. determined from optical astrometry agree far better than has been improperly believed. On the other hand, polar motion results show systematic errors of a few decimeters from the VLBI and SLR results, probably due to large observational errors of individual instruments.

Daily values of the l.o.d. caused by the zonal wind and redistribution of atmospheric mass have been computed since September 1983 with a month delay using the global analysis data compiled at the Japan Meteorological Agency (Naito and Yokoyama 1984). The results are in good agreement with those computed at the National Meteorological Agency of the USA. This series of the l.o.d. will soon be dissiminated regularly.

In accordance with the official adoption of the IAU 1976 System of Astronomical Constants and IAU 1980 Theory of Nutation from January 1, 1984 onwards, the IPMS, in collaboration with the Tokyo Astronomical Observatory and the Japan Hydrographic Department, developed algorithms and computer softwares to calculate apparent places of stars for common use in the reduction of observations by optical astrometry at the individual stations (Yokoyama 1983; Yokoyama 1984b). Magnetic tape versions of the computer softwares were distributed to every country where optical astrometry instruments are in operation. Thus, apparent places of stars are being computed at individual stations based on the same theoretical basis.

In cooperation with the USNO, all PZT observations of Washington and Mizusawa were re-reduced using the system of MERIT Standards and the unified catalog of star positions and proper motions was compiled in the system of FK 4. This work was being extendedly applied to PZT observations of other stations.

In collaboration with the Tokyo Astronomical Observatory, improvement of the computer software for satellite laser ranging, which was originally developed at the SAO using an analytical method, is in progress. Aimed goal is to achieve the precision comparable to that already achieved by other softwares using the method of numerical integration.

IAU/IUGG Joint Working Group on the Rotation of the Earth (MERIT)

The IAU/IUGG Joint Working Group on the Rotation of the Earth has been very active since the IAU General Assembly in Patras in organizing the MERIT campaigns of observation and analysis. It has cooperated closely with the IAU/IUGG Joint Working Group on the Establishment and Maintenance of the Conventional Terrestrial Reference System (COTES). The Second MERIT Workshop was held at the Royal Greenwich Observatory on 1983 May 15 - 19 to review the operational arrangements for the MERIT Main Campaign, which was held from 1983 September 1 to 1984 October 31, and for the COTES Intensive Campaign, which was held from 1984 April 1 to 1984 June 30. In addition to the monitoring of the rotation of the Earth (universal time or length of day and polar motion) by six different techniques, special attempts were made to obtain simultaneous high-precision observations by two or more techniques at selected sites and to determine the position of all sites by the Doppler technique.

Reports on the operations during the period of the main campaign and on the subsequent analyses of the data are published in the MERIT Monthly Circular which is prepared and distributed to participating organizations by the Bureau International de l'Heure. It is clear from the results that have already been obtained that Project MERIT has been successful in stimulating the development and regular use of new high-precision techniques (especially satellite laser ranging and very long baseline radio interferometry) for monitoring the rotation of the Earth; it has also led to an improvement in the quality of the results obtained by the optical astrometry. The project has generated a high level of cooperation, as exemplified by the adoption of MERIT Standards for constants and numerical models for use in the analysis of observations. The improved precision of the earth-rotation parameters has also made possible detailed comparisons with the variations in the angular momentum of the atmosphere that are of great interest to meteorologists. The extension of the MERIT campaign to provide additional data for COTES has emphasized the importance of monitoring earth-rotation for geodetic purposes, and the dense data sets obtained during the intensive campaign are likely to be valuable in many other studies.

The principal disappointment has been that the lunar laser ranging observations have been much fewer than had been hoped, and so it has been decided to hold a second intensive campaign in 1985 May to July so that the potential of LLR observations for monitoring earth-rotation can be properly assessed and so that the relationships between the reference systems can be better determined. The third MERIT Workshop will be held at Columbus, Ohio, on 1985 July 29-30 and will be followed immediately by a conference for the presentation of the preliminary results from the MERIT-COTES campaigns. The implications of the results for the international services will be discussed at the IAU General Assembly in Delhi. In the meantime the operational arrangements developed for the campaigns will be continued, although the level of observation will not be so high.

IAU/IAU Joint Working Group On The Establishment And Maintenance Of a Conventional Terrestrial Reference System (COTES)

The International Association of Geodesy, IUGG/International Astronomical Union Joint Working group on the Establishment and Maintenance of a Conventional Terrestrial Reference System (COTES) was established in 1981 following a resolution at the IAU Colloquium #56 on Reference Coordinate Systems for Earth Dynamics, where it was recommended:

> that a working group be established by the Presidents of IAU Commissions 4, 19 and 31 and the President of the IAG to prepare a proposal for the establishment and maintenance of a conventional terrestrial reference system.

The membership of COTES is as follows: E. M. Gaposchkin (USA), B. Guinot (France), B. Kolaczek (Poland), J. Kovalevsky (France), D. D. McCarthy (USA), W. G. Melbourne (USA), P. Melchoir (Belgium), K. Yokoyama (Japan), Y. S. Yatskiv (USSR), I. I. Mueller (USA), Chairman.

In 1981/82 COTES prepared a proposal entitled "Reference Frame Requirements and the MERIT Campaign" for the purpose of planning the observations during Project MERIT in such a way that they contribute everything possible to studies leading to the establishment and maintenance of the new conventional terrestrial reference system to be proposed. In particular, the necessity to identify precisely the coordinate system being used implicitly by each of the MERIT networks was noted. It would then be necessary to establish relationships between these systems.

The proposal in a somewhat modified form was carried out through the Main Campaign of Project MERIT between September, 1983, and October, 1984, especially during an intensive campaign in the Spring of 1984, carried out especially for the purposes of COTES. The results of the observations are being analyzed and are to be reported on at the MERIT Workshop in July, 1985, to be held at The Ohio State University in Columbus, Ohio, where a proposed resolution will be prepared if at all possible for adoption at the IAU General Assembly in India in November, 1985, and subsequently at the next General Assembly of the IUGG.

IAG/COSPAR International Radio Interferometric Surveying (IRIS) Sub-Commission

The International Radio Interferometric Surveying (IRIS) joint sub-commission of the International Association of Geodesy (IAG) and the Committee on Space Geodesy (COSPAR) implemented an Earth rotation monitoring service, based on Very Long Baseline Interferometry (VLBI). Since January 1984 the IRIS observations have been made regularly at five-days intervals, using three-stations in the United States (Westford, Massachusetts; George R. Agassiz Station, Texas; and Richmond, Florida) and the Wettzell Observatory, in Bavaria, Federal Republic of Germany (FRG). The Onsala Space Observatory, Sweden, participates in one IRIS session each month. The observations are correlated at the Haystack, Massachusetts or Max Planck, FRG, correlators, and the final reductions and analysis to derive the x and y components or polar motion and UT1 are performed at the National Geodestic Survey, Rockville, Maryland. The polar motion and UT1 time series are distributed monthly through the Bureau International de l'Heure and by direct mailing in the form of the IRIS Bulletin A. The Formal Standard errors of the time series are typically 1 to 2 milliseconds of arc in x and y, and 0.05 to 0.1 milliseconds in UT1. Experiments with the Westford-Wettzell interferometer have indicated that it is feasible to monitor UT1 to the 0.1 millisecond level on an hourly, or even shorter, basis. A 90-day series of such measurements were made in support of the MERIT intensive campaign, during April - June, 1984.

REFERENCES

Bernacca P. L. (Ed.), 1983 : Processing of Scientific Data from the E. S. A. Astrometry Satellite HIPPARCOS. F.A.S.T. held in Asiago, Italy.

Bondareva M. B., Yurkina M. I., 1984 : Geodesiya i Kartografiya, 5, p.9-13 (in Russian).

Boucher C., Feissel M., 1984 : Paper presented to International Symposium on Space

Feissel, M., & P. Wilson (eds), 1983. MERIT Campaign: Connection of reference frames: Implementation Plan. Paris: Bureau International de l'Heure

Kolaczek B., Brzesinski A., Kosek W., King R. W., 1984 : Paper presented to International Symposium on Space Techniques for Geodynamics, Sopron, Hungary. Techniques for Geodynamics, Sopron, Hungary.

Lehmann, M., 1982: The Program of Observations with the Danjon Astrolabe at Astromonical Latitude Observatory in Borowiec, Mitteilung des Lohrmann, Observatoriums der Technischen Universitat Dresden.

Melbourne, W. (ch), 1983. Project MERIT Standards. U. S. Naval Observatory Circular No. 167.

Molodenskij S. M., 1984 : Prilivy, nutatsiya i vnutrenneye stroeniye Zemli. Moskwa, 213 pp. (in Russian).

Parijskij N.N., 1985 : In: Povtornye gravimetriche nablyudeniya. Moskwa (in Russian).

Wilkins, G. A. (ed.), 1984. Report on the Second MERIT Workshop and on other activities in 1983 Herstmonceux, UK: Royal Greenwich Observatory.

Yokoyama, K. 1983, Report of Second MERIT Workshop, Greenwich, GBR, May.

Yokoyama, K. 1984, Guidelines for the Reduction of Optical Astrometry Observations during the Main Campaign in Project MERIT. Publ. Int. Latit. Obs. Mizusawa, 17, 2.

20. POSITIONS AND MOTIONS OF MINOR PLANETS, COMETS AND SATELLITES (POSITIONS ET MOUVEMENTS DES PETITES PLANETES, DES COMETES ET DES SATELLITES)

PRESIDENT: E Roemer
VICE-PRESIDENT: C J van Houten
ORGANIZING COMMITTEE: Yu V Batrakov, M P Candy, Y Kozai, Ľ Kresák, L K Kristensen, B G Marsden, H Scholl, G E Taylor, D K Yeomans, Y Z Zhang

I. INTRODUCTION

Interest in minor planets and comets continued to grow during the triennium, sparked in part by the highly successful 1983 mission of the Infrared Astronomy Satellite (IRAS) as well as by activities associated with the return of P/Halley and the first spacecraft missions to comets. A trial of the Astrometry Network of the International Halley Watch (IHW) on P/Crommelin was quite successful. Yet with an increasing need for precise ephemerides, there continued to be concern for acquisition and timely reporting of astrometric observations of even the best known comets.

May 1983 saw the closest approach of a comet to the Earth since that of P/Lexell in 1770. With the recovery of 1009 Sirene and the demonstration that the 1892 observations of 330 Adalberta were not valid, the number of lost numbered minor planets was reduced to six.

On the recommendation of Commission 20's Satellite Nomenclature Liaison Committee (SNLC), chaired by K Aksnes, and the Working Group on Planetary System Nomenclature (WGPSN), the IAU Executive Committee approved permanent designations and names for three more of the recently discovered small satellites:

1979 J1	J XV	Adrastea
1979 J3	J XVI	Metis
1980 S28	S XV	Atlas

Although some decisions have been made by WGPSN concerning names for the F Ring guardians, 1980 S26 and 1980 S27, the orbits of these bodies are not yet considered by SNLC to be sufficiently well known to support assignment of permanent designations. Thus names for these objects remain provisional.

The Committee on Magnitude Ephemerides of Minor Planets and Satellites, chaired by E Bowell, has consulted broadly concerning the most suitable two-parameter equation for calculation of magnitudes of minor planets and satellites. Because of closely related requirements involved in the reduction of photometric observations of minor planets made by IRAS, interaction with the group at the Jet Propulsion Laboratory that is managing the IRAS minor planet data reduction has been particularly important. Among the decisions taken was that the two parameters are to be the absolute magnitude at zero phase angle and a quantity that describes the slope of phase curves. The definitive IRAS data reduction will be based in part on recommendations of the committee.

The Comet Nomenclature Committee appointed at Patras made recommendations concerning naming of comets discovered from spacecraft (e.g., SOLWIND and IRAS), and for temporary designations of comets beyond the 26th in a given year.

A number of meetings of interest to members of the Commission have taken place during the triennium, and proceedings from most of them, as well as from some earlier meetings, have been published. These include the Sixth European Meeting on Astronomy: Sun and Planetary System (Dubrovnik, Yugoslavia, October 1981); Workshop on the

Motion of Planets and Natural and Artificial Satellites, II (Embre, Brazil, December 1981); Workshop on Interstellar Comets (Edinburgh, April 1982); IAU Colloquium No. 74, Dynamical Trapping and Evolution in the Solar System (Chalkidiki, Greece, September 1982); IAU Colloquium No. 75, Planetary Rings (Toulouse, September 1982); Conference on Cometary Exploration (Budapest, November 1982); Colloquium on Asteroids, Comets and Meteors (Uppsala, June 1983); IAU Colloquium No. 77, Natural Satellites (Ithaca, New York, July 1983); IAU Colloquium No. 83, Dynamics of Comets (Rome, June 1984); and the IHW Astrometry Network Workshop (Garching, June 1984).

The passing of Harley W Wood (1911-1984) and of Gerald E Merton (1893-1983), both former members of Commission 20, is noted with regret. Wood was a former director of the Sydney Observatory, an outstanding astrometrist, and a leader of the astronomical community in Australia. Merton was associated for many years with the observation of comets and the determination of comet orbits, serving from 1945 to 1958 as director of the Comet Section of the British Astronomical Association.

The organization of this report follows that of earlier ones, with sections devoted to minor planets, comets, satellites, and prediction of occultations by minor planets and satellites. Information for the several sections has been compiled by the director of the Minor Planet Center and by the chairmen of the Working Groups of the Commission. I owe deep appreciation to B G Marsden, Ľ Kresák, Y Kozai, and G E Taylor for their fine work. I am also indebted to D W Dunham and P D Maley of the International Occultation Timing Association (IOTA) for additional information concerning prediction and observation of occultations by minor planets. V A Shor and N A Belyaev provided information about work done in the USSR on minor planets and satellites, and on comets, respectively. Thanks are also due to many individual members of the Commission who sent information about their own work and about other work being done at their institutions. As in the past, references generally refer to citations in *Astronomy and Astrophysics Abstracts*.

II. MINOR PLANETS

a) GENERAL

The dramatically increased activity in observational and orbital work on minor planets, documented in the last triennial report, has been maintained during 1981-1984. The two most direct indications of the activity are the *Minor Planet Circulars*, issued under the direction of Marsden at the Minor Planet Center, Smithsonian Astrophysical Observatory, Cambridge, Massachusetts; and the *Efemeridy Malykh Planet*, published under the direction of Batrakov at the Institute for Theoretical Astronomy, Leningrad. Bardwell and Green also participate in the production of the monthly batches of *Circulars*; and Ashkova, Izvekov, Khanina and Shor are involved in the preparation of the annual volumes of *Efemeridy*.

The production of *Minor Planet Circulars* seems to have reached a plateau at around 1000 pages per year. The annual number of astrometric observations published has been close to 25 000. Beginning in 1981, the observations have also been made available on magnetic tape. Updated versions of the tape were issued in 1982 and 1984; the latter contained 305 636 entries, of which 69 416 involved minor planets that were then unnumbered. Perhaps 20 percent of the new entries can be utilized in the computation of single-opposition elliptical orbits, and preliminary and improved such orbits are being published at a rate of around 1000 per year. In some 10 percent of these cases observations are subsequently also being identified at other oppositions, and with the inclusion of older identifications the annual number of new multiple-opposition orbits published is in excess of 400, roughly half of which involve minor planets that are as a result being permanently numbered. In addition to the printed publication in the *Minor Planet Circulars*, beginning in January 1984 orbits were also available over a "dial-in" computer service. On contacting the Smithsonian Astrophysical Observatory's computer over a modem a user can obtain (essentially therefore in a form for subsequent machine-readable use) orbital elements for numbered minor planets more recent than those in the current *Efemeridy* volume and unnumbered minor planets that have either been observed recently or at more than one opposition in the past. There is also a feature for the automatic computation of eleven points of an ephemeris at a step-size from 0.1 to 20 days.

In April 1982 the Minor Planet Center issued a *Catalogue of Orbits of Unnumbered Minor Planets* (31.098.108). The catalogue contained orbits for 2471 unnumbered objects, 211 of which were known to have been observed at more than one opposition, and fully half of the orbits were from observations made since September 1978. Except for the single-opposition orbits from the Palomar-Leiden survey (which would have doubled the number of entries), the catalogue was essentially complete for all cases where elliptical orbits could be determined. By October 1984, however, the number of unnumbered objects for which elliptical orbits were available had increased to 4397! This contrasts with the 3131 numbered minor planets, a total that had just doubled since the Minor Planet Center began operation in Cincinnati in 1947. There were fewer than 600 elliptical orbits in the previous tabulation of data on unnumbered objects, published by the Minor Planet Center in 1961.

A companion publication, *Catalogue of Discoveries and Identifications of Minor Planets* (31.098.109), included references to the discovery observations of the 29 157 objects (again omitting the Palomar-Leiden objects) given provisional designations since 1925 (or pre-1925 'A'-type designations). Although necessarily in abbreviated form, this was intended to be an update of Strobel's *Identifizierungsnachweis* of 1963. The 8550 known identifications with numbered minor planets were also listed, identifications with other unnumbered minor planets having been given instead in the orbit catalogue. By October 1984 the number of provisional designations had increased to 36 871, and 10 975 involved numbered minor planets.

The Institute for Theoretical Astronomy prepared and published the *Efemeridy* volumes for 1983, 1984 and 1985, and work on the 1986 volume was well underway. During the triennium, orbit improvements were made for more than 300 numbered minor planets. As in previous years, valuable contributions have been made by Dirikis at the Latvian State University Observatory, Riga. The increased rate at which new minor planets are now being numbered means that each *Efemeridy* volume is about 7 percent larger than its predecessor, and that for 1985 contains 320 pages. The largest increase was for the 1984 volume, which included 308 newly-numbered objects. Several extended ephemerides are given for bright and unusual objects, but 60 percent of each volume is devoted to the standard eight-date opposition ephemerides. The annual updating of the orbital elements to a new osculation epoch is a feature that continues to gain in importance, as increased numbers of users are thereby able to produce their own ephemerides. On the other hand, the Institute for Theoretical Astronomy continues to provide daily ephemerides to the observatories participating in the special Leningrad program of observations of 20 selected minor planets.

b) OBSERVATIONS AND ORBITS

In connection with the Leningrad program, the Institute for Theoretical Astronomy reported the receipt during the triennium of some 2500 observations made at 30 observatories around the world. Among the Soviet observatories involved in this program are those at Abastumani, Alma-Ata, Dushanbe, Goloseevo, Kazan, Kharkov, Kitab, Moscow, Nikolaev and Tashkent. Commission 20 members elsewhere who have specifically mentioned in connection with this report their involvement with the program during this triennium are Quijano at San Fernando, Lomb at Sydney and Protitch-Benishek at Belgrade. Beginning in January 1983 observations were made at San Fernando of 55 minor planets for the Hipparcos project. Most regrettably, the Sydney Observatory was formally closed in late 1982; its astrometric camera and other astrometric equipment are being transferred to Macquarie University.

Among other projects devoted to bright minor planets is the continuing work by Kristensen at Aarhus on 51 Nemausa and the fundamental reference system; an important impetus was provided by the occultation of 14 Psc in 1983 (34.096.015). Johnston, Seidelmann and Wade (32.098.072) succeeded in obtaining high-quality astrometric observations from the thermal emission from 1 Ceres and 2 Pallas detected at microwave frequencies using the Very Large Array in New Mexico; they remark in a follow-up paper that an extended program of observations of this type--measurements have already been made also of 4 Vesta and 10 Hygiea--could allow the equinox to be located and provide a truly fundamental coordinate system for positions of radio sources.

Duma and Kisjun (37.098.010) have refined the orbits of the first four minor planets from meridian observations and discussed the effect of the inclination of the true equator to that of the star-catalogue system. Other specific orbital investigations not included in the publications sponsored by the Commission are a computation on 4 Vesta by Imnadze (34.098.012) and a relativistic examination of the motion of 1566 Icarus by Shefer (33.098.042, 37.098.014).

Wide-field programs involving photographic observations of the fainter minor planets and discoveries of new main-belt minor planets are in progress at Brorfelde (Fogh Olsen and colleagues), El Leoncito (C U Cesco), Flagstaff (Bowell), Kleť (Mrkos), La Silla (Debehogne, de Freitas Mourão, De Sanctis, Lagerkvist, Zappalà and several others), Nanking (J-x Zhang), Nauchnyj (Chernykh), Perth (Candy) and Zimmerwald (Wild); at the Lincoln Laboratory's New Mexico installation further real-time discoveries have been made using a television device (Taff); and several real-time discoveries (with rather crude astrometry) were made from the IRAS satellite (Davies, Green; 37.103.014). The Flagstaff (Lowell) project, which involves the use of a microdensitometer to derive accurate positions, typically yields more than 5000 positions annually. The Nauchnyj (Crimean Astrophysical Observatory) program produces perhaps half as many positions but sets the useful general magnitude (about 18.0) limit for programs of this type; more than 500 numbered planets and a comparable number of discrete unnumbered planets are observed each year. The program with the 0.46-m Palomar Schmidt (E Shoemaker, C Shoemaker) generally concentrates on fast-moving objects, and the deeper limiting magnitude of the 1.2-m Schmidt imposes even greater selectivity toward objects of unusual motion (separate programs by Helin and Kowal) or is advantageous to the follow-up and recovery of specific objects (Gibson). Useful observations of faint, new objects are also obtained from time to time with the 1.3-m Tautenburg (Börngen) and 1.05-m Kiso (Kosai, Hurukawa) Schmidts. The photographic follow-up and recovery programs for faint objects with the large to moderate conventional reflectors at Oak Ridge (McCrosky) and Mt. John (Gilmore, Kilmartin) and by Japanese (Furuta, Seki) and Italian (Colombini, Baur) amateurs also yield a fair number of new discoveries. There have recently been concerted attempts at CCD astrometry with conventional reflectors, notably with the Flagstaff 1.7-m reflector (Bowell), the Palomar 1.5-m reflector (Gibson) and the Steward Observatory's 0.9-m reflector (the "Spacewatch" project) on Kitt Peak (Gehrels).

There is much fruitful interaction between all the above faint-object programs and the Minor Planet Center, which assists with identifications and attempts to check for errors before observations are published. The Crimean program also operates in close association with the Institute for Theoretical Astronomy, which puts considerable effort into calculating ephemerides for both numbered and unnumbered minor planets and has involved during the triennium the computation of some 250 preliminary orbits by Kastel'. Kristensen makes orbit computations for the Brorfelde program; Williams works closely with Palomar observers; and the amateurs Nakano, Oishi, and Urata provide substantial computational support for all the Japanese observing programs. Some of this interaction involves the suggestion and subsequent demonstration of identifications, and this is an area in which Bowell, Bowman (Cincinnati), Kippes (Würzburg), Kretlow (Wiesbaden), Landgraf (Lindau) and Schmadel (Heidelberg) have also made useful recent contributions.

Much effort has been put into the measurement and analysis of positions of minor planets from old plates. Bowell has generously provided measurements on request from the older Flagstaff plates (1929-1975), and Indiana students (under the direction of Edmondson and Owings) have continued to provide measurements from the Goethe Link plates (1949-1966). Oterma has in fact been able to complete the measurement of all the unnumbered minor planets found in the course of the valuable Turku program (1935-1957), and she continues to measure the positions of numbered minor planets when there is a good reason for doing so. Remeasurement of Budapest plates from the Kulin program (1936-1942) is to begin at Belgrade in December 1984.

The Palomar-Leiden survey (1960) has also finally been completed with the publication by van Houten, Herget and Marsden (1984, *Icarus* 59, 1) of 1198 orbits that supplement or amend those in the 1970 version of the survey. The completion of the survey included an extension to 170 new objects found in the field originally used only for calibration purposes, and in several instances the earlier class 4 orbits (generally based on only four-day arcs) could be improved with the identification of observations a month later. Most of the class 4 orbits were in fact rederived using positions measured from the second plate of each blink pair. The total number of objects in the survey is 2403, of which 2256 were still single-opposition objects as of October 1984. Van Houten-Groeneveld has continued her work in Leiden on the 1977 Trojan program (the plates for which were also taken by Gehrels with the Palomar 1.2-m Schmidt). All fields have been blinked, and 80 percent of the fields have been measured for positions. Orbit computations will be possible for some 1600 objects, of which 32 are evidently Trojans or Hildas; in a collaboration with the Minor Planet Center, Bardwell has so far computed orbits for 166 objects. Bus, now located in Flagstaff, has completed the main phase of the March-April 1981 "UCAS" survey (organized from the California Institute of Technology) of faint minor planets with the U K Schmidt. Orbits have been calculated, mainly at the Minor Planet Center, for 1202 objects, of which some 200 have so far been identified at other oppositions. For the remainder, the observed arcs range from nine to 40 days. Except for one fast-moving object, the minor planets followed for fewer than nine days are felt to yield orbits that are too uncertain to be useful. On the other hand, the important next phase in this work will involve the extension of the existing orbits to observations in February and May 1981; it is hoped that this will yield several hundred high-quality orbits, with arcs up to 85-90 days, and there will then be a very real possibility of making deliberate searches for many of the objects at additional oppositions.

With the help of long-term numerical integrations made a few years ago by Scholl for some 2300 of the first 2357 numbered minor planets, the Minor Planet Center has calculated the residuals for all the observations of the numbered minor planets. Thorough analysis of the residuals is proving to be very time-consuming, but the process of deleting, correcting or otherwise identifying observations showing large residuals has so far been completed for minor planets 901 and onward.

In the last report it was remarked that the number of lost numbered minor planets had dropped from 23 in 1978 to eight in 1981. Gibson recovered 1009 Sirene in 1982 quite close to the prediction by Kristensen. A careful investigation by West, Madsen and Schmadel (31.098.105) demonstrated that the (two) original observations of 330 Adalberta by Wolf in Heidelberg in 1892 were certainly imaginary, so this number and name have been applied instead to another object, A910 CB, also discovered by Wolf. Since 1982 there have been no further successes with the recovery (or elimination) of lost objects, and they remain as follows: 473 Nolli, 719 Albert, 724 Hapag, 878 Mildred, 1026 Ingrid and 1179 Mally.

c) THEORETICAL INVESTIGATIONS

Garfinkel (31.098.038, 32.098.053, 32.042.063, 34.042.013) summarized and extended his theoretical treatment of the orbits of Trojans. Also continuing his earlier work on Trojans, B Érdi (30.098.008) used an asymptotic solution for the cylindrical coordinates to obtain perturbations in the orbital elements and made a comparison (33.098.027, 34.098.023) with Garfinkel's theory; with Varadi (034.098.059), he investigated perihelion motion in the non-planar case. Using numerical integration and Labrouste's procedure for isolating periodic effects, Bien and Schubart (34.098. 021, 34.098.060) studied the long periods in the motion of Trojans and pointed out the importance of secular resonances for special Trojan orbits. A simple theory of Trojans was devised by Giacaglia (34.098.136), and Hori (30.098.097) treated the 1:1 (mutual) commensurability case using canonical perturbation theory. D'yakov and Reznikov (32.042.028, 37.098.012) integrated hypothetical Trojan orbits for various mass ratios of the primary and secondary bodies. Zagretdinov (37.098.061) remarked on some regularities in Trojan motion, and Agafonova and Drobyshevskij (37.098.043) discussed the capture on Trojan orbits of objects ejected from the vicinity of Jupiter.

Schubart (32.098.019, 32.098.054) found a numerical way of determining a proper inclination for a Hilda-type minor planet and characterized the orbits of the librating Hildas by this and two other parameters. Froeschlé and Scholl examined the stochasticity of peculiar orbits in the 2:1 Kirkwood gap (29.042.038) and systematically explored three-dimensional motion near the gap (32.098.004). Scholl (31.098. 098) also reviewed earlier results on resonant motion and the possibilities of using resonant motion for mass determinations. In a general study of stability near resonance, Hadjidemetriou and Ichtiaroglou (31.098.106, 34.098.020) found that small stability regions of doubly-symmetric periodic orbits may be present near the generally unstable second-order resonances and that two separate regions, one stable and the other unstable, exist near the first-order resonances. The stability of real 2:1 and 3:2 objects was also discussed by Kiang (31.098.099, 34.098.057). Gerasimov (32.098.081, 33.098.002) made a qualitative analysis of the 2:1 resonance and developed short-period terms using the von Zeipel method. Kazantsev and Sherbaum (37.098. 030, 37.098.033, 37.098.105) examined the boundaries of the a-e zones and the question of the asymmetry of the Kirkwood gaps. The main resonance features of the minor-planet belt showed up in a computer simulation by D'yakov and Reznikov (37.098.011). Veres (32.042.065) compared with a numerical integration the results of a twice-averaged solution from the restricted three-body problem. Perozzi (34.098.058) suggested that the reported increasing mean values of eccentricities and inclinations away from the gaps is due to a lack of highly eccentric and inclined objects near the gaps.

Torbett and Smoluchowski (31.042.113, 34.042.065) examined the sweeping effect on the 2:1, 3:1, 5:2 and 7:3 resonances induced by dissipation of an accretion disk; in the case of the 3:2 resonance the resulting accumulation of objects and ejection of objects nearby resembles the real situation with regard to the Hildas. Dermott and Murray (33.098.015) also concluded that the Kirkwood gaps were formed after the minor planets had dispersed from the near-coplanar disk in which they accreted, while Henrard and Lemaître (34.098.015) discussed in detail the production of Kirkwood gaps from a displacement of the Jovian resonances following the removal of an accretion disk in the early stages of the solar system. Colombo and Franklin (31.098.039) examined motion around the second-order resonances 3:1 and 5:3. Wisdom (31.042.038, 34.042.066) developed a mapping for motions near the 3:1 commensurability and suggested that there occur sudden increases in eccentricity that could easily account for the Mars crossers and possibly even objects that come close to the Earth. The general problem of stability regions and the capture of minor planets was evaluated by Szebehely, Vicente and Lundberg (34.042.119, 34.098.018, 34.098.019).

Analysis by Milani and Nobili of the depletion problem (34.042.076, *Astr. Ap.* in press) showed the insignificance of collisions in the outer part of the minor-planet belt, confirming the absence there of families and small objects generally. Recognizing the destabilizing effect of Jupiter's eccentricity, they discovered three cases of minor planets with mean distances near 3.7 AU where the difference in perihelion direction of the object and Jupiter librates (in the planar problem) and noted the clear influence of coupling of perihelion direction and eccentricity for objects down to a distance of 3.4 AU.

In the course of a numerical integration of the Lagrange secular equations for 2 Pallas, Taylor (31.098.001) noted the dominance of periods of 15 000 and 190 000 years and the retrograde circulation of the longitude of perihelion in a period of 544 000 years. Chen and Tong (33.098.119, 34.042.043) developed an expansion for the gravitational potential of the minor-planet belt. Williams (37.098.006) pointed out that some three dozen minor planets perturb Mars by up to 5 m and thus that some of their masses should be determinable from Viking lander ranging data.

Three papers by Zhuravlev and Kiryushenkov (30.098.090, 31.098.004, 31.098.073) were concerned with the discovery of faint minor planets and comparison of the statistics of the population of bright and faint objects. Taff (34.098.110) discussed the optimization of searches for minor planets. Michel (30.098.100) made a computer simulation of the dispersion of orbital inclinations. Knezević (32.098.016, 34.098.095) discussed observational selection effects, particularly those involved with the inclinations. Beck (32.098.076) compared the eccentricity distribution of minor planets with an empirical rule by Plummer.

Carusi and Valsecchi (32.098.043) stressed the need for consideration of the classification of Hirayama families in terms of both dynamical and other physical parameters. Kozai (34.098.017) applied to the numbered minor planets his criterion for the definition of families and drew a distinction between compact families and loose families bounded by secular and mean-motion resonances. Yuasa (34.098.026) used secular-perturbation theory to estimate the ages of the families.

Farinella, Paolicchi and Zappalà (30.098.069, 31.098.092, 31.098.093) studied the distribution of rotation rates of minor planets in terms of a complex and size-dependent collisional history and concluded (33.098.010, 33.098.037, 33.098.116, 33.098.117, 34.098.024) that all objects have undergone catastrophic collisional events; they paid particular attention (30.098.106) to a class of elongated, rapidly rotating objects and enunciated (34.098.041) the open problems concerning Hirayama families. Fujiwara (30.098.096, 33.098.011) attempted to reconstruct the parent bodies of the Themis, Eos and Koronis families using data from laboratory impact experiments and concluded that the parent bodies of the families were completely fragmented. The dependence of rotation rates of main-belt objects on both size and type was also studied by Dermott, Murray and Harris (31.098.011, 37.098.007), and some of their results were considered by R. Halling (34.098.052) in terms of gravitational compaction of porous matter and Alfvén-Arrhenius accretion theory. Dobrovolskis and Burns (37.098.064) proposed that the relatively slow rotation rates of intermediate-sized objects could be due to angular momentum drain from impacts. Barricelli (32.098.080, 33.098.131) discussed the tendency for the smaller bodies to have slower axial rotation and allowed for the effect of possible satellites. On the basis of elongated photographic images, Wang and others (30.098.011) proposed that 9 Metis possesses a satellite, but a careful examination by many astronomers (32.098.059, 33.098.062) at a subsequent favorable opposition apparently disproved the satellite's existence.

Batrakov (31.021.047) reviewed methods of computation of perturbations, and Khotimskaya (31.042.036) demonstrated the application of a combination of the Myachin-Sizova method and a uniform polynomial approximation process for the positions of the perturbing planets. Sitarski (34.042.124) adapted his recurrent power-series method to incorporate general relativistic effects.

Duncombe and Hemenway (31.031.529, 34.098.135) continued their examination of the combination of Space-Telescope and ground-based observations to probe the systematic accuracy of the fundamental reference system, and the use of minor planets in the Hipparcos astrometric project was discussed by Söderhjelm and Lindegren (31.041.034).

Using results from a restricted circular n-body study, Vashkov'yak (30.098.035) constructed families of integral curves describing the orbital evolution of 944 Hidalgo, 2060 Chiron, as well as Apollo-type and other unusual objects. Landgraf (33.100.019) determined the masses of Saturn and Uranus from observations of Hidalgo and Chiron.

Levin and Simonenko (30.098.075, 33.098.036, 37.098.013) continued their work on interrelations of minor planets, comets and meteoroids and the inconclusive evidence for the cometary origin of most Apollo-Amor objects. Yamada (32.098.011) considered the influence of encounters with the Earth and Venus on possible cometary progenitors. Galibina and Kastel' (33.098.004) examined over an interval of 10 000 years the particularly plausible case of relationship of 2212 Hephaistos and P/Encke. Fox, Williams and Hughes (37.098.062) investigated the orbital evolution of the Geminid parent 1983 TB. Hedervari (33.098.088) made a statistical comparison of the orbital inclinations of Apollo-Amor objects and short-period comets. Sitarski (34.098.061) and Ziolkowski (34.098.062) have tended to conclude that nongravitational effects are present in the motions of some minor planets, but this claim has been disputed by both Landgraf and Marsden. Drummond (31.098.037) published a list of theoretical meteor radiants for Aten, Apollo and Amor objects, while Babadzhanov and Obrubov (33.098.092, 34.098.063, 37.098.101) considered variations in radiants due to secular perturbations.

III. COMETS

a) DISCOVERIES, RECOVERIES AND ASTROMETRIC OBSERVATIONS

During the interval 1981 July 1 - 1984 June 30, 10 long-period comets and 7 new short-period comets were discovered, and 27 previously known short-period comets were recovered. Two of these were accidental rediscoveries of poorly observed one-apparition comets which had been classified as hopelessly lost since 1846 (P/Peters-Hartley) and 1945 (P/du Toit-Hartley). The latter consisted at recovery of two components separated by nearly 1°, with major changes in relative brightness resembling the behavior of P/Biela in 1852. Two additional comets were found on the exposures with the SOLWIND coronagraph from 1981 January 27 and 1981 July 20 (Sheeley, Howard, Koomen and Michels, 32.102.060). Both comets apparently belonged to the Kreutz group and may have collided with the Sun, like 1979 XI. Two comets were found on the Southern Sky Survey plates from 1975 May 4 and 1978 March 5 by Eberst and Tritton (31.103.007), but the images were only sufficient to determine the approximate direction of apparent motion. Two discoveries that received provisional designations 1983a and 1984b turned out to be plate defects. A comprehensive search by Nakano (1984 *Japan Astr. Circ.* 409) revealed that 10 objects observed in 1940-1981 and provisionally designated as asteroids were in fact known short-period comets. In one case, 1967 EU = P/Smirnova-Chernykh, the observation predates the discovery by 8 years.

Many discoveries of long-period comets still come from systematic visual searches in which amateur astronomers play a leading role. A majority of the discoveries, and many ephemeris-aided recoveries of short-period comets, are due to large Schmidt cameras: 5 of the discoveries or independent rediscoveries of short-period comets to the 1.2-m Schmidt at Siding Spring (Russell, Hartley) and as many as 13 routine recoveries to the 1.2-m Schmidt at Palomar (Gibson). Many critical observations from the southern hemisphere are made with the 0.6-m reflector at Mt. John (Gilmore, Kilmartin). The only consistently conducted program with a large long-focus reflector (1.5-m at Oak Ridge, McCrosky et al.) continued to produce astrometric observations of comets down to magnitude 19-20, including many last observations important for orbit determinations.

Two modern search techniques, which may affect comet discoveries very substantially in the years to come, were implemented for the first time in 1982 and 1983. Replacement of a photographic plate by a CCD detector made it possible to increase the effective sensitivity by a factor of 100, and to shift the detection limit to 24-25 mag. This technique was successfully applied to the recovery of P/Halley (by Jewitt et al. at Palomar; confirmed by Belton and Butcher at Kitt Peak and by Baudrand et al. at Mauna Kea: 1982 *IAU Circ.* 3737, 3742, 3753) and of P/Giacobini-Zinner by Djorgovski, Spinrad, Will and Belton at Kitt Peak (1984 *IAU Circ.* 3937). An early recovery of both of these comets was especially desirable in connection with the planned spacecraft missions Vega, Giotto, Planet A and ICE.

The other novel search technique was implemented by the Infrared Astronomy Satellite (IRAS) launched on 1983 January 26 (Davies, Green, Stewart, Meadows and Aumann 37.103.014; IRAS *Asteroid Workshops* 2 and 3, JPL, 1984). During the 9 months of operation of the Fast Moving Asteroid Detector, 6 new comets (4 of long period and 2 of short period) were discovered, and 5 known comets were observed. The evaluation of the IRAS data files is still in progress, and other comets may await identification.

An effective method of ground-based faint comet searches, making use of a CCD in a scanning mode, was proposed and discussed by Gehrels (30.031.061, *MPC* 9198). The selection effects controlling comet discoveries and the observed distribution of comet orbits were analyzed by Hughes (31.102.007, 34.102.002), Radzievskij and Tomanov (33.102.005), Potapov (33.102.090) and by Kresák (32.102.013), who also suggested a method of optimizing the searches for newly captured short-period comets.

Two long-period comets passed remarkably close to the Earth in May and June 1983. The encounter distance of 1983d IRAS-Araki-Alcock, 0.031 AU on May 11, is the smallest on record since P/Lexell in 1770, and that of 1983e Sugano-Saigusa-Fujikawa, 0.063 AU on June 12, the fourth smallest in this century. Owing to the exceptional configuration, the former became the brightest comet of the triennium ($m_1 < 2$), but without any conspicuous tail. Two other comets, 1982g Austin and 1984i Austin, were barely visible with the naked eye ($m_1 \simeq 5$).

A large-scale worldwide astrometric program was initiated as a part of the International Halley Watch (Rahe and Newburn, 32.013.057). For more detailed information, see *Proceedings of the Astrometry Network Workshop*, ESO, Garching, 1984, and *IHW Newsletter* Nos. 1-5. The network, coordinated by Discipline Specialists Yeomans, West, Harrington and Marsden, consists of 176 members from 35 countries. P/Halley remained accessible mainly to large telescopes equipped with CCD detectors, a total of 44 positions having been obtained (mainly at La Silla) from the recovery up to conjunction with the Sun in 1984 June. The preparatory work included production of a special reference star catalogue, including a total of 16175 stars from various sources. During a very successful trial run on P/Crommelin, 283 astrometric observations were received from 34 observatories. In contrast to this, only eleven observations of the unique P/Encke have been reported from between the 1980 and 1984 perihelion passages, in spite of its observability all along its orbit (1984 *IAU Circ.* 3934, *MPC* 9175, 9254). Another comet of special interest was P/Giacobini-Zinner, in connection with the planned rendezvous of the ISEE 3/ICE spacecraft.

The astrometric observations of comets published in the *MPC*s, together with the observations published in the *IAU Circs.* and in a number of other publications since 1964 (and in some cases earlier), have now been included in the machine-readable files issued by the IAU Minor Planet Center. The 1982 version of the magnetic tape contained 13 385 cometary observations, while the 1984 version contained 16 567. Some 10 000 additional pre-1964 observations are also available in machine-readable form at the Center, but they require some attention before they can be incorporated with the more recent observations. Observations appearing on the *IAU Circs.* can now be otained directly in machine-readable form, as the current *Circulars* can themselves be acquired by "dialing in" over a modem to the new CBAT computer service. A brief statistical analysis of the astrometric observations of comets from 1964 to 1984 was presented by Marsden (IHW Astrometry Workshop, 1984). Helpful information on the use of the telegraphic code for reporting new discoveries and on the use of plate constants in astrometric reductions was summarized by Marsden and Roemer (32.012.009).

b) ORBITS AND EPHEMERIDES

The fourth edition of Marsden's *Catalogue of Cometary Orbits* (33.002.052) includes orbital information for 1109 apparitions of 710 different comets, with a number of auxiliary tables. Preliminary and improved orbits and ephemerides of comets have been computed and published routinely by the Central Bureau for Astronomical Telegrams/Minor Planet Center. The *MPCs* also regularly contain predicted orbits and ephemerides for returning periodic comets; a reduced-precision tabulation of preliminary predictions from 1982 to 1999 was published by Marsden and Roemer (32.012.009-Appendix). The same paper includes a listing of "original" and "future" 1/*a* values for 220 long-period comets, classified according to the anticipated accuracy. More details on the recent computations of 28 "original" and "future" orbits were given by Everhart and Marsden (33.102.014). A catalogue of Tisserand invariants of comets with respect to Jupiter and Saturn was compiled and analyzed by Artem'ev, Kitaeva, Mamedov, Markova and Radzievskij (31.021.045, 31.021.047), and a catalogue of their perihelion directions and nodal positions by Tomanov (33.102.003). The distribution of the perihelion directions of short-period comets was discussed by Kazantsev and Sherbaum (34.102.068).

From among the orbit improvements for short-period comets, the following link a number of revolutions: P/Grigg-Skjellerup and P/Wolf-Harrington by Sitarski (32.103.121 and 32.103.801), P/Ashbrook-Jackson and P/Wolf by Kazimirchak-Polonskaya (32.103.363 and 32.103.901), P/Schwassmann-Wachmann 3 by Belyaev and Shaporev (34.103.701). The orbit of P/Halley was very carefully updated by Yeomans (IAU Coll. 83, 1984), Savchenko (33.103.928), Akim, Savchenko and Stepan'yants (37.103.910) and Landgraf (IHW Astrometry Workshop, 1984), and its ephemeris uncertainty estimated by Sukhanov and El'yasberg (*Kosm. Issl.* 21: 6, 1984) and by Zadunaisky (IAU Coll. 83, 1984). Emel'yanenko (33.102.007) has analyzed the uncertainties in the time of perihelion passages of long-lost short-period comets. For one of them, P/Schorr, the predictions have been improved by de Vegt, Kohoutek and Marsden (32.103.661) by remeasuring all available photographic plates from 1918 and recomputing the orbit.

Nongravitational effects in the motion of individual short-period comets were determined by Landgraf (33.102.047), Forti (34.103.003), Todorovic-Juchniewicz (34.103.006) and others. Gyrdymov and Evdokimov (33.103.001) and Batrakov and Chernetenko (IAU Coll. 83, 1984) have suggested a correlation between these effects and solar activity. Marsden (IAU Coll. 83, 1984) reviewed all the previous work in the theory and numerical evaluation of nongravitational effects, and presented a classification of short-period comets according to the character of these effects. Some of the implications were also discussed by Belyaev and Ivanovskaya (IAU Coll. 83, 1984).

Carusi, Kresák, Perozzi and Valsecchi have integrated the motion of all known short-period comets back to 1585 and forward to 2406. The main results are summarized in an atlas of long-term orbital evolutions (*IAS Int. Report* 12, Rome, 1984); the methods are explained and a preliminary general evaluation is presented in two separate papers (IAU Coll. 83, 1984). One of the interesting conclusions is that P/Van Biesbroeck and P/Neujmin 3 are apparently fragments of a single comet which split apart a few years before a close encounter with Jupiter in 1850 (*IAU Circ.* 3940, 1984). Contrary to the earlier views of Öpik (06.102.025) and others, similar genetically associated pairs or groups of detached components seem to be totally absent among the known long-period comets, as demonstrated by Kresák (31.102.052) and Lindblad (IAU Coll. 83, 1984). The only exception is the Kreutz group, with 11 members known at present. The dynamical criteria of orbital similarity between different types of interplanetary objects were discussed by Kresák (31.102.012).

The motion of seven comets that spent some time near the 2:1 resonance with Jupiter (P/Lexell, P/Tempel-Swift, P/Wirtanen, P/Kohoutek, P/West-Kohoutek-Ikemura, P/Haneda-Campos and P/Wild 2) was integrated by Vaghi and Rickman (31.102.048) to evaluate the role of a repeated encounter mechanism. Integrations of the motion of the two comets librating around the 1:1 resonance (P/Boethin and P/Slaughter-Burnham) extending over two millennia have been carried out and analyzed by Benest, Bien and Rickman (34.103.009). The same authors have also studied the capture of P/Boethin by Jupiter (31.103.741, 34.103.240). By backward integration of the motion of all known short-period comets with high values of the Tisserand invariant, Carusi and Valsecchi (31.102.026, 31.102.046) have identified a number of temporary satellite captures, which apparently represent a rather frequent phenomenon. Carusi, Kresák and Valsecchi (31.042.130) have shown that a limiting value of the invariant, $2\sqrt{2}$, splits the comets into two families, those that can be ejected on hyperbolic orbits but cannot be captured into satellite orbits, and those for which the situation is reversed.

Rickman and Karm (31.102.047, 32.103.016) have developed a method for identifying close planetary encounters using unperturbed Keplerian orbits. An application to all known short-period comets enabled them to estimate the frequency of such events. A quick method for estimating the perturbing effects at close encounters, suited to maximize the yield of Monte Carlo simulation experiments, was devised by Rickman and Froeschlé (33.102.042). An interesting possibility of assessing the previous orbital evolution of Earth-approaching comets from the observed structure of meteor streams associated with them was proposed by McIntosh and Hajduk (34.104.003) and by Hajduk (IAU Coll. 83, 1984).

The current status of orbit determinations was reviewed by Kazimirchak-Polonskaya and Belyaev (IAU Coll. 83, 1984). The RADAU integrator, which has proven very efficient and accurate for the integration of comet orbits, was tested and compared with other integrators by Everhart (IAU Coll. 83, 1984). The local and propagated errors of numerical integrations were discussed by Milani and Nobili (IAU Coll. 83, 1984), and the subtle dynamical and observational effects inherent in the underlying data by Sitarski (34.098.061). Todorovic-Juchniewicz (34.102.026) has determined the perturbing effects of the three largest minor planets on 53 comets. Optimum targets for future space missions from among the short-period comets were identified and evaluated by Yeomans (IAU Coll. 83, 1984).

c) MODEL COMPUTATIONS, EVOLUTION AND ORIGIN

Potential evolutionary tracks from different reservoirs into the Jupiter family of comets were reviewed by Carusi and Valsecchi (IAU Coll. 83, 1984). Transitions between different types of cometary orbits starting in the Oort cloud were discussed by Everhart (32.102.032) with emphasis on the connecting links between individual regimes of chaotic motion, and by Kresák (31.102.042) with emphasis on the stepwise capture at low-velocity encounters with the outer planets. Modeling experiments on low-velocity encounters with Jupiter revealed distinct families of evolutionary patterns of widely different occurrence rate (Carusi, Kresák and Valsecchi, 31.042.130; Carusi and Valsecchi, 31.102.045), which are substantially different from the symmetrical patterns appearing at higher velocities (Carusi, Kresáková and Valsecchi, 32.103.831, 34.103.221). A family of temporary satellite captures by Jupiter, of the type of P/Gehrels 3, was investigated in detail by Rickman and Malmort (31.102.049). Statistical results of computer simulations of 1000 close encounters of minor bodies with each of the four outer planets were presented by Carusi and Valsecchi (31.102.046).

While the concept of stepwise capture clearly implies temporary binding of cometary perihelia to individual outer planets, there is little consensus as to the irregularities in the distribution of their aphelion distances. Some authors attribute them to the existence of comet families produced by eruptions on these planets, their satellites, or even transplutonian planets (Vsekhsvyatskij, 33.102.064; Vsekhsvyatskij and Demenko, 32.102.070, 33.102.010; Vsekhsvyatskij and Guliev, 33.101.026, 33.102.027; Guliev, 32.102.066, 34.102.063); but others present dynamical and statistical arguments against (Kresák, 33.101.026; Tomanov, 33.102.001, 33.102.045, 33.102.063, 34.102.049).

Bogart and Noerdlinger (31.102.051) and Khanna and Sharma (34.102.043) have re-examined the distribution of the orientations of the orbits of long-period comets, and concluded that there is a significant relation to the galactic plane and to the solar apex. Lüst (IAU Coll. 83, 1984) has shown that there are major differences in this respect between dynamically old and new comets. All of these authors and Kresák (32.102.013) have pointed out local anisotropies, which limit the use of an ellipsoidal analysis. The aphelion clustering of new comets is attributed by Biermann, Huebner and Lüst (34.102.072) to recent star tracks through the Oort cloud, and Delsemme (IAU Coll. 83, 1984) considers two additional possibilities: the presence of an unseen companion of the Sun, and a grazing incidence of a giant molecular cloud 3 to 30 million years ago. Babenko (34.102.009) finds the observed distribution of binding energies of new comets inconsistent with the expected stellar perturbations in perihelion distance. Nakamura (37.102.061) has checked the random character of the sequence of perihelion passages of long-period comets over two centuries.

Shteins and Salitis (37.102.012) have developed the theory of diffusion of comets, including a method of computation of the distribution of velocity perturbations by passing stars (Salitis, 33.102.006). Numerical simulations of this process have been performed and analyzed by Yabushita, Hasegawa and Kobayashi (32.102.002); Scholl, Cazenave and Brahic (32.102.004); Weissman (32.107.008, 33.102.034, IAU Coll. 83, 1984); Mignard and Remy (IAU Coll. 83, 1984). Galactic perturbations on the Oort cloud of comets were investigated by Byl (34.102.042). Fernández (31.102.041, 31. 102.043) and Weissman (Protostars and Planets II, Tucson, 1984) have followed the dynamical evolution from different birthplaces to check the plausibility of different theories of comet origin. The possible processes of comet formation were reviewed by McCrea (32.012.059) and Greenberg (IAU Coll. 83, 1984).

The main arguments against the classical concept of the Oort cloud as a primordial part of the solar system concern its potential survival time, and its adequacy to replenish the inner cometary zone. The long-term stability of the cloud was questioned especially due to the tidal forces experienced during encounters with molecular clouds, and these clouds were also suggested as a source of episodic replenishment by new comets (Napier and Staniucha, 31.107.001; Napier, 31.102.044, 32.102.084, IAU Coll. 83, 1984; Clube, 32.102.082, 34.102.036, IAU Coll. 83, 1984; Clube and Napier, 33.102.033). This, however, would require dense clouds of extremely low encounter velocity and small velocity dispersion (Valtonen and Innanen, 31.102. 006), or assistance of an unseen companion of the Sun (Valtonen, 33.102.023). Present quantitative data about the molecular clouds are not sufficient to be sure about the destruction rate of the Oort cloud; Weissman (33.102.034) finds it low, and estimates the total depletion of the Oort cloud over the history of the solar system at 85%. Some authors agree with the disruptive effects of the molecular clouds but not with the replenishing ones, and suggest a substantial inner concentration of comets in the region of the outermost planets or not far beyond. This would be sufficiently stable against disruption and could regenerate the outer envelope (van den Bergh, 32.107.038; Bailey, 33.102.019, 34.102.003). The inner concentration may include 99% of the total mass of the cloud and feed the inner planetary zone only at the time of close star passages (Hills, 30.102.049). Bailey (34.102.038) also suggests three methods by which the concentration could be detected: by the total gravitational effects on other bodies, by the infrared emission, and by monitoring of selected stars for occasional occultations. The presence of an inner concentration would alleviate the problem of a source from which enough objects could be perturbed into short-period orbits (Fernández and Ip, 33.102.058, 34.102.039; Fernández, IAU Coll. 83, 1984). The nature of the concentration, together with the limited knowledge of the size distribution of comets, is the main source of uncertainty in present estimates of the total mass of the comet population (Biermann, 31.102.054; Weissman, 33.102.013, IAU Coll. 83, 1984).

Another open question is the effect of limited physical lifetimes of comets on the distribution of their binding energies. Current physical theories of cometary nuclei predict relatively slow fading rates, which cannot account for the $1/a$ distribution anomalies (Yabushita, 34.102.005, 34.102.051; Bailey, IAU Coll. 83, 1984; Weissman, IAU Coll. 83, 1984). Observational identification of the cases of extinction of comets during their apparitions (Kresák, 37.103.029; Sekanina, 37.102.073) suggests rather short lifetimes of long-period comets. There is also a major discrepancy between the observed and expected rate of brightness decrease of short-period comets (Hughes and Daniels, 33.102.021) and a significant difference between the mass distributions of long- and short-period comets (Hughes and Daniels, 31.102.003). Unfortunately, the underlying data are strongly biased by observational selection. The problem of comet fading and survival times was reviewed by Kresák (IAU Coll. 83, 1984).

Until recently it was widely believed that the ultimate fate of almost all comets is a total disintegration at the time of extinction. A real breakthrough was presented by the recent discovery of several Apollo and Amor minor planets moving in comet-like orbits (1982 YA, 1983 LC, 1983 SA, 1983 VA, 1983 XF, 1984 BC, 5025 P-L), and of one (1983 TB) which may be the missing parent body of the Geminid meteor stream (Whipple, 34.104.033; Williams, Fox and Hunt, IAU Coll. 83, 1984; Fox, Williams and Hughes, 37.098.062). In the latter case the meteoric debris have physical properties quite similar to those of active comets, in spite of a definitely asteroidal aphelion distance of only 2.40 AU. While the orbits of the new comet-like planets are not yet known with an accuracy that permits a large extrapolation, perturbation integrations do not reveal any significant differences from active comets (Benest, Bien and Rickman, IAU Coll. 83, 1984). Comparative long-term integrations for the minor planets 2212 Hephaistos and P/Encke were carried out by Galibina and Kastel' (33.098.004), and the existence of a family of Encke-like orbits was discussed by Clube (32.102.082) and Napier (34.102.040). Sitarski (34.098.061) and Ziolkowski (34.098.062) have found indications of weak nongravitational effects in the motions of three Apollo planets (1566 Icarus, 1685 Toro, 1862 Apollo) and one main-belt object (1162 Larissa), but the findings have been disputed by both Landgraf and Marsden. Various aspects of the comet-minor planet relationship were discussed by Degewij and Tedesco (32.102.033), Hughes (32.102.083), Kresák (32.098.039, 33.098.035) and Rickman (IAU Coll. 83, 1984).

Close encounters and collisions of comets with the Earth were statistically evaluated by Sekanina and Yeomans (37.102.003). Their estimate of 2 to 3 impacts per 10^7 years is only 20% higher than the previous estimate by Kresák (21.098.042), the difference being apparently due to updating the statistics by the inclusion of the recent encounters with comets 1983d and 1983e. They also confirm the cutoff in the absolute magnitude distribution of long-period comets; this may be responsible for considerable overestimates of the total population of the Oort cloud through over-correction for undiscovered faint objects. The collision rate of comets is similar to the repetition rate of episodic terrestrial catastrophism, as revealed by the global-scale geochemical anomalies and mass extinctions (Shoemaker, 33.105.230). However, the latter also agrees with the frequency of solar system passages through the mid-plane of the Galaxy and its encounters with the giant molecular clouds (Clube and Napier, 31.107.003, 37.102.047; Clube, 34.102.036). The impact rates of comets are appreciably smaller than those of the minor planets. Weissman (*Geological Soc. of America Spec. Paper* 190, 15, 1982) estimates the share of Earth cratering at 10% for long-period comets, 20% for short-period comets and 70% for Apollo-type minor planets. Zimbelman (37.102.004) concludes that 3% to 9% of the large craters ($d >$ 10 km) on terrestrial planets are due to long-period comets, and that the crater production rate is quite similar throughout the planetary system. The influx of comets and minor planets on the Earth is also discussed by Hughes (31.102.023).

A study of cometary impacts on the Sun by Weissman (34.102.011) has shown that there is no dynamical explanation for the very small perihelion distance, $q = 0.35\ R_\odot$, of comet 1979 XI, and suggests that the problem may be errors in the measured positions or in the solution for the orbital elements. Whipple (33.102.031) suggests that the violent outbursts of P/Holmes in 1892 and P/Tuttle-Giacobini-Kresák in 1973 were due to collisions of a cometary satellite with its primary. His scenario presumes a grazing encounter followed by destruction of the satellite at the next pericomet passage.

The cloud of particulate material discovered by IRAS around Vega and some other main sequence stars may represent an analogy to the inner Oort cloud, and the lack of warm dust closer to each star suggests that planets may have formed and swept out the dust, as suggested by Weissman (*Science* 224, 987, 1984). The possibility of comets revolving in double and multiple star systems was discussed by Dvorak (IAU Coll. 83, 1984).

IV. SATELLITES

a) ASTROMETRIC OBSERVATIONS

Rohde, Ianna, Stayton and Levinson (31.099.089) published astrometric observations for satellites of the outer planets made in 1978-1979, 1980 and 1981 at the Leander McCormick Observatory, a total of 1032 pairs of spherical-equatorial coordinates and 1316 intersatellite positions being reported. Debehogne and Houziaux (32.099.111) gave positions for the Galilean satellites observed in April 1979 at ESO. Ianna's group is continuing their astrometric observations of brighter satellites of Jupiter, Saturn, Uranus and Neptune at the McCormick Observatory and with the Yale-Columbia refractor at Mt. Stromlo. Bocşa started astrometric observations of Jupiter and its Galilean satellites, photographing the five bodies with the 0.38-m astrograph at Bucharest Observatory; observations made in 1982 have been reported.

Harrington and Walker (37.101.021) reported positions of satellites of Uranus and Neptune relative to their respective planets during oppositions of 1979, 1981, 1982 and 1983. Their plates were taken with the 1.55-m reflector in Flagstaff, the planet being occulted with neutral density spots for magnitude compensation. The positions of the satellites of Uranus were sent to the Jet Propulsion Laboratory for use in planning for the 1986 flyby mission.

Pascu made photographic observations of satellites of Mars, the Galilean satellites of Jupiter, and Saturnian satellites I through VII with the 0.66-m refractor and with the 1.55-m reflector at Flagstaff. Pascu, Seidelmann, Baum and Schmidt (32.031.546, 34.091.048) observed Jupiter XIV, Saturn XII, XIII and XIV and the satellites of Uranus and Neptune using the Space Telescope Wide-Field Planetary Camera Ground-Based CCD system on the 1.55-m reflector.

Veillet and Martins (*IAU Circ.*3940, 1984) reported their observations of Saturn XII and the Tethys Lagrangian satellites, S XIII and S XIV on 1984 April 12-19 with the Danish 1.5-m reflector at ESO and gave the following data for the orbital separations: S XII-Dione = +74$^\circ\!.$6 (36 measurements), S XIII-Tethys = +61$^\circ\!.$3 (11 measurements) and S XIV-Tethys = -65$^\circ\!.$0 (25 measurements), the standard deviations being roughly 0$^\circ\!.$6.

The following observations made in the USSR were published: observations of the Galilean satellites in 1976-1978 with the normal astrograph at Pulkovo by Bronnikova and Kiseleva (32.099.151); observations of the Galilean satellites and bright satellites of Saturn in 1976-1979 at the Nikolaev Observatory by Voronenko and Gorel' (32.099.007); observations of the Galilean satellites in 1977-1978 at the Engelhardt Astronomical Observatory by Nefed'ev and Chugunov (37.099.072); observations of the Saturnian satellites in 1977-1978 at the Engelhardt Observatory by Chugunov (37.100.060), and those in 1980 and 1981 by Kitkin and Chugunov (32.100.012, *Izv. AEO* No. 48, 96, 1984); observations of the Galilean satellites in 1975 at the Goloseevo Observatory (Kiev) by Levitskaya (31.099.051); observations of Deimos in 1978 at Goloseevo Observatory by Major and Sereda (*Ref. Zh. Astr.* 3.51.115, 1984).

Several attempts have been made to separate Charon, a possible satellite of Pluto, from the planet. For example, Reitsema, Vilas and Smith (34.101.019) reported that images obtained with a CCD detector attached to the 1.54-m Catalina telescope of the University of Arizona on 1980 February 3 showed an elongation caused by the satellite. Analysis of the data separated the planet and satellite components, yielding a Pluto/Charon brightness ratio of 5.5. Hege et al. (32.101.002) and Hege and Drummond (*IAU Circ.* 3986, 1984) have applied the technique of speckle interferometry to observations of the Pluto-Charon system and reported corrections to the ephemeris of Harrington and Christy (29.101.011). Speckle interferometric observations have also been reported by Hetterich and Weigelt (34.101.007). Mulholland and Binzel (37.101.020, *A.J.* 89,1759, 1984) discussed the expected eclipse phenomena, which had not yet started in 1983.

b) MOTIONS OF SATELLITES

Mutual phenomena of Galilean and Saturnian satellites have been used to improve their orbital elements. Aksnes et al. (37.099.008) gave astrometric data derived from photoelectric observations of mutual occultations and eclipses of the Galilean satellites in 1973 and 1979/1980 and of four Saturnian satellites in 1980, with accuracy approaching 0".01. Orbital elements based on these data and derived by using radii of satellites measured by the two Voyager spacecraft are in generally good agreement with other recent analyses. Some of the observations used were reported by Arlot et al. (32.099.001). Dourneau (32.100.007) observed two mutual events of Saturnian satellites, S III partially occulting S IV on 1980 April 5, and S V partially eclipsing S III on 1980 April 20, finding that the two events occurred, respectively, about 1 min and 3 min in advance of the predictions by Aksnes and Franklin (1978). It was concluded that a small difference could exist between observed and theoretical longitudes for S III, Tethys, which was involved in both events.

Aksnes and Franklin (*Ctr. for Ap. Preprint* No. 1987, 1984) gave predictions for nearly 300 observable mutual eclipses and occultations of the Galilean satellites expected between 1985 May and 1986 April. Predictions for the 1985-1986 events were given also by Arlot (*Astr. Ap.* 138, 113, 1984).

Arlot, Morando and Thuillot (*Astr. Ap.* 136, 142, 1984) rediscovered an important set of old observations of the Galilean satellites that form a part of the eclipse observations collected by Delambre; some 845 mutual phenomena of Io between 1775 and 1802 are included in their list. Lieske (34.099.028) examined the Delisle manuscript and the Pingré book and recorded and reduced the observations listed there. He now has more than 6800 eclipse observations of the Galilean satellites before 1800 and over 16 000 prior to 1982 and has shown the usefulness of such observations for the adjustment of the theory, particularly for the correction of the mean motions.

Tsuchida, Ferraz-Mello and Biancale (31.099.090) discussed the accuracy of photographic observations of the Galilean satellites of Jupiter in the interval 1913-1928 and showed that the residuals are generally very small, 0".06 to 0".08 in the mutual distances. Arlot (31.099.038) calculated new constants for the Sampson-Lieske theory of motion of the Galilean satellites by using a set of 8856 photographic observations; comparison was made with observations of mutual phenomena in 1979.

Thuillot and Vu (34.099.112) reported the development of a first approximation of an analytical theory for the Galilean satellites using Sagnier's method, the results being reduced to numerical tables of longitude, radius vector and latitude for easy comparison with Sampson's or Lieske's tables.

Rocher (33.099.052) computed ephemerides of Jupiter's satellites, J VI and J VII, for the years 1981-1990 by a numerical integration method and by fitting observational data analogous to that used by Bordovitsyna and Bykova (1978).

Boronenko and Schmidt (Tomsk University) continued to work on a theory of the motion of distant satellites using the Lie transformation technique; they plan to apply it to Jupiter VI, VII, X and Saturn IX.

Bec-Borsenberger and Rocher (32.100.083) collected the topocentric observations of Phoebe (S IX) from 1904 to 1981 and obtained initial integration constants adjusted with these observations. They then made a numerical integration up to 1990, taking into account perturbations by the Sun and Titan, and computed an ephemeris for the ten years 1981-1990. Bykova and Shikhalev (37.100.019) analyzed observations of 1898-1981 and derived new orbital elements of Phoebe which represent observations with a mean error of $1''.5$.

Zhang and Liu (32.100.104) surveyed the theory of motion of Hyperion (S VII). Taylor (*Astr. Ap.*, in press, 1984) used astrometric observations of Hyperion from 1967 to 1982 to derive values of parameters of Woltjer's theory for a best fit to the modern data. He found, however, that there are significant inadequacies in the theory. Accordingly, Sinclair and Taylor have tried to fit a numerical integration of the motions of Titan, Hyperion and Iapetus to the modern data, obtaining a much better fit to the Hyperion data and a good determination of various physical parameters of the system.

Yoder et al. (33.100.047) presented a simple analytical theory describing the 1:1 orbital resonance and applied it to the Saturnian co-orbiting pair, 1980 S1 (S X) and 1980 S3 (S XI). These small satellites can approach within 15 000 km, but they are prevented from passing each other by their mutual gravitational interactions. Long-term stability was also discussed, and a tie was established between the 1966 and 1980 observations of the two satellites.

Chugunov (32.100.016, 33.100.060, 33.100.061, *Ref. Zh. Astr.* 11.51.65, 1983; *Ref. Zh. Astr.* 12.51.73, 1983; *Ref. Zh. Astr.* 4.51.266, 1983; *Ref. Zh. Astr.* 4.51.120, 1984; *Ref. Zh. Astr.* 6.51.179, 1984) determined new orbital elements and masses of Saturnian satellites S I - S VI as well as the mass of Saturn and the values of J2 and J4 by using 28 500 observations.

Thuillot (34.099.032) proposed a new method to derive compact sets of ephemerides of various phenomena of the Galilean satellites by calculating a Chebyshev representation; 15 coefficients give a representation of each type of phenomenon for one satellite for one year with a precision better than 51 seconds.

The Bureau des Longitudes, Paris, publishes each year in the *Connaissance des Temps* the positions of the planets and the four Galilean satellites of Jupiter developed into Chebyshev polynomials, and in its *Supplements* the configurations and phenomena of the Galilean satellites, the positions of the first eight satellites of Saturn, and the positions of J VI, J VII, J VIII, J IX and S IX developed also into Chebyshev polynomials.

Veillet (33.101.004) gave 112 positions of Miranda relative to Uranus in 1980-1981 and confirmed an inclination which agrees with the value found by Whitaker and Greenberg from observations in 1948-1972. However, the eccentricity derived is very small. A more extended study, using Van Biesbroeck's data from 1948-1949 as well as all published positions of Miranda (300 points), confirmed all the orbital elements obtained except for those linked to the eccentricity, for which the value found was 0.0021±0.0005. The masses of Ariel and Umbriel were derived by using the values of J2 and J4 derived from the apsidal precession rates of the rings and the nodal precession rate of Miranda.

Veillet (32.101.005) used a series of observations of Nereid made with the Danish-ESO 1.5-m reflector in 1981 April, together with two unanalyzed plates taken at the McDonald Observatory in 1977 and 1978 to determine orbital elements of Nereid and the mass of Neptune.

c) THEORETICAL STUDIES OF SATELLITE MOTIONS

Most theoretical studies of satellite motions have been devoted to the question of dynamical evolution of the orbits, considering such topics as capture mechanisms and the effect of resonances.

Tanikawa (33.042.024) proved the impossibility of the capture of retrograde satellites in the frame of the circular planar restricted problem of three bodies. Huang and Innanen (34.042.037) explored numerically the stability and capture region for retrograde jovicentric satellites in the frame of the restricted three-body problem and derived the value of the Jacobian constant for the greatest probability of temporary capture.

Szeto (34.097.008) examined the role of tidal dissipation within the Martian satellite system and assessed theories of origin through calculations of collision probabilities between Phobos and Deimos in the distant past. An accretion model is preferred over capture, although no such model consistent with the likely carbonaceous chondritic composition of the Martian satellites has yet been established.

Henrard (33.099.020) reexamined Yoder's scenario (25.099.047) for the capture into resonance of the first three Galilean satellites by introducing a more refined dynamical model for resonance and tidal effects. Gailitis (32.099.099) showed that the limited energy of the radial motion has preserved Io from a runaway type of melting as the energy-momentum balance relates tidal heating to orbital expansion as in Yoder's secular equation.

Sinclair (34.100.023) critically reviewed the hypothesis that the origin of the resonances among the Saturnian satellites is due to orbital evolution caused by tidal dissipation within the planet. He concluded that the hypothesis provides a plausible explanation for the origin of the Mimas-Tethys resonance, but it is unsatisfactory for Enceladus-Dione, since their resonance now has little effect on the relative evolution rate of these satellites.

Poirier, Bolah and Chambon (34.100.014) investigated tidal dissipation in a viscoelastic homogeneous sphere having the orbital and physical characteristics of the icy inner satellites of Saturn and found that tidal dissipation with current orbital eccentricity cannot account alone for the surface activity observed on Enceladus if it is composed of water ice. Lissauer, Peale and Cuzzi (37.100.051) argued that the angular momentum transfer between Enceladus and Janus could have sufficiently enhanced the eccentricity of Enceladus' orbit for tidal heating to have melted the interior of Enceladus if Janus were ever locked into a stable 2:1 orbital commensurability with Enceladus. The rapid time scale for dynamical evolution of the ring and inner satellites as presently situated remains a major problem.

Farinella et al. (33.100.089) proposed that the peculiar orbital motion of Hyperion, characterized by a strong orbit-orbit resonance with Titan, was responsible for the ineffectiveness of the process of repeated reaccretion from narrow rings of collisional fragments proposed for the other satellites. The irregular shape of Hyperion thus stands in contrast to the regular figures of the other small satellites of Saturn.

Pauwels (34.100.005) studied the Rhea-Titan secular resonance as a special case of secular orbit-orbit resonance. Taking into account both masses, it was shown that in the Rhea-Titan case, the resonance is dominated by the effect of the large difference between the proper precession rates of the lines of apsides of the two satellites.

Sinclair (*Astr. Ap.* 136, 161, 1984) examined the effects of orbital resonances on satellites in tadpole or horseshoe orbits relative to Mimas, Enceladus, Tethys or Dione and concluded that a tadpole companion of Enceladus would be the most highly perturbed. Although the perturbations would not cause instability, they might prevent the initial formation of a satellite in such an orbit. Tadpole companions of Mimas would have large forced eccentricities, which would probably prevent the simultaneous existence of horseshoe orbits. He also examined the effects of tidal forces, which were found to cause a small displacement of the L4 and L5 equilibrium points.

d) DYNAMICS OF RINGS

Study of the dynamical processes that cause the complex structural features of rings, such as those manifested by the narrow elliptic rings of Uranus and the F ring of Saturn, as well as the spokes in the Saturnian rings, has been an extremely active field during the triennium. Reviews have been published by Goldreich and Tremaine (32.091.025) and by Cuzzi (37.091.052). IAU Colloquium No. 75, held in Toulouse in September 1982, was devoted entirely to the subject of planetary rings. The proceedings of the colloquium, containing contributed papers and also the discussion, have been edited by Brahic and published in 1984 under the title *Anneaux des Planetes/ Planetary Rings* by Cepadues-Editions on behalf of CNES, the French space agency. A second book, based on a set of review papers selected from among those presented at Colloquium No. 75, has been edited by Greenberg and Brahic and published in 1984 under the title *Planetary Rings* in the Space Science Series of the University of Arizona Press. The material in this second book has been carefully reviewed and refined, and reflects the results of investigations published in 1983. Since full lists of literature citations are given in both books, mention is made here of only a few of the most recent studies.

Borderies, Goldreich and Tremaine (34.042.039) extended their investigations of the evolution of eccentric rings under the influence of differential precession due to the oblateness of the planet, self-gravity, viscous forces due to interparticle collisions, and eccentricity excited by shepherding satellites. They concluded that uniform precession can be enforced by self-gravity, the resulting configuration being both secularly and dynamically stable, that due to viscous forces the line of the apse at the inner ring edge is not exactly aligned with the line of apsides at the outer edge, the apse shift being detectable in the α- and β-rings of Uranus, and that the mean eccentricity is determined by a balance between viscous damping and excitation by shepherding satellites.

Lissauer (37.100.002) has studied ballistic mass transport in Saturn's rings, developing an analytic model that includes the effects of angular momentum advection. He showed that net material movement due to angular momentum advection is comparable to that caused by direct ballistic transport.

Seidelmann, Harrington and Szebehely (37.100.052) studied the dynamics of Saturn's E ring, which extends from three Saturn radii, the orbit of Mimas, to at least eight Saturn radii, just inside the orbit of Rhea. Although the brightness profile has a peak at the orbit of Enceladus, there are apparently not corresponding peaks at the orbital distances of Tethys or Dione. It is known that there are satellites at the Lagrangian points of Tethys and Dione within the E ring. Thus the E ring is not constrained by satellites, but rather co-exists at the same orbital distances as satellites, which have other satellites at their Lagrangian points. A discussion of preliminary numerical and analytical investigations of the motions of the ring particles is given.

V. PREDICTION OF OCCULTATIONS

a) IDENTIFICATION OF EVENTS

The number of minor planets whose orbits have been investigated by Taylor for occultations has risen to approximately 200. Ephemerides for Ceres, Pallas, Juno and Vesta were provided by H M Nautical Almanac Office, and an ephemeris of Nemausa was supplied by Kristensen and Møller. Ephemerides of all other minor planets were generated at the Royal Greenwich Observatory from osculating elements supplied by Shor, Institute for Theoretical Astronomy, Leningrad. All these ephemerides have been searched against a combined AGK3 + SAO star catalogue by Taylor, who issued predictions for suitable occultations in *Bulletins* circulated to members of the Working Group. Predictions up to the end of 1985 have been issued in this manner. Predictions for 1984 and 1985 have also been issued by Wasserman, Bowell and Millis (34.096.012) and by Millis et al. (37.096.017). Supplementary calculations have been carried out by D W Dunham, International Occultation Timing Association, and issued, along with other data, in IOTA's *Occultation Newsletter*. Tabulations of possible events involving minor planets have appeared annually in the *Handbook of the British Astronomical Association*, in the January issue of *Sky and Telescope*, and in several other publications.

Searches for occultations of faint stars (not AGK3 or SAO) have been intensified, particularly for occultations by Uranus, Neptune and Pluto (Mink and Klemola, *A.J.*, in press). Searches for occultations by comets are now being made on a routine basis, particularly those involving P/Halley. Predictions are given by Taylor (37. 103.914), and by Bowell and Wasserman (*IHW Newsletter*, No. 5, 1984).

Since star positions based on the AGK3 and SAO catalogues commonly involve errors of 0.6-0.8 arcsecond and more at the current epoch, these catalogues are not suitable as a basis for precise predictions. It has been the practice to engage in "last-minute" astrometry as the basis for refinement of the expected ground tracks. However, in a substantial number of cases, the actual paths of occultation events are far from contact with the Earth. At least semi-precise (to 0.3-0.5 arcsecond) measurements of positions of stars involved in events predicted on the basis of the AGK3 and SAO star catalogues would permit weeding out of the obvious non-events at an early stage. This work could be done at leisure, requires relatively modest equipment, and would have the valuable effect of conserving resources for the high-precision last-minute astrometry for the more certain events.

b) "LAST-MINUTE" PREDICTION IMPROVEMENT

Experience with "last-minute" astrometry during the report period has tended to reinforce the view that only the AGK3R and SRS (or Perth 70) catalogues, each with a density of about one star per square degree, contain stellar data accurate enough at current epochs to provide a suitable reference frame for this demanding work. The astrograph field must be large enough to include several of these stars, yet also have a plate scale adequate for accurate measurements. Schmidt cameras do not yield satisfactory results because of the number of stars needed to determine the higher-order plate constants; the AGK3R and SRS catalogues are not dense enough to give the needed coverage. Good results have been obtained with the 13-inch astrograph at the Royal Greenwich Observatory (Taylor). But the best and most consistently satisfactory results during the report period were those obtained with the 0.5-m Carnegie double astrograph at the Lick Observatory (Klemola), due in part to the wide field and favorable plate scale of that telescope. Long-focus instruments, such as the 1.55-m astrometric reflector at the U S Naval Observatory, Flagstaff, and the 1-m reflector and 26-inch Clark refractor at the University of Virginia (McNamara), also gave good results when the positions of secondary reference stars were measured from Lick plates. In general, better accuracy was achieved when several exposures were measured.

The photoelectric transit telescope at Bordeaux, France, has proved to be an alternate source of accurate data for improving predictions of minor planet occultations. The observations can be made farther in advance of events, since star and minor planet are not limited to the same field. Both objects must be brighter than 13th magnitude, and both must cross the meridian in a dark cloudless sky during several nights of the few weeks preceding the event.

Arrangements have been made with Candy at Perth for astrometry involving stars in the southern hemisphere, but at least one more southern astrometric station is needed.

c) OBSERVATIONS AND DIAMETERS

An historical review and report on the status of observations of occultations by minor planets through the end of 1981 was given by Maley (33.098.020). Millis et al. (34.098.104) have given a brief overview of recent progress and of prospects for the future.

Events well enough observed since those listed in the previous report of the Commission to permit the derivation of useful diameter information include the following:

375 Ursula	1982 Nov. 15	6 chords: Texas, Utah R L Millis et al., 37.098.032
93 Minerva	1982 Nov. 22	20 chords: Arizona, Michigan L H Wasserman et al., 34.098.106
106 Dione	1983 Jan. 19	10 chords: The Netherlands, Germany, Denmark L K Kristensen, 37.098.100
52 Europa	1983 Apr. 26	6 chords: Texas, New Mexico, Arizona
2 Pallas	1983 May 29	130 chords: USA D W Dunham et al., 34.098.105
51 Nemausa	1983 Sept. 11	55 chords: USA E W Dunham et al., *A.J.* 89, 1755, 1984
9 Metis	1984 Feb. 19	10 chords: Denmark, Germany L K Kristensen, *Astr. Raumfahrt* 1984 May-June, p. 76
230 Athamantis	1984 May 11	4 chords: Australia
47 Aglaja	1984 Sept. 16	13 chords: USA

Papers presenting results from earlier occultation events were published by Kristensen (29.098.073) and Millis et al. (33.098.009). A progress report on accurate determinations of diameters of minor planets was given by Taylor (33.098.109).

E Roemer

President of the Commission

21. LIGHT OF THE NIGHT SKY (LUMIERE DU CIEL NOCTURNE)

PRESIDENT: R.H. Giese
VICE PRESIDENT: K. Mattila
ORGANIZING COMMITTEE: R. Dumont, Yu.I. Galperin, M.S. Hanner, A.C. Levasseur-Regourd, F. Sanchez-Martinez, H. Tanabe, J.L. Weinberg

I. INTRODUCTION

The light of the night sky consists of atmospheric components (airglow, light scattered in the atmosphere) and - even in the case of spaceborne observations - of zodiacal, galactic and extragalactic light. Although all components are of similar importance, investigations on zodiacal light have profitted most by the space age since their object of research, the interplanetary dust cloud, became accessible to direct in-situ measurements. Lunar samples and measurements by micrometeoroid detectors provide individual and eventually detailed information on impact events, which however are limited in number and therefore restricted in statistical significance. Zodiacal light investigations involve scattered light of many particles in large volume elements and therefore provide global information about physical properties and spatial distribution of interplanetary dust grains, however just in terms of average values. Therefore both sources of information are complementary and a synthesis can only be achieved by synoptic interpretation of zodiacal light, micrometeoroid, and meteoroid investigations also including dynamical aspects. Measurements of zodiacal light (and emission) from rockets, manned or non manned spacecraft, and deep space probes gained drastically in importance compared to ground based observations. On the other hand investigations on airglow have become more and more a topic of geophysics (aeronomy). They remain relevant however to astronomy as far as photometric features are concerned. These general trends continued in the last triennium and have influenced the activities of our commission.

Although the light of the night sky involves three rather different fields of interest (aeronomy, physics of the planetary system, galactic and extragalactic astronomy) the observational methods and problems are closely related and the results (surface brightness and separation of components) are of equal importance for all investigators, independently of their preferred field of interest. Therefore the commission provided an excellent platform for exchange of results and experience on common problems concerning the light of the night sky which is understood not exclusively in terms of visible light but involving more and more the domain of ultraviolet (UV) and infrared (IR) radiation.

Activities in the field of the commission were concentrated during the past triennium mainly in rocket and satellite experiments and in thorough interpretations of the results obtained by earthbound spacecraft (e.g. D2B, Salyut) and especially by the great deep space probe missions Helios and Pioneer. A highlight was the latest results of the IRAS team contributing much to our field. On the other hand cancellation of the US-Spacecraft for the Out-of-Ecliptic Solar Polar Mission (ISPM) caused considerable damage, since with the Zodiacal Light - Background Starlight Instrument (ZLE) scheduled for ISPM NASA spacecraft we lost the unique opportunity to obtain photopolarimetric measurements from a position out of the ecliptic plane.

At the end of the triennium preparations were completed for a meeting summarizing the present state of the art concerning the zodiacal dust cloud in a synoptic way: the IAU Colloquium No 85 at Marseille (July 9 to 12, 1984). The proceedings of this meeting ("Physical Properties and Interactions of Interplanetary Dust", Giese and Lamy eds., to be published 1985 by Reidel, Dordrecht) will include about 70 reviews and contributed papers. Books covering topics relevant to our commission were published by Hodge (1981: Interplanetary Dust), by Fishkova (1983: The Night Airglow of the Earth Midlatitude Upper Atmosphere, in Russian), and by Megrelishvili (1981: Regularities of the Variations of Scattered Light and Emission of the Earth Twighlight Atmosphere, in Russian). A review article on "Origin and Evolution of Interplanetary Dust Grains" (33.106.029) was presented by Lamy (1983).

This report was prepared with the help of Drs. Dumont, Mattila, and Tanabe, taking into account comments and valuable information kindly provided by Drs. Galperin, Lamy, Lauche, Levasseur-Regourd, and Weinberg.

II. AIRGLOW AND ATMOSPHERE

Numerous papers on the airglow have been published in this triennium. However, since most of them are included in the category of geophysics, we report here only papers on photometric features of astronomical interest.

<u>557.7 nm [OI], NaD, and OH</u> A faint pattern visible to the naked eye was observed by Armstrong 1982 (31.082.037) in the northern region of the night sky at the CSIRO Solar Observatory (30°3S, 149°6E), similar to the phenomenon previously observed by Peterson (26.082.143) in Hawaii. From his photographic and photometric observations, Armstrong concluded that the emission responsible for the visible pattern is 558 nm airglow enhanced by the effect of a gravity wave passing through the emitting layer.

Analysing airglow data obtained at Cachoeira Paulista (22.7°S, 45.0°W) in 1977-82, Takahashi et al. 1984 ( ) concluded that (1) monthly averaged variation of 558 nm emission has revealed a significant semidiurnal oscillation with a seasonal change of the phase of maximum, (2) the 558 nm and NaD emissions show a predominant semiannual variation, but no significant seasonal variation was found for the OH(8,3) emission, (3) a possible correlation was observed in long term variation between the 558 nm emission and the solar activity.

Gegawale and Tillu 1983 (33.082.011) found that the intensity ratios of OH(9,4)/OH(7,3) are systematically different between two types of the nocturnal intensity variations; "continuous decrease" type (observed on 70% of nights) and "increase followed by decrease" type (30% of nights) , observed at Poona, India.

The first photographs of the OH airglow taken by Peterson and Adams 1982 (34.082.072) at Albuquerque during the 5-6 July 1982 total lunar eclipse revealed large (20-30km) east-west waves from horizon to horizon which drifted north-ward at 10-15 m/sec. These waves were not typical of the ripples they have found to be well-correlated with lower lunar transit.

Observations of the nocturnal 558 nm glow in Ashkhabad were used for the study of internal gravity waves by Gavrilov and Sved 1982. Height measurements of the E layer were related to the intensity of the 558 nm glow by Gorbunova et al. 1982 (31.083.029). Several papers on the height of the 558 nm emission layer, characteristics of atmospheric waves according to OH observations, and aeronomical processes were published by Megrelishvili 1982 (31.003.090), 1983

(37.082.088) and Megrelishvili and Fishkova 1983 (37.082.087), 1984 (37.082.032) and Toroshelidze 1983 (37.082.089).

630 nm[OI] Orographic sources were related to emisions (Shefov et al. 1983 (34.002.085) such as to hydroxyl emission temperatures and 630 nm intensity.

The north-south aligned equatorial 630 nm depletions were observed by Mendillo and Baumgardner 1982 (32.082.093) with a new all-sky imaging system on Ascension Island (8.0°S,14°W), and by Carman 1983 (33.082.010) with photometers at Vanimo (2.7°S, 141.3°E). Mendillo and Baumgardner reported that the airglow depletions occur most often before midnight, often cover one-fourth of the sky and are usually not aligned with magnetic meridians. Instrumentation and results of the tropical airglow works during the last few years were reviewed by Kulkarni 1983 (37.082.023)

Space borne Observations From the airglow data measured by the AE-E satellite, Abreu et al. 1982 (32.082.051) developed global pictures of the 630 nm emissions at low latitudes between 1800 and 0400 LT for solstice and equinox conditions. The latter map is more symmetric than the one for solstice condition, in which the winter hemisphere is brighter than the summer hemisphere.

Torr 1984 (preprint) reported first results of atmospheric spectral imaging from Spacelab 1 covering the atmospheric emissions between 30 nm and 1270 nm.

Photographic pictures of Sodium D and airglow emissions (80-100 km) were obtained from space during the Salyut 6 and 7 mission (cf. Koutchmy and Nikolski 1983 (33.106.012). Furthermore noctilucent clouds were observed by Salyut 7.

Atmosphere Color and brightness of the night sky in the crimea derived from observations between 1973-79 were presented and compared to other observation sites by Lyutyi and Sharov (1982).

Gadsden 1982(32.082.085) reviewed the observational facts about the noctilucent clouds, which appear during the summertime at high latitudes near the top of the mesosphere and seem to be ice particles nucleated by either meteoric smoke or by atmospheric ions.

III. ZODIACAL LIGHT

Measurements

General Observations of the Zodiacal Light Most observations have been space observations, especially with the two Helios probes. The general picture of the zodiacal cloud arising from 4 papers (Leinert and Planck 1982, 30.106.026, 31.106.005, Leinert et al. 1982, 30.106.031 and 30.106.032), and summarized in a very useful table (Leinert et al. 1982, 31.106.039) by the Heidelberg group, is that of a very regular, smooth and constant distribution of the dust. No dissymmetry in the polarization state with respect to the antisolar direction has been found, and the only changes (both with short time scale and over the solar cycle) deal with interplanetary plasma, not with the dust. The negative polarization on the borderline of the gegenschein is confirmed. The invariance of the brightness and degree of polarization over 6.5 years including a minimum and a maximum of the solar cycle is remarkable (less than 2% and 0.01, respectively).

Observations, from the atmospheric Explorer C, D, and E, yielded some galactic and zodiacal light surface maps between 732 and 429 nm (Abreu et al. 1982, 31.106.028).

Verlhac (1983) gave preliminary results of a polarimetric study of the zodiacal light far from the sun, as observed photographically in the blue (460 nm) and in the near IR (800 nm) from Salyut 7. The polarization degree seems to be somewhat higher in the near IR.

Preliminary results of a Doppler-Fizeau spectrometry at high ecliptic latitudes by Robley et al. 1982 (31.106.034) are in favour of high radial velocities.

Ground-based observations are reported by Asaad 1982 (31.106.040 and 31.106.37) from Abu-Simbel, Egypt, with special attention to possible intrinsinc variations. Pfleiderer and Leuprecht 1983 (34.155.099) reported about results of the analysis of Pfleiderers surface photometry which is meanwhile completed (isophote maps in V: zodiacal light and Milky Way).

A contribution to the problem of separating the zodiacal light from other components of the night skylight from ground has been given in his thesis by Alvarez Martin (1981).

<u>Inner Zodiacal Light</u> Photographic observations at rather low elongations have been made from the Salyut manned spacecraft. From Salyut 6 (Grechko et al., 1982 31.106.003) have been reported a strong N-S dissymmetry of the cone and an elongation gradient from 25 to 40° steeper than previously found. However, a telluric interpretation, due to the upper airglow emissions, is suggested by the authors for both effects. Pictures have also been taken from Salyut 7 by Russian and French cosmonauts, down to 22.5° elongation (Koutchmy and Nikolski, 1982 32.106.043), adding their contribution to knowledge of the inner zodiacal cloud. During the solar eclipse July 31, 1981 Shestakova and Shcheglov 1983 obtained velocity distributions of the inner F-corona grains at 3-7 solar radii by Fabry-Perot interferometry, which were the basis of a model (Shestakova 1983). A balloon observation of the total solar eclipse June 11, 1983 was carried out by a Japanese-Indonesian team involving photopolarimetry of the F-corona in the visual and near infrared wavelength region.

<u>Ultraviolet Observations</u> An important study of the zodiacal light in the UV domain has been made by Cebula and Feldman 1982 (32.106.045) in two directions at small elongations (21 and 30°), near the ecliptic. Down to 210 nm the colour was found slightly redder than the sun. The upper limit found for the zodiacal brightness at 180 nm, although being 4 times higher than expected from the solar flux, is lower than previous determinations and cannot be considered a true indication of any brightening of the zodiacal light in the UV. The elongation gradient is only slightly steeper than usually obtained in the visible domain. Fahr, Ripken and Lay 1981 (30.106.20) investigated plasma-dust interactions in the solar vicinity which lead to observational consequences in UV. Their corresponding sounding rocket experiment scheduled for 1985 is described by Neumann et al., 1983 (34.035.010).

<u>Infrared Observations</u> Hauser et al. 1984 reported first results of the IRAS satellite. They reveal bright emission from interplanetary dust which dominates the celestial background at 12, 25, and 60 μm except near the galactic plane. At 100μm zodiacal dust emission is dominant only near the ecliptic plane.

<u>The plane of symmetry</u> The IRAS infrared data of Hauser et al. 1984 show that the angular variation in the plane (ε =68.6 to 103.3°) and at a plane ε =90°

including annual variation at the pole are generally consistent with previously determined dust distributions and with the previously determined symmetry plane ($i=3°$, $\Omega=87°$ from visual observations (Misconi and Weinberg (22.093.001), Leinert et al. (27.106.003).

From numerical integrations of pertubation equations considering the gravitational forces of the planets, radiation pressure, ion drag, and the Poynting-Robertson effect, Gustafson and Misconi (1983) showed that the long-term dynamical evolution of interplanetary dust is strongly affected by the inner planets, and the perturbations may explain the observed inclination of the symmetry plane.

On the other hand, Winkler et al. 1982 (31.106.010) showed from their photographic observation in $140° \leq \varepsilon \leq 180°$ that the brightness profile of the Gegenschein perpendicular to the ecliptic can be represented by a superposition of three components of Gaussian form. The symmetry plane they determined from the centroids (not the peak brightness) of the brightness profiles has $\Omega = 14° \pm 10°$, $☋ = 206° \pm 6°$ and $i = 1.0° \pm 0.4°$.

Burkhardt 1982 (31.091.006) revised the position of the invariable plane of the solar system using new sets of masses and orbital elements of the planets yielded by recent spaceflights. Inclination and ascending node of the calculated invariable plane at the new standard epoch J2000.0 are 1°35'13".86 and 107°36'30".8, respectively.

Interpretation of Measurements

Scattering properties of dust grains from laboratory work and theory. Interpretation of optical or infrared measurements of the zodiacal cloud needs information about the differential scattering cross sections and absorption efficiency of typical dust grains including a possible wavelength dependence. Using laser (663 nm) light scattered on single particles in the size range 10 to 200 µm Weiss-Wrana 1983 (33.022.050) obtained scattering functions for terrestrial and meteoritic materials. These investigations are being extended to different wavelengths. Microwave analog measurements were continued at Bochum (8.6 mm) and at Gainesville (3.18 cm) for linear configurations of two (30.063.062) and multiple spheres and in a systematic study for dielectric ellipsoids (Schuerman et al. 1981, 30.063.061). Theoretical studies of rough grains using geometrical optics approaches were continued by Perrin and Lamy (1983), Mukai et al. (1982), Schiffer and Thielheim (1982, 32.106.040). They were in agreement with laboratory results. Colour effects of microroughness were taken into account by Schiffer and Thielheim (1983, 34.106.033). Generally angular dependence of scattered intensity and linear polarization of absorbing irregular grains resemble the volume scattering functions derived from zodiacal light observations. The problem of the exact definition of albedo and its application to irregular particles was treated by Hanner et al. (30.106.034) on the basis of microwave results obtained at Bochum. A workshop on Small Particles and Light Scattering took place at Kiel (Proceedings Thielheim ed., 1984).

Dust Properties from Zodiacal Light Observations To achieve information on scattering properties of interplanetary grains and on the spatial distribution two types of work were published during the triennium:

1. Modelling of the optical properties and of the sizes of interplanetary dust grains mainly from the empirical scattering function and polarization curve, with reference to laboratory experiments (microwaves, laser: Greenberg and Gustafson 29.106.001, Weiss-Wrana 33.022.050 or to theoretical work (Schiffer and Thielheim 32.106.040). There are still disagreements on the nature (dielectric vs. absorbing) and on the shape (irregular, fluffy particles, bird's nests) of the

scatterers. An unsolved question seems to be the degree of confidence in the empirical volume scattering functions and polarization curves due to the unavoidable increase of their error bars when the scattering angle decreases.

In a paper devoted to the F-corona, Beard (35.160.) showed that infrared coronal polarization measurements could bring rather accurate determinations of the size and albedo of the interplanetary dust, and reveal the heliocentric distance at which the dust vaporizes.

2. Remote sensing of the zodiacal cloud in or near the ecliptic. A valuable survey of the in-situ inversion concept, mainly applied to space probes and especially to Pioneers 10-11 has been given by Schuerman and Weinberg 1983 (33.106.028). As shown by Dumont and Pelletanne (30.106.029), local information at given points on the line of sight can be retrieved from the brightnesses and the polarization degrees at the two intersections of that line with the terrestrial orbit. Two "nodes of lesser uncertainty" exist, one inside and one outside the orbit, the locations of which are favourable for disentangling the heliocentric from the angular variations of the scattering coefficient and polarization (Pelletanne 1982, Dumont 1983 34.106.018). The work has meanwhile been extended to locations far from the Earth's orbit and to topics like Doppler and infrared measurements. The method avoids the increasingly doubtful assumption of a homogeneous cloud which in fact is invalidated by its results (significant radial gradient of polarization). This corroborates the conclusions resulting from earlier works of Leinert et al. (30.106.026) and Schuerman et al. (28.106.029).

Spatial Distribution of Dust Attempts have been made by Buitrago et al. 1983 (33.106.011), Buitrago et al. (31.106.038), Mujica et al.(34.106.005) to derive the off-ecliptic distribution of the dust, either by a new inversion process based on integral equations of Volterra types, or by an extension of previous works valid in the ecliptic (28.106.002). The isodensity surfaces are found to be fan-like with a possible, theoretically complete, depletion of the dust near 45° heliographic latitude. Unfortunately, due to the difficulty of the problem, the model-dependence is heavy. Both the ecliptic and the off-ecliptic cases, with less model-dependence in the former case, have been developed in a thesis by Mujica (1981). More conventional models (fan, ellipsoid) were discussed by Kinateder (1983) with respect to diagnostic signatures observable by an Out-Of-Ecliptic Mission. Since there is no such opportunity in the space program this investigation is now being applied at Bochum to all sky observations from the ecliptic plane, including low elongations. Preliminary results seem to provide signatures which allow one to exclude multiple fan models as proposed by Buitrago et al.

IV. GALACTIC COMPONENTS

A review article and summary of the workshop on the ultraviolet background radiation including all different components from airglow to extragalactic has been presented by Henry, 1981 (32.142.082). A computer program for estimating the different components of light of the night sky as observed from a space platform has been presented by Barbieri and Nota (1983).

Optical An extensive photographic UBVR surface photometry of the whole Milky Way has been carried out by Schmidt-Kaler et al. 1983 (33.155.005) using a superwide angle camera with a spherical mirror. Detailed analysis of the data have already been published of the Coalsack region by Seidensticker et al. 1982 (32.155.026), the central area of the Galaxy by Pröll et al. 1983 (33.155.006) and the Northern Milky Way in U colour by Winkler et al. 1982 (31.106.010). Using the Helios 1 and 2 space probes Leinert and Richter 1981 (33.155.033) have measured the UBV intensities and colours along eight strips across the Milky Way. The calibration appears to be very reliable and the results have been used to test previous all sky photometric maps of the Milky Way. Leinert and Richter 1983 (33.167.067) have further used the Helios data in search for polarization of the Milky Way light at l=150° and l=223° and obtain some marginal indication for a polarized component perpendicular to the galactic plane. Toller 1981 (32.155.018), 1982 (33.131.184), 1983 (34.155.096)) has used the Pioneer 10 measurements for a background starlight study in B and R colours with the intention to derive information on large scale galactic structure and the scattering properties of the dust. For the albedo and asymmetry parameter at 440 nm Toller gives values of 0.61 ± 0.07 and 0.6 ± 0.2, respectively, which are in agreement with the previous results. Reynolds 1983 (33.155.109) has analyzed the diffuse Hα emission along galactic equator from l=0° to 240°, also with respect to Hα being scattered by dust.

Near Infrared Near Infrared surveys of surface brightness near the galactic plane were performed by Hayakawa et al. 1981 (30.156.001), at 2.4 and 3.4 μm. The interpretation of near IR surface brightness is similar to the optical wavelength region, including the model for the distribution of volume emissivity (i.e. stars) and for the extinction. Galactic mid- and far IR emissions are dominated by a different source (thermal emission by dust) and are not covered by this review.

Ultraviolet Zvereva et al. 1982 (32.157.012) and Severny and Zvereva 1983 (34.142.043) have presented 110-185 nm surface brightness measurements carried out on board of the satellite Prognoz-6. At low latitudes $|b|\leq 30°$, DGL is observed and interpreted in terms of dust scattering models. At high galactic latitudes a good correlation has been found with soft X-ray background brightness and neutral hydrogen column density. The authors suggest that hot (10^5 - 10^6K) interstellar gas is responsible for the observed emission in some high galactic regions. A line emission component of the hot coronal gas is suggested by the observations of Feldman et al. 1981 (30.158.161). Deharveng et al. 1982 (31.131.162) have investigated the contribution of the warm ($\sim 10^4$K) intercloud gas to the UV background and conclude that it is in general unimportant. Joubert et al. 1983 (34.142.044) in a continued analysis of the D2B-Aura satellite data at 169 and 220 nm confirm the existence of a correlation of the UV sky brightness with the HI column density and the interstellar extinction. The asymmetry factor of the scattering phase function is determined to 0.6-0.7. Further small scale I(UV) vs. N(HI) correlation studies have been made by Jakobsen et al. 1982 (33.142.090). Andersson et al. 1982 (32.157.001) and Henry 1981b (29.157.005), 1982 (32.162.188), 1982 (32.142.082) using Apollo 17 data, have found no DGL from dusty moderate galactic latitude fields at λ =118-168 nm and conclude that either the grains are extremely strongly forward scattering (asymmetry factor g = 0.9) or their albedo is low ($\leq$ 0.2). Jakobsen 1982 (31.131.067) suggests that the UV background intensity vs. HI column density correlation may be partly due to rapid H_2 photodissociation in addition to the scattering.

V. EXTRAGALACTIC COMPONENT

Using the Pioneer 10 photopolarimeter data Toller 1983 (33.161.030) has derived an upper limit of 3.9 $S_{10}(V)_{G2V}$ at the 2σ level for the brightness of the extragalactic background at 440 nm. This measurement was free of airglow and zodiacal light contamination but a large correction was needed to account for the galactic foreground starlight. The UV background level, after substraction of the galactic components, has been found to be in the range 160-360 units at 220 nm and 150-700 units at 160 nm (Feldman et al. 1981, 30.158.161, Joubert et al. 1983, 34.142.044, Tennyson et al. 1982, 32.142.064) (unit= photons $cm^{-2}s^{-1}Å^{-1}sterad^{-1}$). Weller 1983 (33.142.103) using 122-150 nm measurements from the Solrad 11 satellite gives the upper limits of 180 and 280 units near the north and south galactic poles, respectively. Wulf-Mathies et al. 1983 (33.0335.056) have presented observations of the extreme UV background and give as upper limit to the extragalactic contributions the value 3.90×10^4 units in the wavelength regime 75-94 nm, and 9.7×10^3 units in the band 104-100 nm. In addition, EUV measurements have been reported by Kimble and Bowyer 1982 (33.142.089). Paresce 1983 (34.161.294) has presented a review on the attempts to measure the UV background radiation of cosmological origin.

Theoretical calculations on the contribution of galaxies to the ultraviolet and optical background radiation have been presented by Code and Welch 1982 (31.158.094) for nonevolving and evolving galaxies. They used the UV energy distribution curves as obtained from OAO 2 measurements. They conclude that normal galaxies can account for anything from a few percent to all of the measured background radiation. Bruzual 1981 (32.066.034) has carried out similar calculations. Ceccarelli et al. 1983 (34.161.148) have proposed for the measurement of the far IR extragalactic background a new method, which utilizes the dipole anisotropy due to the observer's motion.

REFERENCES

Alvarez Martin 1981, Thesis, University of Canary Islands
Barbieri C and Nota A 1983, Preprint University of Padova (in Italian)
Fishkova L M 1983, "The night airglow of the Earth mid-latitude upper atmosphere" (in Russian) 'Metsniereba' Publ. House Tbilisi
Gavrilov M M, Shved G M 1982, Ann. de Geophys. 38 N6, p.789
Gustafson B A S and Misconi N Y 1983, pg.121 in T.I. Gombosi (ed) "Proc. Internat. Conf. on Cometary Exploration", Budapest
Hodge P W 1981, "Interplanetary Dust", Gordon and Breach New York, London, Paris, Montreux, Tokyo
Kinateder G 1983, Thesis Ruhr-Universität Bochum
Lyutyj V M and Sharov A S 1982, Astron J Acad Sci USSR Moskow 59, p.174 (in Russian)
Megrelishvili T G 1981, "Regularities of the Variations of the Scattered Light and Emission of the Earth Twilight Atmosphere", (in Russian) 'Metsniereba' Publ. House, Tbilisi
Mujica A 1981, Thesis, Univ. of Canary Islands
Mukai S, Mukai T, Giese R H, Weiss K, Zerull R H 1982, The Moon and the Planets 26, p. 197
Pelletanne B 1982, Thesis, University of Bordeaux I
Perrin J M and Lamy P L 1983, Optica Acta 30, p. 1223
Shestakova L I, Sheglow P V 1983, Astronomic Circular N 1286, p. 6
Takahashi H, Sahi Y and Batista P P 1984, Planet. Space Sci. 32, p.897
Thielheim K O (ed) 1984, Workshop "Kleine Teilchen und Lichtstreuung", Verhandlungen Deutsche Physikalische Gesellschaft 7
Verlhac A 1983, Thesis, University of Paris 6
Toller G N 1983, Thesis, State University of N.Y. at Stony Brook

R.H. Giese

22. METEORS AND INTERPLANETARY DUST (METEORES ET LA POUSSIERE INTERPLANETAIRE)

PRESIDENT: O.I. Belkovich
VICE-PRESIDENT: D.E. Brownless
ORGANIZING COMMITTEE: P.B. Babadzhanov, W.J. Baggaley, W.J. Elford, H. Fechtig, M.S. Hanner, J.A.M. McDonnell, J. Stohl, K. Tomita

I. INTRODUCTION

Following past practice, a number of authors have contributed reviews of their own field. The contributions have been edited by the President in order to avoid some overlaps, to reduce the length of the reviews an to add some publications. When possible reference numbers in the Astronomy and Astrophysics Abstracts are used. As editor, the President accepts responsibility for any shortcamings in the report.

The next Proceedings of a major meetings and books were published during the review period:

" Meteor Matter in the Interplanetary Space". Proc. of Symp. held at the Kazan University, Sept. 9-II, I980, ed. O.I.Belkovich, P.B.Babadzhanov, V.A.Bronshten, N.I.Suleymanov, Moskva, I982 (Russian).

"Asteroids, Comets, Meteors". Proc. of a meeting held at the Astron. Obs. of Uppsala University, June 20-22, I983, ed. C.-I.Lagerkvist and H.Riekman, Uppsala, Sweden, I984, (34.0I2.042).

"Aerosol in the Upper Atmosphere and Interplanetary Dust". V.N.Lebedinets, Gidrometeoizdat, Leningrad, I98I (Russian).

"Distribution of Meteor Bodies near the Earth's Orbit". Yu.I.Voloshchuk, B.L.Kashcheev, Nauka, Moskva, I98I (Russian), (30.003.I56).

"Results of Photographic Observations of Meteors on the Programms of the International Geophysikal Year, International Quiet Sun Year and International Active Sun Year. Catalogue". Materials of the World Data Center B. E.N.Kramer, I.S.Shestaka, Moskva, I982.

"Radio Meteor Investigations in Obninsk. Catalogue of orbits. Sept-Dec. I967". Materials of the World Data Center B. V.N.Lebedinets, V.N.Korpusov, A.V.Manokhina. Moskva, I98I.

"Radio Meteor Investigations in Obninsk. Catalogue of orbits. Jan.-Aug. I968". Materials of the World Data Center B. V.N.Lebedinets, V.N.Korpusov, A.V.Manokhina, Moskva, I982.

"Meteor Matter in the Earth's Atmosphere and the Circumsolar Cosmic Space". E.N.Kramer, I.S.Shestaka, Nauka, Moskva I983 (Russian).

" Physics of Meteoric Phenomena". V.A.Bronsten, Reidel, Holland, I983, (34.003.040).

"BMS Meteor Observer's Handbook. Vol.I. Naked Eye Work". R.A.Mackenzie, Dover, England, I983, (32.003.079).

It is with sorrow that we record the deth of members of the Commission 22, C.L.Hemenway, L.A.Katasev, A.Kizilirmak, E.L.Krinov, A.N.Simonenko.

II. PHOTOGRAPHIC AND TV METEORS P.B.Babadzhanov.

The Canadian MORP and Europian EN camera networks continue in operation during the review period. Systematic photographic observations are carrying out in the USSR by small cameras in Dushanbe, Kiev and Odessa. Geometric, dynamic, photometric and orbital data on all 29 fireballs photographed within the European Network in the year I977 have been published by Ceplecha et al (34.I04.006). In two cases meteorite impact points and areas are given.

Similar data on the Traunstein Fireball photographed at two Czech station of EN are given by Cplecha et al (33.I04.050). Multiple meteorite fall of 3 to 4 bodies of comparable mass close to I kg was predicted. Indirect data on fragmentation of the type-II-fireball body were presented and carbonaceous chondrite fall proposed. The determination of reliable orbits for meteors from extende camera networks have been reviewed by Halliday et al (33.I04.0I8). It appears that typical orbits of earth-crossing objects that deposed meteorites are direct orbits of low inclinations with perihelia between Venus and the Earth and with aphelia in the asteroid belt. In August I982 the bright fireball $-I3^{m}.8$ of the Upsilon Pegasid meteor shower was observed at five Czech stations of EN. The velocity of the meteor body near the earth is 5I.5 km s^{-1}, the perihelion distance 0.2 AU, and the inclination of the orbit 85° (Povenmire and Ceplecha, I983). Another bright fireball $-I9^{m}$ accompanied with sound phenomena was observed by the cameras of EN Oct 2, I98I (Rentdel, I982).

Orbits, trajectories and light curves of 44 meteors obtained in Dushanbe in I968-I977 have been published by Babadzhanov et al (33.I04.036). There are 33 Perseids, I Geminid, I Alpha Capricornid and 9 sporadic meteors. The instantaneous photographs with exposures 5.6 x IO^{-4}s permit to investigate the disruption process of meteoroids in the atmosphere, particularly, the disruption of a body into plenty of fragments. It was shown that in most cases the short meteor trails (L< 200 m) are caused by fragments with masses of IO^{-5} to IO^{-7}g while the long tails are caused relatively large fragments with masses of IO^{-3}g and more (Babadzhanov and Konovalova, I983, Babadzhanov et al , 33.I04.033, Babadzhanov, 34.I04.023). Results of photographic observations of meteors on programms IGY, IQSY, IASY are given in the book of Kramer and Shestaka (I982) in form of catalogue trajectories, light curves, geometric and physical characteristics of meteor flares, radiants and orbital elements of Perseids.

Photographic and visual observations of meteor showers have been carried out by amateur astronomers in Great Brittain. 24 photographs of Geminids were obtained in I980 (Spalding, 32.I04.002). A comparison of simultaneous photographic, radar and visual observations have realized by Czech amateur astronomers. Selections of the methods and real differences in distributions of the orbit elements for different masses of meteoroids have been found (Novakova, 32.0I3.076).

Kramer and Shestaka (30.I04.03I, 30.I04.040), Kramer et al (I983) have obtained new values of masses of meteor bodies from the photographic observations in Odessa and Kiév using the new luminous efficiency coefficient that led to increase masses of faint meteors (nearly by the order) and decrease masses of bright meteors. Estimations of terrestrial influx of the large meteoric bodies have been made by Halliday et al (I984), Baggaley (34.I04.003) and Kovshun (3I.I04.0I5). Seven doubly photographed meteors are shown by Drummond (32.I04.02I) to be members of the Delta Aurigids wich now appear to extend from Sept. 29 to Oct. I8. The mean orbital elements of the stream point to an unknown short period (II5 years) retrograd comet as parent. Andreev et al (I983) have derived the formulae for the accounting selectivity of photographic observations of meteors on the basis of geometrical conditions of observations and the physical theory.

The problem of automatisation of TV meteor observations have been discussed by Meleev et al (33.I04.026). Hapgood and Rothwell (3I.I04.023) have presented a study of meteor trails, which were imaged with low light level TV cameras simultaneously from two stations. They determined the characteristics of 98 meteor trails, 39 recorded in the South of England during the Perseid meteor showers of I977 and I978, and 59 recorded in

Northen Norway in Oct./Nov. I977.

III. RADAR METEORS O.I.Belkovich, C.S.Keay.

Analyses of physical processes in ionized meteor trails were given by Levitskij and Abdrakhmanov (30.I04.002), Levitskji (I983) and Kolomiets (I983). Kolmakov (33.I04.004)have proposed a new theory of the initial radius of a meteor trail. A method of determination of the initial radius in the characteristic height from forward scattering radio observations was given by Gajdaev et al (30.I04.0II). Baggaley has found from the measurements of line densities of non-shower meteors that there is a difference in ablation characteristics according to height, indicating the presence two distinct meteoroid components. The component consisting of meteoroids which ablate at height above I05 km appears to be associated with retrograde orbits. Baggaley (32.I04.007) has also found that reported diurnal variations in observed meteor mass distribution indices may be explained as a consequence of the characteristics of the height-magnitude distribution which is in turn dependent on meteoroidal structure and ablation characteristics. He also analysed irregularities in meteor trails (33.I04.0I0). The dependance of average height on meteor velocity from radar observations was found by Olejnikov (30.I04.027). A catalogue of ionization curves, masses and densities of 276 meteor bodies according to radar obsevations from 5 points is presented by Gartman and Chebotarev (33.I04.002).A comparison with results of meteor TV observations was given. Some parameters of meteor trails wre obtained by Chumak and Mojsja (I984) from two-frequency radar observations. Descriptions of meteor radar systems and methods of observations were given by Baggaley (34.I04.026), Malynyak (I982), Kashcheev and Nechitajlenko (I983), Nechitajlenko and Voloshchuk (I983) and Sakhibgareev (I983).

Effects of the resonance, polarisation and physical processes in meteor trails on characteristics of radio-echoes were discussed by Chumak and Moysja (I983), Soljanik and Tkachuk (I983), Gorovenko (I984), Novikov and Zygankov (I982) and Abdrakhmanov and Bajrachenko (33.I04.0I5). A study of diffractions of radio waves on meteor trails has been carried out by Kostylev and Kostylev (30.I04.007), Kostylev (3I.I04.009, I983). A number of authors have used the diffraction pictures for the determination of some characteristics of meteor trails (Kostylev, 32.I04.0II, I982, Andrianov, 3I.I04.0I0, Kolmakov, 33.I04.003). The problem of the meteoroid velocity measurement was discussed by Tkachuk (33.I04.006) and Andrianov and Sidorov (I983). The selektivity of meteor radars have been considered by Voloshchuk (I982) and Voloshchuk et al (32.I04.009).

Morton and Jones (3I.03I.502) have presented an entirely new method for obtaining maps of meteor radiant distributions obtained from a new microprocessor-based radar system. Another method of statistical analysis of density variations of radiants of sporadic meteors over the celestial sphere using an ordinary radar with a rotating aerial have been proposed by Pupyshev (30.I04.008). The maps of the radiant density over the whole celestial sphere obtained by the method have been published (Pupyshev et al, 30.I04.009). Pecina (3I.I04.002 and 003, I984) developed his method of the flux density determination from radar observations. He shows that the usual method of finding the mass distribution index from log N vs log T curve for overdence echoes can give incorrect results. On the contrary, Belkovich et al (I982a) show that the values of the mass exponent S obtained for the Geminid meteor shower by different methods are nearly the same. The problem also discussed by Bibarsov (I982), Kostylev and Svetashkova (I982), Sviridov and Kshcheev (33.I04.0I6), Voloshchuk and Kashcheev (I982) and Šimek (I984).

A number of papers have analysed results of radar meteor observations:

Malynyak (30.I04.026), Voloshchuk and Malynyak (30.I04.025), Voloshchuk and Nazarenko (34.I04.025), Bibarsov and Kolmakov (I984), Lebedinets and Manokhina (3I.I04.0I3).

Meteor wind determinations continue from several groups. Results of radar observations of meteor showers are given in the next section.

IV. METEOR SHOWERS B.A.Lindblad.

Observations of the major meteor showers have been carried out by visual, photographic, radar and television techniques. Data from long series of radar observations are being used to determine the cross-sections of meteor streams and the distributions of particles along a stream. Many of these studies are listed under the individual showers heading below. For a general discussion of meteor stream structure see Tkachuk (33.I04.0I7). For a historical account of meteor shower observations see Hughes (34.I04.036). Several studies of the dynamical effects associated with planetary perturbations on the orbits of meteor streams and associations have appeared in the period I982-84. See papers by Kramer and Timchenko-Ostroverkhova (29.I04.022), Sherbaum and Kazantsev (29.I04.0II), Babadzhanov and Obrubov (3I.I04.029 , 33.I04.020, 34.I04.0I5), Andreev and Babadzhanov (32.I04.0I2, I982), Zausaev and Galimova (33.I04.035), Williams and Fox (34.I04.0I8), Emelyanenko (I984), Kruchinenko (I984) and Kramer and Shestaka (I984). The secular perturbations of meteor orbits over very long time intervals have been studied by Galibina and Terentjeva (33.I04.007). The investigation details the mechanism producing North- South branches of meteor showers.

The theoretical radiant distribution expected for ecliptical meteor orbits is discussed by Štohl and Hajdukova (34.I04.037 and 049). the determination of telescopic meteor radiants from one station is investigated by Porubčan and Hajdukova (34.I04.046). A theoretical discussion of the zenith attraction correction for meteor radiants is presented by Andreev (33.I04.040). The dependance of shower hourly rates on zenith distance of the radiant is studied by Zvolankova (33.I04.0I3). Various geometrical factors affecting the detection probability of radiants in photographic and radar surveys are discussed by Tkacuk and Matsenko (30.I04.029). A comparison of some methods of the meteor flux density determination from radar observations of meteor showers have been carried out by Andreev et al (I984). Sidorov and Karpov (I984) and Andreev et al (3I.I04.0II) have used statistical methods for investigations of some flux structures. Lebedinets (I984) has discussed a possibility of meteor shower observations at the background of the zodiacal light. Statistical analisis of catalogues of orbits of the main meteor streams was carried out by Lebedinets and Manokhina (I982a). They show that the accuracy of the best of them is unsufficient for the investigation of the structure of streams.

The relation between comets and specific meteor streams - in particular the association of Comet Halley with Eta Aquarids and Orionids - is the subject of numerous studies. Predicted meteor radiants from earth-orbit approaching comets are listed by Drummond (30.I04.022, 3I.098.037, 3I.I04.0I7) and comparison is made with published lists of photographic and radar determined stream radiants. Predicted radiants from earth-orbit approaching asteroids are listed by Drummond (3I.098.037) and Babadzhanov and Obrubov (33.098.092, 34.098.063).

The detection of new meteor streams and/or unexpected recurrences of old streams depends almost exclusively on the efforts of amateur meteor observers. For references on amateur groups and observations see the previous report of Commission 22 and Sky and Telescope (I983, 66, 63). Visual observations by large supervised teams of amateur astronomers have been carried out in Czechoslovakia (32.104.006, 33.104.032), USSR (33.104.023, 34.104.008) and elsewhere. Observer's handbooks for visual and telescopic work have been published by Mackenzie (32.003.079, 34.104.029).

The relation of the visual magnitude scale to the color index of meteors is discussed by Hajdukova and Štohl (34.104.045). The observational errors in magnitude estimates of meteors by experienced visual observers have been discussed by Lindblad and Stohl (34.104.025).

Quadrantids. Mathematical models for the formation of the Quadrantid stream and studies of the nodal regression have been presented by Hughes et al (29.104.010), Murray (31.104.016), Fox (34.104.010) and Zausaev and Pushkarev (1984). Porubchan and Hajdukova (34.104.048) have studied the cross-section of the Quadrantid stream using Ottava radar observations 1963-68. McIntosh and Šimek (1984) have combined radar data from Ottava and Ondrejov to obtain a continuous twenty five year sequence. This study indicates that the activity curve of the shower is asimmetric with a very fast decay after shower maximum which occurs at solar longitude 282$^\circ\!.$6 (1950.0). Variations of the flux density and the mass exponent S for the Quadrantid stream are obtained by Belkovich et al (1982b, 1984) from radar observations at the Engelhardt Astronomical Observatory for the eight year period. The stability of the shower from year to year is found out. The standard of random variations of the maximum flux density do not exceed 10%. The activity curve of the Quadrantid stream is also discussed by Tkachuk (1983a), Svetashkova (1982), Andreev et al (1982), Isamutdinov (1983), Zausaev and Pushkarev (1984). Quadrantid observations are discussed in (31.104.022) and by Schippke (32.104.015). Visual Quadrantid recordings in 1971-80 are analyzed by Levina and Martynenko (33.104.008).

Lyrids. Visual recordings of Lyrids in 1945-52 from Skalnate Pleco observatory have been analysed by Porubčhan and Stohl (1983). The magnitude distribution of Lyrid meteors is studied and mean activity curve of the shower is derived. Maximum activity occurs at solar Longitude 31$^\circ\!.$5. Lyrid observations are also discussed in references (31.104.021) and (31.104.038). An exceptional Lyrid display was reported by visual observers in 1982 (Meteor News, No.58).

Eta Aquarids. The orbital evolution of Comet Halley and Eta Aquarid and Orionid streams under the influence of planetary perturbations is discussed in several papers. For references see Orionid stream.

From Ottava radar observations in 1958-67 Hajduk (30.104.014) has derived the activity curve of the Eta Aquarid shower. Hajduk and Buhagiar (32.104.008) have combined northern hemisphere radar observations and southern hemisphere visual observations to derive Eta Aquarid activity vs solar longitude.

Arietids. Radio observations in Tomsk in 1965-66 of the Perseids, Orionids, Ursids and daytime Arietids are analyzed by Svetashkova (33.104.029).

Delta Aquarids. This major southern hemisphere meteor shower has largely been neglected by visual and radar observers. It deserves detailes study. Observations of the Delta Aquarid shower by television technique are described by Sarna and Jones (28.104.038).

Perseids. Computer studies of the dynamics of the Perseid stream are presented by Fox (34.104.010). The stability of the node of the stream is discussed by Hughes and Emerson (31.104.008). A study of the Perseid stream structure and radiant motion based on a compilation of photographic data is presented in the monograph by Kramer and Shestaka (34.003.093). Onsala radar recording of the Perseid shower in 1956-78 are analyzed by Lindblad and Šimek (34.104.021). The activity curve of the shower is studied. Peak shower activity for long duration radar echoes occurred at solar longitude 139°.21 (1950.0). Doublestation TV observations of Perseids in 1978 are discussed by Hapgood and Rothwell (29.104.002, 31.104.023) and by Hapgood et al (32.104.013). Perseid beginning and end heights were obtained and the results were compared with dustball ablation theory.

The expected perihelion passage in 1980-84 of Comet 1862 III and its association with the Perseid stream is discussed in (34.103.261). Although no reports of the return of the parent comet have appeared(as of September 1984) high Perseid activity in 1980-82 has been reported. See Russel (31.104.024), Martynenko and Levina (32.104.005), Savrukhin (32.104.019) and Sky and Telescope (1981, 62, 624). A detailed analysis of the 1980 Perseid shower by visual, photographic and radar techniques is presented by Mason and Sharp (29.104.021). Observations in 1980 have been described by Eltry and Stomeo (31.104.007). Large scale visual observations in 1980 and 1982 have been reported by Martynenko et al (33.104.023, 34.104.008). Teams of visual observers spread out over 100° in geographical longitude were employed in 1982 to obtain 16 hrs of continuous recordings of Perseid activity around shower maximum.

Draconids. The orbit of this stream has been considerably perturbed by the major planets. Numerical modelling of the stream has been performed by Reznikov (30.104.018). The geometrical conditions for observing the strong showers of 1926, 1933, 1946 and 1952 are analyzed.

Taurids. The association between Comet Encke and the Taurids is studied by Katasev and Kulikova (33.104.022). Detailed computations of comet and stream orbits are reformed and the results are compared with original investigation by Whipple and Hamid. A similarity between the orbits of Comet Encke, the Taurids and certain asteroids is discussed by Napier (33.102.080). See also Galibina and Kastel' (33.098.004).

Orionids. The relation of the orbital parameters of the Orionid stream (and Eta Aquarid stream) to those of Comet Halley have been discussed by a number of authors. See papers by Babadzhanov and Obrubov (33.104.034), Kramer and Shestaka (33.104.054), Simonenko (33.104.009) and Hajduk (33.104.044, 34.104.020). McIntosh and Hajduk (34.104.030) present a new model of the stream based on the orbital history of the parent comet and historical accounts of Orionid showers. The total mass of the combined Eta Aquarid-Orionid stream is estimated by Hajduk (34.104.027) to be 5×10^{14} kg. Radar observations of the 1978 and 1979 returns of the Orionids have been reported by Hajduk and Cevolani (30.104.006) and Hajduk et al (1984). Radar observations in 1980 and 1981 are discussed by Jones (34.104.002). The activity curve and magnitude distribution of the Orionid stream is investigated by Stohl and Porubchan (30.104.013). The data base for this study is an extensive series of visual observations carried out at the Skalnate Pleso observatory in 1944-50. Observations of the Orionids also reported in (31.104.036).

Leonids. A detailed study of Comet Temple-Tuttle and the evolution of the Leonid meteor stream is presented by Yeomans (30.104.021). Geometrical conditions for observing an intence Leonid shower are discussed. The orbit of the stream is studied by Kondrat'eva (33.104.039) and Kondrat'eva and Reznikov (34.104.040). Prediction for future encounters of the Leonid stream with the Earth are given. Leonid radar observations in 1966-68 are reported by Andreev et al (29.104.026).

Geminids. Model calculations of stream evolution are presented by Kazantsev and Sherbaum (3I.I04.0I8, I982) and Fox et al (3I.I04.03I, 34.I04.03I). The evolution of the Geminid radiant under the influence of secular perturbations is investigated by Terent'eva and Galibina (34.I04.037). The spatial structure of the shower is analyzed by Pupyshev et al (30.I04.039). See also a review by Tkachuk (1983b). The activity of the stream is discussed by Jones (3i.I04.00I) and model calculations are compared with observational results.

The newly discovered asteroid I983 TB moves in an orbit resembling that of the Geminid meteor stream (34.I04.033).A genetic assosiation between the asteroid and the Geminid stream has been proposed (34.098.065, 34.098.089), Fox et al (I984).

Radar observations of the Geminid stream have been reported by Andreev et al (30.I04.032, 33.I04.028, I982a) and Jones and Morton (3I.I04.030), Isamutdinov and Chebotarev (I984). Variations of the meteor flux density and mass exponent S vs the solar longitude were obtained from radar observations by Belkovich et al (I982a). A stability of the flux curve for different years of observation is confirmed. The good agreement with the results of photographic and visual observations was ascertained. Radar observations at the stations in Ottava, Ondrejov and Dushanbe have been compiled and analysed by Šimek et al (32.I04.0I8). Visual, photographic and radar Geminid observations in I980 are summarized by Spalding (32.I04.002, I984).

Minor streams and possible new streams. Radar studies of a minor shower observed on I7-20 Nov. I967 are presented by Andreev et al (29.I04.027). Newly detected minor meteor showers are reported by Korpikiewicz (30.I04.038), Nesterov(32.I04.0I7), Macek (34.I04.043) and in reference (30.I04.04I). A moderately strong visual display in I982 of meteors associated with Comet Grigg-Skjellerup has been reported by Simmons (I982). In an analysis of visual and telescopic meteors observed in August Znojil (32.I04.006) identified a large number of minor meteor streams. Most of these occur in previous photographic and radio lists and they should therefore be considered as significant.

A possible new minor meteor shower the Upsilon Pegasids is discussed by Povenmire (3I.I04.005), Povenmire and Ceplecha (I983), McRobert (I984) and McLeod (I984). Possible meteor stream associated with Comet Lexell are discussed by Carusi et al (32.I03.83I, 34.I04.0II).

V. METEOR ORBITS

P.B.Babadzhanov.

A Meteor Data Centre had been set up by Dr. B.A.Lindblad at Lund observatory in Sweeden. The purpose of the Centre is to collect and store data obtained by photographic and radar methods on atmospheric trajectories and heliocentric orbits of meteors and fireballs. The Executive Committee of the IAU had allocated SwFr I800 per annum for the three calendar years I983-85 forvards the cost of maintaining the Centre.

Two catalogues of radar meteor orbits (Lebedinets et al, I98I,I982) and two catalogues of orbital elements of bright meteors (Kramer and Shestaka, I982, Babadzhanov et al, 33.I04.036) were published during the review period. The problem of the reliability of orbital catalogues was discussed by Lebedinets and Manokhina (I982b).

A number of papers were devoted to the problem of the "cosmic weight". Kramer and Shestaka (I983) have found the probability of the detection of meteor streams that have the certain values of the orbital elements. Modifications of the Öpik's formula were done by Bronshten (I983a, I983b). Belkovich (34.I04.0I3) has presented an exact solution of the problem both for the set of orbital elements and for the transformation of the flux of sporadic meteors in the gravitation field of a moving body. The influence of the form of the cosmic weight formula on the distributions of the orbital elements of the sporadic meteors has been treated by Lebedinets and Manokhina (I982c, 3I.I04.025). Andreev V.V. et al (I982) and Stohl (34.I04.0I9) have analysed the distribution of sporadic meteor orbits in the solar system.

Physical processes affecting the motions of the meteor bodies in the solar system have been discussed by Mendis (30.09I.066), Tokhtasjev (I982a), Mukai and Yamamoto (3I.I06.009), Grimshaw (32.I06.026), Kapisinsky (33.I06.057) and Williams (34.I04.009).

Work related to the modelling of the orbital evolution of meteor streams has been carried out by Andreev and Sukhotin (I982), Babadzhanov et al (I982), Kulikova (I982), Kazantsev and Sherbaum (30.I04.005), Zausaev and Galimova (33.I04.035). Babadzhanov and Obrubov have shown that a number of features of meteor showers is explaned by joint influence of planetary perturbations, the Pointing-Robertson effect and its corpuscular analogue, light pressure and ejection velocities of different mass meteoroids from cometary nuclei. The dynamical evolution of meteor particles ejected from Comet Lexell was computed by Carusi et al (34.I04.0II) to investigate the propeties of that part of the swarm wich remained in earth-crossing orbits.

The problem of hyperbolic orbits was discussed by Matsenko and Tkachuk (I982), Kazantsev and Sherbaum (32.I04.003) and Tkachuk et al (33.I04.0I4). Studies interrelations among asteroids, comets and meteors by Levin (I982), Simonenko and Levin (34.098.036) have shown that most promising to regard the asteroid belt as the primary supplier of Apollo and Amor type objects. Their numerous fragments of all sizes can produce meteorites. It is possible that the asteroid belt is the supplier of short period meteor streams. The method of search of objects geneticaly related to Apollo, Aten and Amor asteroids was proposed by Simonenko and Terentjeva (I984). It was based on the principle of the spatial comparison of the orbits with taking into account a similarity of physical or chemical parameters of objects.

VI. PHYSICAL THEORY OF METEORS AND FIREBALLS D.O.ReVelle.

In this section both recent advances in physical modelling and related observations of effects produced by meteors and fireballs are considered.

A major contribution to the literature of meteors, fireballs and meteorites is the publication in English of "The Physics of Meteoric Phenomena" by Bronshten (34.003.040). This treatment of the subject is an attempt to meaningfully present the current state of the art and its evolution using the latest scientific advancements in the fields of radiation gas dynamics and plasma physic especially. All things considered, Bronshten has written an exellent monograph for the training of the next generation of researchers in this field.

Work related to meteor modelling has been carried out by Bronshten (29.I04.028, 34.I04.039), Kalenichenko (32.I04.0I0, 33.I04.02I, I982), Apshtein, Pilyugin and Vartanyan (32.I04.0I4), Hapgood, Rothwell and Royrvik (32.io@.0I3), Hawkes, Jones and Ceplecha (I984), Kovshun (33.I04.025, I984), Smirnov and Kovshun (I982), Blokhin and Novikov (I982), Benyukh (32.I04.004) and Smirnov (34.I04.038).

Bronshten's paper used elastic interaction theory combined with fireball observations to deduce the tangential momentum accomodation factor. The range of values determined compare well with values obtained by other methods. The method used can also be applied to meteor data, but for both type of data, the theory is only suitable near the beginning of the trajectory, i.e., for conditions of weak shielding (near-free molecular flow). In a series of papers Kalenichenko considered a quasicontinuous fragmentation process of meteoroid destruction during entry to the atmosphere. Using this model he identified the modes of ablation of meteors using both light curve and velocity data. Apshtein et al applied their earlier model which includes shape change to calculate the mass removal rate, body temperature, light curve and electron line density in meteor trail. These results have been compared with the simple physical theory of meteors and important differences were found under certain conditions. Hapgood et al compared the dust-ball meteor model of Hawkes and Jones with two-station TV observations of Perseid meteors. From their results the observed Perseids were deduced to be even weaker structurally than that predicted using the model Hawkes and Jones. In the paper by Hawkes, Jones and Ceplecha theoretical beginning heights are calculated considering simultaneously the effects of thermal conduction into the meteoroid, thermal radiation emitted by the body and deceleration. It should be noted that as far back as I974 the physical model of the ablation of meteor bodies in mass range of IO^{-6} - I g was developed by Tokhtasjev (Belkovich and Tokhtasjev, I2.I04.05I) taking into account the same effects. The model have been successfully applied for decade for the interpretation of meteor shower observations (Belkovich et al, I982a, I982b) and for the study of densities of meteoroids in meteor showers (Tokhtasjev, I982b). The results obtained by Hapgood et al (32.I04.0I3) confirm this model completely. Finally, Kovshun demonstrated that his model B of the light curve produced better agreement with the observations for the portion of the trail at altitudes above the maximum magnitude altitude than did the earlier predictions using Model A. The problem of fragmentation of meteor bodies was discussed by Babadzhanov et al (I982), Novikov et al (I984). Obrubov and Konovalova (33.I05.229) considered the effect of the rotation of a meteor body in the Earth's atmosphere.

Work related to fireball modelling has been carried out by Wetherill and ReVelle (30.I04.042, 32.I02.0I9), Lebedinets (29.I04.024), Korobeinikov, Chushkin and Shurshalov (33.I05.238), Zelenin, Konstantinov, Mikheev and Salimov (33.I05.I66), Sekanina (34.I05.0I0), Pecina and Ceplecha (33.I04.0I2, I984), Halliday, Griffin and Blackwell (33.I04.0I8), Ceplecha (33.I04.043, I983), Babadzhanov (33.I04.042), ReVelle (33.I04.047, I983), Padavet (34.I04.032), Bronshten (29.I04.025, 34.I04.0I6, 34.I04.039), Kalenichenko (I984). Work related to fireball modelling has also been carried out by Tiwari and Subramanyan (I980).

The papers by Wetherill and ReVelle adress the problem of identifying meteoritic and "cometary" types of fireballs respectively. The theoretical single-body approach is used to estebilish four relative criteria which are applied to the US Prairie Network fireball observations using data from the Lost City and Innisfree Meteorite falls as a calibration. It is deduced up to one third of the fireballs are of the meteoritic type. Also, "cometary" fireballs have a much wider range of structural properties than the meteoritic material studied. These results are very similar to those deduced earlier by Ceplecha and McCrosky methods. Lebedinets considered a model of quasi-continuous fragmentation of meteorites during atmospheric entry. Unlike the work of Kalenichenko (see above), Lebedinets considered a constant value of coefficient of heat transfer, etc. On the other hand, Lebedinets developed his theory using the classical approach as modified by Jaccia in the I950's, using the progressive fragmentation parameter. Both workers reached similar conclusions, however, namely that the low density friable meteoroid hopothesis is not necessary to explain the behavior of photographic meteors and that many observed meteors can be modelled assuming ablation occured predominantly in the solid phase, i.e. as a quasicontinuous fragmentation process.

Two papers have appeared in an attempt to better understand the Tunguska (1908) event. Korobeinikov et al considered the effects of a nonzero lift to drag ratio on the trajectory of the fireball. Also the visible and infrared radiation energy losses were computed as a function of the horisontal range of the shock wave from the explosion center. Sekanina in a comprehensive review of all reliable "data" through 1983 concluded that Tunguska could not have been composed of low density material typical of a comet. Insted he concluded that Tunguska was probably a small Apollo-type asteroid. Its behavior in the atmosphere consistent with that of a type II fireball using the notation of Ceplecha and McCrosky.

Pecina and Ceplecha have shown the impotance of atmospheric variables in deducing meteoroid parameters in two impotant papers. Their work has shown that the form of deceleration assumed previousli in deducing values from observations is not a good approximation for fireballs. Using single-body theory they have deduced an expression for distance along the meteor trajectory as a function of time which can be used to reliably retrieve ablation coefficients, etc, if the instantaneous atmospheric density profile is used. Halliday et al have used the results of Wetherill and ReVelle and data from the MORP and EN networks to identify meteoritic fireballs from data from all three fireball networks. They considered the orbital parameters of the fireballs and concluded that typical meteoritic orbits are direct and of low inclination with perihelia between Venus and the Earth and with aphelia in the asteroid belt. Ceplecha recomputed bulk density, ablation parameter, etc, for group I and II fireballs after correcting for the effects of using a seasonal air density profile rather than simply using a standard atmosphere profile. Ablation coefficients (mean values) of Group I fireballs now agree quite well with the ablation modelling of ReVelle. Ceplecha also presented a review paper summarizing the connections that are currently recognized between meteoroids and meteorites. Babadzhanov analysed fireball data taken using the method of instantaneous exposures in order to study fireball fragmentation processes. This technique is one that has great potential for developing a better understanding of both the fragmentation process and of reliably determining fireball compositions from photographic data. ReVelle considered the modelling of the light production of fireballs in paper that has yet to appear. Modifications of the classical theory are needed at both very high and very low altitudes in order to obtain reasonable agreement between theori and observations. In a separate paper ReVelle formulated a model of porous meteoroid completing a model developed earlier by Liu. Application of the model to the US Prairie Network observations revealed very similar mean ablation coefficients and bulk densities as compared to those obtained previously by Ceplecha and McCrosky for the three fireball groups. Padavet developed a theory for fireballs in the continuum flow regime but relied on perfect gas relations among the atmospheric variables behind the bow shock wave. The author also hypothesizes that the fireball groups identified by Ceplecha and McCrosky correspond only to the known meteorite types. Bronshten has considered the problem relating the green atomic oxigen line emission in meteor spectra and the meteoric head echo phenomena observed using radar. This topic is also briefly considered in his recent monograph. He has also considered the problem of explaining electrophonic noises from very bright fireballs in a separate paper. A theoretical mechanism is proposed and tested which relies on the action of magnetohydrodynamic processes along the trail as suggested qualitatively in 1980 by Keay. The first communication on the instrumental detection of electromagnetic waves from meteors has come from Epifanova (32.082.080). She has observed radiopulses produced by faint meteors at the wavelength 3 cm. Classification of meteor acustics was given by Clarence (31.105.014).

Finally the work of Tivary and Subramanian considered the effects of shape change on the continuum flow field during Jovian entry conditions. Parameters such as shock-standoff distance, radiative heating rate, etc can be cignificatly changed by the shape change of the body.

Consistent with the previous report we now consider new areas research which rely heavily on the results from improved physical modelling of meteors, fireballs and meteorites:

a) The problem of understanding the Moon and Mars as source regions for meteorites found on the Earth. Specifically the case for the SNC meteorites being from Mars is considered in detail by Wetherill (I984);

b) Entry of meteors into the atmosphere of a comet. This is considered for example in Dobrovolskij and Ibadov (33.I04.005);

c) Entry of meteors into the atmospheres of Mars and Venus and the inverse problem of remotely sensing these planetary atmospheres using meteors as in situ probes. The latter is considered in the works of Levin and Simonenko (I982) and Apshtein, Pilugin and Vartanyan (32.I04.0I4); and

d) Entry of Sun grazing comets into the solar atmosphere. Weissman (I983) has considered the early stages of the latter problem. The question of the nature of the cometary nucleus and the relation of comets to carbonaceous chondrites and the various fireball groups has a direct bearing on the solution of the problem.

VII. METEORS AND AERONOMY — W.J.Baggaley.

Work has continued on the relationship of meteor products to noctilucent clouds, nacreous clouds and stratospheric aerosols.

A comprehensive review (30.082.025) of the meteoritic influence on clouds and aerosols included reviews of observations, models of the various channels followed by meteoric species - micrometeorites, smoke particles (2 - I0 Å size) and vapours (considering gas phase reactions and absorption into meteoric dust). Scenarious were presented of the nucleation growth to the particle sizes observed in mesopause (83 km) noctilucent clouds. Particle orientation, ice crystal form, and sedimentation times were considered. The interaction of meteoric products with stratospheric aerosols was also reviewed. A model of ice crystal nucleation growth in noctilucent clouds and the role of metal ion products has also been given (30.082.046).

In an account of the current status of noctilucent cloud theories (32.082.067) the authors presented the first numerical scheme to couple together particle nucleation, condensation, evaporation, coagulation, gravitational coalescence, sedimentation and diffusion with ion-electron processes. The comprehensive study considered ice crystals of various shapes and concluded that meteoric dust acts as effective nucleation centres with large water cluster ions (with cores of meteoric species) being effective for mesopause less than I30 K.

A later report (33.093.0I6) compared the role of meteoric debris as a nucleation agent in the Venus and Earth atmospheres, while mass spectrometer measurements of condensation nuclei have been made by Björn and Arnold (I98I).

For a more definitive model of meteoric species in noctilucent clouds, work is needed on rocket sampling of the particles size distribution and composition during different stages of cloud formation.

Several studies (for example Kirchhof, Clemeska and Simonich, 30.082.0I3, Kirchhof and Clemeska, I983a, Kurbanmuradov and Mukhamednazarov, I982) have been devoted to the aeronomy of meteoric sodium in the lower thermosphere. Models (Kirchhof and Clemeska, I983b) of time variations in the 90 km sodium layers calculated the dinamical and chemical lifetimes of Na atoms and response to changes in the meteor influx rate.

Using the two dimentional model employing mass spectrometer measurements of Si^+ and $SiOH^+$, H_2O mixing ratios have been inferred (Soloman, Fergusson, Fahey and Crutzen, I982). The authors find that Si^+ loss is dominated by O_2 association rather than as a result of H_2O reaction and also discuss the question of the extraterrestrial H_2O sourse in meteoric dust. A series of positive ion mass spectrometer measurements (Kopp and Herman, I984) detailed the height distributions of proton hydrates, NO^+ cluster ions, meteoric metals Fe^+, Mg^+, Al^+, Na^+, as well as the usual thermospheric components O_2^+ and NO^+. The authors consider the role of atomic oxygen in controlling the transition layer and the chemistry of meteoric metals. Comparisons between mass spectrometer measurements of meteoric ions with lidar Na observations and resonance line estimates of Fe and Mg atoms concentrations have been used by Swider (I984) to model the variations of meteoric ion layer during aurorae in terms of charge transfer from auroral produced O_2^+ and NO^+. Measurement of phase path changes and absorption using I6 kHz and 60 kHz one hop links have been used by Rumi (I982) to determine the influence of increased meteor activity on the nocturnal D region. Enhanced meteoric ionization deposited during the Geminid shower yield changes in the LF radio path which permitted the determination of the electron density and collision frequency and height of the ionization ledge at about 83 km. The results were compared with models of ionic processes affecting deposited meteoric ionization.

Although rocket measurements reveal that sporadic-E ionization irregularities (90 - II5 km) are composed mainly of positive ions of meteoric origin, inconsistent conclusions have been found in attempts to associate E_s events and enhanced meteor influx. Using nearly four decades of E_s data and four years of radio-meteor rates for two southern hemisphere stations the seasonal characteristics of both phenomena were examined at different diurnal intervals. No significant changes in medium intensity E_s occurence were found to result from major meteor shower ionization influx (Baggaley, I984a, I984b).

The influence of height variations of atmospheric density scaleheights on the interpretation of fireball trajectories has been considered by Pecina and Ceplecha (I984). Radar meteor rate have been found to vary with solar activity (Prikryl, 33.I04.0II), while meteor heights have been found to depend on atmospheric changes near sunrise (Porubčan and Cevolany, 34.I04.007, Porubčan et al, 34.I04.024). The relationship between visual meteor magnitude and meteor trail ionization emploing aeronomical models has been determined in a simultaneous visual-radio programme by Znojil et al (I982).

The dependence of the ambipolar diffusion coefficient of a meteor trail on the geomagnetic field has been studied theoretically and experimentally by Abdrakhmanov (33.I04.027) and Levitskij et al (I982).

VIII. TEKTITES

B.P.Glass.

Stauffer (33.I05.I92) has researched tektite localities in Indonesian Borneo and he concluded that the only authentic localities are at Martapura Pelaihari and the Sungei (river) Riam Kanan which are all within 40 km of each other in the southeastern corner of the island. O'Keefe (I983a) discussed new tektite localities in the Czechoslovakian strewn field and finds that the strewn field trends SSE to NNW. Han-chang et al (I982) reported three new deep-sea microtektite locations in the western equatorial Pacific Ocean which they suggested may be from a previously unidentified strewn field (or strewn fields). However, the locations are all within or very close to the boundary of the Australasian strewn field as defined by Glass (1982), and the composition they give is similar to the composition of Australasian microtektites.

The question of the origin of tektites has long been a subject of debate. Although most researches support a terrestrial impact origin for tektites, there are some who favor a lunar volcanic origin. O'Keefe (1983b) is the major proponent of a lunar volcanic origin. Yuan (1981) also feels that tektites must have a volcanic rather than impact origin; and since the tektites do not have compositions similar to local volcanic matarial, he supports the lunar volcanic origin for tektites. Rost (1981) pointed out that the irghizites were formed by accretion of numerous fibres, rags and tiny spherules. Heide et al (31.105.163) compared the surface structures of irghizites with volcanic ejecta (pyroclastic material). Bouška et al (30.105.028) concluded that the irghizites were formed by melting of sedimentary rocks at the Zhamanshin area and that they were enriched in less mobile elements of the impacting meteorite. According to Florensky and Dikov (1981) the conditions necessary for tektite formation include: very high temperature, low pressure and short time. They state that only impact can produce those conditions and they propose that tektites are formed by the fall of a giant meteorite into silica-rich rocks. Additional evidence for the impact origin of tektites comes from the discovery of an iridium anomaly associated with the North American microtektite layer (Asaro et al, 1982, Ganapathy, 1982, 31.105.216, Alvarez et al, 1982a, 1982b). A study by Glass et al (1982) shows that the late Eocene iridium anomaly is actualy associated with a layer of crystal-bearing glass spherules that occurs below the normal North American microtektites. However, the authors argued that the crystal-bearing spherules were formed at the same time.

An earlier suggestion that Elgygytgyn crater in northern Siberia might be the sourse for the Australasian strewn field appears to be ruled out because of its old age (Komarov et al, 1983). Shaw and Wasserburg (1982) found that their Sm-Nd and Rb-Sr isotopic data are consistent with the Ivory Coast tektites being derived from the Bosumtwi crater in Ghana. They also found that the Sm-Nd and Rb-Sr systematic support the hypothesis that the Czechoslovakian tektites were derived from the Ries crater in Germany. Hörz (1982) argued that the Czechoslovakian tektites were derived from Tertiary freshwater sediments that capped the Ries area prior to the impact that formed the crater. The source crater for the North American tektites has still not been found.

Some attempts have been made to identify the impacting body that produced the tektites. Bouska et al (30.105.028) belive that the impacting body that produced the Zhamanshin crater and associated irghizites was probably a chondrite. Glass et al (1983) concurred with that conclusion. Based on ratios of siderophile elements, Koeberl and Kiesl (1983) concluded that Czechoslovakian tektites (moldavites), like the Ries crater, were contaminated by an achondrite and that the Australasian tektites were contaminated by a chondrite projectile. The North American tektites may have been formed by the impact of an undifferentiated meteorite (Ganapathy, 1982, 31.105.216).

One of the most interesting aspects of tektite studies is the possibility that the impact events that produced the tektites may have also been responsible for climatic and biological changes as well (Glass, 1982). The North American microtektite layer appears to be associated with the extinction of several species of one-celled marine organisms (Glass, 1982). Alvarez et al (1982a) pointed out that the North American tektite event is also within 4 m.y. of the mass extinction of terrestrial mammals. Ganapathy

(I982, 3I.I05.2I6) suggested that a massive ($>$ 3 km, $>$50 billion tons), chemically undifferentiated meteorite collided with the Earth $\sim$34 m.y. ago producing the North American tektites and microtektites and causing worldwide extinctions.

IX. INTERPLANETARY DUST - DISTRIBUTION AND DYNAMICS H.Fechtig.

Grün et al (I984) have complited a study of the dynamics of dust grains within I AU sun distance. It has been shown that the collisions of small dust particles ($<$ IO^{-6}g) with larger ones ($>$ IO^{-6}g) play a major role. The question now to maintain the spatial distribution of interplanetari dust has been investigated by Leinert et al (33.I06.004). The authors come to the conclusion that from the primary sourse of solid particles a reservoir of radio meteors is preserved. A continuous input of $\geq$ 200 kg/s to the interplanetary dust cloud is created by catastrophic collisions.

Bishop and Searl (33.I04.022) and Haranyi and Kecskemety (34.I02.030) have considered analytically the possibiliry of power-law mass distributions of meteoroids as asymptotic solutions to the rate equations govering both accretion and fragmentation of particles that interact only on collision. The influence of the Lorentz force on the spatial distribution of the interplanetary dust grains has been investigated by Mukai and Giese (I984). Particularly the orbital inclinations are influenced by electromagnetic perturbation. As a result, the authors could explain the dependance of heliocentric distance r as $n(r) \propto r^{-I.3}$ on the basis of the influence of the Pointing-Robertson effect and the Lorentz force.

Zook et al (I984) have showed that impacts at low impact angles (I5°) produce up to 2 orders of magnitudes more ejecta than normal impacts. As a result one has to expect many secondary impact craters in the submicron-sized crater range on lunar samples which can not be differentiated from primary impact craters. The estimations of the flux density of interplanetary bodies of different masses were made by Kresak (30.098.I05), Zacharov et al (30.I04.003) and Lebedinets (I98I). The nature, origin and evolution of interplanetary dust grains have been discussed by Lamy (33.I06.029).

Progress has been made in another aspect: the structural and dynamical change of interplanetary dust grains with time. The Helios Dust Experiment (Grün et al, 27.I06.0I6) has registered two types of dust grains: compact dust grains orbiting the sun on ellipses with low eccentricities $e < 0.6$ and fluffy ones orbiting the sun on ellipses with high eccentricities $e > 0.6$. One has to consider the comets as the only significant source. According to Greenberg (I983, I984) the cometary particles are conglomerates of core/mantle particles. The cores are believed to be silicates as were directly observed from infrared observations by Ney (32.I02.020). The mantles consist of the light elements H, C, N, O and, after being photoprocessed, from more or less complex organic molecules as shown by laboratory simulation experiments. Just after release from a comet nuclei the organic mantles may be coated by ices. Conglomerates of these building blocks are the cometary particles and one would like to call them "Greenberg-particles". Particles larger than I μm in diameter orbit the sun first on the comet's trajectory. That means, however, that they warm up - as the comet nucleus itself - during the perihelium passages and they cool down during aphelium. The corpuscular irradiation sputters the mantle material and above a certain temperature limit the organic mantles may slowly sublime. Mukai and Fechtig (33.022.II4) have calculated the situation for a Comet Halley dust grains and found that with the mass loss of approximately IO^{-13}g/cm^2 s all mantle material is lost in a time comparable to the time needed to change the high eccentric (cometary) orbit into a quasicircular orbit due to the Pointing-Robertson effect. By the loss of the mantle material the Greenberg-particles of rathe low density (< I g/cm^3) change into Brownlee-particles of higher densities (~ I g/cm^3) and, possibly, by rearranging the naked cores into compact grains of normal densities (~ 3 g/cm^3). And this is exactly what was observed by the Helios Dust Experyment. The very similar result was obtained independently by Tokhtasjev (I982b, I984) from radar and photographic observations of meteor streams. He has found the dependence of meteoroid densities on their orbital elements.

Considering the existance of these Greenberg-particles leads us to anothe interesting consequence: organic mantles become black when exposed to high energetic corpuscular irradiation (Johnson, I984). This means the albedos ofthe Greenberg-particles are considerably lower than the albedos of silicates. This could possibly explain the discripancy of the results of the Pioner IO and II dust experiment: the penetration experiment sees constant fluxes between I and 20 AU (Humes, 28 I06.079) while the zodiacal light experiment registers a decreasing dust frequency between I and 3.3 AU sun distance but no scattering sun light beyond 3.3 AU (Hanner et al, I7.I06.056). If the albedos of the Greenberg-particles are sufficiently low, the optical experiment may only see the Brownlee-particles and all compact particles. In the picture given above, the Greenberg-particles are the youngest ones if one counts the time after release from comet nuclei. These young particles orbit on highly eccentric ellipses, as analyzed by Humes (22.I06.079). Most of the time they stay outside the inner solar system. Only during the fast perihelium passages they are within I AU sun distance. The older the particles, the higher their albedos: first as Greenberg-particles, later as Brownlee-particles and finally as compact grains. The eccentricities of their orbits become smaller and smaller, meaning that the old particles finally always stay within the inner solar system (Fechtig,I984).

The current dust experiments, using space shuttle techniques and particularly the Long Duration Exposure Facility of NASA are primarily of interest for the chemical analysis of dust. Dynamical aspects can only play a minor role since the experiments are passive. However, impact craters, as recently recorded on a shuttle experiment (McDonnell et al, I984) and similar experiments of LDEF may give detailed insight into the near Earth dust inviroment. In particular, the space debris dynamics is of interest because artifical dust is expected to stay in Earth orbit. As recently discussed on a COSPAR Workshop on Space Debris (Graz,I984) collisional fragments are continuously produced and build up an Earth's dust belt in the future. The natural dust belts around the Earth were found from the new processing of the meteor experiment data carried out on the space probes "Electron -I" and "Electron-3" (Barsukov and Nazarova, I983). The theory of the dust belt formation have been developed by Gulak (34.I04.004, I983).

X. INTERPLANETARY DUST - PHYSICAL CHARACTERISTICS AND SOURSES

Hughes (34.I02.028) has presented a brief review of our knowledge of the source and characteristics of cometary dust. He has considered that all particles responsible for meteor showers, Brownlee-particles in the stratosphere etc originate in the cometary nucleus. Singer and Stanley (33.I04.024) have thought that submicron particles observed with Explorer 46 in I974-75 and affilated with meteor streams to be produced by comets during perihelion passage. On the other hand Parkin et al (34.I05.057) have argued that rarely occuring white stony cosmic spherules in deep-sea sediments together with ordinary black spherules originate in asteroidal collisions. Evidence was adduced by Greenberg (34.I02.029) to show that comets and carbonaceous meteorites have a common origin via aggregation of interstellar dust. A possible relation of the interplanetary dust with comets and presolar interstellar particles discussed by Brownlee (I982).

Samples of interplanetary dust have been succesfully collected for laboratory study from the stratosphere and from the sea floor. Elemental and isotopic abundances of He, Ne and Ar suggest a space exposure of small particles in the inner solar system for at least tens of years (Fraundorf et al, 32.106.017). Unique crystal morphologies and microstructures of enstatite whiskers and platelets observed in dust particles strongly suggest that they are primary vapor phase condensates which could have formed either in the solar nebula or in prosolar inviroment (Bradley et al, 33.106.013). Dust agregates are black, fine grained material and exhibit significant differences in composition, morphology and mineralogy (Fraundorf, 29.106.022, Brownlee, 30.106.053, Papanastassion et al, 34.106.026).

XI. METEOR SPECTROSCOPY

A summary of the major results of the meteor spectroscopy in the recent years was given in the review papers by Millman (33.104.019) and Bronshten (1982). Other results on the topic see Russel (29.104.005), Babadzhanov and Getman (1982) and Smirnov (31.104.014).

References

Alvarez, W., Asaro, F., Michel, H.V. and Alvarez, L.W.: 1982a, Science, 216, p.886.
Alvarez, W., Asaro, F., Michel, H.V. and Alvarez, L.W.: 1982b, Geol. Soc. Am. Abstracts, 14, p.431.
Andreev, G.V. and Babadzhanov, P.B.: 1982, MMIS, p.261.
Andreev, G.V. and Ryabova, G.O.: 1982, MMIS, p.129.
Andreev, G.V. and Sukhotin, A.A.: 1982, MMIS, p.175.
Andreev, G.V., Lazarev, R.G., Rubtsov, L.N. and Ryabova, G.O.: 1982, MMIS, p.118.
Andreev, G.V., Belkovich, O.I., Ryabova, G.O. and Svetashkova, N.T.: 1984, Astron. Vestn. 18, p.74.
Andreev, V.V., Belkovich, O.I. and Zabolotnikov, V.S.: 1982, MMIS, p.8.
Andreev, V.V., Belkovich, O.I. and Tokhtasjev, V.S.: 1983, Astron. Vestn. 17, p.244.
Andrianov, N.S. and Sidorov, V.V.: 1983, Meteorn. Rasprostr. Radiovoln, 18, p.25.
Asaro, F., Alvarez, L.W., Alvarez, W. and Michel, H.V.: 1982, Geol. Soc. Am. Spec. Paper, 190, p.517.
Babadzhanov, P.B., Bibarsov, R.Sh., Narziev, M. and Chebotarev, R.P.: 1982, MMIS, p.190.
Babadzhanov, P.B. and Getman, V.S.: 1982, MMIS, p.225.
Babadzhanov, P.B., Zausaev, A.F. and Obrubov, Yu.V.: 1982, MMIS, p.131.
Babadzhanov, P.B. and Konovalova, N.A.: 1983, Doklady Akad. Nauk Tadzh. SSR, 26, p.494.
Baggaley, W.J.: 1984a, Bull. Astron. Inst. Czech. in press.
Baggaley, W.J.: 1984b, Planet. Space Sci. in press.
Barsukov, V.L. and Nazarova, T.N.: 1983, Astron. Vestn. 17, p.238.
Belkovich, O.I., Sulejmanov, N.I., and Tokhtasjev, V.S.: 1982a, MMIS, p.88.
Belkovich, O.I., Sulejmanov, N.I. and Tokhtasjev, V.S.: 1982b, MMIS, p.121.
Belkovich, O.I., Sulejmanov, N.I. and Tokhtasjev, V.S.: 1984, Bull. Astron. Inst. Czech. 35, p.123.
Bibarsov, R.Sh., and Chebotarev, R.P.: 1982, MMIS, p.63.
Bibarsov, R.Sh. and Kolmakov, V.M.: 1984, Astron. Vestn. 18, p.71.
Björn, L.G. and Arnold, F.: 1981, Geophys. Res. Lett. 8, p.1167.
Blokhin, V.A. and Novikov, G.G.: 1982, MMIS, p.236.
Bronshten, V.A.: 1982, MMIS, p.208.
Bronshten, V.A.: 1983a, Astron. Vestn. 17, p.175.
Bronshten, V.A.: 1983b, Astron. Tsirkular, No. 1266, p.5.
Ceplecha, Z.: 1983, Meteoritics, 18, p.278.
Chumak, Yu.V. and Mojsya, R.I.: 1983, Vestn. Kiev. Univ., Phys. 24, p.23.

Chumak, Yu.V. and Mojsya, R.I.: 1984, Meteornye Issledovaniya, Kiev, 9, p.69.
Emeljanenko, V.V.: 1984, Pis'ma v Astron. Zhurnal, 10, p.315.
Fechtig, H.: 1984, XXV COSPAR Meeting Graz.
Florensky, P.V. and Dikov, Yu.P.: 1981, Geochemistry International, 18, p.92.
Fox, K., Williams, I.P. and Hughes, D.W.: 1984, MNRAS, 208, p.1155.
Ganapathy, R.: 1982, Geol. Soc. Am. Spec. Paper, 190, p.513.
Glass, B.P.: 1982, Geol. Soc. Am. Spec. Paper, 190, p.251.
Glass, B.P., Dubois, D.L. and Ganapathy, R.: 1982, J. Geophys. Res. 87, p.A425.
Glass, B.P., Fredriksson, K. and Florensky, P.V.: 1983. J.Geophys.Res.88, p.B319.
Gorovenko, A.P.: 1984, Komety i Meteory, 35, p.47.
Greenberg, J.M.: 1984, Scientific American, June 1984, p.96.
Grün, E., Zook, H.A. and Giese, R.H.: 1984, Icarus, in press.
Gulak Yu.K.: 1982, Astron. Tsirk. No.1273, p.2.
Gulak, Yu.K.: 1983, Astron. Vestn. 17, p.232.
Hajduk, A., Cevolani, G., Formiggini, C., Babadzhanov, P.B. and Chebotarev, R.P.: 1984, Bull. Astron. Inst. Czech. 35, p.1.
Halliday, I., Blackwell, A.T. and Griffin, A.A.: 1984, Science, 223, p.1405.
Han-chang, P., Kui-huan, Z. and Sui-tian, C.: 1982, J.Geophys.Res. 87, p.5563.
Hawkes, R.L., Jones, J. and Ceplecha, Z.: 1984, Bull.Astron.Inst.Czech.35, p.46.
Hörz, F.: 1982, Geol. Soc. Am. Spec. Paper, 190, p.39.
Isamutdinov, Sh.O.: 1983, Bull.Astrophys.Inst.Akad.Nauk Tdzh.SSR, 74, p.20.
Isamutdinov, Sh.O. and Chebotarev, R.P.: 1984, Komety i Meteory, 35, p.35.
Johnson, R.E.: 1984, XXV COSPAR Meeting Graz.
Kalenichenko, V.V.: 1982, MMIS, p.199.
Kalenichenko, V.V.: 1984, Astron. Vestn. 18. p.151.
Karpov, A.V., Sidorov, V.V. and Stepanov, A.M.: 1984, Astron. Vestn. 18, p.44.
Kashcheev, B.L. and Nechitajlenko, V.A.: 1983, Meteor Res. 8, p.5.
Kazantsev, A.M. and Sherbaum, L.M.: 1982, MMIS, p.111.
Kirchhoff, V.W.J.H. and Clemeska, B.R.: 1983a, J. Geophys. Res. 88, p.442.
Kirchhoff, V.W.J.H. and Clemeska, B.R.: 1983b, Planet. Space Sci. 31, p.369.
Koeberl, C. and Kiesl, W.: 1983, Meteoritics, 18, p.326.
Kolomiets, A.R.: 1983, Probl. Kosm. Phys. 18, p.65.
Kopp, E. and Herrmann, V.: 1984, Annales Geophys. 2, p.83.
Kostylev, K.K.: 1982, MMIS, p.252.
Kostylev, K.V.: 1983, Meteorn. Rasprostranenie Radiovoln, 18, p.15.
Kostylev, K.V. and Svetashkova, N.T.: 1982, MMIS, p.61.
Kovshun, I.N.: 1983, Astron. Vestn. 17, p.183.
Kramer, E.N. and Shestaka, I.S.: 1982, Materials World Data Center B, Moskow.
Kramer, E.N. and Shestaka, I.S.: 1983, Problemy Kosm. Phys. 18, p.48.
Kramer, E.N. and Shestaka, I.S.: 1984, Astron. Vestn. 18.p.147.
Kramer, E.N., Shestaka, I.S., Kruchinenko, V.G., Markina, A.K. and Musij, V.I.: 1983, Vestn. Kiev. Univ., Astron. 25, p.53.
Kruchinenko, V.G.: 1982, MMIS, p.193.
Kruchinenko, V.G.: 1984, Vestn. Kiev. Univ., Astron. 26, p.98.
Kulikova, N.V.: 1982, MMIS, p.177.
Kurbanmuradov, K. and Mukhamednazarov, S.: 1982, MMIS, p.231.
Lebedinets, V.N.: 1984, Astron. Vestn. 18, p.35.
Lebedinets, V.N. and Manokhina, A.V.: 1982a, MMIS, p.104.
Lebedinets, V.N. and Manokhina, A.V.: 1982b, Trudy Inst. Exp. Meteorologii, No. 12/96, p.49.
Lebedinets, V.N. and Manokhina, A.V.: 1982c, Trudy Inst. Exp. Meteorologii, No. 12/96, p.55.
Levin, B.Yu.: 1982, Izvestiya Akad. Nauk USSR, 6, p.25.
Levin, B.Yu. and Simonenko, A.N.: 1982, MMIS, p.257.
Levitskij, S.M.: 1983, Problemy Kosm. Phys. 18, p.67.
Levitskij, S.M., Abdrakhmanov, N. and Timchenko, V.P.: Izvestiya Vuzov, Radiophysika, 25, p.1240.
Malynjak, M.I.: 1982, MMIS, p.248.
Matsenko, S.V. and Tkachuk, A.A: 1982, MMIS, p.36.

McDonnell, J.A.M., Carey, W.C. and Dixon, D.G.: 1984, Nature, 309, p.237.
McIntosh, B.A. and Šimek, M.: 1984, Bull. Astron. Inst.Czech. 35, p.14.
McLeod, N.: 1984, Sky and Telescope, 67, p.108.
McRobert, A.: 1984, Sky and Telescope, 68, p.148.
Mukai, T. and Giese, R.H.: 1984, Astron. and Astrophys. 131, p.355.
Nechitajlenko, V.A. and Voloshchuk, Yu.I.: 1983, Meteor Researches, 8, p.77.
Novikov, G.G., Lebedinets, V.N. and Blokhin, A.V.: 1984, Pis'ma v Astron. J. 10, p.71.
Novikov, G.G. and Tsygankov, S.F.: 1982, Izvestiya Vuzov, Radiophys. 25, p.239.
O'Keefe, J.A.: 1983a, Trans. Am. Geophys. Union, 64, p.257.
O'Keefe, J.A.: 1983b, Plenary Lecture Intern. Conf. on Glass in Planet. and Geol. Phenomena, New-York.
Pecina, P.: 1984a, Bull. Astron. Inst. Czech. 35, p.5.
Pecina, P.: 1984b, Bull. Astron. Inst. Czech. 35, p.120.
Porubčan, V.and Štohl, J.: 1983, Contrs Astron. Obs. Skalnate Pleso II, p.169.
Povenmire, H. and Ceplecha, Z.: 1983, Sky and Teleskope, 66, p.174.
Rentdel, J.: 1982, Astron. und Raumfahrt, 20, p.150.
ReVelle, D.O.: 1983, Meteorites, 18, p.386.
Rost, R.: 1981, Chem. Erde, 40, p.265.
Rumi, G.C.: 1982, J. Atmos. Terr. Phys. 44, p.773.
Sakhibgareev, D.G.: 1983, Trudy Inst. Exper. Meteorologii, No. 6/107, p.84.
Show, H.F. and Wasserburg, G.J.: 1982, Earth Planet. Sci. Lett. 60, p.155.
Šimek, M.: 1984, Bull. Astron. Inst. Czech. 35, p.126.
Simmons, K.: 1982, Meteor News, 58,p.7.
Simonenko, A.N. and Terentjeva, A.K.: 1984, Astron. Vestn. 18, p.215.
Smirnov, V.A. and Kovshun, I.N.: 1982, MMIS, p.235.
Solyanik, O.A. and Tkachuk, A.A.: 1983, Meteor Researches, 8, p.67.
Solomon, S., Fergusson, E.E., Fahey, D.V. and Crutzen, P.J.: 1982, Planet. Space Sci. 30, p.1117.
Spalding, G.H.: 1984, J. Brit. Astron. Assoc. 94, p.109.
Svetashkova, N.T.: 1982, MMIS, p.102.
Swider, W.: 1984, Planet. Space Sci. 32, p.307.
Terentjeva, A.K.: 1983, Astron. Tsirk. No.1256, p.5.
Tiwari, S.N. and Subramanjan, S.V.: 1980, Acta Astronaut. 7, p.583.
Tkachuk, A.A.: 1982, MMIS, p.67.
Tkachuk, A.A.: 1983a, Meteor Researches, 8, p.51.
Tkachuk, A.A.: 1983b, Probl. Kosm. Phys. 18, p.53.
Tokhtasjev, V.S.: 1982a, MMIS, p.162.
Tokhtasjev, V.S.: 1982b, MMIS, p.184.
Tokhtasjev, V.S.: 1984, "Meteor Bodies in Interplanet. Space and Earth's Atmosphere, Theses of the Conf., Dushanbe, p.53.
Voloshchuk, Yu.I.: 1982, MMIS, p.32.
Voloshchuk, Yu.I. and Kashcheev, L.B.: 1982, MMIS, p.45.
Weissman, P.R.: 1983, Icarus, 55, p.448.
Wetherill, G.W.: 1984, Meteorites, 19, p.1.
Yuan, B.: 1981, Scientia Geologica Sinica, 4, p.329.
Znojil, V., Simek, M., Grygar, J. and Hollan, J.: 1981, Bull. Astron. Inst. Czech. 32, p.1.
Zook, H.A., Lange, G., Grün, E. and Fechtig, H.: 1984, Lunar Planet. Sci. Conf. XV, p.9652.

Note: MMIS = "Meteor Matter in the Interplanetary Space". Proc. of Symp. held at the Kazan University, Sept. 9 - 11, 1980. Moskva,1982.

24. PHOTOGRAPHIC ASTROMETRY (ASTROMÉTRIE PHOTOGRAPHIQUE)

PRESIDENT: Wilhelm Gliese
VICE-PRESIDENT: Arthur R. Upgren
ORGANIZING COMMITTEE: T.E. Corbin, H.K. Eichhorn, G. Gatewood, I.I. Kanaev, T.E. Lutz, J.D. Stock, W.F.Van Altena

I. Introduction

The question of the title of Commission 24, obviously, offers a difficult problem as already mentioned in recent reports. Photographic Astrometry no longer describes the whole scope of the commission. This problem has continued during the last three years especially in view of the preparations for the astrometric tasks of the NASA Space Telescope and of the ESA satellite HIPPARCOS.

A review of the field of astrometry was published in Ann.Rev.Astron.Astrophys. (van Altena, 34.041.014). Murray edited a volume entitled "Vectorial Astrometry" (33.003.083).

Many of the activities of Commission 24 coincide with the interests of other commissions, especially with Commission 8. Surveys of new topics were made by Fredrick and Smith: "Photoelectric Astrometry" (Instrumentation in Astronomy 4, S.P.I.E. Proc.312, 111), Fredrick: "Methods for Photoelectric Astrometry" (S.P.I.E. Proc.331), Johnston: Radio Interferometric Astrometry; Techniques and Results" (IAU Symp. No.109, Gainesville), McAlister: "Speckle Interferometry in Astrometry" (IAU Symp.No.109, Gainesville).

In the past three years several meetings were held which were attended by members of Commission 24 who discussed topics and problems in the fields of activities of our commission:
ESA Colloquium "Scientific Aspects of the ESA Space Astrometry Mission HIPPARCOS", 22-23.02.1982, Strasbourg, France (31.012.011).
Workshop on "Calibration of Stellar Luminosities from Astrometric Data", 4-5.05.1983, Herstmonceux Castle, England.
IAU Colloquium No.76 "The Nearby Stars and the Stellar Luminosity Function", 13-16.06.1983, Middletown, Conn., USA (34.012.065).
IAU Symposium No.109 "Astrometric Techniques", 9-12.01.1984, Gainesville, Florida, USA.
IAU Symposium No.111 "Calibration of Fundamental Stellar Quantities", 24-29.05.1984, Como, Italy.

II. Data

TRIGONOMETRIC PARALLAXES

The new edition of the Yale Parallax Catalogue is nearing completion with an expected publication date of 1985. Van Altena gave preliminary reports at Middletown (34.002.125) and at Gainesville (IAU Symposium No.109). He also discussed the question "Can a system be defined for the new edition of the YPC ?" (37.002.086).

Trigonometric parallaxes of individual objects were published by Harrington et al.(34.111.002 - VB8, VB10), Harrington et al.(34.111.034 - LHS 292), Russell et a (31.111.007 - Wolf 294, 39 com, Case 2, Vega), Ianna et al.(32.111.005 - HD 149499), Kiselev (32.111.018 - Vyssotsky Nos. 5, 233, 28, 29, 191), Wang (34.111.021; 37.111.018 - BD +2°348), Guinan and Ianna (33.119.013 - R CMa).

Dahn and Monet reported about CCD Astrometry at KPNO and USNO-Flagstaff (IAU Symposium No.109). They gave preliminary results from the KPNO 4m-CCD parallax program (34.111.010; 34.111.032 - VB10, LP750-28, LP750-29, LHS 205a, 1405, 1625, 1625a, 1742a, 3481, 5356, VB8 and LHS 2924).

Lists of newly measured parallaxes: Van Vleck: Upgren and Breakiron (29.111.018), Upgren et al.(31.111.005, 33.111.005, 34.111.012); USNO: Dahn et al. (31.111.006); Sproul: Lippincott (30.031.562, 33.111.012), Heintz and Borgman (Astron.J.89, 1068); McCormick: Ianna and Whitman (37.111.009); Herstmonceux: Scales and Zhao (37.111.011); Yale Southern Station: van Altena and Sawada (34.111.011).

Further observations and reductions of trigonometric parallax series are in progress at Yerkes Obs. (Vilkki), Roy. Greenwich Obs. (Murray and Argyle, plates taken with the UK Schmidt-Telescope), South African Astr.Obs.(Churms), Yale-Columbia refractor at Mt.Stromlo (Ianna and Culver, giant stars, 37.111.005, and at IAU Symposium No.111), CTIO 4m-telescope (van Altena).

Murray et al. reported on "Trigonometric parallaxes obtained with the UK Schmidt Telescope" (IAU Symposium No.109).

Astrometric perturbations, search for "unseen companions": Lippincott (30.118.008 - Zeta Her, 85 Peg; 30.118.009 - Mu Cas; 32.118.004 - 7 visual binaries; 33.118.010 - 10 double or suspected multiple systems; 34.122.081 - EV Lac; 34.118.041 - 36 UMa), Hershey and Lippincott (31.118.034 - Lalande 21185, BD+5°1668), Borgman and Lippincott (33.111.004 - White Dwarfs), Lippincott et al. (33.118.025 - BD+41°328), Harrington et al. (30.118.004 - quadruple system G107-69/70), Harrington and Kallarakal (32.111.006, abstract - G208-44, L850-62, W922, Barnard's star), McNamara (Bull.A.A.S. 16, 507 - BD+66°34). Lippincott reported on the status of the Sproul astrometric plate series on the nearest stars in the research for planetary systems (32.118.016).

Also speckle interferometry in astrometry concerns searches for unseen planetary companions (Papers presented at IAU Symposium No.109 at Gainesville by McAlister and by McCarthy). At IAU Colloquium No.76, D.W. McCarthy gave a report and a list of astrometric companions confirmed by infrared speckle interferometry (34.118.045).

Gatewood and Stein described the high presision of electronic detectors which reveal parallactic motion more quickly than previously possible (34.036.206).

An error analysis of the USNO absolute parallaxes was published by Wang (37.041.029).

Halliwell gave an abstract of an orbital and trigonometric parallax comparison (32.111.008).

In two papers Lutz et al.(30.111.011) and Hanson and Lutz (33.111.001) investigated systematic effects in trigonometric parallaxes.

NEARBY STARS

Gliese and Jahreiss are preparing a third edition of the Catalogue of Nearby Stars which will be extended to a limit of 25 pc. Basic data will include the trigonometric parallaxes of the new Yale Catalogue but again supplemented by spectroscopic and photometric distance determinations. The latter are outside the scope of Commission 24, but the calibrations of spectral type-luminosity relations and color-luminosity relations are mostly made from trigonometric parallaxes and the difficulties arising from such procedures depend on statistical problems encountered in trigonometric parallaxes (Lutz, 34.115.026).

At Gainesville (IAU Symposium No.109) Gliese read a paper on "Astrometric Desiderata for Nearby Stars".

PROPER MOTIONS

The measurement of trigonometric parallaxes includes the determination of relative proper motions in the publications cited above.

Klemola reported (IAU Symposium No.109) on "Proper Motions with respect to Galaxies". The Lick proper motion program with respect to galaxies (B.F. Jones, Klemola, Hanson) has completed second-epoch photography for 93% of the 1246 fields north of -23° declination. Hanson and Sirk have completed proper motion and position reductions from -3° to +68° declination. The Lick "Input Catalogue of Special Stars" (Klemola) contains over 65 000 entries.

The Yale Southern Observatory has now extended the region of overlap with the Lick proper motion survey to the +25° declination zone (van Altena, 72 additional fields).

Investigating the reductions of relative to absolute proper motions of stars obtained from galaxies, Fatchikhin found that the accuracy of the measurements of galaxies at Pulkovo proved to be smaller than that of 15^m stars by 2.4 times (32.031.671).

Rybka gave the systematic differences between four catalogues of proper motions with respect to galaxies (Pulkovo, Tashkent, Moscow, and Goloseevo). The size of the differences depend on the stellar magnitude (31.111.009).

Rakhimov and Umarova investigated the Pulkovo Catalogue of Proper Motions of Stars relative to galaxies for determining precession corrections and inaccuracies (34.043.005; 34.111.027).

Absolute proper motions of faint blue objects near the galactic north pole were measured with reference to 45 galaxies. Plates were taken with the Tautenburg Schmidt telescope (Schilbach, 32.111.015).

Luyten has continued analysing pairs of Palomar Survey plates and published several hundred new proper motions for faint stars: Large proper motions (31.111.031); BD+12°584 (31.111.032); another 800 double stars with common proper motions (37.111.020), thus bringing the total number of his double stars (LDS) to more than 6 000. Luyten and H.S. Hughes published the "First Supplement to the NLTT Catalogue" (31.111.030) and proper motions for 374 and 280 faint stars (32.111.021, 34.111.023).

Herbig and B.F. Jones published proper motions of additional Herbig-Haro objects measured on Crossley and 120-in. plates (34.121.003).

Proper motions of ultraviolet-excess objects were determined by Noguchi et al. (32.111.014).

Cudworth gave a report "Astrometry of Star Clusters" (IAU Symposium No.109). Such observations and measurements of proper motions have been done at several observatories. Some of the Commission members have reported their work to IAU Commission 37 (van Altena), others informed Commission 24 on proper motion work in globular clusters, open clusters and nearby galaxies: B.F. Jones and Cudworth have published a study of Praesepe membership (33.153.008). Currently under study at Yerkes Observatory are the globular clusters M71, M22, M13, Pal.11, 47 Tuc, NGC6712, the old open clusters NGC2158 and NGC6791, and the UMi dwarf spheroidal galaxy. A several-year project obtaining proper motions, photometry and spectra for large samples of stars to V = 16.5 at intermediate galactic latitudes is initiated (Cudworth, Kron, Rybski); Adler and Ianna: Membership in the open cluster NGC2287 (37.153.078); Brosche and Geffert: NGC5466 (32.154.016; 34.154.034) and Brosche et al.(34.154.095); Brosche et al. determined the absolute motion of NGC4147.

Stock reports on the activities in Venezuela: Positions and proper motions for the Coma cluster and for Tr 10 have been published (Cova, 32.111.012), NGC129 and NGC752 are ready for publication and h Per is being measured. The plates were taken with the CIDA refractor.

A catalogue of 7120 proper motions for faint stars in the region of the Hyades was published by Luyten et al.(32.111.019) and another similar list of 6 300 such motions is in press.

A catalogue of relative proper motions and photographic UBV magnitudes in the region of the sky with the open star cluster NGC7092 (M39) was published at Leningrad by Platajs (34.002.037).

PHOTOGRAPHIC POSITIONS AND CATALOGUES

The last of the southern Yale Zone Catalogues was published as "Catalogue of Positions and Proper Motions of Stars between declinations -60° and -70° (Fallon, Hoffleit, 37.002.073). It includes 14 597 stars.

King and Lomb produced the Sydney Southern Star Catalogue (37.002.068 - SSSC) which gives 26 926 positions in the zone -51° to -63°.5, available in a tape version. Plates have been taken from -36° to the South Pole.

Nicholson et al. published the first volume of the Second Cape Photographic Catalogue 1950.0, CPC2 as provisional catalogue of positions of stars in the Cape zone -40° to -52° (37.002.024). It contains 51 018 stars, printed in microfiche and available also on magnetic tape. A total of 5820 plates covering the whole southern hemisphere was taken at Cape 1962 - 1972 and measured on GALAXY at Herstmonceux. A new block adjustment program version which can handle very large global plate overlap patterns is under development. This system will be used for the final reduction of the whole CPC2 (de Vegt).

The Pulkovo Bolivian Expedition (26.041.025) has taken already 1650 plates of the southern sky; measuring work is continued (Kanaev).

Murray gave a report on Astrometry with Schmidt telescopes (37.041.035).

In September 1983 the "VI Lohrmann-Kolloquium Geodätische Astrometrie" was held; five papers on new results using the 2m-Schmidt at Tautenburg, GDR for photographic astrometry were read (in press: Mitt. Lohrmann-Obs., Stange).

At IAU Symposium No.109 Stock presented a paper on "Astrometric positions from objective prism plates".

Stock and Cova published a catalogue of 2911 stars of the Carte du Ciel (equ. 1950.0, epoch 1901.1: 11^h39^m to 12^h30^m, +26° to +32°) using recently developed block adjustment methods (33.041.036).

RADIO SOURCES

The Working Group on Optical/Radio Astrometric Sources for the Establishment of an Inertial Reference Frame has produced an interim Progress Report (Argue and de Vegt, Abh. Hamburger Sternw., Band 10, 101-176, 1982) and summarized the first step of this progress publishing a list of suitable objects for future work: A catalogue of 234 strong compact extragalactic radio sources for the construction of an extragalactic radio/optical reference system (Argue et al., 37.043.001).

Optical positions of radio sources were reported by West and Walter resp.: Walter and West (30.141.121, 32.141.014), Harrington et al.(34.041.005), Clements and Argyle (Monthly Notices Roy.Astron.Soc.209, 1, 1984), Brosche and Geffert (30.141.126), Geffert and Richtler (33.041.003). Plates for sources were taken with the double refractor at Bonn (Brosche), with the Mexican 2.12m telescope (Lentes) and at Kazan (Rizvanov).

SOLAR SYSTEM OBJECTS

Plates for determining positions of Solar system objects (minor planets, comets, major planets and their satellites) were taken at Zelenchuk Mountain Station of the Kazan University, at Engelhardt Observatory (Rizvanov), at the Zvenigorod Tracking Station of the Astron. Council of the USSR Acad. of Sciences (Lozinsky), at San Fernando (Quijano), and at Bucharest (Bocşa).

III. Astrographic Catalogues

We should call attention to the recent revival in Astrographic Catalogue work which was discussed at a meeting of the Rêunion Scientifique du Conseil Scientifique du Centre de Données Stellaires at Grasse, France, 10-11.05.1984. For details and desiderata for the future see Bull. d'Inf. du CDS 27, 1984.

IV. Satellite Based Astrometry

A detailed description of the HIPPARCOS-TYCHO project for the period 1981-1984 is given by Høg in a report to Commission 8. It includes the work in progress concerning the Input catalogue preparations (Turon, Ref.: Turon-Lacarrieu 34.051.044; Turon and Réquième at IAU Symposium No.109; Turon at IAU Coll.No.78).

The work of the HIPPARCOS Input Catalogue Sub-Group of Double Stars is reported by Dommanget. A specific Catalogue of the Components of Double and Multiple Systems (C.C.D.M.) giving positions to an accuracy of ±1" and all needed identifications is in work; several papers have been published by Bacchus (34.002.115, 37.118.038), Dommanget (34.002.016), Nys (34.002.011, 34.002.116, 37.002.028), Souliê (34.118.016, 37.118.036).

Linking the HIPPARCOS system to an extragalactic frame by observations of radio stars and of stars in the neighborhood of radio sources is within the scope of the Working Group on Optical/Radio Astrometric Sources (Argue, IAU Symp. No.109; see also Walter, 34.041.038).

Duncombe and Hemenway reported at IAU Symp. No.109 about "The use of the Space Telescope to tie the HIPPARCOS reference frame to an extragalactic reference frame".

V. Instrumentation and Techniques

The 1m Telescope at the Roque de los Muchachos Observatory on La Palma is now being commissioned. This telescope is to be known as the "J.C. Kapteyn Telescope". Time on this telescope has been assigned to a team led by D.H.P. Jones to carry out a five-year astrometric programme, including trigonometric parallaxes, radio source positions and proper motions of cataclysmic variables.

Ianna reported on astrometric applications of the PDS 1010A microdensitometer (Astronomical Photography 1984, Occassional Reports of the Royal Observatory, Edinburgh, ed. Sim and Ishida).

The PDS 2020G Microdensitometer of the Yale Observatory has been fitted by van Altena with a Hewlett-Packard dual axis laser transducer system and is now routinely producing astrometric positions whose accuracy appears to be limited only by the photographic emulsion and the telescope optics.

At Yale a theoretical analysis of the positional and photometric properties of photographic emulsions by J.F. Lee has shown that the IIIa-J emulsions should be capable of producing positional accuracies of ±0.2µm (Astron.J.88, 1683, 1983).

During the last few years special investigations were made in reduction and handling plates of the 2-m-Schmidt at Tautenburg by Böhme, Sandig, and Schilbach (29.041.051) and Böhme (29.041.014).

Stock has developed an automatic program which efficiently reduces simultaneously a system of overlapping plates. It can also be used to reduce existing catalogues like the CdC no matter whether these are given in x and y coordinates or in equatorial coordinates. He also developed an automatic program intended particularly for comet or asteroid positions. It can start out with only two identified stars and will automatically search for all additional reference stars on the plate. These programs have been presented at the IHW meeting at Munich in June, 1984 and will be published as part of the proceedings.

Due to the decision of the New South Wales Government, research at Sydney Observatory has been terminated. The astrographic telescope and other astrometric equipment will be transferred to Macquarie University in Sydney (Lomb)

The growth of the city around the observatory (Observatorio de Marina de San Fernando) is seriously disturbing the work; the 2nd epoch of the Pulkovo program of Extragalactic Nebulae is abandoned (Quijano).

Funding of the GALAXY machine at Herstmonceux ceased in March 1984 (Murray).

VI. Reduction Methods

Eichhorn and Standish derived a transformation which reduces the most general class of least-squares problems to the classical standard case (29.021.007). Eichhorn gives an algorithm to avoid underestimation of internal errors of estimated parameters (34.021.037). Eichhorn developed rigorous formulae for the dependence of the location and velocity of stars on these values themselves, and also for computing parallaxes rigorously from non-uniform proper motions (30.008.026;34.111.024).

VII. Machine-readable Data

Machine-readable astrometric data are available at Centre de Données Stellaires, Strasbourg, France (Bull. d'Information du CDS) and at Astron. Data Center of the NASA Space Flight Center (NSSDC/WDC-A-R&S), Greenbelt, Maryland, USA.

VIII. Acknowledgements

I am very grateful to all who have assisted by submitting information on their activities. I am deeply indebted to the Vice-President A.R. Upgren for reading and supplementing this report.

Wilhelm Gliese
President of the Commission

COMMISSION 25. STELLAR PHOTOMETRY AND POLARIMETRY
PHOTOMETRIE ET POLARIMETRIE STELLAIRES

PRESIDENT: J. Tinbergen.
VICE-PRESIDENT: F.G. Rufener.
ORGANISING COMMITTEE: M.S. Bessell, P.F. Chugajnov, J. Dachs, P. Edwards, I.S. Glass, J.A. Graham, B. Hauck, D. Kilkenny, A.U. Landolt, I.S. McLean, V. Piirola, N.M. Shakhovskoj.

I. Introduction

This triennium has in general not been one of spectacular new developments in photometric of polarimetric techniques; rather, existing techniques have been improved and are being exploited by more observatories than before. For this reason, the previous report (31.113.097) should be consulted for a broad survey of the subject, while the present has the character of a progress report.

Memorable events were the succesful launch and operation of the infrared point source survey satellite IRAS (Ref. 1); the publication of massive photometric catalogues in the Strömgren system (34.002.029), Geneva system (30.002.008 and 31.002.064), DDO system (31.002.003) and the ANS ultraviolet system (32.002.001); the appearance of a Standard Star Newsletter (Ref. 2) and the distribution of Supplement 3 of the Catalogue of Photometric Sequences (Ref. 3).

Space telescope is beginning to make its influence felt (32.012.013 and Ref. 4): apart from stimulating ground-based calibration work at high precision, the most immediate benefit to general practice of our craft will probably be the experience gained with CCD detectors at a precision useful for general photometry. Some of the work elsewhere is of similar photometric quality, but implementing the array detectors for precise spectrophotometric work is in general proceeding more slowly than we hoped for 3 years ago. Progress in this field is often reported first in the series "Instrumentation in Astronomy, published as separate volumes of the Proceedings of the Society of Photo-Instrumentation Engineers (SPIE).

Major observatories are beginning to accept the premise that the polarization option should be present on all suitable instruments. However, "clean" polarization treatment as an important design aim for the future Very Large Telescopes has not yet become fully integrated into the design process for these telescopes. It will be up to our Commission (among others) to press for this.

The present report is based mainly on the classical optical wavelength region; the limits to our interest are vague, but for this report we have concentrated on those spectral regions where classical technology is used. The report was prepared jointly by Rufener and Tinbergen, with input from the Organising Committee. In one or two special cases, we refer to literature more recent than June 1984. To allow each section to be read independently, a small amount of redundancy has been allowed to remain, both in the text and in the references. As far as possible, references are in the system of Astronomy and Astrophysics Abstracts.

REFERENCES

1. Astrophys. J. 278, L1, 1984.
2. L.E. Pasinetti (ed.) Standard Star Newsletter (Osservatorio Astronomico di Brera, 22055 Merate, Italy).
3. A.N. Argue, E.W. Miller and H. Warren, Supplement 3 of the Catalogue of Photometric Sequences, 1983 (Dept. of Physics, Ricks College, Rexburgh, Idaho).

4. Space Telecope Optical Calibration Target Working Group Report (in press; Space Telescope Science Institute).

II. Advances in photometry techniques (F.G. Rufener)

The interpretation of a photometric measurement is dependent not only on its accuracy but also on the exact definition of the instrumental passband. Present-day techniques may often bring about satisfactory results regarding the measurement's precision but are we so sure about the conformity of the passband? Ensuring this conformity is certainly one way to progress.

UBVRI PHOTOMETRY

A number of previous reports have already pointed to the increasingly frequent use of photomultipliers with good sensitivity over the whole domain from the near UV to the near IR. Development of such photometry with only one detector has been made possible by these photomultipliers. Apart from the S-20 or ERMA photocathodes it is the GaAs (Gallium Arsenide) photocathode with a high quantum yield which is mostly employed, the RCA 31034 and HAMAMATSU 943-02 photomultipliers being equipped with it. This technical improvement led to a new definition of the UBVRI system, the $UBV(RI)_C$ variant introduced by A.W.J. Cousins (17.113.005), which is widely applied in the southern hemisphere. The following references concern on the one hand the definition of the instrumental system and on the other hand the extension of the standard star network (showing a wide range of V magnitudes) which is spread over nine Harvard E-regions ($\delta = -45^o$) and distributed over the equatorial band. For example: M.S. Bessell (26.113.045), A.W.J. Cousins (28.113.035), N. Vogt et al. (30.113.016), J.A. Graham (31.113.084), A.W.J. Cousins (South African Astron. Obs. Circ. 8, 69).

The work of A.L. Landolt (33.113.054) deserves particular attention. It covers an extended series of broad-band $UBV(RI)_C$ celestial equatorial photoelectric standard stars mostly in the Selected Areas. The UBV values are on the Johnson system. The RI are tied into Kron-Cousins' standards. The system is thoroughly related to Cousins' superb bright E-region standard stars. The new standard star magnitude range is $7 \leq V \leq 12.5$ and the colour range is $-0.3 \leq (B-V) \leq +2.0$. The mean error of the mean of Landolt's magnitudes is $\pm$ 0.003 magnitudes. Landolt also has published (33.113.054) an internally consistent set of photometrically measured stars ranging from -75° to $+30^o$ declination at several hour circles. These stars are useful to check for orientation effects in photometric equipment mounted at the telescope. Furthermore he has an ongoing program to extend the standard stars to fainter limits ($V \sim 17.5$) in these same equatorial regions. This work is being done at Cerro Tololo Inter-American Observatory. He also is extending the equipment orientation check stars into the northern hemisphere to $+80^o$ declination via an observational program at Kitt Peak.

So far there is no equivalent to this extension in the northern hemisphere, where Johnson's original UBVRI system (using the S-1 photocathode response) is still being used as the standard. The resulting risk of confusion and degree of insecurity have been clearly and sharply criticised by M.S. Bessell (34.113.020) and more indulgently by J.D. Fernie (34.113.055). In a recent discussion A.W.J. Cousins (South African Astron. Obs. Circ. 8, 59) shows how difficult it really is to obtain a precise standardization. It is to be hoped that the use of "modern" instrumental passbands equivalent to those of the $UBV(RI)_C$ definition of the southern sky becomes widespread also in the northern sky. Simultaneously, the standardization efforts by A.V. Landolt should be exploited and extended towards positive declinations.

A.W.J. Cousins reports that a "revised catalogue of $UBV(RI)_C$ photometry of E region stars is being prepared under the direction of John Menzies using additional material from Sutherland".

The development and frequent use of bi-dimensional photoelectric detectors, including the CCD, has opened a new era for photometry in general and for $UBV(RI)_C$

photometry in particular (for example, R.E. Schild, P.A.S.P. 95, 1021). However, we should be cautious and keep in mind the difficulties that could limit its range of action (see section IV). It is stressed that:

a) care must be taken with regard to the preservation and conformity of the definition of the instrumental passbands;
b) these observations must be connected via secondary faint standard sequences that are sufficiently well known.

These two requirements are necessary but not sufficient. Careful study of the performance of each "chip" is necessary which could lead to the maximum adaptation of application and reduction methods. S. Djorgovski exposed these problems in an excellent article (Proceedings of the NASA/SDSU Workshop on Improvements in Photometry, San Diego, June 1984, W. Borucki & A. Young, eds.).

STRÖMGREN PHOTOMETRY

This photometric system has found wide application in the study of galactic populations. The introduction to the comprehensive catalogue of E.H. Olsen (34.002.029) sets out the problems linked to the homogenization of observational data. The difficulties encountered arise both from measurements obtained with instrumental systems defined either by filters or by slots on the spectrum and from the heterogeneity of the selection of standard stars. This remarkable work underlines the importance of the stability of instrumental definition and the choice of the sequence of standard stars for maintaining the significance of the photometric parameters. Using synthetic photometric simulations, J. Manfroid illustrates these problems in the study "On photometric standards and colour transformation" (IAU Symposium 111, 1984). A. Ardeberg and H. Lindgren have prospected the aptitude of this system for the study of stars later than F (31.113.096; 32.113.025).

Very important is that uvby,β filters are proposed for the High Speed Photometer built for the Space Telescope. This will require standardization of very

The difficulties encountered arise both from measurements obtained with instrumental systems defined either by filters or by slots on the spectrum and from the heterogeneity of the selection of standard stars. This remarkable work underlines the importance of the stability of instrumental definition and the choice of the sequence of standard stars for maintaining the significance of the photometric parameters. Using synthetic photometric simulations, J. Manfroid illustrates these problems in the study "On photometric standards and colour transformation" (IAU Symposium 111, 1984). A. Ardeberg and H. Lindgren have prospected the aptitude of this system for the study of stars later than F (31.113.096; 32.113.025).

Very important is that uvby, β filters are proposed for the High Speed Photo-

OTHER SYSTEMS

Photometric systems with intermediate or narrow passbands (Walraven, Geneva system, Vilnius, DDO etc.) have been used actively during the last three years. Efforts are now being made to obtain photometric measures for stars which are presently candidates appearing in the Input Catalogue of the future European astrometric satellite Hipparcos.

The last issue available (raw computerprint) of the "Catalogue of stars measured in the Geneva Observatory photometric system" contains 150.000 measurements in seven colours for approximately 23.000 stars which have been collected over the last 25 years. A comparative study is being made of the [V] magnitude scale of this catalogue with Johnson's V-band definition, which is very close.

I.A. Glass reports that B.C. Carter of the South African Astronomical Observatory has obtained a new set of 156 JHKL standards of very high quality. These have been compared to previous standards published by Elias et al. (A.J. 87, 1029, 1982) on the C.I.T. - Cerro Tololo system, with those of Allen & Cragg (M.N. 203, 777, 1983) on the A.A.O. system and with Engels et al. (A. & Ap. Suppl. 45, 5, 1981) as well as Johnson's and his own much older lists. It is found that the consistency of the C.I.T. and Carter standards is the best, the r.m.s. differences being 0.01 or less at JHK. The A.A.O. system is only slightly worse, and the Engels et al. system much worse, the r.m.s. differences being .043 at J and .023

each at H and K (all comparisons after taking care of systematic transformations).

SYNTHETIC PHOTOMETRY

This subject has developed due to the particular interest of a precise confrontation between the fluxes predicted by the most accurate models, the photometric observations and the latest spectrophotometric data tied to the energy distribution of a black body. In this context the spectrophotometric atlas by J.E. Gunn & L.L. Stryker (33.002.042) and the list by I.N. Glushneva (34.113.009) are particularly useful. The following could be undertaken in the course of these numerical experiments:

a) search for and control of the best description of the pass bands defining each photometric system;
b) estimate of the effects caused by deviating natural (instrumental) systems;
c) calibration of photometric parameters from models and confrontation with empirical calibrations;
d) detection and analysis of inconsistencies;
e) application to the computation of appearances for stellar populations considered globally.

This enumeration is far from being exhaustive and shows the large variety of possible applications. A "joint meeting" on this technique is planned during our next assembly.

MISCELLANEOUS

The extensive use of computer-controlled microdensitometers exploring photographic plates whose sensitivity was enhanced by physico-chemical treatment gives access to an enormous amount of information. Development and refinement of software codes for evaluation are in continuing progress. The nature of the source is determined by shape-recognizing algorithms (stars, globular clusters, galaxies etc.) Even the most refined photometry will remain dependent on the definition of the instrumental systems and on the homogeneity of the emulsions and their treatment (an important reference: 33.012.059).

In recent years several new photometers have been described (29.034.031; 30.034.027; 30.034.054; 30.034.068). Special mention should be made of differential photometers which were conceived to make accurate measurements of the magnitude difference between two stars even if the sky is not of photometric quality (30.034.015; 34.034.087). A very interesting attempt in this direction is the development of RADS (rapid alternate detection system) which was tested by E.F. Milone et al. (32.031.675; 34.036.104). This instrument was conceived to measure successively during a rapid sampling cycle not only the two stars to be compared but also their neighbouring sky background. If we disregard the sensitivity variation caused by the shifting of the pupil's image on the detector, this method has the advantage of taking into account the variations of the sky brightness which are often anti-correlated to extinction variations. The statements and remarks made by J.F. James (30.031.002) gave rise to a pointed and justifiable reaction by A.T. Young et al. (Quart. Jour. Royal Astr. Soc. 25, 56).

A new way of detecting variability has been found by performing rapid sampling (29.034.050; 29.113.013; 33.034.066) or by lowering the detection threshold by a maximum increase of the "signal-to-noise" ratio of the measurements (D.W. Kurtz, 32.122.022). In this respect see also the article by F.L. Deubner & J. Isserstedt (34.113.012). Also related to this field are the measurements of D. Dravins (32.031.642) on photon statistics with nanosecond resolution.

Several articles have been written concerning statistical discussions for a better way of measuring photometric quentities. M. Claudius & R. Florentin-Nielsen (30.031.505) discuss the optimum choice of integration, I.R. King (33.036.017) studies the accuracy of measurement of star images on a pixel array. P. Bartholdi et al. (Astron. Astrophys. 134, 290) show that the application in real-time of statistical evaluation criteria in association with measurements obtained by rapid sampling lead to objective quality criteria, thus allowing filtering of accidental anomalous recordings.

The systematic application of computer processing has brought about a new approach to problems of reduction outside the atmosphere and the relation to standard stars. General procedures have been proposed and criticized (W.E. Harris et al., 30.031.558; D.M. Popper, 31.031.555; J. Manfroid & A. Heck, 33.036.025).

The eruption of the Mexican volcano El Chichon in April 1982 contaminated the stratosphere with a huge quantity of dust and particles of sulfuric acid. Called "the dust cloud of the century", it has significantly increased atmospheric extinction over the entire planet. The following increases were recorded by the Geneva Observatory photometric station at La Silla (ESO): Δk_U=.070; Δk_B = .055; Δk_V = .045. The decrease in this additional extinction has been slow and lasted until 1984.

IMPORTANT PUBLICATIONS

The "proceedings" of several meetings contain interesting descriptions and numerous references in direct relation with our Commission's activities:

-- Instrumentation for astronomy with large optical telescopes, IAU colloquium 67 (32.012.059).
-- Second ESO Infrared workshop, Garching (32.012.060).
-- Solid State Imagers for Astronomy, SPIE 290 (32.012.098).
-- Proceedings of the second New Zealand symposium on photoelectric photometry, Auckland (33.012.049).
-- Proceedings of the workshop on astronomical measuring machines, Edinburgh (33.012.059).
-- Instrumentation in Astronomy IV, SPIE 331 (33.012.085).
-- Proceedings of the NASA/SDSU Workshop on improvements in photometry, San Diego, 1984 (W. Borucki and A.T. Young, ed.)

Some recent book are useful for consultation:

-- Software for Photometric Astronomy, by S. Ghedini (32.003.041)
-- Astronomical photometry, by A.A. Henden and R.H. Kaitchuck (32.003.056)
-- Low light level detectors in astronomy, by M.J. Eccles, M.E. Sim, K.P. Tritton (34.003.061).
-- Microcomputers in Astronomy (R.M. Genet, A.M. Genet, ed.) (34.003.069).
-- Advances in photoelectric photometry. Vol. 1 (R.C. Wolpert, R.M. Genet, ed.) (34.003.160). Completed by vol. 2, published in 1984.

The last two references were published under the auspices of the very active group I.A.P.P.P. (International Amateur-Professional Photoelectric Photometry Group) which also publishes interesting communications.

Mention should be made of the remarkable activity of the data centres which collect and make available the most recent catalogues in computerized form: "Astronomical Data Center at NASA Goddard Space Flight Center, Greenbelt, Maryland, USA" and the "Centre de données stellaires (CDS), Strasbourg, France". The updating of these catalogues is indicated in the "Bulletin d'information du CDS". In this field, B. Hauck draws our attention to:

-- "Homogenous catalogue of red and infrared magnitudes in the photoelectric photometric system of Kron (magnetic tape), G. Jasniewicz (31.002.066).
-- "Le catalogue général des données photométriques", B. Hauck (31.002.022).
-- "Early-type MK standard stars in the Strömgren 4 color and Geneva photometric systems", A.G. Davis Philip, B. Hauck, Proceedings at the MK classification workshop, Toronto, June 1983.
-- "UBV data 1976-1982; a complete supplement to the General Catalogue", J.-C. Mermilliod (34.002.121).
-- "Photometric data", B. Hauck, Proceedings of colloquium on astronomical catalogues, Tbilisi, September 1984.
-- "Photometric data files", B. Hauck, J.-C. Mermilliod, M. Mermilliod, Proceedings of International Course on Data Handling in Astronomy and Astrophysics, Trieste, July 1984.
-- "Collection of data for stars in open clusters", J.-C. Mermilliod, Proceedings of International Course on Data Handling in Astronomy and Astrophysics, Trieste, July 1984.

-- "uvby,β photoelectric photometric catalogue new version with 60044 records concerning 40487 stars", B. Hauck, M. Mermilliod, to be published.

III. Advances in polarimetry techniques

Polarization modulation has become the accepted technique for adding polarimetric capability to an instrument with photoelectric detectors. The highest precision reported for this technique is about 0.003%, used in phase-locked linear polarimetry for detecting non-linear pulsation in a Beta Cephei star (29.122.165). Several new modulation polarimeters, with various precision levels, are described in the literature of the last three years (30.031.521, 30.034.023, 30.032.617. 32.034.155, 32.034.073 or Ref. 1, 34.034.030, 30.034.008, 32.034.004, 34.034.107 and Ref. 12, p. 238; the last four of these are for the infrared, the fifth last is a scanning spectropolarimeter).

The modulation technique is particularly necessary and exacting for array polarimeters (whether area or spectro-polarimeters), where calibration of all polarimetric effects for every elementary detector would be particularly cumbersome. An imaging polarimeter is described in Ref. 13. The Lick Cassegrain spectropolarimeter is in full production (Ref. 2); a high-dispersion Coudé spectropolarimeter has been implemented on the same telescope (Ref. 12, p. 62). Two new array polarimeters are described in Ref. 3; results are reported in 34.125.001, 34.158.002 and Ref. 4; precision approaches 0.1%.

The polarimeter in the Space Telescope's Faint Object Spectrograph is described in Ref. 14. Best precision of a few tenths of a percent is predicted. Both of the imaging cameras and the High Speed Photometer on Space Telescope also have polarimetric capability.

Photographic and electronographic area polarimeters are described in 29.034.001 and 34.034.019 respectively; a precision of a few percent is reported for these on objects only a little above sky background.

Solar spectropolarimetry with a resolution of 500000 is reported in Refs. 5 and 6. Though such resolutions may never be applied to stars, the demonstration that there are no fundamental problems with polarimetry using a Fourier transform spectrometer is valuable for Very Large Telescopes, which after all is where our future lies. Another point of great interest for precision spectropolarimetry using the Very Large Telescopes is the design of the Large European Solar Telescope, which has an intermediate focus for mounting the polarization modulator (Ref. 7), ensuring that polarization capability is not compromised by transfer optics.

A very worthwhile discussion of future instrumental possibilities is given in Ref. 8, which also has its roots in solar polarimetric practice.

Polarised spatial interferometry is discussed in Ref. 9.

An X-ray polarimeter of the "classical" Behr 2-beam type is described in 30.032.509; the best precision reported is 0.2%

Instrument-theoretical considerations are the subject of 33.034.026 (polarimeter design), 34.036.035 (Stokes parameter statistics), Refs. 10 and 11 (retarder design).

Observations suitable for polarization standards are reported in 32.034.004 (UBVRIJHK), 32.116.016 (UBVR, 0.75; very precise degree and angle of polarization), Ref. 15 (UBVRI) and 33.131.147 (UBVRIJ). These are valuable additions to the bibliography in 29.031.614 and 31.113.097 sec IIIb.

Discussions of polarimetric accuracy required for particular scientific objectives, though valuable for polarimetric practice, are not mentioned here; they are considered to be the province of other Commissions.

REFERENCES

1. A.M. Magalhães, E. Benedetti and E.H. Roland, P.A.S.P. 96, 383, 1984.
2. R.R.J. Antonucci, Astrophys. J. 278, 499, 1984.
3. I.S. McLean, Proc.S.P.I.E. 445, (Instrumentation in Astronomy V), p. 547, 1984.

4. I.S. McLean, Astron. Astrophys. 134, 333, 1984.
5. J.O. Stenflo, D. Twerenbold, J.W. Harvey and J.W. Brault, Astron. Astrophys. Suppl. 54, 505, 1983.
6. J.O. Stenflo, J.W. Harvey, J.W. Brault and S. Solanski, Astron. Astrophys. 131, 333, 1984.
7. LEST Foundation Annual Report 1983, p. 16.
8. J.O. Stenflo, Applied Optics 23, 1267, 1984.
9. F. Vakili, Proc. S.P.I.E. 445 (Instrumentation in Astronomy V), 484, 1984.
10. J.H. Apfel, Applied Optics 23, 1178, 1984.
11. A.M. Title and W.J. Rosenberg, Proc. S.P.I.E. 307 (Polarizer Applications), 120, 1981.
12. Bull. Amer. Astron. Soc. 16, 1984.
13. J.A. Tyson and R.W. Lee, Proc. S.P.I.E. 290 (Solid State Imagers in Astronomy), 144, 1981.
14. R.G. Allen, J.R.P. Angel, Proc. S.P.I.E. 331 (Instrumentation in Astronomy IV), p. 259, 1982.
15. T. Korhonen, V. Piirola and A. Reiz, ESO Messenger no. 38, December 1984, p. 20.

IV. Detector developments

There is little doubt that array detectors of some kind will eventually take over from the single-pixel detectors now used for precise work. The take-over, however, is a gradual process; more and more problems are encountered as the required precision is raised. Very broadly speaking, the best array detector practice has now reached 0.01 mag in photometry and 0.1% in polarimetry (0.001 and 0.005% respectively for single detectors). Whenever this precision of the array detectors is sufficient and the multiplex advantage vital to the programme contemplated, the array detectors are the indicated path to progress (see, for instance, Ref. 1).

Imaging a field rather than a single point with photometric precision introduces extra problems such as ghosts (Ref. 2); recording such images is further complicated by pixel-to-pixel variation of both gain factors and offsets. On the other hand, the availability of many parallel detectors offers the possibility of intercomparing the registrations of the same source by different element detectors and thus constructing an internally consistent calibrated image; progress in this field is considerable, but does not seem to have approached any fundamental limits yet and often suffers from ad hoc simplifications.

The problem, of course, is that in general both the zero offset and gain of elemental detectors are not identical and are functions of position of the element, wavelength, temperature, polarization, etc., and that some of these functional dependences are coupled. Decoupling and calibrating them in a relative sense are the keys to succesful photometry. The most "decoupled" situations would seem to be sky images in narrow bands excluding night-sky lines and spectrophotometry of single point objects, while "objective prism' images constitute the most complex situation and are least likely to yield reliable precision results (both these and broadband direct images are of course of great interest for their efficient acquisition).

Calibration sources are sometimes artificial, at other times the sky itself; in both cases it needs to be demonstrated that the spectral content and evenness of illumination are appropriate to the problem. From the point of view of our Commission, reports of array detector work should be read with the above considerations in mind.

To get an impression of the problems involved and the complexity of the solutions, good "student texts" are 34.002.043, Ref. 7 and Ref. 8 (the advanced student may proceed to Ref. 15, p. 2, Refs. 4, 5 and 6, probably to his complete bewilderment). The fundamental solution to our problems will be homogeneous array detectors, but for the moment this is wishful thinking at almost all levels of precision.

The largest collection of papers on electronic area detectors is to be found in Refs. 4, 5, 6 and 15; modern photographic measuring machine practice is also very relevant (e.g. 33.012.059). The CCD for the Space Telescope's Wide Field & Planetary Camera is one of the most fully documented devices (Ref. 3, 33.035.098 or 33.035.122, 34.157. 148 and lately Ref. 10, p. 477, where detection of 26th magnitude on the Hale 5-metre telescope is reported). For photometry, one of the promising ways of using a CCD is the drift-scan method (Ref. 7, and Ref. 9 or Ref. 5, p. 137). In this technique, pixel inhomogeneities are averaged along a column and the sky itself is used for gain calibration; the limits to the technique and its inherent assumptions need to be systematically investigated for photometric use (as opposed to mere detection) and the merits compared to those of, for instance, a series of strongly overlapping discrete exposures.

Much work remains to be done before stellar photometry with array detectors becomes an everyday reality at the precision we are used to from single-detector photometry. However, the potential gains for statistical investigations and for spectro-photometric systems tailored to the application are enormous, so that the effort involved in raising the precision is fully justified. For the present, the focus of attention is on CCDs, because they are superior in faint object detection. For more precise work on brighter objects, they may not be the ideal type of array detector.

The following developments, though not yet "photometric", are of interest for their future promise:

-- A new photon-counting detector called Mepsicron is reported in Ref. 11. It has high uniformity of sensitivity and a large number of picture elements, coupled with adequate quantum efficiency and maximum count rate. It, and similar devices, may become an attractive alternative to the ubiquitous CCD.
-- Infrared array detector developments are reported in 34.034.006, 34.034.061, Refs. 13, 14, a number of papers in Ref. 6 and in Ref. 10, p. 36, 232 and 238. The largest size reported so far is 64 x 64 elements.
-- Ref. 10, p. 304 reports encouraging developments of an X-ray CCD, with spectral as well as spatial resolution and high quantum efficiency.
-- At Kitt Peak a comprehensive, transportable, array detector software system called IRAF is being developed, which includes photometric programmes (Ref. 10, p. 497). We shall all experience this system frequently in the future.

The problems of polarimetry with array detectors are different from those of photometry. Photometric precision is not of paramount importance in most applications; what is important is the ability to operate with polarization modulation at a few Hz or faster. Array polarimeters in the literature are mentioned in sec. III of this report; Ref. 3 of that section describes experience with the on-chip storage of the 2 charge images corresponding to the 2 polarised optical images. This promising method (Ref. 5, p. 76) has not yet been brought to full operational use and may in fact be limited in scope (about 0.1% precision). An alternative possibility would seem to be a modified drift-scan camera; no developments in this direction have been reported.

The Mepsicron (Ref. 11) and similar high-readout-rate photon-counting detectors are attractive for polarimetry, although even 100 counts/sec from a pixel barely allows 0.1% polarimetric precision without further smoothing. A developmental detector Rubicon, based on the Digicon principle, may become important for spectropolarimetry (Ref. 12).

REFERENCES

1. A.J. Penny, Gemini no. 7, p. 5, 1983 and no. 12, p. 1, 1984.
2. C.G. Wynne, S.P. Worswick, C.M. Lowne and P.R. Jorden, Observatory 104, 23, 1984.
3. M.M. Blouke, J.R. Janesick, J.E. Hall, M.W. Cowens, P.J. May, Opt. Eng. 22, 607, 1983.
4. Solid State Imagers for Astronomy, Proc. S.P.I.E. 290, 1981.

5. Instrumentation in Astronomy IV, Proc. S.P.I.E. 331, 1982.
6. Instrumentation in Astronomy V, Proc. S.P.I.E. 445, 1984.
7. P. Hall and C.D. MacKay, M.N.R.A.S. 210, 979, 1984.
8. K.U. Ratnatunga and E.B. Newell, Astron. J. 89, 176, 1984.
9. J.T. McGraw, H.S. Stockman, J.R.P. Angel, H. Epps, T. Williams, Opt. Eng. 23, 210, 1984.
10. Bull. Amer. Astron. Soc. 16, 1984.
11. C. Firmani, L. Gutierrez, E. Ruiz, G.F. Bisiacchi, L. Salas, F. Paresce, C.W. Carlson, M. Lampton, Astron. Astrophys. 134, 251, 1984.
12. Th. Schmidt-Kaler, R. Rudolph, H. Tüg, I.A.U. Coll. 79 (Very Large Telescopes, their Instrumentation and Programs), 659, 1984.
13. J.F. Arens, G.M. Lamb, M.C. Peck, H. Moseley, W.F. Hoffmann, R. Tresch-Fienberg, G.G. Fazio, Astrophys. J. 279, 685, 1984.
14. E.L. Dereniak, J.P. Britt, A.M. Fowler, R.R. Joyce, G. Boreman, W.S. Ewing, Applied Optics 23, 889, 1984.
15. State-of-the-Art Imaging Arrays and Their Applications, Proc. S.P.I.E. 501, 1984.

J. TINBERGEN
President of the Commission

26. DOUBLE AND MULTIPLE STARS (ETOILES DOUBLES ET MULTIPLES)

PRESIDENT : M. G. Fracastoro

VICE PRESIDENT : K. D. Rakos

ORGANIZING COMMITTEE : J. Dommanget, O. G. Franz, W. D. Heintz, H. A. McAlister, A. Poveda, C. D. Scarfe.

Commission 26 has changed its name. The new denomination, warmly promoted by Dommanget, represents more adequately the actual field of the Commission itself and appears as a good start for extending its interest beyond the traditional aspects (morphology and statistics) and plunging into problems connected with the origin, evolution,mass spectrum and dynamics of small star associations.

Commission 26 did not organize official meetings during the triennium. However, it was very actively participating to some Colloquia and Symposia.

On the occasion of the 60 th anniversary of the Bosscha Observatory, a Colloquium (I A U Nr. 60) was held in Bandung, Java (June 1983). A. H. Batten, J. Dommanget, O.G. Franz, E. P. J. van den Heuvel, Z. Kopal and B. Warner were among the members of the Organizing Committee. During the Colloquium, the role of double and multiple systems received special emphasis. Prof. Kopal presented the V.Bappu Memorial Lecture on "Evolution of Binary Systems". Subsequent sessions dealed with: Physical properties of Double and Multiple Systems; Occurrence of Binaries in different stellar systems; Wide systems and their evolution; Close binaries, new observations and evolution; Generic relations between wide and close systems.

At the end of the Colloquium, Dommanget spoke about the role of double stars within the Hipparcos project (see below).

At the I A U Symposium Nr. 111 in Como, Italy (May 1984) P.Couteau has illustrated a modified HR diagram for 164 visual binaries with measured orbits, by applying Kepler's third law, Pogson scale and mass-luminosity relation, and plotting M_* (magnitude for unit mass) against B - V. At the same Symposium, W. D. Heintz has discussed the problem of mass determination in double stars.

Not far from Couteau's research, K. D. Rakos has examined 147 visual binaries for a parallax independent calibration of the mean sequence in the interval $-0.15 < B-V < +0.80$. Furthermore, he has calculated the luminosity distribution for close binary systems and finds that the most probable mass ratio in a close binary system is 0.5. Finally, it seems that the number of triple systems is larger than expected.

In June 1984 an International Conference was held in Bamberg, FRG, in honor of F. W. Bessel 200 anniversary. Z. Kopal (chairman), Dommanget, Fracastoro, van de Kamp were among the members of the Organizing Committee. Significant space was assigned to double star problems and a crowded session was dedicated to astrometric binaries.

Fracastoro spoke on current state of Astronomy of visual binaries, P. L. Bernacca, Dommanget and Söderhjelm illustrated some aspects of the interaction between the double star population and the astrometric mission of the satellite Hipparcos.

A very important work has been accomplished at the US Naval Obser= vatory, in Washington D.C. USA. The revised and updated version of the IDS, known as the Washington Catalog of Visual Double Stars, 1984 (W D S) has been completed by Ch. Worley and Douglass. They have com= pletely edited, corrected, added to, and revised the Index and Obser= vations Catalog. This new version, which virtually has all published observations up to January 1, 1984, has been transmitted to the Data Centers, where machine-readable copies are available for distribution.

Worley is also performing an assiduous micrometer work with the 61-cm refractor at the U S N O and with various instruments at Cerro Tololo Inter American Observatory in Chile. The photographic program at Washington has been formally discontinued, however J. Hershey did get a few plates of Sirius during 1984.

Worley and Heintz have coedited the 4 th Catalog of Visual Binary Orbits (Publ. of the USNO, vol. XXIV, part VII; Washington, 1983).

From 1981.5 to 1984.5 Heintz has made some 7200 double star mea= sures (5400 by micrometer at the 61-cm refractor in Swarthmore, 1100 at Cerro Tololo; plus 700 plates). 130 new pairs were found and some 150 corrections of I D S entries were recorded (identifications, revi= sed positions, etc.). Orbits have been determined by Heintz for 42 binaries, trigonometric parallaxes for 39,and photographic mass ra= tios of 18 pairs.

The 2nd Catalog of Ephemerids, containing the relative radial ve= locities of the components of visual binaries with known orbit, has been published at the Observatoire Royal de Belgique, Uccle, by J. Dommanget et O. Nys. In this catalog 862 curves of relative radial velocities are plotted for 814 systems, covering the period 1970 - 1990. Furthermore, the geometric elements are reported for 102 stars, for which the ascending node is well established from the radial ve= locity observations.

E. van Dessel has obtained several hundred spectrograms at the Haute Provence Observatory, St. Michel, France, and at the European Southern Observatory in Chile, of the orbital doubles B 1909, Rst 3352, Kpr 7, φ363, φ326, ADS 8119, I 365, ADS 9247 BC, ADS 9979, ADS 12096, I 1416, φ307 and Kpr 114, as well as for several other systems showing some peculiarities or being suspected for duplicity.

Still at Uccle, Dommanget has oriented the activity of his group towards the problems raised by the presence of double and multiple stars (not necessarily forming physical systems) on the mission of the astrometric satellite Hipparcos. The necessity to provide it with the absolute positions (within 1 arcsecond) of all the individual com= ponents has obliged the compilation of a special catalog, which is now under preparation.

Furthermore, Dommanget has organized an international cooperation for the observation and research which are preliminary to the Hippar= cos mission in relation to these systems. To this purpose, several thousands of DM identifications, omitted in the Index 1976.5, have been accomplished and several hundred others, which were erroneous, have been corrected by P. Bacchus and O. Nys.

Orbits have been calculated by O. Nys (γ Cir), E. van Dessel (ADS 9453), A. Norro (B 594) and L. Zimmermann (ADS 1102).

Other orbits have been published by R. R. de Freitas Murão, in Rio de Janeiro. He is also an active observer.

At Flagstaff, Arizona, USNO Station, R. L. Walker is preparing the 5th list of 1270 visual measurements made with the 91-cm refractor at Lick Observatory. He has discontinued his visual observations to devote more time to photoelectric photometry of close visual binary and multiple systems.

At the Dominion Astrophysical Observatory, Victoria B.C. Canada, Batten, Fletcher and Scarfe (University of Victoria) have continued to obtain radial-velocity observations of binary stars, using the 1.2 m telescope and coudé spectrograph. Use of the radial-velocity spectrometer has permitted several additional systems to be added to the observing program.

Observations of Σ 2367 (HD 172865) obtained when the spectra were resolved near the late-1980 periastron passage have been published. Similar observations, but fewer in number, were obtained of Ho 296 (HD 214850). Periastron passage was a few months earlier than expected, but no other major modification to the visually-determined elements seems necessary. The brighter component had the more positive velocity. Attempts to resolve the spectra of Kui 114 (HD 214810) were unsuccessful; it appears that nodal passage occurred when the star was too close to the sun for observation, or that it occurred much earlier than expected.

A published study of coudé radial velocities of ζ Her A, obtained during more than half of the 34-year orbit, indicates less-than-perfect agreement with the astrometric parallax and mass ratio. A reinterpretation of π Cep, based on a new spectroscopic orbit for the primary, has also been published.

Observations of 70 Oph (both components) have been continued, and no evidence for a third component has been found. A paper on the parallax and mass of the system is in press. Work continues on the occultation-spectroscopic triple 64 Ori, the 15-year binary ε Hya, the triple system 1 Gem, and both components of 61 Cyg.

From 1981.5 to 1984.5 P. Couteau has discovered, with the two refractors of the Nice Observatory (50- and 74-cm aperture), 246 new binaries, more than one third of them being closer than 0.5 arcsec. During the same interval of time, 1757 measures have been made of different couples, 155 of them with the 91-cm refractor of the Lick observatory in 1983.

Still at Nice, P. Muller has measured 1514 couples, most of them being Mlr's binaries discovered at Nice. During the survey of the polar region ($\delta > 52°$) started in 1969, Muller has examined 5150 stars and discovered 79 new couples, the total number being at present 682. According to the AGK2, the region contains something more than 38 thousand stars; therefore, about 3500 of them have still to be examined.

On his side, Couteau has patrolled more than 102 000 out of 169 243 BD stars contained in his zone ($+17° < \delta < +52°$), discovering 2200 new couples. The situation for the northern hemisphere is now the following:

separation	number of couples
s < 0"25	1066
0.50	3119
1.0	5454

Since 1983, Couteau has succeeded to P. Muller as Editor of the *Circulaires d'Information*, where the orbits, the newly discovered binaries and all information required by the various authors are assembled and published.

To remain in France, at CERGA, Grasse, Bonneau, Foy and Koechlin are using speckle interferometry techniques: a further list of measurements is expected after the one published in 1980.

Finally, Dr Baize has published some 45 orbits during the triennium, as a demonstration of his indefatigable activity inside our commission.

P. van de Kamp communicates that current analysis of Sproul measurements of the Barnard star yields 32 yearly mean residuals, and for one orbit a perturbation with a period of 12 years. An improvement is made by an additional perturbation with a period of 20 years. The radii of the two perturbation orbits are 0.010 and 0.008 arcseconds.

S. L. Lippincott has published astrometric orbits, mass ratios and masses for ζ Her, 85 Peg and other visual binaries. She also analyses astrometric data on μ Cas and other double or suspected multiple systems. A comparison is made between the Sproul data and the Infrared Speckle ones for the unresolved binary BD +41° 328 .

According to Lippincott, the majority of the unseen faint companions to nearby stars, discovered at Sproul, have been detected by IR speckle interferometry by D. Mc Carthy.

The problem of the existence of substellar companions lies in a boundary territory between the astrometric search of double stars (limiting masses slightly below 0.1 $M_\odot$) and planetary bodies (the so called "jupiters", with masses of the order of 0.001 $M_\odot$). The general feeling is that a certain and definitive answer about the existence of these companions, their masses, orbital periods and so on, depends on new techniques, applied to opportunely selected stars.

The Georgia State University Speckle Camera, Atlanta, USA, has been used at the KPNO 4-m telescope by H. A. Mc Alister, W. I. Hartkopf, O. G. Franz and D. J. Hutter to obtain approximately 5000 series of binary star observations. These data are being reduced and analysed by means of a hardwired vectorautocorrelator and interactive software that removes the seeing component from the autocorrelograms and cen=

troids the binary star peaks. Systems as faint as V = +15 can be easily observed, but the gains in astrometric accuracy and the potential for photometric calibration are higher priority goals. A list of results from the new camera is expected to be submitted for publication in early 1985. The camera also continues to be used monthly on the 1.8-m Perkins telescope in Flagstaff for a long-term program searching for low mass companions within a sample of some 65 nearby binaries having separation in the range 0.1 to 1.5 arcseconds. Repeatability tests are indicating astrometric accuracies of 0.1 to 0.3 milliarcseconds, obtained from a single one-minute series of video speckle data.

Mc Alister and Hartkopf published a catalog of interferometric measurements of binary stars as the first contribution from the newly formed Center for High Angular Resolution Astronomy.

In Yugoslavia, visual observations were carried out by G. M. Popović and D. J. Zulević (Zeiss refractor 65/1055; 1780 measurements altogether). The results are published on the Bull. Obs. Astron. of Belgrad, Nrs. 132 - 3 - 4, or will appear shortly. This research is going on.

Popović has found 24 new pairs among BD stars ($+34° < \delta < +44°$). He has calculated orbits for ADS 3174, 5959 13028 AB and 1786. Six orbits are due to Zulević (ADS 1371, 7758, 11989, 16877, 4 and 2446) and ten orbits to V. Erveg (ADS 674 AB, 1345, 1393, 2111, 2609, 3058, 3686, 5707, GLE 1, Hu 1566). Popović is also analysing the triple systems contained in the IDS.

Djurković, who died at the beginning of 1981, has left a theoretical research about double stars and the newly discovered galaxy-satellite of the Milky Way, and a paper on the use of Dembowsky's method of double star measurements as a criterion for site selection.

Further visual and photographic measurements of double and multiple systems are planned at Belgrad (Popović et al.) as well as a statistical analysis of the orbital elements as a function of the galactic coordinates (Zulević).

At the Astronomical Observatory of Torino, Italy, G. Massone started (spring 1982) a program of visual double star observations with the 30-cm refractor. 480 measures are under publication. Successively, the observing program is continued with the new 42-cm refractor. It includes mostly double stars with known orbits, to improve their elements, and pairs with quick relative motion, in view of future orbit determinations. Furthermore, from the new IDS, the program will be extended to recently discovered couples or long disattended pairs with quick relative motion and presumably short periods.

A long term program of photographic observations is carried along by R. Pannunzio and R. Morbidelli with the 105-cm astrometric reflector. Beside possible identifications of perturbations due to low mass companions, the observations will continue selecting, from the initial list of pairs, those showing a significant relative motion. To this purpose, the accuracy of the plates is carefully controlled.

With the same instrument, parallaxes and proper motions of faint and late type nearby double stars are the subject of a long term pro=

gram.

In view of the Hipparcos astrometric mission, a statistical mo= del of a binary star has been elaborated from the IDS Catalog by Pan= nunzio and Scaltriti.

Still in the frame of the Hipparcos mission, Fracastoro has ana= lysed the information contained in the Yale Catalog of Bright Stars, examining the trend of Δm's with separation, and comparing the num= ber of observed couples to the number expected from a random distri= bution. The results are in agreement with the findings of Rakos (see above).

R. H. Wilson Jr. has 115 micrometer measures of double and mul= tiple stars, done with the 26-inch Innes telescope at Johannesburg.

N. P. Wieth-Knudsen has made 70 measures of double stars with se= paration smaller than 2.5 arcsec (30-cm reflector, Muller type micro= meter).

At Lisbon, Portugal, the restoration of the old 38-cm refractor is planned, in view of systematic observations of visual binaries.

ACKNOWLEDGMENTS

I am very grateful to those who have supplied information and collaborated to the preparation of this report.

M. G. Fracastoro
President of the Commission

Addendum:

Report on the work of Soviet astronomers in the field of visual double stars in 1981-1984

1. Observations

Photographic observations of visual double stars in the region of the sky to the north of 30° declination were continued systematically at the Pulkovo Observatory. More than 1000 plates were taken and studied for about one hundred double stars ($m < 12.0$, $\Delta m < 1.0$, $\rho \geq 3''$) not observed for a long time.

A catalogue of relative positions and motions of two hundred visual double stars containing the results of 20 years of double star observations in Pulkovo was prepared for publication. 23 visual double stars were identified which are possibly situated at distances not exceeding 25 parsec from the Sun.

Perturbations in the orbital motion of the double star ADS 11632 were studied on the base of a 20-year observational series: N.A. Shakht, Astron.Zhurnal Pis'ma, 10, 1984.

A.A. Tokovinin (P.K. Sternberg Astronomical Institute, Moscow) has developed

instrumentation for speckle-interferometric observations of close double stars ($\rho > 0''.06$, m > 9.0). The device was used with a 1-m reflector in Crimea: A.A. Tokovinin, Astron.Zhurnal Pis'ma, 8, No. 1, 1982,
8, No. 3, 1982,
9, No. 9, 1983.

Duplicity of the stars ιLyr and ζ Dra was found: G.A. Starikova, A.A. Tokovinin, Astron.Zhurn.Pis'ma, 7, No. 11, 1981. The interferometric method for measuring close double stars has been calibrated: A.A. Tokovinin, Astron.Zhurnal Pis'ma, 9, No. 10, 1983. The orbits of the following double stars were determined: Cou 292, β CrB, η Vir, ADS 10360, 12973, 14787, 15281; the orbits of the first three stars were determined for the first time: A.A. Tokovinin, Astron.Zhurnal Pis'ma, 10, No. 4, 1984.

Observations of double stars using the polarisation micrometer at the 30-cm refractor in Moscow were continued at the Sternberg Institute: G.A. Starikova, Astron. Circular, No. 1249, 1983.

Orbits of 98 visual binaries were determined: G.A. Starikova, Astron.Zhurnal Pis'ma, 7, No. 4, 1981; 8, No. 5, 1982; 9, No. 6, 1983; Sternberg Institute, Trudy, 57, 1984; Astron. Circular No. 1325, 1984.

The Special Astrophysical Observatory of the USSR Academy of Sciences (Zelenchukskaya) has introduced the digital speckle-interometric technique of measuring close double stars. The equipment is installed at the primary focus of the 6-m telescope. Since 1983 about 200 stars were measured with the resolution $0''.027 - 0''.818$. The r.m.s. error of relative positions is $\pm\ 0''.003$. The observational programme includes:

- speckle interferometric pairs first resolved by McAlister and by Labeyrie;
- spectral binaries observed with the 6-m telescope;
- stars with composite spectra;
- known "rapid" visual stars;
- close double stars recently discovered by P. Couteau.

The results of these observations are published: Y. Balega, D. Bonneau, R. Foy, Astron. and Astrophys. Suppl., 1984, 57, No. 1, 31.

2. Theoretical studies

Studies of the method of derivation of orbits for visual binaries using the observations of a short arc were continued at the Pulkovo Observatory. The orbits of ADS 48, 7251, 14636 (61 Cyg) were determined using this method: A.A. Kiselyov, O.V. Kiyayeva, Astron. Zhurnal, 57, No. 6, 1980.

Studies of the dynamical evolution of triple systems were carried out at the Leningrad University. By means of a numeric experiment with a computer it was shown that in 90% of cases triple systems were destroyed into a binary and a single star. The mean lifetime for triple systems in the solar neighbourhood is 10^6 - 10^7 yrs. 10% of cases lead to quasistable hierarchic systems. Dynamic processes of formation, evolution, and dissolution of double systems in a gravitational irregular field were also investigated. The results of the study are given in the review: Zh.P. Anosova, V.V. Orlov, "Dynamic evolution of triple systems", Astron. Observ. Leningrad Univ. Trudy, 40, 1984.

A statistical study of motions of visual double stars was carried out at the Sternberg Astronomical Institute: G.A. Starikova, "On the kinematics of visual double stars and multiple stars of the FK4 catalogue", Astron.Zhurnal, 58, No. 2, 1981.

A.A. Kiselyov

27. VARIABLE STARS (ETOILES VARIABLES)

PRESIDENT: N. Baker
VICE PRESIDENT: B. Szeidl
ORGANIZING COMMITTEE: M. Breger, A.N. Cox, J.D. Fernie, A.M. van Genderen, R.E. Gershberg, M. Jerzykiewiez, L.N. Mavridis, L.N. Mirzoyan, J.R. Percy, B. Warner

1. Introduction

The field of variable-star research is so broad that no report of this nature could possibly mention all the papers that have appeared in the last three years. It is hoped, however, that the reviews below include the most important work and identify the most significant trends. This report comprises ten sections on as many different research topics, each written by a different member of Commission 27. In addition there are (in Section 12) three short reports about ongoing activities of the commission. The commission president is very grateful to the authors of the individual contributions who have worked so conscientously.

A number of international conferences on one aspect or another of variable stars have been held during the period covered by this report. These are mentioned in the appropriate sections below.

There are two very active and important fields on which one might expect to find reports here, but which are mentioned only incidentally. The first is the subject of *Catacylsmic Variables*. An impressive amount of work, both observational and theoretical, has been done on these fascinating objects in the last few years, and there have been a number of major conferences and review articles. The topic is reviewed in the report of Commission 42 (Close Binary Stars), and in order to avoid duplication is not covered here.

The second important omission is *Solar Oscillations*. It is by now clear that the Sun belongs to the ranks of variable stars, and the study of solar oscillations has become a very active field. From the theoretical side stellar and solar oscillations now constitute a single field of research. Since the discovery of stellar variations similar to the solar five-minute oscillations, the essential unity of stellar and solar oscillation phenomena is becoming obvious from the observational viewpoint as well. Variable-star astronomers should be aware of the opportunities now arising in this type of research. Observational work on solar oscillations is discussed in the contribution of T. Duvall to the Report of Commission 12 (Radiation and Structure of the Solar Atmosphere). The theoretical side of the subject is covered by J. Christensen-Dalsgaard in the Report of Commission 35 (Stellar Constitution).

In the following a uniform and very terse style has been adopted for citations to the literature. Volume numbers appear in bold face, other numbers are page numbers unless otherwise identified. In general the abbreviations used in *Astronomy and Astrophysics Abstracts (AAA)* have been used. Frequently-cited journals, however, have been given very succinct designations: these are identified by their *AAA* abbreviations in the following key.

AA	*Astron. Astrophys.*
AA Supp	*Astron. Astrophys., Suppl. Ser.*
AJ	*Astron. J.*
AN	*Astron. Nachr.*
Ann Rev AA	*Annu. Rev. Astron. Astrophys.*
ApJ	*Astrophys. J.*
ApJ Lett	*Astrophys. J., Lett. Ed.*
ApJ Supp	*Astrophys. J., Suppl. Ser.*
Ap Sp Sc	*Astrophys. Space Sci.*

BAAS	*Bull. Am. Astron. Soc.*
Bull Ast Soc India	*Bull. Astron. Soc. India*
Bull Crimean Ap Obs	*Bull. Crimean Astrophys. Obs.*
DAO Publ	*Publ. Dom. Astrophys. Obs.*
IAU Coll 00	*Proc. IAU Colloquium No. 00*
IAU Symp 00	*Proc. IAU Symposium No. 00*
IBVS	*Inf. Bull. Variable Stars*
JAAVSO	*J. Am. Assoc. Variable Star Obs.*
J Ap Ast India	*J. Astrophys. Astron.*
JRAS Can	*J. R. Astron. Soc. Canada*
MN	*Mon. Not. R. Astron. Soc.*
MVS	*Mitt. Veränderliche Sterne*
Obs	*Observatory*
PASP	*Publ. Astron. Soc. Pac.*
Per Zv	*Perem. Zvezdy*
Per Zv Supp	*Perem. Zvezdy, Prilozh.*
Rev Mex AA	*Rev. Mex. Astron. Astrofis.*
SAAO Circ	*S. Afr. Astron. Obs., Circ.*
SAO Spec Rep	*Smithson. Astrophys. Obs., Spec. Rep.*
Sov Ast	*Sov. Astron.*
Sov Ast Lett	*Sov. Astron. Lett.*

2. T Tauri Stars
(Martin Cohen)

An unusual amount of attention has been directed toward the T Tauri stars (TTS) since the previous Commission 27 report. This is demonstrated by the number of major reviews of TTS, or of Herbig-Haro (HH) objects, that have appeared recently, by Bertout (*Rep Prog Phys* **47**, No2), Schwartz (*Ann Rev AA* **21**,209), Cohen (*Phys Rep*, in press).

The primary foci of recent studies of the TTS have been attempts to define the geometry of circumstellar shells; major observational efforts with the VLA at centimeter wavelengths; progress in theoretical modelling of their chromospheres and outer atmospheres; the recognition of highly variable phenomena in all spectral regions; the likelihood that episodic and anisotropic mass loss characterizes at least the most active (youngest?) TTS; and the potential relationship between TTS and the exciting stars of HH nebulae.

Several of these recently developed areas bear upon the geometry of the circumstellar envelopes of TTS, particularly the prospects for disks, and the likely nature of the dusty component. Bastien's (*AA Supp* **48**,153,513) major optical polarimetric survey of the TTS showed the rarity of high degrees of visible polarization, and highlighted the importance of the outstandingly polarized stars (e.g., HL Tau and DG Tau, with 13% and 6%, respectively). Both Bastien and Hough *et al* (*MN*, in press) noted a correlation between near-infrared color indices and visual polarization, which they interpreted as a consequence of flattened dust shells. Hough *et al* (*MN* **195**,429) saw evidence specifically for the presence of silicate particles in asymmetric shells, on the basis of the $p(\lambda)$ curves which they extended as far as 2.2 and sometimes 3.5 μm. From 8-13 μm spectrophotometric surveys of almost 50 young stars, Cohen (*MN* **191**,499) and Cohen and Witteborn (*ApJ*, in press) concluded that silicate materials are definitely found around TTS. They felt that the circumstellar envelopes are likely to be disklike rather than spherical in order to explain the correlations between silicate optical depth and visual extinction, near-infrared color indices, and optical polarization, as well as the infrequency of absorption, as opposed to emission, features. Cohen (*ApJ Lett* **269**,L70) has interpreted the unique properties of HL Tau in terms of our viewing this star through an edge-on circumstellar dust disk. Likewise, DG Tau has shown a resolved biconical radio continuum emitting zone (Bieging, Cohen, & Schwartz *ApJ* **282**,699) with "waist" oriented in the plane of the electric vector of polarization. Rydgren & Cohen (*Protostars and Planets II*, Univ Ariz Press, in press) too have discussed all the diverse observations that are relevant to the dust geometry in TTS and have decided in favor of disklike dust distributions.

It is possible that some of the circumstellar dust grains around TTS either were created in the stellar winds or at least represent thermally-modified prestellar grains. Certainly there are grains present with temperatures rather close to the condensation temperature for silicates (Rydgren, Schmelz, & Vrba *ApJ* **256**,168). These silicates seem to be more crystalline than typical interstellar silicates and their optical depths may correlate with wind activity in the TTS (Cohen & Witteborn, *ApJ*, in press). Variable CaII absorption features toward some TTS could also represent the condensation of gas into grains, locking up Ca atoms; these same data strengthen the case for episodic rather than continuous mass loss from TTS (Bertout *et al*, *AA Supp* **47**,419; Mundt *ApJ* **280**,749).

Centimetric radio emission, detected with interferometers which yield very high spatial resolution, characterizes some TTS (Felli *et al*, *AA* **107**,354; Bertout & Thum *AA* **107**,368; Cohen, Bieging, & Schwartz *ApJ* **253**,707), although an unbiased, luminosity-limited survey at 6 cm of the nearby Taurus-Auriga complex suggests that it is extremely rare to detect typical, older TTS, as opposed to those vigorous, young stars that excite HH objects (Bieging, Cohen, & Schwartz, *ApJ* **282**,699). The nature and origin of the radio emission is rather unclear. A variety of spectral indices is seen, from definitely non-thermal ($\propto \nu^{-0.4}$) through very flat spectra, to steeply rising ones ($\propto \nu^{1}$). The surprising radio strength of V410 Tau, a fast rotator, its recently-discovered strong radio variability and the flat spectrum in the off-state but steep spectrum in the on-state (Bieging & Cohen, *Proc. Boulder Worshop on "Radio Stars"*, D. Reidel, in press) suggest a possible kinship with the flarelike activity seen in RS CVn stars. T Tau itself was found to be a double source at 6 cm, with some 100 AU projected separation (Cohen, Bieging, & Schwartz *ApJ* **253**,707), which prompted a very accurate redetermination of its optical position (De Vegt *AA* **109**,45; Hanson, Jones, & Lin *ApJ Lett* **270**,L27). This probable binarity was also seen in the near-infrared, using the speckle technique in declination (Dyck, Simon, & Zuckerman *ApJ Lett* **255**, L103). It is the optically unknown companion that contributes the dominant radio emission, which has been attributed variously, either to an outflow from a very young, low-mass object, or to accretion onto a protostar (Bertout *AA* **126**,L1) or perhaps even onto a giant protoplanet (Hanson *et al*, *ApJ Lett* **270**,L27). These observations are a disturbing reminder of how little we understand even the prototype, and it raises again the rather puzzling issue of the paucity of known binaries among these young stars.

Theoretical progress in understanding the chromospheres of TTS has been discussed in detail by Bertout (*Rep Prog Phys* **47**, No2), summarized by Calvet (*Rev Mex AA* **7**,169). Recently Calvet, Basri, & Kuhi (*ApJ* **277**,725) have studied, semi-empirically, the effects of chromospheric temperature structure on diagnostic spectral features. A very detailed model of the physical and excitational conditions in the outer atmosphere of T Tau itself has been presented by Brown *et al* (*Nature* **290**,34) and by Brown, Ferraz, & Jordan (*MN* **207**,831), who propose a two-component atmosphere: one to yield the X-ray emission, the other the mass loss. Bertout (*Proc 3rd Eur IUE Conf*, 89) favors a second transition zone too, between the wind and the circumstellar medium. Ulrich *et al* (*ApJ* **267**,199) have advanced a model for DR Tau in which the 3200-11000 Å energy distribution is modelled by emission from an optically thin, very hot (65,000 K) gas. This they term a "turbosphere". It is powered by turbulence dissipated by shocks generated by the accretion of cometary-sized clumps onto the star.

Variability of the TTS is legion. Of late, rapid behavior has been detected in the ultraviolet, where an ensemble of solar-type flare events may be responsible (Worden *et al*, *ApJ* **244**,520), and in the optical, for example in RW Aur (Mundt & Giampapa *ApJ* **256**,156). In the X-ray region, a dramatic individual flare event was seen in its entirety in AS 205 (Walter & Kuhi *ApJ*, in press). Repeated flaring of several pre-main-sequence stars in ρ Oph, with a power-law dependence of frequency of occurrence on amplitude strongly reminiscent of solar flares, has been described by Montmerle *et al* (*ApJ* **269**,182). Longer-term optical monitoring (Herbst, Holtzman & Klasky *AJ* **88**,1648) has also been fruitful, especially for CO Ori (and UX Ori) which showed occasional "aperiodic minima", incompatible with mere variable circumstellar obscuration. Most bizarre has been the singular behavior of SY Cha (Schaefer *ApJ Lett* **266**,L45) which underwent periodic (∿6 days), large-amplitude optical variations for 3 years, after which it reverted to its former, low amplitude, irregular variability. If this were caused by a giant, bright starspot, modulated by stellar rotation, then the stability and longevity of this putative spot would be truly remarkable. Given this general framework, it is not surprising to find radio variability (see Cohen *Phys Rep*, in press) and X-ray variability of TTS.

The phenomenon of anisotropic mass loss from TTS (Cohen *PASP* **94**,266) has also been addressed by Jankovics, Appenzeller, & Krauter (*PASP* **95**,883), who found a strong tendency for forbidden emission lines in some TTS to be blue-shifted relative to the stars. If these lines represented an interaction between a bipolar stellar outflow and the ambient medium, then the red-shifted lines might be hidden by an extended circumstellar dust disk (Appenzeller *Rev Mex AA* **7**,151). Microwave CO emission provides a ready probe of the existence of systematic, high-velocity outflows from TTS, for example, that near T Tau itself (Kutner *et al*, *ApJ Lett* **259**,L35). These high-velocity wings to the cloud CO line profiles also occur toward other TTS (Edwards & Snell *ApJ* **261**,151; Calvet, Canto, & Rodriguez *ApJ* **268**,739). However, it is important to recognize that these bipolar molecular flows characterize primarily those TTS associated with HH nebulae (e.g., Edwards & Snell *ApJ* **281**,237), rather than isolated stars, a point little-noted in the literature.

It is important to understand the generic relationship, if any, between stars that excite HH objects and typical TTS. From near- and airborne far-infrared studies, Cohen and his colleagues (*ApJ* **265**,877;**278**,671;**281**,250) have discovered a number of optically invisible HH-exciting stars with rather low (<100 $L_\odot$) bolometric luminosities (except for the much more luminous R Mon, at around 800 $L_\odot$). These stars have been identified as extremely early (<10^5 yr) precursors of visible TTS (Cohen *Rev Mex AA* **7**,241), whose mass loss is highly non-spherical and which sometimes possess substantial dust disks (e.g., Cohen *Phys Rep*, in press). These same objects are sometimes associated with well-collimated radio and optical emission "jets" (Cohen, Bieging, & Schwartz *ApJ* **253**,707; Bieging & Cohen, *ApJ Lett*, in press; Mundt & Fried, *ApJ Lett* **274**,L83), and at least one (L1551IRS5) may be a binary (Bieging & Cohen, *ApJ Lett*, in press).

Along more global, evolutionary lines, Appenzeller, Jankovics, & Krautter (*AA Supp* **53**,291) have presented low-resolution optical spectroscopy for 60 southern TTS, in Cha and in Lup, sufficient to permit spectral classification. These were supplemented with a number of medium-resolution spectra and with near-infrared photometry for 28 stars. Their spectra are displayed, and classified, in the identical format and by the same criteria as in the massive Cohen-Kuhi sample (*ApJ Supp* **41**,743), although HR diagrams were not constructed. Stahler (*ApJ* **274**,822) has investigated when TTS first become visible objects [extending the earlier spherical protostellar collapse calculations by Stahler, Shu, & Taam (*ApJ* **241**,637;**242**,226) to other masses]. He found a remarkable coincidence between the "birthline" - the locus in the HR diagram along which stars of different mass first become visible - and the upper envelope of all the TTS observed by Cohen and Kuhi. Previously determined ages for the youngest stars (∿10^5 yr), based on isochrones, were underestimated and should be augmented by the duration of the accretion phase (about 10^5 yr for a 1 $M_\odot$ star). One would expect to find a large population in dark clouds of objects with similar, but less extreme, properties to the TTS, representing the "post-T Tauri" phase of evolution. The bulk of this group is yet unknown, although the optical characteristics of a few serendipitous X-ray sources suggest their post-TTS nature (Mundt *et al*, *ApJ* **269**,229). These may differ from the TTS principally because they lack extensive circumstellar envelopes (Walter & Mundt *Rev Mex AA* **7**,195), which may have been dissipated at the end of the true TTS phase (Walter *Rev Mex AA* **7**,196). However, there remains confusion as to the distinction between weak-lines TTS and post-TTS; perhaps some low-mass stars can entirely bypass the T Tauri phase?

3. Early-Type Variable Stars
(John R. Percy)

INTRODUCTION

This report contains selected highlights in the study of early-type (primarily pulsating) variable stars, mid-1981 to mid-1984. The reader should consult *Astronomy and Astrophysics Abstracts* for a more complete bibliography.

The following reviews and conference proceedings have recently been published: *Workshop on Pulsating B Stars*, ed. Auvergne *et al* (Obs de Nice, 1981), *IAU Symposium 98* on "Be Stars", Jaschek & Groth, eds (D Reidel 1982), "Rapid Variability of Early-Type Stars", Harmanec & Pavlovski, eds (*Hvar Obs Bull* **7**,1) and *B Stars with and without emission lines*, Underhill & Doazan, eds (NASA SP-456 & 457, 1982). A conference on "The Origin of Non-radiative Heating and Momentum in Hot Stars", with several papers relevant to variability, was held at NASA Goddard Space Flight Center in 1984. A comprehensive review of the photometric variability of Be stars was published by Harmanec (in *Advances in Photoelectric Photometry*, Wolpert and Genet, eds, Fairborn Obs 1983).

CLASSICAL β CEP STARS

Searches continued, both in the field (Sterken & Jerzykiewicz *Acta Astr* **33**,89) and in clusters: NGC 3293 (Balona & Engelbrecht *MN* **202**,293.; Shobbrook *MN* **205**,1215) and NGC 6231 (Balona *MN* **203**,1041; Balona & Shobbrook *MN* **205**,309). In NGC 6231, β Cep stars have been found near the main sequence. Early-type variables are also being discovered through surveys such as the UBV photometric campaign on bright Be stars co-ordinated by Harmanec and colleagues, and the ESO southern survey co-ordinated by Sterken.

An indirect result of the study of β Cep stars in clusters is the conclusion (supported by other studies) that the absolute magnitudes of early B stars are 0.6 magnitude fainter than previously thought. The new absolute magnitudes for β Cep stars lead to the conclusion that most of them pulsate in the fundamental radial mode. In 16 Lac, this conclusion is independently reached using the mass and radius derived from the spectroscopic-eclipsing binary nature of the star (Garrido *et al*, *AA* **122**,193).

Waelkins & Rufener, *AA* **121**,45 have noted a deficiency of β Cep stars in *close* binary systems. Agrawal *et al* (*MN* **208**,845) have studied the X-ray emission from β Cep stars, and found it to be normal for stars of their luminosity. Percy (*JRAS Can*, **78**,241) has emphasized the continuing need to confirm the variability and properties of suspected β Cep stars. The value of intensive, co-ordinated observations of early-type variables has also been stressed and demonstrated.

INDIVIDUAL OBJECTS

Published observations of β Cru and ν Eri by van Hoof have been reanalyzed by Cuypers (*AA* **127**,186) and Cuypers & Goossens (*AA Supp* **45**,487) respectively, using more effective period determination techniques. The interpretation of BW Vul, the largest-amplitude β Cep star, is further complicated by the discovery of cyclic (O-C) variations, ascribed to a binary companion or a second pulsation mode (Odell *PASP* **96**,657). In β Cen, an unusual phase relation between light and radial velocity has been noted (Kubiak & Seggewiss *Acta Astron* **32**,371). The amplitudes of some of the pulsation modes in 16 Lac have decreased considerably (Garrido *et al*, *AA* **122**,193; Le Contel *et al*, *AA* **118**,294). Waelkins & Rufener (*AA* **119**,279) have discovered a β Cep star at a moderately high galactic latitude. The 6.02 (or 12.04) day variability in the UV spectrum of β Cep remains unexplained (Fischel & Sparks in *The Universe at UV Wavelengths*, Chapman, ed,

NASA CP-2171, Goddard Sp Fl Cent, 1981, p 217; Fahey *et al* in *Advances in UV Astronomy*, Kondo, Mead, & Chapman, eds, NASA CP-2238, Goddard Sp Fl Cent, 1982, p 442).

LINE PROFILE VARIATIONS IN EARLY-TYPE STARS

Line Profile variations have been observed in the spectra of OB stars with a variety of temperatures, luminosities and rotational velocities. These have generally been ascribed to non-radial pulsation, though in some cases, it is not possible to rule out other interpretations. The observations require high signal-to-noise ratios, and good wavelength and time resolution, and have therefore been made at only a few suitably-equipped observatories. Photometric variations have been observed in some of these same kinds of stars, but progress in understanding the relation between the two forms of variability has been hampered by the lack of extensive simulataneous spectroscopic and photometric observations of the same stars. Statistics of the occurrence of these variations are also incomplete, because the stars which have been observed have not always been chosen in unbiased fashion. The time scales of the variations (generally 0.3 to 2 days) create a practical problem which can best be solved by contemporaneous observations from different longitudes. Successful spectroscopic-photometric campaigns on ε Per and several Be stars have recently been carried out.

Line profile variations were first studied intensively in several sharp-lined OB stars, notably 53 Per (Smith *ApJ* **215**,574; Smith *et al, ApJ* **282**,226), and were attributed to low-order prograde non-radial pulsations. The technical feasibility of observing similar variations in broad-lined stars was not fully appreciated at the time, but variations have now been observed in the lines of moderately-rotating B stars such as ε Per (Penrod, 1984 priv comm; Smith *ApJ* **288**, in press), η Lep, δ Sco and α Vir (Walker *et al, PASP* **94**,143; Smith 1984 preprint), early-type supergiants such as γ Ara and θ Ara (Baade *AA* **124**,211; Baade & Ferlet, *AA* **140**,72), rapidly-rotating Be stars such as 28 CMa (Baade *AA*, **105**,65), η Cen (Baade *AA* **124**,283), μ Cen (Baade *AA* **135**,101), λ Eri (Bolton in *Be Stars*), EM Cep (Hilditch *et al, MN* **200**,1153) and others (Baade *AA* **134**,105), and Oe stars, notably ζ Oph (Vogt & Penrod *ApJ* **275**,661). These line profile variations have been ascribed to low, medium and high-order non-radial modes (often present at the same time); these are primarily retrograde in the rapidly-rotating stars. The variations are often complex, and variable with time, and it is premature to try to discuss their nature in this brief report. Several groups, particularly those of Baade, Bolton, Harmanec, Penrod, Smith & Walker, are presently active in this area.

Photometric variations have been observed in stars of these same kinds: 53 Per (Smith *et al, ApJ* **282**,226) and similar stars (Burki *AA* **121**,211; Waelkins & Rufener *AA Supp* **52**,21), Bn stars (Jerzykiewicz & Sterken in *Workshop on Pulsating B Stars*, Auvergne *et al*, eds, Obs de Nice 1981), B supergiants (most recently by Percy & Welch *PASP* **95**,491, and in numerous Be stars [see Percy (*IAU Working Group on Be Stars, Newsletter* No 6,8) for a reasonably complete list]. In some Be stars such as LQ And (Percy *AJ* **88**,427), the photometric variations are stable, but in others, they change with time, often in a complex way. For instance, Baade (*AA* **110**,L15) found no significant photometric variations in 28 CMa at the period of the line profile variations (1.37 days), but he found small variations at 0.435 day. Later, Stagg (1984 preprint) found larger, significant variations at both of these periods. Harmanec (*IBVS* 2506) found a permanent spectroscopic and photometric period of 1.57 days in o And, but the amplitude varies with time. A spectroscopic-photometric campaign on o And and four similar Be stars has revealed a wealth of interesting (but as yet incompletely-understood) behavior. Stagg has completed a major survey of rapid photometric variability in bright southern Be stars.

In some of these Be and Oe stars, changes in line profile and photometric variations appear to be related to emission episodes, notably in λ Eri (Bolton in *Be Stars*, D Reidel; Penrod, 1984 priv comm) and ζ Oph (Vogt & Penrod *ApJ*

275,661). A possibly similar phenomenon has been observed in the B supergiant ρ Leo (Smith & Ebbets *ApJ* **247**,158). The search for such a relation is hampered by the lack of long-term systematic spectroscopic (and photometric) observations.

The relation between pulsation and mass loss in stars in general has been discussed by Willson & Bowen (*Nature* **314**,429). In early-type stars, the pioneering work of Smith and McCall, who observed a "bouncing shell" phenomenon in many β Cep stars and who proposed that a small fraction of the shell is lost to the star in each bounce, has been followed by studies by Burger *et al* (*AA* **107**,320; *AA* **109**,289) and Le Contel and Morel (*AA* **107**,406), which suggest that matter is being ejected from the (radially-pulsating) β Cep stars. A detailed study of shell ejection in non-radially pulsating stars has not yet been pursued, but Smith (*ApJ* **288**, in press) and Penrod (1984 priv comm) have both noted that the non-radial pulsations observed in stars such as ε Per are primarily radially-directed, and frequently approach supersonic velocities.

Narrow components have recently been observed in the UV spectra of several early-type stars including ε Per (Henrichs *et al*, *ApJ* **268**,807; Gry *et al*, *AA* **137**,29), and have been ascribed to discrete "puffs" of material ejected from the stars into the circumstellar environment - perhaps driven by pulsation. For a different interpretation, however, see Howarth (*MN* **206**, 625).

THEORETICAL STUDIES

A picture is emerging in which non-radial pulsation is found in a wide assortment of early-type stars, and radial pulsation is found only in a very restricted group - the classical β Cep stars. The cause of pulsation is still unknown. Stellingwerf's (*AJ* **83**,1184) ionization-zone mechanism, subsequently discussed by Lee & Osaki (*PAS Japan* **34**,39) certainly contributed to destabilization in a narrow region of the H-R diagram. To produce more widespread destabilization, Osaki's (*ApJ* **189**,469) convection-rotation-pulsation mechanism remains an interesting and viable one. Several recent studies have dealt specifically with the interaction of rotation and pulsation (Ando *MN* **197**,1139; *PAS Japan* **35**,343; Carroll & Hansen *ApJ* **263**,352; Cox *PASP* **96**,577), and this appears to be a promising direction for study. There is serious need to bring interested theoreticians and observers of all kinds together, to begin to make sense of the rapidly-accumulating mass of data. Other interesting developments include the discovery of a possible relation between pulsation and tidal-rotational effects in Vir (Smith 1984 preprint) and the studies of supergiant pulsation by Lovy *et al* (*AA* **133**,307) and de Jager (*AA* **138**, 246).

MISCELLANEOUS STUDIES OF OTHER EARLY-TYPE VARIABLES

The ultra-short-period B type variables originally proposed by Jakate appear not to exist (Balona *IBVS* 2120; Shaw *et al*, *IBVS* 2288). The mid-B to early A type "Maia variables" remain hypothetical (as Struve called them), but they may possibly be very elusive non-radial pulsators. Pulsations have been discovered and studied in a variety of degenerate stars, most recently the DB white dwarfs, driven by ionziation zones of helium, and the very hot "PG 1159-035 stars", driven by ionization zones of heavier elements. Studies of the pulsating helium star BD+13°3229 continue to serve as a useful probe of its structure and evolution (Hill *et al*, *MN* **197**,81). There is some evidence for pulsations in Wolf-Rayet stars, with a time scale of 0.5 day (Vreux, 1984 preprint). Observational (e.g. Humphreys & Davidson *Science* **223**,243) and theoretical (e.g. Stothers & Chin *ApJ* **264**,583; de Jager *AA* **138**,246) studies of S Dor (Hubble-Sandage) variables suggest that these are luminous early-type stars which, due to hydrodynamic instabilities, are undergoing extensive mass loss. Long-term monitoring of bright Be stars is being carried out in a systematic way, spectroscopically by Barker, Shore and others, and photometrically by Harmanec and his colleagues and collaborators. Progress and results from these campaigns are reported regularly in the *Newsletter* of the IAU Working Group on Be Stars. See papers by Doazan *et al* (*AA* **115**,138; *AA Supp* **50**,481; *AA* **128**,171), Horn *et al* (*Bull Astron Inst Czechoslovakia* **33**,308) and Harmanec *et al* (*IBVS* 2447) for representative results. Magnetic fields have been observed in a number of Bp stars (Borra *et al, ApJ Supp* **53**,151) and an oblique magnetic rotator model for these stars has been developed (Shore & Bolton, 1982 preprint; Bolton *Hvar Obs Bull* **7**,1).

4. Cepheids
(J.D. Fernie)

INTRODUCTION

The following survey of the cepheid literature covers approximately the period August 1981 to September 1984. It omits most theoretical topics since these are covered elsewhere in the Commission 27 Report, and, in the interests of space, it also omits what in my judgement are minor investigations of a routine nature, e.g. a few observations of an already well-observed star.

IAU Colloquium 82 on *Cepheids: Observation and Theory* was held in May 1984. Its proceedings, edited by B.F. Madore and published by Cambridge University Press, should appear in early 1985. In it will be found major review articles and contributed papers that expand on many topics which necessarily receive only very brief mention here.

Each subsection below starts with a listing of papers followed by a short commentary in which I have tried to summarize the advances in that area over the past three years.

SURVEYS

Caldwell & Coulson (*So Afr Ast Obs Circ* No 8,1); Dean (*MN* **197**,779), Diethelm (*AA* **124**,108); Diethelm & Tammann (*AA Supp* **47**,335); Eggen (*AJ* **88**,361,998); Gieren (*ApJ* **260**,208; *ApJ Supp* **46**,287;**47**,315); Harris (*AJ* **86**,1192;**88**,507); Kwee & Diethelm (*AA Supp* **55**,77); Moffett & Barnes (*ApJ Supp* **55**,389); Serrano (*Rev Mex AA* **8**,131); van Genderen (*AA Supp* **52**,423); Wayman, Stift, & Butler (*AA Supp* **56**,169); Welch *et al* (*ApJ Suppl* **54**,547).

The period has been marked by the production of much new and generally high-precision data for cepheids. Notable among such work is the BVRI photometry of Moffett and Barnes for northern cepheids, similar data (with radial velocities) from Gieren for southern cepheids, and the work of Caldwell & Coulson, Dean, Harris, van Genderen, and Wayman *et al* mainly on Magellanic Cloud cepheids. New ground was broken by Welch *et al* in providing JHK photometry for many Galactic cepheids, by Dean in providing DDO photometry, and by Harris in providing Washington system photometry. Most of the remaining papers, notably those of Eggen and Diethelm were concerned with the photometric properties and classification of cepheids.

NEWLY-DISCOVERED CEPHEIDS

Balona (*MN* **201**,105); Berthold (*IBVS* 2192); Eggen (*IBVS* 2106 & 2331, *AJ* **88**,379).

There seems to have been relatively little activity in this area; in particular, I am unaware of any searches (other than perhaps Balona's) specifically aimed at discovering new cepheids.

THE P-L-C RELATION

Caldwell (*Obs* **103**,244); Clube & Dawe (*AA* **122**,255); Eggen (*AJ* **88**,386); Fernie & McGonegal (*ApJ* **275**,732); Gieren (*ApJ* **282**,650); Ivanov (*Ap Sp Sc* **79**,107); Ivanov & Nikolov (*Ap Sp Sc* **94**,191); Karimova & Pavlovskaya (*Sov Ast* **25**,56); Madore (*ApJ* **253**,575); McAlary *et al* (*ApJ* **273**,539); McAlary, Madore, & Davis (*ApJ* **276**,487); McGonegal *et al* (*ApJ Lett* **257**,L33;*ApJ* **269**,641); Opolski & Ciuria (*IBVS* 2528); Stift (*AA* **112**,149); Stothers (*ApJ* **274**,20); van Genderen (*AA* **124**,223); Wayman, Stift, & Butler (*AA Supp* **56**,169); Zsoldos (*AN* **305**,33).

A number of problems and controversies continue to plague the calibration (indeed the form) of the P-L-C. The value of the coefficient multiplying (B-V) continues to be debated. While most workers still accept a value of about 2.7, others such as Madore and Opolski & Ciurla arrive at values of up to 6. At the other extreme Stift, Clube, & Dawe, and Wayman *et al* have argued strenuously that there is no real color term at all and that discussion should be conducted in terms of the P-L relation alone. The purely observational aspects are unclear; Fernie & McGonegal were unable to find any color term from twenty-seven Galactic cepheids in clusters and associations, but earlier work by Martin *et al* on LMC cepheids definitely indicated the presence of a color term in the P-L-C.

While there is now relatively little argument over the slope of the P-L relation, its zero-point may still be uncertain by as much as half-a-magnitude. A series of four-color investigations by Schmidt on clusters containing cepheids suggests the previously determined moduli were too large by up to 0.5 mag, requiring a concomitant lowering of cepheid luminosities by this amount. This result has received some support from the application to cepheid clusters of a new β-index/M_V calibration by Balona & Shobbrook, using, however, Schmidt's measurements of β. Balona & Shobbrook caution that even a small systematic error in the observed β could lead to significant changes in distance moduli. Reduced cepheid luminosities lead to aggravation of difficulties in reconciling pulsation and evolutionary masses of cepheids, not to mention, of course, the implications for the distance scale of the universe. The verification of Schmidt's results is therefore of fundamental importance.

INDIVIDUAL CEPHEIDS IN CLUSTERS AND ASSOCIATIONS

Eggen (*AJ* **88**,197,379, *ApJ Supp* **50**,199); Fitzgerald & Miller (*PASP* **95**,361); Forbes (*AJ* **87**,1022); Hodge & Lee (*ApJ* **276**,509); Schmidt (*AJ* **88**,104); Turner (*AJ* **88**,650, *JRAS Can* **77**,31, *PASP* **94**,422,655,1003); Turner & Evans (*ApJ* **283**,254); van den Bergh, Brosterhus, & Alcaino (*ApJ Supp* **50**,529); van den Bergh & Brosterhus (*ApJ Supp* **53**,765).

Systematic searches for new cases of cepheids in clusters or associations, notably the searches of van den Bergh and collaborators, have been generally unsuccessful, but more or less accidental discoveries by Eggen and Turner have revealed a number of new possible cases. Forbes' discovery that GY Sge is a member of an association is of particular importance because of the long (51 days) period of that star.

CEPHEID BINARIES

Balona (*Obs* **103**,163); Coulson (*MN* **203**,925;**205**,1135); Eichendorf *et al* (*AA* **109**,274); Evans (*ApJ* **272**,214;**281**,760), Gieren (*ApJ Supp* **49**,1); Harris, Olszewski, & Wallerstein (*AJ* **89**,119), Lloyd Evans (*MN* **199**,925, *Obs* **104**,26); McNamara & Feltz (*IAU Coll 59*,389); Russo, Sollazzo, & Coppola (*AA* **102**,20); van Genderen (*AA* **100**,175).

Most of these papers deal with individual cases, although Lloyd Evans, Gieren, and Russo *et al* present the results of surveys. Lloyd Evans found 18% of the cepheids he surveyed are cepheids, based on radial velocity data, while Russo *et al* found a 25% rate from VBLUW photometry. Gieren's results supported a binary rate in the range of 20% - 40%.

Individual cases of particular interest included Evans' discovery from IUE data that X Cyg is not a binary despite several other indicators to the contrary, and the work of Harris, Olszewski, & Wallerstein on AU Peg, a Pop II cepheid which appears to have a compact object as companion and to have undergone a fairly traumatic evolution.

COLOR AND REDDENING SCALES

Coppola, Russo, & Sollazzo (*Ap Sp Sc* **86**,157); Harris (*AJ* **86**,1192); Madore (*ApJ* **253**,575); van Genderen (*AA* **119**,192).

Activity in this area of interstellar reddening corrections to cepheid colors has been surprisingly low compared to earlier but still recent years. The Coppola, Russo, & Sollazzo paper appears to be the only one to address Galactic cepheids, the other three being concerned with Magellanic Cloud cepheids.

ATMOSPHERES AND CHEMICAL COMPOSITION

Becker & Cox (*ApJ* **260**,707); Benz & Mayor (*AA* **111**,224); Böhm-Vitense & Parsons (*ApJ* **266**,171); Boyarchuk (*Bull Crimean Ap Obs* **64**,1); Giridhar (*J Ap Ast India* **4**,75); Harris (*AJ* **86**,1192;**88**,507); Harris, Olszewski, & Wallerstein (*AJ* **89**,119); Harris & Pilachowski (*ApJ* **282**,655); Henden, Cornett, & Schmidt (*PASP* **96**,310); Rautela & Joshi (*J Ap Ast India* **4**,1); Schmidt & Parsons (*ApJ* **279**,202,215, *ApJ Supp* **48**,185); Wallerstein (*PASP* **95**,422).

Abundance studies, both spectroscopic and photometric, have shown that in general Galactic cepheids have metallicities within about 0.3 of the sun's value, the true range probably being less than this. There is, however, a discernible decline in metallicity with increasing galactocentric distance that is in reasonable agreement with that found from other objects. Likewise, photometric abundances found for cepheids in the Magellanic Clouds are in good agreement with those determined from other sources.

Schmidt & Parsons have used IUE data to study chromospheric emission in cepheids, which they find to be strongly dependent on period and with mean flux levels if anything somewhat lower than in similar non-variable stars. The multimode cepheid, TU CAS, shows no chromospheric emission, which is consistent with other short-period cepheids. Böhm-Vitense & Parsons report an unsuccessful search for X-ray emission from cepheids.

MASSES AND RADII

Burki (*AA* **133**,185), Coppola, Russo, & Sollazzo (*Ap Sp Sc* **86**,157); Evans (*ApJ* **272**,214); Fernie (*ApJ* **282**,641); Gieren (*ApJ* **260**,208; *PASP* **94**,960); Imbert (*AA Supp* **53**,85); Ivanov (*Ap Sp Sc* **79**,107); Magee, Merts, & Huebner (*ApJ* **283**,264); Russo (*Ap Sp Sc* **90**,147); Simon (*ApJ Lett* **260**,L87).

The problem of the cepheid mass anomaly is not yet resolved to everyone's satisfaction, and may have worsened if Schmidt's new distance scale based on cluster cepheids is upheld. An interesting suggestion by Simon that the problem may be caused by incorrect opacities in theoretical models was rejected by Magee, Merts, & Huebner.

The determination of cepheid radii also continues to present problems. A major survey by Fernie has revealed serious discrepancies between theoretical radii and those deduced from luminosities and temperatures for long-period cepheids in clusters and associations, as well as discrepancies between both these and other methods. No solution to the problem is presently apparent.

DOUBLE-MODE CEPHEIDS

Aikawa (*MN* **204**,1193); Antonello & Mantegazza (*AA* **133**,52 & *IBVS* 2411); Balona (*Obs* **103**,163); Barrell (*MN* **196**,357;**200**,127;**200**,139;**204**,1); Fuhrmann & Schult (*IBVS* 2456); Mantegazza (*AA* **118**,321).

Barrell has made a specific search for new cases of double-mode cepheids without finding any. The only recent new case is that of CO Aur, proposed by Antonello & Mantegazza. It does, however, have unusual properties, and although Fuhrmann & Schult have announced confirmation, Balona has queried whether any period other than the fundamental really exists.

Other work by Barrell continues to show the double-mode cepheids as being no different in all other properties from single-mode cepheids, with the possible exception of Hα emission, which remains to be checked. The main challenges of double mode cepheids remain the reconciliation of their theoretical masses with evolutionary masses, and a theoretical explanation for their existence at all.

TYPE II CEPHEIDS

Bond, Carney, & Grauer (*PASP* **96**,176); Carson & Stothers (*ApJ* **259**,740); Harris, Olszewski, & Wallerstein (*AJ* **89**,119); Harris & Wallerstein (*AJ* **89**,379); Lloyd Evans (*Obs* **103**,276); Luck & Bond (*ApJ* **279**,729); Petersen & Hansen (*AA* **134**,319); Zinn & King (*ApJ* **262**,700).

Harris & Wallerstein have confirmed a long-standing suspicion that the so-called Population II field cepheids in fact form a decidedly inhomogeneous group. In particular, they include members that are relatively metal-rich (probably from the old disk population) that have no counterparts among the globular cluster variables. Thus Harris has suggested use of the term Type II cepheid rather than Population II cepheid to emphasize that not all Type IIs are Population IIs. An interesting sidelight is that truly metal-poor cepheids are underrepresented in the field compared to clusters.

Carson & Stothers report considerable success in fitting theoretical models to the observed properties of BL Her stars.

CEPHEID-LIKE STARS

Bond, Carney, & Grauer (*PASP* **96**,176); Fernie (*ApJ* **265**,999); Luck & Bond (*ApJ* **279**,729); Luck, Lambert, & Bond (*PASP* **95**,413); Sasselov (*IBVS* 2314,2387 & *Ap Sp Sc* **102**,161); Takeuti (*Obs* **103**,292).

There is a growing interest in variable stars which are similar to cepheids in luminosity and temperature, but which are more erratic in behavior and probably lie outside the usual instability strip. Foremost among these is a group called either 89 Her or UU Her stars. They are typically high-latitude F-supergiants of long period (>40 days), low amplitude ($\sim$0.1 mag), and fluctuating lightcurve shapes; at least some of them appear to have solar metallicities. Their pulsation mechanisms, evolutionary histories, and basic nature (e.g. high- or low-mass) are uncertain at present.

MISCELLANEOUS

Arellano Ferro (*ApJ* **274**,755); Efremov (*Sov Ast Lett* **9**,51); Fernie (*IAU Symp 105*,441); Grivnev (*Sov Ast Lett* **9**,287); Joshi (*Bull Ast Soc India* **10**,217); Kamper, Evans, & Lyons (*JRAS Can* **78**,173); Karimova & Pavlovskaya (*Sov Ast* **25**,56); Opolski (*IBVS* 2425); Simon & Teays (*ApJ* **265**,996); Szabados (*Ast Sp Sc* **96**,185).

These investigations cover a wide range of topics from period changes in cepheids, to Fourier decomposition of their velocity curves, to galactic structure and kinematics. Arellano Ferro has made the intriguing discovery that Polaris's amplitude of variability has been steadily diminishing for many decades.

CONCLUDING REMARKS

In reviewing these mainly observational aspects of cepheid research over the past three years I would say that there has been a great improvement in available data, both photometric and spectroscopic, and that there is now a need to compile all this information into one uniform database for general application. Among the more specific problems that confront us is the zeropoint of the period-luminosity relations, the size of any color term in the P-L-C relation, and the continuing anomalies in mass and radius determinations.

5. δ Scuti Stars
(D. W. Kurtz)

The nomenclature for δ Scuti stars remains confusing. In the literature of the last three years these stars are referred to as δ Scuti stars, dwarfs Cepheids, RRs stars, RR Lyrae stars (erroneously), AI Vel stars, and Ultra-short period Cepheids (USPC stars). Part of the reason for this confusion is that there are both Population I and Population II δ Scuti stars. In general, known Population I stars are called δ Scuti stars while known Population II stars are called either dwarf Cepheids or AI Vel stars. Note, however, that AI Vel itself, and many stars labelled dwarf Cepheids, are actually Population I δ Scuti stars. Both populations are called USPC stars. The RRs designation is now archaic. Caution is therefore advisable in making inferences about the properties of a star based on a particular author's choice of nomenclature.

No new reviews of the δ Scuti stars have been written in the last three years. Students new to the field are advised to start with the 1979 reviews of Breger (*PASP* **91**,5) and Eggen (*ApJ Supp* **41**,413).

Most of the literature of the last three years presents *brief* observational studies of the light variations of δ Scuti stars. It is clear that some δ Scuti stars pulsate in a single radial mode, some pulsate in more than one radial mode, many pulsate in several radial and non-radial modes, and some *may* switch modes. Observationally, *extensive* observation sets of selected multiperiodic δ Scuti stars are needed now, rather than short studies of known δ Scuti stars or the discovery of new variables. Stellar seismology (Christensen-Dalsgaard in *Proc Workshop on Sp Research Prospects in Stellar Activity and Variability* Mangeney, ed, Obs de Paris 1984) is an exciting new field which promises the direct study of the interiors of the sun and stars. For δ Scuti stars to be a part of this field we need *complete* frequency solutions for multiperiodic δ Scuti stars.

Jorgensen (*AA* **108**,99) and Jorgensen & Hansen (*AA* **133**,165) have discovered three Population II δ Scuti stars in ω Cen. These stars are of considerable interest both for the study of Population II δ Scuti stars (of which there are only a few confirmed examples) and for the study of blue stragglers.

Antonello *et al* (*IAU Coll 69*, 27), Antonello (*IAU Coll 69*, 33), Antonello & Conconi (*Ap Sp Sc* **88**,185), and Antonello *et al* (*Ap Sp Sc* **78**,435) have looked at relationships among amplitude, period, luminosity, temperature, and radius for δ Scuti stars. Wiertz & van Genderen (*AA* **121**,35) have obtained VBLUW photometry of some δ Scuti, δ Del, and Am stars. Verma *et al* (*Ap Sp Sc* **92**,173) have shown that 8 δ Scuti stars have normal JHK colors. Yang *et al* (*PASP* **94**,317) obtained radial velocities of β Cas accurate to ±0.2 km s^{-1}. Fracassini & Pasinetti (*AA* **107**,326) have found emission features in IUE observations of the Mg II h and k lines in several δ Scuti stars. Peniche *et al* (*PASP* **93**,735) and Ortega *et al* (*Rev Mex AA* **8**,45) have continued their spectral classification of δ Scuti stars. Tsvetkov (*Sov Astr* **26**, No 6) has discussed radial pulsation in 83 δ Scuti stars. Saez *et al* (*AA* **101**,259) have discussed the nature of the δ Del stars.

Smith (*ApJ* **254**,242) has looked at line profile variations in nine δ Scuti stars. He determined pulsation modes consistent with those independently found in photometric studies of the same stars.

Theoretical studies of evolutionary sequences and pulsation parameters have been made by Fitch (*ApJ* **249**,218), Tsvetkov (*Ap Sp Sc* **89**,435), Andreasen (*AA* **121**, 250), and Andreasen, Hejlesen & Petersen (*AA* **121**,241). Further discussion of the theory of pulsating stars can be found elsewhere in this volume.

For information on individual δ Scuti stars a good starting point is the list of Halprin & Moon (*Ap Sp Sc* **91**,43). The following references partially update that list. In addition, many notes on individual δ Scuti stars can be found in the *IBVS*. The usual caveat applies here: the following list is not complete.

Studies of δ Scuti stars are: HD 432 (HR 21, β Cas) Yang *et al*, *PASP* **94**,317; HD 2628 (HR 114) Ibanoğlu *et al*, *Rev Mex AA* **5**,261; HD 24832 (HR 1225) DuPuy, Collins, & Swingler *PASP* **94**,177; HD 26322 (HR 1287, 44 Tau) Ibanoğlu *et al*, *Rev Mex AA* **5**,261; HD 27397 (HR 1351, 57 Tau) McNamara *Proc SW Regional Conf Astron Ap* **8**,43; HD 28024 (HR 1392) Bossi *et al*, *AA Supp* **53**,395; HD 50420 (HR 2557) Bossi *et al*, *Ap Sp Sc* **89**,429; HD 55595 (HR 2724) Baade & Stahl *AA* **114**,131; HD 64191 (AD CMi) Balona & Stobie *SAAO Circ* **7**,19; HD 66260 Helt, *AA Supp* **56**,457; HD 67523 (HR 3185, ρ Pup) Fracassini *et al*, *Ap Sp Sc* **97**,323, Balona & Stobie *SAAO Circ* **7**,19; HD 69213 (AI Vel) Sturch & Wu *PASP* **95**,211; HD 73576 Bossi *et al*, *Ap Sp Sc* **85**,25; HD 73857 (VZ Cnc) Balona & Stobie *SAAO Circ* **7**,19, Sturch & Wu *PASP* **95**,211; HD 110411 (HR 4828, ρ Vir) Antonello & Mantegazza *AA Supp* **49**,703; HD 110951 (HR 4847, 32 Vir) Bartolini *et al*, *AA Supp* **53**,139; HD 115308 (HR 5005) Peña *et al*, *AA Supp* **53**,81; HD 115604 (HR 5017, 20 CVn) Bossi *et al*, *AA Supp* **53**,399, Peña & Gonzalez *AJ* **86**,1679; HD 116994 (V743 Cen) Balona & Stobie *SAAO Circ* **7**,19; HD 124675 (HR 5328/9, κ Boo) Ibanoğlu *et al*, *Rev Mex AA* **5**,261; HD 129798 (HR 5492) Bossi *et al*, *Ap Sp Sc* **79**,463; HD 143466 (HR 5960, CL Dra) DuPuy & Burgoyne *PASP* **95**,61; HD 158741 (HR 6522) Waelkens & Bartholdi *AA Supp* **52**,1; HD 160589 (V703 Sco) Balona & Stobie *SAAO Circ* **7**,19; HD 174553 Guerrero & Mantegazza *Ap Sp Sc* **86**,139; HD 177392 Schoneich & Želwanowa *AN* **302**,181; HD 181333 (HR 7331) Ibanoğlu *et al*, *Rev Mex AA* **5**,261; HD 197461 (HR 7928, δ Del) Fernandes *BAV Rundbrief* 30.

Jahrg,42; HD 199908 (DQ Cep) Pena *et al*, *AA Supp* **51**,71; HD 208664 Kurtz *MN* **200**,497; HD 211336 (HR 8494, ε Cep) Fernandes *BAV Rundbrief* 30.Jahrg,42; HD 223338 (BS Aqr) Kilambi *Bull Astron Soc India* **10**,234; BD +12°3028 (DY Her) Szeidl & Mahdy *Commun Konkoly Obs* No 75; BD +28°1494 Broglia & Conconi *AA* **100**,201; BD +37°2635 (YZ Boo) Szeidl & Mahdy *Commun Konkoly Obs* No 75; BD +43°1894 Yamasaki *et al*, *PASP* **95**,447; SX Phe, Coates, Halprin & Thompson *MN* **199**,135, Sturch & Wu *PASP* **95**,211; XX Cyg, Szeidl & Mahdy *Commun Konkoly Obs* No 75, Jones *PASP* **94**,289.

RAPIDLY OSCILLATING Ap STARS

The Rapidly Oscillating Ap Stars are cool magnetic Ap Stars which oscillate with periods in the range of 4 to 15 minutes with peak-to-peak light variations of $\Delta B \leq 0.016$ mag. Their oscillations have been described as high overtone non-radial p-modes of low degree which have the axis of oscillation aligned with the magnetic axis of the star which is oblique to the rotation axis. Kurtz (*MN* **200**, 807) has given an introduction and review of these stars. A list of nine Rapidly Oscillating Ap stars with references to the rest of the literature can be found in Kurtz (*MN* **209**,841).

The Rapidly Oscillating Ap stars lie within the δ Scuti instability strip, but they may or may not be δ Scuti stars. The problem of their very short periods (high pulsational overtone) and pulsation axis orientation have led Shibihashi (*ApJ Lett* **275**,L5) to suggest that their oscillations are excited by overstable magnetic convection. Cox (*ApJ* **280**,220) has given further discussion of Shibihashi's ideas. If this model is correct, then the Rapidly Oscillating Ap stars are phenomenologically different from the δ Scuti stars whose instability is due to the κ-mechanism.

There are also pulsating Ap stars with longer periods which may belong to either the δ Scuti class or the Rapidly Oscillating Ap class (presuming those are different). These stars are (probably magnetic) Ap stars which show normal Scuti light variations. Weiss (*AA* **128**,1) gives a discussion of them and further references.

6. RR Lyrae Stars
(B. Szeidl)

Ephemerides of RR Lyrae type variables have been compiled by Tsesevich, Firmanyuk & Kreiner for the years 1982, 1983 and 1984 (Rocznik Astr Obs Krakow). Elements have been derived, revised or refined for a great number of known RR Lyr stars and for some accidentally discovered ones. The results have been mostly published in *Astron Tsirk*, *IBVS*, *JAAVSO* and *Per Zv Suppl*. Eight new RR Lyrae stars have been found in a field around γ Aql (Gessner *MVS* **10**,35) and 24 new ones in a field around M92 (Meinunger *MVS* **10**,1). Liller (*AJ* **88**,1463) found ten new variables in a field of NGC 6681 and concluded that at least some of them belong to the galactic bulge population. Forty-two new RR Lyrae variables have been discovered and periods have been determined for 40 of them in a field surrounding the globular cluster NGC 6304 (Barlow & Hesser *AJ* **86**,1044).

Photoelectric photometry has been reported on RZ Cep (Garbuzov *Astron Tsirk* **1200**,5), BC Dra (Szabados & Stobie *AA Supp* **47**,541) UW Gru (Bernard *PASP* **94**,700), AV Peg (Alaniya *IBVS* 2558), DH Peg (Alaniya *IBVS* 2558; Garbuzov *Astron Tsirk* **1200**,5; Hopp *As Sp Sc* **79**,239) and two new discoveries (Grauner *PASP* **96**,84; Tan Huisong *IBVS* 2533). The photometry of BD+16°2356 revealed that this star is the brightest representative of c-type variables with periods longer than 0.40 day (Oja *AA* **103**,339). High-phase resolution uvbyβ photometry of SU Dra, RX Eri and RR Lyr were used to find the variations of effective temperatures and surface gravities with phase. The Baade-Wesselink method was applied in order to find their absolute magnitudes (Siegel *PASP* **94**,122). Simultaneous photometric and radial velocity data of VY Ser were obtained and were used to solve for the star's radius and distance. A phase difference $\Delta\phi \sim 0.075$, between the photometric and spectroscopic radii has been found. It may be a consequence of the star's strong convection, which has not been properly taken into account (Carney & Latham *ApJ* **278**,241). A theoretical relation between stellar surface brightness and V-R color has been calculated from model atmospheres for parameters appropriate to RR Lyrae stars (Manduca & Bell *ApJ* **250**,306). In a subsequent paper this relation has been combined with new VR photometry to determine distances, radii and absolute magnitudes for X Ari and RR Lyr (Manduca *et al, ApJ* **250**,312). Photometric observations of RR Lyr in the ultraviolet have been obtained using the ANS. The observations have been compared with theoretical light curves calculated using photometry at longer wavelengths and a good agreement has been found. A bump in the observed UV light curves in the phase range 0.6 to 0.8 supports the existence of a shock (Bonnel *et al, PASP* **94**,910).

Fourier decompositions have been made of the light curves of a large sample of RR Lyrae field stars (Simon & Teays *PASP* **93**,550; *ApJ* **261**,586). Although the period of the small-amplitude pulsator, XZ Cet is 0.823 day, the Fourier decomposition coefficients of its light curve fall far from the regions occupied by the ab pulsators (Simon & Teays *BAAS* **15**,655).

Wisniewski's conclusion that the RRc variable RW Ari is an eclipsing binary simultaneously has been questioned (Goranskij & Shugarov *Per Zv* **21**,211).

Abundance analysis has been reported on VY Ser (Carney & Jones *PASP* **95**,246) and MT Tel (Prybylski *Acta Astron* **33**,141. [Fe/H] values and mean temperatures have been derived for a sample of field RRc-type stars and the Oosterhoff effect was studied for this group (Kemper *AJ* **87**,1395). Kinman, Mahaffey, & Wirtanen (*AJ* **87**,314) extended the Lick survey to three fields in the galactic anticenter (28 RR variables in 84 deg^2). Δs indices and [Fe/H] values have been determined for these stars and compared with the corresponding values for RR Lyrae stars in the north galactic poles. The entire sample shows no direct evidence for an abundance gradient (Butler *et al, AJ* **87**,353). The carbon abundances for a large number of field stars have been studied and it was found that the RR Lyrae stars have carbon-to-iron ratios which are similar to those of unevolved stars (Butler *et al, AJ* **87**,640). A number of papers have been dealing with the helium content and metal abundance of RR Lyrae stars in globular clusters (Caputo, Castellani & Tornambe *IAU Coll 68*, 309; Smith *ApJ* **250**,719; *ApJ* **281**,148; Smith & Perkins *ApJ* **261**,576; Smith & Manduca *AJ* **88**,982). The pulsational properties of field RR Lyrae stars have been studied by Castellani, Maceroni & Tosi (*AA* **102**,411;**128**,64). The Oosterhoff period shift for RR Lyr stars in globular clusters and in the galactic disk has been discussed and it has been established that it correlates with metal abundance (Sandage *ApJ* **252**,553;**252**,574). The latest review of the pulsational properties of RR Lyraes in globular clusters and the Oosterhoff problem has been given by Dickens (*Pulsations in Classical and Cataclysmic Variable Stars*, p.182). Metal abundance parameters have been determined for RR Lyrae stars in the Magellanic Clouds and population, mass, and evolutionary ages have been discussed (Butler, Demarque & Smith *ApJ* **257**,592). Some RR Lyraes have been investigated in the direction of the LMC (Connolly & Smith *BAAS* **15**,927).

Stothers (*ApJ* **274**,20) redetermined the visual absolute magnitude of metal poor RR Lyraes using five independent methods. Catalogues of proper motions, space velocities and absolute magnitudes of RR Lyrae stars have been compiled (Wan, Mao & Ji *Ann Shanghai Obs* No 2,1;No 3,110; *Proc Shanghai Astrometry Symp*, p 326).

The period changes of a few RR Lyrae stars have been investigated (IM Aql by Lada & Belserene *JAAVSO* **10**,81; XZ Cyg by Blasberg *IBVS* 2361; RR Gem by Goranskij *Astron Tsirk* **1226**,1; RR Lyr by Romanov, Fedotev & Movchan *Astron Tsirk* **1205**,4; HK Pup by Lysova *Astron Tsirk* **1261**,6). The characteristics of period behavior of field RR Lyraes have been studied by Firmanyuk (*Astron Tsirk* **1118**,1;3). A model has been given for the explanation of fluctuations of period in multiply periodic variables (Marik *Commun Konkoly Obs* No 83,225).

The Blazhko-effect has been detected in a few RR Lyrae stars. The Blazhko periods of DM Cyg (Lysova & Firmanyuk *Astron Tsirk* **1122**,3) and V5 in M3 (Panov *Per Zv* **21**,391) have been determined. Changes in the shape and amplitude of the light curves of V672 Aql (Tsesevich *Astrometr Astrofiz* **43**,3), RR CVn (Belserene &

Larson *JAAVSO* **9**,61), El Com (Wheatley *JAAVSO* **11**,17), TV CrB (Alaniya & Abuladze *Abastumani Bull* **55**,71) and V802 Cyg (Ventura *JAAVSO* **10**,90) have been found. On the other hand Glownia (*AN* **304**,45) showed that the secondary variability of RZ Cen found earlier was spurious. Photometric observations of RR Lyr were published and short discussions were given on its light curve variations (Alaniya & Abuladze *Abastumani Bull* **53**,13;**55**,53; Murnikova *Per Zv Suppl* **4**,1; Mironov, Moshkalikov & Kolykhalova *Per Zv Suppl* **4**,7; Kazimirskij *Astron Tsirk* **1204**,7). Tayler (*JAAVSO* **9**,57) noticed that the amplitude of secondary cycle magnitude maxima of XZ Cyg appeared to have decreased abruptly, coincident with the increase in its primary period. From a comparison of calculated and observed light curves of XZ Cyg, Gadun, Zajkova & Romanov (*Astrometr Astrofiz* **46**,23) determined the star's mean radius at two phases of the Blazhko-cycle. Polarization observations of RR Lyr have been carried out by Piirola (*Per Zv Suppl* **4**,31). Low dispersion spectroscopy has been done on XZ Cyg (Romanov & Fenina *Per Zv Suppl* **4**,35) and RR Lyr (Fenina & Romanov *Per Zv Suppl* **4**,13; Romanov, Fenina & Vasileva *Astrometr Astrofiz* **43**,43) in order to investigate the variation of spectral characteristics with the phase of the Blazhko effect. Smith (*PASP* **93**,721) discussed the period distribution of irregularly variable RR Lyrae stars in M3, M5, M15, ω Cen and the Draco dwarf galaxy, and found that irregular variability is more frequent among shorter period ab-type RR Lyraes. These period distributions are consistent with some of the mode mixing explanations for the Blazhko effect. On the other hand, an attempt was made to discover what effects the coupling of the magnetic field and the pulsation have (Biront *et al, Liege Coll* **23**,337), and in a subsequent paper (Cousens *MN* **203**,1171) it was found that if rotation is slow enough, the dynamics of the oscillating magnetic star are altered very little, but the kinematic effect of the differing aspects of the oscillation presented to an observer at different epochs leads to modulated light and velocity curves. It has been concluded that the oscillating oblique magnetic rotator is a plausible model for the Blazkho effect in RR Lyr.

The recent discovery of many double-mode RR Lyrae variables in three globular clusters, in the Magellanic Clouds and even in the nearby Draco galaxy (Hodson, Cox & Clancy *BAAS* **13**,870; Goranskij *IBVS* 2007; *Astron Tsirk* **1216**,5; Nemec *IAU Symp 105*, 465; Nemec, Liller & Hesser *IAU Symp 108*, 39) has spurred interest in these stars. A detailed analysis of ten double-mode pulsators in M15 (Cox, Hodson & Clancy *ApJ* **266**,94; Cox *Pulsations in Classical and Cataclysmic Variable Stars*, p 157) provided period ratios, and masses, luminosities, effective temperatures, radii and He content have been derived for these stars. They appear in the instability strip between the pure fundamental and the overtone pulsators, and it is suspected that these stars undergo mode switching as they are evolving (Hodson & Cox *Pulsations in Classical and Cataclysmic Variable Stars*, p 201). The masses of RR Lyrae stars determined from double-mode pulsation range from 0.55 $M_\odot$ to 0.65 $M_\odot$ (Cox *IAU Symp 105*, 421). This is a very large range of mass and it has some constraints on stellar evolution (Cox & Hodson *BAAS* **16**,190). The only known field double-mode RR Lyrae star AQ Leo has been revisited. Its mean B-V color index places AQ Leo close to the transition line between the c- and the a-type pulsators in the instability strip (Jerzykiewicz, Schult & Wenzel *Acta Astron* **32**,357; *Commun Konkoly Obs* No 83,22).

7. Variable Stars in Globular Clusters and Related Systems

(Amelia Wehlau)

INTRODUCTION

This review is the first in many years that has not been prepared by Helen Sawyer Hogg whose work in the field spans more than 50 years and still continues. The present author wishes to acknowledge Dr. Sawyer Hogg's invaluable assistance in making available her extensive catalogue of references on the subject. The addition of the phrase "and related systems" in the title reflects another change, as studies which were once limited to globular clusters have now been extended to dwarf spheroidal galaxies.

Although an effort has been made to be comprehensive, because of space limitations not all references are included. Readers are also referred to the Commission 38 Report on globular cluster research and to the report of the Frascati Workshop on Population II Variables held in September, 1984 at which a number of interesting papers on the subject were presented.

NEW VARIABLES, NEW PERIODS, AND NEW OBSERVATIONS

This section is given in tabular form in order of IAU cluster designation.

C0512-400 (NGC 1851) Following the discovery of new variables by Stetson (*AJ* **86**,687) periods were found for 4 of these and variability confirmed for the new red variable by Wehlau *et al* (*AJ* **87**,1295). With the exceedingly short period of 0.017 day, one may be a field W UMa star of twice that period. No variation was detected in another star on the edge of the instability strip which appears to be a member (Hesser *et al, AJ* **87**,1470).

C1003+003 (Pal 3) Three RR Lyrae variables and a possible Population II Cepheid have been found by Gratton and Ortolani (*AA Supp* **57**,177).

C1339+286 (NGC 5272) Meinunger has observed several variables (*MVS* **9**,117,119, 122,146). A search by Kadla *et al* (*Izv Glav Astron Obs Pulkovo, Astrofiz Astrometr* No 199) led to discovery of 15 variables bringing the total number of known RR Lyraes in M3 to 231.

C1343-511 (NGC 5286) Fourcade *et al (Bol Asoc Argent Astron* No 20-24,249) have determined periods for six variables.

C1403+287 (NGC 5466) Buonanno *et al* (*AA Supp* **56**,79) have found several new variables including some suspected by earlier observers.

C1452-820 (IC 4499) Fourcade *et al* (*Bol Asoc Argent Astron* No 20-24,156) have constructed a color-magnitude diagram and measured magnitudes for 171 variables.

C1514-208 (NGC 5897) Spectroscopy by Smith (preprint) of SK120, a possible non-variable in the variable star gap, indicates metallicity and radial velocity consistent with cluster membership.

C1531-504 (NGC 5946) Four new variables within the cluster tidal radius and 7 new field variables have been found by M. Liller (*AJ* **88**,404).

C1614-228 (NGC 6093) Three new short period variables have been found by Wehlau *et al* (*IBVS* 2586).

C1639+365 (NGC 6205) In M13 two new variables have been found by Russeva *et al, Per Zv* **21**,175; *IBVS* 2223).

C1659-262 (NGC 6273) Samus (*IBVS* 2555) has identified the Population II Cepheid V2 as FK Oph and confirmed its period.

C1707-265 (NGC 6293) Three new variables have been found by Clement *et al* (*AJ* **87**,1491).

C1716-184 (NGC 6333) Clement *et al* (*AJ* **89**,1707) have found that V12 in M9 is a BL Her star with a period of 1.34 days.

C1736-536 (NGC 6397) El-Worfaly and Budding (*Ap Sp Sc* **94**,253) found only two possible field variables in a careful search for main sequence variables. They also discuss observations of the three previously known variables.

C1821-249 (NGC 6626) Wehlau and Sawyer Hogg present observations of 24 variables in and around M28 (*AJ* **89**,1005).

C1840-323 (NGC 6681) Photometry is presented for 5 RR Lyrae variables including three newly discovered by M. Liller (*AJ* **88**,1463).

C1936-310 (NGC 6809) Eight new possible field variables near M55 were found by Irwin and Trimble (*AJ* **89**,83).

C2003-220 (NGC 6864) Pinto *et al* (*AJ* **87**,635) present photometry of 14 variables in M75 including 3 new RR Lyrae stars.

C2127+119 (NGC 7078) Chu *et al* (*Acta Astron Sin* **23**,118) report on the photometry of three RR Lyrae variables in M15.

MAGELLANIC CLOUD CLUSTERS

Graham & Nemec in a paper presented at IAU Symposium 108, *Structure and Evolution of the Magellanic Clouds*, reported on their search for RR Lyrae variables in old clusters of the Magellanic Clouds which was successful for NGC 1786 and NGC 2210. RR Lyrae stars have now been found in 7 Magellanic cloud clusters, all of them being of Type VII, the oldest and most metal-poor classification. Nemec *et al* (*ApJ Supp*, 1985) have studied 38 RR Lyrae variables in NGC 2257 and Walker (*MN*, 1985) has made CCD observations of RR Lyraes in NGC 2210. For the much younger (80±30 million years) blue LMC globular cluster, NGC 1856, Hodge & Lee (*ApJ* **276**,509) confirmed two previously known Cepheids and found two more probable Cepheids.

STUDIES OF PERIOD CHANGES, DOUBLE MODE PULSATION AND LIGHT CURVE PARAMETERS

There has been a considerable expansion of work in this field, both observationally and theoretically, in the last three years and several papers on the subject were given at *IAU Colloquium 82*, *Cepheids: Theory and Observation*, held in Toronto, Canada in May, 1984. Period changes have been determined for RR Lyrae variables in M4 (NGC 6121,C1620-264) by Sujarkova & Shugarov (*Per Zv* **21**,505) and M92 (NGC 6341,C1715+432) by Kukarkin & Kukarkina (*Per Zv* **21**,365). Barlai (*Commun Konkoly Obs* No 85) has obtained O-C diagrams for 49 RR Lyraes in M15 and also reports on the interesting behavior of V15 in that cluster (*Commun Konkoly Obs* No 83,223). Nemec *et al* (*ApJ Supp*, 1985) have studied the period changes of 38 RR Lyrae stars in the LMC Cluster NGC 2257. Studies of period changes for longer period variables include the investigation by Wehlau & Bohlender of 12 BL Her stars in galactic globular clusters (*AJ* **87**,780), by Wehlau *et al* (*IAU Coll 82*) of two variables in M56 (NGC 6779,C1914+300) and by Clement *et al* (*IAU Coll 82*) for two Cepheids in M10 (NGC 6254,C1654-040).

Double-mode pulsation has been found for RR Lyrae stars in several clusters. There have been two studies on these stars in M15 by Cox *et al* (*ApJ* **266**,94) and Nemec (*AJ*, 1985). Another probable double-mode pulsator has been found by Clement *et al* (*AJ* **89**,1707) in M9. Three such stars found in Draco by Goranskij (*Astron Tsirk* No 1216,5) have been confirmed by Nemec (*AJ*, 1985) who found 10 in all in Draco and five probable candidates in the Ursa Minor dwarf spheroidal galaxy. However preliminary analysis by Nemec & Norris of 54 variables in ω Cen (NGC 5139,C1323-472) shows no evidence for double-mode pulsators in that cluster. Smith (*PASP* **93**,721) has studied the period distributions of irregularly variable RR Lyrae stars and has found behavior consistent with mode-mixing.

Petersen (*IAU Coll 82*) has Fourier analyzed the light curves of 75 RRab and 55 RRc variables in ω Cen and finds strong evidence of the progression of Fourier parameters with pulsation period. Hodson & Cox (*BAAS* **14**,900) have similarly analyzed the light curves of M15 variables and find this progression with period is seen more clearly for a sample of variables for one cluster than for field variables.

ABUNDANCE DETERMINATIONS IN RR LYRAE VARIABLES

Low resolution spectra of three RR Lyrae variables in Pal 13 (C2340+124) have enabled Zinn & Diaz (*AJ* **87**,1190) to compare its metallicity to that of other clusters. Determinations by ΔS spectroscopy of the metal abundances of RR Lyrae variables in a number of clusters have been made by Smith (*AJ* **88**,1762; *ApJ* **281**,148), Smith & Perkins (*ApJ* **261**,576) and Smith & Manduca (*AJ* **88**,982). The subject is reviewed for both field and cluster variables by Smith (*PASP* **96**,505). Kemper (*AJ* **87**,1395) compares metal abundances of field and cluster RRc variables.

COLOR STUDIES AND THE OOSTERHOFF DICHOTOMY

A three-color study of 47 variables in M5 (NGC 5904,C1516+022) has been carried out by Goranskij (*Astron Tsirk* No 1207,4) who finds stars with positive period changes are concentrated in the region which corresponds to the fast phase of horizontal branch evolution. Cacciari (*AJ* **89**,231,1082) has studied RR Lyrae variables in NGC 3201 (C1015-461) and confirmed the similarity of this cluster to M3. A mean value of M_V=0.60±0.08 was derived for the RR Lyrae stars in this Oosterhoff Type I cluster. Bingham *et al* (*MN* **209**,765) present BV photometry for 56 RR Lyrae stars in M15. Their new period-color relation shows less overlap between c and ab type variables than that found by earlier observers. For this Type II cluster a mean value of M_V=0.39±0.15 is obtained. Multimode variables lie close to the (ab/c) transition color.

The problem of the underlying physical mechanisms responsible for the observed differences in RR Lyrae properties between the Oosterhoff groups has been discussed by Goranskij (*Astron Tsirk* No 1234,1) who points out the basic tendency of the horizontal branch to become bluer with decreasing [Fe/H] reverses for the most metal poor clusters. A careful review of the subject by Renzini (*Mem Soc Astron Ital* **54**,335) suggests the Oosterhoff effect can be explained by a combination of this non-monotonic behavior and the period shifts of RR Lyraes of given effective temperture with decreasing cluster metallicity. Caputo *et al* (*AA* **123**,141) and Sweigart *et al* (*BAAS* **16**,526) have suggested differences in [CNO/Fe] instead of an anticorrelation of metallicity and helium content could be a contributing factor. However synthetic horizontal branches constructed by Rood (*IAU Symp 105*) with varying CNO do not appear to support this idea.

POPULATION II CEPEHIDS

There have been two recent reviews of the properties and evolutionary status of population II variables. Harris (*IAU Coll 82*) lists 46 cluster members, consisting of 23 short-period BL Her stars, 17 longer-period W Vir stars, and 6 probable RV Tauri stars, all found in lower metallicity clusters with blue horizontal branches. He compares these with other low mass Cepheids in the galaxy and uses the period changes of Wehlau & Bohlender to discuss their evolutionary status. Wallerstein & Cox (*PASP* **76**,677) also review both field and cluster Population II Cepheids, derive limits on their masses and discuss their relationship to RV Tauri and Mira variables in globular clusters. They interpret the fact that these stars are only found in clusters with blue horizontal branches in terms of greater mass loss before the Helium flash in these clusters.

STUDIES OF RED VARIABLES

Substantially more work is being done on red variables. There has been a series of papers by Lloyd Evans (*SAAO Circ* No 7,82,96; *MN* **204**,945,961;**209**,825) reporting on photometric and spectroscopic studies of red variables and other red giants in a dozen clusters. A four color study of the red variables of 47 Tuc (NGC 104,C0021-723) by Fox (*MN* **199**,715) has yielded new periods and evidence of overtone pulsation. Frogel (*AJ*, 1985) presents infrared photometry of 15 red giants, including 5 variables, in the metal rich cluster NGC 6712 (C1850-087) while in an earlier paper (*ApJ* **272**,167) he discusses the evolutionary state and pulsation characteristics of red globular cluster variables in general. A number of radial velocity or spectroscopic studies of stars on the red giant branch have included red variables. For instance see Pilachowski *et al* (*ApJ* **263**,187), Frogel *et al* (*ApJ Supp* **53**,713) and Mayor *et al* (*AA Supp* **54**,495). The studies of long period variables in the Magellanic Clouds by Wood *et al* (*ApJ* **272**,99) also include a number of globular cluster stars.

UNUSUAL VARIABLES

A third dwarf Cepheid, E39, in ω Cen has been found by Jorgenson & Hansen (*AA* **133**,165). Chu *et al* (*IAU Coll 82*) have found two low amplitude variables, K64 and K152, which lie on the mid-giant branch of M15.

BINARIES AND X-RAY VARIABLES

There has been considerable work on the problem of close binaries in globular clusters. Da Costa (*PASP* **94**,769) has shown that UV5 in NGC 1851 is probably not a binary and radial velocity studies in 47 Tuc by Mayor *et al* (*AA* **134**,118) and in M3 by Pryor *et al* (*BAAS* **16**,111) have turned up only two possibilities in the latter cluster. However a study by Harris & McClure (*ApJ Lett* **265**,L77) of a random sample of K giants indicates that M3 may not be deficient in binaries. There is also some positive evidence for binaries. Zinn & King (*ApJ* **262**,700) found that the anomalous Cepheid, V19, in NGC 5466 which has been shown to be a cluster member by Brosche & Geffert (*AA* **127**,415) is pulsating in the first harmonic and has twice the mass of horizontal branch stars, consistent with the theory that such stars are formed by mass transfer in close binary systems. Margon & Downes (*ApJ Lett* **274**,L31) have suggested that V19 in M30 (NGC 7099, C2137-237) is a second candidate for a U Gem star in a globular cluster while Wehlau & Sawyer Hogg (*AJ* **89**,1005) believe V7 in M28 may be similar.

Reviews of X-ray globular cluster sources by Grindlay (*Adv Space Res* **2**,133) and Hertz & Grindlay (*ApJ Lett* **267**,L83) and theoretical discussions by Hut & Verbunt (*Nature* **301**,587) and Krolik (*Nature* **305**,506) predict globular clusters should contain substantial numbers of close binaries and low-mass contact binaries while Van der Woerd & Van den Heuvel (*AA* **132**,361) compare the expected incidence of close binary systems to the observations. X-ray sources have now been found within the regions of 18 galactic globular clusters (Hertz & Grindlay, *ApJ* **275**,105). Among these, Lawrence *et al* (*ApJ* **267**,301) report partially simultaneous X-ray, radio and infrared observations of the rapid burster MXB 1730-335 in Liller 1 (C1730-333) for which Singh & Duorah (*Ap Sp Sc* **92**,143) have suggested a model, while Stella *et al* (*BAAS* **14**,618) have analyzed the observations of the X-ray source 4U1820-30 in NGC 6624 (C1820-303).

8. Mira Variables
(M. W. Feast)

It has long been known that Mira variables in our Galaxy obey a period-(visual) luminosity relation, at least roughly. Glass & Lloyd Evans (*Nature* **291**,303) discovered the first known Miras in the LMC and obtained infrared photometry of them. A rather precise period-M_{Bol} relation was found. New LMC Miras have subsequently been found (Wood *et al*, *ApJ*, in press; Glass & Reid *MN*, submitted) and these confirm the P-L relation with very small scatter (Feast *MN*, in press). Carbon and non-carbon Miras fit the same relation at K though there is a small separation of the two groups in m_{bol}. The small scatter about a P-L relation is confirmed by observations of Miras in the Galactic Center (Glass & Feast *MN* **198**,199; see also van den Bergh *PASP*, in press). The zero point of the relation may be obtained either by adopting a modulus for the LMC (from Cepheids, say) or from Miras in galactic globular clusters adopting some distance scale for these objects (*eg* from RR Lyrae variables, Menzies & Whitelock *MN*, in press). Miras in the LMC and in galactic globular clusters show the same period-T_{BB} (=temperature of a black body fit to IR data) relation as those in the general galactic field (Feast *1984 Frascati Workshop*, in press). The small amount of (infrared) occultation data suggests that T_{BB} is close to T_{eff} (Glass & Feast *MN* **199**,245) though further work is desirable on this crucial point and some workers have preferred other T_{eff} calibrations (*eg* Frogel *ApJ* **272**,167). Assuming $T_{eff} = T_{BB}$ and the P-L relation indicates near-solar masses and first overtone pulsation (*cf* also Fox & Wood *ApJ* **259**,198). Omitting a few known supergiant objects, the galactic type II OH/IR variables extend the Mira period-M_{Bol} relation to longer periods (Engels *et al*, *AA* **124**,123; Feast *Obs*, submitted). The scatter about the relation can be accounted for by expected errors in the kinematic distances. This makes unlikely theories (which involve amongst other hypotheses an evolution from a Mira to an OH/IR source with a change from overtone to fundamental mode pulsation) based on the supposed isolation of low luminosity OH/IR sources using kinematic distances (*eg* Jones *et al*, *MN* **197**,413; *ApJ* **253**,208; *ApJ* **273**,660; *ApJ* **273**,669). Evolution from Mira to OH/IR source has been suggested on other grounds (Baud & Habing *AA* **127**,73) though it is not clear that this is consistent with the well known relationship of Mira kinematics to period or the small range of Mira periods within a given globular cluster and the dependence of this period on cluster metallicity (*cf* Feast in *Physical Processes in Red Giants*, Iben & Renzini, eds. D Reidel 1981). Present evidence suggest that the OH shells of OH/IR (Mira) variables are at least roughly spherical (Bowers *et al*, *ApJ* **274**,733; Herman *et al* 1984 ESA preprint). If so distances of improved accuracy should be obtained geometrically by combining the angular size of the shell with phase lag measurements of the diameter in the line of sight (Herman *et al*). Such phase lag measurements yield (OH) shell radii of 0.8±0.7 (10^{16} cm) for optical Miras with OH emission and 5.5±0.9 (10^{16} cm) for the (longer period) OH/IR Miras (Herman Thesis, Leiden 1983). However some of these shells are evidently quite complex [*eg* U Ori (Chapman & Cohen *MN* in press) where amongst other things a rotation of the envelope seems to be occurring].

New speckle studies of Miras (Mariotti *et al*, *AA* **120**,237; Bonneau *et al*, *AA* **106**,235) give amongst other results a visual diameter for Mira Ceti of 28±6 milliarcseconds. Wayall & Cahn (*ApJ* **275**,225) have attempted a new analysis of the space motions of Mira variables based on an adopted period-mass-luminosity relation and assuming fundamental model pulsation. It may be noted that the theoretical scheme adopted by these workers implies a substantial change in period of a given star during its Mira phase which seems inconsistent with the evidence on the relation of period to other properties (see above). Basic observational data on Mira variables recently published include: newly discovered variables (including spectroscopic discoveries), Margoni & Stagni *AA Supp* **56**,87; Bidelman *IBVS* 2054; Bidelman & MacConnell *AJ* **87**,792; Spectral types, Crowe *JRAS Can* **78**,103; Celis S. *AJ* **89**,527; UBVRI photometry, Celis S. *AJ* **87**,1791; Hill *et al*, *Publ Dominion Astrophys Obs* **15**,339. A KOIII and Mira binary was studied by Menzies *et al* (*Obs* **103**,195); Variable polarization in Mira Ceti by Hayes (*IBVS* 2064) and the variable periods of R Aql and R Hya by Davis (*JAAVSO* **11**,27). Bolometric corrections are discussed by Bessell & Wood (*PASP* **96**,247) and the effects of H_2O absorption on JHKL colors by Frogel *et al* (*ApJ Supp* **53**,713; see also Iyengar *et al*, *AA* **128**,255). The infrared properties of Mira variables in the solar neighborhood were established by Feast *et al* (*MN* **201**,439) on the basis of extensive JHKL observations. Infrared observations leading to period determinations for OH/IR variables have been given by Engels (*IBVS* 2301; *Verröff Astron Inst Bonn No 95*). New IR counterparts of type II OH sources are reported by Epchtein & Nguyem-Quang-Rieu (*AA* **107**,229) and by Willems & de Jong (*AA* **115**, 213) and H_2O masers detected from further Mira variables (Crocker & Hagen *AA Supp* **54**,405). Accurate optical positions and proper motions have been derived for Miras showing SiO masing. These may be valuable in connecting the HIPPARCOS system to the VLBI extragalactic reference system (Soulie & Baudry *AA Supp* **52**,229).

A number of papers relating to OH/H_2O masing in late type variables appear in *IAU Symposium 110*. New OH/IR sources have been found near the Galactic Center and used to study the mass distribution; the velocity dispersion is 140-160 km/sec (Habing *et al*, *AA* **128**,230). Fix & Mutel (*AJ* **89**,406) give infrared observations of a number of OH/IR sources and (in common with several other workers) fit the data to cool black body energy distributions. However, Whitelock (to be published) shows that these are very unlikely to be real temperatures and that the observed near IR colors are largely determined by circumstellar reddening. A significant number of IRAS sources are likely to be OH/IR variables or (perhaps) dust enshrouded carbon-Miras (Whitelock & Feast *MN* **210**,25P); Herman *et al*, *ESA* preprint; Whitelock to be published). Masing by H_2O in an excited vibrational state has not been found (Myers & Barrett *ApJ* **263**,716) in Mira Variables. SiO masing is found in many OH/IR sources (Jewell *et al*, *AA* **130**,L1). Huggins & Glassgold (*AJ* **87**,1828) develop the Goldreich-Scoville idea of OH production in the outer parts of circumstellar shells by ambient u-v radiation, whilst Clegg *et al* (*MN* **203**,125) discuss the effects of stellar u-v (chromospheric and shock-wave induced) on circumstellar silicon chemistry. Polarization studies of maser emission lines are powerful probes of circumstellar envelopes (Claussen & Fix *ApJ* **263**,153; Western & Watson *ApJ* **274**,195; *ApJ* **275**,195). The expansion velocity of the circumstellar shells of Miras may be correlated with the 1 μm amplitude (Ukita *AA* **112**,167). The correlation betwen stellar flux and SiO maser flux has been confirmed (Cahn *AJ* **86**,1935) and the general theory of SiO masing reinvestigated (Elitzur *et al*, *ApJ* **274**,210). Both masing and non-masing SiO emission from Miras has been studied (Wolff & Carlson *ApJ* **257**,161) and a number of differences (in relative intensities of transitions etc) have been found. The variations in the polarization of the SiO maser emission in R Cas place the SiO just above the stellar photosphere and allow the radial thickness of the line-forming region to be estimated (Clark *et al*, *ApJ* **261**,569). Similar results were obtaied for R Leo (*ApJ* **276**,572; *AJ* **87**,1803) and a transient increase in the mass of SiO involved was observed.

There exists a group of short-period semi-regular variables which show H_2O masing. Since these objects tend to lie at high galactic latitudes they may belong to a distinct type of maser (Dickenson & Dinger *ApJ* **254**,136). In one of these stars, R Crt, rapid variations in both 22 GHz H_2O and at optical wavelengths have been suggested (time scale ∿1 hour) Livi & Bergmann *AJ* **87**, 1783). The unusual microwave and infrared source OH 0739-14 (= OH 231.8+4.2) is an (infrared) long period variable (period = 648 days) (Feast *et al*, *MN* **203**,1207). There is some doubt as to whether it is an OH/IR Mira type variable or a more luminous supergiant object. Bowers & Morris (*ApJ* **276**,646) consider it to be a Mira on the basis of luminosity estimates. However in view of the complex bipolar nature of the object, the flux is unlikely to be isotropic so that the problem has not been finally settled.

Tuckman (*MN* **208**,215) predicted Mira masses of 2 $M_\odot$ or greater (P>300 days) from a study of the termination of the AGB by helium shell flashes. Such masses are greater than estimated either from pulsation theory or galactic kinematics. A general review of the evolutionary state of type II OH/IR sources is given by de Jong (*ApJ* **274**,252).

Further studies of atmospheric kinematics in Miras (including shock induced Hα emission) have been interpreted in terms of fundamental mode pulsation (Wilson *et al*, *MN* **198**,483; Gillet *et al*, *AA* **128**, 384). Hinkle (*PASP* **95**, 550) gives a general review of spectroscopy of circumstellar shells. An important study of the stellar atmosphere and circumstellar envelope of χ Cyg was made by Hinkle *et al* (*ApJ* **252**, 697) using high dispersion infrared spectra. The result are interpreted in terms of photospheric shock waves and a circumstellar envelope which is dominated by a stationary layer at 800 K and is ∿10 stellar radii from the star. This layer built up rapidly and then steadily dissipated over the next three pulsation cycles. In Mira itself evidence for infalling material has now been found (Ferlet & Gillet *AA* **133**,L1). SiO absorption bands in the 4μ region have been observed in Miras (Rinsland & Wing *ApJ* **262**,201). The CO J = 2→1 line at 230 GHz has been shown to be a powerful tool in studies of circumstellar envelopes and the mass loss from Miras (Knapp *et al*, *ApJ* **252**,616). Radio continuum measurements (Spergel *et al*, *ApJ* **275**, 330) suggest that Mira Ceti has an optically thick chromosphere extending several stellar radii. Tielens (*ApJ* **271**,702) has developed a model for a stellar wind driven by radiation pressure on dust grains in Mira circumstellar envelopes whilst Berruyer & Frisch (*AA* **126**, 269) discuss the degree of coupling between gas and dust in various parts of the envlope. Recent work continues to support the view that OH masers are pumped by 35 μm radiation (with the possible exception of R Aql) (Hagen *PASP* **94**,835; Le Bertre *et al*, *AA* **132**,75).

It is not possible to give a general account here of work on symbiotic systems though several of them have been shown to contain Mira components (*MN* **202**,951; **203**,351; **203**,363; **203**,373; **208**,161; *cf* also *PASP* **96**,321). These systems show dust emission with blackbody temperatures near 1000K, much higher than in single Miras. Dust obscuration episodes have been observed in R Aqr and some similar systems and BI Cru (Whitelock *et al*, *MN* **205**,1207) shows circumstellar CO emission at 2.3 μm. R Aqr is double in the radio continuum (Spergel *et al*, *ApJ* **275**,330). The secondary source is believed to be a dense region of the stellar wind. Brugel *et al* (*PASP* **96**,78) have discussed evidence that reddening accompanied the 1977 obscuration phase of R Aqr. BX Mon has sometimes been regarded as the (optical) Mira of longest known period (1374 days). However this periodicity is more likely to be associated with orbital motion in a symbiotic system (Whitelock & Catchpole *IBVS* 2296). SY For is similar to Mira Ceti in having a secondary heated by accretion (Feast *et al*, *Obs* **104**,217).

The pulsation of carbon Miras was discussed with particular reference to a dust shell model (Bergeat & Sibai *AA* **119**,207). As already noted carbon Miras in the LMC follow accurately a P-L relation. The carbon Mira R For became unusually faint in the visible and infrared in 1983 and this is attibuted to dust obscuration (particle radius ~0.15 μm) (Feast *et al, MN* in press). Molecular H_2 has been detected in the 2μm region of several carbon Mira variables though it is missing from some non-Mira carbon stars possibly indicating hydrogen deficiency in these latter stars (Johnson *et al, ApJ Lett* **270**,L63; Goebel & Johnson *ApJ Lett* **284**,L39). Work on the complex circumstellar chemistry of dust enshrouded carbon Mira variables has continued (Jewell & Snyder *ApJ Lett* **255**,L69; Nejad *et al, AA* **134**,129; Huggins *et al, ApJ* **279**,284; Johansson *et al, AA* **130**,227). CO millimeter emission has been detected in some of these objects (Thronson & Mozurkewich *ApJ* **271**,611). In IRC+10216 the CO fundamental band at 4.8 m has been spatially resolved by speckle interferometry. A review of the distribution and motions of C and S type Miras is given by Catchpole & Feast (1984 Strasbourg Meeting, "Cool stars with excesses of heavy elements", in press). The carbon Mira T Lyn may have a uv bright companion (Sanduleak *IBVS* 2548). R Vol which is undoubtedly a carbon star (Thé *PASP* **80**,104; Feast 1983, unpublished) has been suspected of showing a circumstellar 10.6 μm (silicate) emission feature (Hagen *PASP* **94**,835). In contrast to oxygen rich objects, carbon stars with shells tend to show non-spherical geometry (*cf* Bowers *et al, ApJ* **274**,733) and some progress has been made modelling these structures (*cf* Jura *ApJ* **275**,683). The infrared spectra of S, SC and CS stars (including Mira variables) show (a) the sensitivity of some absorption bands to C/O ratio; (b) the unusual strength of CO 2.3 μm absorption in superlithium rich stars (Whitelock & Catchpole *MN*, in press; 1984 Strasbourg Meeting, in press).

9. Flare Stars

(R.E. Gershberg and N.I. Shakhovskaya)

The Proceedings of IAU Colloquium No 71, *Activity in Red-dwarf Stars* (Bryne & Rodonò, eds, Reidel 1982), hereafter referred to as BR, contain detailed reviews on physical parameters of FSs, on their radiation in optical, UV, X-ray and radio wavelengths during flares and in the quiescent state, and on theoretical interpretations of FS activity.

Several new FSs in the solar neighborhood have been discovered: Asteriadis (*AA* **113**,165) has registered a flare on Gliese 487, Pettersen (*IBVS* 2141) has confirmed the flare activity of G 9-8 = CU Cnc and discovered such activity on G 51-15 (*AA* **95**,135), V 780 Tau (*AA* **120**,192), G 141-29 (*AA* **97**,199), G 9-38, G 119-62, Gliese 171.2A and Gliese 890 (in preparation), Sandmann (in preparation) has discovered flare activity on DK Leo, Torres *et al (Proc 3rd Regional Latin American Astron Mtg*, Buenos Aires, 1983) have observed a flare on Gliese 425 AB, Egge and Pettersen (BR, 481) registered a flare on the contact binary VW Cep, Patkos (*Mitt Astron Ges* **55**,82) found several short flares on the eclipsing binary SV Cam. Sanduleak (*IBVS* 2433) and Gavin (*IAU Circ* 3573) have suggested flare activity on an anonymous star and the Parenago 1644 stars. BY Dra type brightness variability is found by Bopp *et al* (*ApJ* **249**,210;**275**,691) on Gliese 171.2A, Gliese 410, Gliese 233, HD 143313, HD 175742, by Dorren & Guinan (*SAO Spec Rep* **392**,49) on HD 149661, HD 152391 and 61 Cyg A and by Torres *et al* (BR, 175) on DK Leo, V 914 Sco, V 4046 Sgr, FK Ser, Gliese 425 AB, Gliese 879 = TW PsA; they suggested the BY Dra syndrome on AD Leo, V 1216 Sgr, AT Mic = Gliese 799, AY Ind, Gliese 566 and 567. Young *et al*, in *Cool Stars, Stellar Systems, and the Sun* (Baliunas & Hartmann, eds, Springer 1984) (hereafter BH), p 112 have found for the single dMe star Gliese 890 the BY Dra syndrome, a very short period (10^h20^m) and large $v \sin i$ (70 km/sec). Results of photoelectric monitoring of known FSs have been published by Mavridis *et al* (*IBVS* 2022,2133,2174,2209), Asteriadis *et al* (*IBVS* 2183,2210,2408), Panov *et al* (*IBVS* 2128,2220,2358,2359; *Commun Konkoly Obs* No 83,208; *CR Acad Bulgare des Sciences* **37**,557), Tsvetkov *et al* (*IBVS* 2340, 2437),

Contadakis *et al* (*IBVS* 2088), Melikian & Jankovics (*IBVS* 2038), Ilyin (*IBVS* 2484), Pettersen *et al* (*ApJ Supp* **54**,375), Andrews (*IBVS* 2252,2253), Herr & Frank (*IBVS* 2426), Allen & Edwards (*South Stars* **30**,159), Sanwal (*IBVS* 2143), Soliman (*IBVS* 2193), Andersen (BR, 203), Ichimura & Shimizu (*Tokyo Obs Bull*, in press), Rojzman & Shevchenko (*Sov Astron Lett* **8**,163). Statistical analysis of patrol observations of FSs has been carried out by Mahmoud (*IBVS* 2294), Andrews (*IBVS* 2254), Pettersen *et al* (*ApJ Supp* **54**,375), Pettersen (BH, 194). Gershberg & Shakhovskaya (*Ap Sp Sc* **95**,235) have analyzed energetics of different manifestations of FS activity and shown that the sum of the energy of short-lived flares and permanent radiation of stellar chromospheres and coronas is close in order of magnitude to the radiation deficit of starspots on FSs.

Results on the search for FSs in regions of stellar clusters have been published by Melikian (*IBVS* 2018,2352), Melikian *et al* (*IBVS* 2019), Oganian *et al* (*IBVS* 2149), Chavushian *et al* (*IBVS* 2339), Parsamian *et al* (*Astrofizika*, in press), Natsvlishvili (*IBVS* 2062,2231), Lopez (*IBVS* 2279), Hojaev (*IBVS* 2412), Tsvetkov *et al* (*IBVS* 2067,2132,2224,2338), Tsvetkova *et al* (*IBVS* 2131,2365). A catalogue of FSs in the Pleiades has been published by Haro *et al* (*Bol Inst Tonantzintla* **1**,3). Parsamian (*Astrofizika*, in press) has found increased number of visual binaries amongst FSs in the Pleaides and has shown that in the Orion cluster the fraction of FSs amongst Orion variables increases for variables with larger brightness amplitudes: she has considered the correlation between stellar cluster age and flare activity characteristics of FSs that belong to the clusters (*Proc 3rd regional Latin American Astron Mtg*). Korotin & Krasnobabtsev (*Izv Krymskoj Astrofiz Obs* **72**) have carried out a statistical analysis of FSs in 4 clusters, confirmed the power character of energy spectra of flares in all cases and found a constancy of spectral index for FSs of different luminosities in each cluster and a dependence of this index on cluster age: the flare energy spectrum becomes flatter for older clusters and the relative proportion of stronger flares increases.

Rojzman (*Sov Astron Lett* **9**,41;**10**,279; *Astron Zh* **61**,500) has confirmed the BY Dra type photometric period of EV Lac, suggested the existence of an unseen eclipsing companion of this FS and found that as a rule a slow increase of brightness of $0^{m}.1$-$0^{m}.2$ and of 1-2 hours duration precedes a flare. Therefore during short preflare dips the stellar brightness decreases only to its normal level but not below this level.

Avgoloupis (PhD Thesis, University of Thessaloniki, 1984) and Mavridis & Avgoloupis (in preparation) have considered the distribution of 184 light curves of flares on EV Lac along 4 Oskanian curve types and characteristics of flares within each type.

Chugainov (*IBVS* 2471) has suggested that there are repeating oscillations in the brightness of BY Dra on time scales of 1-3 hours. Cristaldi (*AA Supp*, in press) has found a correlation between the period of BY Dra brightness variations and the mean level of the brightness: the period decreases when mean brightness increases. This may be due to latitude migration of starspots.

Pettersen (BR, 239; *Lund Obs Rep* **18**,114) has carried out synchronous UBVR and narrow band $H\alpha$ and $H\beta$ observations of flares on G 141-29, V 577 Mon and UV Cet. He confirmed much slower evolution of flares in lines compared with the continuum and noted an absence of structures having time scales shorter than seconds in light curves in lines. He also found that the color index V-R may be a good approximation to a pure continuum flare index and that it varies during a flare, while U-B and B-V indices are strongly affected by line contributions and show only small variations during flares. Pettersen *et al* (*AA* **123**,184) have analyzed observations of YZ CMi and EV Lac and did not find a significant correlation of frequencies and energies of flares with visibility of spotted regions of stars.

Clayton & Martin (*AJ* **86**,1518) carried out high-precision polarimetric measurements of 5 FSs and detected no linear or circular polarization in any of the stars. Pettersen & Hsu (*ApJ* **247**,1013) have carried out multifilter polarimetric measurements of 19 dM0e-dM6e FSs and did not find polarization although single and double spotted stars and stars undergoing flare activity were in their observational program.

Bryne *et al* (*MN* **206**,907) have found normal frequency of flares on Gliese 182, confirmed the existence of BY Dra type brightness variations for this FS, and in IUE spectra found chromospheric emission that is typical for its spectral class. Bryne *et al* (*MN*, in press) have carried out optical photometry and UV spectrometry for FSs Gliese 229 = HD 42581 and Gliese 735 = V 1285 Aql.

Worden *et al* (*ApJ* **276**,270) observed 6 flares on YZ CMi with simultaneous photometry and high-dispersion, time-resolved spectroscopy and found significant increase in Hα and Hβ emission without noticeable broadening of the lines. Chugainov *et al* (BR, 237) carried out synchronous photometric and spectral observations of AD Leo, DT Vir and YZ CMi and registered an Hβ increase of 10-50% at 10-20 minutes before flare maximum and red asymmetry of this line in maximum, when broadening up to ±15 Å were observed; a search for He II emission was not successful. Ambaryan (*Astrofizika* **18**,654) has obtained a YY Gem flare spectrum in the photographic region. In one of a series of 5 min spectrograms of YZ CMi, Haisch & Giampapa (*PASP*, in press) have found two strong emission lines, at 4007 Å and 4276 Å, that are stronger than H and Ca II lines but have not yet been identified.

Bromage *et al* (BR, 245) from $17^h.5$ patrol observations with the IUE have detected 4 flares on UV Cet, AT Mic, EV Lac and EQ Peg; averaged over 30-35 min exposures, the flux increases in the C IV line at 1550 Å were from 1.5 to 9 times; in the AT Mic flare a strong continuum which increases monotonically from 1250 to 2000 Å and can be represented by a black body at T = 13000 K has been found; in the EQ Peg flare the strongest line emissions were found; the UV Cet flare with $\Delta U = 2^m.4$ was not accompanied by noticeable variations of the EUV line spectrum; during the EV Lac flare with $\Delta B = 0^m.7$ the variations were very faint. Butler *et al* (BR, 249) have found in an AU Mic flare that the C IV emission decays much faster than do the C I, C II, Si II and He II emissions. Baliunas & Raymond (*ApJ*, **282**,728) have reported UV and visible observations of a flare on EQ Peg B: for several minutes the Hα and Hβ emission increased significantly, then returned to normal states for an hour; simultaneously in the EQ Peg AB Spectrum the C IV, He II and C II emissions were enhanced; similarity to large two-ribbon solar flares is noted.

Butler *et al* (*MN* **197**,815) have described IUE spectra of FSs Gliese 867 A and AU Mic in the quiescent state and a flare spectrum of Gliese 867A with a strong continuum. In review by Bryne *et al* (*Ir Astron J* **14**,219) a description of this flare spectrum is given and characteristics of Mg II lines for FSs and non-active stars are presented. Butler *et al* (*Proc 4th Eur IUE Conf*, Rome, 1984) have observed BY Dra in different phases but did not find flux modulation of C I, O I, C II, Si IV and C IV lines by stellar rotation. They discovered an anticorrelation of stellar brightness in the Mg II doublet and in a band near 2620 Å containing a blend of Fe II lines with optical radiation that may be due to bright chromospheric plages above dark photospheric spots. Rodonò *et al* (*Proc 4th Eur IUE Conf*) have given preliminary results of cooperative observations for FSs YZ CMi, Gliese 182 and AD Leo in EUV, optical, IR and radio ranges. During a strong - $\Delta U \approx 3^m.8$ - flare in YZ CMi a sharp increase in radio flux took place at 6 cm, EUV lines of low and high excitation were enhanced noticeably, and a rather flat continuum increased; radio flux increased at optical maximum but the radio flare had a much longer duration than optical one had. During the Gliese 182 flare registered with IUE an excess pulsating radio emission at 2 cm was observed. A strong ($\Delta U = 2^m.1$) and long (about $0^h.5$) flare on AD Leo was studied in the wavelength range from 2000 Å to 20 cm: near the maximum a noticeable enhancement and broadening of Balmer lines were found, several minutes after maximum short-lived dips in the K band (2.2 micron) were noted, and in an IUE spectrum obtained 20 min after the maximum, a 20-fold increase in the Mg II-Fe II blend near λ2620 Å and in the continuum near 2500 Å were seen.

Bopp *et al* (*ApJ* **249**,210;**275**,691) surveyed spectroscopically with moderate dispersion about 20 KM dwarfs, and in 5 cases discovered faint Hα emission; all these objects turned out to be spectroscopic binaries with periods less than 10^d and with BY Dra type photometric variability. 5 km/sec is found to be a critical equatorial rotation velocity to manifest BY Dra type variability and flare activity on red dwarfs. The system BD +26°730 proved to have the most interesting UV spectrum, with emission line surface fluxes in He II, C IV, and N V 50-200 times the values for quiet sun. However, Torres *et al* (*Proc 3rd Regional Latin American Astron Mtg*, Buenos Aires, 1983) pointed out the BY Dra syndrome for more slowly rotating stars - with rotation velocities less than 4 and even 3 km/sec - and did not find Hα emission in high dispersion spectra of the BY Dra type star TW PsA. Pettersen & Coleman (*ApJ* **251**,571) studied with a spectral resolution of 0.45 A line profiles for Hα, Na, He and Ca II lines in the red for the quiescent state of FSs AD Leo and GX And, which have similar photospheres but very different levels of flare activity; chromospheric emission lines are discussed in detail and correlations between spectral line features and flare activity are investigated. Since the single FS EV Lac and the component of a binary system EQ Peg A are very similar but the component of the short period system V 1054 Oph that should have fast rotation has a low level of activity, Pettersen *et al* (*ApJ*, in press) have concluded that stellar age as well as rotation velocity must be an important factor in determining flare activity level; they noted that the IR Ca II triplet can dominate in chromospheric radiative losses, being more effective than hydrogen lines. Giampapa *et al* (*ApJ* **246**,502; *ApJ Supp* **46**,159) have studied Ca II line profiles with resolution 0.142 A in spectra of KM dwarfs and found that when stellar effective temperatures decrease absolute radiative losses of chromospheres also decrease. Using these high dispersion profiles of Ca II line Giampapa *et al* (*ApJ* **258**,740) have constructed single-component, homogeneous semi-empirical models of quiet chromospheres for 3 dMe and 2dM stars. Vogt *et al* (*ApJ* **269**,250) have determined v sin i from high-resolution line profiles for 17 KM dwarfs, and concluded that all stars with the BY Dra syndrome rotate faster than stars without this feature. Kodaira & Ichimura (*Proc Astron Soc Japan* **34**,21) have shown that the heterogeneity of surface brightnesses of the YY Gem components leads to distortion of the ratio of the amplitudes of radial-velocity curves. They confirmed the stability of a double sectorial distribution of active regions for the primary and changes of the distribution for the secondary. Ayres *et al* (*ApJ Lett* **270**, L17) obtained a high-dispersion IUE spectrum for FS AU Mic in the quiescent state from 1150 to 3000 Å and found that emission lines are symmetrical, do not show differential shifts and have widths (FWHM) from 38 to 74 km/sec. Linsky *et al* (*ApJ* **260**,670) carried out quantitative comparison of UV and optical emission spectra of dM and dMe stars and found that Mg II lines are the strongest chromospheric lines for G-M dwarfs but that for dMe stars Ca II and Fe II emissions are comparable and the Balmer series is more intensive than Mg II. Chromospheric radiative losses for dMe stars are 5 times larger than for dM stars and decrease quickly for later types, but the contributions of these losses relative to total stellar luminosity are approximately the same for a wide range of effective temperature for dM stars and increase by 5-10 times toward lower temperature for dMe stars. Relative magnitudes of radiative losses in chromosphere-corona transition zones and in coronas for dMe stars are 100 times higher than for active regions on the Sun. Torres *et al* (in preparation) have obtained reticon spectra for about 90 M and K dwarfs in the solar neighborhood, in the Pleiades and in the Hyades, and found a rather wide range in Li I 6707 line intensities. They concluded that Li destruction may be slower than was previously thought.

Kahler *et al* (*ApJ* **252**,239) have observed in the radio, optical and X-ray ranges a significant flare on YZ CMi which in character of development, X-ray plasma temperature, and a marked delay of the radio flare relative to the optical one suggest strong similarity to solar flares. X-ray emission fading is found to occur much more slowly than in the optical flare and the main relaxation mechanism is radiation. Haisch *et al* (*ApJ* **267**,280) during a 5 hour patrol with the IUE and the *Einstein* Observatory have recorded a strong flare on Prox Cen and obtained detailed data on the development of this event. In soft X-rays the total duration of the flare was about 2 hours, its luminosity at maximum reached 2×10^{28} ergs/sec and temperature 27×10^6 K; EUV flare emission was about 10% of the X-ray emission. A definite difference in intensity of X-ray emission from the quiet corona of this FS between 1979 and 1980 was found but the coronal temperature was estimated to be the same, namely 4×10^6 K; an analysis of coronal emission in the framework of the loop model leads to the conclusion that active X-ray regions cover about 2% of the stellar surface. Ambruster & Wood (BH, 191) have found fast transients of X-ray emission from EQ Vir and EV Lac. Swank & Johnson (*ApJ Lett* **259**,L67) studied the corona of FS Wolf 630 AB and found strong main emission with $T \cong 6.5\times10^6$ K and an additional emission component - 30% in power - with $T > 10^7$ K; similar two-component emission has been found from the corona of the FS AD Leo.

Caillault (*AJ* **87**,558) observed with the *Einstein* Observatory 5 BY Dra type stars and in all cases noticeable X-ray fluxes were registered; since one of these objects was almost certainly single, this result lends strong support to the suggestion that rapid rotation and not duplicity is the determining factor in defining the level of activity in such stars. Surveying the Pleiades stars in the soft X-ray region with the *Einstein* Observatory, Caillault & Helfand (*ApJ*, 1985 in press) have registered 2 flares: one on a cluster member, the 12^m K star HZ 1136, and one on the 15^m background M star HZ 1733. Within the framework of the loop model for coronal plasma they estimated: temperature, $10\text{-}50\times10^6$ K; luminosity, $10^{29}\text{-}10^{32}$ ergs/sec; emission measure, $10^{51}\text{-}10^{54}$ cm^{-3}; electron density, 10^{12} cm^{-3}; linear size, $10^{10}\text{-}10^{11}$ cm; and magnetic field strength, (1-2 kG for the flare structures. Stern *et al* (*ApJ Lett* **264**,L55) have observed in the Hyades a strong X-ray flare in the eclipsing system HD 27130 consisting of a G and a K dwarf and having a period $5^d.6$. During the flare the X-ray luminosity increased by 35 times and reached 10^{31} ergs/sec, the flare plasma temperature was about 4×10^7 K and the entire radiation in the soft X-ray range was 3×10^{34} ergs. Stern (BH, 150) has considered a general evolutionary scheme for coronal activity on late-type stars of the main sequence. Agrawal *et al* (*Proc Symp on X-ray Astron*, Bologna, 1984) have found for FSs clear correlations between X-ray luminosities of quiet coronas and luminosities of chromospheric and transition-region lines.

Slee *et al* (*Nature* **292**,220) have observed in the optical and at 6 cm a flare on AT Mic. They found a rather good correlation between light curves for these different wavelength ranges and find support for theories in which non-thermal electrons are responsible for both emissions. Fisher & Gibson (*SAO Spec Rep* No 392,109) using the VLA have monitored FSs YZ CMi, CN Leo, BD +16°2708, UV Cet, Wolf 424 AB and EQ Peg AB and 6 and 20 cm, simultaneously, in part, with optical observations. They found that radio flares had typical durations of about 20 min, significant or even very high circular polarization and no correlation with optical flares; they also observed radio emission at 6 cm from YZ CMi and at 20 cm from UV Cet in the quiescent state of these FSs. Gary *et al* (*ApJ Lett* **263**, L79) have recorded at 6 cm a flare on L 726-8 A with 100% circular polarization, intensity oscillations with a quasiperiod of about 56 sec, and maximum brightness temperature higher than 10^{10} K. Lang *et al* (*ApJ Lett* **272**,L15) have observed on AD Leo at 20 cm a rapid sequence of highly polarized spikes during the gradual rise of a longer lasting event; the rise time for the spikes was shorter than 0.2 sec and the brightness temperature was higher than 10^{13} K. The spikes are interpreted as a coherent emission at the second harmonic of the gyrofrequency in a longitudinal magnetic field of 250 G. Mechanisms that may be responsible for the slow component of this event have been discussed by Holman *et al* (*Proc Workshop on Stellar Continuum Radio Astronomy*, Boulder, 1984).

Topka & Marsh (*ApJ* **254**,641) have registered at 6 cm emission from each of the components of EQ Peg AB in the quiescent state. Linsky & Gary (*ApJ* **274**,776) at the same frequency have discovered emission from FSs UV Cet, YY Gem and Wolf 630 AB outside of flares; the emission was highly polarized and is interpreted as a gyroresonant emission from thermal electrons spiraling in magnetic fields of about 300 G or as synchrotron emission from a relatively small number of electrons with effective temperature greater than 10^8 K. The radio emission maximum of YY Gem corresponds to the phase of best visibility of a large dark spot on the surface of the faint component of the system, which may be evidence in favor of the magnetic nature of the radio emission. To explain strong and variable polarization of radio emission from FSs Gibson (BH, 197) has suggested a model of two starspots of different polarity that cross the stellar disk and are responsible for a variable component of magnetic field along the line of sight in the upper atmosphere of the star. Mullan (*Proc Workshop on Stellar Continuum Radio Astronomy*, Boulder 1984) has carried out a comparison of radio emission from UV Cet and RS CVn type variables and concluded that radio emission from FSs is due to coherent mechanisms which are related to impulsive release of magnetic energy during large-scale reconnection.

Continuing to elaborate the gas dynamical model of red dwarf flares that was reviewed by Katsova & Livshits (BR, 617), Livshits (*Astron Zh* **60**,964) has shown that the character of short-lived continuous optical radiation in a stellar flare is determined by the energy balance within a region compressed by a downward propagating shock wave and is weakly dependent on assumptions used in the numerical calculation of a rather complex gas dynamical scheme.

Cram (*ApJ* **253**,768) has shown that powerful X-ray emission from dMe star coronas should noticeably affect the energy balance in quiet chromospheres of these stars. Cram & Woods (*ApJ* **257**,269) have studied effects of stellar atmosphere structure variations during flares on spectral characteristics of stellar flares. They concluded that increased pressure and temperature in the flaring stellar chromosphere and changes in the stellar flare area, both with time and with depth in the atmosphere, play an important role in producing the observed flare spectrum.

Mullan (*ApJ* **279**,746) has pointed out that recent observations in the visible, radio and X-ray portions of the spectrum do not exclude the existence of starspots with magnetic field strengths of order 10 kG or more on cool dwarfs. Considering expected variations of coronal heating efficiency along the main sequence, Mullan (*ApJ* **282**,603) has shown that Ionson's electrodynamic coupling theory predicts a maximum in this efficiency for early M dwarfs and a rapid decrease for cooler M dwarfs, in agreement with X-ray observations of lower main sequence stars.

Bohn (*AA* **136**,338) has revised the theory of generation of acoustic energy from convection zones of late type stars: including the effect of absorbtion by dust and molecules in the opacity tables and improving the calculation of thermodynamic quantities by including molecular dissociation yields larger convective velocities; then, including monopole - and dipole-source terms for the calculation of the sound generation by the Lighthill-Proudman theory, the total expected generation of acoustic energy for very late-type stars turns out to be 5 orders of magnitude more than in previous theoretical calculations.

10. Observations of Pulsating Degenerate Stars
(Edward L. Robinson)

The pulsating degenerate stars include the pulsating DA white dwarfs - known as the ZZ Ceti stars, the pulsating DB white dwarfs such as GD 358, and the pulsating pre-white-dwarfs, also known as the PG1159-035 stars after the prototype of the group. Recent reviews of the ZZ Ceti stars have been given by Robinson (*IAU Coll 53, White Dwarfs and Variable Degenerate Stars* Van Horn & Weidemann eds, Univ Rochester, p 343) and Winget & Fontaine (in *Pulsations in Classical and Cataclysmic Variables*, Cox & Hansen eds, JILA, p 142); no general reviews of the pulsating DB white dwarfs or the PG1159 stars have been published.

Nineteen ZZ Ceti stars have been found. The seven found since the review by Robinson (*IAU Coll 53*) are GD 385 (Fontaine *et al*, *ApJ* **239**,893), G255-2 (Vauclair, Dolez & Chevreton *AA* **103**,L17), G185-32 and G191-6 (McGraw *et al*, *ApJ* **250**,349), G226-29 (Kepler, Robinson & Nather *ApJ* **271**,744), GD 66 (Dolez, Vauclair & Chevreton *AA* **121**,L23), and G238-53 (Fontaine & Wesemael *AJ* **89**,1728).

The importance of temperature in determining whether a DA white dwarf pulsates has been shown much more clearly in the past few years. Earlier studies of the temperatures of the ZZ Ceti stars used the (B-V) or (b-y) colors of the white dwarfs to measure their temperatures (see Robinson *IAU Coll 53*). These colors are not perfectly correlated with temperature, so the effects of temperature were blurred and difficult to investigate. The (G-R) colors of white dwarfs are more perfectly correlated with their temperatures than are the (B-V) or (b-y) colors because the wavelength baseline of the (G-R) index is larger and the G and R passbands are not strongly affected by hydrogen absorption lines. Two recent studies of the temperature dependence using (G-R) colors have shown that all ZZ Ceti stars have (G-R) colors between -0.38 and -0.45 on Greenstein's 1969 calibration. Furthermore, there are few and possibly no constant stars within that range of colors (Greenstein *ApJ* **258**,661, Fontaine *et al*, *ApJ* **258**,651). It appears likely, therefore, that temperature is the sole criterion for pulsation of DA white dwarfs. Depending on the exact calibration of the (G-R) color system, the ZZ Ceti instability strip runs from 10,600 K to 11,800 K. The multichannel observations by Greenstein (*ApJ* **258**,661) also showed that the gravities of the ZZ Ceti stars lie in the range log(g) = 8.0±0.2, and thus are normal for white dwarfs.

It is generally believed that the pulsations of the ZZ Ceti stars are non-radial g-mode pulsations. This belief is based on the long periods of the ZZ Ceti pulsations, 109 s for G226-29 to 1186 s for GD 154, and on the theoretical investigations by Winget and his colleagues that have shown that models of DA white dwarfs near 12,000 K are unstable to g-mode pulsations (Winget *et al*, *ApJ Lett* **252**,L65). Nevertheless, calculations by Saio (*ApJ* **256**,717) suggested that DA white dwarfs with temperatures near 12,000 K should also pulsate in r-modes, another form of non-radial pulsation. To test this suggestion, Robinson, Kepler & Nather (*ApJ* **259**,219) and Kepler (*ApJ*, 1985 in press) calculated the observable photometric and radial-velocity variations of white dwarfs undergoing g-mode and r-mode pulsations. Their observations of the two ZZ Ceti stars R548 and G117-B15A agreed with the calculated variations of g-mode pulsations and not r-mode pulsations.

Much time and effort has been invested in measuring the periods and the rates of change of the periods of the ZZ Ceti stars because these quantities can be used to test theories of the structure and evolution of white dwarfs. The measurements are arduous because the ZZ Ceti stars are all multiperiodic; even the simplest one, R548, has four pulsation periods simultaneously present in its light curve. Nevertheless, the number of ZZ Ceti stars for which accurate periods have been established has jumped from one, R548, to five in the last two years. In addition to R548, we now have L19-2 (O'Donoghue & Warner *MN* **200**,563), G117-B15A (Kepler *et al*, *ApJ* **254**,676), GD 385 (Kepler *ApJ* **278**,754), and G226-29 (Kepler, Robinson & Nather *ApJ* **271**,744). None of these stars has shown measurable period changes. The most stringent upper limit on the rate of change of any of the periods in these stars is for the 215 s period of G117-B15A for which $|dP/dT| < 1.4\times10^{-14}$ s/s at the 68% confidence level (Kepler, Robinson & Nather in *Pulsations in Classical and Cataclysmic Variables*, p 73).

For completeness, we note that Liller & Hesser (*PASP* **92**,319) measured the long term variability of 12 ZZ Ceti stars on Harvard patrol plates and found no evidence of variability - except for their pulsations - on time scales up to 90 years. Also, Angel, Borra & Landstreet (*ApJ Supp* **45**,457) observed several ZZ Ceti stars during their survey for circular polarization in white dwarfs. They did not find any evidence for strong magnetic fields in the stars they observed.

The basic correctness of the theory of pulsating white dwarfs has been confirmed by the discovery of the pulsating DB white dwarfs. Based on their theoretical studies of the ZZ Ceti stars, Winget *et al* (*ApJ Lett* **252**,L65) predicted that the DB white dwarfs should also pulsate in the nonradial g-modes. The pulsations of the DB white dwarfs should be driven by the helium partial ionization zone instead of the hydrogen partial ionization zone that drives the ZZ Ceti pulsations, and the instability strip for the DB white dwarf should be hotter than about 20,000 K. A survey for variable DB white dwarfs made as a result of this prediction has now found two pulsating DB white dwarfs, GD 358 and PG1654+160 (Winget *et al, ApJ Lett* **262**,L11; Robinson & Winget *PASP* **95**,386; Winget *et al, ApJ Lett* **279**,L15). The pulsations of these two stars are similar to the pulsations of the ZZ Ceti stars, leading to the conclusion that they are, indeed, pulsating in g-modes. Furthermore, the temperature of GD 358 is about 30,000 K, in agreement with the theoretical prediction (Winget *et al, ApJ Lett* **268**,L33; Koester, Weidemann & Vauclair *AA* **123**,L11). This appears to be the first time in the history of astronomy that an entirely new class of variables stars was successfully predicted before it was discovered.

Faith in the accuracy of the theory of pulsating white dwarfs should not be boundless, however. Using the best available white dwarf models and pulsation codes, two groups independently predicted that DA white dwarfs should pulsate in radial modes as well as nonradial modes (Starrfield *et al, ApJ* **269**,645; Saio, Winget & Robinson *ApJ* **265**,982). An exhaustive search for radially pulsating white dwarfs by Robinson (*AJ* **89**,1732) failed to find any. Some modifications of the pulsation codes or the white dwarf models will be necessary to resolve this discrepancy.

The star PG1159-035 was found to be a multiperiodic variable by McGraw *et al* (*IAU Coll 53*, p 377). The morphology of its light curve and the presence of broad He II absorption in its spectrum suggested that PG1159-035 was a hot, high gravity pulsating star unlike any other known variable. Three other stars similar to PG1159-035 have now been found: the central star of the planetary nebula K1-16 (Grauer & Bond *ApJ* **277**,211), PG1707+427, and PG2131+066 (Bond *et al, ApJ* **279**,751). Analyses of optical and IUE spectra have shown that all four stars have temperatures greater than 100,000 K and high surface gravities, supporting the suggestion originally made by McGraw *et al* (*IAU Coll 53*, p 377) that they are highly evolved stars rapidly cooling to the white dwarf sequence (Wesemael, Green & Liebert *ApJ Supp*, 1985 in press; Bond *et al, ApJ* **279**,751). The dominant pulsation periods range from 414 s for PG2131+066 to 1700 s for the central star of K1-16. Periods this long suggest that, like the pulsations of the white dwarfs, the pulsations of the PG1159-035 stars are g-modes. Only the light curve of PG1159-035 has been fully decomposed into its component pulsation modes (Winget, Kepler, Robinson, Nather & O'Donoghue *ApJ*, 1985 in press). The period of its largest amplitude pulsation, at 516 s, is changing at the rate $dP/dt = (-1.2\pm0.1)\times10^{-11}$ s/s. The magnitude of this rate - but not its sign (!) - is consistent with theoretical calculations of the g-mode pulsation periods of a cooling and contracting pre-white-dwarf (Winget, Hansen & Van Horn *Nature* **303**, 781). With further theoretical studies it will be possible to convert this rate of period change into a direct determination of the rate of evolution of stars from the region of the planetary nebulae nuclei to the white dwarfs sequence in the H-R diagram.

11. Theory of Stellar Pulsation

(W. Dziembowski)

11.1 INTRODUCTION

Several major international conferences have been held in which papers on stellar pulsation theory are included. The most important were:

Problems of Solar and Stellar Oscillations (IAU Colloquium 66), Crimean Astrophysical Observatory, Sept. 1981. Proceedings in *Solar Physics* **82**,1-498 (1983), D.O. Gough, ed.

Pulsations in Classical and Cataclysmic Variable Stars, Boulder, Colo., June 1982. Proceedings (Boulder: JILA, 1982), J.P. Cox and C.J. Hansen

Oscillations as a Probe of the Sun's Interior, Catania, June 1983. Proceedings in *Mem. Soc. Astron. Ital.* **55**,1-384 (1984), G. Belvedere and L. Paterno, eds.

Theoretical Problems in Stellar Stability and Oscillations, Liège, July 1984. Proceedings (Liège: Universite de Liège, 1984), A. Noels and M. Gabriel, eds.

The Catania meeting, though devoted to the Sun, included many papers of interest for stellar oscillations generally. Papers from the first three conferences in the list are cited in this report. The Liège meeting was too recent to be included at the time this review was written.

11.2 GENERAL THEORY AND NUMERICAL METHODS

Effects of Convection

Investigation of these effects is still in the exploratory phase. A novel method was developed by Stellingwerf (*ApJ* **262**,330). Its superiority over previously used time-dependent mixing-length formalisms consists in a nonlocal treatment of the convection. An autonomous system of equations for stellar pulsation was derived in which convection is described by its mean characteristics. The relative simplicity of the equations makes them suitable for widespread use but the derivation, though physically plausible, is very nonrigorous. Some results obtained with this new method will be described in section 11.3.

Stellingwerf's formalism is not able to describe resonant energy input from the convective motion to pulsations. Methods to study this problem were described by Poyet (*Sol Phys* **82**,267) and by Dolez, Legait & Poyet (*Mem Soc Astron Ital* **55**,293) but not developed to a tractable level. Christensen-Dalsgaard & Frandsen (*Sol Phys* **82**,469) used an approximate method developed earlier by Goldreich & Keeley to predict amplitudes for stellar analogues of the solar "five minute" modes.

The same authors (Christensen-Dalsgaard & Frandsen *Sol Phys* **82**,165) pointed out an important and hitherto ignored effect of the departure from radiative equilibrium in the subphotospheric convective zone on stability of high frequency modes. Taking it into account reverses the conclusion about overstability of the "five minute" modes.

Oscillations in Nonspherical Stars

Effects of a slow nonuniform rotation on frequency spectra of the poloidal modes were studied by Vorontsov (*Sol Phys* **82**,379), Gough & Taylor (*Mem Soc Astron Ital* **55**,227), and by Dziembowski & Goode (*Mem Soc Astron Ital* **55**,185). In the last two papers some effects of a magnetic field were also calculated. Consequences of the singular nature of the magnetic perturbation were discussed by Biront *et al* (*MN* **201**,619). Effects of the tidal force were calculated by Mohan & Singh (*Ap Sp Sc* **85**,83) and by Smeyers & Martens (*AA* **125**,193).

In all these papers a perturbation approach has been used. Effects of rotation were calculated including quadratic terms in the angular velocity Ω; for the other effects only the lowest order terms were calculated. Two distinct numerical methods suitable for treating the case of rapid rotation were developed by Clement (*ApJ* **249**,746;**276**,352) and applied to uniformly rotating polytropic models.

Fully nonadiabatic, first order (in Ω) equations for poloidal modes in uniformly rotating stars were derived by Caroll & Hansen (*ApJ* **263**,352). Numerical solutions for main-sequence star models were obtained.

Studies of toroidal (r-) mode properties have continued. A generalization of the adiabatic equations including perturbation of the gravitational potential and higher order terms in Ω was obtained by Smeyers, Craeynest & Martens (*Ap Sp Sc* **78**,483). Saio (*ApJ* **256**,717) derived the equations for nonadiabatic r-mode oscillations and solved them using realistic stellar models.

Nonlinear Mode Coupling

Direct numerical approaches to the nonlinear radial pulsation problem have been very successful in explaining many properties of the classical pulsating variables. They failed, however, to explain the multiperiodic variability observed in some of these stars and to account for amplitude limitation in delta Scuti variables. This created considerable interest in the perturbation approach. While limited in its applicability it has advantage of being more revealing and more economical. Moreover, it is the only available tool for studying nonlinear effects in nonradial oscillations.

A rigorous and elegant perturbation method based on the two-time formalism was developed by the team at University of Florida (Regev & Buchler *ApJ* **250**,679; Buchler & Regev *ApJ* **250**,776; Regev, Buchler & Barranco *ApJ* **257**,715; Barranco, Buchler & Regev *Ap Sp Sc* **84**,463; Buchler *AA* **118**,163; Buchler & Regev *AA* **123**,331). A less general (adiabatic approximation for the coupling term) but more explicit method was used by Dziembowski (*Acta Astron* **32**,147), by Aiakawa (*MN* **204**,1195;**206**, 833) and by Dziembowski & Kovacs (*MN* **206**,497).

The critical role of the resonant mode coupling in the finite amplitude development of linear instabilities was clearly demonstrated. In particular, it was shown that it may cause amplitude limitation to a very low level, as encountered in delta Scuti stars, and lead to complicated unsteady limit cycles. It may be expected that in the near future more specific results will be obtained with this method.

11.3 APPLICATIONS

Variable Degenerate Dwarfs

The opacity mechanism acting in the H ionization zone was shown to be responsible for gravity mode instability in ZZ Ceti variables (Dolez & Vauclair *AA* **102**,375; Winget *et al, ApJ Lett* **252**,L65). In contrast to the previously suggested driving in the He^+ ionization zone with this hypothesis, the restriction on the mass of hydrogen rich envelope is acceptable. The strongest support for it, however, is a confirmed prediction of the gravity-mode excitation occurring under exactly analogous conditions in DB white dwarfs at higher effective temperatures (Winget *et al, ApJ Lett* **252**,L65). Although the temperatures of the discovered variables turned out to be higher than predicted, it was shown (Winget *et al, ApJ Lett* **268**,L33) that agreement may be reached upon a reasonable rescaling of the mixing length formalism.

Variability in very hot ($T \gtrsim 10^5$ K) degenerate dwarfs was suggested to be caused by gravity mode excitation due to the opacity mechanism acting in the carbon and oxygen ionization zones (Starrfield *et al, ApJ Lett* **267**,L27; *ApJ* **281**, 800). The excitation occurs only if there is essentially no helium left in subphotospheric layers and if the relative amount of oxygen is larger than predicted by stellar evolution theory.

Studies of radial mode instability in degenerate dwarfs made by Saio, Winget & Robinson (*ApJ* **265**,958) and by Starrfield *et al* (*ApJ Lett* **267**,L27) revealed that the high overtones are unstable in the same models that have unstable gravity modes. So far, however, there is no observational evidence for the presence of high frequency oscillations in white dwarfs.

Identification of the instability mechanism for oscillations observed in degenerate dwarfs is certainly an outstanding achievement of the theory. It may be expected that future effort will focus on a detailed interpretation of the complicated patterns of the observed power spectra.

Delta Scuti Stars

These objects owe their variability to the same driving mechanism that causes Cepheid pulsation. It is believed that differences in pulsational behavior are due to the greater role of resonant mode coupling in amplitude limitation of delta Scuti stars than in the classical pulsating stars. Quantitative results supporting this view are not yet available.

In recent years efforts have concentrated on the determination of physical parameters for these stars with use of pulsation data. Fitch (*ApJ* **249**,218) performed an extensive survey of pulsation periods for models calculated with a standard stellar evolution code assuming population I chemical composition. The calculation included nonradial modes of a low spherical harmonic degree ($l \leq 3$). He found an overall consistency of standard stellar evolution theory with the observational data. A similar conclusion was reached by Andreasen, Hejlesen & Petersen (*AA* **121**,241) and Andreasen (*AA* **121**,250) who considered only the radial modes but included both population I and II models and studied effects of helium depletion in the outer envelopes. They stressed, however, the problem posed by the high masses ($> 1\ M_{\odot}$) they found for population II objects. Models with helium depletion are neither supported nor rejected by the observations.

Rapidly Oscillating Magnetic Stars

The theoreticians' response to the discovery of this new type of variable star characterized by excitation of high order acoustic modes was a search for the driving mechanism. Dolez & Gough (in *Pulsations in Classical and Cataclysmic Variable Stars* Cox & Hansen eds, JILA, 1982, p 248) showed that in a normal main sequence star model with an appropriate effective temperature such modes are unstable if effects of convection are accounted for in a certain way. They failed, however, to explain the role of magnetic fields in the excitation. An effect of the magnetic field consisting in converting convective modes into overstable gravity modes was discussed in this context by Shibahashi (*ApJ Lett* **275**, L5) and by J.P. Cox (*ApJ* **280**,220). In their models the connection between the latter modes and observed acoustic ones remains to be clarified.

Beta Cephei Stars

The cause of their variability is still unknown. Lee & Osaki (*Publ Astron Soc Japan* **34**,39), who conducted standard nonadiabatic pulsation calculations, failed to confirm Stellingwerf's suggestion that the helium opacity bump at $T \sim 10^5$ K, if calculated with a fine grid, might turn out sufficiently large to cause an instability. More exotic models were considered by Cox & Hodson (*Pulsations in Classical and Cataclysmic Variable Stars*, p 332) who suggested an excitation due to the overshooting from the convective core and by Ando (*MN* **197**,1139) who discussed the possibility that the oscillations result from the Kelvin-Helmholtz instability in the differentially rotating interior. Osaki (*Pulsations in Classical and Cataclysmic Variable Stars*, p 303) reviewed all hypotheses concerning the driving mechanism in these variables.

Classical Pulsating Variables (Population I and II Cepheids, RR Lyrae)

Work by Whitney (*ApJ* **274**,830) clarified the nature of bumps in light curves for Cepheids by combining Christy's idea of pulse echo with Simon & Schmidt's explanation in terms of the 2:1 resonance between the fundamental and second overtone modes. The resonance hypothesis offers a natural interpretation for the Hertzsprung progression and a tool for mass determination. Cepheid masses determined in this way come out significantly less than those following from stellar evolution theory.

A similar bump progression is observed in short-period Population II Cepheids (BL Her stars). Theoretical modelling of their light-curves by Hodson, Cox & King (*ApJ* **253**,260) indicates that the masses of these stars are close to $0.55\ M_{\odot}$. This number is not in conflict with the stellar evolution models.

Major problems posed by double-mode pulsation phenomena remain open. Dziembowski & Kovacs (*MN* **206**,497) suggested that it arises when the mode that would otherwise be the only one present in the nonlinear regime is in the 2:1 resonance with a higher order mode. Their arguments are based on a simplified theory and therefore the idea should be verified by means of a direct numerical approach. A dramatic discrepancy between Cepheid masses derived from the observed period ratios and the evolutionary masses has not been resolved. Simon's

(*ApJ Lett* **260**,L87) suggestion that the Cepheid mass anomalies could be rectified if the heavy element opacities are increased by factor two does not find support from the Los Alamos opacity team (Magee, Merts & Huebner *ApJ* **283**,264). To pursue further his own idea that the period ratio problem may be solved by allowing for a strong tangled magnetic field, Stothers (*ApJ* **255**,227) calculated nonlinear pulsations for models with such a field. The results are not really encouraging.

The masses derived from period ratios for double-mode RR Lyrae stars fall into the range 0.55 - 0.65 $M_\odot$, which is in a reasonable agreement with stellar evolution theory (A.N. Cox in *Pulsations in Classical and Cataclysmic Variable Stars*, p 157; Cox, Hodson & Clancy in *IAU Coll 68, Astrophysical Parameters for Globular Clusters* Philip & Hayes eds, L. Davis Press, 1982, p 337). It should be stressed, however, that as long as the double-mode Cepheid problem remains unsolved the whole diagnostic value of pulsation data remains questionable.

Stellingwerf (*ApJ* **277**,322) showed that inclusion of convection in RR Lyrae models causes a blueshift of the blue edges of the instability strip. This surprising effect is caused by changes in the structure of the H ionization zone which enhance the local driving effect. Sensitivity of the blue edge location to the helium content is significantly reduced in comparison with purely radiative models.

Long Period Variables

Fox & Wood (*ApJ* **259**,198) carried out a new survey of the linear nonadiabatic pulsation properties of stellar envelope models relevant to these stars. They found, in particular, that a periodic modulation of light curves seen in some Miras may be due to a simultaneous excitation of the fundamental and first overtone modes. An up-to-date review of theoretical work covering all types of long period variables was published by Wood (*Pulsations in Classical and Cataclysmic Variable Stars*, p 152).

Variability in Early and Medium Spectral Type Supergiants

Small-amplitude variability is a common phenomenon in all supergiants. No instability mechanism was found that could account for this phenomenon. Shibahashi & Osaki (*Pub Astron Soc Japan* **33**,427) confirmed that there are unstable nonradial modes in stars in a range of the H-R diagram to the left of the Cepheid instability strip. However, the l-values of such modes are uncomfortably large.

Lovy *et al* (*AA* **133**,307) calculated periods of the fundamental radial mode in models of supergiants. A comparison with the observed periods revealed that in some cases the latter are significantly longer. The authors suggest gravity mode excitation in such cases.

12. Commission Activities

A. GENERAL CATALOG OF VARIABLE STARS (N.N. Samus)

The *New Catalogue of Suspected Variable Stars* (*NSV*) appeared as a book in the "Nauka" Publishing House in early 1982. This was the first stage of the work on the 4th edition of the *General Catalogue of Variable Stars* (*GCVS*). The magnetic tape of *NSV* is now available from the astronomical data centers.

The magnetic tape of the first volume of *GCVS IV* is ready. It will be sent to the Strasbourg Center of Stellar Data by the end of 1984. The book of the first volume (Andromeda - Crux, about 7900 stars, with a foreword containing the description of the catalogue and of the classification system) is expected to be available by March, 1985. The second volume (Cygnus - Orion, about 10240 stars) is also almost ready and probably will be available, as a book and as a magnetic tape, by the XIXth General Assembly.

The 66th Name-list of variable stars printed in November 1981 (*IBVS* 2042) was the last one belonging to *GCVS III*. With this name-list the number of named variables reached 28457. These stars enter *GCVS IV*, the stars designated later will enter the Supplements to it. The 67th Name-list is now compiled (*IBVS*, in press). It contains 648 stars. This name-list, the first one in the system of *GCVS IV*, differs considerably in its form from earlier name-lists. In particular, it gives, along with the usual information, a list of stars ordered by right ascension (1950.0) as well as data on the range and type of variability.

During the period of this report the number of variable stars discovered in the USSR and designated in the "SVS" (Soviet Variable Star) system increased from 2493 (late 1981) to 2681 (September 1984).

B. INFORMATION BULLETIN ON VARIABLE STARS

The *Information Bulletin on Variable Stars* (*IBVS*) is designed to provide astronomers with rapid communication of important announcements and information about observational work. It has since its inception (1961) been edited at the Konkoly Observatory. From 1974 to 1984 the sole editor was Dr. B. Szeidl. In 1984 Dr. L. Szabados, who had assisted in this work for some time, joined Dr. Szeidl as co-editor.

During the three-year period covered by this report, the number of issues that have been published grew from 1985 (July 1981) to 2543 (June 1984). At present about 200 issues are prepared each year.

The editors strive for promptness of publication, which is of importance to authors and users alike. To this end it is required that manuscripts (one or two typewritten pages) be submitted in camera-ready form. Lengthy tabular material and visual observations are not accepted. A more detailed description of the requirements can be obtained from the editors.

The *IBVS* is sent to more than 500 astronomical institutes, observatories, and astronomy departments free of charge in exchange for their own publications or related scientific material. All correspondence concerning the *IBVS* should be addressed to:

The Editors, *IBVS*
Konkoly Observatory
H-1525 Budapest, P.O. Box 67
Hungary.

C. ARCHIVES OF UNPUBLISHED OBSERVATIONS (Michel Breger)

The Archives of Unpublished Photoelectric Observations of Variable Stars was created to provide permanent archives in different parts of the world. The Archives can replace lengthy and expensive tables in scientific publications by a single reference to the archival file number. Furthermore, many valuable observations are never used for scientific publications, and the Archives makes such observations available to other astronomers at a time when they might become very important.

Since the previous report to the IAU, the number of assigned file numbers has grown from 91 to 141. This represents another welcome acceleration of growth. Rapid retrieval of past files, often at no cost, has been reported from all three depositories.

The *Publications of the Astronomical Society of the Pacific* has kindly agreed to publish, on a regular basis, the summaries of recent files. Detailed reports can be found in *PASP* **91**,408, *PASP* **93**,528 and in an early 1985 issue of the *PASP*. Other summaries and announcements can be found in the *Information Bulletin on Variable Stars*. At present, astronomers who wish to *obtain* unpublished photoelectric measurements on variable stars may do so by requesting whole files (*not* partial files) from one of the three archives:

P.D. Hingley, Librarian
Royal Astronomical Society
Burlington House
London, W1V 0NL
Great Britain

Dr. C. Jaschek
Centre de Données Stellaires
Observatoire de Strasbourg
11, Rue de l'Université
67000 Strasbourg
France

Dr. E. Makarenko
Odessa Astronomical Observatory
Shevchenko Park
Odessa 270014
U.S.S.R.

There is no charge for short files. Astronomers who wish to *submit* unpublished observations should send *three* copies as well as a brief descriptive cover sheet to the Coordinator (address listed below). The Coordinator will assign file numbers and forward the observations to London, Strasbourg and Odessa. New files must be printed or handwritten in *black ink*. The printed part should be larger than 8.5 by 11 inches (21.6 by 28 cm). If a new file number is required for scientific publications (in place of extensive tables of measurements), the file number can be assigned by the Coordinator before receipt of the actual measurements. The address is:

Professor Dr. M. Breger
Institut für Astronomie
Türkenschanzstr. 17
A-1180 Wien
Austria.

Since our previous report to IAU Commission 27, the following files have been assigned and completed:

92. 16 Algol Systems: XZ And, KO Aql, RZ Cas, U Cep, U CrB, SW Cyg, W Del, AI Dra, TT Lyr, RW Mon, RV Oph, ST Per, U Sge, RW Tau, X Tri, and TX UMa - E.C. Olson.
93. HD 8357 - D.S. Hall
94. 29 Dra - D.S. Hall
95. HD 26337 - D.S. Hall
96. HD 136905 - D.S. Hall
97. RS Ind - D.S. Hall
98. Algol - G.A. Bower
99. 10 Ap Stars: HR 4082, HR 4109, HR 4327, HR 4552, HR 4817, GC 17563, HR 4965, HR 5158, HR 5269, and GC 19369 - P. Renson and J. Manfroid.
102. HR 2724 - D. Baade and O. Stahl
103. ST Ind - V.J. Mugherli
104. 16 Lac - R. Garrido
105. δ Pic - C.C. Wu
106. HD 1833, 37847, 185510, 190540, 205249, - D.S. Hall
107. DK Dra - D.S. Hall
108. IM Peg - D.S. Hall
109. λ And - D.S. Hall
110. MM Her - D.S. Hall
111. SS Boo - D.S. Hall
112. FZ CMa - N. Vogt
113. SS Psc - G. Garbusov
114. FK Com - S. Morris & E. Milone
115. CL Dra - D.L. Du Puy
116. HR 1225 - D.L. Du Puy
117. V711 Tau - D.S. Hall
118. V444 Cyg - J.A. Eaton
119. 39 AJ Cet - D.S. Hall
120. σ Gem - D.S. Hall
121. 16 RS CVn-type stars - J.A. Eaton
122. 4 W UMa binaries - J.A. Eaton
123. W Gru - M.A. Cerruti
124. 44 i Boo - E.F. Milone & R.M. Robb
125. 1 Cam, HR 8105, V358 Per, V568 Cyg, 2 ES Vul, V819 Cyg - C. Stagg
126. CC Com - D.H. Bradstreet
127. FG Sct - D.H. Bradstreet
128. BI Vul - D.H. Bradstreet
129. FS CrA - D.H. Bradstreet
130. VZ Psc - D.H. Bradstreet
131. Automatic telescope results: 4th quarter, 1983 - L.J. Boyd, R.M. Genet and D.S. Hall
132. Eta Ori - R.H. Koch
133. XY Leo - B.J. Hrivnak
135. CG Cyg - R. Robb
136. Automatic telescope results: 1st quarter, 1984 - L.J. Boyd, R.M. Genet and D.S. Hall
138. BD +28°1494 - P. Broglia
141. RS Cep - D.S. Hall

N. H. BAKER
President of the Commission

28. GALAXIES

PRESIDENT: V. C. Rubin
VICE-PRESIDENT: P. C. van der Kruit
ORGANIZING COMMITTEE: S. d'Odorico, J. Einasto, J.A. Graham, P.W. Hodge, I.D. Karachentsev, D. Lynden-Bell, A. Toomre, M.-H. Ulrich, K.-I. Wakamatsu, B.E. Westerlund

The present Report covers the period 1982-1984. As with previous Reports, it has not been possible to cover all publications or even every field of research. Some subsections of this Report, the Magellanic Clouds for example, report a body of work as extensive as that in some commissions.

This Report is divided into eleven sections, and has been prepared by the President, Vice-President, members of the organizing committee, and chairpersons of the working groups. References are given by commonly used abbreviations for journals, or by reference numbers in Astronomy and Astrophysics Abstracts, in which summaries of the papers are found. The symbol + is used in place of et al.

1. GALAXIES IN GENERAL
(B.E. Westerlund)

I. Survey work; Catalogues

The "ESO/Uppsala Survey of the ESO(B) Atlas" is completed and published (Lauberts, ESO). It contains 18,438 non-stellar objects, of which 16,000 are galaxies south of $-17^{\circ}.5$. Magnitudes and radial velocities are given for 2102 galaxies. B and R magnitudes are being measured for all galaxies. A "Merged Catalogue of Galaxies" has been compiled on magnetic tape at the Abastumani Astrophysical Observatory, and is to be permanently updated (Bull Inf CDS, Strasbourg, No. 25). It contains data on 31,917 galaxies north of -33° and brighter than $15^{m}.1$. The data have been collected from the MCG, the UGC, the RCBG, and the CGCG catalogues and include positions, magnitudes, dimensions, morphological types, radial velocities and radio continuum fluxes.

Dubois+ report the extension of the data base SIMBAD to galaxies. Three catalogues are included: the Uppsala General Catalogue of Galaxies (UGC), the Morphological Catalogue of Galaxies (MCG) and the ESO/Uppsala Survey of the ESO(B) Atlas. Positions (1950), magnitudes and dimensions for 45,317 galaxies are included. The next steps will be to add the Catalogue of Markarian Galaxies, the Second Reference Catalogue of Bright Galaxies, and a list of quasars.

Longo+ (Univ Tex Monogr Astron No. 3) have presented a general catalogue of 3578 galaxies brighter than $V = 16^{m}$. It contains information on photoelectric magnitudes and UBV colors published during 1962-1982. E. Sadler has presented new optical and radio data for 248 southern E and S0 galaxies (AJ 89, 23, 34, 53). Radial velocities, photoelectric magnitudes (169 new), and the result of a radio continuum survey at 2.7 and 5.0 GHz (38 detections) are included. A list of the "Second Byurakan Spectral Survey" has been given by Markarian and Stepanyan (Astrofiz 19, 639). 109 galaxies and QSO candidates brighter than 20^{m} were found in a region 4° by 4° centered on $\alpha = 11^{h}30^{m}$, $\delta = +59^{o}$. A bibliography of the surface photometry of galaxies has been published by Davoust and Pence (AA Suppl 49, 631). For 650 galaxies from more than 300 references, it contains methods of observation and reduction, photometric information and limiting surface brightness. A catalogue of 2251 quasars with published redshifts and of 554 Seyfert and related galaxies known prior to Sept. 1983, has been published by Veron-Cetty and Veron (ESO Sci Rep

No. 1). It includes positions, redshifts, photometry and 6 and 11 cm flux densities.

Images in the Gunn r-band have been obtained with the Palomar 1.52 m telescope of fields around 108 bright quasars (ApJ Suppl 54, 495; ApJ 280, 79). The catalogue gives positional offsets, magnitudes and classifications into stars or quasars. An analysis of the quasar and control fields suggests that quasars are associated with galaxies at their cosmological redshifts. Brosche reports on a "List of Elite Galaxies" (Tartu Teated No. 68). The aim is to provide a first approximation of a generally agreed data base and to make the "black holes" in that data obvious.

Rudnicki reports that at the Jagellonian University Observatory the general catalogues of galaxies are surveyed. Atmospheric extinction, visibility of galaxies on plates, and the effect of interstellar extinction in terms of galactic and supergalactic coordinates are examined. A search for intergalactic obscuring clouds in selected areas is in progress. Intergalactic clouds may be located in volumes of space with a scarce population of galaxies. Magnitudes, colours, ellipticities and position angles have been compiled for 6724 galaxies in a 2°63 area centered on the Coma cluster (MN 202, 113). Kraan-Korteweg has presented a catalogue of a complete sample of 180 Virgo cluster galaxies (AA Suppl 47, 505), within 6° of M87 and brighter than $B_T = 14^m.0$. Positions, morphological types, magnitudes and radial velocities are given. Lucey+ (MN 203, 545) present redshifts, morphological classifications, orientations and ellipticities for 129 galaxies within a $6^o \times 6^o$ region in Horologium-Reticulum. The prominent clusters in the region form a supercluster with a mean redshift of 18,000 km s^{-1}. Crane+ (The Birth of the Universe, p. 325) are surveying galaxies in Horologium for radial velocities and photoelectric photometry.

Pesch and Sanduleak are producing a low-dispersion objective prism survey of the northern galactic polar region. Two parts have been published (ApJ Suppl 51, 1711; 55, 517). 196 galaxies and 179 unresolved objects (possible QSO candidates) are reported, all being blue and/or emission line objects with $11^m < B < 18^m$. Kinman reports on an objective-prism survey for emission-line galaxies with the CTIO Curtis Schmidt and the KPNO Burrell Schmidt (MN 202, 53). Particular care is taken to ensure completeness and to get confirming spectroscopic IIDS scans (8A resolution) and BV photoelectric photometry. This confirming work is about 75% complete. Moss and Whittle are using the Burrell Schmidt to conduct an Hα-survey for emission-line galaxies in nearby Abell clusters (cz < 9500 km s^{-1}). Data for four clusters have so far been obtained. 10-15% of the galaxies listed in the Zwicky CGCG catalogue are detected in Hα-emission.

An attempt to find the optimal two-dimensional classification system for spiral galaxies has been described by Whitmore (ApJ 278, 61) on the basis of a large number of observational parameters. Studies of polar-ring galaxies have been presented by Wakamatsu+ (PASP 95, 2983; ApJ 273, 167; AJ 87, 980).

Courtes reports that a balloon survey in UV (2000A) has been carried out for the nearest galaxies as a joint program of the Laboratoire d'Astronomie Spatiale of Marseille and the Geneva Observatory. Excellent photographs of M33, M31 and M81 reveal the hot stellar populations and show close correlations with the faint diffuse Hα general emission, as well as the dust emission observed by IRAS. A new UV survey will start in Sep. 1984 with a new 40 cm telescope (FOCA-1000).

A near-IR photographic atlas of 54 spiral galaxies has been presented by D.M. Elmegreen (ApJ Suppl 47, 229). The sample was subdivided into categories of grand design and flocculent spirals. A 12-division morphological system has been developed by D.M. and B.G. Elmegreen (MN 201, 1021) and applied to the classification of 305 nearby galaxies. The system has been applied to 829 spiral galaxies in 22 clusters (MN 201, 1035) and to 261 low-inclination galaxies in groups (ApJ 267, 31). "A Catalogue of Infrared Magnitudes and HI Velocity Widths for Nearby Galaxies" has been published by Aaronson+ (ApJ Suppl 50, 241). It supplies integrated H(1.6 μ) magnitudes, HI velocity widths and velocities, positions, types, inclinations, diameters, HI flux density integrals and distances.

Some results from the IRAS survey have been published. The IRAS observations of 165 galaxies in the Revised Shapley-Ames Catalogue led to the detection of 108 predominantly spiral galaxies (ApJ 278, L67). In the IRAS mini-survey 86 galaxies were detected at 60 μ (ApJ 278, L71), again many spirals.

A new "Catalogue of Galaxy Redshifts" (CGR) has been compiled containing known redshifts for ~4000 galaxies (ApJ Suppl 49, 111). Davis reports that a redshift survey has been completed at the Harvard-Smithsonian Center for Astrophysics. The catalogue contains radial velocities for a magnitude-limited sample of 2400 galaxies within a solid angle of 2.7 steradians (IAU Symp 104, 167). The sample is complete to Zwicky magnitude = $14^{m}.5$.

A catalogue of rotation curves for 271 galaxies is given by Baiesi-Pillastrine+ (AA Suppl. 53, 373) with bibliographical information from 332 references from before 1982.

II. Measurements in the Radio and X-ray Regions

Heeschen reports observations of 226 extragalactic radio sources made at 9 and 6 cm at Green Bank (IAU Symp 97) and of 34 sources at 20 cm and 11 cm made at Arecibo. This is a part of an investigation of the global properties of variability of an almost complete sample of extragalactic radio sources. Compact sources exhibit variability on scales of 2 to 20 days with a mean amplitude of 2%. Whether this is intrinsic or the result of interstellar or circumsource scintillation is unclear. Observations to investigate variations on time scales of 30 seconds to 2 days and 20 days to 6 years are under way.

Sherwood reports that 464 Markarian galaxies with UV-excess from lists 7-11 have been observed at 11 cm with the Effelsberg 100 m telescope. 13 detections were made with flux densities ≥ 30 mJy, a number appreciably less than expected from earlier observations of galaxies from lists 1-5.

Total power and polarization maps of NGC 253 have been obtained at 10.7 GHz (AA 127, 177) according to a report from Urbanik. He also reports mappings of NGC 520, NGC 2276, NGC 3448, NGC 3267 and four bright Virgo cluster galaxies (NGC 4254, 4231, 4501 and 4569). Among these only the interacting galaxy NGC 2276 showed an extremely high disk emission, indicating very intense star formation.

A resurvey of the $\pm 4^{o}$ declination zone of the Parkes 2700 MHz survey has been made with the Parkes 64 m telescope. The catalogue of 822 sources is now complete to 0.25 Jy and all but one source have been measured at 5 GHz. 59 new optical identifications have been made (Austr J Phys 35, 177, 207). Results for 294 bright galaxies observed with the Molonglo Radio Telescope at 408 MHz are given in Austr J Phys 35, 321. Radio luminosity functions are derived as well as spectral indices for 73 galaxies observed at other frequencies.

The Very Large Array (VLA) has been extensively used. 602 radio sources from the 5 GHz MIT-Green Bank Survey have been observed (ApJ 278, L95) and their radio structures have been classified. A deep survey in a field at $\alpha = 8^{h}52^{m}15^{s}$ and $\delta = +17°16'$ has been carried out at 1.4 GHz. 159 sources stronger than 84 μJy were found and 45 have been identified with optical objects brighter than B = $21^{m}.5$. Another deep survey is described in Science 225, 23. Many VLA observations study QSO's, BL Lac objects and nuclei of galaxies (MN 208, 231; ApJ 279,93; 280, 476; AJ 89, 5, 189, 195, 203, 753, 932; AA 134, 207).

VLBI observations with the NASA Deep Space Network (Goldstone-Madrid and Goldstone-Tidbinbilla) at 2.29 GHz surveyed 257 extragalactic radio sources within 10° of the ecliptic (AJ 89, 336), to find compact radio sources which form an accurate reference frame for planetary spacecraft navigation. Components with sizes smaller than 2.5-3.2 milliarcsec and with flux densities > 0.1 Jy were found in 144 sources.

The Bologna Northern Cross radio telescope investigated at 408 MHz the low-frequency spectral index distribution of a complete sample of 308 extragalactic sources north of $\delta = -10°$ (AJ 89, 323). From a sample of 132 bright elliptical galaxies a strong correlation between radio activity and flattening is found (MN 206, 899). No ellipticals with axial ratios below 0.65 are found to be radio emitters.

The Westerbork Radio Synthesis Telescope has observed a large sample of S0 galaxies (AA 106, 183) at 1415 MHz. Continuum properties were compared with those of S and E galaxies. Many investigations at the HI 21 cm line wavelength have been published. "A General Catalogue of 21 cm Line Data" (AA Suppl 47, 171) includes 1207 galaxies. Corrected 21 cm line widths at three different levels are given together with HI fluxes and radial velocities. "A General Catalogue of HI Observations of External Galaxies" (ESO Sci Prepr No. 250, 176) has been compiled from pre-1983 data. It contains information for 4000 galaxies and is more than 95% complete for references and more than 98% complete for entries up to the deadline. A 21 cm line survey of 359 galaxies of types S0 to Im, south of $\delta = -18^{o}$, has been carried out with the Parkes 64 m telescope. 196 galaxies were detected and their spectra and integrated properties have been presented (AA Suppl 50, 451).

Mirabel reports on HI surveys of 91 Seyfert galaxies (ApJ 277, 92) and 80 spirals with strong nuclear radio sources (ApJ 260, 75; 270, L35) undertaken with the Arecibo 305 m telescope. In Seyfert galaxies, redshifts from nuclear optical emission lines are systematically smaller than systemic velocities derived from HI line profiles. A 21 cm line survey of a magnitude-limited sample of interacting and isolated galaxies has been carried out at NRAO and Arecibo (ApJ Suppl 53, 269). 166 interacting and 113 isolated galaxies were observed. The radio data were combined with optical data on blue diameters, axial ratios and magnitudes. A HI survey has been made of galaxies in UMa and CVn using the Mk IA 76 m telescope at Jodrell Bank (MN 201, 1073). New accurate systemic velocities, HI masses and total masses are given. The region contains possibly four associations of galaxies.

Benn+ (MN 200, 747; 209, 683) have conducted a deep optical/radio survey near the North Galactic Pole, the 5C 12 survey: at 408 and 1407 MHz with the Cambridge One-Mile telescope, at 4.85 GHz with the 100 m Effelsberg telescope, and in the optical with the 48" Schmidt telescope of Hale Observatories. 299 radio sources were detected at 408 MHz within a 4° diameter circle and 65 at 1407 MHz within a 1° diameter circle. The optical part (MN 206, 475) presents 1200 objects brighter than $m_v = 22\overset{m}{.}5$ in 1' fields centered on the radio positions.

HI and optical parameters are presented for 151 southern low surface brightness galaxies (LSB) selected from UK Schmidt plates and observed with the Parkes 64 m telescope (MN 200, 325). The 21 cm line was detected in 100 of the galaxies. The data are consistent with the LSB being of systematically lower mass than brighter galaxies of same type and linear dimensions. Observations have been made at Arecibo of 325 isolated galaxies to define a standard sample for the comparison of the HI content of galaxies in different intergalactic environments (AJ 89, 758). HI emission was detected in 288 cases with a definite increase of the non-detection rate for the earliest types.

A catalogue of HI diameters of 238 galaxies has been compiled from 98 new measurements at Nancay, and 140 previously published results from Nancay and

Arecibo (AA Suppl 55, 55). The catalogue has been used to study HI expansions and relations between gaseous and stellar extensions in galaxies (AA 122, 273). HI fluxes, optical and infrared luminosities and broad-band colors of 300 spiral galaxies in clusters are presented by Bothun+ (preprint) together with a discussion of a number of derived parameters. The Tully-Fisher relation has been the subject of a number of studies (Bottinelli+, AA 118; 4; ApJ 254, 8; 280, 34; and Aaronson+ ApJ 265, 1; 279, 475).

A review of molecular clouds in external galaxies appears in Ann Rev Astron Astrophys 20, 517. The proceedings of a workshop on "Extragalactic Molecules" (Publications Div., NRAO) held at Green Bank is published. D.M. and B.G. Elmegreen report on ^{12}CO 2.6 mm observations of 23 spiral galaxies with the NRAO 36 ft radio telescope (AJ 87, 626). 4 SAB galaxies, NGC 157, 2903, 4321 and 5248, and the Seyfert galaxy NGC 1068 were detected. For further investigations in the radio region the reader is referred to the Report of Commission 40.

An X-ray survey with the Einstein Observatory of a complete sample of 40 3CR radio galaxies has been made by Fabbiano+ (ApJ 277, 115). 26 galaxies were detected in X-rays, having X-ray luminosities between 10^{41} and 10^{45} ergs s^{-1}. The authors found a correlation between X-ray luminosity, 5 GHz nuclear source luminosity, strong optical emission line spectra and the presence of double radio morphology. They suggest a nuclear origin of the X-ray emission.

Long has discussed X-ray properties of normal galaxies as derived from observations with the Einstein Observatory of more than 50 nearby normal galaxies (Adv Space Res Vol 2, 177). X-ray luminosities were ~2×10^{-4} of optical luminosities and showed no strong correlation with morphological type. Nuclei of some normal galaxies were found to be luminous X-ray sources. Many X-ray observations have been devoted to galaxies with active nuclei and QSO's. See the relevant subsection in this Report and also the Reports of Commissions 44 and 48.

2. STRUCTURE AND EVOLUTION OF GALAXIES
(P.C. van der Kruit)

One recent IAU meeting was devoted to the structure and evolution of galaxies, IAU Symp 100: "Internal Kinematics and Dynamics of Galaxies" (33.012.004). However, relevant discussions took place in other meetings: "Extra-galactic Molecules" (31.012.048), "Morphology, and Dynamics of Galaxies" (Saas-Fee; 32.012.032), "Origin and Evolution of Galaxies" (33.012.023), "Kinematics, Dynamics and Structure of the Milky Way" (33.012.016), "Evolution in old Stellar Populations" (in 33.012.028), "The Nearby Stars and the Stellar Luminosity Function" (34.012.065) and "The Milky Way Galaxy" (IAU Symp 106; to be published). In this Report, nuclei and active galaxies are not included. These subjects are covered at other places in this commission Report and in the Reports of commissions 40 and 48.

I. Elliptical Galaxies

DYNAMICS. Models of early type galaxies have been reviewed in detail by Binney (Ann.Rev.20, 399; 32.151.048; 33.151.052). Emphasis is placed on calculation of individual stellar orbits in building models and in studying their properties; this follows the work by Schwarzschild (see his older chapter 30.158.103). Models are usually tri-axial, such as that of Wilkinson and James (31.151.012) or de Zeeuw and Merritt (33.151.093). A numerical simulation of Schwarzschild's model shows it to be "rugged and robust" (31.151.058). Petrou has introduced a third integral to construct oblate models with reasonable rotation curves (33.151.014/15). Pismis and Moreno (preprint) have found that an initially oblate stellar system oscillates between oblate and prolate form.

Lake and Norman (34.151.001) considered both stellar orbits and gas motions and showed how stable orbits of the gas may exist in planes tilted with respect to the symmetry axis of the triaxial potentials. Such motions may explain "spindle" systems like NGC 2685. In ordinary E and S0 galaxies the gas motions appear different and the velocity dispersions lower than in the stars (AJ 89, 356).

Tonry (33.151.012) has used models of elliptical galaxies with constant M/L ratio but with anisotropic velocity distributions to show that their projected properties on the sky fit actual observations. Binney and Mamon (31.158.320) show that the observed kinematics of M87 can be fit with a model of constant M/L. Although the models suggest a tight star cluster in the inner regions, there is no need to suppose the presence of a black hole of 5×10^9 $M_\odot$. Efstathiou+ (32.158.267) model their observations of NGC 5813 with an increasing velocity anisotropy with radius combined with a very modest increase in M/L.

One interpretation of the X-ray observations of M87 points to a dark halo with a mass of a few times 10^{13} $M_\odot$ within 200 kpc (33.158.109). The opposite conclusion of no dark halo and $5 \times 10^{11} M_\odot$ within 100 kpc has however been reported previously (30.158.007) and in that case matter beyond 200 kpc does exist but is associated with the Virgo cluster. The dark mass is associated with M87 rather than with the cluster.

Davies+ (33.157.029) have measured rotation and central velocity dispersion in faint ellipticals. These rotate more rapidly compared to the velocity dispersion than bright ellipticals do and are almost consistent with the predictions provided by oblate models supported by rotation (see also 33.157.097). This agrees with expectations based on formation by merging or fragmentation in pancakes. Lake (33.151.006) finds that tidal torques make the same prediction.

For analytical purposes Jaffe (33.157.041) has proposed a simple formula to describe the three-dimensional light distribution in spherical galaxies. Van Albada (32.151.065) has given stellar dynamical collapse models that produce luminosity profiles like observed galaxies; abundance gradients are probably conserved through violent relaxation and so do not force us to conclude that the collapse was dissipational. Colour gradients have been studied by Wirth and Shaw (30.158.063; 33.157.037) and Boroson+ (34.157.123), who find rounder isochromes than isophotes at small radii, probably supporting dissipationless collapse or merger as the formation process.

GLOBULAR CLUSTERS. Frogel (ApJ 278, 119) has measured infrared colours of clusters in NGC 5128 and found that these are old; some have higher metallicities than the sun. Comparison with Virgo ellipticals shows that the mean metallicity may be a function of the parent galaxy's luminosity. Hesser+ (ApJ 276, 491) also observed clusters in NGC 5128. One conclusion is that the velocities indicate a mass of 1.6×10^{12} $M_\odot$ within 36 kpc.

SHELLS. Discovery of shells in the light distribution in the outer parts of elliptical galaxies has led to a major southern hemisphere survey, a catalog of 137 galaxies by Malin and Carter (34.157.147). Photometry of one shell in NGC 1344 shows that it is made of stars and is consistent with a (young) disk population (31.158.058). Schweizer (33.157.105), who calls these features ripples, argues that they arise from a merger of an elliptical and a disk galaxy. Quinn (ApJ 279, 596) has modeled such encounters with numerical simulations and derives information on the form of the potential. This suggests a massive dark halo.

GAS AND STELLAR CONTENT. Mapping of HI in large elliptical galaxies is now revealing its kinematics and distribution (ApJ 246, 708; Knapp and Raimond, preprint). The gas appears to have originated externally via accretion or merging rather than resulting from mass loss in stellar evolution or ejection from the nucleus (Knapp, 33.157.100). UV observations have revealed the presence of a component of hot stars (32.158.228). These are most easily, but not unambiguously, interpreted as a small population of recently formed stars (32.158.192/194; Veron and Veron-Cetty, preprint).

CCD's make it possible to study the stellar content of dwarf ellipticals in the Local Group. Observations of NGC 147 (34.157.027) and NGC 205 (ApJ 278, 575) are reported. Besides improved metallicities and distances, these studies give information on possible intermediate-age components from a study of the asymptotic giant branch. No such component is detected in excess of 10% of the stars observed. The majority of stars have ages similar to globular clusters. This contrasts with a similar study in BV of the Corina dwarf spheroidal (34.157.108) for which the bulk of stars indicate an intermediate age of about 7 Gyr. There may actually be no population II at all.

Clusters of galaxies at large redshift have more blue galaxies than present day clusters. This Butcher-Oemler effect has recently been confirmed by its discoverers in a cluster at z = 0.38 (Nature 310, 31). Recent deep counts in two colours by Shanks+ (MN 206, 767) provides evidence that this effect also occurs outside of clusters.

II. Disk Galaxies

PHOTOMETRY OF BULGES. Wirth and Shaw (30.158.063; 33.157.037) measured (B-V) colour gradients in bulges as a function of luminosity and morphology. Pure spheroids have small gradients but show less correlation with luminosity of the bulge. Griersmith+ (32.158.018) have found gradients In the near infrared in (V-K) and (J-K) as well as in (U-V) but these are smaller than those in E and SO galaxies. Colors are similar in spiral bulges and ellipticals, suggesting similar populations. The strong bulge Sab system NGC 7814 has a measurable abundance gradient; its HI rotation curve (31.158.309) indicates an M/L gradient.

KINEMATICS OF BULGES. Dramatic developments have occurred in this field. The main conclusions by Kormendy and Illingworth (31.158.1551; 33.157.024), Whitmore and Kirshner (30.151.051) and Schechter (34.157.041) are that in contrast to giant ellipticals but similar to ellipticals of comparable luminosities, bulges appear flattened mainly by their rotation. Like ellipticals they exhibit a relation in which the bulge luminosity is proportional to a power between 4 and 8 of the velocity dispersion. Box-shaped bulges are cylindrical rotators and bulges in barred galaxies appear to be rapidly rotating triaxial structures (31.158.193). Jarvis (preprint) has constructed dynamical models in which bulges are flattened by rotation; these models represent the photometry and observed kinematics in bulges. The SO-galaxy NGC 3115 has a spiral-like bulge flattened by rotation (31.158.156). Dressler and Sandage (33.157.026) find the same result in a sample of 32 mostly SO-galaxies. These follow the luminosity-velocity dispersion relation. SO's have central velocity dispersions intermediate between those of ellipticals and spirals; this argues against recent stripping of present-day spirals to form SO-galaxies. The diffuse component of the spindle galaxy NGC 4650A has a higher rotation velocity than central velocity dispersion (ApJ 277, 526). Reviews of much of these matters have been presented at IAU 100 (33.151.051/52). A detailed dynamical model of the bulge of M31 has been presented by McElroy (34.157.028) and Lawrie (34.157.111) finds the bulge velocity dispersion from planetary nebulae to be similar to that of the stellar population.

GLOBULAR CLUSTERS. Searches for globular clusters have been made in a number of spirals: NGC 891 and 4565 (31.154.015); M33 (32.158.032); NGC 55 (32.154.015; 33.157.002), NGC 253 (33.157.014), NGC 4594 (ApJ 89, 216) and NGC 2403 (AA 130, 162). The analysis of NGC 4594 by three Harrisses indicates that spirals have fewer globular clusters per unit spheroid light than ellipticals (see also 30.151.064).

Cohen+ (ApJ 281, 141) find that the clusters in M33 have an age range similar to the LMC and unlike larger spirals.

ROTATION CURVES AND MASSES. Observations, systematics and analyses of optical and radio rotation curves have been presented and reviewed by Rubin (32.158.125; 33.157.058; 33.157.246; ApJ 289) and Bosma (30.158.239/40; 33.157.059); see also Lequeux (AA 125, 394). There must be dark matter of at least the mass of the luminous material and maybe constituting 90% of all mass in the universe. Interpretation of the rotation curves needs to take the disk and spheroid contributions into account (Bahcall+, 31.151.078).

The shape of the rotation curve out to about the Holmberg radius can be used to estimate the halo distribution. Carignan and Freeman (preprint) use pure-disk systems and show that their rotation curves can be modelled excellently by a disk with constant M/L plus an isothermal sphere. The latter contains at least half the mass within the last measured point. Analyses of rotation curves of NGC 5907 by Casertano (33.157.118) and NGC 891 by Bahcall (33.157.119) give comparable results. A different approach is to use both stellar and gaseous kinematics in the z-direction in the disks to estimate the disk density distributions. Van der Kruit and co-workers (33.157.071; ApJ 278, 81; AA 132, 20) show that out to the optical edge the disks contain no more than 1/3 to 1/2 of the total mass.

Total masses out to larger radii are derived by sampling the gravitational field by the following methods: (i) The HI rotation curve of NGC 3198 has now been extended to more than 2 Holmberg radii (Bergeman+ preprint) and still fails to decline; (ii) HI absorption at a similar distance from NGC 3067 against the QSO 3C232 would suggest a still rising rotation curve according to Rubin+ (31.158.141); (iii) Polar rings may be used to estimate the potential (33.157.246; AJ 89, 618, 975); (iv) The velocities of companions or in binary galaxies can be examined (e.g. van Moorsel, 34.157.038/39). All studies indicate large amounts of non-luminous matter; only the study of the barred system NGC 3992 (32.158.094) comes to a different conclusion.

The nature of the dark matter is not clear. Several studies (30.158.176; 31.158.330; 34.157.114) conclude that faint M stars would provide too much luminosity or are locally too scarce to provide sufficient mass. The search for dark matter in dwarf galaxies (22.157.032/033), and the velocity dispersions of globular clusters in Fornax (34.157.030) or carbon stars in Draco (33.157.031) do not yet allow a definitive conclusion. Lake and Schommer (ApJ 279, L19) claim large masses in binary dwarf galaxies.

Burstein (31.158.032) has critically discussed the Tully-Fisher relation between luminosity and rotation speed, while Tully+ (31.158.212) have discussed a colour-magnitude relation. Burstein and Sarazin (33.157.008) propose a simple model that explains the distribution of global properties among spirals. A model for the formation of disk galaxies including a dark halo has been given by Jones and Wyse (33.151.065).

STRUCTURE OF THE DISK. Boroson (30.158.059) determined radial luminosity profiles for 26 spirals and finds that the bulge-to-disk ratio is related to Hubble type with a scatter larger than observational uncertainty. His sample displays a constant extrapolated central surface brightness of the disk but with a larger scatter than Freeman's original sample. Meisels and Ostriker (preprint) argue that the Hubble type is independent of disk but is only related to the absolute magnitude of the bulge.

The vertical structure of the stellar disk was further studied by van der Kruit and Searle (31.158.038) in terms of their three-dimensional model with constant thickness with radius. The observational result was confirmed in NGC 4565 by Jensen and Thuan (32.158.307). A very modest increase of thickness with radius (~10%) is consistent with the data (32.151.008). Hamabe and Okamura (31.158.248; 32.158.211) also decompose edge-on galaxies into bulge and disk. Van der Kruit and Freeman (ApJ 278, 81) measure high stellar z-dispersions in the central parts of two disks in agreement with the predictions of constant thickness. Mechanisms to explain the age-velocity dispersion relation in the solar neighbourhood have been investigated (31.151.018; 33.151.070; 33.151.103; 34.115.006/089), but no definitive conclusion regarding the actual process has appeared.

The thickness of the HI layer can provide information on the mass distribution in the disk. The HI velocity dispersion of 7-10 km s^{-1} measured in face-on systems by van der Kruit and Shostak (31.158.079, AA 132, 20, AA 134, 258) leads to an M/L of the old disk population (33.157.071, AA 134, 258). Van der Kruit and Freeman (preprint) have observed that the vertical velocity dispersion in the old disk of NGC 5247 declines in accordance with the constant thickness. M/L in the old disk is 6±2 in B.

ABUNDANCE GRADIENTS AND CHEMICAL EVOLUTION. Radial gradients in chemical abundances have been deduced from observations in many systems (31.158.037; 30.158.135; 31.158.065). McCall's extensive observations for 20 galaxies (33.157.256) have significantly added to the available data. Chemical evolution has been reviewed by Pagel (30.151.022). Noteworthy is also the deduction by Rayo+ (31.158.065) that the pregalactic He abundance was $Y_p = 0.216 \pm 0.01$ (3σ).

SPIRAL STRUCTURE AND STAR FORMATION. Toomre (30.151.021) has described the swing amplifier as the basis of spiral structure in galaxies. Some aspects of density wave theory have been studied by Bertin and Haas (31.151.040, see also 34.151.038). Self-propagating star formation as the basis of spiral structure has been reviewed by Seiden and Gerola (31.151.084). Observationally, Elmegreen and Elmegreen (32.158.268/9; ApJ Suppl 54, 127) have concentrated on "flocculent" spiral galaxies. Studies in the near IR show in the latter an absence of structure in the older disk populations, and these galaxies appear not to display a density wave. Observationally the situation is the opposite in grand-design spirals (e.g. M51), while M33 and M101 are examples of intermediate cases which display evidence for both phenomena. Freedman and Madore (ApJ 280, 592) stress the non-uniqueness of self-propagating spirals in their model simulations. A comprehensive review of all aspects involving spiral structure has been written by Athanassoula (Physics Report, in press).

Kennicutt has studied the relation of spiral arm structure and morphology with classification schemes and model predictions (30.158.241; 31.158.026/71). A significant addition to models for the star formation rate in galaxies has been provided also by Kennicutt (34.157.068) who includes integrated Hα emission together with UBV photometry as input. These constraints better disentangle the effects of star formation rate and initial mass function. Spirals in the Virgo cluster indicate a lower star formation rate (33.157.111). A similar approach by de Giola-Eastwood+ (ApJ 278, 564) estimates the radial dependence of the present star formation rate in NGC 6946; it turns out strongly correlated with the H_2-surface density. Correlations between physical properties in HI-HII associations have been discussed in detail by Viallenfond+ (32.158.249). Hodge and Kennicutt (33.157.163) have studied the radial distribution of HII regions as a function of Hubble type and luminosity of the galaxy. Wevers (Thesis, Groningen) found that colour indices in spiral disks are approximately constant with radius. This implies that the time-dependence of the star formation rate has not been a strong function of galactocentric distance.

MOLECULES. The last three years have seen an enormous increase in the study of CO in external systems (see 33.157.064/65). This compound traces the distribution of H_2. Many major spirals have been studied (31.158.340/45; 32.158.011;

33.157.013/051/239, ApJ 276, 476). The CO follows the distribution of blue light with radius as first pointed out by Young and Scoville (32.15.011).

BARRED GALAXIES. The gas exhibits strong streamings in most systems (31.158.011/14; 32.158.007; 34.157.048; MN 207, 9). The narrow dust lanes in bars are coincident with ridges of radio continuum emission (33.157.242) and thus are almost certainly strong shocks. The general morphology and stellar kinematics has been reviewed and observed by Kormendy (30.158.104; 331.57.089; 34.157.187). Many studies examine orbits of stars and gas flows in barred potentials (31.151.033; 32.151.033/35/80; 34.151.040). Strong deviations from circular rotation are predicted and have been confirmed in general. Tight spiral features in nuclei are described as the loci of matter ejected from the rotating nucleus (Pismis and Moreno, Astrofizica 20, 7; preprint).

STABILITY. Sellwood (33.151.039) has reviewed various criteria for disk stability; prominent among these is that of Toomre for local stability. Global stability can be provided by a dark halo if it contains sufficient mass or if the central disk has a large velocity dispersion (also 33.151.040). The halo masses reported above are sufficient to provide stability for the disks according to the criterion of Efstahiou+ (31.151.081). Van der Kruit and Freeman (preprint) find that the central velocity dispersion in NGC 7184 is also high, about twice that required as minimum by Toomre's local criterion out to about 2 scalelengths. This agrees with numerical simulations by Sellwood and Carlberg (ApJ 282). Kormendy (preprints) has observed the stellar kinematics in the disks of an SO and SBO system. These are very hot and there seems a clear correlation between high velocity dispersions and absence of structure in disks.

RADIO EMISSION. Radio continuum emission from galaxy disks contains a thermal (HII regions) and non-thermal component. Observations show in many cases a steeper radio spectrum with increasing radius. This cannot be entirely explained by a varying thermal/non-thermal ratio (31.158.016/069/111; 34.157.074/125; AA 135, 213) as was thought earlier. Polarization to map magnetic fields is now feasible (31.158.085; AA 135, 213). SO galaxies usually have much fainter radio disks than spirals (31.158.086). Individual sources can often be related to HII-regions or supernova remnants (e.g. 31.158.018; 33.157.006). The general distribution of non-thermal emission does not show concentrations towards regions of recent star formation (34.157.074; 31.158.016/111). However, the integrated Hα-flux correlates well with the non-thermal flux density and Kennicutt interprets this to mean that the old disk population is not important in the generation of relativistic electrons and cosmic rays (33.157.114).

Edge-on spirals show much thicker disks in radio continuum than the HI or young population (AA 133, 1) so cosmic rays may diffuse in the z-direction. In NGC 891 Allen and Hu (preprint) find a close correspondence with the optical surface brightness. Parker's instability may play an important role in establishing the z-structure of the radio disks (see also 34.157.105).

WARPS. The theories of warps have most recently been reviewed by Toomre (33.151.036). Several models involving hydrodynamic phenomena related to interactions between a rotating disk and non-rotating halo (30.151.059; 31.151.025), corrugations (30.151.0031), triaxial potentials (33.151.090) or self-gravity (MN 206, 691) have been published. A particularly simple suggestion is that warps have to do with the shape and tilt of a massive dark halo (33.151.037, ApJ 280, 117).

IRAS. Infrared observations with IRAS have been published in ApJ 278 Letters (March 1, 1984). M31 has emission associated with the dust, HI, HII and radio continuum in a ring of radius about 10 kpc. Most Shapley-Ames galaxies are detected. However in an (infrared) flux limited sample the infrared to blue luminosity ratio varies by a factor 100.

ENVIRONMENTAL INFLUENCES. Schweizer has reviewed the observational evidence for mergers among galaxies (31.158.007; 33.157.105). Villumsen (31.151.041; 34.151.008) and White (33.151.059, also 34.151.057) have presented numerical simulations of merging galaxies. The final remnants show some characteristics similar to actual galaxies. The effects of cluster environment on the gas content of galaxies has been studied by Sullivan and co-workers (30.158.060/61; 31.158.323/24). Most clusters do not contain preferentially gas-deficient spirals. One example of a major spiral outside large clusters that appears presently to undergo stripping is NGC 1961 (32.158.248). HI mapping in S0-galaxies has shown that current stripping is not a major process among these galaxies (33.157.077).

3. GROUPS AND CLUSTERS OF GALAXIES
(J.A. Graham)

This field has been very active during the past three years; stimuli have come from a variety of sources. New photographic sky surveys are available, such as the ESO/SRC Atlas which provides a first deep view over the whole area of the southern sky. New detectors, such as the CCD, allow us to carry out area photometry of galaxies to a greater precision than possible previously (Schneider+, ApJ 264, 337; 268, 476), and multiple galaxy integrated photometry within a single frame is being explored (Bautz+, ApJ 255, 57). Optical fibre devices (Hintzen+, AJ 87, 1656) enable spectra to be obtained of many objects within a small field thus raising the efficiency of data collection during valuable large telescope time. Larger and faster computers have added to our theoretical understanding of gravitational interactions within groups and clusters of galaxies. Perhaps the principal difficulty confronting workers in the field is the need to find time to digest and interpret the vast amount of new data which is coming in.

I. The Local Group

Galaxies within the Local Group are easily resolved into individual stars and are basic to the calibration of the distance scale. Cepheid variables have been studied in M33 (Sandage and Carlson, ApJ 267, L5) and in NGC 6822 and IC 1613 (McAlary+ ApJ 273, 539; 276, 487). CCD detectors are especially useful for accurate stellar photometry (Walker, MN 209, 83). Hoessel and collaborators (ApJ 254, 68; 271, 65; 274, 577) and Christian and Tully (AJ 88, 934) have applied the technique with success to several dwarf galaxies of the Local Group. These data will be essential for interpreting the integrated spectra of more distant galaxies. Low dispersion spectroscopic studies have led to the identification of carbon stars in the Local Group (Richer and Westerlund, ApJ 264, 114) while Moffet and Shara (ApJ 273, 544) as well as Massey and Conti (ApJ 273, 576; BAAS 16, 12) have detected and studied individual Wolf-Rayet stars. The mass and dynamics of the Local Group have been examined by Einasto and Lynden-Bell (MN 199, 67). The motion of the Local Group with respect to a backdrop of distant galaxies has been determined by Hart and Davies (Nature 297, 191).

II. Nearby Groups

A useful all-sky catalog of nearby galaxy groups has been published by Huchra and Geller (ApJ 257, 423), based on a parameter-space search for density enhancements. A remarkable one-to-one correspondence is found between this computerized search and the earlier catalog of de Vaucouleurs. A second catalog, by Geller and Huchra (ApJ Suppl 52, 61), uses a similar search procedure on the deeper, northern hemisphere CFA redshift survey. A search of the Palomar Sky Survey by Hickson (ApJ 255, 382) has led to the identification of 100 compact groups. Optical velocities have been obtained by Huchra, integrated 21-cm profiles by B.

Williams and Rood, and rotation curves by Rubin, Ford, and Schechter.

The galaxy group in Sculptor, which is the nearest to our own, is important to the distance scale calibration. Its member galaxies extend over a distance of 1 Mpc or more along the line-of-sight (Graham, ApJ 252, 474). Webster and Smith (MN 204, 743) have observed HII regions to determine abundance gradients over the faces of the Sculptor galaxies NGC 55, 253, 300 and 7793 as well as for the Centaurus galaxies M83 and NGC 5253. Carignan (ApJ Suppl, in press) presents three color surface photometry for NGC 247, 300 and 7793 as part of a larger study of the properties of disk galaxies. Photometric sequences have been published by Alcaino and Liller (AJ 89, 814).

A survey for new dwarf galaxies in the M81/M82 group has been carried out by Borgnen and Karachentseva (AN 303, 189). Borgnen+ also give photometric data for 39 dwarf galaxies in the area (AN 303, 287). Recent work by Sandage (AJ 89, 621; 89, 630) shows that the galaxies which have been regarded as members of the group may also spread over a large radial distance range. More than one group of galaxies may be involved.

Elmegreen and Elmegreen (MN 201, 1021; ApJ 267, 31) study the relative frequency of grand design spiral structure features among galaxies in groups and in the general field. They find a significant trend towards higher fractions of spirals with grand design features when the galaxy has companions or has a bar feature. However it seems that the presence or lack of bars or companions do not in themselves determine uniquely the occurrence of grand design features.

Correlations between luminosities, velocity dispersions and virial radii for 49 Huchra-Geller groups have been analysed by Menzett+ (AA 125, 368). A similar sample has been examined by Mardirossian+ (AA 126, 86). They find no evidence for merging. Danks and Materne (AA, in press) suggest that the Klemola 25 group forms a ring structure which resembles model simulations of a material ring undergoing bead instabilities. The mass of the system is equivalent to that of a giant elliptical galaxy. Schneider+ (ApJ 273, L1) have discovered an intergalactic HI cloud in the western part of the Leo group of galaxies. Its nominal mass is about 10^9 $M_\odot$. The NGC 1961 group of galaxies has been studied by Gottesman+ (MN 202, 21P). They conclude that a large amount of unseen matter is associated either with NGC 1961 itself or with the galaxy group as a whole.

Sulentic reports that a sample of 196 companion galaxies in 60 spiral-dominated groups show a significant positive redshift excess. A second sample of 62 companions in 21 E/S0 dominated groups show no excesses. A third study with Arp was of a redshift sample obtained in a 21 cm survey (AJ 88, 489). An analysis by Sulentic and Lorre of VV172 and Seyfert's Sextet (AA 120, 36) reveals evidence for physical and unusual structural and color properties. 21 cm mapping of Stephan's Quintet (AJ 88, 2671) suggests that the discordant redshift galaxy in that group interacts with its neighbors. A statistical analysis of the frequency of interacting quartets contradicts the findings of Rose (ApJ 211, 311) and suggests that the above three groups are not chance optical configurations (ApJ 270, 417).

III. Clusters of Galaxies

Clusters of galaxies are the largest known bound aggregates of matter in the Universe and much of observational cosmology depends on our understanding of their basic properties. The subject was well covered at IAU Symposium 104, Early Evolution of the Universe and its Present Structure. Superclustering is covered in the Report of Commission 47.

Abell's cluster catalog (ApJ Suppl 3, 211) continues as a fundamental reference. At the time of his premature death, Abell was working with Corwin on a southern extension using the ESO/SRC Sky Atlas. The Abell-Corwin survey is based on the deep J survey photographs. A search for distant, highly red-shifted clusters is being carried out by West and collaborators (Rev Mex Astron Astrophys 6, 35;IAJ, 15, 25). Deep CCD frames are being obtained for photometric calibration and for the identification of faint cluster members. A useful table of redshifts for 339 Abell clusters is given by Sarazin+ (AA 108, L7). Two color photometry for 17 rich clusters of galaxies has been published by Butcher+ (ApJ Suppl 52, 183). The CFA redshift survey by Huchra+ (ApJ Suppl 52, 89) contains the best available radial velocities for 2401 galaxies brighter than M_B = 14.5. Geller reports that several studies are in progress to examine the role of substructure in rich clusters (PASP 94, 421; ApJ 257, 23). Substructure can provide clues to the cluster's dynamical age and can affect the derived mass/light ratios. A complete velocity survey revealed that the Cancer cluster is not real but consists of an unbound collection of groups.

The nearest clusters of galaxies reveal the most about content, structure and evolution. An ESO Workshop, "The Virgo Cluster of Galaxies", was taking place as this Report was being prepared. In Virgo, we can study the dwarf galaxies in detail and estimate their contribution to the mass of the cluster. Reeves (ApJ Suppl 53, 375) has published a new catalog of Virgo dwarf galaxies. He feels (BAAS 15, 946) that an appreciable mass contribution from the dwarfs cannot be ruled out. Reeves is also investigating whether dust, swept out of the now dustless E and SO galaxies, can be detected from the reddening of background objects. A major series of papers on the Virgo cluster has been initiated by Sandage and collaborators (AJ 89, 64; 89, 919). Particular attention has been given so far to the morphology of the dwarf galaxies. Caldwell (AJ 88, 804) and Caldwell and Bothun (ApJ 280, 528) have been studying structure and stellar content of the Virgo dwarfs. A complete sample of 180 Virgo cluster galaxies has been isolated by Kraan-Kortweg (AA Suppl 47, 505). These lie within 6° of M87 and are brighter than B = 14.0. Positions, morphological types, magnitudes and velocities are provided. Radial velocities for 100 Virgo galaxies are given by Karachentsev and Karachentseva (Pis'ma Astron Zh 8, 198) who include many faint blue objects which appear to be members of the Virgo cluster. Dressler (ApJ 281, 512) has used new, high quality spectra for 23 galaxies in Virgo and 30 in Coma to derive luminosity-velocity dispersion and luminosity-metallicity relations. These are then used to estimate the difference in distance modulus between the Virgo and Coma clusters.

The extent and deficiency of HI in spiral galaxies in Virgo has been studied by Giovanelli and Haynes (AJ 88, 881) and by Giovanadi+ (ApJ 267, 35). Stauffer (ApJ 264,14) finds no difference between emission line strengths for the Virgo galaxies and a field sample.

The dynamics of nearby galaxy clusters are being analysed with the improved data base. Tully and Shaya (ApJ 281, 31) have studied the infall of galaxies belonging to the Virgo southern extension into the main body of the cluster. They find that the influx of gas-rich systems is sufficiently large that all spirals and ellipticals presently observed in Virgo appear to have arrived on a time scale of only 1/3 to 1/2 of the age of the Universe. Rich clusters of galaxies which have attained a high degree of symmetry, such as Coma, are ideal objects for studying structure and long-term evolutionary trends. Godwin+ (MN 207, 113) have completed a catalog of magnitudes, colors, ellipticities and position angles for 6724 galaxies in the Coma field. Kent and Gunn (AJ 87, 945) have derived values for projected surface density and velocity dispersion out to a radius of 3° from the cluster center. Constraints on the anisotropy in Coma are determined; see also Pryor and Geller (ApJ 278, 51).

Luan Ling and Li Jing of the Beijing Astronomical Observatory are analyzing a $9^{\circ} \times 9^{\circ}$ field using r and J UK Schmidt plates to count galaxies to magnitude 20 with a PDS. The globular cluster NGC 1851 is in a corner of the field. They are constructing luminosity functions and making statistical analyses of the structure and substructure within clusters.

Bright galaxies with extended halos and sometimes with multiple nuclei are often found in the centers of clusters. Beers and Geller (ApJ 274, 491) show that such galaxies usually lie on significant peaks in cluster galaxy distributions. Despite suggestive circumstances, there are doubts as to whether galactic cannibalism is always taking place (Tonry, ApJ 279, 13; Merrit, ApJ 280, L5). The majority of multiple nucleus systems may only be transcient phenomena. Optical line emission has been found surrounding bright galaxies in the centers of clusters. In addition to M87, Fabian+ (MN 201, 17P) detected gaseous filamentary structure around NGC 4696 in the Centaurus cluster. Van Bruegel+ (ApJ 276, 79) find similar emission around the radio galaxy 4C 26.42 in A1795. Accretion flows from intracluster gas are thought to be responsible. In some cases, star formation may be a consequence (Fabian+ MN 201, 933; Sarazin and O'Connell, ApJ 268, 552). Materne reports work at the Technical University of Berlin on clusters of galaxies. Studies have been carried out for the Antlia cluster, STR 0627-544 (AA, in press) as well as for Coma and Hydra (AA 13, 85).

At the Beijing Astronomical Observatory, Jia-sheng has derived a power law form of the two-point correlation function, in terms of a natural cut-off of BBGKY heirarchy equations. The evolutionary history of the correlation function and the effects of peculiar motions and backgrounds are described.

More distant galaxy clusters with substantial look-back times are important for studying the early evolution of galaxies. Efforts have been made to confirm the Butcher-Oemler effect which finds distant clusters of galaxies to contain a high proportion of blue and presumably younger galaxies. The original Butcher-Oemler result was apparently exaggerated in at least one cluster by insufficient correction for foreground and background galaxy contamination (Dressler and Gunn, ApJ 270, 7) and by errors in the photometry (van den Bergh, ApJ 265, 606). Osterbrock (ApJ 280, L43) notes that selection effects can influence conclusions in this type of study. The general consensus seems to be that the effect is real but that careful observations of distant clusters are still needed. Segal (AA 123, 151) deals with the use of clusters as cosmological test objects.

The study of X-ray radiation from clusters of galaxies has reached new maturity during the last three years as many of the results from the Einstein Observatory were analysed and published. An excellent review by Forman and Jones appeared (Ann Rev Astron Astrophys 20, 547). Recent work is discussed by Jones and Forman (ApJ 276, 38). Hard X-ray emission in the energy range 20-100 keV is reported in 3 clusters of galaxies by Bazzano+ (ApJ 279, 515). The X-ray luminosity function of Abell clusters at a redshift of 0.25 has been compared with a low-redshift, nearby sample by Henry and Lavery (ApJ 280, 1). There seems to be no evidence of any difference between the two distributions, so evolutionary effects must be small. Two-dimensional spectrophotometry of the cores of X-ray luminous clusters by Cowie+ (ApJ 272, 29) has provided further examples of optical line emission, either from extended 20-100 kpc systems of long filaments or more compact homogeneous regions associated with a dominant central galaxy.

It has long been recognized that head-tail sources found in high resolution continuum maps give unique information about the intracluster gas. Several surveys have been carried out recently (Fanti+, AA Suppl 52, 411; Burns and Gregory, ApJ 87, 1245). Einstein X-ray observations for two clusters, A1446 and A1220, which both contain radio tail galaxies are discussed by Burns and Balonek (ApJ 263, 546). The X-ray data are used to explore mechanisms which may distort radio jets and tails. Slee and Siegman (Proc Astr Soc Aust 5, 114) have used the Culgoora circular array to make 80 and 160 MHz measurements of a selection of cluster radio sources. Owen+ (AJ 87, 1083) have surveyed 1478 Abell clusters at 1400 MHz with the NRAO

91 m telescope. 485 sources were found but optical identifications are needed before the data can be fully interpreted.

4. QUASARS AND RELATED OBJECTS
(M.H. Ulrich)

Conferences related to topic have been held: IAU Symposium No. 104 "Early Evolution of the Universe and its Present Structure", G.O. Abell and G. Chincarini editors; IAU Symposium No. 110 "VLBI and Compact Radio Sources", R. Fanti, K. Kellermann and G. Setti editors; XI Texas Symposium on Relativistic Astrophysics, D.S. Evans editor, New York Academy of Sciences, Volume 422; and "Quasars and Gravitational Lenses": 24th Liège International Astrophysical Colloguium. In general, no references to articles published in the Proceedings of these conferences are made in this Report.

I. Continuum of the Central Source in Radio, Infrared, Optical, and X-ray

VLBI observations of large samples of active nuclei have revealed an elongated structure on the milliarcsec scale in a large fraction of nuclei, up to 80% in the case of quasars (ApJ 269, 387). More cases of superluminal motions have been found including 3C 120, a relatively nearby Seyfert galaxy (ApJ 257, 56). In 3C 345 a small substructure perpendicular to the jet raises the interesting possibility that it could be synchrotron emission from an optically thick disk (ApJ 265, 107). Other evidence for relativistic bulk motions has been provided by the burst observed at 1mm in 3C 273 (AA 117, L5). Sensitive searches for radio emission from radio quiet objects have yielded a number of positive detections, suggesting a fairly continuous, rather than bimodal, distribution of the luminosity ratio, L(optical)/L(radio), for active nuclei (except for the BL Lac objects, Kellermann+, preprint).

Infrared/optical observations of high-redshift quasars show that the energy distribution of high and low redshift quasars are similar in their respective rest frame (ApJ 265, 18; ApJ 281, 76) providing circumstantial evidence for the absence of intergalactic dust. Infrared observations of a number of quasars, radio galaxies and Seyferts reveal the fraction of the energy emitted by a non-thermal continuum, a normal stellar population and radiation by dust (AA 107, 276; MN 199, 943 and 953; MN 207, 35; ApJ 252, L53; ApJ 257, L57; ApJ 257, 570; ApJ 272, 400; ApJ Suppl 52, 341). Some quasars show pure non-thermal IR continuum while some Seyferts produce the IR by dust re-radiation. The 3.3μ feature which is thought to be a dust indicator is present in IC 4329A and NGC 5506 (AA 100, L16). It has also been reported in the spectrum of 3C 273 and NGC 4151 but it is not confirmed by new observations (Nature 295, 214).

The flux density of NGC 4051 was measured in a ballon experiment at 130μ and 170μ (ApJ 274, 571). Preliminary IRAS results show that the strongest far IR emitters are the starburst galaxies and the Seyfert galaxies. The latter have a power-law IR spectrum well represented by $f_\nu \propto \nu^{-1}$, and the far IR flux lies close to the intrapolation of the near IR-radio spectrum. In contrast, starburst galaxies have a steeper far IR spectrum, $f_\nu \propto \nu^{-3}$, and the far IR flux lies well above the intrapolation of the near IR-radio spectrum. Still it is possible that starburst galaxies and Seyfert galaxies are generically related (ApJ 259, 96; AA 124, 89; AJ 88, 1077; ApJ 268, 602). Much effort has been devoted to the measurement and the interpretation of the "3000 Å bump" in quasars and Seyfert galaxies (ApJ 255,25; ApJ 254, 22; ApJ 257, 487; ApJ 275, 445; ApJ 277, 64).

The X-ray and γ-ray end of the spectrum is most important in distinguishing among various emission mechanisms, since the bulk of this emission arises close to the active nuclei. All active nuclei are strong X-ray emitters. Efforts have been devoted to revealing correlations between the luminosity ratios L_X/L_O and L_R/L_O and other properties such as the absolute luminosity, the redshift, the mm flux, the radio structure, etc. Among quasars there is a strong correlation between optical and X-ray luminosities. This correlation is different for radio quiet and radio loud quasars in the sense that for a given optical luminosity the radio quasars are stronger X-ray emitters. The region which produces X-ray and optical luminosity in radio quiet quasars may be present also in radio loud quasars, but in addition radio quasars may have a separate source (possibly a jet) emitting in the radio and X-ray ranges (MN 198, 921; AA 108, 153; ApJ 260, 437; ApJ 257, 499). The cosmic evolution of the X-ray properties of quasars and active nuclei has also been investigated (ApJ 268, 60; ApJ 283, 479 and 495). These objects probably contribute 80% of the 2 keV diffuse X-ray background (ApJ 283, 486).

Very rapid variations of the X-ray fluxes have been found in NGC 4051 and NGC 6814 (ApJ 269, L31) although such events seem to be rare among Seyfert galaxies (ApJ 264, 92). Variations on the same time scales have not been observed in the UV or optical ranges, possibly because the observations have not been conducted with the appropriate time resolution. The spectrum of a number of Seyfert galaxies shows soft X-ray absorption indicating a higher covering factor of the X-ray source than in quasars. The absorber has sometimes been identified with the gas producing the broad emission lines (ApJ 262, 564). A sharp X-ray absorption feature has been discovered (ApJ 278, L99) at 600 eV in the BL Lac object PKS 2155-304 but a definite model has not yet been established (ApJ 278, L103). Simultaneous observations of BL Lac objects from X-ray to the radio range show the overall energy distribution to be consistent with the synchrotron self-compton model in a beamed source (ApJ 253, 19; Advances in UV Astronomy, ed. Y. Kondo, NASA CP-2238, p. 181, 185; ApJ 253, 38). One week continuous observations of the spectrum of BL Lac show that the time variations of the optical polarization angle and of the total optical flux is consistent with a random walk spectrum (ApJ 260, 415).

POLARIZATION. Several processes, sometimes simultaneously, produce optical polarization in galactic nuclei: intrinsic polarization of the non-thermal continuum, dust, electron scattering by a thin or thick disk. Bright BL Lac objects and violently variable quasars show large intensity variations of the optical/UV flux, and large variations of the degree and the angle of polarization (MN 198, 1; ApJ 265, 625; ApJ 279, L39). They also have VLBI sources and in those cases where the optical polarization has a preferred angle, the polarization vector is aligned with the VLBI structure (IAU Symposium 110, p. 73).

A simple, but not unique, model fitting all these properties is one in which the energy is emitted by a small number of independent non-thermal sources (ApJ 260, 415), each emitting a fully polarized synchotron radiation of random orientation and varying on a time scale of a few days. If the emitting material is beamed towards us aberration exaggerates the effects and small changes in the source geometry are amplified by relativistic effects. Alternatively, if there is no beaming effect then the large intensity outbursts must be due to large increases in the average luminosity of the "independent sources" (ApJ 260, 415). The polarization outbursts would originate from sudden large scale re-orderings of the magnetic field (MN 198, 1).

Most Seyfert 2 galaxies and radio galaxies (ApJ 271, 59) show weak polarization with the angle of polarization perpendicular to the radio axis. This suggests the interesting (and speculative) possibility that the polarization is caused by electron scattering in a physically thick disk. Yet, even for well studied objects (NGC 1068, ApJ 271, L7) the interpretation of the observed polarization is ambiguous (Nature 304, 609).

II. The Emission Line Spectrum

THE BROAD EMISSION LINES. Physical properties of the dense clouds forming the broad line region in Seyfert 1 and quasars are now understood through the detailed analysis of increasingly accurate line intensity ratios (AA 106, 362; MN 200, 271; Astrophysics 17, 1; ApJ 273, 36; ApJ 254, 489; ApJ 260, 44). The discrepancy between observed and calculated intensities of optical multiplets of FeII suggests that energy might be provided to the broad line region under the form of mechanical energy in addition to the radiative energy of the UV/X-ray spectrum (AA 106, 362). Remaining discrepancies between observed and calculated ratios are probably due to the attempt to reproduce an "average quasar" spectrum from well defined physical conditions, while, in fact, line intensity ratios vary from quasar to quasar and, moreover, in a given active nucleus the lines come from different clouds which most likely have different physical conditions.

Two types of data provide clues on the kinematics of the clouds in the broad line region, the emission line profiles and the change of the line profiles in response to the variations of the central continuum source (see below). Neither set uniquely constrains the type of motions of the clouds in the broad line region. There is as yet no observational evidence for or against the clouds being gravitationally bound to the central object.

Differences in profiles of emission lines suggest that not all clouds have the same physical conditions (MN 206, 221; ApJ 259, 48). The symmetry of Lyα (MN 201, 645) can be due to pure rotation motions (a not very likely situation) or by emission by optically thin clouds. Additional evidence for both optically thin and optically thick clouds is provided by the difference in redshift between the lines of high and low ionized atoms (ApJ 263, 79; MN 207, 73). Finally, it is interesting to note the discovery of two quasars with narrow emission lines (ApJ 252, 69; PASP 95, 117).

Observations with IUE reveal spectroscopic differences between intrinsically weak Seyfert nuclei and the intrinsically bright quasars: (i) the CIV line has a larger equivalent width (100-200Å) in Seyfert galaxies than in quasars (~30-80Å); (ii) line intensity ratios CIV λ 1550/Lyα and CIV λ 1550/CIII λ 1909 are both larger in Seyfert nuclei. This suggests a higher degree of ionization in Seyfert galaxies than in quasars and possibly also a higher covering factor of the UV source by the broad line gas in Seyferts (ApJ 266, 28).

The degree of ionization in the broad line region and the absolute intensity of the ionizing flux at λ ~ 1200 Å (measured from the ground in high redshift quasars and with the IUE in Seyfert galaxies) lead to a characteristic linear radius of the broad line region of $\sim 10^{16}$ cm to 10^{18} cm. The small linear dimensions of the broad line region in nearby Seyferts (intrinsically weak active nuclei) is confirmed by line variations on time scales of a few weeks found in a number of Seyferts: Akn 120, NGC 3516, NGC 3783, NGC 4151, NGC 4593, NGC 5548, NGC 7469 and others (AJ 88, 926; ApJ 267, 515; MN 202, 455; MN 206, 221; MN 204, 189; Soviet Astronomy Letters Vol 9; ApJ Suppl 49, 149; AA 132, 136). These observations clearly show the lines varying in response to the continuum variations. In NGC 4151 the variations can be explained with a time delay of 10 days between variations of the continuum and of the CIVλ1550 line, providing, for the first time, a lower limit to the size of the broad line region (MN 206, 221).

THE NARROW EMISSION LINES. Typically, these lines have FWHM intensity of ~500 km/s. Intensity ratios are consistent with photoionization by the central source (AA 135, 341). Classical reddening indicators show that narrow line gas is mixed with dust. Asymmetrical profiles are often interpreted as evidence for radial motions and obscuration (ApJ 281, 525; and M. Whittle 1984, preprint). Evidence for a gradual increase of density and velocity of gas closer to the nucleus is provided by the relation between the critical density of a transition and the width of the corresponding emission line (ApJ 275, 493). Lines of very highly ionized species (e.g., Fe IX) also give information on the transition between broad narrow line regions (MN 204, 1231; MN 207, 347; ApJ 273, L31). Improved measurements and analysis of emission line intensities in apparently normal elliptical galaxies show that observed ratios can be produced by a power law spectrum illuminating a gas with a gradient of different densities (AA; ApJ 266, L89) instead of the gas being excited by shocks as previously thought. Intrinsically weak non-thermal continuum sources may be present in nuclei of apparently normal galaxies. The discovery of a Seyfert nucleus in M81 (ApJ 260, 495) suggests that other weak Seyfert line nuclei remain unknown.

III. The Absorption Line Spectrum

THE NARROW ABSORPTION LINES. Spectra of distant quasars show narrow absorption lines at different redshifts which are believed to be caused by gas clouds located along the line-of-sight. Some of these clouds are metal rich and are located in galaxies. Others are detected only in the Lyman lines. These two types of clouds differ by their cosmological evolution, metal abundances, and velocity structure. There appears to be no or only weak cosmological evolution of the number of metal rich absorption line systems (MN 198, 91; ApJ 252, 10; Nature 298, 427; ApJ Suppl 48, 455; Quasars and Gravitational Lenses, p. 589; Woolf+, 1983, ApJ 268, L1) whereas the number of Lyα clouds appears to increase with redshift above what is expected from a uniform space distribution of comoving absorbers. Observations with a resolution of 0.25 Å reveal that in Q1101-264 almost all the Lyα lines are resolved and have a b-parameter larger than 10 km s^{-1} (ApJ 278, 486). Space Telescope should give the distribution of the absorption system redshifts at $z < 1.6$. The upper limit to the abundances in the gas producing the "Lyα only" absorptions depends on the model for the Lyα clouds but it is equal to or below the Population II abundances (ApJ 273, 450; Chaffee+, preprint).

Narrow absorption lines of CIVλ1550 at the redshift of the emission lines are detected in about half of the quasars. Of particular interest are quasar pairs: the presence in spectra of both quasars of an absorption line at the emission redshift of the lower redshift quasar could indicate an extended gaseous halo associated with the lower redshift quasar. However, there exists the alternative possibility that this line is caused by a galaxy at the same redshift as the lower redshift quasar (ApJ 261, L7; MN 204, 698; MN 199, 81).

THE BROAD ABSORPTION LINES. Observations of broad absorption line quasars in the optical, X-ray and radio ranges have been made (ApJ 280, 51; 277, 51; 276, 423; 282, 33; 280, 476). Q1414+089 has the broadest absorption lines, with the absorbing gas moving relative to the gas producing the emission lines at up to 66000 km s^{-1} (PASP 95, 341). Models for the formation of the broad absorption lines are discussed (ApJ 265, 73; MN 201, 27).

NEW REDSHIFTS. Quasars, radio galaxies, and Seyferts are discovered through new or continuing surveys (Astrophysics 19, 1; Astrophysics 19, 2; Soviet Astronomy Letters Vol 8; Astron Circ 17; ApJ 269, 73; ApJ 266, 485; MN 200, 191; MN 202, 703; ApJ 266, 485; 273, 478). Redshifts of several very distant radio galaxies (up to z = 1.819 for 3C 256; Spinrad, 1984, preprint) have been measured through the intense emission lines which they exhibit (PASP 94, 397).

IV. Extended Nebulosity - Gravitational Lenses

EXTENDED GASEOUS EMISSION AND UNDERLYING GALAXIES IN QUASARS. An extended envelope of ionized gas has been found around several quasars and radio galaxies. In the most extreme examples, MR2251-178 (MN 202, 125) and PKS 2158-380 (MN 201, 991) ionized gas has been detected up to 170 kpc and 15 kpc from the center, respectively (H_o = 50 km s^{-1} Mpc^{-1}). The gas is probably ionized by the non-thermal continuum from the nucleus (see also MN 208, 589 and 955) .

Centaurus A is an example of ionized gas detected at large distances from the center. Gaseous filaments ~30 kpc NE of the center are associated with one of the radio lobes and with an extended source of soft X-rays. Optical observations have revealed differential velocities of ≥ 800 km s^{-1} within this group of filaments; they will disperse over a time of $\lesssim 10^7$ years (ApJ 269, 440). An extended region of broad-line high excitation gas has also been detected (Nature 310, p.554; ApJ 273, 154). Additional optical and radio associations are 3C 293 (ApJ 277, 82), 3C 305 (ApJ 262, 529) and NGC 1068 (ApJ 275, 8).

Morphology of Seyfert galaxies has been studied to see how frequent are such peculiarities as condensations and double nuclei (Soviet Astron Lett Vol 8; Astrofizika Tom 18, 192 and 227; AA 118, 166). Direct imaging of quasars at redshift $\lesssim 0.6$ shows a surrounding nebulosity with an absolute luminosity and a brightness distribution similar to those of a galaxy (ApJ 262, 48; 257, L1; 278, 11; 280, 41). The nebulosity spectrum is in some cases dominated by a blue continuum and emission lines and in other cases consistent with that of a normal stellar population (ApJ 265, 4; ApJ 265, L1). Quasar fields have also been imaged in order to see if quasars are more frequently found in tight groups than are average galaxies (ApJ 257, 33; ApJ 271, 471). Evidence has been presented that nuclear emission line activity is more frequent in interacting spirals (ApJ 279, L5).

GRAVITATIONAL LENSES. In addition to 0957+561A,B and 1115+081A,B,C, a third case has been reported, 2345+007 (ApJ 255, L5). Optical and UV observations study the light variations (ApJ 255, 20; Nature 296, 415) and search for deflecting galaxies (Quasars and Gravitational Lenses, p. 182 and p. 192). Images of the triple quasar 1115+081A,B,C have been analyzed to find the distribution of mass of the object producing the lensing (MN 200, 941; Quasars and Gravitational Lenses, p. 78). Several authors have questioned if all quasars are the result of gravitational lens magnification of a distant Seyfert nucleus by a foreground galaxy (AA 118, L1; IAU Symp 104, p. 375; Quasars and Gravitational Lenses, p. 139 and p. 143). In 2345+007 several Lyα absorption lines are not in common or are of substantially different strengths. If confirmed this will yield a size of 5-25 kpc for the characteristic size of the Lyα clouds at z = 2 (ApJ 281, L1).

V. Models and Evolution

The black hole and spinar models (ApJ 269, 57) for energy generation have continued to be actively investigated. Spherical accretion (ApJ 267, 18), thin and thick accretion disks have been investigated (MN 203, 323). The key feature of thick disks is their capacity to produce, collimate and accelerate jets (Acta Astr 32, 285; ApJ 266, 502; Astron Gesell Mitterlanger 57, 27). Extraction of energy from the black hole and production of the observed overall continuous spectrum remains one of the most puzzling and difficult problems (ApJ 267, 371; ApJ 273, L59; ApJ 265, 620; Astrophys Space Science 93, 277). Models of quasars as binary black holes (Quasars and Gravitational Lenses, p. 473) or as supermassive stars (Nature 305, 407) have been proposed to explain the broad emission line spectrum. An attractive possibility is that the broad line gas comes from the stars in the nucleus (ApJ 264, 446; ApJ 272, 390).

The search for quasars with z > 3.5 (ApJ 253, 28) has produced evidence for an apparent decrease in the space density of quasars at z > 3.5 (see however: AA 113, 336); samples have increased with the completion of the bright quasar survey. The analysis of these samples leads to the conclusion (ApJ 269, 352) that the increase of space density with redshift depends strongly on absolute luminosity, being close to zero for low luminosity quasars. Detailed information about the variation of the luminosity function with redshift is still lacking.

VI. The Distribution of Quasars

A distance criterion unrelated to redshift is to associate quasars with other objects of known distance. Reported association of quasars with nearby galaxies (ApJ 271, 479; A+A 138, 408.) have been criticized (MN 200, 47P; ApJ 263, L7); have partially supported (MN 204, 675) or strongly supported associations of quasars with low redshift galaxies (Comptes Rendus 297, 259; ApJ 265, L49 and ApJ 285, 44). One investigation associated quasars with galaxies of similar redshift (ApJ 269, 29; see also PASP 96, 148). The subject of alignment, clustering of quasars, and periodicity in redshifts has received continuing attention (Quasars and Gravitational Lenses, 374; Occas Rep R Obs Edinb 10, 243; MN 201, 179 and ApJ 260, 469; ApJ 277, L27; Journ Astrophys Astron 5, 31; Nature 302, 397; A+A 134, 234; Nachrichten 305, 339).

5. GALAXY REDSHIFTS
(J. P. Huchra)

The last several years have seen both a marked improvement in the instrumentation for small to intermediate diameter telescopes as well as new instrumentation on large telescopes. Photon counting detectors (Shectman and collaborators), CCD based spectrographs (e.g. the "PFUEI" developed by Gunn and collaborators), "Medusae" (fiber optically coupled multiaperture spectrographs; Angel and collaborators), aperture plates (in use at KPNO), plus moveable multislit spectrographs (developed by Huchra and Oke and collaborators) have increased our ability to obtain medium to high quality galaxy redshifts dramatically.

The gathering of high quality redshifts has been spurred on by interest in two major problems: the unraveling of the complexity of the large scale structure in the universe (Early Evolution of the Universe and Its Present Structure, I.A.U. Symposium No. 104); and the structure and evolution of clusters of galaxies (Clusters and Groups of Galaxies, Mardirossian, Giuricin and Mezzetti, Dordrecht, D. Reidel Co.). Rather than slowing down, the growth of the number of redshifts is continuing. The Catalogue of Radial Velocities of Galaxies, compiled by Palumbo, Tanzella-Nitti and Vettolani in 1980 contains 8,250 galaxies with redshifts. The current tally of the CfA Redshift Catalog (Huchra+) is approaching 20,000 galaxies with measured redshifts. This corresponds to a doubling time of less than five years! Not only is the number of redshifts increasing, so is their quality. 21-cm velocities are routinely measured with errors smaller than 20 km/sec and optical velocities (especially those made with photon counters) also have errors in that range. Older and usually poorer velocities have been replaced and discrepant velocities remeasured.

I. The Field Galaxies

Since the last Report, several major field galaxy velocity surveys and catalogs have been published: "Neutral Hydrogen Observations of a Large Sample of Galaxies," Fisher and Tully, 1981, ApJ Suppl. 47,"The CfA Redshift Survey," Huchra, Davis, Latham, and Tonry, 1983, ApJ Suppl. 52 "General Catalog of 21 cm line Data," Bottinelli, Gouguenheim, and Paturel, 1982, Ast Ap. 106 "A Catalog of HI Measurements of Galaxies," Huchtmeier, Richter, Bohnenstengel, and Hauschildt, 1983, ESO preprint 250 "A Catalog of Galaxies South of Declination -30° That Have Been Observed Spectroscopically," Fairall, Lowe, and Dobie, 1983, Pub. of the Dept. Ast. Univ. Cape Town, No. 5. "A Deep Survey of Galaxies," Kirshner, Oemler, Schechter, and Shectman, 1983, AJ 88, 1285. as well as the ESO/Uppsala Survey of the ESO(B) Atlas (Lauberts 1982, ESO, Munich) with many previously unpublished velocities for southern galaxies. These have been supplemented by spectroscopic surveys of peculiar galaxies from wide field objective prism surveys (by Markarian and Kazarian, and by Arakelian, Petrossian, Lipovetskii and collaborators at Byurakan and at the Special Astrophysical Observatory (6m) and published in a series of papers in Astrofizica).

Large bodies of unpublished redshifts are associated with new and/or continuing surveys. Giovanelli and Haynes are continuing their HI measurements of galaxies from Arecibo. The Center for Astrophysics survey is being pushed to a magnitude limit of B = 15.5 (above Dec = 0 and b > 40° or b < -30°). When completed, this will be a magnitude limited survey that will reach a mean almost 10,000 km/sec and contain over 8000 galaxies. Kirshner+ are continuing to survey the Void in Bootes. The analogue of the original CfA survey (magnitude limit of approximately B = 14.5) is now underway in the Southern Hemisphere with several groups (Da Costa and collaborators in Brazil, W.L.W. Sargent and collaborators at Las Campanas and also SAAO, and Fairall at SAAO) surveying different areas of the sky.

Several groups are also surveying very small regions to faint limiting magnitudes (Ellis and collaborators, Kron and Koo, and Geller and collaborators). A few groups are also pursuing the use of objective prism surveys (Palmer 1983 in Clusters and Groups of Galaxies) and galaxy colors (Koo+ in the same volume; Loh, June 1984 A.A.S. meeting) for large scale surveys, but the best redshifts available from the prism techniques are only +/- 3000 km/sec, while multicolor work seems to be reliable only for studying clusters of galaxies.

II. Binaries and Groups of Galaxies

An exhaustive reanalysis of the dynamics of Turner's binary galaxies was done by White+ (1983, MN 203, 701). With significantly better velocities they found that mean M/L is approximately 50 to 100 h (in solar units), essentially the same answer as Turner. Tifft has continued his investigations of double galaxies (ApJ 257, 442; ApJ 262, 44; ApJ 268, 56; and following papers) and continues to find evidence for the quantization of redshifts. This effect is neither well understood nor generally accepted although the velocity measurements are of very high quality. Karachentsev and Karachentseva are also continuing their studies of interacting double and triple galaxies.

Two new major compilations of galaxy groups have been published. Hickson and collaborators are studying a sample of compact groups of galaxies (1982, ApJ 255, 382; and in Clusters and Groups of Galaxies). Geller and Huchra have identified groups with complete dynamical information from the Revised Shapley Ames Catalog (Sandage and Tammann, 1981 Carnegie Institution, Washington) and also from the CfA Redshift Survey (Huchra and Geller 1982, ApJ 257, 423; Geller and Huchra 1983, ApJ Suppl. 52, 61). A summary of this work can be found in Clusters and Groups of Galaxies. Theoretical work on the interpretation of group and cluster positional and velocity information is available from Bahcall and Tremaine and collaborators (Bahcall and Tremaine 1981, Ap. J. 244, 805) and Mezetti, Mardirossian and Giuricin (1983, AA 125, 368; AA 126, 86).

III. Clusters of Galaxies

Interest in the internal dynamics of clusters of galaxies has grown with the advent of powerful X-ray telescopes in space (Einstein and HEAO-A) and with improved dynamical and evolutionary models for clusters. "Deep" surveys of clusters have revealed that these objects are physically very complex (Beers+ 1982, ApJ 257, 23; Bothun+ 1983, ApJ 268, 540; Huchra+ 1983, in Clusters and Groups of Galaxies). Extensive velocity surveys of most nearby northern cluster of galaxies are nearing completion as part of the CfA Redshift survey. More distant clusters and southern clusters are now being surveyed by several groups: Chincarini and collaborators

(Hercules and Perseus), Melnick and Quintana (southern Hemisphere), Dressler and Shectman (selected clusters) and Geller and Huchra and collaborators (CfA overlap - Pegasus, Cancer, Virgo, Coma, A1367).

Significant surveys published in the last two years include the Coma Survey of Kent and Gunn (1982, AJ 87, 945), the Hydra I survey of Richter+ (1982, AA 111, 193), the Perseus survey of Kent and Sargent (1983, AJ 88, 489). Huchra has over 600 velocities for galaxies in the core of the Virgo cluster (1985 ApJ in press). It is also in the study of medium distant galaxy clusters that multi-object spectroscopy can really come to the fore (e.g. Hintzen+ 1982, AJ 87, 1656).

Spinrad and Longair and collaborators have continued the use of radio sources to identify high redshift galaxies (usually brightest cluster members). There are now more than 25 galaxies (clusters) with measured redshifts greater than $z = 0.75$, with the largest galaxy emission line redshift at $z = 1.819$. Several clusters with $z > 0.15$ have even had their velocity dispersions measured (Abell 115: Beers+ 1983, ApJ 264, 356; 3C295, Cl0024+1654: Dressler and Gunn 1983, ApJ 270, 7; preprint).

IV. Superclustering

Studies in the large scale structure of the universe have increasingly used clusters of galaxies as tracers of that structure. Oort has compiled an excellent review of earlier work (1983, Ann. Rev. 21, 373). Over 600 redshifts are now available for Abell clusters (Postman+ 1985, ApJ in press). Several groups are working on surveys of cluster velocities including Karachentsev and Kopylov in the Soviet Union, Burns and Batuski at Kitt Peak, Ford and Harms and collaborators at Lick and Kitt Peak (1983, ApJ 273, 24), Shectman at Las Campanas and Palomar and Huchra and Henry at Mt. Hopkins and Hawaii.

Nearby superclusters (including our own) have also been extensively studied as well as Coma, Perseus, Hercules and the newly uncovered Lynx-Ursa Major chain (Giovanelli and Haynes 1982, AJ 87, 1355). According to some authors, the magnitude of the infall of the Local Group into Virgo is approximately 250 km/sec (Aaronson+ 1982, Ap. J. 258, 64). A next major step will be more detailed mapping of the distortion in the velocity field caused by the flattening and non-uniformity of the Local Supercluster mass distribution.

6. EXTRAGALACTIC RESEARCH IN THE U.S.S.R
(I.D. Karachentsev)

Af - Astrofizika, Erevan	Crim - Izvestia Crimean Astrophys. Obs.
AZ - Astron. Zhurnal	LGU - Trudy Astron. Obs., Leningrad Univ.
LAZ- Letters to AZ (Pis'ma)	AfI - Trudy Astrophys. Inst., Alma-Ata
AC - Astron. Circ.	DArm - Doklady Acad. Nauk Armenia
Tartu - Publ. Tartu Obs.	SAO - Communic. Special Astrophys. Obs.
Kazan - Trudy Kazan Astron. Obs.	BAO - Communic. Byurakan Astrophys. Obs.
MGU- Trudy Sternberg State Astron. Inst., Moscow	

I. Surveys and Lists of Galaxies

Markarian and Stepanian (Af 19, 639; 20, 21) began a new survey of objects with UV-excess on the Byurakan 40" Schmidt telescope with an objective prism. Using combinations of three prisms and IIIaF, IIIaJ emulsions, they reached the survey limiting magnitude ~19^m.5. In two areas 0950+55 and 1130+59, 4° x 4° each, they have found 252 blue objects. The relative number of QSO and Seyfert galaxies among them is expected to be ~20%. Kazarian and Kazarian (Af 18, 512; 19, 213) have presented a list of 228 blue galaxies which were found at the same telescope with a 1°5 prism applying a softer UV-excess criterion.

Abdulova and Denisjuk (AfI 40, 41) have compiled a catalogue of radial velocities for 408 emission galaxies observed on the 0.7-m telescope with an image tube. For ~400 interacting galaxies a reference list of radial velocities is prepared by Vorontsov-Velyaminov and Arkhipova (MGU 53, 3). Redshifts for 130 faint blue galaxies located in the Virgo cluster center direction were published by Karachentsev+ (LAZ 8, 198; 9, 387). Kostjuk and Kopylov (LAZ 8, 520) measured radial velocities for 53 ring-like galaxies from Kostjuk's list. "Catalogue of 603 isolated pairs of galaxies" by Karachentsev is now recorded on magnetic tape. The initial version of the catalogue is supplemented by new data on radial velocities, spectral types and orbital masses for all double galaxies. Kogoshvili (Af 19, 584) studied the inhomogeneity of the data in the Morphologic Catalogue of Galaxies by Vorontsov-Velyaminov.

II. Active Galactic Nuclei

Markarian+ (Af 19, 221; 20, 213; AC 1233, 1) and Denisjuk+ (Af 19, 229; 18, 329) continued spectral observations of Markarian galaxies from Byurakan surveys. Slit spectra have been obtained by Kazarian and Tamazian (Af 18, 192,; LAZ 8, 454) and Kazarian (Af 19, 411; 20, 35; LAZ 9, 648) for many galaxies having UV-excess. Markarian galaxies, especially with double-nuclei structure, have been studied spectrally by Petrosian (AC 1231, 3; 1266, 1; 1277, 2) and Khachikian+ (Af 19, 239; 19, 393,; LAZ 8, 270). Spectral observations and emission line profile classification have been made for 38 Seyfert galaxies by Doroshenko and Terebizh (MGU 55, 64), Doroshenko+ (Af 19, 823) and Afanasjev+ (Af 18, 329). Velocity fields in the central regions of Mkn 538 and Mkn 617 have been investigated by Afanasjev and Rassokhin (LAZ 8, 515; 9, 389). Morphological properties of Markarian galaxies have been considered by Petrosian+ (Af 19, 619; 171; 20, 51), Petrosian (Af 18, 548; LAZ 9, 339; AC 1277, 5) and Eghiazarian+ (Af 19, 631; 18, 541). Integral characteristics of superassociations and of galaxies with double nuclei are presented.

The UBVR observations of galaxies with active nuclei have been continued. Doroshenko and Terebizh (Af 19, 5) have investigated 22 Seyfert galaxies in the UBVR system and in the narrow band filters. Dibay+ (AZ 61, 1) have measured U,B,V,R-magnitudes for 37 Markarian and Arakelian galaxies. Scanning 17 Seyfert and 7 normal galaxies with the slit spectrophotometer Afanasjev+ (LAZ 8, 579,; SAO 33, 5) have confirmed the interconnection of Seyfert properties with a volume luminosity at their circumnuclear regions. Photometry of a number of UV-excess galaxies has been made by Tamazian (BAO 54, 80; Af 20, 43), Kalloglian and Börngen (Af 19, 403) and Hagen-Torn+ (Af 19, 199). Optical variability has been detected in Seyfert galaxies Mkn 699 (BAO 58, 99) and IV Zw29 (LAZ 9, 82). Pronik studied the emission line variablity in NGC 1275 and NGC 3227 (Af 19, 245; AC 1209, 1; Crim 68, 93).

Radio observations of ~400 active nuclei galaxies from Byurakan lists have been made at 327 MHz, 102 MHz, and from 3.9 to 31 cm by Sanamian+ (Af 18, 651; 19, 429) and by Artyukh+ (Af 18, 215; LAZ 9, 86). Other such objects were also studied (Af 18, 343; AC 1247, 1). The radio-luminosity function of Seyfert galaxies has been discussed by Kandalian (Af 18, 580). A dependence of flux and radio-spectrum of a galaxy on the type of its nucleus according to Byurakan classification was noted (Af 19, 251; 18, 25; 19, 655). Physical conditions in the nuclei of Seyfert galaxies and the nature of their activity have been considered by Dibay+ (AZ 60,441; 61,209; 61,10; AC 1216,1; 1269,1; LAZ 9,707), Zentsova (AZ 59,1089; 60,34; 846; 1081; LAZ 8,535), Komberg (AZ 59,1062) and by others (Af 19,15; AfI 39,21; 39,83; LAZ 8,263; AC 1264,1).

III. Compact Galaxies and Quasars

Börngen and Kalloglian (Af 18, 17) have measured the brightness distribution for 262 compact galaxies. Arkhipova and Savel'eva (MGU 54, 33) have made photoelectric UBV observations for 49 compact and 70 interacting galaxies. The spectrophotometry of 4 Arakelian galaxies is presented (LAZ 8, 334). Hagen-Torn+ (LGU 410, 104; 413, 43) continued photometry of compact extragalactic objects. Malymian+ (Af 18, 324; AZ 59, 1075) have carried out a survey of Arakelian galaxies, having high surface brightness, at frequencies of 3.66 GHz and 102 MHz.

In the Byurakan survey, Markarian+ (Af 19, 29; AC 1233, 2; 1265, 1) have found 37 new quasars. Spectral and photometric observations of OQ 172 and 1136+486 are reported (LAZ 10, 83; AC 1241, 1). A search for optical variability was provided (Af 18, 657; 18, 245). Artyukh (AZ 61, 240) has observed quasars from Parkes survey at 102 MHz. Bochkarev and Antokhin (AC 1238, 1; AZ 60, 448) have proposed an interpretation of a possible variability of emission lines of quasars. Komberg and Sheffer (LAZ 9, 529), Katysheva (Af 19, 55; AZ 61, 35), and Grinin (AZ 59, 1070) investigated energetics and dynamics of emitting gas in quasars. Levshakov and Varshalovich (Comments Astrophys 9, 199; Af 18, 49), Waltz and Komberg (AZ 59, 844; Af 19, 449) have undertaken new attempts to identify absorption lines in quasar spectra. Evolution changes in the number of absorption lines for quasars has been estimated by Nazarov (AC 1284, 1). Komberg (Af 20, 73) has assumed the existence of two populations of quasi stellar objects. Lipunov and Sazhin (LAZ 8, 649) connect quasar activity phenomena with gravitational waves. QSO luminosity evolution was considered (AZ 60, 840).

IV. Extragalactic Radio Sources

With the RATAN-600, Amirkhanian+ (SAO 36, 50) have made a sky survey within the declination range $4^{\circ}00'$ to $5^{\circ}50'$ at 3.9 GHz. In a total area of 0.17 steradian there have been found 585 radio sources, including 329 new ones. Using the same radio telescope, Malymian (LAZ 10, 90) observed 15 radio galaxies from Bologna survey, and Ohanian, Panajian (BAO 54, 10) have measured fluxes and spectral indices at 968 MHz and 3.66 GHz for 20 objects from UTI-lists, drawn up from lunar occultations. Artyukh+ (BAO 54, 3) have determined angular dimensions of 52 UTI-sources, using interplanetary scintillations at 102 MHz.

Efanov+ (Crim 66, 205) have obtained high-frequency spectra for 143 radio sources and estimated their optical depth at 8-22 GHz. Observing 8 variable sources, Efanov+ (Crim 68, 104) have shown that the most intensive flux variations occur at λ = (1.3-2.8) cm. Soboleva+ (LAZ 8, 205; 10, 585) discovered extended structures around 3C 84, 3C 120, and 3C 273 radio sources. Apparent superluminous motions have been analyzed by Kyril'chik (AZ 59, 657) in the fine structure components of 3C 120.

V. Structure and Kinematics of Galaxies

An extensive investigation of the system of globular clusters in M 31 was undertaken by Sharov and Ljuty (AZ 60, 1; 666; 61, 245) and by Sharov+ (LAZ 8, 451; 9, 393). UBV magnitudes are measured for several hundred clusters at the 2.6 m and 6 m telescopes. Conclusions based on modern data are presented in Sharov's monograph "Andromeda Nebulae" (Nauka, Moscow, 1982). Efremov (LAZ 8, 339; 585) studied stellar complexes in M 31. Efremov and Ivanov (Astrophys Space Sci 86, 117) have found a relation between a volume density of the brightest stars in an M 31 arm and a gradient of age for these stars.

Karachenstsev and Mineva (LAZ 10, 253) measured rotation curves, V(R), for 21 isolated galaxies and for 37 double galaxies (LAZ 10, 563), and obtain the mean mass-to-luminosity ratio 7 $M_{\odot}/L_{\odot}$. Zasov and Kyazymov (AZ 60, 656) have compiled a list of ~100 normal galaxies with V(R) determination out to large R. They noted that Keplerian rotation curves are relatively rare (~20%) and usually occur among galaxies having close neighbours. Determinations of V(R) for the interacting galaxies VV48, VV219, and for the irregular galaxy NGC 2814 have been made (AZ 60, 20; Af 18, 337). Spectral and photometric investigations of NGC 2685, 2184, 2968, 3034, and 3718 are presented (Af 19, 599; 19, 45; 19, 575; 18, 31; LGU 413, 73). Zasov (AC 1232,1) considered a galaxy bulge influence on its rotation curve properties.

Pasha+ (Astrophys Space Sci 86, 215; MGU 53, 15) and Korchagin (AC 1221, 1) continue the study of regularities of spiral patterns. Nazarova (Kazan 47, 66; 48, 47) investigated the distribution of hot spots in the central parts of NGC 2903, 3310, and 6951. Iskudarian (DArm 74, 122; 217), Vorontsov-Velyaminov (AC 1257, 7), and Fessenko (Af 18, 37) considered some relations between integral properties of galaxies of different structural types. A SN burst frequency was estimated for spiral galaxies (Af 18, 563; LAZ 9, 478; AZ 61, 234).

Sil'chenko with Zasov and Tarasova (AZ 59, 855; 60, 1063; 61, 45; LAZ 10, 19) have performed evolution simulations of colors and spectra of galaxies, and also of M/L for different type galaxies. These calculations define the star formation intensity using its integral properties. Estimates of the temperature of gas sweeping out of a galaxy due to radiative pressure from active nuclei, and also as a result of galaxy motion through intergalactic medium, have been made by Gorbatsky (Af 18, 234) and Kritsyk (Af 19, 471).

VI. Systems of Galaxies

Karachentsev (LAZ 9, 67) has finished radial velocity measurements for the double galaxies in his catalogue. Individual masses have been determined for 227 members of pairs. A joint analysis of orbital and individual masses confirms two main conclusions: galaxy orbital motions are nearly circular; and the essential part of the mass of members of the pairs is concentrated within their standard radii R_{25}. Karachentsev and Shcherbanovsky (LAZ 9, 332) have investigated the double galaxy centers distribution and found considerable difference of the two-point correlation function, $\xi(r)$, for them from that for galaxies. Arakelian (Af 19, 666; 673) and Petrosian (AZ 61, 441) studied a luminosity function for isolated pairs of galaxies and the role of different selection factors. Dostal (Af 18, 201) and Tovmassian (Af 18, 227) discussed observable manifestations of galactic nuclear activity and of interaction phenomena in tight double systems. Bysnovaty-Kogan (LAZ 10, 181) analyzed rotation of a galaxy in a pair taking into account tidal forces.

Karachentseva and Karachentsev continued radial velocity measurements for triple system members. Virial mass estimates have been made for 48 triplets of galaxies (Af 18, 5; 19, 613). Vennik+ (Af 18, 523; 533) measured redshifts for 110 galaxies in the vicinity of nearby groups. Karachentseva+ (AN 303, 189; 305, 53; 305, 59) have discovered many new dwarf members of M 81 group, resolved into stars. Typical luminosity profiles follow the exponential law. Arkhipova (AZ 59, 209) made UBV photometry of 80 "nests" of interacting galaxies. Kuz'min+ (Tartu 49, 140) and Jaaniste (Tartu 49, 170) studied orientation of galaxies in groups. The statistics of structural types and of nuclear classes of galaxies in groups have been considered by Magtesian (BAO 53, 102).

Shcherbanovsky (SAO 31, 21) and Joeveer (AC 1239, 1) investigated the space structure of Virgo cluster of galaxies. The virial mass of this cluster and its luminosity function down to $m = 12^m$ have been determined by Karachentsev (Af 18, 501). Zasov (LAZ 9, 327) has supposed that Virgo cluster has an extended envelope populated by blue dwarf galaxies. Kalloglian+ (Af 19, 183) have made two-color photometry of the cluster A 1185. Gubanov (Af 18, 177; AZ 60, 1051) and

Dagkesamansky+ (MN 200, 971; AZ 60, 9) have made radio observations of Abell clusters at 102 MHz. Interaction of intergalactic gas with cluster galaxies was studied by Syunjaev (LAZ 8, 323), Gorbatsky (Af 20, 61), Volkov and Ivanov (Af 19, 91). Sholomitsky (LAZ 10, 260) considered the possibility of the submillimeter spectroscopy to measure peculiar velocities of clusters of galaxies.

VII. Large Scale Structure of the Universe

Joeveer (Tartu 49, 161) studied the space structure of the Local Supercluster. Einasto+ (Highlights Astron 6, 757, Patras) have presented evidence for existence of two local giant voids with the diameter about 200 Mpc. Using the radial velocities of 6775 galaxies with $V < 17000$ km/s, Einasto+ (MN 206, 529) investigated the Local Supercluster geometry. Applying methods of correlation functions, cluster analysis, and percolation theory, they found that 95% of the galaxies belong to the filament system, which occupy only 5% of the volume under consideration.

Tago+ (MN 206, 559) have studied the space distribution of Zwicky clusters in the wide vicinity of the Coma supercluster. They detected chain-like structures at 80 Mpc, which connect Coma with the Local supercluster and Hercules supercluster. Guberman+ (Af 19, 97) have made a geometric analysis of different catalogues of galaxies to verify the consequences of the "pancake" theory of galaxy system formation. Pavlova (AfI 40, 34) studied the orientation of galaxies in Perseus and Pisces regions. Fessenko (LAZ 8, 458; Af 17, 719; AZ 60, 1077) has analyzed different effects of observable biases, which play an important role in selection of superclusters of galaxies. The space correlation function for rich clusters of galaxies were derived by Klypin and Kopylov (LAZ 9, 75) on the scale of 5 to 100 Mpc. Its amplitude exceeds that for galaxies. According to Karachentsev (LAZ 10, 175), the correlation function for blue emission galaxies has negative amplitude within the range 30 to 60 Mpc and a secondary peak at ~90 Mpc.

7. WORKING GROUP ON THE MAGELLANIC CLOUDS
(M.W. Feast)

A complete bibliography of MC research is found in the special section on the MC in A and A Abstracts. An overview of much current research on the MC is given in the volume of IAU Symposium 108 (Structure and Evolution of the Magellanic Clouds = S108). Both IAU Symposium 105 (Observational Tests of Stellar Evolution Theory) and IAU Colloquium 82 (Cepheids, Observation and Theory = C82) contain much important MC data.

I. 30 Doradus, HII Regions, Interstellar Extinction

In 1981 it was suggested that R136, the central object of the 30 Doradus HII region was a supermassive object (~3000 $M_{\odot}$). The arguments involved the high brightness of the object (if single), the supposed uniqueness of the ultraviolet spectrum and the possible light variability. Many papers have subsequently appeared (S108). It now appears that the ultraviolet line spectrum is consistent with the presence of one or more O3 stars and that evidence for variability is dubious (ApJ 273, 597). Moreover the object has been split into a number of components. The most detailed work is that of Walker and O'Donoghue (Ast Exp, in press). The upper limit to the mass of the brightest component is 200-300 $M_{\odot}$ and this component may be a small group of normal stars. The mass loss rate of R136 (AA 120, 269) and the infrared excess (ApJ 272, 123) probably need reassessing in the light of these results.

McGregor and Hyland (ApJ 250, 116) use JHK photometry to deduce that there has been at least two bursts of star formation in 30 Dor during the last 5×10^7 years.

Evidence for metal deficiency in 30 Dor supergiants is also given. Work on WR stars in 30 Dor is discussed in IAU Symposium 99. Further studies have been carried out on the large scale gas motions in 30 Dor (AA 117, 265; MN 196, 19p). The halo of 30 Dor consists of expanding ionized shells. 1410 MHz observations (AA 133, 93) reveal a low density component extending over 2100pc. Huchra+ (ApJ 274, 125) discuss star formation in 30 Dor in relation to that in blue galaxies. Caplan and Deharveng (S108, AA and AA Suppl submitted) have made absolute Hα, Hβ photometry of 75 LMC and SMC HII regions. The data yield reddenings and indicate dust within the HII regions. Regions of high excitation have been found in LMC N11 and N159A (AA 118, 116; 111, L11). ^{12}CO (J = 2-1) has been detected in three LMC HII regions (ApJ 262, 100). The SMC appears to be significantly deficient in CO with respect to the Galaxy whilst the LMC is at most marginally deficient. In MC HII regions the observed depletion of carbon is consistent with most of this element being produced in stars in the mass range 4-10 $M_\odot$ (ApJ 252, 461). The frequency distribution of HII region diameters in the MC is discussed by van den Bergh (AJ 86, 1464). The S/O abundance ratio in MC HII regions and planetaries is the same as in our Galaxy (ApJ 244, 54; AA 118, 234). Other abundances and temperatures are discussed by Mathis (ApJ 261, 195). Lequeux+ (AA 103, 305) and Hunter (ApJ 260, 81) relate the properties of MC HII regions to the history of star formation in these regions. The velocity field of supergiant shell LMC 2 shows that it may result from combined action of stellar winds and supernova explosions (AA 110, 185). Line ratios and velocities have been derived for a number of giant LMC bubbles. New velocity components in several giant interstellar shells have been detected (MN 206, 705; ApJ 255, 447). Theoretical arguments suggest that giant LMC gas clouds are unstable (ApJ 252, 481).

The MC is a fruitful region for the study of very early stages of star formation. H_2O masering from a number of MC HII regions has been reported (AJ 87, 528; MN 205, 275) and although the one in N159 (LMC) was not detected at a later date (MN 200, 19p) both this region and N105 (LMC) (MN 197, 23p) show type I OH masering. Infrared work revealed a very red object in N159 which is believed to be similar to galactic infrared protostars (MN 197, 17p).

Extensive work on the nature of the interstellar particles in the MC has continued. In both clouds the far uv extinction law differs from that in our Galaxy with the 2200A feature being weak or absent in MC. The effect, more pronounced in the SMC than LMC, appears associated with a deficiency of graphite grains in MC (cf. ApJ 247, 860; 255, 70, S108; AA 132, 389; MN 201, 1p; 204, 29p). Nevertheless some SMC stars (particularly SK 143) have a normal galactic type uv extinction (AA 113, L15, S108). The gas/dust ratio for this star is similar to typical galactic values whereas for most MC stars it is much higher. Furthermore, some galactic objects (the Plèiades star HD23512, ApJ 244, 199) have a weak 2200A feature as in MC stars. The LMC extinction law has been put in analytical form (MN 203, 301). Nandy+ find that the strength of the diffuse band at 4430A in LMC stars is not related to the population of small grains responsible for the far uv extinction (Ast Sp Sc 85, 159). Fortunately for the practical problem of making absorption corrections, infrared photometry has been used to show that in both clouds the ratio $R = A_V/E_{B-V}$ is close to the galactic value (3.05) (MN 199, 979; AA 107, 247; Obs 104, 193). Accurate BVI reddening for 77 MC Cepheids give mean values of E_{B-V} = 0.074 (LMC) and 0.054 (SMC) each with small dispersion (Caldwell and Coulson, MN, in press). The wavelength dependence of interstellar polarization (0.36 - 0.84 μm) in the LMC fits the Serkowski galactic empirical relation well and a similar range of λ (max. polarization) is found (ApJ 265, 194).

II. Supernova Remnants, X-ray Sources and Planetary Nebulae

The MC have proved exceptionally important in the study of SNR's (IAU Symposium 101 and S108). Remnants have been identified from optical, radio and X-ray surveys (ApJ Sup 51, 345; 55, 189; MN 200, 1007). Their frequencies are estimated as one in ~275 years (LMC) and ~800 years (SMC). The first remnants of SNI outside our Galaxy have been identified in the LMC (ApJ 261, 473) and the first oxygen-rich, young SNR in the SMC has been found (ApJ 248, L105). X-ray spectra of 4 LMC SNR are consistent with thermal emission whilst two others are probably of synchrotron origin (ApJ 255, 440). Detailed study of N63A and N49 show these SNR's to be evolving in a heterogeneous, cloudy interstellar medium (ApJ 275, 592, 611). 30 Dor B has radio and X-ray properties similar to the Crab Nebula (MN 202, 927). Absolute photometry of several SNR confirms the model of optical emission originating in dense cloudlets embedded in a tenuous medium (AA 111, 171). N185 (LMC) probably originated in an SN explosion (AA 115, 61) whilst N70 (LMC) is a mass loss bubble within a massive collapsing HI Cloud (ApJ 250, 103). The gamma ray burst from the LMC SNR N49 may have been due to the nonexplosive collapse of a white dwarf which exceeded its mass limit (ApJ 261, L71).

Individual binary X-ray sources in the MC are discussed by Commission 42 and in S108. A significant result is the identification of LMC X-3 as a black hole candidate (ApJ 272, 118; 273, L81; 274, L65; AA 126, 265; 132, 240). The Einstein satellite has detected 75 X-ray sources in the LMC, concentrated around 30 Dor. At least 25 of the sources are SNRs (ApJ 248, 925).

Work on MC planetaries is discussed in S108 and in IAU Symposium 103. Carbon abundance is similar to that in galactic planetaries, in contrast to the MC interstellar medium in which carbon is very under-abundant (ApJ 253, L43). Further studies of abundances in LMC PN show (ApJ 273, 590) He, N and O varying with object, but with Ne and Ar remarkably constant. IUE observations of the central stars of 3 MC PN's lead to $M \sim 1 M_\odot$, $L \sim 4.5 \times 10 L_\odot$ and $T \sim 1 \times 10^5$K. The progenitors were probably $4 M_\odot$ stars and the upper limit to the PN nuclei luminosities gives upper limits to the MC distances (ApJ 262, L41). Planetary Nebula formation in various galaxies including the MC is discussed by Papp+ (ApJ 268, 145). New results on excitation classes of MC PNs confirm earlier work (MN 208, 633).

III. Cepheids, RR Lyrae Stars, Blue Supergiant Variables, Novae

Only work subsequent to S108 and C82 are referenced here. The Cape photoelectric BVRI photometry of SMC Cepheids (SAAO Circ 8, 1) and the Dunsink photographic observations of LMC Cepheids (AA Sup 56, 169) are now published. Extensive radial velocity work on MC Cepheids using CORAVEL is in progress. Observations of long period "pseudocepheids" is reported by Eggen (AJ 88, 1458). The first accurate magnitudes and colours of MC RR Lyrae variables have been obtained by Walker (MN, in press). He observed the LMC cluster NGC 2210 with a CCD camera. If $\langle M_V(RR) \rangle = 0.6$ then $(m-M)_0 = 18.42 \pm 0.10$. Graham has found additional RR Lyrae variables in this cluster (PASP 93, 546).

Work on S Dor variables and related objects has continued both because of their importance for understanding stellar evolution in massive stars and as possible distance indicators (AA 127, 49; 110, 272; 112, 61; ApJ 273, 177; IAU Symposium 98). Some but not all S Dor variables show circumstellar dust emission (MN 209, 759; AA 120, 287; 131, L5). Amongst suggested mechanisms for these objects is sudden massive outflow of matter possibly coupled with radial oscillations of the stellar envelope (ApJ 264, 583). A remarkable ~1 mag outburst of R127 was recorded (IAU Circ 3767).

Data on novae in the MC were reviewed at S108. No new novae were reported during the period under review: the importance of a continuing survey for MC novae needs stressing. Photometric and spectroscopic observations of Nova LMC 1981 were reported (AA 124, 143; IAU Circ 3648) and the spectrum of Nova LMC 1977b was described (PASP 93, 581). An eruptive variable in the LMC was reported (IAU Circ 3662).

IV. Red Stars (including variables)

The study of red stars continues to be very active (S108 and IAU Symposium 105; review by Catchpole and Feast, Strasbourg meeting on Cool stars with excesses of heavy elements, in press). Spectra and infrared photometry of red giants in MC clusters (MN 202, 59) are particularly important in understanding the evolutionary connections of M, C and S type stars. In intermediate age clusters C stars first appear on the AGB at M_{bol} = -4.3 to -4.6 (though there is a separate fainter class of ^{13}C-rich C stars). Observations in clusters confirm the association of S stars with helium shell-flashing in a transition phase from M to C types. The M to S and S to C transitions occur at higher luminosity and lower surface temperature in the younger (more massive) stars (MN 204, 985). IR photometry and visual spectrophotometry has been compared with theory to deduce the evolutionary history of M and C stars in MC clusters and in the general field (ApJ 253, 580). Differences between M star populations of LMC and SMC clusters is attributed to age and/or metallicity effects. The effect of C stars on the integrated light of MC clusters was studied (ApJ 266, 105) on the basis of extensive infrared photometry. Lack of high luminosity C stars is notable. Integrated infrared colours of clusters were analysed as a function of cluster age. Use of infrared luminosities of M supergiants as extragalactic distance indicators was suggested (ApJ 249, L55). The high proportion of C to M giants in the SMC compared to the Galaxy has now been found to extend to the cool components of symbiotic systems in the two galaxies (MN 203, 25; see also ApJ 275, 584). New Mira variables in the LMC have been discovered by Wood, Bessell and Paltoglou (ApJ, in press) and by Reid and Glass (to be published). These show a well defined period-luminosity relation with carbon and non-carbon Miras following the same relation at K (Feast, MN, in press). The more luminous red variables seem to fall into at least two distinct groups which have been interpreted as core helium (or carbon) burning supergiants and AGB stars. The amplitudes appear to be smaller in the first group than the second (ApJ 272, 99; 273, 116). The large amplitude variables of high luminosity seem to show a P-L relation which is steeper than that of the Mira variable (Lloyd Evans, MN, in press). Catchpole and collaborators (SAAO) are continuing infrared observations of the more luminous MC red variables.

V. Clusters

The large young clusters in the MC are of great importance in the study of the evolution of massive stars. NGC 1866 in particular with its Cepheids is a vital cluster with which to compare evolutionary predictions. Such comparisons (ApJ 270, 155; PASP 95, 122) suggest (Y, Z age) = (0.27, 0.016, 86 x 10^6 yr). NGC 1856 is of similar age (ApJ 276, 509) and also contains Cepheids, NGC 1847 is ~25 x 10^6 years old (PASP 95, 5). Ages of ~10^8 years have been derived for NGC 2133, 2134, 2058 and 2065 (PASP 96, 28; 94, 894) and [Fe/H] ~ -1.0 was found for the first two clusters. For a number of late-type supergiants in young MC clusters, infrared photometry shows that the LMC ones are metal deficient by ~3 compared to the Galaxy with the SMC supergiants more deficient. End-main-sequence masses of ~14$M_\odot$ are estimated (ApJ 277, 149). There is a suggestion of radial colour variations in LMC young populous clusters possibly due to delayed star formation in the cluster cores (AA 110, 348). Data suggest that the cluster main sequences are much wider than predicted by conventional theory (AA 108, 145). Strömgren photometry has been used to derive metallicities of stars in one MC young cluster. The young cluster NGC 330 (SMC) has [Fe/H] = -1.8. Results in the LMC suggest a range of abundances at a given age (AA 119, 75).

The current situation concerning intermediate and old clusters in the MC may be summarized as follows. c-m diagrams of 6 LMC clusters all indicate ages less than 10^9 years and a significant population of populous clusters older than this is unlikely (ApJ 278, 582). Analysis of integrated spectra leads to similar conclusions (ApJ 261, 85; see also Astrophys Sp Sc 83, 143). For one cluster in the above sample an estimate of [Fe/H] = -1.3 was made (ApJ 275, 15). Faint photometry indicates that NGC 1466, NGC 2210 and NGC 2257 (LMC) are genuinely old clusters (like galactic globular clusters) (ApJ 266, 82; PASP 95, 967; see also Walker, MNASSA, in press).

The age and metallicity of old and intermediate age MC clusters has been studied on the basis of JHK photometry of cluster members (ApJ Sup 48, 161; ApJ 263, 629). The age distribution of LMC clusters peaks at ~4 x 10^9 years. Also, in the LMC, there is an age-metallicity relation though with an (apparently significant) scatter. Spectrophotometry of cluster stars yields similar results (ApJ 258, 143). Simple models of chemical evolution are adequate to fit the observed LMC age-abundance correlation (see also ApJ 256, 447). Attempts to bring all the age estimates onto the same system (Hodge, ApJ 264, 470) suggest that ages based on bolometric luminosities of C star members may be in error due to uncertainties in the theory. There may be different radial distributions of HB stars, giants and subgiants in the old LMC cluster NGC 1806 (Ast Sp Sc 84, 133).

The kinematics of clusters in the LMC younger than 10^9 years are similar to those of young objects (HI gas HII regions, etc.). Older clusters appear to belong to a different system which however is also disc-like. There is no evidence for a kinematic halo amongst LMC clusters (ApJ 272, 488; 259, 89). Velocities and metallicities for 3 SMC globular clusters have been derived (ApJ Sup 55, 45). Radii, structural parameters and luminosity functions for SMC clusters are published (AA Sup 49, 1; 52, 143; AA 108, 344). Integrated magnitudes and colours of 55 MC clusters have been determined (PASP 95, 461). Bright clusters in the LMC have greater ellipticities than fainter ones. A correlation between ellipticity and age has been suggested but this needs to be disentangled from the correlation with luminosity (cf. S108, PASP 95, 839; MN 199, 565; AA 125, 359; Obs 102, 228).

VI. Stellar Content

Cataloguing of the contents of the MC is basic to MC research. The following catalogues have been published: (1) About 200 late K and early M supergiants in the SMC (AA Sup 53, 255); (2) 839 late type supergiants in the LMC (with B, V, R) (AA Sup 51, 277) (in (1) and (2) the frequencies of variables are estimated); (3) 524 SMC members (with UBV) (AA Sup 50, 291); (4) UBV for 245 LMC stars (AA Sup 50, 7); (5) an addendum to non-supergiants near MC (AA Sup 57, 169); (6) Identification charts for LMC stars (AA Sup 48, 1); (7) a comparison of the catalogue of Stock+ (LMC) with other work (Bull Cen Don Stell No. 22); (8) objective prism radial velocities of 418 LMC members and 1127 foreground stars (AA Sup 48, 409) (extensive CORAVEL velocities have also been obtained for MC red supergiants). Photoelectric sequences are: (1) A UBVRI sequence in the SMC bar (AA Sup 50, 119); (2) extension of the LMC Bok-Tifft LMC sequence (AA 114, 213); (3) three LMC cluster sequences (MN 197, 693).

Blanco and McCarthy summarized results of their major survey for C and M stars in the MC. The numbers of C stars in the LMC and SMC are in proportion to the masses of the systems (AJ 88, 1442). The C/M ratio has been discussed as a metallicity index (ApJ 263, 259). The existence of M giants on two distinct AGB's in the LMC suggests two discrete epochs of star formation (ApJ 274, L57). Infrared scans show that objects in the LMC with M_{bol} <-6.4, T_{BB}> 700K must be very rare (ApJ 275, 84). Faint photographic photometry suggests that even 9 kpc from the LMC bar the main population is of intermediate age (ApJ Sup 55, 127), and similarly for the western periphery of the SMC (MN 198, 935) and in the LMC bar (ApJ 278, 592, Ardeberg+, in progress). Courtes+ have obtained MC isophotes at 8500A from the Salyut-7 satellite. From Spacelab-1 experiments they find a UV stellar cloud which appears to belong to the LMC/SMC bridge (Science 225, 179; COSPAR 1984).

Rocca-Volmerange+ (AA 104, 177) discuss MC star formation rates. They find a similar IMF as in the solar neighbourhood and (contrary to some other work) a smooth rate of evolution. However evidence for stochastic star formation has been given (AA 131, 347; 123, 121; 110, 54). Star formation in 30 Dor may have been triggered by the large scale compression of gas moving around the LMC bar. Studies of large gas "bubbles" and associated OB associations in the LMC lead to an age-diameter correlation (AA 127, 113).

Humphreys has analysed the differences between the distribution of the brightest supergiants in the HR diagrams of the SMC and LMC (ApJ 265, 176). The infrared properties of normal and peculiar A supergiants in the SMC are similar (Obs. 104, 193). Further evidence for metal deficiencies in SMC supergiants has been given (ApJ 265, 716; AA Sup 48, 375).

Magellanic Cloud supergiants are important sources of information on stellar winds and mass loss rates. Mass loss rates in excess of 10^{-4} M⊙/yr were found for LMC F-G supergiants (PASP 93, 567). In MC early type supergiants the strength of the stellar wind is generally less than in comparable galactic stars (ApJ 255, 70; 256, L49; ApJ Sup 55, 1). Observations over a wide wavelength range have been used to derive effective temperatures and radii for early type MC supergiants (MN 200, 55; 205, 231; ApJ 255, 70; ApJ Sup 55, 1). The effects of circumstellar absorption on MC O-type stars is small (AA 91, 53; cf. also AA 115, 97; 117, 53) though some examples of circumstellar dust have been found (ApJ Sup 55, 1).

Current results on WR stars (IAU Symposium 99; ApJ Sup 51, 505) do not indicate major global differences between MC WR stars and those in the Galaxy (AA Sup 54, 229; ApJ 268, 228). Despite similar spectral appearances LMC early WN stars span 3 mags in Mv (ApJ 264, 126; cf. also PASP 95, 411). Masses have been derived from several WR binaries (ApJ 257, 110; 257, 116; see also AA 103, 83; 102, 401). Ring nebulae round WR stars may perhaps be regarded as supergiant planetaries (ApJ 256, 452; 255, 79; 269, 202). Sanduleak's star has a nitrogen over-abundance suggesting a similarity to η Car (ApJ 275, 584). S22 (LMC) may have an accretion disk (AA 126, 427). Shell ejection from the carbon star HV2379 has been suggested (MN 202, 31p) though the interpretation is not entirely straightforward (Feast+, MN, in press).

VII. Distance, Structure, Magellanic Stream

The distances of the MC are important in their own right and also for extragalactic distance scale determinations (S108, C82). Feast (MN, in press) has summarized recent evidence for the LMC (true) modulus and finds 18.5. An estimate of 18.3+0.2 has been obtained from G supergiants (AJ 88, 1187). A value of 19.1 was found from OB stars for the SMC (PASP 94, 31). SMC Cepheids indicate a very large tilt of the SMC to the line of sight (BVI, Caldwell and Coulson; JHK, Stobie and Laney) with a resulting large range of distances. The suggestion (Mathewson and Ford S108) of a division of the SMC into two distinct bodies may be too simple a picture. This structure may be related to the Magellanic Stream (see also M31, MN 205,773). Structure of Magellanic type galaxies is summarized by Feitzinger (Ast Sp Aci Rev 27,35).

The possibility of a gaseous halo in the direction to the MC has been discussed. Some workers believe that the observations indicate a halo to the MC whilst some believe the results should be interpreted differently. Others have suggested that this line of sight through the halo of our Galaxy is not typical of the entire galactic halo (ApJ 243, 460; 248, 945; 257, 587; 267, 93; 248, 956; MN 204, 317; 205, 1191). Fehrenbach and Terzan (AA Sup 55, 103) find the density of foreground high velocity stars towards the MC twenty times greater than in other directions of similar galactic latitude. Perhaps this is connected with Saha's result from RR Lyraes that the galactic halo does not have spheroidal symmetry (Frascati Workshop on Pop II Variables, 1984).

8. WORKING GROUP ON INTERNAL MOTIONS IN GALAXIES
(S.T. Gottesman)

Studies of internal motions in galaxies are tabulated below; R,O, and T refer to radio, optical, and theoretical investigations. Three general references are of interest: <u>IAU Symposium No. 100 (Reidel, 1983)</u>; <u>Morphology and Dynamics of Galaxies</u> (Geneva Observatory, 1982); <u>Instrumentation in Astronomy V</u> (SPIE, 1983).

Object	First Author	Reference	Object	First Author	Reference
NGC 157	Zasov	O 29.158.044	NGC 185	Johnson	R 34.157.188
NGC 205	Johnson	R 34.157.188	NGC 221	Pellet	O 31.158.087
NGC 221	Bendinelli	T 33.157.160	NGC 224	Sawa	RT31.158.101
NGC 224	Linke	R 31.158.348	NGC 224	Sofue	O 32.158.158
NGC 224	Sawa	R 32.158.051	NGC 224	Bajaja	R 32.158.005
NGC 224	Lawrie	O 34.157.111	NGC 224	McElroy	O 34.157.028
NGC 224	Unwin	R 34.157.078	NGC 224	Unwin	R 34.157.079
NGC 224	Boulanger	R 29.158.004	NGC 225	Haud	T 29.158.078
NGC 253	Pence	O 30.158.008	NGC 315	Dressel	R 33.158.050
NGC 596	Williams	O 29.158.084	NGC 598	Boulesteix	O In Prep.
NGC 613	Blackman	O 29.158.130	NGC 613	Peterson	O In Prep.
NGC 628	Shostak	R AA 132,20	NGC 628	Briggs	R 32.158.092
NGC 672	Carozzi - Meyssonier	O 31.158.050	NGC 891	vanderKruit	T 29.158.285
NGC 891	Solomon	R 31.158.342	NGC 925	Marcelin	O 31.158.110
NGC 936	Kormendy	O 34.157.187	NGC 1055	Barabanov	O 29.158.139
NGC 1058	vanderKruit	R AA 134,258	NGC 1068	Scoville	R 34.158.049
NGC 1068	Galletta	OT32.158.044	NGC 1079	Gallagher	OR33.157.035
NGC 1097	Wolstencroft	O 29.158.112	NGC 1097	Meaburn	O 29.158.033
NGC 1097	Blackman	O 29.158.130	NGC 1275	van Gorkom	R 33.158.108
NGC 1313	Marcelin	O 31.158.014	NGC 1313	Blackman	O 29.158.130
NGC 1316	Bosma	O In Press	NGC 1365	Phillips	O 33.158.066
NGC 1365	Jörsäter	O Preprint	NGC 1365	Phillips	O 32.158.287
NGC 1365	Blackman	O 29.158.130	NGC 1553	Kormendy	O In Press
NGC 1566	Comte	O AA 114,7	NGC 1569	Young	R ApJ276,476
NGC 1961	Shostak	R 32.158.248	NGC 1961	Gottesman	T MN 202,21p
NGC 2525	Blackman	O 31.158.011	NGC 2655	Huchtmeier	R 31.158.305
NGC 2685	Richstone	O 32.158.040	NGC 2712	Krumm	R 32.158.286
NGC 2715	Huchtmeier	R 31.158.305	NGC 2814	Andreasyan	O 33.157.131
NGC 2841	Young	R 32.158.105	NGC 2903	Marcelin	O 34.157.084
NGC 3031	Bash	T 30.158.010	NGC 3031	Pellet	O 31.158.087
NGC 3031d	Sargent	R 33.157.027	NGC 3034	Stark	R 31.158.347
NGC 3034	Olofsson	R AA 136,17	NGC 3067	Rubin	OR31.158.141
NGC 3115	Illingworth	O 31.158.156	NGC 3200	Rubin	O 33.157.127
NGC 3310	Galletta	OT32.158.044	NGC 3359	Gottesman	R 31.158.327
NGC 3504	Peterson	O 32.158.007	NGC 3628	Tacconi	R 31.158.344
NGC 3690	Fischer	I 34.157.106	NGC 3726	Meysonnier	O In Press
NGC 3729	Schwarz	R In Press	NGC 3783	Pelat	O 29.158.147
NGC 3810	Meysonnier	O In Press	NGC 3893	Meysonnier	O 33.157.113
NGC 3938	vanderKruit	R 31.158.079	NGC 3992	Gottesman	R 32.158.094
NGC 3992	Gottesman	R In Press	NGC 3998	Blackman	O 33.158.035
NGC 3998	Knapp	R In Press	NGC 4151	Schulz	O In Press
NGC 4203	Burstein	RO30.158.206	NGC 4278	Gallagher	O 29.158.245
NGC 4359	Meysonnier	O In Press	NGC 4594	Bajaja	O In Press
NGC 4650A	Schechter	O ApJ 277,526	NGC 4730	Pellet	O 31.158.087
NGC 4731	Gottesman	R Preprint	NGC 5055	Johansson	R In Prep.
NGC 5128	Möllenhoff	O 29.158.00	NGC 5128	Graham	O 30.158.033
NGC 5128	Graham	O 33.158.175	NGC 5128	Marcelin	O 33.158.062

NGC 5128	van der Hulst	R 33.158.011	NGC 5173	Knapp	R In Press
NGC 5194	Hjalmarson	R In Prep.	NGC 5194	Rydbeck	R In Press
NGC 5194	Monnet	O 30.158.137	NGC 5194	Scoville	R 31.158.343
NGC 5194	Scoville	R 33.157.013	NGC 5236	Comte	O 29.158.236
NGC 5236	deVaucouleurs	O 34.157.048	NGC 5301	Krumm	R 32.158.286
NGC 5363	Haynes	R 29.158.248	NGC 5383	Duval	OT33.157.128
NGC 5457	Solomon	R 31.158.345	NGC 5457	Solomon	R 33.157.051
NGC 5457	Bosma	R 29.158.006	NGC 5728	van Gorkom	R 34.157.054
NGC 5813	Kormendy	O In Press	NGC 5813	Efstathiou	O 32.158.267
NGC 5905	van Moorsel	R 31.158.092	NGC 5907	Casartano	RT33.157.118
NGC 5908	van Moorsel	R 31.158.092	NGC 5963	Romanishin	O 31.158.295
NGC 6081	Williams	R 34.157.126	NGC 6118	Meysonnier	O In Press
NGC 6221	Pence	O 33.157.257	NGC 6240	Fried	O AA 118,166
NGC 6503	deVaucouleurs	O 32.158.114	NGC 6503	Kyazumov	O 33.157.215
NGC 6946	Young	R 32.158.011	NGC 6946	Young	R 31.158.340
NGC 6946	Winnberg	R In Prep.	NGC 6946	Peton	O 32.158.168
NGC 7013	Knapp	R AA 133,127	NGC 7252	Schweizer	O 31.158.007
NGC 7331	Young	R 32.158.105	NGC 7479	Peterson	O In Prep.
IC 342	Young	R 32.158.011	IC 342	Young	R 31.158.340
IC 342	Lo	R ApJ 282,L59	IC 694	Fischer	I 34.157.106
IC 1727	Carozzi-Meyssonier	O 31.158.050	IC 2082	Carter	O 29.158.037
IC 5063	Danziger	O 30.158.014	UGC 7576	Mould	O 32.158.104
UGC 7576	Schechter	R MN 208,111	UGC 12423	Bothun	R 32.158.017
VV 48	Arkhipova	O 33.157.005	VV 219	Arkhipova	O 33.157.005
VV 574	Arkhipova	O 34.157.021	II ZW 73	Schechter	R MN 208,111
VII ZW403	Tully	OR29.158.231	Hydra	Ekers	RO33.158.012
3C 218	Ekers	RO33.158.012	3C 293	Baan	R 29.158.074
Mark 10	Schulz	O AA 115,209	Mark 79	Schulz	O AA 115,209
Mark 266	Petrosyan	O 29.158.027	Mark 348	Heckman	R 31.158.143
Mark 538	Afanas'ev	O 32.158.085	Mark 744	Afanas'ev	O 31.158.128
Vela Ring	Taylor	O MN 208,601	Cartwheel	Davies	O 32.158.271
Fornax	Cohen	O 34.157.030	AM064-741	Few	O 31.158.145
Pks0349-27	Danziger	O MN 208,589			

Seyfert Galaxies

Tohline	T 31.158.009	Afanas'ev	O 30.158.038
Mirabel	R ApJ 277,92	Bieging	R 30.158.004
Shuder	O 32.158.028	Dibaj	O 31.158.045

Active and Radio Galaxies

Bieging	R 33.158.033	Jenkins	O MN 207,361
Shostak	O 33.158.054	Jenkins	O 30.158.018
Heckman	R 33.158.115		

Elliptical Galaxies

Sharples	O 33.157.020	Brosche	T 33.157.109
Copaccioli	T MN 209,317	Brosche	T 33.158.117
Illingworth	O 33.157.029	Davies	O 33.157.046
Dressell	R 32.158.029	Knapp	R 32.158.336
Tonry	O 30.158.213	Bertola	O 30.158.101
Tonry	O 29.158.243		

Spheroidal/SO Galaxies

Hawarden	OR29.158.282	Dressler	O 33.157.026
Schechter	O 34.157.041	Giovanardi	R 34.157.124
Lentes	T P Stat Meth ESA 1983, 73		

Early Type Galaxies

Davis	O 29.158.010	Mirabel	R 34.157.029

Disk Galaxies

Kormendy	O 33.157.024	Kormendy	O ApJ 256,460
Binney	OT12th Course SaasFee	Kormendy	O ApJ 265,632

Kormendy	O 31.158.155			
Spiral Galaxies				
Bosma	R 30.158.240		Lequeux	T 34.157.015
Rubin	O 33.157.246		Young	R 32.158.098
Barred Galaxies				
Kormendy	O 30.158.104		Kormendy	O 31.158.143
Kormendy	O ApJ 257,75		Pfenniger	T AA 134,373
Martinet	T AA 132,381		Athanassoula	T AA 132,349?
Double, Multiple, and Interacting Galaxies				
van Moorsel	R 34.157.007		van Moorsel	R 34.157.038
van Moorsel	R 34.157.039		Helou	R 32.158.284
Gallagher	R 30.158.238		Arkhipova	O 29.158.241
Clusters				
Schommer	R 30.158.061		Sullivan	R 30.158.060

Nuclear Emission	Vernon	O	30.158.021
cD Galaxies	Valentijn	R	32.158.170
Irregular Galaxies	Hunter	O	32.158.097
Superthin Galaxies	Goad	OR	30.158.173
Face On Galaxies	Rickard	R	30.158.167
Galaxy Kinematics	Athanassoula	ORT	30.012.004
Galaxy Dynamics	Illingworth	O	30.158.102
HI in Galaxies	Sancisi	R	30.158.106
Sa Galaxies	Rubin	O	In Press
Sab Galaxies	Elmegreen	R	31.158.322
Sb Galaxies	Rubin	O	32.158.215
Sc Galaxies	Burstein	T	31.158.024

9. THE WORKING GROUP ON GALAXY PHOTOMETRY AND SPECTROPHOTOMETRY (J.-L. Nieto)

This triennium (the period mid-1981 to Oct. 1984) ended with a Colloquium "New Aspects of Galaxy Photometry" held in Toulouse on Sept. 1984, covering recent technological improvements and their application to star formation, galactic dynamics and galactic evolution. Related to galaxy photometry and spectrophotometry is IAU Symposium 110 "Internal kinematics and dynamics of galaxies" (33.012.004) and IAU Colloquium 78 "Astronomy with Schmidt-type telescopes" (Aug. 1983).

Three major books have been published: "The Revised Shapley-Ames Catalogue of Galaxies by Sandage and Tammann (30.002.091); "The ESO/Uppsala Survey of the ESO (B) Atlas" (Lauberts, 31.002.098); "The Atlas of Selected Galaxies" by Takase, Kodaira and Okamura (University of Tokyo Press), and a useful tool: "Detailed Bibliography on Surface Photometry" by Davoust and Pence (32.002.003) which is being updated.

I. Optical, UV, and IR Surface Photometry

LUMINOSITY STANDARD NGC 3379. Ratcliff and Burstein (32.158.337), Nieto (34.157.006 and 34.157.014), Capaccioli and de Vaucouleurs (34.157.042) for profiles, Nieto and Vidal for geometry (AA 135, 190) and center (MN 209, 21P), Bendinelli+ (AA 140, 174) for nucleus.

NEARBY GALAXIES. M31 (Hodge and Kennicutt, 31.158.072; Hiromoto+, 34.157.087; Ann and Yu (31.158.158 with also M106 and M33); Innanen+, 31.158.044), M51 (Monnet+, 30.158.137; Smith, 32.158.217), Leo II DG (Hodge, 32.158.338), NGC 205 (Price and Grasdalen, 34.157.189), DDO 155 (de Vaucouleurs and Moss, 34.157.033).

EARLY-TYPE GALAXIES. M87, NGC 2300, NGC 2768 with a use of a CCD in the scanning mode by Boroson+ (34.157.123), NGC 3379 (see above), NGC 404 and NGC 524 (Barbon+, 32.158.250), NGC 1600 and 4261 (Barbon+, AA 137, 166), S0s with polar rings (A0136-081, Schweizer+, 34.157.001; NGC 3384, Davoust+, MN 209, 503), several Es and S0s (Souviron, thesis, University of Paris 7), NGC 5253 (Morwood and Glass, 32.158.245), NGC 3557 (fast rotating elliptical, Capaccioli+, AA 135, 89), ellipticals with shells (a catalogue by Malin and Carter, 34.157.147; NGC 1344, Carter+, 31.158.058; NGC 5128, Malin +, 34.158.077); dust lane ellipticals (survey by Longmore and Sharples, 32.158.149; ellipticals with ionized gas disks and search for [OII] by Caldwell (ApJ 278, 96 and PASP 96, 287); NGC 4696, Jorgensen+, 33.157.219; IC 4320, Warren-Smith and Berry, 34.157.080; IC 5063, Appenzeller and Goida, 30.158.139; NGC 5128, Harding+, 30.158.219, and Adams+, 33.158.019; NGC 6251, Nieto+, 33.158.067; Cygnus A, Thompson (ApJ 249, L47)), dark halo around NGC 1199 (Walker and Andersen, preprint), halo of M87 (Boughn and Saulson, 33.157.028), and correlation between photometric parameters (Thomsen and Frandsen, 33.157.228).

SPIRAL GALAXIES. 25 objects by Boroson (30.158.059), 34 non-barred and 15 barred by Elmegreen and Elmegreen (ApJ Suppl 54, 127 and in preparation, with a near infrared atlas, 30.002.073), several barred spirals by Baumgart (thesis, U. of Missouri), 16 galaxies by Wevers, Ph.D. thesis (University of Groningen), NGC 5963 and 16 low surface brightness spirals (Romanishin+, 31. 158.295 and 34.157.049), NGC 253 (Pence, 30.158.008), NGC 1313 (Marcellin and Athanassoula, 31.158.014), Sculptor Group Galaxies (Carignan, preprint), NGC 5383 (Duval and Athanassoula, 33.157.128), 109 field galaxies (Kent, preprint), NGC 7479 (Blackman, 33.157.021 and Benedict, 31.158.056), NGC 1566 (Hackwell and Schweizer, 33.157.025), NGC 4501 (Send, 32.158.042), NGC 7331 (Telesco+, 32.158.258), NGC 7814 and 5746 (Jones+, 34.157.024), NGC 4156 (Nieto and Tiennot, AA 131, 291), NGC 1079 (Gallagher and Bushouse, 33.157.035), face-on spirals by van der Kruit and Shostak (NGC 3938, 31.158.079; NGC 628, AA 132, 20; NGC 1058, AA 134, 258), inner region of NGC 3351 (Alloin and Nieto, 32.158.274), of NGC 2997 (Meaburn and Terrett, 31.158.315), halo of NGC 253 (Beck+, 31.158.084), a comparative study of spiral bulges with pure spheroidal galaxies (Wirth and Shaw, 30.158.063, 33.157.037); galactic disks as exponentials (Seiden+, ApJ 282, 95).

EDGE-ON GALAXIES. NGC 5907, 5023, 4217, 4013 (van der Kruit and Searle, 31.158.308), NGC 7814 (31.158.309), NGC 4762 and other galaxies by Hamabe, Okamura and colleagues (31.158.099, 31.158.284, 32.158.211), NGC 4762 by Hamabe and Wakamatsu (in press), UGC 7576 (Mould+, 32.158.104), NGC 4565 (Boughn +, 30.158.176, Jensen and Thuan, 32.158.307), NGC 7172 (Sharples+, MN 208, 15), discussion on the width of the dust by Hacke+ (32.158.167).

PECULIAR GALAXIES. NGC 7252 (Schweizer, 31.158.007), NGC 6240 (Fried and Schulz, 33.158.016), NGC 4656-4657 (Stayton+, 33.157.168), UGC 6697 (Gavazzi+, AA 137, 235), NGC 1316 and 1052 (Carter+, 34.158.055), NGC 3448 (Bertola+, AJ 89, 350), Mrk 325 (Coupinot+, 31.158.144; Dettmar+, AA 130, 424), nebulosities surrounding active nuclei (Weistrop+, 30.158.152 and 34.158.063; Hutchings+, 32.141.059, 32.141.131 ApJ 280, 41); Malkan+ (ApJ 280, 66); Wyckoff+, 30.141.023, 33.159.008; McKenty and Stockton (ApJ 283, 64); Mrk 421 and 521, Hickson+, 31158.293; 20 Mrk Seyferts, (Yee, 34.158.087), interacting objects (NGC 5426-5427, Blackman, 32.158.035; IC 5174/75, West and Barbier, 31.158.083; IIZw40, Baldwin+, 31.158.012; AM 064-741, Few+, 31.158.145; VV371, Scholl and Grayzeck, PASP 96, 216; Arp 91, Jenkins, ApJ 277, 501 and others, Joseph+, MN 209, 111), Haro galaxies (Kinman and Hintzen, 31.158.123), others (Brosch, 32.158.046; Barbieri+, 32.158.081; Wakamatsu, 33.157.237; Wakamatsu and Arp, 33.157.103; Sulentic, 33.158.026; Talent+, 32.158.102; Bertola and Sharp, MN 207, 47; Romanishin, PASP 96, 24; Kinney+, PASP 96, 398; Morwood and Glass, AA 135, 281), Mrk or UV-excess objects (often with spectrophotometry) (Mitchell, 32.158.347; Takase +, 34.157.141; Petrosyan+, 34.158.093, etc. See reports on active galaxies and on research in USSR).

DWARF ELLIPTICAL OR COMPACT GALAXIES. Virgo cluster (Caldwell, 34.157.229; Bothun and Caldwell, ApJ 280, 528; Vigroux+, preprint; a catalogue by Reaves, 34.002.052), in the M81/M82 group (Börngen and Karachentseva, 32.158.165 and 32.158.244), near NGC 1023 (Davies and Kinman, MN 207, 173), with near IR photometry (Gondhalekar, MN 209, 59), a JHK survey (Thuan, 33.157.122), central region of NGC 1510 (Eichendorf and Nieto, AA 132, 342), others (Börngen and Kalloglyan, 32.158.013).

cD GALAXIES. 7 X-ray galaxies by Valentijn (33.157.019), giant ellipticals in clusters (Thuan and Romanishin, 30.158.049, Morbey and Morris, 34.160.102), VZw311 (Schneider and Gunn, 32.160.058), cluster brightest members (Murphy +, 33.160.019; Schneider+, 33.160.007 and 33.160.082).

GALAXIES IN CLUSTERS. 20 galaxies in Virgo and 261 in Virgo and Ursa major (Watanabe+, 32.160.026; 33.157.255; 34.157.150), 84 Virgo galaxies (Fritze, 32.158.166), 81 Virgo disk systems (Boroson+, 34.160.097), 109 galaxies in Virgo cluster center to serve as standards (Bingelli+, AJ 89, 64), as well as a classification scheme and an Atlas of dwarf galaxies (Sandage and Bingelli, AJ 89, 919), IR survey of 53 virgo spirals (Scoville+, 34.157.046), edge-on spirals in Ursa Major and Cancer (Weekes, 30.158.169), E's and S0's in Fornax, Virgo and Sersic 129-1 (Griersmith, 31.160.033), in Abell 370 and Cl 2244-02 (Bautz+, 31.160.016), in 0016+16 (Koo; 30.160.048), 17 intermediate redshift clusters (Butcher+, 34.160.006), Abell 1367 (Godwin and Peach, 32.158.037), Abell 1775 (Richter and Schnell, 30.158.156).

LINE PHOTOMETRY. Spirals in three clusters by Kennicutt+ (AJ 89, 1279), narrow-band photometry of dominant cluster galaxy by Heckman+ (30.158.002), cores of X-ray clusters by Cowie+ (34.160.065).

NUCLEI AND/OR CENTRAL REGIONS. Center of M31 (Kent, 33.157.049), of M32 (Tonry, ApJ 283, L27), nucleus of M33 (Nieto and Auriere, 31.158.319; Gallagher+, 32.158.257), central region of NGC 3379 (see above), elliptical galaxies (Lauer, preprint), nuclear structure of M87 (Zavatti+, AA 134, 99), cDs (Kron and Albert, 32.158.350), (near) IR photometry of the nuclei of NGC 5128 (Lepine+, AA 131, 72), NGC 253 (Beck and Beckwith, MN 207, 671), of normal spirals (Willner+, PASP 96, 143), barred spirals (see above).

ULTRAVIOLET PHOTOMETRY. 13 ellipticals with ANS (Astronomical Netherlands Satellite) (de Boer, 32.158.228), of M101 (Stecker+, 31.158.076 and Hill+ ApJ 277, 542), of the nucleus of M83 (Bohlin, 34.157.151), of Virgo galaxies (Smith and Cornett, 32.160.025) and energy distribution of spiral and irregular galaxies (Code and Welch, 31.158.094) and of M31 and M33 (Davis+, 31.158.327).

IRAS results on M31 (Habing+, ApJ 278, L59), Shapley-Ames galaxies (de Jong+, ApJ 278, L67), infrared galaxies (Soifer+, ApJ 278, L71), the Hercules cluster (Young+, ApJ 278, L75), Arp 220 (Soifer+, ApJ 283, L1). IR survey by Lee (31.158.059), by Heckman+ (34.158.086), two-band photometry of 3664 faint galaxies by Hutchins (33.157.174).

OPTICAL POLARISATION. M 82 (Axon, 30.158.046).

II. Geometry, Morphology and Technical Aspects

Elliptical galaxies (Leach, 30.158.051), variation of geometry (Gamaleldin and Issa, 33.157.052), orientation of the plane of M33 (Maucherat+, AA 133, 341), isophotometry of the Leo group (Dickson and Hodge, 30.160.009), inclination and brightness (Arreguine and Sersic, 34.157.145), diameters of rings of spirals (Pedreros and Madore, 30.158.024), standard D_{25} for coma galaxies (Paturel, 31.158.052), for 237 galaxies (Fouque and Paturel, 34.157.013), 500 southern late-type galaxies (Corwin+, 31.158.055), objective and/or quantitative classifications systems (Whitmore, ApJ 278, 61; Okamura+, ApJ 280, 7).

Discussions on Zwicky's magnitudes (Bothun and Schlommer, 31.158.067; Auman+, 31.158.077; Giovanelli and Haynes, AJ 89, 1; de Vaucouleurs and Corwin, in preparation), on faint galaxy photometry from ground-based telescopes and ST (Tyson, PASP 96, 566), estimates of reddening for galaxies in catalogs (Burstein and Heiles, ApJ Suppts 54, 33), influence of seeing (Ellis+, 34.036.038; Peterson and Collins, 34.158.003; Disney and Phillips, 34.157.113; Djorgovski, J Astr Astr 4, 271), calibration of photographic plates (Feitzinger+, 34.036.037), automatic classification of spiral galaxies (Schlosser and Musculus, 33.036.088; Thonnat, 34.036.204), shape analysis (Pasian and Santin, preprint), digital treatment (Sulentic and Lorre, 33.160.027), deconvolution of photographic images from PSF (Bendinelli+, AA 138, 337), detection of faint objects with photographic and computer techniques (Couch+, MN 209, 307), systematic and accidental errors (Capaccioli and de Vaucouleurs, 34.157.042), influence of ellipticity on profiles (Nieto, 33.158.053), accuracy from quick derivation of profiles (Whitmore and Kirschner, 31.158.142); see also workshops on Astronomical Measuring Machines (Sept 1982, 33.012.059) and on Astronomical Photography (April 1984).

III. Multiaperture Photoelectric Photometry

A major event is the publication of the general catalogue of photoelectric magnitudes and colors in the UBV system of 3578 galaxies brighter than the 16th V magnitude (1936-1982) by Longo and de Vaucouleurs (34.002.060). A similar catalogue for infrared bands is in preparation (1600 galaxies). Also appeared the third catalogue of spectroscopic and photometric observations of galaxies from the ESO/Uppsala list (West+, 30.002.046), a similar catalogue for galaxies from the Horologium region (Chincarini+, AA Suppls Ser 57, 1) and an extensive survey of 248 E and SO southern galaxies (Sadler, AJ 89, 34).

UBV PHOTOMETRY. 36 Mrks and SOs (Peterson+, 30.158.055), 61 bright galaxies (Stone, 31.158.289), disk galaxies near clusters (Bothun+, 31.158.323 and 31.158.324), 46 Es and SOs (Michard, 32.158.003), early-type galaxies with known line widths (de Vaucouleurs, Corwin and Frueh, in preparation), galaxies in the direction of the Hercules supercluster (Buta and Corwin, in preparation), six possible Seyfert (Kinman, 33.158.017), 163 DDO dwarf galaxies (de Vaucouleurs+, 30.158.170 and 33.157.227), 8 Es and SOs by Dottori+ (33.157.154), 105 double galaxies (30.002.049), 74 southern galaxies (32.1580.006), brightest ellipticals by Jackson (Ph.D. thesis, U. Calif. Santa Cruz); 6724 galaxies in the field of Coma (Godwin+, 33.160.013).

BVR of early-type cluster galaxies (Lugger, ApJ 278, 51) UBVRI of 28 Mrk galaxies (Hutter+, 30.158.225), 300 nearby galaxies (Corwin and Smyth, in preparation), VRI of Virgo and field spirals (Visvanathan, 30.158.030), g and r of 48 galaxies in clusters (Melnick and Quintana, AJ 89, 1288), UBVJKH of early type spirals (Griersmith+, 32.158.018), intermediate band filter spectrophotometry with 10 filters for 56 galaxies (Soleim+, 31.158.300 and 32.158.272), IR multiaperture photometry for isolated galaxies (Brosch and Isaacman, 32.158.082).

IV. Spectrophotometry

This technique has been applied extensively to active galaxies, whose report is presented elsewhere. We shall concentrate here on normal or mildly active galaxies (galaxies studied also photometrically are included in section A).

NUCLEI. M51 (Rose and Searle, 31.158.033; Rose and Cecil, 33.157.047), nucleus of M33 (O'Connell, 33.157.120), M101 (as well as 5 HII regions) (Rayo+, 31.158.065), NGC 7793 (Diaz+, 32.158.224 and 33.158.023), 93 remote galaxies (Keel,

34.157.005), 116 nuclei (Feldman+, 31.158.153).

HII REGION NUCLEI. NGC 1097 (Talent, 31.158.179), NGC 1097 and NGC 1598 (Phillips+, MN 210, 701), NGC 1365 (Alloin+, 30.158.135; Edmunds and Pagel, 31.158.019) and five others (Phillips+, MN 207, 25).

ELLIPTICAL GALAXIES. Population models (Gunn+, 30.158.154), NGC 4388 (Phillips and Malin, 31.158.311), NGC 3344 (McCall+, 30.158.054), IC 5063 (Danziger+, 30.158.014), NGC 6946 (de Gioia-Eastwood+, ApJ 278, 564), 12 metal poor dwarf galaxies (Kunth and Sargent, 34.157.252), IZw36 (Viallegond and Thuan, 33.157.240), clumpy irregulars (Boesgaard+, 31.158.008 and Duflot-Augarde and Alloin, 32.158.043), irregular galaxies (Hunter+, 31.158.290 and ApJ 276, L35), NGC 3660 (Kollatschny+, 33.157.056), NGC 3690 (Gehrz+, 33.157.162), NGC 5128 (Graham and Price, 30.158.033), NGC 1275 (Wirth+, 33.158.170), NGC 4449 (Sabbadin+, AA 131, 1), interacting galaxies (Kennicutt and Keel, ApJ 279, L5; Bushouse and Gallagher, PASP 96, 273).

SPECTROPHOTOMETRY OF HII REGIONS IN GALAXIES. We restrict the list to the studies of Zinn (Ursa Major DG, 30.158.212), of Boulesteix+ (M33, 30.158.216), of Kennicutt and Kent (200 field and Virgo cluster galaxies, 34.157.003). Mapping of line emission in some galaxies can be found in Fabian+ (32.158.152), Comte and Duquennoy (32.158.169), Marcellin and Gondoin (33.157.055), Graham (30.158.196), Courtes+ (30.158.031), Williaams+ (ApJ 281, 571).

IUE OBSERVATIONS. UV data on the following (non active) objects: M31 (Deharveng+, 31.158.082; Welch, 32.158.030), M33 (Ciani+, AA 137, 223), NGC 4649 (Bertola+, 31.158.041), M81 (Bruzual+, 32.158.103), NGC 4594, 3031, 5194, 4258 (Ellis+, 32.158.150), clumpy irregulars (Benvenuti+, 31.158.017), NGC 4214 and 4670 (Huchra+, 34.157.146), isolated galaxies (Brosch+, AA 135, 330).

10. THE WORKING GROUP FOR SPACE SCHMIDT SURVEYS
(K. Henize)

This Working Group was formed at the 1976 Grenoble IAU General Assembly to coordinate the interests and efforts of those astronomers who feel the growing need for a large-aperture (~one meter), wide-field, fast-focal-ratio space telescope. The basic objectives of such a telescope are: 1) to conduct a Palomar-quality sky survey in the far UV, 2) to detect and study very faint extended sources which, because of field diameter and focal ratio constraints, are very difficult to study with presently planned space telescopes, and 3) to permit objective prism/grating spectroscopy in the far UV. In particular, it is expected that this telescope would serve the Space Telescope in the same way the Palomar Schmidt served the 200-inch telescope - by detecting those stars and galaxies with special or anomalous characteristics in UV wavelengths which would be of interest to observers using the Space Telescope.

Studies directed at the design and use of a Space Schmidt telescope during the past three years have been carried out primarily at the Asiago Observatory/Padova University in Italy, at the Observatoire de Mareseille/CNRS Laboratoire d'Astronomie Spatiale in France, and at the Naval Research Laboratory (NRL) in the USA.

IAU Colloquium 78, "Astronomy with Schmidt-type Telescopes" (L. Rosino, chairman) was held at Asiago Observatory in August 1983. This colloquium reviewed the many uses of Schmidt telescopes and included a session pointing out the needs for a large-aperture Schmidt telescope in space. There were 120 participants from 27 countries in attendance. In early 1984, F. Bertola, in collaboration with M. Golay and R. West, submitted to ESA a statement of the Space Schmidt concept in response to a call for mission concepts for future ESA planning.

Lemaitre at the Observatoire de Marseille continues optical studies of the all-reflecting Schmidt telescope and maintains a laboratory capable of producing large aspheric mirrors of the type required as the corrector for all-reflecting Schmidts. He presented a paper entitled "Optical Design with the Schmidt Concept: 1. Ground-based Development, 2. The Space Schmidt Project for the 1990's?" at IAU Colloquium 78.

Carruthers and Page have completed their analysis of data obtained with the S201 electrographic Schmidt camera operated on the moon during the Apollo 16 mission. A summary, "The S201 Far Ultraviolet Imaging Survey: A Summary of Results and Implications for Future Surveys" (PASP 96, 447, 1984), contains much information related to the conduct of UV surveys. The NRL group is preparing to utilize the electrographic Schmidt camera on two future space missions. The Mark I camera (used for the S201 program), modified to utilize a microchannel intensifier stage, will be used in the Space Test Program to study diffuse nebulae and to measure the galactic UV background. It is estimated that this camera will be 100 times more sensitive to diffuse sources than the original S201 camera. A related instrument, the Mark II version of the electrographic Schmidt camera, will be flown on the Spartan 3 Shuttle sub-satellite. This instrument has an aperture of 15 cm, a resolution of 30 arcsec over an eleven-degree field and is expected to reach a limiting UV magnitude (M_{1400}) of 15.

Carruthers and Heckathorne have continued studies of large-format, high-resolution electrographic cameras suitable for use as the detector in a Space Schmidt telescope. Henize, using data from the Skylab S019 UV spectrograph, has derived a spectrum classification system for O and early-B stars using the very strong C IV and Si IV lines in the far ultraviolet (AJ 86, 1658, 1981). This system gives sensitive discrimination of luminosity among the O-stars and of temperature for the late-O and early-B stars. Since it utilizes the strongest absorption lines in the UV/optical region of the spectrum, such a system should be of interest in studying the O and early-B stars in nearby galaxies via objective-prism spectroscopy with a Space Schmidt telescope.

11. SUPERNOVA WORKING GROUP
(V. Trimble)

The Supernova Working Group currently consists of a mailing list of about 150 professional and amateur astronomers from 18 countries, who engage in supernova searches, observations, and interpretation, either through their own research or through their institutional positions or IAU commission responsibilities. Not all are IAU members. An official chairman and membership list will be established at the XIVth General Assembly. The Working Group communicates via a semi-annual newsletter. Contributions from anyone who has items of interest to those engaged in supernova research would be welcome.

SEARCHES. IAU Circulars carried 83 reports of new SNe between 1981 January 1 and 1984 September 1, including a handful of false alarms and simultaneous independent discoveries. Of the reports, 36 came from Maza and his colleagues at Cerro el Roble, Chile; 12 from Terebizh+ at the Sternberg Crimean Station; 8 from Rev. Robert Evans at Maclean, NSW; 6 each from Wild+ at Berne and Lovas+ at Konkoly Observatory, Budapest; 3 from ESO; 2 each from Asiago and Okazaki at Yamagata, Japan; and 1 each from the Harvard Agassiz Station, Palomar (Arp), Westerbork, Moore Observatory, Abastumani (Kimeridze), Itajuba Brasil (Pellegrini), CIT (Kowal), and Las Campanas (Tammann).

Of the automated searches that have been on the horizon for several years, only that of Cawson (Cambridge UK; U. Arizona) using rejected UK Schmidt plates and the Cambridge Automatic Plate Measuring system has found any supernovae. Three other projects at U. Arizona (McGraw), UC Berkeley (Muller and Pennypacker), and New Mexico (Colgate and Pearce) have hardware and software substantially complete and have begun recording reference images. All expect to begin searching regularly during winter 1984-85.

OBSERVATIONS. Recent new kinds of data include (a) the discovery of radio emission from the Type I event 1983n in M83 (Sramek, Panagia and Weiler, 1984 ApJ Lett, in press; radio emission had previously been seen only from Type II events), which has been interpreted by Chevalier (1984 ApJ Lett, in press) as requiring a binary progenitor of initial masses 8 + 6.5 $M_\odot$ and (b) the first pre-maximum spectrum of a Type II event, 1983k in NGC 4699 (Niemela, Ruiz, and Phillips 1984, preprint), which shows high-ionization NIII and HeII emission lines, atop a strong blue continuum, somewhat reminiscent of Wolf-Rayet stars, a possible class of SNII progenitor. Finally, Cadonau and Tammann (Basel) are engaged in a study of SNI light curves that will probably rule out a division into fast and slow subclasses.

PROGENITORS AND MODELS. Most workers continue to bet on core collapse in massive single stars as the underlying energy source for Type II's and some form of nuclear detonation or deflagration following mass transfer onto a white dwarf in a close binary system for Type I's. The most serious problem for the SN II's has been transferring enough energy to the stellar envelope to eject it at the observed speeds. In the 8-10 $M_\odot$ range, recent models by Nomoto (Tokyo) and Hillebrandt (Munich) indicate that a shock wave set off by core bounce can be fed enough energy by burning of oxygen in intermediate, infalling layers to keep going and eject material. In the 10-30 $M_\odot$ range, Wilson (Lawrence Livermore) finds that significant neutrino heating occurs at late times (after a second rather than a millisecond) and so re-energizes stalled shocks, yielding the requisite ejection.

For the SN I's, problems remain in matching the products of the nuclear explosions to the compositions implied by observed spectra and in identifying sufficient numbers of progenitor systems. For instance, Branch (Oklahoma) finds that a standard deflagration model (Nomoto, Thielemann, and Wheeler, 1984, ApJ 279, L23) reproduces observed abundances only if products from all the layers of the exploding star are well mixed before we see them, so as to yield roughly solar composition (apart from the total absence of H). The problem of identifying enough SNI progenitors (Observatory 102, 133, 1982) becomes rather worse if white dwarfs in typical novae eject more than they accrete and so erode away into AM CVn-like objects (Nather, Texas). Both difficulties are considerably alleviated if the immediate progenitor is a close pair of degenerate dwarfs that spiral together in something like the Hubble time (Paczynski in "Seventh North American Workshop on Cataclysmic Variables and Low-Mass X-Ray Binaries", January 1983). Detailed discussions of the scenario have been given by Webbink (1984 ApJ 277, 355) and Iben and Tutukov (1984 ApJ Sup 54, 335).

DISTANCE INDICATORS. The 45 year history of the use of supernovae as standard candles (Wilson 1939 ApJ 90, 634) has been a patchy one. Recently, Hershkowitz, Linder, and Wagoner (Stanford, preprint) have concluded that the proper allowance for scattering in the low-density atmospheres of Type II events reduces the emergent flux (for a given, measured color temperature) sufficiently below black body levels to imply a Hubble constant, $H_o \geq 104$ km s^{-1} Mpc^{-1}. For Type I events, on the other hand (Sutherland and Wheeler 1984, ApJ 280, 282; Arnett, work in progress), simultaneous matching of light curves and spectra by nuclear explosion models can only be achieved for H_o in the range 36 to 72 km s^{-1} Mpc^{-1}.

29. STELLAR SPECTRA (SPECTRES STELLAIRES)

PRESIDENT: J. Jugaku
VICE-PRESIDENT: G. Cayrel de Strobel
ORGANIZING COMMITTEE: Y. Andrillat, W. K. Bonsack, P. S. Conti, C. R. Cowley, C. O. R. Jaschek, O. H. Levato, M. A. Smith, R. Viotti, S. C. Wolff.

During the triennium under review, Commission 29 has sponsored or cosponsored the following IAU meetings: Symposium 102, "Solar and Stellar Magnetic Fields," Zurich, Switzerland, August 1982; Symposium 108, "Structure and Evolution of the Magellanic Clouds," Tübingen, FRG, September 1983; Symposium 111, "Calibration of Fundamental Stellar Quantities," Como, Italy, May 1984; and Colloquium 82, "Cepheids: Observation and Theory," Toronto, Canada, May-June 1984. Commission 29 has also supported or sponsored several IAU meetings proposed for 1985 and 1986. They include "Luminous Stars and Associations in Galaxies," Porto Heli, Greece, May 1985; "Upper Main Sequence Stars with Anomalous Abundances," Crimea, USSR, May 1985; "Astrochemistry," Goa, India, December 1985; "Hydrogen Deficient Stars and Related Objects," Bangalore, India, December 1985; "Circumstellar Matter," Heidelberg, FRG, June 1986; and "Be Stars," Boulder, USA, August 1986.

This report has been prepared by members of the Organizing Committee and several other members of the Commission and collaborators. Reports on Be stars and CP stars have been written in close contact with the Working Groups (cosponsored by Commission 45) on these subjects. Similarly a short report on the Working Group (cosponsored by Commissions 30 and 45) on standard stars has been prepared.

1. O, Of and Wolf-Rayet Stars -- P. S. Conti

Dr. Underhill and I are completing a monograph volume on O and W-R stars for the NASA series of special publications on stars. Since the review there is much more complete than I can discuss here, I will only consider the highlights of work in this field in the last few years. This short review should be considered a highly personalized account of recent efforts. Interested readers should also see Dr. Abbott's discourse of problems of W-R star's atmospheres which is carried in the reports of Commission 35.

Several recent IAU Colloquia and Symposia have been concerned with O and W-R type stars. Among these are Symposia No. 99 on Wolf-Rayet Stars, edited by de Loore and Willis (1982, Reidel); No. 105 on Observational Tests of Stellar Evolution Theory, edited by Maeder and Renzini (1984, Reidel); Colloquium No. 59 on The Effects of Mass Loss on Stellar Evolution, edited by Chiosi and Stalio (1981, Reidel). Kitchin (1982, Adam Hilger) has published a monograph on Early Emission Line Stars.

A review of the properties of the most luminous stars, in our galaxy and elsewhere in the Local Group, and beyond, has been provided by Humphreys and Davidson (1984, Science 223, 243). Extensive spectrophotometry of faint W-R stars in our galaxy has been provided by Lundstrom and Stenholm (1984, AA Suppl 56, 43). These data supplement the W-R catalogue of van der Hucht et al. (1981, Space Sci Rev 28, 227).

There has been considerable recent activity in discovery of W-R stars in other galaxies. Massey (1985, PASP in press) has written a very nice review paper concerning this topic. There are some 100 W-R stars now known in M33 (Massey and Conti 1983, ApJ 273, 576), some 17 in M31 (Moffat and Shara 1983, ApJ 273, 544), although Massey and Conti have readily discovered another dozen or so, and a half dozen or so each in NGC 6822 and IC1613 (Armandroff and Massey 1985, preprint). The exciting objects of giant H II regions sometimes are found to be W-R like in their spectra. Conti and Massey (1981, ApJ 249, 471) and D'Odorico and Rosa (1981, ApJ 248, 1015) found this to be the case for several objects in M33; similar results were found for a few H II regions in M101 and other galaxies (see D'Odorico et al. 1983, AA Suppl 53, 97). The centers of some distant galaxies also appear to contain W-R like objects, often hundreds or thousands in number (e.g. Tololo 3: Kunth and Sargent 1981, AA 101, L5; NGC 6764 and Mkn 309: Osterbrock and Cohen 1982, ApJ 261, 64; NGC 5430: Keel 1982, PASP 94, 765). Apparently, star formation in some galaxies may have proceeded in bursts, in which many massive stars were born in a short time scale; the presence of short-lived W-R stars illustrates this concept.

Bertelli and Chiosi (1982, IAU Symp No. 99, 359), Garmany et al. (1982, ApJ 262, 777), and Conti et al. (1983, ApJ 274, 302) have discussed the distribution of O and W-R stars near the sun. These authors have pointed out the galactic distribution of O stars has a gradient in that fewer are seen towards the anti-center direction compared to the galactic center. Such a gradient is presumed to be similar to the overall stellar mass distribution: however, they find that the brightest O stars, those with masses greater than some 35-40 solar masses have an even steeper gradient. This observational anomaly is shared by the W-R star distribution: none are observed in the anti-center region indicating a steep gradient in the numbers. The similar galactic distributions of massive O stars and W-R stars suggest they are a related population. No other spectral type of star shows the steep gradient in numbers that is indicated by these massive stars. Since the W-R stars have masses, on the average, appreciably smaller than 40 solar masses (Massey 1981, ApJ 246, 153), it is clear from these arguments that they are decendant from the massive O star population. Such a result was also being found from evolutionary models of massive stars with mass loss and mixing processes at work (e.g. Maeder 1984, IAU Symp No. 105, in press; Doom et al. 1984, AA in press; Conti 1983, IAU Symp No. 99, p. 3), and W-R composition arguments (e.g. Smith and Willis 1982, MNRAS 201, 451). The overall evolutionary picture that emerges for O and W-R stars is that the massive O stars are the progenitors for the highly evolved decendants, the helium burning W-R objects. The WN subtypes are seen as showing CNO equilibrium products (enhanced helium and nitrogen) on their surfaces due to mass loss and mixing; the WC stars, generally more highly evolved, have the products of helium burning (enhanced helium, carbon and oxygen) visible. Thus the W-R stage is a natural occurrence in the evolution of massive stars. These stars can then be used as probes of the massive star population in other galaxies (Massey 1985, PASP in press).

Considerable notoriety has been associated with the central exciting object of the 30 Doradus complex in the LMC: R 136a. Its spectrum appears to be a mix of O and WN type (Vreux et al. 1982, AA 113, L10; Ebbets and Conti 1982, ApJ 263, 108) and has been claimed to be a single supermassive star (Cassinelli et al. 1981, Science 212, 1497). During IAU Symposium No. 108 on the Magellanic Clouds (proceedings edited by van den Berg and de Boer, 1984) a panel discussion was held on the nature of R 136a, and a review prepared by Walborn on the stellar content of the surrounding star field. The consensus that seems to be emerging is that the central object is made up of several very luminous components with individual masses greater than some hundred solar masses (Chu 1984, preprint). There is also a surrounding star cloud which contains a number of very luminous stars, some of O3 spectral type. These very bright stars provide the excitation for the 30 Dor complex, a "hypergiant" H II region.

Important spectroscopic work on W-R eclipsing binaries, particularly V444 Cyg has been carried out by Cherepashchuk and collaborators (e.g. Cherepashchuk et al. 1984, ApJ 281, 774; Eaton et al. 1984, ApJ in press). They have combined optical and IUE data to arrive at a reasonably consistent picture of the eclipse phenomena, both in the continuum and in several of the emission lines. An important finding is that the radius of this well known WN5 star is dependent upon the continuum wavelength under consideration and is small, thus leading to a relatively high "effective" temperature near 10^5 K. Strong phase dependent behavior is found for various emission lines, both in the optical and in the UV.

Interesting spectroscopic work has been reported for a few of the very luminous, often variable, blue supergiants readily visible in our galaxy and others. The results of a mini-workshop on P Cygni was published by de Groot and Lamers (1983,

Irish Astr J 16, 162). High dispersion studies of P Cyg (Lamers et al. 1983, AA 128, 289), and the LMC stars R66 (Stahl et al. 1983, AA 120, 287), and R127 (Stahl et al. 1983, AA 127, 49) are noteworthy. P Cygni stars may be intermediate in evolution between red supergiants and W-R stars (Lamers et al. 1983, AA 123, L8) as well as the S Doradus stars (Stahl et al. 1983, AA 127, 49) or they may represent a stage which never becomes a red supergiant but turns directly into a W-R object. Walborn (1982, ApJ 256, 452) has noted the similarity of some very luminous Of stars to some WN9 subtypes in their spectroscopic properties.

I would be amiss were I not to mention the "alternative" models for O and W-R stars, particularly, suggested by Underhill (1983, ApJ 265, 933). In this type of model, magnetic fields, analogous to what is seen on the sun, play a major role in the heating and support of the stellar winds. Their resultant interaction with the low density plasma also alters atmospheric properties of the stars themselves, along with the deduced effective temperatures and compositions. More detailed summaries of these suggestions are completely discussed in the forthcoming NASA monograph. My personal feeling is that magnetic fields may play some role, but not to the extent claimed by Underhill.

Ending on a very speculative note, Vreux et al. (1985, AA in press) have found a short period, 0.4 day, in the line profile variation of the WN6 star HD 192163. Vreux (1985, PASP in press) has reexamined the periods of several previously identified W-R plus compact systems found by Moffat and associates and concludes very short periods might also be present in them. He suggests these periods, if confirmed by further work, might be evidence of non-radial pulsations. Such pulsations appear to be present in the well known O-type star, ζ Oph according to Vogt and Penrod (1983, ApJ 275, 661). Perhaps these non-radial pulsations are a key feature of the dynamics of stellar winds in luminous early type stars. Certainly more observations will be necessary on this fascinating new topic.

2. Stars of Types O-F -- H. Levato

2.1 O-type stars. The satellite UV data was the base of several studies as has happened in previous periods. Carrasco et al. (1981, AA 100, 183) reported changes in the resonance line profiles in the far-UV spectra of HD 175754. They interpreted these changes as variations in the dynamics and density-ionization structure of the star's wind. Gathier et al. (1981, ApJ 247, 173) determined empirical mass-loss rates for 25 O and B stars (range O4-B1) using ultraviolet line profiles of O IV, N V, P V, Si IV, Si III and C III lines observed with Copernicus. Garmany et al. (1981, ApJ 250, 660) derived mass-loss rates for 31 O-type stars in clusters and associations with IUE observations. Lamers et al. (1982, ApJ 258, 186) found narrow shifted absorption features superposed on the wide P Cygni profiles of the UV resonance lines, in 17 out of 26 OB stars. They considered several explanations, among them, one in which the wind consists of a two-component gas. Franco et al. (1983, AA 122, 9) monitored with IUE in the far UV five O-type stars to study resonance line profile variability. Walborn and Panek (1984, ApJ 280, L27) performed an extensive survey of ultraviolet O-type spectra using high resolution data from IUE. They found a strong correlation between the photospheric and stellar wind features and the optical spectral types. They described the behaviour of the Si IV resonance doublet and reported a new emission feature at 1574 Å. Olson and Ebbets (1981, ApJ 248, 1021) matched the observed Hα profiles of 10 stars ranging from O4f to B3 Ia against computed profiles to derived mass-loss rates.

Fundamental parameters of the stars were the subjects of some important papers. Massa and Conti (1981, ApJ 248, 201) found that the ultraviolet energy distributions of O-type stars in the open cluster NGC 2264 have cooler ultraviolet color temperatures than the cluster early B-type stars. They considered the possible origins of this effect. Underhill (1982, ApJ 263, 741) determined temperatures, radii and luminosities of 24 O3, O4 and O5 stars. She used ultraviolet and visible photometry. The effective temperatures range from 24,800 K to 63,000 K and a typical radius for an O star of the main sequence is 13.4 solar radii. The errors of the determination are discussed.

Underhill (1984, ApJ 285, 668) published information about the irregular, small amplitude ultraviolet light changes of four O and B supergiants. One possible interpretation is in terms of a rotating spotted disk. Underhill (1985, ApJ in press) studied line strengths and profiles in 10 Lac, S Mon, HD 46223 and ζ Pup demonstrating that although these 4 stars have about the same effective temperature and similar photospheres, their mantles differ widely. This observation raises the problem of how stars can have similar photospheres but different mantles.

Kudritzki et al. (1983, AA 118, 245) performed a non-LTE analysis of the O4f ζ Pup. They derived effective temperature, gravity and helium abundance. Simon et al. (1983, AA 125, 34) analyzed three O3 stars (HD 93128, HD 93129A and HDE 303308) using non-LTE models. They derived effective temperatures and surface gravities, masses and radii. Sakhibullin and Solověva (1983, Tr Kazan Gorod Astr Obs Vyp 48, 27) performed a detailed non-LTE analysis of the intensities of the triplet lines of twice ionized carbon in the spectra of O stars.

2.2 B-type stars. Underhill (1981, AA 97, L9) demonstrated observationally that the $3s3p3d^2\,F^\circ$ levels of Si II autoionize in the atmospheres of B-type stars.

UV observations were used by several investigations on B stars. Hamann (1981, AA 100, 169) fitted the UV resonance lines of τ Sco using an empirical wind model and derived a mass loss rate $\log(\dot{M}/M_\odot/yr) = -8.9 \pm 0.5$. Burki (1982, AA 107, 205) using IUE observations studied the variation of the shapes of the UV lines of the spectrum of ξ^1 Sco and the UV flux variability.

Rogerson and Upson (1982, ApJ Suppl 49, 353) presented an ultraviolet spectral atlas of β Ori (B8 Ia) between 999 and 1561 Å. Holberg et al. (1982, ApJ 257, 656) provided observations of some B-type stars in the spectral region between 912 Å and 1200 Å obtained with Voyager 1 and 2.

Malagnini et al. (1983, AA 128, 375) compared the UV and visual spectral distributions of a group of B5-A0 stars with Kurucz's models. They used an automatic procedure first applied to A and F stars. They derived T_e values which are in good agreement with other determinations.

Underhill and Fahey (1984, ApJ 280, 712) reviewed the observations of narrow absorption components of the UV resonance lines in the spectra of O and B stars. They presented new observations for ρ Leo (B1 Ib) and outlined a theory to account for the observations. They also made some suggestions on how to make progress with understanding the spectra of early type stars, which concern basically with modelling the mantles in terms of closed and "open" magnetic loops. Underhill (1983, ApJ 268, L127) showed that supergiants with similar T_{eff} can have winds widely differing terminal velocities. This observation suggests that more than one accelerating force acts on the wind.

Adelman and Pyper (1983, ApJ 266, 732) presented the energy distribution in the optical region for 13 normal stars B, A and F. They derived effective temperatures. Adelman (1984, AA Suppl 55, 479) in an attempt to derive more useful indices to describe the peculiarities of the energy distributions of CP stars in the optical region, used the energy distributions of normal main sequence stars (B and A types) to derive the behaviour of the indices as a function of photometric colour indices.

Didelon (1982, AA Suppl 50, 199) collected information on the equivalent widths of spectral lines of B-type stars published in the literature.

M. Smith and Ebbets (1981, ApJ 247, 158) studied Reticon spectra of ρ Leo (B1 Iab) and found variations of a few percent in the absorption lines of Si III λλ4552-74. The time scale is three hours or less. There is also present a variable Hα incipient emission. They suggested an intimate link between subphotospheric processes and the outer atmosphere in this supergiant. Walker et al. (1982, PASP 94, 143) studied line profile variations in the double line spectroscopic binary Spica (B1.5 IV, V). They used high signal-to-noise Reticon spectra covering from Hα to He I λ6678. They found features which move linearly through He I and O II λ6721 lines of the primary. They found also the width of the primary lines variable. Meisel et al. (1982, ApJ 263, 759) provided profiles of He I λ10830 lines for 65 early type stars (O6 to A1). They compared the results with model calculations.

Dufton et al. (1981, AA 97, 10) have shown that the method of determining microturbulent velocities in B-type stars from the requirement that abundances derived from individual lines show no systematic effect with line strength provides results that overestimate the microturbulent velocity. Sadakane (1981, PASP 93, 587) derived abundances in 21 Peg (B9.5 V). Adelman (1984, MNRAS 206, 637) made an abundance analyses of six B-type stars with moderately sharp lines (π Cet, 134 Tau, HR 2154, HR 5780, 21 Aql and ν Cap). All of them present abundance values similar to those of the Sun.

Casinelly and Swank (1983, ApJ 271, 681) presented X-ray observations of the three stars in the Orion's Belt. The observations are tested against different models.

Several papers were published concerning with the rotation of B-type stars. The distribution of rotational velocities of late B-type stars in galactic clusters was analyzed by Guthrie (1982, MNRAS 198, 795) and found it clearly bimodal. Levato and Morrell (1983, Astrophys Lett 23, 183) determined rotational velocities of OB stars members of the open cluster NGC 6231. García and Levato (1984, Rev Mex Astr Astrofis 9, 9) determined V sin i values for 207 stars of types B, A and F. Guthrie (1984, MNRAS 210, 159) compared the distribution of rotational velocities of B0-B5 stars for young and older subgroups of OB associations and the field. He discussed the excess of slow rotators in the older distribution in connexion with the problem of star formation.

Barsukova et al. (1984, Bull Spec Astrophys Obs North Caucasus 16, 30) published an Atlas of the Spectra of eight supergiants of spectral types in the range O9.5-A2.5 and in the wavelength region 3819-4927 Å.

2.3 A-type stars. Among the investigations on individual stars we noted that using the UV region Saha and Oke (1982, PASP 94, 802) combined IUE and optical data of the supergiant BD 39°4926 and they compared it with model atmospheres. They obtained temperatures and gravity. Kunasz et al. (1983, ApJ 266, 739) analyzed the Mg II line 2802 and the profile of H in the supergiant HR 1040 (A0 Ia). They derived the mass loss rate.

Vega and Sirius were the subject of some papers. Michelson (1981, MNRAS 197, 57) compared high resolution near-ultraviolet observations of α Lyr and β Ori with synthetic spectra. Model atmospheres in LTE were used. The basic parameters of the stars' atmospheres are derived. The velocity fields in the atmosphere of β Ori are discussed. Bell and Drilling (1982, PASP 94, 50) compared theoretical spectra with observations of Sirius and Vega and Lambert et al. (1982, ApJ 254, 663) derived the abundances of C, N and O in the atmospheres of Vega and Sirius. Vega has normal abundances, while Sirius is C and O defficient with respect to Vega. Dragomiretskij et al. (1983, Probl Kosm Fiz Vyp 18, 101) made an absolute calibrations of the energy distribution of Vega in the region 0.8-2.3 μm.

Among the general investigations on A-type stars based on UV data, we mention the paper by Crivellari and Praderie (1982, AA 107, 75) who obtained observations of the region 1400-1800 Å in late A-type stars with IUE. After looking for emission lines formed in the transition region they proposed a schematic model for the outer layers in A-type stars.

Glushneva (1984, Sov Astr 28, 193) determined effective temperatures, radii and bolometric corrections and surface gravities for a sample of stars in the range A-G. Glushneva (1984, Sov Astr 28, 305) determined electron temperatures of the mantles of the A2 supergiants α Cyg and ν Cep, on the basis of an analysis of the enery distributions.

Among the optical studies C. Cowley et al. (1982, ApJ 254, 191) examined high dispersion spectra of a sample of superficially normal late B and early A-type stars with low V sin i values by the methods of wavelength coincidence statistics. They used around 70 strong and weak Fe I lines. They suggest that several stars in the sample may be deficient in iron and they mentioned some possibilities that may be responsible for the peculiar chemistry of these stars. Sadakane (1981, PASP 93, 587) derived abundances in HR 7338, an A0 III star.

C. Cowley and Adelman (1983, QJRAS 24, 393) discussed a variety of new techniques to determine more accurate abundances for stars near spectral type A. They gave some suggestions for avoiding the most common sources of error in abundance work.

2.4 F-type stars. During this period the work of Gray applying Fourier analysis of spectral lines has continued. He performed (1981, ApJ 251, 152) a Fourier analysis of 34 individual lines in the spectrum of Procyon. He found a projected rotation rate of 2.8±0.3 km/s. Gray (1981, ApJ 251, 583) measured asymmetries in the lines of the spectrum of Procyon. He found a blueshifted core in all of the lines. Finally, Gray (1982, Ap J 258, 201) looking the Fourier transforms of the line profiles of seven F-type stars found evidence of rigid rotation.

Among the abundance studies Clegg et al. (1981, ApJ 250, 262) determined CNO abundances in a sample of 20 F and G stars. They used Reticon spectra and they also obtained Fe and S abundances.

Brown and Jordan (1981, MNRAS 196, 757) obtained IUE observations of Procyon. They discussed the range of electron pressure and calculated several models. The maximum temperature for a model with uniform emission is deduced to be $3x10^5$ K, which is lower than that in the solar corona.

Desikachary and Hearnshaw (1982, MNRAS 201, 707) obtained the atmospheric parameters of the F-type supergiant Canopus using spectrophotometry and Balmer line profiles. They derived abundances for 33 elements and found tentative evidence that nitrogen in Canopus may be five times more abundant than in the sun. Boyarchuk and Lyubimkov (1984, Bull Crimean Astrophys Obs 64, 1) performed a detailed analysis of the F supergiants γ Cyg and α UMi.

3. G- and K-type Stars -- G. Cayrel de Strobel

G- and K-type stars are important not only because our Sun is among them, but because they are indicators of galactic and extragalactic evolution and are used to study implications on nucleosynthesis imposed by their chemical composition. The spectroscopic analysis of these stars not only gives information on effective temperatures, spectroscopic gravities, chemical abundances but also on velocity fields, rotations, magnetic fields, chromospheric activities, mass loss etc. in stars. More than 190 papers were analyzed for this report. In the past three years the spectroscopic researches on G and K stars rely more and more on high resolution, high S/N spectra taken with solid state detectors. The increase in accuracy in spectroscopic measurements has sometimes been by a full order of magnitude. References are given as abstract numbers in Astronomy and Astrophysics Abstracts.

3.1 G and K field stars. The researches of Gray in the field of turbulence and rotation of G and K stars have been remarkable during these three years. Gray (30.116.041) has studied rotation and turbulence in G giant stars and made an essential discovery in finding along the giant sequence a rotational discontinuity at G5. He proposed a dynamo generated magnetic brake as the cause (32.116.015). Gray (32. 116.006) extended somewhat his discovery to main-sequence stars. He (31.114.304) discovered and measured stellar granulation in 27 F, G and K stars of luminosity classes III, IV and V. Another essential result of Gray (37.116.045) is that he is now able to discriminate between lower and higher rotational velocity scales for G and K dwarfs. Gray (37.116.009) has measured Zeeman broadening in G and K dwarfs. Vaughan et al. (30.116.029) have measured stellar rotation in lower main sequence stars from time variations in H and K emission line fluxes. Durney et al. (30.116. 035) have interpreted in terms of dynamo concepts stellar rotation measurements combined with Wilson's stellar cycle observations. Middelkoop and Zwaan (30.116.003) and Middelkoop (31.116.013) have studied the relation between rotation and Ca II H and K emission in main sequence stars. A possible detection of oscillations or running waves in Aldebaran and Arcturus has been made by M. A. Smith (33.114.008). Very interesting results are those of M. A. Smith (33.116.017) who gives the values of rotational velocities for a sample of F and K dwarfs with X-ray emissions. Rengarajan (38.114.-) has studied the relationship: age-rotation in late type main-sequence stars. Marcy (34.116.004) presented the results of measurements of magnetic fields in a sample of 29 G and K main-sequence stars, and found 19 detections of magnetic field. Booth and Blackwell (34.114.012) discuss the effect of hyperfine structure on abundance analyses.

The Fe-content of G and K Ib supergiants has been found by Luck (26.114.015) enhanced by a factor of 2 in respect to the Sun. Luck (31.114.043) reanalyzed a sample of 19 of these supergiants and found that their Fe-abundance decreased from +0.3 to about +0.1 dex of the solar value. Hardorp et al. (31.114.047) and Hardorp and Tomkin (34.114.042) proposed HD 44594 and 16 Cyg B as solar analogs. A sample of solar type stars of known age is presented by Duncan (37.113.039). Tomkin et al. (39.114.-) have determined the abundances of Na, Mg, Al, Si, Ca, and Sc in 20 F, G and K disk stars and discussed them in terms of even-Z and odd-Z nucleosynthesis. Cayrel de Strobel et al.(39.114)have produced a renewed version of their [Fe/H] catalogue, which contains 1035 field and cluster stars analyzed in detail on high resolution material.

3.2 Population II field stars. Bessel and Norris (31.114.092) have discovered the most metal deficient star known up to now: CD -38°245, with [Fe/H] = -4.60. Griffin and Griffin have discovered also one of the most metal poor stars: HD 115444 with [Fe/H] = -2.95. R. Gratton (34.114.005) has studied the abundance in 11 metal poor inner-halo stars. The range of Fe-abundances of these stars goes from -0.4 to -2.5 dex. In a new detailed analysis of the halo dwarfs HD 19445 and HD 140283 Magain (37.114.029) finds them much more metal poor than previously thought. Arpigny and Magain (34.114.041) have determined the Al/Mg abundance ratio in the halo dwarfs HD 19445 and HD 140283. M. Spite et al. (34.114.055) have given an upper limit to the deuterium abundance in a few halo dwarfs.

3.3 G and K stars in open clusters. Two K giants in the old open cluster MEL 66 and NGC 2243 have been analyzed by R. Gratton (31.153.035). The four stars have been found quite metal poor by ~ -1.01. Duncan and Jones (34.153.015) combined new Reticon spectrographic observations with spectrometer and image tube spectrogram measurements, in order to obtain independent information on the age of the Pleiades G and early K stars by determining their Li-abundances and comparing it with the Hyades stars.

Brown and Twarog (33.153.027) have investigated on Reticon spectra the photometric CN anomaly of the Hyades, indicating possible CN variation among main-sequence stars. No anomaly in the Hyades in respect to normal stars of similar temperature was found.

3.4 G and K stars in globular clusters and galaxies. A rather large range in chemical composition has been found by Bell et al. (30.114.105) in analyzing a sample of subgiants belonging to ω Cen. The inhomogeneities in the chemical composition of this cluster have been confirmed by R. Gratton (32.114.145). Pilachowski et al. (34.154.005) have analyzed in detail stars in seven globular clusters bringing to 23 the number of globular clusters having their chemical composition determined by means of detailed analyses.

Bessell (34.154.017) has determined the Fe abundance relative to H and those of Ti, V, and Zr relative to Fe on IPCS spectra of a giant in 47 Tuc and 2 giants in M 71. The abundances found for these stars are in very good agreement with the photometric ones and are definitively smaller than some recent echelle derived abundance analyses. Metal abundances of some giants in four distant globular clusters were determined by Pilachowski et al. (34.154.050). Pilachowski et al. (34.154.083) found for the Fe-abundances in M71 a value of -1.0 dex. R. Gratton et al. (37.114.028) have also presented new data on Hα emission in 24 globular cluster giants. Suntzeff (30.154.037) has analyzed low resolution spectrophotometric scans of 29 giants in M3 and 35 giants in M13 in order to determine their C and N abundances. A G-type supergiant, AZ 369 analyzed by Foy (30.114.136) in the Small Magellanic Cloud on high dispersion echelle electronographic spectra, exhibits a metal deficiency of -0.4 ± 0.3 dex. Withford and Rich (34.114.102) have determined the metal content of K-giants in the nuclear bulge of the Galaxy, and found a very important result, namely that the majority of the stars are metal rich with a mean of [Fe/H] ∼ +0.35. Available data on the metal abundances of globular clusters are combined by Pilachowski (37.154.086) to examine the question of the existence of metallicity gradient in the halo of the Milky Way. Although the gradient is weak, its existence implies that the clusters did participate in a general, homologous collapse of the Milky Way and did not form separately.

3.5 Ba stars. The discovery by McClure et al. (27.117.045) and by McClure (33.120.019) that Ba stars appear to be members of binary systems where the companions might be white dwarfs has in these three years reanimated the interest in Ba stars. McClure (37.114.069) dedicated to the Ba stars a review paper in which he emphasizes the impact of these stars on nucleosynthesis and on advanced stellar evolution researches.

3.6 Li in field stars. In 1981, F. and M. Spite (30.114.104, 31.114.090, 31.162.122, 32.114.146, 34.114.020, 34.114.070, 38.114.--) have discovered the Li-resonance line at 6708 Å in extreme Pop II turn-off dwarfs. This discovery is of paramount importance because it provides information about the primordial Li-abundance in the Galaxy. The preservation of Li in Pop II turn-off dwarfs is due to the shallowness of their convection zone, and to the smaller temperature at its bottom than for normal metal-abundance dwarfs of the same effective temperature. Duncan (30.114.022) has studied in over 100 field dwarfs and subgiants their Li-abundances. Duncan has discussed different mechanisms which may explain Li-depletion in the Sun and in solar type stars.

The relatively low luminosity giant HD 112127 exhibits an extremely strong Li resonance line in its spectrum; Wallerstein and Sneden (31.114.035) have carried out a detailed analysis of this star. Luck (32.114.208) has discovered strong Li-lines in two early G-type supergiants.

Steenbock and Holweger (37.114.004) have searched for non-LTE effects in the Li resonance lines of the Sun, a K3 Ib supergiant, halo dwarfs and halo giants. R. Cayrel et al. (37.114.145, 38.114.-) have put into shape an observational test on stellar interior mixing in studying the Li-depletion in twelve Hyades solar type dwarfs and later. The authors have found in analyzing high quality Reticon CFHT spectra of Hyades G and K dwarfs that their Li-abundances decline more rapidly with decreasing temperature than heretofore realized.

3.7 C, N, O, in field stars. The synthetic spectrum technique was employed by Barbuy (30.114.094) to obtain C abundances in some very old extremely metal deficient Pop II stars. For unevolved stars she found [C/M] = 0, whereas for the evolved ones C-depletion appears as a signature of mixing effects. Red and near-infrared high resolution Reticon spectra have been analyzed by Clegg et al. (30.114.128) in order to determine C, N, O and S abundances in a sample of 20 F and G stars. Barbuy (33.114.096) has derived N abundances for 8 Pop II stars and O abundances for two of them. Sneden (34.114.117) has analyzed in detail several CH stars and Lambert (37.114.078) has determined N abundances in disk and halo dwarfs covering a metal deficient range $-2.3 \leqslant$ [Fe/H] $\leqslant -0.3$. Langer and Kraft (37.114.141) reviewed C and N abundances of nearly 300 old, very metal poor stars [Fe/H] $\leqslant -1.3$.

3.8 Chromospheric activity: Optical and IUE observations of field and cluster stars. Fernandez-Figueroa et al. (32.114.096, 32.114.219, 33.112.013) and Rego et al. (33.114.022) have studied and interpreted UV and far-UV emission in solar type stars observed with IUE. M. Spite (30.114.137) interpreted in terms of mass loss far-UV emission in two Pop II stars. Boesgaard and Simon (32.114.215) have analyzed and interpreted IUE spectra of several young solar type stars and have compared with T Tauri stars. Ayres et al. (32.114.205) compared IUE spectra of α Cen A and Arcturus. Signatures of chromospheric activity in cool stars are discussed by Dupree (32.114.211). Evidence of high chromospheric activity in the Hyades has been found by R. Cayrel et al. (33.114.097). Cram (37.114.031) has outlined contemporary ideas regarding the physical phenomena responsible for stellar chromospheric structure.

Campbell (38.114.-) correlates color anomalies in Hyades solar type dwarfs with star spots. Herbig (39.114.-) has compared Hα emission in G-type dwarfs with Hα emission in T Tau stars. Rose (38.114.-) has studied spectral line anomalies in the Hyades, Pleiades and field stars.

4. M, S, C Stars -- P. C. Keenan

The spectra of many of the groups of cooler stars are discussed and interpreted in two books:

1. V. Straižys, Zvezdy s defitsitom metallov (Metal-deficient stars), 1982; Institut Fiziki AN Litovskoj SSR. Molkslas, Vil'nyus.
2. C. and M. Jaschek, Stellar taxonomy, in press; Cambridge Univ. Press.

A general review of infrared spectroscopy of late-type stars was given by Ridgway (in Galactic and extragalactic infrared spectroscopy, ed. M. F. Kessler and J. P. Phillips, p. 309, 1984; Reidel) and an atlas of the 2400-2778 cm^{-1} region in K, M, C and S stars was prepared by Ridgway et al. (1984, ApJ Suppl 54, 177).

Bessell (1982, Proc Astr Soc Australia 4, 417) obtained spectra of faint high-velocity M dwarfs, and found that they divided into two groups: 1. Similar to ordinary old-disk dwarfs, but about 1 mag fainter, and 2. Having weak TiO but very strong CH, MgH, CaH bands. These lie 2-4 mag below the old-disk dwarfs. Computations show that the spectra can be explained by the very low content of metals.

Liebert et al. (1983, IAU Colloquium 76, Van Vleck Obs Contr No. 1, 361), from CCD spectrograms of the extremely red proper-motion star, LHS 2924, concluded that it is the coolest M-dwarf known.

Linsky et al. (1982, ApJ 260, 670) used IUE spectra of 10 dM stars to study their chromospheric emission lines. At higher resolution Ayres et al. (1983, ApJ Lett 270, L17) examined the UV spectrum of the dM1.6e star AU Mic, and Worden et al. (1981, ApJ Suppl 46, 159) obtained echelle spectrograms of 17 M dwarfs, studying emission in the cores of Na D and the Balmer lines.

For the stars of higher luminosity two atlases based on IUE ultraviolet spectra have been published:

1. C. C. Wu et al., The IUE ultravioelt spectral atlas, 1983; NASA Newsletter No. 22. $\lambda\lambda$2400-3400 Å and resolution 7 Å.
2. R. F. Wing et al., Atlas of high-resolution spectra of late-type stars, 1983; Perkins Obs Sp Pub, No. 1. $\lambda\lambda$2500-3230 Å and resolution 0.3 Å.

Also in the ultraviolet, Stickland and Sanner (1981, MNRAS 197, 792) identified lines between $\lambda\lambda$1296 Å and 1915 Å in M giants. Other studies of the UV spectra include that of Johansson and Jordon (1984, MNRAS 210, 239), who used new laboratory spectrograms to identify lines previously observed in β Gru (M3 II), including the strong emission feature at 1870 Å, as due to Fe II levels excited by Lyα pumping. Earlier work on Fe II fluorescence was reported by Engvold (1983, Inst Theor Astrophys Blindern-Oslo, Rep No. 59, 65).

Changes in the line spectrum of the peculiar M-type variable, CH Cyg, have been followed at 30 Å/mm between 3600 Å and 6950 Å by Galkina (1983, Izv Crimean Obs 67, 33) and at higher resolution by Yoo (1984, Ann Tokyo Astr Obs 20, No. 2). Tomov (1984, IBVS No. 2610) concluded that the active phase beginning in 1977 had ended by 1984.

Another star with variable emission, 24 Ser (HD 139216, M5: II-III) was monitored spectroscopically by Yamashita et al. (1982, Ann Tokyo Astr Obs 19, No. 1).

Among Mira variables, R Cyg (Type S) experienced its faintest recorded maximum (9.9 mag) in 1983, and Wallerstein et al. (Univ. of Washington preprint) found that the velocity of the emission lines correlated with the magnitude at maximum, their negative displacement being less at faint maxima. In another Mira variable (R Aqr, M7e at max) H. M. Johnson (1982, ApJ 253, 224) studied the UV spectrum between 1300 and 3100 Å obtained by the IUE satellite. In Mira itself, line-weakening in the period 1965-1980 was reviewed by Yamashita et al. (1981, Ann Tokyo Astr Obs 18, No. 3).

In M giants the λ6707 Li doublet has been measured by Luck and Lambert (1982, ApJ 256, 189), Hänni (1983, Tartu Obs Publ No. 49, 62), and Boyarchuk et al. (1983,

Astrofiz 19, 265). A large project to measure Li, as well as Na and TiO bands, is under way at the South African Observatory by Evans and collaborators. Also, a correlation between lithium excess and UV excess was found by Melik-Alaverdyan (1981, Astrofiz 17, 225).

Among individual M-giants β Peg (M2+ II-III) has been studied most extensively. Orlov et al. (1983, Astrometr Astrofiz 50, 3) found sodium to have nearly solar abundance, while the heavy elements Mo and Ru were underabundant by a factor of 20 (1984, Pisma Astr Zh 10, 135 and Sov Astr Lett 10, 53).

Much work has been done on their molecular spectra, particularly in the infrared. Hänni (1981, Tartu Obs Publ 48, 138) measured the relative abundance of ^{46}TiO and ^{47}TiO in 15 stars, mostly M giants. First overtone SiO bands, in the 4μ region were found by Rinsland and Wing (1982, ApJ 262, 201) to strengthen with advancing type in M stars. The weak but important H_2 quadrupole line, observed by Hinkle, Ridgway and others, were compared with synthetic spectra by Tsuji (1983, AA 122, 314).

The ratio of $^{16}O/^{18}O$ was found by Harris and Lambert (1984, ApJ 285, 674) to be essentially solar in 7 stars, including β And (M0+ IIIa), β Peg (M2+ II-III), μ Gem (M3 III) and α Her (M5 Ib-II). The $^{16}O/^{17}O$ ratios, however, ranged from ~1000 for β Peg to 160 for β And. The IR observations of α Her by Connes have been analyzed by Tsuji (1984, Colloquium on Cool Stars with Excesses of Heavy Elements, Strasbourg, in press).

An extensive analysis of M and MS giants is under way by V. Smith and Lambert (Texas).

Turning to the cool supergiants, we find measurements of mass loss in α Sco by Bernat (1982, ApJ 252, 644) and in α Ori by Stencel et al. (1982, in Advances in UV Astronomy, NASA CP-2238, p. 259). In these two stars Harris and Lambert (1984, ApJ 281, 739) measured the abundances of the oxygen isotopes. Lambert et al. (1984, ApJ 284, 223) also determined the abundances of C, N and O in α Ori. For this same supergiant Carpenter (1984, ApJ 285, 181) studied the velocity field through the chromosphere by means of Fe II lines observed by the IUE. Mass loss from the supergiant binary system TV Gem was estimated at from 10^{-3} to 10^{-5} $M_{\odot}$/yr by Michalitstanos and Kafatos (1982, in Advances in UV Astronomy, NASA CP-2238, p. 213) by means of IUE spectra. Details of the blended spectra of the B3.5 and M1 components were studied by Underhill (1984, PASP 96, 305).

High-resolution spectrograms of six objects near the galactic center were used by Lebofsky et al. (1982, ApJ 263, 736) to show that they are all supergiants. On the other hand, Luck (1982, ApJ 263, 215) analyzed echelle data of 27 stars of types F to M within 1 kpc of the sun.

Among stars in the sequence M, MS, S, SC and C, Murty (1982, Astrophys Space Sci 88, 189; 1982, Astrophys Lett 23, 7; 1983, Astrophys Space Sci 94, 295) identified new bands of YO and CeO in R Cyg (Se) and π^1 Gru (S5). The presence of Nb in the atmospheres of S stars was discussed by Locanthi (1984, Colloquium on Cool Stars with Excesses of Heavy Elements, Strasbourg, in press), while abundances of Nb and Te were measured in R CMi (SCe) and CY Cyg (SC2/7.5) by Smith and Wallerstein (1983, ApJ 273, 742).

Technetium lines in these stars continue to receive much attention. A systematic search for them is being carried out by Little-Marenin. The abundance of Tc has been measured by Orlov and Shavrina (1983, Astr Tsirk No. 1271, 7) in MS and C stars, and a program to study the lines in Mira and χ Cyg has been started by Dominy and Wallerstein.

Dominy and Wallerstein analyzed 4 SC stars in the 2μ region (in press), finding low nitrogen abundances. In the same spectral region Catchpole and Whitelock (in press) found that the 3.1μ absorption feature is a sensitive indicator of the C/O ratio when it is near 1.0. They observed also the 2.3μCO band, finding a correlation with the lithium 6707 Å lines.

A spectacular change in the spectrum of the SC Mira variable BH Cru was observed by Evans (1984, Colloquium on Cool Stars with Excesses of Heavy Elements, Strasbourg, in press). In the early 1970's the spectrum showed little evidence of either ZrO or C_2 at any phase, but the recent spectra are dominated by strong C_2 bands.

A review of the spectra of carbon stars forms chapter 1 in the book "Carbon stars," by Z. K. Alksne and Y. Y. Ikaunieks (1981, translated and edited by J. H. Baumert, Pachart Publ. House). Alksnic (1983, Izv Acad NAUK Latvia SSR No. 4, 93 and 98) has continued the Latvian work.

Carbon stars have received particular attention in Japan. Fujita (1983, Proc Japan Acad 59B, 145) identified bands of HCN in HD 182040 (C2,2 Hd) and W Ori (C5,4) and studied also H_2O lines in the near IR (1984, Proc Japan Acad 60B, 217) and criteria for the ratio $^{12}C/^{13}C$ (1981, Proc Japan Acad 57B, 169). Hirai used high-dispersion spectra from Palomar to resolve the double lines in the spectrum of U Cyg near maximum light (1982, Bull Fukuoka Univ 31, 81; 1984, ibid in press). Tsuji (1981, AA 99, 48) has applied new models to predict the infrared flux and determine the effective temperatures of carbon stars.

Extensive spectrographic investigations of carbon stars are under way at the University of Washington, where Dominy has obtained about 50 image-tube spectra of cool carbon stars, and at the South African Observatory, where Catchpole is working on the southern R stars. At Indiana University the UV spectra of R stars are being studied by Eaton et al., while abundances in 11 R stars were estimated by Dominy (1984, ApJ Suppl 55, 27).

Among individual carbon stars, the obscured object IRC +10216 was searched unsuccessfully in the 3μ region by Clegg et al. (1982, MNRAS 201, 95) for evidence of interstellar HCl. In HD 218875 (C1,3:) Oganesyan (1982, Astrofiz 18, 460) observed the variations of the absorption bands. Goebel et al. (1983, ApJ 270, 190) measured C_2H bands at 2.9 and 3.6μ in HD 19557 (C4,5 J), but did not detect any SiC emission. Evidence that hydrogen may be deficient in late carbon stars was found by H. R. Johnson et al. (1983, ApJ Lett 270, L63) by measuring the quadupole 1-OS(1) line at 4712.9 cm^{-1}, and by Goebel and H. P. Johnson from airborne spectrophotometry of the 1.6μ flux peak due to H^-.

Variations of C_2, CN and rare-earth ions were followed in X Cnc (C5,9) and UU Aur (C6,4) by Eglitis (1983, Nauchnya Inform No. 57, 142), who noted that the lines of neutral metals did not change with phase. In the ultraviolet, both rapid and slow changes in the emission lines in the spectrum of TW Hor (C7,2) have been observed by M. and F. Querci (1985, AA in press), and similar work on 19 TX Psc (C7,2) by H. R. Johnson et al. is under way.

5. Stars in the Late Stage of Evolution -- R. Viotti

This review covers the spectroscopic observations of highly evolved stars. The systematic study, especially in the ultraviolet with the IUE satellite during the last three years has provided a large amount of new results, which have been discussed during the ESA and NASA IUE Conferences held in 1981, 1982 and 1984. The spectra of highly evolved stars are also discussed in the Proceedings of the IAU Symposia 103 (33.012.005) and 105 (37.012. 078), and of the IAU Colloquia 59 (30. 012.015), 70 (32.012.045), 72 (33.012.019), and 80 (37.012.031).

5.1 Hot subluminous stars. Hot subdwarf stars are normally faint in the visual, but because of their high effective temperature and low interstellar extinction, they can be easily observed at high resolution in the IUE ultraviolet. For this reason they are also good spectrophotometric standards for the UV as, for example, discussed by Code in the IAU Symposium 111. Several investigations have been devoted to these stars. The evolutionary scenario was discussed among others by Wasemael et al. (31.126.010) and Bergeron et al. (37.113.037). The high resolution UV spectrum of sdO contains a large number of lines, mostly of metals in different ionization stages, whose identification is still a problem. Extensive line identifications were made of sdO by Bruhweiler et al. (30.126.021) and Rossi et al. (37.126.014), and of the sdOB star HD 149382 and Feige 66 by Baschek et al. (31.126.022; 32.126.008). In sdO Fe V is by far the ion with the largest number of lines. Narrow absorption lines of Fe IV and Fe V were also found in the UV spectrum of WR stars by Fitzpatrik (32.114. 106) and Wills and Stickland (33.117.050), and of the central star of the planetary nebula IC 3568 by Harrington and Feibelman (33.134.007). Heber et al. (37.114.002) derived T_e, gravity and helium content from the study of the optical and UV spectra of eight sdB stars, and concluded that they are He-main-sequence stars with a mass of about 0.5 $M_⊙$. The UV flux distribution of 13 extreme helium stars was investigated by Drilling et al. (37.114.067) who found T = 8500 to 32000 K. The IUE low resolution spectra of eight very hot sdO were used by Schönberner and Drilling (37.114.091) to derive their temperatures much higher than 60000 K. Three turned out to be central stars of planetary nebulae and four near the WD stage. Detailed studies of stellar winds from sdOs, extreme He stars and central stars of planetary nebulae by Hamman (37.112.106) revealed that these stars fit well into the general picture of stellar winds from early type stars. An upper limit to the X-ray flux from the mass-losing sdO BD -3°2179 was derived by D'Antona et al. (33.126.029) from HEAO-2 observations.

Two workshops were held in Paris (1981) and Kiel (1984) entirely devoted to the white dwarfs (no published proceedings). The main result is the discovery of the carbon absorption features in the ultraviolet spectrum of the DC white dwarfs G33-49 by Vaclair et al. (30.126.002) and G218-8 by Wegner (30.126.007). These features were later identified in a number of white dwarfs by Vauclair et al. (31.16.029) and Wegner (33.126. 021). A high resolution study of the UV spectrum of eight white dwarfs with IUE was undertaken by Bruhweiler and Kondo (33.126.033) who found lines of ionized heavy elements in three DA and indication of mass loss in two of them. Greenstein et al. (34.126.010) analyzed the optical spectrum of the new WD binary Sanduleak-Pesch, and found that they are young Pop I stars with masses of 0.80 and 0.43 $M_⊙$. Extensive multicolour spectrophotometry of white dwarfs was performed by Greenstein (37.126.027) and Oke et al. (37.126.085).

Photospheric X-ray emission has been detected by Kahn et al. (37.126.049) in four DA white dwarfs.

A number of studies were devoted to the determination of the physical parameters of the central stars of planetary nebulae (CPN) mainly based on IUE observations. The main result is the discovery of stellar winds with velocities from 1000 to 3000 km/s and mass loss rates of 10^{-10} to $10^{-8} M_⊙$/yr (33.134.043; ESA SP-218, p. 333)

Non-LTE model atmospheres were used to determine T_e, log g and helium abundance in many CPN by Mendez et al. (1981, AA 101, 323; IAU Symp 103; and AA in press), and by Kudritzki et al. (1981, AA 94, L15).

5.2 Symbiotic stars and related objects. A great progress in the knowledge of the nature of symbiotic stars was made in recent years especially from observations outside the visual region and with satellites.

Two meetings held in 1981 on this subject (Boulder Workshop and IAU Colloquium 70) have given fresh ideas for new investigations both observationally and theoretically, and several research papers were published in the following years. Excellent reviews on the previous works on the subject can be found in the Proceedings of the IAU Colloquium 70 (32.012.045), in the Ph. D. dissertation of Kenyon (1983, Univ Illinois), and in a review by Allen (37.117.131). Blair et al. (34.117.142) and Kenyon and Webbink (37.117.176) presented a broad spectroscopic survey of symbiotic and related objects. An extensive catalogue of the optical spectra of symbiotic stars was made by Allen (1984, Proc Astr Soc Australia 5, 369).

Special efforts were devoted to the detailed study of several individual objects, mostly based on IUE observations, such as: Z And (29.122.058), CH Cyg (32.122.021, 040; 37.117.046), CI Cyg (31.117.017; 1984, AA 140, 91), V 1016 Cyg (30.114. 006; 32.114.172), V 1329 Cyg-HBV 475 (34.117.034), AG Dra (33.117.042), YY Her (37.117.220), SY Mus (37.117.081), RX Pup (31.117.060), HM Sge (Mueller and Nussbaumer 1984, AA in press), RR Tel (33.117.024). Additional IUE observations of symbiotic stars are discussed by Michalitsianos et al. (32.117.015), and Sahade et al. (37.117.084). Observation of these objects in other galaxies may provide a new understanding of their nature. Three symbiotics were discovered in the SMC (33.117.028). The IUE and optical spectrum of the LMC object S63 is described by Kafatos et al. (34.114.137). Lutz found that HDE 330036 is a dense planetary nebula with a central binary system (37.134.046). Allen (34.117.017) established the presence of a variable M giant in the high excitation symbiotic H1-36. From the detailed analysis of the optical spectrum of the Kuwano object (PU Vul) a binary model was suggested consisting of an M giant and of a variable low luminosity star. Additional results include the study of the symbiotic stars CI Cyg and AR Pav during eclipse (31.117.017; 34.117.233) and the spectral variations of CH Cyg (31.117.032).

It is noticeable to recall that two objects, AG Dra and Z And, underwent an outburst which was followed with IUE (Viotti et al. 1984, ApJ 283, 226; EAS SP-218, p. 403). In general these works give support to a binary model where the mass exchange is supported by the stellar wind from the cool component or in a few cases by Roche lobe outflow [as suggested by Iijima for CI Cyg (32.117.199)]. Various binary models are discussed by Kenyon and Webbink (37.117.176). A colliding winds model is developed by Wallerstein et al. (37.117.093) to explain the emission line spectrum of V1016 Cyg and HM Sge. Friedjung et al. (34.117.053) give evidence for the presence of a warm wind from the red component of symbiotic systems. Johansson (34.117.130) proposed a fluorescence mechanism for the Fe II emission lines in RR Tel and V1016 Cyg.

Kenyon and Truran (34.124.007) proposed a model of thermonuclear burning episodes onto accreting white dwarfs to explain the nova-like outbursts. A bipolar mass outflow was found by Solf (33.112.011; 1984, AA 139, 296) in V1016 Cyg and HM Sge from very high spatial and spectral resolution observations. Actually, VLA radio maps of many symbiotics revealed the presence of compact asymmetric nebulae (29.116.034; 32.116.017; 1984, ApJ 284, 202). Fundamental information on the nature of the cool component of the symbiotic systems was derived from observations in the infrared. Systematic monitoring of northern objects was carried out by Taranova and Yudin (30.113.061; 30.117.065; 31.113.003,007,024; 32.122.168; 33.117.004), who found large optical and infrared variability in several symbiotics. Large amplitude Mira-type oscillations were found in many southern D-type symbiotics by the South Africa Astronomical Observatory group (33.117.024, 030, 031, 032). Roche et al. (34.117.033) discussed the IR photometry of 20 objects, mostly dust-rich. Kenyon and Gallagher (33.117.119) and Allen et al. (31.112.059) studied the spectroscopic features in the infrared.

6. Pre-main-sequence Stars -- I. Appenzeller and B. Wolf

The period covered by this report (1982-1984) was characterized by a significant improvement of the spectral resolution and signal-to-noise ratio which could be achieved in the spectroscopy of faint stellar objects. This progress resulted from the introduction of new efficient spectrographs and new linear detectors (notably CCDs) at all major observatories. Due to these technical advances much new information and important new insights into the physics of PMS stars and their circumstellar envelopes and gas flows could be derived.

Comprehensive reviews and many original papers on the spectra of PMS stars can be found in the proceedings of the following recent meetings: IAU Colloquium No. 71 (Activity in red dwarf stars) held 1982 in Catania, Italy; "Symposium on Herbig-Haro Objects, T Tauri Stars, and Related Phenomena" held 1983 at Mexico City and published in Rev Mex Astr Astrofis Vol. 7, 1983; "Third Cambridge Workshop on Cool Stars" held 1983 in Cambridge, Massachusetts and published as Springer Lecture Notes in Physics Vol. 193; "Second Meeting on Protostars and Planets" held 1984 in Tucson, Arizona (Proceedings in preparation). General reviews on PMS objects have been presented by Appenzeller (1982, Fund Cosmic Phys 7, 313) and Mirzoyan (1982, Astrofiz 18, 463).

Progress on the various classes of PMS objects can be summarized as follows:

6.1 T Tauri stars. An excellent overview of our present knowledge on the T Tauri stars has been published by Bertout (1984, Rep Prog Phys 47, 111). Recent developments in the T Tauri star research were also reviewed by Appenzeller, Calvet, Kuhi, and by S. E. Strom in the proceedings of the 1983 Mexico City Symposium.

Spectroscopic searches for new T Tauri stars were carried out successfully by Ogura and Hasegawa (1983, PAS Japan 35, 299) and by Duerr et al. (1982, ApJ 261, 135). The three puzzling and disputed emission line objects TW Hya, BM Cha, and LkHα 324 were confirmed to be T Tauri stars (cf. Rucinski and Krautter 1983, AA 121, 217; Krautter and Mouchet 1983, AA 125, 378; Chavarria et al. 1983, AA 118, 189). Appenzeller et al. (1983, AA Suppl 53, 291) published a catalog and spectral atlas of the T Tauri stars south of $\delta = -35°$, which were not included in the Cohen-Kuhi catalog.

With the IUE satellite additional UV spectrograms of T Tauri stars were obtained. The corresponding results were reported and discussed by Penston and Lago (1983, MNRAS 202, 77), Giampapa (1984, Proc Third Cambridge Workshop on Cool Stars, p. 14), Imhoff and Giampapa 1982, Smithsonian Astrophys Obs Spec Rep 392, 175), and by Brown et al. (1984, MNRAS 207, 831; 1983, Proc IAU Coll No. 71, p. 509).

Gahm and Krautter (1982, AA 106, 25) determined upper limits for the coronal line emission of T Tauri stars. They could prove that, compared to normal late type stars, in T Tauri stars the coronal emission (if present at all) is much less enhanced than the chromospheric emission.

Detailed spectroscopic investigations of individual T Tauri stars were carried out for RU Lup by Lago and Penston (1982, MNRAS 198, 429 and 495; 1984 MNRAS 210, 323) and by Boesgaard (1984, AJ 89, 1635), for S CrA and GQ Lup by Bertout et al. (1982, AA Suppl 47, 419) for AS 353, DI Cep, and RY Tau by Krasnobabtsev (1982, Bull Crimean Astrophys Obs 65, 100), for RW Aur by Appenzeller and Wolf (1982, AA 105, 313) and by Aiad et al. (1983, PASP 95, 656) and for DR Tau by Ulrich et al. (1983, ApJ 267, 199).

Papers concerning the theoretical interpretation of the T Tauri spectra and the problem of line formation in T Tauri atmospheres and circumstellar envelopes were presented by Bastian (1982, AA 109, 245), Bertout (1982, Third European IUE Conference, p. 89), Chevalier 1983 (ApJ 268, 753), Edwards and Strom (1984, Proc Third Cambridge Workshop on Cool Stars, p. 1), by Hartmann (1984, Proc Third Cambridge Workshop on Cool Stars, p. 60), Jordan (1983, Obs 103, 129), Katyesheva (1981, Astrofiz 17, 301), and Wagenblast et al. (1983, AA 120, 6).

Several extended observing programs were devoted to studies of the spectral variability of T Tauri stars on time scales between minutes and years. Aiad et al. (1984, AA 130, 67) organized an international observing campaign to monitor the spectral variations of several YY Ori stars continuously over a four day time interval. Making use of the long winter nights north of the arctic circle Appenzeller et al. (1983, AA 118, 75) used the Skiboth Observatory of the University of Tromsö in North Norway to obtain an almost continuous record of the spectral variations of RW Aur during a 12 day period. Mundt and Giampapa (1982, ApJ 256, 156) observed line profile variations of RW Aur on time scales as short as 10 minutes. The spectral variability of this star on longer time scales (days to months) was studied extensively by Bukach et al. (1982, Sov Astr Lett 8, 172) and Grinin et al. (1983, IAU Coll No. 71, 513). Efimov et al. (1980, Perem Zvesdy 21, 247) studied the spectral variability of DI Cep. Very interesting information on the variability of the emission line spectrum of T Tauri stars (variations of the Hα line emission) resulted from the <u>photometric</u> monitoring program of Herbst et al. (1982, AJ 87, 1710) which included Hα intensity measurements.

Hartmann (1982, ApJ Suppl 48, 109) and Mundt (1984, ApJ 280, 749) used high resolution absorption line profiles to derive fundamental physical properties of the T Tauri stellar winds. Jankovics et al. (1983, PASP 95, 883) and Appenzeller et al. (1984, AA 141, 108) found the forbidden lines in the spectra of the T Tauri stars to be systematically blueshifted, with line profiles indicating highly anisotropic mass flows already in the nonresolved close circumstellar envelopes of these stars.

Meyer and Ulrich (1984, ApJ 283, 98) used the diffuse interstellar bands in the spectra of T Tauri stars to study the properties of their circumstellar dust.

6.2 <u>Post-T Tauri stars (PTT)</u>. The evaluation of Einstein X-ray observations of star forming regions led to the discovery of PTTs which seem to be PMS stars in an evolutionary stage between the T Tauri phase and the ZAMS (cf. Mundt et al. 1983, ApJ 269, 229; Walter 1984, Third Cambridge Workshop on Cool Stars, p. 75). Additional PTTs were observed spectroscopically by Feigelson and Kriss (1983, AJ 88, 431) and by Smith et al. (1983, ApJ 271, 237).

6.3 <u>Herbig Ae/Be stars (HBeS)</u>. Finkenzeller and Mundt (1984, AA Suppl 55, 109) and Finkenzeller and Jankovics (1984, AA Suppl 57, 285) obtained high quality spectrograms and collected fundamental data on all known HBeS. Their data support the PMS nature of these stars. Talavera et al. (1982, Third European IUE Conference, p. 99) and Tjin et al. (1984, AA 134, 273) investigated the UV spectra of HBeSs. Harvey (1984, PASP 96, 297) obtained IR spectrograms of nine HBeSs. Detailed spectroscopic studies were carried out for HD 200775 by Baschek et al. (1982, AA 105, 300) and for AB Aur by Praderie et al. (1982, ApJ 254, 658), Felenbok et al. (1983, Rev Mex Astr Astrofis 7, 239), Finkenzeller (1983, AA 124, 157), and Catala et al. (1984, AA 134, 402).

6.4 FU Ori objects. New spectroscopic observations of V 1057 Cyg were obtained by Kolotilov (1983, Sov Astr Lett 9, 622) and by Chalonge et al. (1982, Astrofiz 18, 263). Kolotilov and Petrov (1983, Sov Astr Lett 9, 171) also observed V 1515 Cyg. Levreault (1983, ApJ 265, 855) used molecular line spectroscopy to study the wind from the FU Ori object Elias 1-12. Graham and Frogel (1985, ApJ in press) discovered and investigated a new FU Ori star. Interestingly this fifth known FU Ori object appeared on the edge of a known HH object (No. 57).

6.5 Herbig-Haro objects (HH); jets. The last three years brought decisive progress in the understanding of the HH objects. It now seems to be generally accepted that these line emitting gaseous nebulae are caused by collimated mass ejection from very young low mass stars or protostars. Recent general reviews of the HH objects have been given by Schwartz (1983, Ann Rev Astr Astrophys 21, 209; 1983, Rev Mex Astr Astrofis 7, 27). The stars exciting the HH objects were studied by M. Cohen and Cohen et al. (1982, PASP 94, 266; 1983, ApJ 265, 877; Rev Mex Astr Astrofis 7, 241; 1984, ApJ 278, 671) and by Mundt and Hartmann (1983, ApJ 268, 766).

New spectroscopic data on individual HH objects were collected by Dopita et al. (1982, ApJ 263, L73), Meaburn and White (1982, MNRAS 199, 121), Böhm et al. (1983, AA 125, 23), Graham and Elias (1983, ApJ 272, 615), Axon and Taylor (1984, MNRAS 207, 241), and by Krautter et al. (1984, AA 133, 169). IUE-UV spectrograms were obtained by Böhm et al. (1982, ApJ 262, 224; 1982, ApJ 263, L35; 1984, ApJ 277, 216), Brugel et al. (1982, ApJ 262, L35), and by Meaburn (1982, AA 114, 367). Brown et al. (1983, ApJ 203, 785) and Schwartz (1983, ApJ 268, L37) observed molecular hydrogen in the spectra of HH objects. A highly interesting new phenomenon discovered during the period covered by this report is strongly collimated HH-like gas flows (or "jets") close to young stars (cf. Mundt et al. 1983, ApJ 265, L71; 1983, ApJ 274, L83; 1984, AA 140, 17; Strom et al. 1983, ApJ 271, L23). An excellent review of this exciting new topic has been prepared by Mundt (1984) and will be published in the proceedings of the Second Meeting on Protostars and Planets.

7. Be Stars -- A. Slettebak

The field of Be-star research has been very active since the last Commission 29 report. Earlier research was summarized in the proceedings of IAU Symposia No. 70 (1976, "Be and Shell Stars," ed. A. Slettebak) and No. 98 (1982, "Be Stars," ed. M. Jaschek and H.-G. Groth). More recent general reviews include those by Doazan (1982, in "B Stars With and Without Emission Lines," ed. A. Underhill and V. Doazan, NASA-CNRS, NASA SP-456), Harmanec (1983, in Proc Workshop on Rapid Variability of Early-Type Stars, Hvar Obs Bull 7, No. 1), Kogure and Hirata (1982, Bull Astr Soc India 10, 281), Hirata and Kogure (1984, Bull Astr Soc India 12, 109), and Thomas (1983, in "Stellar Atmospheric Structural Patterns," NASA-CNRS, NASA SP-471).

Several catalogues of Be-star data, very useful for researchers in this field, appeared during this report period. The Catalogue of Stellar Groups by M. Jaschek and D. Egret (1982, Publ Spec CDS No. 4) lists some 1000 Be stars. Photometric data for these objects have been collected by B. Hauck and M. Mermilliod (1984, Bull Inf CDS No. 26). Both catalogues are available through the Centre de Données Stellaires (CDS) of the Strasbourg Observatory. Also, Page of the Mount Tamborine Observatory (Australia), has made his Be Spectrum and Magnitude Databank (1981, Be Star Newsl No. 4, 2) available in several forms through both the CDS and ADS (NASA), and completed a revision in 1984.

Of special importance is the Be-Star Newsletter, edited by M. Jaschek and distributed twice annually by the Strasbourg Observatory. This very valuable publication makes reports such as this one somewhat superfluous. Thus, during the three-year period covered by this report, the Newsletter included 44 research contributions and 382 references to Be-star research, plus regular reports on international observing campaigns, announcements of meetings, etc. Since it is neither possible nor very meaningful to attempt to summarize this vast amount of work into a short report, only some highlights will be discussed in what follows.

It should be stated at the outset that we still do not understand why or how Be stars throw off shells, nor is their evolutionary status clear. This seemingless slow progress is due in part to the nature of Be-star observations in earlier years: sporadic "looks" at individual stars, usually in visual or blue light only. The past several years have seen a dramatic change in this respect, however: increasingly, long series of observations of a given star are being made, often simultaneously in different wavelength regions. Such observations should be very fruitful toward understanding Be stars which are, after all, (on various time scales) variable stars.

Mention should be made, especially, of two international observing campaigns started several years ago. One, a spectroscopic campaign, was initiated by Barker and concentrates on Hα observations of Be stars (cf. Baker 1983, PASP 95, 966). Other large programs of Hα observations include those of Andrillat and Fehrenbach (1982, AA Suppl 48, 93), Andrillat (1983, AA Suppl 53, 319), Dachs et al. (1981, AA Suppl 43, 427), and Fontaine et al. (1982, ApJ Suppl 49, 259; 1983, AJ 88, 527).

The other international observing campaign is a photometric one and was initiated by Harmanec, Horn, and Koubsky, with many participating observers. A successful mini-campaign on five rapid Be variables was organized by Percy and Stagg in 1983. Among the many papers on Be star photometry and variability written during this report period are those by Alvarez and Schuster (1982, Rev Mex Astr Astrofis 5, 173), Ashok et al. (1984, MNRAS 211, 471), Bhatt et al. (1984, AA Suppl, in press), Dachs and Wamsteker (1982, AA 107, 240), Goraya (1984, AA 138, 19), Huang (1984, Publ Beijing Obs, in press), Kogure (1984, Publ Beijing Obs, in press), Mazeh and Brosch (1981, AA 95, 3), Mendoza et al. (1983, PASP 95, 48), Mermilliod (1982, AA 109, 48), Neto and Pacheco (1982, MNRAS 198, 659), Schuster and Alvarez (1983, PASP 95, 35), Sharov and Lyuty (1984, Astr Tsirk No. 1333, 1), and Zorec et al. (1983, AA 126, 192). An excellent review of Be star variability was presented by Harmanec at the Hvar Workshop on Rapid Variability of Early-Type Stars (1983, Hvar Obs Bull 7, No. 1).

Long series of observations of a number of bright Be stars now exist, as well as simultaneous observations in different wavelength regions, which permit modelling of these objects in a number of cases. Recent papers include studies of o And (Pastori et al. 1982, Astrophys Space Sci 86, 179; Poeckert et al. 1982, PASP 94, 87; Harmanec 1984, IBVS No. 2506), V 923 Aql (Ringuelet et al. 1984, AA 131, 9), 28 CMa (Baade 1982, AA 105, 65; 1982, AA 110, L15), γ Cas (Doazan et al. 1983, AA 128, 171; Galkina 1981, Bull Crimean Astrophys Obs 64, 72; Henrichs et al. 1983, ApJ 268, 807; Lowe et al. 1985, ApJ, in press), μ Cen (Baade 1984, AA 135, 101; Peters 1984, PASP, in press), θ CrB (Doazan et al. 1984, AA 131, 210; Harmanec 1984, Bull Astr Inst Czech 34, 324; Underhill 1984, AA, in press), 59 Cyg (Barker 1982, ApJ Suppl 49, 89; 1983, ApJ 88, 72; Hubert-Delplace and Hubert 1981, AA Suppl 44, 109; Tarasov and Shcherbakov 1983, Sov Astr Lett 9, 350), 88 Her (Doazan et al. 1982, AA 115, 138; Barylak and Doazan 1984, Fourth European IUE Conference; Barylak et al. 1985, AA in press), EW Lac (Kogure and Suzuki 1984, Publ Astr Soc Japan 36, 191), ω Ori (Hayes and Guinan 1984, ApJ 279, 721), ϕ Per (Poeckert 1981, PASP 93, 297), ζ Tau (Dawanas and Hirata 1984, Astrophys Space Sci 99, 139; Harmanec 1984, Bull Astr Inst Czech 35, 164), HR 2142 (Peters 1983, PASP 95, 311), and HR 5223 (Dachs and Hanuschik 1984, AA 138, 140).

The above papers generally support the picture, which was already developing at IAU Symposium No. 98, of the Be envelope consisting of a cool, low-velocity, flattened Balmer-emitting region plus a hot, high-velocity, spherically-symmetrical stellar-wind region. See also the papers by Marlborough and Zamir (1984, ApJ 276, 706) and Oegerle and Polidan (1984, ApJ 285, 648). Not all researchers accept this picture, however (cf. Doazan and Thomas 1982, "The B Stars With and Without Emission

Lines," ed. Underhill and Doazan, NASA SP-456), and the relationship between the cool and hot regions is not yet clear. Various possible scenarios were discussed by Poeckert (1982, IAU Symp No. 98, p. 480).

An exciting development, already suggested in papers by Baade and by Bolton at IAU Symp No. 98, is the presence of nonradial pulsations in Be stars. Recent work includes papers by Baade (for example, 1984, AA 135, 101), Percy (1983, AJ 88, 427) and Vogt and Penrod (1983, ApJ 275, 661). The latter suggest that sporadic release of pulsation energy during mode switching provides a plausible explanation for Be-type outbursts in ζ Oph, and for the occurrence of the Be phenomenon in general.

An especially interesting aspect of ultraviolet observations of Be stars is the discovery of narrow, high-velocity absorption components in the spectra of γ Cas (Hammerschlag-Hensberge 1979, IAU Circular No. 3391), 59 Cyg (Doazan et al. 1980, ApJ 235, L17), and ω Ori and 66 Oph (Peters 1982, ApJ 253, L33). The observations and possible models were reviewed by Henrichs (1984, Fourth European IUE Conference), who points out that while 50 percent of luminous stars show narrow absorptions, their occurrence in less luminous stars is strongly connected with the Be phenomenon.

Another interesting ultraviolet result, in this case from Voyager in the far-UV (910-1700 Å), comes from Polidan et al. (1985, in preparation), who find that the energy distributions of Be stars in the ultraviolet are identical to those of non-Be stars of similar ground-based spectral type. This result (no gravity darkening) implies that Be stars do not rotate at the critical velocity.

In addition to the above and to the ultraviolet studies of selected bright Be stars already referenced, the following investigators were among those who wrote papers based on IUE and Copernicus observations of Be stars: Barker et al. (1984, NASA Goddard Conference, "Six Years of IUE Research"), Bernacca et al. (1983, AA 122, 17), Codina et al. (1984, AA Suppl 57, 239), Hubert-Delplace et al. (1983, AA 121, 174), Kitchin (1982, MNRAS 198, 457), Molaro et al. (1983, AA 119, 150), Oegerle et al. (1983, PASP 95, 147), Selvelli (1984, Fourth European IUE Conference), Slettebak and Carpenter (1983, ApJ Suppl 53, 869), and Tarafdar (1983, MNRAS 204, 1081).

The association of low-luminosity X-ray sources with Be stars was established with the detection of X-ray emission from X Per and γ Cas. Among papers written during this report period on these and other X-ray sources are those by Charles et al. (1983, MNRAS 202, 657), Corbet and Mason (1984, AA 131, 385), Grindlay et al. (1984, ApJ 276, 621), Kelley et al. (1983, ApJ 274, 765), Koenigsberger et al. (1983, ApJ 268, 782), Kriss et al. (1983, ApJ 266, 806), Mazeh et al. (1982, ApJ 256, L13), Peters (1982, PASP 94, 157), Sanwal et al. (1984, Astrophys Space Sci, submitted), Steiner et al. (1984, ApJ 280, 688), van der Klis et al. (1983, MNRAS 203, 279), Weisskopf et al. (1984, ApJ 278, 711), and White et al. (1982, ApJ 263, 277). Mention should also be made of the Bologna-Frascati Be-star working group (Giovannelli, Bartolini, Bianco, Guarnieri, Piccioni, Giangrande, Ferrari-Toniolo, Persi and others), who published a number of papers on Be X-ray sources during the report period.

Several investigators have attempted to measure shell sizes for Be stars. Thus, Schmidtke and Africano (1984, AJ 89, 663) observed a lunar occultaion of the B1 IV shell star, ζ Tau, in Hα light to determine an upper limit to the shell size of about four stellar radii. Granes, Vakili, and collaborators have undertaken an interferometric program to determine sizes of Be stars. They obtain an upper limit to the diameter of γ Cas of about 46 solar radii (Vakili et al. 1984, Publ Astr Soc Japan 36, 231).

Magnetic fields have been suggested as playing a role in Be star models. A recent search for magnetic fields in Be stars by Barker et al. (1984, ApJ in press) finds these objects to be magnetically indistinguishable from normal upper main-sequence stars. The authors point out, however, that large surface fields could escape detection, depending upon field geometry and orientation, and that circumstantial evidence suggests that many Be stars may indeed be magnetic.

Rotation in Be stars has been studied by several investigators during the report period. Thus, Ruusalepp and Sapar (1983, Publ Tartu Obs vol 50; 1984, Publ Tartu Obs vol 51) have attempted to separate V from sin i for rotating stars, while Slettebak (1982, ApJ Suppl 50, 55) and Gao (1984, Publ Beijing Obs, in press) estimated V sin i for a number of bright Be stars.

Abt and Cardona (1984, ApJ 285, 190) investigated Be stars in binaries, confirming the known near-absence of Be binaries with periods less than 0.1 year, but finding no difference in binary frequencies of Be and B stars for longer periods, up to the 10^4 AU limit of their survey.

A principal activity of the Working Group on Be Stars (IAU Commissions 29 and 45) was the planning of a conference on the Physics of Be Stars. This has been designated as IAU Colloquium No. 92 and will be held in Boulder, Colorado, U.S.A. during the period August 18-22, 1986. The undersigned is Chairman of the Scientific Organizing Committee and T. P. Snow, Jr. is Chairman of the Local Organizing Committee for this colloquium.

Finally, I would like to thank my many colleagues who sent reprints, preprints, and reports of their work -- I regret that there is not more space available to include all of the research done during this report period. In particular, since this section has been restricted to work done on "classical" Be stars, research dealing with Herbig Ae/Be (pre-main-sequence) stars, emission-line supergiant stars, and interacting binary systems will be found in other sections of this Commission 29 report.

8. CP Stars of the Upper Main Sequence -- D. S. Leckrone and S. J. Adelman

A new monograph by S. C. Wolff, "The A-Type Stars: Problems and Perspectives," (1983, NASA SP-465) provides a comprehensive review of research on the various subclasses of CP stars. A major review (in Russian) of magnetic stars is given by Khokhlova (1983, in Astrophysics and Space Physics, ed R. A. Syunyaev, p. 233).

Radiatively driven diffusion remains the dominant model for the production of anomalous elemental abundances in CP star photospheres. Alecian and Vauclair (1983, Fundam Cosmic Phys 8, 369) and Vauclair (1983, in Astrophysical Processes in Upper Main Sequence Stars, ed B. Hauck, A. Maeder, Geneva Obs p.167) presented reviews of the theory. Diffusion computations for specific ions suggest important observational tests of the theory. Megessier (1984, AA 138, 267) computed the time variations of the silicon abundance and distribution over the surface of magnetic CP stars with effective temperatures of 12000 K and 14000 K. Silicon overabundances occur over the entire surface of a young magnetic star; in middle age the equatorial regions become Si-free; and Si will be concentrated at the poles of the oldest magnetic stars. Alecian and Grappin (1984, AA 140, 159) studied the temporal building of density inhomogeneities due to the diffusion of a single ion in CP stars. A transient phase occurs prior to the final equilibrium state during which time significant temporal and spatial fluctuations occur. Borsenberger et al. (1984, AA 139, 147) investigated the radiative acceleration of Be, Mg, and Ba in HgMn stars. They predicted overabundances of Be and Ba and underabundances of Mg. A possible merger of the diffusion and magnetic accretion theories was proposed by Havnes and Goertz (1984, AA 138, 421). Neutral diffusion may lead to elemental segregation and overabundances of several heavy elements in some parts of the stellar magnetosphere, with sporadic violent distortions and re-combination of magnetic field lines accelerating particles to the stellar surface.

Moss (1984, MNRAS 209, 607) developed time-dependent models for the structure of rotating magnetic stars, based on a two-dimensional time-dependent solution of the MHD equation. Many of the restrictions previously found for static models disappear. Carpenter (1984, ApJ in press), continuing work begun by Stepien and Muthsam, explored the influence of magnetic field lines on the structure and predicted spectra of model atmospheres with normal composition. Some effects attributed to overabundances may be partly due to Zeeman broadening of the individual lines analyzed, Zeeman broadening of UV lines contributing to line blanketing, and contributions of magnetic pressure to the hydrostatic equilibrium of the atmosphere.

A substantial body of data now support the existence of a new type of variability among the CP stars - rapid oscillations of the surface layers. Kurtz (1984, MNRAS 209, 841 and references therein) has found nine cool, magnetic CP stars which vary photometrically with periods of 6-14 minutes and with amplitudes in the B-band of 0.002-0.016 mag. These stars appear to pulsate in non-radial modes of very high overtone, with the pulsation and magnetic axes co-aligned. Kurtz suggests an "oblique pulsator" model from which he derives angles of inclination of the magnetic and rotational axes and rotational periods independent of the spectroscopic and magnetic oblique rotator models. Shibahashi (1983, ApJ Lett 275, L5) and Cox (1984, ApJ 280, 220) propose an explanation of the rapid oscillations in terms of magnetic overstability in a superadiabatic shell in the magnetic polar regions of the star. Weiss (1983, AA 128, 152) discusses the existence of a second group of rapidly varying magnetic CP stars with about five members - low overtone pulsators of the δ Scuti type with periods of 1-2 hours.

Adelman and Pyper continued their extensive scanner measurements of CP stars (cf. 1983, AA 118, 313) and attempted to relate them to Stromgren and other photometric systems (cf. 1984, AA Suppl 55, 479). Hauck and North (1982, AA 114, 23) investigated the photometric properties of CP stars in the Geneva system. The peculiarity parameter Δ(V1-G), the rotational velocity, and the effective magnetic field strength are found to correlate with one another. Maitzen (1984, AA 138, 493) discovered that the magnetic field strength and the λ5200 feature appear to be correlated in He-weak stars. Efforts continue to reduce and improve photometric periods of CP stars (e.g. Manfroid and Renson 1983, AA Suppl 51, 267).

Lane and Lester (1984, ApJ 281, 723) derived effective temperatures for several classical Am stars using Kurucz line-blanketed model atmospheres and ultraviolet energy distributions. Their results are systematically cooler than previous determinations. They also noted that blanketed models with solar composition are uniquely able to match simultaneously the observed optical and UV flux distributions of numerous Am stars, suggesting that these stars may be Ca-deficient rather than metal-rich. Nicolet (1983, AA Suppl 51, 245) used the Geneva photometric parameter m to select Am star candidates. Wiertz and van Genderen (1983, AA 121, 35) discussed the properties of Am stars in the Walraven photometric system.

Infrared measurements of magnetic CP stars were made by Groote and Kaufmann (1983, AA Suppl 53, 91; 1984, AA 130, 184). But neither Bonsack and Dyck (1983, AA 125, 29) nor Odell and Lebofsky (1984, AA 140, 468) support their finding of infrared excesses at 5μm in these stars.

With the extended life of the IUE satellite, ultraviolet spectroscopy of CP stars moved rapidly from qualitative surveys to detailed quantitative analyses - a trend that will surely continue with the expected launch in 1986 of the Hubble Space Telescope and its High Resolution Spectrograph. Jacobs and Dworetsky (1982, Nature 299, 535) identified in the extreme HgMn star HR 7775 numerous lines of Bi, whose only stable isotope is the heaviest in the period table and measured its abundance to be $\simeq 10^6$ x solar. Bi was not found in the 12 other HgMn stars they examined. Ultraviolet lines of Pt II in IUE spectra of HgMn stars were analyzed by Dworetsky et al. (1984, Phys Scr T8), who found a Pt abundance log A $\gg$ 6. Leckrone (1984, ApJ 286, 725) studied the ultraviolet resonance lines of Hg II in IUE spectra of CP stars and found abundances for five HgMn stars in the range log A = 6.0-6.4, consistent with values obtained from Hg I λ4358. The Hg II λ3984 line was found to be poorly suited to abundance studies. Ultraviolet resonance lines of Al II and Al III in 35 CP and 7 normal stars were observed by Sadakane et al. (1983, ApJ 274, 261). Aluminum is, in general, underabundant in CP stars, in apparent disagreement with predictions of diffusion theory. Sadakane (1984, PASP 96, 259) reported anomalously strong C IV and Si IV resonance lines, and an anomalous ratio of C IV to Si IV line strengths, in the IUE spectra of the Bp star, 36 Lyn. IUE data for the remarkable CP star HD 101065, Przybylski's star (Wegner et al. 1983, ApJ 272, 646), demonstrate the clear presence of singly-ionized iron-peak elements. This observation, combined with the ultraviolet flux distribution, implies that HD 101065 has an effective temperature between 7500 and 8000 K, which is hotter than found in some previous studies. Application of the technique of wavelength coincidence statistics by Bord and Davidson (1982, ApJ 258, 674; 1982, AA 111, 362) and Chojnacki et al. (1984, ApJ 286, 736) to IUE spectra showed promise of revealing the presence of interesting species and establishing statistical abundances. Lanz (1984, AA 139, 161) used IUE low dispersion spectra to derive semi-empirical bolometric corrections for CP stars whose flux distributions are distorted by enhanced UV blanketing and backwarming effects.

Interest continues in the remarkable long period variable HR 465, which according to Cowley and Rice (1981, Nature 294, 636), was returning to rare earth maximum when observed in late 1981. Guthrie (1982, Astrophys Space Sci 85, 315) found Cr to vary out of phase with the rare earths and also with respect to Sc, Ti, V, and Ni. The Cr/Fe ratio is abnormally large at rare earth minimum and abnormally small at rare earth maximum. Scholz (1983, Astrophys Space Sci 94, 157) established HR 465's magnetic period as 23.3 years and found that the extrema in the absorption line strengths occur at magnetic cross-over.

The current University of Michigan program to reclassify all the HD and numerous fainter stars on the MK system, using an objective prism all-sky survey, is yielding an abundance of new candidate CP stars as well as other interesting objects (Bidelman 1983, AJ 88, 1182).

Didelon's (1983, AA Suppl 53, 119) new catalogue of published magnetic field measurements covering the years 1958-1981 will be valuable to workers desiring data on specific stars or interested in the statistics of magnetic fields. Borra and Landstreet (1983, ApJ Suppl 53, 151) detected fields in 12 He-weak CP stars which appear to be hot extensions of the magnetic Ap stars. But they were unable to detect fields at the 100-400 gauss level in any of the P- and Ga-rich B stars such as 3 Cen A. These latter stars may be hot extensions of the HgMn group.

Walborn (1983, ApJ 268, 195) presented a homogeneous list of He-rich stars, to visual magnitude 10-11. The frequency of slowly rotating He-rich stars is similar to that of normal early B stars, so that slow rotation does not appear to be a fundamental parameter of the He-rich phenomenon.

Cowley and Adelman (1983, QJRAS 24, 393) discussed the status of abundance analyses of CP stars as well as the prospects for substantial improvements. Fine analyses of specific CP stars which attempt to implement some of their suggestions include those by Adelman et al. (1984, MNRAS 206, 640), Adelman (1985, AA and AA Suppl, in press), and Cowley and Muthsam (1984, AA 130, 348). Boesgaard et al. (1982, ApJ 259, 723) determined beryllium abundances from observations of the Be II resonance doublet near λ3130 in normal and HgMn stars. 75% of the HgMn stars, particularly the hotter ones, showed values at least 20 times solar. Guthrie (1984, MNRAS 206, 85) surveyed the abundances of 14 elements in the photospheres of 20 sharp-lined HgMn stars. By using mainly weak lines, he hoped to better define the abundance anomalies of this class. Johansson and Cowley (1984, AA 139, 243) identified numerous Fe II with lower excitation potentials of at least 10 eV in several CP stars. Many other high level transitions are probably present in their spectra. Cowley (1984, Phys Scr T8, 28) reviewed the rare earth abundance determinations of CP stars and evaluated the status of rare earth partition functions.

Goncharskij et al. (1982, Astr Zh 59, 1146; 1983, Astr Zh 60, 83) and van Rensbergen et al. (1984, AA 136, 31) have attempted to map the surface distributions of elements from time dependent observations. Higher signal-to-noise observations with good phase coverage are needed to resolve ambiguities in the results.

Maitzen and his colleagues (cf. 1984, AA 138, 189) as well as Borra et al. (1982, AA 111, 117) continued the search for CP stars in clusters and in the field by obtaining of the measured Δa index. North (1984, AA Suppl 55, 259) has determined periods for 25 CP stars in clusters or associations from Geneva photometry. One of these stars, Cox 38, in NGC 2516 is an eclipsing binary. Studies of CP stars in binary systems included orbital determinations by Dworetsky (1982, Observatory 102, 138) and Stickland et al. (1984, Observatory 104, 74). Abt and Cardona (1984, ApJ 276, 266) classified stars with Ap or Am primaries and found an apparent excess of Am secondaries. According to Lamb et al. (1983, ApJ Lett 274, L71) the magnetohydrodynamic torque due to electric currents produced by the asynchronous rotation of magnetic stars in close binary systems can lead to rotational synchronization in binaries with a magnetic CP star component.

Cowley et al. (1982, ApJ 254, 191) suggested that weak-lined stars including λ Boo stars may represent a sizable fraction of the late B and early A stars. Hauck and Slettebak (1983, AA 127, 231) proposed a new spectroscopic definition of the λ Boo class and showed how their intermediate band colors could be used to distinguish them from other types of stars with similar temperatures. Baschek et al.

(1984, AA 131, 378) observed several classical and suspected λ Boo stars with IUE and found inconsistencies between ultraviolet and optical classification criteria.

9. Working Group on Standard Stars -- L. E. Pasinetti

This report covers the period 1 January 1982 - 1 October 1984.

The Working Group deals with the problem of identifying "normal" standard stars and providing standard reference data for various problems in stellar astronomy. A Standard Star Newsletter (SSN) edited by Pasinetti at the Department of Physics of the Milan University, is published twice a year since November 1982. SSN also serves as an official voice of the IAU Working Group on Standard Stars; it is currently distributed to members and, on request, to interested colleagues and institutes.

The Working Group participated in organizing the IAU Symposium No. 111 "Calibration of Fundamental Stellar Quantities." By the time this report is distributed, the Proceedings of the Symposium will be published, edited by Hayes, Pasinetti and Philip. The colleagues can find in the 20 review papers reported in them, recent data and exhaustive discussions on the topics. Other problems are also reported in the 65 contributed papers. As the Symposium No. 111 has provided a detailed overview of the subject, I will not report all the results appeared in the last three years, but I will give some additional information.

Calibrational lists are published in The Astronomical Almanac. Preliminary lists of standard stars were published in Bull Inf CDS (Strasbourg) by Breger and Hsu (1982, No. 23, 51) Batten, Glushneva, Keenan, Fracassini et al. (1983, No. 24, 3, 7, 19, 31, respectively). A microfiche concerning standard stars in various systems, to be included with the Proceedings of the Symposium No. 111, has been prepared by Philip and Egret.

Comparison stars which turn out to be variable are regularly reported in the SSN. The listed references are also sent to IBVS.

SSN is covering regularly the literature of all the countries since October 1984. Besides the papers already quoted in the past Newsletters, the following ones are relevant for our Working Group:

9.1 Standard stars and standardizations. Lists of standard stars (also for use on large telescopes) for JHKL, H_2O and CO photometry are given by Elias et al. (1982, AJ 87, 1029, 1983). Taylor (1982, ApJ Suppl 50, 391) has presented new spectrophotometric data in the range λλ4040-6180 for the standards of the Spinrad-Taylor scanner photometric system and has made a calibration to absolute units. Koornneef (1983, AA Suppl 51, 489) has made a homogenization of near-infrared data from southern bright stars and gives a list of 290 stars and data suitable for standardization purposes. Landolt (1983, AJ 88, 439) has provided new UBVRI standard stars, available to telescopes of a variety of sizes in both hemispheres. Batten et al. (1983, Publ Dominion Astrophys Obs 16, 143) observed most of the 60 stars selected as velocity standards in 1955 by the IAU. A list of stars which are not suitable as standards is given. McAlister and Hartkopf (1983, PASP 95, 775) have proposed a list of 26 binary-star systems as standards for interferometric observations of binary stars. Relationships between the Johnson and Kron-Cousins VRI photometric systems have been investigated by Fernie (1983, PASP 95, 782) using the new Johnson UBVRI standards appeared in The Astronomical Almanac. Trefzger et al. (1983, AA 117, 347) report on an empirical calibration of the RGU system and give colors of a first set of standard stars.

9.2 Energy distributions. New absolute energy distributions for seven standard stars in the region λλ3200-7600 Å are derived by Glushneva and Ovchinnikov (1982, Sov Astr 26, 548) from spectrophotometric comparison with α Lyr. Moreover, for the same stars and α Lyr, Glushneva (1983, Sov Astr 27, 326) gives effective temperatures, bolometric corrections, radii and surface gravities. The following papers are devoted to studies of bright and faint (6-8 mag) standard stars: Ipatov 1982, Astr Zh 59, 607; Ipatov 1981, Astr Zh 58, 569; Depenchuk and Komarov 1982, Astrometr Astrofiz No. 46, p. 15. A determination of the absolute monochromatic flux from Vega at λ2.20

and 3.80 μm has been made by Selby et al. (1983, MNRAS 203, 795) by comparison with a furnace. Oke and Gunn (1983, ApJ 266, 713) give absolute fluxes of four F type subdwarfs, intended as secondary standards for absolute spectrophotometry. Energy distributions for 18 stars intended as flux standards for large southern hemisphere telescopes are given by Stone and Baldwin (1983, MNRAS 204, 347).

9.3 Polarization. Bailey and Hough (1982, PASP 94, 618) present UBVRIJHK polarization measurements of standard stars with large interstellar polarizations and give normalized interstellar polarization curves. Hsu and Breger (1982, ApJ 262, 682) made linear polarimetric measurements to improve the system of standard polarized stars. Polarimetric variability was found in three stars.

9.4 Variability in "standards." Some evidence for variation of the spectrophotometric standard 109 Vir is reported by Taylor (1982, PASP 94, 663). Philip and Hayes, on the contrary, have not found evidence for any variations over the period 1969-1980. McClure (1983, PASP 95, 201) gives results on HD 184467, a double-line spectroscopic binary, which was an IAU velocity standard.

Abbreviations

AA .... Astronomy and Astrophysics
AA Suppl .... Astronomy and Astrophysics Supplement Series
AJ .... Astronomical Journal
ApJ .... Astrophysical Journal
ApJ Suppl .... Astrophysical Journal Supplement Series
Astr Tsirk .... Astronomicheskij Tsirkulyar
Astr Zh .... Astronomicheskij Zhurnal
Astrofiz .... Astrofizika
Astrometr Astrofiz ...... Astrometriya i Astrofizika
Astrophys Space Sci ..... Astrophysics and Space Science
Bull Inf CDS .... Bulletin d'Information du Centre de Données Stellaires
Fund Cosmic Phys .... Fundamentals of Cosmic Physics
IBVS .... Information Bulletin on Variable Stars
Izv Crim .... Izvestiya Ordena Trudovogo Krasnogo Znameni Krymskoj Astrofizicheskoj Observatorii
MNRAS .... Monthly Notices of the Royal Astronomical Society
PAS Japan .... Publications of the Astronomical Society of Japan
PASP .... Publications of the Astronomical Society of the Pacific
Perem Zvesdy .... Peremennye Zvezdy
Phys Scr .... Physica Scripta
Pisma Astr Zh .... Pis'ma v Astronomicheskij Zhurnal
QJRAS .... Quarterly Journal of the Royal Astronomical Society
Rev Mex Astr Astrofis ... Revista Mexicana de Astronomia y Astrofisica
Sov Astr .... Soviet Astronomy

J. Jugaku
President of the Commission

30. RADIAL VELOCITIES (VITESSES RADIALES)

PRESIDENT: A. G. Davis Philip
VICE-PRESIDENT: J. Andersen
ORGANIZING COMMITTEE: A. Batten, M. Duflot, D. Hube, M. Mayor, J. Sahade

1. INTRODUCTION

This report covers the period June 1981 to June 1984 and includes some material from IAU Colloquium No. 88, October, 1984). The field of radial velocities has undergone a renaissance in the last few years as the new radial velocity machines have come into use. Not only can stars of much fainter magnitudes be reached, but the precision of the measured radial velocity has been increased by orders of magnitude. Instead of speaking of velocities accurate to kilometers per second it is now possible to measure velocities to tens of meters per second. Research programs, involving these new techniques are now underway involving the study of hundreds, and in some cases, thousands of faint stars.

At the Patras General Assembly Commission 30 meetings Philip proposed that an IAU Colloquium on Stellar Radial Velocities be held to discuss the new techniques and the results obtained with them. This proposal was subsequently ratified by the Executive Committee of the IAU as Colloquium No. 88 and was held at Union College, Schenectady, New York in October, 1984.

At the 16th General Assembly in Grenoble the commission set up two working groups, one dealing with the problems of compiling a Mean Radial Velocity Catalog from heterogeneous sources of material and the other dealing with the problems of standard radial velocity stars. The working groups met in Schenectady, at IAU Colloquium No. 88. M. Barbier reported that work was progressing well on the Bibliographic Catalog of Stellar Radial Velocities. A magnetic tape containing the catalog will be available in the next year. Barbier and Duflot have been working on the Catalog of Mean Radial Velocities and they are currently working on radial velocities published during 1970 - 75. As is reported in the section on Standard Stars, Batten recommends that the present system of IAU radial velocity standards be revised completely.

2. INSTRUMENTATION

Campbell and Walker (1984, IAU Coll. No. 88) list stellar precision Velocity Techniques in their Table 1 (shown on the next page).

Work continues with Griffin's original radial velocity spectrometer (Cambridge and Palomar Observatories). (30.120012 31.114008 32.120007 32.120009 32.120018 32.120025 34.120010 34.120011 34.120012 34.120026 88.09 Papers 54-60 in Observatory, 10-11 in J. Astrophys. Astr.).

Major additions to the numbers of stars with measured radial velocities have been made by CORAVEL (Haute-Provence Observatory and ESO, La Silla) and by the stellar radial velocity program at the Center for Astrophysics, (MMT Observatory, Whipple Observatory and Oak Ridge Observatory), the LPL Radial Accelerometer, the McDonald observatory Radial Velocity Spectrometer and radial velocities from CCD detectors (Kitt Peak National Obs.). The radial velocity program at Haute-Provence Observatory, using the objective prism continues, with a present focus on measuring the radial velocities of Hipparcos stars.

TABLE 1

Technique	Investigators	Velocity Range m/s
Telluric Lines + Phot. Spectra	Griffin and Griffin (1973)	50
Telluric Lines + Reticon	M. Smith (1982, 83)	6-11
PEPSIOS (3 Fabry-Perots)	Traub, Mariska, Carleton (1978)	3-46
Fabry-Perot Interferometer	Reay, Ring and Pietraszewski (1983)	4-20
Fiber image scrambler + Fabry Perot + Echelle + CCD	McMillan et al. (1984)	(15)
Fabry-Perots + Reticons	Cochran and Young (1984)	(1)
N20 absorption cell + FTS	Hall and Hinkle (1981)	20
HF absorption cell + Reticon	Campbell and Walker (1978)	12
Vacuum echelle spectrog. + interferometer	Connes (1984)	(?)
Optical fiber + Coude spectrog.	Heacox (1984)	(50)

Development work has been started towards a combined observing instrument and reduction package which will give on-line reduced velocities for a wide range of stars of spectral types OBAF (J. Andersen, Copenhagen and G. Hill, DAO, Victoria).

3. GALACTIC STRUCTURE

Programs on galactic structure are being conducted in a collaboration among the Geneva, Mareseille, Copenhagen, ESO and Lund observatories, based on the two CORAVEL scanners in both hemispheres. Throughout this and following sections, the persons named are the proposers/coordinators for each program.

The radial velocities of stars from the catalog by Woolley et al. (Stars less than 25 pc from the Sun) are being determined by M. Imbert (Marseille) with CORAVEL. Ardeberg (Lund) and Lindgren (ESO) are determining radial velocities of faint metal-deficient F, G, and K stars.

Andersen and Nordstrom (Copenhagen) have observed the remaining 790 late-type stars in the Bright Star Catalog with CORAVEL and another 740 B - F4 stars have been observed photographically at ESO. The results have been published (Andersen and Nordström 1983 33.0020045 33.111009 33.111010 34.111003 88.14 Andersen et al. 1984 Astron. Astrophys., in press). Preliminary data for some 700 stars were supplied for inclusion in the 4th edition of the Bright Star Catalog.

Current CORAVEL programs (Andersen, Nordström and Olsen, Copenhagen; Mayor, Geneva) include some 4000 F stars in the whole sky with known metal abundances (and ages) in order to study the evolution of the dynamical properties of this group.

Florsch, Marcout and Valbousquet have extended the search for high velocity stars, using objective prism plates taken with the Schmidt telescope at La Silla. In two 2X2 degree fields near the SMC only one high velocity star was found, to a limiting magnitude of 11.5 (88.15). This indicates that foreground high-velocity stars are probably rare among stars found in the direction of the Magellanic Clouds.

At Haute-Provence Observatory measures of radial velocities of stars in fields at galactic latitude +30° were continued with the objective prism on the Schmidt telescope. A study of the precision of the radial velocities obtained

(Fehrenbach, Burnage and Agniel 1982 32.111001) shows that with at least three plates per field, the precision was 4 km/s for all spectral types; these radial velocities were in the same system as that of the IAU standards. These measures are continuing in a program to measure radial velocities for Hipparcos stars. This accuracy is sufficient for the statistical study of stellar movements in the galaxy.

Fehrenbach and Duflot (88.18 88.19) have obtained 260 plates of 4X4 degrees with the small objective prism (PPO) of the OHP (magnitude 10 obtained in 2 X 10 min) and 132 plates of 4X4 degrees (magnitude 10 obtained in 2 X 15 min) taken with the large objective prism on the Franco-Liegois telescope (SPO). This work has given about 12,000 spectra or the equivalent of 3,000 radial velocities, with four measures for each star.

For peculiar stars (Ap and Am) and isolated stars which do not justify the taking of a Schmidt plate, a program is underway to obtain individual spectra at 80 Å/mm with the 120 cm telescope at OHP. 2,500 spectrograms have been obtained. A study of 55 stars of large radial velocity in the direction of the Magellanic Clouds (Fehrenbach and Terzan 1984) found four stars to be subdwarfs and fourteen stars had large (U-B) excesses.

At the Copenhagen Observatory, Andersen and Nordström have been working on the completion of the radial velocities for the Bright Star Catalog. The B and F5 - M stars have been published and the measurements of the A0 - F4 stars have been completed and will be submitted for publication by the end of 1984. All together some 1,500 southern stars have been completed on this program, 740 stars having been observed photographically and 790 with CORAVEL.

Philip (1984 88.43) has combined four-color photometric data with radial velocity data for early-type stars at high galactic latitudes and finds he can classify stars as members of Pop I, Pop II and an intermediate population and confirm these photometric classifications by means of the velocity dispersions for each group. Similar trends of the velocity dispersion increasing with later population type have been obtained by Ardeberg and Lindgren (1984 88.13) and Adamson et al. (1984 88.42).

4. STAR CLUSTERS

Griffin and Gunn have had one observing run with their spectrometer on the Palomar 5 m telescope each year to continue a survey of radial velocities of candidates for Hyades membership. Observing time for this project has also been granted to Griffin at the DAO.

240 radial velocities measured with CORAVEL for 169 stars in the globular cluster 47 Tuc have been published (Mayor et al. 1983 34.154056) These data were used to show that the cluster rotates (Mayor et al. 1984 Astron. Astrophys. 134, 118). Thanks to complementary observations of stars near the center of the cluster, the velocity field V(r,z) and the radial variation of the velocity dispersion have been established.

Similarly, the dynamics of Omega Cen have been investigated, on the basis of about 400 observations for 300 stars (Meylan and Mayor 1984 IAU Symposium No. 113, Mayor and Meylan 1984 88.52). The mean precision of the radial velocity measurements is 0.6 km/s in the case of 47 Tuc and 0.9 km/s for Omega Cen.

R. Peterson (1984 88.44, 88.45) has determined that the velocity dispersion for nine BHB stars in M 4 is ±1.9 km/s. This low value has implications that the mass-to-light ratio in M 4 may be less than solar. Spectra, from 3,800 - 5,200 Å have been obtained of stars in 20 globular clusters with the R-C spectrograph on the KPNO 4 m telescope. Eight remote globular clusters are included in the sample. Radial velocities for each star are being measured on a single axis Grant engine.

J. Mermilliod and Mayor have continued their survey of red giants (800 stars) in the fields of 150 open clusters, primarily for membership determination. Approximately 150 binaries have been discovered and 50 stars are suspected variables. Fifteen orbits have been obtained. They also pursued the observations of about 500 F5 - KO dwarfs in several open clusters (Pleiades, Praesepe, Coma Ber, Alpha Per, IC 2391, IC 2602, NGC 6675 and Beta Scl). Twenty-five orbits have been determined among the numerous binaries discovered. As a byproduct of the survey, v sin i values have been determined in the Hyades, Pleiades, Praesepe and Coma Ber clusters. The discussion of these data and the comparison of the distributions have been published (Benz, Mayor and Mermilliod 1984 Astron. Astrophys. 138, 93).

About 100 stars in the corona of Praesepe, selected from the proper motion studies by Artjukhina (1966) have been tested for membership, with an appreciable rate of success. Membership has also been confirmed by photometric data obtained by E. Weis (Van Vleck Obs.).

In the Orion Ic association, stars with low v sin i values from the sample of Smith et al. (1983 Astrophys. J. 271, 237) have been observed for membership determination. Many stars with low v sin i are non-members and the v sin i distribution of the remaining stars is notably different from that obtained by Smith et al.

The orbits derived for stars in open clusters (dwarfs and giants) have been used to determine the initial eccentricity distribution (Mayor and Mermilliod 1983 V'eme Journee de Strasbourg, p. 45). This distribution differs completely from that of the visual binaries. The orbital parameters have also been used to test the circularization timescale by tidal forces (Mayor and Mermilliod 1984 IAU Symposium No. 105).

At the Harvard Smithsonian Center for Astrophysics echelle spectrographs have been used to measure the radial velocities of more than 10,000 stars with accuracies better than 1 km/s. Part of these data have been employed in the study of the dynamics and evolution of open and globular clusters.

Harris and McClure (1984 88.26) (DAO) have made a study of the radial velocities of about 50 giant-branch stars in NGC 188. Multiple observations of about 40 of these stars have revealed that more than 15% are binaries.

5. GALAXIES

A major accomplishment at the Harvard Smithsonian Center for Astrophysics was the completion of the the first phase of the CfA Redshift Survey, with results published for 2,400 galaxies brighter than a B magnitude of 14.5 and at high galactic latitudes (Huchra, Davis, Latham and Tonry 1983). The data from the CfA Redshift Survey and its extensions were used for a variety of applications, such as a study of the large-scale structure of the universe within 50 megaparsecs of the Sun (Davis, Huchra, Latham and Tonry 1982 Astrophys. J. 253, 432); several studies of groups and clusters of galaxies (e.g. Huchra and Geller 1982 Astrophys. J. 257, 432; Geller, Beers, Bothun and Huchra 1984 Astron. J. 89, 319; Beers, Geller, Huchra, Latham and Davis 1984 Astrophys. J. 283, 33; and a reanalysis of the Turner binary galaxies (White, Huchra, Latham and Davis 1982 MNRAS 203, 701). In collaboration with the Observatorio Nacional de Brasil, work was begun on the extension of the northern redshift survey to cover the southern skies (da Costa, Pellegrini, Nunes, Willmer and Latham 1984 Astron. J. 89). Altogether the facilities at

Mt. Hopkins have now been used to measure more than 6,000 galaxy redshifts.

Late-type supergiants in the direction of the Magellanic Clouds are being observed with CORAVEL. In the SMC, 232 F-M type stars have been measured, 207 of which are most probably SMC members (Maurice et al. 1984 Astron. Astrophys. Suppl. 57, 275). For the LMC, the corresponding numbers are 451 and 415 (Prevot et al. 1984 Astron. Astrophys. Submitted). 17 Cepheids in both clouds are being observed in order to determine their radial velocity curves (See section 7.)

V. S. Popov and G. D. Polyakova (Pulkova Observatory) report on radial velocities determined for 15 galaxies (photographic magnitudes 11.0 - 15.4) with the 6 meter telescope of the Special Astrophysical Observatory. Relative to the velocities in de Vaucouleurs's Second Reference Catalogue their velocities were about 275 km/s larger. A large difference was found in the case of NGC 7385 (88.53).

6. BINARIES

Griffin has continued to determine the orbits of spectroscopic binaries; about 70 orbits have been published during the triennium.

About 200 northern F-M supergiants have been observed with CORAVEL during the past 7 years. Nineteen new spectroscopic binaries have been discovered. The rate of detected SB among northern F-M1 stars is at present 21% without any dependence on spectral type or luminosity class. The total rate of binary systems among the same stars is in the range of 31-38% (Burki and Mayor 1984 IAU Symposium No. 105). A similar program for the southern supergiants has been undertaken from La Silla. Forty spectroscopic and eclipsing binaries are being observed continuously with CORAVEL (M. Imbert)

Using the techniques developed for measuring galaxy redshifts, the Harvard Smithsonian Center for Astrophysics have obtained radial velocities of stars in order to study the frequency and characteristics of binaries in a variety of populations, such as clusters, the disk, and the halo. The discovery of wide binaries with separations as large as 0.1 parsec is reported by Latham, Tonry, Bahcall, Soneira and Schecter (1984 Astrophys. J. Lett. 281, L41), while observational evidence that all the blue stragglers in M 67 are binaries is given by Peterson, Carney and Latham (1983 Astrophys. J. 279, 237) and Beers, Geller, Huchra, Latham and Davis 1984 Astrophys. J. 283, 33). Latham, Hazen-Liller and Pryor (1984 88.29) report on the discovery of a spectroscopic binary in the globular cluster M 3. The radial velocities for von Zeipel 164 varied with an amplitude of over 18 km/s and a period of a few years.

7. VARIABLE STARS

M. Imbert (Marseille) is observing 21 northern galactic Cepheids with B magnitudes brighter than 11th mag. using CORAVEL.

Two very small amplitude Cepheids (HD 9250 and HD 14462) have been discovered and the mean radii have been calculated (Burki and Benz 1982 32.122081). The monitoring of the strange classical Cepheid HR 7308 continues with CORAVEL at Haute-Provence Observatory. This star, which has the shortest period known for a classical Cepheid and whose amplitude varies by over a factor of 16 in 1200 days, is probably pulsating in the second radial overtone (Burki et al. 1982 31.122189). On the basis of new radial velocity light and color curves, the mean radii have been derived for 9 short period Cepheids (Burki). CORAVEL radial velocity curves have been obtained for four RR Lyrae stars: RR Cet, DX Del, BS Aqr and DY Peg. By using simultaneous photometric observations the mean radii of these stars have been calculated (Burki et al. 1984 IAU Colloquium No. 82).

CORAVEL radial velocity curves have been completed for six Cepheids in the LMC and their radius variations have been determined (Imbert et al. 1984 Astron. Astrophys. Suppl. Submitted).

8. STANDARD STARS

Batten and others at Victoria have continued to observe IAU standard-velocity stars to check their constancy. A first discussion has been published (Batten et al. 1983 Pub. DAO 16, 143). Batten also discussed the whole concept of standard stars (with particular reference to standard-velocity stars) at IAU Symposium No. 111 (1984) and proposed that the Commission should consider a radical revision of the system of velocity standards (Batten 1984 88.36).

Scarfe has continued to make radial velocity observations using the DAO radial velocity spectrometer which has lead to a concern with the system of standard stars (1982 32.034158, 1983 Pub. DAO 16, 143).

IAU and potential high radial-velocity standard stars are being observed from both OHP and La Silla with CORAVEL. At La Silla about 1,000 measures of IAU standards have been obtained. The analysis of these data resulted in bringing to light a zero-point difference between the brighter and fainter IAU standard subsets:

$$V_r \text{ (IAU bright)} - V_r \text{ (IAU faint)} = 0.8 \text{ km/s}.$$

In addition, four standards, (HD 36673, 44131, 115521 and 156014) showed important variations during the period from 1981 to 1984 (Mayor and Maurice 1984 88.33).

Prevot and Rebeirot have made a study of h and Chi Per and of the Perseus Arm. Anderson and Nordström (1984) have studied the standardization of radial-velocity observations and the influence of rotation.

9. ABSTRACT NUMBERS OF PAPERS 1981 -1983

In this section of the report the abstract numbers of all Radial Velocity papers in volumes 30 -34 of Astron. Astrophys. Abstracts are listed by topic. The papers to appear in the proceedings of IAU Colloquium No. 88 are indicated by 88.xx. (The abstract numbers and categories appear on the next two pages.) A more compete bibliography appears in the appendix of the proceedings of IAU Colloquium No. 88.

A. G. Davis Philip
President, Commission 30

31. TIME (L'HEURE)

PRESIDENT: G. Hemmleb VICE-PRESIDENT: D.D. McCarthy

ORGANIZING COMMITTEE: N.S. Blinov, C.C. Costain, H.F. Fliegel, B. Guinot, S. Iijima, Y.R. Miao, A. Orte, P. Paquet, I. Pilkington, E. Proverbio

INTRODUCTION

This report will try to review briefly the work achieved from 1982 to 1984 in different "subjects to be considered by Commission 31 Time" as adopted in Grenoble 1976. It contains also information provided by Commission members, for which hearty thanks are to be given. The limitation of space required the abbrevation of some institution reports.

MERIT

At IAU Symposium No 82 on "Time and the Earth's Rotation", organized by Commission 19 and 31 at Cadiz, Spain, May 1978 the Project MERIT was initiated. MERIT is a special programme of international collaboration to Monitor Earth-Rotation and Intercompare the Techniques of observation and analysis.

The project includes after an initial short campaign of observation (during 1980) the main campaign of observation from 1983 September 1 to 1984 October 31, the period of data analysis, and a review phase during which the results are to be discussed. It has been proposed that at IAU General Assembly 1985 a review of the scientific results should be presented which will provide the scientific background for the IAU/IUGG recommendations concerning future Earth orientation services.

A great number of observatories with different techniques (Astrometry, Doppler, SLR, LLR, VLBI, CERI) took part in the Project MERIT. The results enable better knowledge on the behaviour of Earth's rotation and consequently also of UT.

FUNDAMENTALS

At the IAU Symposium No 109 on "Astrometric Techniques" in Gainsville, Jan. 1984, there was a strong feeling among participants that the exact definitions of some of the entities which define (astronomical) space-time are in need of attention. The majority of these entities was conceived at a time when the precisions and accuracies which we can achieve at present were not even dreamed of, and most of these depend intimately on the kinematics of the Earth, whose recently uncovered complexities may make it desirable to redefine some of the current concepts (H.K. Eichhorn).

For all astronomers discussions on general concepts of reference systems are significant. Guinot has proposed quasi-ideal theory-independent reference frames for both space and time. As a result of the discussions in Gainsville it has been decided to create a Working Group on the "Fundamental Concepts of Astronomical Space-Time and their Realization".

Last year consultations took place between interested organisations (BIH, CIPM, IAU, IUGG, URSI, CCIR, BIH Directing Board) on the transfer of the activities concerning TAI from BIH to the BIPM. For this decision it is to be taken into consideration that TAI is based solely on physical measurements independent of astronomy, that TAI and UTC are of general interest as a basis for numerous technical and public applications and that therefore it is natural to entrust the intergovernmental organization in charge of the metrology of basic physical quantities, CGPM and its bodies (CIPM, BIPM) with TAI and UTC.

OBSERVATION OF UT AND ET ON A ROUTINE BASIS

Astronomical observations for UT0-UTC have been continued on a routine basis in participating observatories to BIH. The BIH has continued the regular computations of UT1 on the basis of a solution by optical astrometry, from the observations of 88 stations in 22 countries. Moreover, provisional raw and smoothed values of UT1-UTC are computed from the combination of the optical astrometry solution with the VLBI results obtained by the U.S. National Geodetic Survey and partly with the CERI results obtained by the U.S. Naval Observatory.

In order to increase the precision and the accuracy of optical astrometric observations in some observatories efforts have been made for the automatization of observation and for other improvements of the instruments (e.g. Chinese observatories, Mizusawa, San Fernando, Potsdam). USNO has completed a new 20 cm PZT and has installed it on the original site of PZT1.

Results on ET2 - TAI were obtained by the Hydrographic Department of Japan (JHD) for the epoch 1981.5, 1982.5, and 1983.5 as 30.60, 30.71, and 30.68 s respectively with the mean error of $\pm$0.04 s by use of the occultation timing data observed by JHD. The services of the International Lunar Occultation Center at JHD have been continued since 1981. Total data reported from 34 countries amount to 43,000.

CLOCKS, T-F STANDARDS AND TIME SCALES

In the last few years an increasing number of observatories with industrial cesium clocks have participated in the formation of TAI (33 laboratories in July-August 1984, operating 140 clocks), but laboratories with primary standards contributing to TAI have not increased. In some laboratories investigations on frequency standards have been made and new primary frequency standards have been constructed.

Highly accurate primary clocks (NRC, PTB) seem to be especially qualified for establishing TAI because of their accuracy and their long term stability. Most of the individual time scales based on sets of industrial c sium clocks show a seasonal variation with respect to the primary clocks of NRC and PTB. This effect seems to be produced partly by the LORAN-C time comparison, partly by environmental effects on the industrial clocks.

Since its introduction on 1972, January 1 the UTC system has gained widespread acceptance as the reference time scale throughout the world not only for practical and scientific application but also as the basis of legal time.

TIME SYNCHRONIZATION AND DISSEMINATION

Different methods have been used for the comparison of time scales: TV, LORAN-C,portable clocks, satellite techniques, VLBI. These methods differ in accuracy and precision, coverage, operational convenience, cost, and availability. Therefore no single system can satisfy all requirements. The necessary accuracy of world-wide time synchronization now is in the 10 to 1 ns level, in the next decade nano-second accuracy or better is required also for intercontinental time synchronization and comparison.

Most of the regular time comparisons between the laboratories contributing to the TAI were carried out by LORAN-C. For seven months in 1982 measurements NRC-OP and NRC-PTB via satellite Symphonie were used for the atlantic time link between North America and Europe. In 1983 GPS time comparisons using GPS time receivers have been introduced in some laboratories (NBS, USNO, OP/BIH, VSL, PTB, TUG, TAO and laboratories in Australia). The GPS time comparison permitted the start of an active participation of laboratories in Japan and PR China in the establishment of TAI.

Many experiments of time transfer or time synchronization via satellites have been made in last few years. New experiments are planned for the near future. One-way and two-way techniques are used, each of them offering certain advantages. For the one-way method simple equipment can be used but the accuracy is lower than for the two-way technique.

In 1982 the planned LASSO experiment with an estimated accuracy of time synchronization of 1 ns or better failed because of the loss of the carrier rocket. A new LASSO experiment is planned for June 1986.

For the purpose of calibrating propagation time and instrument delay when requested (e.g. LORAN-C, TV), the comparison of time scale by using portable atomic clocks is the most precise operational method over short and long distances. Commercial cesium clocks can give accuracies of about 10 ns.

RELATIVITY

In the field of time scales, time dissemination, and time synchronisation relativistic effects are of importance. The state-of-the-art navigation and communication systems, the nano-second accuracies of clocks need relativistic corrections. Rate changes because of the velocity of satellite- and aircraft transported clocks and of the gravitational potential must be taken into account.

REFERENCES

Because of the limitation of space only a list of major references concerning the work of the commission is given. Complete references to the reports of BIH, observatories and laboratories can be furnished separately on request.

1) Proc. of the 14th (15th) Annual Precise Time and Time Interval (PTTI) Planning Meeting 1982 (1983)

2) Advance Program of the 16th Annual PTTI Meeting 1984

3) BIH Annual Reports 1981, 1982, 1983

4) CCIR, XVth Plenary Assembly, 1982, Recommandations and Reports, Vol. VII, Standard Frequencies and Time Signals

5) BIPM, Comptes Rendus 17e Conférence Générale des Poids et Mesures, 1983

6) Documents on Commission A -"Electromagnetic Metrology", XXI URSI General Assembly, Florence 1984 (to be published)

7) Papers on the IAU Symposium 109, Astrometric Techniques, Gainsville, USA, Jan. 1984

Report of the Director of the Bureau International de l'Heure

The BIH is organized in two sections: (a) Time, (b) Rotation of the Earth, respectively under the scientific responsibility of B. Guinot and M. Feissel.

The following report will cover the activities on atomic time and some administrative and operational matters common to both sections. The BIH activities on the Rotation of the Earth and related problems are presented to Commission 19.

ATOMIC TIME SCALES

(a) Time comparisons and participating laboratories.

The International Atomic Time (TAI) and the Coordinated Universal Time (UTC) are reference time scales. But they can also be regarded as a worldwide system of coordination of clocks: to this respect, the time comparisons of remote clocks are crucial.

During the last three years, the major event was the implementation of the time comparisons by the Global Positioning System, GPS. The operational link established between NBS and USNO in June 1981 demonstrated that the accuracy of 10 ns on daily averages was realized, as announced by Allan and others. In 1983/84, many other laboratories acquired GPS time receivers (10 laboratories equipped in Nov. 1984) and, for the first time, the network of permanent and precise time comparisons extended to Far East and Australian laboratories. On July-August 1984, over 140 clocks, which entered in the TAI computations, 78 are directly linked by GPS. Most of the remaining clocks are linked to a GPS-equipped laboratory by common reception of the same LORAN-C station and thus benefit from the GPS advantages.

The following table shows the improvement of the worldwide coordination of time.

	July-Aug. 1981	July-Aug. 1984
Nr of clocks participating in TAI	105	140
Nr of Laboratories for which UTC-UTC(lab) is published in BIH Circular D	33	38
Nr of Laboratories the clocks of which enter in TAI computations	26	33
Nr of Laboratories for which TAI-TA(lab) is published in BIH Circular D	8	10

However some difficulties remain.

The potential accuracy of 10 ns by GPS can be realized by the common view mode, which is well organized by NBS. But this requires also that the delays are correctly handled. In practice, discrepancies of the order of 100 ns can be observed due to the lack of precise definition of UTC(lab), to misunderstandings concerning the cable delays, to the lack of precise measurements of the differential delays between GPS receivers. That shows:

- the possible need of upgrading the time equipment of laboratories in order to make it compatible with a few ns-seconds accuracy,
- the importance to continue clock transportation,
- the interest of a portable GPS calibration system (planned by USNO),
- the need of a good coordination concerning the adopted delays.

Concerning the last item, the BIH has begun to collect informations on the adopted delays. The BIH also suggested that the GPS coordination be a topic for a future meeting of the Consultative Committee for the Definition of the Second, to be held in 1985.

In the current work the BIH uses either its own evaluations of the UTC(i)-UTC(j) from the GPS common views, either evaluations from other laboratories. In this domain also a coordination is desirable in order to avoid duplication of work.

Studies of the statistical properties of the GPS time comparisons have been made, especially thanks to the loan to BIH of GPS time receivers by NBS and by Stanford Telecommunications Inc. (these instruments being operated in cooperation with the LPTF at the Paris Observatory).

(b) Computation of TAI and UTC.

No changes have been made in the BIH algorithms. The free running time scale EAL is derived from all the clocks. Then the second of EAL is compared to the realization of the second by the primary standards and a steering process is used to derive TAI from EAL, implying frequency steps of 0.2×10^{-13} when needed.

The primary clocks PTB CS1, NRC Cs V and VI-A, -B, -C, are used both as clocks with an input in EAL, and as frequency standards for the steering. The frequency standards functioning in a discontinuous mode, NBS-5 and RRL Cs1 are used for the steering only.

The long term stability of the participating clocks still continues to improve, as shown by the percentage of clocks reaching the (unchanged) maximum weight accepted by the BIH algorithm: 35 % over January - August 1984, instead of 28 % over January - August 1981.

A number of studies on the time scale algorithms has been performed, but did not enter in the practice of the TAI computations. The possibility of modeling annual frequency variations of the clocks is being considered and will be tested when a sufficient amount of past data obtained by the GPS will be availabe.

The alignement of BIH results with the GPS data and the clock transportation in 1983 was attempted, as explained in a note "Use of GPS time comparisons for establishing TAI in 1983 and related problems", which is available on request.

The operation of the UTC system raised no special problems.

DATA PROCESSING AND DISSEMINATION OF RESULTS

Much effort was devoted to the simplification of the current work of both BIH sections. With the help of USNO, the General Electric Mark III system is used intensively. A set of mini-computers has been installed in the BIH offices at Paris Observatory; they are used for automated data transfers, local computations, and as terminals on several computation centers or systems. The edition of Circular D and of most of the pages of the BIH Annual Report has been fully automated.

The main series, Circular D and BIH Annual Report are sent to about 750 addresses. Circular D and other BIH results are available through the GE Mark III System.

BIH ORGANIZATION

In 1985, the Time section of BIH will be moved from the Paris Observatory to the Bureau International des Poids et Mesures (BIPM) in the outskirts of Paris. Computing facilities are being installed at BIPM, and arrangements are being made so that this move will affect neither the contributors to the BIH work, nor the users of its results. The Section of the Rotation of the Earth remains at the Paris Observatory where it receives an increased support.

ACKNOWLEDGEMENTS

There are so many participants to the BIH work, who not only contribute data, but help the service by constructive exchange of ideas, by sending visiting scientists, that it is not possible to thank individually each of them. However, we would like to mention some organizations to which we are especielly indebted for their help.

. USNO for loan of equipment, the administration of the GE Mark III catalog, the distribution in USA of BIH publication.
. NBS for loan of a GPS receiver and for the coordination of GPS common views.
. Stanford Telecommunications Inc., for the loan of a GPS receiver.

B. Guinot
Director

<u>Reports of Observatories and Laboratories</u>

A. BELGIUM. OBSERVATOIRE ROYAL DE BELGIQUE, UCCLE/BRUSSELS (ORB)
(Reported by P. Paquet)

The astronomical observations for the determination of UT are discontinued since January 1979.

The Doppler tracking station integrated into the DMAHTC Polar Monitoring Service has continued permanent operation. For the time scaling of the Doppler tracking network the satellite time marks are recorded during all passes. The standard deviation is of the order of 10 microseconds.

The Belgian time service is still provided with 2 HP 5061 A Cesium Standards, a visual Loran-C receiver operating on Sylt, and a TV receiver used for daily time comparisons between the Observatories of Brussels and Paris.

According the method used to realize the time synchronisation: Transit satellites, Loran-C or TV signals the precision is respectively 10, 0.20, 0.05 microseconds. In 1984 a receiver for time signals synchronisation via the GPS system has been ordered. Delivery is expected in late 1985.

B. BRAZIL. OBSERVATORIO NACIONAL, RIO DE JANEIRO (ONRJ)
(Reported by P. Mourilhe Silva)

Time keeping for legal time transmissions in UTC and all other national applications of Time and Frequency, is achieved with two HP 5061 A commercial Cesium Standards, (one standard tube and one high performance tube). Two Rubidium Standards TRACOR 304D and HP 5065 A are also kept at Rio de Janeiro installations of the Time Service. A third Rubidium, HP 5065 A is at a new station in Brazilia.

By line-10 TV method and clock transportation, the Cesiums and Rubidiums of other institutions are intercompared.

For global synchronization, some clock trips were made to Rio de Janeiro by USCG, ONBA and PTB.

The Time Service was qualified by INMETRO (National Institute of Metrology, Standardization and Industrial Quality), as the National Primary Laboratory for Time and Frequency measurements and basic researches.

The time reference was supplied during the last experiments of VLBI, from December 1983 to February 1984, between Itapetinga Radio Observatory (INPE) and foreign Observatories. For the future time transfer by this way is foreseen.

C. FEDERAL REPUBLIC OF GERMANY, DEUTSCHES HYDROGRAPHISCHES INSTITUT, HAMBURG (DHI)
(Reported by H. Enslin)
Observations of UT0 - UTC with the PZT - made from 1957.7 onwards without interruptions worth to be mentioned - have been continued; the results have been communicated weekly to the BIH and the IPMS.

The time-scale UTC(DHI) has been produced by a HP 5061 A Cesium Standard combined with a micro-stepper. It has been compared regularly with the time-scales of other institutions via Loran-C and TV. The DHI was included yearly in portable clock trips of the USNO. Time-signals have been transmitted regularly on 4 frequencies (2 MF, 2 HF).

D. FEDERAL REPUBLIC OF GERMANY. PHYSIKALISCH-TECHNISCHE BUNDESANSTALT, BRAUNSCHWEIG (PTB)
(Reported by K. Dorenwendt)
The PTB primary Cesium beam standard CS1 has been operated continuously generating the time scales UTC(PTB) and TA(PTB).

After two and a half year of duration the two-way time transfer between PTB, OP (France) and NRC (Canada) via the Symphonie satellite had to be stopped in August 1982. It could be shown that the primary standards at PTB and NRC agreed in frequency within the claimed uncertainties of 5.10^{-14}. A comparison of the time transfer across the Atlantic via Loran-C and via the Symphonie satellite revealed seasonal variations of the order of 1 μs in the Loran-C link.

From February to December 1982 one-way time transfer via the test satellite OTS-2 were executed between different timing institutes in Europe. Since July 1983 the satellite system GPS has been used for intercontinental time comparisons with an uncertainty of the order of 10 ns.

In June 1983 experiments with a pseudo-random phase noise (phase modulation ± 10°) on the frequency of the transmitter DCF77 have been started. First results showed that an attribution of the transmitted second markers to a period of the carrier can thus be achieved.

An experimental Cesium beam standard using atoms with an average velocity of 400 m/s was put into operation in 1982. The apparatus allowed to determine the phase gradient in a Ramsey resonator and the resulting uncertainty contribution to the frequency of a primary standard. The influence of field inhomogeneities on the frequency of Cesium beam standard will be investigated next. The work on three new primary standards is continued.

E. GERMAN DEMOCRATIC REPUBLIC. ZENTRALINSTITUT FÜR PHYSIK DER ERDE, POTSDAM (ZIPE)
(Reported by G. Hemmleb)

Regular time and latitude determinations are being continued with one astrolabe Danjon and one PZT. The results are being transmitted to the international centres BIH and IPMS and published in two-monthly booklets.

Special efforts were made to obtain a large number of observations during the MERIT Main Campaign. Beginning with 1984 January 1 all observations were reduced using the MERIT Standards. Also the observations during the period from 1983 September through December were re-reduced using the MERIT Standards.

The efficiency of the PZT 2, which is working since 1980, was estimated using the observation results from 1981. The results confirm that the PZT 2 is expecially suited for precise time and latitude determinations in geodynamic observatories.

The investigations for the development of a Photoelectric Zenith Tube (PEZR) were continued.

An independent time scale TA(DDR) is computed by the Amt für Standardisierung, Meßwesen und Warenprüfung (ASMW) using the data of 4 cesium standards of ASMW and ZIPE. Time scale UTC(ZIPE) is based on 1 HP cesium clock 5061 A. This scale is compared daily with UTC(ASMW) using TV method. UTC(ASMW) on its part is based on 3 atomic clocks (1 Ebauches B 5000 and 2 Oscilloquarz 3200).

For time comparisons with time services BIH, TP, PTB, and AOS the TV method is used. Receptions of LORAN-C station Sylt and phase tracking of VLF transmitters were also carried out.

F. ITALY. ASTRONOMICAL OBSERVATORY, CAGLIARI (CAO)
(Reported by E. Proverbio)

The Time Service of the Cagliari Observatory is based on 2 commercial cesium standards and 2 quarz clocks. During the period 1981-1984 the reference local time scale has been continuously compared by mean VLF and Loran-C techniques. The UTC (CAO) scale was also compared via television pulses with IEN (Turin) and ISPT (Rome).

The accuracy of the UTC(CAO) scale versus UTC is about $1\text{-}3\times10^{-13}$. The result of time and frequency comparisons are published in the Monthly Bulletin of Cagliari Astronomical Observatory and in Circular D of the BIH.

Relativistic effects on physical time scales has been studied and particular techniques for time dissemination & synchronisation were investigated.

G. JAPAN. TOKYO ASTRONOMICAL OBSERVATORY (TAO)
(Reported for all Japanese establishments by Sh. Aoki)

Astronomical observations for time and latitude have been made regularly with the PZT. From the beginning of 1984, astronomical observations have been referred to the new system of IAU astronomical constants.

UTC(TAO) has been kept with a Master Clock, selected out of 5 HP cesium clocks, controlled with a phase-microstepper.

Domestic time comparisons of UTC clocks have been continued by a cesium portable clock of TAO regularly twice a year against the ILOM, NRLM, RRL, and GSI (Geodetic Survey Institute), and KGO (Kanozan Geodetic Observatory). Also TAO clocks have been linked with TV-signals. Clock comparison between the TAO and the USNO with a flying clock of the USNO has been made regularly once a year with an accuracy of 0.1 micro-seconds by courtesy of the USNO.

The receptions of Loran-C signals from the Okinawa slave station (Northwest Pacific Loran-C chain) was commenced on 1 September 1983. The measurements of Loran-C from the Iwo-Jima Master Station (9970-M) and from the Okinawa Station (9970-Y) with Loran-C receiver (Austron 2000C) have been continued with an accuracy of the order of 0.1 micro-seconds.

From 1 April 1983, the TAO has started formal receptions of the Global Positioning System (GPS) timing signals with a GPS receiver (STI Model TTS 502) and has continued regular time comparisons among the TAO and six laboratories: OP, PTB, TUG, USNO, VSL, and NBS with an accuracy of the order of a few 10 nano-seconds.

The results of time comparison via the GPS and the Loran-C time transfer are transmitted every week by the General Electric (GE) Mark III system. By these means, the TAO clocks and the Asian clocks, which had been hitherto excluded from contribution to the International Atomic Time scale (TAI), have effectively participated in TAI from the beginning of 1984.

H. JAPAN. INTERNATIONAL LATITUDE OBSERVATORY OF MIZUSAWA (ILOM)

1. Observational Activities: Time and latitude observations have been made with the PZT No. 2 and Danjon astrolabe. The fully automated electronic astrolabe has been subjected to a preliminary observation since July 1983 and will join to a regular observation after the MERIT main campaign. UTC(ILOM) has been maintained with a selected HP cesium clock out of four.

Loran-C signals from Iwo-Jima and Hokkaido stations and Omega signals from Hawaii have been received for the clock comparison. The comparison by TV has been made against clocks in NRLM, RRL, and TAO. Loran-C signals from the secondary station Hokkaido has been received at ILOM since April 1982. Delay times of the receiving systems and propagation times of the Loran-C signals were investigated to examine a consistency between these two Loran-C data and also the portable clock data. Horiai et al. (1983) have pointed out the discrepancy in these and institutes located around Tokyo area. The portable clock of USNO visited ILOM in November 1983 and showed a discrepancy of around 4 μs between portable clock data and the Loran-C data. An experiment is planned by ILOM to determine the propagation time of Loran-C signals over land this year and forth.

2. Astronomical Reference Frame: Yokoyama (1983, 1984) presented a full description of the method for reducing the optical astrometry observations at each station, as well as at the international centers, in accordance with the adoption of the FK5 system and the IAU (1976) System of Astronomical Constants for 1984 Jan. 1 onwards. Yokoyama and Tanikawa (1983) completed a standard FORTRAN program for computing apparent places of stars based on the new astronomical constants and circulated the program to all countries where optical

astrometry instruments are in operation.

3. Reference Frame: Fujii and Murakami (1982) derived the corrections of the PZT star places of Mizusawa using the data of the Mizusawa new PZT, and compared their results with those of the Tokyo PZT observations and the NPZT 74 based on an improvement of the Tokyo PZT Catalogue (Fujii, 1981).

4. Astronomical Instrumentation: A fully automated pendulum astrolabe (Automatic Electronic Astrolabe) was designed, constructed, and tested at the ILOM. The telescope is controlled by the mini-computer which is installed at 170-meter distance, and is drived by seven pulse (stepping) motors. Star-image motions along both azimuth and altitude directions are photoelectrically detected by ETD (Elektronic Transit Detector). ETD output signals are used for tracking, guiding, and correction for observed time. An instrumental accuracy is expected to be about 0".001. The automatic astrolabe can be used for observations of two different almucantars with two astrolabes. This gives a way for a precise determination of the star place and EOP simultaneously.

5. Reduction Method: A new method for the reduction of PZT observations was proposed (Manabe and Kitago, 1981). In this method, the UTO-UTC and latitude for individual stars, and the plate parameters are simultaneously estimated by solving a non-linear least-squares problem. Linearized equations of observation are solved to give approximate but explicit expressions of the estimates and their dispersions in terms of the measured coordinates. It was shown that the scale value is determined mainly from the diurnal motion. The method is applied to the data of the Mizusawa PZT.

An algorithm has been developed at the IPMS to provide a frequent series of the Earth orientation parameters (EOP). Vondrak's method was generalized for calculating smoothed EOP, directly from individual values of observed time and/or latitude. Statistically optimum smoothing factors can be determined with the aid of the procedure in which a Bayesian information criterion is minimized. This method was applied to the astrometric observations which have been accumulated at the IPMS, and the estimated EOP for every 0.5 days was compared with the EOP from the new techniques such as VLBI, LLR and SLR. Agreement was generally close with one another, and this means that observations with optical astrometry could reveal short periodic or irregular variations of the Earth's rotational speed, whose amplitude is a few milliseconds of time (Manabe et al, 1982).

I. JAPAN. HYDROGRAPHIC DEPARTMENT OF JAPAN (JHD)

For the purpose of establishing the relation between ET2 and TAI the observations of occultations of stars by the Moon have been continued at the head office in Tokyo and branch observatories. Around 1000 timing data including 550 photoelectric data have been obtained every year.

Results on ET2-TAI based on the data obtained above for the epoch 1981.5, 1982.5 and 1983.5 were: 30.60, 30.71 and 30.68 s reaspectively with the mean error of ± 0.04 s. Details are published in Data Report of Hydrographic Observations, Series of Astronomy and Geodesy and Japanese Ephemeris.

The services of the International Lunar Occultation Center have been continued since 1981. Total data reported from 34 countries amount to 43,000.

K. SPAIN. INSTITUTO Y OBSERVATORIO DE MARINA DE SAN FERNANDO CADIZ (IOM)

(Reported by A. orte and J. Benavente)

Series of observations with the Danjon Astrolabe for the determination of UT0 and polar coordinates have continued up to May 10, 1983. At this date, the instrument has been disassembled in order to proceed to its modernization by incorporating the "full pupil" feature. Some difficulties with micrometer movement system have impeded resuming the series of observations.

Timekeeping activities, based upon the operation of six Cs beam atomic clocks have continued along the present term; to improve the environmental operating conditions of the clocks, thermal control chamber, capable for three or four standards, has been specified and ordered. Solid state, electronically controlled cooling, will provide a high degree of higrometric and thermal stability. This chamber is to be fully operative in the last quarter of 1984.

San Fernando has been included in USNO flying clock trips for global synchronization.

Operation of the mobilde t/f laboratory was continued to satisfy national needs, and four calibration and synchronization trips have been achieved. Link with INTA/NASA space station at Robledo de Chavela (Loran C) includes now a GPS timing receiver at Robledo along with periodical comparisons with our time scale by means of clock transport.

Funds have been appropiated for the purchase of a GPS receiver to be operated at San Fernando, hopefully by mid 1985.

L. U.K. ROYAL GREENWICH OBSERVATORY, (RGO)

(Reported by J. Pilkington)

During this period the RGO has been obliged to reduce both the number of staff employed and its expenditure on equipment and operations. The effects have been felt particularly strongly in the fields covered by Commissions 19 and 31 although, since the satellite laser ranging facility at RGO commenced regular operation in 1983 October, the RGO has been contributing more accurate data than ever before to the international programmes for the determination of Earth-orientation parameters. It has, however, been necessary to stop observations with the Herstmonceux PZT, and unless additional sources of funding can be found the work of the atomic-time service will soon be directed almost exclusively towards meeting the requirements of the RGO's observational programme.

Throughout 1982, 1983 and 1984 the RGO has continued to form the independent atomic timescale TA(RGO) from the readings of up to eight commercially-produced cesium-beam clocks on site. Two of these, models HP 5060A which were purchased before 1968, are no longer in service and several of the others are known to be working at a level of predictability substantially below the best that may be achieved by such clocks. Although TA(RGO) has often been explicitly based on the readings of only two clocks from the ensemble, the process of clock evaluation requires that several more clocks must be kept in undisturbed operation and so limits the opportunities for diagnostic investigation.

The Time Service has also continued to publish results obtained by using the related UTC(RGO) scale as a reference in monitoring four Loran-C and two VLF signals. These results have provided the main link between the clocks and timescales of RGO and those of other establishments; they are used by the BIH in the formation of TAI and they contribute to the accuracy and effectiveness with which the systems being monitored may be used elsewhere for position-fixing or as sources of precise time and frequency.

M. USA. U.S. NAVEL OBSERVATORY, (USNO)
(Reported by D.D. McCarthy)

Optical observations were made with Photographic Zenith Tubes (PZT's) in Washington., D.C. and at the Time Service Substation in Richmond, near Miami, Florida. Two PZT's were operated at each site: a 20 cm (PTZ3) and a 65 cm (PZT7) in Washington, and two 20 cm instruments (PZT2 and PZT6) in Richmond. A new 20 cm PZT at Washington (PZT8) has been completed and is being installed on the original site of PZT1. Stars previously identified as supplemental stars have been added to the regular observing programs at Washington and Richmond, and at the Richmond Substation, several photographic plate and filter arrangements were tested in order to find a low-cost combination which would yield satisfactory PZT observations. The Kodak "M" plate plus a light yellow filter was adopted for routine use.

A new radio antenna was constructed for the Green Bank connected Element Interferometer (CEI). Located in Monterville, West Virginia, it is 14 meters in diameter, and forms one end of the new 1-3 baseline.

Radio observations were obtained daily with the CEI throughout the period. Measurements were made of two 35-km-baseline vectors, to determine Earth orientation parameters.

A new atmospheric model and new source coordinates were adapted in the CEI reduction on 1 January 1984. An improved estimate of UTO based only on the variations of the East-West components of the 1-4 baseline was made available to users.

Plans are underway to install a VLBI Mark III Correlator at the Naval Observatory. This is a joint effort with Naval Research Laboratory (NRL), the National Geodetic Survey (NGS) and the National Aeronautics and Space Administration (NASA).

VLBI time transfer experiments have been carried out in this period with VSL and Onnsala using the NRL antenna at Maryland Point. Studies of the use of the Global Positioning System (GPS) for time transfer indicate that remote scale information is possible to obtain with an accuracy better than 50 nanoseconds.

Two Smithsonian Astrophysical Observatory (SAO) Hydrogen masers were installed in 1983.

Until June, 1984 two cesium reference clocks were maintained to within 50 nanoseconds of UTC(USNO, MEAN). Since June, one of the cesium reference clocks was replaced by a hydrogen maser reference clock, and new computer algorithms to steer the maser to a preliminary estimate of UTC(USNO, MEAN) were developed.

The Data Acquisition and Control System (IBM Series 1 and HP 1000) collected data from approximately 70 cesium clocks about 20 of which define UTC(USNO, MEAN).

Seven cesium clocks contributed to the time scale at Richmond, Florida. This time scale was synchronized with USNO (MC) to within 150 nanoseconds.

All available modern techniques have been uncorporated into the USNO algorithm to determine polar motion and Universal Time. The accuracy of these estimates is $\pm\ 0''.005$ in polar motion and $\pm 0^{s}.0014$ in UTI-UTC.

The services available on the General Electric Mark II Computer Network have been expanded. Many Time Service publications and data obtained from systems such as the Global Positioning System and Loran-C are exchanged on the RC28 Catalog, which is composed of laboratories and observations concerned with timing matters.

Regular publications of astronomical time and polar motion information continued in the weekly Earth Orientation Bulletin, Time service Announcement Series 7. Predictions of polar motion and UT1-UTC were continued and published in this bulletin. The predictions were also disseminated through the GE Mark III and the Time Service Automatic Data Service (ADS), accessible by telephone through the HP 1000 at Time Service headquarters.

Timing information continued to be recorded from eleven VLF stations (U.S. Navy, OMEGA and GBR), eleven LORAN-C chains, television stations, five GPS satellites, and six transit satellites.

Investigations were carried out on past values of the length of day, the nature of the variations of the length of day, possible use of a super conducting gravimeter for the determination of Earth orientation parameters.

N. USSR.

(Reported by N. Blinov)

Astronomical determination of the Universal Time was carried out at 13, and since August, 1984 at 14 time services with photoelectric passage instruments (12), visual passage instruments (2), Danjon prismatic astrolabes (4) and photographic zenith tubes (2).

For reductions of passage instrument observations the merged catalogue of the USSR time services (CTS) was used. For reductions of photographic zenith tube and prismatic astrolabe observations, catalogues were used which were improved by original observations. Right ascensions are being calculated since January 1, 1984 in the new FK-5 system using the new IAU reduction constants and the new computational methods. Older observations (September 1983 - January 1, 1984) were re-reduced in the new system.

The results of the astronomical determinations of time were telexed weekly to the International Time Bureau (ITB) and sent by post to the InternationalarPole Motion Service.

The Universal Time was calculated at the Main Metrological Centre (MMC) of the State Service of Time and Frequency (SSTF USSR) by joint reduction of the results of astronomical observations made by all the time services of the USSR as well as by 10 time services of Czechoslovakia, Poland, GDR, Romania, Mongolia, and Yugoslavia taking part in the operation of the SSTF USSR.

The total number of astronomical determinations of time available for the joint reduction during a calendar year amounted to about 6000. For urgent (preliminary) calculations of the Universal Time, preliminary values of the co-ordinates of the pole were used which were telexed weekly from the ITB to the SSTF USSR. Preliminary values of the Universal Time and of the co-ordinates of the pole were published weekly in the A series of the SSTF bulletins. Accurate values of the Universal Time calculated using the final values of the co-ordinates of the pole according to the ITB data were published monthly in A series bulletins and in the three-month bulletins of the E series (Universal Time) along with accurate values of the co-ordinates of the pole and with the results of astronomical determinations of time.

The national scales of the atomic and co-ordinated time of the USSR were based on the State Primary Standard of Time and Frequency consisting of a metrologic cesium reper, hydrogen storages of the dimensions of time and frequency units, hydrogen, cesium, and rubidium clocks - storages of time scales.

To intercompare the scales of time services, to compare them with State Standards and with the ITB scale, different canals of communication were used: TV, meteor canals, receiving signals of LW and SLW bands, as well as transportation of a movable clock.

G. HEMMLEB

President of the Commission

33. STRUCTURE AND DYNAMICS OF THE GALACTIC SYSTEM (STRUCTURE ET DYNAMIQUE DU SYSTEME GALACTIQUE)

PRESIDENT: R. Wielen
VICE-PRESIDENT: W.B. Burton
ORGANIZING COMMITTEE: L. Blitz, W. Iwanowska, E.K. Kharadze, G.G. Kuzmin, D. Lynden-Bell, G. Lynga, M. Mayor, M. Miyamoto

I. Introduction

Several authors have contributed to this report: L. Blitz (Section V), W.B. Burton (Sections IIIB and IVB), J. Einasto (Section VII), B. Fuchs (Sections VIC and VID), W. Hermsen (Section VIF), G. Lynga (Sections IIIA and IVA), M. Mayor (Section II), M. Miyamoto (Sections VIB and VIE) and R. Wielen (Sections I, VIA, and editing). The layout of this report follows previous practice. The galactic center is included in Sections IV and V. The references are, as far as possible, coded by their numbers (VV.CCC.NNN) in the bibliography "Astronomy and Astrophysics Abstracts"(AAA). VV identifies the volume of AAA, while CCC.NNN gives the subject category and the serial number within that volume.

The field of galactic research has been very active in the last three years. During the triennium under review, many conferences of importance for galactic research have taken place. Among the IAU Symposia or Colloquia, we mention IAU Symposium No. 100 "Internal Kinematics and Dynamics of Galaxies" (33.012.004), IAU Symposium No. 105 "Observational Tests of the Stellar Evolution Theory" (37.012.078), IAU Symposium No. 106 "The Milky Way as a Galaxy", IAU Symposium No. 113 "Dynamics of Star Clusters", IAU Colloquium No. 76 "Nearby Stars and the Luminosity Function" (34.012.065), and IAU Colloquium No. 78 "Astronomy with Schmidt-type Telescopes" (37.012.077).

Among other relevant meetings were "Giant Molecular Clouds in the Galaxy" (29.012.012), "Galactic Evolution" (29.012.024), "Galaktische Struktur und Entwicklung" (29.012.039), "Structure and Evolution of Normal Galaxies" (30.012.010), "The Phases of the Interstellar Medium" (30.012.018), "Scientific Aspects of the HIPPARCOS Space Astrometry Mission" (31.012.041), "Morphology and Dynamics of Galaxies" (32.012.032), "The Galactic Center" (32.012.076), "Kinematics, Dynamics and Structure of the Milky Way" (33.012.016), "Evolution of Old Stellar Populations in Galaxies" (33.012.028), "Theoretical Aspects on Structure, Activity, and Evolution of Galaxies" I (33.012.048) and II (37.012.034), "Galactic Astrophysics and Gamma-Ray Astronomy" (34.012.016), "Surveys of the Southern Galaxy" (34.012.023), "Star Clusters and Associations and their Relation to the Evolution of the Galaxy" (34.012.068), "Formation and Evolution of Galaxies and Large Structures in the Universe" (37.012.002), "Statistical Methods in Astronomy" (37.012.003), and "Population Synthesis" (37.012.081).

A text book on the Galaxy, "Bau und Physik der Galaxis", has been written by Scheffler and Elsässer (32.003.108). Subvolume 2c of the Landolt-Börnstein (32.003.055) contains articles on our Galaxy by Gliese, Scheffler and Wielen. Among relevant general review papers, in addition to those published in the proceedings of the conferences listed above, are those of Bok (34.155.056), Pagel and Edmunds (30.155.039), Mould (32.155.027), York (32.155.028), Binney (32.151.041), Bignami and Hermsen (34.143.025), Kron (32.151.066), Güsten and Mezger (34.155.081), Oort (32.155.025), Delhaye (32.155.050), Downes and Güsten (32.155.053), Buser (32.155.054), Neckel and Klare (32.155.055), and Contopoulos (34.151.103).

II. Basic Data and Calibrations

A. BASIC DATA

It would hardly be possible to present in this summary an exhaustive list of all the important data for studies of galactic structure that were collected during the corresponding triennium. However, we tried to keep this report as complete as possible.

The Centre de Données Stellaires (CDS) has continued to publish basic information for galactic research. Their information bulletin summarizes the available information as well as the catalogues (Bull. Inf. Cent. Données Stellaires 21 to 25) (30.002.030, 30.002.095, 31.002.037, 31.002.033, 32.002.114, 34.002.021, 34.002.026, 34.002.124, 34.002.024). Many catalogues are available on magnetic tape or microfiche. Direct telephone connection is now available between observatories and the CDS data base. In the US, astronomical data can be obtained from the Astronomical Data Centre; information on available data are given in 30.002.021, 30.002.023, 30.002.034, 30.002.043, 31.002.032, 31.002.051, 31.002.081 to 31.002.095, 32.002.082 to 32.002.088, 34.002.020, 34.002.087. See also "Automated data retrieval in astronomy", IAU Colloquium No. 62 (32.012.002).

A large set of basic data for galactic structure is expected from the future HIPPARCOS/TYCHO space mission: star positions, parallaxes, proper motions, magnitudes ("The scientific aspects of the HIPPARCOS Space Astrometry mission", 34.012.070). During the last triennium, intensive preparative work has been done in relation with the input catalogue (see INCA News letters, contact C. Turon, Obs. de Meudon).

Schmidt-Kaler in Landolt-Börnstein (31.003.064) presents a survey, a critical appraisal of relevant work (up to the end of 1981) and tabulations of "Physical Parameters of the Stars".

Among new photometric catalogues or investigations we can mention: "The third catalogue of stars measured in the Geneva Observatory photometric system", giving 7-colour photometry for 14633 stars, presented by Rufener (30.002.008); "An infrared astronomical data base and catalogue of infrared observations" (1 μm - 1000 μm) by Schmitz, Gezari and Mead (30.002.037); McClure and Forrester have prepared a catalogue of homogeneous photometry of 2196 bright G and K stars in the DDO systems (30.002.077), and Meylan has established a compilation of measurements in the DDO photometric system, available on magnetic tape for 2911 stars (31.002.003); Upgren, Dawson and Lu have obtained DDO photometry for a number of dK-M stars (30.113.051).

25654 photoelectric measurements in the UBV photometric system published from 1976 to 1982 have been collected by Mermilliod (34.002.121). All photometric data were collected by Hauck in a general catalogue of photometric data (31.002.022); red and infrared photoelectric magnitudes in the Kron photometric system have been compiled by Jasniewicz in a homogeneous catalogue on magnetic tape (31.002.066). Positions and uvbyβ photometry are presented for about 1000 O-F8 stars towards the north galactic pole by Hill, Barnes and Hilditch (33.002.066, 33.002.067). UBVr magnitudes of more than 13000 stars at high galactic latitude have been prepared on magnetic tape by Weistrop (34.002.017). Olson has measured uvbyβ photometry of 14816 A5 to G0 stars brighter than 8.3 (34.002.029). Reed and Fitzgerald have collected UBV photometry of 610 stars in Puppis (34.002.028). The colours, magnitudes and parallaxes of 1077 nearby stars belonging to the 1969 edition of Gliese's catalogue have been collected by Grenon and Rufener with an up-to-date spectral classification (30.113.017).

RGU photometry has been measured and analyzed: of fields in the direction of the Scutum cloud, Karaali (31.113.073); in an anticentre field around NGC 2129, Topaktas (31.113.074); in fields near the galactic centre, Wiedemann (32.113.018), Fenkart et al. (33.113.019), Becker, Fang (31.155.048); in a field of the galactic bulge, Spaenhauer et al. (33.113.027); in a field in Cassiopeia, Fenkart, Kandemir (34.113.035); in a field in the large Sagittarius Cloud, Topaktas (30.155.032), and in a field in Centaurus, Spaenhauer, Fang (31.155.008). As tests of galactic models, the importance of these colour and magnitude counts is certainly increasing (see also 33.002.044). From RGU photometry of 1810 stars down to the limiting magnitude G = 19, density gradients for disc and halo stars have been derived by Becker and Karaali, as well as the luminosity function of m.s stars of population I within 400 to 800 pc (33.155.068).

Vilnius photometric determinations have been made for halo stars and stars at high galactic latitudes by Speranskas, Bartkevicius, Zdanavicius and Bartasinté (31.113.079, 31.113.080, 31.113.081). uvbyβ photometry was used in many investigations, for example: by Degewij in McCormick proper motion fields (31.113.089), by Westin for 355 O-A0 type stars near the galactic equator (32.113.003), by Hill and Barnes for about 700 AF stars in the region of the north galactic pole (33.113.070, 33.113.071), by McFadzean, Hilditch and Hill for 572 O-F stars near the south galactic pole (34.113.023). UBV and Hβ photometry of luminous stars between $l = 335^o$ and 6° was carried out by Dachs, Kaiser, Mikolov and Sherwood for 316 OB stars (32.113.031). UBV photometry has been determined by Pier for 234 AB stars in the southern halo (32.113.046), by Oja for 320 FK4 and FK4 suppl. stars (33.113.028).

A set of 223 stars was observed to provide new UBVRI standard stars (33.113.003); V-R observations and effective temperatures are given for 90 extremely metal-deficient red giants by Stone (33.113.011). New UBVRI photometry is presented by Fernie for more than 900 supergiants (33.113.042). Near-infrared photometry (JHKLM) of 203 bright southern stars from various sources has been transformed by Koornneef into one homogeneous data set (33.113.025).

Schuster and Alvarez have observed Be and shell stars with the 13-colour photometric system (33.113.012), and Mendoza et al. have carried out a narrowband study for 228 O4-B8-type stars (33.113.013).

A list of almost 2000 near-infrared sources in the galactic plane brighter than 6.5 in the K-band has been published by Kawara et al. (34.133.019). An analysis of the stellar component of the Galaxy as seen by the AFGL Infrared Sky Survey has been reported by Grasdalen et al. and Kleinmann, Gillet and Joyce (34.155.065, 30.133.008).

Depenchuk and Komarov, discussing the homogeneity of spectrophotometric standards, have stressed the necessity of a new absolute calibration of α Lyr in the UV and blue spectral regions (31.113.023). Secondary standard stars for absolute spectrophotometry based on an adopted absolute spectral energy distribution for the primary standard α Lyr are proposed by Oke and Gunn (33.113.017). A determination of the absolute monochromatic flux from Vega at $\lambda = 2.20$ and 3.80 μm through comparison with a furnace has been carried out (33.113.031, 34.113.039). Gunn and Stryker have presented a spectrophotometric catalogue of 175 stars covering a complete range of spectral types and luminosity classes for the wavelength region $3130 < \lambda < 10800$ Å (33.002.042).

Preliminary results on the calibration of the RGU photometric system have been presented by Thévenin, Spaenhauer and Foy (34.113.006); the relationship between the Johnson and Kron-Cousins VRI photometric systems can be found in (34.113.055).

Basic data for all calibrations are certainly also the colours and absolute magnitude of the Sun. A new direct measurement gives B-V = 0.636, U-B = 0.183, and M_V = -26.74. uvby measurements show that the Sun seems to be somewhat metal-poor (Tüg and Schmidt-Kaler, 31.031.515).

Newly determined radial velocities, magnitudes and colours are given for 339 stars in the southern hemisphere (31.113.072), and spectral classification and photometry have been established for 221 barium stars (33.002.043); Alexander, Jones and Sinclair have presented narrow-band photometry of faint red stars (34.002.066).

The "Catalogue bibliographique de vitesses radiales stellaires", a continuation of the Abt-Biggs catalogue, is done by Barbier and Petit and is available at the CDS (32.002.035).

The Bright Star Catalogue has seen its radial velocities completed thanks to 106 southern stars observed by Beavers and Eitter (30.113.002), 139 B-type stars and 53 late-type stars measured by Andersen and Nordstroem (33.111.009, 33.111.010), and 790 late-type HR stars determined in the southern hemisphere by Andersen et al. (37.111.016). In a study of late-type dwarfs, radial velocities have been obtained by Dawson for 20 of the stars in Sanduleak's catalogue. The velocity dispersion and the space density appear to be "normal", i.e. there is no evidence for an excess density of low-velocity M dwarfs (30.111.007).

Radial velocities of 617 stars belonging to four stellar fields and 1127 galactic stars in the direction of the LMC have been measured with an objective prism technique by Fehrenbach and collaborators (32.111.001, 31.002.063). Radial velocities for 1246 stars measured at Herstmonceux and Kottania have been published by Woolley et al. (31.002.054). Radial velocities have also been determined by Stetson for 124 southern early-type stars possibly having large space motions (34.111.005). Of interest is also a compilation of radial velocities of 524 galactic planetary nebulae (34.002.034), and a list of radial velocities for 116 southern red stars published by Jones and Fisher (37.111.012).

Absolute proper motions of blue objects near the galactic pole have been derived by Schilbach with reference to 45 galaxies (32.111.015). Luyten and collaborators have produced numerous lists of proper motions in the last three years: with Morris and Hill, a list of 7120 stars in the Hyades region (32.111.019), and a catalogue of 929 possible candidates for Hyades membership (31.111.033); with Hughes a sample of 374 faint stars near the north galactic pole (32.111.021), 280 faint stars (34.111.023), and a first supplement of 398 stars to the NLTT catalogue (31. 111.030).

Proper motions have been determined in the Orion association by Qian and Xu (30.111.018, 30.111.019), for 117 O-type stars as well as for 211 stars in the north polar cap by Karimova and Pavlovskaya (30.002.050, 33.111.015), for AGK3R stars by Kalinina (33.002.023), in the field of the open cluster NGC 6913 by Gerts (31.111.010), in the area of the galactic cluster Coma Ber by Cova (32.111.012), and in the Hyades region by Stelmakh (33.111.016). A catalogue of proper motions from the Moscow PZT program (33.111.014) by Tochilina and a catalogue of relative proper motions and photographic UBV magnitudes in the region of M 39 by Platajs (34.002.037) have also been published. Wan, Mao and Ji have collected proper motions of RR Lyrae variables (30.002.081), and Fu and Gong have determined proper motions in R.A of 904 bright stars (30.002.080). See also proper motions studies 34.111.008 and 33.111.014.

Spectroscopic, photometric and kinematic data for 488 white dwarfs were collected by Agayev et al. (31.002.012). A compilation prepared by Mennessier and collaborators of the most reliable kinematical data for stars of the HD and HDE, including positions and spectroscopic, photometric data is now available from the CDS (31.002.025).

The space distribution of stars in the region of Centaurus-Crux has been investigated by McCuskey (31.002.025).

The galactic poles and centre-anticentre direction have been and continue to be directions of great interest for galactic studies. We should also notice surveys in course of progress to study stellar kinematics in the main meridional section of the Galaxy (see for example 34.155.018 and 33.155.060).

The relative parallaxes determined by five major general catalogue observatories have been intercompared to investigate their systematic differences by Lutz and collaborators (30.111.011). New trigonometric parallaxes appear in a few studies: by Breakiron et al. (31.111.005), from the US naval faint stars programme (31.111.006), by Russel et al. (31.111.007), by Kiselev (32.111.018), by Borgman and Lippincott (33.111.004), by Upgren and Kuzma (33.111.005), by Lippincott and Hershey (33.111.012), by van Altena and Sawada (34.111.011), and by Weis et al. (34.111.012).

We should further mention some important sources of basic information: The systematic reclassification of the HD stars on the MK system is progressing. Houk published the third volume of the Michigan catalogue of two-dimensional spectral types(declinations -40° to -26°) (31.002.126). Spectral types and luminosity classes have also been determined by Bartaya for 10396 stars in the Kapteyn Areas NN 2 - 43 (34.002.018). The fourth revised edition of the Bright Star catalogue has been published by Hoffleit and Jaschek (31.002.077), as well as the fifth general catalogue of MK spectral classification (31.002.078), and a first supplement of the catalogue of star clusters and associations by Ruprecht et al. (31.002.111).

A new catalogue of stars suspected of variability containing data on 14810 variable stars which have not received definite nomenclature until 1980 has been published by Kukarkin and collaborators (32.002.039, 32.002.090).

Of considerable interest for galactic structure studies are the numerous CO or OH data: A catalogue of CO radial velocities towards galactic HII regions by Blitz, Fich and Stark (32.002.004), CO observations of southern HII regions by de Graauw et al. (30.132.035), CO observations around $l = 45^\circ$ by Israel (31.131.059), CO observations of southern galactic-plane HII regions by White and Phillips (31.132.046), a CO survey covering $294^\circ \leq l \leq 354^\circ$ by McCutcheon, Robinson and Whiteoak (32.131.032), CO observations towards southern HII regions (33.132.040), towards $l = 50^\circ$ and 72° by Solomon, Stark and Sanders (33.155.031).

Recently acquired general information on OH, CO, HI and gamma ray surveys is available in "Surveys of the southern Galaxy" (34.012.023) or in "Regions of recent star formation" (31.012.002). In particular: a CO survey of 372 optical HII regions by Blitz et al. (31.132.043), and HI and CO observations of distant HII regions in the galactic anticentre by Grayzeck et al. (31.132.044). See also 34.155.024, 34.155.025, 34.155.038, 34.155.039, 34.155.043, 34.155.044, 34.155.045 .

From a southern survey of OH masers at 1612 MHz, the velocity and spatial distributions of the emission sources have been discussed with special reference to their kinematic properties and population type by Caswell et al. (30.131.107).

Let us also note numerous works of interest dealing with the distribution of molecular clouds and HII regions in the Galaxy and their kinematics (33.131.123,

33.131.124, 33.131.126, 33.132.036, 33.132.037, 33.132.038) presented in "Kinematics, dynamics and structure of the Milky Way" (33.012.016).

High velocity HI streams, high velocity clouds or gas infalling on the Galaxy are detected and discussed in various studies: by Cohen (30.131.006, 32.131.021), Giovanneli (30.132.042), Mirabel (31.132.022, 32.131.150), Morras and Bajaja (32. 131.151, 32.131.264, 33.131.017), de Boer (33.131.138), and Lo and Clausen (34.155. 059).

The following HI surveys have been published: by Olano, Pöppel and Vieira ($320^\circ \le l \le 345^\circ$, $+18^\circ \le b \le +26^\circ$ and $346^\circ \le l \le 350^\circ$, $+18^\circ \le b \le +20^\circ$), by Burton and Liszt ($349^\circ \le l \le 13^\circ$, $-10^\circ \le b \le 10^\circ$), for 21 cm absorption at low latitudes towards 88 compact sources by Dickey et al. (34.155.066), and in the halo of the Galaxy (34.155.057) by McGee, Newton and Morton. The 21 cm line radiation of neutral hydrogen in the central region of the Galaxy has been measured by Braunsfurth and Rohlfs (29.156.016), the region covered is $-1^\circ\!.5 \le l \le 1^\circ\!.5$, $-1^\circ\!.5 \le b \le 1^\circ\!.5$. These authors have also carried out (in press) a survey of the innermost 21° in longitude. Rohlfs, Kreitschmann and Siegman are preparing the publication of a 21 cm survey in the ranges $315^\circ \le l \le 340^\circ$, $-5^\circ \le b \le 5^\circ$.

We should further mention a catalogue of galactic supernovae remnants (34.002. 035), a catalogue of O-type stars (34.002.101), the Lund catalogue of open-cluster parameters (34.002.127), the Geneva photometric-boxes catalogue on magnetic tape (31.002.064), a catalogue of [Fe/H] determinations (30.002.001), and a catalogue of Hα observations (30.002.002).

Sakhibov and Smirnov (31.125.023) presented a complete survey of distance estimations for 57 galactic SNR's.

Weis has started a programme to determine photometric parallaxes of all stars of colour class m from the NLTT Catalogue which are brighter than V = 14 and north of the Celestial Equator. Results in VRI colours for 413 of these stars have already been published by Weis (37.111.014).

B. INTRINSIC COLOURS AND INTERSTELLAR REDDENING

Infrared properties of Mira-type variable and other cool stars as determined from JHKL photometry have been used by Feast and collaborators to establish period-colour relations and to determine interstellar reddening (32.122.072).

Ultra-violet intrinsic colours of early-type stars have been presented from satellite observations and compared with a grid of stellar model atmospheres (30. 113.003). For UV intrinsic colours, see also Carnochan (32.113.044), Wesselius (32. 113.059), Galecki et al. (33.113.050), and Zorec et al. (34.113.011).

Cramer and Maeder, applying Geneva photometry, have classified 3600 B and Ap stars (30.113.032), and Cramer has determined intrinsic colours of B stars in the Geneva photometry (32.113.014). A comparison between the observed intrinsic colours $(b-y)_o$ and the calculated theoretical index (b-y) has been carried out by Burkhart et al. for Ap stars (30.113.039). Wide-band photometric systems in the red and near infrared have been discussed by Straizys and collaborators in relation with the influence of reddening (30.113.056). Mean intrinsic U-B and B-V indices for population II stars as a function of temperature have been studied by Malyuto and Traat. Based on the method of "Geneva photometric boxes", Nicolet has determined a reddening of $E_{B-V} = 0.04$ in the directions of both galacic poles (31.113.008). However, from uvbyβ photometry of 572 O-F stars in the south pole direction, Fadzean et al. have estimated a zero value for the interstellar extinction (34.113.023).

Becker and Wooden have studied the space density of stars and interstellar extinction near h and χ Persei, using RGU photometry (30.115.025).

E(b-y) variations on scales smaller than 30' have been investigated by Knude, using uvbyβ observations in intermediate and low-latitude areas SA 128 and 156 (31.113.090). With infra-red photometry, Castor and Simon constructed reddening-independent normal colours for O stars (33.113.010).

Calibrations of photometric indices of red giants in the old disk population for the determination of reddening, luminosity and metal abundance were discussed by Eggen (33.113.052) and are applied to determine the reddening, metal abundance and luminosity of high-luminosity G-type stars (34.113.002).

Intrinsic colours and absolute calibration values have been established between one and five microns by Koornneef (34.113.025).

Ultraviolet interstellar extinction towards 1367 stars has been measured by Meyer and Savage in application of the ANS narrow-band photometry. Many stars have been isolated with significant deviations from the mean galactic extinction curve (30.131.023). Deviations from the normal absorption law are also presented from UBVRI photometry measurements in some dusty HII regions (30.132.029).

A high-resolution optical survey of interstellar absorption lines, including line-of-sight measurements towards 3C 273, the nucleus of M 87 and the cores of globular clusters allowed an estimation by Songaila, Cowie and York of the galactic halo column density (30.131.027). The distribution of gas in the halo has also been studied by Blades et al., using the nuclei of two Seyfert galaxies as background probes (32.131.043, 32.131.186), by de Boer and Savage, using a blue star in M 3 (33.131.022), and by using distant early-type stars (32.131.082, 33.131.168).

For UV absorption by interstellar gas at large distances from the galactic plane, see Savage and de Boer (30.131.043), and Pettini and West (32.131.082). The distribution of the interstellar gas in the galactic halo has been studied by Hobbs et al., using 21 cm emission lines and Lyα absorption lines towards high-latitude OB stars (32.155.060). From a high-latitude HI cloud with optical emission, Goerigk et al. found the ratio of HI column density to extinction to be about a factor of 10 smaller than the standard value found from UV observations (33.131.101).

Deviations of galactic reddening from the cosecant relation have been discussed by Noonan (30.131.037).

Maps of reddening in the Galaxy, derived from HI and galaxy counts, have been established by Burstein and Heiles; these cover almost the whole galactic sphere with $b \geq 10^{\circ}$ (32.131.005). Based on uvbyβ photometry of about 3450 bright A-F stars, a map of interstellar reddening within 300 pc was done by Perry and Johnston (32.131.296). For interstellar reddening within 100 pc, see 32.131.309.

Interstellar extinction within 5 kpc from the Sun in the direction of the galactic centre has been investigated by Ichikawa et al. (32.131.037). Interstellar light absorption in the Galaxy has also been investigated from the galactic latitude distribution of quasars by Nazarov and Khodyachikh (33.155.065), and of extragalactic objects by de Vaucouleurs and Buta (34.155.002). Hilditch, Hill and Barnes have established the interstellar reddening for the North Galactic pole region and uvbyβ intrinsic colour calibration from a sample of 1000 stars with $b > 75^{\circ}$ (34.131.023).

Golay et al. (1984) have used 123 6° fields near the galactic plane, photographed by balloon in the medium ultraviolet ($\lambda \approx 2000$ Å), to determine the distribution of interstellar extinction.

Frisch and York have composed a synthesis map of ultraviolet observations of neutral interstellar gas from satellite observations of 140 stars, 10-3000 pc distant (34.131.091).

A catalogue gives equivalent widths of the interstellar 2200 Å band obtained from TD-1 spectrophotometric data (34.131.012).

Finally, we can mention reddening determinations in specific directions (34. 113.035; Orion 30.131.028; Taurus dark clouds 32.131.169; Cep OB 3 32.131.174; Cas OB 6 32.131.175; Cygnus 32.131.306; Aur OB-2, Sco OB-1 32.131.318; Carina neb. 34.131.028).

C. ABSOLUTE MAGNITUDES

Using Strömgren photometry for Orion B stars, Anthony-Twarog has rediscussed the Hβ distance scale for B stars (32.111.004).

Systematic effects in trigonometric parallaxes have been searched for by Lutz et al. through inter-observatory comparisons (30.111.011) or through comparisons with spectroscopic and cluster parallaxes (33.111.001).

Absolute magnitudes analysis have been carried out for Ap stars (30.115.001) by Grenier et al., Carbon stars (30.115.014) by Cohen et al., M supergiants (30. 115.015) by Elias et al., δ Scuti and δ Del-type stars (30.115.017) by Grenier et al., OB stars (30.115.024) by Vega and Muzzio, G5-M3 stars near the giant branch (31.115.006) by Egret, Keenan and Heck, non-supergiant Be stars (31.115.012) by Kozok, O and B stars (31.115.035) by Divan and Zorec, F, G, K and M-type stars (32.115.020) by Mikami and Heck, M giants in the nuclear bulge of the Galaxy (32. 155.003) by Frogel and Whitford, for nearby stars (30.113.017, 34.155.136) by Grenon and Rufener, B and Ap stars (30.113.032) by Cramer and Maeder, B-type supergiants (33.113.053) by Zhang, O3 to O5 stars (32.115.022) by Underhill.

Grenier et al. (1984) have extented their absolute magnitude calibration to normal stars.

Bolometric corrections have been determined by Habets and Heintze for the main-sequence stars (32.115.019).

The period-luminostiy relation for cepheids has been rediscussed in relation with infrared photometry by Madore et al. (30.122.040), McGonegal et al. (30.122. 111, 31.122.089, 33.122.147), and by Welch et al. (34.122.121); in relation with its zero point by Karimova and Pavlovskaya (30.122.081), and by Caldwell (34.122. 061); in relation with the surface brightness - colour relation by Ivanov (30.122. 061). For the cepheid PL relation, see also 32.122.128, 32.122.129, 34.122.032, and 34.122.220.

Stift has argued that the canonical cepheid PLC relation leads to an incorrect cosmic distance scale (32.122.017). In his classification of intrinsic variables, Eggen has discussed the cepheid domain (33.122.018). For a study of cepheid properties in the photometric Walraven system, see Coppola et al. (32.122.060).

Clube and Dawe followed some considerations arising from the faint absolute magnitude of halo RR Lyrae variables and proposed a modification of the PLC relation (33.122.139).

The surface brightness relation has been applied by Manduca et al. for the determination of absolute magnitudes of RR Lyrae stars (30.122.100). Luminosities of

RR Lyrae variables in Osterhoff-group clusters have been studied by Cox et al. (31.122.073) (see also 30.154.011). Glass and Feast derived distances for 70 Mira variables in the Galactic Centre Windows by using the period-luminosity relation for Mira variables (31.122.001).

It was shown by Cousins that the surface-brightness parameter F_V can be obtained from VRI_C (30.113.053). A new calibration of the Vilnius photometric system in spectral types and absolute magnitudes has been proposed by Straizys and collaborators (30.113.054, 30.113.055, 32.113.064), as well as for the Vilnius-Geneva system (32.113.066).

D. THE STELLAR LUMINOSITY FUNCTION

The degree of completeness of nearby stars and the stellar luminosity function have been analyzed by Upgren and Armandroff (30.111.017). Probst and O'Connell, from an infrared search for very low mass stars in binary systems with white dwarf primaries, found that the IMF declines steeply for $M_V > 14$, and it appears unlikely that very low mass stars make a substantial contribution to the mass of either the disk or the massive halo components of the Galaxy (31.126.005, 34.126.015, 34.126.020). Areas around 21 nearby stars have been scanned by Jameson et al. (34.126.008) but no black dwarfs were found. The luminosity function in the solar neighbourhood was analyzed by Reid and Gilmore (32.115.008, 32.115.009). Form a purely photometric derivation of the stellar luminosity function to $M_V = +19$ (magnitude comparable to the minimum mass for thermonuclear burning), they conclude that the missing mass in the solar neighbourhood, if any, does not reside in main-sequence stars brighter than $M_V \approx +17$ mag.

From a complete sample of about 12500 stars brighter than I = 18 towards the south galactic pole, Gilmore and Reid (33.155.015) have derived stellar luminosity function and density laws with distance from the galactic plane. They identified the 300 pc-scale-height old-disk component and a 1350 pc-scale-height "thick disk" component.

Kipp derived the stellar luminosity function in the direction of the south galactic pole by a statistical method using proper motions (32.031.551). Bisiacchi et al. have studied the mass function for massive stars (33.115.003). From UBV photometry in the anticentre direction, Morales Duran showed that the slope of the luminosity function at about 1 kpc is the same as for the solar neighbourhood (31.113.012). The space density and the luminosity function for the halo and disk populations in the test field SA 54 of the Basel halo program have been derived by Fenkart and Esin-Yilmaz. See also density and luminosity function in Cassiopeia by Fenkart and Kandemir (34.113.035), and a field in Centaurus by Spaenhauer and Fang (31.155.008).

From a proper motion study, a new determination of the halo luminosity density of the Galaxy was established by Richstone and Graham (30.155.007). From a deep blink survey of faint red objects towards the south galactic pole, Staller et al. obtain a surface density for M dwarfs per square degree that is approximately twice as large as would be expected from Luyten's (1968) luminosity function (adopting a scale height of 260 pc) (31.155.003). The density distribution of faint stars in the direction of the north galactic pole has been analyzed from star counts in three-colour-photometry by Yoshii (32.155.034). The initial mass function (IMF) for stars more massive than 20 $M_\odot$ was derived by Garmany, Conti and Chiosi from a set of 750 O stars. Substantial variation in the IMF of the most massive stars has been found with varying distance to the galactic centre (32.155.061).

In an analysis of the observed colour distribution of faint stars, Bahcall, Schmidt and Soneira place strong upper limits on a possible intermediate population with $5 \leq M_V \leq 8$ in the galactic spheroid (33.155.012).

Mezzeti et al. have estimated the influence of binarity on the local stellar mass density determination (33.155.102).

The IAU Colloquium No. 76 "The nearby stars and the stellar luminosity function" (34.012.065) offers an excellent review of the subject. In particular, we should mention the studies on the luminosity function of white dwarfs by Dahn et al. (34.126.040), by Liebert et al. (34.126.041), by Ishida (34.126.042), and by Shipman (34.155.133); further, the search for brown dwarfs by Probst (34.126.043), a list of one thousand new K and M dwarfs brighter than the 13th magnitude by Stephenson (34.155.103), and the new limits on the surface density of M dwarfs by Booshaar and Tyson (34.155.106). From the Two Micron Sky Survey, the space density distribution of late-type giants and supergiants in the solar vicinity (34.155.107) was derived by Ishida. In the context of an extensive survey of trigonometric parallaxes and proper motions in the region of the south galactic cap, Murray discussed the populations of red stars and the space density and kinematics of K dwarfs (34.155.108). Stellar populations at higher galactic latitudes were discussed by Burer and Kaeser (34.155.111) and by Upgren and Philip (34.155.135). Schmidt studied the stellar luminosity function based on selected proper motion stars (34.155.112), Wielen et al. the luminosity function of nearby stars (34.155.133), and Pritchet the luminosity function from star counts (34.155.114). Robertson discussed the luminosity function and kinematics of late M dwarfs from an objective-prism survey (34.155.131). Smethells, from a sample of 185 red dwarfs selected in an objective-prism survey, has reanalyzed the luminosity function of early M dwarfs (34.155.134). Tables and plots of luminosity functions were prepared by Philip and Upgren (34.155.137). Gilmore reported on an extensive analysis of the stellar distribution in the galactic spheroid and the faint end of the local luminosity function currently under way, using automated photographic photometry. Results concerning the luminosity function of main sequence stars are given (34.155.118).

Stellar populations and luminosity functions are discussed by Reid (34.155.115) and Crézé (34.155.116). A comparative study of the Bahcall-Sonaira Galaxy model with various observations has been carried out (34.155.119); all of the data are derived by a two-component model with an exponential disk and a de Vaucouleurs spheroid and do not seem to confirm the fat-disk model proposed by Gilmore and Reid.

The compilation of a "Third catalogue of nearby stars" is in progress. Trigonometric parallaxes exceeding 0".039 of about 1900 stars are known in 1984 (Gliese and Jahreiß). Luminosity relations, mainly for red dwarfs, were calibrated for the data of Smethell's stars (Gliese, 29.111.009), for Eggen's photometry of large proper motion stars (Gliese, 31.111.004), and for the late Kuiper types (Jahreiß and Gliese, 32.036.022).

E. REFERENCES

Braunsfurth, E., Rohlfs, K.: 1984, Astron.Astrophys.Suppl. (in press).

Golay, M., Cramer, N., Huguenin, D., Nicolet, B., Blecha, A.: 1984, Astrophys. Space Sci. (submitted).

Grenier, S., Gomez, A.E., Jaschek, C., Jaschek, M., Heck, A.: 1984, Astron.Astrophys. (submitted).

III. Local Galactic Structure

A. OPTICAL STUDIES

This section contains work dealing with the Milky Way characteristics in the solar neighbourhood and supported by observations with wavelengths shorter than 1 mm. Thus, results are collected from gamma-ray, X-ray, UV, optical and IR observations in the different subsections. These are (1) the volume closest to the sun, (2) more distant objects at low galactic latitudes, and (3) objects at high galactic latitudes. Overall properties of the Galaxy and evolutionary effects are presented in Section IV.

1. The Volume Closest to the Sun

The nearby stars and the stellar luminosity function are discussed in detail in Section IID of this report.

Bahcall (37.155.014) used the distribution of F stars to determine the amount of matter near the sun. Unobserved disk material is at least 50 % of total disk material in the models considered.

Space densities for late-type giants have been determined by Ghosh and Tandon (32.155.063), and for Am stars by Ochsenbein (33.155.010). The mass spectrum in the solar neighbourhood was studied by Vereshchagin (34.155.062), and space densities for nearby stars were derived by Guseva et al. (33.155.019). Baggett (31.155.023) studied the stellar distribution in Gould's belt. Ishida and Mikami (31.155.014) have calculated the emissivity of the solar neighbourhood from density distributions of luminous stars.

Upgren and Chabotte (37.118.001) have examined about 200 K and M dwarfs with accurate parallaxes, proper motions and radial velocities. They find that the frequency of wide pairs can be accounted for by random space motions. Mezzetti et al. (33.155.102) point out the importance of taking unresolved binaries into account when deriving luminosity functions.

UV radiation from the sky back-ground was observed by Zvereva et al. (32.157.012) and compared to interstellar radiation at other wavelengths. Paresce (34.155.072) shows that the state of the interstellar matter near the sun is consistent with a stellar wind or a supernova shock originating in the Sco-Cen association.

2. Objects at Low Galactic Latitudes

The effect of interstellar dust on observations of galaxies has been reviewed by de Vaucouleurs and Buta (34.155.002) who derive a csc-b-law. Nazarov and Khodyachikh (33.155.065) study the galactic extinction law from the apparent distribution of quasars.

In a series of papers (32.155.026, 32.157.011, 33.155.005, 33.155.006, Winkler et al., 1984), the Bochum group present their photographic surface photometry of the Milky Way with a super-wide-angle camera. Reynolds (33.155.109) has mapped the Hα-emission-line profiles along the galactic equator. Feitzinger and Stüwe (1984) have catalogued dark nebulae and globules in longitudes $l = 240^{\circ}$-360°.

Optical observations of interest for galactic structure work include UBVβ photometry of luminous stars in the region $l = 335^{\circ}$-6° (Dachs et al., 32.113.031) and 8-colour, Wing-system photometry of M supergiants with $l = 228^{\circ}$-304° (MacConnell, Michigan State University). MacConnell also reports of a low-dispersion infrared objective-prism survey of $l = 210^\circ-30^\circ$, $b = \pm 6^\circ$. Vega (31.113.092) searched for Hα-emission-line objects in Vela; Bassino et al. (32.113.043) studied faint OB stars in the Southern Milky Way; and Rabolli (33.113.058) made a UBV photometry of early-type stars near the HII region RCW 99. The spiral structure of the Sagittarius arm at $l = 305^\circ$ was studied with radio, infrared and optical techniques by Danks et al. (37.155.023).

The Basel group has investigated luminosity and density functions based on RGU photometry in several Milky Way regions: $l = 290^\circ$, $b = -0^\circ.3$ (Becker, Hassan, 31.155.007); a Centaurus field (Spaenhauer and Fang, 31.155.008); $l = 9^\circ.3$, $b = 12^\circ.1$ (Becker and Fang, 31.155.048); $l = 277^\circ.0$, $b = 0^\circ.8$ (Becker and Marsoglu, 32.113.013); $l = 3^\circ.7$, $b = -8^\circ.8$ (Fenkart et al., 33.113.019); towards Cassiopeia (Fenkart and Kandemir, 34.113.035); in the anticentre direction (in preparation). Two more parts (VIII, 1982, and IX, 1983) of the Basel photometric catalogue of stars in selected areas and other fields have also appeared.

Optical spiral structure at $l = 30^\circ-70^\circ$ was discussed by Forbes (33.155.051, 37.155.051). Neckel and Klare (32.155.055) discussed spiral structure out to 3-4 kpc from the sun. The space distribution of stars in the direction of NGC 6823 was studied by Kuznetsov and Metreveli and space densities around the direction $l = 298^\circ$, $b = +1^\circ$ were derived by McCuskey (34.155.003). Burns et al. (37.155.012) studied the distribution of B5-A5 stars at $l = 253^\circ$, $b = -7^\circ$. Loden (Uppsala) has preliminarily concluded his study of galactic fine structure, particularly analysing the high clustering frequency among stars hotter than F stars.

During the triennium, much attention has been given to galactic structure as observed in the infrared. In the near-infrared, the SRC atlas of the galactic plane was presented by Hartley and Dawe (32.156.003). Sources were counted in the galactic plane by Kawara et al. (33.155.123). There is also an objective-prism survey (Fuenmayor, 32.155.023) towards the galactic centre and the anticentre.

Presenting observations in the near-infrared from a rocket-borne telescope, Noguchi et al. (31.156.009) conclude that most of the sources are late-type stars. The 2.4 μm radiation enhancement at $l = 355^\circ$, $b = -1^\circ$ is related to newly detected M stars by Hamajima et al. (31.156.010), who conclude that extinction is low in that direction. Grasdalen et al. (34.155.065) used the AFGL infrared sky survey to study statistical properties of the stellar sources. Ananth and Nagaraja (32.156.005) find that high fluxes at 2.2 μm coincide with areas of strong correlation between far-infrared radiation and CO emission, indicating active star formation. Eaton et al. (37.155.060) observed stars at 2.2 μm; counts at different longitudes define an exponential disk.

Sources of far-infrared radiation have been studied by Jaffe et al. (31.156.001) at $l = 10^\circ-16^\circ$, and by Price et al. (31.156.003) in the direction of Cygnus X. By utilising lunar-occultation methods, Stacey et al. (33.155.111) were able to directly estimate fine-structure emission at 157 μm, i.e. [CII], from galactic longitude $l = 8^\circ$. Bussoletti et al. (31.156.002) observed diffuse, far-infrared emission from the galactic plane at longitudes 233°, 230°, 222° and 110°, and concluded that it originates in dust of around 10 K.

The γ-radiation from the galactic anticentre was observed by Graser and Schönfelder (32.157.013), and interpreted as originating partly in the Crab Nebula. The diffuse γ-radiation between $l = 100^\circ$ and $l = 200^\circ$ is, according to Schlosser and Feitzinger (33.155.021), due to the central shingle of the local spiral arm.

3. Objects at High Galactic Latitudes

Becker and Karaali (33.155.068) investigated a field near NGC 7006 using RGU photometry and determined density functions for both populations. The Basel halo program also includes several papers by Fenkart and collaborators (33.113.027, 34. 113.036, and Astron. Astrophys. (in preparation)). Staller et al. (31.155.003) found a large number of red objects near the South galactic pole and discussed the M-dwarf distribution. Trefzger et al. (1984) used Walraven photometry to show a metallicity gradient for stars in SA 141 at the South galactic pole. The density function of faint stars towards the North galactic pole was derived by Yoshii (32. 155.034). Early-type stars near IC 4665 were investigated by Paparó and Balázs (33. 155.124). Kron, Cudworth, and Rybski (Yerkes) are beginning a programme of photometry, spectroscopy and proper motions for complete samples of faint stars in about twenty fields at intermediate latitudes. Tobin (37.113.034, 1984) has continued the studies by Tobin and Kaufmann (37.114.030) of high and intermediate-latitude B stars; uvbyβ photometry shows that some of these are more than 500 pc from the galactic plane.

The globular cluster system was studied by Frenk and White (31.155.001), who concluded that the sun's distance to the galactic centre is $R_o = 6.8 \pm 0.8$ kpc.

4. References

Feitzinger, J.V., Stüwe, J.A.: 1984, Astron. Astrophys. Suppl. (in preparation).
Tobin, W.: 1984, Astron. Astrophys. Suppl. (in press).
Trefzger, Ch.F.: 1984, IAU Symposium No. 106 (in press).
Winkler, Chr., Schmidt-Kaler, Th., Schlosser, W.: 1984, Astron. Astrophys. Suppl. (submitted).

B. RADIO STUDIES

Several regional surveys of southern HI made with the Argentine 100-m telescope were published (30.155.031). HI and other data were used by Dolidze (30.155. 044) to study the local distribution of gas and star forming regions. The Perseus arm region was the subject of an analogous study by Gerasimenko (34.155.139). Vallée (34.155.005) used rotation measure data from extragalactic sources in the direction of the Perseus arm to determine the magnetic field structure there and to constrain gravitational collapse theories of magnetic compression. HI in the direction of the Puppis window was surveyed by Stacy and Jackson (32.131.277) and used to study the turbulent characteristics of the interstellar medium. HI related to Gould's Belt was studied by Pöppel and Olano (32.155.024, 33.155.030). Salter (34. 131.014) reviewed radio and other observations pertaining to Loop I, the North Polar Spur, and considered the influence which this structure might have on the local medium.

Several regional studies were carried out in the radio continuum. Kononov and Pyatunina (34.155.017) surveyed the galactic plane in the region of Mononceros at λ7.6 cm. The extended component of the radio continuum at 408, 820, and 1420 MHz from the Cassiopeia-Perseus region was analyzed by Kallas et al. (34.155.069). Kanbach (34.155.020) reviewed the nonthermal radio emission from the local (< 2 kpc) region of the Galaxy and separated the emissivity into thin- and thick-disk components.

Shell-like structures in the interstellar medium received much attention during the triennium. Hu (30.155.006) searched for HI shells at $|b| > 10^o$, using filtering in velocity to enhance the shell features; she found correlations of the HI shells with radio continuum loops. Gosachinskij (31.132.048) saw little correlation of the HI shell phenomenon with supernova remnants, but Velden and Hirth (32.131.058) found a well-defined HI shell apparently associated with the nearby pulsar PSR 0809+74. Bochkarev (37.131.106) argued that the large-scale bubble structure of the interstellar medium is due to the combined operation of stellar winds from Wolf-Rayet, supernovae, and other stars. Heiles (33.155.045) found an explanation for some of the galactic noncircular motions in the HI shells, and stressed (34.131.106) the need for homogeneous and complete observational studies of the shells.

IV. Overall Galactic Structure

A. OPTICAL STUDIES

This section contains work appertaining to observations at wavelengths shorter than 1 mm and giving information on large scale properties of the Galaxy. The different subsections contain work regarding (1) the galactic disk, (2) the galactic centre, (3) the halo and the spheroidal system and (4) the evolution of the Galaxy.

Several conferences during the triennium have discussed our Galaxy as such. The Vancouver workshop 1982 (33.012.016) discussed the Kinematics, Dynamics and Structure of the Milky Way. At the General Assembly in Patras, a joint discussion was held on the subject "Evolution in Old Stellar Populations in Galaxies" (33.012.028). The Frascati workshop 1982 on the first stellar generations (34.012.057) contains a number of papers on the early evolution of the Galaxy. IAU Symposium No. 106 in Groningen was entitled "The Milky Way Galaxy"; its proceedings are being edited by H. van Woerden and published by Reidel. A conference was held in Prague 1983 (34.012.068) on the topic "Star Clusters and Associations and their relation to the Evolution of the Galaxy". The European regional meeting in Florence, December 1983, contained several contributions on galactic structure and about the centre of the Galaxy.

Review papers of interest in this section include the discussion of stellar populations by Mould (32.155.027), the presentation of galactic gamma-ray sources by Bignami and Hermsen (34.143.025) and the Henry Norris Russell Lecture by B.J. Bok (34.155.056) on "Current Trends in Milky Way Research".

1. The Galactic Disk

The galactic distribution of WR stars was discussed by Hidayat et al. (31.155.042, 32.155.040), that of WR run-away stars by Vanbeveren (32.155.011); Bertelli and Chiosi (31.155.029, 32.155.041) analysed the reasons for a gradient in the ratio of supergiants to WR stars in the galactic disk. Efremov (33.155.002) analysed the difference between cepheid distribution at 10-12 kpc in our Galaxy and in M31. The relation of giant clouds and HII regions to spiral structure was discussed by Elmegreen and Elmegreen (33.157.042), and again by Elmegreen (1984). The common origin of a group of open clusters was discussed by Lynga and Wramdemark (37.153.008), and the distribution of open clusters in the disk was reviewed by Lynga (1984). Fich and Blitz (37.155.064) studied the distribution of optical HII regions in the outer Galaxy. The distribution of interstellar sodium has been mapped by Ardeberg et al. (1984) for the southern sky. Three-dimensional structure of spiral arms has been discussed by Kolesnik and Guseva (31.155.019) and by Voroshilov et al. (33.155.118).

Metal abundances in distant RR Lyrae stars towards the galactic anticentre have been observed and discussed by Kinman et al. (31.122.029) and by Butler et al. (31.122.030). There is no direct evidence of an abundance gradient. Photometry and ephemerides for RR Lyrae stars in the field RRI are given by Kinman et al. (1984).

Abundance determinations of the ab type variables in this field show (Kraft, Kinman, Suntzeff) that these stars at a galactocentric distance of 6 kpc are more metal-rich than those at the North galactic pole or in the anticentre direction.

Garmany et al. (32.155.061) studied the IMF of massive stars and found more massive stars inside than outside the solar circle. Meylan and Maeder (34.155.004) determined stellar density functions for various types of massive stars. Considering absolute magnitudes of galaxies and the distribution of HII regions, Hodge (34. 155.098) concluded that our Galaxy is of Hubble type Sc or SBc.

Pritchet (34.155.022) used the deep star counts of Jarvis and Tyson (29.031. 530) to study the stellar population of the disk. Some problems of stellar statistics were discussed by Buser (32.155.054).

Sources of near-infrared radiation were detected and their stellar origin investigated by Mikami et al. (32.155.009) and by Kawara et al. (32.155.035). The origin of the far-infrared and sub-millimeter galactic emission is to be found in dust; models were proposed by Mezger et al. (31.156.006), by Gispert et al. (31. 156.007), by Puget (32.156.009), and by Hauser et al. (33.155.048).

Observations of γ-radiation from the Galaxy have made a great impact on our knowledge about the structure of the disk. A large survey was made by the ESA satellite COS-B (Mayer-Hasselwander et al., 31.157.002, 32.157.003), and observations towards the anticentre region were made by the Natalya-1 telescope (Iyudin et al., 33.155.098). The relations between γ-ray features, interstellar gas and other sources were studied by Strong and Wolfendale (31.157.012), by Haslam et al. (31.157. 014), by Salvati and Massaro (31.157.001), and by Bloemen and collaborators (33. 155.035, 34.155.031, 34.155.034, 37.155.065).

The emission of γ-rays as the result of interaction between cosmic rays and interstellar gas was analysed by Bignami (31.157.013), by Massaro (33.155.016), by Mayer-Hasselwander (33.155.052), by Riley and Wolfendale (34.155.033), and by Lebrun (34.155.032). The mass of H_2 as inferred from γ-radiation is much lower than that given by CO data (Li et al., 33.155.017; Bhat et al., 1984). A different view is presented by Korchagin et al. (33.155.100) who interpret the γ-ray emission in terms of galactic shocks.

The structure of the galactic disk in X-ray was studied by Worrall et al. (31. 157.003) and by Nousek et al. (31.157.010). Hertz and Grindlay (37.142.034) published the statistical analysis of the Einstein galactic plane X-ray survey.

The electron scale height of the disk was studied through a statistical analysis of pulsar dispersion measures by Harding and Harding (31.155.028).

2. The Galactic Centre

Observations and theories regarding the galactic centre were presented at the Caltech conference in 1982 (32.012.076) and at IAU Symposium No. 106 in 1983.

Use of CCDs and other detectors for observations in the infrared with large telescopes have made possible detailed mappings of galactic centre sources. Active groups have been Storey and collaborators (31.155.010, 32.155.068, 32.155.069, 34. 155.012), Biretta et al. (32.155.042, 32.155.070), and Ricker et al. (32.155.071). IR spectra of sources have been investigated by Wollmann et al. (32.155.001), Hall et al. (32.155.048), and Herter et al. (33.155.032). Several surveys of the centre of the Galaxy were presented, notably by Becklin et al. (31.156.019), by Dent et al. (32.156.018), by Stier et al. (32.156.019), and by Matsumoto et al. (32.156.020).

The high-resolution IR survey by Allen et al. (34.155.011) was interpreted in terms of stellar distribution. Genzel et al. (37.155.025) mapped the far-infrared radiation around the galactic centre and discussed the physical conditions in the nucleus.

The galactic centre is a source of continuum γ-radiation (Paciesas et al., 31. 157.015, 32.157.017; Matteson, 32.157.014). Several other groups have also attempted to observe the 511 keV line of electron-positron annihilation (Jacobson, 32.157. 006, 32.157.015; Leventhal and MacCallum, 32.157.005, 32.157.016; Ramaty et al., 32.157.007; Gardner et al., 32.157.018; Jardim et al., 33.155.029). γ-ray spectroscopy by Mahoney et al. (32.157.010) placed limits on emission in the 60-Fe, 26-Al and 22-Na decay lines. Implications of the γ-ray observations for the models of the galactic centre were discussed by Lingenfelter and Ramaty (32.157.019) and by Forrest (32.157.020). Stellar kinematics in the centre were studied by means of OH/IR stars (Baud, 34.155.042; Habing et al., 34.155.048, and IAU Symposium No. 106).

There have been a number of attempts to make a model of the galactic centre and to compare it with other galaxy nuclei. Oort (32.155.025, 32.155.073, and IAU Symposium No. 106) has investigated which models are possible in view of observed data. He finds the evidence for a central black hole "not compelling"; there has probably been a burst of star formation in the nucleus about one million years ago. Gatley (32.156.017) made a model of the galactic nucleus based on observations of infrared emission. Lacy et al. (32.155.038), Lacy (32.155.066), Townes et al. (33. 155.022), and Rees (32.155.072) also examine arguments for and against the existence of a massive black hole in the centre of the Galaxy. Rieke and Lebofsky (32.155. 074) compare the galactic nucleus with nuclei of other galaxies. Oda (34.155.021) related the infrared features near l = 356° to the γ-ray hump. Evolutionary scenarios involving shock-wave generation in the galactic nucleus and subsequent star formation are discussed by Loose et al. (31.155.012) and by Bhattacharyya and Basu (31.155.045). Shklovskij (34.155.010) gives arguments for explaining galactic-centre phenomena as caused by a recent supernova outburst.

The spatial distribution of RR Lyrae stars near the galactic centre is being studied by Wesselink (Nijmegen). M giants near the galactic centre appear to be super-metal-rich according to results by Frogel and Whitford (32.155.003).

3. The Halo and the Spheroidal System

The chemical evolution of the galactic halo was discussed at the IAU General Assembly Joint Discussion on "Evolution in Old Stellar Populations in Galaxies" (33.012.028), particularly by Spite (33.155.062), and also at the Vulcano workshop (34.012.057), particularly by Castellani (34.155.078). Busso (33.155.066) and Busso and Gallino (34.155.014) have studied heavy-element enrichment processes in the halo. Hartwick (34.155.075) has compared [Fe/H] distributions, space densities and kinematics for various halo components. Abundance gradients in the halo were studied by Daido (33.155.077). Smith (37.154.055) finds that globular clusters with galactocentric radii < 9 kpc are more metal-rich than outer ones but that there is no pronounced gradient. Extensive counts of faint stars have been used by Bahcall and Soneira (31.155.013, 33.155.050), by Gilmore and Reid (33.155.015) and by Gilmore (33.155.107, 37.155.016, IAU Symposium No. 106) to study the stellar distribution in the galactic spheroid. Gilmore and collaborators are completing their photometric survey in eight directions of the galactic spheroid. First results show a population that is old and relatively metal-rich. Blanco et al. (37.155.071) picked out and observed late-type giants in Baade's window confirming an earlier finding of an extremely small ratio of C to M stars in the nuclear bulge. B.M. Blanco (CTIO) reports from her blink survey of RR Lyrae stars in Baade's window that the majority of RR Lyraes are relatively metal-rich. Edmunds (34.155.001) and

Edmunds and Phillipps (37.155.017) estimate the luminosity and mass of the galactic spheroid, finding it a major component similar to an elliptical galaxy of $M_V \sim -20$.

The outer ranges of the halo have been discussed by Ratnatunga (33.155.057) and, particularly concerning population II giants, by Cayrel and Boulon (34.155.076). Hawkins (37.122.003) studied the galactic halo from RR Lyrae star distribution finding a power law out to r = 6o kpc.

4. The Evolution of the Galaxy

The chemical evolution of the Galaxy has continued to be an active area of research. Chemical abundances, abundance gradients and isotope ratios have been compared to model predictions assuming rates of star formation and of infall of matter. A review of the field was given by Güsten and Mezger (34.155.081). Particular aspects have been discussed by Tosi (31.155.005, CNO isotope ratios and evolution), by Twarog and Wheeler (32.155.033, heavy element production), by Audouze et al. (34.155.046, Li isotope evolution), by Yokoi et al. (33.155.003, ^{187}Re - ^{187}Os chronology) and by Yoshii (1984, a two-zone model of chemical evolution). Abundance gradients have been studied by Marsakov and Suchkov (33.155.058), by González (33.155.114), and by Binette et al. (32.155.043). The IR excess at 5 kpc galactocentric radius has been introduced in models of galactic evolution by Rocca-Volmerange and Guiderdoni (31.155.030, 31.155.049). Selected evolutionary models were examined by Lee and Ann (31.155.021), by Lee and Hong (34.155.080), by Meusinger (34.155.082), and by Yoshii and Arimoto (33.155.078). Comparisons of evolutionary models of the solar neighbourhood with photometry of edge-on galaxies were made by Lacey and Fall (34.155.006). Díaz and Tosi (37.157.159) compared models of chemical evolution applying them to our own Galaxy and to other spiral galaxies.

B. RADIO STUDIES

Several reviews pertinent to galactic morphology were published during the triennium. Bok's Henry Norris Russell lecture (34.155.056) and his other reviews (33.013.022, 37.155.087) summarized many of the current problems in Milky Way studies. Blitz (33.155.115) emphasized work on the outer reaches of the Galaxy based on new information on the galactic rotation curve. Downes and Güsten (32.155.053) reviewed radio investigations of the large-scale structure and kinematics of our system, including those dealing with the various phases of the interstellar gas. Reviews specific to the core of the Galaxy were given by Oort (32.155.025, 32.155.073) and by Kraus (33.155.020).

Two workshops were held whose proceedings contain contributions on the morphology of the Milky Way. The Green Bank workshop (31.012.048) dealt with the molecular content of our and other normal, nearby galaxies; the Leiden workshop (34.012.023) dealt with surveys of the southern Galaxy. IAU Symposium No. 100 on the Internal Kinematics and Dynamics of Galaxies (33.012.004) was co-sponsored by Commission 33. The proceedings of IAU Symposium No. 106 on the Milky Way Galaxy, held in Groningen, are scheduled for publication in early 1985. Bajaja (34.155.029) gave a compilation of HI surveys made with the Argentine radio telescope; Xiang (37.002.023) summarized galactic CO surveys.

1. Observational Surveys

a) Atomic Hydrogen. Most of the HI work done during the triennium was directed toward specific astrophysical problems, rather than toward general surveying, although a new survey of the southern galactic plane made with the Parkes 64-m telescope was reported by Riley (34.155.030) and by Strong et al. (32.156.010). Bregman et al. (33.155.009) spatially resolved the 21-cm HI Zeeman effect in the Cas A

Perseus arm absorption features. Quiroga (33.155.112) studied turbulent motions in HI structures and their relations with galactic structure. Observations of galactic HI absorption against the continuum spectra of 69 radio sources were published by Mebold et al. (30.141.147). Dickey et al. (34.155.066) used the VLA to measure 21-cm absorption toward 88 sources at $|b| < 15^{o}$. Lockman (33.155.055) reported evidence for HI gas in the inner Galaxy more than 500 pc from the plane that corotates with the material in the plane but has a larger velocity dispersion than the HI confined closely to $b = 0^{o}$.

Bania and Lockman (37.155.074) used the Arecibo telescope to measure closely-sampled HI spectra which reveal numerous self-absorption features that arise in cool HI clouds, for which they were able to specify size parameters. Peters and Bash (37.155.021) sought correlations between CO clouds and HI self-absorption features in the galactic plane at large distances from the Sun.

b) Carbon Monoxide. The triennium saw the publication of several surveys of emission from CO; for the first time, extensive coverage of the galactic disk is available for this important tracer of the density of the interstellar medium. The CSIRO telescope extended the coverage in the ^{12}CO line to the southern portions of the galactic equator. Publications by Manchester, Robinson, Whiteoak, and collaborators deal with various aspects of the latitude and longitude distribution (32. 131.032, 33.131.126, 34.155.027, 34.155.039, 34.131.150). Southern observations of the 2→1 CO transition at λ1.3 mm were made at ESO by Israel, De Graauw, and collaborators (34.155.025, 37.155.066).

The northern hemisphere ^{12}CO data bank was substantially extended by the FCRAO survey (34.155.025, 37.155.015); the analysis of that material included discussion of the axisymmetric distribution of H_2. Conversion from CO intensities to H_2 densities was discussed by Xiang et al. (37.131.169), who also gave closely-sampled observations of ^{13}CO along the galactic equator. Comparison of the ^{13}CO and ^{12}CO aspects of a selected galactic-disk region was given by Stark et al. (34.155.045).

The Bordeaux Observatory telescope was used to observe ^{13}CO emission in the 1st and 2nd galactic quadrants (30.155.060, 34.155.043, 34.155.044, 37.131.095). Particular emphasis was paid to the mass spectrum of the observed clouds. The cloud-to-cloud velocity dispersion of the molecular clouds in the inner Galaxy was discussed by Liszt and Burton (33.131.125) on the basis of new ^{13}CO data and by Stark (33.131.320).

The 1st quadrant was searched by Verter et al. (34.131.009) to reveal over 100 directions in which no molecular cloud of consequence occurs along the lines-of-sight, constraining cloud size and distribution statistics. Evans et al. (33. 155.024) searched for ultra-cold ($T_{ex} < 5$ K) molecular gas but found no evidence for a large amount of such material.

c) OH and Other Radio Lines. Turner (31.156.005, 31.156.021, 33.155.047) observed widely extended, weak OH emission at 1720 MHz which is purportedly a good tracer of galactic structure. Southern galactic plane surveys of OH, H_2O masers, and other lines were reviewed by Caswell and Haynes (34.155.026) and by Whiteoak (34.155. 027). Hart et al. (34.155.028) reported on a southern survey of H166α emission from the galactic plane.

d) Continuum Emission. Haslam et al. (31.141.036) published their major 408 MHz all-sky continuum survey combining data from telescopes at Jodrell Bank, Effelsberg, and Parkes as sets of contour maps of radio brightness in α,δ as well as l,b coordinates. Reich (33.155.008) and Wielebinski (37.155.075) reviewed the various uses of this survey. Reich (31.141.085) published the part of the Stockert 1420 MHz survey of the northern sky at $\delta > +20^o$ in the form of contour maps. The combined 408 and 1420 MHz surveys yielded the distribution of the spectral index across the Galaxy to Reich et al. (34.155.100). The spectrum of galactic nonthermal radiation between 200 and 375 MHz was studied by Belyaev et al. (37.155.055).

2. Structural Properties of the Inner Galaxy

The extensive new CO surveys were the focus of several attempts to derive the morphology of the gas distribution. Analyses of the CO data in terms of spiral structure were made by Stark (33.131.124) and by Dame (37.131.091). Bash, Leisawitz, and collaborators used the ballistic-particle approach to cloud motions in several analyses of the CO and other data (31.131.277, 32.155.002, 33.132.038, 34.155.035). Kimura and Tosa (37.151.058) predicted the distribution of molecular clouds on the assumption that they move on ballistic orbits and confronted these predictions with the CO data. Sawa et al. (33.155.071) analyzed the CO distribution in terms of three alternative kinematic models. The kinematics of HII regions was investigated by Grivnev (30.132.018) supposing the existence of density waves in the Galaxy. Seiden (33.131.127) used the stochastic self-propagating star formation model to account for the observed radial distribution of atomic and molecular gas.

The distribution of free electrons in the inner Galaxy was the subject of work by Harding and Harding (31.155.028) based on dispersion measures of pulsars. Gusejnov et al. (30.155.037) studied the distribution of electrons using H_2O maser, radio recombination line, and other data. Those aspects of the overall structure which involve comparison of γ-ray data with nucleon densities from CO and HI observations are mentioned in Section VIF of this report.

3. Galactic Center

Several investigations were carried out dealing with scales of order 100 pc. Rohlfs and Braunsfurth (32.155.012) published closely-sampled HI data for the inner few degrees and discussed the material in terms of galactic rotation. Burton and Liszt (33.155.027) published a survey of the region $349 \leq l \leq 13^o$, $|b| \leq 10^o$, $|v| \leq 350$ km s^{-1} made with the NRAO43-m telescope and gave summarizing remarks concerning the anomalous central features. Surveys in the CO lines were reported by Inatani (32.131.289) and by Heiligman (30.155.024). Burton and Liszt (34.155.040) described a CO structure near the center with strong positional and kinematic gradients. Surveys in the OH lines were published by Cohen (32.131.228), who emphasized the tilted nature of the gas distribution, and by Cohen and Dent (33.155.025, 34.155.041). OH/IR stars and the information they reveal on the kinematics and mass in the galactic bulge were the subject of observational and interpretive work by Isaacman and Oort (30.155.042), Habing et al. (33.112.019, 34.155.048), and Baud (34.155.042). Ammonia in the neighbourhood of the galactic center was mapped by Güsten et al. (30.155.045), and its temperature distribution derived.

4. Structural Aspects of the Outer Reaches of the Milky Way

The rotation parameters of the outer Galaxy, found in studies mentioned elsewhere in this report, are now known firmly enough that renewed attention could be given to the properties of the warped HI disk at large radii. Henderson, Kerr, and Jackson (32.155.045, 33.155.043, 34.155.036) used the Parkes 18-m and Hat Creek 26-m telescope surveys to derive, assuming a flat rotation curve, the HI surface density, mean layer positions, and thickness as a function of galactocentric radius and angle. Kulkarni, Blitz, and Heiles (32.155.014, 33.155.044, 34.155.037) used the CO-derived rotation curve and the Hat Creek data to get the HI structural parameters to distances of 30 kpc from the center. The z-extent of the HI found in these studies implies that the large invisible mass in the outer Galaxy does not reside in a thin disk.

Kutner and Mead (30.155.036, 31.131.271, 34.131.105) reported detection of extensive low-level CO emission from molecular clouds well outside the solar circle, whose distribution apparently followed the HI warp. Solomon et al. (33.155.031) were unable to reproduce some aspects of those observations.

Work on the high-velocity cloud phenomenon involved several detailed studies of selected regions. Evidence for interaction of a high-velocity stream with the galactic disk was found by Cohen (30.131.006) in Jodrell Bank observations and by Mirabel (31.132.022) in Arecibo observations of the anticenter region. A survey by Cohen (32.131.021) of the Cetus region led to the conclusion that the high-velocity HI there is debris from the tidal interaction between our Galaxy and the Magellanic Clouds. Mirabel and Morras (37.155.063) published results from a search for high-velocity HI in a large area of the sky around the direction to the galactic center.

C. REFERENCES

Ardeberg, A., Lindgren, H., Maurice, E.: 1984, IAU Colloquium No. 81 (in press).
Bhat, C.L., Issa, M.R., Houston, B.P., Mayer, C.J. and Wolfendale, A.W.: 1984, (in preparation).
Elmegreen, B.G.: 1984, IAU Symposium No. 106 (in press).
Kinman, T.D., Wong-Swanson, Wenz, Harlan: 1984, Astron. J. (in press).
Lynga, G.: 1984, IAU Symposium No. 106 (in press).
Yoshii, Y.: 1984, Astron. J. (submitted).

V. Kinematics

A. STARS

1. Galactic Center

Wollman et al. (32.155.001) present high-resolution 2 μm spectra of three red giant stars in the core of the Galaxy. The properties of OH/IR stars were studied by Baud et al. (29.131.023), Olnon et al. (29.155.032), Habing et al. (34.155.048), and Baud (34.155.042). Vanderspeck and Ricker (34.041.001) determined absolute coordinates for 19 stars within 2' of the galactic center for epoch 1958.3 and 1976.3 and calculated their proper motions. Mould (33.155.013) obtained the radial-velocity dispersion for M stars in the nuclear bulge.

2. Disk and Solar Vicinity

Stetson (30.111.009) published a list of 371 A-F population I stars of high velocity determined from proper motion studies. Kibblewhite et al. (33.111.018) analyze proper motions of disk stars to obtain measures of the galactic differential rotation. Murray (34.155.108) discusses the kinematics of dwarf stars and obtains histograms of proper motion for 900 stars with $B < 14$ near the south galactic cap. An observational program to measure absolute proper motions of stars with respect to galaxies was described by Kharchenko (34.155.018).

Zentelis (34.111.004) measured the radial velocities of 353 B0-A0 stars with $6.5 < V < 10.8$ in six galactic plane regions with a velocity uncertainty of 1.4 km/s. Fehrenbach and Burnage (29.111.003) and Fehrenbach et al. (32.111.001) published radial-velocity catalogues each containing four $4^o \times 4^o$ fields. Hartkopf and Yoss (32.155.062) obtained radial velocities for 302 polar giants.

A statistical study of 403 OB stars at $r < 630$ pc was performed by Quiroga and Tarsia (34.151.039) which shows that z motions are predominantly hydrodynamic. Robertson (34.155.031) has made a comparison of the kinematics of M stars having M > 10.5 with K and M dwarfs selected from Gliese's catalogue. Lotkin (34.155.079) investigated the kinematics of 614 red giants in the solar vicinity and found differences between components of peculiar space velocities and the normal distribution. He also examines, in another work (33.155.101), the radial velocities of 369 B stars in the solar neighborhood and concludes that they exhibit equipartition of energy per unit mass. Kharachenko (33.155.060) uses absolute proper motions of stars to obtain kinematic characteristics of stellar groups. Yoshizawa (33.155.076) presents comments on stellar proper motion analyses and the velocity field in the solar neighborhood. An analysis of the kinematical properties of dwarfs grouped by spectral type was performed by Lee (31.155.022); Jahreiß and Wielen (34.155.123) used Gliese's catalogue to investigate the kinematical characteristics of stars in the solar neighborhod. Suchkov (33.155.001) examined the velocity ellipsoid of nearby dwarf stars. Gradients of the ellipsoid of the residual velocities in the galactic plane in the Π and Θ directions were calculated by Oblak (34.111.001). The correlation between X-ray luminosity and space velocity of nearby K dwarfs was analyzed by Johnson (34.155.127). Gliese (34.155.102) in a review of stars in the solar neighborhood, discussed a number of problems related to stellar velocity distributions.

Equations of the centroid and velocity disperion have been derived by Eelsalu (32.155.006) connecting the stellar-statistical kinematic data with the corresponding descriptive variables. He also considered (32.155.007) the special case of dispersion equations corresponding to high galactic latitudes. Nunez and Torra (31.155.050) calculated velocity momenta through the fourth order for different samples of Gliese's catalogue. Nunez and Figueras (34.155.125) calculated central momenta of the local velocity distribution to fourth order. Ogorodnikov and Osipkov (1980) considered the influence of high-velocity stars on estimating the dispersion of stellar velocities. Shuter and Goulet (34.155.126) analyze the radial velocities of 260 A V stars and show that they exhibit significant departures from pure circular motion. King (33.155.037) discusses possible reasons for the dip in the K_z curve derived from stellar kinematics.

Byl and Ovenden (30.155.001) performed numerical experiments of fictitious proper motions with and without spiral arms. Ardeberg and Maurice (29.155.038) studied radial-velocity data for young stars and gas on the inner border of the Carina arm. Using five samples totaling 500 stars, Berman and Mishurov (30.155.038) obtained the spiral structure parameters from a non-linear description of the stellar motion. Pavlovskaya and Suchkov (31.155.006) study statistical fluctuations in a sample of ∿ 150 stars within 3-4 kpc. Balázs (33.155.116) uses moderately old open clusters to determine the angular velocity of the local spiral arm pattern. Pellegatti Franco (34.155.052) finds that velocities of OB stars agree well with non-linear density-wave theory. The motions of 55 Cepheids were traced back in time by Grinev (34.155.067) who finds consistency with density-wave theory. Clube (34.155.124) reviews current attempts to establish peculiar motion of the local spiral arm. Gerasimenko (34.155.139) locates the inner edge of the Perseus arm from the velocity field of O-M stars.

Wyatt and Cahn (34.155.144) investigate the kinematics of 124 Mira variables and find the largest velocity dispersions in the oldest Miras. Baud et al. (29.131.022) observed the OH emission from 114 OH/IR stars and showed that there are two distinct kinematic groups related to the ages of the stars.

The slope dV/dl of the radial velocity of a feature close to $l = 0^o$ of $l = 180^o$ is shown by Rohlfs (30.155.030) to be a good measure of the distance to the galactic center. Palous (34.155.143) estimates the A,B,C and K constants from gradients in the U,V components in the X,Y plane. Vader and de Jong (30.155.002) constructed models of the galactic disk near the Sun by calculating the chemical and kinematic evolution of the gas. Lindblad (33.155.038) discusses the local expansion of gas and stars in terms of star formation and giant molecular clouds. Rohlfs and Kreitschmann (30.155.040) make a three component mass model of the Galaxy to, among other things, reproduce the rotation curve. Bahcall et al. (31.151.078) examine how various mass distributions can produce flat rotation curves out to 60 kpc. A review of the recent literature on the age distribution, kinematic properties and distribution over composition of the stellar component of the Galaxy was made by Mould (32.155.027).

3. Halo

Hartwick (34.115.075) finds that the kinematics of halo dwarfs and giants are similar to the RR Lyrae stars. Ratnatunga (33.155.057) discusses the kinematics and metal abundances of stars in the halo with respect to the formation and evolution of the Galaxy. Keenan and Dufton (34.114.031) find OB stars at high galactic latitude to be normal population I objects requiring ejection velocities > 100 km/s to be consistent with their lifetimes. A distant galactic RR Lyrae star was discovered by Hawkins (33.155.074) with a high negative radial velocity implying a lower galactic mass limit of 1.4×10^{12} $M_\odot$.

Reviews of the kinematical properties of globular clusters were given by Freeman (33.154.021, 33.154.026). Ninkovic (34.155.083) calculated orbital eccentricities of globular clusters. Innanen et al. (33.155.014) examine the shapes of typical globular cluster orbits. Lynden-Bell and Frenk (30.155.053) determine the circular velocity of the Galaxy (= 212 ± 16 km/s) from globular clusters. White and Frenk (33.154.028) discuss the kinematics of the globular cluster system. Lynden-Bell et al. (34.155.013) remeasured carbon stars in dwarf spheroidal galaxies and argue that the mass of the Galaxy is $2.6 - 10 \times 10^{11}$ $M_\odot$.

B. INTERSTELLAR MATTER

1. Central Regions

Reviews of the kinematic structure of the galactic center were presented by Oort (32.155.025, 32.155.073) and Townes et al. (33.155.022). Shklovskii (34.155.010) suggests that the kinematics and other features seen toward the galactic center can be explained by a supernova outburst in a massive binary ∿ 100 years ago. Rohlfs (34.155.070) shows that expanding features near the center can be explained either by large-scale expansion or by gas in closed orbits of approximately elliptical shape. Blitz et al. (29.155.005) proposed that the bending-wave theory of Bertin and Mark can provide the dynamical explanation of the tilt of the inner gas disk.

Becker and Sramek (32.141.053) made the first measurements of the proper motion of the compact non-thermal radio source at the galactic center by VLBI. Arguments for a massive black hole at the center were reviewed by Lacy et al. (32.155.038) and by Lacy (32.155.066). Brown (32.155.037) suggests that the radio continuum brightness distribution is due to collimated gaseous outflow along twin opposing beams. Lo and Claussen (34.155.059) identify three streams that could be due to infalling molecular gas. Hall et al. (32.133.005) present the discovery of a helium feature with a line width of ∿ 1500 km/s.

Lester et al. (30.132.012) detected the 63 μm line of OI toward Sgr A and find rotational and radial motions similar to ionized gas in the core of the Galaxy. Nadeau et al. (29.155.016) show that the spectra of Bγ toward the center has central velocities and line shapes similar to those reported for [NeII]. Bα and [NeII] lines were observed toward the center by Geballe et al. (32.133.022) showing evidence for ionized gas associated with IRC 16 and with $-500 < V < 450$ km/s. Bregman and Schwartz (32.156.004) present Westerbork data on the H110α recombination line toward Sgr A west. Caswell and Haynes (31.132.005) report the discovery of 3 HII regions near the center with high velocities.

Molecular gas at large negative velocities is detected in absorption against the galactic center by Linke et al. (29.155.013). Güsten and Downes (29.131.267) observed H_2CO and HI toward the galactic center and find weak high velocity absorption extending to $V = -210$ km/s. Güsten (32.131.323) summarizes detections of high-velocity gas toward the center and the observations and mapping of the central 15' in lines of NH_3 and H_2CO. CO observations of the inner 1° of the Galaxy are discussed by Inatani (32.131.289). Fukui et al. (34.131.299) detect molecular emission HCN which is shown to be in the central 10 pc of the Galaxy.

Cohen and Dent (34.155.041) make a large-scale survey of OH in the galactic center region which indicates a bar-like structure. Burton and Liszt (34.155.040) find an inclined CO feature in the quadrant where l, b are positive with a velocity gradient of ∿ 100 km/s over ∿ 1° and with a large velocity dispersion. Sandqvist (32.131.324) observed H_2CO toward Sgr A west. The large-scale spatial and kinematic data for the overall system of molecular clouds at the center were reviewed by Scoville (29.131.129). Cohen and Few (29.131.010) made an H_2CO survey within a few degrees of the center and showed that the H_2CO is strongly concentrated toward the center.

HI observations from the central regions of the Galaxy were reported by Burton and Liszt (33.155.027). Rohlfs and Braunsfurth (32.155.012) investigate the kinematics of HI within 1.5^o of the center. Quiroga (29.155.004) analyzed HI toward the center and finds systematic oscillating motions in the nucleus.

2. Galactic Rotation

Bok in his Russell Prize lecture (34.155.056) reviewed the recent progress made in unraveling the large-scale structure of the Milky Way and the determination of the outer Galaxy rotation curve. Knapp (33.155.053) reviewed determinations of the rotation law of the Galaxy. Schneider and Terzian (34.155.087) used radial velocities and a Σ-D relation for planetary nebulae to derive the rotation curve for $R > R_0$. Blitz (33.155.023) discusses the implications of a rising rotation curve on the mass distribution of the outer Galaxy. Blitz and Fich (33.155.046) examine the errors in the determination of the CO rotation curve for $R > R_0$. Gill and Shuter (34.155.058) made sinl - v plots in the first and fourth quadrants to analyze the HI kinematics and derive the rotation curve. Shuter (29.155.001) uses published 21 cm and CO data to reanalyze the rotation curve.

Ovenden et al. (33.155.040) and Pryce (33.155.041) examine the kinematics of stars and gas in the solar neighborhood. Shuter (33.155.042) uses a second-order Taylor series expansion for radial velocity to determine AR_0 from O and B stars and CO observations. Pismis (32.155.022) reviews the determination of the galactic rotation curve; Bektasova and Petrovskaya (34.155.019) and Petrovskaya (30.155.003) determine the rotation law from 21 cm profiles. Heiles (33.155.045) argues that motions of gas due to HI shells and supershells are the reason for much of the non-circular gas motions found toward the galactic anticenter. Yuan (33.155.036) discusses predictions of density-wave theory to argue that the LSR has a net motion with respect to young objects. Shuter (31.155.004) argues against the conventional LSR for studies of galactic rotation and suggests using a rotational standard of rest.

3. Spiral Structure

Blitz et al. (33.155.115) and Blitz (34.155.037) use kinematics of the outer Galaxy to determine the properties of the HI disk beyond the solar circle. Turner (31.156.005, 31.156.021, 33.155.047) finds widely extended 1720 MHz OH emission and argues that it is a good tracer of spiral structure. Bash and Turek (33.132.038) compared l-v diagrams of HII regions and 1720 MHz OH emission and find that both suggest the presence of spiral arms. Gordon and Burton, however, (29.155.034) discuss the kinematics of molecular clouds and find the existence of clustering which is not satisfactorily explained solely by postulating spiral arms in the Milky Way. Clube (33.155.026) reports observations that may support the possibility that spiral arms are recurrently produced, short-lived material bodies.

Dolidze (33.155.069) tries to explain peculiarities of local spiral structure by the coexistence of leading and trailing arm systems of different power and development. Downes and Güsten (32.155.053) review investigations of large-scale galactic structure. Grinev (30.132.018) examines the kinematics of HII regions under the assumption of density waves in the Galaxy. Using three different kinematic models, Sawa et al. (33.155.071) show that the distribution and kinematics are consistent with the density-wave model and that the CO clouds form in spiral arms.

4. Disk and Solar Vicinity

Quiroga (33.155.112) examines the hierarchy of motions using the 21 cm line to study turbulence in the galactic disk. Lousto and Muzzio (33.155.033) use published photometric and spectroscopic results to show that turbulent motions in the Sagittarius arm which have been invoked to explain interstellar-line observations for $l > 340^o$ extend to $l < 340^o$ also. Schwartz et al. (31.155.051) present a Westerbork synthesis of the HI seen in absorption toward the galactic center. Pöppel et al. (32.155.024, 33.155.030) analyze HI profiles in Ophiuchus, Scorpius and Lupus. In a model of the local gas related to Gould's belt, Olano (32.155.008) considers the interaction between the expanding ring and the stationary gas.

De Graauw et al. (34.155.025) using their J = 2-1 CO survey of the southern galactic plane, measured the cloud-to-cloud velocity dispersion of the molecular gas. Liszt and Burton (33.131.125) examine the kinematics of CO in the inner Galaxy and find a velocity dispersion of 4.2 ± 0.5 km/s. Stark (33.131.124) uses the scale height of molecular clouds in the inner Galaxy as an indication of their kinematics. Liszt et al. (30.131.128) compare ^{13}CO terminal velocities with those of HI to show that the kinematics of dense molecular clouds are essentially the same as that of HI. Liszt and Burton (29.131.052) present synthetic l-v diagrams of CO for comparison with observations and include perturbations associated with galactic spiral structure. Liszt et al. (29.131.217) show that cold residual HI in molecular clouds is likely to be responsible for small-scale scatter in measured HI terminal velocities.

Lacey and Fall (34.155.006) examine the kinematic and chemical evolution of the galactic disk making detailed comparisons with observations. Bash (34.155.035) uses the ballistic particle model to predict what should be found in CO surveys of the southern galactic plane. Hilton and Bash (31.151.022) use the ballistic particle model to predict the velocities of young stars near the Sun and the vertex deviation toward $l = 320^o$.

5. Halo

Lockman (33.155.055) presents evidence that high z (> 500 pc) HI in the inner Galaxy shares in the rotation closer to the midplane. A complete set of radial velocities for all l, b for gas in a corotating and differentially rotating halo are presented by Feitzinger and Kreitschmann (32.155.005). McGee et al. (34.155.057) find low-intensity HI in the galactic halo in three distinct velocity ranges. A number of studies of high-velocity clouds have been made by Mirabel (29.131.252, 30.159.017, 31.132.022, 32.131.150). De Boer (33.131.128) reviews the progress in understanding the high-velocity component of the interstellar medium.

York (32.155.028) reviews data on gas in the halo including gas motions for comparison with quasar absorption lines. York et al. (34.131.194) observed SiIV absorption toward 3C273 which showed halo gas centered at zero velocity with no high-velocity components. De Boer and Savage (33.131.022) find an absorbing cloud toward M13 in an IUE spectrum and conclude that the cloud's motion is incompatible with halo gas corotating with the disk. York et al. (31.131.058) detect narrow interstellar absorption lines toward MK 509 and Fairall 9 that are consistent with a corotating halo.

VI. Dynamics

A. STELLAR ORBITS

1. Reviews

During the last years, work on stellar orbits concentrated on two areas: Firstly, the existence of non-classical integrals ("third integral") and the occurence of "stochastic orbits" has been investigated for various systems of two degrees of freedom (motions in the comoving meridional plane of axisymmetric galaxies and in the equatorial plane of spiral or barred galaxies). Secondly, three-dimensional orbits have been studied in detail for triaxial systems (galactic bars, prolate elliptical galaxies).

Reviews have been presented by Binney (32.151.041, 32.151.048) and by Contopoulos (34.151.103, 1984a). Many relevant papers can be found in the proceedings of IAU Symposium No. 100 (33.012.004). Relativistic stellar dynamics has been summarized by Contopoulos (34.151.034).

2. General Problems

Binney and Spergel (31.151.003, 37.151.008) developed a spectral description of orbits by using Fourier transforms and action integrals. Ratcliff et al. (37.151.080) have applied this technique to two-dimensional orbits in triaxial systems. Carnevali and Santangelo (37.151.097) used the phase-space correlation function for determining the effective number of isolating integrals. Binney (32.151.034) discussed the applicability of Jeans's theorem.

The relation between the (infinite Feigenbaum-type) bifurcations of orbit families and the onset of stochasticity of orbits has been investigated by Contopoulos (1981, 34.151.011, 34.151.073, 1983a,b, 1984b).

General problems of non-classical integrals or stochastic behaviour of stellar orbits have been discussed by Antonov (31.151.045, 32.151.082), Barbanis (1984), Binney et al. (1984), Genkin and Genkina (32.066,013, 34.042.097, 1984), Gerhard (1984a,b), and Malasidze (30.151.089).

Orbital behaviour in systems of three degrees of freedom in general has been studied by Contopoulos et al. (32.151.087), Contopoulos (34.042.080), Magnenat (32.151.064, 34.151.048), Martinet and Magnenat (29.151.039), and Martinet et al. (30.021.015).

Binney (30.151.002) studied the resonant excitation of motion perpendicular to galactic planes. Osipkov and his colleagues investigated numerically galactic orbits of various types of objects (30.153.009, 30.152.032, 30.153.035, 31.153,015). For relaxation of stellar orbits due to the irregular part of the galactic gravitational field, see Section VID3 of this report.

3. Spiral and Barred Galaxies

There is continuing interest in the behaviour of orbits in the equatorial plane of spiral or barred galaxies. Contopoulos (31.151.095) discussed the trapping of orbits around stable periodic orbits. Periodic orbits in barred galaxies have been investigated by Contopoulos (34.151.109), Petrou (33.151.045), and van Albada and Sanders (32.151.035). Regular and stochastic motion in barred or spiral galaxies has been studied by Athanassoula et al. (33.151.043, 34.151.040), Binney (32.151.033), Contopoulos (33.151.003), and Teuben and Sanders (33.151.044). The effect of resonances in barred galaxies is discussed by Contopoulos (30.151.040, 30.151.058), Martinet (37.151.020), and Papayannopoulos and Petrou (33.151.023).

A number of investigations aim at obtaining the response of a galactic disk to an imposed spiral, oval or barred perturbation of the potential from the studies of orbits. We mention Levinson and Roberts (29.151.058), Roberts et al. (33.151.031), Schwarz (29.151.090), Sanders et al. (33.151.047), Polzin and Thielheim (30.151.038), Spreckels and Thielheim (31.151.039), and Thielheim and Wolff (29.151.052, 31.151.011). See also Section VID of this report.

4. Oblate Ellipticals

Richstone studied orbits in oblate scale-free potentials (31.151.004) and used Schwarzschild's self-consistent field method to produce solutions to the collisionless Boltzmann and Poisson equations for oblate ellipticals (37.151.096); see also Miller (31.151.017). Other models for oblate spheroidal systems have been constructed by Petrou (33.151.014, 33.151.015) on the basis of an approximate third integral. Periodic orbits in elliptical galaxies have been investigated by Caranicolas (32.042.075, 34.151.033, 1984a,b), Caranicolas and Barbanis (32.151.046), Caranicolas and Maropoulou (34.151.059, 1984), Caranicolas and Varvoglis (1984), and Davoust (34.151.012).

5. Triaxial Systems

Schwarzschild and his coworkers (32.151.001, 33.151.132) have continued to study orbits in triaxial systems (prolate ellipticals or three-dimensional bars). Rotation has now been included (32.151.078). Special attention has been paid to the existence of semi-stochastic orbits with long evolution times (29.151.060). The collective stability of Schwarzschild's triaxial galaxy model has been studied by Smith and Miller (31.151.058).

Many other interesting investigations on stellar orbits in triaxial systems have been carried out. We mention here those by Caranicolas (37.151.037), de Zeeuw (37.151.003), de Zeeuw and Merritt (33.151.093), Magnenat (31.151.035), Magnenat and Martinet (33.151.057), Martinet and de Zeeuw (33.151.054), Mulder (33.151.090), Mulder and Hooimeyer (37.151.047), Pfenniger (37.151.068), and Wilkinson and James (31.151.012).

The orientation and kinematics of gaseous disks may provide clues concerning the true three-dimensional shape of triaxial systems. Such preferred orbital planes in triaxial systems have been studied by David et al. (34.151.090), Durisen et al. (33.151.005), Lake and Norman (34.151.001), Merritt and de Zeeuw (33.151.069), and Steiman-Cameron and Durisen (37.151.012).

6. References

Barbanis, B.: 1984, Celest. Mech. 33, 385.
Binney, J., Gerhard, O.E., Hut, P.: 1984, "Structure of Surfaces of Section" (preprint).
Caranicolas, N.: 1984a, Celest. Mech. 33, 209.
Caranicolas, N.: 1984b, Astron. Nachr. (in press).
Caranicolas, N., Maropoulou, J.: 1984, Z. Naturforsch. (submitted).
Caranicolas, N., Varvoglis, H.: 1984, Astron. Astrophys. (in press).
Contopoulos, G.: 1981, Lett. Nuovo Cimento 30, 498.
Contopoulos, G.: 1983a, Physica 8D, 142.
Contopoulos, G.: 1983b, Lett. Nuovo Cimento 38, 257.
Contopoulos, G.: 1984a, ESO Sci. Prepr. No. 342.
Contopoulos, G.: 1984b, in Singularities and Dynamical Systems, Ed. S. Pneumaticos, North Holland (in press).
Genkin, I.L., Genkina, L.M.: 1984, Tr. Astrofiz. Inst. Alma-Ata 43,82.
Gerhard, O.E.: 1984a,b, Astron. Astrophys. (both papers in press).

B. MODELS OF THE GALAXY

1. Basic Data for Constructing Galactic Mass Models

Reviews of implications of the recent observational data concerning the galactic structure were given by Bok (33.013.022) and Blitz et al. (33.155.115).

The recent data of the rotational velocity in the Galaxy were discussed by Blitz (33.155.023) and Knapp (33.155.053). A method to derive the circular velocity at the sun was proposed by Lynden-Bell and Frenk (30.155.053).

Analyzing the structure of the globular-cluster system, Frenk and White (31. 155.001) estimated the distance of the Sun from the galactic center as $R_o \sim 6.8$ kpc, while Ovenden and Byl (33.155.039) reported that the direct analysis of the radial velocities of distant OB stars gives $R_o \sim 10.4$ kpc. From the local velocity field of stars, Shuter (33.155.042) derived again the IAU values of $2AR_o$ = 300 km/s and R_o = 10 kpc.

Shuter et al. (33.155.040) discussed the velocity fields of gas and stars within five kpc of the sun. The local standard of rest was re-examined by Shuter (31.155.004) and Yuan (33.155.036). Oblak (34.111.001) analyzed the gradients of the velocity ellipsoid for the nearby stars.

Gilmore, Reid and Hewett (32.115.008, 32.115.009, 33.155.015, 34.155.118) have carried out an extensive photometric survey of low-luminous main-sequence stars with absolute magnitudes above the thermonuclear burning limit. They found a broad maximum near M_V = +13 in the luminosity function, and the total luminous mass density of about 0.04 $M_\odot/pc^3$ at the sun. They identified the old thin disk with the scale height of about 300 pc and a thick disk with the height of about 1500 pc, and excluded low-luminosity stars as candidates for the missing mass in the solar neighbourhood; see also the review paper 34.157.114, Gilmore (37.155.016), Reid and Gilmore (37.115.004), and Gilmore et al. (1984). Based on the data given by Gilmore and Reid , Edmunds and Phillipps (37.155.017) suggested that the galactic spheroid represents a major component of the Galaxy and may be equivalent to an elliptical galaxy with $M_V \sim -20$.

Bahcall et al. (33.155.012) evaluated the stellar content of the galactic spheroid, and fitted the galactic rotation curve with a four-component galactic mass model of an exponential disk, a de Vaucouleurs spheroid, a massive halo, and a central component. Bahcall and Soneira (33.155.050, 34.155.119, 34.155.120) discussed galactic parameters based on the available data of star counts, and Bahcall (34.155.122, 37.155.014) derived a total mass density at the sun of 0.185 $M_\odot/pc^3$.

2. The Dark Galactic Halo

The true nature of the dark galactic halo remains still an embarrassing problem in modelling the overall distribution of the galactic matter.

A total mass of the Galaxy out to about 100 kpc of 2.6 x 10^{11} $M_\odot$ was derived by Lynden-Bell et al. (34.155.013) from the best modern data on all the nine galactic companions. Hawkins (33.155.074) discovered a candidate for a very distant galactic RR Lyrae star with high negative radial velocity, from which a lower limit of the total galactic mass of 1.4 x 10^{12} $M_\odot$ was derived.

Under the condition that the boundaries of globulars and dwarf companions are tidally limited, Wakamatsu (31.155.002) and Innanen et al. (33.155.014) discussed the galactic mass distribution. They came to different conclusions on the existence of the massive galactic halo.

In determining the logarithmic slope of the halo density law, Bahcall et al. (31.151.078) have illustrated the importance of taking account of the interior luminous mass components of the Galaxy, and concluded that, on the basis of the observations, there is little direct evidence for a R^{-2} halo density law.

3. Models of the Overall Mass Distribution

Based on currently existing galactic-structure data, three-component mass models for the Galaxy, which consist of an exponential disk, a spheroid with the Hubble or the de Vaucouleurs density law, and a pseudo-isothermal spherical halo, were proposed and compared with previous models by Caldwell and Ostriker (30.155.051, 33.155.054), and Rohlfs and Kreitschmann (30.155.040). Schmidt (1984) reviewed published mass models and examined a set of test models. He showed that even with a given rotation curve and given density laws for the three components of the Galaxy, there is a variety of acceptable solutions.

The implication of the outer flat rotation curve was discussed by Shuter (29.155.001) and Rohlfs (31.155.011).

4. Theoretical Models

A family of exactly self-gravitating, collisionless stellar systems with axial symmetry was constructed by Toomre (32.151.017). Two-component versions of these models yield disk-halo stellar system with flat rotation curves. Hunter et al. (37.151.045) found a class of the self-gravitating truncated disks by generalizing the Mestel and the Tommre disks.

Binney and Spergel (34.151.111) proposed a method to construct a self-gravitating stellar system that takes full account of the third integral. The combined Boltzmann and Poisson equations for the isothermal distribution of stars perpendicular to a galactic disk was solved by Bahcall (37.151.016) in the presence of a spherical halo. He suggested a method for measuring the ratio of halo to disk mass distribution.

Based on a stability criterion, Morozov (34.155.138) tried to avoid an arbitrary choice for the mass in the disk and halo in galactic mass modelling.

5. References

Gilmore, G., Reid, N., Hewett, P.C.: 1984, Mon. Not. R. Astr. Soc. (in press).
Schmidt, M.: 1985, IAU Symposium No. 106.

C. SPIRAL STRUCTURE

1. Reviews

The spiral structure of galaxies is probably related to density waves, primarily governed by gravitational forces. Density waves may result from an inherent instability of galaxies against spiral perturbations as conceived in the conventional density-wave theory or may be forced by other internal or external perturbations of the gravitational field, such as neighbouring galaxies or oval distortions and bars in the inner regions of the galaxies. Reviews of recent developments on the various aspects of gravitational theories of spiral structure have been given by Ambastha and Varma (30.151.029), Athanassoula (33.151.051, 1984), Contopoulos (32.151.021, 34.151.103), Donner (30.151.085), Hunter (34.151.053), James and Wilkinson (29.151.023), Kalnajs (33.151.024), Kormendy (32.151.049), Lin (32.151.040, 33.151.025, 33.151.071), Lin and Bertin (30.151.068, 1984), Lin and Roberts (30.151.045), Martinet (30.151.043), McElroy (34.157.160), Norman (33.157.088), Pasha and Tsitsin (34.151.042), Sorensen (29.151.024), Thonnard (31.158.357), and Toomre (30.151.021). Seiden and Gerola (31.151.084) reviewed the theory of formation of spiral structures by stochastic self-propagating star formation in galaxies.

2. Spiral Modes, Origin and Maintenance of Spiral Structure

The conventional density-wave theory has been extended by Bertin (33.157.192), Bertin and Haas (31.151.040), Dzigvashvili and Malasidze (1984), Korchagin and Korchagin (37.151.001), Marochnik and Suchkov (30.151.049), Tang (34.151.076), Terzides (31.151.029), Ting and Zheng (32.151.053), Tong, Wu and Peng (34.151.072), Tong, Zheng and Peng (31.151.088, 31.151.089), Xu (29.151.063, 31.151.074), and Yue (34.151.074).

Global spiral modes in stellar disks have been investigated by Ambastha and Varma (32.151.043, 32.151.044, 32.151.058, 33.151.007, Bertin (33.151.026), Haass (33.151.027, 33.151.072), Iye, Ueda, Noguchi and Aoki (33.151.020), Korchagin and Shevelev (33.151.077), Litvintzev (1984), Pannatoni (33.062.113), and Yue (31.151.085, 31.151.086, 31.151.087).

Further aspects of the excitation and maintenance of spiral density waves, especially regarding to non-linear phenomena have been discussed by Abramyan and Arutyunyan (37.151.041), Ando (37.151.054), Berman and Mark (32.151.023), Chalov (29.151.005, 37.151.039), Eliseev and Stepanov (37.151.078), Fridman, Palous and Pasha (29.151.003), Grivnev (1984), Korchagin and Korchagin (31.151.038, 31.151.069), Litvintsev and Maksumov (34.151.071, 34.151.108), Louise (32.151.059), Maksumov (1982, 33.151.130), Meinel (33.151.078), Renz (31.151.054), and Sugiyama and Mark (33.151.109).

Barred spirals have received much interest in recent years. The formation and properties of stellar bars in the inner regions of galaxies have been studied by Combes and Sanders (29.151.040), Contopoulos (30.151.040, 30.151.058, 30.151.086), Sellwood (29.151.094, 33.151.080), Terzides and Michalodimitrakis (33.151.122), Tohline, Durisen and McCollough (37.151.026), Vandervoort and Welty (30.151.014, 32.151.080), and Weinberg and Tremaine (34.151.029, 34.151.030). The response of stellar disks and interstellar gas to bar forcing has been discussed by van Albada and Sanders (32.151.035, 33.151.049), Chen, Weng and Xu (29.151.044, 32.151.052, 37.151.099), Colin and Athanassoula (33.151.050), Combes and Gerin (33.151.048), Davoust (37.151.092), Elmegreen and Elmegreen (1984), Huntley and Gerola (30.151.006, 32.151.031), Karchagin and Shevelev (31.151.036), Marcelin and Athanassoula (31.158.014), Matsuda and Isaka (29.151.064), Nishida, Watanabe, Fujiwara and Kato (33.151.096, 37.151.049), Oldershaw (33.151.125), Polzin and Thielheim (30.151.038), Prendergast (33.151.046), Sanders, Teuben and van Albada (33.151.047), Schempp (31.151.077), Sorensen and Matsuda (31.151.008), Thielheim (29.151.073, 31.151.092, 32.151.061, 34.151.114), Thielheim and Wolff (31.151.011, 33.151.042, 37.151.015), Wolff (34.151.105), and Yuan (37.155.124).

The formation and lifetime of spurs in the disks of galaxies have been studied by Byrd, Smith and Miller (34.151.100), and Feitzinger and Schwerdtfeger (32.151.063).

Non-gravitational theories for the formation of spiral structures in galaxies have been pursued as well. The role of stochastic self-propagating star formation on the large-scale structure of galaxies has been investigated by Chiang and Elmegreen (32.151.025), Comins, Statler and Smith (31.151.075, 34.151.002), Freedman and Madore (37.151.090), Feitzinger and Seiden (29.151.088, 33.151.033), Schulman and Seiden (30.151.012, 32.151.037), and Seiden, Schulman and Feitzinger (31.151.010). Ikeuchi and Nozakura (33.151.100, 34.151.116, 34.151.117, 37.151.062, 37.151.067) have developed a model of galactic spiral structure as a self-organized dissipative structure in galaxies. Further alternatives have been suggested by Balbus (37.151.033), Cowie and Rybicki (32.151.020), McCanney (29.151.015), Korchagin and Rakhimov (31.151.070), Morozov (30.151.081), and Pismis and Moreno (31.151.074).

3. Gas Flow and Shocks

Shock calculations for the gas flow in spiral galaxies have been performed by Bogdan (33.151.068), Brand and Heathcote (31.151.005), Hu and Ao (29.151.045, 32.151.039), Ishibashi and Yoshii (33.151.099, 37.151.050), Matsuda (32.151.054), Matsuda, Hachisu and Sawada (37.151.060), Mishurov (31.151.060), Nelson, Matsuda and Johns (1984), Shore (30.151.046), Tang (29.151.043), Tosa (33.151.166, 37.131.159), Yuan (33.151.110), and Yuan and Wang (31.131.010). In particular, the influence of supernovae and shock-induced star formation on the structure of the shock fronts and the multi-component nature of the gas in galaxies have been considered. The occurence of shocks in particle models simulating the large-scale behaviour of interstellar gas clouds has been investigated by Bash (33.151.032), Fukunaga (34.151.004), Kwan and Valdes (34.151.031), Leisawitz and Bash (32.155.002, 37.131.308), Roberts (33.131.128), Roberts and Hausman (34.151.095, 34.151.096, 34.151.098, 37.151.034), Roberts, Hausman and Levinson (33.151.031), and Schwarz (37.157.111). Feitzinger and Spicker (33.151.034), and Nelson, Matsuda and Johns (33.151.035) studied the vertical extension of gaseous spiral arms and spiral shocks.

Van Albada and Roberts (29.151.086) performed a high-resolution study of the gas flow in barred spirals. Continuum models for gas in disturbed galaxies have been developed by Scalo and Struck-Marcell (33.151.112, 37.151.011, 37.151.032).

4. Observational Aspects

Bash (30.155.049) reconsidered the question whether the Galaxy has two or four spiral arms. Hilton and Bash (31.151.022) investigated the vertex deviation of young stars near the sun in the framework of their ballistic particle model for molecular clouds. Haass, Bertin and Lin (32.151.026, 32.151.055) presented a dynamical interpretation for the classification of spiral galaxies. Various measured spiral arm properties have been compared by Kennicutt and Hodge (31.158.026) and by Louise (31.151.030) with model predictions.

5. References

Athanassoula, E.: 1984, Physics Reports (in press).
Dzigvashvili, R.M., Malasidze, I.A.: 1984, Abastumanskaya Astrofiz. Obs. Byull. (in press).
Elmegreen, B.G., Elmegreen, D.M.: 1984, Astrophys. J. (in press).
Grivnev, E.M.: 1984, Astron. Zh. (in press).
Lin, C.C., Bertin, G.: 1985, IAU Symposium No. 106.
Litvintzev, S.I.: 1984, Byull. Inst. Astrofiz. 76 (in press).
Maksumov, M.N.: 1982, Dokl. Akad. Nauk Tadzh. SSR 25, 720.
Nelson, A.H., Matsuda, T., Johns, T.: 1984, Mon. Not. R. Astron. Soc. (in press).

D. STABILITY AND EVOLUTION

1. Stability of Galactic Disks

The global stability of galactic disks against various kinds of instabilities and oscillations have been analyzed by Abramyan (32.151.085), Aoki (37.151.053), Caimmi and Dallaporta (32.151.012), Chalov (32.151.083), Durisen and Bacon (29.151.059), Hachisu (32.151.047), Ishibashi and Ando (37.151.055), Iye (37.151.028), Iye, Ueda, Noguchi and Aoki (33.151.028, 1984), Morozov (29.151.047), Nishida, Yoshizawa, Watanabe and Inagaki (30.151.084), Nishimoto (33.151.095), Nuritdinov (29.151.083), Tajima (29.151.095), Vandervoort and Welty (29.151.077), Vandervoort (31.151.044, 32.151.022, 34.151.055), Watanabe, Inagaki, Nishida and Tanaka (30.151.083), Wiegandt (31.151.023, 31.151.024), and de Zeeuw, Franx, Meys, Brink and Habing (33.151.053). Local stability criteria which take into account the multi-component nature of galactic disks have been developed by Abramyan (33.151.076), Balbus (33.151.111), Churilov and Shukhman (29.151.048), Grishchuck and Zeldovich (29.151.082), Grivnev, Ivannikova and Maksumov (1984), Jog and Solomon (37.151.013, 37.151.014), Morozov (29.151.009), Ogorodnikov and Osipkov (1980), and Polyachenko and Fridman (29.151.011).

Bar and spiral instabilities of galactic disks under the stabilyzing influence of a nuclear bulge or halo have been investigated by Athanassoula (33.151.051), Athanassoula and Sellwood (33.151.040), Berman and Mark (32.151.023), Combes and Sanders (29.151.040), Efstathiou, Lake and Negroponte (31.151.063, 31.151.081), Nishida (37.151.056), Robe and Leruth (37.151.043), Sellwood (29.151.094, 33.151.039, 33.151.080), Sellwood and Carlberg (1984), Terzides and Michalodimitrakis (33.151.122), and Tohline, Durisen and McCollough (37.151.026).

2. Warps of Galactic Disks

Recent developments of the theory of warps of galactic disks have been reviewed by Toomre (33.151.036). Detailed studies on the warps of galaxies, tilting of galactic disks and vertical disk oscillations have been presented by Blitz, Mark and Sinha (29.155.005), Dekel and Shlosman (33.151.037), Fleck (34.155.015), Johns and Nelson (1984), Nelson (30.151.003), Mark (33.151.073), May and James (37.151.010), Papp and Innanen (32.151.079), Simonson and Tohline (33.151.127), Sparke (37.151.084, 1984), and Yoshii and Fujimoto (30.151.059).

3. Secular Evolution of Galaxies

Reviews on recent progress in understanding the evolution of galaxies have been given by Martinet (30.151.027), and Strom and Strom (31.151.059). The secular evolution of galactic disks due to angular momentum transport and matter accretion has been investigated by Bertin (34.151.038), Gorbatskij and Serbin (34.151.045), Mayor and Vigroux (29.151.065), Norman (37.151.005), and Simonson (37.151.027). The evolution of perturbed gaseous disks has been studied by Sorensen (33.151.030). Biermann (29.151.072) has presented a new code to calculate the evolution of galaxies.

Recently, the interaction of stars with fluctuations of the galactic gravitational field due to massive objects or short-lived spiral perturbations, as reviewed by Wielen and Fuchs (1984), has found special interest. The resulting secular evolution of galaxies has been studied by Carlberg (37.151.006), Carlberg and Sellwood (33.151.029), Carlberg, Freedman and Sellwood (34.151.052), Icke (31.151.018), Kamahori and Fujimoto (37.151.061), Lacey and Fall (34.155.006, 1984), Lacey (37.151.007, 37.151.076), Rohlfs and Wiemer (32.151.008), Villumsen (34.151.089, 37.151.102, 1984), Wielen and Fuchs (32.151.072, 32.151.073, 33.151.070, 34.151.104).

4. References

Grivnev, E.M., Ivannikova, E.I., Maksumov, M.N.: 1984, Astron. Zh. (in press).
Johns, T.C., Nelson, A.H.: 1985, IAU Symposium No. 106.
Lacey, C.G., Fall, S.M.: 1985, IAU Symposium No. 106.
Ogorodnikov, K.F., Osipkov, L.P.: 1980, Byull. Abastumansk. Astrofiz. Obs. 52, 37.
Sellwood, J.A., Carlberg, R.G.: 1984, Astrophys. J. (in press).
Sparke, L.S.: 1984, Astrophys. J. (in press)
Ueda, R., Noguchi, M., Iye, M., Aoki, S.: 1984, ESO Sci. Prepr. No. 310.
Villumsen, J.V.: 1985, IAU Symposium No. 106.
Wielen, R., Fuchs, B.: 1985, IAU Symposium No. 106.

E. COMPUTER SIMULATIONS

Computer simulations provide a powerful tool for investigating the evolution of gravitating systems.

1. Clustering of Galaxies

Using the Monte-Carlo simulation technique, Dodd et al. (31.151.028) and MacGillivray and Dodd (31.160.074, 31.160.075, 32.160.055, 32.160.056) investigated the geometric properties of positions, orientations, shapes, etc. of galaxies in clusters.

Ishizawa et al. (32.151.022), Carnevali et al. (30.151.048), and Guiricin et al. (37.160.078) examined the dynamical evolution of a group of galaxies. Miller (34.151.018) showed that the observed structure of galaxy clusters and superclusters is most easily described in terms of matter being swept away from growing empty regions in an expanding universe. A two-dimensional simulation of the gravitational superclustering of collisionless particles was carried out by Melott (33.161.016). Bhavsar et al. (29.151.085) supported the gravitational instability picture for clustering, in which the multiplicity function depends strongly on the initial density fluctuation spectrum index.

Saslaw and Aarseth (31.162.015) examined the evolution of the velocity distribution of galaxies as they cluster in the expanding universe and showed that the velocity dispersion of extreme field galaxies is a good cosmological indicator of Ω. To estimate Ω, Miller (1984) made also numerical experiments on galaxy clustering in open universes. Shaya (37.160.113) made Monte-Carlo simulations to determine the probability distribution of peculiar velocities and shear velocities of superclusters.

The merging rate of galaxies in an expanding universe for explaining the galaxy mass function was derived by Roos (29.151.035). The evolution of rich clusters of galaxies with halo was simulated by Roos and Aarseth (32.160.042). Farouki et al. (34.151.020) examined the collapse and violent relaxation of N-body systems, and found the mass segregation and the secondary density maximum similar to those observed in some cD clusters of galaxies. Malumuth and Richstone (37.160.024) simulated the evolution of galaxy clusters with different richness, and confirmed that the central region of the cD clusters is produced by mergers while the halo is produced by stripping. N-body simulations of galaxy clustering was compared with observations by Zieba et al. (31.151.007).

2. Interacting Galaxies

Simulations of galaxy mergers were reviewed by White (33.151.059). A new technique for simulating galactic collisions was proposed by James (30.151.061).

Villumsen (31.151.041, 34.151.008) carried out N-body simulations of mergers of two spherical non-rotating systems with isothermal or Hubble density profile, and showed that the density profiles of the remnants are inconsistent with the de Vaucouleurs law. Farouki and Shapiro (29.151.018) simulated the passage of a flat disk galaxy through the core of a cD cluster, and suggested that only the flattest of E-galaxies could be created from the disk systems through repeated collisional heating.

The collision and merging of two disk-halo galaxies were investigated by Gerhard (30.151.018), Smith and Miller (33.151.063), and Negroponte and White (34.151.057). The collision of two spherical isothermal systems with a central core was studied by Carlberg (31.151.082). Farouki and Shapiro (32.151.004) found a criterion for the merging of two identical disk-halo galaxies.

Quinn (37.151.079) showed that the interaction between disk and elliptical galaxies creates shell-like features in the outer regions of E-galaxies. The tidal interaction between galaxies was simulated by Chatterjee Ahmed (29.151.022), and Korovyakovskaya and Korovyakovskij (33.151.010). Duncan et al. (34.151.021) applied N-body simulations to galactic cannibalism. The decay of satellite galactic orbits was simulated by Lin and Tremaine (33.151.004) and White (34.151.077).

3. Gravitational Collapse of Protogalaxies

Based on a numerical integration of the collisionless Boltzmann equation, Fujiwara (34.151.063) studied the evolution of the distribution function during the collapse of a homogeneous sphere. Fujiwara (33.161.125, 1983) examined also the collapse of massive neutrinos which yields a core-halo structure. Collapses of stellar protogalaxies which lead to the elliptical or bar deformation were investigated by Polyachenko (29.151.012), and Miller and Smith (29.151.029). Collapses of pressureless inhomogeneous spheroids were examined by Goodman and Binney (33.151.012), and Noguchi (1984).

The formation of galactic halos during the secondary infall was simulated by Dekel et al. (30.151.055), and Pryor and Lecar (33.151.133). Van Albada (32.151.065), and May and van Albada (1984) showed that a variety of irregular initial conditions produce density distributions in agreement with the $r^{1/4}$ law. The formation of E-galaxies was also simulated by Doroshkevich and Klypin (29.151.046), and McGlynn (37.151.095). Among remnants of merging simulations, Gerhard (33.151.013, 33.151.018) found out quasi-stable principal axis twists in models of elliptical galaxies.

In numerical experiments on the protogalactic collapse, Miller and Smith (29.151.036) investigated the effect of star formation, and Carlberg and Hartwick (30.151.036) considered an instantaneous mass loss.

4. Evolution of Galactic Structures

Taking into account of the scattering of stars out of the plane due to giant molecular clouds, Villumsen (34.151.089) simulated the evolution of the vertical structure of galactic disks. Integrating numerically the Boltzmann and Poisson equations, Watanabe et al. (30.151.083), Nishida et al. (30.151.084), and Nishida (33.151.096) investigated the evolution of the flat stellar disks. Inagaki et al. (1984) investigated the reliability of two totally independent numerical methods in simulating the evolution of disk galaxies. Numerical experiments on the evolution of a rotating stellar bar were made by Miller et al. (32.151.018), and Carnevali (33.151.011). Self-gravitating warps in disk galaxies were examined by May and James (37.151.010). Thielheim and Wolf (37.151.015) simulated the development of bar and spiral structure in a self-consistent way.

The evolution of the spiral structure in flat galaxies was also investigated by computer simulations: The stellar and gas responses to oval distortions in disk galaxies were studied by Thielheim and Wolff (29.151.052), and Sorensen and Matsuda (31.151.008), respectively. Levinson and Roberts (29.151.058) investigated the evolution of the system of clouds in response to a background spiral density wave of a model galactic disk. The process of stochastic self-gravitating star formation was simulated by Comins (29.151.001), Seiden and Gerola (31.151.084), and Statler et al. (34.151.002).

5. Evolution of Star Clusters

Based on extensive orbital calculations of elliptic and circular three-body systems, Keenan (29.151.033, 29.151.034) investigated the galactic tidal limits and outer structure of star clusters. Monte-Carlo simulations of the evolution of galactic nuclei and star clusters with a central black hole were carried out by Duncan and Shapiro (31.151.016, 33.158.163).

Binaries in star clusters are important for the energy budget, absorbing energy from and injecting energy into the clusters. Hut and Bahcall (33.151.118) and Hut (34.151.037) carried out numerical orbit calculations for binary-single star scattering, and showed that soft binaries become softer while hard binaries become harder through the heat exchange process. Fullerton and Hills (31.151.021), and Hills (34.151.003) simulated also encounters between binary and single stars.

Encounters between pairs of binary to obtain various cross-sections were simulated by Mikkola (32.151.077, 33.151.102, 34.151.041) and Hoffer (34.151.014); see also Dokuchaev and Ozernoj (29.151.010, 29.151.013, 31.151.067, 31.151.068, 32.151.003) for the dynamical role of binaries in star clusters.

Based on a N-body simulation, the relaxation processes of one-dimensional self-gravitating system were investigated by Wright et al. (32.151.014), Luwel and Severne (33.151.017), and Severne et al. (1984). Smith (29.151.008) and Farouki and Salpeter (31.151.014) carried out N-body simulations to determine the timescales of relaxation and segregation, respectively.

6. References

Fujiwara, T.: 1983, Theor. Phys. 70, 603.
Inagaki, S., Nishida, M.T., Sellwood, J.A.: 1984, Mon. Not. R. Astr. Soc. (in press)
May, A., van Albada, T.S.: 1984, Mon. Not. R. Astr. Soc. 209, 15.
Miller, R.H.: 1984, Astron. Astrophys. 138, 121.
Severne, G., Luwel, M., Rousseeuw, P.J.: 1984, Astron. Astrophys. 138, 365.

F. MAGNETIC FIELDS, PULSARS, X-RAY AND GAMMA-RAY EMISSION

1. Magnetic Fields

During the triennium under review many papers reported on studies of the structure of the galactic magnetic field. Andreasyan used rotation measures (RM) of large samples of extra-galactic radio sources and pulsars (29.156.001) or radio sources (32.156.002), and Inoue and Tabara (31.156.011) used in addition optical polarization of stars to investigate the direction of the large-scale regular magnetic field. Thomson and Nelson analyse the RMs of 459 extragalactic sources (32.161.001) to determine the best fit parameters for a galactic magnetic-field model, and find agreement with their earlier work using pulsars (27.156.009). Similarly, Sofue and Fujimoto (33.155.011) show that the characteristic features of the RM distribution on the sky are well reproduced by a model in which the magnetic field is in a bisymmetric, two-armed logarithmic spiral configuration. Finally, Welter, Perry and Kronberg (37.159.096) present a statistical analysis of the (Galaxy-corrected) residual rotation measure (RRM) of 116 QSOs.

In the triennium, large-scale magnetic field structures in nearby galaxies were discussed, e.g., M31 by Beck (31.158.085), and NGC 6946 and NGC 253 by Klein et al. (31.158.111, 34.157.071). The fields in these nearby galaxies are found to be substantially higher than had been previously thought to exist. Ordered fields of up to 10 μG, over kiloparsecs appear frequently.

Rumaikin and Shukurov (30.156.007, 31.156.017) evaluate again the large-scale and small-scale galactic magnetic fields, generated and maintained due to the turbulent dynamo-action. Vallee and colleagues (32.156.006, 34.131.110, 37.131.179) discuss the discovered "magnetic bubbles" of sizes ranging from about 0.1 to 1 kpc; in analogy to the two known magnetic bubbles, loop I and loop II, they propose the areas of the Gum Nebula and the Monogem X-ray ring (30.142.013) to be new 'magnetic bubbles' near the Sun. A quantitative estimate of the apparent effects of the four large interstellar magnetic bubbles near the Sun on the RM data of distant radio sources has recently appeared (Vallee, 1984): The mean RM contribution is $\sim$ 90 rad m^{-2}. Finally, the large-scale structure of the magnetic field in the Perseus spiral arm is derived by Vallee (34.155.005).

Phillips et al. (29.156.008, 30.156.010) interpret the new all-sky radio continuum map at 408 MHz (Haslam et al., 30.141.017) to derive a large-scale emissivity distribution. Sarkar (32.156.015) evaluates the possibility that the galactic synchrotron radio background originates in old supernova remnants. Brown (33.155.049) and Kanbach (34.155.020) discuss in detail the galactic distribution of cosmic-ray electrons and magnetic fields, reflected in the galactic non-thermal radio background. Brown stresses the intimate relation between interpretations of the radio and gamma-ray backgrounds and the distribution of the thermal material in the Galaxy. Earlier, Haslam et al. (29.157.001) showed the large-scale similarity of the gamma-ray and synchrotron distributions in the galactic plane. Spoelstra (37.155.086) discusses the polarization characteristics of Galactic radio emission at frequencies 408, 465, 610, 820 and 1411 MHz.

2. Pulsars

The proceesings of IAU Symposium No. 95 on "Pulsars" (29.012.037) provide a complete review of the topic up to 1981. Three contributions are concerned with the galactic distribution of pulsars (29.141.593, 29.141.594, 29.141.595) and are of special interest for this report. Lyne (29.141.593, 32.141.535), Guseinov, Kasumov and Yusifov (29.141.594, 31.141.514, 31.141.526) and Harding (29.141.595) arrive at different parameters on the details of this distribution, but agree on the large-scale appearance, such as a wide (∿ 8 kpc) distribution above and below the galactic plane, a ring-shaped (or spiral-arm-like) structure reaching a maximum at ∿ 5 kpc and a pulsar birth rate of one pulsar every 20 to 50 years. Morini (30.141.546) performed Monte-Carlo simulations to derive the galactic distribution.

Vivekanand and Narayan (30.141.555) and Huang, Huang and Peng (33.141.009) study the P-$\dot{P}$ relation. The latter authors propose under a new pulsar spindown configuration ($\dot{P} \propto P^2$) that there might be two types of pulsars which could be correlated with two types of supernovae. Anderson and Lyne (33.141.085) study a sample of 26 pulsars and point out that their velocities are closely correlated with either the magnetic dipole moments, or their moments of inertia. They find no relation between the directions of the proper motions and the pulsar spin axes.

3. Diffuse Galactic Gamma Radiation

In the context of the XVIII general assembly of the IAU, a meeting was organised on "Galactic Astrophysics and Gamma-Ray Astronomy". The readers are referred to the proceedings (34.012.016) for a complete review of the field of gamma-ray astronomy up to the end of 1982. Most new results are derived from the COS-B data base for energies $\gtrsim$ 70 MeV (e.g. 31.157.002). At lower gamma-ray energies, new results have been derived from balloon measurments by Graser and Schönfelder (32.157.013), Sacher and Schönfelder (34.143.021), and Bertsch and Kniffen (34.143.003).

For this report, the most important progress made during this triennium concerned correlation studies between the gamma-ray distributions and the distributions of gas tracers; gamma radiation being a tracer of the product of the cosmic-ray density and the total interstellar gas density, integrated along the line of sight. The contribution from Inverse-Compton gamma rays (e.g. 30.157.015, and Bloemen, 1984) and point sources (e.g. pulsars, 31.142.528, 29.157.002, 30.157.001, 34.141.076) has been ignored in most works.

Gamma rays and the local interstellar medium. The SAS-2 detection of the structure of Gould's Belt in gamma rays has been confirmed in detail using COS-B data (31.157.002). A second local disc-shaped structure similar to Gould's Belt, but independent of it (Dolidze, 27.155.037, 28.155.046), has been shown to be visible in the COS-B data by Bignami (30.157.012) and Bignami et al. (32.012.070). Lebrun et al. (31.142.507) and Strong et al. (32.131.267) study the relation between galaxy counts and gamma rays measured by COS-B for $10^{\circ} < |b| < 20^{\circ}$. They show that a close bidimensional correlation exists between gamma-ray intensities and total-gas column densities as estimated from galaxy counts. Analyses of this type applied to SAS-2 data by Lebrun and Paul (1979, 33.131.050) and Strong and Wolfendale (30.157.011) reached similar conclusions. Strong et al. (32.131.267) and Bloemen (33.155.035) present for this latitude range a map of the distribution of molecular-hydrogen column density, $N(H_2)$, based on the excess gamma-ray flux over that predicted from atomic hydrogen, HI, alone. The map includes both hemispheres, and establishes the H_2 angular distribution on scales $< 10^{\circ}$. The Orion (28.131.145) and ρ-Ophiuchus (34.143.013, 34.155.031) cloud complexes have been resolved in gamma rays. Bloemen et al. (1984) find for the large complex of interstellar clouds in Orion and Monoceros a good correlation between the gamma-ray emission and the total-gas distribution, using a Columbia CO survey to trace the H_2. This correlation is used to calibrate the ratio between H_2 and the integrated CO line intensity: $N(H_2)/W_{CO} = (2.6\pm1.1)\ 10^{20}$ molecules $cm^{-2}\ K^{-1}\ km^{-1}s$.

Gamma rays from atomic and molecular gas in the galactic plane. Bloemen, Blitz and Hermsen (33.155.086, 37.155.065) present a method by which the radial distribution of galactic high-energy gamma-rays (70 MeV-5 GeV) can be determined. In a follow-up paper (Bloemen et al., 37.155.072) the radial distribution of the gamma-ray emissivity in the outer Milky Way is derived, using HI column-density maps in various galacto-centric distance ranges in the outer Galaxy in combination with COS-B gamma-ray data. Their results are interpreted as a steep gradient in the cosmic-ray electron density and a near constancy of the nuclear component out to large ($\sim$ 20 kpc) galacto-centric distances. The variation of the electron component, with effective pathlength $\sim$ 4 kpc, is shown to be consistent with low-frequency radio continuum observations (e.g. 27.156.001, 29.156.008). In this triennium, earlier studies on cosmic-ray gradients are presented by Li Ti Pei and Wolfendale (30.155.011) and Issa et al. (30.155.012). Arnaud et al. (32.142.523) and Lebrun et al. (34.131.199) compare the COS-B gamma-ray survey with a Columbia CO survey and Berkeley HI surveys for the inner galaxy. Lebrun et al. found that a simple model in which uniformly distributed cosmic rays interact with the interstellar gas, as traced by HI and CO, can account for almost all the gamma-rays (E > 300 MeV) observed from the first quadrant. They calibrated the ratio $N(H_2)/W_{CO}$ = (1-3) 10^{20} molecules cm^{-1} K^{-1} km^{-1} s, fully consistent with the value derived from Orion-Monoceros data. Korchagin, Korchagin and Suchkov (33.155.100) and Fichtel and Kniffen (37.155.056) model the interstellar gas distribution, using recent considerations of this distribution from radio and mm data, to predict the galactic gamma-ray distribution.

4. References

Beuermann, K., Kanbach, G., Berkhuijsen, E.M.: 1984, Astron. Astrophys. (in press).
Bloemen, J.B.G.M.: 1984, Astron. Astrophys. (in press).
Bloemen, J.B.G.M., Caraveo, P.A., Hermsen, W., Lebrun, F., Maddalena, R.J., Strong, A.W., Thaddeus, P.: 1984, Astron. Astrophys. 139, 37.
Lebrun, F., Paul, J.A.: 1979, Proc. 16th Int. Cosmic Ray Conf. (Kyoto), 12,13.
Vallee, J.P.: 1984, Astron. Astrophys. 136, 373.

VII. Galactic Environment

The galactic environment comprises of five distinct populations: hot gaseous corona, high-velocity clouds of neutral hydrogen (HVCs), dwarf companion galaxies and extended distant globular clusters, and an invisible massive corona of still unknown origin.

The presence of the hot gaseous corona has been demonstrated principally by IUE observations of distant stars, galaxies and quasars. Absorption features of highly ionized atoms (CIV, CVI, SiIV) have been discovered. Redshifts indicated that absorption originated partly in the Galactic disk and in the Magellanic Clouds, but partly in the gaseous corona (30.131.027, 30.158.161, 31.131.058, 32.131.082, 32.155.060; reviews are given in 32.158.075, 32.155.028, 33.131.022, 33.131.138).

Mirabel and Morras (37.155.063) have performed a new sensitive survey of neutral hydrogen in the velocity interval -1000 to +1000 km/s; 120 clouds have been discovered. In the galactic anticenter region, all clouds have negative galacto-centric velocity; in the galactic center region, 15 % of the clouds have positive velocities. Cohen (30.131.006) discovered a 25° long, high-velocity stream colliding with gas in the galactic disk. Detailed studies of individual HVCs have been continued (30.131.112, 31.159.005, 32.131.089, 32.131.264, 33.131.017, 33.131.173). The current observational status of HVCs and the Magellanic Stream is reviewed by Mirabel (30.159.017, 32.131.150) and by Mathewson and Ford (37.156.032).

The dominance of negative galactocentric velocities of HVCs is explained in the galactic fountain model (Shapiro and Field, 17.131.112). Cold and dense gas of the galactic disk will be locally heated by supernova explosions up to a temperature 10^6 K. The heated gas flows away from the galactic plane, where it cools and condenses slowly. The cooled clouds fall back toward the disk. Dynamics of the galactic fountain has been studied in detail (29.131.110, 30.155.052, 30.158.146, 31.131.012, 31.131.029, 32.131.261). HVCs can also be tidal debris of the Magellanic Clouds falling to the Galaxy. Chevalier and Fransson (37.155.073) suggested a cosmic-ray supported galactic corona with an equilibrium temperature 10^4 K. Jaaniste (37.156.035) proposed a method of determining the distances of HI clouds using the equilibrium condition between clouds and surrounding hot coronal gas.

Modern sensitive detectors and new generation telescopes have been extensively used for the detailed study of distant globulars and dwarf companion galaxies. The range of metallicity of these objects is surprisingly large, several clusters exhibit anomalously red horizontal branches for their metallicity (30.154.006, 30. 158.212, 32.154.005, 32.154.006, 32.154.046, 34.157.032, 37.154.076). Globular Clusters in the Local Group dwarf elliptical galaxies tend to be bluer than the halos of their parent galaxies (33.154.010). The lack of luminous red supergiants in the Pegasus dwarf irregular indicates the absence of recent star formation (31. 158.036). Isochrone fitting of the Carina dwarf elliptical yields an age estimate of 7.5 Gyr and the absence of an extremely old population II component. Photometric properties and improved distances have been determined for a number of companion galaxies and their clusters (30.158.189, 32.154.003, 32.154.010, 32.158.338, 33. 157.040, 34.154.030, 34.157.036, 37.154.005). Intrinsically bright carbon stars have been discovered in almost all companion galaxies (31.158.004, 31.158.042, 31. 158.043, 33.157.003). New data suggest that the Sculptor dwarf galaxy has a radial velocity of +20 km/s, not -149 km/s as obtained earlier (33.157.003).

Lynden-Bell stresses that all dwarf ellipsoidal satellites of the Galaxy belong to one of two streams of tidal debris and hydrogen clouds (32.158.185, 33.157. 074). Draco and Ursa Minor galaxies are oriented along the stream they belong to (31.159.006). Searches for stars associated with the Magellanic Stream have given negative results (32.159.015, 34.156.002).

The motion of companions and gaseous streams has been used to determine the total mass of the Galaxy including its massive corona. Fujimoto and Murai (37.156. 031) and Lin and Lynden-Bell (31.159.002) have calculated a consistent model of the dynamics of the Magellanic Stream; the mass within a sphere of 50 kpc is 7×10^{11} $M_\odot$; see also 30.159.007. Other test bodies with less accurate results have also been used (33.155.074, 34.155.013). Einasto and Lynden-Bell (31.160.008) have calculated orbits of the Galaxy and M31 supposing that both galaxies have equal and opposite angular momenta as seen from the barycentre of the Local Group. The total mass of the Local Group lies between 4 to 7×10^{12} $M_\odot$. Tidal radii of dwarf spheroidal companions have also been used to estimate the mass of the Galaxy with controversial results (31.155.002). Miller and Smith (37.151.104) demonstrated by numerical experiments that companion envelopes become expanded by tidal encounters, rather than truncated. Thus, tidal radii cannot be used for a mass determination of the parent galaxy. Environmental influences on companion and parent galaxies have been studied in detail (32.151.081, 33.151.004, 33.151.022, 32.151.050, 33.151.059).

Shuter (29.155.001) suggested that a flat rotation curve can be produced by a disk-like population. Rohlfs (31.155.011), however, demonstrated that in respective models, the vertical gravitational force must be larger than the radial one (in the galactic plane), contrary to observations.

Carbon stars have sharp emission lines in their spectra and can be used to derive very accurate radial velocities of individual stars in dwarf ellipticals. Respective observations by Aaronson (33.157.031) indicate for these ellipticals higher M/L than that found in galactic globulars. This is used as an argument suggesting the presence of dark halos in these ellipticals (33.157.032, 33.157.033). Some Magellanic irregulars have a rising rotation curve which also suggests the presence of dark halos (Freeman, 37.156.030). The possibility that Magellanic irregulars may possess hot gaseous coronas is discussed by de Boer (37.156.061). Velocity dispersion of globular clusters in the Fornax dwarf galaxy indicates a normal mass to light ratio (34.157.030).

The possibility that dwarf galaxies may possess massive coronas can solve a dispute concerning the nature of the dark matter around galaxies. Other available evidence strongly supports a non-baryonic character of the dark matter. Non-baryonic dark matter can be made of neutrinos or particles like axions, gravitinos or photinos. In the first case, dark matter can form only aggregates comparable in mass with clusters of galaxies; in the second case, it can concentrate also around smaller objects, including dwarf galaxies and even globular clusters (Primack and Blumenthal, 37.161.268, and Peebles, 37.161.105).

R. WIELEN
President of the Commission

34. INTERSTELLAR MATTER

(MATIERE INTERSTELLAIRE)

PRESIDENT: M. Peimbert
VICE-PRESIDENT: J. Lequeux
ORGANIZING COMMITTEE: S. D'Odorico, B.G. Elmegreen, E.B. Kostyakova, J.S. Mathis, U. Mebold, V. Radhakrishnan, P.A. Shaver, Y. Terzian, P.G. Wannier.

I. Introduction
(M. Peimbert)

It has become more evident during the last three years that the study of interstellar matter is paramount to understand the evolution of the universe and its constituents. From observations of the present state of the interstellar medium, in our galaxy, in other galaxies, and between galaxies, it is possible to test theories of: evolution of the universe, formation and evolution of galaxies, formation and evolution of stars and of the evolution of the interstellar medium itself. The amount of information on the interstellar medium that has been gathered during the 1982-1984 period has been very large and the theoretical models that have been ellaborated to explain these observations have been very numerous, these facts show that the subject of our Commission constitutes a very active field of astronomical research.

This report covers the period 1982-1984 and is divided in self-contained sections. References are given by commonly used abbreviations (see the 1979-1981 report of the Commission). It should be noted that due to the space allotted for this report it was not possible to cover all the papers relevant to the work of our Commission produced during this period; moreover for those papers considered, only very brief summaries are presented here. A detailed list of articles on the physics of the interstellar medium and gaseous nebulae carried out in the Soviet Union in the 1981-1984 period was prepared by N.G. Bochkarev and G. Rudnitskij; only a small fraction of these articles are discussed in this report; copies of this list are available from the office of the President of Commission 34.

The presentation of this report was made possible by the generous reviewing efforts of its writers and by the cooperation of the many members of Commission 34 and others who submitted relevant information. We list below a selection of books, conference proceedings, catalogues and review articles published since 1981; many others, particularly the more specialized, are mentioned in other sections of this report.

1. MONOGRAPHS AND OTHER BOOKS

Belov, K.P.: 1983, "Magnetism on the Earth and in Space", Nauka, Moscow. (In Russian)
Duley, W.W., Williams, D.A.: 1984, "Interstellar Chemistry", Academic Press, London.
Goudis, C.: 1982, "The Orion Complex: A Case Study of Interstellar Matter", D. Reidel Publishing Co.
Gurevich, L.E., Chernin, A.D.: 1983, "Origin of Galaxies and Stars", Nauka, Moscow. (In Russian)
Hellwege, K.H., et al. (eds.): 1982, "Landolt-Börnstein, New Series, VI Vol. 2, Subvolume C: Interstellar Matter-Galaxy-Universe", Springer, Heidelberg.
Kostyakova, E.B.: 1982, "Physics of Planetary Nebulae", Nauka, Moscow. (In Russian)
Pottasch, S.R.: 1984, "Planetary Nebulae", D. Reidel Publishing Co.
Rudnitskij, G.M.: 1983, "Molecules in Astrophysics", Institute of Scientific and Technical Information, Moscow. (In Russian)
Sharov, A.S.: 1982, "The Andromeda Nebula", Nauka, Moscow. (In Russian)

Syunyaev, R.A. (ed.): 1982, "Astrophysics and Space Physics", Nauka, Moscow. (In Russian)

Vajnshtejn, S.I.: 1983, "Magnetic Fields in Space", Nauka, Moscow. (In Russian)

Zel'dovich, Ya. B., Ruzmajkin, A.A., Sokoloff, D.D.: 1984, "Magnetic Fields in Astrophysics", Gordon and Breach, London.

2. SYMPOSIUM REPORTS AND CONFERENCE PROCEEDINGS

Audouze, J., et al. (eds.): 1983, "Diffuse Matter in Galaxies", D. Reidel Publishing Co.

Battrick, B., Mort, J. (eds.): 1980, "Second European IUE Conference", ESA Special Pub. 157.

Brahic, A. (ed.): 1982, "The Formation of Planetary Systems", Cepadues Editions.

Bruhweiler, F.C., et al.(eds.): 1984, "The Local Interstellar Medium", IAU Coll. 81, in press.

Cantő, J., Mendoza, E.E. (eds.): 1983, "Symposium on Herbig-Haro Objects, T Tauri Stars and Related Phenomena", RevMexAA, 7.

Danziger, J., Gorenstein, P. (eds.): 1983, "Supernova Remnants and their X-ray Emission", IAU Symp. 101, D. Reidel Publishing Co.

Dickey, J.M., et al. (eds.): 1981, "The Phases of the Interstellar Medium", NRAO, Green Bank, WV (U.S.A.)

Flower, D.R. (ed.): 1983, "Planetary Nebulae", IAU Symp. 103, D. Reidel Publishing Co.

Glassgold, A.E., Huggins, P.J., Schucking, E.L. (eds.): 1982, "Symposium on the Orion Nebula to Honour Henry Draper", AnnNYAcadSci, 395.

Heidmann, J. (ed.): 1983, "Giant H II Complexes Outside our Galaxy", in Highlights of Astronomy, 6, 609.

Kessler, M.F., Phillips, J.P. (eds.): 1984, "Galactic and Extragalactic Infrared Spectroscopy", D. Reidel Publishing Co.

Kondo, Y., Mean, J.M., Chapman, R. (eds.): 1982, "Advances in Ultraviolet Astronomy", NASA, CP-2238.

Radhakrishnan, V. (ed.): 1983, "The Violent Interstellar Medium", in Highlights of Astrononomy, 6, 655.

Rees, M.J., Stoneham, R.J. (eds.).: 1981, "Supernovae: a Survey of Current Research", D. Reidel Publishing Co.

Rolfe, E., et al. (eds.): 1982, "Third European IUE Conference", ESA Special Pub. 176.

Roger, R.S., Dewdney, P.E. (eds.): 1982, "Regions of Recent Star Formation", D. Reidel Publishing Co.

Shaver, P.A., Kunth, D., Kjär, K. (eds.): 1983, "Primordial Helium", European Southern Observatory.

van den Bergh, S., de Boer, K. (eds.): 1984, "Structure and Evolution of the Magellanic Clouds", IAU Symp. 108, D. Reidel Publishing Co.

Wolstencroft, R.D. (ed.): 1984, "Star Formation Workshop", Royal Observatory, Edinburgh.

Wolstencroft, R.D., Greenberg, J.M. (eds.): 1984, "Laboratory and Observational Infrared Spectra of Interstellar Dust", Hilo Workshop, Royal Observatory, Edinburgh.

3. CATALOGUES AND SURVEYS

Alfërova, Z.A., et al.: 1983, "Catalogue of Positions, Angular Sizes, and Flux Densities of Galactic Radio Sources at 21 cm, Astrofiz. Issled. Izv. Spec. Astrophys. Observ., 17, 67.

Blitz, L., Fich, M., Stark, A.A.: 1982, "Catalogue of CO Radial Velocities Toward Galactic H II Regions", ApJSuppl,49, 183.

Gulyaev, S.A., Sorochenko, R.L.: 1983, "Catalogue of Recombination Lines", Prepr. Phys.Inst.Acad.Sci.USSR, Nos: 145,146,168.

Hodge, P.W., Kennicutt, R.C.Jr.: 1983, "An Atlas of H II Regions in 125 Galaxies", AJ, 88, 296.

Maciel, W.J.: 1984, "A Catalogue of Distances of Planetary Nebulae", AASuppl, 55, 253.
Milne, D.K., Aller, L.H.: 1982, "Radio Observations at 14.7 GHz of Southern Planetary Nebulae", AASuppl, 50, 209.
Schneider, S.E., et al.: 1983, "Radial Velocities of Planetary Nebulae", ApJSupp, 52, 399.

4. REVIEW ARTICLES

Czyzak, S.J., et al.: 1982, "The Formation and Properties of Grains in the Interstellar Medium", Vistas Astron, 25, 337.
Ginzburg, V.L., Ptuskin, V.S.: 1983, "Origin of Cosmic Rays", Sum.Sci.Tech.Ser. Astron.,Inst.Sci.Tech.Inf., Moscow, 24, 94.
Hildebrand, R.: 1983, "Cloud Masses and Dust Properties from Submillimeter Thermal Emission", QJRAS, 24,267.
Ho, P.T.P., Townes, C.H.: 1983, "Interstellar Ammonia", AnnRevAA, 21, 239.
Lequeux, J.: 1982, "The Chemical Composition of Gas and Grains in Dense Interstellar Clouds", Geochim.Cosmochim.Acta, 46, 777.
Lozinskaya, T.A.: 1983, "Supernova Remnants: Observational Data Evolution in the Interstellar Medium", Sum.Sci.Tech.Ser.Astron., Inst.Sci.Tech.Inf., Moscow, 22,33.
Mallik, D.C.V.: 1982, "Planetary Nebulae", BullASIndia, 10, 73.
Morris, M., Rickard, L.J.: 1982, "Molecular Clouds in Galaxies", AnnRevAA, 20, 517.
Nagirner, D.I.: 1983, "Theory of Transfer of Radiation in Spectral Lines", Sum.Sci. Tech.Ser.Astron.,Inst.Sci.Tech.Inf., Moscow, 22, 220.
Pagel, B.E.J., Edmunds, M.G.: 1981, "Abundances in Stellar Populations and the Interstellar Medium in Galaxies", AnnRevAA, 19, 77.
Peimbert, M. et al.: 1984, "Interstellar Matter and Chemical Evolution", Science, 224, 345.
Raymond, J.C.: 1984, "Observations of Supernova Remnants", AnnRevAA, 22, 75.
Schwartz, R.D.: 1983, "Herbig-Haro Objects", AnnRevAA, 21, 209.
Shore, S.N. et al.: 1984, "Stochastic Processes in Astrophysics: Stellar Formation and Galactic Evolution", Adv.Chem.Phys., in press.
Shull, J.M., Beckwith, S.: 1982, "Interstellar Molecular Hydrogen", AnnRevAA, 20,163.
Stein, W.A., Soifer, B.T.: 1983, "Dust in Galaxies", AnnRevAA, 21, 177.
Torres-Peimbert, S.: 1984, "Nucleosynthesis and Planetary Nebulae", in Chiosi and Renzini (eds.), "Stellar Nucleosynthesis", D. Reidel Publishing Co., p. 3.
Trimble, V.: 1983, "Supernovae. Part II. The Aftermath", Review of Modern Physics, 55, 511.
York, D.G.: 1982, "Gas in the Galactic Halo", AnnRevAA, 20, 221.

II. Diffuse Interstellar Medium

(J. Lequeux, U. Mebold)

This review covers the literature from July 1981 to mid -84, but does not pretend to be complete. Emphasis has been put on observations, but some theoretical work is also cited. Overlaps with other sections are unavoidable. Amongst topics not or very incompletely covered one finds: atomic and molecular processes, interstellar chemistry, acceleration and interaction of cosmic rays with interstellar matter, topics related to galactic structure.

The last 3-year period has mainly seen a series of high quality observations or of reanalysis of earlier observations (i.e. Copernicus) aiming at precising physical parameters and at checking and refining ideas which emerged in the previous period: multi-component, non-stationary structure of the interstellar medium (ISM), role of stellar winds, supernova explosions, ionization fronts and shock waves in general, origin and structure of the gaseous halo, etc. A wealth of efforts has also been devoted to the local ISM. The reader should look at the introduction for a list of general references and review papers, some of which cover the present topics.

1. THE DIFFUSE DISK ISM: PHYSICAL PARAMETERS, CHEMICAL COMPOSITION

Survey-like HI observations supplemented the available information about the galactic structure. Blitz (1983, in Surveys of the Southern Galaxy, ed. Burton & Israel, 117) and Blitz et al. (1981, BullAAS, 13, 539) deal with the spiral structure of the outer parts of the galaxy. Kerr, (1983, in Kinematics, Dynamics and Structure of the Milky Way, ed. Shuter, 91) studied the HI distribution in the outer galaxy and the thickness and the total mass of the HI Layer. Westerhout and Wendlandt (1982, AASuppl, 49, 143) finally published their atlas of the HI distribution near the galactic plane (2°>b>-2°). A survey at 31 <ℓ<64°, -3° <b<3° with the Arecibo telescope by Bania and Lockman (1984, ApJSuppl, 54, 513) is used to study HI self-absorption features. The results of the Pulkovo sky HI line survey are presented in various forms (Bystrova, 1981, Nauka; Bystrova, 1983, IzvSAO, 17, 63; Bystrova & Tselovalnicov, 1983, in VINITY, No. 5036-83D; Bystrova, 1984, IzvSAO, 18, 94). Special effort went into completing the coverage of the southern galactic plane (Bajaja, 1983, in Surveys of the Southern Galaxy, ed. Burton & Israel, 49; Kerr, l.c. 113; Riley, l.c.55; Olano et al. 1981, AASuppl, 46, 41). Other publications treated the central part of the galaxy (Rohlfs and Braunsfurth, 1982, AA, 113, 237; Lockmann, 1983, BullAAS, 15, 661). HI absorption line surveys were published and discussed by Mebold et al. (1981, AASuppl, 46, 389; 1982, AA, 115, 223) and Dickey et al. (1983, ApJSuppl, 53, 591). Payne et al. (1982, ApJSuppl, 48, 199) report on emission and absorption observation of extended background sources.

New Zeeman measurements were discussed by Troland and Heiles (1982, ApJ, 252, 179) and Heiles et al. (1981, BullAAS, 13, 828). Liszt et al. (1982, ApJ, 261, 102) carried out HI absorption measurements in the direction of 3C123. Statistical analyses of HI clouds were presented by Anantharamaiah et al. (1984, AA, 138, 131) and by Belfort and Crovisier (1984, AA, 136, 368).

The association of HI to specific galactic objects is frequently discussed. Rodríguez and Moran (1982, Nature, 299, 323) find an association of an HI component at -40 km s^{-1} with the nebula 6302. Braunsfurth (1983, AA, 117, 297) investigated the HI-gas in the Cas OB6 association. Near the molecular cloud in W3 and W4 Hasegawa et al. (1983, AJ, 88, 658) found a fragmentary cold cloud. Bania (1983, AJ, 88, 1222) found HI self-absorption towards an anonymous OB association (ℓ = 55°, 6, b = 0). Pöppel et al. (1982, RevMexAA, 5, 223) observed HI emission related to Gould's belt, and Bystrova (1984, in Structure and Evolution of the Magellanic Clouds, ed. van den Bergh & de Boer, 139) searched for low velocity HI in the Magellanic Stream.

The association of HI, CO and occationally OH was studied in a number of investigations. Wannier et al. (1983, ApJ, 268, 727) compared CO line emission with HI observations and found indications for warm HI halos around molecular clouds. Kazès and Crovisier (1981, AA, 101, 401) investigated the correlation of HI emission and absorption with CO emission and OH absorption lines. The association of CO emission with cold HI is studied by Peters and Bash (1982, in Regions of Recent Star Formation, ed. Roger & Dewdney,469). King et al. (1982, MN, 198, 255) studied the relation between HI and dust near the south celestial pole.

The general ionization of the disk ISM has been studied using the dispersion of pulsar radio signals (Vivekanand and Narayan, 1983, JAA, 3, 399) combined or not with Hα diffuse emission (Reynolds, 1983, ApJ, 268, 698 and in preparation): it appears that 10 - 30% of the IS volume is filled with ionized gas at T ≅ 10^4 K with n_e ≅ 0.1 - 0.2 cm^{-3}. O stars and perhaps planetary nebulae are responsible for this ionization (Reynolds, 1984, ApJ, 282, 191). This gas has a z-extent of at least 500 pc (Reynolds, in preparation). It may be responsible at least in part for the CIV and SiIV absorption lines in the disk (Cowie et al. 1981, ApJ, 248, 528), although CIV sometimes comes from hotter gas. The presence of very hot gaseous components with 8×10^4<T< 10^7 K has been confirmed by many observations at IS lines in the UV (cf. Hartquist and Snijders, 1982, Nature, 299, 783), and in the soft X-ray continuum (Garmire, 1983, in IAU Symp. 101, 347; Mc Cammon et al. 1983, ApJ, 269, 107) and lines (Schnopper et al. 1982, ApJ, 253, 131; Rocchia et al. 1982, AA, 130, 53). The contribution of the warm and hot medium to the far UV background through line or 2 photon emission is controversial (Feldman et al. 1981, ApJ, 249, L51; Deharveng et al. 1982, AA, 109, 179; Paresce et al. 1983, ApJ, 266, L107; Zvereva et al. 1982, AA, 116, 312; Paresce et al. 1983, AA, 124, 300; Joubert et al. 1983, AA, 128, 114). A large fraction of the hot medium is contained inside bubbles, shells and supershells (see later).

The abundance of elements in the ISM has been the subject of many papers. Extensive series of data obtained using Copernicus, or optical observations can be found in Rogerson and Upson (1982, ApJSuppl, 49, 353), Chaffee (1982, ApJSuppl, 50, 169), Bohlin et al. (1983, ApJSupp, 51, 277) and Hobbs (1984, ApJSuppl, in press). New discussions of the abundance of deuterium yield a small value $D/H \cong 5\ 10^{-6}$ (Laurent, 1983, in Primordial Helium, Shaver et al. ed., ESO, p. 335; Vidal-Madjar, 1983, in Diffuse Matter in Space, Audouze et al. eds., Reidel, p. 57; Vidal-Madjar et al. 1983, AA, 120, 58). Lithium appears to be not much depleted with respect to solar abundances (Snell and Vanden Bout, 1981, ApJ, 250, 160; Hobbs, 1984, ApJ, in press), but the $^{7}Li/^{6}Li$ ratio is >25 (Ferlet and Dennefeld, 1983, in Primordial Helium, ed. Shaver et al. ESO, p. 373). Na, K, Rb, Mg, Zn are not much depleted either (Jura and Smith, 1981, ApJ, 251, L43; York and Jura, 1982, ApJ, 254, 88; Chaffee, 1982, op.cit.; Harris et al. 1983, MN, 203, 1225; Murray et al. 1984, ApJ, 282, 481; Phillips et al. 1984, MN, 206, 337). This is also the case for C, N and O (Hobbs et al. 1982, ApJ, 252, L21; Vidal-Madjar et al. 1982, ApJ, 260, 128; York et al. 1983, ApJ, 266, L55). Conversely, many elements are known to be considerably depleted, and there is a general consensus that depletion increases with column density, hence presumably volume density. This is the case for Si (Snow and Joseph, 1981, AJ, 86, 1916), Al (Barker et al. 1984, ApJ, 280, 600), Ca (Snow et al. 1983, ApJ, 265, L67, and others), Cl (Harris and Bromage, 1984, MN, 208, 941) and possibly B (York et al. 1982, ApJ, 255, 524); see also the more general studies by Phillips et al. 1982, MN, 200, 687; Cardelli and Böhm-Vitense, 1982, ApJ, 262, 213; and Tarafdar et al. 1983, ApJ, 267, 156. Fe, also very depleted, poses a special problem since a number of oscillator strengths had to be determined through astrophysical observations (Lugger et al. 1982, ApJ, 257, 135; Shull and van Steenberg, 1983, ApJ, 271, 408). However Nussbaumer et al. (1981, AA, 102, 351) have recalculated f values for FeII, and more recently Nussbaumer and Storey (1984, AA, in press) have done the same for CI. An interesting old observation has been confirmed by many new ones: the depleted elements (Ca, Ti, etc.) are less depleted in high velocity components (Martin and York, 1982, ApJ, 257, 135; Hobbs, 1983, ApJ, 265, 817, etc.). Several authors see in this property the effect of sputtering of grains by relatively high energy particles (Trivedi and Larimer, 1981, ApJ, 248, 563; Duley, 1982, ApSpSc, 85, 221).

A number of studies of the physical conditions in the diffuse clouds have used optical and UV lines of atoms and molecules: Federman and Hobbs, 1982, ApJ, 265, 813; Snow, 1983, ApJ, 266, 576; Welsh, 1983, ApSpSc, 90, 437; Jenkins et al. 1983, ApJ, 270, 88; Younan and Dufton, 1984, MN, 209, 123, and others.

A large number of papers are concerned with molecules in diffuse clouds. Tentative detections of CS^{+} (Ferlet et al. 1983, AA, 125, L5), and of H_2O (Snow et al. 1981, ApJ, 250, 163), have been reported. H_2O^{+} could not be found (Hayden Smith et al. 1984, ApJ, 277, 196), as well as C_3 (Clegg and Lambert, 1982, MN, 201, 723). Many new observations concern C_2, an interesting tracer of temperature but which may be affected by radiative pumping (Hobbs and Campbell, 1982, ApJ, 254, 108; van Dishoeck and Black, 1982, ApJ, 258, 533; Danks and Lambert, 1983, AA, 124, 188; Hobbs et al. 1983, ApJ, 271, L95; Lutz and Crutcher, 1983, ApJ, 271, L101, Dishoeck and de Zeeuw 1984, MN, 206, 383). CO and ^{13}CO have been observed and discussed by Wannier et al. 1982, ApJ, 254, 100, and Tarafdar and Krishna Swamy, 1982, MN, 200,

431, and CH by Federman (1982, ApJ, 253, 601 and 257, 125), Danks et al. (1984, AA, 130, 62) and Bouloy et al. (AA, 130, 380). The abundances of C_2, CO, CH and CN are reasonably well predicted by standard gas-phase chemistry (Dickman et al. 1983, ApJSupp, 53, 55; Federman et al. 1984, BullAAS, 16, 484), although formation on grains cannot be excluded (Pickles and Williams, 1981, MN, 197, 429; Mann and Williams, 1984, MN, 209, 33). The exception remains CH^+, although its rate of formation by radiative association has been revised upwards by Graff et al. (1983, ApJ, 269, 796): the general consensus is that it is formed in shocks (see e.g. Federman, 1982, op. cit.; Dickman, 1983, op. cit.; Mitchell and Watt, 1984, BullAAS, 16, 464). New analyses of the formation of H_2 have been published by Snow (1983, ApJ, 269, L57) and Jenkins et al. (1983, ApJ, 270, 88).

Diffuse interstellar bands may be due to isolated molecules, but they will be treated in the section on grains.

2. THE LOCAL ISM

A comprehensive review of the local interstellar medium is given in the proceedings of the IAU Coll. 81 (1984). New analyses of the H and He lines scattered by the ISM streaming into the solar cavity show that the hydrogen in this gas is 20-30% ionized, has a density of the order of 0.1 cm^{-3} and a temperature of 10^4 K (Burgin, 1981, Comments Ap, 9, 157; Blum, 1981, AdvSpRes, 1, 197, COSPAR; Ripken and Fahr, 1983, AA, 122, 181; Kunc et al. 1983, Planet SpSc, 31, 1157; Wallis, 1984, AA, 130, 200; Dalaudier et al. 1984, AA, 134, 171). Partial indications suggest that this gas extends to a few parsecs and is photoionized (Crutcher, 1982, ApJ, 254, 82; Oegerle et al. 1982, ApJ, 252, 302; Bobroff et al. 1984, AA, 277, 678). However, the density decreases further away and the temperature rises sharply, suggesting that we are dealing with a bubble or superbubble interior (Frisch, 1981, Nature, 293, 377; Bruhweiler and Kondo, 1981, ApJ, 248, L123; 1982, ApJ, 259, 232 and ApJ, 260, L91; Innes and Hartquist, 1984, MN, 209, 7). However the existence of absorption lines at various radial velocities in the spectra of numerous nearby stars suggest the presence of cold or warm discrete clouds (York, 1983, ApJ, 264, 172; Paresce, 1983, Nature, 302, 806). Frisch and York (1983, ApJ, 271, L59; 1984, BullAAS, 16, 411) have mapped these clouds using Lyman α and NaI, CaII lines.

A review of the situation including the information from the 21-cm line has been given by York and Frisch (1984, IAU Coll. 81).

3. SHELLS, HOT GAS AND HALO

The phenomenon of the shells and supershells first detected in the 21-cm line and recently reviewed by Heiles (1983, IAU Symp. 101, 367) has been studied with special emphasis on understanding its ubiquity (Stacy, Jackson, 1982, AASuppl, 50, 377) and its astrophysical nature (Gosachinskij, 1982, SovietAstrL, 8; Heiles, 1981, BullAAS, 13, 507).The number of detected objects has been increased (Hu, 1981, ApJ, 248, 119) by employing various criteria including a search list derived from radio continuum data (Sofue and Nakai, 1983, AASuppl, 53, 57; Sufue, 1983, PASJapan, 35, 91) and a search in M31 (Brinks and Bajaja, 1983, IAU Symp. 100, 139). Individual objects studied are those in the Rosette nebula and the Monoceros OB2 association (Gosachinskij and Khersonskij, 1983, Soviet Astr, 26, No. 2; Lockman and Ganzel, 1983, ApJ, 268, 117), the Lupus Loop (Colomb and Dubner, 1982, AA, 112, 141) and a shell close to the pulsar PSR 0809+74 (Velden and Hirth, 1982, AA, 113, 340).

The shells and supershells seen in the 21-cm line and more generally the interaction zones between stellar winds, supernovae and HII regions and the ISM have often been detected in interstellar absorption lines, mainly of CaII,AlIII, CIV, SiIV and NV. See for individual supernova remnants: Jenkins et al. 1981, ApJ, 248, 977; Hobbs et al. 1982, ApJ, 252, L17; Phillips and Gondhalekar, 1983, MN, 202, 483; Jenkins et al. 1984, ApJ, 278, 649; and Blair and Raymond, 1984, BullAAS, 15,

930. For the Carina region, see Walborn and Hesser (1982, ApJ, 152, 156), Walborn (1982, ApJSuppl, 48, 145); Laurent et al. (1982, ApJ, 260, 163), Welsh and Thomas (1983, ApLett, 23, 103) and Walborn et al. (1984, ApJ, 276, 524). Other specific regions have been discussed by Phillips and Gondhalekar (1981, MN, 196, 533) (I Per OB association); Phillips et al. 1984, MN, 206, 55 (I Cyg OB association); de Boer and Nash (1982, ApJ, 255, 477) (HD 46402 in the LMC); Bates et al. (1983, AA, 122, 64) (Loop II); Chanot and Sivan (1983, AA, 121, 19), Bruhweiler et al. (1983, CommentsAp, 10, 1) and Reynolds (1984, in preparation), (Gum Nebula); Königl (1983, MN, 205, 471) (W50); Welsh (1983, MN, 204, 1203) (M8); Cox and Deharveng (1983, AA, 117, 265), Goudis and Meaburn (1984, AA, 137, 152 and 138, 57) (shells in the LMC). Hobbs (1984, ApJ, 280, 132) has even succeeded in discovering [FeX]6375 in absorption in front of λ Cep, but the [FeXIV]5303 line in HD 9352 is variable hence probably circumstellar (Hobbs and Albert, 1984, ApJ, 281, 639).

Many successful attempts have been made to detect interstellar lines in high-latitude distant stars, globular clusters or extragalactic objects. These attempts have confirmed the existence of a galactic halo and have precised its properties. 21-cm line observations show that the Galaxy contains corotating HI that extends to $/z/\sim$ 1500 pc from the plane (Lockman, 1983, BullAAS, 15, 661). The detection of HI counterparts of ionized and neutral heavy elements in the galactic halo is reported by McGee et al. (1983, MN, 205, 1191). This could be produced in the warm interface between hot gas and the far side of molecular clouds in the halo (Hartquist et al. 1982, ApJ, 259, 591). A high-latitude HI-cloud with optical emission (Goerigk et al. 1983, AA, 120, 63) located at $0.5 \lesssim /z/ \leqq 1.5$ kpc above the plane and observed in CO and H_2CO molecular lines (Mebold et al. 1983, AA, in press) shows clear evidence for an interaction with high velocity clouds (HVCs) at velocities $\leqq$-140 km s^{-1} (Kalberla et al. 1984, in IAU Coll. 81, in press). This result, a possible interaction of a high velocity stream at -110 km s^{-1} with local gas (Cohen, 1981, MN, 196, 835) and a study of Mirabel and Morras (1984, ApJ, 279, 86) show that HVCs are halo objects flowing toward the galactic plane. Further studies iterate the discussion whether or not HVCs are intergalactic (Giovanelli, 1981, AJ, 86, 1468), produced by tidal interaction of our Galaxy with the Magellanic Clouds (Cohen, 1982, MN, 199, 281 and 200, 391; Morras, 1982, AA, 115, 249; Morras and Bajaja, 1983, AASuppl, 51, 131) or are galactic objects (Shchekinov, 1980, Astrophys, 16, No. 2). Possibly soft X-ray emission (Bone et al. 1983, ApSpSc, 89, 173; Kalberla et al. 1984, in IAU Coll. 81, in press) and star formation (Dyson and Hartquist, 1983, MN, 203, 1233; Mebold et al. 1984, AA, in press) occurs during interaction of high and low velocity clouds.

The halo gas is visible in low ionization (NaI, CaII, etc.), as well as in high-ionization (CIV, SiIV) lines. One of the difficulties is that it is very inhomogeneous and that the halo lines are not visible in every direction. The statistics of optical absorption lines of Songaila et al. (1981, ApJ, 248, 956) shows that 1/3 to 1 component only can be detected per 100 kpc. The lines of sight to the Magellanic Clouds (Songaila, 1981, ApJ, 248, 945), where the halo has been discovered, might even be atypical (Blades and Morton, 1983, MN, 204, 317). The kinematics of the halo is somewhat controversial: York et al. (1982, ApJ, 255, 467; 1984, ApJ, 276, 92) find a corotation with the disk while de Boer and Savage (1983, ApJ, 265, 210) find a slower rotation at large distances z above the galactic plane. This is embarrassing since many authors use radial velocities of halo lines to locate them in z. de Boer and Savage (1984, AA, 136, L7) find that the halo gas in the direction of the N galactic pole falls on the plane, in agreement with the galactic fountain model.

The halo contains several types of gaseous components with different physical properties, e.g. temperature: Hartquist and Tallant (1981, MN, 196, 527). The warm phase ($T < 2.4\ 10^4$ K) is seen in the lines of low ionization species. Observations of these lines (Songaila et al. 1981, op.cit.; Keenan et al. 1981, MN, 197, 799; York et al. 1982, 1984, op. cit.; Jenkins et al. 1984, ApJ, 281, 585) show that this component is rather concentrated on the galactic plane, although Pettini et al. (1982, MN, 199, 409) suggest a larger extent, based on the hypothesis of corotation. The most thorough study is by Albert (1983, ApJ, 272, 509) who uses IS lines in stars at different distances from the plane; he finds that the warm phase consists in an extension of the disk gas, with low velocities and large amounts of metal depletion, and in a much less depleted, high velocity gas extending at higher z-distances, which makes at least 24% of the mass of the halo gas (see also Albert, 1982, ApJ, 256, L9 and Keenan et al. 1983, MN, 203, 963). The hot halo gas seen in the CIV and SiIV lines extends at even larger distances from the plane with a scale height of at least 3 kpc (Pettini and West, 1982, ApJ, 266, 561; Hartquist et al. 1984, ApJ, 276, 519). Apparently these observations give a better fit with the model of a halo supported by cosmic rays (Hartquist et al. 1983, MN, 203, 1233; Chevalier and Fransson, 1984, ApJ, 279, L43) than with the galactic fountain model. Other theoretical studies bearing on special points are by Hartquist et al. (1982, ApJ, 259, 591), Dyson and Hartquist (1983, MN, 203, 1233). See also the review on the galactic fountain model by Kahn, 1981, in Investigating the Universe, ed. Kahn, Reidel, p. 1. The existence of a halo in the SMC is still controversial (Fitzpatrick and Savage, 1983, ApJ, 267, 93).

4. THEORETICAL STUDIES

Apart from the previous studies concerning the halo, theoretical work has been performed on a number of topics concerning the diffuse medium. A detailed general review is by Lazareff (1983, in Diffuse Matter in Space, ed. Audouze et al. Reidel, p. 141). The determination of the "Spin-Temperature", T_{spin}, from observed quantities was discussed by Payne et al. (1983, ApJ, 272, 540). Kalberla et al. (1983, Mitt. Astron. Ges., 60, 392) related the spin temperature to the clumping in the HI gas. Liszt (1983, ApJ, 275, 163) considered the correlation of T_{spin} with the optical depth and discussed a core-halo model for the HI clouds.

A further topic is the turbulence in the galactic HI-gas. Crovisier and Dickey presented a statistical study of the small-scale structure based on a power-spectral analysis; cf. also Kalberla and mebold (1983, Mitt. Astron. Ges, 58, 101). Alferova and Gosachinskij (1983, Astrofiz. Issl. Izv. Spets. Astrofiz. Obs, 17, 77) made a computer simulation of the HI-emission from spherically symmetric gas clouds.

Some studies are concerned with the evaporation of clouds in a very hot gas: Balbus and McKee, 1982, ApJ, 252, 529; Giuliani, 1984, ApJ, 277, 605; Draine and Giuliani, 1984, ApJ, 281, 690; Balbus, 1984, BullAAS, 16, 444. The fate of clouds in shocks has been treated by Nittmann et al. 1982, MN, 201, 833; Heathcote and Brand, 1983, MN, 203, 67 and Krebs and Hillebrandt, 1983, AA, 128, 411. The dynamics of shells driven by stellar winds, supernovae, or even radiation pressure from field stars (Elmegreen and Chiang, 1982, ApJ, 253, 666) has been discussed by Huang and Weigert (1982, AA, 116, 348), McKee et al. (1984, ApJ, in press) and Wolf and Durisen (1984, BullAAS, 16, 465). Tenorio-Tagle, Rozycska and collaborators have written a large series of papers on stellar wind bubbles and related topics. Theoretical studies on the mass spectrum of clouds are due to Hausman (1982, ApJ, 261, 532) and Pumphrey and Scalo (1983, ApJ, 269, 531). Abbott (1982, ApJ, 263, 723) and Tarrab (1983, AA, 125, 308) have discussed the deposition of energy and momentum in the diffuse ISM of our Galaxy and other galaxies, and Dupree and Raymond (1983, ApJ, 275, L71) have emphasized the interaction of white dwarfs with the ISM. Global models of the ISM have been discussed by Ikeuchi and Tomita (1983, PASJapan, 35, 77), Satoru et al. (1984, MN, 207, 909). Hydrostatic equilibrium with magnetic field has been reviewed by Kuznetsov and Ptuskin (1983, ApSpSc, 94, 5), and Gosh and Ptuskin (1983, Soviet Astr, 9, 49).

III. Molecules and Molecular Clouds
(P.G. Wannier)

1. ISOTOPE ABUNDANCES

Molecular transitions have continued to provide valuable abundance information with only small interpretational errors. A major topic of study has been to determine precisely how large these errors can get.

Carbon isotopic ratios have been made using the 6-cm lines of formaldehyde yielding a value (25) of $H_2{}^{12}CO/H_2{}^{13}CO$ in the galactic center and values near or slightly lower than terrestrial (90) in other galactic sources (Wilson and Jaffe, 1981, ApJ, 245, 866; Gardner and Whiteoak, 1982, MN, 199, 23P; Henkel et al. 1982, AA, 109, 344; 1983, AA, 127, 388). These results are generally consistent with observations of $H_2C^{18}O/H_2{}^{13}CO$ which are somewhat smaller than 90 in the galactic plane and half as large in the galactic center (Gardner and Whiteoak, 1981, MN, 194, 37P; Machnik et al. 1982, ApJ, 254, 538). Observations of $H^{13}CO^+$, $H^{12}CO^+$ and $HC^{18}O^+$ yielded somewhat lower ^{13}C abundances in all sources, in better agreement with previous CO surveys and with a UV determination of $^{12}CO/^{13}CO \sim 55$ in the local interstellar medium (Stark, 1981, ApJ, 245, 99; Wannier et al. 1982, ApJ, 254, 100; Crutcher and Watson, 1981, ApJ, 244, 855).

Four observational projects and one theoretical paper sought to determine the effects of isotopic fractionation and line saturation for CO and concluded that there is fractionation in the outer boundary layers of dense clouds and in the wings of the line profiles (Wilson et al. 1981, ApJ, 243, L47; Bally and Langer, 1982, ApJ, 255, 143; Young et al. 1982, ApJ, 261, 513; Chu and Watson, 1983, ApJ, 267, 151; Penzias, 1983, ApJ, 273, 195). A general search for the unstable ^{14}C variant of CO failed to detect any in seven sources, in keeping with generally accepted age estimates for molecular clouds.

Measurements of the oxygen isotopic ratios used OH, HCO^+ and CO. A survey of galactic sources in CO indicates that $^{18}O/^{17}O$ is very constant (within 5 percent) throughout the galactic disc including the galactic center, but is significantly (40 percent) lower than the terrestrial value of 5.5 (Penzias, 1981, ApJ, 249, 518). Measurements using HCO^+ and OH are in excellent agreement with the CO results (Guelin et al. 1982, ApJ, 263, L89; Bujarrabal et al. 1983, AA, 128, 355). A measure of $^{16}OH/^{18}OH$ is about half the terrestrial one in the galactic center (Whiteoak and Gardner, 1981, MN, 197, 39P).

Measurements of the nitrogen isotopes were made with the J = 1-0 transition of HCN in 16 galactic sources (Wannier et al. 1981, ApJ, 247, 522), yielding $^{14}N/^{15}N$ equal to the terrestrial value with no evidence for a gradient except for a deficiency of ^{15}N in the galactic center, consistent with extensive processing in stars. Measurements of HNN^+ were consistent (Linke et al. 1983, ApJ, 271, L85). A new measurement of the silicon isotopes using SiO has yielded results equal to terrestrial values, contrary to an earlier report (Penzias, 1981, ApJ, 249, 513).

Interstellar D/H has the lure of providing cosmological information, but suffers from fractionation and contamination with newly processed material. UV observations including atomic and molecular hydrogen yielded D/H $\sim 2\text{-}5\ 10^{-5}$, but have shown some evidence for contamination by stellar winds (Bruston et al. 1981, ApJ, 243, 161; Vidal-Madjar et al. 1983, AA, 120, 58). The deuterated species of heavy molecules in dense clouds continue to provide information about fractionation rather than about abundances (Beckman et al. 1982, MN, 201, 357; Herbst, 1982, AA, 111, 76).

2. MOLECULAR LINES

2.1 New Lines and Undetected Lines.

New molecular detections have continued to provide valuable information about conditions in dense clouds, though the surprises have come from circumstellar envelopes and dense, cold clouds instead of from active star-forming regions as in the past. Eleven new molecular species have been detected and/or confirmed.

Lab spectra and astronomical observations confirmed tentative detections of C_4H and C_3N, species especially prominent in cold clouds (Gottlieb et al. 1983, ApJ, 275, 916; Guelin et al. 1982, AA, 109, 23; Bell et al. 1982, ApJ, 255, L75; Irvine et al. 1981, ApJ, 248, L113). Both C_4H and C_3N give rise to methylated forms which have also been detected in dense clouds. These are CH_3C_4H (Walmsley et al. 1984, AA, 134, L11; Loren et al. 1984, ApJ, 286, L23; MacLeod et al. 1984, ApJ, 282, L89) and CH_3C_3N (Broten et al. 1984, ApJ, 276, L25). Two new ions, HCS^+ and $HOCO^+$, have been detected (Thaddeus et al. 1981, ApJ, 246, L41; Irvine et al. 1983, AA, 127, L10) and C_3H has been detected and confirmed (Johansson et al. 1984, AA, 130,227) as has C_3O. The first molecular ring (albeit a small one), was discovered in IRC+10216: SiCC (Thaddeus et al. 1984, ApJ, 283, L45). A theoretical discussion of $^{16}O^{18}O$ (Black and Smith 1984, ApJ, 277, 562) has led to several unpublished searches and a possible detection. In the near-IR, SiH_4 has been detected in IRC+10216 (Goldhaber and Betz, 1984, ApJ, 279, L55) as has C_2H_4 (Betz, 1981, ApJ, 244, L103). Most recent is a detection of HCl (Blake et al. 1985, ApJ, submitted). Improved spectroscopic information was obtained from observations of CCH (Ziurys et al. 1982, ApJ, 254, 94) and from lab measurements and observations of H_2 (Jennings et al. 1984, ApJ, 282, L85). A tentative detection of HOC^+ has been made on the basis of one weak line toward the SagB2 and upper limits toward 13 other clouds (Woods et al. 1983, ApJ, 270, 583). Frequencies were reported for the diatomic molecules MgO (Steimle et al. 1984, ApJ, 277, L21) FeO (Endo et al. 1984, ApJ, 278, L131) and LiH (Plummer et al. 1984, ApJ, 282, L113).

Several explorations of frequency bands have been made which have uncovered a host of new lines, but almost entirely of species already familiar. One survey of the 3 mm region toward SgrB2 and Orion A yielded 274 new lines (Turner, 1983, ApJ, 23, 217) and another yielded 170 lines in Orion and 45 toward IRC+10216 (Johansson et al. 1984, AA, 130, 227). A survey at 215-247 GHz has detected 544 molecular lines, all but 27 or which are identified as lines of previously detected molecules (Sutton et al. 1985, ApJ, submitted). New physical probes are provided by high frequency lines such as those of CO at 434μ, 153μ and 168μ (Koepf et al. 1982, ApJ, 260, 584; Goldsmith et al. 1981, ApJ, 243, L79; Stacey et al. 1982, ApJ, 257, L37; Stacey et al. 1983, MN, 202, 25P), NH_3 at 524μ (Keene et al. 1983, ApJ, 271, L27) and the 1.2 mm lines of CH_3OH (Boland et al. 1983, ApJ, 271, 183).

Null detections, though less compelling, are nontheless important. A useful set of null detections for practitioners of millimeterwave astronomy is a published set of 100 locations near the galactic plane with no detectable CO emission (<0.5K) (Verter et al. 1983, ApJSuppl, 52, 289). Molecular searches which have not been successful are ones for NaH (Plambeck and Erickson, 1982, ApJ, 262, 606), H_2O^+ (Smith et al. 1984, ApJ, 277, 196), SiN (Ziurys et al. 1984, ApJ, 281, 219). FeO (Merer et al. 1982, ApJ, 256, 151), N_2O (Wilson and Snyder, 1981, ApJ, 246, 86), H_3^+(Oka 1981, Phil.Trans.R.Soc.Ser, A, 303, 543), C_3 (Clegg and Lambert, 1982, MN, 201, 723), $C_3N_2H_4$ and $CH(CN)_3$ (Rydbeck et al. 1981, AA, 97, 192), glycine (Snyder et al. 1983, ApJ, 268, 123), HOCl, CH_3CH_2CCH and $HCOOCH_3$ (Hollis et al. 1983, AA, 126, 393) and HCP and H_2SO_4 (Hollis 1981, ApJ, 251, 541).

2.2 Molecular Excitation

Almost all studies involving molecular spectroscopy require some knowledge of excitation. Calculations have been made where the energy input is from shocks (Draine and Roberge 1984, ApJ, 282, 491), from collisions with hot grains (Krugel and Walmsley 1984, AA, 130, 5), from IR radiation (Carroll and Goldsmith 1981, ApJ, 245, 891) and star formation (Stenholm 1983, AA, 124, 247). The gas temperature is dependent, in turn, on molecular line cooling and studies have been made of radiation from CO (Gilden 1984, ApJ, 283, 679), H_2O (Takahashi et al. 1983, ApJ, 275, 145) and OH (Flower et al. 1982, MN, 200, 55P). The numerical techniques of radiative transfer in molecular clouds have been scrutinized (Sobolev 1982, Nauchn. Inf., 50, 47; Stenholm 1983, AA, 117, 41; Deguchi and Kwan 1982, ApJ, 260, 579). Analyses have been made of HCO^+ and HCS^+ (Montiero 1984, MN, 210, 1) and of CH (Lien 1984, ApJ, 284, 578). Several observational and theoretical papers have addressed the anomalies of the hyperfine brightnesses of HCN (Sandell et al. 1983, AA, 118, 306; Walmsley et al. 1982, ApJ, 258, L75; Zinchenko and Khersonskij 1982, Astron.Zh.,59, 676). Observations and analysis of ortho-NH_3 indicate an inversion and possible weak maser (Guilloteau et al. 1983, AA, 124, 322). Studies of ortho/para ratios of H_2CO (Kahane et al. 1984, AA, 137, 211) and H_2 (Flower and Watt 1984, MN, 209, 25) determine under what conditions the statistical ratios do not apply. H_2 line excitation has been modelled (Chernoff et al. 1982, ApJ, 259, L97; McKee et al. 1982, ApJ, 259, 647).

2.3 Lines as Probes in Clouds

New probes of physical conditions in clouds have been tried. The existence of weak radiative coupling (metastable states) is desirable for temperature probes and the CH_3C_2H (Askne et al. 1984, AA, 130, 311; Kuiper et al. 1984, ApJ, 276, 211), CH_3CN (Loren and Mundy 1984, ApJ, 286, 232; Andersson et al. 1984, AA, 136, 243) and the metastable states of NH_3 (Walmsley and Ungerechts 1983, AA, 122, 164; Ungerechts et al. 1982, AA, 111, 339) have been so used in molecular clouds (but see Stutzki et al. 1984, AA, 139, 258, for discussion of NH_3). Density probes require molecules with many allowed transitions. HCN (White et al. 1982, MN, 199, 375), HC_3N (Vanden Bout et al. 1983, ApJ, 271, 161), HC_5N (Jennings and Fox 1982, ApJ, 254, 111) and HNCO (Jackson et al. 1984, ApJ, 280, 608) have been used to make such measurements. CO, because of its stability and low dipole moment has been the workhorse to trace the general distribution of molecular material, but its lines are often saturated and ^{13}CO has been used to obtain better statistics of cloud sizes and distribution (Liszt et al. 1981, ApJ, 249, 532).

3. CHEMISTRY

Since the last IAU Report, significant progress has been made in the understanding of the molecular chemistry in interstellar clouds. Reactions critical to the modeling of the chemistry have been measured in the laboratory and studied theoretically. Additional observations of both diffuse clouds and dense clouds have provided additional constraints for the theoretical models. Moreover, the models have become more sophisticated by the inclusion of time dependent phenomena. Many relevant references are given in the section on molecular lines and are not repeated here.

3.1 Molecular Processes

Isotope exchange reactions have been measured in the laboratory (Adams and Smith 1981, ApJ, 248, 373; Smith et al. 1982, ApJ, 263, 123). The reactions involving O^+ and H, as well as the reactions between CH^+ and H or H_2, have been investigated experimentally (Federer et al. 1984, Phys.Rev.Lett., 52, 2084). Laboratory measurements relate to the synthesis of complex molecules (Adams and Smith, 1981, ApJ, 247, L123; Herbst, Adams and Smith 1983, ApJ, 269, 329; Smith et al. 1983, ApJ, 272, 365). Lab measurements have been undertaken at the characteristic temperatures of interstellar clouds (Barlow et al. 1984, Phy.Rev.Lett., 52, 902; Smith and Adams 1984, ApJ, 284, L13), including the recombination of H_3^+, which affects the determination of the electron abundance. The measurements confirm theoretical predictions (Michels and Hobbs 1984, ApJ, 286, L27).

The radiative association of C^+ and H has been studied (Graff et al. 1983, ApJ,

269, 796), as has the association of C^+ and H_2 (Herbst 1982, ApJ, 252, 810). More general discussions have been reported (Bates 1983, ApJ, 267, L121; 1983, ApJ, 270, 564). The endothermic reactions $C^+ + H_2$ and $O + H_2$ have also been treated (Herbst and Knudson 1981, ApJ, 245, 529). The photodissociation rates in the presence of the interstellar ultraviolet radiation field for C_2 (Pouilly et al. 1983, J.Phys.B., 16, 437), OH (van Dischoeck and Dalgarno 1983, ApJ, 277, 576), and CN (Lavendy et al. 1984, J.Molec.Spectros., 106, 395) have been estimated theoretically. Recent laboratory data on photoabsorption cross sections have been obtained, indicating that in some cases the quantal estimates may be uncertain by a factor of five or so (Lee 1984, ApJ, 282, 172; Nee and Lee 1984, ApJ, in press). Other related theoretical investigations include the selective photodissociation of CO isotopes (Bally and Langer 1982, ApJ, 255, 143; Chu and Watson 1983, ApJ, 267, 151) and the transport of UV radiation (Roberge et al. 1981, ApJ, 243, 817).

3.2 Diffuse Clouds

Several important observations of the diffuse gase are given in section 4b and are not repeated here. Measurements of CO absorption with IUE (Tarafdar and Krishna Swamy 1981, MN, 196, 67; 1982, MN, 200, 431; Tarafdar 1983, MN, 204, 1081; McLachlan and Nandy 1984, MN, 207, 355) have substantiated an earlier finding that CO correlates with H_2. New measurements of CH and CH^+ (Federman 1982, ApJ, 257, 125) constrain chemical models. CN observations indicate that the reaction $C_2 + N$ is crucial to the formation of CH (Federman 1984, ApJ, in press).

Time independent calculations (Pickles and Williams 1981, ApSpSc, 80, 337) and time dependent calculations (Mann and Williams 1984, MN, 209, 33), as well as those including the effects of shocks (Mitchell and Deveau 1983, ApJ, 266, 646) have been performed.

3.3 Dense Clouds

New observations of the cyanopolyynes (Bujarrabal et al. 1981, AA, 99, 239; Snell et al. 1981, ApJ, 244, 45; Benson and Myers 1983, ApJ, 270, 589; Cernicharo et al. 1984, AA, 138, 371), the detection of other acetylene-based carbon chained molecules (see section 2a on molecular lines) the detection of CH_3CN (Matthews and Sears 1983, ApJ, 267, L53), and the detection of C_3O (Matthews et al. 1984, Nature, 310, 125) have been obtained. The distribution of C_2H has been measured also (Huggins et al. 1984, AA, 133, 347). These and earlier observations have been modeled with gas phase reactions playing the dominant role (Freeman and Millar 1983, Nature, 301, 402; Herbst 1983, ApJSupp, 53, 41; Suzuki 1983, ApJ, 272, 579; Millar and Freeman 1984, MN, 207, 405; 1984, MN, 207, 425; Herbst et al. 1984, AA, 138, L13; Stahler 1984, ApJ, 281, 209), with grain surface reactions (Millar 1982, ApSpSc, 87, 435), and with shocks included (Mitchell 1983, MN, 205, 765; 1984, ApJSuppl, 54, 1; Duley and Williams 1984, MN, 211, 97). A time dependent model with gas phase chemistry (Leung et al. 1984, ApJSuppl, 56, 231) has illustrated that the large observed abundance of neutral carbon (Phillips and Huggins 1981, ApJ, 251, 533; Wootten et al. 1982, ApJ, 256, L5) may be the key to hydrocarbon production. The role shocks play in the neutral carbon abundance has also been analyzed (Williams and Hartquist 1984, MN, 210, 141).

The chemistry of sulfur-bearing compunds has been studied both observationally and theoretically. Emission from CS (Goldsmith and Linke 1981, ApJ, 325, 482), SiS (Dickinson and Rodriguez Kuiper 1981, ApJ, 247, 112), and SO_2, as well as HCS^+ has been analyzed. These measurements have been compared to the results of theoretical models based on gas phase reactions (Prasad and Huntress 1982, ApJ, 260, 590) and grain chemistry (Millar 1982, MN, 199, 309). The gas phase models predict that O and O_2 are not present in large quantities because of their rapid reactions with sulfur.

The ionization fraction in dense clouds has been analyzed in terms of DCO^+/ HCO^+ (Guelin et al. 1982, AA, 107, 107; Wootten et al. 1982, ApJ, 255, 160), although the new recombination measurements for H_3^+ (Smith and Adams 1984, ApJ, 284, L13) may alter the interpretations. The ionization fraction may also be determined by HCS^+/CS (Millar 1983, MN, 202, 683). The chemistry involving CO^+ has been studied in the laboratory (Huntress et al. 1982, AA, 114, 275), theoretically (Defrees et al. 1984, ApJ, 279, 322), and through observations of the isomers HOC^+ and HCO^+. Observational studies of CN (Churchwell and Bieging 1983, ApJ, 265, 216; Crutcher et al. 1984, ApJ, 283, 668) and the isomers HCN and HNC (Goldsmith et al. 1981, ApJ, 249, 524) have been made. Deuterated cyanopolyynes have been detected and have been analyzed in terms of gas phase versus grain formation (MacLeod et al. 1981, ApJ, 251, L33; Schloerb et al. 1981, ApJ, 251, L37); a recent theoretical study of surface reactions has been presented (Tielens 1983, AA, 119, 177). Deuterated water has been observed as well (Olofsson 1984, AA, 134, 36). Comprehensive theoretical models for the simpler species have been carried out (Graedel et al. 1982, ApJSuppl, 48, 321; Watt 1983, MN, 205, 321; Boland and de Jong 1984, AA, 134, 87; Langer et al. 1984, ApJ, 277, 581). The chemistry of phosphorus has been analyzed (Thorne et al. 1984, ApJ, 280, 139). The formation rate of H_2 on grain surfaces has been studied (Snow 1983, ApJ, 269, L57), and the effects of the formation on cloud evolution have been investigated (Suchkov et al. 1982, Astrofizika, 4, 629; El Shalaby et al. 1983, JASEgypt, 5, 1). Shock formation in the cloud near IC443 has been studied observationally (DeNoyer and Frerking 1981, ApJ, 246, L37) and theoretically, where it was found that HCO^+ is not formed in the shock but through enhanced ionizing radiation, (Elitzur 1983, ApJ, 267, 174). The collision-induced dissociation of molecules behind shocks has been analyzed theoretically (Roberge and Dalgarno 1982, ApJ, 255, 176; Lepp and Shull 1983, ApJ, 270, 578).

3.4 Circumstellar Shells

Long-chain carbon chemistry was studied by means of a survey of HC_3N (Jewell and Snyder 1982, ApJ, 255, L69; 1984, ApJ, 278, 176) and by observations of HC (nC_2)N out to n = 5 ($HC_{11}N$)(Feldman et al. 1982, Nature, 295, 389). Observations of CN (Wootten et al. 1982, ApJ, 257, 151), of HCN (Bieging et al. 1984, ApJ, 285, 656) and of SiS (Nguyen-Q-Rieu et al. 1984, ApJ, 286, 276) in IRC+10216 indicate the importance of photodissociation in the outer shells (Huggins et al. 1984, ApJ, 279, 284) and have led to the realization that self-shielding from ambient UV photons is important for the abundant species, CO and H_2 (Wootten et al. 1982, ApJ, 256, L5; Morris and Jura 1983, ApJ, 264, 546; Huggins and Glassgold 1982, AA, 87, 1828). The chemistry of C_3H has been modelled (Nejad et al. 1984, AA, 134, 129).

4. CLOUD MORPHOLOGY, DYNAMICS AND EVOLUTION

4.1 Magnetic Fields

The role of magnetic fields in the evolution of interstellar clouds remains an important, but largely unsolved problem. This is not for lack of effort, but is due to the difficulty of measuring the weak fields which may nontheless dominate. Zeeman splitting in the 21-cm line has indicated magnetic energy densities unexpectedly insensitive to density, and smaller than energies of macroscopic motions (Troland and Heiles 1982, ApJ, 252, 179; Brown and Chang 1983, ApJ, 264, 134; Fleck 1983, ApJ, 264, 139). Zeeman splitting of the OH maser line has led to mixed results in dense regions (Crutcher and Kazes 1983, AA, 125, L23; Hansen 1982, ApJ, 260, 599; Crutcher et al. 1981, ApJ, 249, 134). A prediction of several percent polarization of millimeterwave lines toward dense clouds (Goldreich and Kylafis 1981, ApJ, 243, L75; Kylafis 1983, ApJ, 267, 137) led to a search using CO, ^{13}CO, HCN and CS toward 14 dense clouds which yielded no detections to levels as small as 0.5 per cent (Wannier et al. 1983, ApJ, 267, 126). Nontheless, magnetic fields have been invoked to explain the curious kinematics of the dark cloud B5 (Dudorov and Sazonov 1982,

Nauchn.Inf, 50, 98; Young et al. 1981, ApJ, 251, L81). The role of fields in collapsing clouds has been studied theoretically by several groups with a view, especially, to determining whether or not the fields can escape (Paleologou and Mouschovias 1983, ApJ, 275, 838; Mouschovias and Paleologou 1981, ApJ, 246, 48; Black and Scott 1982, ApJ, 263, 696; Nakano, 1982, PASJapan, 34, 337; Scott 1983, ApJ, 275, 836; Nakano 1983, PASJapan, 35, 209; Dorfi 1982, AA, 114, 151).

4.2 Diffuse Molecular Gas and Dense Cloud Boundaries

Several studies have focused attention on how dense clouds interact with other components of the interstellar medium. It is by such interactions that dense clouds are formed and, presumably, often destroyed.

HI and CO emission lines correlate in velocity, but not in line intensity (Kazes and Crovisier 1981, AA, 101, 401), consistent with a conclusion from observations of CH that most diffuse clouds may actually be the outer portions of dense clouds (Federman and Willson 1982, ApJ, 260, 124). The first direct observations of diffuse gas at the boundary layers of dense clouds were made with high-resolution strip maps across projected edges of CO clouds, indicating the presence of warm (> 100 K) HI halos 0.5 to several pc thick (Wannier et al. 1983, ApJ, 268, 727). The detection of absorption by HCO^+, CS, C_2 and HCN millimeter lines toward HII regions indicates the presence of molecules in low density material, probably surrounding dense clouds (Nyman 1983, AA, 120, 307; 1984, AA, 141, 323). Absorption line studies of H_2CO (Goss et al. 1984, AA, 139, 317) and NH_3 (Batrla et al. 1984, AA, 136, 127) toward CasA find similar cloudlets, some also present in HI observations.

Studies concentrating on diffuse molecular gas include optical absorption line studies of C_2 (Danks and Lambert 1983, AA, 124, 188; Hobbs and Campbell 1982, ApJ, 254, 108; Cosmovici and Strafella 1981, AA, 98, 408; van Dishoeck and DeZeeuw 1984, MN, 206, 383; Hobbs et al. 1983, ApJ, 271, L95; Crutcher and Lutz 1983, ApJ, 271, L101), CH (Danks et al. 1984, AA, 130, 62). Studies toward the region of the Pleiades have been made using millimeter emission lines (Federman and Wilson 1984, ApJ, 283, 626) and optical absorption-lines (White 1984, ApJ, 284, 685; 1984, ApJ, 284, 695). Observations of interactions of dense clouds with energetic, ionized gas have been made toward bright optical rims (Wootten et al. 1983, ApJ, 269, 147), the Orion Bright Bar (Omodaka et al. 1984, ApJ, 282, L77), compact HII regions (Ho et al. 1981, ApJ, 246, 761; Forster and Boland 1982, AA, 114, 109) Planetary Nebulae (Storey 1984, MN, 206, 521), and supernova remnants (DeNoyer 1983, ApJ, 264,141); Huang et al. 1983, ApJ, 272, 609; Wootten 1981, ApJ, 245, 105).
Recent calculations of dense cloud evaporation in the presence of hot gas (Giuliani 1984, ApJ, 277, 605; Bedijn and Tenorio-Tagle 1984, AA, 135, 81) have characterized the resulting winds and shocks and bright rim formation. Generally, molecular observations have not proved to be sensitive probes of such interactions.

Three surveys have studied small clouds in the solar neighborhood (Stark 1984, ApJ, 281, 624; Lebrun and Huang 1984, ApJ, 281, 634; Blitz, Magnani and Mundy, 1984, ApJ, 282, L9) and conclude that these high-latitude objects are young, low-mass ($\sim 100\ M_{\odot}$) objects, associated with HI clouds but not gravitationally bound. However, another search for high latitude CO indicates a lack of molecular material outside of dense clouds (Wilkinson and Stark 1983, AJ, 88, 1832). A compilation of diffuse cloud data has been made (Dickman et al. 1983, ApJSuppl, 53, 55). A careful study of CO and its rare isotopic variants have yielded a new relation between visual extinction and CO column density, (Frerking et al. 1982, ApJ, 262, 590).

4.3 Dense Clouds

A survey of ~ 100 dense cloud cores has been carried out using lines of CO and NH_3 (Myers et al. 1983, ApJ, 264, 517; Myers and Benson 1983, ApJ, 266, 309; Myers 1983, ApJ, 270, 105) and concludes that cores are smaller, denser and have narrower lines in regions with active star formation. Two surveys, using lines of H_2CO and NH_3, indicate that cloud motions do not increase markedly toward dense cores (Bieging et al. 1982, AASuppl, 49, 607; Macdonald et al. 1981, MN, 195, 387). Two surveys have been made using CO J = 2-1 and J = 1-0 especially toward giant dense clouds with self-absorbed line profiles and have found the physical properties of the absorbing envelopes and have corrected previous values for the core gas temperatures (Huggins et al. 1981, ApJ, 245, 512; Loren et al. 1981, ApJ, 245, 495). Surveys were made of 80 southern hemisphere dark clouds in CO J = 2-1 (de Vries et al. 1984, AASuppl, 56, 333) and toward 100 clouds in H_2CO absorption (Gardner and Whiteoak 1984, MN, 210, 23). Autocorrelation analyses of CO emission can determine the structure of clouds (Scalo 1984, ApJ, 277, 556; Stenholm 1984, AA, 137, 133). Surveys have also been made using the millimeter lines of CN (Crutcher et al. 1984, ApJ, 283, 668) and CS (Snell et al. 1984, ApJ, 276, 625).

Studies of individual star forming regions include the Orion complex (White and Phillips 1981, MN, 194, 947; Johnston et al. 1983, ApJ, 271, L89; Harris et al. 1983, ApJ, 265, L63). CepA (Hughes and Wouterloot 1984, ApJ, 276, 204; Gusten et al. 1984, AA, 138, 205), the Serpens object (GGD29) (Ungerechts and Gusten 1984, AA, 131, 177), the Taurus complex (Henkel et al. 1981, AA, 99, 270; Baudry et al. 1981, AA, 104,101; Walmsley et al. 1981, AA, 95, 143), W33 (Stier et al. 1984, ApJ, 283, 573) W3 (Wright et al. 1984, ApJ, 281, L71; Thronson et al. 1984, ApJ, 284, 597; Zeng et al. 1984, AA, 140, 169; Dickel et al. 1984, AA, 135, 107), NGC 2024 (Thronson et al. 1984, ApJ, 280, 154; Black and Willner 1984, ApJ, 279, 673), MonOB1 and MonOB2 (Wouterloot 1984, AA, 134, 244), M17 (Jaffe and Fazio 1982, ApJ, 257, L77; Martin et al. 1984, MN, 208, 35), NGC 7538 (Dickel et al. 1981, ApJ, 250, L43; Henkel et al. 1984, ApJ, 282, L93; Campbell and Thompson 1984, ApJ, 279, 650), eight IR-selected star-formation regions (Jaffe et al. 1984, ApJ, 281, 237), southern HII regions (Brand et al. 1984, AA, 139, 181), Heiles Cloud 2 (Schloerb and Snell 1984, ApJ, 283, 129), Rho Oph (Zeng et al. 1984, AA, 141, 127) TMC (Goldsmith and Sernyak Jr., 1984, ApJ, 283, 140; Irvine and Schloerb 1984, ApJ, 282, 516; Gaida et al. 1984, AA, 137, 17), the Ophiuchus complex near the upper Scorpius association (Wouterloot 1984, AA, 135, 32), B335 (Menten et al. 1984, AA, 137, 108), S88 (Harvey et al. 1981, ApJ, 250, 200), S235 (Nordh et al. 1984, AA, 131, 221; Evans and Blair 1981, ApJ, 246, 394), S128 (Ho et al. 1981, ApJ, 243, 526) and various dark clouds (Phillips and White 1981, MN, 194, 1981).

The galactic center region has continued to fascinate molecular astronomers. Lines of NH_3 and CH_3CN probed gas temperatures with varying results (Cummins et al. 1983, ApJ, 266, 331; Wilson et al. 1982, AA, 115, 185; Morris et al. 1983, AJ, 88, 1228). H_2CO absorption line studies have focused on both low velocity and high velocity features, the latter indicating the ejection of gas from the galactic nucleus (Whiteoak et al. 1983, MN, 202, 11P; Gusten and Downes 1981, AA, 99, 27). Additional studies have been made with the CO J = 1-0 line and the main lines of OH (Inatani 1982, PASJapan, 34, 515; Venger et al. 1981, Pis'maAZh, 7, 295) and with the 2μ lines of shocked H_2 (Gatley et al. 1984, MN, 210, 565).

5. CIRCUMSTELLAR MOLECULAR ENVELOPES

A key event in post-main-sequence evolution is the rapid loss of material from red giants in dense winds. As with the advanced stages of star formation, the dense envelopes are best studied by means of molecular lines, and recent advances have been made from thermal microwave, infrared and maser observations. Chemistry in circumstellar shells is particularly interesting and is covered in the section on chemistry.

5.1 Millimeterwave Observations

Surveys of millimeterwave CO (and ^{13}CO) emission have been carried out by several groups using infrared sky surveys as guides to the source selection process

(Zuckerman 1981, AJ, 86, 84; Knapp et al. 1982, ApJ, 252, 616; Knapp and Morris 1985, ApJ, in press; Wannier and Sahai 1985, ApJ, in press). These surveys indicate that there are about equal proportions of carbon-rich and oxygen-rich object among stars with massive winds (10^{-7} to 10^{-4} M_o/yr) and that the mass loss rates of carbon-rich stars are larger than for oxygen-rich stars. The $^{13}C/^{12}C$ ratio is larger in the oxygen-rich stars, consistent with the formation of carbon-rich envelopes by injection of fairly pure ^{12}C into envelopes rich in ^{13}C from CNO processing. $^{17}CO/^{18}CO$ in IRC+10216 is far larger (∿5 as opposed to 1/5) than terrestrial or interstellar values (Penzias 1981, ApJ, 249, 518).

Detailed observations of the carbon-rich object IRC+10216 have led to models with larger mass-loss rates than previously calculated (Kwan and Linke 1982, ApJ, 254, 587) though interferometric observations of HCN indicate that the rate may be decreasing with time (Bieging et al. 1984, ApJ, 285, 656). Extensive observations of the millimeter-wave spectrum of IRC+10216 (e.g. Johansson et al. 1982, AA, 107, 128) have refined our understanding of chemistry and radiative transfer.

5.2 Infrared Spectroscopy

Infrared molecular lines have yielded important information about the inner envelopes, where mass-loss is initiated, and about the outer photospheres, where there is interaction with the Mira shocks. Most observations have used CO, either the $\Delta J = 1$ fundamental vibration-rotation band or the overtone bands. A survey of nine giants indicated episodic mass-loss (Bernat 1981, ApJ, 246, 184). Time sequence observations of variable stars in the CO overtone bands traced the dynamics of the shocks and have detected warm (∿ 1000 K), quiescent layers of gas which may represent where outward acceleration is initiated (Hinkle et al. 1982, ApJ, 252, 697; 1985, ApJSuppl, in press). Techniques of spatial mapping have been combined with high-resolution CO spectroscopy. Combined speckle/spectral observations of IRC+10216 and Mira have shown CO emission (Dyck et al. 1983, ApJ, 271, L79; Beckwith et al. 1985, ApJ, submitted). Larger scale mapping has used annular observing apertures to isolate weak CO emission features out to 6 arcsec (Sahai and Wannier 1985, ApJ, in press). A survey of oxygen isotope ratios in nine stars has yielded mixed results, but indicates that the standard assumptions about late stellar evolution are in need of some revision (Harris and Lambert 1984, ApJ, 285, 674). Observations of 2.1μ H_2 in the bipolar nebulae CRL 2688 and 618 (Beckwith et al. 1984, ApJ, 280, 648) and in NGC 7027 (Isaacman 1984, AA, 130, 151) indicate that shocks may result from interaction of a fast wind with an older, slower dense wind remnant.

5.3 OH Maser Observations

The most fruitful maser lines for observations of red giants are those of OH. Most, but not all 1612 MHz sources are associated with late-type variable stars (Engels 1982, Veroff.Astron.Inst.Bonn. Nr. 95, 91; Jones et al. 1983, ApJ, 273, 669) and a survey of 114 OH masers has been made (Baud et al. 1981, AA, 95, 156,171). The most dramatic results have come by use of radio interferometers, including the VLA, MERLIN and VLBI. Two surveys using the VLA indicate that the OH masers are in shells distributed spherically around the central stars (Reid et al. 1981, AJ, 86, 897; Bowers et al. 1983, ApJ, 274, 733; Chapman et al. 1984, MN, 207, 149). These results agree with MERLIN observations in two sources, but are at odds with results from IRC+10420, an F8 supergiant, which seems to contain a bipolar flow (Norris et al. 1982, Nature, 299, 131; Diamond et al. 1983, AA, 124, L4; Booth et al. 1981, Nature, 290, 384) and from NML Cyg which has a complex structure (Diamond et al. 1984, MN, 207, 611). VLA and time sequence observations of another 1667 MHz source, OH 231.8+4.2, indicate bipolar flow (Morris et al. 1982, ApJ, 259, 625; Bowers and Morris 1984, ApJ, 276, 646). VLBI observations of the VYCma 1665 and 1667 MHz masers indicate a rotating disc, and Zeeman splitting indicates a ∿1 milligauss field (Benson and Mutel 1982, ApJ, 253, 199). A survey of stokes polarimetry was carried out in IRC+10420 and in seven late-type variables yielding information about their magnetic fields (Claussen and Fix 1982, ApJ, 263, 153). Models of OH masers involve photodissociation of water by ambient UV photons (Huggins and Glassgold 1982, AJ, 87, 1828).

5.4 Other Masers

SiO has maser and thermal emission lines. Maser emission has been studied using the v = 1, J = 1-0, J = 2-1, J = 3-2 and J = 5-4 (Clemens and Lane 1983, ApJ, 266, L117; Clark et al. 1984, ApJ, 276, 572) lines of SiO as well as the v = 2, J = 2-1 transition (Olofsson et al. 1981, ApJ, 247, L81). There is strong evidence from a survey of six objects, of variable intensity correlating with the IR fluxes of periodically variable stars (Wolff and Carlson 1982, ApJ, 257, 161) and a survey of 54 v = 1, J = 1-0 masers has yielded a statistically complete survey to approximately 300 pc. Maser emission was found toward four objects near dense clouds, objects which seem to be red super-giants (Ukita and Goldsmith 1984, AA, 138, 194). Modelling of SiO maser emission has also been improved (Elitzur et al. 1983, ApJ, 274, 210; Langer and Watson 1984, ApJ, 284, 751).

A systematic survey of short-period semi-regular stars has yielded six new water masers, indicating the presence of a different class of maser star (Dickenson and St. Clair Dinger 1982, ApJ, 254, 136). An SiS maser (v = 0, J = 1-0) was detected in IRC+10216 (Grashoff et al. 1981, AA, 101, 238; Henkel et al. 1983, ApJ, 267, 184), understood by observations of many rotational transitions of SiS (Sahai et al. 1984, ApJ, 284, 144).

6. HIGH-VELOCITY OUTFLOWS

One of the most striking results of recent years is the discovery of numerous high velocity, generally bipolar flows near dense cores and embedded IR sources in molecular clouds, apparently arising from outflow along the poles of stellar accretion discs (Bally and Lada 1983, ApJ, 265, 824). The best studied object is the high-velocity source in Orion, though the general importance of such outflows can only be appreciated from the studies of many other occurrences.

6.1 The Orion High-Velocity Source

The first direct indication of the importance of the IR source, IRc2, came from VLBI observations of the water masers, yielding proper motions of several arcsec/yr, centered on the deeply embedded source. (Genzel et al. 1981, ApJ, 244, 884). Observations of the J = 3-2 transition of CO and of SiO v = 0, J = 2-1 with the Hat Creek interferometer further indicated the importance of IRc2 by detecting red-shifted and blue-shifted jets to the east and west of IRc2 respectively (Erickson et al. 1982, ApJ, 261, L103; Wright et al 1983, ApJ, 267, L41). Interferometry of J = 1-0 CO (Masson et al. 1984, ApJ, 283, L37) and of HCO^+ (Vogel et al. 1984, ApJ, 283, 655) have further refined our understanding of the interaction between the high velocity material and the quiescent gas. Interferometry located the centers of OH (VLA) and SiO (Hat Creek) maser emission around IRc2 (Hansen and Johnston 1983, ApJ, 267, 625; Wright and Plambeck 1983, ApJ, 267, L115) and MERLIN observations of OH indicate a rotating torus (Norris 1984, MN, 207, 127. IR polarization convincingly showed that IRc2 and the Becklin-Neugebauer sources are the illuminating sources in the entire region, with other IR sources merely being reflection nebulae (Werner et al. 1983, ApJ, 265, L13). Polarimetry of the SiO maser indicates its origin in a rotating, expanding disc (Barvainis 1984, ApJ, 279, 358). High spectral resolution IR spectroscopy at 2-5μ indicated the existence of several components associated with the outflow region and provided a detailed look at the central, ultra-dense wind and surrounding dense molecular disc (Scoville et al. 1983, ApJ, 275, 201).

The region of molecular gas shocked by the outflow has been extensively studied by means of vibration/rotation lines of molecular hydrogen, leading to estimates of the mass (.02-.05 $M_{\odot}$), temperature (2000-3000 K) and line-of-sight extinction (1-4 mag at 2μ) (Scoville et al. 1982, ApJ, 253, 136; Beckwith et al. 1983, ApJ, 264, 152; Knacke and Young 1981, ApJ, 249, L65; Beck et al. 1982, ApJ, 253, L83; Nadeau et al. 1982,ApJ, 253, 154; Beck and Beckwith 1983, ApJ, 271, 175; Beck 1984, ApJ, 281, 205). Other molecular observations which have focused on the highly excited gas near the origin of the outflow have used high excitation lines of HCCCN (Loren et al. 1981, ApJ, 244, L107; Goldsmith et al. 1982, ApJ, 260, 147), the 153μ and 168μ rotational lines of CO (Stacey et al. 1982, ApJ, 257, L37; 1983, MN, 202, 25P), CH_3CN (Goldsmith et al. 1983, ApJ, 274, 184) and CH_3OH (Lovas et al. 1982, ApJ, 253, 149). New Herbig-Haro objects have been discovered (Axon and Taylor 1984, MN, 207, 241).

Several other molecular transitions have been used to further define the nature of the outflow and shocked gas including CO, H_2CO, H_2O, HCO^+, SO, SiO, NH_3 and HCN (van Vliet et al. 1981, AA, 101, L1; Solomon et al. 1981, ApJ, 245, L19; Olofsson et al. 1982, AA, 113, L18; Phillips and White 1983, MN, 202, 1093; Plambeck et al. 1982, ApJ, 259, 617; Kuiper et al. 1984, ApJ, 283, 106; Hasegawa et al. 1984, ApJ, 283, 106; Friberg 1984, AA, 132, 265; Kuiper et al. 1981, ApJ, 251, 88; Harris et al. 1983, ApJ, 265, L63; Rydbeck et al. 1981, ApJ, 243, L41; Wright et al. 1981, ApJ, 245, L87; Wootten et al. 1984, ApJ, 277, 189).

6.2 Other High-Velocity Sources

Observations of NGC 2071 show its bipolar flow to be orthogonal to a dense, rotating disc, presumably associated with a forming star (Lichten 1982, ApJ, 253, 593; Bally 1982, ApJ, 261, 558; Schwartz et al. 1983, ApJ, 267, L109; Persson et al. 1981, ApJ, 251, L85; Bally and Stark 1983, ApJ, 266, L61; Takano et al. 1984, ApJ, 282, L69; Wootten et al. 1984, ApJ, 279, 633). CO observations in several objects indicate that molecular emission arises in the outer boundaries of a cavity swept out by the outflow (Snell et al. 1984, ApJ, 284, 176). A survey of high-velocity sources reviews the properties of known outflows (Bally and Lada 1983, ApJ, 265, 824). Individual sources observed are W3 (Claussen et al. 1984, ApJ, 285, L79). AFGL2591 (Lada et al. 1984, ApJ, 286, 302), HH 7-11 (Snell and Edwards 1981, ApJ, 251, 103), L1529 (Lichten 1982, ApJ, 255, L119), L1551 (Kaifu et al. 1984, AA, 134, 7), HH 24-27 and HH 1-2 (Snell and Edwards 1982, ApJ, 259, 668), NGC 7538 (IRS 1-3) (Campbell and Thompson 1984, ApJ, 279, 650; Campbell 1984, ApJ, 282, L27), GL437 (Arquilla and Goldsmith 1984, ApJ, 279, 664), CepA (Cohen et al. 1984, MN, 210, 425), GL490 (Lada and Harvey 1981, ApJ, 245, 58; Kawabe et al. 1984, ApJ, 282, L73), G35.2-0.74 (Matthews et al. 1984, AA, 136, 282), W49 and W51 (Ho, Genzel and Das, 1983, ApJ, 266, 596; Downes et al. 1982, ApJ, 252, L29; Beckwith and Zuckerman 1982, ApJ, 255, 536; Genzel et al. 1981, ApJ, 247, 1039), S156 (Joy et al. 1984, ApJ, 284, 161), RCW97 Brand et al. 1984, AA, 139, 181; R Mon (HH 39) (Canto et al. 1981, ApJ, 244, 102), GGD 12-15 (Rodríguez et al. 1982, ApJ, 260, 635), GGD 27/28 (de Vries et al. 1984, AASuppl, 56, 333), PV Cephei (Levreault 1984, ApJ, 277, 634), Elias 1-12 (Levreault 1983, ApJ, 265, 855), HH objects 19-27, 1-3, 35,36 and 63 (Edwards and Snell, 1984, ApJ, 281, 237) and in four dark clouds (Frerking and Langer 1982, ApJ, 256, 523). Observations were made using millimeterwave CS (Thronson and Lada 1984, ApJ, 284, 135).

Interpretations of the widespread occurrence of high-velocity outflows generally involve massive discs (Torrelles et al. 1983, ApJ, 274, 214) but sometimes also involve magnetic fields (Draine 1983, ApJ, 270, 519; Pudritz and Norman 1983, ApJ, 274, 677). It has also been suggested that most outflowing regions will result in stars of less than 4 $M_{\odot}$ (Beckwith et al. 1983, ApJ, 267, 596).

7. MASERS

During the past four years maser research has benefitted from widespread application of new radio interferometers and from the completion of a number of complete surveys. Much important maser work has focused on observations of material around late-type stars and bipolar flows, and the references given in those sections are not repeated here.

OH masers have been studied in a set of three surveys of mainline emission covering the galactic longitude range 326°-60° including the galactic center region (Caswell et al.1980, AustJPhys,33, 639;Caswell and Haynes 1983, AustJPhys,36,361; 1983,AustJPhys,36,417) yielding 144 maser sources, almost entirely associated with HII regions. Interferometry of OH was done with the VLA (Ho et al.1983,ApJ,265,295; Dickel et al.1982, MN, 198,265), MERLIN (Norris et al.1982,MN,201,209) and using VLBI techniques (Haschick et al. 1981,ApJ,244,76;Fouquet and Reid 1982,AJ,87,691; Forster et al. 1982,MN,201,7P). Generally the OH masers are located in shells at the edges of HII regions, indicating shock excitation of the masers.

Water masers have been used in surveys made according to various selection criteria: toward known OH masers (Caswell et al. 1983, AustJPhys, 36, 401; 1983, AustJPhys, 36, 443), radio continuum peaks (Braz and Scalise, 1982, AA, 107, 272), infrared sources (Jaffe et al. 1981, ApJ, 250, 621; Crocker and Hagen 1983, AASuppl, 54, 405), bright nebulosities (Sandell and Olofsson 1981, AA, 99, 80) and Herbig-Haro objects (Haschick et al. 1983, ApJ, 265, 281). Of these, IR sources and OH masers proved to be the most reliable signpost of H_2O maser emission.

Several studies have focused on the 1979 maser outburst in Orion, associated with action of an emerging proto-star (Abraham et al. 1981, AA, 100, L10; Matveenko et al. 1982, Pis'ma AZh, 8, 711;Matveenko et al. 1983, Pis'ma AZh, 9, 456; Strel' nitskij 1982, Pis'ma AZh, 8, 165). Another time-variable maser has been studied in the direction of the supernova remnant Kes 45 (Sandell et al. 1983,AA, 124, 139) and three interferometric studies have been made toward W51 (Downes et al. 1981, ApJ, 249, 124), W49 (Walker et al. 1982, ApJ, 255, 128) and toward 7 other HII regions (Lada et al. 1981, ApJ, 243, 769).

VLA maps have been made of the only two known H_2CO masers, toward SgrB2 (Whiteoak and Garner 1983, MN, 205, 27P) and NGC 7538 (Rots et al. 1981, ApJ, 245, L15) and further studies have been made of the equally unusual SiO maser in Orion (Ukita et al. 1981, PASJapan, 33, 341; Olofsson et al. 1981,AA, 100, L30). Calculations have modelled SiO, H_2O and methanol masers in star-forming regions (Varshalovich et al. 1983, Pis'ma AZh, 9, 395; Grinen and Grigor'ev 1983, Pis'ma AZh, 9, 463; Sobolev and Strel'nitskij 1983, Pis'ma AZh, 9, 26; Strel'nitskij 1984, MN, 207, 340) and have modelled maser pumping in the presence of cold grains (Deguchi 1981, ApJ, 249, 145).

IV. Interstellar Grains
(J.S. Mathis)

For reasons of space, this overview cannot cover some very intersting work on circumstellar and interplanetary grains, both of which have bearing on interstellar dust.

Continuum extinction has been determined much more accurately than before in the near-infrared (1 - 10μm) spectral region (Koornneef 1983, AA, 128, 84; Castor and Simon 1983, ApJ, 265, 304; Willner and Pipher 1983, ApJ, 265, 760; Leitherer and Wolf 1984, AA, 132, 151; Landindi et al. 1984, AA, in press; Rieke and Lebofsky 1984, ApJ, in press). The derived values of $A(\lambda)/A(V)$ differ by up to 50% in the 2 - 4 μm range. The extinction law has been investigated in several dark clouds (Herbst et al. 1982, AJ, 87, 98; Vrba and Rydgren 1984, ApJ, 283, 123; Chini and

Krügel 1983, AA, 117, 289). As is well known, R, defined as A(V)/E(B-V), is larger in dark clouds than for diffuse dust. A large R and a large wavelength of maximum polarization, λ(max), are usually taken to imply a larger-than-average grain size. However, Chini and Krügel (op.cit.) show from models that this is not necessarily the case; they can produce many counterexamples, except that large λ(max) is always correlated with the size of silicate cylinders. Extinction in dark clouds is apparently rather similar to that in diffuse regions for $\lambda < 0.5$ μm, and the large value of R in dark clouds arises from the extinction between V and B increasing more slowly than normal (Hyland et al. 1982, MN, 201, 1095). The Taurus dark cloud has a normal value of R (Straizys et al. 1982, ApSpSci, 85, 271). The IR/visual extinction is also the same in the LMC and SMC as in the Galaxy (Koornneef 1982, AA, 107, 249; Morgan and Nandy 1982, MN, 199, 979; Nandy et al. 1984, MN, in press).

Maps of the visual extinction in various directions have been determined from galaxy counts (Burstein and Heiles 1982, AJ, 87, 1189; Strong 1983, MN, 202, 1015) and Strömgren photometry (Perry and Johnston 1982, ApJ, 50, 451). Knudel (1983, AA, 126, 89) suggest that about 5% of the sky at high latitudes is covered by clouds. The results of the IRAS satellite on IR "cirrus" (Low et al. 1984, ApJ, 278, L19) should clarify the picture.

Ultraviolet extinction has been studied by means of the IUE satellite. Variations in extinctions from star to star put powerful constraints on the nature of grains. A large catalog of measurements of the strength of the λ2175 feature ("the bump") is compiled in Guertler et al. (1982, AstrNach, 303, 105) and in Friedemann et al. (1983, AstrNach, 304, 237).

Krelowski and Strobel (1982, ActaAstr, 32, 331) and Greenberg and Chlewicki (1983, ApJ, 272, 563) find that the bump and visual extinction are correlated, but the extinction in the 0.17-0.13 μm region (hereafter called the "FUV rise") varies independently of the other two. However, de Boer (1983, AA, 125, 258) finds a star with a normal FUV extinction and an extinction excess from 0.5 μm through the bump. Variations of UV extinction from star to star have been discussed in 1367 stars of type B3 or earlier by Meyer and Savage (1981, Apj, 248, 545) using ANS data. They find that there are wide variations in the strength of the bump and in the FUV rise. They show examples of stars which are not embedded in dark clouds in which the visual extinction, FUV rise, and bump very independenly. In detailed IUE studies, Massa et al. (1983, ApJ, 266, 662) and Witt et al. (1984, ApJ, 248, 545) confirmed the result that deviations for stars in the general diffuse ISM are not uncommon. The implications of this fact is that the estimation of the reddening of an object by removal of the λ2175 bump is rather dangerous. A most interesting finding is that the maximum of the bump is at almost exactly the same wavelength in all stars, independent of its strength.

Ultraviolet extinction within the Cep OB3 Association has been studied by Barsella et al. (1982, AA, 111, 130), who concluded that the dust is graphite-rich from the strength of the bump, and by Massa and Savage (1984, ApJ, 279, 310), who found a regular decrease of the FUV rise as the line of sight progresses into successively denser parts of the associated molecular cloud. The "graphite richness" may be a "small-silicate poverty" within the cloud, assuming that indeed the bump is related to graphite (see the following discussion on theories). The dust in front of Herschel 36 in M20 (Hecht et al. 1982, ApJ, 263, L391) is similar to that in the Orion Nebula (Bohlin and Savage 1981, ApJ, 249, 109). The star HD 29647 in the Taurus cloud has a very peculiar extinction, with a weak bump (Seab et al. 1981, ApJ, 246, 788) possibly caused by icy coatings as implied by the observed (but weak) 3.01 μm absorption band (Goebel 1983, ApJ, 268, L41). Peculiar extinction for the Wolf-Rayet stars has been suggested by Garmany et al. (1984, ApJ, 278, 233).

Extinction in the LMC is like that in the Milky Way for the IR and visual regions (Nandy and Morgan 1982, MN, 200, 979; Koornneef 1982, AA, 107, 247), but with an N(H)/E(B-V) four times larger than the Galactic average. This ratio presumably reflects the gas/dust ratio, by mass, in the two objects. The UV extinction in the LMC is well known to differ from Galactic, with a weak bump and a steep FUV rise (Nandy et al. 1981, MN, 192, 955; Koornneef and Code, 1981, ApJ, 247, 860). The extinction has been fitted by a simple analytical expression (Howarth 1983, MN, 203, 301). The bump is stronger than the LMC average towards the core of the giant HII region 30 Doradus (Fitzpatrick and Savage, ApJ, 279, 301). The difference in extinction between two stars of the same spectral type but different reddenings suggests a Galactic extinction law, rather than LMC, for the dust between them.

Ultraviolet extinction deviates even more from Galactic in the SMC than in the LMC. There is almost no bump (Nandy et al. 1982, MN, 201, 1p; Prevot et al. 1984, AA, 132, 389). The dust/gas ratio is even higher than in the LMC, but must be regarded as highly uncertain because of the very low extinctions involved (see also Fitzpatrick 1984, ApJ, 282, 436). At least one star in the SMC, Sk 143, has a "normal" Galactic UV extinction law (Lequeux et al. AA, 113, L15).

Infrared absorption and emission features have been recently reviewed by Willner (1983, 16th ESLAB Symp.). The 3.07 μm "ice" absorption feature has been studied extensively observationally (Knacke et al. 1982, ApJ, 260, 141; Joyce and Simon 1982, ApJ, 260, 604). The feature is fitted well by amorphous water and ammonia ices. Léger et al. (1983, AA, 117, 164) suggest that the long-wavelength wing is caused by large ice particles; Hagen et al. (1983, AA, 117, 132) feel that it is due to complex molecular ices, perhaps by formaldehyde ice as well (Hagen et al. 1980, AA, 86, L3). Extensive laboratory studies of ices of water and other substances (Hagen et al. 1983, AASuppl, 51, 389; Moore et al. 1983, Icarus, 54, 388) show that the C-O stretch absorption at about 5 μm is strong. It is not seen in many sources with strong 3.07 μm bands, which suggests that ammonia is the more likely cause of the longward wing of that band. Whittet et al. (1983, Nature, 303, 218) have detected the 3.07 μm band in several stars in the Taurus cloud which have only modest amounts of extinction (3-6 mag at V), whereas it was formerly believed that about 20 mag were required for the ice band to be detectable.

Roche and Aitken (1984, MN, 208, 481) find that the shape of the 9.7 μm band in six Wolf-Rayet stars is better fitted by the band profile in μ Cephei than by the usually assumed profile, taken from the Orion Nebula. The ratio of visual extinction to optical depth at 9.7 μm is 18.5, larger than usually quoted. A very peculiar 9.7 μm profile has been observed in Mark 231 by Breger and Witteborn (1984, ApJ, 281, L17).

Kitta and Krätchmer (1983, AA, 122, 105) have considered the 12 μm feature of water ice in various mixtures of water and other materials. They conclude that the 12 μm ice band must be present in objects which show the 3.07 μm band. However, Hagen et al. (1983, AASuppl, 51, 389) suggest that the feature may be shifted to longer wavelengths where there is a strong telluric band. Of course, the 12 μm feature coincides with the "silicate" feature and may be hidden if the continuum is improperly drawn.

The 3.4 μm absorption feature of the C-H stretch, observed in IRS 7 near the Galactic center (Allen and Wickramasinghe 1984, Nature, 294, 239) is especially interesting. There are many other complex absorption features in the 2.9-3.3 μm region of this object, but no convincing sign of the 3.07 μm ice band. This fact is important because it shows that C-H bonds exist on grains which are not deeply embedded in dark clouds. Which molecules give rise to the 3.4 μm absorption band is not clear. There is a rather weak absorption at 6.0 μm (Willner et al. 1982, ApJ, 253, 174), well correlated with the 3.07 μm strength, which is presumably the bending mode of water ice. There is also an unidentified band at 6.8 μm which is correlated with the short-wavelength side of the ice band, but not with the long. Strazzulla et al. (1984, AA, 133, 77; 1983, MN, 204, 59p) have studied synthesized polymetric residues produced from proton bombardment of hydrocarbons. The 3.4 μm absorption band is produced independently of the hydrocarbon used in the bombardment.

The 4.67 μm absorption band has been detected in several molecular cloud sources by Lacy et al. (1984, ApJ, 276, 533). It has a structure which can be well explained by solid CO and some molecule containing the C-N triple bond. In one source, gaseous CO bands were observed instead of the solid-CO feature.

The 3.2-3.6 μm emission bands are very complex, with different profiles from source to source (Allen et al. 1982, MN, 199, 1017). The presence of many telluric bands makes the 3-4 μm region a difficult one for the determination of accurate profiles. The unidentified emission features at 3.3,6.2,7.6,8.8 and 11.4 μm are observed in many objects (Aitken and Roche 1984, MN, 208, 751). Emission features near 3.5 μm have been identified with formaldehyde (Bass et al. 1983, ApJ, 265, 290). Most of the emission features have been suggested by Leger and Puget (1984, AA, 137, L5) to be produced by warm coronene, a polycyclic aromatic with a structure of linked benzene rings, and H on the edges. They suggest that a molecule of about 50 carbon atoms arranged in a matrix of linked benzene rings might agree rather well with observations. Dwek et al. (1981, ApJ, 243, 677) show that a very high efficiency of fluorescence is required if the emission bands are to be produced by that process.

Scattering by grains has been investigated in the visual reflection nebulae in the Pleiades (Cottrell and Witt, 1983, AJ, 88, 418) and NGC 1999 (Smith 1983, MN, 205, 349). In the UV, Morgan et al. (1982, MN, 199, 399) and Tanaka et al. (1984, ApJ, 280, 213) have observed the Orion Loop Nebulosity. Witt et al. (1982, ApJ, 261, 192) have studied NGC 7023 extensively. Unfortunately, the determination of albedo and phase parameter from the measurements depends critically upon the assumptions regarding the geometry of the reflecting grains and illuminating star.

The most interesting finding of the studies of reflection nebulae is that there is an excess emission in the 2-5 μm range in NGC 2023, 7023 and 2068 (Sellgren et al. 1983, ApJ, 271, L13; Sellgren 1984, ApJ, 277, 623). This excess emission extends into the red spectral region (Witt et al. 1984, ApJ, 281, 708). At present, it is not clear whether the excess represents the thermal radiation from a single tiny grain following its absorption of a single photon, or alternatively some fluorescence process. If the emission extends to wavelengths as short as 0.65 μm, the required grain temperatures are improbably high. All of these nebulae show a strong 3.4 μm emission band.

Interstellar Polarization has been studied in molecular clouds (Werner et al. 1983, ApJ, 265, L13; Cudlip et al. 1982, MN, 200, 1169; Moneti et al. 1984, ApJ, 282, 508; Joyce and Smith 1982, ApJ, 260, 604), in the Pleiades (Breger 1984, AA, 137, 145), and in field stars (Tinbergen 1982, AA, 121, 158; Wolstencroft and Smith 1984, MN, 208, 461). The last paper shows an interesting ripple structure in the wavelength dependence of the polarization. An improved empirical fit to the wavelength dependence of polarization has been given by Wilking et al. (1982, AJ, 87, 695). The width of the $p(\lambda)$ curve increases with $\lambda(max)$. Clarke and Al-Roubaie (1983, MN, 202, 137) have discussed the effects of noise on the $p(\lambda)$ law. Polarization in the LMC is very similar to that in the Galaxy, both in the form for $p(\lambda)$ and in the values for $\lambda(max)$ (Clayton et al. 1983, ApJ, 265, 194).

The size distribution of the grains affects the x-ray scattering from the heavy elements in the grains. Observations of halos around x-ray sources are consistent with a power-law distribution of particles sizes (Catura 1983, ApJ, 275, 645). The chemical composition is also uncertain. Sakata et al. (1984, Nature, 301, 493) show that a λ2175 feature is produced by the residue from discharges in gaseous hydrocarbons. It has been suggested (Duley et al. 1979, ApSpSc, 65, 69) that the bump is caused by surface defects on MgO grains. MacLean et al. (1982, ApJ,

256, L61) show that MgO produces more blue light by fluorescence when irradiated by UV at λ2200 than at other wavelengths, and interpret this observation to mean that MgO absorbs more at λ2200 than elsewhere. However, Hoyle et al. (1983, ApSpSc, 92, 433) show that the observed extinction is about an order of magnitude too much to be produced by these grains, and Duley (1983, ApSpSc , 92,45) states that CO absorption onto small MgO grains shifts the wavelength of the bump to shorter wavelengths, which is not observed in several stars with CO absorption lines in their IUE spectra. Murray et al. (1984, ApJ, 282, 481) show that almost half of Mg is in the gas phase, which makes the abundance constraints on the MgO-bump connection even more severe. Millar (1982, ApSpSc, 86, 497) has claimed that SiO has enough stability to exist in the interstellar chemically reducing environment, so that the 9.7 μm bump could be caused by SiO, but Nuth and Donn (1983, ApSpSc, 95, 175) point out major additional problems with the SiO interpretation of the 9.7 μm feature. Duley and Najdowsky (1983, ApSpSc, 95, 187) have given an extensive explanation of interstellar extinction using oxide grains.

Grains form in the upper photospheres of O-rich M giants and supergiants because of the rapid drop in temperature caused by the spherical extension of the atmosphere (Schmid-Burgh and Scholz 1981, MN, 194, 805). The formation of graphite, as opposed to amorphous carbon, in C-rich atmospheres has been discussed (Woodrow and Auman 1982, ApJ, 257, 247; Gail and Sadlmayr 1984, AA, 132, 163; Gail et al. 1984, AA, 133, 320). The carbon star R Cor Bor does not have the bump in its UV spectrum (Hecht et al. 1984, ApJ, 280, 228), so presumably makes amorphous C grains. Novae are C-rich, but it is not clear whether the grains which emit the observed IR are formed within the expanding shell or were in the ejecta from the red giant progenitor (Bode and Evans 1981, MN, 197, 1055; 1982, ApJ, 254,263). A similar situation applies to supernovae (Dwek et al. 1983, ApJ, 274, 168 and 175).

The chemical composition of mantles of ices on grains have been predicted theoretically by Thielens and Hagen (1982, AA, 114, 245). The trapping of molecules on grains was discussed by Burke and Hollenbach (1983, ApJ, 265, 223). Leger (1983, AA, 123, 271) finds that CO should condense onto cold grains in only modestly dark clouds, while observations show that it does not. Johnston et al. (1983, AA, 123, 343) show that mantles can be eroded surprisingly efficiently by low-energy cosmic rays. Processed icy mantles, when warmed, produce a refractory "yellow stuff" residue (Greenberg 1982, in Comets, ed. L. Wilkening, U.Ariz.Press, p. 131; Moore et al. 1983, Icarus, 54, 388). The destruction of more refractory mantles on grains in molecular clouds by shocks (Draine et al. 1983, ApJ, 264, 485) is reasonably efficient and is heavily influenced by the presence of a magnetic field, which softens the shock and increases the importance of collisions of gas and dust.

Dust and gas can possibly be gravitationally separated within a molecular cloud (Williams and Bhatt 1982, MNRAS, 199, 465; Bhatt and Desai 1982, ApSpSc, 84, 163). The IRAS satellite also detected dust emission from regions of the sky in which there seems to be little HI (Low et al. 1984, ApJ, 278, L19), and there is a separation of gas and dust as seen in a particular high-latitude position (King et al. 1982, MN, 198, 255).

Possibly grains have a biological origin (Jabir et al. 1982, ApSpSc, 86, 321; Hoyle and Wickramasinghe 1982, ApSpSc, 86, 341). However, Duley (1984, QJRAS, 25, 109) and Whittet (1984, MN, 210, 479) have shown that the biological model of interstellar grains requires too much phosporous by more than an order of magnitude.

Depletion studies of ions in the gas phase are surveyed by Bohlin et al. (1983, ApJSuppl, 51, 277). Depletions increase with increasing gas density, even in the diffuse ISM (Harris et al. 1984, ApJ, 284, 157). There are only modest depletions of O and N, about the same factor for each (York et al. 1983, ApJ, 266, L83), and in C (Hobbs et al. 1982, ApJ, 252; Jenkins et al. 1983, ApJ, 270, 88). Zinc is mostly gaseous and is a good tracer of the gas column density (York and Jura

1982, ApJ, 254, 88), while Si, Mn and Fe are severly depleted (Lugger et al. 1982, ApJ, 259, 67; Phillips et al. 1982, MN, 200, 687), and Ca within a dark cloud has the greatest depletion yet measured (Snow et al. 1983, 265, 267). There is strong evidence that the depletion is less at large distances from the galactic plane (Albert 1982, ApJ, 256, L9); Hobbs 1983, ApJ, 265, 817), and is greater within dark clouds (Tarafdar et al. 1983, ApJ, 267, 156).

It is still an open question whether or not most interstellar chemistry takes place on grain surfaces or in the gas phase. The arguments are summarized in papers in the 16th ESLAB Symposium, Galactic and Extragalactic Infrared Spectroscopy, 1983. It is clear that molecular hydrogen is produced on grain surfaces, apparently very efficiently (Jenkins et al. 1983, ApJ. 270, 88). Grain-surface chemistry has been studied (Millar 1982, MN, 199, 309; 1982, ApSpSc, 87, 435; Jones et al. 1983, ApSpSc, 96, 141). The results look promising. However, so do the predictions from gas-phase chemistry (Graedel et al. 1982, ApJSuppl, 48, 321; Freeman and Millar 1983, Nature, 301, 402; Herbst 1983, ApJSupp, 53, 41;Dickman et al. 1983, ApJSupp, 53, 55), both in equilibrium and following shocks (Draine et al. 1983, ApJ, 264, 485; Mitchell,MN, 1983, 205, 765). Interstellar molecules within dark clouds may be the products of reactions within grain mantles after being warmed (Allamandola et al. 1982, AA, 109, L12). The amount of gaseous C in clouds suggests rapid circulation within these clouds (Boland and de Jong 1982, ApJ, 261, 110).

The far-infrared emission ("FIR"; $\lambda > 80\mu m$) observed from the Galaxy is emitted following absorption of starlight. The FIR has been discussed, with references to observations, by Mezger et al. (1982, AA, 105, 372) and Mathis et al. (1983, AA, 128, 212) in terms of the bare-grain silicate and graphite ("MRN") model. Other observations are in Jaffe et al. (1982, Apj, 252, 601) and Campbell et al. (1984, ApJ, 284, 566). The results are, of course, dependent upon the FIR emissivities of the assumed constituents.

The optical constants of graphite and silicates have recently been discussed by Draine and Lee (1984, ApJ, 285, 89). Tanabe et al. (1983, PASJapan, 35, 397) have given constants for graphite, amorphous C, and SiC; Millar (1981, ApSpSc, 78, 505) for oxide grains; Kitta and Krätschmer (1983, AA, 122, 105) and Léger et al. (1983, AA, 117, 164) for amorphous ice.

V. STAR FORMATION

(B.G. Elmegreen)

1. INTRODUCTION

The rate of progress in understanding star formation and pre-main sequence (PMS) objects has been higher in the last three years than it has at any previous time. Radio and speckle interferometric observations reveal the clumply structure of clouds and the disks around embedded stars. The structure of the Kleinmann-Low nebula in Orion is nearly understood. The gravitational collapse phase of star formation has been observed in a few sources, and embedded stars have been found to possess strong winds, or, in some cases, jets. Numerical computations of cloud collapse can now simulate fragmentation in some detail, although a shift in emphasis toward analytic methods for simple geometries has occurred.

Star formation has also been observed to influence the surrounding medium over distances as large as a kiloparsec, leading to the speculation that star formation can propagate over these scales, and to the concept of feedback control between star formation and interstellar matter. Cloud formation mechanisms are also beginning to be recognized on kiloparsec scales, and several observations of star and cloud complexes of this size have been made. Models of propagating star formation in galaxies are beginning to recognize the predictions of these models.

Reviews of star formation may be found in the proceedings of "Regions of Recent Star Formation" (1982, ed. Roger and Dewdney, Reidel), "The Formation of Planetary Systems" (1982, ed. A. Brahic, Cepadues Editions), "Symposium on the Orion Nebula to Honor Henry Draper" (ed. Glassgold, Huggins and Schucking, 1983, AnnNYAcadSci, Vol. 395), and "Proceedings of a Symposium on Herbig-Haro Objects, T Tauri Stars and Related Phenomena" (1983, ed. Cantó and Mendoza, RevMexAA, Vol.7).

2. OBSERVATION OF STAR FORMATION ON SMALL SCALES

The contraction of gas onto a protostar has been observed in W3(OH) by Rickard et al. 1982, AJ, 87, 1806) and Wouterloot (1984, AA, 135, 32), and in a few other sources by Forster and Boland (1982, AA, 114, 109). OH associated with possible young protostars has been found by Caswell and Haynes (1983, AustJPhys, 36, 361). The contraction of small molecular clouds was observed by Myers, Linke and Benson (1983, ApJ, 264, 517), and of larger clouds by Riley et al. (1982, MN, 199, 197). Star formation in the smallest clouds may proceed by the quasistatic contraction of a dissipating turbulent envelope onto a thermal-pressure-supported core (Myers, 1983, ApJ, 270, 105). Some globules may contract from filaments (Leung et al. 1982, ApJ, 262, 582). These contractions are so slow that many globules appear to be nearly in equilibrium (Myers and Benson 1983, ApJ, 266, 309; Dickman and Clemens 1983, ApJ, 271, 143).

Small-scale clumpy structure in clouds was observed using a variety of techniques. NH_3 shows clumps in the following sources: Orion (Ziurys et al. 1981, AA, 104, 288; Zuckerman et al. 1981, ApJ, 250, L39; Harris et al. 1983, ApJ, 265, L63; Pauls et al. 1983, AA, 124, 23; Batrla et al. 1983, AA, 128, 279), W51 (Ho et al. 1983, ApJ, 266, 596); Taurus (Ungerechts et al. 1982, AA, 111, 339), G35.2-0.74 (Brown et al. 1982, MN, 201, 121) and DR21 (Matsikis et al. 1981, ApJ, 250, L85). Clumps are inferred from H_2CO in W3 (Arnal et al. 1982, MN, 201, 317), from HC_3N in TMC1 (Schloerb et al. 1983, Ap, 267, 163), from HNCO in several clouds (Jackson et al. 1984, ApJ, 280, 608), from CS in W51 and DR21(OH) (White et al. 1983, MN, 204, 1117), in M17, S140 and NGC 2024 (Snell et al. 1984, ApJ, 255, 149) and in other clouds (Snell et al. 1982, ApJ, 255, 149), from several molecules in Taurus (Baudry et al. 1981, AA, 104, 101), and from near-infrared observations in Orion (Smith et al.1983, AJ, 88, 469) and far-infrared observations in Orion (Drapatz et al. 1983 AA, 128, 207), W3 (Jaffe et al. 1983, ApJ, 273, L89), Cha I (Wesselius et al. 1984, ApJ, 278, L37) and S235 (Nordt et al. 1984, AA, 131, 221). Line profiles of clumpy clouds were discussed by Martin et al. 1984, MN, 208, 35).

PMS circumstellar disks or rings have been found in the following sources: IRc2 in Orion (see Orion discussion below), HL Tau (Cohen and Schmidt 1981, AJ, 86,1228; Cohen 1983, ApJ, 270, L69, FU Ori (Smith et al. 1982, ApJ, 258, 170, near the Herbig-Haro objects HH 24-27 (Matthews and Little, 1983, MN, 205, 123), S106 (Gehrz et al. 1982, ApJ, 254, 550); Staude et al. 1982, ApJ, 255, 95; Bally and Scoville 1982, ApJ, 255, 497; Harvey et al. 1982, ApJ, 258, 568; Churchwell and Bieging 1982, ApJ, 258, 515; Stutzki et al. 1982, AA, 111, 201; Solf and Carsenty 1982, AA, 113, 142; Bally et al. 1983, ApJ, 272, 154, N2071 (Bally 1982, ApJ, 261, 558; Schwartz et al. 1983, ApJ, 267, L109), N2261 (Gething et al. 1982, MN, 198, 881), NGC 1333 (Schwartz et al. 1983, ApJ, 267, L109), N6334V (Harvey and Wilking 1984, ApJ, 280, L19), L1455 and L1551 (Davidson and Jaffe 1984, ApJ, 277, L13), Cha I (Cohen and Schwartz 1984, AJ, 89, 277), M8 (Lightfoot et al. 1984, MN, 208, 197), LkHα208 (Shirt et al. 1983, MN, 204, 1257), and in a few other sources (Torrelles et al. 1983, ApJ, 274, 214; Turner and Matthews 1984, ApJ, 277, 164). These disks are usually perpendicular to bipolar flows from centrally located stars. Dense, unresolved material has also been found associated with embedded A and B stars by Harvey et al. 1984, ApJ, 278, 156.

Bipolar outflows are commonly observed in the vicinity of embedded PMS stars. Survey for broad pedestals in the line profiles from molecular clouds were made by

Loren 1981,(ApJ, 249, 550), Frerking and Langer (1982, ApJ, 256, 523), Bally and Lada (1983, ApJ, 265, 824), Plambeck et al. (1983, ApJ, 266, 321), and de Vries et al. (1984, AASuppl, 56, 333). Surveys for such high velocity molecular emission near T Tauri stars were made by Kutner et al. (1982, ApJ, 259, L35), Edwards and Snell (1982, ApJ, 261, 151), Levreault (1983, ApJ, 265, 855), and Calvet et al. (1983, ApJ, 268, 739). Surveys near Herbig-Haro objects were made by Snell and Edwards (1981,ApJ, 251, 103; 1982, ApJ, 259, 668), Rodríguez et al. (1982, ApJ, 260, 635), Edwards and Snell (1983, ApJ, 270, 605; 1984, ApJ, 281, 237); and Haschik et al. (1983, ApJ, 265, 281). Surveys for high velocity H_2O masers were made by García-Barreto et al. (1981, RevMexAA, 5, 87) and Jaffe et al. (1981, ApJ, 250, 621). An optical search for bipolar nebulae was made by Neckel and Staude (1984, AA, 131, 200). Other evidence for PMS winds is the excited H_2 emission from shocked gas. Studies of this emission were made by Persson et al. (1981, ApJ, 251, L85); Beckwith and Zuckerman (1982, ApJ, 255, 536; Bally and Lane (1982, ApJ, 257, 612); Simon and Joyce (1983, ApJ, 265, 864); Brown et al.(1983, MN, 203, 785), and for Orion (see below). Surveys for radio continuum or line emission from embedded stellar winds were made by Simon et al. (1981, ApJ, 251, 552; 1983, ApJ, 266, 623); Thompson (1982, ApJ, 257, 171); Bally and Predmore (1983, ApJ, 265, 778); Thompson et al.(1983, ApJ, 266, 614), and McGregor et al. (1984, PASP, 96, 315). HI emission was found in a region of mass outflow by Bally and Stark (1983, ApJ, 266, L61). Anisotropic outflow from T Tauri stars was discussed by Cohen (1982, PASP, 94, 266).

Other studies of molecular outflow refer to the following sources: the Carina nebula (White and Phillips 1983, MN, 202, 255), G35.2-0.74 (Little et al. 1983, MN, 203, 409), N2071 (Lichten 1982, ApJ, 253, 593; Wootten et al. 1984, ApJ, 279, 633), S206 (Pişmiş and Hasse 1982, RevMexAA, 5, 209), the Cha T association (Schwartz and Henize 1983, AJ, 88, 1665), W49N (Downes et al. 1982, ApJ, 252, L29; Walker et al. 1982, ApJ, 255, 128), W51 (Downes et al. 1982, ApJ, 252, L29; Ho et al. 1983, ApJ, 266, 596). L1529 (Lichten 1982, ApJ, 255, L119), L1551 (Phillips 1982, AA, 116, 130; Nagata et al. 1983, AA, 119, L1; Emerson et al. 1984, ApJ, 278, L49), S106 (Pişmiş and Hasse 1981, RevMexAA, 5, 79; Felli et al. 1984, AA, 135, 261), the Serpens cloud near S68 (Nordh et al. 1982, AA, 115, 308; Ungerechts and Gusten (1984 AA, 131, 177, M8E (Simon et al. 1984, ApJ, 278, 170), Cep A (Ho et al. 1982, ApJ, 262, 619), Sh2-269 (Heydari-Malayeri et al. 1982, AA, 113, 118), AFGL 961 (Lada and Gautier 1982, ApJ, 261, 161), NGC 7538 (Campbell and Thompson 1984, ApJ, 279, 650), GL 437 (Arguilla and Goldsmith 1984, ApJ, 279, 664), and in a distant dark cloud (Bruck and Godwin 1984, MN, 206, 37). Some of these PMS flows may have been eruptive (Mundt and Hartmann 1983, ApJ, 268, 766), and some may be clumpy (Phillips 1982, AA, 116, 130; and see Orion paragraph, below).

PMS jets were found in regions of mass outflow by Mundt et al. 1983, ApJ, 265, L71), Strom et al. 1983, ApJ, 271, L23) and Mundt and Fried (1983, ApJ, 274, L83). Other Herbig-Haro (HH) objects were discovered in M16 (Meaburn and White 1982, MN, 199, 121), on the periphery of the Gum Nebula (Dopita et al. 1982, ApJ, 263, L73), in NGC 7129 (Magakyan 1983, Pis'ma AZh, 9, 155), in Orion (see below), and in a number of regions by Gyul'budagyan (1982, Pis'ma AZh, 8, 232). Studies of exciting stars near HH objects were made by Axon et al.(1982, MN, 200, 239), Dyck et al. 1982, ApJ, 255, L103), Graham (Surveys of Southern Gal 229 1983); Cohen and Schwartz (1983, ApJ, 265, 877); Reipurth and Wamsteker (1983, AA, 119, 14); Cohen et al. (1984, ApJ,278, 671; 1984, ApJ, 281, 250) and Schwartz et al. (1984, ApJ, 280, L23). Searches for exciting stars near H_2O masers were made by Moorwood and Salinari (1981, AA, 102, 197; 1983, AA, 125, 342) and Braz and Epchtain (1982, AA, 111, 91); stars near S106 were found by Lacasse et al. (1981, AA, 104, 57 and Hofmann and Larson (1982, AA, 116, 179). Spectra of HH objects were analyzed by Cohen and Schmidt (1981, AJ, 86, 1228), Brugel et al. (1981, ApJSuppl, 47, 117), Bohm-Vitense et al. (1982, ApJ, 262, 224; Brugel et al. (1982, ApJ, 262, L35), Schwartz (1983, ApJ, 268, L37);Graham and Elias (1983, ApJ, 272, 615); Bohm et al. (1983, AA, 125, 23), Krautter et al. (1984, AA, 132, 169); Hartmann and Raymond (1984, ApJ, 276, 560), and Bohm and Bohm-Vitense (1984, ApJ, 277, 216). Proper motions of HH objects were detected by Herbig and Jones (1981, AJ, 86, 1232); 1982, AJ, 87, 1223; 1983, AJ, 88, 1040).

The Kleinmann-Low Region of the Orion cloud complex is the most thoroughly studied site of massive star formation. It appears to be comprised of a large cavity (Wynn-Williams et al. 1984, ApJ, 281, 172) swept out by a bipolar wind from IRc2 (Knapp et al. 1981, ApJ, 250, 175; Phillips et al. 1982, MN, 199, 1033; Genzel et al. 1982, ApJ, 259, L103; Erickson et al.1982, ApJ, 261, L103; Wright et al. 1983, ApJ, 267, L41; Wright and Plambeck 1983, ApJ, 267, L115). Two other nearby PMS stars, BN and IRc9, also have winds, although weaker than that from IRc2 (Moran et al. 1983 ApJ, 271, L31; Scoville et al. 1983, ApJ, 275, 201; Beck 1984, ApJ, 281, 205). IRc2 is surrounded by a disk or torus of dense neutral matter (Werner et al. 1983, ApJ, 265, L13; Chelli et al. 1984, ApJ, 280, 163) that may be rotating and expanding (Plambeck et al. 1982, ApJ, 259, 617; Barvainis 1984 ApJ, 279, 358) and, possibly, unbound (Norris 1984, MN, 207, 127). The wind from IRc2 is clumpy (Kuiper et al. 1981, ApJ, 251, 88; Wootten et al. 1984, ApJ, 277, 189) and it interacts with the cavity wall by driving a hydromagnetic shock wave into the surrounding cloud (Draine and Roberge 1982, ApJ, 259, L91; Chernoff et al. 1982, ApJ, 259, L97). The shock produces vibrationally and rotationally excited H_2 emission (Knacke and Young 1981, ApJ, 249, L65; Joyce and Simon 1982, MN, 200, 39p; Scoville et al. 1982, ApJ, 253, 136; Nadeau et al. 1982, ApJ, 253, 154; Beck et al. 1982, ApJ, 253, L83; Davis et al. 1982, ApJ, 259, 166). The wind itself may produce HH objects (Harvey et al. 1983, AnnNYAcadSci, 395, 199; Axon and Taylor, 1984, MN, 207, 241) and high velocity H_2O maser emission (Abraham et al. 1981, AA, 100, L10). Low mass embedded stars are in the region (Lonsdale et al. 1982, AJ, 87, 1819; Hyland et al. 1984, MN, 206, 465), and low mass visible stars are in the Orion Nebula (Walker 1983 ApJ, 271, 642; McNamara and Huels 1983, AASuppl, 54, 221; Herbig 1983, AnnNYAcadSci, 395, 64).

Observations (not mentioned elsewhere) of other regions of massive star formation concern W51 (Genzel et al. 1981 ApJ, 247, 1039; and 1982, ApJ, 255, 527; Jaffe et al. 1984, ApJ, 279, L51), W51N (Schneps et al. 1981, ApJ, 249, 124), W58 (Vogel and Welch 1983, ApJ, 269, 568), W33 (Haschick and Ho 1983, ApJ, 267, 638), W3 (Neugebauer et al. 1982, AJ, 87, 395; Dyck and Staude 1982, AA, 109, 320; Dickel et al. 1983, AA, 125, 320; Wright et al. 1984, ApJ, 281, L71), W3(OH) (Norris et al. 1982, MN, 201, 209), W3/4/5 (Thronson and Price 1982, AJ, 87, 1288), IC 1805 (Joshi and Sagar 1983, JRAS Can, 77, 40), M16 (Mufson et al. 1981, ApJ, 248, 992; McBreen et al. 1982, ApJ, 254, 126; Walsh and White 1982, MN, 199, 9p, M17 (Jaffe and Fazio 1982, ApJ, 257, L77), S128 (Mampaso et al. 1984, MN, 207, 465), NGC 6334 (Rodríguez et al. 1982, ApJ, 255, 103; Gezari 1982, ApJ, 259, L29; Harvey and Gatley 1983, ApJ, 269, 613), NGC 2024 (Thompson et al. 1981, ApJ, 249, 622; Jiang et al. 1984, AA, 135, 249; Thronson et al. 1984, ApJ, 280, 154), NGC 2071 (Calamai et al. 1982, AA, 109, 123), IC 5146 (McCutcheon et al. 1982, ApJ, 256, 139; Wilking et al. 1984, AJ, 89, 496), OH 351.78-0.54 (Fix et al. 1982, ApJ, 259, 657), ON-1 (Israel and Wootten 1983, ApJ, 266, 580), Cep OB III (Sargent et al. 1983, AJ, 88, 1236), Cep OB IV (Rossano et al. 1983, AJ, 88, 1835), Vul OB I (Turner 1981, JRAS Can 75, 252), Mon OB I and NGC 2264 (Ogura 1984, PASJapan, 36, 139; Sargent et al. 1984, AA, 135, 377), Sgr B2 and G34.3+0.2 (Benson and Johnston 1984, ApJ, 276, 181), CrA (Cruz-Gonzâlez et al. 1984, ApJ, 279, 679) and Mon R2 (Willson and Folch-Pi 1981, AJ, 86, 1084; McCarthy 1983, ApJ, 257, L93). Additional observations of embedded young clusters were made by Ho and Haschick (1981, ApJ, 248, 622), Sargent et al. (1981, ApJ, 249, 607), Hackwell et al. (1982, ApJ, 252, 250), Ho et al. (1983, ApJ, 265, 295), Krassner et al. (1983,AJ, 88, 972) and Jaffe et al. (1984, ApJ, 281, 225). Infrared spectra of young stars were studied by Willner et al.(1982, ApJ, 253, 174) and Harvey (1984, PASP, 96, 297), and infrared speckel interferometry of protostars was studied by Dyck and Howell (1982, AJ, 87, 400).

Low mass stars have been found in globules with dense NH_3 cores (Benson et al. 1984, ApJ, 279, L27), in B5 (Beichman et al. 1984, ApJ, 278, L45), B18 (Myers 1982, ApJ, 257, 620), B335 (Keene et al. 1983, ApJ, 274, L43), the Cha I cloud (Hyland et al. 1982, MN, 201, 1095; Baud et al. 1984, ApJ, 278, L53), and the Ophiuchus cloud (Wilking and Lada 1983, ApJ, 274, 698; Sargent et al. 1983, AJ, 88, 88). Star formation in Bok globules was discussed by Reipurth (1981, Messenger, 26, 2; 1983, AA, 117, 183). A search for Hα emission line stars in Bok globules was made by Ogura and Hasagawa (1983, PASJapan, 35, 299).

Observations of magnetic fields in regions of stars formation include polarization studies of the ρOph cloud (Hong, 1981, JKoreanASoc, 14, 37), M17 (Chesterman et al. 1982, MN, 200, 965), the Serpens nebula (King et al. 1983, MN, 202, 1087) and Per OB 2 (1981, Markkanen Proc. 3rd Finnish-Soviet Astron.Symp. 17). Zeeman measurements of molecular cloud field strengths were made for the HI line in the Orion molecular cloud (Heiles and Troland 1982,ApJ, 260, L23), and for the OH line throughout the Orion region by Hansen (1982, ApJ, 260, 599), Crutcher and Kazes (1983, AA, 125, L23), Hansen and Johnston (1983, ApJ, 267, 625), and in several other sources by Norris et al. (1982, MN, 201, 191). Magnetic field strengths were inferred from Faraday rotation in HII regions by Heiles et al. (1981, ApJ, 247, L77) Field strength upper limits were reported for several sources by Crutcher et al. (1981, ApJ, 249, 134) and Silverglate (1984, ApJ, 279, 694).

Regions where star formation appears to have been stimulated include: W4 (Braunsfurth 1983, AA, 117, 297), W5 (Wilking et al. 1984, ApJ, 279, 291), W28 (Odenwald et al. 1984, ApJ, 279, 162), W51 (Lightfoot et al. 1983, MN, 205, 653), M8 (Lightfoot et al. 1984, MN, 208, 197), G134.2+0.8 (Hughes and Viner 1982, AJ,87, 685), the Carina Nebula (de Graauw et al. 1981, AA, 102, 257), NGC 7538 (Campbell and Thompson 1984, ApJ, 279, 650), S128 (Heske and Wendker 1982, AA, 113, 170), Stock 16 (Turner 1983, JRAS Can, 77, 261), a cometary globule (Brand et al. 1983, MN, 203, 215), OH maser clusters near cloud edges (Ho et al. 1983, ApJ, 265, 295), OH 342.01+0.25 (Sandell et al. 1983, AA, 124, 139) and Orion (Isobe 1983, AnnNYAcad Sci, 395, 165).

Star formation persists in a region longer than the collapse time of the molecular cloud. Extended formation times for star clusters have been suggested by Jakobsen (1981, PASP, 93, 547), Guetter (1982, AJ, 86, 1057), Herbst and Miller (1982, AJ, 87, 1478), Duncan and Jones (1983, ApJ, 271, 663), Adams et al. 1983, ApJSuppl, 53, 893), and Stauffer et al. (1984, ApJ, 280, 202). There may be a problem with the determination of ages from Li abundances, however (Giampapa 1984, ApJ, 277, 235), and from theoretical pre-main sequence tracks on an HR diagram (Mercer-Smith et al. 1984, ApJ, 279, 363), although Stahler (1983, ApJ, 274, 822) identifies a "birthline" on this diagram.

3. THEORY OF STAR FORMATION ON SMALL SCALES

For non-magnetic clouds, I-D numerical collapse calculations were performed by Wang and Qian (1981, Ann Shanghai ObsAcadSci, 3, 166), and 1-D semi-analytical calculations were made by Stahler et al. (1981, ApJ,248, 727) and Ballone and Fofi (1982, ApSpSci, 85, 293). Collapse models for Bok globules were studied by Villere and Black (1982, ApJ, 252, 524). Two-dimensional cloud collapse was followed by Narita et al. (1983, MN, 203, 491), Virgopia and Ferraioli (1981, ApSpSci, 78,211, and 1981, ApSpSci, 79, 129) Boss and Haber (1982, ApJ, 255, 240), Boss and Black (1982, ApJ, 258, 270); Cassen and Moosman (1982, Icarus, 48, 353), and Boss (1984, ApJ, 277, 768). Fragmentation during 3-D collapse was studied numerically by Boss (1981, ApJ, 250, 636; 1983, Icarus, 55, 181), Gingold and Monaghan (1981, MN, 197, 461; 1983, MN, 204, 715), Bodenheimer and Boss (1981, MN, 197, 477), Wood (1982, MN, 199, 331), Rozyczka (1983, AA, 125, 45), Miyama et al. 1984, ApJ, 279, 621), and analytically by Tohline (1981, ApJ, 248, 717), Bernstein and Book (1981, ApJ, 251, 271), Silk (1982, ApJ, 256, 514), and Boss (1982, ApJ, 259, 159).

The equilibrium of clouds was studied by Hayashi et al. (1982, Prog.TheorPhys, 68, 1949), Izotov and Kolesnik (1982, AstrometrAstrofizVyp, 46, 3), Kolesnik and

Kravchuk (1982, Pis'ma AZh, 8, 234), Stahler (1983, ApJ, 268, 155 and 165) and Schmitz (1984, AA, 131, 309). The effect of the stellar velocity dispersion on cloud collapse was discussed by Kegel and Volk (1983, AA, 119, 101). The stability of cylinders was studied by Karnik (1982, ApSpSc, 83, 209) and Bastien (1983, AA, 119, 109); the stability of rings by Qian and Wang (1981, Ann ShangaiObsAcadSci, 3, 161), and the stability of layers by Welter (1982, AA, 105, 237), Tomisaka and Ikeuchi (1983, PASJapan, 35, 187), and Vishniac (1983, ApJ, 274, 152). Stability of the Taurus clouds was discussed by Nurmanova (1982, AstronZh, 59, 61). Gravitational torques in protostellar disks were discussed by Larson (1984, MN, 206, 197), magnetic fields in disks were discussed by Ray (1983, IrishAJ, 16, 39), and accretion on binary stars in disks was studied by Artymowicz (1983, ActaAstr, 33, 223).

The self-gravitational collapse of magnetic clouds was followed numerically by Phillips (1982, ProcAstrSocAustr, 4, 371) and Dorfi (1982, AA, 114, 151). An analytical study of field topology was made by Dudorov and Sazonov (1982, NauchnIfnVyp, 50, 98). Stability criteria for 3-D media with rotation,magnetic fields, finite conductivity and various other properties were studied by Sharma and Thakur (1982, ApSpSc, 81, 95), Sharma (1982, ApSpSc, 85, 263; 1983, ApSpSc, 94, 419), and Chajlani and Purohit (1983, ApSpSc, 94, 425). The collapse of magnetic cylinders was investigated by Karnik and Talwar (1981, ApSpSc, 79, 379), Karnik (1982, ApSpSc, 82, 283) and Nuzhnova (TrAstrofizInstAlma-Ata, 39, 114). Time dependent 1-D and 2-D collapse calculations with magnetic diffusion were performed by Nakano (1983, PAS Japan,34, 337; 1983, PASJapan, 35, 87; 1983, PASJapan, 35, 209), Black and Scott (1982, ApJ, 263, 696), Scott (1983, ApJ, 275, 836; 1984, ApJ, 278, 396), Shu (1983, ApJ, 273, 202), and Paleologou and Mouschovias (1983, ApJ, 275, 838). Magnetic rotational braking was studied by Dudorov and Sazonov (1983, NauchnInfVyp, 52, 29).

The triggering of star formation in a pressurized cloud or cloud clump was studied theoretically by several groups. The ionization and compression of inhomogeneities in a cloud was discussed by Klein et al. (1983, ApJ, 271, L69), and the compression of isolated globules was calculated by Nittmann (1981, MN, 197, 699), Sandford et al. (1982, ApJ, 260, 183), LaRosa (1983, ApJ, 274, 815), and Krebs and Hillebrandt (1983, AA, 128, 411). Rayleigh-Taylor instabilities in supernova-compressed regions were discussed by Baierlein et al. (1981, Icarus, 48, 49) and Baierlein (1983, MN, 205, 669). Triggering by cloud-cloud collisions was discussed by McCrea (1982, Prog. in Cosmol. 239), Struck-Marcell (1982, ApJ, 259, 127), Suchkov et al. (1982, Astrofizica, 18, 629), Yoshii and Sabano (1983, Theor Aspects on Struct Activity and Evol of Gal, 76), Dyson and Hartquist (1983, MN, 203, 1233), and Gilden (1984, ApJ, 279, 335). Such triggering mechanisms may produce self-regulated star formation inside a cloud core (1983, ApJ, 264, 508). Self-regulation by X-ray ionization was considered by Silk and Norman (1983, ApJ, 272, L49). The role of the magnetic field in propagating star formation was discussed by Welter and Nepveu (1982, AA, 113, 277).

Models for the formation of bound galactic clusters were considered by Elmegreen (1983, MN, 203, 1011) and of globular clusters by Suchkov et al. (1981, Pis' ma AZh, 7, 617), and Smith (1982, ApJ, 259, 607). Theoretical models for the origin of the initial stellar mass function were made by Hong (1981, J Korean ASoc, 14, 89), Zinnecker (1982, MittAstronGes, 55, 160), Hunter and Fleck (1982, ApJ, 256, 505) Ferrini et al. (1983, MN, 202, 1071), and Elmegreen and Mathiew (1983, MN, 203, 305). Differences in the stellar mass fucntion for large and small clouds were noted by Larson (1982, MN, 200, 159). A discussion of star formation efficiencies for galactic clusters is in Mathieu (1983, ApJ, 267, L97) and for OB associations, in Duerr et al. (1982, ApJ, 261, 135).

Models for winds around PMS stars were discussed by Barral and Cantó (1981, RevMexAA, 5, 101), Hartmann and MacGregor (1982, ApJ, 259, 180), Konigl (1982, ApJ, 261, 115), Beckwith et al. (1983, ApJ, 267, 596), Draine (1983, ApJ, 270, 519),

Pudritz and Norman (1983, ApJ, 274, 677), Dyson (1983, AA, 124, 77), Torbett (1984, ApJ, 278, 318) and Kundt (1984, ApSpSc, 98, 275). A model for the H_2O maser environment was studied by Strelnitskij (1984, MN, 207, 339), and a numerical hydrodynamical model of HH objects was made by Sandford and Whitaker (1983, MN, 205,105).

4. STAR FORMATION ON LARGE SCALES

Evidence for coherent cloud structure or star formation on scales larger than several hundred parsecs comes from the observation of giant star complexes (Efremov 1982, Pis'ma AZh, 8, 585 and 663; Smith 1982, ApJ, 261, 463), spiral arm and interarm superassociations (Dolidze 1981, Pis,ma AZh, 7, 666; Condon 1983, ApJSuppl, 53 , 459), stellar streams and moving groups (Latyshev 1980, AstronTsirk, 1141, 7; Eggen 1983, MN, 204, 377,391 and 405), the simultaneous formation of multiple galactic clusters (Barkhatova and Pavlovskaya 1981, AstronTsirk, 1155,4; Lynga and Wramdemark 1984, AA, 132, 58), and the giant HI clouds and star formation patches in the Large Magellanic Clouds (Page and Carruthers 1981, ApJ, 248, 906), M101 (Blitz et al. 1981, ApJ, 249, 76; Viallefond et al. 1981, AA, 104, 127; 1982, AA, 115, 373) and M31 (Nakai and Sofue 1982, PASJapan, 34, 199). The molecular cloud mass distribution is dominated by the largest masses (Casoli et al. 1984, AA, 132, 99).

Theoretical studies of cloud and star formation on a kiloparsec scale were concerned with self-gravitational instabilities in the interstellar gas (Elmegreen 1982, ApJ, 253, 634 and 655; Viallefond et al. 1982, AA, 115, 373; Elmegreen and Elmegreen, 1983, MN, 203, 31; Jog and Solomon 1984,ApJ, 276, 114 and 127), and the growth of large clouds from colliding small clouds (Casoli and Combes 1982, AA, 110, 287; Kwan and Valdés 1983, ApJ, 271, 604; Roberts and Hausman 1984, ApJ, 277, 744). Such cloud formation operates faster in a density wave spiral arm, so these mechanisms may explain how density waves trigger star formation. Large-scale triggering appears necessary because density wave arms may contain the largest cloud complexes and HII regions in a galaxy (Rumstay and Kaufman 1983, ApJ, 274, 611).

Supernovae and other pressures from young stars may disturb the environment of a star-forming region over a distance of several hundred parsecs or more. In some cases such disturbances have led to the formation of new stars (Dixon et al. 1981, ApSpSc, 78, 189; Pişmiş and Hasse 1982, RevMexAA, 5, 161; Gosachinskij and Khersonskij 1982, AZh, 59, 237; Elmegreen 1982, in Submm-Wave Astr, 1; de Boer and Nash 1982, ApJ, 255, 447; Braunsfurth and Feitzinger, 1983, AA, 127, 113). Two-way interactions between star formation and interstellar matter may produce self-regulated star formation, as discussed by Shore (1981, ApJ, 249, 93; 1983, ApJ, 265, 202), Tomisaka et al. (1981, ApSpSc, 78, 273), Loose and Fricke (1982, MittAstron Ges, 55, 100), Loose et al. (1982, AA, 105, 342), Cox (1983, ApJ, 265, L61), Franco and Cox (1983, ApJ, 273, 243), Comins (1983, ApJ, 274, 595), and Ikeuchi et al. (1984, MN, 207, 909). Cloud-fluid/star formation models were simulated numerically by Prendergast (1982 in Extragalactic Molecules), and Struck-Marcell and Scalo (1984, ApJ, 277, 132).

Models of stochastic self-propagating star formation (SSPSF) in galactic disks included a two-component interstellar medium to study grand design spirals (Seiden et al. 1982, ApJ, 253, 91) and gas distributions (Seiden 1983, ApJ, 266, 555). SSPSF simulations also modeled spiral arm spurs (Feitzinger and Schwerdtfeger 1982, AA, 116, 117), thick disks (Statler et al. 1983, ApJ, 270, 79) and long-time galaxy evolution (Freedman and Madore 1983, ApJ, 265, 140). Spiral arm generation by stochastic inhibition of star formation was discussed by Freedman and Madore (1984, ApJ, 280, 592), and spiral arm shapes were derived analytically by Cowie and Rybicki (1982, ApJ, 260, 504) and Balbus (1984, ApJ, 277, 550). Observations related to SSPSF predictions were made by Kaufman (1981, ApJ, 250, 534), Janes and Adler (1982, ApJSuppl, 49, 425), Isserstedt (1984, AA, 131, 347) and Schlosser and Musculus (1984, AA, 131, 367).

Enhanced star formation in interacting galaxies has been suggested for Arp 299 (Bushouse and Gallagher1984, PASP, 96, 273), NGC 5253 (Moorwood and Glass 1982, AA, 115, 84), N5430 (Keel 1982, PASP, 94, 765), Arp 91 (Jenkins 1984, ApJ, 277, 501), and several galactic nuclei by Kennicutt and Keel (1984, ApJ, 279, L5). Star formation in cluster galaxies that accrete gas has been studied by Fabian et al. (1982, MN, 201, 933) and Sarazin and O'Connell (1983, ApJ, 268, 552). Variations of the initial stellar mass function within a galaxy or from galaxy to galaxy were discussed by van den Bergh (1982, The Most Massive Stars, 253), Lequeux (ibid.,261), Puget et al. (1982, Regions of Recent Star Form. 249), Berkhuijsen (1982, AA, 112, 369), Garmany et al. (1982, ApJ, 263, 777), and Sil'chenko (1984, SovietAstrL, 10, No. 7).

VI. HII Regions
(P.A. Shaver)

1. INTRODUCTION

HII regions are now being studied at wavelengths spanning almost the entire electromagnetic spectrum. Progress is particularly notable in high-resolution radio mapping, in millimeter wave spectroscopy, and infrared spectroscopy and imaging. Some of these areas are reviewed in preceding sections (e.g. molecular clouds,star formation), and the following discussion is largely restricted to representative papers on the properties of well-evolved, conventional H II regions.

Recent reviews have been written on the Orion Nebula (Peimbert 1982, AnnNYAcad Sci, 395m 24; Goudis, The Orion Complex: A Case Study of Interstellar Matter), on far-IR line observations (Harwit 1982, AnnNYAcadSci, 395, 56), on ultraviolet observations (Peimbert 1981, The Universe at UV Wavelengths, p. 557; Patriarchi and Perinotto 1983, MemSAItal, 54, 529), on abundances (Pagel and Edmunds 1981, AnnNY AcadSci, 19, 77; 1984, Stellar Nucleosynthesis, Reidel 341), and on extragalactic HII regions (1983, Highlights of Astronomy, 6, pp. 609-635). Two recent books contain several relevant papers, and will be referred to frequently: Regions of Recent Star Formation, 1982, eds. Roger & Dewdney, Reidel (hereafter RRSF), and Primordial Helium, 1983, eds. Shaver, Kunth & Kjär, ESO (hereafter PH). Two surveys of galactic HII regions were made recently by Wink et al. (1982, AA, 108, 227) in the radio continuum and recombination line, and Blitz et al. (1982, ApJSuppl, 49, 183) in the CO line. Fich and Blitz (1984, ApJ, 279, 125) have studied the distribution and properties of HII regions in the outer galaxy.

2. STRUCTURE

A great deal of effort has gone into mapping the structure of HII regions over a wide range in wavelength. At the lowest radio frequencies they are observed in absorption against the galactic background (Abramenkov and Krymkin 1982, SovietAstr, 26, 160; 1983, SovietAstr, 27, 32). At higher radio frequencies the emphasis has been on high-resolution mapping, particularly with interferometers like the VLA (e.g. 1981, AA, 100, 28 and 42; 1981, AA, 101, 39; 1981, ApJ, 248, 622; 1981, ApJ, 250, 227; 1981, AA, 103, 50; 1982, ApJ, 255, 103 and 527; 1982, ApJ, 256, 127; 1982, RRSF, 39 and 181; 1982, AA, 108, 412; 1982, ApSpSc, 84, 143; 1982, AASuppl, 48,345; 1982, AJ, 87, 685; 1982, AA, 115, 164; 1983, AJ, 88, 972; 1983, AA, 120, 322; 1983, AA, 124, 116; 1983, AA, 125, 320; 1983, AA, 126, 433; 1983, AJ, 88, 1470; 1983, ApJ, 267, 638; 1983, ApJ, 272, 154; 1983, SovietAstr, 27, 628; 1984, ApJ, 278, 170). At the highest resolution and sensitivity, compact HII regions are often resolved into shell-like structures (Turner and Matthews 1984, ApJ, 277, 164), and ultracompact HII regions, presumably the youngest, are found embedded in more diffuse nebulae (e.g. 1982, AnnNYAcadSci, 395, 204; 1983, ApJ, 271, L31; 1984, ApJ, 277, 181), and in dense molecular clouds (e.g. 1983, ApJ, 265, 778; 1983, ApJ, 266, 623; 1984, ApJ, 276, 204; 1984, ApJ, 281, 225). Main-line OH masers are invariably found to be associated with such ultracompact HII regions (Garay et al. 1984, ApJ, in press). Radio recombination line maps have also been made, using both single dish and interferometer radio-telescopes (1982, RRSF, 39; 1982, AASuppl, 48, 345; 1982, AA, 115, 164; 1983, AJ, 88, 1470; 1984, AJ, 89, 95; 1984, ApLett, 24, 1).

Far-infrared broadband maps (1982, ApJ, 254, 126; 1982, ApJ, 258, 568; 1983, MN, 205, 653; 1984, ApJ, 279, 679; 1984, MN, 207, 659; 1984, AA, 135, L14; 1984,MN, 208, 197) have been used to study the distribution and temperature of dust in and around the HII regions and the overall energy balance, and near-infrared maps (1982, ApJ, 252, 250; 1982, ApJ, 254, 550; 1982, AJ, 87, 1288; 1983, AA, 120, 1; 1984,MN, 206, 465; 1984, MN, 207, 465) have been used to study the hot dust component and the exciting stars. Several HII regions have been mapped in infrared emission lines (1981, ApJ, 250, L35; 1982, RRSF, 73; 1982, ApJ, 255, 510; 1983, ApJ, 264, 538;1983, MN, 202, 859; 1983, ApJ, 265, L7; 1983, ApJ, 268, 721; 1983, ApJ, 275, 130) to determine density and ionization distributions, and in some cases abundances.

Narrow-band optical images have been obtained for several HII regions (1981, ApSpSc, 78, 235; 1981, AA, 101, 397; 1981, AA, 102, 316; 1982, AA, 105, 329; 1982, PASP, 94, 453; 1982, ApSpSc, 85, 405; 1982, ApSpSc, 88, 477; 1983, AA, 124, 1; 1983, ApSpSc, 89, 407; 1983, AASuppl, 53, 403; 1983, AA, 125, 320; 1983, MN, 203, 977; 1983, Astron Tsirk, 1252, 3). Münch and Hippelein (1982, AnnNYAcadSci, 395, 170) found that the [CI]9850A distribution is flatter than that of Hα in the Orion Nebula, and that the [CI] emission is particularly strong in the dust lane of NGC 2024. Cesarsky (1982, AA, 113, L7) suggests that this line originates in regions where radio Cnα emission is found, and that the excitation is recombination, not collisions. Walsh (1982, MN, 201, 561) found a complex structure south of θ^2 Orionis in [SII] maps of N_e and velocity.

3. PHYSICAL CONDITIONS

New techniques are being used to study the properties of HII regions.Distances to some HII regions have been obtained using H_2O maser proper motions (Genzel et al. 1981, ApJ, 244, 884; 1981, ApJ, 247, 1039; Schneps et al. 1981, ApJ, 249, 124). Magnetic field strengths in HII regions are being explored using Faraday rotation of polarized background sources (Heiles et al. 1981, ApJ, 247, L77) and Zeeman splitting of recombination lines (Silvergate 1984, ApJ, 279, 694). Far-UV (1981, ApLett, 22, 135), X-ray (1982, Science 215, 61) and γ-ray (1981, PhilTransRSocLondon, 301, 569) observations have been made of the Orion Nebula.

The measured electron densities in HII regions now span almost eight decades. At one extreme, very low-density HII regions ($N_e \sim 0.1 - 10$ cm^{-3}) have been found around high-latitude O stars (Reynolds and Ogden 1982, AJ, 87, 306) and star clusters (Cersósimo 1982, ApLett, 22, 157). At the other extreme, densities as high as 10^7 cm^{-3} have been found in ultracompact HII regions using high-resolution interferometer observations (e.g. Moran et al. 1983, ApJ, 271, L31); some of these objects have EM $\sim 10^9$ pc cm^{-6}, M $\sim 10^{-6}$ $M_\odot$, R $\sim 10^{-4}$ pc, and turnover frequencies above 20-30 GHz. Smirnov et al. (1984, AA, 135, 116) studied Stark broadening in RRLs from Orion, and found it to be consistent with current models of that HII region. Infrared lines are increasingly being using to determine electron densities (e.g. 1981, ApJ, 250, 605; 1982, ApJ, 259, L109; 1983, MN, 202, 859; 1983, ApJ,268, 721), and optical lines continue to be used for this purpose (e.g. 1982, Astron Tsirk 1228, 3 and 6; 1983, Astron Tsirk 1301, 1 and 3). A comparison of radio (rms) and optical densities yields the filling factor (e.g. 1983, MN, 204, 53).

Electron temperatures have been measured in more HII regions using optical lines (1983, MN, 204, 53; 1984, ApSpSc, 100, 451) and radio recombination lines (e.g. 1981, MN, 196, 889; 1983, MN, 204, 53; 1983, MN, 205, 719; 1983, SovietAstrL, 9, 341; 1983, ApJ, 266, 263; 1983, AA, 127, 211). Radio and optical temperatures agree well, indicating that temperature fluctuations are small. The relative effects of T_{eff}, Z and Ne on electron temperature have been studied observationally (1983, MN, 204, 53) and theoretically (Rubin 1984, ApJ, in press). Maciel and Pottasch (1982, AA, 106, 1) studied the contribution to heating by electrons ejected from dust grains after absorption of UV photons.

Dust within HII regions has been studied in a number of ways: infrared maps and spectroscopy (above), extinction (1981, AASuppl, 45, 451; 1984, AA, 133, 313), polarization (1982, PASJapan, 33, 313; 1982, ApJ, 255, 95; 1982, MN, 200, 965;1983, MN, 202, 11; 1983, Astrophys 19, No. 1), and UV scattering (1981, ApJ, 249, 99; 1981 ApLett, 22, 5; 1982, ApJ, 255, 87). Williams and Zealey (1983, MN, 203, 433) showed that the 3.28 μm feature is associated with dust. Nakano et al. (1983, ApSp Sc, 89, 407) concluded from Hα and V-band maps that there is a dust depletion zone in the centre of some HII regions. Mathis (1983, ApJ, 267, 119) showed that a wavelength-dependent dust albedo affects the extinction determined from the Hβ/Hα ratio, and suggested methods for dereddening emission-line objects.

The excitation and energy balance have been examined by identifying and studying the exciting stars and their relationship with the HII regions (e.g. 1981, Ap Lett, 22, 5; 1982, AnnNYAcadSci, 395, 64; 1982, Science, 215, 61; 1982, ApJ, 263, 130; 1983, ApJ, 270, 169; 1983, AA, 124, 273). Chanot and Sivan (1983, AA, 121,19) have measured the total Hα flux of the Gum Nebula, and find it to be in ionization balance with the known exciting stars; they conclude that it is a normal HII region, except for an unusual [SII]/Hα ratio indicative of shocks. Franco and Savage (1982, ApJ, 255, 541) studied CIV and SiIV absorption lines in IUE spectra of stars in the Orion Nebula, and considered the possible cause of the high excitation: X-rays, collisions, or UV photons. Herter et al. (1983, AASuppl, 51, 195) outlined a method of determining excitation conditions in HII regions independent of chemical abundance, using IR fine structure lines.

Increasingly sophisticated model HII regions have been produced. Extensive grids of model HII regions, covering a wide range in stellar effective temperatures, stellar and nebular abundances, and nebular densities have been published by Stasinska (1982 AASuppl, 48, 299), Köppen et al. (1983, AA, 118, 203), and Rubin (1984, ApJ, in press). Various authors have studied the effects of stellar abundances in NLTE atmospheres (Borsenberger and Stasinska 1982, AA, 106, 158), HII regions ionized by associations (Copetti and Bica 1983, ApSpSc, 91, 381), HII regions with arbitrary density distributions (Oliva and Panagia 1983, ApSpSc, 94, 437), far-IR lines from dusty HII regions (Aannestad et al.(1984 AA, in press), the effects of opacity due to heavy elements in the HII region (Rubin 1983, ApJ, 274, 671), and radiative transfer processes (Elitzer 1984, ApJ, 280, 653). McCall (1984, MN, 208, 253) has given revised emission coefficients for several prominent lines as a function of temperature and density. Detailed models of some individual HII regions have also been produced (e.g. 1982, RRSF, 53,; 1982, ApJ, 255, 497; 1983, ApJ, 265, 239; 1983, AA, 124, 77; 1983, ApJ, 274, 650; 1984, ApJ, 279, 650).

The internal kinematics of several HII regions have been studied using radio and optical emission lines (e.g. 1981, MN, 196, 995; 1981, ApJ, 248, 992; 1981, ApJ, 250, 615; 1981, ApSpSc, 78, 189; 1982, MN, 200, 771; 1982, AA, 113, 118; 1982, RevMexAA,5, 161 and 209; 1982, RevMexAA, 6, 269; 1982, ApSpSc, 86, 331; 1982, ApSp Sc, 87, 13; 1982, AA, 115, 164; 1982, ApJ, 252, 156; 1983, ApJ, 269, 164; 1983, RevMexAA, 8, 51 and 83; 1983, ApJ, 273, 639; 1983, ApJ, 274, 650; 1984, ApLett, 24, 1). Optical and UV absorption-line spectroscopy of stars in HII regions provide an independent view of these kinematics from a variety of ions (1982, ApJ, 252, 156; 1982, ApJ, 260, 163; 1983, ApSpSc, 90, 437; 1983, MN, 204, 1203; 1984, ApJ, 276, 524; 1984, MN, 207, 167); Walborn (1982, ApJSuppl, 48, 145) has found a systematic expansion of 15-20 km/s and smaller scale features with velocities of hundreds of km/s in the Carina Nebula, and Franco and Savage (1982, ApJ, 255, 541) found velocities of -25 km/s in the Orion Nebula. Fich et al. (1982, RRSF, 201) conclude from a study of 151 HII regions that typically the bulk of ionized gas flows away from the parent molecular cloud at 5-10 km/s.

4. EVOLUTION

Radio and infrared observations can now discern the earliest stages in the evolution of an HII region as it evolves from the protostellar cocoon (e.g. 1982, ApJ, 257, 171; 1982, RRSF, 83; 1983, AA, 125, 342; 1983, ApJ, 265, 778; 1983, ApJ, 266, 614 and 623; 1983, ApJ, 271, L31; 1984, ApJ, 276, 204; 1984, ApJ, 277, 164 and 181; 1984, ApJ, 281, 225).

Theoretical calculations of the subsequent evolution of model HII regions allow for an increasing complexity of conditions, as computer codes evolve. Tenorio-Tagle and co-workers have modelled the evolution of HII regions in collapsing molecular clouds, around evolving OB associations, and into a hot phase III (e.g. 1982, RRSF, 1 and 15; 1982, AA, 108, 25; 1982, AA, 112, 1 and 104; 1982, AA, 115, 207). Ionization front interactions and the formation of bright rims and globules have been studied by Brand (1981, MN, 197, 217), Sandford et al. (1982, RRSF,129), and Bedijn and Tenorio-Tagle (1984, AA, 135, 81). Icke (1981, ApJSuppl, 45, 585) and Yorke et al. (1983, AA, 127, 313) have produced theoretical radio continuum and far-IR maps of a variety of model HII regions. Smith (1982, ApJ, 259, 607) has modelled HII regions formed in globular and open cluster protoclouds, and McKee et al. (1984, ApJ, 278, L115) have studied the evolution of photoionized stellar wind bubbles. The observations indicate that no single model is universally applicable (e.g. 1982, RRSF, 25; 1984, ApJ, 277, 164), presumably a reflection of the wide variety of initial conditions and evolutionary stages that actually exist.

5. ABUNDANCES AND GALACTIC GRADIENTS

Further work has been done to derive chemical abundances in HII regions from radio, infrared, and optical spectroscopic data, and their distribution across the galactic plane. Some papers critically examined methods of abundance determination for helium(Lockman 1982, ApJ, 256, 543; Stasińska 1983, PH, 255) and sulfur (Dennefeld and Stasińska 1983, AA, 118, 234), and others suggested new approaches to abundance determination (Mathis 1982, ApJ, 261, 195; Simpson and Rubin 1984, ApJ, 281, 184). Rood et al. (1984, ApJ, 280, 629) have detected the cosmologically important 8.7 GHz hyperfine line of $^3He^+$ in a few HII regions. Silvergate (1984,ApJ, 278, 604) found gas phase Si, Mg, and Fe to be depleted near HII regions, using radio recombination line observations.

Shaver et al. (1983, MN, 204, 53) combined radio and optical data on HII regions distributed over the galactic disk to determine the gradients in electron temperature and abundances (O,N,S,Ne,Ar). The temperature gradient, which is thought to be caused by the abundance gradient, has also been measured in other recent radio studies (1981, MN, 196, 889; 1983, ApJ, 266, 263; 1983, AA, 127, 211). Radio observations have also been used to search for a helium abundance gradient, but it is not yet clear if such a gradient exists (1981, MN, 196, 889; 1982, ApJ, 259, 595; 1983, MN, 204, 53; 1983, PH, 281 and 299).

Several papers have been written on abundances and gradients from IR spectroscopy (e.g. 1981, ApJ, 250, 186; 1982, ApJ, 262, 153 and 164; 1983, AA, 127, 383); Lester et al. (1983 ApJ, 271,618) used the ratio of the [NIII] 57 μm and [OIII] 52 μm lines to determine a model-independent estimate of the N/O ratio in several HII regions.

6. GALACTIC CENTRE

A new VLA map reveals Sgr A to have a spiral structure, with the non-thermal point source close to the peak in the thermal emission (Ekers et al. 1983, AA, 122, 143). The cause of this morphology is unclear, but tidal distortion of infalling material is a strong possibility. Interferometric radio recombination line observations with the WSRT (Bregman and Schwarz 1982, AA, 112, L6) and the VLA (van Gorkom et al. in preparation) should assist in the interpretation. A 76α survey of the galactic centre region has been made (1982, Proc ASAust, 4, 453), and several infrared lines have been detected in Sgr A (1981, ApJ, 248, L109 and 524; 1983, ApJ, 267, L37; 1984, ApJ, 276, 551). Lacy et al. (1982, ApJ, 262, 120) modelled the central parsec in terms of clouds generated by stellar collisions.

7. EXTRAGALACTIC HII REGIONS

HII regions in other galaxies have been reviewed and catalogued by Hodge (1982 AJ, 87, 1341) and Hodge and Kennicutt (1983, AJ, 88, 296). New catalogues have been given of HII regions in NGC 1313 (1983, AASuppl, 51, 353) and M83 (1983, ApJ, 274, 611). Radio, infrared, and optical maps have been made of several extragalactic HII regions (e.g. 1982, AA, 105, 229; 1982, AA, 111, L11; 1983, AA, 118, 116; 1983, AA, 119, 185; 1984, AA, 133, 93); Georgelin et al. (1983, AASuppl, 54, 459) and Lawrie and Kwitter (1982, ApJ, 255, L29) find filamentary and ringlike nebulae in the LMC and M101, and show that their [SII]/Hα ratio is intermediate between normal HII regions and SNRs. Van den Bergh (1981, AJ, 86, 1464) derived an analytical expression for the frequency distribution of HII region diameters, and this has been confirmed for irregular galaxies by Hodge (1983, AJ, 88, 1323).

The excitation and internal kinematics of extragalactic HII regions have received much attention. Several observations were made of R136a, the "supermassive star" in the 30 Doradus Nebula (e.g. 1981, Science, 212, 1497; 1983, ApJ, 273, 597); Worley (1984, ApJ, 278, L109) and others conclude that it is not a single object. Massey and Hutchings (1983, ApJ, 275, 578) find spectroscopically similar objects in M33 HII regions. Internal motions of extragalactic HII regions have been studied in more detail (1981, ApSpSc, 78, 443; 1983, AA, 117, 265); they appear to be chaotic, and no single model accounts for them all (1984, ApJ, 280, 580). Gallagher and Hunter (1983, ApJ, 274, 141) and Rosa and Solf (1984, AA, 130, 29) disagree with the suggestion of Terlevich and Melnick (1981, MN, 195, 839) that the internal motions are largely gravitational. Several papers have been written on W-R stars in extragalactic HII regions (1981, ApJ, 248, 1015; 1981, ApJ, 249, 471; 1982, AA, 105, 410; 1982, AA, 108, 339; 1983, AASuppl, 53, 97).

Abundances have been measured in several more extragalactic HII regions (1981, AJ, 86, 989; 1982, ApJ, 252, 461 and 594; 1982, ApJ, 255, 1; 1983, J KoreanAS 16, 1). Blair et al. (1982, ApJ, 254, 50) find the M31 abundance gradient to be similar to that of other spirals. Rosa (1983, PH, 317) has pointed out possible systematic observational effects in such abundance determinations. Kunth (1983, PH, 305) reviewed the determination of Y_p from extragalactic HII regions.

Other studies of extragalactic HII regions include the following: 1981, AA, 102, 245; 1981, ApSpSc, 80, 267; 1981, ApJ, 249, 76; 1981, ApJ, 250, 103; 1983,ApJ, 266, 568; 1983, AA, 127, 395; 1984, ApJ, 279, L1 and 578; 1984, MN, 207, 25 and 801. Lequeux et al. (1981, AA, 103, 305) considered the properties of several extragalactic HII regions, and concluded that they mainly result from massive bursts of star formation of very short duration.

VII. Supernova Remnants
(S. D'Odorico)

The status of supernova remnant research as in late 1982 is comprehensively reviewed in the proceedings of the IAU Symposium No. 101, "Supernova Remnants and their X-ray Emission" (1983, J. Danziger and P. Gorenstein, eds., R. Reidel Publ. Co., herewith referred to as IAU 101). In this report, the papers published in the Proceedings are quoted unless published elsewhere in a more extended version. A review of the subject is also given by Raymond (1984, AnnRevAA, 22, 75).

1. X-RAY OBSERVATIONS

As it is indicated by the title of Symposium 101, the more interesting body of new information on SNR came in the X-ray band, mostly through the observation of the Einstein satellite and its predecessor, HEAO-1. There are now 135 galactic radio SNR, out of which 40 are seen optically and 33 have been detected in X-rays (van den Bergh 1983, IAU 101, 597). X-ray maps with inferences on the physical conditions and/or the nature of the progenitor have been published for the remnants G296.1-0.7 (Markert et al. 1981, ApJ, 248, L17), Puppis A (Petre et al. 1981, ApJ, 258, 22), G292.0+1.8, the second galactic remnant with oxygen rich filaments (Tuohy et al. 1982, ApJ, 260, L65; Braun et al. 1983, IAU 101, 159), W28 (Andrews et al. 1983, ApJ, 266, 684), G29.7-0.3 (Becker et al. 1983, ApJ, 268, 93), G78.2+2.1 (Higgs et al. 1983, IAU 101, 281), the Cygnus Loop (Ku et al. 1983, ApJ, 278, 615), IC 443, W44 and W49B (Watson et al. 1983, IAU 101, 273), IC 443 and Puppis A (Petre et al. 1983, IAU 101, 289), HB3 (Venkatesan et al. 1984, MN, 208, 251), and W49B (Pye et al. 1984, MN, 207, 649). In some cases (W49B, W28 and G29.7+0.3) the X-ray data were coupled with radio observations of comparable resolution. Detection of X-ray emission from 3C 400 and MSH 15-56 has been reported by Agrawar et al. (1983, ApSp Sc, 89, 279). Ballet et al. (1984, AA, 133, 357) interpret the X-rays and the coronal lines observations in the Cygnus loop.

Young remnants, being powerful X-ray sources, have been the subject of detailed investigation. The X-ray data do provide information on the distribution and total mass of the hot gas and some indication on the composition of the ejecta. Fabian et al. (1981, Nature, 295, 508) have discussed the remnant of SN 1006AD. Tycho has been studied by Reid et al. (1982, ApJ, 261, 485), Seward et al. (1983, ApJ, 266, 287) and Gorenstein et al. (1983, IAU 101, 1). Evidence on a halo around Cas A is reported by Stewart et al. (1983, IAU 101, 59). Pisarski et al. (1983, ApJ, 277, 710) infer from the X-ray map of RCW 86, the remnant of SN 185AD, that it was most likely a type I event. White and Long (1983, ApJ, 264, 196) present the HRI map of Kepler. Remnants with center filled morphologies (plerions) are established as a separate category in the X-ray domain as well. Becker (1983, IAU 101, 321) presents a review of X-ray observations while Weiler summarizes our present understanding of these objects (1983, IAU 101, 299). MSH 15-52 is discussed by Seward et al. (1984, ApJ, 281, 650), 3C 58 and CTB 80 by Becker et al. (1981, ApJ, 255, 557) and G21.5-09 by Becker and Szymkowiak (1981, ApJ, 248, L23).

X-ray maps of remnants have been carefully scanned for the detection of compact sources, possible remnants of the SN progenitor or ejecta of the explosion. Seward et al. (1983, ApJ, 267, 698) discuss two compact sources in MSH 15-17 (RCW 89), Gregory and Fahlman (1983, IAU 101, 429) and Gregory et al. (1983, IAU 101, 437) the X-ray pulsar in the remnant G109.1-1.0.

Spectroscopic studies of the hot gas at X-ray frequencies give information on the physical conditions and the high ionization stages of iron and other elements. Manzo et al. (1983, AA, 122, 124) report on the spectrum of Cas A between 2 and 10 KeV, Kahn et al. (1983, ApJ, 269, 212) on the HEAO-1 soft X-ray observations of Vela. A review of the Einstein results with the low resolution Solid State Spectrometer on young remnants is given by Holt (1983, IAU 101, 17), while Canizares et al. (1983, IAU 101, 205) discuss the high resolution observations of bright remnants with the Focal Plane Crystal Spectrometer. The velocity structure of Cas A from the X-ray emission lines is discussed by Markert et al. (1983, ApJ, 268, 134). Winkler et al. (1983, IAU 101, 245) derive the properties of a bright X-ray knot in Puppis A from FPCS data. Galas et al. (1981, ApJ, 250, 216) discuss the low energy X-ray spectrum of IC 443. While these spectroscopic data do provide information on the relative abundances, it appears still rather difficult to derive accurate values both for lack of spectral resolution and of accurate models.

2. THEORY

Theory of the SNR emission and evolution has however made considerable progress in this period with focus on the production of X-rays. Non-ionization equilibrium modelling of X-ray emission has been done by Shull (1982, ApJ, 262, 308 and 1983, IAU 101, 99) for young remnants and by Hamilton et al. (1983, ApJSuppl, 51, 115) for adiabatic SNR. Non-equilibrium effects have been discussed also by Groneschild and Mewe (1981, AASuppl, 84, 305) and Teske (1983, ApJ, 277, 832). Jones et al. (1981, ApJ, 249, 185) discuss X-ray production in the pre-blast wave phase. The effect of the reverse shock has been studied by Hamilton and Sarazin (1984, ApJ, 281, 682), in particular for the case of pure heavy elements SN ejecta (1984, Princeton preprint). Reynolds and Chanan (1984, ApJ, 281, 673) studied the processes to produce X-ray emission from Crab like SNR. Still centered on the theoretical interpretation of X-ray emission are the works by McKee (1983, IAU 101, 87), Pravdo and Nugent (1983, IAU 101, 29), Fabian et al. (1983, IAU 101, 119), Yorke et al. (1983, IAU 101, 393), Long et al. (1982, ApJ, 260, 202) for the long term effects of ejecta, Falle and Garlick (1982, MN, 201, 635) for the case of the Cygnus Loop, and by Berman and Kahn (1983, MN, 205, 303) for the X-ray emission shortly after the SN event.

Strictly related to the previous papers are those which deal in general terms with the interaction of the SN explosion with the interstellar or circumstellar medium. Several scenarios and problems have been considered: The evolution of a remnant in an inhomogeneous, three-phase medium (Cowie et al. 1981, ApJ, 247,908; McKee 1982, in "Supernovae, a Survey of Current Research", 433), the interaction of SN of type II and I with the surrounding medium (Chevalier 1981, ApJ, 251, 254; 1982, ApJ, 259, L85 and IAU 101, 71; Itoh and Fabian 1984, MN, 208, 925), the spherization of the remnant of a non-symmetrical explosion (Bisnovaty-Kogan and Blinnikov 1982, AstZh, 59, 876), the evaporation of clouds in a hot gas (Balbus and McKee 1982, ApJ, 252, 529), the production of acoustic waves (Spitzer 1982, ApJ, 262, 315), the heating of the IM by SNR (Cox,IAU 101, 385), the effect of the explosion in a plane-stratified medium (Falle et al. 1984, MN, 208, 925). The interaction of two remnants (Bodenheimer et al. 1984, AA, in press), the interaction with a molecular cloud (Tenorio-Tagle et al. 1984, Max Planck, MPA preprint 117) and the evolution in a six-phase IM (Ikeuchi et al. 1984, MN, 207, 909).

The mechanisms for acceleration of relativistic electrons have been discussed by Fedorenko (1981, AstZh, 58, 790; 1983, ApSpSc, 96, 25), Cavallo (1982, AA, 111, 368), Blandford and Cowie (1982, ApJ, 260, 625), Hartquist and Morfill (1983, ApJ, 266, 271), Boydan and Völk (1983, AA, 122, 129). The properties of the cooling filaments in old SNR W50, the SNR which contains SS 433, have been discussed by Panagia and Weiler (1981, Vistas in Astr, 25, 87) and van den Bergh (1981, Vistas in Astr, 25, 104). A possible mechanism to produce radio emission from very young remnants is discussed by Bandiera et al. (1983, AA, 126, 7). A number of papers are related to the Crab: The probable mass of the progenitor (Nomoto 1982, IAU 101, 139), the origin of the spur (Kundt 1983, AA, 121, L15), the evolution as driven by the pulsar (Reynolds and Chevalier 1984, ApJ, 278, 630; Chevalier 1984, ApJ, 280, 797). The transition between radio SN and plerions is discussed by Bandiera et al. (1984, ESO preprint 325). Modelling of oxygen rich ejecta was computed by Itoh (1981, PASJapan, 33, 521) and by Dopita et al. (1984, ApJ, 282, 142). A word of caution to the theoreticians comes from a laboratory simulation of a SN explosion (Borowksy et al. 1984, ApJ, 280, 802): no sweep-up of the ambient plasma was detected.

3. STUDIES IN OPTICAL, ULTRAVIOLET AND INFRARED FREQUENCIES

There has been no breakthrough or new major discoveries in this spectral range, but a large body of new information has been collected. Radial velocity and/or spectrophotometric studies of the emission lines have been carried out for W63

(Rosado and González,1981, RevMexAA, 5, 93), G65.2+5.7 (Rosado 1981, ApJ, 250,222), the Cygnus Loop, Cas A and Kepler in the near infrared (Dennefeld and Andrillat 1981, AA, 103, 44 and Dennefeld 1982, AA, 112, 215), the Cygnus Loop (Doroshenko et al. 1982, AstrZh, 59, 699; Fesen et al. 1982, ApJ, 262, 171), and Hα filament in the Cygnus Loop (Treffers ApJ, 250, 213), 3C 400.2 (Rosado 1983, RevMexAA, 8,59), RCW 103 and Milne 23 (Ruiz 1983, AJ, 88, 1210), RCW 86, 103 and Kepler (Leibowitz and Danziger 1983, MN, 204, 273), 3C 58 (Fesen 1983, ApJ, 270, L53), IC 443 (Fesen 1983, ApJ, 281, 658), and G292.0+1.8 (Dopita and Tuohy 1984, ApJ, 282, 135). Pskovsky (1981, AstrZh, 58, 1022) has rediscussed the data on Cas A. Danziger (1983, IAU 101, 193) has presented a summary of results and open problems on the interpretation of spectroscopic data. Studies based on imagery in the light of emission lines have been carried out for HB3 (Fesen and Gull 1983, PASP, 95, 196), G126.2+1.6, G206.9+2.3, CTB 13 (Rosado 1982, RevMexAA, 5, 127), G65.3+5.7, G126.2+1.6 (Sitnik et al. 1983, AstrZh, 50, 503), W28 (Zealey et al. 1983, IAU 101, 267) and the Cygnus Loop (Hester et al. 1983, ApJ, 273, 219).

The Crab is still the subject of many investigations. Murdin and Clark (1981 Nature, 294, 543) discuss the optical halo, McLean et al. (1983 Nature,304, 243) the optical polarization, Fesen and Gull (1983, IAU 101, 145) the jet, Dennefeld and Pequignot (1983, AA, 127, 42), Henry et al. (1984, ApJ, 278, 619) and Henry (1984, ApJ, 281, 644) the optical near infrared spectrum.

Proper motions have been studied in the remnant of SN 1006 (Hesser and van den Bergh 1981, ApJ, 251, 549) and in Cas A and RCW 89 (van den Bergh and Kamper 1983, ApJ, 268, 129; 1984, ApJ, 200, L51). The velocity dispersion in single filaments of the Cygnus Loop, IC 443, Vela and Puppis A has been studied by Shull (1981 ApJ, 253, 682; 1983, ApJ, 269, 218).

Studies of the interstellar absorption have been carried out with IUE in stars in the direction of the Monoceros Loop (Vidal-Majar et al. 1981, 3rd Eur. IUE Conf., 412), of S 147 (Phillips and Gondhalekar 1983, MN, 202, 483), SN 1006 (Wu et al. 1983, ApJ, 269, L5) and Vela (Jenkins et al.1984, ApJ, 278, 649). A review of UV observations of SNR is given by Raymond (1983, AdvSpRes, 2, 145). New data are presented for Cygnus and Vela (1981, "The Universe at UV Wavelengths", ed. Chapman, NASA, 595), the Crab Nebula (Davidson et al. 1981, ApJ, 253, 696) and again the Cygnus Loop (Raymond et al. 1983, ApJ, 275, 636).

There is a lack of good infrared data on SNR, despite theoretical predictions on possible emission (Dwek 1981, ApJ, 247, 614, and Dwek and Werner 1981, ApJ, 248, 138). Dinerstein et al. (1982, ApJ, 255, 552) report an unsuccessful search in Cas A, D'Odorico and Moorwood (1982, ESO Messenger 28,24) a possible detection in N49 in the LMC and Marsden et al. (1984, ApJ, 278, L29) a detection of an infrared excess due to dust in the Crab with IRAS. There are also unpublished reports of detection in other remnants with the same satellite.

4. STUDIES AT RADIO WAVELENGHTS

Since most of the remnants are detected at radio frequencies only, the number of new investigations in this domain is still quite large.

Studies of a general character have been carried out on the following topics: on the radio morphology of SNR (1981, Shaver AA, 105, 306), on SNR classification (Sakhibov and Smirnov 1981, Pis'ma AZh, 8, 281), on the status of radio studies (Dickel 1983, IAU 101, 213), on the association with pulsars (Radhakrishnan and Srinavasan 1983, IAU 101, 487), on the statistical determination of the SN birth rate (Srinivasan and Dwarakanath 1983, JAA, 3, 351), on SNR evolution from an updated catalogue (Green 1984, MN, 209, 449), on their collective properties (Cowsik and Sarkar 1984, MN, 207, 745) and on the Σ/D relationship (Allakhverdiyev et al. 1983, ApSpSc, 97, 287). Studies of, or related to, individual remnants are generally based on improved angular resolution or on the extension of the known spectrum to other frequencies. They are listed below in order of increasing galactic longitude: Map of Sgr A East (Goss et al. 1983, IAU 101, 65), HI near W28 (Venger et al. 1982, AstrZh, 59, 20), G9.8+0.6, a newly identified SNR (Caswell 1983, MN, 204, 833), the plerionic SNR G24.7+0.6 and G27.8+0.6 (Reich et al. 1984, AA, 133, L4), 1.4 GHz maps of G15.9+0.2, G27.4+0.0, G41.1-0.3 (Caswell et al. 1982,MN, 200, 1143), 2.7 and 4.7 GHz observations of G33.2+0.6 (Reich 1981, AA, 106, 314), HI near W44 (Venger et al. 1981, AstrZh, 58, 1187), 1.7 and 2.7 GHz fluxes and polarization of W50 (Downes et al. AA, 103, 277), maps at two frequencies of G57.1+1.7, a newly identified SNR (Gómez-González and del Romero 1983, AA, 123, L5), the possible plerionic remnant G65.7+12 (Landecker and Caswell 1983, AJ, 88, 1810), 10 GHz observations of G68.8+2.6 (Sofue et al. 1983, PASJapan, 35, 437), map of the Cygnus Loop at 34.5 MHz (Sastry et al. 1981, AA, 103, 393) and at high resolution (Green and Gull 1983, IAU 101, 329), maps of G93.3+6.9 at 4.75 GHz (Lalitha et al. 1984, AA, 131, 196), multi-frequency observations of G109.2-1.0 (Hughes et al. 1983, IAU 101, 455; Downes 1983, MN, 203, 695; Sofue et al. 1983, PASJapan, 35, 447), Cas A at 20 and 25 MHz (Bovkoon et al. 1981, ApSpSc, 81, 221), HI and continuum from CTB 1 (Landecker et al. 1982, AJ, 87, 1374), multi-frequency study and polarization of G119.5+9.8 (Sieber et al. 1981, AA, 103, 393), 11 and 6-cm observations of G126.2+1.6 (Fürst et al. 1984, AA, 133, 11), morphology and spectrum of G127.1+0.5 (Pauls et al. 1982, AA, 112, 120; Geldzahler and Shaffer 1982, ApJ,260, L69; Fürst et al. 1984, AA, 133, 11), on the distance of 3C 58 (Green and Gull 1982, Nature, 299, 606), observations of HB9 at 25 and 34.5 MHz (Abranin and Bazelyan, 1982, ApSpSc, 88, 199; Dwaratakanath et al. 1982, JAA, 3, 207), on the structure of VRO 42.05.01 (Landecker et al. 1982, ApJ, 261, L41), 69.8-cm observations of S 147 (Angehofer and Kundu 1981, AJ, 86, 1003), 21-cm map of Puppis A at improved resolution (Milne et al. 1983, MN, 204, 237), on the single nature of G296.05-0.50 (Caswell and Barnes 1983, ApJ, 271, L55), G320.4-1.2 at 843 MHz (Manchester and Durdin 1983, IAU 101, 421), radio maps at 1' resol. of G327.6+14.6 (Caswell et al. 1983, MN, 204, 915), of G330.2+1.0 (Caswell et al. 1983, MN, 204, 915) of G340.4+0.4, G340.6+0.3, G341.9-0.3, G342.0-0.2, G352.7-0.1 (Caswell et al. 1983, MN, 203, 595), 21-cm continuum and line maps of G347.+22, a possible new SNR (Colomb et al. 1984, AA, 130, 294) and high resolution maps of G349.7+0.2 and G357.7-0.1 (Shaver et al. 1984, Nature, in press).

Changes of the radio structure with time have been studied in young SNR and a review is given by Strom (1983, IAU 101, 37). Cas A has been studied by Tuffs (1983, IAU 101, 49) and Tycho by Strom et al. (1982, MN, 200, 473). The decline of radio fluxes from Cas A, Crab and SN 1572 is discussed by Ivanov et al. (1981, Pis'ma AZh, 8, 83).

5. EXTRAGALACTIC SNR

I consider the progress made in the study of SNR in the Magellanic Clouds through the Einstein results and the new radio and optical observations as the most significant achievement in this period. There are now 32 SNR in the LMC and 11 in the SMC and the sample is judged to be complete to 40 pc diameter. Papers of general character on the MC SNR concern spectroscopy and imaging (Lasker 1981, PASP, 93, 422), radio obs. of SNR candidates in the SMC (Mills 1982, MN, 200, 1007), optical studies (Mathewson et al. 1983, ApJSuppl, 51, 344), X-ray properties from the LMC SNR (Long 1983, IAU 101, 525) and the SMC SNR (Inoue et al. 1983, IAU 101, 535), radio maps and general properties (Mills et al. 1984, AustJPhys, 37, 321), a complete catalogue with a general discussion and its updating (Mathewson et al. 1984, ApJSuppl, 55, 189 and 1984, ApJSuppl, in press), the interpretation of the Σ/D relationship (Fusco-Femiano and Preste Martinez 1984, IAU Symp. 108, 315), an optical, spectroscopy study of 9 remnants with discussion of the abundances (Danziger and Leibowitz 1984, MN, submitted).

Detailed investigations have been carried out on 1E 0102.2-7219 in the SMC

(Dopita et al. 1981, ApJ, 248, L105; Tuohy and Dopita 1983, ApJ, 268, L11), on the X-ray spectra of 6 SNR in the LMC (Clark et al. 1981, ApJ, 255, 440), on the Balmer lines SNR (Tuohy et al. 1982, ApJ, 261, 473), on multicolor photometry of remnants (Greve et al. 1982, AA, 111, 171), on N175B in 30 Dor B (Gilmozzi et al. 1983, MN, 202, 927), on the optical spectra of N132D and 0540-69.3 in the LMC (Dopita and Tuohy 1984, ApJ, 282, 135), on N49 as the possible site of γ-ray and optical bursts (Pizzichini et al. 1983, IAU 101, 573; Pedersen et al. 1984, Nature, 312, 46), and on high resolution spectra of N63A and N49 (Shull 1983, ApJ, 275, 592 and 611).

There are only a few new results on other extragalactic SNR. D'Odorico and Dopita (1983, IAU 101, 517) and Dopita et al. (1984, ApJ, 276, 653) present the optical spectra of SNR in IC 1613 and NGC 6822 and discuss the dependence of line intensities of old SNR on the abundances in the ISM. New data on the powerful SNR in NGC 4449 are given by de Bruyn (1983, AA, 119, 301), Blair et al (1983, ApJ, 271, 84 and 1984, ApJ, 279, 708), Bignell and Seaquist (1983, ApJ, 270, 140). A possible SNR in NGC 185 is identified by Gallagher et al. (1984, ApJ, 281, 263) but not confirmed at radio frequencies (Dickel et al. 1984, AJ, submitted).

SNR in M31 have been further studied optically (Blair et al. 1981, ApJ, 247, 879; Dennefeld and Kunth 1981, AJ, 86, 989), at 21 and 6-cm (Dickel et al. 1981, ApJ, 252; Dickel and D'Odorico 1984, MN, 206, 351). Berkhuijsen (1983, AA, 120, 147) discuss the Σ/D correlation for SNR in M31 and M33.

The detection at radio frequencies of powerful SNR near the nucleus of M82 is reported by Kronberg and Biermann (1983, IAU 101, 583).

VIII. Planetary Nebulae
(Y. Terzian)

1. GENERAL STUDIES

The IAU held Symposium No. 103 on Planetary Nebulae in 1982 in London, U.K. The proceedings were edited by D. Flower and appeared in a volume Planetary Nebulae (1983, D. Reidel Publ. Co.). The proceedings report on: observations of planetary nebulae; physical processes; chemical abundances;origin; and central stars.

A book, Planetary Nebulae, by S.R. Pottasch was published in 1984 (D. Reidel Publ. Co.). This book surveys the history; morphology and evolution, distribution, spectra and abundances, line and continuum emission, distances, mass loss, central stars, and the influence of planetary nebulae on the interstellar medium. Kaler (1984, preprint) has completed a general review on "Planetary Nebulae and their Central Stars", where he presents up-to-date information on the nebulae, their central stars, and their place in stellar evolution.

Several comprehensive studies of planetary nebulae were undertaken. Maciel (1984, AASuppl, 55, 253) compiled a catalogue of distances of 486 galactic planetary nebulae. Milne and Aller (1982, AASuppl, 50, 209) presented radio data for 397 southern nebulae at 14.7 GHz. Schneider et al. (1983, ApJ Suppl, 52, 399) published a comprehensive catalogue of planetary nebulae radial velocities, and Schneider and Terzian (1983, ApJ, 274, L61) used this catalogue to derive the galactic rotation curve of our galaxy beyond the solar system. Gathier (1984, Ph.D. Thesis, Groningen) studied the distances and physical properties of planetary nebulae. Ultraviolet studies of planetary nebulae and their central stars were reported by Adams (1983, Ph.D. Thesis, London).

A number of unusual planetary nebulae have been investigated. F.G. Sagittae was found to have unusual brightness variations (Arkhipova et al.1983, Astron.Circ. USSR 1250, 1). Arkhipova (1983, Variable Stars USSR, 22, 25), and Kostjakova and

Arkhipova (1983, Astron. Circ. USSR 1270, 3) have also reported photometric variations in V1016 and IC 4997. Dufour (1984 ApJ, in press) determined that if NGC 2818 is associated with a Population I star cluster, then it is of exceptional size, mass, expansion velocity and age. Feibelman (1984, ApJ, in press) studied the Butterfly nebula, M2-9; and the bipolar nebula NGC 2346 was studied by Feibelman and Aller (1983, ApJ, 270, 150). Recillas-Cruz and Pişmiş (1984, MN, 210, 57) have studied the kinematics and morphology of NGC 650-651. Luminosities and masses of three planetary nebulae nuclei in the Magellanic Clouds were determined from IUE observations by Stecher et al. (1982, ApJ, 262, L41). These authors suggest that the progenitors of these three nebulae had a mas of $\sim$4 $M_\odot$.

Adams et al. (1984, MN, 207, 471) have presented a detailed study of the planetary nebula K648, which is a member of the globular cluster M15.

2. YOUNG PLANETARY NEBULAE

A number of extensive studies of young planetary nebulae have been undertaken. Issacman (1984, MN, 208, 399) observed 62 compact nebulae at λ20 and λ6-cm and determined that compact nebulae are both younger and more distant than those which are more resolved. Turner and Terzian (1984, AJ, 89, 501) also observed compact nebulae at λ12 cm including CRL 618, HMSge, and He2-459; and Kwok (1984, preprint) observed small nebulae at λ6 cm. Sabbadin et al. (1984, AA, 136, 200) studied the internal motions of fourteen compact nebulae. Rodríguez and García-Barreto (1984, RevMexAA, 9, 153) concluded that the HI absorption lines in five of seven nebulae are due to line of sight clouds unrelated to the planetary nebulae, the exceptions are NGC 6302 and NGC 2440.

Goodrich and Dahari (1984, ApJ, in press) studied M4-18 and reported the coolest central star temperature of $22{\pm}2{\times}10^3$K. They suggest that M4-18 is a young nebula about 4000 years old. Feibelman (1983, ApJ, 275, 628) has reported IUE observations of the compact nebula M1-2, and Sabbadin (1984, MN, 209, 889) obtained high dispersion spectra of fourteen compact nebulae.

3. MASS LOSS AND NEBULAR EXPANSION

Harrington and Feibelman (1983, ApJ, 277, 716) reported on the remarkable nebula A30 and its central star. They find that the central star is losing mass via a 4000 km s^{-1} stellar wind, which must produce a shock upon encountering the nebulosity.

O'Dell and Ball (1984, ApJ, in press) have measured the velocities of material in NGC 2392 and found very high velocities of about 93 km s^{-1} for the inner shell. High velocity material is streaming outwards at places with a velocity of ±190 km s^{-1}. This material is probably accelerated by a strong stellar wind. Kaler et al. (1984, preprint) have discussed a new method for determining planetary nebulae distances by using the P Cygni profiles which are produced by strong stellar winds.

4. CENTRAL STARS

Reay et al. (1984, AA, 137, 113) determined Zanstra temperatures for nine central stars using a method which involves imaging the nebulae. They find stellar temperatures between 10^5 and $2{\times}10^5$ K. Méndez et al. (1984 AA, in press) report on the spectra of central stars of planetary nebulae and derive physical parameters. Kaler and Feibelman (1984 ApJ, in press) discuss the central star of the nebula Abell 78, and Feibelman and Kaler (1983 ApJ, 269, 592) report on the binary central star of the nebula LT5(339+88°1). Kondratjeva (1984 Trudy AstrophysInstUSSR, 44,30) obtained the continuous spectral distribution of eleven central stars of planetary nebulae and derived Zanstra temperatures and luminosities.

5. GENERAL OBSERVATIONS AND COMPOSITION

Carrasco, Serrano and Costero (1983, RevMexAA, 8, 187) have reported the Hβ absolute fluxes of fifty-five planetary nebulae. Pritchet and Grillmair (1984 PASP, 96, 349) presented high-sensitivity spectrophotometry of several planetary nebulae. Whitelock (1984, preprint) has made infrared JHK photometric measurements of eighty-two planetary nebulae. Pottasch et al. (1984 AA, 278, L33) report infrared results from IRAS and discuss the N III and S IV spectral lines. Shure et al. (1984 ApJ, 281, L29) present the [Ne III] 36.02 μm line in NGC 6543.

Feibelman (1983 AA, 122, 335) has studied the profiles and intensity ratios of the C IV λ1548, 1550 emission lines in eleven planetary nebulae.

The chemical abundances of twenty-eight planetary nebulae have been computed by Kondratjeva (1983 Trudy AstrophInstUSSR, 42). Barker and Cudworth (1984 ApJ, 278, 610) derive the chemical abundances of a new halo planetary nebula 61+41°1.

Aller and Czyzak (1983 ApJSuppl, 51, 211) used theoretical models to interpret observations of forty-one planetary nebulae and to derive chemical abundances.

6. DUST

Pottasch et al. (1984 AA, 138, 10) have reported IRAS measurements of 46 bright planetary nebulae. Dust temperatures were computed and the dust in the younger nebulae was found to be hotter. Pnillips et al. (1984 AA, 133, 395) made near infrared measurements of planetary nebulae and concluded that small grain of radius $\sim 10^{-3}$μm, with low thermal capacity, may explain their observations.

Gee et al. (1984 MN, 208, 517) have performed submillimeter observations of the cold dust associated with NGC 7027 and derive a dust temperature of ∿ 20 K. Arens (1984 ApJ, 279, 685) explain the λ10μm emission from NGC 7027 by assuming two distinct grain populations. They also suggest that the warm dust component is well mixed with the nebular ionized gas.

Bentley et al. (1983, ApJ, 278, 665) found that the infrared dust emission in BD+30°3639 spacially co-exists with the emission of the ionized gas.

7. MODELS OF PLANETARY NEBULAE

Yorke et al. (1983, SovietAstrL, 9, 156) have computed evolutionary sequences of spherically symmetric numerical models simulating planetary nebulae, covering the period from envelope detachment to dispersal. Phillips (1984 AA, 137, 92) has discussed the dynamics, evolution and formation rate of planetary nebulae and finds that the formation rate is similar to that for white dwarfs, $\sim 2\times10^{-3}$kpc^{-3}yr^{-1}.

Sabbadin et al. (1984 AA, 136, 181) try to interpret the internal motions of planetary nebulae with two possible models, a fast stellar wind, and a sudden ejection by the central star. Also, Sabbadin et al. (1984 AA, 136, 193) interpret their observations of NGC 1535 and NGC 2022 by the model of the sudden mass ejection from a central star.

M. PEIMBERT
President of the Commission

35. STELLAR CONSTITUTION (CONSTITUTION DES ÉTOILES)

PRESIDENT: A.N. Cox (USA)
VICE-PRESIDENT: D. Sugimoto (Japan)
ORGANIZING COMMITTEE: P.H. Bodenheimer (USA), C.S. Chiosi (Italy), D.J. Faulkner (Australia), A. Maeder (Switzerland), J.-L. Tassoul (Canada), R.J. Tayler (UK), A.V. Tutukov (USSR), V. Weidemann (F.R. Germany), and J.-P. Zahn (France).

I. INTRODUCTION

This report of Commission 35, as in past reports, consists of some details of only a few selected topics. This is necessary because a survey of the entire field of stellar formation, structure, stability, evolution, pulsation, and explosions for the three year period from mid-1981 to mid-1984 would be excessively long. Our topics here, in order from the most massive stellar classes to the least are: Massive Stars (R.M. Humphreys), Rotation in Late Type Stars (W. Benz), Helioseismology (J. Christensen-Dalsgaard), Planetary Nebula Central Stars (E.M. Sion), Pulsations in Hot Degenerate Dwarf Stars (A.N. Cox and S.D. Kawaler), and White Dwarfs (V. Weidemann). There is some overlap in the reviewing of these last three reports because the topics are very closely related. Concentration in this dying stage of stellar evolution seems appropriate because of the great current interest in these matters.

In the course of our considerations of these topics, the Organizing Committee had two other topics that we could not get reviewed in time. Nevertheless, we hope for appropriate reviews in the next report on star formation and the new infrared observations of them as well as the comparisons of surface abundances that are observed and those predicted from stellar evolution calculations.

References are given as their number in the Astronomy and Astrophysics Abstracts when possible. In many cases, the conventional references needed to be given because the 1984 volumes were not available in December 1984.

Interest in stars continues at a high level. During the reporting period five Symposia and five Colloquia were held on topics of interest to Commission 35. These are: Symposium 102 Solar and Stellar Magnetic Fields, Zurich, Switzerland, August 2-6, 1982; Symposium 103 Planetary Nebulae, London, England, August 9-13, 1982; Symposium 105 Observational Tests of the Stellar Evolution Theory, Geneva, Switzerland, September 12-16, 1983; Symposium 108 Structure and Evolution of the Magellanic Clouds, Tubingen, GFR, September 5-8, 1983; Symposium 111 Calibration of Fundamental Stellar Quantities, Como, Italy, May 24-29, 1984; Colloquium 68 Astrophysical Parameters for Globular Clusters, Schenectady, New York, USA, October 7-10, 1981; Colloquium 71 Activity in Red Dwarf Stars, Catania, Italy, August 10-13, 1982; Colloquium 72 Cataclysmic Variables and Related Objects, Haifa, Israel, August 9-13, 1982; Colloquium 76 Nearby Stars and the Stellar Luminosity Function, Middletown, Connecticut, USA, June 13-16, 1983; and Colloquium 82 Cepheids: Observation and Theory, Toronto, Canada, May 29-June 1, 1984.

From the following brief reports and the above listed international meetings, one can see that there are still many frontiers for Commission 35. More detailed and accurate observations of a variety of observables and more elaborate calculations of how stars are born, evolve and die make this field still very exciting.

[I. MASSIVE STARS (R.M. Humphreys, University of Minnesota)

The evolution of the most massive stars ($\geq$15 $M_\odot$) is a subject that has progressed very rapidly in the past ten years. Much of this progress has been stimulated by observational advances including ultraviolet measurements of mass loss rates in hot stars via IUE and other satellites and our ability with the modern, large relectors and new detectors to observe the luminous star populations in other galaxies. These observations have led to the realization that mass loss plays a major role in the lives of massive stars across the HR diagram. Consequently, the models of massive star evolution now routinely include mass loss as well as other new physical processes such as internal mixing and convective overshooting, plus differences in metallicity to explain the observed features of the HR diagram in our galaxy and others.

There have been a number of meetings, symposia and reviews on massive stars, their evolution and related topics, including the ESO Workshop on The Most Massive Stars (31.012.034), IAU Symposium #105, Observational Tests of the Stellar Evolution Theory (1984, ed. Maeder and Renzini), IAU Symposium #99, Wolf-Rayet Stars: Observations, Physics, Evolution (32.012.042), and The Effects of Mass Loss on Stellar Evolution (30.012.015). De Jager's book The Brightest Stars (28.003.060) is a comprehensive review of all aspects of massive star research up through about 1979, and the recent review article by Humphreys and Davidson (Science, 223, 243, 1984) gives a brief history and summarizes the most recent ideas and developments. Interested scientists are referred to these publications for numerous reviews and papers on massive stars and massive star evolution.

In this brief review of the current literature I will summarize the models and the comparison with observations plus three subjects that have recently received considerable attention; the evolution to Wolf-Rayet stars, luminous blue variables, and the possibility of supermassive stars.

1. Models and Evolutionary Tracks

The observed HR diagram for massive stars in our galaxy and other nearby galaxies cannot be adequately explained by conservative, non-mass loss evolution. We now realize that stellar winds and mass loss occur in massive stars of all types and may alter their evolution. There is also increasing evidence from the chemical abundances in the atmospheres and ejecta of the most massive stars for mixing and turbulence in their interiors.

Models showing the effects of mass loss have been studied by de Loore et al. (21.065.079), Chiosi et al. (21.065.008, 25.065.025), Stothers and Chin (26.065.014), Maeder (28.065.068, 29.065.112, 30.065.066, 33.065.025), Noels and Gabriel (30.065.008), Falk and Mitalas (30.065.024, 33.065.004), Sreenivasan and Wilson (31.065.010), Brunish and Truran (31.065.015, 32.065.006), Doom (32.065.092, 32.065.093), and Klapp (33.065.087). Inclusion of mass loss in the models results in evolution across the HR diagram at essentially constant luminosity but at a lower luminosity than without mass loss (the star is overluminous for its mass), an increase in the mass fraction of the core, an increase in the main sequence lifetime, and a widening of the main sequence. Mass loss also increases the length of time the star spends in the red supergiant stage in better agreement with the observations; although, higher mass loss rates reduces the lifetime of the most luminous M supergiants. In contrast, the models by Brunish and Truran (31.065.015) predicted an absence of red supergiants not supported by the observations of supergiants in associations and clusters by Humphreys (22.002.058, 26.115.001, Humphreys and McElroy Ap.J., 284, 565, 1984).

Overall, comparison between the mass loss models and the observations is reasonably good except for the excess of luminous stars outside or to the right of the hydrogen burning main sequence band on the HR diagram. Meylan and Maeder (31.153.008) have shown that this main sequence widening is very likely a real effect in young clusters in our galaxy and the Magellanic Clouds.

Models with convective overshooting by Bressan et al. (30.065.047), Stothers and Chin (30.065.004), and by Doom (32.065.092, 32.065.093) produce a somewhat broader main sequence, but are still insufficient to explain the observations. Bertelli et al. (A.&A., 130, 279, 1984), have suggested that a combination of mass loss, convective overshooting and a moderate CNO opacity enhancement produces satisfactory agreement with observations. Indeed, with these combined effects they find that hydrogen burning may extend all the way to the red supergiant region. Maeder (31.065.002) has shown that mixing by turbulent diffusion will also broaden the main sequence. The effects of internal mixing on the most massive stars will be discussed later.

Metallicity variations should appear more frequently in future models to examine the possible dependence of massive star evolution on chemical composition gradients in our galaxy and for comparison with massive star populations in other galaxies, especially the Magellanic Clouds. A few evolutionary tracks for massive stars have been computed with the chemical abundances of the LMC and SMC by Maeder (28.065.068), Hellings and Vanbeveren (30.065.027), and Brunish and Truran (32.065.006).

2. Evolution to Wolf-Rayet Stars

Wolf-Rayet stars and their evolution were the subject of IAU Symposium #99 (32.012.042). Four main scenarios have emerged as the most probable explanations, whether separately or in concert, for the origins of the WR stars: 1) mass loss in massive main sequence stars (the Conti scenario), 2) mass loss from red supergiants, 3) mass transfer in binaries, and 4) mixing.

The initial masses of the WR stars are particularly important for understanding their evolutionary origins. A lower limit to their initial masses close to 20 $M_\odot$ was proposed by Chiosi (31.012.034), Maeder (30.065.066, 31.065.002), and Fermani (32.012.042); however, Conti et al. (34.010.061), suggested that WR stars evolve from stars more massive than 40 $M_\odot$ based on their analysis of the galactic distributions of O- and WR-type stars. Recently Schild and Maeder (A.&A., 136, 237, 1984) from an analysis of WR stars in clusters and associations found lower limits to their initial masses of 18 $M_\odot$ for WN stars and 35 $M_\odot$ for WC stars, but concluded that the majority probably originate from stars >40 $M_\odot$. In similar studies Vanbeveren (A.&A., 1984, in press) found that WR stars descend from stars with initial masses >35 $M_\odot$. and Humphreys, Nichols, and Massey (A.J., 1985, in press) concluded that while WR stars have a lower limit to their initial masses >30 $M_\odot$, 80% of them have initial masses $\geq$ 50 $M_\odot$. There is a consensus emerging from these different studies that most WR stars have a lower limit to their initial masses near 40 to 50 $M_\odot$. This implies that most M supergiants, with initial masses of 15-30 $M_\odot$, are not progenitors of WR stars. Only the most massive, most luminous red supergiants would have the potential of becoming WR stars. A lower limit of 40 $M_\odot$ also implies a much greater number of WR stars compared to massive main sequence stars than predicted by the theoretical models.

There is a very pronounced lack of WR stars in the outer parts of our galaxy; for all the known WR stars within 2.5 kpc of the sun (34.115.018) the ratio is about 6 to 1 between the regions interior and exterior to the solar orbit. This difference can be largely explained by fewer massive star progenitors in the outer parts. Recent studies of the IMF for massive stars (Garmany et al., 32.155.061, Humphreys and McElroy, Ap.J., 284, 565, 1984) show that the number of massive stars per unit area is greater in the inner regions, and the IMF for the interior region would predict about three times as many massive stars in the inner region compared to the observed ratio for WR stars of about six. The remaining difference of a factor of two may be due to metallicity effects; higher metallicity leads to higher mass loss rates which may lead more of the most massive stars to the WR stage, as suggested by Maeder (31.065.002).

3. Luminous Blue Variables

A group of very luminous, hot stars with peculiar emission line spectra which are also known to be unstable and undergoing unsteady mass loss provide important clues to the evolution of the most massive stars and their eventual fate. Well known examples of this group of stars include η Car, P Cyg, S Dor and the Hubble-Sandage variables in M31 and M33. Information on their temperatures, luminosities, and mass loss rates have recently been determined for many of these stars primarily as a result of ultraviolet (IUE) and infrared observations. [η Car: Pagel (2.122.090), Davidson (6.114.064), Neugebauer and Westphal (1.113.014), Robinson et al. (9.114.025), Gehrz et al. (09.114.030), and Hyland et al. (26.133.013); P Cyg: Cassatella et al. (26.114.059), Underhill (26.114.088), and Lamers et al. (A.&A., in press, 1984); S Dor etc. in the LMC and SMC: Wolf et al. (28.122.024, 30.122.057, 30.122.103), Appenzeller and Wolf (31.012.034), and Shore and Sanduleak (Ap.J. Suppl., 55, 1, 1984); and for the Hubble-Sandage Variables in M31 and M33; Humphreys (14.114.322, 21.158.012), Gallagher et al. (30.114.101), and Humphreys et al. (Ap.J., 278, 124, 1984)]. Their measured mass loss rates range from 10^{-5} to 10^{-3} $M_\odot$/yr with a possible high of 10^{-1} $M_\odot$/yr for η Car (22.064.080). These stars occupy a critical position on the HR diagram with luminosities between M_{bol} = -10 and -12, on or near the observed upper luminosity boundary for normal hot stars (see Science, 223, 243, 1984 and Humphreys et al., Ap.J., 278, 124, 1984). η Car, P Cyg, some of the H-S variables (Var A in M33) and most recently R127 in the LMC by Walborn (in Workshop on MK Spectral Classification: Criteria and Applications) and by Stahl et al. (34.122.067), are thought to suffer spectacular episodes of mass ejection. These stars may represent a stage in the evolution of the most massive stars when they are very near an 'instability limit' (Humphreys and Davidson 26.115.001).

The famous star η Car is probably the most massive and most extreme member of this group and has undergone more than one episode of large mass ejection. Davidson, Walborn, and Gull (31.112.009) found that the ejecta from η Car is nitrogen-rich but carbon and oxygen-poor. Many of the other variables are also known to be nitrogen-rich [Shore and Sanduleak, (Ap.J. Suppl., 55, 1, 1984) and Walborn (31.159.014)]. This suggests that these hot, luminous and often unstable stars are on their way to becoming Wolf-Rayet stars of the late WN type. A possible evolutionary sequence for stars >60 $M_\odot$, O → Of → LB Var. → WN7-9 has recently been proposed by several authors. However, Bath (26.064.026) has suggested an alternative model for these luminous stars in which they are really accretion disks around moderately massive stars in binary system.

The exact cause of the instability in these stars is not known but there are several alternatives as discussed by Stothers and Chin (33.122.005). An internal vibrational instability is possible, but very likely surface radiation pressure is also involved. Maeder (33.065.025) has proposed that a deep convective zone develops in the outer layers of very massive stars as they evolve to cooler temperatures and gives rise to a turbulent pressure gradient which can oppose gravity if the star's luminosity is high enough. The de Jager limit (28.003.060) for stability is reached when the turbulent pressure gradient equals the acceleration due to gravity. This halts the star's evolution to lower temperatures accompanied by enhanced mass loss. Maeder's models show that the star then evolves to warmer temperatures at essentially constant luminosity on short time scales. This corresponds to the observations of R71 and S Dor (Appenzeller and

Wolf 31.012.034) and Var 83 in M33 (Humphreys et al., Ap.J., 278, 124, 1984) between maximum and minimum.

In a very recent paper de Jager (A.&A., 138, 246, 1984) has presented a more quantitative discussion of the observed luminosity boundary for luminous supergiants and the stability limit confirming many of the ideas originally put forward by Humphreys and Davidson (26.115.001) and by Maeder (33.065.025).

4. Supermassive Stars

Radcliffe 136a, the central object of the giant nebula 30 Doradus in the Large Magellanic Cloud has received considerable notariety as a possible supermassive star. Several years ago Schmidt-Kaler and Feitzinger (27.132.017, 31.012.034) suggested that it accounts for most of the ultraviolet radiation that ionizes the nebula. After obtaining UV spectra with IUE, Cassinelli et al. (30.132.008), concluded along with Schmidt-Kaler and Feitzinger that R136a is a single star with a luminosity nearly 10^8 $L_\odot$ and a temperature > 60,000°K corresponding to a mass near 3000 $M_\odot$ although additional data on the UV continuum by Savage et al. (34.115.011), reduced the mass estimate to about 2000 $M_\odot$.

Such extreme mass and luminosity estimates for R136a naturally made it a very controversial object. Years ago Walborn (9.131.168) suggested that R136a is analogous to the very compact group of luminous stars in NGC 3603. Numerous astronomers have recently discussed the possible nature of R136a and an entire session at IAU Symposium #108 Structure and Evolution of the Magellanic Clouds (1984, ed., van den Bergh and de Boer) was dedicated to the 30 Doradus nebula and R136a including a lively panel discussion on whether R136a is a single star or a compact group.

It is now realized that R136a is definitely a binary (Innis Catalogue of Double Stars, 1927, Worley, Ap.J., 278, L109, 1984, and Chu, Cassinelli and Wolfire, Ap.J., 283, 560, 1984) and very likely a multiple system (G Weigelt, unpublished speckle interferometric measurements) similar to NGC 3603. These considerations will reduce the mass estimate for the primary to a few hundred solar masses; still very high, comparable to or greater than that of η Car.

Massey and Hutchings (34.157.191) have suggested that several objects in M33 with Wolf-Rayet-like spectra may resemble R136a, but M33 is much further away than the LMC and it is not possible to tell whether any of them are single stars.

The formation of these very massive stars and especially the large groups of very massive stars represented by the η Car group (Tr 14, 15, 16) and NGC 3603 in our galaxy, the R136 complex in the LMC and NGC 604 in M33 is an important problem for future work on massive evolution.

III. ROTATION IN LATE TYPE STARS (W. Benz, Geneva Observatory and Los Alamos National Laboratory)

1. Introduction

The tremendous development of observational techniques in the visible as well as in the UV and X-ray spectral domain in the last years has brought a great rebirth of interest in the study of rotation and its effects in late type stars. Rotation was found to play an important role in so many fields that a complete review is far beyond the scope of this brief survey of the past three years literature. I will therefore only review some aspects of the subject: rotational measurement techniques, temporal variation of the rotation, and chromospheric and coronal activity.

There have been several meetings about these topics and related ones: the second and the third Cambridge workshop on cool stars, stellar systems and the sun (1981, eds. M.S. Giampapa, L. Golub, 32.012.065; 1983, eds. S.L. Baliunas, L. Hartmann, Lecture Notes in Physics, Vol. 193), the IAU Symposium #102 on Solar and Stellar Magnetic Fields: Origins and Coronal Effects (1982, ed. J.O. Stenflo, 34.012.017) and the IAU Colloquium #71 on Activity in Red Dwarf Stars (1982, eds. P.B. Byrne, M. Rodono, 34.012.026). Interested scientists are referred to these publications for numerous reviews and papers as well as for further references.

2. Rotation velocity measurements

Rotation and age correlation studies obviously require high precision velocity measurements that for slow rotators are difficult to achieve. In the last years two different spectroscopic techniques have shown to be able to challenge that goal. The first one is based on a Fourier transform of the spectral line and was proposed several years ago by Smith and Gray (19.031.402) and the second one, based on a correlation technique, was proposed by Benz and Mayor (29.031.502). Comparison papers have been written to check the coherence between all determined rotational velocities, Soderblom (32.116.018), Benz and Mayor (1984, A&A 138,183). Even though there is a general agreement between all determinations, there are still unexplained differences for individual stars even between authors using the same determination technique. Recently Bruning (1984, Ap.J., 281,830) has proposed that the differences between users of the Fourier transform technique may exist because of the breakdown of some basic assumptions of the method.

Observations of periodic variations either in the Ca II emission reversals, Middelkoop et al. (29.116.013), Vaughan et al. (30.116.029), Baliunas et al. (34.116.042), or in luminosity, Van Leeuven and Alphenaar (32.153.043) have provided an indirect method of deriving rotational velocities that has the advantage of being independent of line profile analyses. In a study of K stars in the Pleiades, Van Leeuven and Alphenaar have found very short periodic luminosity variations that imply unusually high rotational velocities for this type of stars. Soderblom et al. (34.153.051), spectroscopically confirmed these high velocities concluding therefore that they are really spotted stars and definitively not ellipsoidal stars or close binaries. However, Benz et al. (1984, A&A, 138,93), do not find such a close agreement in five stars they had in common with the list of Van Leeuven and Alphenaar. Finally Stauffer et al. (1984, Ap.J., 280,202), show that the rotation distributions of the Pleiades K dwarfs is a superposition between two components, one that seems unusual for this spectral type. They conclude that the fast rotators are stars only recently arrived on the main sequence implying an age spread of at least 3×10^7 yrs in the Pleiades. Smith et al. (34.121.012), found the same two components structure in a study of G stars in the Orion Ic cluster (despite some possible membership problems).

Macroturbulence is a major problem in determining rotational velocities in late type stars as its broadening effect may be as important as the rotational one. The Fourier transform technique allows a separation of both turbulence and rotational broadening. Gray (30.116.041) has given macroturbulence velocities for five giants whereas Soderblom (32.116.018) and Gray (1984, Ap.J., 281,719) list macroturbulence values for dwarfs. The latter show that macroturbulence decreases rapidly with advancing spectral type and that Zeeman broadening may have a significant effect. This was confirmed by the study of Benz and Mayor (1984, A&A, 138,183) who also suggest an increase of turbulence with equatorial rotational velocity.

3. Rotation and age

Several attempts to determine the decay of the rotation with age in main sequence stars have been made by using either field stars or open clusters stars. The first stars have the advantage of being brighter than the second, but cluster stars are much more easily dated. Soderblom (32.116.018, 34.116.006) has shown that rotation in solar type stars is well fitted by the $t^{0.5}$ relation proposed a long time ago by Skumanich (7.114.020). However, Smith et al. (34.121.012), and Benz et al. (1984, A&A, 138,93), using rotational velocities of nearby clusters stars, do not confirm this relation. They both show that the rotation decay must be smaller for stars younger than the Hyades. Benz et al., show that the same decay is to be expected for all stars in the spectral range F8 to K3. Gray (32.116.006) by looking at rotation rates as a function of spectral type for F, G, and K stars even suggests that besides the slow spindown, there must be a strong braking phase before these stars reach the main sequence. Endal and Sophia in numerical models of the evolution of a solar-type star indeed found that a phase characterized by a strong decrease of angular momentum was followed by another one with a much lower variation of the angular momentum (29.080.016). Using rotation periods deduced from Ca II emission variations, Catalano and Maritti (33.116.010, 34.012.026) showed that for stars of spectral type between F8 to K5 the rotation period only depends on mass and age of the star. Geroyannis and Antonakopoulos (30.065.063) constructed polytropic models with differential rotation in order to study the decay of the rotation. The characteristic decay time they got is consistent with observations. A theoretical review of angular momentum loss produced by stellar winds may be found in Roxburgh (34.012.017).

The spindown in giant stars was discussed by Gray (30.116.041, 32.116.015, 33.116.004, 34.012.017). He showed that there exists a powerful mechanism bringing the rotation down from 25 km/s to 5 km/s and that all giants leave this phase with about the same velocity. Gray and Endal (31.153.002) have shown that this slow rotation rate might be explained either by a braking mechanism or differential rotation in the convectives zones. Endal (34.012.017) suggests that a dynamo mechanism is responsible for the sudden drop of the rotational velocity at G5 III.

4. Rotation and stellar activity

Ca II, Mg II and X-ray emission have been widely studied as they are believed to be a signature of chromospheric or coronal activity. The correlation between rotation and emission intensity got particular attention as it may give the clue to the physical mechanism driving the stellar activity. Knobloch et al. (30.116.036), proposed a theoretical explanation of the correlation between magnetic activity and rotation in terms of a transition between convection patterns as the angular velocity becomes smaller. Further discussions of the connection between activity and rotation might be found in Noyes (32.012.065), Dupree (32.012.65), and Rosner (34.012.017).

4.1 Chromospheric activity

Variation of Ca II emission with rotation was found by Middelkoop (30.116.015, 31.116.013), Pallavicini et al. (32.012.065), and Noyes et al. (1984, Ap.J., 279,763). Noyes et al., also show that the decrease of Ca II emission with rotation is the same for stars above and below the Vaughan-Preston gap, thus casting doubts on explanations of the gap in terms of a discontinuity in dynamo characteristics. That the gap might be an artificial structure was suggested by Hartmann et al. (1984, Ap.J., 276,254), who showed that a similar structure is obtained with a smooth function for chromospheric decay.

Böhm-Vitense by looking at Mg II and transition layer emission lines (32.012.006, 32.114.002, 32.012.065) in F stars did not find an increase of emission with increasing rotation for stars rotating faster than 30 km/s whereas Hartmann et al. (1984, 279,778), present IUE observations that correlate well with

Ca II emissions. However, a definite conclusion that Mg II lines exhibit the same feature as Ca II lines could not be made because of the small number of data points.

4.2 Coronal activity

The Einstein X-ray Observatory satellite program of mapping the occurrence of X-ray emissions in late type stars (Ayres et al., 30.115.018) has shown that these stars may have a hot coronae. Walter (29.116.018, 29.116.019) shows that, similarly to chromospheric activity, coronal activity is a function of angular velocity and depth of the convective zone. Later by reexamining the rotation-activity relation, Walter (31.116.006, 34.012.017) suggests that a gap at 12 days rotation period exists in the L_x/L_{bol} versus rotation period relation. This gap was confirmed later by Smith (33.116.017). Walter also found that a correlation between X-ray emission and rotation appear only for stars redder than B-V=0.45. He speculates that this limit represents the turn-on of a solar-like dynamo. Further confirmations of the existence of a relation between rotation and emission was given by Smith et al. (34.153.018), in a study of G stars in the Orion Ic cluster and by Stern and Skumanich (33.118.014). Rucinski however showed that X-ray luminosities for M dwarfs are too large to be due to rotation, they must derive ultimately from nuclear energy sources.

A theoretical estimate of the dependence of the X-ray coronal flux from late-type stars on various stellar parameters is given by Paterno and Zuccarello (34.064.083). They show that an approximate relationship between X-ray surface flux and mass, angular velocity and depth of the convection zone may be derived. Finally, Mangeney and Praderie (1984, A&A, 130,143) suggested that the convection-rotation-interaction can be represented by an effective Rossby number. Interesting relations are presented between L_x and F_x and this Rossby number which are valid for almost all spectral types, whereas using L_x/L_{bol} no general relation could be found.

IV. HELIOSEISMOLOGY (J. Christensen-Dalsgaard, Nordita and University of Aarhus)

1. Introduction

The study of the solar interior structure and dynamics by means of observations of solar oscillations has matured very considerably during the last three years. In 1981 the existence of global solar oscillations had been firmly established, although in most cases the modes of oscillation had not yet been definitely identified. Now a direct determination of the sound speed in a large part of the solar interior has become possible, and the equatorial rotation rate has been measured in most of the Sun.

Several conferences have been concerned with this subject. These include IAU Colloquium No. 66 on "Problems of Solar and Stellar Oscillations" in Crimea in 1981 (published in Solar Phys., vol. 82; ed., D. O. Gough), the conference on "Pulsations in Classical and Cataclysmic Variable Stars" at JILA, Boulder in 1982 (eds., J. P. Cox & C. J. Hansen; JILA), the EPS study conference on "Oscillations as a Probe of the Sun's Interior" in Catania in 1983 [published in Mem. S. A. It., vol. 55; eds., G. Belvedere & L. Paterno: in the following referred to as (Catania)], the workshop on "Solar Seismology from Space" in Snowmass in 1983 [ed., R. K. Ulrich; JPL: (Snowmass)], and the International Astrophysical Colloquium on "Theoretical Problems in Stellar Stability and Oscillations" in Liège in 1984 [published by Institut d'Astrophysique, Liège; (Liège)]. The proceedings of these conferences contain reviews on various aspects of the subject. Reference may also be made to the reviews by Deubner & Gough (Ann. Rev. Astron. Astrophys., 22, 593), by Christensen-Dalsgaard (33.080.020), and by Gough (Adv. Space Res., in press, 1984).

The advances in helioseismology have in large measure been caused by improvements in the observations of solar oscillations (see also the section by Duvall in the report from Commission 12). Particularly important were the observations by Duvall & Harvey (33.080.16) which for the first time closed the gap between the 5 min modes of high and low degrees, thereby establishing the radial orders of the latter. In addition, observations from two widely spaced stations [Claverie et al., (Catania), p. 63] have resulted in the measurement of frequencies of low-degree 5 min modes with an accuracy which exceeds that with which the mass and radius of the Sun are known. There have also been continued observations of oscillations at longer periods (e.g., Scherrer & Wilcox, 33.080.22; Delache & Scherrer, 34.080.049; Severny et al., Nature, 307, 247), although the modes responsible have not been definitely identified yet. Finally, observations of the atmospheric behaviour of the oscillations are becoming increasingly detailed [e.g., Staiger et al., (Catania) p. 147; Andersen, (Liège) p. 220].

2. Inverse theory

The frequency of a given mode of solar oscillation is determined by the structure of that part of the Sun where the mode has appreciable amplitude (as measured, e.g., in terms of the energy density). The location of this region depends on the nature of the mode. The 5 min modes are standing acoustic waves which are trapped between the surface and an inner turning point. For modes of low degree ℓ[1] the trapping region extends over most of the solar interior, whereas at high degrees the modes are confined close to the surface. Thus, by combining frequencies of modes of different degree, localized information about the structure and dynamics can in principle be obtained. The long-period oscillations must be high-order gravity modes (cf. Christensen-Dalsgaard et al., 33.080.009) which are concentrated towards the centre of the Sun; their frequencies therefore give information about the deep interior.

The observed quantities are mean frequencies, which (for adiabatic oscillation) are determined by the density ρ and the adiabatic exponent γ in the Sun, and the frequency splittings induced by rotation, which are linear functionals of the rotation rate. The simplest way of interpreting the observed frequencies is by comparing with frequencies computed for solar models, to determine the model which best fits the data. To determine the rotation rate one may attempt to specify it in terms of a small set of parameters (e.g., as a piece-wise constant or piece-wise linear function, or a polynomial), the parameters being determined by least squares fitting. More systematic inversion procedures have been developed in geophysics (Parker, 19.081.035) and may be applied to the solar data [Gough, (Snowmass) p. 49].

To test the methods it is very useful to apply them to fictitious data, such as rotational splittings for a given, imposed rotation law, to see how well they recover the input. Studies of this nature have been carried out for density inversions by Cooper & Gough (cf. Cooper, Ph.D. Dissertation, University of Cambridge, 1981), and for rotational inversions by Gough [(Snowmass) p. 49] and Christensen-Dalsgaard & Gough [(Snowmass) p. 79]. Broadly speaking, from observations of 5 min modes over a sufficiently large range of degrees, localized information can be obtained about the region scanned by the turning points of the modes. Alternatively, the interior of the model can be resolved with only modes of low degree, provided all acoustic modes, and a few low-order gravity modes, are

1. ℓ is the degree of the spherical harmonic Y_ℓ^m which describes the variation of the mode over the solar surface.

included; however, it must be emphasized that low order modes have not been definitely identified on the Sun.

3. Results of solar structure

The frequencies of 5 min modes are largely determined by the variation in the sound speed in the solar interior. From observations of modes of high degree it has been inferred that the depth of the convection zone is about 0.3 $R_\odot$ (e.g., Berthomieu et al., 27.080.068; Lubow et al., 27.080.067). This result is unlikely to be affected by atmospheric non-linearities of the oscillations (Belvedere et al., 33.080.043). However, it may depend somewhat on the physics of the model or the oscillations in the layers close to the surface.

Low-degree 5 min modes provide information about global properties of the Sun. Among models with different heavy element abundances Z, the agreement between the observed and the computed frequencies is generally best for Z ≅ 0.02 [Christensen-Dalsgaard & Gough, 30.080.064; Shibahashi et al., 34.080.008; Christensen-Dalsgaard, (Liège) p. 155]; this is very close to the spectroscopically determined value (Grevesse, Proc. 7th Europ. Regional Meeting in Astr., Mem. S. A. It., in press, 1984). Such models have a helium abundance Y ≅ 0.25 (e.g., Gough, 34.080.023). Models with low Z are ruled out by the oscillation observations. However, even for the models which best fit the observations, the differences between theory and observation are considerably higher than the observational, or the intrinsic computational, errors.

These differences are often discussed in terms of parameters based on the asymptotic theory of low-degree high-order acoustic modes (e.g., Tassoul, 28.065.044). Thus the mean separation $\Delta\nu_\ell$ between the frequencies of adjacent modes of the same degree is a measure of the inverse sound travel time between the centre and the surface of the Sun. In most solar models (e.g., Christensen-Dalsgaard & Gough, 28.080.062; Ulrich & Rhodes, 33.080.007; Noels et al., A. & A., 130, 389) this quantity is 0.5 - 1 μHz higher than the observed value of about 135.2 μHz; but for Z = 0.02 Shibahashi et al. (34.080.008) obtained a value quite close to the observed one. A second important characteristic of the spectrum is the average $\delta\nu_\ell$ of the small separation $\nu_{n,\ell} - \nu_{n-1,\ell+2}$ between frequencies of modes differing by 1 in radial order n and 2 in degree; this is largely a measure of the properties of the solar core (e.g., Gough, Phys. Bull., 34, 502). For normal solar models $\delta\nu_\ell$ is generally consistent with the observations, whereas it is about 50 percent too large in models with substantial mixing in the interior [e.g., Christensen-Dalsgaard & Gough, loc. cit.; Ulrich & Rhodes, loc. cit.; Cox & Kidman, (Liège) p. 259]. Thus observations of low-degree 5 min modes constrain the extent to which the core of the Sun can have been mixed.

Analysis of low-degree 5 min modes, taken in isolation, provides little information about the origin of the discrepancies between theory and observations. With observations of 5 min modes at all degrees, however, one should be able to locate the causes of the discrepancies. A preliminary comparison (Christensen-Dalsgaard & Gough, [Snowmass] p. 193) of the observations of Duvall & Harvey with a normal solar model indicated that the errors in the model were predominantly close to its surface, with a possible additional contribution from near the base of the convection zone; most of the radiative interior seemed to be consistent with the observed frequencies. Qualitatively similar results are obtained from the published frequencies for most other normal solar models (e.g., Ulrich & Rhodes, loc. cit.; Noels et al., loc. cit); the model of Shibahashi et al., agrees considerably better with the observations. In this model the convection zone is only very slightly deeper than for the model considered by Christensen-Dalsgaard & Gough, but the central temperature and density are higher by about 9 and 23 percent, respectively. The reasons for these differences, and their relation to the computed frequencies, require further study.

The frequencies, degrees and radial orders of the 5 min modes satisfy a simple, approximate relation, first found by Duvall (32.080.029) from observational data on high-degree modes. Asymptotic theory (e.g., Gough, Phil. Trans. R. Soc. A., in press, 1984) shows that this relation depends on the variation of sound speed c with radius r; it may be inverted to yield c(r) purely in terms of observed quantities. Using this technique, Christensen-Dalsgaard et al. (Nature, in press) found that c(r) could be determined with an accuracy of about one percent on the interval $0.4 < r/R_\odot < 0.9$. The result was very close to a normal solar model, although there were indications that beneath the convection zone c was slightly higher in the Sun than in the model.

The frequencies of g modes are much more sensitive to conditions in the solar core than are frequencies of 5 min modes. From such data Berthomieu et al. (Nature, 308, 254) found evidence for partial mixing of the solar core. However, it was pointed out by Gough (Proc. Kun Ming Workshop on Solar Physics, ed., B. Chen & C. de Jager; 1984) and Gabriel (A. & A., 134, 387) that the observed frequencies could be identified with modes in normal, unmixed solar models. Thus definite information about solar structure from observations of g modes must await identification of the modes observed.

4. Results on the internal solar rotation

Observations of oscillations in the solar limb intensity (Bos & Hill, 33.080.027) have yielded rotational splittings for about 7 p and g modes of relatively low order. The inversion of these data was far from unique, but it indicated that much of the Sun was rotating at 4 - 6 times the surface rate (Hill et al., 32.066.238; Gough, 32.080.008). However, the observed power spectra had a very high density of peaks which made the mode identification difficult. Data on 5 min modes that did not suffer from this problem were obtained by Duvall & Harvey (Nature, 310, 19). Inversions of these data using a variety of techniques [Duvall et al., Nature, 310, 22; Gough, Adv. Space Res., in press, 1984; Leibacher, (Liège), p. 298] showed that the equatorial rotation rate in most of the Sun is at, or slightly below, the surface value, except perhaps for a small region in the core that may be rotating substantially faster. Although the conflict between the limb data and the observations by Duvall & Harvey should be resolved, the relative simplicity and extensive nature of the latter set of data give credence to the resulting rotation curve. From this the rotational flattening of the Sun may be calculated; it has a negligible effect on present observations of planetary motion (e.g., Shapiro et al., 17.066.072), which are therefore in accordance with Einstein's theory of General Relativity.

The rotation in the solar convection zone is clearly of major interest for studies of convection zone dynamics. So is information about large-scale, time-varying velocity fields that may also be obtainable from observations of oscillation frequency splittings [Hill et al., (Catania) p. 153]. There is some evidence for an increase of the equatorial rotation rate with depth in the upper part of the convection zone (see also Deubner et al., 25.080.003), and variations in the velocity field from day to day may have been detected, possibly associated with convective giant cells. However, better observations data are required to confirm and further study these phenomena.

5. Atmospheric behaviour and excitation

The oscillations are very nearly adiabatic in most of the solar interior. However, non-adiabatic effects are strong just below the surface and in the atmosphere. Here the details of the interaction between the oscillations and the radiation field may affect the frequencies and damping rates of the modes, as well as the variation of the oscillations with the height in the atmosphere. The latter is required to interpret the observed amplitudes and phases.

The general problem of linear oscillations in a radiating atmosphere is quite complex, if the variation of the radiation field with angle and with the wavelength of the radiation is consistently taken into account. Christensen-Dalsgaard & Frandsen [33.080.034; (Catania) p. 285] computed numerical solutions where either the angle variation or the wavelength variation were included. They showed that the grey Eddington approximation is quite accurate for the calculation of frequencies and damping rates. The treatment of radiation had some effects, of the order of 10 - 20 percent, on the temperature perturbation in the atmosphere; this in turn may affect the amplitudes and phases of the computed intensity oscillations in spectral lines [cf. Frandsen, (Liège) p. 303]. In contrast Hill & Logan (Ap. J., 285, 386) found very substantial effects of including non-grey radiation; however, their results were based on a number of approximations which may be responsible for the discrepancy.

A major uncertainty in the computation of nonadiabatic solar oscillations is the treatment of the perturbation of the convective flux. This could well have a substantial effect on the computed atmospheric properties of the oscillations, and on the relation between the observed velocity and intensity amplitudes.

The excitation of the observed oscillations is still badly understood. The instability found in some calculations of 5 min modes (e.g., Ando & Osaki, 14.080.058; Antia et al., 31.080.043) may depend on the (erroneous) assumption of thermal equilibrium in the equilibrium model (33.080.034). Also the perturbation in the convective flux (e.g., Gough, 27.080.066) or the inclusion of turbulent viscosity (Goldreich & Keeley, 19.080.007) seem to stabilize these modes. Stochastic excitation by convection (Goldreich & Keeley, 19.080.007) can possibly account for the observed amplitudes (Gough, loc. cit.; Christensen-Dalsgaard & Frandsen, 33.065.044), although the theoretical treatment of this process leaves a great deal to be desired.

In normal solar models the high order gravity modes responsible for the long period oscillations are almost certainly stabilized by radiative damping (e.g., Saio 28.065.036). They could be destabilized in models with a mixed core, where the amplitudes of selected modes are enhanced by mode trapping (Gavryuseva et al., 33.080.036); but such models may not be consistent with the observed frequencies of the 5 min modes. Stochastic excitation in the convection zone is totally inadequate to provide the observed amplitudes (Keeley, 27.080.063). Thus at present there seems to be no satisfactory explanation for these oscillations.

6. Conclusion

The initial phase of helioseismology consisted of discovering and establishing beyond reasonable doubt, the presence of large-scale modes of oscillation on the Sun. For the 5 min modes this was completed around 1980; the study of longer period modes is still in this phase. This was accompanied by some preliminary attempts at theoretical interpretation and use of the data for probing the solar interior. In the second phase the first reasonably adequate sets of data were assembled, and some definite results concerning the solar interior obtained. This phase may be reaching its culmination for the 5 min oscillations, with the determinations of the internal rotation rate and sound speed. Thus we are now entering the third phase, where helioseismology is to be used as a tool for understanding the processes going on in the solar interior. The results obtained on solar rotation are not easy to rationalize; and the discrepancies on the structure, while fairly subtle, have proved difficult to eliminate. To proceed we need better and more extensive data, which will enable resolution of the solar interior in increasing detail; such data are promised by the new observing schemes being set up or planned. But we also need a parallel theoretical effort, to devise methods for dealing efficiently with the very large amount of data that is becoming available, to make sure that the computations of solar models and frequencies are sufficiently accurate, to understand the relationship between solar models and their frequencies, and to integrate the results in the broader picture of solar and stellar evolution. There is little doubt that the development of helioseismology will continue to profit from the close interaction between observations and theory, which has been one of its distinguishing, and deeply satisfying, characteristics.

V. PLANETARY NEBULA CENTRAL STARS (Edward M. Sion, Villanova University and Arizona State University)

1. Introduction

Planetary nebula nuclei (PNN), as transition objects, hold the essential keys to understanding the details of late stellar evolution from the nature of their asymptotic giant branch (AGB) progenitor stars to the domain of white dwarfs. PNN can shed important light on a diverse array of poorly understood physical processes: nebula ejection and hot stellar wind mass loss phases, convective dredge-up efficiency and dilution, diffusion, late thermonuclear shell flashes, formation of binary nuclei, non-radial oscillations and the detailed origin of white dwarf spectroscopic subgroups. Recent comprehensive discussions of PNN can be found in two conference proceedings (IAU Symp. 103; 34.012.005 and IAU Symp. 99; 32.012.042) and in two up-to-date monographs by Pottasch (1984, Planetary Nebulae, Reidel: Dordrecht; hereafter PN84) and Kaler (34.134.010). This report will cover work completed since 1981 concerning PNN spectral subgroups temperatures, mass loss, mass distribution and evolutionary links to the hottest DA and non-DA degenerates, topics which have stirred considerable controversy recently.

2. Spectroscopic Subgroups of PNN

Various spectroscopic subgroups of PNN are generally defined by the nature of their line spectra at visual wavelengths. Given the observational difficulties posed by their faintness and nebular contamination (only ~ 75 central stars out of 1400 known planetary nebula are amenable to quantitative analysis), the following spectral categories are recognized (cf. Heap 32.114.133; Pottasch PN84):

1) Wolf-Rayet stars. Very broad emission lines of H, He, C N and O (see Heap 32.114.133 for sub-classes);
2) Of stars. Very similar to young Of stars having emission lines of H, HeII, NIII, and usually CIII;
3) WR + Of. These stars have spectra which are a combination of the above types and do not have a counterpart in normal main sequence stars;
4) OVI stars. The spectra are very similar to WR or Of stars except for the fact that they show very prominent emission lines due to O^{+5} at λλ3811 and 3834 Å. These are among the hottest PNN known;
5) Continous spectra. With high enough resolution these objects show lines in their visual spectra;
6) O type. The absorption lines of an O star, especially H and HeII are observable. The absorption line λ4542 of HeII is often broader than the same line in an Of star, indicating higher gravity;
7) sd O type. Sub dwarf O stars with still broader absorption lines. Often the absorption lines of H are blended by H 12, indicating high gravity.

These spectral categories correlate with the state of ionization in the nebula. While spectral differences may be due, in first order, to the thermal evolution of the PNN, different compositions, surface gravities and atmospheric structure may also have a key role (Kaler 34.134.010; Kaler and Shaw 1984 Ap. J., 278, 195; Heap 32.114.133; Pottasch PN84).

3. Surface Temperatures of PNN

Observations of the continous and line spectra of PNN are seriously hampered by nebular line and continuum emission as well as by uncertainties in interstellar reddening corrections to the observed continuum fluxes. Some alleviation of these problems has occurred due to International Ultraviolet Explorer (IUE) observations in the far ultraviolet where better resolution can help eliminate nebular line contamination and where hot central stars usually dominate the continuum.

Black body fits from the ultraviolet to optical yield good first approximations to the effective temperatures of PNN (e.g., Pottasch PN84). Direct model atmosphere fits to PNN line and continuum observations, where possible, provide the most reliable determinations of effective temperatures and gravities. Recent non-LTE models and line profiles due to Kudritzki (1981; unpublished models) have been applied extensively (Mendez et al., 34.114.024, Mendez, Kudritzki and Simon (1984 Astr. Ap., in press). LTE model atmospheres due to Hummer and Mihalas (1970, JILA report No. 101, University of Colorado) and Wesemael et al. (28.064.011), though less realistic, have also been applied. For $\lambda > 1300$ Å the model fluxes are consistently lower than the black body fluxes and models have identical slopes which agree very closely with the black body slope. The non-LTE model atmosphere analyses have yielded effective temperatures, gravities and helium abundances for several PNN which are absorption line objects (Mendez et al., 34.114.024). Approximately, 60 percent of PNN can be analyzed in this way. However, caution must be exercised in selecting only those lines that form deep enough in the photosphere that the assumption of hydrostatic equilibrium is not violated. When direct model fits are difficult or impossible, Zanstra methods, nebular ionization equilibrium models and 'Stoy' energy balance methods are widely used (cf. Pottasch PN84; Kaler 34.134.010; and references therein). The Stoy method is particularly useful when the visual magnitude of the PNN is poorly known and the nebula is optically thin to its ionizing radiation (cf. Gathier 1984, Ph.D. Thesis Groningen, hereafter G84; Pottasch PN84). Kaler (34.134.010) has calculated hydrogen and HeII Zanstra temperatures and luminosities for 82 PNN.

4. Mass distribution of PNN

What is the true distribution of PNN masses? This question has raised considerable controversy. Schönberner (30.065.067) Schönberner and Weidemann (29.115.023; 33.115.004) and Weidemann and Koester (34.065.055) claim a very narrow mass range ($0.55 < M/M_{\odot} < 0.65$) for PNN based upon a comparison of nebula radii (ages) and absolute visual magnitude with those predicted from theory. They used Shklovsky distances derived by Cahn and Kaler (5.133.017). They also found no correlation between PNN brightness and helium abundance. In sharp contrast to these results, Kaler (34.134.010) finds a greater (less narrow) mass range $0.5 \lesssim M/M_{\odot} \lesssim 0.8$ and correlates higher N/O nebula abundances with higher initial (progenitor) masses and thus higher core masses. This is expected in part because higher than average nitrogen and helium abundances should manifest the higher efficiency of dredge-up processes for higher initial masses.

A less narrow mass distribution than Schönberner and Weidemann (33.115.004) is supported by Gathier (G84) who shows that the mass range should extend to lower ($0.5\ M_{\odot}$) mass, in better agreement with the mass range of DA white dwarfs. Gathier demonstrates that the use of Shklovsky distances can artificially create a narrow distribution in the $M_{\odot}$ versus nebular radius plane. Pottasch (PN84) and Gathier (G84) find that higher luminosity nebulae generally have higher progenitor masses, higher helium and nitrogen abundances and generally higher ionized masses than lower luminosity planetary nebulae. Disagreements about the PNN mass range are likely to persist until proper account is taken of significant differences in distance scales, evolutionary tracks and the still uncertain effects of observational selection. In a later section, however, new evidence is presented which appears to lend some support to a larger PNN mass range.

5. Mass Loss

Stellar wind mass loss has been observed in several types of PNN ranging from subdwarf O nuclei like NGC 6543 (Castor Lutz and Seaton 29.135.002) to Wolf-Rayet-OVI nuclei like Abell 78 (Heap 32.114.133, Kaler and Fiebelman 1984, Ap. J., in press) to PNN with "continuous" visual spectra like NGC 3242 (Hamann et al., 1984, Astr. Ap., 139, 459). It is generally believed that unless mass loss rates are $\gtrsim 10^{-7}\ M_{\odot} yr^{-1}$ the evolution of the PNN will not be greatly affected (cf. Schönberner 34.065.037). The exact role of wind mass loss and "superwind" phases in post-AGB evolution (cf. Iben 1984, Ap. J., 227, 333) is still poorly understood and further theoretical and observational investigations are clearly needed. PNN winds may also have a role in formation of the nebula and their study is highly important to general theories of mass loss from early type stars, for example the radiation-driven wind theory of Castor, Abbot and Klein (13.064.004).

Far ultraviolet observations, at both high and low resolution, with the International Ultraviolet Explorer (IUE) satellite have proven invaluable in studies of PNN winds. The analysis of P Cygni profiles at low IUE resolution is facilitated by the highly useful techniques for extracting wind parameters by Castor, Lutz and Seaton (29.135.002). There have been extensive studies of PNN mass loss by Benvenuti and Perinotto (Proc. 2nd Europ. IUE Conf., ESA SP-157, p. 187), Perinotto (34.134.043), Kahn (34.134.038) and Hamann et al. (1984, Astr. Ap., 139, 459). The paper by Hamann et al., treats the hottest PNN (NGC 3242) for which mass loss has been analyzed. The derived rate is in the range $-10.1 \lesssim \log \dot{M}(M_{\odot}\ yr^{-1}) \lesssim -8.1$ and the wind parameters bear the same relation to stellar parameters (e.g., $\log T_e$, $\log g$) as that established for winds of more massive, early type stars. Further observational and theoretical work on PNN wind mass loss at different post-AGB phases is badly needed.

6. Binary PNN

The discovery of short period, Roche lobe-detached dK/dM + DA/sdO binary PNN, such as UU Sge (Bond et al., 21.121.048) and Abell 41 (Grauer and Bond 34.134.001) established these objects as the products of common envelope evolution (Paczynski 18.117.051) and the immediate precursors of a substantial fraction of the cataclysmic variables. The growing number of such systems (cf. Table 1, Sion et al., 1984, Ap. J., 279, 761) together with the large number of short period DA + dM binaries to be found in the Palomar-Green Survey (Green, Schmidt and Liebert 1985, in preparation; Wade, Liebert and Green 1985, in preparation) provides a sample of post-common envelope binaries large enough to help achieve a deeper understanding of this poorly known phase of evolution. While the spiralling-in of the secondary to form a double core red giant or supergiant cannot yet be followed accurately with numerical models, more accurate determinations of binary periods, component dynamical masses and effective temperatures and gravities for the white dwarf/ subdwarf companions should yield valuable insight into the efficiency of the common envelope ejection process itself and the termination of or coalesence from the spiralling in process. The central star of the low surface brightness planetary nebula Abell 41 was discovered to be a binary with a period of 2h 43m by Grauer and Bond (34.134.001). Green, Liebert and Wesemael (1984, Ap. J., 280, 177) found the primary to be a 50,000 K subdwarf, one half the effective temperature derived by Bond and Grauer from the reflection effect due to the hot subdwarf O star heating the facing hemisphere of the dM star.

Following further less violent loss of orbital angular momentum the binary central stars should evolve to a cataclysmic variable through a combination of gravitational radiation, tidal friction and magnetic braking via flares of winds (e.g., Eggleton 33.117.082; Verbunt and Zwaan 30.117.005; Taam 33.117.171).

7. PNN-White Dwarfs: Final Evolutionary Links

The obvious PNN-white dwarf general evolutionary connection has long been established by the position of white dwarfs in the H-R diagram at the older, low luminosity extension of the Harmon-Seaton sequence, by the heretofore comparable formation rates of planetary nebulae and white dwarfs and the near-degenerate physical properties of the post-ABG cores (cf. Weidemann 68.819, 32.126.025). However, important questions remain concerning (1) how many distinct evolutionary paths with or without planetary nebula ejection lead to white dwarf formation; (2) which types of AGB giant progenitors and descendant PNN's evolve into which spectroscopic subgroups of white dwarfs?

Significant progress in both theory and observation has been made in the last three years.

On the theoretical side, detailed post-AGB evolutionary tracks have been computed by Schönberner (30.065.067, 34.065.037); Renzini (32.065.076); Iben and Renzini (34.065.032) and Iben et al. (34.065.002). In the last two papers, it is argued that the high excitation planetaries, Abell 30 (Hazard et al., 27.134.033) and Abel 78 (Jacoby and Ford 34.134.009), with pure helium inner zones, may be due to ejection of a second nebular shell as a PNN suffers a final helium shell flash as it evolves along the declining luminosity portion of its pre-white dwarf track. The second ejection phase rids the PNN of its remaining hydrogen and most of its CNO-processed envelope material, which can be removed quickly via a hot stellar wind. Indeed Heap (32.114.133) has shown that nitrogen rich planetaries correlate with carbon-rich, H deficient PNN's. The percentage of PNN which undergo a final thermal pulse is largely determined by the precise thermal pulse cycle phase at which "superwind" mass loss begins or stops. Iben (1984, Ap. J., 277, 333) followed the evolutionary behavior of a model PNN as a function of the phase in its nuclear burning cycle when its progenitor leaves the asymptotic giant branch for the first time. He identified several distinct groups of models whose members are expected to evolve into non-DA white dwarfs and suggests that 25% of all PNN should fall into these groups. He argues that winds of at least two types are responsible: a low velocity or slow wind which operates when and if the PNN returns to the AGB after experiencing a final helium thermal pulse; and a high velocity or fast wind which operates when the PNN is at high enough surface temperatures that emitted photons will excite nebular emsssion. Iben's estimate that 25% of all PNN become non-DA white dwarfs is rougly consistent with the empirical non-DA/DA ratio for the proper motion-selected white dwarf sample down to $T_e \gtrsim 10^4$K (Sion 1984, Ap. J., 282, 612). Further post-AGB calculations are now in progress including the effects of stellar wind mass loss phases, diffusive separation and diffusion-induced burning (cf. Iben and MacDonald 1984, preprint).

On the observational side, links between specific classes of PNN and white dwarfs have recently been established. First, two PNN, NGC 7293 and Abell 7, both appear to contain DA01 white dwarfs with helium abundances a factor of ten lower than solar (Mendez et al., 34.114.024; Mendez, Kudritzki and Simon 1984, Astr. Ap., in press). If this underabundance is due to gravitational diffusion, the time scale for helium depletion is comparable to or shorter than the time scale for nebular dissipation. Hence PNN's like Abell 7 and NGC 7293 should be important producers of DA white dwarfs. Second, oxygen has recently been discovered in the photospheres of the hottest known white dwarfs, the PG1159 DOZ1 degenerates (Sion, Liebert and Starrfield 1985, Ap. J., in press). The presence of high n OVI absorption features at λλ3311, 3434, 3811, 3834, some with sharp emission reversals, together with CIVλλ3689, 3934, 4658, all associated in strong emission with the OVI PNN class (cf. Heap 32.114.133; Kaler and Shaw 1984, Ap. J., 278, 195) suggests a direct evolutionary link. That such a link is correct, is strengthened by the OVI nucleus of K1-16 which is itself a PG 1159, non-radially pulsating DOZ1 degenerate (Grauer and Bond 1984, Ap. J., 277, 211). The presence of oxygen in significant abundance at the surfaces of the PG 1159 stars also provides tentative confirmation of the cyclical ionization of oxygen as driving the oscillations as predicted by Starrfield, Cox, Kidman and Pesnell (1984, Ap. J., 281, 800). Furthermore, the direct association of the hottest, most exotic PNN's, the OVI central stars, with the hottest ($T \gtrsim 10^5$K), fully or near fully degenerate stars, the PG1159 class, has far reaching implications for the mass range of PNN progenitors and hence the masses of DA and DO/DB white dwarf descendants. The PG1159 stars appear to be evolving from the highest excitation, helium-rich nebulae, which are generally associated with PNN of relatively high masses (Kaler 34.134.010; Pottasch PN84). Noting the (a) PNN's like Abell 7 are at degenerate gravities but are cooler ($T_e < 10^5$K) then OVI PNN's and (b) no DA white dwarfs have a temperature approaching 10^5K despite being 4 times more numerous, Sion, Liebert and Starrfield (1985) argue that the OVI/PG1159 progenitors may have higher than average ($> 0.6\ M_\odot$) masses. Sion, Liebert and Starrfield (1985) claim that either (1) the PG1159 stars (and hence their OVI parents) have higher masses than the apparent PNN progenitors if DA white dwarfs generally have very thin H envelopes (cf. Dolez and Vauclair 30.126.024) or (2) there is no significant difference in the masses of the hottest DO and DA degenerates (and thus their PNN progenitors) if the DA stars have hydrogen envelopes which are quite thick as suggested by the results of Iben and Tutukov (1984, Ap. J., 282, 615).

On the other hand, Schönberner and Weidemann (33.115.004) and Mendez, Kidritzki and Simon (1984, Astr. Ap., in press) favor an evolutionary path to non-DA white dwarfs originating from isolated sdO stars which have masses on the low side of the PNN mass distribution and that these stars, already having He-rich atmospheres, evolve into DO/DB white dwarfs while the PNN's evolve primarily into DA white dwarfs.

A recent re-determination by Fleming, Liebert and Green (1984, preprint) of the DA white dwarf luminosity function, based upon 353 confirmed DA white dwarfs from the Palomar-Green sample (complete to B = 16.5) yields good agreement with the Iben and Tutukov theoretical luminosity function which is based upon DA models having thick H envelopes ($M_H \gtrsim 10^{-4}\ M_\odot$) with H shell burning as the dominant luminosity source. Assuming a constant white dwarf formation rate, Fleming et al., derive a DA white dwarf birth rate of $5.9 \times 10^{-13}\ pc^{-3}\ yr^{-1}$ which is only 30% of the currently accepted formation rate of planetary nebulae. These new results reveal for the first time, a wide discrepancy between the formation rate of PN and DA white dwarfs, that cannot be reasonably explained by further PN ejections from "born again" AGB stars via the Iben et al. (34.065.002) scenario.

More refined determinations of the formation rates of different classes of PNN are badly needed, together with a more detailed understanding of the extent of ongoing mass loss as PNN begin their descent into the white dwarf domain. Comparisons with formation rates of the hottest DA and non-DA white dwarfs would then be more fruitful.

VI. PULSATIONS IN HOT DEGENERATE DWARF STARS (A.N. Cox, Los Alamos National Laboratory and S.D. Kawaler, University of Texas)

Over the past five years, four extremely hot degenerate dwarf stars have been reported to be multiperiodically pulsating variable stars. The first of these was PG1159-035, initially reported by McGraw, Starrfield, Angel, and Carleton (26.122.137). These discoveries have been made by using high time resolution photometry on the very blue stellar objects found in the Palomar Green (1977, thesis Caltech), PG, catalogue. Seven stars plus the planetary nebula central star K1-16 display hydrogen-free spectra that all the known variables in this class show. Now two more of these stars have been found to pulsate with similar periods (near 500 seconds): PG1707+427 and PG2131+066. These last two variables have been observed by Bond, Grauer, Green, and Liebert (1984, Ap.J., 279, 751). The planetary nebula nucleus star in Kohoutek 1-16, with periods between 1400 and possibly 4500 seconds (Grauer and Bond, 1984, Ap.J., 277, 211) also seems to be in this class according to Starrfield, Cox, Kidman, and Pesnell (Ap.J. Lett., submitted). The gravity for this latter star is probably lower than for the others in this class according to model atmospheres. Surface temperatures of these objects are extremely high, with estimates for PG1159-035 ranging from 80-150 × 10^3K (Wegner, Barry, Holberg, Forrester, and McGraw 33.114.076). Kaler (34.134.010) has provided a lower limit of 90,000K for K1-16. Model atmosphere analyses suggest a surface gravity of $7<\log g<8$ for the PG1159 stars according to Wesemael, Winget, Cabot, Van Horn, and Fontaine (31.126.010). Spectroscopically these stars show He II lines in absorption with emission cores, and absorption lines of C IV and other heavy elements including oxygen are present. See Green and Liebert (27.126.039), Liebert, Green, and Wesemael (33.120.018), and Sion, Liebert, and Starrfield (1984 Ap.J., in press). The absence of surface hydrogen, coupled with high surface gravities, implies an advanced evolutionary stage for these pulsating stars. Their approximate location in the H-R diagram between the planetary nebulae regime and the white dwarfs suggests that these PG stars are most likely rapidly evolving and cooling down into the white dwarf region. K1-16 is probably evolving to become hotter before it reaches this cooling stage.

The observational and theoretical situation for these stars and their relation to the cooler white dwarfs has been briefly reviewed twice by Cox (1984, Nature, 303, 752 and 1984, Nature, 111, 509).

The interior structure of pre-white dwarf stars may be inferred from what is known of their possible progenitors and their descendents, the white dwarfs. One attractive possibility is that the progenitors of these PG stars are the nuclei of planetary nebulae, just as K1-16 is observed to be. These nuclei stars have been studied from a stellar evolution standpoint recently by Iben (1984 Ap.J., 277, 333), Iben and Tutukov (1984 Ap.J., 282, 615), Kovetz and Harpaz (29.065.010), and Schonberner (30.065.067, 30.065.037). They are assumed to be the hot cores of asymptotic branch stars which have ejected most of their hydrogen rich envelopes as red giants and during the planetary nebula stage. By altering the amount of hydrogen and helium rich material remaining in the outer layers following the nebula ejection, Iben (1984 Ap.J., 277, 333) demonstrated a variety of evolutionary possibilities. His models go through various phases of shell helium and hydrogen burning before reaching the pre-white dwarf configuration. Schonberner has also followed the evolution of these late asymptotic giant branch stars by including a Reimers-like stellar wind. A superwind phase seems to be necessary to remove enough envelope material to place the model on the horizontal planetary nebula nucleus track. The Schonberner models also experience episodes of shell burning following nebular ejection while evolving through the planetary nebula stage to the pre-white dwarf region.

If the pulsating PG stars are hot degenerates, then the observed pulsation periods are much longer than the expected radial mode periods of 20-50 seconds. In an early study, Starrfield, Cox, and Hodson (28.122.200) found that models at 10,000 $L_\odot$ were unstable to radial oscillations for very specific circumstances. Later observations have indicated higher surface gravities, implying a lower luminosity of only 100 $L_\odot$. This, along with the multiperiodic nature of the light curve, strongly suggests that these stars are undergoing nonradial g mode pulsations such as are seen in the pulsating white dwarfs, in many other classes of stars, and even in our sun. Nonadiabatic analyses of the PG stars on the cooling track, and of K1-16 on the planetary nebula nucleus track, have been made by Starrfield, Cox, Hodson, and Pesnell (33.126.023), Starrfield, Cox, Kidman, and Pesnell (1984 Ap.J., 281, 800), and Starrfield, Cox, Kidman, and Pesnell (Ap.J., submitted). These studies show that static stellar envelopes constructed along the evolutionary tracks display unstable nonradial modes due to the cyclical partial ionization of carbon and oxygen. Too much surface helium will dilute the κ and γ effects of the CO ionization and stabilize at least the cooler models of these stars.

McGraw, Starrfield, Liebert, and Green (27.126.070) pointed out that, if PG1159-035 is indeed a pre-white dwarf star, then its rapid evolutionary changes in structure might be indirectly observable through observations of secular period changes. This suggestion was elaborated upon by Winget, Hansen, and Van Horn (33.065.063) who estimated that these stars should exhibit period decreases with an e-folding time of about a million years. This means that in only a few years such changes could be observed. These authors made a rough estimate of the period decrease. Now Winget, Kepler, Robinson, Nather, and O'Donoghue (Ap.J., preprint) have measured such a period decrease of -1.2 ± 10^{-11} sec/sec for the dominant (516 second) period of PG1159-035. The best current theoretical predictions by Kawaler, Hansen, and Winget (Ap.J., preprint) show period increases of the same order of magnitude as the observed value. They suggest that rotational spinup from contraction can provide an additional negative contribution to the theoretically derived period change. Reasonable rotation rates can account for the observations. These issues are at the current frontier of stellar evolution at the dying stages of stars.

VII. WHITE DWARFS (V. Weidemann, University of Kiel)

The second edition of the "Catalogue of Spectroscopically Identified White Dwarfs" has now been published by McCook and Sion (Villanova Press, 1984) and contains relevant information for about 1500 stars - more than tripled compared to the earlier edition, essentially due to the addition of objects from the Palomar Green North Galactic Pole Survey of hot white dwarfs by R. Green, M. Schmidt and J. Liebert which have been spectroscopically identified as degenerates mainly by J. L. Greenstein at the 5 m Palomar telescope. For about 500 stars multichannel spectrophotometry is available, the results of which are summarized by J. L. Greenstein (Ap. J., 276, 602, 1984) who has also investigated correlations to Strömgren colors (P.A.S.P., 96, 62, 1984). Strömgren photometry has been obtained for 182 stars by G. Wegner (33.126.006) the results are also found in the McCook - Sion Catalogue. The latter contains valuable enlarged cross-references lists as well as a section on stars which have been shown not to be white dwarfs but subdwarfs. The stars in the Catalogue are classified according to the new system proposed by Sion et al. (33.126.030) which subdivides DA stars according to temperatures, assigns DZ to white dwarfs with metal lines (instead of the former types DF and DG) and DQ to those which show carbon features in their spectra (former DC/C_2, λ4670 types). Improved spectroscopy has enabled reclassification of many white dwarfs formerly considered to be of type DC (G. Wegner, 30.126.019, 31.126.012, 34.126.003) but now showing faint features of H_α, HeI, C_2 or broad unidentified shallow absorption. Results of the proper motion survey with the 48" Palomar Schmidt telescope on white dwarf statistics in the NLTT catalogue have been published by Luyten (University of Minnesota, 1982) whereas photometric investigations of faint blue stars in the Giclas lists were made by Rupprecht and Bues (32.126.030) and astrometric studies by Borgmann and Lippincott (33.111.004). A blue object survey was carried out by Ishida et al. (32.126.021) to a limiting magnitude of m = 18 for a meridian belt through the galactic poles and center; for m < 16 more than 50% are white dwarfs. Ishida has also estimated the space number density of white dwarfs (32.126.021). General information can be found in the new (1982) edition of Landolt-Börnstein, where the section on white dwarfs has been prepared by Weidemann (31.003.064).

UV spectroscopy of the most common spectral type, DA, with hydrogen-atmospheres has revealed the unexplained absorption feature at 1400 Å in additional stars (Wegner, 32.126.020, Sion et al., Ap. J., 279, 758, 1984) as well as a feature at 1600 Å both of which have now been ascribed to quasimolecular H_2^+ - or

H_2 - absorption (Koester et al., Wegner, 1984 preprints). Atmospheric parameters have been evaluated by Guseinov et al. (34.126.013); however, their conclusions concerning a wider mass distribution have shown to be incorrect by Koester (Astr. Space Sci., 100, 471, 1984). An improved analysis, using multichannel observations and a least squares technique in the temperature range in which the energy distribution is quite sensitive to surface gravity, 8000 K $\leq T_{eff} \leq$ 16000 K has yielded a mass distribution for 70 DA stars which is remarkably narrow and shows the steep increase at 0.45 $M_\odot$ which is expected from calculations of stellar and galactic evolution (Weidemann and Koester, Astron. Astrophys., 132, 195, 1984). Resulting constraints on evolutionary schemes, especially on the initial-final mass relation have been discussed by Weidemann and Koester (33.065.055). The initial-final mass relation can also be determined emperically by observation of white dwarfs in open clusters with known turn-off mass. Reimers and Koester (32.153.030) succeeded in finding white dwarfs in NGC 2516 whose spectroscopically derived masses are higher than 1 $M_\odot$ with progenitors at least up to 8 $M_\odot$. The newly determined initial-final mass relation thus demonstrates that the intermediate mass single stars do not become carbon detonation supernovae. However, Weidemann (Astron. Astrophys., 134, L1, 1984) has pointed to evidence for differential mass loss leading to the replacement of a single-valued initial-final mass relation by a band of finite width. The absence of luminous red giants in younger Magellanic cloud clusters like NGC 1866 is also evidence of tremendous mass loss for intermediate mass stars, and thus confirms the white dwarf results. An earlier study of white dwarfs in clusters by Anthony-Twarog (32.153.006) puts the upper limit for white dwarf production at 5 $M_\odot$. The newer results and the ensuing constraints are summarized by Weidemann (IAU Symposium 105, "Observational Test of the Stellar Evolution Theory", Eds., A. Maeder, A. Renzini, Reidel, 1984).

The white dwarfs of spectral type DB, with HeI spectra and helium-rich atmospheres have also been restudied in order to improve the determination of atmospheric parameters. Whereas Koester et al. (30.126.023) from spectroscopy and photometry of 25 DB stars and a new set of model atmospheres derive a mean mass of 0.44 $M_\odot$, smaller than that of the DA stars, 0.58 $M_\odot$, Wickramasinghe and Reid (33.126.013) evaluate high resolution line spectra and find log g = 8, corresponding to a mean mass of 0.58 $M_\odot$. Reobservation of all northern DB stars with the multichannel spectrophotometer, and theoretical evaluation by Oke et al. (Ap. J., 281, 276, 1984) have confirmed the higher average mass; however, only if the new calibration of Oke and Gunn (33.113.017) is used. In that case the DB stars delineate for the first time a narrow cooling sequence, too, with a mass range M = 0.55 + 0.1 $M_\odot$. However, on the Oke-Gunn calibration scale, the mean DA mass would be increased to M = 0.62 $M_\odot$, a small but not insignificant difference compared to the mean DB mass. Guseinov et al. (34.126.045) conclude that 70% of the He-rich white dwarfs have M < 0.55 $M_\odot$.

It thus cannot yet be established if DA and DB stars arise from different progenitors, or if the difference in atmospheric composition are due to minor differences in late pre-white dwarf stages of evolution. The birth rate of DB white dwarfs is between 10 and 30% of the total and much higher than the death rate of known He-rich progenitors. Iben (Ap. J., 277, 373, 1984) assumes that the phase in its nuclear burning cycle when a star leaves the asymptotic giant branch determines the final outcome: DB or DA, and estimates that 25% of all central stars of planetary nebulae will fall into groups that become DB stars. This scheme implies that 50% of the observable bright central stars are helium-burners opposite to the conclusions of Schönberner (30.065.067) who finds that the vast majority are hydrogen-burners.

Major progress has been made by further detections of CI lines in the UV spectra of DC and DQ (former C_2 or λ4670) stars between 12000 and 6000 K (Vauclair et al. 30.126.002, 31.126.029, Wegner 30.126.007, 33.126.021, 34.126.012). Atmospheric parameters and abundances have been determined for these stars by

Koester et al. (32.126.029). Carbon abundances are fairly low, of the order of C : He = 10^{-6} to 10^{-7} for DC stars and go up to 2 10^{-3} for DQ stars, i.e., those which show C_2 bands or, at higher temperatures also CI lines in the visible like G268-40 (Koester and Weidemann, 31.126.023). The appearance of small amounts of carbon is explained by upwards diffusion and dredge-up by convection, a mechanism proposed by Koester (32.126.029) and subsequently elaborated by Fontaine et al. (Ap. J., 277, L61, 1984). It is predicted that C abundances reach a maximum near T_{eff} = 10000 K, and fall off towards higher and lower temperatures. The absence of CI UV-lines in IUE spectra of the DB stars (Wickramasinghe, 33.126.014) confirms this picture.

However, the appearance of metal lines in white dwarfs of spectral type DZ and the non-systematic variation of the derived metal abundances still constitutes a puzzle. The hottest DZ, GD40, at T_{eff} = 14000 K has Ca, Mg, and Fe abundances of the order of 10^{-8} (Shipman and Greenstein, 33.126.007) but the relative abundances are such as to rule out a simple diffusion explanation. An analysis of the cooler DZ (former DF) stars L119-34 (9300 K) and L745-46 A (7500 K) by Zeidler (5th European Workshop on White Dwarfs, Kiel, June 1984) yields also abundances of Si/He(10^{-8} -10^{-9}) and C (10^{-6} - 10^{-7}). Accretion of interstellar grains (Wesemael and Truran, 32.064.027) and diffusion might explain the varieties in cool helium rich atmospheres. The former DA, F stars, hydrogen-rich with CaII lines have, however, all been shown to be spurious, with only one exception (Lacombe et al., 34.126.001).

The problem of diffusion in white dwarfs has been recently reinvestigated by Muchmore (Ap. J., 278, 769, 1984) who finds with more careful calculations of transport coefficients that diffusion time scales are longer, but remain still much shorter than cooling times: he concludes that trace metals must be transient. In hot white dwarfs, on the other hand, metals may be kept in the atmosphere by radiative levitation as indicated by the presence of SiII, SiIII, SiIV, CIV, and NV lines in high resolution IUE spectra for three out of seven DA stars (Bruhweiler and Kondo, 33.126.033). There are several hotter white dwarfs of mixed spectral type: DO stars with H and He, and GD323, formerly assigned DB, but with the detection of broad shallow Blamer lines now considered a white dwarf with a stratified atmosphere (Liebert et al., 34.126.016).

It has become clear, however, that stratified envelopes do not have sharp boundaries, but that diffusion tails extend into the neighbouring layers. A very interesting mechanism has been suggested in order to explain the absence of hydrogen in He-rich white dwarf atmospheres: diffusion-induced H-burning by the CNO cycle at an intermediate depth, where downward diffused hydrogen meets upward diffused carbon (Michaud et al., Ap. J., 278, 769, 1984). This mechanism might also reduce the thickness of H-envelopes from the 10^{-4} - 10^{-5} $M_\odot$ left when the pre-white dwarf leaves the asymptotic giant brach to the less than 10^{-7} $M_\odot$ which a ZZ Ceti pulsator must have in order to show instability (Fontaine and Michaud, 5th European Workshop in White Dwarfs, Kiel, June 1984).

ZZ Ceti variables, pulsating DA stars, have been studied extensively, with several new discoveries made (McGraw, Fontaine, 30.126.025, Vauclair et al., 30.122.104, Nather, McGraw, 31.126.011, Schiffer et al., 31.122.153, Donoghue, Warner 32.126.006, Dolez et al., 33.126.009, Kovacz 33.126.035, Kepler et al., 30.126.009, Kepler, Ap. J., 278, 754, 1984). The multicolor, radial velocity and line profile variations are consistent with g-mode pulsation (Robinson et al., 32.126.004). From the statistics of color distribution Fontaine et al. (32.126.002) conclude that probably all DA white dwarfs become ZZ Ceti variables on entering the instability strip 13000 > T_{eff} > 11000 K. Greenstein (32.126.003) derives temperatures from multichannel observations and finds the edges to be at 11700 and 10700 K, whereas Weidemann and Koester (Astron. Astrophys., 132, 195, 1984) find most of the ZZ Cetis between 12500 and 11200 K or 500 K cooler, depending on the calibration used. The surface gravity and mass range is typical for DA stars in general, and the fraction of variable DA stars varies from 50 to 70% depending on the temperature range considered.

A theoretical linear nonadiabatic analysis of g-modes by Dolez and Vauclair (20.126.024) shows that the location of the instability strip depends on the thickness of the remaining hydrogen layer, if this is thin, the driving occurs in the helium partial ionization zone and the strip lies between 13500 and 11500 K, if the layers are thicker, the driving is caused by hydrogen ionization and the strip is located between 11500 and 10000 K. On this basis Brickhill (34.126.005) computed flux variations of finite amplitude and was able to explain the observational red edge. Winget et al. (31.126.004), from full nonadiabatic g-mode calculations, find that the blue edge is at about 11000 K for hydrogen-driving but depends on the remnant hydrogen-layer as well as on the thickness of the transition zone. For an observed edge at 12000 K they find remnant hydrogen masses of $10^{-12} < M_H/M_* < 10^{-8}$. If the hydrogen layer goes to zero instability due to helium ionization is predicted to occur at 19000 K. When indeed a pulsating DB star was found, GD358 (Winget et al., 32.126.023), a corresponding analysis by Winget et al. (33.126.024) showed that the location of the helium instability strip is extremely sensitive to the assumed efficiency of convection and might be between 16000 and 19000 K or between 26000 and 29000 K. The temperature of GD 358 derived from IUE observations (Koester et al., 34.126.002) is 26000 ± 2000 K. In the meantime more pulsating DB stars have been found (Winget et al., Ap. J., 279, L15, 1984, 5th European Workshop on White Dwarfs, Kiel, June 1984) as well as null results reported (Robinson, Winget 34.126.007). Still more exciting was the discovery of a new class of very hot pulsating white dwarfs, called after the first object, PG 1159-035 stars (Winget et al., 33.065.063, Bond et al., Ap. J., 279, 751, 1984) which also includes a nucleus of a planetary nebula, K1-16. These stars show HeII, CIV, and OVI lines partly in emsssion and seem to constitute a link between OVI central stars of PN and white dwarfs. It is the first time that oxygen has been observed in white dwarfs (Sion et al., 5th European Workshop on White Dwarfs, Kiel, June 1984). If these stars pulsate by carbon/oxygen partial ionization, as proposed by Starrfield et al. (33.126.023, and Ap.J., 281, 800, 1984), they need significant amounts of oxygen near the surface which would imply that helium-burning in evolved stars produces much more O than normally assumed. Winget et al. (33.065.063), furthermore show that for these objects the periods should change measurably during a few years and thus constitute a strong testing ground of theoretical models of pre-white dwarf evolution. It must be pointed out that in these hot stages the stars are not yet fully degenerate and have radii which are considerably larger and surface gravities which are lower than those of zero-temperature degenerate configurations (Koester, Schönberner, 5th Europ. Workshop on White Dwarfs, Kiel, June 1984).

Hot white dwarfs with $T_{eff} > 50000$ K have been observed in the visible (Liebert et al., 30.126.018), UV (Sion et al., 31.126.014, Dupree and Raymond 32.126.035) EUV (Malina et al., 32.126.024) and soft X-ray region (Cole et al., 30.126.030, Kahn et al., Ap. J., 278, 255, 1984). Sion and Guinan (33.126.005) determine the Einstein redshift of HD 149499 B, a 55000 K DO star from high resolution IUE spectra and find $M = 0.5 + 0.1\ M_\odot$ for this DB progenitor. For Sirius B soft X-ray observations by Martin et al. (32.126.019) and EUV observations by Holberg et al. (Ap. J, 280, 679, 1984) have shown that the origin of the radiation is probably photospheric even if the effective temperature is < 28000 K, as is now well established. The He abundance must be smaller than 10^{-5}. Wesemael (30.126.001) checked that electron scattering is not important for the EUV in hot DA atmospheres.

A survey of magnetic fields in white dwarfs, summarizing research up to 1981 has been given by Angel, Borra and Landstreet (30.126.003). Linear polarization has been observed by Efimov in about 50% of a sample of 85 white dwarfs and subdwarfs

(30.126.026). New observations were reported by Liebert et al. (33.126.001) for two hot DA stars with fields around 1 Megagauss, and by Downes and Margon (34.126.006), where Zeeman splitting indicates 13 Mgauss. The influence of magnetic fields on the spectra has been theoretically investigated by Martin and Wickramasinghe (32.126.009, M.N.R.A.S., 206, 407, 1984) and by Bues (33.126.012), 34.126.036). Spectacular progress has been made in the interpretation of spectra with unidentified features. The ultraviolet line at 1345 Å observed by Greenstein and Oke (31.126.002) was identified by Greenstein (Ap. J., 281, L47, 1984) as a shifted Lyman α component whose location becomes stationary around 350 Mgauss. This interpretation was possible on account of theoretical investigations of hydrogen atoms in very strong magnetic fields by research groups at Baton Rouge (Henry, O'Connell, Ap. J., 282, L97, 1984) and Tübingen (Rosner et al., Forster et al., J. Phys. B.: Atom. Mol. Phys., 17, 29, and 1301, 1984) which meanwhile were able to identify also the famous Minkowski band at λ4135 Å as due to a normally forbidden 2s0-4f0 transition of H_β as well as the 5855 Å band as a stationary component of Hα, both indicate fields around 300 Mgauss.

Information about cool white dwarfs remains relatively poor. From IR photometry Wickramasinghe et al. (31.126.006) conclude that three objects are hydrogen-rich, perhaps due to accretion not being prohibited below 5500 K, whereas Liebert and Dahn (33.126.031) suggest pressure-broadened and shifted C_2 bands as an explanation for the spectrum of LHS 1126. Greenstein (33.126.015) investigates 4 cool objects for which the entire flux is measured, Liebert and Wehrse (33.126.038) show that Ross 627 is a cool hydrogen-rich star whereas Wegner and Yackovich analyze the cool DC binary Stein 2051 B (34.126.046).

That the deficiency of low luminosity white dwarfs seems to be real and not due to observational selection has been shown by Liebert et al. (34.126.041) which study the luminosity function for over 40 cool degenerates with $M_v > 14$. The falloff at $M_v > 15$ is also confirmed from searches for binary companions. It is not yet settled if this indicates earlier accelerated Debye cooling or a late start of white dwarf formation in the Galaxy: in any case the cooling theory would have to be revised. Cooling ages could for example be increased by a phase separation of carbon and oxygen in the interior [Mochkovitch (33.126.027)].

VII. CONCLUDING REMARKS

In addition to the above detailed reviews, there are several other aspects of stellar astrophysics that deserve a very short mention because of recent progress. For more information on these topics, one must consult Astronomy and Astrophysics Abstracts.

Mixing in stellar interiors continues to be an important question that has consequences for the evolution of a star, especially if nuclear fuel is brought to a hot burning region. Nuclear reactions for the burning of carbon to oxygen and those for neutron capture by the r process have recently been revised. Observations of the radio band emission from stars continue to provide information about stellar surfaces. Magnetic stars, observed for years, are now being used for interpretations of them, in particular, the pulsating Ap stars. How the solar dynamo operates to give its surface magnetic fields has resulted from recent elaborate calculations. Evolution of stars in binary systems has been studied extensively because close binaries can and do have very important effects in significantly changing how evolution occurs. Considerable mass exchange or loss can occur to produce some observed astronomical objects. Self-excited radial and nonradial pulsations have been observed and interpreted for stars all over the H-R diagram, and information about masses, radii, luminosities, and compositions have resulted. Investigations of stellar explosions for the surface thermonuclear runaways of novae and the core collapse event for supernovae have still left unsettled details concerning these violent events. Neutron stars that result from the deaths of stars larger than the maximum white dwarf mass, have still unknown interior structures, and their cooling rates are still being calculated. The entire field of black hole research has brought together the disciplines of stellar structure and relativity theory, yielding new insights to both. Finally there is the question of the ages of the oldest stars as discussed in the last Commission 35 report. If the age of the universe is just over 10 billion years, as some now strongly suggest, how do we reconcile this short time with the apparently older population II stars?

With all the questions brought up in the detailed reports above, and those mentioned briefly here, as well as the discussions scheduled for several upcoming IAU Symposia and Colloquia, one can see that Commission 35 has many further research projects and a very bright future.

Arthur N. Cox
President of the Commission

COMMISSION 36: THEORY OF STELLAR ATMOSPHERES (THEORIE DES ATMOSPHERES STELLAIRES)

PRESIDENT: B. Gustafsson VICE PRESIDENT: K. Kodaira

Organizing Committee: D.F. Gray, A.G. Hearn, W. Kalkofen, R.P. Kudritski, D. Mihalas, A.A. Sapar, M.J. Seaton, G. Traving, A.B. Underhill

I. COMMISSION ACTIVITY, SYMPOSIA

Commission 36 acts as a sponsor or co-sponsor at the following symposia and colloquia: IAU Colloquium No. 90 "Upper Main Sequence Stars with Anomalous Abundances", Crimea, USSR (May 1985), IAU Colloquium No. 89 "Radiation Hydrodynamics in Stars and Compact Objects", Copenhagen, Denmark (June 1985), IAU Symposium No. 120 "Astrochemistry", Goa, India (December 1985), IAU Colloquium No. 87 "Hydrogen Deficient Stars and Related Objects", Bangalore, India (December 1985).

The commission participates jointly with several other commissions in the organization of Joint Discussions at the XIXth General Assembly on the topics "Stellar Activity: Rotation and Magnetic Fields" and "Solar and Stellar Non-Radial Oscillations".

II. RECENT PROGRESS IN THE THEORY OF STELLAR ATMOSPHERES

As in earlier Reports on Astronomy, this report is not intended to be a comprehensive review of all work done in the field covered by our commission, nor does it contain a complete bibliography. (However, in Section III, below, an attempt has been made to list monographs, proceedings from symposia etc. of particular relevance for the activity of our commission.) Rather, the following pages focuses on a few areas in which significant progress has been made in the last three years. These areas represent different lines of direction that the study of stellar atmospheres now follows - the invention of new efficient numerical methods, the attempts to relate the atmospheric phenomena to global properties such as the rotation of the stars, the study of complex dynamical systems where mass flows are decisive such as Wolf-Rayet stars, the application of rather classical model-atmosphere techniques for more "exotic" stellar photospheres such as those of white dwarfs, and the use of stellar-atmosphere theory in the study of non-stellar objects, such as accretion discs.

It is obvious to any reader that new observations in new wavelength regions, or at higher spectral, temporal or spatial resolution, is of key importance today for the study of stellar atmospheres. In particular, one should here emphasize the fundamental value of solar studies. It is also clear that the development of new generations of fast computers has led and will lead to impressive progress in the theoretical simulation of stellar atmospheres and of phenomena in them. However, one must strongly stress the need for a deeper physical *understanding* of the interplay between the different complex phenomena that in interaction form the stellar atmospheres. An inquiry to some experienced astronomers in this field led to suggestions as to what problems are yet unsolved but would be worth attacking: The behaviour of radiation-pressure dominated plasmas; structure and physics of magnetized atmospheres, including interaction with motions in deeper layers; the problems of mass loss, coronae and extended layers; effects of dust and polyatomic molecules in cool stars; problems connected with partial redistribution; physical and numerical instabilities; atmospheres of accretion discs. The list could well be much longer - in fact, most problems beyond the classical treatment of plane-parallel atmospheres are still unsolved. Hopefully, it will be possible in future Reports of Astronomy to report on further significant progress in these respects.

1. Numerical Methods (Wolfgang Kalkofen)

Introduction. Much progress has been made in numerical radiative transfer during the last several years; the subject has been treated recently in some detail in a book entitled *"Methods in Radiative Transfer"*, published by Cambridge University Press, and in a workshop held in Trieste, for which the proceedings are to appear shortly by Reidel.

The principal numerical methods to emerge that are suitable for investigations of time-dependent media in which radiative transfer as well as gas dynamics play a role are based either on the probability of photon escape from a medium or on the perturbation of an operator. Both will be reviewed here. For the older methods which are suitable mainly for static media the reader is referred to these publications, as he is for two new ones that show great promise: one, by Anderson (1985), is based on a treatment of the frequency dependence of the radiation field in analogy to the variable Eddington factor for the angle dependence; it results in a significant reduction of the order of the system of coupled equations that must be solved; and the other, by Wehrse (1985), is a generalization of the formal integral of the scalar transfer equation to that of a matrix transfer equation.

This report reviews mainly the escape probability and operator perturbation methods, their derivation and the underlying principle; for applications the interested reader can be guided by the references to the basic papers. In the escape probability method (reviewed recently by Rybicki 1984, 1985), the mathematical description of the transfer process is much simplified; this results in a major economy, but also in errors over which one has only very limited control. In the operator perturbation method (reviewed by Kalkofen 1984, 1985a), the solution of the complete set of equations is speeded up by the use of an approximate integral operator in the calculation of correction terms; this still permits an accurate solution of the problem, but at the price of iterations.

The Escape Probability Method. In the Escape Probability Method one can distinguish between first- and second-order methods. In the former, the global relation between the radiation field and the source function which is ordinarily expressed either by the differential equation of transfer or its formal integral, $J(\tau) = \int K(\tau,\tau')S(\tau')d\tau'$, is replaced by a local relation, $J(\tau) = S(\tau) \int K(\tau,\tau')d\tau'$, for the mean integrated intensity, J, in a line or a continuum for a finite or semi-infinite plane-parallel atmosphere. This equation is obtained from the formal integral by assuming that the scale of variation of the source function, S, is large compared to that of the integral kernel, K. If this condition is satisfied the formal integral can be written in terms of the escape probability from the medium, $P(\tau) = 1 - \int K(\tau,\tau')d\tau'$, giving $J(\tau) = [1 - P(\tau)]S(\tau)$. Combining this relation with the equations of statistical equilibrium in the form of the source function equation, $S(\tau) = (1 - \varepsilon)J(\tau) + \varepsilon B$ (cf. Athay 1976), results in a local relation for the source function, $S(\tau) = \varepsilon B/[\varepsilon + (1-\varepsilon)P(\tau)]$. Thus, the source function and the mean integrated intensity are known throughout the atmosphere when the escape probability $P(\tau)$ is known; and $P(\tau)$ can be determined very rapidly by the use of asymptotic relations (cf. Canfield et al. 1984 who give asymptotic expressions for $P(\tau)$ for lines broadened by the Doppler, Voigt, Lorentz, or Stark effects, and for bound-free continua; cf. Ivanov 1973 for an approach to obtaining asymptotic results).

The first-order escape probability method has fair accuracy deep inside the medium (a factor of two or three for the net rate coefficient $\rho = 1 - J/S$ and hence for the energy loss rate from large depth). But even when the collision parameter ε and the Planck function B are constant the solution carries a large relative error near the surface of a medium (of order $2\sqrt{\varepsilon}$ for the source function of a line transition with complete redistribution in a semi-infinite static atmosphere), especially when scattering is important (i.e., when $\varepsilon \ll 1$). A modification by Frisch (1984) improves the solution near the surface and extends the method to the case of scattering by resonance lines described by the R_{II} redistribution function. - In contrast to static media, the first-order escape probability method works well in the Sobolev limit of a high velocity gradient. - A typical application of the method (for a static medium) is the discussion by Kwan and Krolik (1981) of the formation of emission lines in quasars.

The accuracy is improved significantly with the second-order escape probability method, often referred to as the probabilistic method. It originated in a suggestion by Athay (1972) to describe the radiative transfer in a spectral line by means of a first-order differential equation for the mean integrated intensity, with the escape probability as the independent variable. A rigorous derivation of this equation was given by Frisch and Frisch (1975); a derivation stating clearly the mathematical assumptions was given by Canfield et al. (1984); and one making explicit the physical assumptions, by Scharmer (1981, 1984).

A first-order differential equation, the probabilistic equation, describes the transport of energy and hence information in one direction only. Thus, in the normal course of the solution from the inside in the outward direction, the structure of the outer layers of a medium cannot influence the state of the gas in the deeper layers; a major shortcoming. Other assumptions that must be satisfied for the equation to be valid are that the profile function for the line absorptions be depth-independent, that the line be formed in complete redistribution (CRD), that differential velocities be negligible, and that the background continuum be weak. When these assumptions are satisfied, the equation yields the exact solution at the outer surface of a semi-infinite medium, a distinct improvement over the first-order method. Although the probabilistic equation appears to yield a fair accuracy throughout an atmosphere when these conditions are satisfied (a typical value is 20%), the problems that call for such fast methods tend to be those where these conditions are not met, resulting in larger errors that are difficult to estimate. A further drawback is that the method does not provide for a way to improve the numerical result; another method must be used if a higher accuracy is desired. - A modification of the probabilistic equation by Canfield et al. (1984) separates the equation into two first-order differential equations, for the inward and outward-directed mean integrated intensities, respectively. This allows the specification of two boundary conditions for the incident intensities in a finite medium.

The Operator Perturbation Method. The Operator Perturbation Method for the solution of the differential equation of radiative transfer subject to integral constraints was originally proposed by Cannon (1973). He applied it to partial redistribution (PRD) problems, perturbing the PRD operator about the isotropic CRD operator; to media with differential flow velocities, perturbing about the static case; and to media with spherical symmetry, perturbing about the plane-parallel case (for references to these and to related papers, cf. Cannon 1984).

Scharmer (1981, 1984) and Scharmer and Nordlund (1982) have expressed the operator perturbation in terms of integral equations. In their formulation, corrections to a solution are based on the error incurred in the conservation equation; the reference to a conserved quantity makes this powerful approach suitable also for non-linear problems.

The essence of the method is the separation of the calculation into two parts: the *approximate* calculation of corrections to a solution using an approximate integral operator, and the *accurate* calculation of the error with which the solution satisfies the conservation equation. The accuracy of the converged solution depends only on the accuracy of the error calculation.

In the integral equation formulation, the transfer problem is expressed in the form $\mathsf{L}S = \varphi$, where L is a linear integral operator, S is the source function or another variable for the state of the gas that is to be determined, and φ is the known inhomogeneous term of the problem; this equation may describe the line transfer of a two-level atom with complete (Cannon 1984, Scharmer 1981, 1984) or partial (Scharmer 1983) redistribution, a multi-level atom (Scharmer and Carlsson 1984), or a model atmosphere (Nordlund 1984, Kalkofen 1985b); and both the linear and the non-linear cases. The procedure by which the integral equation is solved is to expand the unknown function in a series $S^{(n)} = S + \Sigma_{i=1}^{n} s^{(i)}$, $S = S^{(\infty)}$, where the initial estimate S is either the converged solution of the preceding time step in a time-dependent problem, or an estimate, usually S = 0, in a static problem. Given the solution in the nth order, the solution in the $(n+1)^{th}$ order is obtained from the equation $L' s^{(n+1)} = \epsilon^{(n)}$, where L' is an approximate integral operator (approximating L) and $\epsilon^{(n)}$ is the error made by the nth order solution in satisfying the conservation equation, $\epsilon^{(n)} = \varphi - LS^{(n)}$.

The speed of the operator perturbation method, giving it a distinct time advantage over the *direct* solution of the equations, depends on the speed with which the approximate integral operator L' can be constructed and with which the set of coupled equations for the corrections $s^{(n)}$ can be solved. Now, the operator L' can be constructed very rapidly with Scharmer's (1981, 1984) one-point quadrature formula for the intensity in terms of the source function, i.e., essentially the Eddington-Barbier relation applied to the interior of a medium. This results in a matrix of nearly triangular structure, leading to fast solution times (proportional to N^2, not N^3 as ordinarily, where N is the number of depth points for a single variable S, or the product of the number of depth points and the number of levels in a multi-level atom) for the corrections $s^{(n)}$ by means of an $l \times u$ decomposition of the matrix L' ; this decomposition needs to be carried out only once in each interation cycle. - The error term $\epsilon^{(n)}$ is computed by solving the (scalar) differential equation of transfer for known source function S either as the Feautrier equation with second-order accuracy, or in Auer's (1976, 1984) formulation with fourth-order accuracy. Thus the approximate integral operator L' is the only matrix that needs to be constructed.

In line transfer, the large thermalization length due to scattering is contained in the approximate integral operator L'; the error calculation, since it solves the transfer equation for known source function separately for each angle and frequency point, contains only the short-range behavior of the monochromatic transfer equation (Kalkofen 1985a). The long-range behavior must be preserved when the matrix L' is simplified in order to speed up the solution. - The operator perturbation method can be adapted to computers with short word length by cancelling analytically intensity-dependent terms in the equations of statistical equilibrium (Scharmer 1984, Scharmer and Carlsson 1984), allowing the treatment of line transfer problems with very small scattering parameter ε.

The Core Saturation Method. One other approximate method that should be mentioned here is the Core Saturation Method (Rybicki 1972). It is not very fast since it may require a large number of iterations for convergence, but the essential physics on which it is based can be used to advantage in reducing the construction time for the approximate integral operator L' in the operator perturbation method (Kalkofen 1985a). In the core saturation method, the (scalar) transfer equation is solved by means of a Λ-iteration for known source function in the shallow layers, where the monochromatic optical depth is smaller than some value of order unity (for a modification that allows for structure inside the medium such as shocks, cf. Kalkofen and Ulmschneider 1984); at larger optical depth, the intensity is usually assumed to have the same value as the source function. The core saturation assumption permits a very general approach to complex problems; it has been used for media with cylindrical symmetry (Stenholm and Stenflo 1977), the transfer of polarized radiation (Stenholm and Stenflo 1978), and the co-moving frame description of the transfer in a moving medium (Stenholm 1980). The method can also be modified to yield the exact solution of a transfer problem (Rybicki 1984).

References

Anderson, L.: 1985, in *Trieste Workshop on Progress in Spectral Line Formation Theory*, (henceforth *TRIESTE*), D. Reidel Publ. Co., Boston.

Athay, R.G.: 1972, *Astrophys. J. 176*, 659.
Athay, R.G.: 1976, *The Solar Chromosphere and Corona: Quiet Sun*, D. Reidel Publ. Co., Dordrecht.
Auer, L.H.: 1976, *J. Quant. Spectroscopy Radiative Transfer 16*, 931.
Auer, L.H.: 1984, *Methods in Radiative Transfer* (henceforth *MRT*), Cambridge University Press, Cambridge, ed. W. Kalkofen, p. 79.
Canfield, R.C., McClymont, A.N. and Puetter, R.C.: 1984, *MRT*, p. 101.
Cannon, C.J.: 1973, *J. Quant. Spectroscopy Radiative Transfer 13*, 627.
Cannon, C.J.: 1984, *MRT*, p. 157.
Frisch, H.: 1984, *MRT*, p. 79.
Frisch, U. and Frisch, H.: 1975, *Monthly Notices Roy. Astron. Soc. 173*, 167.
Ivanov, V.V.: 1973, *Transfer of Radiation in Spectral Lines*, NBS SP-385 (Washington, D.C.: US Dept. of Commerce).
Kalkofen, W.: 1984, *MRT*, p. 427.
Kalkofen, W.: 1985a, *TRIESTE*.
Kalkofen, W.: 1985b, *TRIESTE*.
Kalkofen, W. and Ulmschneider, P.: 1984, *MRT*, p. 131.
Kwan, J. and Krolik, J.H.: 1981, *Astrophys. J. 250*, 478.
Nordlund, Å.: 1984, *MRT*, p. 211.
Rybicki, G.B.: 1972, in *Line Formation in the Presence of Magnetic Fields*, ed. R.G. Athay, L.L. House and G. Newkirk Jr. (Boulder: High Altitude Observatory), p. 145.
Rybicki, G.B.: 1984, *MRT*, p. 21.
Rybicki, G.B.: 1985, *TRIESTE*.
Scharmer, G.B.: 1981, *Astrophys. J. 249*, 720.
Scharmer, G.B.: 1983, *Astron. Astrophys. 117*, 83.
Scharmer, G.B.: 1984, *MRT*, p. 173.
Scharmer, G.B. and Carlsson, M.: 1984, *J. Comp. Phys.*, in press.
Scharmer, G.B. and Nordlund, Å.: 1982, *Stockholm Obs. Rep. 19*.
Stenholm, L.G.: 1980, *Astron. Astrophys. Suppl. 42*, 23.
Stenholm, L.G. and Stenflo, J.O.: 1977, *Astron. Astrophys. 58*, 273.
Stenholm, L.G. and Stenflo, J.O.: 1978, *Astron. Astrophys. 67*, 33.
Wehrse, R.: 1985, *TRIESTE*.

2. Manifestations of the Interaction between Convection and Rotation in Stellar Atmospheres (Lee Hartmann)

Introduction. According to dynamo theories, the interaction between convection and rotation is responsible for the generation of magnetic fields in solar-type stars. Stellar observations can explore the ways the global parameters, like the surface rotation and internal structure, affect the production of magnetic fields. In this very brief review I wish to concentrate on recent progress in understanding the evolution of stellar rotation, and the dependence of stellar cycles and magnetic activity on rotation, internal structure, and age.

Stellar Rotation. Most of the progress that has been made in understanding the origins of magnetic activity derives from an explosion in the amount of stellar rotation data. The measurement of rotational modulation of the Ca II emission in slowly-rotating main sequence stars (Baliunas et al. 1983) has now provided a large sample of stars for which the rotational periods are accurately known. Modulation of broad-band photospheric light by starspots even makes it possible to determine accurate rotational periods for stars in the Hyades (Lockwood et al. 1983).

One of the biggest surprises in this area is the discovery that late-type dwarfs go through a phase of very rapid rotation (van Leeuwen and Alphenaar 1982; Soderblom, Jones and Walker 1983; Stauffer et al. 1984). Low-mass stars apparently spin up (by a factor of 10!) as they contract toward the main sequence (Stauffer et al. 1984). Furthermore, the observation of rapid rotation among G dwarfs in the α Per cluster (Stauffer et al. 1985), compared with the slow rotation of similar stars in the Pleiades cluster (cf. Benz, Mayor and Mermilliod 1984) shows that many solar-type stars go through a phase of rapid spin-down (from ~50 to ~10 km s^{-1}) between the ages of 5×10^7 and 7×10^7 years. While the implications of this behavior are not fully clear, we now have a sample of rapidly-rotating, single dwarfs to test dynamo theories in an expanded range of rotation.

Solar-Type Cycles. Solar and stellar cycles (Wilson 1978) provide the fundamental evidence for dynamo generation of stellar magnetic fields. The extension of the Mt. Wilson survey work has made it possible to investigate the dependence of cycle periods on rotation and other parameters in detail. Noyes, Weiss and Vaughan (1984) found a relationship between cycle period and rotational period for a sample of 13 slowly-rotating main-sequence stars of the form $P(\text{cycle}) \approx P(\text{rot})^n$, $n = 1.25 \pm 0.5$. They suggest that the range of values found for the exponent n apparently rules out quenching of the α effect or differential rotation as mechanisms limiting dynamo action, but is consistent with the effects of magnetic buoyancy.

These results must be considered preliminary since ten of the 13 stars have not yet been observed for two cycle periods. In addition, this sample does not include the more active, rapidly rotating stars. In a recent study of the Mt. Wilson data, Baliunas et al. (1985) conclude that the range of periods among the active stars is very large, and no relationship between rotation and cycle length is present in the larger sample. Short cycle periods have also been seen in rapidly-rotating RS CVn systems (Dorren and Guinan 1984; Brusso et al. 1984), although the different internal structure of these evolved stars make comparison difficult.

Dependence of Magnetic Activity on Rotation and Internal Structure. One would like to use direct observations of stellar magnetic fields to determine the dependence of field generation on rotation and internal structure. However, it is clear from recent studies that the detection of magnetic fields is extremely difficult (Marcy 1984; Gray 1984; Borra et al. 1984). Measurements of photometric variability can be used to infer some aspects of starspot activity; the results of Radick et al. (1982) and Lockwood et al. (1983) indicate that spot activity in the Hyades cluster is much weaker in F than in G dwarfs.

One of the most useful tracers of stellar magnetic activity is Ca II H and K emission, because fluxes can be measured accurately, the emission has been observed in a large number of stars, and the time baseline of observations is sufficiently long to reduce the effects of variability. Although a number of investigations clearly demonstrate the correlation between chromospheric emission and rotation (Middelkoop 1982a; Catalano and Marilli 1983; Noyes et al. 1984), they unfortunately all differ on the dependence of emission on rotation and spectral type.

This disagreement is produced by a number of theoretical and observational difficulties. Theories of magnetic field generation and the resulting chromospheric emission are not sufficiently well developed to predict the fundamental parameters controlling chromospheric emission. One has to infer the proper parameters to study from observations. Unfortunately, many stellar parameters are correlated, and although techniques exist to find combinations of parameters that minimize the scatter of data points (cf. Schrijver 1983), these analyses usually assume straight-line relationships that are not clearly justified. Finally, there are arguments about the proper way to disentangle chromospheric and photospheric emission (Blanco et al. 1974; Linsky and Ayres 1978; Catalano and Marilli 1983; Hartmann et al. 1984).

Noyes et al. (1984) showed that the dimensionless ratio of chromospheric to bolometric luminosity is determined by rotation and a function of spectral type. Noyes et al. identified this spectral-type dependent function as the convective turnover timescale τ, so that the dimensionless activity parameter is defined as a function of the ratio of τ to the rotational period, otherwise known as the Rossby number (see also Mageney and Praderie 1984). This parameterization is also motivated by the idea that the Rossby number is closely related to the dynamo number (Durney and Latour 1978; Parker 1979).

Although the description of Noyes et al. is attractive, it must be emphasized that this formulation is not unique, and that observations at present do not rule out that other parameterizations could be used which would also provide a tight relation between some aspects of rotation and chromospheric emission. Because of the close relation between chromospheric and coronal activity (Ayers, Marstad and Linsky 1981; Oranje, Zwaan and Middelkoop 1982; Schrijver 1983; Vilhu and Rucinski 1983, Vilhu 1984), it appears that any successful parameterization of chromospheric activity will apply to other magnetic field-related emission.

Evolution of Magnetic Activity. The apparent bimodal distribution of chromospheric emission in the Vaughan and Preston (1980) survey led to suggestions that rapid rotators have a qualitatively different dynamo activity than slow rotators, and that there is a discontinuous change in activity at a critical rotation rate (Knobloch, Rosner and Weiss 1981; Durney, Mihalas and Robinson 1981). In a statistical analysis of the Vaughan-Preston survey, Hartmann et al. (1984) pointed out that a discontinuous change in emission as a function of age is not required by the observations. They did suggest that the survey, and framgmentary observations of the Pleiades, indicated weaker chromospheric activity among young stars than predicted by the Skumanich (1972) relation. Deviations from the Skumamich relation were also found by Barry, Hege and Cromwell (1984). Furthermore, studies of coronal X-ray activity also indicate a weaker dependence of activity on age and/or rotation for young, rapidly rotating stars than for older, more slowly-rotating objects (Stern et al. 1981; Walter 1982; Schrijver 1983; Vilhu 1984; Caillault and Helfand 1984; Walter et al. 1984).

Some progress has also been achieved in the analysis of magnetic activity for evolved stars. Simon (1983) found that the C IV emission in giants drops off extremely rapidly near spectral type G8. These results are very reminiscent of the rapid decrease in Ca II emission among giants in the mass range ~3-1.7 $M_\odot$ observed by Middelkoop (1982b), providing new insights into the decay of dynamo activity as stars evolve.

References

Ayres, T.R., Marstad, N.C. and Linsky, J.L.: 1981, *Astrophys. J. 247*, 545.

Baliunas, S.L., Vaughan, A.H., Hartmann, L., Middelkoop, F., Mihalas, D., Noyes, R.W., Preston, G.W., Frazer, J. and Lanning, H.: 1983, *Astrophys. J. 275*, 752.

Baliunas, S.L. et al.: 1985, *Bull. Amer. Astron. Soc.*, in press.

Barry, D.C., Hege, K. and Cromwell, R.H.: 1984, *Astrophys. J. (Letters) 277*, L65.

Benz, W., Mayor, M. and Mermilliod, J.C.: 1984, *Astron. Astrophys. 138*, 93.

Blanco, C., Catalano, S., Marilli, E. and Rodono, M.: 1974, *Astron. Astrophys. 33*, 257.

Borra, E.F., Edwards, G. and Mayor, M.: 1984, *Astrophys. J. 284*, 211.

Brusso, M., Scaltriti, F., Blanco, C., Catalano, S., Marilli, E., Pazzani, V.and Rodono, M.: 1984, *Astron. Astrophys. 135*, 255.

Caillault, J.P. and Helfand, D.: 1985, *Astrophys. J.*, in press.

Catalano, S. and Marilli, E.: 1983, *Astron. Astrophys. 121*, 190.

Dorren, J.D. and Guinan, E.F.: 1984, in *Cool Stars, Stellar Systems, and the Sun*, eds. S.L. Baliunas and L. Hartmann (Springer, Heidelberg), p. 259.

Durney, B.R. and Latour, J.: 1978, *Geophys. Appl. Fluid Dyn. 9*, 241.

Durney, B.R., Mihalas, D. and Robinson, R.: 1981, *Publ. Astron. Soc. Pacific 93*, 537.

Gray, D.F.: 1984, *Astrophys. J. 277*, 640.

Hartmann, L., Soderblom, D.R., Noyes, R.W., Burnham, N. and Vaughan, A.H.: 1984, *Astrophys. J. 276*, 254.

Knobloch, E., Rosner, R. and Weiss, N.O.: 1981, *Monthly Notices Roy. Astron. Soc. 197*, 45P.

Linsky, J.L. and Ayres, T.R.: 1978, *Astrophys. J. 220*, 618.

Lockwood, G.W., Thompson, D.T., Radick, R.R., Osborn, W.H., Baggett, W.E., Duncan, D.K. and Hartmann, L.W.: 1984, *Publ. Astron. Soc. Pacific 96*, 714.

Margeney, A. and Praderie, F.: 1984, *Astron. Astrophys. 130*, 143.

Marcy, G.W.: 1984, *Astrophys. J. 276*, 286.

Middelkoop, F.: 1982a, *Astron. Astrophys. 107*, 31.

Middelkoop, F.: 1982b, *Astron. Astrophys. 113*, 1.

Noyes, R.W., Hartmann, L., Baliunas, S.L., Duncan, D.K. and Vaughan, A.H.: 1984, *Astrophys. J. 279*, 763.

Noyes, R.W., Weiss, N.O. and Vaughan, A.H.: 1984, *Astrophys. J.*, in press.

Oranje, B.J., Zwaan, C. and Middelkoop, F.: 1982, *Astron. Astrophys. 110*, 30.

Parker, E.N.: 1979, *Cosmic Magnetic Fields*, Clarendon Press, Oxford.

Radick, R.R., Lockwood, G.W., Thompson, D.T., Warnock III, A., Hartmann, L.W., Mihalas, D., Worden, S.P., Henry, G.W. and Sherlin, J.M.: 1983, *Publ. Astron. Soc. Pacific 95, 621.*

Schrijver, C.J.: 1983, *Astron. Astrophys. 127*, 289.
Simon, T.: 1984, *Astrophys. J. 279*, 738.
Skumanich, A.: 1972, *Astrophys. J. 171*, 565.
Soderblom, D.R., Jones, B.F. and Walker, M.F.: 1983, *Astrophys. J. (Letters) 274*, L37.
Stauffer, J.R., Hartmann, L., Soderblom, D.R., Burnham, N.: 1984, *Astrophys. J. 280*, 202.
Stauffer, J.R., Hartmann, L., Burnham, N. and Jones, B.: 1985, *Astrophys. J.*, in press.
Stern, R.A., Zolcinski, M.-C., Antiochos, S.K. and Underwood, J.H.: 1981, *Astrophys. J. 249*, 647.
van Leeuwen, F. and Alphenaar, P.: 1982, *ESO Messenger* No. *28*, 15.
Vaughan, A.H. and Preston, G.W.: 1980, *Publ. Astron. Soc. Pacific 92*, 385.
Vilhu, O.: 1984, *Astron. Astrophys. 133*, 117.
Vilhu, O. and Rucinski, S.M.: 1983, *Astron. Astrophys. 127*, 5.
Walter, F.M.: 1982, *Astrophys. J. 253*, 745.
Walter, F.M., Linsky, J.L., Simon, T., Golub, L. and Vaiana, G.S.: 1984, *Astrophys. J. 281*, 815.
Wilson, O.C.: 1978, *Astrophys. J. 226*, 379.

3. Wolf-Rayet star atmospheres (David C. Abbott)

Introduction. Wolf-Rayet (WR) stars are the helium-burning remnants of massive stars (initial mass $\gtrsim 30\ M_\odot$), which have lost their outer hydrogen-rich layers through the processes of Roche lobe overflow to a companion or mass loss by a strong stellar wind. The characteristic emission-line spectrum which defines the WR spectral type is produced by a stellar wind that is so dense and opaque, that the radiation of all lines and continua arise from material in the wind. Because the wind completely screens any radiation emitted by the hydrostatic core of the star, the spectra of WR stars are nearly impossible to interpret quantitatively, and the basic parameters -- such as mass, luminosity, temperature, and chemical composition -- are poorly determined.

Model Atmospheres for Wolf-Rayet Stars. Classic model atmospheres are clearly inappropriate for WR stars because of the flagrant violation of hydrostatic equilibrium and planeparallel geometry. There is no analog to the "photospheric analysis" of OB stars, which yields T_{eff}, log g, chemical composition, and the continuum radiation field. Spectral lines formed in the winds of WR stars cannot be analyzed empirically like their OB star counterparts (e.g. Castor and Lamers 1979) because: (i) Collisional processes cannot be ignored in the statistical equilibrium, as evidenced by the large observed ratio of emission to absorption in the P Cygni profiles. (ii) Radiative rates are not known a priori because the continuum radiation field depends on the run of both density and temperature in the wind. (iii) Almost all wind profiles overlap in frequency with other wind lines because of the Doppler-shift imposed by the expansion velocities. Photons of a given frequency will scatter multiply with several distinct lines in the wind, so the statistical equilibrium of different elements is interlocked by the radiation field.

Given these formidable difficulties it is no wonder that, to my knowledge, no line profile of a WR star has ever been fitted by a theoretical model to derive quantitative knowledge about the atmospheric structure. The basic means of interpreting WR winds remains the escape probability formalism of Castor and Van Blerkom (1970), in which the statistical equilibrium is solved at a "representative point" in the wind. This method essentially fits the total intensity of the observed lines, but does not utilize any information available in the line shape.

Fortunately, progress in the analysis of WR spectra appears imminent. At least three researchers are developing diagnostic models which are intermediate between present crude models and the prohibitive solution of the general problem. All use a semiempirical approach, in which $T(r)$ and $\rho(r)$ are assumed, and the radiation field and the statistical equilibrium of helium are then solved. Hamann (1984) solves the statistical equilibrium with the "equivalent two-level atom" method and the radiative transfer with the co-moving frame formalism. Hillier (1983) uses the complete linearization method and a hybrid of the co-moving frame and Sobolev formalisms. Schmutz (1984) uses a lambda iteration method and Sobolev formalism. A major difficulty for all these techniques is the extremely high opacity of the Lyman transitions of He II, which evidently can make convergence elusive.

Model-Independent Atmospheric Diagnostics. Until the above theory is refined to the point of application, progress in understanding WR atmospheres resides in observational diagnostics which are model-independent, or at least do not require $T(r)$ or $\rho(r)$ for their interpretation. The following are examples of studies of this type:

Absolute Magnitudes. WR stars identified with either galactic clusters (e.g. Lundström and Stenholm 1984) or with the Magellanic Clouds (e.g. Prévot-Burnichon et al. 1981; Massey and Conti 1983) have yielded values of the absolute magnitude, M_V, for 38 galactic and 38 LMC stars which are thought to be single. There is a definite trend for the later spectral subtypes to be the brightest visually, however, the dispersion in M_V at a given subtype can be as large as three magnitudes.

Line Strengths. A systematic study of the equivalent widths of the optical emission lines of He, C, N, and O is given by Conti, Leep and Perry (1983) for the WN sequence and Torres and Conti (1984, and in preparation) for the WC sequence. The dispersion in line strengths is a factor of 10 for a given WN subtype, while the dispersion for the WC subtypes is smaller, but still real.

Colors. Massey (1984) has acquired absolute spectrophotometry of 55 galactic WR stars. He constructed color-color indices to describe the intrinsic shape of the continuum which should be independent of interstellar reddening. (See, however, the anomalous interstellar extinction found by Garmany, Massey and Conti 1984.) Massey (1984) concludes that the differences in the intrinsic colors of WR stars from subtype to subtype is small compared to the considerable variation present among stars of the same subtype.

Mass. Model-independent estimates of stellar mass are possible in WR stars that are members of spectroscopic binaries. The observation and analysis of such systems is now essentially complete for the bright galactic systems (e.g. Massey 1982). Again, there is a large range in observed stellar masses, ($50\ M_\odot \gtrsim M_{WR} \gtrsim 8\ M_\odot$), with no particular correlation between mass and spectral type.

Rate of Mass Loss. Surprisingly, the mass loss rate $\dot{M}$ of the wind can be measured quite reliably by infrared/radio techniques, despite our ignorance of the underlying properties of the star (e.g. Barlow, Smith and Willis 1981; Bieging, Abbott and Churchwell 1982). There are now 22 WR stars with reliable rates from radio observations (Abbott 1984). The rates for WR stars are confined to the range $10^{-4} \gtrsim \dot{M} \gtrsim 10^{-5}\ M_\odot\ yr^{-1}$ with no correlation between $\dot{M}$ and spectral type, binarity, V_∞, or M_V. At a given spectral subtype the dispersion in $\dot{M}$ is as much as a factor of 5.

The unifying factor in all of these studies is the conclusion that spectral type is a hopelessly inadequate description of a Wolf-Rayet star. This is not surprising, as spectral classification in WR stars is one-dimensional in contrast to the two-dimensional MK scheme used for OB stars. The development of the WR analog to the luminosity class of OB stars is a necessity for the proper description of WR spectra. An interesting possibility is that the differences between stars of the same subtype arise from differences in their evolutionary history (e.g. Maeder 1982), although it is not clear at present how one can parametrize "history". Abbott and Hummer (1984) have shown that the "photospheric" spectrum of OB stars can depend as strongly on $\dot{M}$ as on the traditional parameters of T_{eff} and gravity. Given the overwhelming strength of WR winds, the mass loss rate will doubtless also become an important third parameter for describing WR spectra.

Chemical Composition of WR Atmospheres. One quantity sure to correlate with spectral type is chemical composition. The two classification sequences of WR stars represent atmospheres enhanced with the products of hydrogen burning (WN sequence) and with the products of helium burning (WC sequence). While the composition anomalies of WR stars are obvious from even a cursory examination of the spectra, efforts to derive quantitative values of composition have been hampered by the lack of model atmospheres. The best estimates probably come from the escape probability formalism, which yields [C/N] and [N/He] ratios generally in accord with predictions from stellar evolution calculations (e.g. Smith and Willis 1982).

The hydrogen abundance of WR atmospheres is estimated in a fairly model-independent manner by comparing optical emission-line strengths between lines of the He II Pickering series which do, and do not, overlap with the H I Balmer series. Previous work has established that WR stars are hydrogen deficient, although severe blending makes this method questionable for the WC sequence. Recently, this method was used by Conti, Leep and Perry (1983) to derive quantitative estimates of the |H/He| ratio in 37 galactic and 20 LMC stars of the WN type. Hydrogen is present in roughly 1/3 of the stars, typically of a level of H/He $\approx$ 1-2 by number.

Conclusive evidence for the presence of hydrogen in some WR atmospheres has come from radial velocity studies that prove that the hydrogen absorption lines in some WR + O binaries are intrinsic to the WR star (e.g. Niemela 1979). Interestingly, several "single" WR stars (i.e. radial velocity variations $\lesssim$15 km s^{-1}) also show weak hydrogen absorption lines (e.g. Massey, Conti and Niemela 1981). The lines are either intrinsic or belong to a very distant companion (period $\gtrsim$ 500 days). If intrinsic, these stars would include the first definite proof that hydrogen exists in the WC sequence.

V444 Cygni: The Anatomy of a WR Atmospehere. A fantasy of any astronomer studying WR stars would be to have a movable probe in the atmosphere which could directly measure density, temperature and composition. Probably the closest realization of this dream we will ever have is the eclipsing binary system V444 Cygni (WN5+06). As demonstrated by Cherepashchuk, Eaton and Khaliullin (1984), measurements of the eclipse duration and strength for a wide range of lines and continuum frequencies provide a direct measurement of both the intensity of the radiation and the opacity as a function of radius. Further, the measured period change gives a model-independent diagnostic of $\dot{M}$. With such an analysis, Cherepashchuk et al. conclude that:

(i) The atmosphere is composed of a compact ($R_c \sim 3\ R_\odot$), hot ($T_B \sim 10^5$ K) core surrounded by an extended envelope characterized by a decreasing temperature gradient ($T_e \sim 10^4$ K at 8 $R_\odot$).

(ii) The concept of T_{eff} is ill-defined because the "photospheric" radius is a strong function of wavelength. The bulk of the radiation flow is emitted at small radii, which implies that by any definition T_{eff} exceeds 90,000 K.

(iii) The outflow velocity at the stellar core [$\tau_e(R_c) \sim 1$] is roughly 400 km s^{-1}, which verifies that the continuum is formed in an expanding, spherical region. The wind rapidly accelerates to $\approx$ 2500 km s^{-1} between 3 $R_\odot$ and 8 $R_\odot$.

Do WR Atmospheres Pulsate? Wolf-Rayet stars are at or near the theoretical limit for vibrational instability (e.g. Maeder 1984). The possible consequences of pulsation on the structure of the atmosphere remain largely unexplored, mainly because there has been no observational confirmation that the oscillations actually exist. Recently, Vreux (1984) reanalyzed the time variability of the class of "WR + compact companion" binary systems, and he argues that the observed radial-velocity variations are most likely nonradial oscillations in single stars. If this conclusion is verified by further observations, it raises exciting possibilities for the diagnostics of stellar properties from observed periods and modes, and for the study of the interaction between pulsations and stellar winds.

References

Abbott, D.C.: 1984, in *Radio Stars*, eds. R.M. Hjellming (Reidel: Dordrecht), in press.

Abbott, D.C. and Hummer, D. G.: 1984, submitted to *Astrophys. J.*

Barlow, M.J., Smith, L.J. and Willis, A.J.: 1981, *Monthly Notices Roy. Astron. Soc. 196*, 101.

Bieging, J.H., Abbott, D.C. and Churchwell, E.: 1982, *Astrophys. J. 263*, 207.

Castor, J.I. and Lamers, H.J.G.L.M.: 1979, *Astrophys. J. Suppl. 39*, 481.
Castor, J.I. and Van Blerkom, D.: 1970, *Astrophys. J. 161*, 485.
Cherepashchuk, A.M., Eaton, J.A. and Khaliullin, Kh. F.: 1984, *Astrophys. J. 281*, 774.
Conti, P.S., Leep, E.M. and Perry, D.N.: 1983, *Astrophys. J. 268*, 228.
Garmany, C.D., Massey, P. and Conti, P.S.: 1984, *Astrophys. J. 278*, 233.
Hamann, W.R.: 1984, *Astron. Astrophys*, submitted.
Hillier, D.J.: 1983, Ph.D. Thesis, Australian National University, Canberra.
Lundström, I. and Stenholm, B.: 1984, preprint.
Maeder, A.: 1982, *Astron. Astrophys. 105*, 149.
Maeder, A.: 1984, in *Observational Tests of the Stellar Evolution Theory*, eds. A. Maeder and A. Renzini (Dordrecht: Reidel), p. 299.
Massey, P.: 1982, in *Wolf-Rayet Stars: Observations, Physics and Evolution*, eds. C.W.H. de Loore and A.J. Willis (Dordrecht: Reidel), p. 251.
Massey, P.: 1984, *Astrophys. J. 281*, 789.
Massey, P. and Conti, P.S.: 1983, *Astrophys. J. 264*, 126.
Massey, P., Conti, P.S. and Niemela, V.S.: 1981, *Astrophys. J. 246*, 145.
Niemela, V.: 1979, in *Mass Loss and Evolution of O-Type Stars*, eds. P.S. Conti and C.W.H. de Loore (Dordrecht: Reidel), p. 291.
Prévot-Burnichon, M.L., Prévot, L., Rebeirot, E., Rousseau, J. and Martin, N.: 1981, *Astron. Astrophys. 103*, 83.
Schmutz, W.: 1984, in *Observational Tests of the Stellar Evolution Theory*, eds. A. Maeder and A. Renzini (Dordrecht: Reidel), p. 269.
Smith, L.J. and Willis, A.J.: 1982, *Monthly Notices Roy. Astron. Soc. 201*, 451.
Torres, A. and Conti, P.S.: 1984, *Astrophys. J. 280*, 181.
Vreux, J.M.: 1984, preprint.

4. White Dwarf Atmospheres (Rainer Wehrse)

Introductory Remarks. The main characteristic of white dwarf atmospheres is a pressure that is at least two orders of magnitude higher than in main sequence star atmospheres of the same effective temperature. This is due to the high gravity ($\log g \approx 8$), extreme metal under-abundances ($\Delta \log \varepsilon_M \gtrsim 2.5$), and (in many cases) the replacement of hydrogen by helium as the main constituent. As consequences the atmospheres are very thin ($\Delta R \lesssim$ a few km) and the level populations of all species are given by the Boltzmann distribution (perhaps with the exception of the extreme outer layers, Greenstein 1973, Pilachowski 1984). Thus, models can be calculated under the assumption of plane parallel radiative transfer and local thermodynamic equilibrium , which facilitates the numerics very much; but special care has to be taken of pressure effects (e.g., broadening of spectral lines, quenching of levels, changes in the dissociation-ionization equilibria). In addition, the proper consideration of convection, which is very effective and may reach into the optically thin layers, makes the construction of model atmospheres for white dwarfs rather tedious and costly.

Mainly due to the work of Greenstein (see e.g. his paper on spectrophotometry, Greenstein 1984) and R. Green (see Green, Schmidt and Liebert 1984) the empirical basis for studying white dwarf atmospheres has recently widened very much. The new edition of the McCook-Sion catalogue (1984) contains about 1500 white dwarfs for which photometric, spectroscopic and/or astrometric data are available. Most of them are classified in the new system of Sion et al. (1983).

In the following, I want to review some selected aspects of the progress in the understanding of white dwarf atmospheres following the last extensive review (Liebert 1980). However, due to the lack of space, I will not cover the interesting effects of magnetic fields (for a review see Borra, Landstreet and Mestel 1982) and variability (see Winget and Fontaine 1982), nor discuss white dwarfs in close binary systems (see proceedings of IAU Colloquium no. 72).

Spectroscopy in the Ultraviolet. The use of the IUE satellite has revealed the existence of several interesting and unexpected features in the UV spectra of white dwarfs. The most important seem to be:

(i) Several strong CI lines (Vauclair, Weidemann, Koester 1981, Wegner 1981) in the spectral of DC and DQ stars. The resulting abundances span the wide range from C:He = 10^{-2} to 10^{-7}.
(ii) Weak metal lines (Si II - Si IV, C IV, N V) in the high dispersion spectra of some hot white dwarfs of type DA and DO (Bruhweiler and Kondo 1983, Dupree and Raymond 1982). Since they are well separated from the corresponding interstellar lines, they seem to be formed in or at least very near the photosphere. The most likely explanation for their appearance is that the ions are levitated by the strong radiation field.
(iii) Broad depressions in DA stars around 1400 and 1600 Å (Greenstein 1980, Wegner 1982, Sion, Wesemael and Guinan 1984). It seems that they are formed by satellite lines of Lyα and are caused by the interaction of the radiating atom with protons and other hydrogen atoms (Koester et al. 1984).

High Signal-to-Noise Optical Spectroscopy. The application of high efficiency linear detector systems at several observatories made possible the detection of very weak features and/or the study of small, but for the analysis significant differences from star to star.

As a result, for the first time a cooling sequence for DB stars could be constructed (Oke, Weidemann and Koester 1984). The data showed that the mass distributions of DA and DB stars are very similar so that both types seem to have the same progenitor stars.

By use of the Kitt Peak IIDS Reticon Wegner and Yackowitch (1982, 1984) were able to demonstrate that many DC objects had to be reclassified since they show weak lines of C_2, H or He in their spectra. The authors suspect that after further reduction of the noise in the spectra of all DC stars features would be visible. Of particular interest is the possibility to detect high members of the Balmer series (Liebert and Wehrse 1983) since it enables the simultaneous estimation of the gravity and the helium content, even if no He lines are seen. This method therefore completes the infra-red observations of induced H_2 opacity (Wickramasinghe, Allen and Bessel 1982) to higher temperatures.

The consequences of the recent discovery of Balmer lines in very cool white dwarfs (Greenstein 1984) are still unclear, since these lines are not predicted by existing models.

Theoretical Developments. In addition to the model atmospheres constructed to interpret the observations mentioned above, Wesemael (1982, see also Wesemael et al. 1980, Sion, Guinan and Wesemael 1982) has calculated extensive grids for hot white dwarfs composed of H and/or He that had long been missing. They have successfully been used not only in the analysis of white dwarfs (Wesemael, Green and Liebert 1983), but also for hot subdwarfs (Wesemael et al. 1982).

The atmospheric structure of very cool pure He white dwarfs has been studied by Kapranides (1983, see also Kapranides and Böhm 1982). He shows that the pressure may get so high that at an optical depth $\tau=1$ the matter is already degenerate and conduction transports a large fraction of the energy. However, a corresponding star has not yet been identified. If some metals are present the situation is much less extreme, but the additional difficulty arises that several compositions give the same synthetic spectrum to a very high accuracy, since there are no elements with lines from two ionization stages and several potential electron donors do not show up directly (Liebert, Wehrse and Green 1984).

Several papers have been devoted to the origin of the extreme abundance distributions. Although it has been clear for some time that the metal abundances must be the result of an interplay between gravitational settling and accretion (see, e.g., Vauclair and Vauclair 1982) , it was not understood how the He-rich stars could manage to accrete heavy metals, but essentially no hydrogen. Wesemael and Truran (1982) now suggest that a slowly rotating, weakly magnetized white dwarf may be screened from accreting interstellar H by the propeller mechanism of Illarionov and Sunyaev (1975), whereas metals in grains due to their different charge to mass ratio can penetrate to the stellar surface. Another long-standing question has been whether or not DA stars convert into helium rich objects. A promising clue for a solution seems to be the finding of Michaud and Fontaine (1984) that at least under favourable conditions diffusion may transport hydrogen to so deep layers that it can burn into helium.

References

Borra, E.F., Landstreet, J. and Mestel, L.: 1982, *Ann. Rev. Astron. Astrophys. 20*, 191.
Bruhweiler, F.C. and Kondo, Y.: 1983, *Astrophys. J. 269*, 657.
Dupreee, A.K. and Raymond, J.C.: 1982, *Astrophys. J. 263*, L63.
Green, R., Schmidt, M. and Liebert, J.: 1984, *Astrophys. J. Suppl.*, in press.
Greenstein, J.L.: 1973, *Astron. Astrophys. 25*, 29.
Greenstein, J.L.: 1980, *Astrophys. J. 241*, L87.
Greenstein, J.L.: 1984, *Astrophys. J. 276*, 602.
Illarionov, A.F. and Sunyaev, R.A.: 1975, *Astron. Astrophys. 39*, 185.
Kapranides, S.: 1983, *Astrophys. J. 275*, 342.
Kapranides, S. and Böhm, K.H. 1982, *Astrophys. J. 256*, 227.
Koester, D., Weidemann, V., Zeitler- K.T., E.-M. and Vauclair, G.: 1984, *Astron. Astrophys.*, in press.
Liebert, J.: 1980, *Ann. Rev. Astron. Astrophys. 18*, 363.
Liebert, J. and Wehrse, R.: 1983, *Astron. Astrophys. 122*, 297.
Liebert, J., Wehrse, R. and Green, R.: 1984, *Astron. Astrophys.*, submitted.
McCook, G.P. and Sion, E.M.: 1984, *A Catalogue of Spectroscopically Identified White Dwarfs*, Villanova Press, Villanova.
Michaud, G. and Fontaine, G.: 1984, *Astrophys. J. 283*, 787.
Oke, B., Weidemann, V. and Koester, D.: 1984, *Astrophys. J. 281*, 276.
Pilachowski, C.A.: 1984, private communication.
Sion, E.M., Guinan, E. and Wesemael, F.: 1982, *Astrophys. J. 255*, 232.
Sion, E.M., Greenstein, J.L., Landstreet, J., Liebert, J., Shipman, H.L. and Wegner, G.A.: 1983, *Astrophys. J. 269*, 259.
Sion, E.M., Wesemael, F. and Guinan, E.: 1984, *Astrophys. J. 279*, 758.
Vauclair, G., Weidemann, V, and Koester, D.: 1981, *Astron. Astrophys. 100*, 113.
Vauclair, S. and Vauclair, G.: 1982, *Ann. Rev. Astron. Astrophys. 20*, 37.
Wegner, G.: 1981, *Astrophys. J. 248*, L129.
Wegner, G.: 1982, *Astrophys. J. 261*, L87.
Wegner, G. and Yackowitch, F.H.: 1982, *Astron. J. 87*, 155.
Wegner, G. and Yackowitch, F.H.: 1984, *Astrophys. J. 284*, 257.
Wesemael, F.: 1981, *Astrophys. J. Suppl. 45*, 177.
Wesemael, F., Auer, L., v. Horn, H.M., and Savedoff, M.: 1980, *Astrophys. J. Suppl. 43*, 159.
Wesemael, F., Green, R. and Liebert, J.: 1983, *Bull. Amer. Astron. Soc. 14*, 915.
Wesemael, F., Winget, D.E., Cabot, W., v. Horn, H.M. and Fontaine, G.: 1982, *Astrophys. J. 254*, 221.
Wesemael, F. and Truran, J.K.: 1982, *Astrophys. J. 260*, 807.
Wickramasinghe, D.T., Allen, D. and Bessell,M.: 1982, *Monthly Notices Roy. Astron. Soc. 198*, 473.
Winget, D.E. and Fontaine, G.: 1982, in *Pulsations in Classical and Cataclysmic Variable Stars*, University of Colorado, Boulder, p. 46.

5. Atmospheres of accretion discs (James E. Pringle)

Accretion discs are a popular ingredient among theorists for modelling a number of high energy astronomical objects like quasars, active galactic nuclei (Rees 1984) and galactic X-ray sources (Levin and van der Heuvel 1983). However, the observational evidence (as opposed to the strong theoretical presumption) that accretion discs exist in these objects is weak, and in only one case has some attempt been made to argue the case for a disc on the basis of its spectral properties (Malkan 1983). Indeed the structure of accretion discs is sufficiently ill-understood that any progress in this area must rest on a strong interaction between theoretical modelling and the actual observation of accretion discs in action. The only astronomical objects in which the bulk of the radiation comes clearly from an accretion disc are some of the cataclysmic variables - in particular dwarf novae in outburst, some old novae and the UX UMa stars (Pringle & Wade 1985). In these stars accretion takes place from a faint quasi-main sequence red dwarf, via an accretion disc, onto a white dwarf. Typical accretion rates are in the range 10^{-10} - 10^{-8} $M_{\odot}$ y^{-1} and consequent disc temperatures in the range 5,000 - 100,000 K. The simplest case to consider from a theoretical viewpoint is the case of a steady, optically thick disc, and in this report I shall concentrate on this, and restrict my attention to observations and modelling of the continuum distribution of such discs.

In a steady (thin) accretion disc, around an object of mass M, a steady stream of matter is added to the outer disc radius, R_{out}, at a constant rate $\dot{M}$. The mass flow inwards through each radius is the same $\dot{M}$, down to the inner disc radius R_{in} If $R_{out} >> R_{in}$, then the total luminosity available from accretion is $GM\dot{M}/R_{in}$, which is simply the product of the mass flow rate and the depth of the potential well into which the matter is falling. Similarly, the amount of energy released by matter falling from radius $R+\Delta R$ to radius R is approximately $GM\dot{M}$ $(\Delta R/R^2)$. To estimate the rate of energy release per unit area in the disc we divide this by $2\pi R \cdot \Delta R$ to obtain $GM\dot{M}/2\pi R^3$. In fact, a more careful calculation which takes into account the fact that one half of the potential energy lost is retained as increased kinetic energy, and also the energy transport involved in redistributing angular momentum within the disc yields the energy dissipation rate per unit area within the disc as $D(R) = GM\dot{M}[1 - (R/R_{in})^{-\frac{1}{2}}]/4\pi R^2$ (e.g. Pringle, 1981). In turn this implies that at each radius in the disc we may define an effective temperature T_e by $\sigma T_e{}^4 = \frac{1}{2}D(R)$, where the $\frac{1}{2}$ is because the disc radiates from both sides. Away from R_{in}, this gives $T_e \propto R^{-3/4}$ which has been confirmed by some elegant observations by Horne & Stiening (1985). The one simplifying feature about steady disc theory stems from the fact that all these formulae are derived from simple consideration of conservation of mass, energy and angular momentum. The details of how the angular momentum is transported through the disc, i.e. of the viscous process, do not enter. This means that if such a disc radiates locally at each radius like a black body with temperature T_e, the emergent spectrum can be calculated simply.

However although the spectra of some stars in some wavelength ranges follow an approximate blackbody distribution, in general they do not. The next obvious step to calculating a disc spectrum is therefore to assume that each element of the disc radiates locally like a stellar atmosphere with the same T_e, and some appropriate gravity, g. In the plane of symmetry of the disc the gravity of the central object is exactly balanced by centrifugal force and the net gravity is zero. As one moves perpendicular to the disc plane (in the z-direction at fixed R) there is a restoring force towards the disc plane of approximately $g = -GMz/R^2$. Thus the effective gravity at the $\tau \approx 1$ surface depends on the thickness of the disc H at which that surface occurs. For a disc which is isothermal in the z-direction, hydrostatic equilibrium implies $\rho(R,z) = \rho_o(R)\ \exp(-z^2/H^2)$ where $H \equiv 2c_s^2R^2/GM$ and $c_s^2 = R\ T_e/\mu$. Usually H has been estimated by taking $T \equiv T_e$ (Wade 1984), coupled with the argument that the details of the spectrum are not strongly sensitive to g. Wade (1984) reviews in some detail the difference in model spectra between the blackbody and the stellar atmosphere assumptions. He demonstrates that the disc parameters deduced from model fitting depend sensitively on which assumption is taken.

Since the launch of IUE, a large group of us has been involved in obtaining simultaneous observations of dwarf novae in outburst covering the optical and IUE ranges λλ1200 - 7500. The first result (Bath et al. 1980) seemed to indicate that the blackbody approach might be adequate. Almost all subsequent observations have contradicted this (Pringle & Verbunt 1984, Schwarzenberg-Czerny et al. 1985). At the same time it is evident that the stellar atmosphere approach is not fully satisfactory - one obvious failing being the Balmer discontinuity which is barely seen, but present in absorption in model spectra. Hassall (1985) in her study of EK TrA suggests remedying this by adding an exact amount of Balmer continuum emission and in this way does obtain a good fit to the full spectral range of her observations.

On the other hand, the stellar atmosphere models do display naturally some of the spectral details seen in hot stars, which can be seen in observed disc spectra (Panek & Holm 1984, Verbunt et al. 1984, La Dous et al. 1985).

Although the approaches outlined above have yielded hopeful results, the agreement between theory and observation is not yet adequate to measure disc parameters with any confidence. It is clear that more detailed calculations need to be undertaken, but it is equally clear that the next step introduces a large number of uncertainties and considerably broadens the parameter space to be explored. This is because any further step must introduce knowledge of the viscous process occuring in the disc, about which little or nothing is known. For a given accretion rate $\dot{M}$, the surface density and so optical depth through the disc in the z-direction depend inversely on the viscosity. In outbursting dwarf nova discs the optical depths through the disc are thought to be relatively low, of order $10 - 10^3$. Thus since $g \propto z$, simple constant gravity atmospheres may be inadequate. More seriously, it is completely unknown where, as a function z, the accretion energy is liberated, and the assumption of constant energy flux through the atmosphere is unlikely to hold. Even worse, if one adopts the Shakura-Sunyaev α-parameterization of the viscosity (where α is essentially the Mach number of the putative turbulent viscous process) then indications are that $\alpha \sim 1$. This implies sonic velocities in the atmosphere with resultant strong temperature and density fluctuations on scales of H and less. The net effect of such processes might be to make the temperature structure in the disc photosphere more isothermal than is the case in stellar atmospheres, and so to weaken such features as the Balmer discontinuity and some absorption lines. The effect of the Lyman jump on the overall continuum shape, is, however, still likely to be considerable, although observations shortward of 912 Å would help clarify this point. In any case, even with the observations we already have, the task facing theorists of providing credible and accurate fits to the observed spectra is a considerable one.

References

Bath, G.T., Pringle, J.E. and Whelan, J.A.J.: 1980, *Monthly Notices Roy. Astron. Soc.*, *190*, 185.

Hassall, B.J.M.: 1985, *Monthly Notices Roy. Astron. Soc.*, in press.

Horne, K.D. and Stiening, R.: 1985, *Astrophys. J.*, in press.

La Dous, C. et al.: 1985, *Monthly Notices Roy. Astron. Soc.*, in press.

Lewin, W. and van den Heuvel, E.: 1983, *Accretion Driven Stellar X-ray Sources*, Cambridge Univ. Press.

Malkan, M.A.: 1983, *Astrophys. J.*, *268*, 582.

Panek, R.J. and Holm, A.V.: 1984, *Astrophys. J.*, *277*, 700.

Pringle, J.E.: 1981, *Ann. Rev. Astron. Astrophys.*, *19*, 137.

Pringle, J.E. and Verbunt, F.: 1984, *Proc. 4th IUE Conference, p. 377.*

Pringle, J.E. and Wade, R.A.: 1985, *Interacting Binary Stars*, Cambridge Univ. Press.

Rees, M.J.: 1984, *Ann. Rev. Astron. Astrophys.*, *22*, 471.

Schwarzenberg Czerny, A. et al.: 1985, *Monthly Notices Roy. Astron. Soc.*, in press.

Verbunt, F., Pringle, J.E., Wade, R.A., Echevarria, J., Jones, D.H.P., Argyle, R.W., Schwarzenberg Czerny, A., La Dous, C. and Schoembs, R.: 1984, *Monthly Notices Roy. Astron. Soc.*, *210*, 197.

Wade, R.A.: 1984, *Monthly Notices Roy. Astron. Soc.*, *208*, 381.

III. MONOGRAPHS ETC.

A list of monographs, proceedings from symposia etc, published since last General Assembly and of particular relevance for the activity of our commission is given here. For natural reasons, the list is not complete - further volumes can be found in the *Abstracts of Astronomy and Astrophysics.*

Methods in Radiative Transfer, W. Kalkofen (ed.), Cambridge University Press (1984).
Foundations of Radiation Hydrodynamics, D. Mihalas and B.W.Mihalas, Oxford University Press (1985).
B Stars with and without Emission Lines, A.B. Underhill and V. Doazan, NASA SP-456 (1982).
The A Stars: Problems and Perspectives, S.C. Wolff, NASA SP-463 (1983)
Stellar Atmospheric Structural Patterns, R.N. Thomas, NASA SP-476 (1983).
The Solar Granulation, R.J. Bray, R.E. Loughhead and C.J. Durrant (2nd ed.), Cambridge University Press (1984).
Early Emission Line Stars, C.R. Kitchin, Adam Hilger Ltd, Bristol (1982).
Solar Magnetohydrodynamics, E.R. Priest, Geophys. and Astrophys. Monographs Vol. 21, D. Reidel Publ. Co., Dordrecht (1982).
Trieste Workshop on Progress in Spectral Line Formation Theory, D. Reidel, in press.
Observational Basis for Velocity Fields in Stellar Atmospheres, R. Stalio (ed.), Oss. Astron. Trieste (1982).
Calibration of Fundamental Stellar Quantities, IAU Symp. No. 111, D. Hayes, L.E. Pasinetti and A.G.D. Philip (eds.), D. Reidel Publ. Co., Dordrecht, in press.
The Origin of Non-Radiative Heating/Momentum in Hot Stars, A.B. Underhill and A.G. Michalitsianos (eds.), NASA CP, in press.
Astrophysical Processes in Upper Main Sequence Stars, 13th Adv. Course of the Swiss Society of Astron. Astrophys., B. Hauck and A. Maeder (eds.), Geneva Obs. (1983).
Wolf Rayet Stars: Progenitors of Supernovae?, Proc. of workshop at Obs. de Paris, Meudon (1983).
Wolf Rayet Stars: Observations, Physics, Evolution, IAU Symp. No. 99, C.W.H. de Loore and A.J. Willis (eds), D. Reidel Publ. Co., Dordrecht (1982).
Magnetic and Variable Stars, M. Marik and L. Szabados (eds.), Commun. Konkoly Obs. No. 83 (1982).
Cosmic Magnetic Fields, R. Pallavicini (ed.), Mem. Soc. Astron. Ital. *53*, No. 4 (1982).
Solar and Stellar Magnetic Fields: Origin and Coronal Effects, IAU Symp. No. 102, J. Stenflo (ed.), D. Reidel Publ. Co., Dordrecht (1983).
Active Phenomena in the Outer Atmospheres of the Sun and Stars. J.C. Pecker and Y. Uchida(eds.), Observatoire de Paris, Meudon (1984).
Chromospheric Diagnostics and Modelling, B. Lites (ed.), Sacramento Peak Obs., Sunspot, in press.
Small-Scale Dynamical Processes in Quiet Stellar Atmospheres, ed. S.L. Keil, Sacramento Peak Obs., Sunspot, in press.
Cool Stars, Stellar Systems and the Sun. Proc. of the third Cambridge Workshop, S.L. Baliunas and L.W. Hartman (eds.), Cambridge, Mass. (1983).
Activity in Red-Dwarf Stars, IAU Coll. 71, P.B. Byrne and M. Rodonò (eds.), Astrophys. and Space Sci. Library Vol. *102*, D. Reidel Publ. Co., Dordrecht (1983).
Cool Stars with Excesses of Heavy Elements, M. Jaschek and P.C. Keenan (eds.), D. Reidel Publ. Co., Dordrecht, in press.
Observational Tests of Stellar Evolution Theory, IAU Symp. No. 105, A. Maeder and A. Renzini (eds.), D. Reidel Publ. Co., Dordrecht (1984).
Herbig Haro Objects, T Tauri Stars and Related Phenomena, J. Canto and E.E. Mendoza (eds.), Rev. Mex. Astron. Astrofis. Vol. *7* (1983).
Cataclysmic Variables and Related Objects, IAU Coll. 72, M. Livio and G. Shaviv (eds.), Astrophys. and Space Sci. Library Vol. *101*, D. Reidel Publ. Co., Dordrecht (1983).
Ejection and Accretion of Matter in Binary Systems, J. Tremko, VEDA, Publ. House of The Slovek Acad. of Sciences (1982).

Accretion-driven Stellar X-Ray Sources, W.H.G. Lewin and E.P.J. van den Heuvel, Cambridge University Press (1983).
The Nature of Symbiotic Stars, IAU Coll. No. 70, M. Friedjung and R. Viotti (eds.), D. Reidel Publ. Co., Dordrecht (1982).
Planetary Nebulae, IAU Symp. No. 103, D.R. Flower, D. Reidel Publ. Co., Dordrecht (1983).
Ultraviolet Stellar Classification, A. Heck and B. Battrick (eds.), ESA SP-182 (1982)
Third European IUE Conference, ESA SP-176 (1982).
Fourth European IUE Conference, ESA, in press.
Advances in Ultraviolet Astronomy, NASA Conf. Publ. 2238 (1982).

Bengt Gustafsson

President of the Commission

COMMISSION 37: STAR CLUSTERS AND ASSOCIATIONS
(AMAS STELLAIRES ET ASSOCIATIONS)

PRESIDENT: K.C. Freeman
VICE-PRESIDENT: D.C. Heggie
ORGANISING COMMITTEE: G.L.H. Harris, G. Lynga, P.E. Nissen, C. Pilachowski, G.N. Salukvadze.

1. Introduction

The last three years have been very productive for cluster research. This report, on activities in the field, has been compiled by the members of Commission 37. It begins with sections on recent meetings, and on data catalogs (G. Lynga). Detailed tables of work on associations (P.E. Nissen), open clusters (G.L.H. Harris) and globular clusters (R.E. White) are then given. A section on cluster dynamics (D.C. Heggie) follows, and the final section concerns present trends in cluster research (C. Pilachowski).

2. Symposia and Colloquia

Major recent meetings on star clusters include:
IAU Colloquium 68, "Astrophysical Parameters for Globular Clusters" (31.012.023).
The Prague Conference on "Star Clusters and Associations and their Relation to the Evolution of the Galaxy" (34.012.068).
IAU Symposium 113, "Dynamics of Star Clusters", (Reidel; ed. J. Goodman and P. Hut), in press.

Other meetings, at which the subject of star clusters figured more or less prominently, include 32.012.006 and 065; 33.012.016; 34.012.008, 026, 049, 057,061, 062 and 065, and IAU Colloquium 88 "Stellar Radial Velocities" (L. Davis Press: ed. A.G.D. Philip), in press.

At the 19th General Assembly, Commission 37 is participating in Joint Discussion VI, "Evolution in Young Populations in Galaxies", organised by G. Lynga.

3. Catalogs of Cluster Data

Several new catalogs or updates of old catalogs have been made available by the data centers in Strasbourg and at NASA:

Catalog number	Content	Authors	Year
5041	Open cluster data	Lynga, G	1983
5044	Globular cluster bibliography	Ruprecht, J., Balazs, B. and White, R.E.	1983
5045	Masses and ages of stars in twelve open clusters	Myatzutin, V., Sagar, R. and Joshi, U.	1984

5046	Stars in stellar associations and young clusters	Humphreys, R.M. and McElroy, D.B.	1984

A bibliography of radial velocities for stars in open clusters is presented by Mermilliod (1984: Bull. Inform. CDS, No. 26, 9) and a compilation of proper motion studies of stars in an around open clusters is given by van Leeuwen (1984: IAU Symposium 113, in press).

Mermilliod (1985: "International Course on Data Handling in Astronomy and Astrophysics", Trieste, in press) is updating his catalogs on cross identifications and on UBV data and MK types of stars in open clusters.

Janes and Adler (32.153.005) have re-examined color magnitude diagrams for open clusters, while Philip and White (Steward Observatory) are making a reference catalog of globular cluster color magnitude diagrams.

4. Stellar Associations

Papers on associations in general, and on OB, T and R associations are listed in the following four tables. The abbreviations are

abund = chemical abundance(s)	em = emission	pol = polarization
ass = association(s)	IR = infrared	rv = radial velocity
cl = cluster(s)	ist = interstellar	sp = spectra
d = distance(s)	mol = molecular	UV = ultraviolet
distr = distribution(s)	obs = observations	var = variable
	phot = photometry	vel = velocity

TABLE 1. ASSOCIATIONS IN GENERAL

Reference	Subject
Brinks (29.152.002)	holes in HI distr and ass in M31
Casse et al (29.143.063)	cosmic rays from OB ass
Cowie et al (30.152.004)	search for supershells in OB ass
Garmany et al (30.064.080)	mass loss rates from O stars
Tomisaka et al (30.125.026)	model of superbubble formation
Gull et al (30.131.044)	theory of superbubbles
Olive, Schramm (31.107.022)	abund ejected from supernovae in ass
Efremov (32.152.007)	age and dimensions of OB ass
Salukvadze (32.118.027)	trapezium star systems in T ass
Voshchinnikov, Marchenko (32.131.319)	pol of starlight in R ass
Beltrametti et al (32.132.004)	gas dynamics around OB ass
Tenorio-Tagle et al (32.132.005)	gas dynamics around OB ass
Bruch, Sanders (33.153.029)	mass of OB ass
Elmegreen (33.153.029)	formation of ass and cl
Vanbeveren (34.152.001)	evolution of massive stars in cl and ass
Schramm, Olive (34.152.016)	chemical evolution of OB ass
Tenorio-Tagle (34.152.018)	dynamical evolution of OB ass
Mirzoyan (34.152.019)	instability in stellar ass
Bodenheimer et al (34.125.075)	evolution of OB ass
Klein et al (34.131.092)	star formation in OB ass
Cesarsky, Montmerle (34.143.018)	energetics of OB ass, gamma rays
Nissen (34.153.010)	helium abund in cl and ass
Zhilyaec, Marchenko (34.153.074)	rotation of stars in cl and ass

TABLE 2. OB ASSOCIATIONS

Name	Observer	Type of data
Aur OB2	Aiello et al (32.131.318)	uv extinction
Car OB1	Forte, Orsatti (29.114.033)	obj. prism survey
Cas OB6	Braunsfurth (31.152.002)	HI obs
	Braunsfurth (33.131.014)	HI around HII regions
Cep OB3	Evans et al (29.131.058)	IR obs of mol cloud
	Panagia, Thum (29.131.154)	radio obs of mol cloud
	Perinotti, Panagia (30.112.025)	mass loss of O7n star
	Barsella (32.131.010)	peculiar extinction in UV
	Fabian, Stewart (33.152.001)	X-ray emitting bubble
	Sargent et al (34.131.008)	IR and CO obs, star formation
Cep OB4	Rossano et al (34.131.175)	radio obs, kinematics
Cyg OB1	Turner (33.152.002)	membership of X Cyg
Cyg OB2	Hutchings (29.112.014)	stellar winds
	Abbott et al (30.064.079)	mass loss rates
	Leitherer (30.112.039)	mass loss rates
	Persi et al (30.113.052)	IR phot of X-ray stars
	Iyengar et al (30.133.013)	ist OI line
	Leitherer (31.114.049)	sp, IR phot, mass loss
Lac OB1	Bijaoui (30.111.005)	rv
Mon OB1	Ogura (PAS Jap 36 139 1984)	H-alpha em stars
Mon OB2	Guseva (34.131.284)	distr of dust
Ori OB1	Joncas, Borra (29.152.001)	frequency of Ap stars
	Oganesyan, Gasparyan (29.114.017)	UV sp of B stars
	Parsamyan (29.122.010)	frequency of stellar flares
	White, Phillips (29.131.013)	mol line obs
	Guetter (30.152.002)	MK sp, d, age spread
	Borra (30.116.018)	magnetic fields
	Anthony-Twarog (32.111.004)	d from H-beta phot
	Gieseking (33.153.002)	kinematics from rv
	Isobe (34.152.015)	distr and age of stars
	Cowie (34.131.253)	UV obs, supershells of gas
Per OB1	Phillips, Gondhalekar (30.152.001)	high-vel ist lines
	Krelowski, Strobel (34.152.009)	UV extinction
Per OB2	Markkanen (30.156.011)	pol obs, magnetic fields
Pup OB2	Turner (29.153.003)	membership of Ruprecht 44
Sco OB1	Gieseking (32.120.004)	massive binaries
	Aiello et al (32.131.318)	UV extinction
Sco OB2	Olano, Poppel (29.131.064)	HI feature, formation model
	Herbst, Warner (30.121.001)	phot of stars in dust cloud
	Krelowski, Strobel (34.152.009)	UV extinction
	Lipovka et al (34.152.012)	radio em
Sco-Cen OB1	Borra et al (32.113.008)	search for Ap stars
Vel OB1	Eggen (32.152.003)	phot, d
Vel OB2	Eggen (32.152.003)	phot, d, cepheid members

TABLE 3. T ASSOCIATIONS

Name	Observer	Type of data
Cyg T1	Gol'dberg (34.152.006)	content of B stars
Mon T1	Erastova (31.122.090)	number of irregular var

Pho T1	Cersosimo, Arnal (30.152.003)	HI obs
Tau T1	Nurmanova (30.113.067)	phot of T Tauri stars
	Nurmanova (30.152.010)	new member
	Nurmanova (30.113.067)	phot of T Tauri stars

TABLE 4. R ASSOCIATIONS

Name	Observer	Type of data
CMa R1	Baierlein et al (30.125.069)	stellar clustering
	Herbst et al (31.152.001)	reddening law
Mon R1	Herbst et al (31.152.001)	reddening law
Simeis 188	Herbst et al (31.152.001)	reddening law
Vul R1	Herbst et al (31.152.001)	reddening law
Vul R2	Herbst et al (31.152.001)	reddening law

5. Open Clusters

Papers and projects which refer to individual open clusters are listed in Table 5, where the clusters are ordered according to IAU number. At the end of this section, references are given to survey papers on open clusters. Abbreviations used are:

abund = abundance(s)	lf= luminosity function	RG = reg giant
agesp = age spread	ms = main sequence	rot = rotation(al)
bin = binar(y,ies)	memb = membership	rv = radial velocity
d = distance	pg = photographic	SG = supergiant(s)
distr = distribution	pec = peculiar	sp = spectra(l)
E = color excess	phot = photometry	UV = ultraviolet
em = emission	pm = proper motions	var = variable
IR = infrared	pol = polarization	vel = velocity

TABLE 5. OPEN CLUSTERS

Number	Name	Reference, and type of data
C0039+850	NGC 188	34.153.048 abund 34.153.054 PDS phot
		Baliunas et al (1984) W UMa bin, mass ratios, space vel
C0040+615	NGC 225	34.153.023 UBV pg phot
C0057+636	Be 62	30.153.011 UBV phot, E, d, age
C0115+580	NGC 457	30.153.026 uvbyHβ phot, agesp
		32.153.043 uvbyHβ phot, d, F supergiants
		33.153.013 memb, pre-ms
C0126+630	NGC 559	32.113.048 UBVRI phot
C0129+604	NGC 581	34.153.044 UBV phot, d, age,E
C0140+616	NGC 654	33.153.011 UBV phot, E, d, age
C0142+610	NGC 663	30.153.026 uvbyHβ phot, agesp
C0155+552	NGC 744	30.122.134 carbon star, var
C0154+374	NGC 752	33.153.017 uvby phot, ms bimodality, E, d, age, abund
C0211+590	Stock 2	31.115.004 giants, linear radii
C0215+569	NGC 869	30.153.026 uvbyHβ phot, agespread
	(h Per)	31.153.011 ANS phot, circumstellar matter
		31.153.034 SG, var 33.153.037 memb, catalog

C0218+568	NGC 884	31.153.011 ANS phot, circumstellar matter
	(χ Per)	31.153.034 SG, var 33.153.037 memb, catalog
C0228+612	IC 1805	33.153.013 memb, pre ms 31.112.013 UV sp, mass loss
		33.153.020 UBV phot, var E, d, star formation
		34.153.046 UBV phot, d, E, R, age
C0238+425	NGC 1039	31.153.006 pg phot, white dwarfs
C0311+470	NGC 1245	30.153.015 lf
C0318+484	NGC 1252	30.153.003 kinematics, "supercorona"
	Mel 20	34.153.037 internal structure
		34.153.017 UBV phot, age, d, carbon star
C0341+321	IC 348	34.153.046 UBV phot, d, E, R, age
C0344+239	Pleiades	30.153.007 X-ray phot, variability
		30.121.028 sp, em lines, Li abund
		31.153.005 UV phot, 2 color phot. E, age, Be stars
		31.153.036 X-ray survey 31.153.043 var K stars
		31.113.069 phot, var stars 31.113.076 slow flare curves
		31.113.077 flare stars
		31.122.037 flare stars, spatial distr
		32.153.028 BVRIJHK phot, sp, bin, ms
		32.113.069 phot, variability
		32.114.053 Mg II h and k, chromospheric activity
		33.153.030 X-ray sources 34.153.015 Li abund, agesp
		34.153.038 sp types, vsini, age, d
		34.153.065 lf 34.153.051 sp, rot vel, K
		34.153.076 flare stars, statistics
		34.153.066 sp age criteria, kinematics
		Vandenberg, Bridges (1984) ZAMS models, abund
C0417+368	Hyades	30.153.023 X-ray em, stellar coronae
		30.114.040 UV, optical data, giants
		30.122.149 V697 Tau, memb
		31.153.003 rot vel, macroturbulence, giants
		31.153.016 uvby phot, He abund
		31.153.040 IUE sp, X-ray sources, em
		31.153.037 UBVR phot 31.153.048 VRIK phot, bin
		31.153.049 Hipparchos 31.111.033 pm survey, memb
		31.113.076 slow flare curves 31.114.010 sp
		31.115.004 giants, linear radii
		31.116.013 Ca II H and K em, rot vel
		31.120.004 HD 27130, bin, mass-luminosity relation, d
		31.120.006 bin, rv, memb 32.153.003 UBVRI phot, memb
		32.153.023 uvbyHβ,RI phot, d, agesp
		32.153.029 JHK phot, bin 32.153.039 bin, He abund, d
		32.114.053 Mg II h and k studies, chromospheric activity
		33.153.001 BVRI phot, memb 33.153.021 phot, ms
		33.153.025 d 33.153.027 reticon sp, CN abund
		33.153.034 pm, phot, rv, subluminous stars
		33.153.041 stellar orbits
		33.114.097 chromospheric activity, dwarfs
		34.153.015 Li abund, agesp 34.153.016 IUE sp, giants, em
		34.153.020 X-ray activity 34.153.034 Ca II H and K em
		34.153.026,.027 Hyades problems 34.153.039 structure
		34.153.065 lf 34.113.032 phot, variability
		34.153.058 d, pm 34.153.067 X-ray lf
		34.153.077 A stars, motion 34.153.073 bright stars, distr
		Vandenberg, Bridges (1984) ZAMS models, abund, d
		Upgren et al (van Vleck) BVRI phot
C0443+108	NGC 1662	34.153.012 Ap stars
C0447+436	NGC 1664	30.122.193 carbon star, var star

C0459-058	HR 1614	32.153.006 DDO phot, CN abund
C0532+099	Orion	30.153.005 pm, memb 31.113.077 flare stars
		34.153.002 UV flux distr, extinction curves
		34.153.014 MK types, rv, memb
		34.153.018 Vsini, X-ray phot, rot, T Tauri stars
		34.153.022 pm, memb 34.153.025 sp, HI lines
		34.153.061 low mass stars, IR pg
		Lee et al (Yale) pm, internal motions
C0532+341	NGC 1960	30.153.026 uvbyHβ phot, agesp
		Barkhatova et al (1984)
C0533-01	Cr 70	33.153.002 vel distr
C0540-022	NGC 2023	34.153.003 IR phot, reflection nebulae, star formation
C0540+002	NGC 2068	34.153.003 IR phot, reflection nebulae, star formation
C0546+336	King 8	29.153.037 BV phot, E, d, age, metallicity
C0548+217	Berk 21	32.153.010 slitless sp, automated sp classification, memb
C0549+325	NGC 2099	30.122.194 carbon star, var star
		31.115.004 giants, linear radii 32.153.021 UBV phot
C0604+241	NGC 2158	Smith, Norris (1984) sp, giants, Ca II H and K, CN
		Cudworth (Yerkes) pm, phot
C0605+243	NGC 2168	Cudworth, McNamara (Yerkes) internal motions
C0611+128	NGC 2194	30.153.022
C0613-186	NGC 2204	34.153.050 phot, d, E, age, lf
		Claria (Cordoba) UBV, CMT_1T_2 phot, abund
C0624-047	NGC 2232	34.153.041 DDO phot
		Claria (Cordoba) UBV, DDO phot
C0627-312	NGC 2243	31.153.035 high disp sp, RG, abund, age
C0629+049	NGC 2244	30.114.015 IUE data, energy distr 32.153.027 pm
C0634+031	CV Mon	33.153.004 uvbyHβ phot, d, E
C0635+020	Coll 10	32.153.013
C0638+099	NGC 2264	30.153.009 X-ray phot 30.115.006 theor HR diagram
		33.122.025 UBVRI phot, light curve, W92
		34.153.021 UBV phot, var E, d
		34.153.046 UBV phot, d, E, R, age
		34.153.059 pre ms stars, star formation, model comparison
		Claria (1984) UBV, DDO phot, evolved stars, memb, abund
C0639-480	IC 2395	Claria et al (Cordoba) UBV phot, sp, lf, d, age, abund
C0644-206	NGC 2287	30.153.003 uvbyHβ phot, E, d, age
		34.153.041 DDO phot 34.153.012 Ap stars
		34.153.028 MK sp, star counts, d, E
C0645+411	NGC 2281	30.153.025
C0649+005	NGC 2301	34.153.045 phot, d
C0650+030	Berk 28	Bijaoui et al (1984) UBV phot, E, d, age, lf
C0652-245	Coll 121	30.153.003 uvbyHβ phot, E, d, age 34.153.011 Ap stars
C0658-204	Tomb 1	33.153.019 UBV phot, cepheid, memb, d, age
C0700-082	NGC 2323	Claria, Lapasset (Cordoba) UBV phot, E, age
		34.153.045 UBV phot, d, diameter, E, memb
C0701-207	Tomb 2	32.153.042 pg phot, E, d
C0704-100	NGC 2335	Claria (1984) UBV, DDO phot, evolved stars, memb, abund
C0705-105	NGC 2343	Claria (1984) UBV, DDO phot, evolved stars, memb, abund
C0712-102	NGC 2353	34.153.028 UBV phot, MK sp, d, var E, star counts
C0712-310	Coll 132	33.153.007 uvbyHβ phot, age, space motion
C0715-155	NGC 2360	Mermilliod (Geneva) 7-color phot, RG
		Mermilliod, Mayor (Geneva), rv, RG
C0715-367	Coll 135	33.153.007 uvbyHβ phot, age, space motion
C0716-248	NGC 2362	32.153.024 BV phot, Ap stars
C0722-321	Coll 140	33.153.007 uvbyHβ phot, age, space motion
C0724-287	Ru 20	Claria (Cordoba) DDO, CMT_1T_2 phot, abund
C0724-476	Mel 66	31.153.035 high dispersion sp, d, age, abund

		Claria (Cordoba) UBV, CMT_1T_2 phot, abund
C0735+216	NGC 2420	31.153.035 high dispersion sp, RG, abund, age
		33.153.042 structure, memb, surface densitydistr
C0734-205	NGC 2421	Mermilliod, Mayor (Geneva) rv, RG
C0734-143	NGC 2422	33.153.041 stellar orbits 34.153.012 Ap stars
C0734-137	NGC 2423	33.153.006 phot, var stars, d, age, RG
		Claria (Cordoba) CMT_1T_2 phot, abund
		Mermilliod (Geneva) 7-color phot, RG
C0739-147	NGC 2437	30.153.024 uvbyHβ phot, E, d, age
C0740-354	Ru 31	Claria (Cordoba) UBV phot
C0742-237	NGC 2447	33.153.005 phot, d, age, abund
		Mermilliod (Geneva) 7-color phot, RG
		Mermilliod, Mayor (Geneva) rv, RG
C0743-328	NGC 2451	33.153.007 uvbyHβ phot, age, space motion
		33.114.023 c Pup, companion, E, T_e, angular diameter, d
		34.153.041 DDO phot
		Claria (1984) UBV, DDO phot, evolved stars, memb, abund
		Mermilliod, Mayor (Geneva) rv, RG
C0750-384	NGC 2477	33.153.003 JHK phot Mermilliod, Mayor (Geneva) rv, RG
		33.153.035 sp, blue stragglers, rot, age, abund, rv
		Claria (Cordoba) CMT_1T_2 phot, abund
C0752-241	NGC 2482	34.153.042 CMT_1T_2, DDO phot, RG, abund, homogeneity
		Mermilliod, Mayor (Geneva) rv, RG
C0754-299	NGC 2489	Mermilliod, Mayor (Geneva) rv, RG
C0757-607	NGC 2516	30.114.119 coude sp, HD 66318 32.120.002 rv, bin
		32.153.030 UVR search, white dwarfs, mass limit
		34.153.009 VBLUW phot, E, d
		34.153.012 Ap stars
		Mermilliod, Mayor (Geneva) rv, RG
C0802-461	Coll 173	33.153.007 uvbyHβ phot, age, space motion
C0803-280	NGC 2527	Claria (1984) UBV, DDO phot, evolved stars, memb, abund
		Mermilliod, Mayor (Geneva) rv, RG
C0808-126	NGC 2539	Claria (Cordoba) DDO, CMT_1T_2 phot, abund
		Mermilliod, Mayor (Geneva) 7-color phot, RG
C0810-374	NGC 2546	32.153.024 BV phot, Ap stars, var E
		Claria (1984) UBV, DDO phot, evolved stars, memb, abund
		Mermilliod, Mayor (Geneva) rv, RG
C0809-491	NGC 2547	31.153.004 uvbyHβ phot, E, d, age, lf, RG, var, ms gap
C0811-056	NGC 2548	Claria (1984) UBV, DDO phot, evolved stars, memb, abund
C0816-304	NGC 2567	Claria (1984) UBV, DDO phot, evolved stars, memb, abund
C0837+201	NGC 2632	30.153.010 BVRI phot 31.153.006 pg phot, white dwarfs
	Praesepe	31.113.026 slow flare curves 31.114.197 sp, abund
		31.115.004 giants, linear radii
		32.153.019 BVRI phot, lower ms
		33.153.008 pm, memb 33.153.018 MK sp, d
		34.153.073 bright stars, distr
		Anthony-Twarog (1984) phot survey, white dwarfs
		Vandenberg (1984) ZAMS models, abund, d
C0838-528	IC 2391	33.122.037 var, o Vel, HR 3467
		Levato, Garcia (1984) rot vel, axial rot, bin
		Levato, Malaroda (1984) sp types, pec stars, d
C0840-447	NGC 2659	30.153.024 uvbyHβ phot, E, d, age
C0840-469	NGC 2660	33.153.003 JHK phot
C0843-527	NGC 2669	Mermilliod, Mayor (Geneva) rv, RG
C0847+120	NGC 2682	30.153.002 uvbyHβ phot, E, d, blue stragglers
	M 67	30.113.059 phot
		31.115.004 giants, linear radii
		31.153.035 high dispersion sp, RG, d, age, abund

		32.153.040 pm, memb Lopez (Yale) pm 34.153.048 abund
		Peterson et al (1984) IR phot, sp, blue stragglers, bin
		Janes, Smith (1984) UBV, DDO phot, RG, sp, abund, bin
		Schild (1984) BVRI CCD phot
C0921-770	E 3	34.153.050 phot, d, E, age, lf
C0925-549	Ru 77	32.153.002 RGU phot, d, E, earliest sp type
C0927-534	Ru 78	32.153.002 RGU phot, d, E, earliest sp type
C0931-561	Basel 20	32.153.002 RGU phot, d, E, earliest sp type
C0938-501	NGC 2972	Mermilliod, Mayor (Geneva) rv, RG
C0939-536	Ru 79	33.153.004 uvbyHβ phot, d, E
C0947-543	Ru 83	32.153.002 RGU phot, d, E, earliest sp type
C0947-561	NGC 3033	Mermilliod, Mayor (Geneva) rv, RG
C0949-529	Pis 16	32.153.002 RGU phot, d, E, earliest sp type
C1001-598	NGC 3114	31.153.042 uvbyHβ phot, d, E, age, blue stragglers
		Claria (Cordoba) UBV, DDO, CMT_1T_2 phot, abund
		Mermilliod, Mayor (Geneva) rv, RG
C1019-514	NGC 3228	32.153.024 BV phot, Ap stars
		Mermilliod, Mayor (Geneva) rv, RG
C1025-573	IC 2581	32.153.043 uvbyHβ phot, d, F SG
C1033-579	NGC 3293	32.153.026 UBV phot, agesp, lf, initial mass function
		33.122.011 B phot, β Cep stars
		34.153.031 uvbyHβ phot, ZAMS, d, β Cep stars
C1035-583	NGC 3324	32.114.081 sp types, O stars
C1041-593	Tru 14	32.114.081 sp types, O stars
		34.153.035 UBVRI phot, memb, var E, d, age
C1041-597	Coll 228	31.153.002 MK sp types, d, bin, shell stars 31.153.017
C1042-591	Tru 15	31.153.019
C1043-594	Tru 16	32.153.001 UBVRI phot 32.153.037 sp types, bin, shell star
		32.114.081 sp types, O stars
C1104-584	NGC 3532	32.152.020 UBVRI phot
		Mermilliod, Mayor (Geneva) rv, RG
C1112-609	NGC 3603	32.114.081 sp types, O stars
		34.153.060 SIT Vidicon surface phot
C11115-624	IC 2714	Mermilliod (Geneva) 7-color phot, RG
		Mermilliod, Mayor (Geneva) rv, RG
C1123-429	NGC 3680	34.153.042 CMT_1T_2, DDO phot, RG, abund, homogeneity
		Mermilliod (Geneva) 7-color phot, RG
		Mermilliod, Mayor (Geneva) rv, RG
C1141-622	NGC 3960	30.153.017 BV, DDO phot, E, d, age, CN abund
		Mermilliod, Mayor (Geneva) rv, RG
C1141-622	Stock 14	32.153.018 UBV phot, d, age V810 Cen
		34.153.013 MK sp types, E, cepheid
C1148-554	NGC 3960	30.153.017 BV, DDO phot, E, d, age, Cn abund
		Mermilliod, Mayor (Geneva) rv, RG
C1204-609	NGC 4103	30.153.024 uvbyHβ phot, E, d, Be star, age
C1221-616	NGC 4349	Claria (Cordoba) UBV, CMT_1T_2 phot, abund
		Mermilliod, Mayor (Geneva) rv, RG
C1222+263	Mel 111	31.153.016 uvby phot, He abund 32.111.012 pm
	(Coma)	32.114.053 Mg II h and K studies, chromospheric activity
		34.153.065 lf 34.153.073 bright stars, distr
C1226-605	Coll 258	Mermilliod, Mayor (Geneva) rv, RG
C1227-645	NGC 4463	Mermilliod, Mayor (Geneva) rv, RG
C1232+365	Upgren 1	31.153.041 uvby phot, rv, cluster identification, d, age
C1236-508	Ru 106	31.153.021 HI
C1239-627	NGC 4609	Claria (Cordoba) UBV, DDO, CMT_1T_2 phot, abund
		Mermilliod, Mayor (Geneva) rv, RG
C125+275	NGC 4745	33.153.041 stellar orbits
C1250-600	NGC 4755	Shobbrook (1984) uvbyHβ phot, ZAMS, d, age

C1313+179	NGC 5053	31.153.046 UBV phot catalog
C1315-623	Stock 16	34.153.029 star formation
C1315-669	Coll 268	Mermilliod, Mayor (Geneva) rv, RG
C1324-587	NGC 5138	Claria (Cordoba) CMT_1T_2 phot, abund
		Mermilliod, Mayor (Geneva) rv, RG
C1343-626	NGC 5281	Mermilliod, Mayor (Geneva) rv, RG
C1350-616	NGC 5316	Mermilliod, Mayor (Geneva) rv, RG
C1404-480	NGC 5460	31.153.021 HI Mermilliod, Mayor (Geneva) rv, RG
		Claria (Cordoba) UBV, DDO, CMT_1T_2 phot, abund
C1426-605	NGC 5617	32.153.044 UBV phot, cepheid, memb
		Mermilliod, Mayor (Geneva) rv, RG
C1440+697	Coll 285	32.114.053 Mg II h and k studies, chromospheric activity
	U Ma	33.153.026 Sirius moving group, supercluster, abund
		34.153.005 sp, A stars, memb
		34.153.037 internal structure
		34.153.066 sp age criteria, kinematics
C1445-543	NGC 5749	Mermilliod, Mayor (Geneva) rv, RG
C1501-541	NGC 5822	30.153.024 uvbyHβ phot, E, d, age
		Claria, Lapasset (1984) UBV, DDO, CMT_1T_2 phot, G and K stars, memb, E, d, age, abund, mass of RG
		Anthony-Twarog (Kansas), van Altena (Yale) pm, memb
		Mermilliod (Geneva) 7-color phot, RG
		Mermilliod, Mayor (Geneva) rv, RG
C1502-554	NGC 5823	30.153.017 BV, DDO phot, E, d, age, CN abund
		Mermilliod, Mayor (Geneva) rv, RG
C1601-517	Lynga 6	33.153.004 uvbyHβ phot, d, E
		Mermilliod, Mayor (Geneva) rv, RG
C1609-540	NGC 6067	33.153.009 phot, cepheids
		Claria (Cordoba) DDO, CMT_1T_2 phot, abund
		Mermilliod, Mayor (Geneva) rv, RG
C1614-550	Coll 299	Topaktas et al (1984) RGU phot
C1614-577	NGC 6087	33.153.043 pm memb Mermilliod, Mayor (Geneva) rv, RG
C1622-405	NGC 6124	Mermilliod, Mayor (Geneva) rv, RG
C1636-432	NGC 6192	32.153.017 UBV phot 34.153.007
C1650-417	NGC 6231	33.153.016 uvby, Hγ, Hε phot, sp, abund
		33.122.089 β Cep stars 34.153.068 rv, rot vel
		34.153.032 uvbyHβ phot, d, ZAMS
		34.122.053 β Cep stars, statistics
		Keenan et al (1984) high resolution sp, abund
C1652-394	NGC 6242	Claria (Cordoba) UBV, DDO, CMT_1T_2 phot, abund
		Mermilliod, Mayor (Geneva) rv, RG
C1654-447	NGC 6249	Mermilliod, Mayor (Geneva) rv, RG
C1657-446	NGC 6259	Mermilliod, Mayor (Geneva) rv, RG
C1658-396	NGC 6268	Mermilliod, Mayor (Geneva) rv, RG
C1701-378	NGC 6281	Claria (Cordoba) UBV, DDO, CMT_1T_2 phot, abund
		Mermilliod, Mayor (Geneva) rv, RG
C1715-387		31.153.012 Walraven, VRIJHKL phot, Of stars, WN stars
C1717-360	NGC 6334	32.114.081 sp types, O stars
C1720-499	IC 4651	31.153.007 DDO phot, E, D, abund, blue stragglers
		34.153.042 CMT_1T_2, DDO phot, RG, abund, homogeneity
		34.114.022 sp, CN and CH bands, giants
		Mermilliod (Geneva) &-color phot, RG
		Mermilliod, Mayor (Geneva) rv, RG
C1725-294	Tru 26	Mermilliod, Mayor (Geneva) rv, RG
C1732-334	Tru 27	33.153.015 UBVRI, JHKLM phot, R, d, age
C1734-362	Ru 127	Mermilliod, Mayor (Geneva) rv, RG
C1736-321	NGC 6405	31.153.006 pg phot, white dwarfs
C1741-323	NGC 6416	Mermilliod, Mayor (Geneva) rv, RG

C1743+057	IC 4665	30.153.029 Vilnius phot, MK types
C1743-315	NGC 6425	Mermilliod, Mayor (Geneva) rv, RG
C1753-190	NGC 6494	33.153.010 pm, cluster mass
C1757-442	NGC 6496	33.153.041 stellar orbits
C1800-279	NGC 6520	Mermilliod, Mayor (Geneva) rv, RG
C1801-243	NGC 6530	30.115.012 UBV Hβ phot, d, pre-ms
		33.153.039 pm 33.153.041 stellar orbits
		Feinstein (La Plata) UBVRI phot
C1804-233	NGC 6546	Mermilliod, Mayor (Geneva) rv, RG
C1806-240	Coll 367	Claria, Lapasset (Cordoba) UBV phot, OB association ?
C1816-138	NGC 6611	34.112.058 pol, Be stars
		Feinstein (La Plata), Westerlund (Uppsala) UBVRI phot
		The, Walraven (Amsterdam) WULBV, JHKL phot
C1828-192	IC 4725	31.153.049 uvbyHβ phot, memb, d, E
	(M25)	33.154.041 stellar orbits
C1834-082	NGC 6664	32.153.004 uvbyHβ phot, d, var E, age, blue stragglers
C1836+054	IC 4756	33.153.036 DDO phot, abund
C1848-063	NGC 6705	32.153.012 dynamics
C1849+102	NGC 6709	33.153.014 pm, memb
C1850-204	Coll 394	31.153.021 HI
C1919+377	NGC 6791	30.153.027 UBV phot, E, d
		Cudworth, Anthony-Twarog (Yerkes) pm
C1941+231	NGC 6823	31.114.086 sp types
C2014+374	IC 4996	32.113.063 pg phot, E, d
C2018+385	Berk 86	30.153.011 UBV phot, E, d, age
C2019+372	Berk 87	32.153.036 UBV phot, molecular cloud complex
C2021+406	NGC 6910	34.153.062 pm, memb
C2022+383	NGC 6913	30.114.183 UBVRI phot, energy distr
		31.111.010 pm, memb
		34.153.024 UBV phot, var E, d, agesp
C2032+281	NGC 6940	32.153.033 structure of cluster
		Gotz (1984) FG Vul, light curve
C2042+355	X Cyg	30.153.006 new cluster, cepheid, UBV phot, sp, E, d
C2059+679	NGC 7023	34.153.003 IR phot, reflection nebulae, star formation
C2111+422	NGC 7044	31.153.010
C2130+482	NGC 7092	30.153.029 Vilnius phot, MK types
	(M39)	32.153.038 phot sequence
		33.153.026 Sirius moving group, supercluster, abund
		Platais (1984) UBV phot, pm, memb
C2137+573	Tru 37	32.113.048 UBVRI phot
		Marschall (Gettysburg), van Altena (Yale) pm, memb
C2144+655	NGC 7142	30.153.029 d
C2152+623	NGC 7160	30.153.015 phot 32.113.048 UBVRI phot
C2227+551	Berk 96	Del Rio (1984) UBV phot, d, earliest sp
C2245+578	NGC 7380	33.153.013 memb, statistical methods, pre-ms
C2313+602	Mark 50	34.153.004 UBV phot, d, age, WN star, mass loss
C2322+613	NGC 7654	30.153.012 uvbyHβ phot, E, d, SG
C2354+564	NGC 7789	30.122.192 BVR phot, carbon stars, var stars
		31.153.024 RG 31.153.025 He abund
		31.115.004 giants, linear radii
		32.153.025 pol, memb, blue stragglers
		Stryker and Hrivnak (1984) blue stragglers
		34.153.057 giant branch
Sco-Cen		33.153.028 Bp, Ap stars 34.153.037 internal structure

References to Table 5

Anthony-Twarog, B.J. 1984, Astron.J. 89, 267
Baliunas, S.L., Gurnan, E.F. and Hartmann, L. 1984, "Cool Stars, Stellar Systems and the Sun", Springer Verlag, p. 223
Barkhatova, K.A., Zacharova, P.Y., Orechova, L.K. and Shashkina, L.P. 1984, Astron.Zh., in press.
Bijaoui, A., Lacoarret, M. and Mais, G. 1984, Astron.and Astrophys. Suppl. 55, 393.
Bohm-Vitense, E., Baliunas, S.L. and Hartmann, L. 1984, "Cool Stars, Stellar Systems and the Sun", Springer Verlag, p. 273.
Claria, J.J. and Lapasset, E. 1984, Mon.Not.Roy.Astron.Soc. (in press).
Claria, J.J., Lapasset, E., Levato, H. and Malaroda, S. 1984. Astrophys. and Sp. Sci. (in press).
Del Rio, G. 1984, Astron.and Astrophys. Suppl. 56, 289.
Gotz, W. 1984, IAU Inf.Bull.Var.Stars, No. 2461, p. 1
Janes, K.A. and Smith G.H. 1984, Astron.J. 89, 487.
Keenan, F.P., Brown, P.J.F., Dufton, P.L. and Lennon, D.J. 1984. Astrophys.J. Lett. 279, 11.
Levato, H. and Garcia, B. 1984. Astrophys.Lett. 24, 49.
Levato. H. and Malaroda, S. 1984. Astrophys.Lett. 24, 37.
McGregor, P.J. and Hyland, A.R. 1984. Astrophys.J. 227, 149.
Peterson, R.C., Carney, B.W. and Latham, D.W. 1984. Astrophys.J. 279, 237.
Platias, I.K. 1984. Astron.Zh. 10, 203.
Schild, R.E. 1984. Publ.Astron.Soc.Pacific. 95, 1021.
Shobbrook, R.R. 1984. Mon.Not.Roy.Astron.Soc. 206, 273.
Smith, G.H. and Norris, J.E. 1984. Astron.J. 89, 263.
Stryker, L. and Hrivnak, B.J. 1984. Astrophys.J. 278, 215.
Topaktas, L., Kandemir, G., Boydag, S. and Fenkart, R. 1984. Astron.and Astrophys. Suppl. 56, 11.
Vandenberg, D.A. and Bridges, T.J. 1984. Astrophys.J. 278, 679.

SURVEY PAPERS ON OPEN CLUSTERS

Abt, Cardona (34.153.019) "Confirmation Among Visual Multiples of an Increase of Ap Stars with Age".
Abt, H.A. and Cardona, O. 1984. "The Nature of the Visual Companions of Ap and Am Stars", Astrophys.J. 276, 266.
Balazs, Lynga (34.153.022) The Galactic Distribution of Open Clusters if Different Ages and the Anglular Velocity of the Spiral Structure".
Balona, L.A. and Shobbrook, R.R. 1983. "The Zero Point Calibration of the Cepheid Luminosity Scale from a Calibration of the Luminositires of Early Type Stars", Mon.Not.Roy.Astron.Soc. 205, 309.
Barkhatova, Pavlovskaya (30.153.034) "Probable Multiple Galactic Star Clusters"
Barkhatova, Pylskaya (34.153.069) "On the Scale of Distances of the Galaxy Based on the Study of Open Star Cluster Subsystems".
van den Bergh (31.153.001) "Formation Rate and Decay Time Scales of Open Clusters Near the Sun".
Bruck, Sanders (33.153.022) "The Absolute Masses of 72 Galactic Clusters and 12 OB Associations".
Claria, J.J. 1984. Astrophys. and Sp.Sci (in press).
Cayrel, Cayrel, Campbell (34.153.033) "Steps Towards the Abundance Scale I - the Nearest Open Clusters".
Danilov (31.153.014) "On the Dynamics of the Early Stages of Evolution of Open Star Clusters I".
Danilov (31.151.061) "On the Dynamics of the Early Stages of Evolution of Open Star Clusters II".

Elmegreen (33.153.029) "Quiescent Formation of Bound Galactic Clusters".
Elmegreen (34.153.064) "The Initial Stellar Mass Function as a Statistical Ensemble, and Implications for the Formation of Bound Clusters".
Esin (31.153.038) "Observational Isochrones Determined by Open Clusters and their Comparison with Theoretical Results".
Freedman (34.153.063) "The Effects of Crowding on the Determination of Apparent Luminosity Functions".
Gieseking (31.153.009) "Investigations of the Kinematics of Open Clusters and OB Associations from Radial Velocity Measurements".
Glaspey (32.153.015) "Simulated Rotational Velocity Distributions in Open Clusters".
Gotz (32.153.034) "On the Behaviour of Bright Double Stars and Yellow Giants in Open Clusters"
Gratton, R.G., Pilachowski, C.A. and Sneden C. 1984. "H-alpha Emission in Old Giants", Astron.and Astrophys. 132, 111.
Guthrie (31.116.003) "The Bimodal Distribution of Rotational Velocities of Late B-type Stars in Galactic Clusters".
Haro (34.153.052) "Post T Tauri Stars in Galactic Clusters".
Ivanov (33.153.023) "The Gradient of Age of Open Clusters Across the Spiral Arm of the Galaxy".
Janes, Adler (32.153.005) "Open Clusters and Galactic Structure".
Lundstrom, Stenholm (32.153.022) "Wolf-Rayet Stars in Open Clusters and Associations".
Lynga (30.002.036) "The Lund-Strasbourg Catalogue of Open Cluster Data".
Lynga (31.153.044) "Open Clusters in our Galaxy"
Lynga (31.153.051) "Evolutionary Effects among Open Clusters".
Lynga (32.153.035) "IAU Numbers for some Recently Discovered Clusters".
Lynga (34.153.011) "IAU Commission no. 37; IAU Numbers for some Recently Discovered Clusters".
Lynga (34.002.099) "Open Star Clusters and the Evolution of the Galatic Disk".
Lynga (34.002.127) "The Lund Catalogue of Open Cluster Parameters".
McGregor, P.J. and Hyland, A.R. 1984. "A Photometric Comparison of Late-type Cluster Supergiants in the Magellanic Clouds and the Galaxy", Astrophys.J. 277, 141.
Maitzen (30.113.030) "CP Stars in Open Clusters".
Manduca (34.153.056) "Theoretical Integrated Spectra of Star Clusters".
Mathieu (33.131.156) "Dynamical Constraints on Star Formation Efficiency".
Mayor, Mermilliod (33.120.012) "Duplicity in the HR Diagram".
Mermilliod (31.153.028) "Stellar Content of Young Open Clusters I. Blue Stragglers".
Mermilliod (31.153.029) "Stellar Content of Young Open Clusters II. Be Stars".
Mermilliod (31.153.031) "Analysis of UBV Data in Open Clusters".
Mermilliod (31.115.011) "Composite Color-Magnitude and Color-Color Diagrams for Be Stars in Open Clusters".
Mermilliod (33.120.013) "The Nature of the Binary Stars in the HR Diagrams of Open Clusters".
Mermilliod (34.153.043) "Stellar Content of Young Open Clusters III. The 'Sn' Stars".
Mermilliod (34.153.071) "Age Groups and Composite Diagrams for Young Open Clusters".
Mermilliod, J.C. 1984. "Bibliography of Individual Radial Velocities for Stars in Open Clusters II. NGC and IC Clusters", Bull.Inf.Centre Donnees Stellaires, No. 26, p. 9.
Meylan, Maeder (33.153.008) "Comparisons on the HR Diagrams of the Youngest Clusters in the Galaxy, the LMC, SMC: Evidence for a Large MS Widening".
Myakutin, V.I., Sagar, R. and Joshi, V.C. 1984. "Catalogue of Masses and Ages of Stars in Twelve Open Clusters", Bull.Inf.Centre Donnees Stellaires, No. 26, p. 103.

Nicolet (30.153.028) "Geneva Photometric Boxes III. Distances and Reddenings for 43 Open Clusters:.
Nissen (34.153.010) "Helium Abundances from Young Stars and Open Clusters".
Popova, Krajcheva, Bakoev (34.153.036) "Distribution of Eclipsing and Spectroscopic Binaries in Regions of Open Clusters".
Rastorgriev, Samus (31.011.013) "Star Clusters and Problems of Stellar Evolution".
Roth (34.002.099) "A Catalogue of Open Cluster Ages".
Ruprecht (34.153.069) "Structural Properties of Open Star Clusters and their Relation to the Galaxy".
Sagar, Joshi, Sinvhal (34.153.006) "Integrated Photometric Parameters of Open and Globular Clusters".
Shobbrook (31.122.204) "uvbyHβ Photometry of Young Star Clusters Containing β CMa Stars".
Tarrab (31.153.045) "The Initial Mass Function for Young Open Clusters I".
Tarrab (32.153.008) "Integrated Colors for Young Open Clusters as a Function of Age".
Wramdemark (31.153.058) "Radial Velocities of Open Clusters".
Zakharova (30.153.030) "On Universality of the Zero Age Luminosity Function of Open Star Clusters".
Zhao, Tian (33.153.038) "Determination of Membership in Open Clusters by Means of Statistical Decision Theory".

6. Globular Clusters

In this edition of the triennial report is found a noticeably greater abundance of information concerning clusters in galaxies other than our own. Obviously the impact of high quantum efficiency detectors and large telescopes is being felt. Catalogs of cluster information are given in Table 6. Then Table 7 presents new data for the galactic globular clusters, and Table 8 gives data for globular clusters in other galaxies. Information about variable stars in globular clusters (except that found here which pertains to cluster chemical abundance estimates etc) will be given in the Commission 27 (Variable Star) Report. Abbreviations used are :

abund = abundance(s)
atm = atmosphere(s)
betw = between
BHB = blue horizontal branch
bl = blue
br = brightness
BV = Johnson B and V mag
By = billion years
chem = chemical
cl = cluster(s)
CMD = color-magnitude diagram
col = color(s)
comp = compared
disc = discussion
distr = distribution
el = element(s)
em = emission
evol = evolution(-ary)
gt = giant(s)
HB = horizontal branch
he = heavy elements
integr = integrated
lf = luminosity function
m-p = metal poor
m-r = metal rich
ms = main sequence
no. = number
obs = observation(s)
pe = photoelectric
pd = period(s)
phot = photometry, photometric
res = resolution
RG = red giant(s)
rv = radial velocity
seq = sequence(s)
sp = spectrum (-tra,-tral)
sp-phot = spectrophotometry
st = stars
str = strength(s)
synth = synthetic(-esis)
temp = temperature
theo = theoretical
UV = ultraviolet
var = variation

IR = infrared   w/ = with

TABLE 6. CATALOGS OF CLUSTER INFORMATION

Bica, Pastoriza (33.154.038) integr DDO and UBV phot (91 cl)
Brodie, Hanes (31.154.054) integr sp-phot (27 cl)
Cacciari et al (32.154.019) UV and optical sp-phot w/IUE of Magellanic Cloud cl
Freeman, Illingworth, Oemler (34.156.010) rv, kinematics of 35 LMC cl
Frenk, Fall (31.154.014) ellipticity-age relation (93 cl in Galaxy, 52 in LMC)
Frogel, Cohen, Persson (34.154.102) V-K col, CO, abund (31 cl)
Geyer, Richtler (31.154.044) axial ratios for 25 LMC cl
Grindlay (29.154.017) survey for X-ray sources (30 cl)
Hertz, Grindlay (34.154.100) X-ray survey, 14 new sources in 8 cl (71 cl)
Hodge (31.159.017) ages for 48 LMC and 18 SMC cl, abund for 24 LMC and 5 SMC cl
Huchra, Stauffer, van Speybrock (32.154.014) dynamics, abund of M31 cl
Kontizas, Kontizas (33.156.010) radii and structural parameters (23 SMC cl)
Kontizas, Danezis, Kontizas (31.159.036) radii and structural parameters (23 SMC cl)
Mould, Aaronson (32.154.041) extended giant granch (30 LMC, 12 SMC cl)
Persson et al (33.156.006) integr IR phot of 84 LMC, SMC cl
Sandage (31.154.001) Oosterhoof pd groups, cl ages (30 cl)
Sharov, Lyutyj, Esipov (32.154.011) pe phot of 58 M31 cl
Straizys (31.154.010) 2D classification (75 cl)
van Albada, de Boer, Dickens (29.154.017) far UV phot (27 cl)
van den Bergh (30.159.010) UBV integr phot for 147 LMC and 61 SMC cl; ages
Zdanavicius (33.154.005) integr phot in Vilnius system (39 cl)

TABLE 7. NEW DATA FOR INDIVIDUAL CLUSTERS

C0021-723 (NGC 104, 47 TUC)

29.154.007 ratio of HB to RG st
33.154.017 abund comp to Arcturus
33.154.054 Hα em in RG
31.154.066 CMD turnoff
31.154.058 He abund
31.159.020 age-abund relation
29.154.018 CN and Na anomalies
32.154.008 var in N on ms
32.154.007 abund -0.8 dex
34.154.012 apparent flattening
31.154.025 optical, X-ray positions
31.154.022 possible diffuse X-ray em
32.154.024 TiO band str for RG
31.154.003 anticorrelation of C, N
34.154.036 correlation of Al I, CN
34.154.005 high res sp, abund
33.154.019 Stromgren phot, RG
31.154.051 evol seq, isochrones
33.154.062 sp of 11 dwarfs, 5 subgt
34.154.036 He abund, Iben's R-method
29.154.002 faint BV pe seq
32.154.005 sp abund indicators
31.154.064 disc of cl ages
34.154.002 abund for RG in m-r cl
32.154.001 st counts, ms lf
31.154.021 supra-bl st UV col
29.154.032 IR phot 64 RG
31.154.059 cl lf differences
33.154 061 BV CMD to V = 22
34.154.056 rv for 169 RG
29.154.037 UV energy distr
31.154.034 CN distr in RG
31.154.033 abund, cl near Gal center
32.154.017 integr sp of cl
34.154.061 abund in m-r cl
33.154.013 synth CMD

C0050-268 (NGC 288)

31.154.066 cl CMD turnoffs
31.154.064 disc of cl ages
33.154.055 [Fe/H] = -1.0
33.154.013 synth CMD
32.154.005 sp abund indicators
31.059.020 age-abund relation
33.154.019 Stromgren phot, RG

C0100-711 (NGC 362)
34.154.028 UBV br profiles
33.154.054 Hα em in RG
32.154.005 sp abund indicators
31.154.029 abund in m-r cl
32.154.040 UBV CMD
34.154.005 high res sp, abund
33.154.015 chem inhomogeneity, CN
34.154.036 He abund, Iben's R-method
30.154.033 IUE sp of central region
31.059.020 age-abund relation
34.154.012 apparent flattening
31.154.034 CN distr in RG
32.154.017 integr sp of cl

C0310-554 (NGC 1261)
33.154.059 Anticorrelation between Y, Z confirmed

C0354-498 (AM-1)
33.154.063 BV CMD

C0443+313 (Pal 2)
30.154.001 IR studies

C0512-400 (NGC 1851)
34.154.036 He abund, Iben's R-method
31.154.025 optical, X-ray positions
29.154.037 UV energy distr
31.154.027 radial UBV distr
32.154.032 CN strong st
29.154.030 BV CMD

C0522-245 (NGC 1904, M79)
31.154.059 cl lf differences
33.154.019 Stromgren phot, RG
32.154.017 integr sp of cl

C0911-646 (NGC 2808)
33.154.029 IUE sp of cl nucleus
32.154.026 m-p st abund
34.154.101 CN distr of RG
31.154.027 radial UBV distr
33.154.003 Hα em in RG

C1015-461 (NGC 3201)
30.154.030) BV CMD to ms (1452 st)
33.154.054 Hα em in RG
34.154.002 abund for RG in m-r cl
32.154.037 structure comp to NGC 1806
32.154.026 m-p st abundances
31.154.059 cl lf differences
34.154.036 correlation of Al I, CN
31.154.004 CN distr
33.154.013 synth CMD
34.154.036 He abund, Iben's R-method
31.159.020 age-abund relation
30.154.013 IR phot for 26 RG
34.154.012 apparent flattening
33.154.003 Hα em in RG
31.154.034 CN distr in RG
34.154.005 high res sp, abund
34.154.001 RR Lyr [Fe/H]

C1126+292 (Pal 4)
34.154.011 search for HI

C1207+188 (NGC 4147)
34.154.036 He abund, Iben's R-method
34.154.050 echelle sp of RG, abund
33.154.059 anticorrelation of Y,Z

C1223-724 (NGC 4372)
34.154.012 apparent flattening

C1236-264 (NGC 4590, M68)
33.154.059 anticorrelation of Y,Z
34.154.001 RR Lyr [Fe/H]

C1256-706 (NGC 4833)
34.154.012 apparent flattening
31.154.059 cl lf differences

34.154.005 high res sp, abund

C1320+184 (NGC 5024, M53)

33.154.059 anticorrelation of Y,Z
31.154.059 cl lf differences
29.154.011 synth HB

C1313+179 (NGC 5053)

34.154.011 search for HI
34.154.050 echelle sp of RG, abund
31.154.064 disc of cl ages
33.154.013 synth CMD

C1323-472 (NGC 5139, ω Cen)

33.154.054 Hα em in RG
29.154.002 faint BV pe seq
29.154.008 He abund from RR Lyr
31.154.066 CMD turnoff
30.154.007 abund in RG
34.154.012 apparent flattening
31.154.059 cl lf differences
34.154.020 10 S-st from ZrO bands
30.154.044 BV pe seq
34.154.036 correlation of Al I, CN
34.154.041 integrated br and col
33.154.013 synth CMD
31.154.032 he abund in RG
29.154.003 CMD from 300 faint st
33.154.059 anticorrelation of Y,Z
31.154.064 disc of cl ages
31.154.069 CMD of subgt branch
33.154.003 Hα em in RG
31.154.022 possible diffuse X-ray em
29.154.001 he abund variations in RG
31.154.034 CN distr in RG
30.154.011 RR Lyr, PLA relation
29.154.028 abund gradient of Ca

C1339+286 (NGC 5272, M3)

33.154.041 BV CMD near center
34.154.036 He abund, Iben's R-method
31.154.064 disc of cl ages
32.154.027 isochrone fitting to CMD
31.154.059 cl lf differences
31.154.012 width of subgt, HB seq
30.154.037 Ca, C, N abund for RG
33.154.013 synth CMD
34.154.011 search for HI
31.154.058 He content
34.154.088 rv for bl stragglers
34.154.012 apparent flattening
30.154.011 RR Lyr, PLA relation
33.154.060 br st phot in nucleus
31.154.037 C, N abund for RG

C1353-269 (AM-4)

31.154.077 discovery

C1403+287 (NGC 5466)

34.154.011 search for HI
34.154.034 membership of br st
34.154.050 echelle sp of RG, abund
31.154.016 proper and space motion
31.154.059 cl lf differences

C1500-328 (NGC 5824)

30.154.033 IUE sp of central region
29.154.037 UV energy distr

C1513+000 (Pal 5)

29.154.027 str of sp features for RG

C1516+022 (NGC 5904, M5)

30.154.034 IUE sp
29.154.007 ratio of HB to RG st
33.154.044 N-rich, UV-br st
31.154.072 BV phot, CMD
34.154.036 He abund, Iben's R-method
31.154.058 He abund
32.154.027 isochrone fitting
31.154.036 sp comp w/RG in M71
31.154.034 CN distr in RG
33.154.016 IUE sp, synth models
34.154.011 search for HI
30.154.002 BV phot, CMD
30.154.005 radial col distr
33.154.054 Hα em in RG
31.154.064 disc of cl ages
34.154.082 energy distr for HB A-st
33.154.031 UV sp
34.154.036 correlation of Al I, CN

32.154.004 high res sp of Hα in RG
31.154.033 abund, cl near Gal center
33.154.002 CN distr in RG
33.154.013 synth CMD
33.154.018 errors in high res RG sp
34.154.005 high res sp, abund
34.154.061 abund in m-r cl

C1524-505 (NGC 5927)
34.154.002 abund for RG in m-r cl

C1608+150 (Pal 14)
31.154.090 CMD
32.154.003 st counts, structure

C1614-228 (NGC 6093, M80)
30.154.019 IUE obs
29.154.037 UV energy distr
30.154.033 IUE sp of central region

C1620-264 (NGC 6121, M4)
29.154.007 ratio of HB to RG st
33.154.054 Hα em in RG
31.154.038 high res sp of BHB st
31.154.034 CN distr in RG
31.154.068 faint bl st: 2 WD st ?
34.154.036 He abund, Iben's R-method
33.154.059 anticorrelation of Y,Z
30.154.012 CN distr from sp of 45 RG
34.154.036 correlation of Al I, CN
30.154.011 RR Lyr, PLA relation

C1629-129 (NGC 6171, M107)
34.154.036 He abund, Iben's R-method
34.154.002 abund for RG in m-r cl
31.154.033 abund, cl near Gal center
34.154.001 [Fe/H] for RR Lyr
33.154.059 anticorrelation of Y,Z
31.154.059 cl lf differences
30.154.011 RR Lyr, PLA relation

C1639+365 (NGC 6205, M13)
29.154.007 ratio of HB to Rg st
31.154.058 He abund
31.154.021 UV col for supra-bl st
32.154.027 isochrone fitting
31.154.059 cl lf differences
34.154.082 energy distr of HB A-st
31.154.037 C, N anund for RG
33.154.013 synth CMD
34.154.011 search for HI
31.154.064 disc of cl ages
32.154.007 theo calib of metallicity
34.154.012 apparent flattening
31.154.055 abund, temp of Pop II gt
30.154.037 Ca, C, N abund for 29 RG
30.154.003 far UV phot, energy distr

C1644-018 (NGC 6218, M12)
34.154.036 He abund, Iben's R-method
33.154.056 rv, UV-br st

C1645+476 (NGC 6229)
34.154.050 echelle sp, [Fe/H] for RG

C1654-040 (NGC 6254, M10)
34.154.036 He abund, Iben's R-method
31.154.059 cl lf differences
34.054.074 BV phot, ms turnoff st
33.154.054 Hα em in RG
34.154.005 high res sp, abund

C1656-370 (NGC 6256)
34.154.007 provisional CMD

C1658-300 (NGC 6266, M62)
29.154.011 synth HB
34.154.012 apparent flattening

C1659-262 (NGC 6273, M19)
34.154.012 apparent flattening

C1715+432 (NGC 6341, M92)
34.154.006 st near cl center, BV CMD
31.154.058 He abund
32.154.009 C,N abund for RG st
32.154.027 isochrone fitting
31.154.059 cl lf differences
34.154.082 energy distr of HB A-st
34.154.003 isochrones: age = 18 By
30.154.003 far UV phot, energy distr
34.154.036 He abund, Iben's R-method
31.154.064 disc of cl ages
32.154.007 theo calib of cl abund
34.154.012 apparent flattening
31.154.055 abund, temp for Pop II gt
29.154.037 UV energy distr
34.154.002 width and lf for ms
33.154.013 synth CMD

C1721-484 (NGC 6352)
34.154.002 abund for RG in m-r cl
31.154.076 ms phot
31.154.030 metallicity
31.154.033 abund, cl near Gal center

C1720-177 (NGC 6356)
34.154.012 apparent flattening

C1724-307 (Ter 2)
31.154.025 optical, X-ray positions

C1726-670 (NGC 6362)
34.154.012 apparent flattening
31.154.033 abund, cl near Gal center

C1725-050 (NGC 6366)
30.154.042 B lf

C1728-34x (4U/MXB 1728-34, GX 354+0)
29.154.035 discovery via JHK photometry

C1730-333 (Liller 1)
31.154.025 optical, X-ray positions

C1732-447 (NGC 6388)
29.154.009 BV CMD

C1732-304 (Trz 1)
29.154.036 discovery of new X-ray burster

C1736-536 (NGC 6397)
34.154.039 age, abund for turnoff st, Pop II calibration for uvby system
31.154.071 uvby phot, turnoff st
33.154.054 Hα em in RG
31.154.066 CMD turnoff
33.154.024 IUE sp of 3 BHB st
31.154.024 synth from IUE sp
32.154.026 m-p st abund
31.154.055 abund, temp of Pop II st
31.154.020 age
33.154.013 synth CMD
31.154.102 structure of core
29.154.002 faint BV pe seq
31.154.064 disc of cl ages
32.154.001 st counts, ms lf
31.154.021 UV col for supra-bl st
33.154.003 Hα em in RG
31.154.038 high disp sp of BHB st
30.154.003 far UV phot, energy distr

C1745-247 (Trz 5)
29.154.036 discovery of new X-ray burster

C1746-370 (NGC 6441)
31.154.025 optical, X-ray positions

C1804-250 (NGC 6544)
33.154.023 BV CMD

C1806-259 (NGC 6553)
34.154.002 abund for RG in m-r cl

C1820-303 (NGC 6624)
30.154.020 em line sp from IUE
29.158.062 IUE obs comp to NGC 6624
31.154.025 optical, X-ray positions

C1821-249 (NGC 6626, M28)
29.154.031) BV CMD

C1827-255 (NGC 6638)
34.154.004 BV CMD

C1828-323 (NGC 6637, M69)
34.154.002 abund for RG in m-r cl
34.154.012 apparent flattening
31.154.029 abund in m-r cl

C1832-330 (NGC 6652)
34.154.012 apparent flattening

C1833-239 (NGC 6656, M22)
34.154.021 BV CMD to ms
30.154.007 abund in RG
34.154.012 apparent flattening
33.154.003 Hα em in RG
33.154.011 chem inhomogeneity
34.154.036 correlation of Al I, CN
30.154.003 far UV phot, energy distr
33.154.054 Hα em in RG
31.159.020 age-abund relation
32.154.026 m-p st abund
31.154.022 possible diffuse X-ray em
31.154.034 CN distr in RG
32.154.028 abund in six st

C1838-198 (Pal 8)
30.154.001 IR studies

C1840-323 (NGC 6681, M70)
34.154.012 apparent flattening

C1850-057 (NGC 6712)
33.154.059 anticorrelation of Y,Z
30.154.031 BV CMD
31.154.025 optical, X-ray positions
34.154.061 abund in m-r cl

C1856-367 (NGC 6723)
33.154.059 anticorrelation of Y,Z
34.154.012 apparent flattening
31.154.033 abund, cl near Gal center
31.154.029 abund in m-r cl
31.154.059 cl lf differences
34.154.061 abund in m-r cl

C1902+017 (NGC 6749)
30.154.001 IR studies
30.154.038 BV phot near center

C1906-600 (NGC 6752)
34.154.036 He abund, Iben's R-method
29.154.002 faint BV pe seq
31.154.058 He abund
29.154.018 CN, Na anomalies
32.154.027 isochrone fitting
31.154.055 abund, temp of Pop II gt
29.154.006 bimodal CN distr for RG
34.154.036 correlation of Al I, CN
33.154.054 Hα em in RG
31.154.066 CMD turnoff
31.154.064 disc of cl ages
32.154.001 st counts, ms lf
34.154.012 apparent flattening
29.154.037 UV energy distr
31.154.034 CN distr for RG
34.154.005 high res sp, abund

31.154.006 C, N abund var
33.154.013 synth CMD
30.154.003 far UV phot, energy distr

C1914+300 (NGC 6779, M56)
34.154.011 search for HI
30.154.021 memb via proper motion
33.154.056 rv, UV-br st

C1916+184 (Pal 10)
30.154.001 IR studies

C1936-310 (NGC 6809, M55)
34.154.036 He abund, Iben's R-method
31.154.059 cl lf differences
31.154.004 CN distr
30.154.003 far UV phot, energy distr
34.154.012 apparent flattening
31.154.034 CN distr in RG
34.154.054 CN, CH bands for 2 st

C1951+186 (NGC 6838, M71)
31.154.007 synth col, [M/H] = -0.9
31.154.064 disc of cl ages
32.154.007 [M/H] = -0.8
31.154.034 CN distr in RG
32.154.004 high res sp at Hα in RG
31.154.074 Stromgren 4-col CMD
31.154.033 abund, cl near Gal center
31.154.051 grid of isochrones
33.154.017 abund comp to Arcturus
34.154.002 abund for RG in m-r cl
31.154.036 sp comp w/RG in M5
31.154.035 abund
33.154.018 errors in high res sp of RG
34.154.083 [Fe/H] = -1.0
31.154.004 CN distr
33.154.013 synth CMD

C2050-127 (NGC 6981, M72)
33.154.059 anticorrelation of Y,Z
30.154.011 RR Lyr, PLA relation
31.154.059 cl lf differences

C2003-220 (NGC 6864, M75)
32.154.017 integr sp of cl

C2059+160 (NGC 7006)
31.154.008 CNO not 2nd parameter
29.154.027 str of sp features in RG
32.154.046 C, N, Fe-peak abund

C2127+119 (NGC 7078, M15)
30.154.004 BV CMD near center
31.154.072 BV phot, CMD
34.154.036 He abund, Iben's R-method
32.154.005 abund indicators
31.154.058 He abund
29.154.011 synth HB
32.154.027 isochrone fitting
31.154.059 cl lf differences
34.154.082 energy distr of HB A-st
33.154.030 pop synth, ANS, IUE
32.154.017 integr sp of cl
33.154.037 phot, central core
30.154.003 far UV phot, energy distr
34.154.011 search for HI
33.154.006 UBV phot, CMD
33.154.054 Hα em in RG
33.154.059 anticorrelation of Y,Z
31.154.064 disc of cl ages
31.159.020 age-abund relation
34.154.012 apparent flattening
31.154.025 optical, X-ray positions
29.154.037 UV energy distr
31.154.074 Stromgrn 4-col CMD
34.154.003 isochrones, age = 18 By
33.154.012 C, N abund for RG
33.154.013 synth CMD

C2130-010 (NGC 7089, M2)
33.154.007 BV CMD for central st
31.154.059 cl lf differences
32.154.005 abund indicators from sp
32.154.017 integr sp of cl

C2137-234 (NGC 7099, M30)
32.154.012 UBV phot near center
33.154.054 Hα em in RG

31.154.025 optical, X-ray positions	32.154.017 integr sp of cl
30.154.003 far UV phot, energy distr	33.154.013 synth CMD

C2143-214 (Pal 12)

32.154.005 abund indicators from sp	31.154.064 disc of cl ages
31.154.051 grid of isochrones	33.154.013 synth CMD

C2304+124 (Pal 13)

32.154.046 C,N,Fe-peak abund for RG	29.154.027 str of sp features for RG
32.154.006 [Fe/H] = -1.67	

TABLE 8. NEW DATA FOR CLUSTERS IN OTHER GALAXIES

LARGE MAGELLANIC CLOUD

NGC 1466	29.159.016 cl memb to LMC	32.159.001 [Fe/H], rv of gt
	31.154.021 UV col from IUE sp	31.159.019 new data
	31.159.021 rv, [Fe/H]	32.154.017 integr sp of cl
NGC 1672	31.159.020 age-abund relation	
NGC 1783	31.159.020 age-abund relation	32.154.017 integr sp of cl
NGC 1786	32.159.001 [Fe/H], rv of gt	31.159.018 IUE sp
	31.159.021 rv, [Fe/H]	
NGC 1806	31.154.021 UV col from IUE sp	
NGC 1835	31.159.018 IUE sp	33.154.024 integr IUE sp
	32.159.001 [Fe/H], rv of gt	31.154.021 UV col from IUE sp
	31.159.021 rv, [Fe/H]	
NGC 1841	31.159.020 age-abund relation	32.159.001 [Fe/H], rv of gt
	29.159.016 cl memb to LMC	31.159.019 new data
	31.159.021 rv, [Fe/H]	32.154.017 integr sp of cl
NGC 1846	31.159.020 age-abund relation	32.154.017 integr sp of cl
NGC 1855	31.159.020 age-abund relation	
NGC 1866	33.154.024 integr IUE sp	31.154.021 UV col from IUE sp
	30.159.011 superluminous gt	
NGC 1898	32.154.017 integr sp of cl	
NGC 1978	31.159.018 IUE sp	32.159.001 [Fe/H], rv of gt
	31.154.021 UV col from IUE sp	31.159.021 rv, [Fe/H]
	32.154.017 integr sp of cl	
NGC 1984	31.159.020 age-abund relation	
NGC 1987	31.154.021 UV col from IUE sp	32.154.017 integr sp of cl
NGC 1994	31.159.020 age-abund relation	
NGC 2004	33.154.024 integr IUE sp	31.159.020 age-abund relation
	31.154.021 UV col from IUE sp	
NGC 2019	31.154.021 UV col from IUE sp	
NGC 2100	31.159.020 age-abund relation	
NGC 2121	31.159.020 age-abund relation	34.154.098 age, abund, BV CMD
	32.154.017 integr sp of cl	
NGC 2155	32.159.001 [Fe/H], rv of gt	31.159.021 rv, [Fe/H]
NGC 2173	31.159.020 age-abund relation	32.154.017 integr sp of cl
NGC 2193	31.159.020 age-abund relation	
NGC 2209	31.159.019 new data	32.154.017 integr sp of cl
NGC 2210	31.159.018 IUE sp	31.159.020 age-abund relation
	32.159.001 [Fe/H], rv of gt	31.154.021 UV col from IUE sp
	31.159.021 rv, [Fe/H]	
NGC 2257	31.159.020 age-abund relation	32.159.001 [Fe/H], rv of gt
	29.159.016 memb to LMC	31.159.019 new data
	31.159.021 rv, [Fe/H]	33.156.005 BV CMD to V = 22.4
	31.159.025 BV CMD	

H11	32.159.001 [Fe/H], rv of gt	31.159.021 rv, [Fe/H]

SMALL MAGELLANIC CLOUD

NGC 121	31.159.018 IUE sp	31.159.019 new data
	32.154.017 integr sp of cl	
NGC 339	32.154.017 integr sp of cl	
NGC 411	32.154.017 integr sp of cl	
NGC 416	32.154.017 integr sp of cl	
NGC 419	31.154.021 UV col from IUE sp	32.154.017 integr sp of cl
Kron 3	31.159.019 new data	32.154.017 integr sp of cl
Linds 1	31.159.019 new data	

31.159.012 lf comp to cl in Milky Way: Lindsay 3, 11, 13, 14, 15, 20, 82 and HW 62

OTHER EXTERNAL GALAXIES

NGC 55	33.154.015 3 candidates	33.157.002 51 diffuse objects found
NGC 224	31.154.013 cl search	32.154.016 X-ray cl Bo 158
(M31)	32.154.021 X-ray cl Bo 158	29.158.062 IUE comp to NGC 6624
NGC 253	33.157.014 61 candidates found	
NGC 598	32.158.032, 34.154.099 integr phot and sp of cl candidates	
NGC 891	31.154.011 search for cl, no candidates	
NGC 3226	30.154.040 cl systems detected	
NGC 3311	34.157.082 cl search and distr	
NGC 3377	30.154.040 cl systems detected	
NGC 3379	29.158.062 IUE comp to NGC 6624	30.154.040 cl systems detected
NGC 3607	30.154.040 cl systems detected	
NGC 4278	30.154.040 cl systems detected	
NGC 4374	29.158.109 cl comp to spheroidal pop of elliptical galaxies	
NGC 4406	20.158.109 cl comp to spheroidal pop of elliptical galaxies	
NGC 4472	30.154.040 cl systems detected	29.158.062 IUE comp to NGC 6624
NGC 4486	32.158.285 origin of cl system	29.158.132 system of halo cl
NGC 4565	31.154.011 search for cl: 100 candidates	
NGC 4621	29.158.109 cl comp to spheroidal pop of elliptical galaxies	
NGC 5813	30.154.040 cl systems detected	
NGC 5846	30.154.040 cl systems detected	
Fornax	34.157.030 rv for 4 cl	33.154.010 JHK phot for 3 cl

7. Dynamics of Star Clusters

(In this section, three figure references are to Astronomy and Astrophysics Abstracts and two-figure references are to Physics Abstracts.)

In June, 1984, at the end of the review period for these reports, IAU Symposium 113 on the Dynamics of Star Clusters was held in Princeton. The Proceedings will form the most useful summary of recent research for the next few years. Nevertheless the topics which were emphasised at the symposium do not always coincide with those which are most prominent in the literature of the review period, which is the subject of this report.

Many papers were devoted to the fundamental dynamical process of relaxation. While a number of investigations reman close to the original formulations of Chandrasekhar (29.151.038, 30.151.070-2, 34.151.081-2,

87.72989), there have been some attempts to examine the effects of inhomogeneities, non-rectilinear orbits and so on (31.151.051, 34.151.115). There have been numerical tests of the standard theory (29.151.008, 31.151.014) and consideration of the strong encounters which are usually neglected (34.151.019 and .022). Nevertheless, the Fokker-Planck equation remains the standard tool in applications to the dynamics of star clusters.

Another fundamental process on which much work has been done is the dynamics of binary stars: their interactions with single stars (31.151.021, 34.151.037), which involves consideration of the decay of bound triple systems (29.042.054,1) and their interactions with each other (34.151.014, 87.47653). Which of the observable types of active binary may be formed in globular clusters is a topic which has also received attention from a dynamical point of view (33.154.020, 86.121747, 87.63219).

Now we turn to the application of these processes in the dynamical evolution of star clusters. The main consequence is core collapse, which is now well established, and occurs for some clusters on a short enough timescale to be of importance (32.154.029, 34.154.097; but cf 34.151.032). Therefore much attention has been given to the reversal of the collapse and the subsequent evolution. Binaries are especially important, as can be shown from the very simplest evaporative models (32.151.002-3, 87.22448) or from Monte-Carlo studies (34.154.091). The nature of post-collapse evolution is still in doubt; some studies suggest smooth expansion (eg 34.151.056) but the evolution may be oscillatory (87.77887). Another mechanism, less favoured now, for reversing core collapse is a central black hole (31.151.016, 32.066.070, 34.151.107).

Several of these evolutionary phenomena can be seen as manifestations of certain instabilities, eg the mass-segregation instability (29.151.014, 33.151.008, 34.151.020). Single component collisional systems may also be unstable (29.151.026, 29.151.096), and the role of anisotropy in the stability of collisionless systems has also been studied (34.151.112, 87.13375).

Two other processes are beginning to receive the attention which their importance merits. One is mass loss, which is especially important in the evolution of open clusters (30.151.036, 31.151.061, 32.151.075). The other is the effect of the Galactic tidal field. On one approximation, this can be dealt with by simple generalisations of the Roche formula (32.153.031, 87.47663) but the tide is time-dependent, and the effect of this is still poorly understood (29.151.034, 30.151.030, 87.42505).

Many of these theoretical investigaytions require the use of numerical methods. Several significant new techniques have been devised for the solution of both collisional and collisonless problems (32.151.016, 33.151.067, 34.151.063), and there has been some development of Larson's fluid dynamical method (33.151.084).

Much of this review has been concerned with dynamical evolution, but for some purposes there is a need for equilibrium cluster models which are justified more by their simplicity than on dynamical grounds. There are some new models of single component clusters (30.151.089, 33.154.036, 87.72992), and models containing two components (32.151.015) or a binary (30.154.039).

It remains to mention a number of isolated investigations which cannot be grouped conveniently with any of those described above. These include a numerical investigation of small systems immersed in a uniform medium (34.151.079) and theoretical studies of stellar orbits in clusters (33.153.041), the effect on anisotropy of accretion at the centre of a system

(87.58488), and a modification of the virial theorem due to pair correlations (32.151.089).

To summarise, the review period has seen some excellent work on the dynamics of binaries within clusters, and the realisation of the importance of tidally formed pairs. A fresh start has been made in the attempt to understand post-collapse evolution, while the attention paid to clusters containing a black hole has declined sharply. In the next few years it is possible that more attention will be paid to the effects of mass segregation, to the dynamics of the halo and the tidal boundary, and to the role of rotation, important topics which have recently been rather neglected.

8. Some Trends in Cluster Research

Major advances have been realized during the last three years in star cluster research, and new techniques and detectors are quickly being applied to the study of star clusters. We can anticipate further advances in several areas in the next few years.

Progress in our understanding of the detailed physics of stellar evolution has, until recently, been limited by the uncertainties in the available observational data. The precision of photometric measurements for stars on cluster main sequences and at the main sequence turnoffs has been too poor to offer theoreticians much guidance in constraining their stellar evolution calculations. The effects of many parameters in stellar evolution have been too subtle to distinguish from the observations. The application of CCD detectors to the problem of measurement of cluster H-R diagrams has brought substantial improvement to this situation. Examples of this work are the H-R diagrams of 47 Tucanae recently obtained by Hesser and W. Harris and the H-R diagram of NGC 6752 by Penny and Dickens. For the first time it may be possible to compare the ages of globular clusters to an accuracy better than 10%, to determine if they were formed over an extended period in the early history of the galaxy, or if they formed suddenly in a short time. From the improved theoretical models, the actual ages of clusters may be determined, and the controversy between the globular cluster ages and the age of the Universe determined from H_o may be resolved. We can anticipate significant improvements in the quality and quantity of cluster color-magnitude diagrams during the next several years.

During the last several years, we have also seen much progress in the study of star clusters very distant from the disk of the Milky Way, allowing us to explore very remote regions of the galactic halo. Examples of this work include studies of AM-1 by Aaronson et al, and of the Eridanus cluster by Da Costa.

Investigations of star clusters in other galaxies have also been moving forward quickly during the last three years. Buonanno and his colleagues have surveyed, catalogued, and measured many clusters in M31, and we can anticipate a much more detailed understanding of cluster systems in that galaxy and in our own. Searches for clusters in more distant galaxies have received much attention, and we can expect to begin to understand the formation of cluster systems in galaxies of different morphological types.

A third area in which we can look forward to important progress is the study of stellar rotation and binarism on the main sequences of star clusters. The work of Stauffer has already lead to interesting results on the duration of star formation in open clusters. The combination of the availability of new instrumentation with a renewed interest in solar/stellar cycles and variability has lead us again to use star clusters as laboratories for investigating these phenomena. Similarly, R. Peterson's work on rotational velocities of horizontal branch stars in globular clusters may lead to a solution of the "second parameter" problem in globular cluster research.

The period from 1981 to 1984 has seen continued efforts to resolve the controversy surrounding the metal rich calibration of the globular cluster metallicity scale. Although the question is still not fully resolved, many investigations are consistent with the adoption of metallicities near [Fe/H] = -0.8 for the clusters M71 and 47 Tucanae. The application of new instrumental techniques to the problem may lead finally to a solution.

The primary thrust in studies of cluster abundances has been a detailed examination of star-to-star variations within clusters. Norris, G. Smith, Suntzeff, Kraft and their colleagues have made extensive investigations of molecules and abundances of carbon and nitrogen in star clusters. Their results may lead to a better understanding of mixing and nucleosynthetic processes in stellar evolution.

The ability now to measure accurate radial velocities for relatively faint stars has produced a revolution in observational dynamics of star clusters. In particular, kinematical studies of open clusters and star forming regions are now possible. Recent work in this rapidly expanding area was summarised at IAU Colloquium 88 "Stellar Radial Velocities", in Schenectady. Also, it is now possible to measure velocity dispersions as low as about 4 km/s from the integrated light of globular clusters: Elson recently made a direct measurement of the mass of NGC 1835, in the LMC, with this technique.

COMMISSION 38: EXCHANGE OF ASTRONOMERS (ECHANGE DES ASTRONOMES)

(Committee of the Executive Committee)

President: F. B. Wood Vice President: E. A. Muller

Organizing Committee (Comite d'Organisation): A. A. Boyarchuk, J. Delhaye, D. A. MacRae, P. H. Routly, J. Sahade, F. G. Smith, C. R. Tolbert, H. H. Voight, S. H. Ye

While the chief function and methods of operation of the Commission have remained much the same as in the past, there has been a gradual evolution in the nature of the proposals submitted. A greater fraction now come from countries in which the study of astronomy on a professional basis is as yet very recent and more proposals are being received from relatively young astronomers, although admittedly the distinction between a "young" and an "established" astronomer is not always easy to make. The commission may wish to consider whether or not it is advisable to reconsider its guidelines. Grants awarded during the interval 30 November 1981 and 31 January 1985 were the following.

Grantee	Origin	Destination
C. E. Akujor	Nigeria	India/France
P. Chen	China	USA
E. Corbelli	Italy	USA
M. K. Das	India	England
V. N. Dermendjiev	Bulgaria	France
S. K. Jain	India	England
L. Jing	China	Scotland
M. Kaisig	Germany	USA
S. Manorama	India	Italy
M. V. Mekkanden	India	Germany
M. Melozzi	Italy	USA
R. Morris	Argentina	USA
T. Nagata	Japan	Australia
V. S. Niemela	Argentina	France
R. Nityananda	India	USA
K.-D. Oh	Korea	USA
A. R. Patnaik	India	USA
Y. Sobouti	Iran	USA
C. Sterken	Belgium	New Zealand
K. Subramanian	India	England
J. Trujillo	Germany	USA
R. J. Wainscoat	Australia	Netherlands
Y. Wu	China	USA

Approximately half this number of applications could not be granted because they were not within the guidelines set for the Commission (see IAU Information Bulletin no. 50). Further, a number of preliminary inquiries were not followed up when copies of the guidelines were sent. In a few cases, prospective grantees received funds from other sources and then did not require Commission 38 support. A few applications are still under consideration.

Thanks are due to Prof. Edith Müller, Prof. J. Delhaye and Mrs. Vicki Leino for valuable help in preparing this report.

40. RADIO ASTRONOMY (RADIO ASTRONOMIE)

PRESIDENT: K. I. Kellermann

VICE-PRESIDENT: J. Baldwin

ORGANIZING COMMITTEE: J. G. Ables, N. Broten, G. Dulk, B. Hoglund, N. Kardashev, M. Morimoto, S. Okoye, G. Setti, R. Strom, G. Swarup, L. Weliachew, T. Wilson, S. Wang.

The last triennium marked the 50th anniversary of the paper describing the first observations of cosmic radio emission by Karl Jansky in 1933. Sullivan (82 *Classics in Radio Astronomy*, Reidel) has published a collection of the major historical papers in radio astronomy, and collections of papers discussing the historical development have been published by Sullivan (84 *Early Years of Radio Astronomy*, Cambridge Univ. Press) and by Kellermann and Sheets (84 *Serendipitous Discoveries in Radio Astronomy*, NRAO).

The following symposia of interest to radio astronomers were held during the period under review or their proceedings were published during this period.

IAU Symposium No. 97, Extragalactic Radio Sources, August 1981, (eds.) D. S. Heeschen and C. M. Wade, Reidel, 1982.

IAU Symposium No. 100, Internal Kinematics and Dynamics of Galaxies, Aug. 1982, (ed.) E. Athanassoula, Reidel, 1983.

IAU Symposium No. 104, Early Evolution of the Universe, Aug./Sept. 1982, (eds.) G. O. Abell and G. Chincarini, Reidel, 1963.

IAU Symposium No. 108, Structure and Evolution of Magellanic Clouds, Sept. 1983, (eds.) S. van den Bergh and K. S. de Boer, Reidel 1984.

IAU Symposium No. 109, Astrometric Techniques, Jan. 1984, Reidel.

IAU Symposium No. 110, VLBI and Compact Radio Sources, June 1983, (eds.) R. Fanti, K. Kellermann, G. Setti, Reidel, 1984.

10th Texas Symposium, Dec. 1980, ed. R. Ramaty and F. D. Jones, New York Acad. Sci, 1981.

11th Texas Symposium, Dec. 1982, ed. D. S. Evans, New York Acad. Sci., 1982.

Green Bank Workshop on the Comparative HI Content of Normal Galaxies, April 1982, (eds.) M. Haynes and R. Giovanelli, NRAO.

Green Bank Workshop on Low Frequency Variability of Extragalactic Radio Sources, April 1982, (eds.) W. D. Cotton and S. R. Spangler, NRAO.

Green Bank Workshop on Interference Identification and Excision, Sept. 1982, (eds.) W. Erickson and R. Fisher, NRAO.

Bangalore Winter School on Energetic Extragalactic Sources, Tata Inst. 1983.

Symposium on Astrophysical Jets, (ed.) A. Ferrari and A. Pacholczyk, Reidel,

1983.

Astrophysical Cosmology, Vatican Study Week, Oct. 1981, (ed.) H. Brück, G. Coyne, and M. Longair, Pontifical Acad. Sci., 1982.

Quasars and Gravitational Lenses, 24th Liege Colloquium, June 1983, (ed.) J. P. Swings, Univ. Liege, 1984.

URSI Symposium on Indirect Imaging, (ed.) J. A. Roberts, Camb. Univ. Press, 1984.

IAU Symposium No. 112, The Search for Extraterrestrial Life, (ed.) M. Papagiannis, Reidel, Boston, June 1984.

QUASAT Workshop, Vienna, Austria, June 1984, ESA-SP 213.

Green Bank Workshop on Millisecond Pulsars, (eds.) S. P. Reynolds and D. R. Stinebring.

This report follows the style of recent Commission 40 Reports. The following commission members have contributed to the report: Instrumentation and Techniques (Baldwin, Booth, Cornwell, Shaffer, Weinreb, Weliachew); Solar System (Berge, Dulk); Continuum Research (Bridle, Fanti, Hjellming, Kapahi, Lo, Pauliny-Toth, Strom, Wielebinski); VLBI (Cohen); Pulsars (Backer); Molecular Spectroscopy (Reid, Wilson); HI and HII Research (Haynes, Shaver, van Gorkom); SETI (Papagiannis).

INSTRUMENTATION

Aperture Synthesis Radio Telescopes - The last three years have seen very active developments with several proposals for new telescopes or substantial modifications of existing ones, construction starting on new approved projects, completed instruments being brought into commission, and important improvements in the frequency coverage, sensitivity and image quality of well-established telescopes. The principal new developments not discussed in other sections are summarized below.

Australia - The Australia Telescope (Poulton 84 IAU/URSI Symp Indirect Imaging 67) for which funding was approved in November 1983 will be the most important high resolution radio telescope in the southern hemisphere for many years. It has two main components: (i) a compact array at Culgoora comprising six 22m antennas on a 6 km baseline of which five are movable on a 3 km rail track and (ii) a long baseline array obtained by correlation of signals between the compact array, a new 22m antenna near Siding Spring mountain and the Parkes 64m telescope. It is also planned to use AT in collaboration with other antennas. Important features of the design include rapid changing of the frequencies of operation, very broad receiving bandwidths, and good polarization and spectral capabilities. A large amount of design and development work has been finished and completion is planned for 1988. Initial operation will be at four bands between 1.4 and 10 GHz with extensions to several other bands between 327 MHz and 115 GHz in the longer term.

The Molonglo Observatory Synthesis Telescope (MOST) (Mills 81 Proc Ast Soc Aust 4, 402) has been in operation at 843 MHz giving full synthesis mapping over 23' x 23' cosec δ fields of view with a resolution of 43" x 43" cosec δ and an rms noise sensitivity of 0.2 mJy in 12h.

Regrettably, operation of the Culgoora radio heliograph came to an end in 1983.

Canada - The frequency range of the four-element synthesis telescope at DRAO, Penticton has been extended to cover operation at 408 MHz as well as the original 1420 MHz band.

India - The Ooty Synthesis Radio Telescope (OSRT) at Ootacamund has been commissioned and produced its first astronomical results (Swarup 84 JAA 5, 139). The combination of the Ooty Radio Telescope and seven smaller antennas distributed over an area of 4km x 2km gives a coverage of about ±40° in δ with a resolution of 40" x (50-90)" over a field of view of 3° x 0°.6 and an rms noise of about 3 mJy in 10h integration.

The Gauribidanur T array operating at 34.5 MHz has produced its first astronomical results (Dwarakanath et al. 82 JAA 3, 207) in studies of supernova remnants with a resolution of 26' x 40'.

A proposal for a Giant Metre Wavelength Radio Telescope (GMRT) has been put forward (Swarup 84) with the aim of providing very high sensitivity observations at low frequencies with angular resolutions ranging from 4" at 610 MHz to 75" at 38 MHz. The proposed elements are 34 cylindrical antennas spaced over a 3 km radius with a total collecting area of $65000m^2$. The projected date for completion is 1992.

Netherlands - The Westerbork Synthesis Radio Telescope (WSRT), operational on a 2.7 km baseline since 1980 has extended its frequency range to include 327 MHz with observations starting in December 1983.

United Kingdom - The Multi Telescope Radio Linked Interferometer (MTRLI) at Jodrell Bank, magically transformed into MERLIN in about 1982, has extended its observations to 5 GHz and 22 GHz for three of the antennas. Future plans include operation at 151 MHz in 1984-5 and incorporation of an antenna at Cambridge.

The Cambridge Low Frequency Synthesis Telescope (Baldwin 84 reported at URSI Gen. Assembly), comprising 60 steerable Yagi arrays on a 4.6 km east-west baseline, has been commissioned. Operating initially at 151 MHz it gives 70" resolution over 5°-10° fields of view with an rms noise sensitivity of 20 mJy in 12h. The former 6C array has been operated at 38 MHz to give 16' x 16' mapping of the sky for $\delta \geq 65°$. The 5 km Telescope has made use of small (9m) increments in baselines for studies of large sources at 15 GHz.

U.S.A. - The Very Large Array (VLA) near Socorro, New Mexico (Napier et al. 83 Proc IEEE 71, 1295) has established its position as the most powerful existing radio telescope. The successful development of self-calibration techniques combined with its very high sensitivity has led in the last three years to the most detailed radio maps of sources yet made; r.m.s. noise levels of 100 μJy can be reached in only 10 minutes and sources as weak as 50 μJy have been catalogued. Operation at 327 MHz is planned for 1985, and a 75 MHz system has been proposed.

The Clark Lake Teepee Telescope (Erickson et al. 82 ApJ Sup 50, 403) has been in regular use for solar, galactic and extragalactic programs.

U.S.S.R. - The Siberian Solar Radio Telescope sited near Irkutsk has started astronomical observing. The telescope, composed of 128 2.5m antennas on each of two axes about 600m in length operates at 5.8 GHz with a bandwidth of 120 MHz.

P.R. China - The Miyun synthesis telescope, Beijing Observatory, was put into service in October 1984. This is a 16 x 12 element E-W array spanning 1164 metres, and at present, working at 232 MHz. The synthesis pattern gives a resolution of 3'.8 x 3'.8 csc δ. A polar region "dirty map" obtained in 2 x 12 hours shows a dynamical range of 100:1, where sources of 0.2 Jy are easily detected. The thermal noise-limited sensitivity of the system is about 0.05 Jy. With a field of view of 8° x 8°, it will be used for source survey of sky region $\delta \geq +30$ and a number of other astronomical programmes.

Millimeter Telescopes - After R. B. Leighton's pioneering work on high precision 10.4 m antennas, larger telescopes are now being built or put into operation. The 45 m antenna at Nobeyama, Japan, operates at 3 mm wavelength with an aperture efficiency around 30% and the IRAM 30 m antenna is being tested near Granada, Spain (83 Baars, Infrared and mm Waves 9, 241, Academic Press). A second 30 m antenna is being completed by the Iraq Council for Scientific Research at a 2100 m site on Mt. Koerk and is expected to be operational in 1985.

Progress has been made in surface panel manufacturing, thermal stability or control and surface measurements. Surface panels are now being manufactured with accuracies around 0.05 mm, mostly through molding techniques applied to panels composed of aluminum honeycomb with aluminum or carbon fiber skins. More attention has been given to thermal deformations of surfaces and structures. Greater stability has been achieved by careful design of steel structures and associated thermal control, or through the use of carbon fiber. Examples of later include Nobeyama for the surface panels, and the 15 m antennas of the IRAM interferometer for the surface panels and the back-up structure. Because most radio sources are weak at mm wavelengths, millimeter antennas are made as large as possible, and, thus, have very large apertures in units of wavelengths compared to cm and dm telescopes. For this reason, pointing of their beams - usually less than one arc min - is difficult, especially in the presence of even moderate winds. This may be the ultimate limiting factor in the use of large single dish millimeter telescopes. Pointing with arc sec accuracy requires structures with high resonant frequencies and complex servo-systems. Surface measurements, another critical task, have become more straightforward by the use of radio holography which has been successfully applied to the Green Bank 140-ft, the Kitt Peak 11 m, the Bonn 100 m, the Parkes 64 m, and the RATAN 600 radio telescopes. Measuring and adjustment techniques are often the limiting factor in the surface accuracy of mm telescopes and it is good to have a surface quality test by direct radio astronomical means. For example, with the high accuracy of structure and panels, future improvements of the surface accuracy are probably possible for the Nobeyama 45 m telescope and the 30 m telescope in Spain.

The surface and the back-up structure of the 36-ft telescope on Kitt Peak, Arizona, have been completely refurbished providing an aperture efficiency of 30% at 230 GHz for what has now become a 12 m antenna (84 Gordon, Sky and Telescope 67, 326). Holographic and mechanical surface measurements indicate an rms surface accuracy of 75 microns. Construction proceeds on Mauna Kea of a 15 m telescope funded by the UK and the Netherlands. The design goal of this telescope is to operate it as a diffraction-limited telescope down to 0.8 mm wavelength. Another telescope of 10.4 m will also be erected on Mauna Kea. It will be an improved version of the Owens Valley antennas with acceptable performance in the sub-mm range. A 10.4 m telescope of the same design is being built by the Raman Institute in Bangalore, India. A 13.7 m ESSCO telescope is being constructed at a 3000 m site in the Chaidamu Basin, China for the Nanjing Observatory. Completion is expected by 1986. Finally, a 10 m telescope for the mm and sub-mm range is planned on Mount Graham, Arizona (Baars et al. 84 SPIE 444, 65) as a joint project of the Max-Planck Institute and the University of Arizona.

After the two element interferometer at Hat Creek, the first of its kind, the Owens Valley interferometer has started operating in 1982/83 (Masson et al. 84 ApJ 283, L37). It is composed of three 10.4 m antennas equipped with SIS receivers in the 3 mm band. The surface quality of these antennas allows operation at shorter wavelengths. The interferometer is capable of a few arc sec resolution and is equipped with filter spectrometers. The Nobeyama interferometer, with five 10 m antennas, has been tested at 1.35 cm and will begin operation near 3 mm in the winter 84/85. The three-antenna IRAM interferometer with 15 m antennas is under construction at an elevation of 2550 m in Southern France. The design goal of this instrument is operation in the short mm range. Plans are being developed in the USA for a large millimeter array.

Receivers - Considerable progress has been made over the past years in receiver sensitivity. Low-noise, cryogenically-cooled, gallium-arsenide, field-effect amplifiers (GASFET's) have been widely used for cm-wave radio astronomy receivers. Noise temperatures of < 10K at 1.5 GHz (Weinreb 82 IEEE Trans. on MTT, MTT-30, 849) and 27K at 10.7 GHz (Tomassetti, et al. 81 Electronics Lett. 17, 949) have been reported. Lower noise temperatures are now being measured in various laboratories utilizing high electron mobility transistors (Berenz, et al. 83 IEEE Mic. and Millimeter Wave Monolithic Circ. Symp., Digest of Papers, IEEE #84CH2042, San Francisco). Low noise 22 GHz maser radiometers are now in routine use on the Green Bank, OVRO, Haystack, and Bonn antennas with zenith system temperatures as low as 50K under good sky conditions. The Green Bank 140-ft antenna is equipped with 3 upconverters to give nearly complete coverage of the range 5 to 25 GHz with system temperatures in the range 35 to 80K. A cooled GASFET amplifier gives system temperatures under 25K over the range 1.3 to 1.8 GHz.

Receiver temperatures near 100K DSB have been reached at 3 mm wavelength over large frequency ranges with mixers based on Schottky barrier diodes (Predmore et al. 84 IEEE Trans, MTT-32, 498; Archer and Faber 84 Int J Infrared and mm Waves 5, 1069). Another type of non linear device, the superconducting SIS junction has now come of age (Phillips and Woody 82 Ann Rev AA 20, 285). Here, the tunneling of quasiparticles through a composite thin-film device gives rise to a sharp non-linearity at very small I-V products, that is at very small local oscillator powers. With SIS junction mixers, receiver noise temperatures below 100K DSB near 3 mm (Linke et al. 83 ApJ 271, L85) and below 300K DSB at 1.3 mm (Sutton et al. 83 ApJ 275, L49) have been achieved, both during actual observations on telescopes. Depending on junction fabrication, it may be an advantage to cool those devices down to 2-3 K. This is often done with hybrid cryostats where cooling to such temperatures is obtained from a combination of commercially available refrigerators and refillable He containers (Blundell et al. 82 Int J Infrared and mm Waves 3, 793).

Sources of power for local oscillators are now generally solid state Gunn oscillators followed by frequency doublers or triplers. This is much cheaper in operation cost and more reliable than the previously used klystrons. As a consequence, mm receivers are now progressively becoming pure solid state systems, even at short wavelengths (Archer 82 IEEE Trans., MTT-30, 1247).

Acousto-optical spectrometers (AOS) have now become stable enough to be used in position switching mode on large telescopes (Masson 82 AA 114, 270). Such devices are now routinely used at Yunnan, Nobeyama, RATAN 600 and Owens Valley for single-dish observations. In Nobeyama, a 2 GHz band is analyzed into 16000 elements by eight 250 MHz AOS's and a 160 MHz band can also be analyzed into 4000 channels by four 40 MHz AOS's. At Owens Valley, bands of 100 MHz and 500 MHz are analyzed into 1024 and 512 channels respectively. The millimeter interferometer in Nobeyama has a spectrometer of a novel design (Chikada et al 83 URSI/IAU Symp on Measurement and Processing for Indirect Imaging). This system samples the signals from the individual antennas and Fourier transforms them before performing the antenna products in frequency space. This is in contrast to the usual approach of cross correlation interferometry. This spectrometer uses IC's performing real time FFT's. The present version has 80 MHz bandwidth, and will be extended to 320 MHz. This approach appears to be more cost-effective for the same performance than the usual cross correlation method. Multibit FFT spectrometers with bandwidths up to 10 MHz are in use at JPL, Harvard, and Stanford; and the technology is being extended to 40 MHz with spectrometers under design and construction at NRAO and Cornell.

VLBI Systems and Programs - During the past triennium the dynamic range of VLBI maps has improved dramatically with the use of larger numbers of antennas and more sophisticated calibration procedures. Five to eight antennas are commonly used, and an 18-antenna experiment was successfully conducted in 1984. A dynamic range of 100:1 is now common, and several hundred to one is expected soon. Another major improvement is the inclusion of the VLA, or Bonn 100 m, telescopes with global fringe fitting techniques to dramatically improve the sensitivity. Sources as weak as 20 mJy have been mapped at 5 GHz with a resolution of 0".001. More than 30 different radio telescopes are being used in VLBI observations at wavelength ranges from 3 mm to 1 meter. Major international collaborations are common with more than 16 different countries on all continents being involved in VLBI research.

The major organized VLBI networks which schedule regular observations are those in the USA and Europe. Active telescopes in the U.S. Network are: Fort Davis (Texas), Green Bank (West Virginia), Hat Creek (California, North Liberty (Iowa), Haystack (Massachusetts), and Owens Valley (California). Also used regularly as part of the network are Effelsberg (W. Germany) and the VLA (New Mexico) and Maryland Point (Washington, D.C.). Other affiliate observatories are Algonquin Park (Canada), Arecibo (Puerto Rico), and the NASA DSN stations. The European VLBI Network (EVN) uses: Onsala (Sweden), Effelsberg (W. Germany), Westerbork (Holland), Jodrell Bank (U.K.) and Medicina (Italy). Torun (Poland) and Crimea (U.S.S.R.) are also regularly used. The new 32 m telescope in Medicina is dedicated to VLBI use as will be a similar telescope planned for Noto (Siracusa) in Sicily. Details of the antennas, instrumentation and procedures are published in the Network handbooks: U.S. Network Handbook (ed.) D. C. Backer, University of California, Berkeley, and EVN Handbook (ed.) R. S. Booth, Onsala Space Observatory, Sweden, both available from the editors. Combined Network operation, so-called "Global" Network experiments are conducted on a regular basis. Australian, South African, and US antennas have been linked to study southern hemisphere sources (Preston 82, IAU Symp. 110, eds. Fanti et al.). A Germany-China baseline at 1430 MHz was successfully utilized (Wan 82 Acta Astron. Sinica 23, 376). A 5-station 327 MHz experiment involving Ooty (India) with Torun, Crimea, Westerbork and Jodrell Bank gave fringes. Hartas et al. (83 M.N.R.A.S. 205, 625) have used the Cambridge 3.6 Hectare Array at a frequency of 81.5 MHz together with a portable outstation on 11 baselines from 0.5 to 1500 km to investigate the feasibility of mapping sources with sub arcsec resolution at metre wavelengths. A three-antenna VLBI system is operating in the USSR with 20 m antennas in Crimea and Serpekhev and the new 70 m antenna at Evpatorly (83 Soviet AJ 83, 219). Additional antennas at Medvezhii Osera (64 m), Ulan Ude (25 m), and Usuriisk (70 m) will be added in 1985, and will be equipped with MKII recording systems.

Recording Systems - Although the MKII recording system (Clark 73, Proc IEEE 61, 242) is regularly used, there is considerable pressure on the MKIII system (Rogers et al. 83 Science 219, 51) but its use is limited by the high consumption of tape. This wideband system runs through a single tape in about 13 minutes at 112 Mbps, but a greatly improved recording density is now possible using a narrow track movable headstack to record 28 40-micron wide tracks in each of 16 multiple passes on a single tape. The bit density is approximately 10 Mbits/square inch and represents a more than 10-fold improvement over the original MKIII. Existing recorders will be upgraded to the higher bit density. A Mark III compatible system (K3) has been developed in Japan (Saburi et al. 82, Proc IAG Symp No. 5, 307).

VLBI Processors - MKII VLBI processors are available at NRAO, Charlottesville, and MPI Bonn, both with 3 stations spectroscopic capability and at Caltech, Pasadena (5 stations continuum only). Three-station MKIII processors are in operation at Haystack and Bonn, and Caltech will soon complete a 5-station MKIII processor which will be operational in 1985 and will allow single pass, 16-station MKII processing. A 2-station MKII processor is being built in the USSR.

New VLBI Projects - A start was made in 1984 on a major new project--a dedicated array of ten 25-m télescopes to be built at sites throughout the United States from Hawaii to Puerto Rico spaced so as to provide good u-v plane coverage and hence excellent image quality. This system, the VLB Array (Kellermann 84, IAU Symp 110, 377) will have a wavelength range from 0.7 to 90 cm with full polarization capability. Bandwidths will be at least 56 MHz giving an r.m.s. noise of 0.05 mJy in an 8-hour integration time. For spectroscopic use, up to 512 frequency channels will be available. As many as 19 tapes will be simultaneously replayed, to allow up to 9 non VLBA antennas to be used to further increase the sensitivity and resolution. Studies on the Canadian VLB Array have continued (Legg 84, IAU Symp 110, 383) and the current proposal contains two options: the original array of nine 32 m antennas and a more limited 4-element array (Seaquist 84, Proc. Workshop on QUASAT, ESA SP-213, p. 127). A combination of either array with the US or European Networks will fill important gaps in the baseline plane. A study of a 103-MHz 3-element VLBI Array in India has also been reported (Bhonsle 84, IAU Symp 110, 391).

After the first successful millimeter VLBI experiment at 3.4 mm wavelength, with a baseline of 140 million wavelengths (Readhead et al. 83 Nature 303, 504), more experiments have yielded positive detections even on baselines of 10^9 wavelengths on the quasar 3C273. Coherence times of several hundreds of seconds have been observed. For the first time in astronomy, this practically opens up the possibility of resolutions $\sim 10^{-4}$ arcsec.

Space VLBI - During the past three years much attention has been focussed on a study of a satellite-borne antenna for extending VLBI baseline (Burke 84, IAU Symp 110, 397) and culminating in a design study for a joint ESA-NASA mission - QUASAT (Schilizzi et al. 84, IAU Symp 110, 407, ESA SP-213). QUASAT will be a 15 to 20 m antenna in low Earth orbit and will be used to extend currently available baselines by a factor of about 3 at 1.35, 6 and 18 cm. By observing simultaneously with the ground arrays, the relative motion of QUASAT and the Earth provides for exceptionally dense coverage of the aperture plane of the interferometer system. This system will give superb image quality (Readhead et al. 84, ESA SP-213, 101). Soviet Space VLBI missions are also being planned. Two experiments are under consideration. The first will put a 30 m antenna into an orbit with a height of about 500 km while the second considers a much larger orbit of one million km (Sagdeev, ESA SP-213, 19).

Image Processing Techniques - A timely overview of recent developments in astronomical image processing techniques is available in the proceedings of an IAU/URSI conference on "Indirect Imaging" held in Sydney in 1983 (ed. J. A. Roberts, CUP 1984). Since 1981, the main advance in processing techniques has been the widespread acceptance and use of self-calibration to dramatically improve the quality of images produced from radio interferometer data (Schwab 80 Proc SPIE 231, 18; Cornwell and Wilkinson 81 MN 196, 1097). At the heart of this technique lies the observation that calibration errors in an interferometer array arise mainly at the individual antennas, due to atmospheric or instrumental effects, while data is collected from all possible pairs of antennas. Thus, since effectively one has to solve for the sky brightness anyway by deconvolving the effects of incomplete sampling of the u,v plane, one may as well solve for the antenna based calibration errors as well. The dynamic range of self-calibrated images is limited by those effects not factorizable per antenna, such as receiver noise and bandpass problems (see Cornwell and Wilkinson, in the proceedings of the Sydney meeting, op. cit.). Most arrays now achieve dynamic ranges between 30 and 45 dB (peak/rms) on sufficiently strong sources. There is some contention whether or not instantaneous baseline redundancy is responsible for the further increase in dynamic range now achievable with WSRT (Noordam and De Bruyn 82 Nature 299, 597; Cornwell and Wilkinson op. cit.) but, unfortunately, there is as yet little firm evidence to support either view. We note, however, that the dynamic range of VLA maps, made without redundancy, is beginning to approach that of WSRT maps, made with redundancy.

In the early days of self-calibration, it was generally supposed that a point source somewhere in the field of view was required; subsequent experience with the VLA, MERLIN and VLBI has revealed this to be somewhat conservative and that, instead, sources as complex as SNR may be self-calibrated if the u,v coverage is reasonably complete (see e.g. Cornwell and Wilkinson op. cit.).

In an important extension to self-calibration, Schwab and Cotton (83 AJ 88, 688) have shown that one can also solve for antenna related delay and fringe rate offsets when finding fringes in VLB data. Such "Global Fringe Fitting" lowers the signal to noise required for fringe finding, and will work on sources with considerable structure. In radio-astronomical imaging, self-calibration is now very widely accepted and used, while reconstruction with amplitudes alone, although feasible (see e.g. papers in proceedings of Sydney conference, op. cit.), is too expensive in terms of integration time required to reach a given sensitivity. Reviews of self-calibration have been published by Pearson and Readhead (84 Ann Rev AA) and Cornwell and Wilkinson (op. cit.)

Considerable progress has been made in the deconvolution of images made with radio interferometric arrays. Variants of CLEAN, such as Cotton's MX program and the modified CLEAN algorithm (Steer, Ito and Dewdney 84 AA), adapted to particular circumstances have been introduced. Chen and Frater (Sydney conf. op. cit.) have developed a CLEAN variant ideally suited to parallel processing. Maximum Entropy methods are still quite controversial; new justifications abound (Gull and Skilling, Sydney conf. op. cit.) but the work of Narayan and Nityananda (e.g. Sydney conf. op. cit.) provides a pragmatic interpretation which most people seem to accept. Reasonably sophisticated numerical algorithms for calculating the Maximum Entropy solution in two dimensions have been developed (e.g. Burch et al. 83 Comp. Vision, Graphics and Image Processing 23, 113; Cornwell and Evans 84 AA in press) but an analytic solution would still be very useful. Nityananda and Narayan (83 AA 118, 194) have discussed the extension of MEM to polarization data, and San-Roma and Estalella (84 AA 133, 299) and Shevgaonkar (84 in press) have married MEM and self-calibration.

Image processing for single dishes advanced rapidly during the middle and late seventies (see e.g. the review of Wielebinski, proc. Sydney conf. op. cit.) mainly due to the efforts of the Bonn group, but has progressed very little during the eighties. It is reasonable to expect further refinements to multibeam methods when large multibeam systems are constructed. Also the combination of interferometers and single dishes needed for imaging at millimeter wavelengths should generate new problems and solutions.

The advances in image processing during the last decade have been the result of refinements in hardware, such as computers and correlators, software, in the form both of unified software systems such as AIPS and of specific algorithms, and in the ingenuity of the astronomer. It is to be hoped that all of these factors will continue to improve with time.

Astrometry and Geodetic Applications of Radio Interferometry - Radio interferometry is now routinely providing precise baseline and source position results. Summaries of equipment, techniques, and results from VLBI and connected interferometers appear in several proceedings: IAU Symposia nos. 109 and 110; *High Precision Earth Rotation and Earth-Moon Dynamics* (82 O. Calame, ed., Reidel); *Geodetic Applications of Radio Interferometry* (82 NOAA Technical Report NOS 95 NGS 24); *Proceedings of the Ninth International Symposium on Earth Tides* (82 E. Schweizerbart'sche Verlagsbuchhandlung, Stuttgart); and *Very Long Baseline Interferometry Techniques* (82 CNES Colloquium, Toulouse). Astrometric and geodetic applications of a VLBI array are discussed in *Multidisciplinary Uses of the VLBA* (83, Natl. Acad. Press, Washington, DC).

An important contribution to improved results is the use of the J2000.0 system, which is a more consistent treatment of precession, nutation, aberration, earth rotation, etc. (Kaplan 81 USNO Circular No. 163).

The Mark III VLBI system has greatly advanced precision astrometry and geodesy, and is described by Rogers et al. (83 Science 219, 51). The very best source positions, from VLBI (Ma et al. 81 BAAS 13, 899; Fanselow et al. 84 AJ 89, 987; and Robertson, in IAU 109), have accuracies approaching 0".001, and define an internally consistent reference frame. Numerous optical positions, generally in the FK4 system, of radio sources and/or stars in their vicinity have been determined in efforts to align the optical and radio coordinate systems (Dick et al. 82 AJ 87, 1374; Harrington et al. 83 AJ 88, 1376; Clements 83 MNRAS 203, 861; de Vegt et al. AA 101, 191 and 113, 213; West et al. 81 AA Sup 46, 277; Froeschle et al. 82 AA 116, 89; and Costa et al. 83 AA Sup 51, 425. Argue et al. (84 AA 130, 191) have compiled a recommended list of sources for radio and optical comparison. Efforts are also underway to use radio stars to relate the two systems (Carrasco et al. 83 AA Sup 52, 279). The minor planets also show promise for these purposes (Johnston et al. 82 AJ 87, 1593).

The VLA has produced a large number of modest precision positions (about 0".05: Perley 82 AJ 87, 859) which are in good agreement with VLBI observations. The NRAO radio link interferometer was used by Kaplan et al. (82 AJ 87, 570) to measure 16 sources to about 0".01 - 0".03. These positions were among the first to appear in the J2000.0 system.

Lyne et al. (82 MNRAS 201, 503) used the MERLIN radio link system at 408 MHz to measure proper motions of pulsars at the 0".007 per year level, by referencing a pulsar to another source in the same primary beam. Similar referencing has been used with VLBI to measure the 30" distance between the quasars 1038+52A,B to about 5 and 15 micro-arcseconds at 3.6 and 13 cm respectively (Marcaide et al. 83 AJ 88, 1133).

Robertson (84 Proc. of International Symposium on Space Techniques for Geodynamics) and Shaffer (IAU 110) show the results of repeated VLBI baseline measurements. The Haystack-OVRO baseline is stable to well under 1 cm per year. The Westford-HRAS baseline has been measured almost 200 times, and the length results show a 2 cm rms scatter. Motions of a few cm per year have been measured along the San Andreas fault in southern California. Sovers et al. (84 JGR 89, B9, 7597) present a set of baselines measured between the antennas of the Deep Space Network.

Earth tides, and the Love numbers, have been determined by VLBI (Herring et al., Ninth Int'l Symp on Earth Tides). VLBI measurements of polar motion and

UT1-UTC compare favorably with other modern techniques. VLBI and connected interferometer results are heavily weighted into the BIH averages for these values (Robertson et al. 83 Nature 302, 509; Carter et al. 84 Science 224, 957). The earth rotation measurements show striking correlation with atmospheric angular momentum (the sum of global winds), and suggest that radio astronomy may someday help predict the weather!

SOLAR SYSTEM RESEARCH

The Sun - Radio images of the quiet sun of unprecedented quality have been obtained using Westerbork (Chiuderi Drago et al. 77 AA 60, 277), Arecibo (Lang and Wilson 83 AA 127, 135) and the VLA (Dulk and Gary 83 AA 124, 103). Magnetic fields cause circular polarization of bremsstrahlung radiation from active regions and gyroresonance emission from hot spots (Felli et al. 81 ApJ 247, 325).

Progress has been achieved toward understanding plasma radiation from electron streams, shock waves and coronal transients, both from observations and theory (Stewart et al. 82 AA 116, 217; Vlahos et al. 82 ApJ 258, 812; Kerdroan et al. 83 ApJ 265, L19; Gary et al. 84 AA 134, 222; Bougeret et al. 84 Solar Phys 90, 401; reviews by Melrose 81 Proc ASA 4, 139; Goldman 83 Solar Phys 89, 403). Even so, the immense effort has not produced a complete theory of several aspects of the generation of intense Langmuir waves and their conversion to radio waves.

Westerbork and the VLA have revealed that microwave, gyrosynchrotron radiation from flares usually comes from tops of magnetic loops with scales of $\sim$3" at 15 GHZ and $\sim$30" at 1.5 GHz (Kattenberg et al. 83 Solar Phys 88, 315; Marsh and Hurford 82 Ann Rev AA 20, 497; Kundu 83 Solar Phys 86, 205). Hard X rays from the same, >30 keV electron distribution usually come from footpoints (Hoyng et al. 83 ApJ 268, 865). Temporal variations in X rays and radio lasting <1 s are closely correlated, but sometimes with a small delay (Kaufman et al. 83 Solar Phys 83, 311); similarly with gamma rays which may indicate a small delay in ion acceleration relative to electron (Nakajima et al. 83 Nature 305, 292).

A new radio emission mechanism, a maser operating at a low harmonic of the cyclotron frequency, has been identified in solar and stellar flares (Holman et al. 80 IAU Symp 86; Melrose and Dulk 82 ApJ 259, 844; Gary et al. 82 ApJ 263, L79), the same mechanism which causes Earth's TKR and Jupiter's DAM. Strong arguments suggest that, in the sun and stars, maser energy is produced copiously at the fundamental cyclotron frequency but is reabsorbed by the surrounding plasma, heating it, and thus creating soft X radiation (Melrose and Dulk 84 ApJ 282, 308). If so, this is the first instance in astronomy where radio waves are energetically important.

Planets, Satellites, Comets and Asteroids - High resolution images of the Ishtar Terra region of Venus were obtained with the Arecibo 12.6 cm radar (Campbell et al. 83 Science 221, 644). The number of asteroids studied via radar techniques is increasing (Ostro et al. 83 AJ 88, 565). Radar returns from Comet IRAS-Iraki-Alcock were successfully received at both Arecibo (Campbell et al. 83 BAAS 15, 789) and Goldstone.

Ground-based observations of Jupiter's decameter radiation are still being made routinely, and are being used to attack important questions. For example, Genova et al. (81 Nature 293, 382) reported on a preliminary test of a method to study source location and structure. Dominating the low-frequency observations the last few years, however, has been the Planetary Radio Astronomy instrument on Voyagers 1 and 2. The basic results for Jupiter and Saturn were reported by Steinberg in Vol. XVIIIA, p582-584 of these Transactions. More recently a large amount of work that further organizes and interprets the data has been published. For example, there is a series of papers dealing with the Jupiter studies in 81 JGR 86, Sept. 30 issue, including an overview by Boishot et al. The Saturn observations are summarized by Warwick et al. (81 Science 212, 239 and 82 Science 215, 582). 81 Nature 292, Aug. 20-26 issue, also has a series of papers on Jupiter.

Ford et al. (83 Science 221, 1379) have mapped Venus at 17 cm using the radar receiver on Pioneer Venus. The observed features are interpreted as due primarily to emissivity variations across the surface. Work in progress includes high-resolution mapping at the VLA by Muhleman, Berge, Janssen, Klein, Chapman, and Pettengill. The four VLA wavelengths (20, 6, 2, 1.3 cm) probe to the Venus surface and to various depths in the atmosphere. 3.4 mm interferometer observations at Hat Creek by Good et al. (83 Icarus 53, 538) were used to limit the SO_2 abundance in the Venus atmosphere. Observations begun by Muhleman and Berge with the OVRO interferometer in the 2.6 mm CO(0-1) line show promise in determining the spatial distribution of CO in the upper atmosphere of Venus. Clancy et al. (83 ApJ 273, 829) determined the $^{12}CO/^{13}CO$ ratio in the Venus upper atmosphere at Kitt Peak using the CO(1-2) line.

Possible time variations of CO in the Mars atmosphere have been studied by Good et al. (81 Icarus 47, 166) and by Clancy et al. (83 Icarus 55, 282). Work is in progress by Rudy, Muhleman and Berge to interpret the observed temperature vs latitude of the Mars surface at 2 and 6 cm obtained from high-resolution VLA maps made during the retreat of the north polar cap.

The VLA has been used for high-resolution mapping of Jupiter. de Pater et al. (84 ApJ Sup 54, 405) and Roberts et al. (84 ApJ 282, 345) presented detailed maps, including polarization, of the largely nonthermal emission at 18-20 cm. These observations of synchrotron emission have the potential to help consolidate the various particle and field data for the inner magnetosphere obtained by spacecraft. Not yet published are short-wavelength VLA maps by de Pater et al. showing the Jovian atmospheric thermal emission. Valdes et al. (82 Icarus 49, 17) determined the ammonia abundance in the atmosphere of Jupiter from limb-darkening measurements made at 3.4 mm with the Hat Creek interferometer.

Saturn and the rings have been obvious subjects for VLA mapping at centimeter wavelengths. First results were published by de Pater et al. (82 Icarus 50, 88). As expected from earlier interferometry, the ring brightness was very low, and the rings caused conspicuous absorption where they crossed the disk. A limb-darkening analysis was also done. Later results can be found in de Pater et al. (83 Adv Sp Res 3, 39). Further work is in progress on other VLA maps and OVRO 3 mm interferometer data by Muhleman and Berge.

Additional analysis and interpretation of the time variation in the microwave spectrum of Uranus were presented by Gulkis et al. (83 Science 221, 453). VLA images of Uranus by Jaffe et al. (84 Science 225, 619) showing a subsolar symmetry at 2 cm and a polar symmetry at 6 cm may tie in with the wavelength dependence of the brightness temperature.

VLA measurements of Io at 2, 6, and 21 cm, consistent with thermal emission, were given by de Pater et al. (82 ApJ 261, 396). More recently, de Pater et al. (84 Icarus 57, 93) presented VLA maps of all the Galilean satellites at 2 and 6 cm showing large temperature differences between them and all lower than the IR temperatures. These are interpreted as surface emissivity effects. Similar VLA observations, as well as OVRO 2.6 mm interferometer observations, of Titan and the Galilean satellites have been made by Muhleman, Berge and Rudy. These VLA measurements of the Galilean satellites are also being analyzed for astrometric purposes by Niell et al. The 2.6 mm observations of Titan have also led to a radio detection of CO in the CO(0-1) line (Muhleman et al. 84 Science 223, 393).

Ulich, Dickel and de Pater have made new 1.3 mm planetary observations at Kitt Peak and have determined disk temperatures for Ceres as well as Callisto, Ganymede, and the major planets. Johnston et al. (82 AJ 87, 1593) observed Ceres and Pallas at 6 and 2 cm with the VLA to obtain astrometric data and flux densities.

Radio observations of comets are being vigorously pursued. For a recent review, see Snyder 82 Icarus 51, 1. The OH spectrum of comets is now rather well understood and is proving to be a useful tool to study the velocity field of the material (e.g. Bockelee-Morvan et al. 84 AA 131, 111). Attempts to detect other spectral lines or continuum emission remain frustrating because of the episodic nature of the detections, the difficulty of confirmation, and the large fraction of negative results. For example, see Gibson et al. 81 ApJ 248, 863; Crovisier et al. 81 AA 97, 195; Ekelund et al. 81 Icarus 47, 431; Snyder et al. 83 ApJ 269, L21.

GALACTIC RESEARCH

Surveys - Filled-aperture arrays ranging in size from 1 km to 600 m have been used by Ellis (82 AJP 35, 91) to obtain maps of the southern sky at frequencies of 2.1, 3.7, 4.7, 5.5, 8.3, 13 and 16.5 MHz, with angular resolution ranging from $1\overset{\circ}{.}5$ to $7\overset{\circ}{.}5$, and used to derive the emission measure as a function of galactic coordinates.

An atlas of radio continuum maps of the whole sky, made at 408 MHz with a resolution of $0\overset{\circ}{.}85$ has been published by Haslam et al. (82 AA Sup 47, 1). A description of the survey was previously given by Haslam et al. (81 AA 100, 209). The survey has been used to model the distribution of the galactic synchrotron emission by Phillips et al. (81 AA 98, 286; 81 AA 103, 405) and in a comparison with COS-B γ-ray data by Haslam et al. (81 Nature 289, 470). Spoelstra (84 AA 135, 238) has discussed observations of the linear polarized component of galactic radio emission at 5 frequencies between 408 MHz and 1420 MHz.

At 1420 MHz, Reich (82 AA Sup 48, 219) has surveyed the whole sky north of δ 20° with a beam of 35' arc. The maps are presented in a similar way to those of the 408 MHz survey and reach a full-beam brightness temperature of 50 mK. A survey at the same frequency and with the same resolution by Reich and Steffen (81 AA 93, 27) covers an area of 1450 sq. deg. around the southern part of loop IV ($13^h00^m \leq RA \leq 14^h40^m$, $-32° \leq \delta \leq -12°$) and achieved a similar sensitivity. Hughes (82 AA 111, 358) has mapped a 1 sq. deg. region around the Cep A condensation using the Westerbork synthesis telescope and obtained a catalogue of 20 sources stronger than 3.75 mJy: only 4 of these are thermal.

At 2.7 GHz, Graham et al. (82 AA 109, 145) have mapped a region in Monoceros of 5° x 7°, centered at l = 206°, b = $-0\overset{\circ}{.}5$, with a beam of $4\overset{'}{.}5$ arc, in a study of the Rosette Nebula, the Monoceros nebulosity and the source G206.9 + 2.3.

Two surveys made with the RATAN-600 telescope at 3.95 GHz have been reported. Kononov and Pyatunina (83 AZh 60, 685; 83 Sov Astr 27, 4) have mapped the region of the Monoceros nebulosity ($5^h39^m \leq \alpha \leq 7^h55^m$, $-6°28' \leq \delta \leq -6°08'$) and list 93 sources having flux densities $\geq$ 20 mJy: almost twice as many as the expected number of extragalactic background sources. Pyatunina (84 Pis'ma A Zh 10, 191) surveyed three fields in the range $200° \leq l \leq 230°$ and detected 405 sources $\geq$ 30 mJy in the combined area of 0.025 sr; about 160 of these are estimated to be galactic. At sub-mm wavelengths, Stier et al. (81 Bull AAS 13, 537) have used a balloon-borne telescope to map the region $10° \leq l \leq 45°$, $|b| < 3°$ with a resolution of about 10' arc, at wavelengths of 0.15, 0.25 and 0.35 mm.

Taylor and Gregory (83 AJ 88, 1784) have made repeated surveys of the area ($40° \leq l \leq 220°$, $|b| < 2°$) at 5 GHz in a search for compact, variable sources. They list 534 discrete sources above 10-20 mJy, the completeness limit being 70 mJy. About 30 sources are variable on time scales of days to 1-2 years. One source shows periodic outbursts every 26.5 days (Coe et al. 83 MNRAS 203, 791).

Radio Stars - Radio observations of stellar systems has been a rapidly expanding field since the completion of the VLA in 1980, therefore only a small fraction of the major discoveries can be summarized. Probably the most important observations of radio stars are the VLBI observations (Lestrade et al. 84 ApJ 279, 184; Mutel et al. 84 ApJ 278, 220; Leal 84 ApJL 282, L23; Lestrade et al. 85 *Radio Stars*, eds. Hjellming and Gibson, Reidel:Dordrecht) that have established that Algol and many RS CVn binaries (HR5110, UX Ari, HR 1099) exhibit variable core-halo structures with typical brightness temperatures of (1-3) x 10^{10} K for the cores and (5-10) x 10^8 K for halos with the size scale of the binary systems.

The essential nature of radio flare stars, in doubt for many years because of questions about confusion with interference in earlier observations (Davis et al. 78 Nature 273, 644), has been well established by observations in recent years, particularly with the VLA and Arecibo. The most critical (Lang et al. 83 ApJL 272, L15) observation was a flare in AD Leo observed at 1.4 GHz at Arecibo in which a flare peaking at 120 mJy showed complex, impulsive time structure unresolved on time scales of 0.2 seconds, with almost entirely left circular polarization and brightness temperatures in excess of 10^{13} K. Weaker versions, unfortunately with poorer time resolution, have been found to be common in UV Ceti-type flare stars (Mullan 85 and Gary 85 in *Radio Stars*, eds. Hjellming and Gibson, Reidel:Dordrecht) and impulsive flaring, usually in one hand of circular polarization. A keystone to the interpretation of some of these phenomena was the VLA observations of similar radio flares from the magnetic cataclysmic variable AM Herculis (Chanmugam and Dulk 82 ApJL 255, L107) and their interpretation (Dulk et al. 83 ApJ 273, 249) in terms of optically thick gryosynchrotron emission with occasionally occurrence of cyclotron maser emission at isolated radio frequencies.

Relativistic jets produced by stellar X-ray binaries, first observed as "corkscrews" expanding at 0.26c in SS433 (Hjellming and Johnston 81 ApJL 243, L141) have also been found on two occasions following very large flares in the X-ray binary Cyg X-3, with indications of velocities of the order of 0.3c (Geldzahler et al. 83 ApJL; Johnston et al. 85 ApJ).

VLA observations of OB and WR stellar winds have measured both sizes and temperatures (Newell 81 New Mexico Tech PhD Thesis; White and Becker 83 ApJL 272, L19; Hogg 85 *Radio Stars*, eds. Hjellming and Gibson, Reidel:Dordrecht) of some winds, but simple interpretations as free-free emission from mass loss rates of 10^6 to 10^4 solar masses per year have been questioned (Underhill 84 ApJ 276, 583) with observational support from indications of time variability (Abbott et al. 84 ApJ 280, L71; Lamers et al. 85 and Rodriguez et al. 85 in *Radio Stars*, eds. Hjellming and Gibson, Reidel:Dordrecht).

Flux densities (and upper limits) at 2.3 and 8.4 GHz (S and X band) for 49 radio stars have been obtained by Estalella et al. (83 AA 124, 309). The aim is to provide a list of objects which could be observed by VLBI in order to link the positional radio reference system to the optical reference system which will be provided by Hipparcos.

Results on variability of various types of radio stars are given by Purton et al. (83 AJ 88, 1825), Geldzahler et al. (84 IAU Symp 110, 179), Hughes and McLean (84 ApJ 278, 716) and Abbot et al. (84 ApJ 280, 671). The variability of SS433 has been studied by Seaquist et al. (82 ApJ 260, 220), Bonsignori-Facondi (83, Workshop on "Astrophysical Jets", Reidel p. 161) and Johnston et al. (84 AJ 89, 509). Radio emission from the Nova SU UMa has been observed with the Bonn 100 m telescope (Benz et al. 83 Nature 302, 45).

Pulsars - The discovery of a 1.6 ms pulsar, PSR1937+21, at Arecibo (Backer 82 Nature 300, 615) has rejuvenated research into the final states of isolated and binary stars (Ruderman 83 Comments Ap 10, 15; van den Heuvel 84 JAA 5, 209). Boriakoff announced the discovery of a 6.1 ms pulsar in a 123-day binary (83 Nature 304, 417). Many observational and theoretical developments spawned by these discoveries were reviewed at a Green Bank Workshop in June 84 (proceedings referenced below as GBWS, eds. Reynolds and Stinebring). Millisecond pulsar observations are reviewed by Backer (84 JAA 5, 187).

The sharp pulse of PSR1937+21 and its low rotation noise suggest that ms pulsars may provide a clock for long time scales (Backer 83 Nature 301, 314; Ashworth 83 Nature 301, 313). An accurate pulsar time scale can lead to improved values for the orbit of the Earth and to detection of gravitational waves (Romani 82 ApJ 265, L35; Hellings 82 ApJ 265, L39; Bertotti 83 MNRAS 203, 945; Hogan 84 Nature). Limits to timing accuracy have been analyzed by Blandford (84 JAA Dec). Acquisition of pulsar arrival time data at Arecibo with <100 ns IAT error has required improvements in software, hardware and time transfer (Davis 84 GBWS). Timing measurements also continue at Jodrell Bank.

Binary PSR1913+16 observations continue to confirm the orbit decay predicted by general relativity (Weisberg 84 Phys Rev Let 52, 1348; Boriakoff 82 ApJ 261, L97). The Vela pulsar glitched again (McCulloch 83 Nature 302, 319). The braking index of PSR1509-58, a young X-ray/radio pulsar, is n=2.83 (Manchester 84 GBWS). A $P/\dot{P}$ study by Cowling (83 MNRAS 204, 1237) places a new limit on the existence of a solar companion. New timing program results: Downs (83 ApJSS 53, 169); Manchester (83 MNRAS 202, 209) and Gullahorn (82 ApJ 260, 520). Crab timing continues at Jodrell.

Gwinn (84 Princeton PhD) has determined parallaxes for PSR0950+08 (7.9±0.8 mas) and PSR0823+26 (1.8±0.4 mas) with VLBI. Fomalont's (84 MNRAS 210, 113) comparison of timing and interferometer positions shows unexplained systematic 0".3 discrepancies between the two reference frames. Lyne (82 MNRAS 201, 503) has measured proper motions which imply a curious correlation with inferred surface field (Anderson 83 Nature 303, 597) and allow a comparison with scintillation parameter velocities (Lyne 82 Nature 298, 825; Cordes 84 GBWS).

The search for counterparts of radio pulsars at X and gamma rays (Thompson 83 AA 127, 220; Buccheri 83 AA 128, 245; Graser 83 ApJ 272, 681; Knight 82 ApJ 260, 553) and at optical wavelengths (Lebedev 83 AZh 60, 742) have been successful. The ms pulsars have not been detected (Loredo 84 GBWS; Manchester 84 Nature 310, 569). Xray synchrotron nebulae have been detected for 1055-51 (Cheng 83 ApJ 3=271, 271) and 1509-58 (Seward 84 ApJ 281, 650). Observations of the Crab at high energy continue (Knight 82 ApJ 260, 538; Wilson 83 ApJ 269, 273; Beskin 83 PAZh 9, 280).

Rankin (83 ApJ 274, 359; 274, 333) presented an observational synthesis of pulsar beaming. Narayan (82 AA 113, L3; 83 AA 122, 45) presented a strong case for elongated beams in the unseen coordinate. Others have investigated evolution of beaming (Wu 82 AASin 23, 140; Kuz'min PAZh 9, 149). Stinebring (84 ApJSS) published new polarization data. Pulse nulling and mode changing were studied with emphasis on PSR0809+74 (Fillipenko 82 ApJ 263, 828; Lyne 83 MNRAS 204, 519; Popov 82 AZh 59, 719; Bartel 82 ApJ 258, 776; Davies 84 MNRAS). While Perry at Jodrell Bank reports detection of continuous point-source emission for several pulsars, Kesteven (82 AA 113, 211) and Cohen (83 ApJ 264, 173) do not detect extended emission near pulsars. Important studies of polarization at 39 MHz have been reported by Sulejmanova (83 AZh 60, 554).

The discovery of millisecond pulsars has led to new surveys for fast pulsars employing new techniques. Most are summarized in GBWS: polarized point source (Jodrell, Berkeley/WSRT); interstellar scintillation (Berkeley/GB); scintar/pulsars (many); offline analysis of data (Parkes, Jodrell, Princeton/GB, Princeton/NAIC); online search with pipeline multiprocessor (Berkeley).

Pulsars in the LMC have been found at radio wavelengths (McCulloch 83 Nature 303, 307) and at X-rays; the latter is surrounded by a synchrotron nebula.

Interstellar propagation studies focused on three areas: galactic distribution of turbulence; refractive effects (Seiber 82 AA 113, 311; Roberts 82 MNRAS 201, 1119; Rickett 84 AA; and "applications" to pulsar studies (Lyne 84 Nature 310, 300; Cordes 83 ApJ 268, 370).

The maximum rotation speed of pulsars is discussed by Freidman (83 Phys Rev Let 51, 11). Binary evolution scenarios and neutron star production scenarios are reviewed by van den Heuvel (84 JAA 5, 209) and Chevalier (84 GBWS), resp. Vital to these discussions is a provocative discussion on production of strong magnetic fields by cooling neutron stars (Blandford 83 MNRAS 204, 1025). The relation of pulsars and supernovae remains a puzzle (Radhakrishnan 82 Contemp Phys 23, 207). Lyne (84 MNRAS) has summarized current understanding of galactic distribution of pulsars. X-ray observations of synchrotron nebulae or the neutron star itself are important (Becker 82 ApJ 255, 557).

Supernovae and Supernova Remnants - Supernova (SN) and supernova remnant (SNR) research has grown substantially during the last triennium. This is partly due to the quantum jump in resolution and sensitivity provided by the VLA and partly the result of cross-fertilization from other areas, especially X-ray astronomy, as demonstrated by the large amount of radio data presented at the 1983 IAU Symposium 101 on SNR and their X-ray emission. Progress in the study of SN has been particularly striking with a doubling of the total number detected at radio wavelengths, including the first SN discovered by other than optical techniques. SNR research has been highlighted by radio surveys of several external galaxies with sufficient resolution to reveal the structure of many of the remnants studied, improved maps of many SNR in our Galaxy, and new ideas on their statistics.

Bignell and Seaquist (83 ApJ 270, 140) have studied the luminous SN in NGC 4449 and confirmed variability at 5 GHz. With the detection of radio emission from SN1980k in NGC 6946 (Weiler et al. 83 IAU Symp 101, 171) and the radio discovery of a SN in NGC 4258 (Van der Hulst et al. 83 Nature 306, 566; 83 PASP 95, 607), the number of radio supernovae has increased to four. Three are of Type II (there is insufficient information to classify the SN in NGC 4258) and their characteristics are rather divergent. Nevertheless it is clear that models involving the sudden injection of relativistic particles which subsequently cool adiabatically are inconsistent with the data. Rather, a pulsar-powered source initially shielded by a dense shell of material (Salvati, 83 IAU Symp 101, 177) or particle acceleration by the blast wave plowing through a cocoon of circumstellar matter (Chevalier, 81 ApJ 251, 259) have been suggested. An unsuccessful search for radio emission from recent SN in nearby galaxies was made by Cowan and Branch (83 ApJ 258, 31).

Observations of external galaxies are now contributing significantly to radio studies of SNR. Mills et al., selecting objects on the basis of X-ray data, present maps of SNR in the Small (82 MNRAS 200, 1007) and Large Magellanic Clouds (84 AJP 37, 321)--38 sources in all. Mathewson et al. (83 ApJ Sup 51, 345) also used optical data in studying a similar number of LMC and SMC remnants. More distant galaxies have turned up fewer detections: a handful each in M31 (82 ApJ 252, 582; 84 MNRAS 206, 351) and M33 (82 MNRAS 198, 1059; see also 83 AA 120, 147). In addition, there have been observations of singular SNR candidates in other galaxies, especially NGC 4449 (83 AA 119, 301; 83 ApJ 270, 140) and M82 (83 IAU Symp 101, 583). Statistical studies based on SNR in the LMC, SMC, M31 and M33 and presented in papers cited above suggest that the onset of the adiabatic phase occurs quite late in remnant evolution. This conclusion has been disputed by Green (84 MNRAS 209, 449) who, in presenting and discussing a new inventory of galactic SNR, suggests existing catalogues are quite incomplete. Several authors have commented on the large scatter in the surface brightness-diameter relationship and on its unreliability for determining distances (83 IAU Symp 101).

Morphological studies of galactic SNR have substantially increased the number of high quality maps available of both northern (81 AJ 86, 1003; 81 AA 103, 277; 81 AA 103, 393; 81 J. Astrophys. Astron. 2, 339; 82 AA 105, 176; 82 AA 109, 145; 82 ApJ 261, L41; 82 AJ 87, 1379; 82 J. Astrophys. Astron. 3, 207; 83 MNRAS 203, 695; 83 AJ 88, 1810; 83 Pub.A.S.Jap. 35, 437; 83 Pub.A.S.Jap. 35, 447; 84 AA 131, 196; 84 AA 133, 11) and southern (82 AA 106, 314; 82 MNRAS 200, 1143; 82 MNRAS 203, 595; 83 MNRAS 204, 237; 83 MNRAS 204, 833; 83 MNRAS 204, 915; 83 MNRAS 204, 921; 83 ApJ 271, L55) remnants. There has also been work which combined radio and X-ray data (81 ApJ 248, L23; 83 ApJ 266, 684; 83 ApJ 268, L93; 84 MNRAS 207, 649) and from a consideration of radio morphologies (Shaver (82 AA 105, 306) concludes that shell SNR are brightest on the side nearest the galactic plane. HI studies have been made both for determining SNR distances (82 Reg. of Rec. Star Form. 193; 82 SovA 26, N1; 82 Nature 299, 606) and investigating nearby gas (81 SovA 25, 675; 82 AA 112, 141; 83 A&AS 53, 57; 84 AA 130, 295). There have also been molecular cloud studies of SNR which often show associations (81 Nature 293, 549; 83 AJ 88, 297). The decrease in flux density of several young SNR has been monitored by Ivanov et al. (82 Sov AJ Let 8, 42), while the expansion of Tycho was determined by Strom et al. (82 MNRAS 200, 473) and that of Cas A by Tuffs (83 IAU Symp 101, 49).

SNR resembling the Crab have not been neglected, as shown by the special issue of Observatory (83 Obs 103, 73-116; see especially reviews by Wilson and Weiler). In a search for such remnants, Helfand et al. (84 AJ 89, 819) conclude that few are missing from existing catalogues. Related objects appear in an SS433 conference (81 Vistas 25) and Velusamy and Kundu mapped the center of CTB80 (83 J. Astrophys. Astron. 4, 253).

The Galactic Center - The major new observational results include a series of high-resolution VLA radio images of sources within the central square degree of the Galaxy. The radio source at the center, Sgr A, was shown (Ekers et al 83 AA 122, 143) to consist of (i) a ∿10-pc oval shell of nonthermal emission, Sgr A East, and (ii) a ∿3-pc spiral-like distribution of thermal emission, Sgr A West, (Lo et al. 83 Nature 306, 647; Brown et al 83 ApJ 268, L85). On a larger scale, the "arc-feature" at $\ell=0\overset{\circ}{.}18$ has been resolved into several long (∿50 pc), linear "strands" of emission with curved features merging with the diffuse halo of Sgr A (Yusef-Zaheh et al. 84 Nature 310, 557). These linear features are at the base of a giant Ω-shaped radio loop, straddling the galactic center and extending above the galactic plane by more than 1° (∿200 pc) (Sofue et al. 84 Nature 310, 568). The compact nonthermal radio source at the centroid of Sgr A West was shown to have an elongated structure at 3.6 cm (Lo 84 IAU Symposium 110, 265). HI, CO and HNC observations (Liszt et al. 83 AA 126, 341; Fukui et al. 83 ApJ

275, L55) indicated the existence of neutral gas within 5 pc of the center, perhaps providing the material falling in towards the center--a possible explanation of the ionized gas in Sgr A West (Lo et al. 83; Ekers et al. 83).

HI - Increasing sensitivity, bandwidth coverage and spectral resolution have continued to improve the 21 cm line systems on major single dish telescopes. The preliminary reports of new surveys of the southern hemisphere promise many exciting results, especially in the regions of the Magellanic Clouds and the Magellanic Stream. Much of the detailed 21 cm work on the Milky Way is designed to correlate the HI clouds with other constituents of the interstellar medium. In the region of the South Celestial Pole, King et al. (82 MNRAS 198, 255) find a correlation between gas and dust by examining the HI and optical emission. HI is also seen in association with supernova remnants (Stacy and Jackson 82 Nature 296, 42; Sofue 83 Pub.A.S.Jap. 35, 91; Lockman and Genzel 83 ApJ 268, 117) surrounding CO clouds (Wannier et al. 83 ApJ 268, 727), and in self-absorption towards an OB association (Bania 81 AJ 88, 1222). Absorption spectra have been measured by Dickey et al. (83 ApJ Sup 53, 591) with the VLA and Mebold et al. (82 AA 115, 223) with the Bonn and Green Bank dishes.

Heiles and Troland (82 ApJ 252, 179) have measured the Zeeman splitting in the vicinity of Orion and in expanding HI shells; in eight HI emission regions, the weakness of the magnetic field precludes the alignment of dust grains by the conventional mechanism. Bregman et al. (82 AA 118, 157) have detected Zeeman splitting in absorption toward Cas A.

Detailed studies of the structure and distribution of high velocity HI clouds (Schwarz and Oort 83 AA 101, 305; Mirabel 82 ApJ 256, 112), especially the Magellanic Stream (Cohen 82 MNRAS 199, 291; Morras 83 AJ 88, 62), suggest that most of the phenomena result from the recent tidal interaction of the Milky Way and the Magellanic Clouds (Giovanelli 81 AJ 86, 1468).

HII Regions and Recombination Lines - New surveys of galactic HII regions were made by Wink et al. (82 AA 108, 227) in the radio continuum and recombination line (RRL), and Blitz et al. (82 ApJ Sup 49, 183) in the CO line. At least 70% of all HII regions have associated molecular clouds. Fich and Blitz (84 ApJ 279, 125) have studied the distribution and properties of HII regions in the outer galaxy. Distances have been determined for a few HII regions using H_2O maser proper motions (e.g. Schneps et al. 81 ApJ 249, 124). Much effort has gone into continuum and line mapping of individual HII regions. VLA observations reveal shell structure in compact HII regions (e.g. Turner and Matthews 84 ApJ 277, 164), and detect ultracompact HII regions embedded in more diffuse nebulae and molecular clouds (e.g. Moran et al. 83 ApJ 271, L31); turnover frequencies are as high as 20-30 GHz, and electron densities $\sim 10^7$ cm^{-3}. Garay et al. (84 ApJ, in press) find that main-line OH masers are invariably associated with such ultracompact HII regions.

Magnetic fields within HII regions have been studied using Faraday rotation of the polarized radiation from background sources (Heiles et al. 81 ApJ 247, L77), and the Zeeman effect in RRLs (Silvergate 84 ApJ 279, 694). The cosmologically important 8.7 GHz hyperfine line of $^3He^+$ was detected in a few HII regions (Rood et al. 84 ApJ 280, 629). RRLs were used to determine ^4He abundances and electron temperatures (e.g. McGee and Newton 81 MNRAS 196, 889; Lockman and Brown 82 ApJ 259, 595; Garay and Rodriguez 83 ApJ 266, 263; Wink et al. 83 AA 127, 211), and the galactic abundance gradient (Shaver et al. 83 MNRAS 204, 53).

Low frequency (20-30 MHz) carbon RRLs have been detected in the spectrum of Cas A (Konovalenko and Sodin 80 Nature 283, 360; 81 Nature 294, 135); these are the lowest-frequency spectral lines, and the first RRLs to be observed in absorption. Low-frequency RRLs have also been used to study the properties of the low-density outer parts of HII regions (Anantharamaiah, 84 PhD Bangalore), and the diffuse interstellar medium (Ariskin et al. 82 Sov AJ 26, 23; Payne et al. 84 AJ 89, 668).

Interstellar Molecules - The scientific developments in molecular line radio astronomy are moving away from the discovery phase and toward a consolidation and deepening of our knowledge. Over the next few years, great advances will be made in producing high spatial resolution maps in centimeter lines with the VLA, and in millimeter lines with the Berkeley, Caltech, IRAM and Nobeyama interferometers. The 45-m Nobeyama and 30-m IRAM single dishes, when combined with state-of-the-art receivers, should allow an order of magnitude improvement in sensitivity. The soon-to-be-released IRAS data will provide a starting point for many comparisons of IR properties with molecular cloud parameters. Searches for specific molecules, such as the biologically important species Glycine (Snyder et al. 83 ApJ 268, 123), the reported detection of the long chain carbon $HC_{11}N$ (Bell et al. 82 Nature 295, 389), the detections of CH_3C_3N (Broten et al. 84 ApJ 276, L25) and CH_3C_4H (Walmsley et al. 84 AA 134, L11; MacLeod et al. 84 ApJ 282, L89) have taken place. The assignments of lines to exotic, "non-terrestrial" species have been made by Thaddeus et al. (81 ApJ 246, L41). Searches for HOC^+ (Woods et al. 83 ApJ 270, 583) have been made, and the assignment of a number of unidentified lines to the first ring molecule, SiC_2, in the shell of the carbon star IRC+10216 (see e.g. Olofsson et al. 82 AA 107, 128; Knapp et al. 82 ApJ 252, 616) have been made by Thaddeus et al. (84 ApJ 283, L45). All of this work depends on the availability of laboratory measurements or accurate molecular structure models. Frequency scans give an estimate of the number and intensity of spectral lines in a given range. The best and most complete data are in a 3 mm frequency scan of Orion-KL carried out at the Onsala Space Observatory (Johansson et al. 84 AA 130, 22). These results showed only a few unidentified lines. At higher frequencies, the 200-265 GHz frequency scan of Orion-KL done by the Caltech group shows a similar result.

A great deal of effort has gone into the study of molecular outflows from low mass stars, which in some cases are bipolar (see Snell and Edwards 83 ApJ 279, 668; Bally and Lada 83 ApJ 265, 824). The mechanism for the production of large momentum fluxes deduced from CO outflows is unknown. Lower resolution maps (Bastien et al. 81 AA 98, L4, for Orion-KL; Torrelles et al. 83 ApJ 274, 214, for lower velocity outflows) seemed to indicate that the outflows were focussed by quiescent molecular clouds. However, on scales of 30" this does not seem to be the case (Takano et al. 84 ApJ 282, L69). On a smaller scale, this possibility has been applied to the Orion-KL outflow by Plambeck et al. (82 ApJ 259, 617). There are a rather large number of hot molecular clouds in the galaxy. For the Orion-KL and Sgr B2 clouds, the kinetic temperatures have been estimated by observations of many transitions of CO (Storey et al. 81 ApJ 247, 136; Stacey et al. 82 ApJ 257, L37; Koepf et al. 82 ApJ 260, 584), vibrationally excited HC_3N (Goldsmith et al. 82 ApJ 260, 147), SO_2 (Schloerb et al. 83 ApJ 264, 161); torsionally excited methanol (Lovas et al. 82 ApJ 253, 149) and NH_3 at 1.3 cm (Wilson et al. 82 AA 115, 185). For the Orion region near the KL nebula the kinetic temperature estimates range from $\sim$ 80 K to 10^3 K. However, the centers of the molecular emission are spatially separated. The high velocity wings in J = 3-2 line of CO (Erickson et al. 82 ApJ 261, L103) are $\sim$10" north of the centers of the hot Ammonia peaks seen with the VLA (Genzel et al. 82 ApJ 259, L103; Pauls et al. 83 AA 124, 23).

High quality single-dish maps may allow accurate determinations of the spatial distributions of many species (Olofsson et al. 82 AA 113, L18; Batrla et al. 83 AA 128, 279; Hasegawa et al. 84 ApJ 283, 117). For maps in a single line, however, the trend is toward interferometry (see Forster et al. 81 MNRAS 197, 513; Welch et al. 81 ApJ 245, L87; Harris et al. 83 ApJ 265, L63; Johnston et al. 83 ApJ 271, L89). Although the spatial distributions of the line intensities of various molecules are different, it is not completely clear that the distributions of these molecules differ. High angular resolution maps of warm clouds are also necessary if comparisons are to be made between the location of IR sources or stars and molecular clouds, similar to those made for nearby cold clouds by Myers and Benson (83 ApJ 266, 309). In addition to the VLA results for Orion-KL, VLA observations of NH_3 emission in W51 (Ho et al. 83 ApJ 266, 596) show large kinetic temperatures (from the line brightness temperatures) directly. For the hot clouds in the galactic center (Güsten et al. 81 AA 103, 197), the source of heating is not certain. Emission line measurements toward Orion-KL (Genzel et al. 82 ApJ 259, L103; Pauls et al. 83 AA 124, 23), and absorption line data taken toward the compact HII regions in W31 and NGC 7538 (T.L. Wilson et al. 83 AA 127, L19) show that these hot molecular clouds also appear to have large concentrations of Ammonia. Multi-line analyses also have been used to determine H_2 densities and scale sizes in molecular clouds (Vanden Bout et al. 83 ApJ 271, 161; Schloerb et al. 83 ApJ 267, 163); Plambeck et al. 82 ApJ 266, 321; Ho et al. 81 ApJ 246, 761; Batrla et al. 82 AA 119, 139; Mauersberger et al. 83 Mitt. A.G. 60, 402; Martin-Pintado et al. 83 AA 117, 145; Martin et al. 84 MNRAS 208, 35). The optical depth in many such lines is significant, and the trend is to use a multi-line analysis of rarer isotopic species (Henkel et al. 83 AA 127, 388); in the future, a comparison of different lines of the same species, mapped interferometrically, will become more common (Genzel et al. 82 ApJ 259, L103.

Isotope studies showed that the $^{12}C/^{13}C$) ratios obtained from H_2CO (Henkel et al. 82 AA 109, 344) and $^{13}C^{18}O$ (see R.W. Wilson et al. 81, ApJ 243, L47) give a ratio which is close to terrestrial, for sources near the Sun. Studies of the polarization of absorption lines of molecules such as OH (Crutcher et al. 81 ApJ 249, 134) will lead to better estimates of the magnetic fields, via the Zeeman effect, in what are certainly the denser portions of molecular clouds. In SiO, polarization has also been found (e.g. Clark et al. 82 ApJ 261, 569). This is <u>not</u> caused by the Zeeman effect. Theoretical studies (Goldreich and Kylafis 82 ApJ 253, 606) have shown that, in the general interstellar medium, polarization in the radiation from molecules other than OH might be detectable, and searches for this effect are underway. Frerking et al. (82 ApJ 262, 590) have related the column density of $^{13}C^{16}O$ and $^{12}C^{18}O$ to the visual extinction, A_v, for nearby dust clouds in Taurus and Ophiucus. The extinction was related to the column density of H_2 using the standard gas-to-extinction ratio. Attempts to determine a spiral pattern in our galaxy from CO data can now begin, since a combined north-south data set (taken with different telescopes) has been presented by Robinson et al. (84 ApJ 283, L31). A more uniform data set, from work of the Goddard group, is expected soon. However, the results may be strongly affected by non-circular velocity streaming motions.

<u>Interstellar Masers</u> - Observations and theoretical advances in the study of molecular masers prior to this report have been summarized by Reid and Moran (81 Ann Rev AA 19, 231).

A new application of the technique of statistical parallax has been used to measure distances to H_2O masers in W51 by Genzel et al. (81 ApJ 247, 1039) and Schneps et al. (81 ApJ 249, 124). These studies extend direct astronomical distance measurements by more than an order of magnitude. Walker et al. (82 ApJ 255, 128) used new VLBI fringe-rate analysis techniques and mapped 386 maser components in W49; results based upon these observations indicate for the first time hyperfine splitting in H_2O maser emission.

Abraham et al. (81 AA 100, L10) monitored the dramatic variations in the great Orion H_2O maser flare which exceeded 10^6 Jy while Matveyenko et al. (83 Sov AJ Let 9, 240) have measured the angular size of the flaring region with the Simeiz-Pushchéno interferometer. New H_2O masers associated with far IR sources were discovered by Jaffee et al. (81 ApJ 250, 621) and other new sources were reported by Dickinson and Dinger (82 ApJ 254, 136), Haschick et al. (83 ApJ 265, 281), Scalise et al. (81 AA 104, 166), Gusten and Downes (83 AA 117, 343), Braz and Scalise (82 AA 107, 272), Sandell and Olofsson (81 AA 99, 80) and others.

Accurate positions for 6 cm excited OH masers were determined by Rickard et al. (82 AJ 87, 1806). VLBI synthesis maps of the OH maser emission in NGC 7538 were completed by Forster et al. (82 MNRAS 201, 7P). Benson et al. (82 ApJ 253, 199) and Kent and Mutel (82 ApJ 263, 145) reported main-line OH VLBI results. VLA observations of OH masers associated with compact HII regions have been reported by Ho et al. (83 ApJ 265, 295) and Garay (83 PhD Thesis, Harvard U.). Maps of OH emission from IR stars (e.g., Norris et al. 82 Nature 299, 131; Diamond et al. 84 MNRAS 207, 611) and from Orion (Norris et al. 84 MNRAS 207, 127) were made with the MERLIN telescope. VLA observations of stellar OH sources were published by Baud et al. (81 ApJL 250, L79), Reid et al. (81 AJ 86, 897), and Bowers et al. (83 ApJ 274, 733). An unusual OH source was reported by Morris and Kazes (82 AA 111, 239). The evolution of the peculiar OH star U Ori was monitored by Jewell et al. (81 ApJ 249, 118). The luminosity function and spatial distribution of OH masers in the Galaxy has been studied by Tors et al. (84 Acta Ap Sinica 4, 96).

Polarization properties of SiO masers have been observed (Clark et al. 84 ApJ 276, 572) and simultaneous observations of several transitions have been made (Schwartz et al. 82 ApJL 256, L55). Relative positions of the two SiO maser peaks in Orion were reported by Lane (82 PhD Thesis, U. Mass.) and by Wright et al. (83 ApJL 267, L115). Isotopic maser emission may have been detected in ^{29}SiO by Olofsson et al. (81 AA 100, L30) and Deguchi and Rieu (83 AA 117, 314) from the Orion source. Sun et al. (84 Acta Ap Sinica 4, 104) have modeled the velocity spectra of Mira Variable SiO masers by stellar pulsations. Potential new masing molecular species and transitions were discovered: SiS (Henkel et al. 83 ApJ 267, 184) and SiO J=5-4 (Clemens et al. 83 ApJL 266, L117).

EXTRAGALACTIC RESEARCH

Surveys of Extragalactic Sources - Two new 5C surveys have been completed. The 5C10 survey (Schuch 81 MN 196, 695 and 83 MN 204, 1245) covers the Ursa Major supercluster and lists 265 sources $\geq$ 9.8 mJy and 48 sources $\geq$ 1.7 mJy at 408 and 1407 MHz, respectively. The source counts for this special field appear consistent with those from other 5C surveys. The 5C12 survey (Benn et al. 82 MN 200, 747) covers a region near the north Galactic pole and lists 299 sources $\geq$ 9 mJy and 65 sources $\geq$ 1.4 mJy at 408 and 1407 MHz, respectively. Magnitudes and colors for 1200 objects near the 5C12 sources are given in a separate paper (Grueff et al. 84 MN 206, 475).

At 1.45 GHz, Condon and Mitchell (82 AJ 87, 1429 and 84 AJ 89, 610) have surveyed a region centered at RA = 08 52 15, δ = 17°16'. They find 159 sources $\geq$84 μJy, of which 45 are identified with objects brighter than B = 21.5. They have used the P(D) technique to extend the number-flux relation to 30 μJy.

The RATAN-600 telescope has been used to make two surveys. The first, made at 8.7 and 14.4 GHz (Amirkhanyan et al. 81 AZh 58, 717; Sov Astr 25, 412) covers the region 0° $\leq \delta \leq$ 4° and lists 82 sources $\geq$ 0.36 Jy and 24 sources $\geq$ 0.79 Jy at the respective frequencies. The surveys are believed to be 98 percent complete at 0.45 and 1.01 Jy, respectively. The second, made at 3.95 GHz, covers the region $4°44' \leq \delta \leq 4°54'$ and is 75 percent complete above 0.8 mJy (Berlin et al. 81 Sov AJ Let 7, 161). Source counts down to 0.86 Jy have been derived from the section $13^h \leq RA \leq 14^h$ and down to 7.5 mJy from the rest of the region. The counts converge down to 0.86 mJy, where the source density reaches 2×10^5 sr^{-1}.

Near 5 GHz, previous survey work has been extended with single dish surveys by Wall et al. (82 MN 200, 1123) of a 27 sq. deg. region at RA = 22^h, $\delta = -18°$ down to 32 mJy. Owen et al. (83 AJ 88, 1) list 480 sources $\geq$ 35 mJy in a 0.0691 sr area with $07^h04^m \leq RA \leq 17^h56^m$, $34°09' \leq \delta \leq 35°51'$. In both cases, the surveys have been followed by optical identification work, by Savage et al. (82 MN 200, 1135) and Condon and Ledden (82 AJ 87, 219); in both cases only a small fraction of the fainter sources can be identified with flat-spectrum QSOs. Wall et al. (82 MN 198, 221) have extended the number-flux relation down to ∿ 1 mJy using the P(D) technique and find the convergence suggested by the direct source counts continues. A dramatic step to even lower flux density levels has resulted from VLA surveys at 1.4 and 4.9 GHz.

Little survey work has been done at frequencies above 5 GHz. Seielstad has surveyed five 1 sq. deg. fields at 10.6 GHz down to 50 mJy, but detected only 2 sources (83 PASP 95, No. 563, 32). Seielstad et al. (81 ApJ 244, 717) have used the P(D) technique on 10.6 GHz data to put limits of (0.25 to 3.3) x 10^6 sr^{-1} Jy^{-1} on the source density near 10 mJy. The limited 14.4 GHz survey by Amirkhanyan et al. has been referred to earlier.

Further progress in the definition of the high-frequency source counts can be expected from additional VLA surveys. The application of the P(D) method to VLA data has already extended the counts to ∿ 10 μJy, but the direct counts below 1 mJy are still based on only a small number of sources. With improvements in receivers and multi-beaming techniques, extensive surveys near 15 GHz should become feasible soon.

Cosmological Tests - The period under review has seen a remarkable extension of source counts near 21 and 6 cm λ to the sub -mJy level. At 21 cm, Westerbork surveys down to ∿ 0.6 mJy (van der Laan and Windhorst 82 Astrophysical Cosmology, p349) suggested a possible upturn in the normalized differential counts below a few mJy. Subsequent deeper surveys with the VLA (Windhorst et al. 84 AA Sup 58, 1; Condon and Mitchell 82 AJ 87, 1429) going down to 85 μJy for individual sources and to ∿ 30 μJy by a P(D) analysis (Condon and Mitchell 84 AJ 89, 610) have confirmed a significant flattening of the counts below ∿ 5 mJy. Deep photometry shows a faint blue galaxy population dominating at low flux levels that could be responsible for the upturn (van der Laan et al. 83 IAU Symp 104, 73; Windhorst 84).

Sub-mJy counts near 5 GHz based on VLA surveys have been reported by Bennet et al. (83 Nature 301, 686) and Fomalont et al. (84 Science 225, 23), the latter survey reaching ∿ 60 μJy by direct counting and ∿ 10 μJy by a P(D) analysis. The 6 cm counts show rapid convergence below ∿ 100 mJy and also show the flattening at very low flux levels seen near 20 cm. The convergence in the counts down to ∿ 1 mJy was also inferred from a P(D) observation with the Effelsberg dish by Wall et al. (82 MNRAS 198, 221) who also point out that an earlier discrepant P(D) result with the Parkes dish (Wall 78 MNRAS 182, 381) could be attributed to an instrumental error. Counting "flat" and "steep" spectrum sources separately at 5 GHz shows (Wall and Benn 82 IAU Symp 97, 441; Fomalont et al. 84) that while the former begin to converge below ∿ 1 Jy, the latter do so only below ∿ 100 mJy. This gives the total counts a near "Euclidean" slope over a wide range in flux density. The similar shapes of the subcounts imply evolution for both types of sources, a result that is now supported by the V/V_m tests and also by the multifrequency "free-form" models of the evolution of the Radio Luminosity Function (Peacock and Gull 81 MNRAS 196, 611).

Swarup et al. (82 AA 107, 190; Astrophysical Cosmology, p383) have used the statistics of optical identifications with galaxies to constrain evolutionary models. A multifrequency model using a "parametric" evolution function has also been proposed (Subrahmanya and Kapahi 83 IAU Symp 104, 47). Cavaliere et al. (82 AA 114, L1; 83 ApJ 269, 57) report a physical model for the evolution of luminosity functions based on the conversion of gravitational energy in supermassive black holes.

The suggestion by Hawkins and Steward (81 ApJ 251, 1) that the V/V_m statistics of quasars could be compatible with "no evolution" in view of the selection effect against recognizing quasars at low redshifts has been shown by Avni and Schiller (83 ApJ 261, 1) and by Wills (83 ApJ 270, 48) to have little effect on the conventional interpretation of V/V_m tests.

The possibility of detecting a dipole anisotropy in radio source counts similar to that in the microwave background radiation, in order to test the isotropy of the Universe has been discussed by Ellis and Baldwin (84 MNRAS 206, 377).

The angular size - redshift relation for radio quasars and its interpretation has been investigated by several authors (Ulvestad et al. 81 AJ 86, 1010; Nottale 82 AA 113, 223; Okoye and Onuora 82 ApJ 260, 37; Onuora and Okoye 83 ApJ 270, 360; Hintzen et al. 83 AJ 88, 709; Swarup et al. MNRAS 208, 813) but no definite conclusions with regard to the possible evolution of linear size with epoch has been possible mainly because of the correlation between redshift and luminosity in the existing data.

New determinations of angular sizes down to ∿ mJy level at 408 MHz have confirmed that the median angular size (θ_m) of radio sources appears to saturate at ∿ 10 arcsec (Kapahi and Subrahmanya 82 IAU Symp 97, 401; Downes 82 IAU Symp 97, 393; Downes et al. 81 MNRAS 197, 593; Fielden et al. 83 MNRAS 204, 289). Westerbork surveys at 21 cm in fact show that θ_m remains at this level even down to ∿ 1 mJy (Windhorst 84). The interpretation of the θ_m-S relation is however still controversial because the predicted relation depends on the unknown redshift of weak radio sources.

New limits to small scale fluctuations in the CBR have been published by Fomalont et al. (84 ApJL 277, L23); Uson and Wilkinson (84 ApJL 277, 1); Lasenby and Davies (83 MNRAS 203, 127; and Berlin et al. (82 IAU Symp 104, 121). On larger scales, measurements of the dipole anisotropy have been refined, but no quadrupole moment has been detected (Lubin et al. 83 Phys Rev Let 50, 616; Fixen et al. 83 Phys Rev Let 50, 620). Improved low frequency measurements of the CBR spectrum give a value of 2.79±0.10 K with no evidence of departure from a black body curve over the range 2.5 to 90 GHz.

Flux Densities and Spectra - Flux density measurements of extragalactic sources have been reported over a wide range of frequencies. Over two hundred sources from the 1 Jy catalogue of Kuhr et al. have been measured at 408 MHz with the Bologna Cross by Gregorini et al. (84 AJ 89, 323). Radio sources from the Parkes 2.7 GHz survey have been measured at 5 GHz (Binette et al. 81 AJP 34, 445) and at 8.87 GHz (Shimmins et al. 81 AJP 34, 471). Lawrence et al. (83 ApJ Sup 51, 67) reported 5 GHz flux densities of 2911 sources in the Arecibo 611 MHz survey. Radio sources selected from the NRAO-MPI 5 GHz surveys have been measured at 31.4 GHz by Geldzahler and Witzel (81 AJ 86, 1306) and Geldzahler and Kuhr (83 AJ 88, 1126). Abraham et al. (84 AJ 89, 200) measured flux densities at 22 GHz and at

44 GHz for 30 and 16 sources respectively. Millimeter observations data of several radio sources are given by Ennis et al. (82 ApJ 262, 460) and Gear et al. (84 ApJ 280, 102).

5 GHz flux densities and discussions on the radio spectra of complete samples of radio sources are presented by Machalski and Maslowski (82 AJ 87, 1132) for the GB/GB2 survey and by Slee et al. (82 PASA 4, 278) for the Culgoora survey. Efenov et al. (83 Izv. Krymskoy Astrofiz. Obs., 66, 205) discuss the radio spectra of 143 extragalactic radio sources in a frequency range up to 22 GHz.

Gopal Krishna and Steppe (81 AA 118, 150) and Steppe and Gopal Krishna (84 AA 135, 39) investigated the dependence of median (decimeter-wavelength) spectral index on the 408 MHz flux density and find a significant correlation. Owen et al. (83 AJ 88, 1) made a similar analysis for source samples selected at 5 GHz and confirmed a monotonic decrease in the fraction of flat spectrum sources fainter than 100 mJy.

Variability - Wardle et al. (81 AJ 86, 848) give results on variability of a sample of extragalactic sources at 2.7 GHz. They find evidence for variability as fast as 1-2 months. Abraham et al. (82 AJ 87, 532) reported a search for short term variability in four non-thermal radio sources. None of them showed any short term variability larger than 10%. Heeschen (82 IAU Symp 97; 84 AJ 89, 1111) showed that compact sources display a "flicker" in their cm emission with an amplitude of about 2% and characteristic time scale of a few days. Pustil'nik (82 Soobchch.Spts. Astrofiz. Obs. Vyp. 33, 28) searched for rapid variability ($t>1$ day) in the cm range in fourteen objects with continuous optical spectra. Variability was found in nine cases. Studies of radio variability of BL Lac objects are found in Aller et al. (82 IAU Symp 97) and Altschuler (83 AJ 88, 16). Rudnick and Jones (82 ApJ 255, 39), from an unbiased sample of 40 flat spectrum sources, showed that complex spectrum sources are highly variable, while the simple concave spectrum and the straight spectrum ones show little or no variability. A re-examination of light curves of variable sources has been presented by Legg (84 IAU Symp 110), who showed that bursts from different variable sources, though differing in time scale, have profiles of the same shape.

Detailed studies of the radio variability of specific objects were made by Pustil'nik et al. (Pis'ma Astron Zh 7, 547, trans. Soviet AJ 7: PKS 0757+100), Wills et al. (83 ApJ 274, 62: 1156+295), Kikuchi and Inoue (84 IAU Symp 110, 181: OJ 287), Johnston et al. (84 ApJ 377, L31: BL Lac), Bregman et al. (84 ApJ 276, 454: 0735+178), O'Dea et al. (84 ApJ 278, 89: 3C 84) and Dent et al. (83 Nature 306, 41: 3C 84). Some of these studies include comparison with variability in other wavelength bands.

Ekers et al. (83 AA 120, 297) have published results of a six-year program to monitor the 5 GHz flux of compact radio sources in nearby spirals and ellipticals, and of cores in extended double radio galaxies and quasars.

Studies of the so-called "low frequency variability", typically at $\nu <1$ GHz are given in Fanti et al. (81 AA Sup 45, 61), Dennison et al. (81 AJ 86, 1604), Dennison et al. (84 ApJ 281, L55). Interpretation in terms of intrinsic variability can be found in Fanti et al. (83 AA 118, 171) and references therein; and by slow refraction effects in the interstellar medium by Rickett et al. (84 AA 134, 390) and by Shaperovskaia (82 Astron Zh 59, 246; Sov AJ 26, N.2).

Polarization - Several authors have made measures of integrated linear polarization at several frequencies for extragalactic radio sources (Simard-Normandin et al. 81 AA Sup 43, 19; Simard-Normandin et al. 82 AA Sup 48, 137; Perley 82 AJ 87, 859; Conway et al. 83 MNRAS 202, 813; Vallee 83 AA Sup 52, 125; Bignell and Vallee 83, ApJ Sup 53, 147).

Rotation measures are discussed by Simard-Normandin et al. (82 op.cit.) and Conway et al. (83 op.cit.). By using close-by frequencies, unambiguous rotation measures have been derived by Rudnick et al. (83 AA Sup 52, 317; 83 AJ 88, 518). Aller et al. (82 IAU Symp 97) presented new data on four sources which exhibit large amplitude rotation of the polarization position angle with time.

Circular polarization measures are presented by Ryle and Brodie (81 MNRAS 196, 567), by de Pater and Weiler (82 MNRAS 198, 747) and Komesaroff et al. (84 MNRAS, 208, 409). De Pater and Weiler estimate the magnetic fields by applying uniform synchrotron emission model to the measured circular polarization. Komesaroff et al. discuss different alternatives for explaining the circular polarization variations.

Structure of Radio Galaxies and Quasars - The last three years have seen exciting new synthesis observations from the VLA, MERLIN, Cambridge, Westerbork, Fleurs, Culgoora and Molonglo. The implications of recent synthesis data for source physics generally have been reviewed by Begelman et al. (84 Rev Mod Phys 56, 255) and for radio jets in particular by Bridle and Perley (84, Ann Rev AA 22, 319). Many observations relating to energy transport in extragalactic sources are discussed in the Turin Workshop on "Astrophysical Jets" (Reidel 82, ed. A. Ferrari and A. G. Pacholczyk), though much new work has appeared since this volume.

The traditional distinction between compact and extended sources is vanishing, as many high dynamic range maps have found extended emission around "compact" sources (e.g. Browne et al. 82 MNRAS 198, 673; Browne et al. 82 Nature 299, 788; Perley et al. 82 ApJL 255, L93; Ulvestad et al. 83 ApJ 266, 18; Schilizzi and de Bruyn 83 Nature 303, 26; Ulvestad and Johnston 84 AJ 89, 189; De Pater and Perley 84, ApJ 273, 64; Antonucci and Ulvestad 84, Nature 308, 617). A large fraction of strong extragalactic sources has both compact cores and some kiloparsec-scale or larger structure. The prominence of the cores relative to the larger structures varies by more than 10,000 to 1 from source to source. The success of relativistic flow models for the compact core-jets has encouraged "unified" models wherein this variation is attributed to differing inclinations of different sources to the line of sight.

Jets - Jets are detected in many weak radio galaxies, where they are normally fairly symmetric, i.e. "two-sided" (Bridle and Perley 84), and in a high proportion of sources in well-observed complete samples of extended QSRs, where they are always very asymmetric, i.e. "one-sided" (Owen and Puschell 84 AJ 89, 932; Neff and Brown 84 AJ 89, 195; Burns et al. 84 ApJ 283, 515). They are however relatively hard to detect in strong, edge-brightened radio galaxies (Bridle and Perley 84) though some clear examples are now known (e.g. Linfield and Perley 84, ApJ 279, 60; Perley et al. 84 ApJ 285, L35). Among powerful extended sources, the relative prominence of jets increases with that of their radio cores (e.g. Burns et al. 84; Saikia 84 MNRAS 208, 231). The intensity asymmetries, dominant magnetic field configurations and collimation (spreading rates) of detected jets all correlate with both core and total powers (Bridle 84 AJ 89, 979).

Galaxies and Clusters - Single dish observations have been made at numerous frequencies. Normal nearby galaxies have spectral index which lies in a very narrow range, $\alpha = 0.74 \pm 0.05$ (Gioia et al. 82 AA 116, 164; van der Hulst and Israel 83 AJ 88, 736). The thermal fraction of the radio continuum is low, generally below 40% of the total at 10.7 GHz. A study of samples of extreme blue galaxies (Heidmann et al. 82 AA 105, 188; Klein et al. 83 AA 117, 332) showed that these objects are significantly more radio luminous than the nearby (redder) galaxies.

Maps of galaxies at several frequencies enable the study of the morphology of the radio continuum and the separation of the thermal (free-free) and nonthermal emission components (M31--Beck and Gräve 82 AA 105, 192; Berkhuijsen et al. 83 AA 117, 141; M33--Berkhuijsen 83 AA 127, 395; NGC 253--Klein et al. 83 AA 127, 177; M51--Klein et al. 84 AA 135, 213; NGC 4258--Krause et al. 84 AA 138, 385; NGC 6946--Klein et al. 82 AA 108, 176). Linear polarization observations at high frequencies showed an unusually regular magnetic field structure along the spiral arms of nearby galaxies (M31--Beck 82 AA 106, 121; NGC 6946--Klein et al. 82 AA 108, 176).

Jenkins (82 MNRAS 200, 705) used the Cambridge 5-km array to study a complete sample of 34 bright E and S0 galaxies, showing that weak sources are confined within the optical galaxies while more powerful ones can attain sizes of several hundred kiloparsecs. Ulrich and Meier (84 AJ 89, 203) mapped 18 B2 radio galaxies at 5 GHz to isolate their radio cores, and found that the core luminosity correlated with the stellar optical luminosity. Burns and Gregory (82 AJ 87, 1245) studied a complete sample of 20 4C radio galaxies in poor clusters with the VLA at 1.4 and 5 GHz. They found a high rate of detection of radio jets in weak radio galaxies (cf. above), and found several examples of tailed structures, indicating that this morphology is not restricted to rich cluster environments. They found no clear correlation between the radio morphologies and (low-resolution) X-ray imaging of the clusters. Hanisch (84 AA 133, 192) surveyed 37 poor clusters at 1.4 GHz with the WSRT, but found no further examples of head-tail structure. A study of radio galaxies in 65 rich clusters using the WSRT at 1.4 GHz was begun by Fanti et al. (83 AA 105, 200; AA Sup 51, 179; AA Sup 52, 411).

Quasars - Several samples of extended quasars were mapped with the VLA. Owen and Puschell (84 AJ 89, 932) studied 26 quasars from the Jodrell Bank 966 MHz survey. Hintzen et al. (83 AJ 88, 709) mapped 117 QSRs (mainly at $z < 1.5$). Neff and Brown (84 AJ 89, 195) studied 60 quasars with known VLBI cores. Feigelson et al. (84 AJ 89, 1464) mapped 15 quasars from X-ray samples. Swarup et al. (84 MNRAS 208, 813) studied 31 of the largest quasars. Significant numbers of distorted structures suggesting interactions with ambient IGM have been found in these samples, as well as the one-sided jets mentioned above.

Survey Samples - Peacock and Wall (82 MNRAS 198, 843) completed a structural survey of a sample of 168 sources with 2.7 GHz flux densities >1.5 Jy, and showed that 44% of unresolved (mostly <5") sources have spectral indices >0.5. The VLA was used to map 237 sources from the GB2 survey with 1.4 GHz flux densities >0.55 Jy (Machalski et al. 82 AJ 87, 1150; Machalski and Condon 83 AJ 88, 143) and 126 with 1.4 GHz flux densities between 0.20 and 0.55 Jy (Machalski and Condon 83 AJ 88, 1591). Lawrence et al. (84 ApJ 278, L95) reported statistics of radio structure types and optical identifications with VLA snapshots of 602 MIT-Green Bank survey sources with 5 GHz flux densities >0.05 Jy.

The most detailed information on individual sources comes from high resolution studies of the biggest, brightest and nearest objects, so we highlight such work here.

Jets - Multifrequency studies of the jet in the radio galaxy NGC 6251 by Saunders et al. (82 MNRAS 197, 253) and Perley et al. (84 ApJ Sup 54, 292) showed variations in its spreading rate (implying that it is not free everywhere), lateral oscillations, and knot structure indicating local particle acceleration. Perley et al. also showed that the 3-D magnetic field in this jet is not axisymmetric, and found rotation measure gradients indicating a magnetoionic medium outside the jet but within the optical galaxy.

VLA observations of the jet in M87 at 0.1" resolution (Biretta et al. 83 ApJ 274, L27) discovered a sharp inner edge to Knot A and an overpressure relative to the surrounding X-ray gas over much of the jet, though the variable spreading rate of the jet argues that it is not free. The knot structures suggest that they are shocks in the outflow of M87. There is also evidence for absorption of the knot emission below 408 MHz (Charlesworth and Spencer 82 MNRAS 200, 933).

Multifrequency VLA mapping of the jet in Cen A by Burns et al. (83 ApJ 273, 128) showed that it could be thermally confined by surrounding X-ray gas, and revealed strong similarities between its radio and X-ray intensity distributions. This makes severe demands of particle acceleration mechanisms in the jet if its X-ray emission is interpreted as synchrotron emission. Culgoora and Molonglo maps of Cen A by Slee et al. (83 Proc ASA 5, 247) showed that the lobes have a 327 to 843 MHz spectral index of 0.7 while the jet region has a spectral index of 0.93--this goes against the general trend for jets to have flatter spectra than the lobes they enter, and also questions the synchrotron interpretation of the X-ray jet (though higher resolution observations below 1 GHz are desirable).

Hot Spots - Resolutions better than 0.5" are needed to characterize the hot spot properties of distant powerful sources (e.g., Schilizzi et al. 82 J. Astrophys. Astron. 3, 173). Two main classes of hot spots have been recognized in well-resolved powerful sources--compact flat-spectrum hot spots and more diffuse steep-spectrum spots. These may have similar total fluxes in the same source, but the compact spots have much higher energy densities and much lower total energies (e.g. Lonsdale and Morison 83 MNRAS 203, 833). One-sided radio jets, where these are visible, tend to point toward the most compact hot spot, though exceptions are known. Magnetic fields in hot spots generally lie parallel to the ridge lines of the hot spot emission (e.g. Swarup et al. 84 MNRAS 208, 813).

Some compact hot spots in quasar lobes have internal pressures that are too great to be directly ram confined assuming normal IGM densities and subrelativistic advance velocities (e.g. Lonsdale and Barthel 84 AA 135, 46). If such spots are supplied by jets that flow into them and stagnate, very high jet velocities and conversion efficiencies would be required for energy and momentum balance. The morphologies and spectra of these spots and of their lobes (which often contain larger, secondary hot spots) suggest that they may be emission from oblique shocks at which jet outflows are deflected rather than stopped (see also Lonsdale 84 MNRAS 208, 545), possibly alleviating these difficulties. Lonsdale and Barthel 84 also discuss how plasmoid models may provide acceptable alternatives.

Lobes and Trails - As with jets, the biggest and the brightest continue to reveal new details. Qualitatively new classes of non-relaxed structure have been found in radio lobes using the VLA--polarized filaments (Cyg A--Perley et al. 84 ApJ 285, L35), and highly polarized rings or shells (Her A--Dreher and Feigelson 84 Nature 308, 43; 3C310--van Breugel and Fomalont 84 ApJ 282, L35). The origins of these structures are unclear, but the rings suggest episodic events or instabilities in the plasma outflow from the nucleus to the lobes. The projected magnetic fields run parallel to the ridges of most of the filaments and rings. The prominence of the jets and lack of strong hot spots in the powerful source Her A is unusual, given the general power-morphology trends for extragalactic sources.

Van Breugel (82 AA 110, 225) mapped the spectral index distributions and magnetic field configurations in the jets and outer regions of the radio galaxies 3C31, 3C66B and 3C129 with the WSRT (see also van Breugel and Jaegers 82 AA Sup

49, 529). He found a peculiar Faraday depolarization and rotation at the base of the jet in 3C66B which may be due to foreground magnetoionic material in the parent galaxy (cf. NGC 6251 above). Strom et al. (83 AA 122, 305) also used the WSRT to study the depolarization dichotomy between the North and South lobes of 3C31, suggesting large differences in the thermal densities on the two sides of the active nucleus. They found that the magnetic field, which is predominantly perpendicular to the inner jets, turns through 90 deg. to run parallel to the long axis of the outer lobes. Alexander et al. (84 MNRAS 209, 851) derived the spectral index distribution over the lobes of Cyg A from multifrequency mapping and concluded that if some aspects of this are attributed to synchrotron aging of material flowing away from the hot spots, their velocity of advance should be about 0.05c.

WSRT studies of the largest known rich cluster source (1919+479) by Robertson (84 AA 138, 41) were consistent with buoyant shaping of its large scale "wide angle tail" (WAT) structure, but neither buoyancy, gravitational bending nor dynamic pressure could account for the bent structure observed in 3C465 with the VLA by Eilek et al. (84 ApJ 278, 37) and at Cambridge by Leahy (84 MNRAS 208, 323). There is disagreement over whether motions of the parent galaxies through their clusters are large enough to explain the detailed shapes of this and other WATs. Galaxy velocities of order 500 km/s are required to produce the observed WATs by moving a twin jet away from static trail emission deposited in an earlier stage of its evolution, but it is unclear whether these are consistent with other constraints on the galactic velocities and with the ages of the radio trails estimated from observed spectral steepening (cf. Eilek et al. 84 and Leahy 84).

Symmetries and Orientations - New examples of sources with bends or S symmetry suggesting precessional or rotational motion of the primary collimator have been found, mainly in QSRs (Gower and Hutchings 82 ApJL 253, L1; Hunstead et al. 84 MNRAS 207, 55; Gower and Hutchings 84 PASP 96, 19; Muxlow et al. 84 IAU Symp 110, 141) but also in one radio galaxy (Condon and Mitchell 84 ApJ 276, 472). Specific precessional models fitted to these sources are generally complex and their parameters are not yet obligatory. Lonsdale (84) found morphological and spectral symmetries in 3C196 that are consistent with physical rotation of its primary collimator followed by outflow of material from the most compact hot spots. The total intensity symmetries of the lobes of powerful doubles are often broken at high resolution but it is not clear whether this favors alternating-ejection models for all such sources (Rudnick and Edgar 84 ApJ 279, 74; Ensman and Ulvestad 84 AJ 89, 1275).

Observations of 3C218 (Hyd A) at Fleurs by Simkin and Ekers (83 ApJ 265, 85) showed an edge darkened radio structure (unusual for a source of its relatively high power) elongated along the optical rotation axis of the galaxy, confirming a trend shown by other powerful radio galaxies. Kapahi and Saikia (82 J. Astrophys. Astron. 3, 161) discuss a possibly related alignment between the radio elongations and the optical minor axes in radio galaxies with prominent nuclear cores. Jaegers (83 AA 125, 172) documented three parallel edge-darkened sources in the field of 3C130, all possibly identified with galaxies; he suggested that this parallelism may reflect alignments among the three galaxies resulting from common membership in a cluster.

Steep Spectrum Cores (SSCs) - MERLIN and VLA maps of 3C380, a powerful edge-darkened source (also violating the general power-morphology trend) has been shown to have a complex extended structure (Wilkinson et al. 84 Nature 308, 619). Van Breugel et al. (84 AJ 89, 5) give VLA maps of 23 SSC sources; they find low degrees of polarization and misalignments between the core components and more distant lobes. These results suggest that SSCs manifest strong interactions between radio emitting plasma and dense ISM on a sub-galactic scale, analogous to those seen in radio Seyfert galaxies (e.g. NGC 1068--Wilson and Ulvestad 83 ApJ

275, 8) and studied in detail in some nearby radio galaxies with extranuclear line emission and small distorted radio sources (Heckman et al. 82 ApJ 262, 529; van Breugel et al. 84 ApJ 276, 79 and ApJ 277, 82).

VLBI - The Proceedings of IAU Symposium No. 110, "VLBI and Compact Radio Sources" (1984, eds. Fanti, Kellermann, and Setti) contains many up-to-date accounts of research relevant to this section, as does the earlier Proceedings of IAU Symposium No. 97, "Extragalactic Radio Sources" (1982, eds. Heeschen and Wade), and the review on "Compact Radio Sources" (81 Kellermann and Pauliny-Toth, An Rev Astron Astrophys 19, 373). Matveyenko (81 Radiophysics Institute, Gorki) has published a bibliography of VLBI papers published during the period 1965-1981.

Superluminal Sources - From 7 to 10 superluminal sources now are known; the number depends on the degree of certainty required by different authors (Cohen et al. 84 Proc IAU Symp 110, 95; Porcas 83 Nature 302, 753). Another half-dozen or more are suspected of being superluminal, but more data are needed for confirmation. The best cases are 3C120, 3C273, and 3C345 which are strong and are frequently observed at many wavelengths. Each of these has a bright, marginally unresolved "core" which is optically thick at centimeter wavelengths and whose flux density is constant to within a factor 2. Components move away from the core and travel along a similar track (forming a "jet" or "beam") with apparent velocity $v \sim 5c$ for Ho = 100. The highest well-determined value is $v = 9.5c$ for a component in 3C345 (Unwin et al. 83 ApJ 271, 536). In one source successive components may have somewhat different velocities. The beam is strongly curved and eventually points towards an outer jet (scale 100's of kpc). This is particularly vivid in 3C120 (Benson et al. Proc IAU Symp 110, 125) where there is continuity over most of the range from 1 pc to 100 kpc and a total curvature greater than 90°. The VLBI components evolve with a time scale of a few years; their spectra steepen and they grow weak.

Bartel et al. (Proc. IAU Symp 110, 113) have shown by astrometric measurements that the core of 3C345 is stationary with respect to NRAO 512, and thus that the jet components are indeed moving. The newest component in 3C345 was first detected at about 0.3 mas from the core in 1981, and moved to about 1.0 mas by early 1984. It accelerated in the midst of this journey. Its track has been nearly straight but not radial, although with enough (unseen) curvature at small distances it could have come out of the core (Moore et al. 83 Nature 306, 44; Biretta et al. 83 Nature 306, 42). The jet is best defined in 3C345 because 3C120 and 3C273 have low declination and poor north-south resolution. In 3C345 new jet components have almost overtaken the position old ones held at the beginning of these programs, and in the next few years it will be seen whether the tracks are the same or not. There already is some evidence that the two outer components of 3C345 are on different tracks (Cohen et al. 83 ApJL 269, L1).

4C39.25 is unique in that the centroids of its two components are approaching each other rather than separating. However, it is still possible to interpret this in terms of the birth of an unresolved new component which is separating from its parent, rather than as a contraction (Shaffer 84 Proc IAU Symp 110, 135).

Subluminal or Stable Sources - Only a few compact sources are known to be subluminal or stable. 3C84 is complex but contains a core-jet structure with a component moving at about $v = 0.2c$ (Romney et al. 84 Proc IAU Symp 110, 137). Stable sources include NRAO 150 (Baath et al. 80 AA 86, 364), 2134+004 (Pauliny-Toth et al. 84 Proc IAU Symp 110, 149), and the nucleus of M87 (Reid et al. 84 Proc IAU Symp 110, 145).

Relativistic Beams and Unified Schemes - The most common explanation for the superluminal motion involves a relativistic beam of Lorentz factor γ propagating at a small angle θ to the line of sight (Begelman et al. 84 Rev Mod Phys 56, 255). VLBI measurements of angular size and spectrum, plus Compton X-ray measurements, allow the calculation of a limit to the Doppler (blue-shift) factor δ of the beam, and this plus the superluminal motion sets constraints on θ and γ. In this way Unwin et al. (83 ApJ 271, 536) found $\gamma > 7$, $\theta < 7°$ for 3C345 (H_o = 100). Calculations of beam parameters based on the X-ray flux and a measured diameter are independent of the Hubble constant. Marscher et al. (81 ApJ 249, 406) used these calculations to successfully predict superluminal motion in NRAO 140, but this did not work for 3C147 (Simon et al. 83 Nature 302, 487; Preuss et al. Proc. IAU Symp 110, 29; Simon et al. 83 Proc IAU Symp 110, 111).

In the relativistic beam theory the radiation from a moving component is anisotropic and is boosted by about δ^2. This makes a strong selection effect which explains why such a large fraction of strong core-dominated sources are superluminal. There has been considerable speculation over the identity of the "misdirected" beamed sources, which must be 50 times more numerous than the directed ones. Scheuer et al. (79 Nature 277, 182) discussed the radio-quiet quasars in this context; see also Kellermann et al. (83 Proc Liege Colloq 24, 81) who showed that the radio weak quasars have the distribution expected from randomly oriented beams. Orr et al. (82 MNRAS 200, 1067) discussed the unification of superluminal sources and blazars with the common extended double sources. This point of view is strongly supported by the observation of large low-surface-brightness features (which could be outer lobes of doubles seen end-on), around many core-dominated objects (Schilizzi et al. 83 Nature 303, 26; Browne et al. 82 Nature 299, 788). All the superluminal sources except 3C273 have outer structure which surrounds (or is on opposite sides of) the core, but in 3C273 the outer structure consists solely of a linear straight jet on one side. Two superluminal sources (3C179 and 3C279) have outer structure of the conventional double-lobe type. Other tests of the beaming model have been reported (Kapahi and Saikia 82 J Astroph & Astron 3, 465; Kaikia and Wiita 82 MNRAS 200, 83).

Surveys - Two VLBI mapping surveys of flux-limited samples are underway. Pearson et al. (84 Proc IAU Symp 110, 15) have been making a survey of sources with S(5GHz) $\geq$ 1.3 Jy, $\delta > 35°$, and $b > 10°$. 45 of the 65 sources have VLBI structure and were mapped at 6 cm. These can be divided into 6 categories of which the most numerous (13/45) have the asymmetric core-jet structure. Second-epoch observations are now underway (Readhead et al. 84 Proc IAU Symp 110, 131). Two of 5 objects reobserved at 6 cm show structural variations. Observations at 10 GHz and at optical wavelengths are also being pursued.

Eckhart et al (84 Proc IAU Symp 110, 65) are studying flat-spectrum sources with $\delta > 70°$ and S(5GHz) > 1 Jy. VLBI observations are being made at many wavelengths, and larger-scale maps from MERLIN and VLA are being obtained. Four of the 13 sources were observed twice at 6 cm, and one showed structural variations.

Zensus et al. (84 AA 133, 27) detected 56 of 57 flat spectrum sources at 5GHz and fit simple models to the data. Romney et al. (84 AA 135, 289) have made first-epoch maps at 1.7 GHz of 21 low-frequency variables.

Detection surveys at 2.3 GHz have been made by Preston et al. (83 ApJ 269, 387) on a complete flux-limited sample selected from the Parkes ±4° catalog, by Wehrle et al. (84 AJ 89, 336) on sources in the ecliptic, and by Morabito et al. (83 JPL TDA Progress Report 42-74, 183) on south polar sources. In all these the limiting detectable flux was about 0.1 Jy. Matveyenko et al. (82 Soviet AJ 8, 77) have studied 3C84 and 3C345 at 18 cm wavelength. Fringes were detected in 18 cm observations using the 100 m antenna in Bonn FRG and a 6 m antenna in Shanghai, PRC.

Steep-Spectrum Sources - Many compact steep-spectrum sources show VLBI structure which is severely distorted on scales of 1-10 kpc. Wilkinson et al. (84 Proc IAU Symp 110, 25) show maps of four objects and suggest that the distortions might be due to interaction between a jet and interstellar gas. Fanti et al. (84 Proc IAU Symp 110, 25) show maps of 8 objects.

Compact doubles are a minority of all compact steep spectrum sources with turnover frequency near 1 GHz (Phillips et al. 82 AA 106, 21; Hodges et al. 84 AJ 89, 1327). Maps have been published at 1.6 GHz (Phillips et al. 80 ApJ 236, 89), 2.3 GHz (Phillips et al. 83 ApJ 271, 32), and 5 GHz (Mutel et al. 84 ApJ, in press; Pearson et al. 84 Proc IAU Symp 110, 15; Spangler et al. 83 ApJ 271, 44). Phillips et al. (82 AA 106, 21) suggest that these sources arise when a double-sided beam, advancing into the interstellar medium, is seen near the plane of the sky.

Nearby Galaxies - Studies of nearby galaxies have only recently become possible, because they are weak, with $L \sim 10^{40}$ erg s^{-1}. VLBI detection and/or model-fitting experiments are reported by Bartel et al. (82 ApJ 262, 556), Hummel et al. (82 AA 114, 400), Jones et al. (82 ApJ 261, 422), Preston et al. (83 ApJ 266, L93), and Preuss and Fosbury (83 MNRAS 204, 783). The data have warranted mapping for NGC 4278 (Jones et al. 82 ApJ 261, 422), M87 (Reid et al. 82 ApJ 263, 615), NGC 2911 and NGC 4278 (Schilizzi et al. 83 AA 126, 412), MK 348 (Neff and de Bruyn 83 AA 128, 318), and NGC 1052 (Jones et al. 84 ApJ 276, 480).

Wehrle et al. (84 ApJ 284, 519) made a detection survey for cores in a complete sample of 26 elliptical galaxies with $m_V \leq 16.5$ and S(408 MHz) $\geq$ 0.25 Jy. Thirteen had S_{core}(2.3 GHz) $\geq$ 2mJy.

Lensed Objects - VLBI observations play an essential role in the study of gravitational lensed objects, as they place the tightest constraints on the lens models. Two of the five known lens systems have been studied with VLBI: 0957+561 (Porcas et al. 81 Nature 289, 758; Gorenstein et al. 83 Science 219, 54), and 2016+112 (Gorenstein et al. 84 BAAS 15, 936). The images are particularly weak and require elaborate signal processing with phase referencing.

Extragalactic Molecules - High resolution CO mapping of other galaxies is hindered by the low signal level. High resolution, fully sampled maps of covering many beam areas are just now becoming available (see the review of Morris and Rickard 82 Ann Rev 20, 517 for previous results). The first maps showed little sign of spiral structure (Rickard and Palmer 81 AA 102, L13; Scoville and Young 83 ApJ 265, 148), for which sufficient spatial resolution is available. New CO maps of M51 made with the Onsala telescope (Rydbeck et al. 82 IAU Symp 100, 53) and the Caltech interferometer, however, show such structure. In the case of the Onsala data, a smooth background must be subtracted to clearly show the spiral arms in the disk which are 20%-30% of the spread-out CO emission. It has generally been assumed that the J = 1-0 transition of CO in galaxies (and the Galaxy) is optically thick. For M82, there is conflicting information from comparisons of observed temperatures in the J = 1-0 and 2-1 lines, but isotope ratios (Stark and Carlson 84 ApJ 279, 122) seem to support an effective low optical depth in the J = 1-0 line.

Extragalactic Masers - Extragalactic H_2O maser sources were found in NGC 1068 (M77), NGC 4258 (M106), M82, and NGC 6946 by Claussen et al. (84 Nature 310, 298) and in the Circinus galaxy by Gardner and Whiteoak (82 MNRAS 201, 13P). The NGC

1068 source, the most luminous H_2O maser known, is 350 times more luminous than the sun, assuming isotropic emission. Extragalactic OH masers were discovered in the LMC by Haynes et al. (81 MNRAS 197, 23P) and in IC4553 by Baan et al. (82 ApJL 260, L49). The latter source is more than 10^7 times stronger than the Milky Way source W3(OH) assuming isotropic emission.

Extragalactic Hydrogen - The detection of HI in early type galaxies is a game of patience but positive detections do reward the persistent. HI seems to be more abundant in low luminosity ellipticals (Lake and Schommer 84 ApJ 280, 107) but there is no dependence on environment (Dressel et al. 82 ApJ 259, 55). The availability of new broad bandwidth capabilities promises the detection of very broad HI signals, from massive rotating disks.

The comparative HI content of normal galaxies in varying environments was the subject of a workshop held at Green Bank in April 1982. In the Virgo Cluster, spirals are not only HI deficient (Giovanardi et al. 83 ApJ 267, 35) but the HI disks are also shrunken with respect to their counterparts in the field (Giovanelli and Haynes 83 AJ 88, 881). Virgo continues to provide fertile territory for investigation. In other clusters, comparison with the field produces mixed results (Bothun et al. 82 ApJ 87, 725; Giovanelli et al. 82 ApJ 262, 442).

Huchtmeier (82 AA 110, 12) reported on a survey of HI in Sa galaxies, while Schneider et al. (83 ApJL 272, L1) claim the detection of a massive intergalactic HI cloud.

The 21 cm line is a useful tool in deriving the redshifts for spiral galaxies to a z of about 0.03. The new feed at Arecibo will cover the frequency range to 1300 MHz. The compilation of Bottinelli et al. (82 AA Sup 47, 171) is most useful. The surveys that are currently underway promise clues to the large-scale structure of nearby superclusters, the stability of groups, and the use and calibration of the Tully-Fisher relation. The study of the orientation of spin vectors with the large scale structure (Helou and Salpeter 82 ApJ 252, 75) will constrain theories of galaxy formation. The number of detected highly redshifted absorption lines continues to grow and nearby quasars are also seen to show the emission signatures of rotating disks. The continuing acquisition of luminosity distances for galaxies, especially in the southern hemisphere, will provide increasing accuracy in the determination of the local motion (Hart and Davies 82 Nature 297, 191).

The last three years have been characterized by a steady flow of thorough HI studies with the WSRT and the first spectral line results of the VLA. Some of the highlights are summarized here, many results are more fully described in IAU Symposium 100.

The velocity fields of normal spiral galaxies show large-scale symmetric deviations, interpreted as oval distortions or kinematical warps; rotation curves are flat out to large distances, indicating that M/L increases outward (Bosma 81 AJ 86, 1791,1825). Studies of nearly face-on spirals show that this dark material is distributed in the halo's of the galaxies and not in a flattened disk (van der Kruit and Shostak 81 AA 105, 351; 84 AA 132, 20; AA 134, 258). A study of the rotation curves of binary galaxies shows that 50 to 80% of the mass of spirals is non-luminous (van Moorsel 82 PhD Thesis, Un. of Groningen).

An HI study of M31 with a linear resolution of 80x120 pc has been completed (Brinks 84 IAU Symp 100, 23). Holes are found in the HI distribution in the same area as the HII regions, OB associations, etc. On a large scale the HI layer shows a warped gas distribution similar to our own Galaxy.

With the continued improvement of the sensitivity of the WSRT at 21 cm it has become possible to study galaxies with smaller HI emissivity. The majority of the observed lenticulars has the HI in a ring outside the optical body (van Woerden et al. 83 IAU Symp 100; Knapp et al. 84 AA 133, 127). The HI in 2 spindle-like galaxies coincides with the outer optical components. The velocity fields suggest that these annuli have formed recently (Schechter et al. 84 MNRAS 208, 111).

Since the spectral line system of the VLA has become operational it is possible to observe a larger Declination range. This has resulted in the systematic studies of barred spirals. An interesting example is NGC 3992, which appears to have a flat rotation curve and yet no massive halo (Gottesman et al. 82 ApJ 260, 65). The good instantaneous UV coverage of the VLA makes it possible to map in spectral line snapshot mode. An example is an HI survey of the Virgo cluster (van Gorkom et al. 84, Clusters and Groups of Galaxies, p261, ed. F. Mardirossian), which shows convincingly the effects of ram-pressure sweeping in the center of the cluster.

Both the VLA and WSRT appear to be extremely suitable to study HI in radio galaxies, because of the high spectral dynamic range that can be achieved (van Gorkom and Ekers 83 ApJ 267, 528). The results indicate that in the majority of the cases hydrogen is falling toward the nucleus of these galaxies (Shostak et al. 83 AA 119, L3).

Some further radio studies have been made of HII regions in other galaxies (Viallefond et al. 83 AA 119, 185; Cersosimo and Loiseau 84 AA 133, 93). New searches for RRLs from active galactic nuclei and quasars have been made (Bell et al. 84 AA 130, 1), and further work has been done on the prospects and potential applications (Dravskikh and Parijskij 81 Sov AJ 25, 274; Val'tts 83 Sov AJ 27, 18; Shaver 83 Liège XXIVe, 287; Sarazin and Wadiak 83 AA 123, L1; Wadiak et al. 83 ApJ Sup 53, 351).

SETI

SETI research has made impressive progress in recent years. From the first radio search by Frank Drake in April, 1960, there have been at least 45 different searches (25 in the last six years), with about 130,000 hours of observations, in at least seven different countries (USA, USSR, Australia, Canada, France, Germany, Holland) with Japan getting ready to join.

Besides the SETI directed, short-term radio searches, there are also shared or parasitic SETI projects, such as the SERENDIP program of UC-Berkeley, which share and/or analyze data obtained for other purposes. There are also SETI dedicated facilities which are automated to operate continuously and account for the bulk of the observing hours logged so far. The two major ones are: the Ohio SETI Program (R. Dixon and J. Kraus) at Ohio State University in operation since 1973 and Project Sentinel (P. Horowitz) at Harvard.

A major development program for mega-channel spectrum analyzers and signal recognition algorithms is now in progress with potential uses in many other areas. Horowitz is building a narrow band (350 kHz), 8.4 million channel spectrum analyzer to be incorporated in 1985 to his project Sentinel at Stanford. Peterson and Linscott are building an 8 MHz, 8 million channel spectrum analyzer, which, together with the algorithms now being developed by B. Oliver at NASA-Ames, will become the cornerstone in the long-term SETI program of NASA that will include comprehensive all-sky and targeted radio searches over a wide frequency range.

COMMISSION 41: HISTORY OF ASTRONOMY (HISTOIRE DE L'ASTRONOMIE)

PRESIDENT: O.Pedersen
VICE-PRESIDENT: J.Eddy
ORGANIZING COMMITTEE: H. Eelsalu, E.J. Forbes (representing the Division of History of Science of IUHPS), M.A. Hoskin (Past President), J.-P. Verdet, K. Yabuuti

Commission 41 is one of the largest specialist components of the IAU. At the General Assembly in 1982 at Patras the number of ordinary members was increased from 79 to 86 as a testimony to the growing interest among astronomers in the history of their subject. At the same time the number of consulting members rose from 32 to 40, which shows that an increasing number of highly competent historians of science is engaged in research in the history of astronomy without being professional astronomers. However, many valuable contributions to the subject come from historians who are not members. In consequence, the principal purpose of Commission 41 is to serve as a link between all historians of astronomy whether they be members or not, by initiating research into particular areas, by disseminating new results, or simply by creating possibilities of personal contacts between scholars from different parts of the world. This is mainly achieved by meetings of various kinds.

The Commission as such usually meets on two days during the General Assembly of the IAU, in a number of sessions devoted to general papers, to specific questions of importance to historians, and to a session or joint meeting with another commission. Thus at the General Assembly in 1982 at Patras there was a highly successful joint meeting with Commission 40 on The Early History of Radio Astronomy. A similar session at the forthcoming General Assembly in New Delhi on The History of Cometary Astronomy belongs to the same category. Other subjects discussed at Patras were The Use of Historical Records in Astronomical Research, and The Tools of the Historian of Astronomy. A particular feature of the Patras meeting was the Invited Discourse given in the old amphitheatre by the then president of the Commission, Dr. M.A. Hoskin.

Since many of the consulting members, and other historians of astronomy, are unable to attend the General Assemblies of the IAU a number of other meetings is usually arranged by our Commission in between General Assemblies, often in collaboration with other international or national bodies.

Since Patras several such meetings have taken place. Among these was a very well attended symposium on 1984 July 8-13 at Greenwich called "Longitude Zero" and commemorating the centenary of the international adoption of the Greenwich meridian. No less than sixty papers on astronomy, time keeping, cartography and navigation were presented. The meeting was sponsored by the IAU, the International Union of History and Philosophy of Science, and the International Union of Geodesy and Geophysics, and recognized as the IAU Colloquium N° 84. The organisation of the meeting was superbly handled by professor S. Malin and the staff of the Old Greenwich Observatory. A volume containing the proceedings of the Colloquium is in the press.

Among other international meetings involving either the Commission or a number of its members together with other historians of astronomy we can mention the following, without pretending to provide an exhaustive list.

A conference 1982 August 30 - September 1 to commemorate the 400th anniversary of the Gregorian Calendar was arranged by the Vatican Observatory, the Pontifical Academy of Science, and several members of Commission 41.

The IAU Symposium N° 106 on "The Milky Way Galaxy" on 1983 May 30 - June 3 at Groningen which was the first of its kind to include a historical session.

A colloquium on the history of astronomy held 1983 June 4 in Aarhus, sponsored by the Institute of History of Science, Aarhus University, and the Arnold Willer Foundation, was arranged by the president of the Commission.

The III International Reunion of History of Nautic and Hydrography which took place on 1983 July 3-8 in Lisboa and Sagres.

The First International Ethnoastronomy Conference took place on 1983 September 5-9 at the Smithsonian Institution, Washington D.C.

The III Scientific Instrument Symposium was held 1983 September 26-30 at the Deutsches Museum, München.

Our Commission has noticed with satisfaction a number of historical sections set up by such national bodies as the Royal Astronomical Society of London, The American Astronomical Society, and the Astronomical Society of the Pacific.

The only regular publication of the Commission is a series of informal Information Circulars of which N° 32 (containing names and addresses of all our members) was published in July 1983, and N° 33 in February 1985.

It is a great pleasure to be able to announce that the *General History of Astronomy*, jointly sponsored by the IAU and the IUHPS, has now begun to appear from the Cambridge University Press under the general editorship of Dr. M.A.Hoskin; vol. 4A on *Astrophysics and 20th Century Astronomy to 1950* (ed. by O. Gingerich) was published in the summer of 1984 as a handsomely printed quarto of more than 200 pages.

This is not the place to attempt anything like a bibliography of the works of the members of the Commission, but one cannot help noticing that since the last general assembly a rather impressive array of major works has come to light, among which are *Ptolemy's Almagest*, translated and annotated by G.J. Toomer (London 1984); N.M. Swerdlow and O. Neugebauer, *Mathematical Astronomy in Copernicus's De Revolutionibus, Part 1-2* (New York, Heidelberg 1984); O. Gingerich and Barbara L. Welther, *Planteary, Lunar, and Solar Positions, New and Full Moons, A.D. 1650-1805* (Philadelphia 1983); W. Hartner, *Oriens-Occidens II* (Hildesheim 1984); D.G. Heggie (ed.), *Archaeoastronomy in the Old World* (Cambridge 1982); A.F. Aveni (ed.) *Archaeoastronomy in the New World* (Cambridge 1982); G.V. Coyne, M.A. Hoskin, O.Pedersen (eds.), *Gregorian Reform of the Calendar* (Città del Vaticano 1983).

O. PEDERSEN
President of the Commission

42. CLOSE BINARY STARS (ETOILES BINAIRES SERREES)

PRESIDENT: A.H. Batten
VICE-PRESIDENT: J. Smak
ORGANIZING COMMITTEE: K.D. Abhyankar, J. Andersen, A.M. Cherepashchuk, M. Kitamura, R.H. Koch, Y, Kondo, K-C. Leung, M. Rodonó, G. Shaviv, E.P.J. van den Heuvel, B. Warner

1. Introduction

Research on close binary systems has continued at a high level during the past triennium, although the rate of growth is noticeably slower - probably reflecting the cutbacks in funds to which many of us are subject. There have also been changes of emphasis within the field, which are commented on in the pages that follow. These reflect both changing opportunities for observation and the natural development of the subject. In many areas, the time is ripe for a more critical look at ideas that previously seemed adequate.

No large symposium devoted exclusively to close binary systems has been held in the past triennium. The Commission did sponsor IAU Colloquium 80, *Double Stars, Physical Properties and Generic Relations*, held at Lembang (Java), Indonesia, June 3-7, 1983, to commemorate the 60th anniversary of the Bosscha Observatory. The proceedings have been published in *Astrophysics and Space Science*, Vol. 99. The Commission was a co-sponsor of IAU Symposium No. 111, *Calibration of Fundamental Stellar Quantities* held in Como, Italy, May 24-29, 1984 (dedicated to D.M. Popper) and IAU Colloquium 88, *Stellar Radial Velocities*, held in Schenectady, N.Y., October 23-27, 1984. The proceedings of the first will be published by D. Reidel and of the second by L. Davis press.

Other meetings in our field have been arranged independently of our Commission and the following list is not guaranteed to be complete. Two workshops were held in 1983, one in Baton Rouge, Louisiana, in February, on cataclysmic variables, and one in Cambridge, England, in the summer, on close binaries and dedicated to the memory of J.A.J. Whelan. The Commission on "Physics and Evolution of Stars" of the multi-lateral cooperation of the Academies of Sciences of socialist countries organized two meetings: *Ejection and Accretion of Matter in Binary Systems* (Tatranska Lommica, Czechoslovakia, April 1980, published by VEDA in 1982, edited by J. Tremko) and *Relativistic Objects in Close Binary Systems* (Cluj-Napoca, Romania, Preprint No. 4, Faculty of Mathematics, Babeş-Bolyai University, edited by V. Ureche). A national workshop on binary stars, the first of its kind in the People's Republic of China, was held at the Xinglong observing station of Beijing Astronomical Observatory in September 1983. A Colloquium on *Astrometric Binaries and Related Problems* was held in Bamberg, F.R.G. in June 1984, to commemorate the 200th anniversary of the birth of F.W. Bessel. Proceedings of two meetings mentioned in our last report (*Binary and Multiple Stars as Tracers of Stellar Evolution* and IAU Colloquium 72, *Cataclysmic Variables and Related Objects*) have now been published. Your Organizing Committee hopes to submit a proposal, before the nineteenth General Assembly, for a symposium covering much the same ground as the 1972 Parksville symposium, to be held in 1986 or 1987.

During much of the triennium under review, many observers in our Commission have been interested in the eclipse of ε Aur. References to some preliminary results will be found in this report. A Joint Discussion in New Delhi will be partly devoted to this system, and a workshop on the subject will be held in Tucson, Arizona in January 1985. Most of the results of these observations must wait, however, for the next Commission report. Eleven campaign newsletters have been issued, edited by J. Hopkins and R. Stencel (who provided much of the information in this paragraph) with support from both the IAU and AAS. Collaboration between professional and amateur astronomers, through the International Amateur-Professional Photoelectric Photometry association, has greatly contributed to the completeness of photometric observations. Fourth contact of the eclipse was in May 1984. Reasonably well defined *UBV* light-curves were obtained. Red and near infrared photometric observations, as well as polarimetric (J. Kemp) observations, were also made. Spectroscopic and spectrophotometric observations, both from the ground and from space, are reported. The Mg II feature at 280 nm seems to have been unaffected by the eclipse. The time of egress determined from ultraviolet observations appears to have been several weeks later than that deduced from observations in the visual region of the spectrum.

During the Patras meetings, the Commission discussed the proposal by AAVSO to publish its accumulated observations of cataclysmic variables. We submitted a resolution of support to the General Assembly, which was endorsed. We note with gratitude that the IAU Executive has made a grant towards the cost of publication, which has stimulated private gifts. Although the AAVSO does not yet have all the funds it needs, it is proceeding with its plans for publication.

The Commission has continued to produce its *Bibliography and Program Notes on Close Binaries*. After the distribution of No. 37, Dr. Larsson-Leander relinquished the editorship which he had held since 1968. The high standard and usefulness of the *Bibliography* owes much to Dr. Larsson-Leander's work over 14 years, and to the support given him by Lund Observatory; we express our thanks to both for this important contribution to our work. We are fortunate that Dr. T. Herczeg (University of Oklahoma) has agreed to take over the editorship. Despite unexpected difficulties in financing the printing and distribution, Dr. Herczeg and his collaborators have produced (up to the time of writing) Nos. 38, 39 and 40. Contributors to the *Bibliography* during the last three years have been: K.D. Abhyankar, B. Cester, D.S. Hall, M. Kitamura, H. Mauder, C.D. Scarfe, A. Shul'berg, R.F. Sistero, F. Van't Veer and M. Vetesnik. Other help was given by R.P. Olowin and C. Jascheck (CDS Strasbourg).

In arrangement, this report is similar to previous ones, but some sections have been merged to reflect changing emphases. We propose to discuss further changes in New Delhi. The President acknowledges with thanks those members of the Organizing Committee (and others) who prepared various sections, as identified. Thanks are also due to those members of the Commission who submitted individual reports. Not everything thus submitted could be included, but members may be sure that their reports were helpful. As in previous years, we have found it convenient to use very abbreviated references to which we provide the key here:

AA = *Acta Astron.*
AAp = *Astron Astrophys.*
AAp Sup = *Astron. Astrophys. Suppl. Ser.*
AA Sin = *Acta Astron. Sinica*
AAp Sin = *Acta Astrophys. Sinica*
AJ = *Astron. J.*
AN = *Astron. Nachr.*
Ann Rev A Ap = *Ann. Rev. Astron. Astrophys.*
Ann. Tokyo = *Ann. Tokyo Astron. Obs. Second Ser.*
ApJ = *Astrophys. J.*
ApJSup = *Astrophys.J. Suppl. Ser.*
ApL = *Astrophys. Lett.*
Izv. Krym. = *Izv. Krymskoi Astrofiz. Obs.*
JAAVSO = *J. American Assoc. Variable Star Observers*
JApA = *J. Astrophys. Astron.*
JRASC = *J.R. Astron. Soc. Canada*
Mitt AG = *Mitt. Astron. Ges.*
MN = *Mon. Not. R. astr. Soc.*
MSAI = *Mem. Soc. Astron. Italiana*
Obs = *Observatory*
PASJ = *Publ. Astron. Soc. Japan*
PASP = *Publ. Astron. Soc. Pacific*
Per Zv = *Perem. Zvezdy Bull.*
PisAZh = *Pis'ma v Astron. Zh.*

ApSpSc = *Astrophys. Space Sci.*
ATs = *Astron. Tsirk*
AZh = *Astron. Zh. Akad. Nauk USSR*
BAAS = *Bull. American Astron. Soc.*
BAC = *Bull. Astron. Inst. Czechoslovakia*
BASI = *Bull. Astron. Soc. India*
Bull. Abas = *Bull. Abastamani Astrophys. Obs.*
IAUC = *IAU Circ.*
IBVS = *Inf. Bull. Variable Stars*
Publ. DAO = *Publ. Dominion Astrophys. Obs.*
Publ Tartu = *Publ. Tartu Astrofiz. Obs.*
Rev Mex = *Revista Mexicana Astron. Astrofis.*
Tartu Pub. = *Tartu Astrofiz. Obs. Publ.*
Trudy Kazan = *Trudy Kazan. Gorod. Astron. Obs.*

In addition *IAU* Symposium and Colloquium volumes are referred to by *IAU Sym* or *IAU Coll* and the appropriate number.

2. Methods of Light-Curve Analysis
(K-C. Leung)

There was an outburst of activity in the development of methods of analyzing light-curves in the early 1970s, when several groups developed methods of computation for high-speed computers. In the subsequent decade there seem to have been fewer astronomers participating in this important area of double-star astronomy. In the past triennium, only a few workers have been involved, but there have been significant developments.

R.E. Wilson has continued to make new contributions since the development of the Wilson-Devinney method in 1971. He has been generous in supplying his updated version on request; only recently has the University of Florida instituted a nominal charge for the reproduction of documentation and tapes. His method for simultaneous solution of light-curves and velocity-curves, developed in an earlier paper (*ApJ 234*,1054), is demonstrated for AS Eri and V1134 Cyg in a paper now in press (*AAp*). In another paper in press, Wilson identifies RZ Sct as a good example of a double-contact binary in the sense defined in the *ApJ* paper cited. He has now added to the Wilson-Devinney method a means of including star-spots (cold or hot) in the "light-curve" program. A star-spot is defined by four photoelectric parameters, namely, spot temperature, spot size and two coordinates (longitude and latitude of the spot). This required also a small modification in the differential-correction program.

Yamasaki (*ApSpSc 77*,75) has published 2,700 theoretical light-curves in tabular form, suggesting that they can be used for deriving preliminary parameters for contact systems. Linnell (*ApJSup 54*,17) has developed a program to synthesize light-curves and colour-curves of contact systems. He hopes to use this method to analyze W-type W UMa systems.

Cherepashchuk and Lipunova (*ApSpSc 86*,299, *AZh 599*, 73 *Pis AZh 8*, 242) have discussed methods of solving light-curves of Wolf-Rayet systems. Kopal (*ApSpSc 87*,149,*88*,313) has studied the theoretical light changes to be expected in a system with oscillating components.

3. Observational Data

A. PHOTOMETRIC OBSERVATIONS AND SOLUTIONS (R.H. Koch)

This section was written from materials with publication dates no later than June 30, 1984 and which were held at Pennsylvania by October 30, 1984. The format and content are intended to be consistent with the same section of the 1982 Report.

Table 1 shows that photoelectric observing has continued to increase from the

Table 1. Photoelectric Observing Programs for the Past Three Triennia

	75-78	78-81	81-84
References for photoelectric data	346	564	713
Close binaries observed	209	342	455
Binaries not observed previous triennium	184	240	313
Northern systems ($\delta > +23°$)	88	120	197
Equatorial systems	50	77	135
Southern systems ($\delta < -23°$)	35	76	123

last triennium although at both absolute and relative rates slower than during the previous interval. The second and third entries demonstrate not only the accelerated pace of working on new (and in many cases very faint or very bright objects) but also the commitment to sustain observational coverage of systems that are intrinsically active.

The last three entries of the Table show that sky coverage is approaching uniformity. Naturally, this does not mean that we are yet free of selection bias even for Population I.

Part of the charge toward preparing this section was to include a listing of the light-curve analyses appearing in the last 3 years. The charge was construed in the same way as in 1981 and the list appears in Table 2. This enumeration of models and solutions is not complete. Several Argentine, Chinese, German, Indian, and Russian publications are not held in the Pennsylvania library. From the *Bibliography and Program Notes* it was evident that these journals and serials contain several analyses, but the published material could not be located in the vicinity of Philadelphia or at Goddard Space Flight Center. Additionally, for some of these papers the entries in *Astron. and Astrophys. Abstracts* were uninformative and it seemed best not to guess the contents and conclusions of the papers. In Table 2 the only unconventional abbreviation, *PSB*, refers to *Photometric and Spectroscopic Binary Systems* (E.B. Carling and Z. Kopal, D. Reidel 1981).

Table 2. Photometric Solutions

RT And *AAp 103*,57, *PSB* p313; AB And *PSB* p199, *MN 208*, 123; AD And *AApSup 45*,499; AN And *AAp 114*,74, *Ann Tokyo 19*,361; S Ant *AApSup 47*,211; AE Aqr *A Ap 104*,24; EE Aqr *AAp 107*,197; KO Aql *AApSup 45*,85; OO Aql *PSB* p199; V337 Aql *AApSup45*, 499; V346 Aql *AApSup 45*,85; V603 Aql *MN 195*,51P; V889 Aql *AAp 115*,321; V1182 Aql *AApSup 45*,499; V1343 Aql *AZh 61*,124; V535 Ara *PSB* p199, *ApSpSc 92*,99; SS Ari *ATs 1249*,5; SX Aur *PerZv 21*,445; EO Aur *JAAVSO 10*,13; KR Aur *PASP 95*,265, TY Boo *PSB* p199; VW Boo *PSB* p199; XY Boo *PSB* p199; AC Boo *PSB* p199, *ApSpSc 92*,99; AD Boo *AApSin 2*,232; 44 Boo *AApSup 45*,187; AO Cam *BAAS 13*,514, *JRASC 76*,90; AS Cam *ApSpSc 94*,115; AW Cam *AApSup 45*,85, *52*,311; TX Cnc *MN 196*,305; AC Cnc *PASJ 35*,423, *IAU Coll 72*,17; R CMa *ApSpSc 99*,229; FZ CMa *AAp 120*,278; XZ CMi *AAp 96*,415; X Car *ApSpSc 99*,191; ST Car *AAp 121*,271, *MN 206*,305; GW Car *ApSpSc 99*,191; OY Car *MN 207*,783, *AAp 128*,29, *128*,37; PX Car *MittAG 52*,169; QX Car *AAp 121*,271; TV Cas *AApSup 52*,213; YZ Cas *AApSup 53*,161; AZ Cas *AApSin 2*,143; CW Cas *PSB* p199; HT Cas *ApJ 244*,579, *245*,1035, *ApJSup 45*,517; V368 Cas *IBVS* 2517; V523 Cas *AJ 86*,98, *AApSup 49*,89; RR Cen *PSB* p199; ST Cen *AJ 86*,1546; SV Cen *AAp 110*,246, *IAU Coll 59*,487; BH Cen *AJ 89*,872; V757 Cen *PASP 94*,189; V779 Cen *AAp 124*,294; RS Cep *AJ 89*,562; VV Cep *MSAI 52*,275; VW Cep *PSB* p199; CQ Cep *AAp 134*,45; *244*, CX Cep *AZh 59*, 73, *ApJ 244*,169; EM Cep *IBVS* 2050; GT Cep *AApSup 49*,89, *55*,403; GW Cep *ApSpSc 83*,195; NN Cep *AApSup 51*,27; TV Cet *AJ 86*,102; TW Cet *AApSup 47*,211; Z Cha *ApJ 244*,579, *252*,653; RZ Com *ApSpSc 92*,99, *PSB* p199; CC Com *PSB* p199, *AApSup 49*,123; TZ CrA *AApSup 45*,85; RT CrB *AASin 23*,125; W Crv *BAAS 14*,979; RV Crv *AApSup 49*,89; UZ Cyg *ApSpSc 82*,189; VW Cyg *ApSpSc 82*,189; BR Cyg *AAp 96*,409; DK Cyg *ApSpSc 92*,

99, *PSB* p199; EM Cyg *MN 195*,235; V388 Cyg *MN 203*,235; V444 Cyg *ApJ 281*,774; V463 Cyg *BAC 33*,187; V470 Cyg *AAp 109*,368; V478 Cyg *AApSup 53*,363; V1073 Cyg *PSB* p199, V1357 Cyg *AJ 86*,1259; V1727 Cyg *ApJ 243*,900; RZ Dra *AApSin 2*,144; UX Dra *BAC 35*,65; BH Dra *AApSup 45*,499; BS Dra *ApSpSc 79*,359, *AA 31*,505; BV Dra *AApSup 48*,85, *51*,435, *ApSpSc 88*,433; BW Dra *AApSup 51*,435, *ApSpSc 88*,433; RU Eri *ApSpSc 81*,209, *AApSup 49*,89; UX Eri *PSB* p199; WX Eri *AApSup 45*,85, *52*,311; YY Eri *AApSup 49*,123; BV Eri *ApSpSc 93*,69; BW Eri *ApSpSc 88*,197; EF Eri *MN 195*,155; U Gem *ApJ 246*,215, *258*,572, *MN 204*,1105; RY Gem *AA 32*,411, *Bull.Abas* 55,89; AF Gem *PASP 94*,926; GW Gem *AApSup 46*,185; SZ Her *AApSup 45*,85; TT Her *AAp 126*,94; AH Her *AAp 117*,283; AK Her *ApSpSc 92*,99, *PSB* p199; AM Her *ApJ 247*,984; DI Her *ApJ 254*,203; *ATs 1183*,5; DQ Her *ApJ 244*,579; HZ Her *ApJ 247*,1003, *275*,278; LT Her *AApSup 52*,311; MM Her *ApSpSc 90*,421; u Her *AN 304*,37; TT Hya *JApA 2*,119, *3*,93; WY Hya *AAp 103*,349; KM Hya *AAp 130*,102; RY Ind *PASP 94*,524, *MN 206*,305; SW Lac *AApSup 51*,435, *PSB* p199; AR Lac *ApSpSc 78*,123; AW Lac *AJ 88*,1679; CO Lac *ApSpSc 89*,5; EM Lac *AApSup 51*,435, *PSB* p199; EN Lac *AAp 122*,193; UV Leo *AN 304*,37; UZ Leo *PSB* p199; XY Leo *MN 196*,305; AM Leo *PSB* p199, *ApSpSc 92*,99; RS Lep *AApSup 45*,85; ES Lib *AN 302*,187; FT Lup *MittAG 55*,72, *MN 208*,135; RR Lyn *ATs 1165*,3; UU Lyn *PASJ 35*,131; UV Lyn *PASP 94*,350; MV Lyr *ApJ 245*,644; β Lyr *AAp 95*,328; TU Men *AAp 132*,187; AU Mon *AA 32*,431, *MN 199*, 131; BT Mon *ApJ 254*,646; UZ Oct *AAp 130*,97; RZ Oph *AAp 111*,372, *ApJ 264*,251; V502 Oph *AApSup 49*,123, *PSB* p199; V566 Oph *PSB* p199; V839 Oph *PSB* p199; V2051 Oph *MN 203*,909; VV Ori *ApSpSc 89*,15, *AApSup 51*,111, *ApJSup 49*,531; CN Ori *AAp 115*,190; ER Ori *AApSup 47*,211, *PSB* p199; δ Ori *ApJ 248*,249; θ^1Ori *PerZv 21*,227; BD Pav *AAp 124*,287; U Peg *AApSup 47*,211; *PSB* p199; AQ Peg *ApSpSc 82*,189; BB Peg *AN 302*,285, *AAp 101*,273; BX Peg *AA 32*,131; DI Peg *ApSpSc 81*,283; EE Peg *ApJ 281*,268; II Peg *ApJ 247*,975; RT Per *ApSpSc 75*,329; DM Per *AApSup 54*,193, *Obs 104*,83; β Per *ApSpSc 98*,163; 1 Per *IBVS* 2077; AE Phe *AApSup 45*,187, *PSB* p199; AG Phe *PASP 95*,347; AI Phe *BAAS 15*,877; δ Pic *PASP 95*,319; SZ Psc *ApSpSc 82*,289; UV Psc *AApSup 49*,89, *ApSpSc 99*,239; V Pup *AAp 128*,17; TY Pup *AAp 99*,182, *AApSup 49*,123; VV Pup *MN 201*, 521, *203*,749; PV Pup *AAp 132*,219; TY Pyx *AAp 101*,7, *PSB* p361; VV Pyx *AAp 134*,147; U Sge *PASP 94*,70; UU Sge *PSB* p405; WZ Sge *ApJ 248*,1067; V499 Sco *AApSup 49*,89; V861 Sco *MN 203*,1021, *IAU Coll 59*,481; V884 Sco *AZh 58*,1226; RT Scl *ApSpSc 100*, 117; VZ Scl *ApJ 244*,579; RS Sct *ApSpSc 99*,191; RY Sct *AAp 100*,59, *101*,138; RZ Tau *ApSpSc 92*,99, *PSB* p199; CD Tau *ApSpSc 79*,359; HU Tau *AAp 97*,410; V471 Tau *AA 31*, 37; HO Tel *AApSup 45*,499; RW Tri *ApJ 244*,579, *MN 195*,227, *195*,825; RR TrA *AApSup 45*,85; KZ TrA *ApJ 244*,1001; AQ Tuc *AApSup 45*,187; W UMa *PSB* p199, *ApSpSc 92*,99,*MN 196*,305; TX UMa *AJ 89*,126; TY UMa *AApSup 51*,97; UX UMa *ApJ 244*,579, *MN 195*,505; VV UMa *BAAS 14*,613; AW UMa *ApJ 260*,744; BE UMa *ApJ 272*,202, *PASJ 34*,141; RT UMi *ATs 1165*,1, *AAp 96*,328, *97*,206; RU UMi *BAAS 14*,613; AG Vir *ApSpSc 92*,99; AH Vir *PSB* p199, *ApSpSc 92*,99, *MN 196*,305, *AApSup 53*,13; AZ Vir *AN 303*,311; BF Vir *ApSpSc 78*,141; BH Vir *MN 206*,305; DM Vir *AJ 88*,535; HD5303 *MN 197*,769, *202*,427; HD90657 *ApJ 259*,213; HD 97152 *ApJ 244*,528; HD134518 *AAp 109*,366; HD155638 *ApJ 270*,L79; HR7442 *IBVS* 2552; HR7464 *AAp 135*,194; HR9049 *ApSpSc 74*,83, *MNASSA 40*,48; BV267 *BAAS 14*,613; BD+37°2356 *IBVS* 2191; 2A0311-227(=EF Eri) *MN 199*,801; E2003+225 *ApJ 277*,682; PG1012-029 *ApJ 276*,233; Lanning 10 *ApJ 252*,681.

Table 3 gives a breakdown of analytical procedures used for light-curve study. This is offered with the same caveats and apologies as in 1982. The Russell-Merrill formalisms still survive either as entry-level solutions or because some groups do not yet have available to them the more deeply and physically parameterized methods. The frequency-domain methods have been elaborated and used abundantly by workers affiliated with the Manchester group. Their work has raised anew the question of whether contact and over-contact binaries are really so. At this time, it seems fair to say that results indicating that such binaries are not in contact are not accepted by most students, nor are they construed as evidence that the cycling episodes of, e.g., the theory of thermal-relaxation oscillations have been found (see also section 5C). The past three years have continued the abundant productivity with Wilson's and Wood's codes. Applied retrospectively to many light curves, in the literature, these efforts have certainly removed false and satellite solutions of several binaries, but some confusions and unlikely interpretations do remain.

Table 3. Percentages of Light Curves Studied by Different Computational Methods

Method	%
Budding	2
Kopal	15
Miscellaneous	22
Rucinski	3
Russell-Merrill	11
Spherical and Nelson-Davis-Etzel	3
Wilson-Devinney	22
Wood	22

From Table 4 it is possible to sense the changes of interest in binaries of different configurations. Except at the hottest and coolest ends, there are numerous light curves solved for systems on or near the main sequence and, without a judgment concerning evolution, it is possible to be satisfied with the sample of contact pairs except for the coolest temperatures. Because of their stellar and solar impacts, the RS CVn-like objects and spotted stars have great value but many more light curves are needed. Modelled systems associated with slow mass exchange are plentiful now but we are still limited by only a few showing fast mass transfer. The most obvious trend is associated with binaries having a degenerate component. These attract deserved attention for the light-curve studies support all other investigations and, in many cases, we can hardly believe that we have seen the entire repertoire of a system's photometric behaviour.

Table 4. Percentages of Configurations among Solutions of Binary Light Curves

	75-78	78-81	81-84
Numbers of binaries	173	211	215
Mostly non-contact, ZAMS to TAMS	33%	22%	16%
Near MS contact	21%	29%	37%
Slightly evolved, e.g. RS CVn-syndrome	4%	7%	6%
Substantially evolved but still non-degenerate	37%	34%	24%
Substantially evolved with collapsed component	5%	9%	17%

Space limitations have caused this section to concentrate on photoelectric work. Nonetheless, archival photographic and even visual studies continue to have value as is shown by the exemplary summary concerning GK Per = N Per 1901 in 1983, *AApSup 54*,p393.

B. SPECTROSCOPIC AND SPECTROPHOTOMETRIC OBSERVATIONS

Table 5 shows that spectroscopic observers have been very active. It includes ultraviolet observations, but not radio or X-ray studies, although spectroscopy of the optical counterparts of X-ray sources is included. References to IAU Circulars are not given, neither are references given to abstracts if the subsequent main paper can be identified. A few papers that give only a little information (e.g. spectral classification) about each of many systems have been deliberately omitted. An asterisk indicates that the paper cited contains new information about orbital elements.

A characteristic of Table 5 is the increasing number of spectroscopic observations that are not primarily concerned with the determination or revision of orbital elements. Several studies of stellar rotation and of magnetic fields are included. This reflects the growing realization of the importance of these factors for the understanding of surface phenomena in some kinds of binary. Of particular interest is the application by Vogt and Penrod (*PASP 95*,p565) of Doppler imaging to the study of the distribution of luminous flux over the surfaces of components of RSCVn binaries. The results of this technique lend considerable support to the spot hypothesis as an explanation of the light variations of these stars.

Cross-correlation has been used to determine the velocities of binary components and has proved very useful. Good spectroscopic mass-ratios are at last being obtained for W UMa systems (see e.g. Maclean and Hilditch, *MN 203*,p1), which should help in the definitive solution of their light-curves. Hill (*Publ. DAO 16*,p59) has devised a cross-correlation program that has been found most effective for determining velocities from the spectral lines of weak early-type components, and may lead to more reliable determinations of absolute dimensions for such stars. It has also been used successfully in the measurements of components of composite spectra (see the reference for 93 Leo in Table 5).

The expanding nebulae around old novae have been studied spectroscopically by Williams (*ApJ 261*,p170). Oliversen and Anderson (*ApJ 268*,p250) have made observations of the visual and ultraviolet spectra of symbiotic stars that tend to confirm the binary nature of these objects, since they can be interpreted by a model in which a red-giant star surrounded by a nebula is eclipsed by a hot companion. In this context, Sahade's (*IAU Coll 70*,p1) useful review of symbiotic stars should be noted.

Reticons and similar detectors continue to be used successfully in the detection of faint secondaries of Algol-type systems. An important example is the system of λ Tau in which Fekel and Tomkin (*ApJ 263*,p289) detected the secondary spectrum and confirmed that the system is triple, with a remarkably small ratio of periods. Spectroscopy of the disk of RW Tau with high time-resolution by Kaitchuck and Honeycutt (*ApJ 258*,p224) is also an important development promising a much greater understanding of circumstellar disks in Algol systems. Plavec' (*ApJ 275*,p251) study of the UV spectrum of U Cep is discussed in section 5B.

Table 5. Spectroscopic and Spectrophotometric Studies

RX And *PASP 94*,102*, *BAAS 14*,981; AN And *Ann Tokyo 19*,361*; EG And *ApJ 262*,282, *BAAS 15*,927; KZ And *ApJ 269*,250; α And *AAp 118*,313; λ And *BAAS 15*,650, *ApJ 268*,L121; o And *IBVS 2125*,2284*, *ApSpSc 86*,179*; V337 Aql *ATs* 1212; V496 Aql *RevMex 6*,103; V603 Aql *AA 33*,149, V794 Aql *ApJ 260*,794; V822 Aql *Izv Krym 64*,81*; V882 Aql *ATs* 1212; V1343 Aql (SS433) *PASP 94*,80, *ApJ 256*,222; *BAAS 14*,880, *15*,927; ε Aql *Obs 102*,82*; AE Aqr *MittAG 57*,302*; FF Aqr *BAAS 14*,979; R Ara *ApSpSc 99*,281; AT Ara *MittAG 55*,77*; BF Ara *IBVS* 2286; V359 Ara *AAp 118*,255*; TT Ari *AAp 110*, 281, *BAAS 15*,663; AR Aur *AAp 118*,313; KR Aur *ApJ 267*,922, *PASP 95*,264*; α Aur *BAAS 14*,900*, *ApJ 272*,223,*277*,678, *AZh 60*,267; β Aur *ApJ 258*,369; ε Aur *BAAS 14*,979,*15*, 925, *IBVS* 2326,2534, *PASP 95*,1012, *ApSpSc 99*,269; ζ Aur *AAp 126*, L5,225, *BAAS 15*,925, *AApSup 53*,339; TZ and XY Boo *MN 203*,1*; ZZ Boo *AJ 88*,1242*; AO Cam *JRASC 76*,90; ζ Cap *PASP 96*,226; Y Car *Obs 103*,163*; YZ Car *MN 205*,1135*; QU Car *ApJ 261*,617*; QX Car *AAp 121*,271*; RX Cas *BAAS 15*,916; SX Cas *PASP 95*,364; TV Cas *AAp 127*,297; YZ Cas *AAp 127*,297, *ApJ 251*,291; AR Cas *MittAG 55*,163*; V425 Cas *BAAS 14*,880*; V635 Cas *ApJ 266*,806; MU Cen *MittAG 50*,77*; BV Cen *ApJ 263*,302*; V373 Cen *MittAG 50*,77*; V436 Cen *ibid**, *ApJ 254*,653, V442 *MittAG 50*,77*; Cen XR-3 *ApJ 278*, 266; Cen X-4 *ApJ 278*,270; U Cep *MN 203*,1063, *ApJ 275*,251; VV Cep *AApSup 53*,339; CQ Cep *ApJ 265*,961*; *AAp 134*,45*; EK Cep *ApJ 271*,717*; EM Cep *MN 200*,1153; π Cep *MN 203*,103*; 6 Cet *IBVS* 2313; R CMa *ApSpSc 99*,229*; UW CMa *BAAS 16*,*473; CW CMa ApJ 261*,612*; FZ CMa *AAp 120*,278*; HL CMa *AAp 125*,L1; ξ CMa *IBVS* 2330; TX Cnc *MN 203*,1*; AC Cnc *PASJ 35*,423, *ApJ 280*,235*, *PASP 94*,162,950, *IAU Coll 72*,17; 81 Cnc *Obs 102*,217*; RW Com *BAAS 15*,917*; RZ and CC Com *MN 203*,1*, V691 CrA *ApJ 255*,596*; θ CrB *BAC 34*,324*, *AAp 131*,210; σ CrB *MN 207*,809, *BAAS 15*,665, *16*,473; VY Cru *BAAS 16*,506; SS Cyg *BAAS 15*,982, *ApSpSc 98*,237, *AA 33*,219; BF Cyg *ApJ 268*,250; CH Cyg *PASP 95*,135,399,1006, *AAp 107*,200, *113*,250, *ApJ 262*,282, *Pis AZh 9*,421; CI Cyg *IBVS* 2126,2355,2356, *AAp 116*,210*, *ApJ 268*,250, *AA 33*,403; V367 Cyg *AAp 131*,147;

V386 Cyg *JApA 3*,233*; V1016 Cyg *AAp 122*,343, *133*,137, *ApJ 268*,250; V1341 Cyg (X2) *BAAS 14*,944, *15*,637,663, *AAp 130*,175; V1357 Cyg (X1) *ApJ 260*,240*, *270*,671, *Pis AZh 9*,606*, *AZh 60*,1041, *PASP 94*,149; V1727 Cyg (4U 2129+47) *ApJ 253*,756,*281*,751, *BAAS 15*,925, 31 and 32 Cyg *AAp 126*,225; HR Del *PASP 94*,916*; YY Dra *BAAS 14*,618; AG Dra *ApSpSc 91*,63, *BAAS 15*,665; υ Dra *JRASC 77*,73*; ω Dra *BAAS 15*,665; BM Eri *AAp 125*,391; BV Eri *ApSpSc 93*,69*; CW Eri *AJ 88*,1242*; EF Eri *BAAS 15*,663; U Gem *BAAS 15*,982, *ApJ 277*,700; WY Gem *ApJ 262*,282; χ and 81 Gem *MN 200*,1161*; RZ Gru *MN 206*,819; DI Her *ApJ 254*,203*; *BAAS 16*,505; HZ Her (X-1) *PASP 94*,149, *BAC 33*,122; V624 Her *AJ 89*,1057*; 105 Her *MN 203*,103*; ζ Her *JRASC 77*,126*; TT Hya *BAAS 15*,926, *16*,504; EX Hya *ApJ 258*,576; KM Hya *AAp 130*,102*; χ^2 Hya *PASP 95*,757*; VW and WX Hyi *MN 203*,865; SW Lac *MN 208*,309; AR Lac *AJ 89*,549; T Leo *ApJ 276*,305*; XY Leo *BAAS 15*,916; 93 Leo *PASP 95*,768*; Leo *IBVS* 2188; FT Lup *MN 208*,135; UU Lyn *PASJ 35*,131*, SZ Lyn *IBVS* 2417; MV Lyr *PASP 94*,328; β Lyr *PASP 94*,341, *AAp 126*, 115; TU Men *AAp 132*,187*; T Mon *MN 203*,925*; AU Mon *PASP 94*,113*; BM Mon *JAAVSO 10*,85; BT Mon *ApSpSc 99*,95*; V505 Mon *ApSpSc 88*,115; V616 Mon *PASP 95*,391, *ApJ 226*,L27; HP Nor *MittAG 55*,77*, V442 Oph *PASP 95*,509*; V566 Oph *MN 204*,817*; V1010 Oph *BAAS 15*,1926; V2051 Oph *MN 205*,465; Nova Oph (H1705-25) *ApJ 274*,717; 64 Ori *ApJ 266*, 732; δ Ori *ApSpSc 87*,269*; AR Pav *PASP 94*,107, *ApJ 275*,271; BD Pav *AAp 124*,287; AG Peg *ApJ 262*,282; AU Peg *AJ 89*,119*; BK Peg *AJ 88*,1242*; EE Peg *ApJ 281*,268*; II Peg *IAU Coll 71*, *AA 32*,315; ι Peg *PASP 95*,1000*; X Per *ApJ 256*,L13: AX Per *ApJ 268*,250; KT Per *BAAS 14*,981; 58 Per *MN 204*,927*; β Per *ApJ 259*,719, *AA 32*,379, *AAp 128*,429; AE Phe *ApSpSc 99*,157; AI Phe *BAAS 15*,877*; ζ Phe *AAp 118*, 255*; φ Phe *Obs 102*,145; SS Psc *ApSpSc 98*,237; SZ Psc *AJ 89*,549; UV Psc *BAAS 16*, 473; V Pup *AAp 128*,17*; VV Pup *ApJ 259*,730*, *BAAS 15*,663; CP Pup *ApJ 261*,170; KQ Pup *AApSup 49*,511; PV Pup *AAp 132*,219*; T Pyx *ApJ 261*,170; VV Pyx *AAp 134*,147*; V818 Sco (X-1) *BAAS 15*,663; V856 Sco *IBVS* 2234; V861 Sco *MN 206*,625; RY Scu *ATs* 1246, *AApSup 56*,17; FH Ser *BAC 33*,116; Y Sex *MN 203*,1*; RY Sex *ApJ 258*,209; U Sge *MN 203*,1063; WY Sge *ApJ 264*,560; WZ Sge *AZh 60*,938; HM Sge *BAAS 14*,982; *AAp 133*, 137; δ Sge *AAp 124*,241*; AP and V350 Sgr *Rev Mex 6*,103; V3885 Sgr *ApJ 258*,217; Nova Sgr (1982) *IBVS* 2283, *PASP 95*,506; ν^1 Sgr *PASP 94*,647; RW Tau *ApJ 258*,224, *272*,206; BU Tau *IBVS* 2148; GR Tau *PASJ 36*,175*; V471 Tau *IBVS* 2092*, *MN 202*,587*, *BAAS 15*, 917, *16*,516; V711 Tau *ApJ 254*,L41, *268*,274*, *PASP 95*,565; ζ Tau *ApSpSc 99*,139; λ Tau *ApJ 263*,289*; RR Tel *MN 202*,833; RW Tri *ApJ 267*,239*; R TrA *RevMex 6*,103; SU UMa *BAAS 14*,880*; SW UMa *IBVS* 2354; UX UMa *ApJSup 53*,397*; AN UMa *BAAS 15*,663, *ApJ 254*,232; AW UMa *ApJ 270*,200, *MN 208*,309; BE UMa *ApJ 272*,202*; ε UMi *ApJ 262*,682; Vel X-1 (4U0900-40) *AAp 135*,155*; γ^1 Vel *PASP 96*,88; γ^2 Vel *Obs 103*,154*, *ApJ 276*,281, *ApSpSc 99*,153; FO Vir *IBVS* 2527; α Vir *PASP 94*,143; ER Vul *MittAG 55*,164*; HR 96 *AApSup 53*,29; HR 362 *PASP 96*,179*; HR 753 *ApJ 269*,250; HR 1883 *MN 199*,303*; HR 2142 *PASP 95*,311*; HR 2577 *PASP 94*,169*; HR 2786,2859 *PASP 94*,642; HR 3084 *IBVS* 2242*; HR 3361 *Obs 104*,74*; HR 3626 *PASP 94*,356*; HR 4006 *PASP 94*,557*; HR 4668 *JApA 5*,181*; HR 4896 *Obs 103*,17*; HR 5553, 6524 *PASP 95*,79*; HR 7024 *Obs 102*,27*; HR 7578 *AJ 267*,682*; HR 8580 *Obs 102*,223*; HD 5980 *ApJ 257*, 116*; HD 13725 *Obs 103*,284*; HD 26337 *IBVS* 2110,2323; HD 27130 and 27149 *ApJ 254*, 606*; HD 28475 *JRASC 77*,142*; HD 31487 *ApJ 268*,264*; HD 33708 *Obs 102*,136*; HD 45088 *ApJ 269*,250*; HD 45166 *IAU Symp 99*,447; HD 46407 *ApJ 268*,264*; HD 47129 *ApSpSc 99*,153; HD 47415 *AApSup 52*,293*; HD 50896 *IAU Symp 99*,295; HD 57339 *Obs 103*,252*; HD 58368 *ApJ 268*,264*; HD 64704 *Obs 104*,69; HD 68244 *AJ 88*,1349; HD 69148 *AApSup 54*,187*; HD 77247 *ApJ 268*,264*; HD 80655 *Obs 103*,56*; HD 81817 *AAp 136*,L5; HD 85091 *AApSup 54*,187*; HD 86590 *AJ 89*,683*; HD 90524 *Obs 102*,136*; HD 90657 *ApJ 259*,213*; HD 94305 *ApJ 272*,190*; HD 104328 *AJ 88*,1349; HD 104451 *JRASC 77*,18*; HD 107742 *JApA 4*,19*; HD 116378 *JApA 4*,171*; HD 117064 *JApA 3*,107*; HD 120803 JApA 3,101*; HD 121844 *JApA 4*,171*; HD 123299 *AApSup 53*,215*; HD 128220 *AAp 121*,85; HD 136905 *IBVS* 2111; HD 141458 *JRASC 77*,18*; HD 143414 *AAp 126*,183*; HD 151910,152333,152570 *AApSup 49*,673*; HD 154791 *ApJ 267*,291; HD 157978 *ApSpSc 94*,1; HD 162085 *AApSup 54*,515*; HD 172865 *PASP 94*,860*; HD 177230 *ApJ 269*,596*; HD 181602 *Obs 102*,1*; HD 184467 *PASP 95*,201*; HD 185151 *AJ 87*,1035*; HD 185662 *Obs 103*,145*; HD 186943 *PASP 95*,151; HD 189178 *PASP 94*,515*; HD 190002 *RevMex 5*,285; HD 193077 *ApJ 253*,230*; HD 193793 *ApJ 277*,258*; HD 199547 *Obs 104*,6*; HD 199939 *ApJ 268*,264*; HD 207739 *ApSpSc 99*,281; HD 208095 *PASP 94*,76*; HD 210647 *Obs*

104,148*; HD 210763 *AApSup 52*,293*; HD 211853 *Tartu Pub 49*,84, *PASP 95*,151; HD 224118 *Obs 104*,80*; HDE 245770 (A0535+26) *PASP 95*,391, *96*,312*; HDE 320156 (LSS 4300) *ApJ 276*,229; BD 45°3310 (ADS 14396) *Obs 104*,143*; BD 26°730 *ApJ 269*,250; CPD-48° 1577 *BAAS 14*,978, *ApSpSc 99*,145*; HZ 22 *PASP 94*,815; Case 1 *ApJ 253*,752*; *279*,758; SMC AB6 *ApJ 257*,110*; NGC 1851 UV5 *PASP 94*,769; MWC 17 *IBVS* 2135; 209 BAC *AAp 114*,135*; PS 74 *AAp 114*,L11; PSR 0921-63 *ApJ 256*,605*; H1-36 *MN 204*,113*; Lanning 90 *PASP 95*,206; P2 *ApJ 269*,229*; Abell 41 *ApJ 280*,177; 2S 0114+650 *PisAZh 9*,603; H0139-68 *BAAS 14*,633; 2A0526-238 (TV Col) *MN 200*,1039; A0538-66 *MN 202*,657, *205*,1117, *207*,287; LMC X1 *ApJ 275*,L43*; LMC X3 *ApJ 272*,118*; *281*,354; PG 1012-029 *ApJ 276*, 233*; PG 1030+590 *BAAS 16*,505; CW 1103+254 *PASP 94*,682, *ApJ 271*,725, *BAAS 14*,980; 4U 1223-62 *PASP 94*,541*; 1329-294 *MN 205*,559; E1405-401 *ApJ 264*,575, *BAAS 14*,633; 3A 1431-409 *MN 201*,25P,32P; PG 1550+191 *ApJ 256*,594*; *BAAS 15*,663; MXB 1735-44 *PASP 95*,23; 4U 1735-44 *MN 207*,29P; 2A 1822-371 *MN 200*,793*; E 2003+225 *ApJ 277*,682, *BAAS 14*,980; H2215-086 *AJ 87*,655*; H 2252-035 *ApJ 265*,363, *267*,726.

C. POLARIMETRIC STUDIES (A.M. Cherepashchuk)

The most important results are the discovery of the 294-day photometric and polarimetric period of Cyg X-1 (Kemp *et al.* *ApJ 271*,pL65), the discovery of polarization during the eclipse of Algol (Kemp *et al.* *ApJ 273*,pL85) and the study of circular polarization of the light of white-dwarf pulsars (Krzeminski *et al.* *IAU Coll 69*,p399).

Polarimetric work has been in two directions: the study of circular and linear polarization in individual stars (many of them of the AM Her type) and theoretical studies of the variations in polarization to be expected in a binary system. Work published in the report period includes the following: *MN 194*, pp187, 283; *AZh 58*,p146; *MN 196*,p275; *AAp Sup. 44*,p461; *ApJ 247*,p984, *MN 198*, p787, *199*,p601; *ApJ 256*,p594; *IAU Coll 70*, p139; *PASP 94*,pp692,695; *IAU Coll 72*,pp205,211,217; *ApJ 264*,p237; *Rep. radio Lab. Helsinki Univ. Technol. No. 129*, p46; *ApJ 275*,p709; *BAAS 15*,p665.

D. X-RAY OBSERVATIONS (Y. Kondo)

During the report period, three X-ray satellites have been in operation: the Japanese satellites Hakucha (launched 1979 February) and Tenma (1983 March) and the European satellite EXOSAT (1983 May 26). The American X-ray satellite *Einstein* (HEAO-2) ceased operating on 1981 June 17, before the triennium under review; nevertheless, a brief report on it is included.

Einstein

When the operation of this satellite was terminated by the exhaustion of its gas, a few hundred binaries, including "normal" interacting binaries (e.g. R. Ara, 31 Cyg), RS CVn systems (e.g. ζ And, UX Ari), novae (e.g. U gem, SY Cnc), X-ray binaries containing compact objects (e.g. AM Her, SMC X-1) and pulsars (e.g. PSR 0740-28, PSR 1055-52) had been observed. A complete list can be found in the booklet *A Listing of all Targets Observed by the Einstein Observatory*, available from: Dr. Frederick D. Seward, Head, Einstein Guest Observer Program, Center for Astrophysics, 60 Garden Street, Cambridge, MA 02138, U.S.A., who can slso provide other information about *Einstein* archival data.

Hakucho and Tenma

These two satellites were still in operation at the end of 1984. The X-ray pulsars observed include Her X-1, Cen X-3, 4U 1626-67, OAO 1657-415, 4U 1700-377, A0535 + 26, Vela X-1, 4U 1907 + 09, 4U 1538-52 and GX 301-2. They have also observed the X-ray binaries Sco X-1, 1608-522, GX 349 + 2, GX5 - 1, Cyg X-1 and Cyg X-3, as well as the RS CVn binary HR 1099. The project scientists, both located at the Institute of Space and Astronautical Science, 4-6-1 Komaba,

Meguro-ku, Tokyo 153, Japan, are Dr. Minoru Oda (Hakucho) and Dr. Yasuo Tanaka (Tenma).

EXOSAT

A large number of interesting binaries has been observed with EXOSAT. Amongst the X-ray binaries with compact objects that have been observed are: Cyg X-1, Cyg X-2, Cyg X-3, Her X-1, Sco X-1, LMC X-1, LMC X-4, AM Her and Vela X-1. Cataclysmic binaries observed include U Gem, SS Cyg, SU UMa, WZ Sge and GK Per. Other binaries observed include Algol, W UMa and the RSCVn systems AR Lac and λ And.

The targets observed with this satellite are listed in the newsletter *EXOSAT Express*, available from: Dr. David Andrews, EXOSAT Observatory Manager, ESOC, Robert-Bosch Strasse 5, 61 Darmstadt, Germany, F.R. There is a guest-observer programme and observing proposals are evaluated every three months. Further information may be obtained from EXOSAT Observatory.

E. RADIO OBSERVATIONS (D.M. Gibson)

The interval July 1981 - June 1984 was characterized by solidification of the picture which emerged in the previous decade. Only three new (single-member) classes of objects were detected: AM Her (Chanmugan and Dulk, *ApJ 255*,pL107), YY Gem (Gary, Linsky, and Dulk, *Solar and Stellar Magnetic Fields: Origins and Effects* p387), and W UMa (Hughes and McLean, *ApJ 278*,p716). The latter two exhibit emission similar to that observed in the dMe flare stars (cf. review by Gibson, in *IAU Coll 71*,p273) and the RS CVn binaries (cf. Feldman, *IAU Coll 71*,p429), respectively. In neither case is it believed that the binary nature is important to the observed activity other than through the enforcement of rapid rotation of the active component(s) by synchronization. In the case of AM Her, Dulk *et al.* (*ApJ 273*,p249) interpreted the rapidly-variable, highly-circularly polarized emission as an electron-cyclotron maser associated with the dMe component. For the quiescent emission they propose a new mechanism. The electrons captured and heated during mass transfer radiate thermal gyro-resonance emission in the strong magnetic field of the white dwarf.

Mutel, Lestrade, and their collaborators (cf. Mutel *et al. ApJ 278*,p220; Lestrade *et al. ApJ 279*,p194) made first reports on an ongoing effort to *map* the structure of RS CVn radio sources with the VLBI. They found the stars UX Ari, HR 1099, and HR 5110 to be unresolved within measurement uncertainties, implying that the emission arises from a region of at most a few stellar radii in diameter. The inferred brightness temperatures 10^{10} K and mild circular polarizations ($\pi_c < 10$) are consistent with the working model of a coronal radio source wherein a power-law distribution of mildly relativistic electrons radiates gyrosynchrotron emission in magnetic fields $5 < B < 80$ Gauss. Doiron and Mutel (*AJ 89*,p430) made similar inferences when they did not detect radio eclipses from AR lac. A number of new (primarily southern) RS CVn systems have been detected by Collier *et al.* (*MN 200*, p869) and Slee *et al.* (*MN 208*,p865).

The peculiar Galactic object SS433, a stellar binary displaying twin collimated jets which precess with a period of 164 days, has been the subject of continued systematic monitoring by a number of groups (Johnston *et al. AJ 86*, p1377; Niell *et al. ApJ 250*,p248; Seaquist, *Vistas Astron. 25*,p79; Bonsignori *et al. Vistas Astron. 25*,p91; Johnston *et al. AJ 89*,p509). The radio observations are consistent with continuous injection of relativistic electrons into the optically-thin jets and show no evidence of the periodicities, other than precession, seen at other bands. The bright radio knots seen in VLA (cf. Hjellming and Johnston *IAU Symp. 97*,p187) and *VLBI* (cf. Schilizzi *et al. IAU Symp 97*,p205) are thought to result from synchrotron emission produced by relativistic electrons accelerated in the jets through resonant cyclotron interaction with turbulent waves which arise due to Kelvin-Helmholtz instabilities in the ordered

0.26*c* flow (Bodo *et al.* *AA 133*,p247). *VLBI* observations of Cyg X-3 (Geldzahler *et al.* *ApJ 273*,pL65) have revealed that it too is a twin jet source with a jet velocity near 0.25*c*. The peculiar Be-star LSI+61°303, a binary X-ray and γ-ray source, has been detected (Taylor and Gregory, *ApJ 255*,p210) and confirmed (Coe *et al.* *MN 203*,p791) to have radio outbursts at intervals of 26.52 days. The object is probably similar to Cir X-1.

A number of "wide" binaries have nevertheless been shown to be "interacting" by radio observations. Symbiotic stars (cf. Kwok, *IAU Coll 70*,p17) and related objects such as Antares (Hjellming and Newell, *ApJ 275*,p704) and HM Sge (Kwok *et al.* *ApJ 279*,p188) evidently produce their emission when the hot companion to a late-type giant or supergiant ionizes a portion of its wind. The unusual shapes and symmetries of these circumstellar H II regions, from ellipsoidal cavities to quasi-cones, cause spectral variations significantly different from those expected to be exhibited by early-type mass-loss stars.

Although it falls outside the prescribed interval for this review, I must mention the *Workshop on Stellar Continuum Radio Emission* which was held in Boulder, Colorado, 8-10 August 1984. A wide range of observations and theories on radio-stars in general and binary radio-stars in particular was presented. Members may wish to obtain a copy of the *Proceedings* (ed. R.M. Hjellming and D.M. Gibson, Reidel) when it becomes available in mid-1985.

F. ULTRAVIOLET OBSERVATIONS (G.E. McCluskey Jr.)

High-resolution ($\sim$ 0.02nm) and low resolution ($\sim$0.6nm) spectrometers on the International Ultraviolet Explorer (*IUE*) satellite, operating in the wavelength range 115nm - 320nm, were used for nearly all the ultraviolet observations of close binaries. Important exceptions are the ultraviolet spectrometers on-board *Voyagers* 1 and 2, that have been used in the regions 55nm - 17nm and 50nm - 120nm, respectively. They have obtained spectrophotometric data, at about 2.5nm resolution, for a number of hot stars, including several interacting binaries. In addition, papers based on data obtained by *Copernicus* and *ANS* are still being published.

The *IUE* has provided 37,000 ultraviolet spectra to date and many of these are of close binaries. Important reviews of this work have been published by Dupree (*IAU Symp. 88*,p39), Hack (*2nd Goddard IUE Symp p89*), McCluskey (*ibid.* p102), Plavec (*IAU Coll 69*,p159) and Rahe (*3rd Goddard IUE Symp.*).

Selected studies of particular systems or classes of systems based at least partly on ultraviolet observations can be found in the following references.

<u>Cataclysmic Variables</u>

AAp 108,p243, *112*,p355, *113*,p76; *ApJ 247*,p577, *258*,p217, *271*,p754, *277*,p700, *281*,p194; *MN 203*,p865, *210*,p197.

ε <u>Aur</u>

AAp 130,*p419*; *ApSup 50*,p233; *ApJ 269*,pL97.

<u>Interacting Systems</u>

ApJ 247,p702, *256*,p206, *264*,pL19, *266*,p755, *275*,p251, *283*,p745; *ApJ Sup 47* p333; *ApSpSc 99*,p281; *MN 202*,p1221; *PASP 93*,p621, *94*,p113.

<u>Symbiotic Stars</u>

AAp 126,p407; *AApSup 56*,p17; *ApJ 253*,p735, *257*,p204, *268*,p250, *275*,p271,

283, p226.

Wolf-Rayet Stars

AApSup 47,p257, *54*,p229; *ApJ 276*,p281, *MN 198*,p897.

X-Ray Binaries

AAp 106,p339; *ApJ 265*,p354, *278*,p270, *283*,p249; *MN 204*,p1091.

ζ-Aurigae Stars

AAp 126,p225; *ApJ 248*,p1043, *281*,p751.

Voyager Extreme UV Spectrophotometry

ApJ 257,p656; *BAAS 15*,p982.

4. Derived Physical Data

A. ABSOLUTE DIMENSIONS AND APSIDAL MOTION (D.M. Popper and A.H. Batten)

Masses in Wolf-Rayet binaries have been reviewed by Massey *IAU Symp 99*,p251) and in cataclysmic binaries by Shafter (*PhD dissertation, UCLA*) and by Wargau and Vogt (*MittAG 55*,p77). LMC X-3 may be the most promising Black-hole candidate yet proposed (Cowley *et al. ApJ 272*,p118). Controversy over the possible super-massiveness of R136 in 30 Dor is reviewed by Edwards (*Nature 306*,p5945). Application of a radial-velocity spectrometer may at last give us definitive masses of the components of Capella (Shen *et al. BAAS 14*,p900), while speckle interferometry, by providing the missing inclination for φ Cyg (McAlister *AJ 87*, p563), leads to rare masses of cool giants. Infrared speckle interferometry has found the faint companion of the famous halo astrometric binary μ Cas (McCarthy, *AJ 89*,p433), but the astrometric orbit of μ Cas A is so uncertain that its mass, and consequently the "primordial" helium abundance cannot be evaluated. Use of a radial-velocity spectrometer has also led to the masses of the Hyades eclipsing binary HD 27130 (McClure, *ApJ 254*,p606) and confirmed the larger distance modulus for the cluster. Fekel and Beavers (*Ap.J 267*,p682) have reviewed the evidence on the masses of K-type main-sequence spectroscopic binaries and conclude that they are more massive than given in "standard" tables. An excellent example among the bright stars of the detection of a faint secondary component with a low-noise detector is provided by ι Peg (Fekel and Tomkin *PASP 95*,p1000). Drechsel *et al.* (*AAp 110*,p246) have obtained rough values of the masses in SV Cen, the one-of-a-kind contact B-type binary with rapid period decrease. Popper has removed 2 more B-type binaries, HD 208095 (*PASP 94*,p76) and NY Cep (*ApJ 262*,p641), from the nearly non-existent list of over-massive binaries, and LY Aur (*ApJ 262*,p641) has been elevated to the status of the eclipsing system with the largest directly determined mass (32 $m_{\odot}$).

Following is a selected list of binaries with absolute dimensions determined directly, with a minimum of hypotheses, during the triennium: V539 Ara, ZZ Boo, CW CMa, QX Car, YZ Cas, EK Cep, U CrB, CW Eri, DI Her, V624 Her, KM Hya, BK Peg, ζ Phe PV Pup, VV Pyx and λTau. References for all but U CrB (*Publ. DAO 15*, p419) may be found in Table 5. Three systems (U CrB, V Pup and λTau are semi-detached, all the others detached.)

Apsidal motion in DI Her has proved to be of considerable interest. The relativistic component should dominate in this system but neither spectroscopic observations collected by Popper (*ApJ 254*,p203) nor eclipse timings since 1898, collected by Guinan *et al.* (*BAAS 16*,p505) show any certain evidence of apsidal motion. This result is consistent with a non-symmetric gravity theory recently proposed by J.W. Moffat (*Phys. Rev. Let.* *50*,p709). While it may be too early to hail the result as confirmation of a new theory, it does underline the importance of apsidal-motion studies.

An important discussion of apsidal motion in 27 systems has been published by Jeffery (*MN 207*,p323). He has recalculated values of the apsidal constant k, using Carson's values of opacities, and has greatly improved the agreement between theory and observation (confirming earlier results by Giménez and García-Pelayo *IAU Coll 69*,p37). Only the system CO Lac still shows a significant difference from theoretical expectations. Jeffery also devised a method for estimating the mass and age of a system from the spectral type of the primary and the observed apsidal motion.

Individual systems for which studies of apsidal motion have been published include the following: V889 Aql (*AAp 115*,p221), PV Cas (*AJ 87*,p1233), EK Cep (*AZh 60*,p72), CI Cyg (*JAAVSO 10*,p92), and δ Ori (*ApSpSc 87*,p269). No evidence has been found for apsidal motion in V1347 Cyg (*ApJ 260*, p240) and its presence in β Per has been questioned (*JAAVSO 11*,p1). A new method for computing apsidal constants was published by Giménez and García-Pelayo (*ApSpSc 92*,p203).

B. PROXIMITY EFFECTS AND LIMB-DARKENING (M. Kitamura)

Within the last few years, considerable efforts have been directed towards the study of tidal evolution in close binary systems. Scharlemann (*ApJ 246*,p292; *253*,p298) studied the effect of tidal coupling with particular reference to RS CVn stars with differentially rotating convective envelopes. Hut (*AAp 99*,126; *110*,37) and Verbunt and Hut (*AAp 127*,p161) discussed the problem as applied to X-ray binaries. The tidal evolution of X-ray binaries was fully investigated by Savonije and Papaloizou (*MN 203*,p581; *207*,685) who calculated the response of non-rotating massive stars to a periodic perpurbing tidal potential (see also section 7). Tidally driven circulation in close binaries was also studied by Smith and Smith (*MN 194*,p583) who showed that it can not be responsible for energy transfer between the components.

Saio (*ApJ 244*,p299) investigated the effect of tidal force on the periods of non-radial oscillations with application to β-Cephei binaries. Tidal effects in twin-degenerate binaries were studied by Campbell (*MN 207*,p433) who showed that tidal excitation of non-radial oscillatory modes on the secondary may significantly shorten the synchronization time. Further comments on the importance of tidal effects and magnetic braking are in sections 5C and 7.

Koch and Hrivnak (*AJ 86*,438) examined the sensitivities of Zahn's theory of tidal friction in actual close binaries and statistically concluded that the theory is generally adequate for binaries whose periods are less than about twenty days.

Mathematically thorough investigations of the effects of rotation and tidal distortions on the orbital elements of a close binary system were presented by Zafiropoulos and Zafiropoulos (*ApSpSc 88*,p401). Mohan and Saxena (*ApSpSc 95*,p369) calculated the effect of rotation and tidal distortions on the structure of polytropic components of close binaries by evaluating various parameters on the Roche equipotentials.

The gravity-darkening of highly distorted stars in close binaries was studied by Kitamura and Nakamura (*Ann.Tokyo19*,p413) with second-order treatment, in order to compare with the result previously obtained in the first order. They found serious differences between these treatments in some actual close binaries. Kitamura and Yamasaki (*Ann.Tokyo 20*,p51) calculated photometric proximity effects by taking into account multiple reflections. Their result indicates that the effect of secondary reflection is quite appreciable and should be considered in almost all actual close binaries. The influence of tidal effects upon the structure of accretion disk in dwarf novae was also discussed by Kříž (*MN 199*, p725).

There has been very little work on limb-darkening problems during these three years, but Bogdovov (*A Zh 60*,p529) studied the source function from the limb-darkening of a star with an extended atmosphere in a close binary system, and Zhu *AASin 25*,p118) treated the limb-darkening effect in accretion disks.

C. ATMOSPHERIC ABUNDANCES (M. Kitamura)

One of the most important results on chemical abundances of close binary stars was obtained by Williams (*ApJ 257*,p672; *IAU Coll 72*) who detected spectroscopically a huge over-abundance of He,C,O,N in the accretion disks of cataclysmic variables. Such over-abundance has already been known in the ejecta of a number of classical novae. William's observations of quiescent novae revealed that the similar over-abundance of He,O,N,C definitely exists in the accretion disk as well.

IUE observations of close binaries have offered a number of new physical data for study of their circumstellar and chromospheric structures: e.g., for W UMa stars by Eaton (*ApJ 268*,p800), Capella by Ayres (*ApJ 284*,p784) and symbiotic binaries by Friedjung *et al.* (*AAp 126*,p407). By analysing *IUE* spectra of Ba stars, Dominy and Lambert (*ApJ 270*,p180) studied the current hypothesis that the Ba stars are produced by mass transfer from a companion now present as a white dwarf. McClure (*ApJ 268*,p264) measured radial velocities of such stars and concluded that all Ba II stars should be binaries. A bipolar nebulosity of NGC 2346 has been suggested by Walsh (*MN 202*,p303) to have been formed by mass-loss from a binary.

Luck and Bond (*ApJ 259*,792) determined C and O abundance in giant CH stars from abundance analysis, on the basis of which they also discussed a hypothesis of mass-transfer from a carbon-star binary companion that is now an invisible white dwarf.

5. Structure and Models of Close Binaries

A. EARLY-TYPE SYSTEMS (K.D. Abhyankar)

Evolution of close binary systems has been reviewed by Yungelson and Masevich in *Astrophysics and Space Physics Reviews* Vol. 2, pp29-74, 1983. It is also covered in *IAU Coll 69* and in *IAU Coll 80*.

W-R Binaries

There has been continued interest in the effects of mass loss in early type systems. Vanbeveren (*AAp 105*, p260) showed that mass loss by stellar wind is marginal for stars with $< 40\ M_\odot$ which give rise to OBN stars while stars with $> 40\ m_\odot$ lose 30 per cent of mass by stellar wind and Roche-lobe overflow to produce WR stars. Deloore (*IAU Coll 80*) found that 50 per cent loss of mass and angular momentum is required for bringing observations into agreement with theory. Hidayat *et al.* (*IAU Coll 80*, p175) studied the distribution of WR Stars in the Galaxy and concluded that in the regions where stars are of low metallicity, WR stars can evolve only in binary systems. Doome and De Greve (*ApSpSc 87*,p357) indicated the possible evolution scheme: OB+OB → WR+OB → C+OB → C+WR → WR+WR. Vanbeveren (*AAp 119*,p239) and Doome and De Greve (*AAp 120*,p97) considered the effect of overshooting on the evolution of massive close binaries.

Several binaries with WR components were observed:

(i) For CQ Cep Leung *et al.* (*ApJ 265*,p961) found 50 per cent overcontact: Kreiner and Tremko (*BAC 34*,p341) as well as Antokhina *et al.* (*AZh 59*,p704) interpreted period changes by the process of mass loss and mass transfer while

Walker (*AAp 128*,p394) doubted the period change. (ii) For γ^2 Vel Pike *et al.* (*Obs 103*,p154) derived a new orbit from archival plates and Kondo *et al.* (*ApJ 252*,p208) found that the temperature of the stellar wind increases outward, indicating an additional source of energy for mass flow in all directions, particularly toward L_3. The spectrum of the binary was analyzed by Sahade *et al.* (*ApJ 276*,p281). (iii) Low-mass companions were found in: HD 143414 by Isserstedt (*AAp 126*,p183) and HD 193077 by Lamontagne *et al.* (*ApJ 253*,p230). (iv) WR stars in Magellanic clouds were studied by Moffat (*ApJ 257*,*p110) and by Breysacher et al.* (*ApJ 257*,p116). (v) Other systems included HD 94305 in which Niemela *et al.* (*ApJ 272*,p190) found the minimum masses of 15 and 32 $m_\odot$; CX Cep in which Lipunova and Cherepaschak (*AZh 59*,p73) found the WN5 component to have a core radius of 4.5 ± 2.5 $R_\odot$ and T_b = 750,000; HD 45166 in which Willis and Stickland (*MN 203*,p619) found a stellar wind of 1200 km s^{-1} from UV Spectra; HD 186943 and HD 211853 for which *IUE* observations are reported by Hutchings and Massey (*PASP 95*,p151).

OB Binaries

Popper (*ApJ 262*,p641) found that luminosities of double-line O-type binaries are almost equal and, therefore, published mass-ratios will need revision toward equality of masses. While studying the massive contact system LY Aur Margoni *et al.* (*ApSpSc 79*,p145) found that 11 out of 14 early-type systems are of contact type. Leung (*IAU Coll 80*) found most contact systems to have been in contact at zero age. Contact systems observed during the period of the report include: BH Cen by Sistero *et al.* (*ApSpSc 91*,p427); RY Sct by Milano *et al.* (*AAp 100*,p59); AW Lac by Jiang *et al.* (*AJ 88*,p1679); HR 4975 by Waelkins and Bartholdi (*AAp 108*, p51); SV Cen by Nakamura and Nakamura (*ApSpSc 83*,p103); VV Ori by Chambliss and Leung (*ApJ 49*,p531); V 861 Sco by Howarth (*MN 203*,p1021); V 539 Ara, ζ Phe and QX Car by Andersen (*AAp 118*,p255) and *AAp 121*,p271); and FZ CMa by Moffet (*AAp 120*,p278), who found it to be a massive triple system. Sybesma and Deloore (*AAp 111*,p229) found that the runaway OB stars HD 108 and HD 53975 may have measurable light variations indicating the presence of compact companions.

β Lyrae

A structural model of the disk around the more massive component was given by Wilson (*IAU Coll 69*, p261). Plavec, Weiland and Dobias (*BAAS 13*,p803) found that the flux of the primary agrees with Kurucz's model in the optical region only and shows an excess in UV region. In the BUSS spectrum of the star Hack *et al.* (*AAp 126*,p115) discovered [N II] lines at 214.3nm and 213.9nm; they have suggested a model for the location of the regions of formation of [Si III] and [N II].

Be Stars

There were several studies of Be stars as interacting binaries. Peters (*PASP 95*,p311) has interpreted measurements of broad wings of Balmer and He lines in HR 2142 as binary motion of a 11 $m_\odot$ + 1 $m_\odot$ system, attributing periodic shell phenomena to mass transfer. Charles *et al.* (*MN 202*,p657) find the periodic recurrent X-ray transient A 0538-66 to be an interacting binary (12 $m_\odot$ + 1 $m_\odot$) with Balmer and He I lines in emission and a highly eccentric (e = 0.7) orbit. Antonello *et al.* (*ApSpSc 83*,p381) find evidence for the binary nature of the Be Stars EW Lac, 28 Tau and ζ Tau. Singh (*IBVS* 2284) gives evidence for a short-period and high-eccentricity orbit for the Be star o And.

B. ALGOLS AND RELATED SYSTEMS

Continued observations of the UV spectrum, and other new techniques, have contributed to our understanding of Algol-type systems. Plavec (*ApJ 275*,p251) has found emission lines in the UV spectrum of U Cep, which undergo eclipse. This observation rules out the chromosphere of the late-type component as the place of origin of these emissions. The strengths of the emission lines suggests anomalous abundances of C, N and O which are consistent with at least partial processing of the matter inside a star. Similar results are reported for U Sge and V356 Sgr (Plavec, *JRASC 77*,p283; Plavec and Dobias, *BAAS 15*,p915), and some confirmation is supplied by the synthetic spectra published by Parthasarathy *et al.* (*MN 203*, p1063). This provides important evidence in favour of mass transfer. Kondo *et al.* (*ApJ 247*,p202), also observing in the UV, found further evidence for mass loss from U Cep.

Plavec and Dobias (*ApJ 272*,p206) have also found weak emission lines in the UV spectrum of RW Tau, an observation that tends to confirm the relationship believed to exist between Algols and Serpentids. Spectroscopy of Balmer lines arising in the disk in this system by Kaitchuck and Honeycutt is mentioned in section 3B. They confirm the variability of the emission lines, already suspected by Joy and now known in other systems. They find that the disk is smaller than formerly believed - rarely exceeding 1.5 times the radius of the central star - and the emission lines are about twice as broad as expected from the rotation of the disk. This leads them to question determinations of the masses of stars from observationally measured rotations of accretion disks. Similar spectroscopy of other Algol-type systems should be valuable.

In Algol itself, Sahade and Hernández (*3rd IAU Latin American Regional Meeting*, in press) find evidence from *IUE* observations for two distinct regions of high electron temperature in the system: a suggestion they had already made for AU Mon (also of Algol-type) and for the very different system γ^1 Vel (*PASP 96*, 88). The successful detection of polarization during eclipses of Algol is mentioned in section 3C. Cugier (*AA 32*, p379) and Cugier and Molaro (*AAp 128*, p429) have constructed a stream model for the system, to explain extra components of the UV lines of Mg II. Harrington (*ApJ 277*,pL69) has discussed the relative inclination of the two orbital planes.

Olson (*ApJ 257*,p198; *258*,p702) has examined the light curves of 16 Algol-type systems for transient disturbances. He found evidence of such disturbances in 4 systems, and of non-stellar sources of flux in two. He has also studied the rotation of primaries of Algol systems (*KRTK 58*,p376) finding evidence for possible changes that might be linked with variable rates of mass transfer. McCluskey and Kondo (*ApJ 266*,p755) have found significant variations in the UV continuum level (outside eclipse) of the spectrum of R Ara which indicate that the volume of gas producing continuous radiation is changing. They also report (in press) X-rays from this system. A similarly variable UV continuum is reported in the spectrum of HD 207739 (Parsons *et al.* *ApJ 264*,pL19).

Young and Snyder (*ApJ 262*,p259) have made four-colour photometric and infrared spectroscopic observations of several Serpentids that they believe indicate both accretion within and variable mass loss from the systems. Similar results are reported by Guinan *et al.* (*PASP 95*,p364) for SX Cas. Kondo *et al.* (*ApJ Sup 47*, p33) confirm the presence of hot gases in this system, as well as in RX Cas and TT Hya. Plavec *et al.* (*ApJ 256*,p206) have found emission lines in the UV spectrum of SX Cas and have revised the spectral types of its components, making more obvious its similarity to Algol-type systems. On the other hand, Weiland and Plavec (*BAAS 15*, p916) reclassify both components of RX Cas as late-type giants - a conclusion to some extent confirmed by Taranova and Shenavrin (*PisAZh 9*,p291). In the visual region of the spectrum, RX Cas and SX Cas appear very similar, but these results from the UV make the former look more like RZ Oph, whose 1981 eclipse was studied by Forbes and Scarfe (*PASP 96*,p737) and by van Paradijs *et al.* (*AAp 111*,p372). The system 93 Leonis (Batten *et al.* *PASP 95*,p768), with a period of 72 days, provides an example of a system containing a late-type giant free to expand without yet encountering its Roche lobe.

Apart from ε Aur, the ζ Aur systems have also been studied. With the help of *IUE* observations, the rates of mass loss from ζ Aur, 31 and 32 Cyg have been estimated by Che *et al.* (*AAp 126*,p225) and for δ Sge by Reimers and Schröder (*AAp 124*,p241). In all four systems the rate of mass loss is estimated to be of the order of 10^{-8} $m_{\odot}$ y^{-1}. From *IUE* observations of the 1979-80 eclipse of ζ Aur, Chapman (*BAAS 15*,p925) found evidence for a temperature gradient increasing outwards, reaching values as high as 14,000 K. Hempe (*AAp 115*,p133) and Hempe and Reimers (*AAp 107*,p36) have studied mechanism for the formation of lines in the UV spectra of these systems. Koch and Pfeiffer (*AJ 87*, p1409) found residual polarization in the light of ζ Aur, during the 1979-80 eclipse, which they ascribe to the effects of inhomogeneities in the atmosphere of the late-type component.

Amongst the RS CVn systems HR 1099 (V711 Tau) and AR Lac have probably attracted most attention. In a number of these systems, the migration of the distortion wave of the light curve appears not to be constant, and may be reversible (SS Boo, Wilson *et al. AJ 88*,257, UX Ari, Zeilik *et al. IBVS* 2168 - but see also Sarma *et al. IBVS* 2357, SZ Psc, Eaton *et al. ApSpSc 82*,p289 and possibly RS CVn itself, Blanco *et al. AAp 106*,p311). Several reviews of these systems, up to the beginning of the triennium under review, can be found in *IAU Coll. 71*. Agreement that the basic cause of their phenomena is starspots now seems general and, as mentioned in section 3B, the hypothesis has received considerable support from the results of Doppler imaging. For a dissenting opinion, however, see Kopal (*ApSpSc 87*,149). Middelkoop (*AAp 107*, p31, *113*,p1) has developed a theory of dynamo action in the envelopes of late-type stars, in which the emission flux in the H and K lines is dependent on rotational velocity. Vilhu and Rucinski, in various papers (*AAp 127*,*p5*, *133*, p117, *MN 202*,p1221 and *Physica Scripta T7*,p70) have applied this both to W UMa and RS CVn systems. Dorren and Guinan (*ApJ 252*, p296) have constructed a model for HR 1099 which simulates the light variations by the rotation of two spots on one of the stars. Nha and Kang (*PASP 94*,p496) however, find that neither a spot model nor a shell model fits their photometric observations of AR Lac. This system is known to be exceptional; Kiziloglu *et al.* (*AAp 123*,p17), from their studies of its UV spectrum conclude that both stars are active, about 70 per cent of the activity being on the secondary. The period is known to be decreasing (Evren *et al. ApSpSc 95*,p401; Erten *et al. 87*,p255). The former authors find evidence for long-term photometric variations, and the latter suggest that mass is being transferred by stellar winds. A similar conclusion is reached by Svristava (*BASI 12*,p52).

Mekkaden *et al.* (*JApA 3*,p27) believe that the amplitude of the wave distortion in the light-curve of HR 1099 changes cyclically, increasing sharply and decaying slowly in intervals of 5-6 years. Fraquelli (*ApJ 254*,pL41, *276*, p243) has studied the Hα emission associated with both components and finds that its intensity is correlated with radio flux, but not with the phase of the distortion wave. The Hα emission appears to arise from localized regions and the line profiles are broader than expected from either the orbital motion or the rotation of the stars. Fekel (*ApJ 268*,p274) has derived an excellent set of orbital elements and (different) rotational velocities for each star. He also detected the asymmetries in the line profiles of the more active star which were made the basis of Doppler imaging. Three years of cooperative photometry (Bartolini *et al. AAp 117*,p149) support the spot interpretation and suggest that the spots are migrating on a differentially rotating star. Ayres and Linsky (*ApJ 254*, p18) have detected high-excitation emission lines (e.g. C II, C IV) at quadratures, whose structure changes in about a week. Radial velocities derived from these lines are similar to that of the primary stars; the secondary components are weak or absent. Ayres and Linsky interpret these results as support for the hypothesis of a connection between the rotational activity and the magnetic behaviour of these stars. Both HR 1099 and UX Ari have been observed with the VLBI by Mutel *et al.* (*ApJ 278*,p220) with results that are discussed in section 3E.

Ramsey and Nations (*AJ 89*,p15) find changes in the phase dependence of the Hα intensity in the spectrum of II Peg, and that the flux has decreased. Udalski and Rucinski (*AA 32*,p315) have published a study of the *IUE* spectrum, in which Marstad and Linsky (*BAAS 14*,p866) find enhanced emission between phases 0.45 and 0.95. They ascribe this enhancement to a plage area overlying the spots. Huenemoerder *et al.* (*BAAS 15*,p663) have observed Hα emission in both components of SZ Psc, a system that was very active in 1982-3. There is evidence that the emission is concentrated at particular longitudes of the primary. Giampapa *et al.* (*ApJ 268*,pL121) measured a magnetic field of close to 1300 gauss in the system λ And.

Scaltriti and Busso, together with other colleagues (*AAp 135*,p23,p.255, *139*, p25 and others in press, *IAU Coll 77*,p395 and *Proc. Statistical Methods in Astron. Strasbourg* Symposium, p13) have devised and applied statistical methods for extracting the information contained in the photometric observations of RS CVn stars and are working on detailed models for individual systems. Amongst others reporting observations are Hearnshaw (New Zealand) and Hu (Peoples' Republic of China).

C. W URSAE MAJORIS SYSTEMS (K-C. Leung)

The last decade has been an extremely active period in the development of theoretical models of the structure of contact and near-contact systems. Many papers were published, including Benson's discussion of the possibility of the mass-gaining component swelling into contact, and important advances made by two groups (Lucy, Flannery, Robertson and Eggleton, on the one hand and Shu, Lubow and Anderson on the other). Shu and Lubow (*Ann Rev A Ap 19*,p277) gave a good summary of these developments (up to 1981) and the controversy engendered by them. They questioned whether or not we should accept the swelling into a contact configuration predicted for mass-gaining stars by several investigators. Theories produced by both groups of workers require refinement, and a high level of precision is needed in both theories and computations to solve many problems in this field. Fundamental advances have become much more difficult and more theorists are leaving the field than are attracted to it. It is to be hoped that this situation will change. Nevertheless, theoretical work has been published in the past triennium on e.g., the stability of zero-age contact binaries (Hazlehurst and Refsdal *AAp 113*,p63; Rahunen *AAp 109*,p66) and on magnetic braking in close binary systems (Rappaport *et al. ApJ 275*,p713; Mochnacki *BAAS 13*,p513; Vilhu *ApSpSc 99*, p287 - see also references given in section 7).

Many photoelectric observations of short-period close binary systems have been published and photometric solutions based on the Roche model have been derived from them. References can be found in the *Bibliography and Program Notes on Close Binaries*. More reliable data are now being accumulated for contact and semi-detached systems, which - no doubt - will help us to understand their structure and evolution. It is becoming apparent that there are systems that cannot be readily classified as either contact or semi-detached. There is evidence for two new configurations: almost-contact systems and closely detached systems. In the former, the component of lower mass is in contact with its Roche lobe, and that of higher mass is extremely close to its Roche surface. These systems may be at an evolutionary phase immediately following the evolution into contact. In systems of the latter type, both components are close to their Roche surfaces, but not in contact with them. Some investigators identified these systems with a type described by Lucy that goes in and out of contact. Absolute dimensions are known for some of these systems, however, and the radii of the components of lower mass are considerably larger than those expected for zero-age stars of similar mass. These systems, therefore, are presumably evolved, while those described by Lucy are of zero-age. The closely detached systems, therefore, are not the in-and-out-of-contact systems predicted by thermal-oscillation theory, but, probably, evolved close binary systems. Perhaps they have completed the

Algol phase and the mass-losing components have just detached from their Roche lobes.

While spectroscopic observations of W UMa systems have not been published in the same quantity as photometric ones, important results have been obtained by cross-correlation, as described in section 3B, by McLean and Hilditch (St Andrews) and by Milone, Hrivnak, Hill and Fisher (Calgary and Victoria). Spectroscopically and photometrically determined mass-ratios now agree more closely.

D. NOVAE AND RELATED SYSTEMS (J. Smak)

Many meetings were held devoted entirely or partly to the field of cataclysmic variables (CV's): *IAU Coll Nos. 69* (Bamberg), *72* (Haifa), and *80* (Bandung), three North American Workshops on CV's and related objects (Santa Cruz, 1981; Cambridge, 1983; Baton Rouge, 1984), *Coll. on Pulsations in Classical and Cataclysmic Var. Stars* (Boulder, 1982), *Nato Symp.* on *Interacting Binaries* (Cambridge, England, 1983), Frascati Workshop on *Galactic Accreting Sources* (Vulcano Island, 1984), and the American and European *IUE* Conferences. Their proceedings contain excellent reviews of observational and theoretical aspects. Reviews are also contained in *The Classical Novae* (ed. Bode and Evans, J. Wiley, 1984). Other, individual reviews included: Pringle, (*Ann Rev AAp 19*,p137) and Verbunt (*SpSc Rev 32,p379)* on accretion disks, Smak (*PASP 96*,p5) on dwarf novae, and Truran (in *Essays in Nuclear Astrophysics*, eds. Barnes *et al.*, Cambridge U. Press, 1982) on THR models of nova outbursts. Several useful catalogues of objects and lists of literature were published (e.g. Duerbeck, *ApSpSc 99*,p93; Patterson, *ApJ Sup 54,p443; Ritter, AAp Sup,* in press).

There was a large number of papers based on observations in the visual and UV regions. Particularly important are new results of extensive observations of selected objects; examples are: photometry of Z Cha (Cook and Warner, *MN 207*, p705), photometric and spectroscopic study of OY Car (Vogt, *AAp 128*,p29; Schoembs and Hartmann, *AAp 128*,p37; Krzemiński and Vogt, *AAp* in press), spectroscopic coverage of SS Cyg over outburst cycle (Clarke *et al. ApJ* in press), extensive observations of several dwarf novae in the ultraviolet and visual regions by Cambridge observers (cf. Pringle and Verbunt, and Hassall, *Proc. 4th European IUE Conf.*), and observations of recent novae. Ultraviolet photometry of SS Cyg and U Gem was extended into the extreme UV by Polidan and Holmberg (*Nature 309*,p528). Disk eclipses were analysed by Schwarzenberg-Czerny (MN 208,p57) and Horne (Ph.D. Thesis, Caltech). Studies of CV's belonging to the VY Scl group have shown (Robinson *et al. ApJ 251*,p611; Shafter *et al. ApJ* (in press) that their "low" states are due to considerably reduced mass-transfer rates. The SU UMa subtype was reviewed by Warner (*NATO Symp.*) who finds that all earlier explanations of superhumps are inconsistent with observations and proposes a new, intermediate polar model in which the superhumps result from illumination of asymmetries in the disk.

CV's continue to be favoured targets of X-ray observers. Results of the *Einstein* survey were summarized by Cordova and Mason (*MN 206*,p879) and new data are coming from EXOSAT. The X-ray fluxes provide important constraints on models of the boundary layer; most recent discussions range from nearly alarmist, about the "missing boundary layer" (Ferland *et al. ApJ 262*,pL53) to fairly optimistic, showing reasonable agreement between theory and observations (Patterson and Raymond *ApJ* in press). See also section 5E. The X-ray emission from polar columns in the case of magnetic white dwarfs was discussed by several authors (cf. King, *IAU Coll 72*, p181).

The number of CV's known to contain magnetic white dwarfs is growing. The subtype of intermediate polars (Lamb and Patterson, *IAU Coll 72*, p229; Warner, *IAU Coll 72*,p155; *7th N.American Workshop; NATO Symp)* includes objects ranging from strong X-ray emitters and rapid rotators to (probably) less extreme cases such as the SU UMa variables.

A major breakthrough was achieved in the field of dwarf novae following a pioneering work by Meyer and Meyer-Hofmeister (*AAp 104*,pL10). First time-dependent accretion-disk models, including effects of thermal instability due to hydrogen ionization (Meyer and Meyer-Hofmeister, *AAp 132*,p143; Mineshige and Osaki, *PASJ*, in press; Papaloizou *et al.* *MN 205*,p487; Smak, *AA 34*,p161) reproduce quite well the light-curves and other properties of dwarf novae and demonstrate that their outbursts are due to non-stationary accretion behaviour. There is obvious need to include additional effects, such as the heating of the secondary by radiation from the boundary layer, which appears crucial for the Z Cam (Meyer and Meyer-Hofmeister *AAp 121*,p29) and SU UMa (Osaki, *AAp*, in press) behaviour. Heating effects have already been detected in some dwarf novae (Szkody *et al.* *ApJ 282*,p236; Hessman, *ApJ*, in press). Superoutbursts of SU UMa variables represent a different case. Vogt (*AAp 118*,p95) finds that supermaxima of VW Hyi always develop from immediately preceeding normal maxima and are likely due to increased mass-transfer rate. Time-dependent disk models with mass-transfer bursts (Bath *et al.* *MN 205*,p171; Mantle and Bath, *MN 202*,p151) may be applicable to this case. Dwarf novae, with their outbursts being due to accretion phenomena, are also vulnerable, on a longer time scale, to the TNR phenomena leading to nova type outbursts. Three recorded novae, GK Per, WY Sge, and V3890 Sgr, are indeed showing also the dwarf nova behaviour (Duerbeck, *ApSpSc 99*,p363).

The theory of nova outbursts has reached a new turning point. The TNR models, calculated in spherical case explain satisfactorily many observed properties of fast and slow novae and also of some recurrent novae (Starrfield *et al.* *ApJ*, in press). The predicted CNO overabundance (cf. McDonald, *IAU Coll 72*,77), which was confirmed by observations, can be understood in terms of mixing with the deeper layers of the C-O white dwarf; in the special case of Nova CrA 1981 (Williams *et al.* *MN* in press) observations suggest an O-Ne-Mg white dwarf. The most promising explanation involves mixing due to shear instabilities in the equatorial belt (McDonald, *ApJ 273*,p289); there is obvious need for further work along these lines.

E. X-RAY BINARIES (CATACLYSMIC VARIABLES) (G. Shaviv)

The most important event concerning observations of X-rays from cataclysmic variables is the publication of a catalogue of such variables that are X-ray sources and have been observed by *Einstein* (Cordova and Mason *MN 206*,p879). The catalogue is accompanied with a few detailed studies of particular sources. The 18 close binaries in the catalogue have luminosities in the 1 Kev band of between 10^{30} erg s^{-1} and 10^{32} erg s^{-1}. Most of them show hard spectra (kT > 2 Kev) independently of whether or not they are undergoing an outburst. Of seven dwarf novae observed during optical outbursts, only U Gem exhibits enhanced ultra-soft X-ray emission (kT 10 ev) in addition to weak, hard X-ray emission. The observed X-rays are highly variable on time-scales ranging from tens of seconds to hours.

Observations with EXOSAT have allowed better exploration of possible correlations and predictions. Watson *et al.* (*MN* in press) have discovered a definite X-ray period of 351s in the old nova GK Per. Accompanying ground-based observations (Mazeh *et al.* *AAp* in press) show a bit signal at 380s caused by a long-term transient in the system, observable also in X-rays with a quasi-period of 3000s. These observations confirm the suggestion by King and Shaviv (*MN* in press) that GK Per is an "intermediate polar" - that is, a magnetic white dwarf with a field in the range of 10^5 gauss. The X-ray pulse profile is remarkably constant and has a quasi-sinusoidal shape. The spectrum of the dominant hard X-rays is of a flatness unprecedented among cataclysmic variables. Another important discovery by King *et al.* (*Nature*, in press) is the existence of a correlation between the hard X-ray luminosity L and the temperature T of the bremsstrahlung-emitting plasma.

The first theoretical predictions of X-rays from cataclysmic variables were that they must have a high luminosity in hard X-rays, as a result of either shock or viscous dissipation at the boundary layer. The low observed X-ray flux from cataclysmic variables, however, has led people to look for the "missing boundary layer" or to suspect that the X-rays come from the M-type dwarf or a disk corona. Obviously, the process by which X-rays are formed must be different in each case, as shown by differing properties from one system to another. Recent discoveries by EXOSAT help to clarify the problem. It is natural to suppose that the white-dwarf corona is the hottest source, since it is in the deepest gravitational well. Recent theory by King and Shaviv (*Nature 308*,p519) predicts that boundary-layer instability leads to the formation of a corona at low accretion rates (thus hard X-rays are expected for low masses). For high masses, the boundary layer is stable and only soft X-rays are observed. Exactly this behaviour is observed in SS Cyg. During optical outbursts the hard X-rays decay, while the soft ones increase.

The puzzle of the soft X-rays from AM Her has not yet been settled observationally. Recent observations reported by Heisse at the 1984 North American workshop on cataclysmic variables, seem to indicate that hard X-rays emerge from one pole and soft X-rays from a second pole, since radiations in the two X-ray bands are out of phase. It is not clear how this can happen.

6. Statistical Investigations

A number of wide-ranging statistical studies has been published by Guiricin *et al.* (*AApSup 54*,p211, *ApJSup 54*,p421, *AAp 119*,p218, and *ApJSup 52*,p35) based on samples of up to 1,000 systems. The authors find an underabundance of short-period systems (relative to the numbers of longer periods) and especially of early-type contact systems. Their studies of the angular momentum of Algol-type systems lead them to conclude that considerable amounts of mass and angular momentum must have been lost during the formation of these systems.

The frequency of binaries in both globular and galactic clusters, and among Population II field stars has continued to attract attention. A possible cataclysmic variable has been identified in M30 (*ApJ 274*,pL31). Harris and McClure (*ApJ 265*,pL77) argue from their study of binary incidence amongst late-type giants that it does not necessarily follow from the presently available evidence that M3 is deficient in binaries, while Irwin and Trimble have found some possible W UMa systems in the globular cluster NGC 6809. Krolik (*Nature 305*, p506) published estimates of the number of binaries to be expected in globular clusters; similarly, Herz and Grindley (*ApJ 267*,pL83) predict that an average globular cluster should contain about 200 white-dwarf binaries. Four W UMa systems have been found in NGC 188 (Baluinas and Guinan *BAAS 15*,p924) while Stefanik *et al.* (*ibid*) find that about half the stars in the Hyades are binaries. Levato and Malaroda (*PASP 94*,p807) have identified possible binaries in Trumpler 16. On the other hand, Stryker and Hrivnak (*ApJ 278*, p215) believe that not *all* the blue stragglers in NGC 7789 are binaries. Stryker (*BAAS 16*,p507) also advances arguments for supposing that the binary frequency of Population II stars may be higher than some investigations have suggested, while Carney (*AJ 88*,p623) believes that 20 to 25 per cent of a sample of 71 halo dwarfs may be binaries.

Griffin (*Obs 103*,p280) has published an interesting synopsis of his first 50 orbits from photoelectrically determined radial velocities which indicates how the distribution of (e.g.) orbital periods in a sample can be influenced by selection factors dependent on the method of observation. Halbwachs (*AAp 128*,p399) studied the distribution of binaries in the new edition of the *Bright Star Catalogue* and concluded that 60 per cent of the stars are in binary or multiple systems with a separation greater than 3 A.U. Fofi *et al.* (*AAp 124*,p313) from a multivariate statistical analysis confirm the existence of a period-eccentricity relation and of bimodality in the distribution of mass-ratios (the latter is also confirmed in *AAp 119*,p218). Mezzetti *et al.* (*AAp 122*,p333) have studied the effects of duplicity on estimates of the local luminosity function. Echeverría (*Rev Mex 8*, p109) finds a period-spectral-type relation for cataclysmic variables and deduces that the secondaries are not normal main-sequence stars. Trimble (*Obs 102*,p133) has considered the binary systems likely to produce cataclysmic variables and type I supernovae. She concludes that there is an ample supply for the former but not for the latter. Kenyon and Gallagher (*AJ 88*,p666) have studied a sample of symbiotic stars and conclude that not all the giant components of such stars fill their Roche lobes. Russo (*IAU Coll 69*,p23) believes that upto 25 per cent of Cepheids have companions while Panchatsaram and Abhyankar (*ibid* p.47) have studied period changes in 22 binaries.

Garrison *et al.* (*ApJ 276*,pL13) report the discovery of a probable cataclysmic variable that is, if its nature is confirmed, the brightest member of its class ($V = 9.4$). From the results of a survey of stars to $B = 10.0$, they are able to estimate the space density of cataclysmic variables as 1.3×10^{-6} per cubic parsec.

Popova *et al.* (*ApSpSc 88*,p55) have published a discussion of the properties of 1041 known spectroscopic binaries. Amongst the bright stars, they find a duplicity rate of 40 to 45 per cent and, allowing for selection effects, they estimate that almost all stars belong to binary systems. Related investigations of 482 eclipsing binaries and 333 two-spectra binaries have been published by Kraicheva *et al.* (*PisAZh 7*,p488) and by Popova *et al.* (*PisAZh 8*,p297). From the frequency distributions with respect to mass and separation, the last-named authors are led to postulate the coalescence of some binaries into single stars (see also section 7).

Two useful catalogues are in press: the 14th *Catalogue Complementaire (Binaries Spectroscopiques)* - Pedoussaut *et al.* *AApSup* - and Budding's *Catalogue of Classical (Evolved) Algol-Type Binary Candidate Stars*, to be published by CDS Strasbourg.

7. Origin and Evolution

A review of the evolution of close binaries has been published by Yungelson and Massevich (*Soviet Scientific Review 21*,p27). No attempt will be made to review more than a few of the many papers that discuss the evolutionary status of individual systems. McClure's (*ApJ 268*,p264) discovery of the binary nature of barium stars is clearly of evolutionary interest.

Most of the effort in theoretical studies of evolution seems directed towards understanding systems containing collapsed objects. Kornilov and Lipunov (*AZh 60*, p284,p574) have studied the evolution of close binaries, taking into account the evolution of a rotating neutron-star component. Tutukov *et al.* (*Pis.AZh 8*,p365) have discussed the evolution of dwarf novae, with particular reference to gravitational radiation. Webbink and collaborators (*ApJ 254*,p616) have developed theoretical models of low-mass X-ray binaries, again taking into account gravitational radiation which explains the minimum period found for cataclysmic binaries - a conclusion reached independently by Paczynksi and Sienkiewicz (*ApJ 248*,pL27), *268*,p825). Webbink *et al.* (*ApJ 270*,p678) find that mass-transferring subgiant stars, driven by nuclear evolution, provide a satisfactory explanation for bright-bulge sources and fit very closely the observed parameters of Cyg X-2 and 2S 0921-630. Webbink (*ApJ 277*,p355) has also explored the origin and evolution of close double white dwarfs, suggesting that they may be progenitors of R CrB stars and type I supernovae. The latter suggestion was also made by Iben and Tutukov (*ApJSup 54*,p335), but see the reference to Trimble's work in section 6. Some work by Bath and his collaborators is mentioned in section 5D. Bath and Pringle (*MN 194*,p967, *199*,p267, *201*,p345) have also studied the evolution of viscous accretion disks, finding excellent agreement between predicted outburst evolution and observed photometric and spectroscopic behaviour of dwarf novae and symbiotic systems. A statistical study of outbursts of SS Cyg since 1897 (Bath and van Paradijs *Nature 305*, p33) shows a clear correlation between the structure of the outburst and the light level in the quiescent state.

Campbell (*MN 205*, p1031) has found that the white-dwarf component of AM Her systems cannot be completely synchronized with the orbit while mass is being transferred. As mentioned in section 4B, however, he found that tidal torque can synchronize the white dwarf in a twin-degenerate binary within the lifetime of the system (*MN 207*, p433). Campbell and Papaloizu (*MN 204*, p433) find that synchronization times for convective secondaries can approach the lifetime of the system. Chau and Nelson (*ApSpSc 90*, p245) have discussed non-conservative mass transfer and gravitational radiation in low-mass binaries. Other evolutionary studies of cataclysmic variables have been published by Anzer and Borner (*AAp 122*, p73), Law and Ritter (*AAp 123*, p33), and Miyaji (*IAU Coll 72*, p263). Vilhu (*ApSpSc 99*, p287) has considered magnetic braking of the red star in a cataclysmic variable as a mechanism to drive mass transfer. Ureche (e.g. *IAU Coll 69*, p73) has published several studies of relativistic objects in close binaries.

Studies, by the research group at the Free University of Brussels, on the formation of Wolf-Rayet binaries are reported in section 5A. The same group of authors has studied the structure of early-type contact stars and the origin of Algol-type systems (De Greve and Packet *ApSpSci 99*, p313 and papers in press). Chaubey (*ApSpSc 73*, p503) finds that all semi-detached systems are results of Case B evolution, provided allowance is made for loss of mass and angular momentum. From data for 333 unevolved double-line systems, he deduces that Case A evolution is rare. Guiricin *et al.* (*AAp 125*, p388) find no satisfactory agreement between theoretical calculations and the observed properties of Algols. Kenyon and Gallagher (*AJ 88*, p666) find that the late-type components of many symbiotic stars do not fill their Roche lobes, even though it appears that mass is being transferred. This may have implications for all theories of binary evolution. Other work by Kenyon and his collaborators on symbiotic stars is to be found in *AAp 106*, p109 and *ApJ 273*, p280, *279*, p252.

Vilhu (*AAp 109*, p17) has discussed the formation of W UMa stars by angular-momentum loss from detached systems. See, however, Rucinski's (*Obs 103*, p280) comments. Verbunt and Hut (*AAp 127*, p161) also discuss magnetic braking as an evolutionary mechanism, as mentioned in section 4b. An alternative origin for W UMa systems is proposed by Budding (*AAp 130*, 324), who suggests that they are formed from detached systems with small mass-ratio, by Case B mass transfer. Van Hamme (*AAp 116*, p27) has studied the absolute dimensions and lifetimes of W UMa systems.

Van't Veer (Paris preprint No. 26) has postulated that the components of a binary can coalesce into single stars. Walter and Basri (*ApJ 260*, p735) have suggested that this is happening in FK Com.

Artymowicz (*AA 33*, 233) has studied the formation of binaries by accretion on to protostars from a molecular cloud. Numerical simulations show that the mass-ratio of neighbouring nuclei of condensation tends to equalize during this process, which can thus explain the observed preponderance of binaries with mass-ratios near unity. The possible evolutionary connection between planetary systems and low-mass cataclysmic variables has been discussed by Livio and Soker (*AAp 125*, L12).

Lebovitz (*ApJ 275*,316) has extended his earlier investigation of the possibility of binary and multiple systems being formed by the fission of a single proto-star. When changes in internal energy - ignored in classical treatments in fission theory - are taken into account, it appears that binary systems may be able to be formed by this process.

Alan H. Batten
President of the Commission

44. L'ASTRONOMIE A PARTIR DE L'ESPACE (ASTRONOMY FROM SPACE)

PRESIDENT: M. Oda
VICE-PRESIDENT: Y. Kondo
ORGANIZING COMMITTEE: A.A. Boyarchuk, K. Fredga, M. Grewing, D.C. Morton, L. Peterson, E. Peytremann, A.B. Underhill, B. Valnicek, K.A. van der Hucht, R.J. van Duinen, A.J. Willis

Introduction

M. Oda

A rapid and dramatic change in our views of the Universe which we have witnessed during the past two decades or so is often compared with what happened at the time of Galileo. Revolutionary role of the optical telescope then may be analogized with that of space-astronomy today which has drastically opened the new observational window to the Universe. The revolution is ongoing with a rapid pace or even being accelerated.

Astronomy that is the oldest of the sciences which inspire imagination of mankind has two aspects; one is the exploration of unknown and the other is understanding what is going on in the Universe. Today the frontiers of astronomy extend more and more deeply into the domains of fundamental physics. Space science is now an essential element of astronomy to ask basic questions about the nature of the observational universe.

Recently deep relationships have been developing between astrophysics and elementary particle physics through studies in cosmology and grand-unified-theory. One of the key problemes in this context now is the hidden mass (unidentified matter) in the Universe. It is not an imaginative existence in theoretical cosmologies but now has become a realistic object for observational astronomy to search for.

Today the existence of neutron stars is not anymore the question to be asked but they are subjects of observational studies for physics of extreme environments: extremely strong gravity, high matter density, high temperature, superfluid interior and strong magnetic field. The collapses of stars which produce the neutron stars are origins of dynamic processes in our Galaxy. We now suppose that the interstellar space is the place where such dynamic processes work continuously. We know that in some of the supernova remnants, at least in the Crab Nebula, the spinning neutron star is the cause of all activities of the nebulae.

The black hole is an extremely unusual state of matter. It is of greatest interest to study this in the context of general relativity. X-ray astronomers now believe that they observe black holes in some of the X-ray sources. Intriguing suggestion from observations of active galactic nuclei are that supermassive black holes may sit there.

Infrared astronomy, which had grown its value in all fields of astrophysics over the past decade with relatively modest means, marked a large step by a launch of the Infrared Astronomy Satellite (IRAS). Among fascinating discoveries IRAS has revealed physical processes in active nebulae that are candidate places for star formation and discovered clouds of small particles surrounding stars which are possible origins of the planets.

Among the scheduled space astronomy missions which were listed in the previous commission report, IRAS, ASTRO-B (renamed TENMA), and EXOSAT have been launched. These missions have been extremely rewarding and the results will be presented in this report.

The following missions are to be launched in the near future:

- ROSAT, the German X-ray reflective telescope, scheduled for 1987;
- Space Telescope, scheduled for launch in 1986
- GRO, the gamma-ray observatory to be launched in 1988
- HIPPARCOS, the European astrometry satellite, to be launched in 1988
- ASTRO-C, the Japanese X-ray and gamma-ray burst mission under collaboration with UK and US scientists, scheduled for 1987
- GIRL, the German infrared observatory in SPACELAB.

The missions which have been proposed or are under plan include: XTE (X-ray Timing Explorer), AXAF (Advanced X-ray Astronomy Facility), SIRTF (Shuttle Infrared Telescope Facility), ASO (Advanced Solar Observatory). Mission for the next solar maximum are under consideration in Japan and US. Orbiting VLBI has been discussed on various occasions in USSR, US, ESA and Japan.

The Halley's Comet will be observed in March 1986 under the international cooperation by VEGA 1, 2 (USSR), GIOTTO (ESA), MS-T5 and PLANET A (Japan) and ICE and ASTRO-1 (US).

While the scale of space astronomy programmes tends to increase, it appears that small or moderate missions remain valuable. It looks important to maintain a healthy balance between major programmes and modest ones from viewpoints of generating new and progressive ideas for the following decades and producing a powerful new generations of astronomers.

We observe every year more extensive and deeper international collaboration of a wide variety of modes in various disciplines. Commission 44 should continue to provide opportunities to extend international cooperation and collaboration for ASTRONOMY FROM SPACE.

1. GAMMA RAY ASTROPHYSICS BEYOND THE SOLAR SYSTEM

C. E. Fichtel

A. INTRODUCTION

The promise of high energy ($\gtrsim$ 10 MeV) astrophysics as a valuable new avenue for the exploration of the universe is now beginning to be realized, as the results discussed here will show. Even considering these achievements, γ-ray astronomy is still a young, growing science, and the potential for fundamental contributions to astrophysics in the future is very large. With the study of γ-rays, the forces of change, the formative process in the Galaxy and interstellar clouds, rapid expansion processes, explosions, the largest energy transfers, and very high energy particle acceleration are all examined directly. If the mission opportunities that are discussed here come into being, the results that will be obtained, particularly in combination with those from other areas of astronomy, will provide an entirely new look at the Universe.

This section will concentrate on the γ-ray astrophysics results beyond the solar system with the latter subject being included in the section on the solar system. Gamma ray bursts will be discussed elsewhere and will not be considered here. The remainder of this section will be divided into three parts: our galaxy, extragalactic radiation, and future prospects for γ-ray astrophysics.

B. OUR GALAXY

The γ-ray sky is dominated by radiation from the galactic plane, which is generally assumed to be the sum of diffuse radiation and unresolved point sources. The point source contribution would for the most part appear diffuse to the high energy γ-ray satellite instruments that have flown thus far because the angular resolution of these instruments for individual photons has been only one to a few degrees, or poorer, depending on energy. The source of the true diffuse radiation has been assumed to be cosmic ray interactions since, assuming cosmic rays pervade the Galaxy, they necessarily produce high energy γ-rays as they interact with the interstellar matter and photons. Substantial work has been performed on the calculation of the source functions for these various γ radiations and the intensity to be expected in the vicinity of the solar system. (For a general review see Chapter 5 of Fichtel and Trombka 1981.) Several recent important developments have occurred to aid in the study of this problem, including the detailed results of high energy galactic γ-radiation obtained with the COS-B satellite (Mayer-Hasselwander et al., 1982), the medium energy results of Agrinier et al., (1981), Graser and Schonfelder (1982), and Bertsch and Kniffen (1983), further evaluations of the 21 cm radiation in the galaxy, and hence the atomic hydrogen density distribution, additional CO line observations from which molecular hydrogen column densities are deduced, the high photon density estimate for the inner galaxy which affects the Compton radiation and the electron spectrum in this region, the current longer estimate of the galactic cosmic ray lifetime, further evidence supporting the galactic arm concept, and improved theoretical calculations on the nucleon-nucleon source function. It should be noted that the difficulty in normalizing the molecular hydrogen column density deduced from the CO measurements in an absolute manner remains a problem, but one which can at least be constrained.

The variations of the cosmic ray density in the galactic plane with position in the galaxy is a complicating factor. However, for galactic latitudes where the local contribution may be expected to dominate, $|b|$ greater than 10 or 15, the cosmic ray density as a function of position in the galactic plane presumably does not vary much. For this case, since the scale height of the cosmic rays is expected to be large compared to that of matter, a good approximation for the cosmic ray, matter interaction contribution to the γ-ray diffuse radiation is probably obtained by using a constant cosmic ray density, which allows the direct use of atomic and molecular hydrogen column densities. If the point source contribution is small and if account is taken of the Compton contribution, it should be possible to obtain a good agreement using the matter column densities directly as shown by Strong et al., (1982), and Lebrun et al., (1982). It should also be possible to use this simplified approach successfully at intermediate longitudes (~ 60 to ~ 100 and ~ 250 to ~ 280), where regions which are at galactic radii similar to the Earth are predominantly being viewed as shown, for example, by Arnaud et al., (1982) and Lebrun and Paul (1983), for the ($60 < l < 100$) region. Other attempts to correlate the γ radiation with matter include Issa et al., 1981, Lebrun et al., (1983), and Riley et al., (1983), with generally reasonable results. The more general case wherein the cosmic ray density is assumed to be variable is treated by Fichtel and Kniffen (1984). They show that within uncertainties the present γ-ray results are in agreement with a coupling of the cosmic ray density in the plane with the broad galactic features as predicted by theory. Other analyses which support at least a general galactic radial gradient of the cosmic ray density include those of Dodds et al., (1975), Kniffen et al., (1977), Issa et al., (1981), and Hermsen et al., (1982).

With regard specifically to the galactic center, Blitz et al., (1984) show that the γ-ray flux from the central few hundred parsecs of the Milky Way is nearly an order of magnitude smaller than the value expected from the H_2 masses generally estimated to be present from CO emission data. This result implies that in the galactic center, either the cosmic ray density is very anomalously low or that molecular hydrogen is much less abundant than estimated. These authors argue that circumstantial evidence favors a low H_2/CO abundance as the source of the γ-ray deficiency.

A recent summary of the COS-B γ-ray results by Hermsen (1983) lists about 25 localized excesses as potential point sources, in addition to the Orion and ρ Oph clouds (Bloemen et al. 1984). In spite of considerable effort, there are no new clear identifications of galactic sources as point sources, as opposed to broad localized sources such as clouds, beyond those known a decade ago, namely the Crab and Vela pulsars, Cygnus X-3, and very likely Geminga (195, 5). New observations are needed to determine if these point-like excesses are objects, such as neutron stars, supernovae, or even black holes, or broad features.

New results on the Crab γ-ray source have been reported, e.g., Graser and Schonfelder (1982), Wills et al., (1982), and Mahoney et al., (1984), the latter two showing a definite interpulse flux. The Crab γ-ray results are summarized in detail in an article by Schönfelder (1983).

During the last two years, Samorski and Stamm (1983), Lloyd-Evans et al., (1983), and Dowthwaite et al., (1983), have reported the strongest evidence yet for γ-rays from Cygnus X-3 with energies in the 10^{15} to 5×10^{16} eV range from ground level detector systems. The source shows the same 4.8 hr period reported by Lamb et al., (1977) in the 35 MeV to several hundred MeV range from SAS-2 data. These results suggest a continuous spectrum from the soft γ-ray range to over 10^{16} eV with an integral power-law index of about -1.1. Although the COS-B γ-ray detector did not detect the source (Bennett et al., 1977), those observations were during a period of low X-ray intensity. The positive detection of γ-rays with energies $\gtrsim 10^{16}$ eV from Cygnus X-3 could be very important to the understanding of the acceleration and supply of cosmic rays.

There have been reports of discrete γ-ray absorption or emission lines from Hercules X-1, SS433, and γ-ray bursts and of the half MeV annihilation line from the general direction of the galactic center. (For a summary, see Ramaty 1984). Because of the difficult nature of the experiments, the low intensities, differences between the results, and the sensitivity of the data to analysis assumptions, most of these reports should probably be viewed as needing confirmation. The two most recent attempts to measure the galactic center half MeV line (Leventhal et al., 1982, and Paciesas et al., 1982), set upper limits below any reported positive result (Haymes et al., 1975, Leventhal et al., 1978, Albernhe et al., 1981, and Riegler et al., 1981). There is, of course, the possibility of time variability.

Jacobson (1983) has reported the detection of the 1.809 MeV ^{26}Al line from the more intense central part of the galactic plane with an intensity of $(4.7\pm1.0)\times10^{-4}$ photons $cm^{-2}S^{-1}rad^{-1}$. Because of the long half-life of 7.4×10^{-5} years, this emission would be expected to be diffuse rather than associated with a source region since the nuclei would have had time to spread through the interstellar medium. Ramaty and Lingenfelter (1983) had predicted the presence of this line on the basis of ^{26}Al being synthesized in supernovae and estimated the intensity from the direction of the galactic center region to be 10^{-4} photons $cm^{-2}S^{-1}rad^{-1}$. In view of the uncertainty of some of the parameters and the likelihood that novae make a similar contribution, the agreement between the prediction and observation is reasonable. This observation may thus represent direct evidence of nucleosynthesis in our galaxy within the last million years.

In general, the study of astrophysical γ-ray lines outside the solar system, at present, seems to require more definitive measurements with greater sensitive and lower background before many firm conclusions are possible.

C. EXTRAGALACTIC RADIATION

Beyond our Galaxy, γ-ray emission has already been seen from four active galaxies, two Seyfert galaxies, one radio galaxy, and a quasar. For at least two of these, 3C273 and NGC 4151, more energy is emitted in the γ-ray region (E >0.1 MeV) than in the X-ray, optical, or radio regions. The photon intensities are still relatively low, of course, because of the large energy per photon in the γ-ray region. No normal galaxy, other than our own, has been seen in γ-rays, but this result is not surprising on the basis of the emission level of our own Galaxy. The γ-rays observed from the Seyfert galaxies NGC 4151 and MCG 8-11-11 are in the low-energy γ-ray region (Auriemma et al., 1978; Coe et al., 1980; Gursky et al., 1971; Ives et al., 1976; Meegan and Haymes 1979; Mushotsky et al., 1978; Paciesas et al., 1977; Perotti et al., 1979; Perotti et al., 1981; Schönfelder et al., 1980; White et al., 1980) and only upper limits to the high energy γ-radiation exist (Bignami et al., 1979). The spectra are similar in that both show a very marked increase in the spectral slope in the energy region near 1 MeV. The several measurements in the γ-ray region for NGC 4151 were made at different times, and, assuming no significant errors in the data, clearly show a time variability. For five other Seyferts (and also several other emission-line galaxies) upper limits derived from the SAS-2 γ-ray data (Bignami et al., 1979) are substantially (more than an order of magnitude) below an extrapolation of the power law X-ray spectra (Mushotsky et al., 1979), suggesting that a sharp spectral change in the low-energy γ-ray region may be a general feature of these galaxies. Pollock et al., (1981) come to a similar conclusion.

Turning to quasars, 3C 273 is the brightest X-ray quasar and is the only quasar which has been clearly identified as a source of γ-rays (Swanenburg et al., 1978). The differential energy spectrum of 3C 273 steepens sharply from the X-ray range to the γ-ray region, with the slope of the differential energy spectrum changing from 1.4 in the hard X-ray region to 2.7 in the high-energy (E>50 MeV) γ-ray region. The change in spectral shape between the hard X-ray and γ-ray region seen for 3C 273 is similar to that suggested for the Seyfert galaxies for which data exist. The COS-B instrument has observed 3C 273 in γ-rays in July 1976, June 1978, and June-July 1980, and no significant variation in the γ-ray fluxes among the observation was observed (Bignami et al., 1981).

Centaurus-A (NGC 5128), generally believed to be the closest radio galaxy, is the only radio galaxy that has been seen in γ-rays. It has now been observed in all frequency bands from radio through low-energy γ-rays (e.g., Baity et al., 1981) and, although γ-ray emission is not seen in the 30 to 10^3 MeV region (Bignami et al., 1979 and Pollock et al., 1981), a strong indication of very high energy ($E>3x10^{11}$ eV) γ-ray emission has been found (Grindlay et al., 1975). The observations of the radiation from CEN-A in the X-ray region through the very high energy γ-ray region again suggest a steepening of the spectral slope similar to NGC 4151, MCG 8-11-11, and 3C 273.

A diffuse celestial radiation, which is isotropic at least on a coarse scale, has been measured from the soft X-ray region to about 150 MeV, at which energy the intensity falls below that of the galactic emission for most galactic latitudes. The first indication that diffuse celestial radiation extended from the X-ray region into at least the low-energy γ-ray (1 MeV) portion of the spectrum was reported by Arnold et al., (1962). At energies above 10 MeV, the first measurements related to an extragalactic diffuse radiation were those of Kraushaar and Clark (1962), whose upper limits from Explorer 11 provided an experimental refutation of the steady-state theory of cosmology. The first suggestion of a diffuse high-energy flux came from the OSO-3 satellite experiment (Kranshaar, et al., 1972); however, it was data from the SAS-2 high-energy γ-ray experiment that clearly established a high-energy extension of the diffuse radiation with a steep energy spectrum above 35 MeV (Fichtel et al., 1977). A recent reanalysis of the SAS-2 data which included galaxy counts as a tracer of the interstellar matter has been performed by Thompson and Fichtel (1982) and has added support to the concept of the spectrum being quite steep (having a power low index of about 2.4) in the energy region above 35 MeV.

A large number of theories predicting a diffuse γ-ray background have appeared in the literature over the years. With the measurements of the spectrum and intensity which now exist, most of these seem not to be likely candidates for the majority of the diffuse radiations (see, for example, Fichtel and Trombka, 1981). Two possibilities seem to remain at present. One of these involves a baryon-symmetric universe, containing superclusters of galaxies of matter and others of antimatter. The annihilation of nucleons and antinucleons at the boundaries (Stecker, Morgan, and Bredekamp, 1971) produces the γ-rays. The other possibility is the sum of the radiation from point sources, and particularly active galaxies, integrated over cosmological times (e.g., Strong and Worrall, 1976; Bignami et al., 1978; Schönfelder, 1978; and Grindlay, 1978). Using the data on the few known objects, Bignami et al., (1979) and Fichtel and Trombka (1981) conclude that it is quite conceivable that Seyfert galaxies and Quasars could account for the diffuse γ radiation using conservative evolutionary models. Leiter and Boldt (1982) and Boldt and Leiter (1984) have proposed a model based on supermassive Schwarzschild black holes with accretion disks radiating near the Eddington luminosity limit. The authors believe that, if this theory is correct, there would be detectable variations in the diffuse radiation in small elements of the sky (10 deg^2) over several days in the 1/2 to 3 MeV region, giving a specific test for this theory.

D. FUTURE PROSPECTS FOR GAMMA RAY ASTRONOMY

The major satellite opportunities for γ-ray astronomy in the 1980's are the GAMMA-I and the Gamma Ray Observatory. The information to be obtained from these instruments may be supplemented by results from γ-ray balloon flights and from yet to be approved spacelab and small satellite opportunities. In the more distant future, the NASA Space Station should provide opportunities to fly quite large γ-ray instruments which might be refurbished or reconfigured in space.

I. GAMMA-I

The next γ-ray satellite expected to fly is GAMMA I, which should be launched in about two years on a Soviet satellite. It is similar to SAS-2 and COS-B in the sense that its central element is a multilayer spark chamber system, triggered by a directional counter telescope, and surrounded on the upper end by an anticoincidence system. The upper spark chamber system is a twleve-level wide gap Vidicon system. The directionality of the electrons is determined by a time-of-flight system rather than a directional Cerenkov counter. The sensitive area is about 1600 cm^2 or about 2.7 times that of SAS-2 or COS-B. The area solid angle factor is about the same, because the viewing angle is smaller.

II. The Gamma Ray Observatory

The Gamma Ray Observatory (GRO) is an approved NASA mission with a launch tentatively planned for 1988. There are four instruments covering the energy range from 0.03 MeV to 3×10^4 MeV, with a major increase in sensitivity over previous satellite experiments. It is advantageous to combine the instruments into one mission not only because they place similar requirements on a spacecraft, but also because of the great scientific value of studying the entire γ-ray spectrum of any object at the same time to examine in detail the nature of time variations. The combined compliment of instruments to be incorporated into the Gamma Ray Observatory is expected to have the capability to carry out the following:

i. A survey of γ-ray sources and diffuse emission with sensitivities around 10^{-5} photon cm^{-2} sec^{-1} and energy resolution around 10 percent at energies between 0.1 and 30 MeV.

ii. A survey of high energy γ-ray sources and diffuse emission with a point source sensitivity of 10^{-7} photon cm^{-2} sec^{-1} or better, angular resolution of about 0.1° for strong sources, and energy resolution around 15% at energies above 10^2 MeV.

iii. Detection and identification of nuclear gamma lines with an energy resolution of 4 percent and sensitivity of the order of 5×10^{-5} photon cm^{-2} sec^{-1}.

iv. Observations of γ-ray bursts, including studies of their spectral and temporal behavior.

The Gamma-Ray Observatory will be a shuttle-launched, free-flyer satellite. The nominal circular orbit will be about 400 kilometers with an inclination of 28.5°. Celestial pointing to any point on the sky will be maintained to an accuracy of ± 0.5°. Knowledge of the pointing direction will be determined to an accuracy of 2 arc minutes. Absolute time will be accurate to better than 0.1 milliseconds to allow precise comparisons of pulsars and other time varying sources with observations at other wavelengths from ground observations and other satellites. For further information on the instruments to be flown on GRO, see Kniffen et al., (1981).

III. Space Station

NASA is now considering a Space Station which would be a manned spacecraft permanently orbiting the earth and capable of performing a variety of scientific and operational missions. High energy γ-ray astronomy is certainly among the scientific desciplines which would be able to benefit very significantly from such an opportunity. With the γ-ray sky surveyed in some depth with the GRO, it would, for example, be possible to concentrate on the detailed features of discrete sources and to study carefully limited regions such as clouds, galactic arms, and nearby galaxies.

References

Agrinier, B., et al.: 1981, 17th International Cosmic Ray Conf. 9, p. 72.
Alberne, F., et al.: 1981, Astron. Astrophys. 94, p. 214.
Arnaud, K. Pei, Li Ti, et al.:1982, Mon. Not. R. Ast. Soc. 201, p. 745.
Arnold, J. R., Matzger, A. E., Anderson, E. C., and Van Dilla, M. A.: 1982, J. Geophys. Res. 67, p. 4878.
Auriemma, G., et al., 1978, Ap. J. Letters 221, p. L7.
Bennett, K., et al.: 1977, Astron. Astrophys. 59, p. 273.
Bertsch, D. L., and Kniffen, D. A.: 1983, Ap. J. 270, p. 305.
Bignami, G. F., et al.: 1981, Astron. Ap. 93, p. 71.
Bignami, G. F., Fichtel, C. E., Hartman, R. C., and Thompson, D. J.: 1979, Ap. J. 232, p. 649.
Bignami, G. F., Lichti, G. G., and Paul, J. A.: 1978, Astron. and Astrophys. 68, p. L15.
Blitz, L., Bloemen, J. B. G. M., Hermsen, W., and Binia, T.: 1984, "The Gamma-Ray Deficit Toward the Center", preprint.
Bloemen et al.: 1984, Astron. Astrophys. 139, 37.
Boldt, E., and Leiter, D.: 1984, Ap. J. 276, p. 427.
Coe, M. J., et al.: 1980, M.N.R.A.S.
Dodds, D., Strong, A. W. and Wolfendale, A. W.: 1975, Mon. Not. R. Astr. Soc. 171, p. 569.
Dowthwaite, J. C., et al.: 1983, Astron. Astrophys. 126, p.1.
Fichtel, C. E., et al.: 1977, Ap. J. 217, p. L9.
Fichtel, C. E., and Trombka, J. I.: 1981, "Gamma Ray Astrophysics: New Insight into the Universe", NASA SP-453.
Fichtel, C. E., and Kniffen, D. A.: 1984, Astron. Astrophys. 134, p. 13.
Graser, U., and Schönfelder, V.: 1982, Ap. J. 263, 677.
Grindlay, J. E.: 1978, Nature 273, p. 211.
Gursky, H., et al.: 1971, Ap. J. Letters 165, p. L43.
Haymes, R. C., et al.: 1975, Astrophys. J. 201, p. 593.
Hermsen, W.: 1983, Space Sci. Rev. 36, p. 61.
Hermsen, W., and Bloemen, J. B. G. M.: 1982, Leiden Workshop on Southern Galactic Surveys.
Issa, M. R., Riley, P. A., Strong, A. W., Wolfendale, A. W.: 1981, J. Phys. G. 7, p. 656.
Ives, J. C., Sanford, P. W., and Penston, M. V.: 1976, Ap. J. Letters 207, p. L159.
Jacobson, A.: 1983, Bulletin of the American Physical Society 28, p. 666.
Kniffen, D. A., Fichtel, C. E., and Thompson, D. J.: 1977, Ap. J. 215, p. 765.
Kniffen, D., et al.: 1981, "The Gamma-Ray Observatory Science Plan."
Kraushaar, W. L., and Clark, G. W.: 1962, Phys. Rev. Letters 8, p. 106.
Kraushaar, W. L., et al.: 1979, Ap. J. 233, p. 510.
Lamb, R. C., et al.: 1977, Ap. J. Letters 212, p. L63.
Lebrun et al.: 1982, Astron. Astrophys. 107, 390.
Lebrun et al.: 1983, Ap. J. 274, 231.
Lebrun, F., and Paul, J. A.: 1983, Astrophys. J. 266, p. 276.
Leiter, D., and Boldt, E.: 1982, Ap. J. 260, p. 1.
Leventhal, M., MacCallum, C. J., and Stang, P. D.: 1978, Astrophys. J. 225, p. L11.
Leventhal, M., et al.: 1982, Astrophys. J. 260, p. L1.
Lloyd-Evans, et al.: 1983, Nature 305, p. 784.
Mahoney, W. A., Ling, J. C., and Jacobson, A. S.: 1984, Ap. J. 278, p. 784.
Mayer-Hasselwander, H. A., et al.: 1982, Astron. Astrophys. 105, p. 164.
Meegan, C.A., and Haymes, R.C.: 1979, Ap. J., 233, 510.
Mushotzky, R. F., Boldt, E. A., Holt, S. S., and Serlemitsos, P. J.: 1979, Ap. J. 232, p. L17.
Mushotzky, R. F., Holt, S. S., and Serlemitsos, P. J.: 1978, Ap. J. Letters 225, p. L115.

Paciesas, W. S., Mushotzky, R. F., and Pelling, R. M.: 1977, N.N.R.A.S. 178, p. 23.
Paciesas, W. S., et al.: 1982, Astrophys. J. 260, p. L7.
Perotti, F., et al.: 1979, Nature 282, p. 484.
Perotti, F., et al.: 1981, Ap. J. Letters 247, p. L63.
Pollock, A.M.T., et al.: 1981, Astron. Astrophys. 94, 116.
Ramaty, R., and Lingenfelter, R. E.: 1977, Ap. J. 213, p. L5.
Ramaty, R.: 1984, 4th Moriond Astrophysics Meeting, LaPlagne, France
Riegler, G. R., et al.: 1981, Astrophys. J. 248, p. L13.
Riley, P. A., et al.: 1984, Mon. Not. R., Ast. Soc. 206, p. 423.
Samorski, M., and Stamm, W.: 1983, Ap. J. 268, p. L17.
Schönfelder, V., 1978, Nature 274, p. 344.
Schönfelder, V., Graml, F., and Penningsfeld, F. P.: 1980, Ap. J. 240, p. 330.
Schönfelder, V.: 1983, Adv. Space Res. 3, p. 59.
Stecker, F. W., Morgan, D. L., and Bredekamp, J.: 1971, Phys. Rev. Letters 27, p. 1469.
Strong, A. W., and Worrall, D. M.: 1976, J. Phys. A: Math. Gen. 9, p. 823.
Strong, A. W., et al.: 1982, Astron. Astrophys. 115, p. 404.
Swanenburg, B. N., et al.: 1978, Nature 275, p. 298.
Thompson, D. J., and Fichtel, C. E.: 1982, Astron. Astrophys. 109, p. 352.
White, R. S., et al.: 1980, Nature 284, p. 608.
Wills, R. D., et al.: 1982, Nature 296, 1.

2. X-RAY ASTRONOMY

K A Pounds

A. INTRODUCTION

The period leading up to the last IAU General Assembly was dominated in X-ray Astronomy by the results from the Einstein Observatory. This first application of a large, satellite-borne, high resolution X-ray telescope to the study of cosmic sources had led, by the end of orbital operation in April 1981, to the detection of X-ray fluxes from a wide variety of astronomical objects and the full maturing of X-ray Astronomy. During the past three years a strong scientific output has continued to flow from the analysis of the more than 5600 separate Einstein observations, many of which are now widely available via the Einstein Data Bank. In sharp contrast, the peak of activity and scientific output from Einstein will apparently be followed, in the United States, by almost a decade with no new X-ray satellite. Fortunately, successful launches have been achieved in both the Japanese and European programmes, with Tenma launched in February 1983 (to join Hakucho in orbit) followed by EXOSAT three months later. Both these new missions have attracted world-wide interest and observing time on both is at a premium. Of a range of instruments on board the Tenma satellite the most productive is proving to be the large ($\sim$800 cm^2) array of gas scintillation proportional counters. These detectors, flown for the first time on Tenma (and EXOSAT), provide a factor-of-two increase in spectral resolution over the more conventional proportional counters, an advantage which is particularly valuable in resolving the key emission line and absorption features associated with the K-shell of iron, near $\sim$7 keV. One particularly exciting result reported from this Tenma instrument was an apparent absorption feature at $\sim$4.1 keV in X-ray burst spectra from the source 2S 1636-536 (later supported by EXOSAT observations), the feature being attributed to red-shifted iron absorption in the strong gravitational field close to the neutron star burst source! At the time of writing (December 1984) a power storage problem is limiting the operational efficiency of Tenma, but the scientific payload continues to function well.

The unique orbit of EXOSAT is proving to be a major asset, with continuous observations of more than 70 hours being possible, uninterrupted by Earth eclipse. In addition, the apogee over N. Europe allows real-time operation of EXOSAT, analogous to observations with a ground-based telescope. The scientific payload consists of an array of proportional counters (effective area $\sim$1600 cm^2), a small gas scintillation proportional counter ($\sim$90 cm^2) and 2 imaging telescopes. Good data have been obtained from all the EXOSAT instruments, but problems with some of the focal plane detectors and the mechanical drive are currently preventing the use of the second telescope and of either transmission grating. Given the considerable scientific potential and current scarcity of X-ray spectra of corresponding resolution the latter is particularly disappointing. Nevertheless, to date over 1500 EXOSAT observations have been carried out, the programme continues to be heavily over-subscribed and it is hoped to continue full operation until satellite re-entry in 1987.

An expanding effort in X-ray astronomy is also reported from the Soviet Union with the Salyut-7 orbital station carrying several experiments for X-ray astronomy. Of particular interest is the RT-4M instrument, a grazing incidence telescope with a non-imaging flow gas proportional counter as focal plane detector. The telescope optics are of two-fold nested Wolter type I design, with a maximum entrance aperture 24 cm wide and a focal length of 66 cm. The reflecting surfaces are coated with gold (outer mirror) and nickel (inner mirror) to provide an operating wavelength band between 44 A and 100 A over a field of view of 45' by 3°. The detector entrance aperture is divided into three parallel sectors, viewing different fields (Mandelstam et al, 1983). Results from this and other Salyut instruments are eagerly awaited. In the following section, a (necessarily personal and incomplete) review is made of progress in X-ray astronomy over the period since the Commission's 1982 report.

B. SCIENTIFIC ACHIEVEMENT

(i) Stellar emission.

Einstein observations have provided an outline of the X-ray emission properties of stars throughout the H-R diagram (e.g. Vaiana et al, 1981; Swank et al, 1981; Caillault and Helfand, 1982; Montmerle et al, 1983), but a detailed understanding has so far been limited by the small sample of higher resolution spectra obtained with the Einstein grating or crystal spectrometers (e.g. Mewe et al, 1981; Swank and Cassinelli, 1983; Swank and Johnson, 1982; Vedder and Canizares, 1983). Evidence that stellar surface activity (as an effect of surface convection) does indeed cease along the main sequence is provided by the discovery that the X-ray luminosity decreases from a mean of 10^{29} erg s^{-1} at F5 to upper limits of $\sim 10^{27}$ erg s^{-1} at A0 (Schmitt et al, 1985). Early type stars are readily detectable as X-ray sources but it is not yet clear whether the emission is from coronae, shocked winds, or some other cause. Recent models have been discussed by Waldron 1981), Lucy and White (1980) and Lucy (1982), among others. Of particular value in determining the evolution of stellar X-ray emission is the study of star clusters (of common age), such as those in the Hyades (Stern et al, 1981) and Pleiades (Caillault and Helfand, 1985) Clusters. Preliminary indications are that for solar type stars the fall in X-ray output (at least up to $\sim 10^{9}$ yr) is slower than $t^{-0.5}$.

The significance of RS CVn stars as highly variable X-ray sources was underlined in an analysis of short-lived Ariel-5 transients (Pye and McHardy, 1983), who found 6 of 11 optically identified X-ray transients to be RS CVn systems. The long EXOSAT observing time is well matched to the study of such systems. For example, Barstow (1985) has observed the RS CVn binary HR1099, during which an X-ray outburst was followed through a peak (of 3 x $10^{34.5}$erg s^{-1}, 2 - 6 keV) and subsequent decay over $\sim$ 2 hours. Interesting, non-periodic modulations were also clearly seen. Other recent EXOSAT stellar observations have covered the contact binary VW Cephei, revealing a structured corona (Vilhu and Heise, 1985), the 12 day rotating K2 V star ϵ Eri (Horstman et al, 1985), and the nearby triple star system Algol. Observations of Algol during the secondary optical eclipse (B star in front of K Companion) have provided lower limits to the size of the K star's X-ray corona, while ME and grating spectra revealed a thermal spectrum with Fe XVII and XVIII emission lines and a continuum temperature of 2.4 x 10^{7} K (Culhane et al, 1985). A large X-ray flare was also seen, with an effective temperature of 6 x 10^{7} K and iron emission up to Fe XXV (Parmar et al., 1985). Further detections of soft X-rays from hot white dwarf stars have been reported from Einstein observations (Kahn et al, 1984). EG 187, Gr 288, Gr 289 and LB 1663 have spectra consistent with DA white dwarf photospheric temperatures in the range 3 - 6 x 10^{4} K. Complementary IUE data show the need for some atmospheric opacity, with a H:He ratio in the range (2-600) x 10^{-5}. The implication of these detections is that many more white dwarf stars should be available for study in future XUV surveys.

(ii) Supernova remnants.

The Einstein Observatory has successfully mapped more than 40 galactic supernova remnants and has detected 25 and 6, respectively, in the Large and Small Magellanic Clouds (Mathewson et al, 1983 and references therein). The best data naturally arise for the relatively young galactic remnants and detailed studies have been reported on a number, including Tycho's SNR (Seward et al, 1983), SN 1006 (Pye et al, 1981), Cas A (Stewart et al, 1983) and most recently RCW 86 (=SN185A.D.), observed with Einstein by Pisarski et al (1984). Other high quality X-ray maps have been obtained with Einstein for several older, but bright, remnants, in particular Puppis A (Petre et al., 1982) and Cygnus Loop (Ku et al., 1984). These observations have generally supported the view that the X-ray emission from these remnants is thermal, from shock heated gas (of the supernova itself and the surrounding interstellar medium) at temperatures $\sim$1 - 10 keV. Spatial structures have been compared with radio and optical features. X-ray spectra taken with the Einstein solid state spectrometer (see review by Holt, 1983) have been analysed in terms of thermal models, but, in general, the lack of spatial resolution and uncertainties as to the state of equilibrium of the emitting plasma have limited the reliability of the deduced mass and element abundance distributions. The high resolution crystal spectrometer data from Puppis A (Winkler et al, 1981) have shown the great promise of future, more sensitive observations. EXOSAT X-ray maps of Puppis A (Aschenbach et al, 1985), SN 1006 at higher resolution (Jones et al, 1984) and Cygnus Loop (Arnaud et al, 1985) are now adding to this field.

An increasing number of core-halo remnants have been reported, some with an associated neutron star. Four pulsars have been found in SNR to date, while Einstein observations have added a further six with point sources too weak to allow pulsations to be detected (Seward, 1983). In addition to the Crab Nebula, MSH15-52 has been found to contain an X-ray/radio pulsar with the highest known rotational energy loss rate (Seward and Harnden, 1982) and a further Einstein observation showed CTB109 to contain a long period ($\sim$7 sec) X-ray pulsar (Fahlman and Gregory, 1983). Other remnants found from Einstein observations to have weak point X-ray sources include RCW103, 3C58, CTB80 and W28 (Tuohy and Garmire, 1980; Becker et al, 1982; Andrews et al, 1983). A recent EXOSAT addition to this sub-group is the southern SNR PKS1209-52 (Kellett, 1985).

(iii) Galactic X-ray binaries with compact components.

By their nature, both Tenma and EXOSAT are well suited to detailed study of the bright X-ray binaries containing white dwarf, neutron star or (possibly) black hole components. Over 100 of these sources now have identified optical counterparts (Bradt and McClintock, 1983) and it is conventional to divide them according to system mass, nature of the compact component, or location (e.g. globular cluster). An excellent review of this broad field is now publishd in Lewin and van den Heuvel (1983) while other reviews may be found in a special volume of Physica Scripta (vol. T7, 1984).

Recent progress in the study of massive X-ray binaries (sometimes called class I) where the components are an early type star of $M > 10M_\odot$ and a neutron star has concentrated on the details of mass loss, stellar wind opacity, pulse structure etc., since this class is basically better understood than the low mass systems (most of which are not yet actually established to be binary). Continuing Hakucho and Tenma observations of Vela X-1 have now shown the spin-down, noted in the Commissions's report 3 years ago, to have again reversed, indicating a complex interaction between the accreting material and the neutron star (Nagase et al, 1984 and references therein). The rate of change of the pulsar period in LMC X-4 is also in strong discord with a simple accretion torque model (Keller et al., 1983). Two of the best candidate X-ray binaries to contain a black hole, Cygnus X-1 and GX 339-4 remain 'unproven', though new high quality data has been obtained on the short term X-ray variability (e.g. Maejima et al, 1984; for GX339-4 and Meekins et al, 1984 and Page, 1985; for Cyg X-1). In addition, the analysis of SAS-3 data has revealed an intriguing series of intensity 'dips' generally occurring near superior conjunction of the X-ray source (Remillard and Canizares, 1984). Identification of several 'classical' X-ray transient sources with Be stars (Warwick et al, 1981; Rappaport and van den Heuvel, 1982; Steiner et al, 1984) has been followed by further detailed optical and X-ray studies and it now seems likely that it is the strong equatorial mass loss of the Be star that fuels the neutron star accretion over the large (and probably eccentric) orbits of such long period binaries as 1145-61 (187d; Watson et al., 1981), 4U0115+63 (24d) and the highly luminous LMC transient A0538-66 (16.6d; Charles et al, 1983).

The major recent advance in the understanding of low mass (or class II) systems has been the establishment of a binary period for several cases, typically of a few hours (though there are notable exceptions, e.g. Cygnus X-2 at 9.8 days). Thus 2A 1822-371 has been found to have a binary period of 5.6 h (Cowley et al, 1982 and references therein) and 2A 2129+470 one of 5.2 h (McClintock et al, 1982). The proposition that X-ray bursters are a sub-class of low mass binaries (having a lower mean accretion rate or weaker neutron star magnetic field?) has been justified by the identification of binary periods for 1636-53 (~4h, from optical photometry; Pedersen et al., 1981), 1915-05 (~50m, from X-ray dips; White and Swank, 1982), 1755-33 (4.4h, from X-ray dips; White et al., 1984), 1659-29 (7.1h, from X-ray dips; Cominsky and Wood, 1984) and 1254-69 (~3.9h). The latter result is perhaps the most interesting in that the period is found from both optical photometry (Motch et al., 1984) and X-ray dips (Courvoisier et al., 1985). X-ray burst sources have continued to receive attention in the Tenma and EXOSAT programmes. Use of large detector arrays, with improved spectral and timing resolution, are yielding high quality data which are helping to clarify details of the mechanism and geometry of the emission region. The previously established model for the more common type I bursts, involving thermonuclear fusion on the surface of an accreting neutron star, remains intact, but only now are first attempts being made to calculate the radiation transfer properties in the neutron star photosphere. The intriguing detection of an apparently red-shifted absorption line by Tenma (Waki et al, 1984) and EXOSAT observers (Turner et al, 1984) has been mentioned previously. Correlated optical and X-ray bursts have been detected for 1735-44 (Grindlay et al., 1978), 1636-536 (Matsuoka et al., 1984) and - most recently - for MXB1659-29 (Matsuoka et al, 1984).

Probably the most significant progress in the study of X-ray binaries during the past 3 or 4 years has been of those systems involving accretion onto a white dwarf star, the cataclysmic variables. Because of their proximity (~50-500 pc) cataclysmic variables are excellent objects in which to study the physics of accretion onto a compact object. They are also readily studied by X-ray techniques and some 50 have now been so detected (see review by Cordova and Mason, 1983). Ariel-5, HEAO-1 and Einstein observations have been actively followed up in the EXOSAT programme. An important variable in such systems, which are believed to involve a low mass star and a white dwarf, is the magnetic field strength of the dwarf star. Twenty or so AM Her (or polar) and DQ Her (intermediate polar) type stars have strong fields which, in the former case, lock the white dwarf rotation to the binary period. Most cataclysmics, such as SS Cyg and U Gem, however, show no evidence of magnetic fields. One of the earliest EXOSAT observations showed GK Persei to have a strongly modulated X-ray flux, establishing 351 sec as the white dwarf rotation period and GK Per as the sixth known 'intermediate polar'. Extended observations further showed the depth of modulation to be independent of mean flux (accretion rate), indicating an extended, thin polar cap emission region (Watson et al, 1985a). Repeated EXOSAT observations of SS Cygni between and during optical outbursts have confirmed the anti-correlation of hard and soft X-ray flux (Watson et al, 1985) and are providing detailed information on the emission in such a non-magnetic white dwarf system, which appears to involve thermal radiation (of the soft component) from the surface of the white dwarf and a (harder) component from an overlying hot corona (King, 1985). Rapid variability of the quiescent flux ($\sim$10 sec) was seen during quiescence, consistent with the earlier HEAO-1 observations of SS Cyg and U Gem (Cordova et al, 1984).

(iv) Globular clusters

An important advance in the study of globular clusters has been made using Einstein data which have shown diffuse X-ray emission in the clusters W Cen, M22 and 47 Tuc (Hartwick et al, 1982). The presence of diffuse gas has previously been looked for without success in optical and radio searches, and probably arises from the winds of red giant stars. In the continuing effort to understand the nature of the well established luminous X-ray components, accurate ($\sim$1 arc sec) positions have allowed Grindlay et al (1984) to determine the most probable masses in 8 globular clusters to be $\sim$1.5 $M_{\odot}$, consistent with them being accreting neutron star binaries. More recently a survey of 71 galactic globular clusters (Hertz and Grindlay, 1983) has revealed a bimodal X-ray luminosity function, with a fainter group of sources ($< 10^{34.5}$ erg s^{-1}), believed to be cataclysmic variables (but see Verbunt et al, 1984). The fact that the presence of both high and low luminosity X-ray sources is correlated with the timescale for binary formation by tidal capture is interpreted as evidence for the importance of this process in globular clusters.

(v) Normal galaxies

The Einstein map of M31 contained $\sim$100 discrete sources and was able to distinguish between the X-ray emission from the nucleus, from a bulge population and from a disc population (e.g. Crampton et al, 1984). Repeated observations with EXOSAT have shown substantial changes in the source distribution, presumably due to intrinsic variability and - in part - to the 'softer' response of the EXOSAT telescopes. Although few sources have been optically identified, the high X-ray luminosities strongly suggest a population of X-ray binaries and supernova remnants, as dominate our own Galaxy. The Large and Small Magellanic Clouds were well mapped by Einstein and a programme of source identification in the SMC, based on precise Einstein HRI positions, has now been extended with EXOSAT, with preliminary results yielding 2 foreground stars, an HII nebula, the SNR N 76 and a (background) Seyfert galaxy (Jones et al., 1985). Although the mean X-ray luminosity of the LMC sources is significantly higher than for the Galaxy, the number of luminous stellar X-ray systems per unit mass is similar. Other nearby galaxies which have now been well studied include M82 and M81. Watson et al (1984) found an extended region of emission in M82 with a bright component centred on the nucleus, together with a broader region correlated with Hα filaments and the optical halo, interpreted by the authors in terms of a starburst model. The main X-ray feature of M81 is a luminous X-ray nucleus, a property shared with other nearby spirals M100, M83 and M33 and, as recently reported, M51 (Fabbiano et al, 1985). In summary, the observations of a sample of nearby spiral galaxies suggests a base level of X-ray emission similar to our own Galaxy, probably due to 'classical' X-ray binaries and supernova remnants, but with a level of nuclear activity varying by up to 1000 times that of the Galaxy. Elliptical and lenticular galaxies reveal a greater spread of extra-nuclear X-ray luminosity, with evidence for extended (thermal) emission in some cases, particularly NGC 4636 (Stanger, 1985) and M86. The suspected high mass loss of stars in early-type galaxies and the possibility that this gas is then expelled by galactic winds are fundamental ideas given support by recent detections of hot coronae around galaxies both inside and out of clusters (Nulsen et al, 1984; Forman et al, 1984). In addition, nine extended sources were detected around mainly early-type galaxies in the core of A1367 (Bechtold et al, 1983), the measured temperatures of $\sim$ 1 keV suggesting the ejected gas may have been heated by supernova explosions. More detailed X-ray mapping and spectroscopic observations of early-type ellipticals will be of great interest in determining the amount of hot gas, the heating mechanism, extension, and confining medium (dark halo?) and whether or not (as seems likely in NGC 4636) cooling flows are important in the central regions of the galaxy. On this topic, also see Nulsen et al. (1984).

(vi) Active galactic nuclei (AGN)

Substantial progress in analysing well-exposed Einstein fields (the Medium Deep Survey) has been made and Gioia et al (1984) recently reported the results from 150 fields covering ~40 square degrees of sky. Approximately half of the 112 new X-ray sources are identified with AGN. The rapidly increasing number of AGN identifications will be of great value in improving both the AGN luminosity function(s) and estimates of their contributions to the X-ray background.

Broad-band X-ray spectra have been reported for a sample of the brighter AGN, covering a wide range of luminosity ($\sim 10^{42}$ - 10^{46} erg s^{-1}) and morphological type and Mushotzky (1984), in reviewing these mainly HEAO-1 and Einstein MPC and SSS data, reports a remarkable uniformity of spectral slope. New EXOSAT spectra, reviewed by Pounds et al. (1985), confirm many of these 'canonical' spectra (photon number index ~1.7) but also provide evidence for several steeper spectra and for spectral variability. Two particularly interesting results from this EXOSAT study are the discovery of a soft spectrum and marked variability (on a timescale ~1 hour) for the nearby Seyfert galaxy NGC 4051. The EXOSAT observations of steep, soft X-ray components in several AGN conforms with a result from the re-analysis of Einstein IPC spectra of the brighter QSOs (Elvis, private communication) and may correspond to the spectral break 'required' to join the X-ray and UVOIR continua of AGN. Einstein (Holt et al, 1980) and EXOSAT (Pounds, 1985) observations of NGC 4151 have shown the potential of X-ray spectra for studying the environment of AGN, both analyses providing evidence for a complex absorbing structure in the surrounding gas. A particularly interesting spectral feature was detected in the Einstein OGS observation of the BL Lac object PKS 2155-304. An absorption edge or trough at ~600 eV, probably attributable to ionised oxygen in the vicinity of the AGN or in an (hypothesised) intervening intergalactic medium (Canizares and Kruper, 1984).

(vii) Clusters of galaxies

Analysis of the rich Einstein data base has continued, extending the range of cluster morphologies and demonstrating a continuous distribution of loose groupings to highly evolved clusters (Jones and Forman, 1984). The Einstein observations also show that many of the clusters with a dominant central galaxy probably have central cooling flows, which could contribute substantially to the growth of this central galaxy. From a determination of the central density and temperature of 36 clusters of galaxies, Stewart et al. (1984) estimated cooling times less than the Hubble time and derived mass inflow rates of between 2 and 400 $M_\odot$ yr^{-1}. The search for distant clusters (Z > 0.5) is important in relation to the fundamental questions of cluster formation and evolution. X-ray observations should prove an effective way of discovering very distant clusters, due to their characteristic extended emission and lower confusion level (than in the optical or radio bands). McHardy and Cooke at Leicester have been using Einstein and EXOSAT to search for X-ray emission from suspected distant clusters, selected from a sample of ~150 steep spectrum radio sources with deep CCD optical follow-up, while Harris (CFA) is studying serendipitous Einstein sources for the same purpose.

(viii) Gamma ray bursts

Identification and understanding of the intense (but rare) γ-ray bursts has been brought closer by the determination of several relatively precise source positions. Cline et al (1984) have reported positions accurate to a few (arc min) for the moderately intense bursts of 1978 Nov 24 and 1979 Nov 5 and 16, from the interplanetary network of Soviet, US and European spacecraft. Although all 3 locations are at high galactic latitude none have an obvious steady optical counterpart; however, a search of archived data produced a coincident optical event, confirming the previous γ-ray burst/optical transient association discovered by Schaefer (Schaefer, 1981). Barat et al (1984) report the precise location of the intense 1979 January 13 γ-ray burst, probably coincident with a second of the Schaefer et al (1984) optical flashes and with a possible faint ($m_V \sim 23$) quiescent optical counterpart. A quasi-periodic repetition of γ-ray pulses makes this observation different to most previous detections. On the basis of a total of 3 optical flashes seen near γ-ray burst locations, Schaefer et al (1984) conclude an average recurrence timescale of $\sim$ 1 year and a ratio of γ-ray to optical flux of $\sim$ 900. Although the γ-rays burst sources remain an unsolved puzzle, these new data appear to support the view that they originate within the magnetosphere or at the surface of an isolated or weakly accreting neutron star.

C. FUTURE PROSPECTS

By its nature, the Einstein Observatory examined only a small fraction of the whole sky, an aspect in common with most large ground-based telescopes. The German ROSAT mission, due for launch on the US Space Shuttle in late 1987, will carry a somewhat larger Wolter I telescope, of similar properties to Einstein (see Table), which will be used in the first $\sim$ 6 months in orbit to conduct an all-sky survey with a sensitivity at least comparable to the Einstein medium-deep survey (Trumper 1984). More than 10^5 sources of all astronomical types are anticipated from this survey, greatly increasing the statistical power of many varied studies and - potentially - uncovering major new phenomena. Alongside the ROSAT X-ray telescope will be flown the UK Wide Field Camera, a telescope optimised for the extreme ultraviolet waveband (Pye, 1984). Following a simultaneous all-sky survey (probably the first in the XUV band), the WFC and ROSAT X-ray telescopes will then embark on a programme of observations of selected fields, in the manner of the Einstein Observatory. US astronomers will participate strongly in this phase of the mission.

Several months before the launch of ROSAT, the successor to Tenma is due to be placed in orbit by the Institute of Space and Astronautical Science. This Japanese satellite, ASTRO-C, will carry a 0.5 square metre array of proportional counter detectors, to conduct detailed timing and spectral studies of a wide range of galactic and extragalactic sources. UK astronomers are collaborating with ISAS on the payload development and will share in the scientific operation of ASTRO-C. This mission, complementary in its capabilities to ROSAT, will provide a powerful extension to the studies now being conducted with Tenma and EXOSAT (due to re-enter the atmosphere in mid-1987).

Beyond 1987, NASA plan to fly two satellites in the Explorer programme which cover this field of astronomy. The Extreme Ultra Violet Explorer (EUVE) will carry out a sky survey in the XUV band, similar to the ROSAT WFC, and may also have some additional spectroscopic capability. Later on, the X-ray Timing Explorer (XTE) is due for launch, carrying a powerful array of detectors, including a 1 square metre proportional counter array, for timing and spectral studies of the brighter galactic and extragalactic sources. In Italy, approval has recently been obtained for the SAX mission, an X-ray satellite to be launched on the US Space Shuttle towards the end of the 1980s and designed to carry out a range of spectroscopic and hard X-ray mapping studies. Collaboration of Dutch astronomers is proposed for SAX. Finally, in the present decade, it is hoped that a small Indian X-ray payload will be launched, while further experiments are planned in the Russian programme (though details are not yet known). In summary, it appears that the 1980s should see a total, world-wide activity in X-ray astronomy similar to that of the previous decade, but with a much broader international spread.

TABLE Characteristics of some recent and planned X-ray telescopes

Telescope	Einstein	EXOSAT	ROSAT XRT	ROSAT WFC	AXAF
Type	Wolter I	Wolter I	Wolter I	Wolter-Schwarzschild I	Wolter I
Nested shells	4	2	4	3	6
Focal length (cm)	344	109	240	52	1000
Max. diameter(cm)	58	28	83	58	120
Geom. area (cm)	350	89	1150	511	1500
Energy range(keV)	0.2-4	0.04-1.5	0.1-2	0.015-0.2	0.1-10
On-axis resolution (arc sec)					
FWHM	3	5	5	30	0.5
HER (44 A)	4	10	2.5	60	0.5
HER (8 A)	5	15	2.5	60	0.5

In the 1990s, further Japanese and European X-ray missions (ASTRO-D and XMM, respectively) are firmly planned, very probably with an emphasis on realising the great scientific potential of moderate-to-high resolution X-ray spectroscopy. A major focus will be on the NASA AXAF mission. This natural (US) successor to the Einstein Observatory, with a nested Wolter I telescope of maximum mirror diameter 1.2 metres and sub-arc second resolution, is seen as the X-ray counterpart to the Hubble Space Telescope and will be the most powerful X-ray mission yet conceived.

REFERENCES

Andrews, M., Basart, J., Lamb, R. and Becker, R. (1983) Ap. J. 266, 684.
Arnaud, M., Ballet, J., Rocchia, R. and Rothenflug, R. (1985) Proc. 18th ESLAB Symposium on X-ray Astronomy (Reidel).
Aschenbach, B. et al., (1985) To be published.
Barat, C. et al. (1984) Ap. J. 286, L5.
Barstow, M.A., (1985) Proc. 18th ESLAB Symposium on X-ray Astronomy (Reidel).
Bechtold, J. et al. (1983) Ap. J. 265, 26.
Becker, R., Helfand, D. and Szymkowiak, A. (1982) Ap. J. 255, 557.
Bradt, H. and McClintock, J.E. (1983) Ann. Rev. Astron. and Ap. 21, 13.
Caillault, J.P. and Helfand, D. (1982) B.A.A.S. 13, 811.
Caillault, J.P. and Helfand, D.J. (1985) Ap. J. 289, in press.
Canizares, C.R. and Kruper, J. (184) Ap. J. 278, L99.
Charles, P.A. et al. (1983) Mon. Not. R. astr. Soc. 202, 657.
Cline, T.L. et al. (1984) Ap. J. 286, L15.
Cominsky, L.R. and Wood, K.S. (1984) Ap. J. 283, 765.
Cordova, F.A., Chester, T.J., Mason, K.O., Kahn, S.M. and Garmire, G.P. (1984) Ap. J. 278, 739.
Cordova, F.A. and Mason, K.O. (1983) Ch. 4 in "Accretion driven stellar X-ray

sources" (C.U.P.) Eds. Lewin and van den Heuvel.
Courvoisier, T., Parmar, A. and Peacock, A. (1985) Proc. 18th ESLAB Symposium on X-ray Astronomy (Reidel).
Cowley, A.P., Crampton, D. and Hutchings J.B. (1982) Ap. J. 255, 596.
Crampton, D., Cowley, A.P., Hutchings, J.B., Schade, D.J. and van Speybroeck, L.P. (1984) Ap. J. 284, 663.
Culhane, J.L., White, N.E., Kahn, S., Parmar, A.N., Blissett, R.J. and Kellett, B. (1985) Proc. 18th ESLAB Symposium on X-ray Astronomy (Reidel).
Fabbiano, G., Fransson, C. Palumbo, G.C. and Trinchieiri, G. (1985) Proc. 18th ESLAB Symposium on X-ray Astronomy (Reidel).
Fabian, A.C., Hu, E.M. Cowie, L.L. and Grindlay, J. (1981) Ap. J. 248, 47.
Fahlman, G.G. and Gregory, P.C. (1983) in "Supernova remnants and their X-ray emission" (Reidel) Eds. Danziger and Gorenstein, P445.
Forman, W., Jones, C., & Tucker, W. (1984) in "Large Scale Structure of the Universe" Proc. Trieste Meeting.
Gioia, I.M., et al. 1984) Ap. J. 283, 495.
Grindlay, J.E. et al. (1978) Nature 274, 567.
Grindlay, J.E., Hertz, P., Steiner, J.E., Murray, S.S. and Lightman, A.P. (1984) Ap. J. 282, L13.
Hartwick, F., Cowley, A. and Grindlay, J. (1982) Ap. J. 254, L11.
Hertz, P. and Grindlay, J.E. (1983) Ap. J. 275, 105.
Holt, S.S. (1983) "Supernova remnants and their X-ray emission" (Reidel) Ed. Danziger and Gorenstein, P17.
Holt, S.S. et al. 1980) Ap. J. 241, L13.
Horstman, H., Balunias, S., Cecchini, S., Palumbo, C.G. Ramond, J.C. and Rosso C. (1985) Proc. 18th ESLAB Symposium on X-ray Astronomy (Reidel).
Jones, C. and Forman W. (1984) Ap. J. 276, 38.
Jones, L.R., Pye, J.P. and Culhane, J.L. (1984) Proc. Bologna Symposium on X-ray Astronomy - to be published by ISAS.
Jones, L.R., Pye, J.P. and McHardy, I.M. (1985) Proc. 18th ESLAB Symposium on X-ray Astronomy (Reidel).
Kahn, S.M., Wesemael, F., Liebert, J., Raymond, J.C., Steiner, J.E. and Shipman, H.L. (1984) Ap. J. 278, 255.
Kellett, B.J. (1985) Proc. 18th Symposium on X-ray Astronomy (Reidel).
Kelley, R.L., Jernigan, J.G., Levine, A., Petro, L.D. and Rappaport, S. (1983) Ap. J. 264, 568.
King, A.R. (1985) Proc. 12th Texas Symposium on Relativistic Astrophysics (N.Y. Acad. Sci.)
Ku, W.H.M., Kahn, S.M., Pisarski, R.L. and Long, K.S. (1984) Ap. J. 278, 615.
Lewin, W.H.G. and van den Heuvel, E.P. (1983) Eds. of "Accretion driven stellar X-ray Sources" (C.U.P.).
Lucy, L.B. and White, R. L. (1980) Ap. J. 241, 300.
Lucy, L.B. (1982) Ap. J. 255, 286.
McClintock, J.E. et al. (1982) Ap. J. 258, 245.
Maejima, Y., Makishima, K., Matsuoka, M., Ogawara, Y. and Oda, M. (1984) Ap. J. 285, 712.
Mandelstam, S.L. et al. (193) Adv. Space. Res. 2, 293.
Mathewson, D. et al. (1983) "Supernova remnants and their X-ray emission" (Reidel) Eds. Danziger Gorenstein, P541.
Matsuoka, M. et al. (1984) Ap. J. 283, 774.
Meekins, J.F. et al. (1984) Ap. J. 278, 288.
Mewe, R. et al. (1981) in 'Cambridge Cool Star Workshop' Ed. Dupree.
Montmerle, T., Koch-Miramond, L., Falgarone, E. and Grindlay, J.E. (1983) Ap. J. 269, 182.
Motch, C. Pedersen, H. Ilovaisky, S.A., Chevalier, C. and Pakull, M.W. (1984) Proc. Bologna Symposium - to be published by ISAS.
Mushotzky, R.F. (1984) NASA Tech. Memo. 86071.
Nagase, F. et al. (1984) Ap. J. 280, 259.
Nulsen, P.E.J. et al. (1982) Mon. Not. R. astr. Soc. 199, 1089.

Nulsen, P.E.J., Stewart, G.C. and Fabian, A.C. (1984) Mon. Not. R. astr. Soc. 208, 185.
Page, C.G. (1985) Proc. 18th ESLAB Symposium on X-ray Astronomy (Reidel).
Parmar, A.N., Culhane, J.L., White, N.E. and van den Oord, G.H.J. (1985) Proc. 18th ESLAB Symposium on X-ray Astronomy (Reidel).
Pedersen, H., van Paradijs, J. and Lewin, W.H.G. (1981) Nature 294, 725.
Petre, R., Canizares, C.R., Kriss, G.A. and Winkler, P.F. (1982) Ap. J. 258, 22.
Pisarski, R.L., Helfand, D.J. and Kahn, S.M. (1984) Ap. J. 277, 710.
Pounds, K.A. (1984) Proc. Bologna Symposium on X-ray Astronomy - to be published by ISAS.
Pounds, K.A. Warwick, R.S., Culhane, J.L. and de Korte, P. (1985) Proc. 18th ESLAB Symposium on X-ray Astronomy (Reidel).
Pye, J.P. and McHardy I.M. (1983) Mon. Not. R. astr. Soc. 205, 875.
Pye, J.P. (1984) MPE Report 184 "X-ray and UV emission from AGN". Eds Brinkman and Trumper P261.
Pye, J.P. Pounds, K.A., Rolf, D.P., Seward, F.D., Smith, A. and Willingale, R. (1981), Mon. Not. R. astr. Soc. 194, 569.
Rappaport, S. and ven den Heuvel, E.P. (1982) Proc. IAU Symposium 98 on "Be Stars" (Reidel) Eds. Jaschek and Groth. p327.
Remillard, R.A. and Canizares, C.R. (1984) Ap. J. 278, 761.
Schaefer, B.E. (1981) Nature 294, 722.
Schaefer, B.E. et al. (1984) Ap. J. 286, L1.
Schmitt, J. Golub, L., Harnden, F.R., Maxson, C.W., Rosner, R. and Vaiana, G. (1985) Ap. J. to be published.
Seward, F.D. (1983) in "Supernova remnants and their X-ray emission" (Reidel) Eds. Danziger and Gorenstein, P405.
Seward, F.D., Gorenstein, P. and Tucker, W. (1983) Ap. J. 266, 287.
Seward, F.D. and Harnden, F.R. (1982) Ap. J. 256, L45.
Stanger, V. J. (1985) Ph. D. Thesis, University of Leicester.
Steiner, J.E. et al. (1984) Ap. J. 280, 688.
Stern, R.A., Zolcinski, M.C., Antioches, S.K. and Underwood, J.H. 1981) Ap. J. 249, 647.
Stewart, G.C., Fabian, A.C. and Seward, F.D. (1983) "Supernova remnants and their X-ray emission" (Reidel) Ed. Danziger and Gorenstein.
Swank, J. and Carsinelli J.P. (1983) Ap. J. 271, 681.
Swank, J. and Johnson, H.M. (1982) Ap. J. 259, L67.
Swank, J., White, N.E., Holt, S. and Becker, R. 1981) Ap. J. 246, 208.
Tuohy, I. and Garmire, G. (1980) Ap. J. 239, L107.
Trumper, J. (1984) MPE Report 184 "X-ray and UV emission from AGN". Eds. Brinkman and Trumper.
Turner, M.J.L. and Breedon, L.M. (1984) Mon. Not. R. astr. Soc. 208, 29P.
Vaiana, G.S. et al. et al. (1981) Ap. J. 245, 163.
Vedder, P.W. and Canizares, C.R. (1983) Ap. J. 270, 666.
Verbunt, F., van Paradijs, J. & Elson, R. (1984) Mon. Not. R. astr. Soc. 210, 899.
Vilhu, O. Heise, J. (1985) Proc. 18th ESLAB Symposium on X-ray Astronomy (Reidel).
Waki, I., Inoue, H., Koyama, K. et al. (1984) Pub. Astron. Soc. Japan - in press.
Waldron, W.L. (1981) U. Wisconsin, Thesis.
Warwick, R.S. et al. (1981) Mon. Not. R. astr. 197, 865.
Watson, M.G., King, A.R. and Heise, J. (1985) Proc. 18th ESLAB Symposium on X-ray Astronomy (Reidel).
Watson, M.G., King, A.R. and Osborne, J. (1985a) Mon. Not. R. astr. Soc.
Watson, M.G., Stanger, V.J. and Griffiths, R.E. (1984) Ap. J. 286, 144.
Watson, M.G. Warwick, R.S. and Ricketts, M.J. (1981) Mon. Not. R. astr. Soc. 195, 197.
White, N.E. Parmar, A.N., Sztajno, M. and Zimmermann, H., Mason, K.O. and Kahn, S.M. (1984) Ap. J. 283, L9.
White, N.E. and Swank, J. (1982) Ap. J. 253, L61.
Winkler, P.F., Canizares, C.R., Clark, G.W., Markert, T.H., Kalata, K. and Schnopper, H. (1981) Ap. J. 246, L27.

3. Ultraviolet Astronomy

Stephen P. Maran

INTRODUCTION

During 1982-1984, progress in ultraviolet astronomy continued through extensive observations with the International Ultraviolet Explorer (IUE), continuing programs of sounding rocket and balloon flights, and through observations made from other orbital spacecraft, including the Space Shuttle and ASTRON. In addition, there was important progress in the design and development of facilities for future missions, including the Hubble Space Telescope.

INTERNATIONAL ULTRAVIOLET EXPLORER

As of November, 1984, a total of 554 proposals were received by NASA, ESA, and the UK Science and Engineering Research Council for observations during the 8th annual operating episode of IUE. These requests represent a proposed oversubscription of the available observing time by a factor of about 2.6; those requests that require observations during intervals of low orbital radiation environment represent an oversubscription of about a factor of 2.8. The IUE, clearly of continuing major scientific importance, is operating remarkably well considering its substantial orbital lifetime. Major recent IUE discoveries include the detection of the sulphur dimer in Comet IRAS-Araki-Alcock; this molecule was never previously detected in a celestial object or medium and appears to be a previously unsuspected cometary parent molecule for which it may be necessary to assign a presolar origin.

SOUNDING ROCKETS, SKYLAB, BALLOON EXPERIMENTS

Continuing analysis of data from the S183 experiment on Skylab resulted in new determinations of 2600 A isophotes in the Large Magellanic Cloud and of the integrated ultraviolet fluxes of stellar associations (A. Vuillemin), while 1430 A photographs obtained in a Goddard Space Flight Center sounding rocket experiment revealed what appears to be highly-organized large scale structure in the LMC that resembles forms predicted from the theory of detonation wave propagation in a rotating gaseous disk (A. M. Smith). New developments in ultraviolet sounding rocket astronomy include the completion and first flight of IMAPS, the Interstellar Medium Absorption Profile Spectrograph (E. Jenkins, Princeton University), an objective grating echelle spectrograph that attains spectral resolution R = 200,000 and incorporates an intensified CCD detector. IMAPS operates from 1100 A to the Lyman limit; its principal objective is to study the interstellar medium. Among balloon astronomy results, new photometric and imagery observations were obtained at 2000 A in a joint program of the Laboratoire d'Astronomie Spatiale (Marseille) and the Geneva Observatory. In the SCAP-2000 balloon program, isophotes of the LMC and other nearby galaxies were obtained; the hottest objects were reached down to magnitude 17. The 13-cm SCAP-2000 telescope, equipped with a Fabry-Perot device, will be reflown to make deep detections of sources of 1909 A C III intercombination line emission.

GALACTIKA

The 200 A-resolution scanning spectrometer "GALACTIKA" was built by the Laboratorie d'Astronomie Spatiale and the Crimean Observatory as a component of the France-USSR space program. It detected a systematic increase in background light at 1600 A and determined the functional dependence of brightness on galactic latitude, perhaps due to scattering by high-latitude galactic dust, although other sources have been considered. At apogee (220,000 km) of the highly elliptical orbit, GALACTIKA is beyond the geocorona; accurate corrections for terrestrial Lyman alpha can be made thanks to its varying contribution to observations made along the orbit.

SPACELAB

The FAUST 16-cm, f/1.2 telescope operating at 1500 A (joint project of the Space Science Laboratory, Berkeley and the Laboratoire d'Astronomie Spatiale) flew on Spacelab-1 in November, 1983. Besides Shuttle glow, high altitude airglow, arising well above Shuttle orbit, may have contributed significantly to an interfering background. However, useful photographs were obtained. A Space Shuttle reflight, in which FAUST will be fitted with an electronic detector in place of film, is anticipated on Earth Observation Mission-1, presently manifested for Fall, 1986. The Very Wide Field Camera of the Laboratorie d'Astronomie Spatiale, with a 66° field at f/1.9 and operating at 1650 A, 1930 A, and 2530 A, also flew on Spacelab-1 and is to fly again; a large (1.2 X 2.4 Kpc), hot population stellar group was detected in the neutral hydrogen bridge between the Small and Large Magellanic Clouds.

The first three flights of the Astro mission are manifested for March 6, 1986 (to coincide with the various Comet Halley flyby missions), Fall, 1986, and Summer, 1987. The instruments, Wisconsin Ultraviolet Photopolarimeter Experiment (A. Code, University of Wisconsin), Ultraviolet Imaging Telescope (T. Stecher, Goddard Space Flight Center), Hopkins Ultraviolet Telescope (A. Davidsen, The Johns Hopkins University) are already undergoing thermal vacuum tests and are scheduled for delivery at the Kennedy Space Center for integration into the Spacelab system by March, 1985. A prototype of the Ultraviolet Imaging Telescope was launched on a Black Brant sounding rocket in April, 1982 to photograph globular cluster M 5 and barred spiral galaxy M 83 at 1400 A and 2400 A. In M 5, an unusual ultraviolet-bright star was discovered and the initial helium abundance was determined from the horizontal branch stars. A starburst nucleus was found in M 83.

SPACE STATION ASTRON

The 80-cm UFT telescope, equipped with a ultraviolet spectrometer, provided spectra of stars, galaxies and the interstellar medium in the 1500 A - 3500 A region, at resolutions of 0.2 A - 28 A. In the low resolution mode, the spectrometer, with a single concave grating, is especially efficient for the detection of faint monochromatic emission from galaxies. UFT (Laboratoire d'Astronomie Spatiale and the Crimean Observatory) flew aboard the ASTRON Space Station as a component of the France-USSR space program.

HUBBLE SPACE TELESCOPE

The Hubble Space Telescope has been developed by NASA in preparation for a launch readiness date of June, 1986, with launch likely in the third quarter of 1986. The Optical Telescope Assembly and the five scientific instruments (Faint Object Camera, Faint Object Spectrograph, High Resolution Spectrograph, High Speed Photometer, Wide Field and Planetary Camera) have been delivered, as has the engineering model Fine Guidance Sensor. The instruments were provided respectively by ESA (D. Macchetto), the University of California at San Diego (R. Harms), Goddard Space Flight Center (J. Brandt), the University of Wisconsin (R. Bless), and the California Institute of Technology (J. Westphal). The Space Telescope Science Institute (Director, R. Giacconi) occupied its new quarters in Baltimore, Maryland; the Space Telescope European Coordinating Facility (Head, P. Benvenuti) began activity at the European Southern Observatory premises in Garching, and the Space Telescope Mission

Operations Control Center was inaugurated at Goddard Space Flight Center. Work is underway on the Guide Star Selection System catalog, which involves microdensitometry of survey plates, photometric and astrometric calibration. In the Fall of 1984, NASA issued an Announcement of Opportunity for astronomers to propose second-generation instruments that would be installed in-orbit to replace one or more instruments of the initial payload complement.

STARLAB, FUSE

Studies continued on the international 1-m telescope facility, Starlab. NASA is pursuing definition studies of the Far Ultraviolet Spectroscopic Explorer (FUSE) in response to the Field Committee recommendation that it be accorded highest priority among moderate-scale space astronomy missions. A joint ESA-NASA Scientific Working Group was established to refine common interests in this project.

SPARTAN

Spartan is a major new program to utilize the transportation capability of the Space Shuttle in order to enable short duration (40 hours), free-flying orbital astronomy experiments. Through Spartan, NASA hopes to virtually replicate the previous frequency of sounding rocket flights by the end of the decade. An ultraviolet spectroscopy payload to investigate Comet Halley during the near-perihelion phase in January, 1986 is under development as "Spartan Halley" at the Laboratory for Astronomy and Space Physics in Boulder, Colorado (C. Barth). The Spartan-3 payload (G. Carruthers, Naval Research Laboratory) consists of an electronographic Schmidt camera operating at 1230 A - 1600 A, with 11° field of fiew; it is expected to photograph as many as 30 fields during its first orbital mission. Selection of additional ultraviolet astronomy Spartan payloads was underway in December, 1984.

UVX AND EUVE

The "Ultraviolet Experiment" (UVX) of the University of California at Berkeley (S. Bowyer) and The Johns Hopkins University (P. Feldman) consists of two ultraviolet spectrometers mounted in "Get Away Special" cans; a third can carries the battery and tape recorder. The objectives include definitive measurements of the cosmic ultraviolet background by the two university groups, which have previously differed in the results of sounding rocket measurements of this background. Another objective is to measure the spacecraft glow phenomenon and thus help establish that highly sensitive ultraviolet observations are viable from low earth orbit. Launch aboard the Space Shuttle may occur in 1985.

NASA is continuing work on the Extreme Ultraviolet Explorer (EUVE), with the prospect of mounting it on a leased space platform toward the end of the decade. Equipped with four telescopes, EUVE (S. Bowyer, University of California at Berkeley) will make a complete sky survey in the 100 A - 800 A region, and would also perform low resolution far-ultraviolet spectroscopy and support a guest observer program.

SPACE INTERFEROMETRY

A Workshop on Optical Interferometry in Space was held in Baltimore, Maryland in June, 1984. Among the presentations was a proposal for ultraviolet spectro-interferometry that could be accomplished from the Space Shuttle orbiter bay (S. Synott, Jet Propulsion Laboratory). Abstracts of the Workshop papers were published in the _Bulletin of the American Astronomical Society_, Vol. 16, No. 2.

SPACE SCHMIDT CAMERAS

A Working Group for Space Schmidt Surveys (Chairman, K. Henize, Johnson Space Center) has continued studies. The basic concept is that of a wide field, fast telescope with aperture approximately 1-m, for (1) surveying the sky in the far UV, (2) detecting very faint extended sources, and (3) objective prism or grating spectroscopy. F. Bertola, M. Golay, and R. West have submitted a statement of the Space Schmidt Concept to ESA for future mission planning. G. Carruthers and H. Heckathorn (Naval Research Laboratory) continue studies of electronographic cameras suitable in Space Schmidt cameras and the NRL group has modified an electronographic Schmidt camera to incorporate a microchannel intensifier stage for flight in the U.S. Air Force Space Test Program to study nebulae and the ultraviolet background.

ULTRAVIOLET ASTRONOMY FROM INTERPLANETARY PROBES

Far ultraviolet observations of stellar and interstellar sources continue from Voyager (A. Broadfoot, Lunar and Planetary Laboratory, Tucson, Arizona). Among the ultraviolet instrumentation to be launched on interplanetary probes to investigate Comet Halley in March, 1986 are an electronic camera for the Japanese probe "Planet A" and spectrometers on the USSR Vega probes.

4. Infrared Space Astronomy

Frank J. Low

I. INTRODUCTION

The years 1982 to 1984 have produced the greatest advances in Infrared Astronomy since its beginning. The InfraRed Astronomical Satellite (IRAS) was completed in 1982, launched on its 10 month mission in early 1983 to be followed only one year later by the first extensive release to the public of IRAS data, including the point source catalog of over 245,000 sources and an atlas of the sky in four wavelength bands from 10 to 100 um imaging the absolute sky brightness at 2 arcmin resolution. With its superfluid He cooled telescope system and sensitive IR detectors from the US, its 3-axis stabilized spacecraft from the Netherlands and its ground station in the United Kingdom, this satellite is viewed by many as the most successful example of international cooperation yet to occur in the exploration of Space. In addition, the technological advances and engineering achievements combined with the enormous scientific output from IRAS insure that in the coming years Astronomy will be marked by many new discoveries made around the world as the IRAS data products are analyzed, studied by various techniques and combined with results from other spectral regions; future space missions will build on this new technology and will carry forward the study of the universe even more effectively.

While these dramatic events were taking place in orbit, the more traditional methods of Infrared Space Astronomy were actively pursued using jet aircraft, stratospheric balloons and sounding rockets to reach above the obscuration and thermal background which limits such work from the ground-based observatories. Only a very brief summary of these developments can be given here in order to cover in more detail the results from IRAS.

II. NEAR SPACE ACTIVITIES

The tenth anniversary of NASA's Kuiper Airborne Observatory (KAO), an 85 cm IR telescope carried aloft in a four engine jet transport, was celebrated at the Ames Research Center in June 1984. Among the many new results described at the KAO Symposium and reported in the proceedings were the first measurements of Sub-mm (270 um) polarization of emitted radiation from very low temperature dust grains in the interstellar medium. This work opens up the possibility of studying in some detail the local magnetic fields in regions of active star formation since it is thought that the grain alignment producing the observed polarization is the direct result of the local magnetic field. Hildebrand et al (1984) report 1.7 +/- 0.5 % polarization at two locations within the most active parts of the Orion star-forming region. In addition to these very sensitive airborne measurements by the University of Chicago group using He three cooled bolometers, Harvey et al (1984) from the University of Texas showed that with very long integration times (3000 sec.) and conventional bolometers, sensitivity limits approaching those of the IRAS survey could be obtained at 50 and 100 um for compact extra-galactic sources such as OJ 287 and BL Lac. It is clear that such instruments can be expected to play a vital role in IRAS follow-up studies where it is necessary to extend either the spectral or spatial resolution or to extend the wavelength range in the study of specific objects. In this regard, both the Texas and the Chicago groups quickly confirmed and extended the IRAS observations of a far-IR excess from Vega reported by Aumann et al (1984).

In the past three years a variety of IR instruments were carried into the dry, almost transparent stratosphere by balloons launched in several different countries, including Japan, Italy and India as well as France and the USA. Among the results from these experiments were very long wavelength observations, >100 um, of the thermal emission from the galactic plane (De bernardi et al. 1984). Using He three cooled bolometers operating in two broad bands, 150 to 400 um and 350 to 3,000 um, they measured the large scale emission from the galactic plane over a large range of galactic longitude. When these results are combined with the IRAS data for the same regions at shorter wavelengths it will be possible to construct a more complete spectral energy distribution for a significant part of the galaxy.

Working in the much shorter wavelength range from 1 to 5 um, where it was predicted that the natural background from all known cosmic sources must reach a deep minimum, the Japanese group, led by Matsumoto from the University of Nagoya, began to use sounding rockets to make a series of extremely difficult measurements of the total sky brightness. Their initial results have recently been augmented by even more interesting findings soon to be published (1984).

III. IRAS

The information contained in this section is based on publications of the Joint IRAS Science Team; for more information refer to the catalogs which may be obtained from the various data centers in the USA, Netherlands and the United Kingdom or to the Explanatory Supplement edited by C. Beichman, G. Neugebauer, H. Habing, P. Clegg and T. Chester.

A. Design and performance-The following tables summarize important aspects of the design and inflight performance:

CRYOGENIC SYSTEM
- Cryogen- Superfluid He
- Bath Temp- 1.8K
- Main Tank Cap.- 500 l
- Lifetime- 300 days

TELESCOPE SYSTEM
- Type- f/9.6 Ritchey-Chretien
- Primary Dia.- 57 cm
- Optics Temp.- <5K
- Collecting Area- 2019 cm**2
- Construction- Beryllium

Total Mass- 809 kg (including 73 kg LHe)

FOCAL PLANE INSTRUMENTS

Survey Array- 62 discrete detectors with JFET TIA Amplifiers

Band (um)	Det. Material	# Working Detectors	Field of View (arcmin)	Av. Survey Sensitivity Pt.Sources (1σ, Jy)	Ext.Sources* (1σ, MJy/sr)
12	Si:As	16	0.75x4.5	0.07	0.02
25	Si:Sb	13	0.75x4.6	0.07	0.02
60	Ge:Ga	15	1.5x4.7	0.08	0.02
100	Ge:Ga	15	3.0x5.0	0.3	0.05

* For 30x30 arcmin Synthetic Beam

Chopped Photometric Channel

50	Ge:Ga	1	1.2 Dia.	0.7
100	Ge:Ga	1	1.2 Dia.	0.7

Low Resolution Spectrometer

Band (um)	Det. Material	# Working Detectors	Resolving Power
8-13	Si:Ga	3	14-35
11-23	Si:As	2	14-35

SPACECRAFT AND ORBIT

Type of Orbit- Sun Synchronous, Terminator
Altitude- 900 km
Inclination- 99 deg
Launch Date- 1983Jan26
Inertial Wheels- 3
Attitude Control- Sun Sensor, Gyros, Focal Plane Star Sensors
Pointing Accuracy- <10 arcsec.
Reconstructed Pointing- 2 arcsec
Survey Scan Rate- 3.85 arcmin/sec.

B. Survey Results-The following tables summarize the survey performance and list some of the results of general interest:

Sky Coverage
HCON 1- 96 %
HCON 2- 96 %
HCON 3- 74 %

Reliability- >99.8 %
Completeness- >97 %

Wavelength (um)	12	25	60	100
Limiting Flux Density (Jy)	0.4	0.5	0.6	1.0
b >50 deg. Density (sources/deg*2)	0.6	0.6	0.6	–
Photometric Uncertainty (%)	7	9	11	11

Contents of Point Source Catalog (<1 arcmin dia.)

CLASS	DEFINITION	# IN CAT.	# OF ASSOC.
"STARS"	F (12 um) > F (25 um)	134,092	46,000
"GALCIR"	F (60 um) > F (25 um)	58,295	10,000
"CIRRUS"	100 um only	33,146	–
"OTHER"	None of the above	23,782	–
TOTAL		245,839	71,000
b 50 deg		12,418	
OTHER TYPES			15,000

As indicated in the table above most of the point sources that are brighter at 12 um than at 25 um are stars and positional associations with visible stars have been made for a large number of such sources. At 60 um a very different population of objects dominates the sky at high galactic latitudes, positional association with cataloged galaxies show that these sources are mostly extragalactic; it appears that about 20,000 extragalactic objects have been detected and cataloged. Yet another population dominates the sky at 100 um, the interstellar cirrus; the point source processor could not discriminate between compact features of the cirrus clouds and discrete sources at the resolution of IRAS. In addition to positions, flux densities and associations with previously cataloged sources, the IRAS catalog contains information on the measurement accuracy and on the reliability of sources and in many cases indications of variability. It should be noted that an important astronomical class of infrared point sources, asteroids, have been removed from the catalog and constitute another large data base which is still being prepared for release. At this time the catalog of small extended sources, objects with sizes between 1 and 8 arcmin, is also undergoing further processing and analysis at JPL.

C. Early Scientific Results From IRAS

1. The Cosmic Background-Because of its capacity to measure the total photocurrent from each of the IR detectors in the four wavelength bands and because it was possible, through a special calibration procedure, to determine the inflight values of the electronic offset voltages for each of the channels, accurate data was obtained from IRAS on the absolute sky brightness throughout the mission. As reported by Hauser et al (1984), thermal emission from the cloud of interplanetary dust through which the earth travels on its orbit around the Sun dominates the sky away from the galactic plane at 12, 25 and 60 um (the data given in the tables above reflect the final calibrations and differ somewhat from those published by Hauser et al 1984). It was discovered, however, that the 100 um diffuse background is composed of at least two parts, the much weaker zodiacal emission and a galactic component which is present over the entire sky. The distribution over the sky of this bright, and as yet unexplained, galactic component will be accurately determined by subtraction of the zodiacal emission based on empirically determined models (Good 1984).

2. Solar System Results-One of the most surprising features of the extended emission was the existence of three narrow bands of emission seen in all four wavelengths extending around the sky close to and almost parallel with the ecliptic plane. These so called "zodiacal dust bands" described by Low et al (1984) have been identified with material ejected into specific orbits as the result of multiple collisions between certain families of main belt asteroids.

The discovery of a number of comets, including IRAS-Araki-Alcock 1983d (Walker et al 1984) and pointed observations of several others, provided the first far-IR observations of these primitive objects with their large envelopes of dust and debris. The amount of dust and the sizes of the dust tails greatly exceeded expectations based on optical studies; this explains in part why so many Comets were found in such a short period.

The systematic survey of large numbers of named and unnamed asteroids--perhaps as many as 10,000--has provided a large and rich data base for detailed analysis once the preliminary data reduction has been completed.

3. The Vega Far-IR Excess--As part of their photometry program to provide absolute calibration data for IRAS, Aumann and Gillett discovered that alpha Lyrae (Vega) showed a large, unexpected excess at 60 and 100 um. In their paper, Aumann et al (1984) give the results of special high angular resolution scans which provided data on the size of the 85 K shell of orbiting material, the first direct observational evidence of solid material in orbit around stars other than the Sun. Subsequently, similar phenomena were reported for alpha Pisces and beta Pictor, with indications of perhaps 50 more stars with similar types of far-IR excess. Ground-based photography of beta Pictor has confirmed that the distribution of material is similar in size and shape to that expected for the solid material in a pre-planetary envelope.

4. Absolute Photometric Calibration-Because it was possible to make repeated pointed observations of stars and asteroids to produce photometry accurate to a few percent with IRAS, a new absolute calibration has been developed for each of the four wavelength bands. The results given by Neugebauer et al (1984) extend the ground-based calibration of Rieke et al (1984) at 10 um and 20 um to 25, 60 and 100 um with accuracy adequate for most astrophysical purposes. Detailed studies of the thermal models required to fit the observations of bright asteroids is one immediate consequence of this development.

5. Low Mass Protostars-Surveys of nearby molecular clouds such as those in Taurus, Ophiucus and Chamaeleon show numerous cold unresolved sources with a wide range of temperatures and luminosities (Wesselius et al 1984; Baud et al 1984; Beichman et al 1984). Recently, Beichman has reported objects which have such low luminosities, only 0.1 L(sun), that it is difficult to escape the conclusion that they must be extremely primitive and of very low mass. Thus it appears that the census of infrared stars in dense molecular clouds produced by IRAS includes almost the complete range of masses, luminosities and temperatures expected for the various stages of star formation.

6. Evolved Stars and Supernova Remnants-Circumstellar dust shells are formed by the mass loss mechanisms which play important roles in the evolution of stars as they return much of their mass to the interstellar medium; they are seen as excesses in the color-color diagrams of IRAS sources (Rowan-Robinson et al 1984). As in the case of proto-stars or the pre-planetary systems around nearby stars mentioned above, thermal reradiation of energy from solid particulate matter around stars is a signature that is well imprinted on many of the objects in the IRAS catalog. In these cases, such as the OH/IR stars studied by Olnon et al (1984), it is the final stages rather than the birth of stars and planets that are observed.

The most dramatic form of stars returning their mass to the interstellar medium is the supernova. Marsden et al (1984) reported on the IRAS observations of the Crab Nebula which show a small but significant far-IR excess attributed to dust within the nebula, probably but not necessarily formed from the material ejected by the supernova. Pointed observations of several supernova remnants were made as part of the additional observation program and should provide a better insight into the role of dust in the development of these nebulae as they expand into the surrounding interstellar medium.

7. The Interstellar Medium-Once the solid material formed in the various out-flow processes reaches interstellar space it forms a dilute but optically detectable medium. IRAS has shown a dramatic picture of the intricate patterns of the so called interstellar cirrus clouds. Low et al (1984) reported on the optical depths and temperatures of prominent examples of very high latitude clouds; surprisingly high color temperatures of 30 K at 60 and 100 um were found suggesting that small graphite grains are the chief constituents of these particular clouds and that evidence for variations in temperature from one cloud to another requires a range of material properties. More recently it has been noted by Gautier and Beichman that some clouds have very hot emission, several hundred degrees. If this is the result of grains heated by the normal interstellar radiation field they must have properties very different from the colder grains. Leger et al (1984) have suggested models in which extremely small graphite grains are raised to high temperatures by the absorption of individual energetic quanta. The study of grain properties and the structure of the clouds and their evolution will yield new insights as this work progresses.

8. Galactic Structure-Because even the most dense interstellar clouds are optically thin at long IR wavelengths and because so much of the galaxies energy is radiated in the thermal IR, it was expected that the IRAS survey would reveal details of the structure of the galaxy not seen before. This has been the case. The structure of the inner galaxy, already mapped to some extent by balloon instruments, was revealed in much greater detail and depth. Perhaps the most surprising feature of the galaxy is the central bulge as seen in the distribution of stars bright at both 12 and 25 um. These luminous stars must be short lived which means that they are constantly replaced on a much shorter time scale than expected.

9. Close Galaxies-IRAS has provided quite detailed views of neighboring galaxies such as the Magellanic Clouds and the Andromeda Nebula (Habing et al 1984). Both Magellanic Clouds are found to be quite active sites of star formation and will require much detailed study. On the other hand, when compared to our galaxy M31 seems almost quiescent with its total IR luminosity only 3 % of its total (our galaxy must have a ratio of IR/Optical output of about unity). Clearly the IR emission from spiral galaxies is in general higher than in eliptical galaxies because the star formation rate is much higher and because there is much more dust. Given the new IR results on such a large sample of galaxies out to the distance of the major clusters there is much more to be learned in this area.

10. Active Galaxies and Quasars-With the discovery of extremely luminous galaxies emitting almost exclusively in the IR, such as ARP 220 (Soifer et al 1984) and NGC 6240, it became clear that the space density of these objects is much higher than for quasars which, though of similar energy output emit a much broader non-thermal spectrum (Neugebauer et al 1984). Houck and co-workers (1984) found that many of the 60 um sources with the characteristics of extragalactic sources could not be readily identified on Palomar plates; subsequent studies with more sensitive CCD cameras show that these "unidentified" sources are optically faint systems with very high ratios of IR/Optical luminosity (>50) and very high total luminosity.

It was noted by Lonsdale and others that a high percentage of IR galaxies are morphologically disturbed as the result of interactions between galaxies. A sample of Seyfert galaxies, with their quasar like nuclei and high IR luminosities, were studied by de Grijp et al (1984) who conclude that their nuclear activity is more readily detected by their relatively warm IR spectral energy distributions than by their UV excesses; they predict a very large increase in the number of such objects available for study when the catalog is searched more thoroughly and that many of these will have luminosities as high as quasars. Thus models for the primary energy source in IR galaxies may differ little from those for quasars except for the almost universal presence of extremely high levels of "star burst" activity.

References

Aumann, H. H, *et al.*, 1984, Ap. J. (Letters), *278*, L23
Baud, B. *et al.*, 1984, Ap. J. (Letters), *278*, L53
Beichman, C. A., *et al.*, Ap. J. (Letters), *278*, L45
De Bernardi, P., Masi, S., Melchiorri, F., Moreno, G. 1984, Ap. J, *278*, 150
de Grijp, M., Miley, G. K., Lub, J., and de Jong, T., 1984, Nature, in press
Good, J., 1984, Private Comm.
Habins, H. J., *et al.*, Ap. J. (Letters), *278*, L59
Harvey, P. M., Marshall, J., Lester, D. F., and Wilking, B. A., 1984, Ap. J. (Letters), in press
Hauser, M. G., *et al.*, 1984, Ap. J. (Letters), *278*, L15
Hildebrand, R. H., Dragovan, M., and Novak, G., 1984, in press
Houck, J. R., *et al.*, 1984, Ap. J. (Letters), *278*, L63
Leger, A. and Puget, J. L., 1984, A and A (Letters), in press
Low, F. J., *et al.*, 1984, Ap. J. (Letters), *278*, L19
Marsden, P. L., *et al.*, 1984, Ap. J. (Letters), *278*, L29
Matsumoto, T., Private Comm.
Neugebauer, G., *et al.*, 1984, Ap. J. (Letters), *278*, L83
Neugebauer, G., *et al.*, 1984, Explanatory Supplement, IRAS, JPL
Olnon, F. M., *et al.*, 1984, Ap. J. (Letters), *278*, L41
Rieke, G. H., Lefofsky, M. J., Low, F. J., A. J., in press
Rowan-Robinson, M., *et al.*, 1984, Ap. J. (Letters), *278*, L7
Soifer, B. T., *et al.*, 1984, Ap. J. (Letters), in press
Walker, R. G., *et al.*, 1984, Ap. J. (Letters) *278*, L11
Wesselius, P. R., *et al.*, 1984, Ap. J. (Letters), *278*, L37

5. Solar Research from Space, 1982-1984 (Some Highlights)

E. L. Chupp

I. INTRODUCTION

During this reporting period, the satellites ISEE 3 (now ICE), IMP 8, Helios 1, Hinotori, Nimbus 7 (ERB), P78-1, Pioneer Venus Orbiter (PVO), Spacelab 1 and Solar Maximum Mission (SMM) continue to provide solar observations. We review briefly some recent developments, selected by the author as most significant, concerning: coronal mass ejecta, high-energy neutral emissions, X-ray and microwave bursts, the solar constant, and solar flare periodicity. While the emphasis is on new work, some significant results, obtained in earlier spacecraft experiments and only recently reported, are also mentioned. The repair, in space, of SMM in 1984 April, resulting in reactivation of three of the four pointed experiments is briefly described and a status report of the SMM 2 mission is given. We conclude with a summary of important future programs.

II. RECENT PROGRAMS

1. Coronal Mass Ejecta (CME)

Observations of CME are still being carried out by the SOLWIND coronagraph on the P78-1 satellite (Doschek 1983). The SMM Coronagraph Polarimeter (C/P) was nonoperational from 1980 October until it resumed observing the Sun in 1984 April. Thus, a large CME data base now exists for studying this phenomenon over a large part of the solar cycle.

The SOLWIND data base has led to several correlation studies. CME and type II correlation studies by Sheely et al. (1984) have revealed that "type II bursts without CMEs were associated with short lived soft X-ray events but not with interplanetary shocks at the Helios 1 spacecraft." They also note that "type II bursts with CMEs were associated with longer lived X-ray events (3 h on the average), and interplanetary shocks with CME speeds >400 km s^{-1}." For faster CME (>455 km s^{-1}) without metric type II bursts there was usually an associated interplanetary shock. Sheely et al. (1984) conclude that "shocks without CME, have a relatively impulsive origin, and may die out sooner than many shocks with CME which are piston driven," and that the lack of type II emission is due to formation of the faster CME above the lower corona where metric emission originates or, alternately, that the CME form shocks which (for some reason) cannot excite type II emission.

Kahler et al. (1984a) have also studied flares producing metric type II bursts and CME, and find that both compact and large mass ejection flares are associated with type II bursts and that "CME accompany about 60% of all flares with type II bursts for solar longitudes >30°." On the other hand, for events "with no CME," they suggest that "many of the apparent type II shocks are not piston driven and that there is strong indication that type II shocks are not necessarily produced by large energy releases in the flare's impulsive phase," since small 3 cm impulsive bursts are poorly correlated with occurrence of type II bursts.

The Skylab observations showed that "long duration" (∿ h) "X-ray events were almost always accompanied by mass ejections through the outer corona" (Sheeley et al. 1983). This suggested that the soft "X-ray events fall into two distinct classes," that is, long-duration events with CME and short-duration events without CME. A statistical study of CME observed from 1979-1981 on P78-1 has shed further light on this correlation. In particular, Sheeley et al. (1983) find that there is "no duration that separates X-ray events into two distinct classes depending on whether or not they have associated CMEs." Instead, the X-ray CME association is due to different differential distributions for events with and without CME. Also, a study of associations of CME, Hα flares and solar energetic particle events (SEP) has shown "in 26 of these 27 cases an associated CME was found indicating a high but not perfect association" (Kahler et al. 1984b).

The structure of the solar corona is known to be characteristically different between solar maximum and minimum. This was beautifully demonstrated by C/P observations made in 1980 and in 1984 respectively (Hundhausen et al. 1984a). Other post-repair results are still being analyzed; however, the 1980 C/P data has confirmed the Skylab result "that the type of surface activity most likely to be associated with a CME is an event near the limb with Hα material ejection, such as a prominence erruption" (Sawyer et al. 1984). The CME association with activity in the chromosphere or the low corona is tenuous. An interesting observation by Hundhausen et al. (1984b) is that the occurrence rate for CME at solar maximum as seen by C/P is slightly higher than during the declining phase of cycle 20 as seen by Skylab i.e. 0.9 ± 0.15 and 0.75 per 24 hour day, respectively.

2. High Energy Neutral Emissions

Uninterrupted observations (1980-1984) by the SMM High-Energy X-Ray Burst Spectrometer (HXBRS) and the Gamma-Ray Spectrometer (GRS) and by the Gamma-Ray Detector (GRD) on Hinotori (1981-1982) have made it possible to routinely detect several energetic solar flare neutral emissions. These include: hard X-rays, γ-ray lines and continuum, and direct neutrons at the Earth. Of importance is the strong indication that impulsive ion and electron acceleration occur simultaneously in flares (Chupp 1984 and references therein). Traditionally, the time relation between electron and ion acceleration had been presumed to be determined by the different peak times of X-ray and γ-ray emissions (Ramaty 1980). Such differences have been observed for some flares with time profiles of 4-8 MeV nuclear γ-rays, which are delayed relative to hard X-rays >100 keV (Bai 1982). The inference is that ions are accelerated later than relativistic electrons; however, the time of the peak fluxes in X-ray and γ-ray channels can be simultaneous within instrumental resolution in some events. The starting times of the 4.1-6.4 MeV and the 40-65 keV fluxes were essentially the same (≤ 2 s), even though these two fluxes show a delay in the times of maximum emission (Forrest and Chupp 1983). It is therefore highly probable, that if a single primary accelerator is operable, any apparent delay between X-ray and γ-ray emissions is due to other factors, such as "energy-dependent Coulomb losses in a trap model" (Kundu et al. 1985 and references therein). The GRD on Hinotori sees the same general time relations (Yoshimori et al. 1983a,b).

High-energy neutrons (>50 MeV) which arrive at the Earth after an intense solar flare on 1980 June 21 have been reported by Chupp et al. (1982); also, for a flare on 1982 June 3 (Chupp et al. 1983; Debrunner et al. 1983; Evenson et al. 1983). In the later event, a ground-level response of neutron monitors in Switzerland, Czechoslovakia, and Italy recorded the highest energy neutrons (>500 MeV) through an atmospheric nucleon cascade (Debrunner et al. 1983; Efimov et al. 1983) also, the neutron decay protons (20-100 MeV) were observed in space by detectors on ICE (Evenson et al. 1983). Neutral π^0 mesons were also produced, as expected in the above two events as evidenced by detection of the characteristic γ-ray spectrum (Forrest et al. 1984, Vestrand et al. 1984a). These results can be related to the spectrum of ions at energies up to ∿ 1 GeV.

Evidence for directivity of relativistic electrons has come from a recent study with SMM GRS data of the longitude distribution of the number of solar flares with photons >300 keV. Vestrand et al. (1984b) have shown that the number of such flares is enhanced (>3σ) near the limb (heliocentric angle >65°) relative to the number expected if photon bremsstrahlung emission is isotropic. This observation can be expected if the relativistic electrons are beamed downward toward the photosphere. This interpretation is supported by the observation that the limb enhancement of GRS events with photon emission (>10 MeV) is even stronger (>6σ) as expected since the higher energy bremsstrahlung (>10 MeV) is more anisotropic than for photons <1 MeV (Rieger et al. 1983).

SMM HXRBS has observed more than 7000 flares (1980-1984). Temporal studies have revealed that hundreds of flares have statistically significant hard X-ray spikes of less than 1 s. Further, "rise and decay times as fast as $\sim$20 ms" (Kiplinger et al. 1984) have been seen in X-rays for the first time. These observations, as well as earlier HXRBS data, are consistent with production of hard X-rays ($\sim$100 keV) as due to thick target interaction of electrons in the lower chromosphere. A likely model (Kiplinger 1984) is the "instantaneous injection of electrons into the top of a coronal loop" of height $10^8 < h < 3 \times 10^9$ cm above the interaction region.

ICE and PVO spectroscopic data (Kane 1983) give further information on the spatial structure of the sources of high-energy (>100 keV) photons. Some interesting conclusions are "that (1) the thermal model with adiabatic compression and expansion is not consistent with the observations of a magnetically-confined plasma and the thin target (non-thermal) model; (2) the thick target (non-thermal) model and the dissapative thermal model are partially in agreement with the observations; (3) the emission probably originates in many individual nonthermal sources distributed in altitude," where those at the lowest altitudes are brighter (Kane 1983). In the case of "gradual" or slowly increasing X-ray emissions, the explanation does not seem to be a purely coronal source model, but rather "a partial precipitation model with trapped as well as precipitating electrons." A white light flare on 1981 April 24, observed by the X-ray spectrometer on ICE, has shown a good temporal relationship between these two emissions. Kane et al. (1984) have concluded that the white light emission in this flare "is produced nonthermally in the upper chromoshpere, by the energy provided, by energetic ($\geq$25 keV) electrons if the source in the impulsive phase consists of an over-dense region above the altitude where the density is normally 10^{14} cm^{-3}." Earlier, Ryan et al. (1983) showed that ions (>30 MeV nucleon^{-1}) could <u>not</u> produce the white light emission in a flare on 1980 July 7.

3. Solar Constant

Solar constant observations on Nimbus 7 (Hickey et al. 1980) and SMM Active Cavity Radiometer Intensity Measurement (ACRIM) (Willson 1980) have continued to date. The data from both show a decrease of (0.03-0.05)% per year over the time period 1980-1982. This pattern is similar to that observed form Nimbus 6 over a 10 year period from 1968-1978. Earlier balloon observations apparently showed an increase of $\sim$0.04% per year. On the shorter time scale of days and weeks, both Nimbus 7 and ACRIM observe correlated variations of a few tenths of 1%. Also, it appears that sunspots on the solar disc cause a decrease in the solar constant. A new series of solar constant measurements was initiated on Spacelab 1 (Crommelynck et al. 1984).

4. X-Ray and μ-Wave Correlations

Recent precision comparisons of time histories of 10.6 GHz microwave and <100 keV hard X-ray emission (Cornell et al. 1984) indicate that "microwave emissions are delayed with respect to the X-rays by $\sim$0.2 s." Earlier observations (Takakura 1975; Crannell 1978) with time resolution of $\gtrsim$1 s showed a good correlation; the improved timing accuracy of the recent data now permits a study of the effects of geometry and acceleration time scale of electrons. It is too early to give a definite conclusion on this observation. The results could hold the key to understanding one of the important questions concerning electron acceleration.

At higher microwave frequencies (30 and 90 GHz) correlated observations (Kaufmann et al. 1984) were made with the HXRBS hard X-ray observations (>25 keV) during an impulsive burst on 1984 May 21. In this case the "hard X-ray time structures at energies >25 keV were almost identical to the 90 GHz time structures to <1 s." It is important to note that "The <u>onset</u> of the major 90 GHz burst structure was coincident (with) the hard X-ray <u>structure</u> to better than <u>128 ms</u>." This observation taken with those mentioned above might well indicate that <u>both</u>

the electrons producing the X-rays and microwaves are accelerated much closer in time than previously thought; that is to <128 ms. Thus onset time, rather than peak time is a more significant indicator of the accelerator or injector time scale.

5. Solar Flare Periodicity

From 1980 through 1984 the GRS has recorded 140 events with emission above 300 keV. A time series analysis of the GRS data base has shown that 1/3 of the events recur with a period of 155 days (Rieger et al. 1984). These authors further note that such behavior is also exhibited by GOES X-ray events of class M >2.5 and by the most intense HXRBS (>25 keV) events. It has been shown (Dennis et al. 1984) that the full HXRBS data base of 7000 flares, obtained over 4 years, also exhibits this behavior. Therefore, it appears that this phenomenon is not specific to γ-ray flares but is related to all flares, (i.e. at least X-ray flares). Studies are in progress by several groups to see if other solar phenomena show a similar behavior. It is possible that a connection of solar flares with the internal dynamics of the Sun has been discovered.

III. SPECIAL NOTE ON POST-SMM REPAIR CAPABILITIES

In 1984 April, the SMM satellite was retrieved by the Space Shuttle, repairs were effected in space and the satellite was relaunched. SMM was first launched in 1980 (see previous IAU Report). After 9 1/2 months of operation the precision pointing system and electronics systems for two experiments failed. Subsequently, the satellite attitude was approximately stablized by magnetic torquing, allowing excellent data to be accumulated for the next 3 1/2 years by 3 wide field instruments in X-rays, γ-rays and total solar irradiance. Since the repair, the original instruments with the exception of the Hard X-ray Imaging Spectrometer (HXIS), continue observing the Sun. It is noteworthy that within two weeks of the repair a major solar flare occurred on 1984 April 24. The pointed and wide field instruments obtained joint data of unprecedented quality. A summary of the prospects for the renewed mission has been given by Woodgate (1985) and by Suess et al. (1984).

The results of the first observations on SMM in 1980, the Solar Maximum Year (SMY), have been studied intensively in 1982 and 1983 at a series of NASA sponsored workshops (Kundu and Woodgate 1985).

IV. FUTURE PROGRAMS

The next major set of solar observations from space will take place on Spacelab 2, currently planned for launch in 1985 July. On this mission, payload specialists L.W. Acton and J.D. Bartoe will operate several solar telescopes:

NRL **Solar Ultra Violet Spectral Irradiance Monitor** - Fullsun - 1200-4000 A^o

NRL **High Resolution Telescope/Spectrograph**- (1176-1700) A^o $\Delta\lambda = 0.05$ A^o, spatial resolution 1"

Mullard Space Science Laboratory and Appleton Laboratory **Coronal He Abundance Experiment** - EUV (150-1300) A^o, 15"

Lockheed Scientific Laboratory **Solar Magnetic Velocity Field - Measurement in the form of an Optical Universal Polarimeter & Telescope** (4000-7000) A^o $\Delta\lambda = 0.030$ A^o spatial resolution 0.5"

A detailed description of this second U.S. manned solar mission since Skylab is described by Clifton (1982).

Development of a major solar instrument, the Solar Optical Telescope (SOT) is planned to start in 1985. This instrument will study the Sun on the spatial scale of 73 km or 0.1" of angular resolution using a primary mirror of 0.3 m diameter with diffraction limited operation in the UV, and overall wave length coverage from 1200-10,000 A°.

A European mission known as SOHO is also under development. This spacecraft is a part of a wider international program involving Europe, Japan and the US, which is known as the International Solar-Terrestrial Physics Program (ISTP). The SOHO is a three axis stablized spacecraft which will direct a variety of telescopes to within 10" of the solar center and will study the spectral and total luminosity of the sun, solar oscillations, the chromosphere, transition region and corona at high spectral resolution, outflow velocities in the inner corona and the solar wind shock. The overall wave length coverage is from (66-7000) A°.

The USSR is also planning solar observations from dedicated spacecraft, but no detailed information is available to the author at this time.

References

Bai, T.: 1982, Gamma-Ray Transients and Related Astrophysical Phenomena, eds. R.E. Lingenfelter et al., (American Inst. Physics: New York) p.409
Chupp, E.L.: 1984, Ann. Rev. Astron. Astrophys., 22, 359
Chupp, E.L. et al.: 1982, Astrophys. J., 263, L95
Chupp, E.L. et al.: 1983, Proc., 18th Int. Cosmic Ray Conf., Bangalore, 10, 334
Clifton, K.S.: 1982, NASA TM-82477, SL2 - Expt. Description, Marshall Space Flight Center Cornell, M.E. et al.: 1984, Astrophys. J., 279, 875
Crannell, C. et al.: 1978, Astrophys. J., 223, 620
Crommelynck, D. et al.: 1984, Science, 225, 181
Debrunner, H. et al.: 1983, Proc., 18th Int. Cosmic Ray Conf., Bangalore, 4, 75
Dennis, B.R. et al.: 1984, in EOS Transactions of Am. Geophys. Union, 65, 1067
Doschek, G.A. et al.: 1983, Solar Phys., 86, 9
Efimov, Y.E. et al.: 1983, Proc., 18th Int. Cosmic Ray Conf., Bangalore, 10, 276
Evenson, P. et al.: 1983, Astrophys. J., 274, 875
Forrest, D.J. and Chupp, E.L.: 1983, Nature, 305, 291
Forrest, D.J. et al.: 1984, Bull. Am. Astr. Soc., 16, No. 2, 475
Hickey, J.R. et al.: 1980, Science, 208, 281
Hundhausen, A.J.: 1984a, in EOS Transactions of Am. Geophys. Union, 65, 1067
Hundhausen, A.J. et al.: 1984b, J. Geophys. Res., 89, 2639
Kahler, S.W.: 1984a, Solar Phys., 93, 133
Kahler, S.W. et al.: 1984b, J. Geophys. Res., in press
Kane, S.R.: 1983, Solar Phys., 86, 355
Kane, S.R. et al.: 1982, Astrophys. J., 254, L53
Kane, S.R. et al.: 1984, Submitted to Astrophys. J. (Letters)
Kaufmann, P. et al.: 1984, Nature, in press
Kiplinger, A.L. et al.: 1984, To be published in Astrophys. J. (Letters)
Kundu, M. and Woodgate, B.: Proc. of SMM Workshop, NASA publication expected 1985
Orwig, L.E. et al.: 1980, Solar Phys., 65, 25
Ramaty, R. et al.: 1980, in Solar Flares, ed. P. Sturrock, (Assoc. Univ. Press: Boulder) p. 117
Rieger, E.et al.: 1983, Proc., 18th Int. Cosmic Ray Conf., Bangalore, 10, 338
Rieger, E. et al.: 1984, Nature, in press
Ryan, J.M. et al.: 1983, Astrophys. J., 272, L61
Sawyer, C. et al.: 1984, Submitted to J. Geophys. Res.
Sheeley, N.R. et al.: 1983, Astrophys. J., 272, 349
Sheeley, N.R. et al.: 1984, Astrophys. J., 279, 839
Suess, S. et al.: 1984, in EOS Transactions of Am. Geophys. Union, 65, 1067
Takakura, T.: 1975, in IAU Symposium 68, Solar, Gamma-, X-, and EUV Radiation, ed. S. Kane, p. 299
Vestrand, W.T. et al.: 1984a, Proc., XXV COSPAR Conf., Graz, Austria, p. 55
Vestrand, W.T. et al.: 1984b, Submitted to Astrophys. J.
Willson, R.C.: 1982, J. Geophys. Res., 87, 4319
Woodgate, B.: 1984, Proc., XXV COSPAR Conf., Graz, Austria, in press
Yoshimori, M. et al.: 1983a, Solar Phys., 86, 375
Yoshimori, M. et al.: 1983b, Proc., 18th Int. Cosmic Ray Conf., Bangalore, 4, 85

6. Space Observations of Solar System Objects

J. Caldwell

Solar system objects may be studied in space by two general techniques. Everyone is familiar with the exciting aspects of deep space probes: very high spatial resolution; *in situ* measurements of particles and fields; *in situ* chemistry studies by mass spectrographs and gas chromatographs; unique phase angle and occultation opportunities. However, the Solar system can also be studied to great advantage by observatories in orbit around the Earth. The broader spectral range available above the terrestrial atmosphere is as important for planetary studies as it is for investigations of more distant astronomical targets. Both techniques will be discussed in this brief report.

The period from July 1, 1981 to June 30, 1984, was somewhat of a lull following the extraordinary preceding three years, in which the two Voyager spacecraft had close encounters with Jupiter and Saturn and their rings and satellites. The highlights of those missions include: the measurement of trace gases in the planetary atmospheres and in Titan's atmosphere with concentrations of order parts per billion by means of infrared spectroscopy; the most active volcanic regime in the Solar system on the satellite Io; the rings of Jupiter; the thick atmosphere of the satellite Titan (with a column abundance approximately 10 times thicker than that of the Earth!); and the detailed structure of Saturn's rings. There were so many discoveries in that brief period that they have not yet been completely assimilated into the body of knowledge that is planetary science. The data have occupied many planetary scientists for the years of the current report period, and will do so for many years to come.

The planet which has been explored most actively during the current report period is Venus. This includes the continued activity of the United States' Pioneer Venus orbiter, which first encountered the planet on December 4, 1978. (In fact, this long-lived mission is currently being considered as a unique means to investiate comet Halley near perihelion in February, 1986, when the comet will be most active, but when it will not be visible from the Earth because of the Sun.) No fewer than four Soviet Venera spacecraft encountered Venus during the report period, including both landers and orbiters. Venera 13 and 14 arrived in March, 1982 and Veneras 15 and 16 arrived in October, 1983. A superb book has recently been published (Hunter *et al.*, 1983) summarizing Venus studies up to and including Venera 14. To highlight just one feature of this planet, it has been noted that there has been a recent pulse of SO_2 in the atmosphere, which has decayed, and which suggests that volcanism is currently active there.

Veneras 15 and 16 are in 24 hour, near-polar elliptical orbits, with periapses near 60° N. These orbits are intended to permit the mapping of the northern hemisphere by radar with a resolution of 1.2 km. Preliminary results show a variety of interesting geological features, including ridges and domes. The full import of this exciting mission has not been realized as of the time of writing this report.

At this time, there are a number of missions planned for the near future. The United States is well along with the Galileo orbiter/probe mission to Jupiter, and has begun a Venus Radar Mapper mission. The Voyager flight continues to Uranus (January, 1986) and to Neptune (August, 1989). The Soviet Union will send two probes to comet Halley, with a gravitational assist from Venus. Japan and the European Space Agency will also send probes to this most famous comet, in the first truly international armada of space vehicles with a common target. Arrival will occur in March, 1986.

With regard to Earth orbital astronomy, two United States satellites have made major contributions to planetary studies from space during the report period.

One satellite made broadband infrared measurements and the other did high resolution ultraviolet spectroscopy.

The brief life of the Infrared Astronomy Satellite (IRAS) occurred entirely within the report period. It was cryogen-limited. However, it completely changed the discipline of planetary science by one class of unexpected discoveries and made numerous other contributions.

The major discovery was that many stars near the Sun exhibit infrared emission in excess of that expected for a star of their respective types. The excesses are attributed to dust rings orbiting the stars. More than forty stars exhibit this characteristic, including Vega and Fomalhaut. These were not among any prominently mentioned leading canidates for extrasolar planets. Once again, scientists have been reminded of the importance of real data, despite the presence of very nice theories. It must be emphasized that the infrared emission is from a low density region of matter, not from a planet like Jupiter or the Earth. However, the list of 40 stars provides an excellent starting point for detailed searches at other wavelengths at higher spatial resolution.

IRAS also made observations of small particles orbiting the Sun. It discovered bands of warm zodiacal dust near the inner edge of the main asteroid belt and observed a dust tail on comet Bowell, outside the orbit of Jupiter. It even managed to discover two comets which subsequently approached the Earth more closely than any other comets this century. One was comet IRAS-Araki-Alcock, the first ever named for a spacecraft.

This leads to the other spectral extreme, with the International Ultraviolet Explorer spacecraft, a joint NASA/ESA/United Kingdom venture that has been operational since 1978 and continues even now. Among the discoveries made with the IUE in the report period was the emission of the S_2 molecule in comet IRAS-Araki-Alcock. This short-lived molecule was observable primarily because the comet came so close to the Earth. Other studies with this satellite include the Io-associated plasma torus around Jupiter, aurorae on Jupiter, atmospheric trace gas compositions on the giant planets and on Venus, and the discovery of an aurora on Uranus. Having pushed the ultraviolet frontier for planetary research below 200 mm for the first time, the IUE remains exceptionally valuable for studying the Solar system.

It is expected that the next report of this type, covering the period through June, 1987, will be dominated by the first results of the Hubble Space Telescope, a spacecraft with a primary mirror of 2.4m diameter, diffraction limited optics throughout the ultraviolet and spectrographs with resolving powers up to 2×10^5. The immediate future holds the promise of extraordinary advances in all areas of ultraviolet astronomy with the HST, and planetary science will participate fully in this adventure.

Reference

Hunter, D.M., Colin, L., Donahu, T.M. and Moroz, V.I.: Venus, 1983

45. STELLAR CLASSIFICATION
(CLASSIFICATION STELLAIRE)

PRESIDENT: V. Straižys
VICE-PRESIDENT: R. F. Garrison
ORGANIZING COMMITTEE: R. A. Bell, M. Golay, A. Heck, M. Jaschek, D. J. MacConnell, A. G. D. Philip, A. Slettebak

I. INTRODUCTION

During the report period (1982–1984) several symposia, colloquia, and workshops in the field of stellar classification were held. These include: IAU Symposium No. 111 "Calibration of Fundamental Stellar Quantities" (Como, May 24–29, 1984, ed. D. S. Hayes et al.), IAU Colloquium No. 76 "The Nearby Stars and the Stellar Luminosity Function" (Middletown, June 13–16, 1983, ed. A. G. Davis Philip and A. R. Upgren), IAU Colloquium No. 78 "Astronomy with Schmidt-Type Telescopes" (Asiago, August 30 – September 2, 1983, ed. M. Capaccioli), the workshop "The MK Process and Stellar Classification" (Toronto, June 6–10, 1983, ed. R. F. Garrison), the Colloquium "Cool Stars with Excess of Heavy Elements" (Strasbourg, July 3–6, 1984).

II. CLASSIFICATION USING SLIT SPECTROSCOPY (R. F. Garrison)

(a) O-and B-type Stars.

Wolf-Rayet stars in the Galaxy were the subject of several papers: Silversten (29.114.030), Thevenin et al. (31.114.051), Heap (32.114.133), Mendez and Niemela (32.114.136), Lamontagne et al. (33.114.101), Massey and Conti (34.114.029), Torres and Conti (ApJ 280,181), and Lundstrom and Stenholm (AA Suppl. 56,43). Other peculiar stars studied include He-strong stars (Drilling, 30.114.151; Walborn (33.114.082), subdwarfs (Drilling, 34.126.001), peculiar supergiants (Swings, 29.114.091; Turner and Drilling, PASP 96,292), high-latitude stars (Tobin and Kilkenny, 29.114.008; Hobbs et al., 32.155.060; Walborn, 33.116.007), Bp stars (Hobbs et al., 34.131.095), and white dwarfs (Wegner, 34.126.003). Apparently normal O- and B-type stars were classified by Forte and Orsatti (29.114.033), Walborn (29.114.045, 32.114.081), Drilling and Perry (30.114.079), Bassino et al. (32.113.043), Aab et al. (34.117.163), and Forbes (AJ 89,475). Classification criteria, standards and methods have been discussed by Walborn (31.114.023), Bisiacchi et al. (31.114.045), Barsukova et al. (33.002.001), and Turner and Lyons (33.114.095). Classification of slit spectra for ultraviolet objects was carried out by Dworetsky et al. (32.114.008, 32.114.168). An atlas of yellow-red spectra of OB stars has been published by Walborn (29.002.011).

(b) A- and F-Stars.

Wolff (34.003.159) has produced a comprehensive summary of the classification of normal and peculiar A stars. Bouw (29.114.147) classified supergiants in the infrared, and Böhm-Vitense (32.114.201) has suggested ultraviolet criteria for A-star classification. Gray, working with Garrison, has refined the MK System for the A stars and reclassified most of the bright A stars. Peña et al. (29.122.162, 31.122.015) classified δ Scuti stars, while Hauck and Slettebak (34.113.021) and Abt ("The MK Process") studied λ Bootis stars. The variable A-type supergiant suggested as an x-ray source identification by Garrison and Krzeminski in 1973, has been further studied by van Genderen et al. (34.113.005). Abt et al. have classified HD 164615 as a spotted star (34.122.063). Hauck (32.002.107) has published a supplemental catalogue of Am stars. Rao and Rajamohan (JAA India 3,267) have studied the peculiar star HD 147010. Glaspey (32.115.002) has identified HD 105262 as a possible horizontal-branch star and Garrison and Fernie (ApJ 285,698) have pointed out a serious discrepancy between photometry and spectral type for HD 161796, but reconfirmed Bidelman's type of F3 Ib. Slettebak (32.114.054) has published types for the brighter Be stars and A-F shell stars, and Lutz (ApJ 279, 714) has studied the unusual emission-line object HDE 330036.

(c) Late-type Stars.

Morgan et al. (29.114.049) have reviewed the classification of F and G supergiants and Keenan (34.114.024) has updated his large list of classifications for stars later than G0. Keenan (30.114.143, 31.114.212) has reviewed problems of classification for the red giant stars and Egret, Keenan and Heck (31.115.006) recalibrated the red giant absolute magnitudes. Red dwarfs were classified by Walker (30.114.013, 34.114.082) and by Cowley and Hartwick (31.111.001). Harlan (30.114.177) continued his series on classifications of F-G stars.

While still not using MK spectra or techniques, Hardorp (31.114.010, 31.114.047, 32.114.033) continued to obliquely discuss the MK type of the Sun. Garrison and Zimmerman (33.071.037) have determined integrated MK spectrum of the Solar disk: it was found to be consistent with a type of G2 V and not with a later type. Garrison is classifying all of Hardorp's candidates to the Solar analog with proper MK spectra to provide homogeneous types.

Barium stars were classified by Yamashita and Norimoto (30.114.201) and by Lü et al. (33.002.043). The S star Pi Gruis was studied by Murty (34.114.021). A G star surrounded by nebulosity has been classified by Vogt et al. (29.112.011) and the G-type supergiant HD 179315 has been rediscussed by Fernie and Garrison (30.122.027).

(d) Binaries and Multiples.

During this period, four comprehensive studies have been made: Abt (30.152.002) has classified 865 components of visual multiples, Corbally (ApJ Suppl. 55,657) has classified the components of 170 close visual pairs, Haldebel (PASP 96, 98 and IBVS Nos. 2549 and 2550)has given types for eclipsing systems, and Gahm et al. (33.118.003) classified visual doubles with early-type primaries. Some composite spectra were studied by Naftilan and Milone (30.120.006). Groote and Reimers (33.114.023) have classified the companion of HD 63032 and Menzies et al. (34.118.003) have classified a pair of late-type giants. Wolf-Rayet binaries were studied by Lamontagne et al. (33.114.101), Willis and Stickland (34.117.050), and Turner et al. (34.153.004). A, Am, and Ap binaries have been classified by Hube (30.118.015), Abt (29.118.009), Abt and Cardona (34.153.019 and ApJ 276,266), and Ginestet et al. (31.120.008). Close binaries were analyzed by Young and Schneider (30.117.007), Hutchings and Crampton (30.117.051), and Green et al. (ApJ 280,177). Allen (Proc. ASA 5,369) has compiled a catalogue of symbiotic stars, Huang (32.114.155) has classified the symbiotic star AG Dra, while Andrillat (32.114.148,149) studied the infrared spectra of symbiotic stars.

(e) Variable Stars.

Comprehensive classification studies have been made of Mira variables by Crowe (JRASC 78,103, "The MK Process" and ApJ Suppl., in press), of Cepheids by Garrison and Gauthier ("The MK Process" and ApJ Suppl., in press), of δ Scuti stars by Peña, Peniche et al. (29.122.162, 31.122.015, 33.122.150), and of helium-strong stars by Walborn (31.114.213). Walborn (29.122.022) has also found systematic variations in the spectrum of θ' Ori C.

Garrison et al. (ApJ 276,L13) have used classification techniques to discover the brightest-known cataclysmic variable, and van Genderen et al. (34.113.005) have confirmed Garrison's A2 Iaep classification of the variable supergiant CD-33°12119. Chincarini and Walker (30.117.097) has classified AE Aquarii primary as the K5 V star.

Lloyd Evans (Obs. 103,276) has studied carbon stars among type II Cepheids and Parsons (29.122.048) has classified the probable Cepheid HR 4511 and its hot supergiant companion. Chavarria (30.121.002) has studied a peculiar T Tauri star.

(f) Clusters and Associations.

Levato and Malaroda classified MK spectra in three young clusters (31.153.002, 32.153.037, ApLetters 24,37) and Turner, Forbes et al. (29.153.009, 30.153.011, 32.152.001, 32.153.036, 33.152.002, 34.153.004, ApJ 283,254) classified stars in several young clusters, especially those with an associated Cepheid. Herbst and Warner (30.121.001, 31.152.001) continued their series on classifications in R associations, while Marraco and Rydgren (29.152.005) and Vrba and Rydgren (ApJ 283,123) classified stars in T associations. Orion stars were classified by several, including Geutter (30.152.002), Derviz (34.114.036), and Walker (34.153.014). FitzGerald and Miller (34.153.013) classified stars in Stock 14. Claria has studied NGC 2547 (Bol. Ass. Astr. Arg. No. 27,115) and work is in progress on IC 2395 and several trapezium systems. Slettebak is classifying Be stars in NGC 663, the Perseus Double Cluster, NGC 2421, 2439, 2516, 2581, 3766, and 4755.

In work on intermediate-age clusters, Corbally and Garrison (33.153.018) classified stars in the lower main sequence of Praesepe and recently extended their work to K4 at B = 15.6 with good MK spectra. Arellano, Garrison,and Gray are re-classifying the bright Praesepe stars. Smith and Hesser (33.154.035) classified stars in NGC 2477 and Breger (AASuppl. 57,217) classified Pleiades stars. Rose (AJ 89,1238) classified Hyades and Pleiades stars, but not on the MK System. Gray, working with Garrison, is reclassifying all the A stars in several bright clusters. Common proper motion groups were studied by Lü and Tsay (31.152.010, 34.152.007).

Globular cluster stars were classified by Lloyd Evans (34.154.020, MNRAS 209,825), McClure and Hesser (31.154.036), and Smith and Norris (AJ 89,263). S-type stars in ω Centauri have been detected.

(g) Stars in galaxies.

WR stars in the LMC were classified by Breysacher et al. (31.159.030), Conti and Garmany (32.114.139, 34.114.004), Stahl et al. (34.122.067), and Walborn (31.159.014). WR stars were classified in the SMC by Moffat (31.159.029) and in M33 by D'Odorico and Rosa (31.114.169). Other OB star classifications in galaxies include those in the Clouds by Shore and Sanduleak (34.114.064, ApJ Suppl. 55,1), Walborn (33.159.004), and Crampton and Greasley (31.159.026). Humphreys (33.156.002) classified luminous stars in the SMC. Carbon stars in the Clouds were classified by Richer (29.159.005) and Bessell et al. (33.153.003). A CH star in the direction of the LMC has been described by Fehrenbach and Duflot (30.111.001). In the Clouds Lloyd Evans (34.156.004, MNRAS 208, 447) classified M and S stars and Wood et al. (34.156.008) classified long-period variables.

(h) General

In the volume "The MK Process" Mihalas, Morgan, and Keenan have presented some very important philosophical remarks concerning the distinction between the MK System and the MK Process by which it was established and Gray has discussed the application of new detectors to classification work. Morgan et al. (29.114.049) have discussed the classification of the Hertzsprung gap supergiants. Milone (29.114.129) has discussed classification at 42 Å/mm. The Jascheks (32.114.176) have given a general review of spectral classification. A new system of classification for white dwarfs has been suggested by Sion et al. (32.126.030). Danks and Dennefeld (32.114.197) have extended the MK classification system into the near infrared up to 11000 A. Greidanus and Tinbergen (32.002.100) and Jacoby et al. (ApJ Suppl. 56,257) have published the libraries of digitized spectra of stars classified in the MK System.

Hoffleit (31.002.007) has updated the Bright Star Catalogue with classifications from the literature and has extended (Supplement, 1984) it to include stars up to 7 mag. She has also discussed errata in the La Plata catalogue (29.002.016). Buscombe (31.002.078) continued to publish his catalogues without bothering to give references; his catalogues have been commented upon by Kharadze (32.002.105). Documentation for the Morris-Kennedy (1983) catalogue has been given by Warren (34.002.091). Benest (34.155.104) has discussed the distribution of types in the Solar neighborhood. MacConnell (in preparation) has classified 82 faint, high-velocity stars with a range of types from B to M.

III. OBJECTIVE-PRISM SPECTRAL CLASSIFICATION AND CLASSIFICATION INVOLVING AUTOMATIC METHODS (D. J. MacConnell)

(a) Objective-prism work in the Galaxy.

(1) **General.** The third volume of the Michigan Spectral Catalogue (32.002.126) has appeared containing N. Houk's MK types for 30314 HD stars between -40° and -26°. "A Second Atlas of Objective-Prism Spectra" (Houk and Newberry, U. of Michigan, 1984) has been published and it contains stars over a wider temperature and luminosity range than the first one. Vol. 4 of the Catalogue is to be published early in 1986 and it will have a northern cut-off at -12°. Bidelman (34.114.006) has reported on his survey of the first 100 plates taken for the northern HD reclassification project and listed 175 new peculiar and interesting stars. As of this writing, 50% of the plates necessary to cover the remaining northern sky have been taken. High proper motion stars were classified by MacConnell (34.114.128) and Lee (AJ 89, 702). Bidelman and MacConnell (31.123.033) have presented 27 southern stars of various peculiarity groups, and Jensen (30.114.080) has given MK types for 167 northern HD stars. Bartaya and Kharadze in the Abastumani were working on two-dimensional classification of 12000 stars to mag 13 in SA 44-67. Kuznetsov (31.114.088) has reviewed methods of low dispersion spectral classification.

(2) **Surveys in the Galactic plane.**

Hα-emission stars. Maehara (31.114.201) reports on 19 em. stars found on Kiso and Bosscha plates. Ogura and Hasegawa (34.121.005) searched 15 Bok globules and has found 48 new em. stars, and Ogura (PASJ 36, 139) has found 135 new ones concentrated around NGC 2264. Parsamian (31.114.066) has reported on the total number of em. stars in the Orion Nebula region (~450) including her new findings, and Parsamian and Chavira (34.002.132) have compiled a major catalogue of em. stars in the Orion Nebula region. Tsvetkova and Tsvetkov (31.114.161) has discussed their new em. stars in the region of γ Cyg. MacConnell (31.114.161) has found 107 new em. stars along the southern plane, and Vega (31.114.092) has found 108 in Vela. Kun (32.114.004) has found 110 new em. stars in two fields with several dark clouds. Duerr et al. (32.121.009) have found 21 new em. stars in the region of the λ Ori cloud complex. Arkipova (33.114.084) has published a review of spectral observations of hot em. stars observed with the 50-cm telescope at Crimea, and Kohoutek (34.013.079) reports on the Hamburg/Calar Alto survey of the northern Milky Way. MacConnell and Coyne (Publ. Vatican Obs. 2, Nos. 5, 6, 1983) have revised Parts I–V of the Vatican survey and have given 215 new stars along the northern plane.

Early-type stars. Vega and Muzzio (30.115.024) have discussed the relation between types determined from thin-prism plates and those of higher dispersion for OB stars. Peton-Jonas (30.155.013) gives types for 102 O and B stars in Puppis. The following papers contain classification in the vicinities of the open clusters: Kun (33.152.005) at NGC 7160, Paparo and Balazs (33.155.124) at IC 4665, Gieseking (33.153.002) at Cr 70, Kalandadze and Voroshilov (33.132.046) at NGC 1499, and Christian (32.153.010) at Be 21.

Late-type stars. Maehara (PASJ, in press) has classified carbon stars on the C-system in Cas, Raharto et al. (Ann. Tokyo Obs., 19, No. 4) have published a catalogue of over 2000 M stars, and Hamajima et al. (31.114.202,203) have presented results of their collaborative work using the Kiso and Bosscha telescopes searching for cool stars toward the Galactic center. Mikami et al. (32.155.009) have classified IR sources in Sct. Alksnis et al. (30.114.191, 34.114.047,100; Investig. of the Sun and red stars, 19, 40) have discussed their discoveries of new carbon stars at Baldone. Kurtanidze and Nikolashvili (31.114.065) have reported 40 new carbon stars in Cas. Smriglio and Nandy (31.114.194,195) have discussed spectrophotometry of M stars in Mon and Cep.

(3) Surveys out of the plane.

Stephenson (34.155.103) has completed his visual-red survey north of dec. -25° and further than 10° from the plane and has discovered about 300 new Hα emission stars and over 1000 new K and M dwarfs. Stock (Rev. Mex. A. A. 9, 77) has published the details of his catalogue of over 10000 stars giving positions, types, and indicative radial velocities; the catalogue may be obtained by writing to him. As a result of the Case northern sky survey (33.002.004 and ApJ Suppl. 55, 517), Sanduleak is preparing for publication a list of 184 early-type stars and suspected blue horizontal-branch stars toward the NGP. He also is finding possible new HB members of globular clusters at considerable angular distances from them. Fehrenbach and his collaborators continued their classification and radial velocity work toward the MC (30.111.001 and AA Suppl. 55, 103) and in 4 fields a b $\sim 30^o$ (32.111.001). Schiller (31.155.038) surveyed 840 sq. deg. at the SGP and has found 183 M stars to m_{pg} 14.7. Recillas-Cruz (32.114.108) searched for early-type stars in the direction of the Magellanic Stream. Erculiani (31.141.086) has classified 186 blue objects in the field of 88 Leo. Robertson (34.155.131 and AJ 89, 1229) has presented his work on the kinematics and lum. function of late M dwarfs found on red objective prism plates. Smethells (34.155.134) has given a reanalysis of his southern early M-dwarf survey. Staller et al. (31.155.003) used grism plates to examine the nature of 2600 red stars at the SGP.

(b) Objective-prism and grism work in the Magellanic Clouds and other galaxies.

Sanduleak and Pesch (30.159.008) have reported finding 5 possible new VV Cep stars in the SMC. Prevot et al. (34.156.001) have presented nearly 200 K and M supergiants in the SMC. Rebeirot et al. have detected over 260 new M stars which are presumably members of the LMC. Philip and Sanduleak (33.156.016) have presented their comprehensive list of LMC OB and supergiant stars. Shore and Sanduleak (ApJ Suppl. 55, 1) have given a detailed description of the spectra of 21 early-type, strong emission stars in both Clouds. Mould et al. (31.158.042) have reported finding 8 carbon stars in the Carina dwarf spheroidal galaxy.

There has been much recent activity using "grisms" at the prime focus of major telescopes, and much of this effort was directed toward members of the Local Group. Blanco and McCarthy (34.156.007,012) have discussed their findings on the distribution of C and M giants in the MCs from their near-IR grism spectra, and in IAU Symp. 108 ("Structure and Evolution of the Mag. Clouds", 1983) they have presented their classification scheme. In AJ 89, 636 they and B. Blanco have given the results of their survey for M giants in the Baade's Window. Blanco et al. (30.159.009) have reported the discovery of the first S star in the SMC. Westerlund et al. (33.114.098) have described their program to search for WR stars in galaxies. Several groups have reported their results of surveys for carbon stars in Local Group galaxies: Richer and Westerlund (33.157.003) and Aaronson et al. (31.158.043, 32.158.133,303,304; 33.157.031, 34.157.143). Carbon stars have been found in 7 of the local dwarf spheroidal galaxies.

(c) Automated Classification.

Several groups were active in this area and have reported substantial progress. Schmidt-Kaler (32.031.648) has given a survey report on the goals and methods, and Kurtz ("The MK Process" workshop) has also presented an excellent review. The methods may by divided into three groups: a) quantitative measurements for evaluation of criteria, b) pattern recognition and cross-correlation techniques, and c) combined procedures. Malyuto et al. (30.031.653, 32.114.065, 32.031.655, 33.114.108, and IAU Coll. 78, Asiago, p. 287) are developing the method to determine temperatures, luminosities, and metallicities from low dispersion objective-prism spectra digitized with a PDS. For calibration they use about 150 F-K standard stars and apply a linear polynomial regression model for the fitting. Zekl (31.031.542, 32.031.650) uses Cassegrain spectra for 2-dimensional classification of normal stars in the range B0-K3. Tobin and Nordsieck (30.114.092) have obtained spectra of 26 weak-metal stars with an intensified diode-array scanner over a wide wavelength range and have made mathematical models of the absorption features'dependence on color, luminosity, and metallicity. Kelly et al. (31.114.018) have described their program of microdensitometry of very low-dispersion UK Schmidt plates. Ratnatunga and Freeman (IAU Coll. 78, Asiago) present their automated analysis of plates to locate distant field halo K giants and report a high rate of success. Kurtz (32.031.649) has made important progress in the use of pattern recognition, cross-correlation, and Euclidean metric techniques over the whole MK spectral domain. Bijaoui (30.111.005, 32.031.653, 33.036.129) also applies pattern recognition methods to problems of stellar classification. Simien (32.031.652) uses cross-correlation to find rough spectral classes which are then improved by the evaluation of specific criteria. Luminosity classification presents too many difficulties, however.

Automated classification techniques has been also discussed at the June 1982 meeting of the Scientific Council of the C. D. S. (Bull. Inf. CDS, No. 23) and at the Sept. 1982 "Workshop on Astronomical Measuring Machines" (33.012.059).

IV. CLASSIFICATION FROM EXTRAATMOSPHERIC SPECTRA (A. Heck)

Most of the work in this field during the period was based on spectra collected by the International Ultraviolet Explorer (IUE). A workshop on UV stellar classification has been organized in October 1981 at VILSPA (32.012.062 – ESA SP-182) with contributions of Cucchiaro (32.114.195), Jaschek and Jaschek (32.114.196), Wesselius (32.113.059), Wu et al. (32.002.115), Heck (32.013.095), Benvenuti (32.002.116), Danks and Dennefeld (32.114.197), Nandy (32.114.198), Praderie (32.114.199), Baschek (32.114.200), and Böhm-Vitense (32.114.201), reviewing the classifications carried out from TD1 and ANS data, as well as the work in progress with IUE spectra, together with theoretical considerations. At the Toronto MK workshop, the Jascheks presented their morphological classification from a sample of IUE low-resolution spectra of O–G normal stars, on which the first volume of the IUE low-dispersion reference atlas (Heck et al., 1984, ESA SP-1052) has been built. An objective statistical classification on the same material has been carried out by Heck et al. (32.114.102; ESA SP-218, 1984, p. 257), confirming Jascheks' frame and the fact that one cannot simply transfer MK classifications to the ultraviolet spectral range. Walborn and Panek ("The MK Process" workshop and ApJ, 280, L27) pointed out Si IV as a luminosity criterion for O-type stars from a sample of IUE high-resolution spectra. The resulting frame allows also to identify some anomalies in the CNO features among the ON and OC stars. Walborn et al. (ESA SP-218, 1984, p. 255) also announced an IUE high-resolution atlas of O-type spectra based on these discriminative features. Rountree et al. announced (IAU Symp. 111, Como) a classification of B stars based on IUE SWP high-resolution spectra. Baschek et al. (AA 131, 378) offered a discrimination of λ Bootis stars based on strong identified (CI 1657 Å and CI 1931 Å) and unidentified (1600 Å and 3040 Å) absorption features. Henize et al. (30.114.163) used low-dispersion UV spectra from the S-019 Skylab experiment for a classification of O–B2 stars, particularly by using the Si IV/C IV ratio as a luminosity criterion among O stars, and as a temperature criterion among O9–B2 stars of lower luminosity. Sakhibullin and Solovjev (32.114.177) investigated a possibility of spectral classification of B0-Al stars based on the ratio I(2831 Å)/I(2843 Å) from TD1 observations. Ultraviolet classification was discussed for F-G dwarfs by Garcia-Alegre (29.114.061) and for carbon and S stars by O'Brien and Johnson (32.114.216) and Johnson (Coll. Cool Stars with Excess of Heavy Elements, Strasbourg, 1984). The IUE Ultraviolet Spectral Atlas (Wu et al., 34.002.083) containing a wide representative set of spectral types and luminosities has been published. Wing et al. (34.002.086) has produced an atlas of IUE high resolution spectra of late-type stars, and Benvenuti et al. (34.002.085) of supernovae. Koornneef et al. (31.113.009) have published an atlas of energy distributions for 531 stars in the wavelength range λ 1330–5500 Å based on OAO-2, TD-1, and ANS results.

V. CLASSIFICATION USING MULTICOLOR PHOTOMETRY (V. Straižys)

(a) Wide-Band Systems

Several varieties of the superwide photographic UJFN system intended for crude classification of stars, galaxies and quasars were used (Kron and Chiu, 30.141.074; Koo and Kron, 31.141.016; Couch and Newell, 32.113.005). Schild and Kent (SPIE 290, 186, 1981) have realized a similar system with the CCD. The photometric systems to be used in the Space Telescope and Hipparcos – Tycho experiment were discussed by Westphal (32.032.513), Hog et al. (31.051.019,020), Argue (32.051.078) and Granes and Mignard (34.113.062). The arguments in favour of the revision of the UBV system have been presented by Straižys (34.113.059). Observations in the revised WBVR system have been published by Dorokhov et al. (Problemy kosm. fiziki, Kiev, 19, 102). New intrinsic color indices in the VRIJHKL system have been published by Wing (34.114.045), The et al. (AA 132, 385) and Koornneef (34.113.025). Discovery and investigation of very cool infrared stars in 1983 is related with the IRAS satellite which made an infrared sky survey at 12, 25, 60, and 100 μm (see ApJ Letters 278, No. 1). The RGU system was used to recognize the metal-deficient stars (Buser, 32.155.054; Spaenhauer et al., 31.031.571; 32.155.057; Messenger 36, 6; Thevenin et al., 34.113.006). The Washington system was used for metallicity determination of red giants in open clusters (Claria and Lapasset, 34.153.042; MNRAS, in press). The classification properties of the Washington system were investigated by Straižys and Kurilienė (Bull. Vilnius Obs. 62, 39).

(b) Medium-Band Systems

1. **The uvbyβ system.** Observations and studies of the following objects have been published: field A5-G0 stars (Olsen, 34.002.029; AA Suppl. 56, 229), B-type stars (Westin, 32.113.003; Tobin, AA Suppl. 56, 221; 57, 427, and Zhang, 33.113.053), F-type stars (Eggen, AJ 89, 1606), late-type dwarfs (Olsen, AA Suppl. 57, 443), stars nearer than 100 pc (Perry et al., 32.131,296,309), NGP region (Hill et al., 33.113.070,071; 33.002.067; 34.131.023), SGP region (Degewij, 31.113.089; McFadzean et al., 34.113.023, and Twarog, AJ 89, 523), SA 128 and 156 (Knude, 31.113.090), ten areas at high gal. latitudes (Knude, 29.113.053), open clusters (Stetson, 30.153.024; Schmidt, 31.153.042,047; 32.153.004; 33.153.004; ApJ Suppl. 55, 455; Eggen, 32.153.043; 33.153.005,006,007,009; 33.113.052; Twarog, 33.153.017; Shobbrook, 34.153.031; MNRAS 206, 273; Anthony-Twarog, AJ 89, 655), globular culsters (Richtler and Nelles, 33.154.019 ; Ardeberg et al., 34.154.039), binary stars (Lindroos, 29.118.024; Olsen, 31.064.100; 31.113.020; Wolf and Kern, 34.119.027), high velocity A stars (Stetson, 30.111.016), white dwarfs (Koester and Weidemann, 31.126.023; Wegner, 33.126.006), metal-dificient giants (Ardeberg and Lindgren, 31.113.096; 32.113.025), RR Lyrae-type stars (Siegel, 31.122.079), Cepheids (Eggen, 34.122.006), hot subdwarfs (Bergeron et al., AJ 89, 374), Ap stars (Manfroid and Renson, 29.113.021; 33.113.021). The Strömgren system was used to classify stars in spectral types and luminosities by Oblak and Chareton (30.114.081) and Westin (Stockholm Rep. 25). The method of photometric boxes was used by Philip and Egret (33.113.055). Intrinsic color indices were discussed by Burkhart and van't Veer (30.113.039) and Shulov (Trudy Leningrad AO, 39, 54). Axial rotation effects were investigated by Schmidt and Forbes (MN 208, 83).

2. **The Geneva system.** Rufener (30.002.008) has published the third Geneva photometric catalogue, containing 14633 stars. Nicolet (29.113.023; 30.153.028; 31.113.008; 31.002.064) developed the method of photometric boxes for classification of stars. Egret and Heck (Proc. Statistical Methods in Astronomy Symp., Strasbourg, 1983, 149) have presented a progress report of the prediction of MK classification from photometric indices in the Geneva system. The classification of B-type stars was developed by Cramer (32.113.014 and AA 132, 283), the temperature criteria have been discussed by Meylan and Hauck (30.113.038 and IAU Symp. No. 111, Como). The following types of stars were investigated: the nearby stars (Grenon and Rufener, 30.113.017), the nearby open clusters (Hauck, 29.153.042; Golay and Mauron, 31.153.005), Ap stars (North and Cramer, 30.113.031; Cramer and Maeder, 30.113.032; Hauck and North, 32.113.026; Hauck, "The MK Process" workshop; North, AA Suppl. 55, 259), Am stars (Hauck and Lovy, 30.113.035; Nicolet and Cramer, 33.113.008; Nicolet, 33.113.020), variable stars (Rufener and Bartholdi, 31.123.030), β Cephei stars (Waelkens, 29.113.024), metal-deficient giants (Grenon, 31.154.055).

3. **The Vilnius system.** A number of areas was investigated through two-dimensional photometric classification of stars and datermination of their interstellar reddening: the Taurus dark clouds (Straižys, Meištas et al., 29.131.210; 29.131.243; 29.122.046; 30.131.203; 31.114.027; 32.114.021; 32.131.169; 32.131.901), the area of open cluster IC 4996 (Pučinskas, 32.113.063; 32.131.306; Bull. Vilnius Obs. 65), the area of globular cluster M71 (Janulis and Straižys, Astrophys. Space Sci. 100, 95; Bull. Vilnius Obs. 67, 18), the area of globular cluster M56 (Smriglio, Boyle, Nandy, Straižys et al.), the Scorpio-Ophiuchus dark cloud (Straižys, Bull. Vilnius Obs. 67, 3), the NGP area (Bartašiūtė, Bull. Vilnius Obs. 68, 33). The following types of stars were investigated: metal-deficient giants by Bartkevičius et al. (30.114.161; 32.113.067; Bull. Vilnius Obs. Nos. 63, 66, 68), G-K subdwarfs by Straižys et al. (Astrophys. Space Sci., 104, 219), carbon, zirconium, barium and CH stars by Dzervitis and Paupers (30.113.071,072,075; 34.113.017), Bartkevičius and Šleivytė (Bull. Vilnius Obs. 64, 24) and Šleivytė (Bull. Vilnius Obs. 69), T Tauri-type stars by Meištas (32.121.049). The system was calibrated in terms of spectral types, absolute magnitudes, temperatures and gravities by Straižys et al. (30.113.054; 32.113.064,065).

The Vilnius-Geneva (VilGen) system has been described by Straižys et al. (32.113.066) and North et al. (31.113.018).

4. **DDO system.** Meylan (31.002.003) has prepared the catalogue of DDO observations of 2911 stars on magnetic tape. Other catalogs of the field giants have been published by McClure and Forrester (30.002.077) and Dean (30.113.080). The system was used to investigate the red giants in open clusters (Janes, 30.153.017; Smith, 31.153.007; 33.153.036; Pastoriza and Ropke, 34.153.041; Claria and Lapasset, 34.153.042; Janes and Smith, AJ 89, 487; Schmidt, ApJ Suppl. 55, 455), in moving groups (Lü, 31.152.010; Lü and Tsay, 34.152.007; Smith, 32.153.006; 34.152.011), at the Galactic Poles (Hartkopf and Yoss, 32.155.062), red dwarfs (Upgren et al., 30.113.051), metal-deficient giants (Pastoriza et al., 31.113.088), Cepheids and supergiants (Dean, 30.122.118), barium stars (Lü et al., 33.002.043), RV Tauri and SRd stars (Mantegazza, AA 135, 300), red variables (Bergmann et al., Astrophys. Space Sci. 100, 341).

5. **Walraven system.** The system was used to investigate open clusters (The et al., 31.153.012; Bakker and The, 33.153.015; Verschoor and van Genderen, 34.153.009; Waard et al., AA Suppl. 56, 373), field Cepheids (Coppola et al., 32.122.060), Magellanic cloud Cepheids (van Genderen, 33.122.030, 068; 34.122.014), Magellanic Cloud supergiants (van Genderen et al., 31.113.010; 33.113.007), Ap stars (Deul and van Genderen, 33.113.014; Weiss and Schneider, AA 135, 148), Am, δ Del, δ Sct stars (Wiertz and van Genderen, 33.113.037), metal-deficient giants (Nelles et al., 34.114.122 and AA 139, 220), SGP stars (Trefzger et al., 34.114.121 and Messenger 35, 32).

5. **Arizona 13-color system.** In the system the following objects were investigated: T Tauri-type stars (Chavarria and de Lara, 32.121.018), Be stars (Alvarez and Schuster, 31.122.225; 32.113.024; 33.113.012,034,059), O-type stars (Schuster, 33.113.032), B-type stars (Schuster, Rev. Mex. AA 9, 53).

(c) Narrow-Band Systems

Narrow band Hβ photometry was used together with medium-band uvby photometric system (see b1 section of this report). The measurements of other hydrogen lines have been published by Ducati (30.002.002; 31.002.041), Claria and Escosteguy (30.113.045), Zeuge (31.115.013), Cester et al. (33.113.064). Mendoza et al. (31.113.045; 33.113.013) have used the $\alpha\Lambda$ system to measure the intensities of Hα and OI at λ7774. Maitzen et al. (30.113.020,030; 30.153.004; 32.153.024; 33.113.056; 34.153.012; 34.155.086; AA 138, 189) measured the λ5200 depression in Ap stars.The gnkmfu system was used to investigate G–K type stars at the NGP (Hansen and Radford, 34.113.008) and G dwarfs (Kjaergaard, AA Suppl. 56, 313). The bands of NH, CN, CH, and C_2 in carbon stars were measured by Yorka (34.113.041). TiO bands were measured by Mould et al. (31.113.083; 31.154.100; 32.154.024). Four-color system measuring CaH and TiO bands was used by Jones et al. (29.113.001; 34.002.066). The infrared Wing system was used photographically by Palmer and Wing (32.031.667) for identification of M and carbon stars. Model atmospheres were used to calibrate the Wing system (Johnson et al., 29.064.083; 31.064.099). Wing and Rinsland (31.114.184; 32.113.023; 32.114.114) have investigated the suitability of band strengths in 1–4 μm region for classification purposes. Frogel, Cohen and collaborators continued to measure the infrared bands of CO and H_2O in globular cluster giants (29.154.032; 30.154.013; 34.154.060), M and carbon stars in the Magellanic Clouds and the Galaxy (30.115.014), and K giants in the Galactic central bulge (AJ 89, 1536). Elias et al. (32.113.001) have published a list of CO and H_2O band standards. Greenstein (32.126.003 and ApJ 276, 602) continued to use the narrow-band UBGVRI system for white dwarfs. Similar system was used by Weidemann and Koester (AA 132, 195) and Oke et al. (ApJ 281, 276).

VI. CATALOGUES AND ATLASES (C. Jaschek)

The years covered in this report testimony a rapid growth in the number of catalogues and atlases. Lists of new catalogues are regularly published in the "Bulletin d'Information du Centre de Donnees Stellaires", Strasbourg and the "Astronomical Data Center Bulletin NSSC-NASA", Greenbelt.

Spectroscopic data bases were discussed by C. Jaschek in "The MK Process" workshop and data bases for photometry and other parameters by C. Jaschek in IAU Symp. 111, Como. The compilation of photometric data was discussed by Mermilliod (37.002.025). Ševarlic et al. (34.002.061) have provided a bibliography of photographic catalogues of star positions, which list usually some kind of photometric magnitude.

In what follows, catalogues are divided into (A) observing lists and (B) compilations. The note (T) means that the catalogue is available on magnetic tape from one of the data centers. Lists of type (A) are included if they contain more than about one thousand objects. The section "Atlases" includes spectrophotometric atlases.

(a) Catalogues of Type A; Spectroscopic

N. Houk: Michigan catalogue of two-demensional spectral types for HD stars. Vol. 3. Declinations -40° to -26° (32.002.126). (T)

(b) Catalogues of Type B; Spectroscopic (including lists of standard stars)

A. G. Agayev, O. H. Guseinov: Catalogue of white dwarfs (31.002.012).

C. D. Garmany: Catalog of Galactic O-type stars (34.002.101).

I. N. Glushneva: List of spectrophotometric standards (34.113.009).

P. C. Keenan: Revised MK spectral standards later than G0 (34.114.034).

M. Jaschek, D. Egret: Catalog of stellar groups. Part one: the earlier groups (34.002.079). (T)

G. P. McCook, E. M. Sion: A catalogue of spectroscopically identified white dwarfs (37.002.084).

A. Bartkevičius, J. Šleivytė: CH and metal-deficient barium stars and their color excesses, Bull. Vilnius Obs. No. 64, 24, 1983.

A. Bartkevičius: Metal-deficient giants in the Galactic field. Catalogue and some physical parameters, Bull. Vilnius Obs. No. 66, 1984.

A. Bartkevičius: Catalogue of metal-deficient F–M stars. Part I. Stars classified spectroscopically. Supplement 1, Bull. Vilnius Obs. No. 68, 1984.

(c) Catalogues of Type A; Photometric

J. Dachs et al.: UBV-Hβ photometry of luminous stars between $l = 335^o$ and $l = 6^o$ (32.113.031).

J. R. Pier: Stars of spectral types A and B in the southern galactic halo. I. UBV photometry (32.113.046).

P. R. Wesselius et al.: ANS ultraviolet photometry, catalogue of point sources (32.002.001). (T)

T. N. G. Westin: Four colour and Hβ photometry of O-A0 stars in three regions near the galactic equator (32.113.003).

D. Wiedemann: RGU photometrie eines Sternfeldes in $+23^o$ galaktischer Breite in Richtung zum galaktischen Zentrum (32.113.018).

J. D. Fernie: New UBVRI photometry of 900 supergiants (33.113.042).

G. Hill et al.: Further observations of A and F stars in the region of the North Galactic Pole. IV. A catalogue of uvbyβ photometry and derived quantities (33.002.067).

I. A.Dubyago: Catalogue of magnitudes and colour indices of 2220 stars in Cygnus (37.113.053).

E. H. Olsen: Four colour uvby and Hβ photometry of A5 to G0 stars brighter than $8^{m}.3$ (34.002.029). (T)

E. H. Olsen, C. L. Perry: A catalogue of Hβ photometry of southern A5 to G0 stars brighter than $8^{m}.3$ (37.111.035). (T)

K. Loden: Photometry in the uvbyβ system in some galactic regions (B-type stars; $6^{m}.5-10^{m}.2$; 1700) (in preparation)

A. Pučinskas: Photographic photometry of stars in the region of open cluster IC 4996 in the Vilnius photometric system (32.113.063).

(d) Catalogues of Type B; Photometric

J. Ducati: A catalogue of observations in Hα (30.002.002). (T)

M. Grenon, F. Rufener: The colours, magnitudes and parallaxes of the nearby stars (30.113.017). (T)

R. D. McClure, W. T. Forrester: A catalogue of homogeneous photometry of bright stars on the DDO system (30.002.077). (T)

F. Rufener: Third catalogue of stars measured in the Geneva Observatory photometric system (30.002.008). (T)

J. C. Mermilliod: UBV data (1976–1982): a complete supplement to the general catalogue (34.002.121). (T)

D. Y. Gezari et al.: Catalog of infrared observations (37.002.079). (T)

(e) Atlases (including spectrophotometry)

I. N. Glushneva (ed.): Spectrophotometry of bright stars (33.003.123).

R. F. Wing, K. G. Carpenter, G. M. Wahlgren: Atlas of high resolution IUE spectra of late-type stars, 2500–3230 Å (34.002.086).

E. A. Barsukova et al.: An atlas of the spectra of early supergiants. I. Procedure and its application to the spectra of eight O9.5–A2.5 supergiants in the range λλ 3819–4927 Å (37.114.099)j

A. Heck, D. Egret, M. Jaschek, C. Jaschek: IUE low-dispersion spectra reference atlas. Part I. Normal stars (37.002.045).(T)

N. Houk: A second atlas of objective-prism spectra (37.002.088).

S. T. Ridgway et al.: An atlas of late-type stellar spectra 2400–2778 inverse centimeters (37.114.055).

C.-C. Wu et al.: The IUE ultraviolet spectral atlas (37.002.048).

N. S. Komarov et al.: Spectrophotometry of stars in λ550–900 nm, Kiev, 1983.

J. E. Gunn and L. L. Stryker: Stellar spectrophotometric atlas, λ3130–10800 Å (33.002.042). (T)

G. H. Jacoby et al.: A library of stellar spectra, Astrophys. J. Suppl. 56, 257, 1984. (T)

(f) Miscellaneous

G. Cayrel de Strobel et al.: A catalogue of [Fe/H] determinations (30.002.001).

D. Hoffleit, C. Jaschek: The bright star catalogue. Fourth revised edition (31.002.077). (T)

A. N. Argue: A catalogue of photometric sequences (suppl. 3) (33.002.029). (T)

A. Fernandez et al.: The first dictionary of the nomeclature of celestial objects (solar system excluded) (33.002.035).

D. Hoffleit et al.: A supplement to the Bright Star Catalogue ($m \leqslant 7.01$) (37.002.007). (T)

M. S. Kazanasmas et al.: Atlas and catalogue of photoelectric standards in stellar areas, Kiev, 1981. (30.002.076).

M. S. Kazanasmas et al.: Atlas of photometric standards in stellar areas, Kiev, 1982. (33.002.011).

V. STRAIŽYS
President of the Commission

COMMISSION 46 : TEACHING OF ASTRONOMY - ENSEIGNEMENT DE L'ASTRONOMIE

PRESIDENT : L.N. Houziaux
VICE-PRESIDENTS : M. Rigutti, C. Iwaniszewska

ORGANIZING COMMITTEE : J. Kleczek (Secretary of ISYA), D. McNally (Editor of Newsletter,ICSU CTS representative), D.G. Wentzel (Past-president of Commission, publisher of Newsletter), E. Kononovich, M. Gerbaldi, R.R. Robbins (Astr.Education Materials), S.E. Okoye, A. Sandqvist, S. Ferraz-Mello (Visiting Lecturer Program)

MEMBERSHIP

There has been no change in membership since the Patras General Assembly.

WORK OF THE COMMISSION

The various activities of the Commission have been actively pursued.

1. International Schools for Young Astronomers.

The program has been continued thanks to the financial help of IAU, Unesco, and of local authorities. A school has been held in Lembang, Indonesia from May 16 to June 2, 1983, with students from India, Indonesia, Malaysia, Philippines, and Thailand. The Local Organizing Committee, chaired by B. Hidayat was very efficient, and the school is considered to be one of the most successful held to date. Both teachers and students expressed their appreciation to the School Secretary, Dr J. Kleczek, and to the IAU Executive Committee. Another school, foreseen in Venezuela for September 1983, had to be cancelled due to the lack of support of the Venezuelan Research Council. There are plans to hold several schools during the 1985-1988 period.

2. Visiting Lecturers Projects.

At the last session of the General Assembly, it has been decided in Resolution A3 to support a contract scheme "for Visiting Lecturers with a sponsoring institution in a country newly entering into astronomical work". A subcommittee (set up jointly with Commission 38), under the chairmanship of Dr D.G. Wentzel has reviewed the various possibilities (see IAU information bulletin 50, p.14).
A first contract, initiated by Dr M.L. Aguilar, has been signed in March 1984 between the IAU and the San Marcos University at Lima, Peru. The coordinating committee is chaired by Dr J. Sahade, who, as the first lectured, started the courses in September 1984. A second contract is presently negotiated with the University of Nigeria, for a school in Nsukka, Nigeria. Dr D. Walsh has accepted to chair the coordinating committee for this project, organized locally by Dr S. Okoye.

3. Newsletter.

The Commission Newsletter is issued twice a year, roughly in January and June. During the period under review, issues n° 13 to 17 have been published and distributed. They include over 50 pages of articles, covering a wide variety of subjects, reporting on the current status of astronomy teaching in universities or stating the availability of new teaching material. We recently had the pleasure to welcome a first contribution by our Chinese colleague, Dr Du-Shenyun, from Beijing. The Newsletter Editor is Dr D. McNally ; as in the past, the Newsletter is produced and distributed by Dr D.G. Wentzel.

4. National Reports.

Every three years the Commission publishes (currently as an issue of its Newsletter), a collection of reports from National Delegates. This publication gives an overview of the astronomy teaching activities at general public, school and university levels troughout the world. The report for the period 1983-1985 will appear around May 1985.

5. Astronomy Education Materials.

This series of publications, started in 1970, is brought up to date every three years. It is at present also published as an issue of the Newsletter. The 1982 edition, issued as Newsletter n° 12, met with considerable success. It has been reprinted in 1983 with financial support of the IAU and most of the copies are nowadays distributed to institutions and individuals requiring them. The 1985 update is presently collected by Dr E. Kononovich (material in Slavic languages), Dr R. Robbins (material in English), and Dr M. Gerbaldi (other idioms). It is foreseen that the English part of A.E.M. will be published jointly with the American Association of Physics Teachers.

6. Education sessions at IAU Regional Meetings.

A full day has been devoted to astronomy education during the 8th IAU Regional Meeting, held at Toulouse (France), September 17-21, 1984. In addition to astronomers taking part in the meeting, the sessions were attended by over 80 schoolteachers of all levels, who had the opportunity to present some of their activities : instruments, video and movie films, books and leaflets. Dr L. Gouggenheim (Paris) reported on the activities of CLEA (Comité de liaison enseignants - astronomes), which permit fruitful exchanges of ideas between professional astronomers and schoolteachers.

L. Houziaux

President of the Commission

47. COSMOLOGY (COSMOLOGIE)

PRESIDENT : J. Audouze
VICE-PRESIDENT : G. Setti
ORGANIZING COMMITTEE : J. Gunn, S. Hayakawa, M.S. Longair, I. Novikov, G.A. Tammann, V. Trimble, and G. de Vaucouleurs.

Introduction
(Jean Audouze)

The preparation of a report dealing with such a large domain is almost an impossible task. Because so many different questions, problems and expertises are assembled under the word "Cosmology", my approach has been the following : first to divide this field in a somewhat arbitrary fashion into the following sections : very early universe - elementary particle and cosmology - early nucleosynthesis - cosmological parameters (Hubble constant, deceleration parameter, cosmological constant) - large scale structures, intergalactic gas, missing mass - clusters of galaxies and intercluster gas - anisotropy of the black body radiation - formation of galaxies - quasars and their evolution - cosmological evolution of radiosources. I have then asked to the most knowledgeable specialists to review briefly each of these most important questions on which many exciting and very new results have been obtained not only by the astrophysicists themselves but also by particle physicists, nuclear physicists, theoretical physicists,... This is why the reader will read in section 1 the report on primordial nucleosynthesis written by G. Steigman, in section 2 Anisotropy of the black body radiation by D.T. Wilkinson and F. Melchiorri, in section 3 Clusters of galaxies by J. Einasto, in section 4 Galaxy formation by B.J.T. Jones, in section 5 Quasars and their evolution by M. Schmidt and in section 6 the Cosmological evolution of radio sources by R.A. Windhorst. Let me thank these colleagues for their excellent work in writing these various reviews.

The reader who wishes to obtain supplementary and very recent information on this very fast moving field could consult many recent review articles such as those gathered in Annual Review of astronomy and astrophysics. They could also read many recent books of proceedings, such as those edited by myself and J. Tran Than Van in Editions Frontières which are based upon an annual series of workshops called the Rencontres d'Astrophysique de Moriond - The book of 1981 is entitled Particles and Cosmology, 1982 The birth of the Universe, 1983 Formation and evolution of galaxies and large structures (published by Reidel), 1984 High energy astrophysics. Three other books could also be quoted in this respect : Astrophysical cosmology edited in 1982 by the Scientific Pontifical Academy, in 1984, A tribute to the memory of Mgr Lemaître (edited by A. Berger and published by Reidel) and the same year, the proceedings of the first ESO-CERN conference held in November 1983 at Geneva and edited by L. Van Hove and G. Setti under the title "Large Scale Structure of the Universe, Cosmology and Fundamental Physics". This list is of course like this report quite fragmentary and it is up to the readers to browse in a huge amount of literature to be informed before contributing to any topic of their choice.

1. Primordial Nucleosynthesis - A progress report 1981-1984
(Gary Steigman)

Primordial nucleosynthesis provides a unique window on the early evolution of the Universe. To test the "standard" hot big bang model it is necessary to confront the theoretical predictions with observational data. The last three years have witnessed a wide variety of important observational work of relevance to the goal of deriving the primordial abundances of the light elements.

<u>Deuterium</u> : Since deuterium is destroyed in stars during the course of galactic evolution, the present interstellar abundance of D or that in the presolar nebula provide a <u>lower</u> bound on the primordial abundance; if the D-to-H ratio by number is y_2 then, $y_{2P} \geq y_{2\odot} \gtrsim y_{2ISM}$ is expected. From observations of deuterated methane in the atmosphere of Jupiter, Kunde et al (1982) derive from a comparison of CH_3D to CH_4 : y_2 (Jupiter) $= 3.6^{+1.0}_{-1.4} \times 10^{-5}$. For Jupiter and Saturn, Encranez and Combes (1982) compare observations of CH_3D with those of H_2 and - once a C/H ratio has been adopted - derive : y_2 (Jupiter) $= 1.2 - 3.1 \times 10^{-5}$, y_2 (Saturn) $= 2\text{-}15 \times 10^{-5}$. An indirect approach to the presolar deuterium abundance is to compare the ^3He content of the gas rich meteorites or the solar wind (where presolar D has been burned to ^3He) with that of the primitive, carbonaceous chondrites (presumably ^3He uncontaminated by D). From the data summarized by Geiss (1982) we may derive : $y_{2\odot} \equiv y_{23\odot} - y_{3\odot} = 2.9 \pm 0.5 \times 10^{-5}$. All these results - as well as older data - are consistent with a presolar deuterium abundance in excess of $\sim 1\text{-}2 \times 10^{-5}$ suggesting that $y_{2P} \geq y_{2\odot} \gtrsim 1\text{-}2 \times 10^{-5}$. A cautionary note sounded by Geiss and Reeves (1981) is worth echoing - they point to the enormous enhancement of deuterium in interstellar molecules, especially in dark clouds, and suggest that the presolar D abundance need not have been representative of the D/H ratio in the "average" interstellar gas 4.5 billion years ago. The wide variation in the D/H ratio inferred from the study of interstellar molecules ($10^{-4} \lesssim$ D/H $\lesssim 3$) reminds us that observations of deuterated molecules in the atmospheres of the giant planets may provide more information about the physics and chemistry of planetary atmospheres than about the protosolar abundance of deuterium.

Exciting - and confusing - developments have occured concerning the interstellar abundance of deuterium. Earlier ultraviolet observations of deuterium absorption features in the wings of the much stronger hydrogen Lyman lines had led to estimates of the D abundance which spanned an uncomfortably large range ($4 \times 10^{-6} \lesssim y_{2ISM} \lesssim 8 \times 10^{-5}$). Bruston et al (1981) discussed various mechanisms which may have produced such variations and concluded that the data was consistent with a "universal" interstellar abundance : $y_2(ISM) \approx 2\text{-}2.5 \times 10^{-5}$. Given the uncertainties in the solar system and interstellar abundances, it was unclear if there had been any destruction of D since the formation of the solar system. However, an upheaval occured when Vidal-Madjar et al (1982) had great difficulty in finding a consistent model describing the line-of-sight to εPer; although their data was compatible with $y_2 = 1.5 \times 10^{-5}$, it was, apparently, also compatible with $y_2 \sim 10^{-4}$. More important, they found a high velocity feature which <u>varied</u> by at least a factor of 3 in column density within a few hours. In a follow-up study, Vidal-Madjar et al (1983) argued that the "deuterium" observations in εPer were not D at all but, "wrong" velocity ($\sim -80 km^{-1}$) hydrogen in a stellar wind. They discovered variable features in the spectra towards other stars and concluded that this "contamination" of the previous deuterium abundance determinations would argue for the <u>lowest</u> observed D/H ratio as the best interstellar value; Vidal-Madjar et al (1983) suggest $y_2(ISM) \lesssim 5 \times 10^{-6}$. If their suggestion is correct, it is curious that stars whose line-of-sight hydrogen column densities range over more than two orders of magnitude, must have "wrong" velocity stellar wind hydrogen column densities which - in lock step - also range over two orders of magnitude so that we have been fooled into believing that $y_2 \approx 2 \times 10^{-5}$. If, indeed, D/H in the present interstellar gas is as low as 5×10^{-6}, then deuterium has been destroyed by at least a factor of 2-4 in the last 4.5 byr. In this case, interstellar deuterium observations are of little value to primordial nucleosynthesis but of great value to studies of galactic evolution. An excellent survey of the observational situation is to be found in Laurent's (1983) article.

<u>Helium-3</u> : Since some stars ($M \lesssim 1\text{-}2\ M_\odot$) are net producers of ^3He while others destroy ^3He, the value of He in constraining the "standard" hot big bang model is questionable. However stars do burn deuterium to ^3He, <u>some</u> of which - in the cooler, outer layers, - will survive. Yang et al (1984) exploited this result to place an <u>upper</u> limit on the <u>sum</u> of the primordial abundances of D and ^3He survives as D or ^3He. Using the presolar abundances of D and ^3He, Yang et al (1984)

find : $y_{23P} \lesssim 6\text{-}10 \times 10^{-5}$. It is expected that ^{3}He production in low mass stars would have led to enhanced abundance of interstellar ^{3}He in the time since the formation of the solar system (Rood et al, 1976). After years of heroic effort, Rood et al (1984) reported the detection of interstellar ^{3}He via the hyperfine line (8.7 GHz) of $^3He^+$ in three galactic HII regions; they also found upper limits to ^{3}He/H in three other HII regions. The surprise of the results of Rood et al (1984) is that the ^{3}He abundance ranges over more than an order of magnitude : $y_3 = 4\text{-}40 \times 10^{-5}$ for the three positive detections; $y_3 < 2\text{-}6 \times 10^{-5}$ for the hupper limits. Although none of the upper limits is <u>yet</u> in conflict with the presolar value of $y_{3\odot} \approx 1.5 \times 10^{-5}$, the range and magnitude of the apparent enhancements are a puzzle for galactic evolution.

<u>Lithium</u> : Perhaps the single most important set of observations of relevance to primordial nucleosynthesis in the last three years has been that of lithium in old years of Population II by the Spites (Spite and Spite 1982a,b; Spite et al 1984). Since, in the course of normal stellar evolution, surface lithium is convected to the interior and burned away, no detectable lithium was anticipated in Pop II stars. In observations of more than two dozen Pop II stars, the Spites have found for those with an effective surface temperature in the range T_{eff} = 5500-6250 K, a constant abundance of lithium : y_7(Pop II) = $1.1 \pm 0.4 \times 10^{-10}$ which they interpret as the primordial value ($y_{7P} \approx y_7$(Pop II) $\approx 1 \times 10^{-10}$). This value for Li/H is an order of magnitude lower than that derived for the presolar nebula from meteoritic studies, from the present ISM and for Pop I stars of a variety of ages. Indeed, the apparent constancy of the "Pop I" lithium abundance (y_7(Pop I) 1 x 10^{-9}) over the last 4.5 byr has, until the Spites' observations, led to the natural conclusion that $y_{7P} \approx y_7$(Pop I). If their exciting discovery withstands further scrutiny, the Spites' abondance for Pop II is of immense importance for primordial nucleosynthesis ($y_{7P} \approx y_7$(Pop II)), for stellar structure and evolution (Why haven't the Pop II stars destroyed their lithium ?) and for galactic evolution.

<u>Helium 4</u> : The primordial abundance of ^{4}He is the Rosetta Stone of the hot big bang model. Since the predicted primordial mass fraction of ^{4}He, Y_P, is insensitive to the universal density of nucleons (more precisely, to the nucleon-to-photon ratio $\eta \equiv N/\gamma$) and is only weakly dependent on other parameters (neutron half-life $\tau_{1/2}$; number of light, 2-component neutrino species N_ν), the standard model stands or falls as a result of the outcome of a detailed comparison between theory and observation. Unfortunately, the accuracy of the data required for such a comparison is unprecedented; to truly test or significantly constrain the standard model the primordial abundance inferred from observational data should be determined to better than 2% ($\Delta Y_P \lesssim 0.005$). As Tully et al (1981), Shaver et al (1983), Kunth and Sargent (1983) and, especially, Kinman and Davidson (1984) have emphasized, there are myriad obstacles to deriving highly accurate estimates of Y_P even from observational data of extraordinary quality. Two of the most serious problems are the necessary - and uncertain - correction for neutral helium (Shaver et al 1983) and, the contamination of the primordial abundance by stellar produced ^{4}He - the correction for evolution. Failure to make these corrections and to account for the <u>uncertainties</u> of these corrections leads to conclusions of questionable value. From a very careful study of several HII regions in M101 Rayo et al (1982 derived a ^{4}He abundance for one HII region (NGC5471) which they extrapolated to a (very low) primordial abundance Y_P = 0.216. In deriving this value, Rayo et al (1982) took no account of possible neutral helium nor did they include any uncertainty for the evolutionary correction. In contrast, Rosa (1983) showed that the apparent abundance varies across NGC5471, depending on the aperture used; correcting for neutral helium, Rosa (1983) found $Y_P \approx 0.24$. Steigman (1983) too found that the ionization corrections is unlikely to be negligible ($He^0/He^+ \approx 0.07\text{-}0.13$; Smith (1975) found 0.10-0.16) and derived from the data of Rayo et al (1982) : $Y_P = 0.235 \pm 0.010$.

The most extensive, high accuracy study of extragalactic HII regions has recently been completed by Kunth and Sargent (1983). Having studied 13 low metal abundance HII regions (to minimize the contamination from stellar produced ^{4}He) they derive : Y_p = 0.24±0.003, a result they find in agreement with all previous high quality data. Although Kunth and Sargent (1983) did correct for neutral helium (but assigned no uncertainty to that correction) they argued that no correction was necessary for stellar contamination. With allowance for possible contamination - as well as account for the various uncertainties, Steigman (1983) derives from the Kunth-Sargent data : Y_p = 0.241±0.008. The bottom line - at present - is that it is likely that the primordial mass fraction of ^{4}He is in the range 0.23 $\lesssim Y_p \lesssim$ 0.25; if one were less sanguine about the uncertainties then, 0.22 $\lesssim Y_p \lesssim$ 0.26 is not inconsistent with current data.

For the best recent work on the abundances of the light elements the Proceedings of the ESO Workshop on Primordial Helium (1983) is recommended; the excellent review by Pagel (1982) as well as that by Steigman and Boesgaard (1984) are also recommended. Lest this summary become unbalanced we turn to the theoretical developments of the last three years.

Limits To "New" Light Particles : Olive et al (1981a) considered the uncertainty in the bound on N_ν - the number of light neutrino flavors - from the primordial ^{4}He mass fraction due to the uncertainty in the value of the neutron half-life. Olive et al (1981a) noted that if nucleons dominated the mass on the scale of binary galaxies and small groups of galaxies (so that the nucleon-to-photon ratio $\eta \gtrsim 2 \times 10^{-10}$), then $N\nu \leqslant 4$ if $Y_p \leqslant 0.25$. However, since the dark mass on such scales need not be nucleonic, Olive et al (1981a) found that if only the mass in the luminous parts of galaxies is nucleonic ($\eta \lesssim 3 \times 10^{-11}$) then there is no limit to N_ν.Olive et al (1981b) noted that limits (if they existed!) to N_ν actually provided constraints on all weakly or superweakly interacting light particles. Kolb and Scherrer (1982) considered the effect of neutrino masses to better define "light".Kolb and Scherrer (1982) find that each neutrino species with $M_\nu \lesssim$ 0.1 MeV contributes $\Delta N_\nu = 1$; those with $M_\nu \gtrsim$ 25 MeV, contribute $\Delta N_\nu \lesssim 1$ ($\Delta N_\nu \to 0$ for $M_\nu >>$ 25 MeV). However, for 0.1 $\lesssim M_\nu$ (MeV) $\lesssim$ 10-15 MeV, $\Delta N_\nu \gtrsim 1$. Schramm and Steigman (1984) pointed out that the constraint on N_ν from the width of the Z^0 boson is complementary to that from nucleosynthesis. The Width of the Z^0 increases with the addition of species which couple directly to the Z^0 even if they are heavy ($M_\nu \lesssim$ 45 GeV). In contrast, the constraint from big bang nucleosynthesis is sensitive to all light ($M_\nu \lesssim$ 25 MeV) particles even if they don't couple directly to the Z^0.

The bound $N_\nu \leqslant 4$ was reestablished by Yang et al (1984) who discovered how the bound η from below. Yang et al (1984) noted that D and ^{3}He are overproduced in low η Universes and that some of the primordial ^{3}He (plus primordial D burned to ^{3}He) will survive stellar processing. Using the solar system observations of ^{3}He, Yang et al (1984) argued that $[(D+{}^3He)/H]_p \lesssim$ 6-10 $\times 10^{-5}$. This upper bound to primordial D and ^{3}He leads to a lower bound to the nucleon-to-photon ratio : $\eta \gtrsim$ 3-4 $\times 10^{-5}$ which, for $Y_p \lesssim$ 0.25-0.26 leads to $N_\nu \lesssim 4$.

Consistency Of The Standard Model : Yang et al (1984) have made a detailed comparison of the predictions of the standard model ($N_\nu = 3$, $\tau_{1/2}$ = 10.6 ± 0.2 min.) with the observational data (see also Pagel 1982). They find the standard model is consistent with the data for the nucleon abundance in the range : 3-4 $\lesssim 10^{10}\ \eta \lesssim$ 7-10. With these values for η the ratio of the nucleon mass density to the critical density, Ω_N, is for - $H_0 \approx$ 50-100 kms^{-1} Mpc^{-1} and $T_{\gamma 0}$ = 2.7 - 3.0 K - in the range : 0.01 $\lesssim \Omega_N$, $\lesssim$ 0.14 - 0.19. The lower bound on Ω_N is consistent with that inferred from the dynamics of the luminous parts of galaxies; the upper limit suggests that much - perhaps all - the dark mass observed could be nucleonic.

Beaudet and Reeves (1983) and Yang et al (1984) probed the sensitivity of the predicted abundances to possible variations in various - crucial - nuclear cross sections. While the predicted abundances of D, ^{3}He and ^{4}He should be accurate to better than a few percent, the ^{7}Li abundance is uncertain by a factor of two. Since very detailed comparisons between the predicted and observed abundances of ^{4}He are required to test the standard model, the predicted abundance should be accurate to better than a few percent. In a very detailed study of radiative, finite temperature and density, coulomb and plasma corrections to the weak interaction rates, Dicus et al (1982) found a small but systematic decrease ($\Delta Y_p \approx -0.003$) in the predicted ^{4}He mass fraction; see also Cambier et al (1983) and Johansson et al (1983).

Constraints On Deviations From The Standard Model : The excellent agreement between prediction of the standard model and the observed abundances (Yang et al 1984) leaves little room for deviations from the standard model. The upper limit $N_\nu \leqslant 4$ corresponds to a limit on the speed-up on the expansion rate - during the epoch of nucleosynthesis; the Universe can have expanded no more than 8% faster than the rate given by the standard model. Although this leads to significant constraints on isotropy, Barrow (1984) notes that anisotropic models (even during the present epoch) could yield abundances which are not in conflict with the data.For models which do become isotropic by the present epoch, Rothman and Matzner (1984) find even more stringent limits to anisotropy than found in earlier work.

Nucleosynthesis in inhomogeneous cosmologies has been studied recently by Matzner (1982), Yang et al (1984) and Barrow and Morgan (1983) who find that $\delta\rho/\rho$ is limited to ~ few if ^{4}He is not be overproduced.

The effect of neutrino degeneracy was investigated by Fry and Hogan (1982), Rana (1982), Scherrer (1983) and Steigman (1984). The agreement of the standard - nondegenerate - model is lost if the neutrinos are "too" degenerate; for μ- or τ-neutrinos, $|\zeta_\nu| \lesssim 1.4$; for e-neutrinos, $-0.05 \lesssim \zeta_e \lesssim 0.10$ (Steigman 1984). Also, even for degenerate e- and μ- (or τ-) neutrinos, a Universe closed by nucleons (Ω_N=1) is not allowed (Steigman 1984).

The theoretical and observational activity of the last three years have led to issues and questions which promise to keep us all busy for (at least) the next three years.

REFERENCES

Barrow, J.D., M.N.R.A.S., 211 (In press) (1984).
Barrow, J.D. and Morgan, J., MNRAS 203, 393 (1983)
Beaudet, G. and Reeves, H., In Proc. of ESO Workshop, p. 53 (1983)
Bruston, P., Audouze, J., Vidal-Madjar, A. and Laurent, C., Ap.J. 243, 161 (1981)
Cambier, J-L., Primack, J.R. and Sher, M., Nucl. Phys. B20, 372 (1983).
Dicus, D.A., Kolb, E.W., Gleeson, A.M., Sudarshan, E.C.G., Teplitz, V.L. and Turner, M.S., Phys. Rev. D26, 2694 (1982).
Encrenaz, T. and Combes, M., Icarus 52, 54 (1982).
Fry, J.N. and Hogan, C.J., Phys. Rev. Lett. 49, 1873 (1982).
Geiss, J., Sp. Sci. Rev. 33, 201 (1982).
Geiss, J. and Reeves, H., Astron. Astrophys. 93, 189 (1981).
Harris, M.J., Fowler, W.A., Caughlan, G.R. and Zimmerman, B.A., Ann. Rev. Astron. Astrophys. 21, 165 (1983).
Johansson, A.E.I., Peressutti, G. and Skagerstam, B-S.K., Phys. Lett. B117, 171 (1982).
Kinman, T.D. and Davidson, K., Ap. J. (In press) (1984).
Kolb, E.W. and Scherrer, R.J., Phys. Rev. D25, 1481 (1982).
Kunde, V., Hanel, R., Maguire, W., Gautier, D., Baluteau, J.P., Marten, A., Chedin, A., Husson, N. and Scott, N., Ap. J. 263, 443 (1982).
Kunth, D. and Sargent, W.L.W., Ap. J. 273, 81 (1983).
Laurent, C., In "Proc. of ESO Workshop" p. 335 (1983).
Matzner, R., In "Numerical Astrophysics" (eds., J. Centrella, J. LeBlanc and R. Bowers, Jones & Bartlett, Boston) In Press (1982).
Olive, K.A., Schramm, D.N., Steigman, G., Turner, M.S. and Yang, J., Ap. J. 246, 557 (1981a).
Olive, K.A., Schramm, D.N. and Steigman, G., Nucl. Phys. B180, 497 (1981b).
Pagel, B.E.J., Phil. Trans. Roy. Soc. London A307, 19 (1982).
Rana, N., Phys. Rev. Lett. 48, 209 (1982).
Rayo, J., Peimbert, M. and Torres-Peimbert, S., Ap. J. 255, 1 (1982).
Rood, R.T., Steigman, G. and Tinsley, B.M., Ap. J. (Lett.) 207, L57 (1976).
Rood, R.T., Bania, T.M. and Wilson, T.L., Ap. J. 280, 629 (1984).

Rosa, M., In "Proc. of the ESO Workshop", p. 317 (1983).
Rothman, T. and Matzner, R., Phys. Rev. D30, 1649 (1984).
Scherrer, R.J., M.N.R.A.S. 205, 683 (1983).
Schramm, D.N. and Steigman, G., Phys. Lett. B141, 337 (1984).
Shaver, P.A., McGee, M.X., Newton, L.M., Danks, A.C. and Pottasch, S.R., M.N.R.A.S. 204, 53 (1983).
Smith, H.E., Ap. J., 199, 591 (1975).
Spite, F. and Spite, M., Astron. Astrophys. 115, 357 (1982a).
Spite, F. and Spite, M., Nature 297, 483 (1982b).
Spite, M., Maillard, J.P. and Spite, F., Astron. Astrophys., In Press (1984).
Steigman, G., In Proc. of Yerkes Conf., In Press (1983).
Steigman, G., "How Degenerate Can We Be " (In Preparation) (1984).
Steigman, G. and Boesgaard, A.M., Ann. Rev. Astron. Astrophys. 23, In Press (1984).
Tully, R.B., Boesgaard, A.M., Dyck, A.M. and Schempp, W.V., Ap. J. 246, 38 (1981).
Vidal-Madjar, A., Ferlet, R., Laurent, C. and York, D.G., Ap. J. 260, 128 (1982).
Vidal-Madjar, A., Laurent, C., Gry, C;, Bruston, P., Ferlet, R. and York, D.G., Astron. Astrophys. 120, 58 (1983).
Yang, J., Turner, M.S., Steigman, G., Schramm, D.N. and Olive, K.A., Ap. J. 281, 493 (1984).

Note added by J. Audouze : Like many other fields of cosmology, the consequences of primordial nucleosynthesis on the actual present density of the Universe and the number of lepton flavours might not be as firmly established as written in the report of Professor Steigman. There is now a friendly debate between (1) the Chicago-Bartol school who claims as written e.g. in Yang et al. 1984 that the baryonic density of the Universe is necessarily equal to about 5% of the critical density and that canonical primordial nucleosynthesis implies also a number of three lepton flavours (in agreement with the Grand Unification theory scheme) and (2) the "french" school. Our group has attempted in several publications to write some words of caution about these conclusions.

As shown by Vidal-Madjar and Gry (1984) and by Audouze (1984) the baryonic density deduced from ^{4}He and D primordial abundances seem to be inconsistent to each other. This is why Delbourgo-Salvador et al. (1985) have shown that in order to restore such a consistency, one must invoke some specific models of chemical evolution of galaxies leading to a large destruction of D (by about a factor 10) during the history of the galaxy. Moreover Audouze et al. (1985) and Schaeffer et al. (1985) are proposing some mechanisms by which the results of primordial nucleosynthesis could be consistent with a closed ($\Omega > 1$) Universe.

In the first scenario, if the baryonic density is very large, ^{4}He and ^{7}Li could be synthetized but not D and ^{3}He. Those two last nuclear species could be formed by a secondary process like partial photodestruction of ^{4}He induced by high energy photons coming from the decay of hypothetical massive neutrinos and/or gravitinos. In the second the baryonic density would be still low but the bulk of the matter density could be made of nuggets of quark matter. Finally I do concur with Professor Steigman on his last statement according which our community will kept busy during the next coming years.

Additional references

Audouze, J. : 1984, in Primordial Helium, ed. by P.A. Shaver et al, ESO p.3.
Audouze, J., Lindley, D. and Silk, J. : 1985, Ap. J. (in press).
Delbourgo-Salvador, P., Gry, C., Malinie, G. and Audouze, J. : 1985, Astron. Astrophys. (in press).
Schaeffer, R., Delbourgo-Salvador, P. and Audouze, J. : 1985, submitted to Nature.
Vidal-Madjar, A. and Gry, C. : 1984, Astron. Astrophys. 138, 285.

2. Anisotropy of the blackbody radiation
(D.T. Wilkinson and F. Melchiorri)

INTRODUCTION

The 2.7 K microwave background radiation provides a sensitive probe of the universe in the interesting, but poorly understood, epoch around $z \sim 1000$. At this time (age $\sim 10^6$yr) the universe has cooled to $T \sim 4000$ K, the plasma combines, Thomson scattering ceases, and matter and blackbody radiation decouple. Subsequently, the radiation freely propagates to us, carrying the imprint of temperature fluctuations on the $z \sim 1000$ surface. The temperature fluctuations could have been caused by primordial density fluctuations, anisotropy in the expansion of the universe, or inhomogeneity in the initial temperature distribution; the $z = 1000$ surface we see was not causally connected at the time the radiation was released. Interpretation of the anisotropy measurements is complicated by the possibility that the matter may have been reionized (e.g. by massive stars), so the radiation may have been rescattered, possibly as late as $z \sim 7$.

This picture follows the standard hot big-bang cosmological model [1], and indeed measurements of the spectrum and isotropy of the 2.7 K radiation furnish much of the support for the model [2].

Remarkably, no anisotropy has been detected at a level of $\Delta T/T \sim 10^{-4}$ on angular scales from 1 arcminute to 90 degrees. The "dipole" effect is mostly, perhaps completely, due to the peculiar velocity of the sun through the radiation. Observational techniques and theoretical interpretations logically divide the subject into three regimes of angular scale.

SMALL SCALE ANISOTROPY (< 1°)

The large-scale structure seen in the current universe - clusters of galaxies, voids, strings, etc. - should have been forming at $z \sim 1000$ and had an angular size (as seen now) of a few arcminutes. If reionization is negligible, current estimates of the magnitude of perturbation of the background radiation based on the standard hot big-bang model are in the range $\Delta T/T \sim 10^{-4}$ to 10^{-5} [3]. The observational upper limits on this angular scale are approaching 10^{-5}, placing important constraints on the models for structure formation, the magnitude of initial density fluctuations, and the cosmological model.

Early searches for small scale anisotropy in the 2.7 K background are discussed by Partridge [4] and the theoretical predictions are reviewed by Hogan, Kaiser and Rees [5]. The current best limits are discussed below.

To achieve angular resolution of a few arcminutes, large radio telescopes are needed, so the observational work in this angular regime has been done from the ground. To achieve sensitivity of 10^{-5} K the lowest noise receivers ($\Delta T_{rms} < 10^{-2}$ K $s^{1/2}$) and the most uniform atmospheric conditions are resuired. Since many hours of integration time are required on each sky position, stable telescope and receiving systems, and careful control of ground radiation are essential.

The Green Bank 140' telescope and maser system has been used to place a limit of $\Delta T_{rms}/T \leq 2.1 \times 10^{-5}$ (95% confidence) at angular scales of 1-5 arcminutes [6]. At smaller angular scales the NRAO Very Large Array has produced the best limits : $\Delta T_{rms}/T \leq 10^{-3}$ at 18 arcseconds and 5×10^{-4} at 1 arcminute (95% confidence) [7, 8]. This limit was not dominated by system noise, so further improvement is expected. Parijski and his coworkers have reported early results from the RATAN-600 telescope [9] at a wavelength of 7.6 cm. The smallest sky signal, $\Delta T_{rms} \leq 3 \times 10^{-4}$ K (one sigma) is seen at an angular scale of about 7 arcminutes where Galactic radiation and point source confusion reach a minimum. Correlation with signals at other wavelengths, which are observed simultaneously, allows a tenfold reduction to a (1σ) level of 10^{-5} on scales from 4.5 to 9 arcminutes.

The complicated issue of reducing and interpreting anisotropy data is carefully addressed by Lasenby and Davies [10] and by Partridge [4]. Because of the variety of spatial patterns used by the observers, each upper limit must be carefully interpreted in terms of the particular theoretical model being tested [6]. The actual detection of small scale anisotropy will be an important milestone in contemporary cosmology.

INTERMEDIATE SCALE ANISOTROPY ($\sim 1^{o}$)

Regions of the photon barrier at $z \sim 1000$ which are separated by a few degrees were not in casual contact at the time the 2.7 K photons left there. Therefore, even in the case of strong reionization this angular scale will provide useful informations because statistical mass clumping can cause anisotropy because the 2.7 K photons from different regions suffer different gravitational redshifts [11, 12]. This effect can grow with angular scale as $\theta^{1/2}$: most effects get weaker with θ.

Atmospheric inhomogeneities are a serious problem for ground-based observations at intermediate angular scales. Few observations have been attempted, the best being done from balloons with broadband bolometric detectors [13]. Thermal emission from patchy Galactic dust is the main problem for the broadband mm detectors, but multiband observations and correlation with radio and infrared sky maps can be used to separate signals due to dust and 2.7 K anisotropy [14].

LARGE SCALE ANISOTROPY

If the universe were anisotropic or inhomogeneous on large scales (in violation of Einstein's Cosmological Principle), an anisotropy should be seen in the 2.7 K radiation. No such intrinsic effect has been seen with amplitude greater than 0.1 mK [15, 16]. These results were obtained by high-sensitivity microwave radiometers (λ = 3 mm and 1.2 cm, respectively) carried above atmospheric water vapor by balloons. Most of the sky has been observed. The 1.2 cm observations are the most sensitive (ΔT_{rms} = 4 mK $s^{1/2}$), but residual Galactic radio emission is a problem. A wavelength of 3 mm seems about optimum, located between Galactic dust and radio emissions.

A strong dipole signal is observed in the 2.7 K radiation : amplitude 3.3 ± 0.3 mK, declination - 6° ± 3°, right ascension 11.2^{h} ± 0.2^{h} [15, 16]. The dipole pattern measured by balloon-borne bolometers [2, 17] is less accurate, due to limited sky coverage and Galactic dust emission, but the results are in agreement with the radiometer measurements.

Most of the observational dipole anisotropy is due to the motion of the sun through the radiation, although one cannot rule out a small contribution from intrinsic cosmic anisotropy. If the sun's motion due to Galactic rotation is subtracted from the observed solar motion with respect to the 2.7 K radiation, the peculiar velocity of the Galaxy (or Local Group) is obtained : V_{LG} = (610 ± 50) km/sec, towards $\alpha = 10.5^{h} \pm 0.4^{h}$ and $\delta = - 26^{o} \pm 5^{o}$ [18]. This velocity vector is directed 49° away from the Virgo cluster and 17° away from the velocity vector measured with respect to Sbc galaxies [19].

The small dipole signal due to the Earth's motion has been detected (7σ level) [18] and the spectrum of the dipole signal [20] is in agreement with that expected for backbody radiation. The dipole's spectrum has recently been used as a sensitive test for photon mass [21].

SUNYAEV - ZEL'DOVICH EFFECT

When the 2.7 K photons travel through the hot plasmas in the central regions of some clusters of galaxies, they are Compton scattered upwards in frequency by the energetic electrons. The result is a distortion of the blackbody spectrum which diminishes the intensity at longer (cm) wavelengths but increases the appaparent temperature at short (mm) wavelengths [22]. The effect is estimated to be about 10^{-3} K for clusters having plasmas with strong, smooth x-ray emission [23,24].

Since the plasma clouds are about an arcminute in diameter, large ground-based telescopes must be used for observations, and the problems are similar to those of searches for small-scale anisotropy. Years of work, and much telescope time, have been devoted to this problem [25, 26, 27, 10] with some marginal detections reported. Based on this experience, recent observations at Owens Valley [28] and at Greenbank [29] have detected the Sunyaev-Zel'dovich effect with good statistical accuracy and (more importantly) small systematic corrections. The clusters 0016 + 16, A2218 and A665 show clear, reproducible decrements at their central positions. Attempts [30] to detect the excess signal expected at millimeters wavelengths have been unsuccessful, so far. Atmospheric inhomogeneity is a very severe problem even at mountaintop observatories.

REFERENCES

1. For references and discussion of alternative models see : Ellis, G.F.R. 1984, Ann. Rev. Astron. Astrophys. 22, p. 157.
2. For reviews see : Weiss, R. 1980, Ann. Rev. Astron. Astrophys. 18, p. 489; Sunyaev, R.A. and Zel'dovich, Ya. B. 1980, Ann. Rev. Astron. Astrophys. 18, p. 537; Melchiorri, B., Melchiorri, F. 1982, Alta Cosmologica DCLXXIV, p. 27.
3. Bond, J.R. and Efstathiou, G. 1984, Inner Space/Outer Space Conference Proceedings (Fermilab); Vittorio, N. and Silk, J. 1984, Inner Space/Outer Space Conference Proceedings (Fermilab).
4. Partridge, R.B. 1980, Phys. Scripta 21, p. 624; 1980, Ap. J. 235, p. 681; 1983, "The Origin and Evolution of Galaxies", VIIth Course of the International School of Cosmology and Gravitation (Eds. B.J.T. Jones and J.E. Jones), p. 121.
5. Hogan, C.J., Kaiser N. and Rees, M.J. 1982, Phil. Trans. R. Soc. London A, 307, pp. 97-110.
6. Uson, J.M. and Wilkinson, D.T. 1984, Nature (in press); 1984, Ap. J. Letters 277, L1; 1984, Ap. J. 283, pp. 471-478; 1982, Phys. Rev. Letters 49, p. 1463.
7. Fomalont, E.B., Kellerman, K.I. and Wall, J.V. 1984, Ap. J. Letters 277, L23.
8. Knoke, J.E., Partridge, R.B., Ratner, M.I. and Shapiro, I.I. 1984, Ap. J. 284, p. 479.
9. Berlin, A.B., Bulaenko, E.V. Vitkovsky, V.V., Kononov, V.K., Parijski, Yu. N. and Petrov, Z.E. 1983, "Early Evolution of the Universe and Its Present Structure" IAU Symposium N° 104 (Eds. Abell and Chincarini), p. 121.
10. Lasenby, A.N. and Davies, R.D. 1983, M.N.R.A.S. 203, pp. 1137-1169. Also, see this paper for discussion and early results of a promising small-scale anisotropy program at $\lambda = 6$ cm.
11. Wilson, M.L. and Silk, J. 1981, Ap. J. 243, p. 14.
12. Peebles, P.J.E. 1981, Ap. J. Letters 243, L119-L121.
13. Melchiorri, F., Melchiorri, B.O., Ceccarelli, C. and Pietranera, L. 1981, Ap. J. Letters 250, L1.
14. Ceccarelli, C., Dall'Oglio, G., Merchiorri, B., Melchiorri, F. and Pietranera, L. 1982, Ap. J. 260, p. 484.
15. Lubin, P.M., Epstein, G.L. and Smoot, G.F. 1983, Phys. Rev. Letters 50 p.616.
16. Fixson, D.J., Cheng, E.S. and Wilkinson, D.T. 1983, Phys. Rev. Letters 50, p. 620.
17. Fabbri, R., Guidi, I., Melchiorri, F., Natale, V. 1980, Phys. Rev. Lett. 39 p. 898.
18. Wilkinson, D.T. 1983, "Early Evolution of the Universe and Its Present Structure", IAU Symposium N° 104 (Eds. Abelland Chincarini), p. 143.
19. Hart, L. and Davies, R.D. 1982, Nature 297, p. 191.
20. Boughn, S.P. Cheng, E.S., Wilkinson, D.T. 1981, Ap. J. Letters 243, L113.
21. de Bernardis, P., Masi, S., Melchiorri, F. and Moleti, A. 1984. Ap. J. Letters 284, L21.

22. Sunyaev, R.A. and Zel'dovich, Ya. B. 1972, Comments Ap. Space Phys. 4, p 173.
23. Boynton, P.E., Radford, S.J.E., Schommer, R.A. and Murray, S.S. 1982, Ap. J. 252, p. 473.
24. White, S.D.M., Silk, J., Henry, J.P. 1981, Ap. J. Letters 251, L65.
25. Lake, G. and Partridge, R.B. 1980, Ap. J. 237, p. 378.
26. Birkinshaw, M., Gull, S.F. and Moffet, A.T. 1981, Ap. J. Letters 251, L69; Birkinshaw, M., Gull, S.F. and Northover, K.J.E. 1981, M.N.R.A.S. 197, p.571; birkinshaw, M. and Gull, D. F. 1984, M.N.R.A.S. 206, pp. 359-375.
27. Andernach, H., Schallwich, D., Sholomitski, G.B. and Wielebinski, R. 1983, Astron. Astrophys. 124, pp. 326-330.
28. Birkinshaw, M., Gull, S.F. and Hardebeck, H. 1984, Nature 309, p. 34.
29. Uson, J. and Wilkinson, D.T. 1984, Bull. Am. Astron. Soc. 16, p. 513.
30. Meyer, S.S. Jeffries, A.D. and Weiss, R. 1983, Ap. J. Letters 271, L1.

3. Clusters of Galaxies

(Jaan Einasto)

Abell and Corwin (34.160.040) have continued the search of southern clusters on photographs taken with the U.K. Schmidt telescope at Siding Spring. A catalogue is in preparation, it continues the Abell catalogue toward southern declinations. West (32.160.019, 32.160.036, 34.160.044) has initiated a study of distant southern clusters with automatic identification of clusters and objective evaluation of their properties. Noonan (30.160.006) and Sarazin et al. (31.160.030) have listed clusters with published redshifts. Numerical simulations have demonstrated the presence of bias in observed clusters as difined by Abell (33.160.006), ranked absolute magnitudes of galaxies are distance dependent (33.160.081), some clusters may be due to projection effects (34.160.009).

Kraan-Korteweg has presented a complete sample of bright members of the Virgo cluster and derived the luminosity function and kinematical parameters (30.160.049 31.160.013). Binggeli, Sandage and Tarenghi (1984) have started a program to study faint galaxies in the Virgo cluster using der Pont 2.5-m reflector. Near infrared photometry of 81 disk galaxies has been done in (34.160.097), structural parameters of 20 galaxies are derived from Fourier and momentum analyses of luminosity profiles (32.160.026). Rocket ultraviolet photometry has been done in (32.160.025). The effective radius and central surface brightness decrease with decreasing luminosity, but there is no difference between S0 and later Hubble type galaxies of same luminosity. Kinematics of the Virgo clusters has been studied also in (31.160.072, 31.160.011, 33.160.036, 33.160,039).

Extensive 21-cm HI observations in Westerbork and Arecibo (30.160.044, 33.160.003, 33.160.032, 33.160.076, 34.160.005, 34.160.051) have demonstrated that galaxies in the Virgo cluster core have smaller HI extent and are HI deficient by a factor 2 to 2.7 in respect to galaxies of the same morphological type outside the core. The HI deficiency depends on the velocity of a galaxy in respect to the cluster suggesting a ram pressure sweeping of the intracluster gas. It is shown that Virgo cluster galaxies are dust poor compared to field galaxies (van den Bergh 1984).

The distribution of radial velocities of galaxies in the vicinity of M87 has demonstrated the presence of background clusters (and probably superclusters) of galaxies at redshifts V = 20200 and 26200 km/s (31.160.029, 34.157.034, Huchra and Brodie 1984).

An extensive radio and optical survey of galaxies in 9 nearby clusters (30.158.060, 30.158.061, 30.158.323, 31.158.324, 32.160.027, Bothun, Schommer, Sullivan 1984) shows that most clusters do not contain gas-deficient spirals. The Coma cluster is the only cluster in the sample exhibiting a large fraction of HI-deficient spirals and having other characteristics consistent with present day spiral-to-S0 conversion. In general initial conditions of formation and variations in star formation histories have been more important than environmental influences. The key parameter may be the amount of neutral hydrogen remaining after star formation in the bulge has ceased. 21-cm observations of X-ray galaxies in A1367 show that neutral gas may coexist with the hot coronal X-ray gas. Such galaxies are HI deficient (33.160.052). For the Pegasus I cluster also no HI-deficiency has been observed (31.160.047), but in the Hydra I (A1060) and A262 clusters an increase of the HI-deficiency inwards to the cluster center is found (32.160.053, 34.160.012). The effects of viscosity, thermal conduction and turbulence can exceed that due to ram pressure in stripping of clusters of galaxies (31.160.007).

The stripping of galaxies of gas is most effective in central regions of clusters. The combination of kinematical and HI-deficiency information suggest that Coma cluster velocity dispersion is isotrope (32.160.001, Pryor, Geller 1984). The total mass of the Coma cluster and its M/L has been estimated from kinematical and X-ray data and tidal radii or nearby groups of galaxies. For the Coma cluster mass-to-light radio estimates lie between 100 h and 400 h (30.160.019, 30.160.041, 31.160.003, 32.160.016, 32.160.035, 34.160.019, 34.160.28). Similar arguments suggest for the Perseus cluster the preference of radial orbits and $(M/L)_V$ = 600 h (33.160.080).

Radio observations of the Coma A1367 and Perseus clusters have been reported in (30.160.011, 32.160.006, 33.160.024). Equilibrium distribution of the elements for an isothermal hot plasma in a gravitational potential well suggests that elements with $Z > 2$ have abundances proportional to their standard cosmic values but reduced by a factor of 20 (30.160.020).

To study photometric properties of galaxies in clusters photographic (31.160.078, 34.160.074), photoelectric (31.160.033) methods and CCD detectors (33.160.007, 33.160.019, 33.160.082) were used µ automatic measuring of photographic plates was also used (33.160.013). CCD photometry of the brightest three galaxies in cores of 83 Abell clusters yields for the dispersion of the absolute magnitudes of first, second, and third brightest galaxy 0.35, 0.55 and 0.65 respectively (33.160.082). Core radii were determined for 11 rich southern clusters (33.160.037). A new class of clusters that contain a large number of binary galaxies among their brightest members was reported (30.160.045).

The remote cluster of galaxies surrounding 3C 295 has been extensively studied photometrically (30.160.046, 33.100.017, 34.160.009) and spectroscopically (32.160.068, 34.160.001). The presence of a large number of blue galaxies is confirmed, but a high fraction of them are likely to be field galaxies. Color distribution of member galaxies is typical for local BM type I clusters. Spectral data demonstrate that 6 blue member galaxies are all unusual, including 3 Seyfert galaxies and 3 galaxies which have bursts of star formation. A high percentage of Seyfert galaxies is expected (Osterbrock 1984), since among brightest nearby field galaxies the fraction of Seyfert galaxies and optically selected quasars is large.

Photometric and spectral (preferentially CCD) studies of other remote and intermediate redshift clusters (30.160.047, 30.160.048, 31.160.016, 34.160.006, 34.160.044, 34.160.079) demonstrate that a high fraction of blue galaxies in the 3C 295 cluster may be unique. In the $z = 0.54$ cluster 0016+16 virtually all galaxies are intrinsically very red, thus star formation must have ceased in the S0 galaxies of this cluster more than 9 Gyears ago (30.160.047). Some blue galaxies found in intermediate redshift clusters correspond to normal Sbc galaxies (34.160.079).

Internal dynamics of clusters has been studied by conventional spectroscopy (30.160.007, 30.160.021, 30.160.051, 32.160.018, 34.160.008, Dressler 1984, Geller et al. 1984) and Medusa spectroscopy (32.160.074, Ellis et al. 1984). Rough redshifts (accurate to 1800 km/s) can be obtained from objective prism spectra (30.160.002, 33.160.079). Tully-Fisher relation has been used to study the expansion of nearby clusters (31.160.046). Dynamical modeling suggests that Coma has a static core surrounded by a still collapsing envelope of a radius 10 Mpc (32.160.050, 34.160.029). Kinematical data have been used to estimate the masses of clusters by various methods (31.160.028, 32.160.012, 32.160.021). The relations between cluster velocity dispersion and other cluster parameters (central density, X-ray luminosity, radius, and infrared surface brightness) have been investigated (30.160.012, 32.160.002, 33.160.014, 34.160.011).

Spectrophotometry can also be done with a CCD camera through narrow filters

(34.160.065). A number of X-ray luminous clusters show bright optical emission from extended filaments (30.160.042). Very bright emission-line systems seen towards N1275 arise from a high-velocity impact of a foreground galaxy upon the accretion flow of cooling gas in the center of the Perseus cluster (34.160.118). Accretion-driven star formation may explain the unusual blue color of N1275 (33.160.085). Accretion rate into the central galaxy is estimated to be 20 - 100 M_0 yr_{-1} (34.160.071).

Radio observations of clusters at various wavelengths have been continued in Arecibo (30.160.055, 32.160.009), in Westerbork (31.160.006, 31.160.014, 33.160.023, 33.160.050, Hanisch 1984a, b) and many other radio observatories (30.160.003, 30.160.005, 30.160.008, 30.160.010, 31.160.036, 32.160.005, 32.160.015, 33.160.005, 33.160.054, 34.160.022, 34.160.089). Clusters possessing radio halo (Coma, Perseus, A2255, A2256, A2319) have a number of striking similarities in their optical, radio, and X-ray properties (32.160.059). These also are due to synchrotron radiation by electrons, leaked out of the cluster radio sources (Valtaoja 1984).

VLA 1465 MHz survey of clusters with redshifts of 0.25 to 0.9 shows strong evolution of the radio luminosity function during this period (32.160.051).

Reviews of X-ray observations of clusters of galaxies are given by Giacconi (32.160.039) and Forman and Jones (32.160.043). Data reduction of earlier Einstein observations has been continued (30.160.013, 30.160.016, 30.160.017, 30.160,053, 31.160.010, 31.160.041, 31.160.050, 31.160.080, 32.160.007, 32.160.040, 33.160.022, 33.160.084, 34.160.017, 34.160.024, 34.160.025, 34.160.082, Rothenflug et al. 1984, Harris et al. 1984, Henry and Lavery 1984, des Forets et al. 1984). An X-ray survey of clusters at moderate and high redshifts suggests that up to redshift of 0.5 the slope of the luminosity function is constant (32.160.049).

Surface brightness profiles and a comparison with optical features suggests that X-ray clusters of galaxies can be devided into two classes : XD have small core radii and the X-ray emission is centered on a central dominant galaxy, nXD clusters have larger core radii and emission is not centered on a stationary galaxy (Jones and Forman 1984). This suggests that the formation of a central galaxy occurs in the early stage of the cluster evolution. Poor clusters with central dominant galaxies have X-ray structure similar to rich clusters with dominant galaxies, thus the origin of the dominant galaxy in poor and rich clusters should be similar (34.160.070).

Clusters of galaxies with dominant cD galaxies have been studied in detail (32.160.058, 34.160.102, 34.160.103, van Breugel, Heckman and Miley 1984, Lugger 1984). D and cD galaxies lie on peaks in the cluster galaxy distribution, whether or not they are the brightest cluster members (34.160.101). cD galaxies lie at the kinematical centers of their parent clusters (31.160.015). The stellar velocity dispersion in the individual nuclei of galaxies with multiple nuclei yield $(M/L)_B$ = 10 - 20 (H_0 = 100), but relative velocities are typically 1000 km/s (Merritt 1984a, Tonry 1984). These observations are incompatible with the cannibalism hypothesis. Dynamical friction causes galaxy orbits to decay and produce substantial increase in the number of galaxies which pass near the center without lowering the mean orbital velocity. Thus multiple systems so produced would be transient phenomena. Simulations of the dynamical evolution of clusters of galaxies under the influence of galaxy mergers, dynamical friction, two-body relaxation and tidal stripping have been performed (31.160.012, 32.160.042, 33.160.004, 33.160.072). These studies demonstrate that cluster morphology is probably fixed during the cluster collapse and changes little afterward (Merritt 1984b). Specific globular cluster frequency in central galaxies of nearby clusters (Virgo and Hydra I) is unusually high, thus central galaxies were very special already at an early stage of the evolution of their parent clusters (33.160.074). The luminosity stripped from galaxies and forming the halo of cD galaxies is proportional to the number of cluster galaxies squared, thus the relative importance of the cD halo is a strong function of the cluster richness (Malumuth and Richstone 1984).

A number of clusters have substructures and secondary maxima in the galaxy distribution (31.160.017, 31.160.044, 32.160.003, 32.160.010, 32.160.013,

33.160.008, 34.160.035, 34.160.045, 34.160.095). Some of these clusters may be unbound collections of groups of galaxies, as the Cancer cluster is (33.160.073). Numerical simulations show (McGlynn and Fabian 1984) that two clusters can pass through one another without destroying the individual optical components. In some clusters an encounter may now be underway, as in the Centaurus cluster. High-resolution observations of the cluster A1367 indicate the presence of 8 point sources and 13 extended features associated with cluster galaxies (33.160.010). The structure of clusters with bimodal or clumpy X-ray morphology is consistent with the idea that clusters form by coalescence and merging of subclusters (32.160.004).

Clusters of galaxies are strongly elongated and nearby clusters point with their long axis to each other (31.160.024). Galaxies in clusters also show a tendency to point toward their nearest neighbours (34.160.098). Monte-Carlo simulations of the orientations and shapes of galaxies in clusters (31.160.074) demonstrates that phenomenon may be due to the history of the cluster formation via adiabatic scenario or to subsequent evolution via tidal forces, the first explanation is more plausible (34.160.094, 34.160.098, Dekel, West and Aarseth 1984).

Sunayev and Zeldovich (34.160.081) published a review on the intergalactic gas in clusters, the microwave background radiation and cosmology. Significant depression in the temperature of the microwave radiation at 10.6 GHz is found only in several clusters (A576, A665, A2218, 0016+16), observations at 20.3 GHz confirm the reality of the effect in most cases (30.160.038, Birkinshaw and Gull 1984). The motion of the cluster can produce two-sided brightness pattern in the microwave background and used to test for transverse motion of clusters (33.160.025). Studies of the intracluster gas have been reported in (31.160.001, 32.160.011, 33.160.069, Nottale 1984). Parameters of the intergalactic medium can be determined by using Lyman-alpha absorbing systems in quasars spectra (33.160.087), diffuse UV-radiation (34.160.015) and by X-ray observations (Forman, Jones and Tucker 1984). Allowed density of the hot intercluster gas is 2×10^{-8} cm^{-3}, allowed temperature is 300×10^{6} K.

Counts of faint galaxies in the vicinity of quasars indicate that quasars are situated in first ranked galaxies of groups and clusters of galaxies (Gehren et al. 1984). Active galaxies have been discovered in clusters A484 and A2534 (32.160.023).

REFERENCES

Binggeli, B., Sandage, A., Tarenghi, M. : 1984, Astron. J. 89, 64.
Binkinshaw, M., Gull, S.F. : 1984, Mon. Not. R. Astron. Soc. 206, 359.
Bothun, G.D., Schommer, R.A., Sullivan III, W.T. : 1984, Astron. J. 89, 466.
Dekel, A., West, M.J., Aarseth, S.J. : 1984, Astrophys. J. 279, 1.
Dressler, A. : 1984, Astrophys. J. 281, 512.
Ellis, R.S., Gray, P.M., Carter, D., Godwin, J. : 1984, Mon. Not. R. Astron. Soc. 206, 285.
des Forets, G., Dominguez-Tenreiro, R., Gerbal, D., Mathez, G., Mazure, A., Salvador-Sole, E : 1984, Astrophys. J. 280, 15.
Forman, W., Jones, C., Turner, W. : 1984, Astrophys. J. 277, 19.
Gehren, T., Fried, J. Wehinger, P.A., Wyckoff, S. : 1984, Astrophys. J. 278, 11.
Geller, M.J., Burs, T.C., Bothun, G.D., Huchra, J.P. : 1984, Astron. J. 89, 319.
Hanish, R.J. : 1984a, Astron. Astrophys. 139, 276.
Hanish, R.J. : 1984b, Astron. Astrophys. 133, 192.
Harris, D.E., Costain, C.H., Dewdney, P.E. : 1984, Astrophys. J. 280, 532.
Henry, J.P., Lavery, R.J. : 1984, Astrophys. 280, 1.
Huchra, J., Brodie, J. : 1984, Astrophys. J. 280, 547.
Jones, C., Forman, W. : 1984, Astrophys. J. 276, 38.
Lugger, P.M., : 1984, Astrophys. J. 278, 51.
Malumuth, E.M., Richstone, D.O. : 1984, Astrophys. J. 276, 413.
McGlynn, T.A., Fabian, A.C. : 1984, Mon. Not. R. Astron. Soc. 208, 709.
Merritt, D. : 1984a, Astrophys. J. Lett. 280, L5.
Merritt, D. : 1984b, Astrophys. J. 276, 26.

Nottale, L. : 1984, Mon. Not. R. Astron. Soc. 206, 713.
Osterbrock, D.E. : 1984, Astrophys. J. Lett. 280, L43.
Pryor, C., Geller, M.J. : 1984, Astrophys. J. 278, 457.
Rothenflug, R., Vigroux, L., Mushorzky, R.F., Holt, S.S. : 1984, Astrophys. J. 279, 53.
Tonry, J.L. : 1984, Astrophys. J. 279, 13.
Valtaoja, E. : 1984, Astron. Astrophys. 135, 141.
van Breugel, W., Heckman, T., Miley, G. : 1984, Astrophys. J. 276, 79.
van der Bergh, S. : 1984, Astron. J. 89, 608.

4. Galaxy formation
(Bernard J.T. Jones)

This article surveys the literature from July 1981 to June 1984. It is neither possible nor desirable to refer to all papers on the subject, and accordingly only papers that are generally representative of some particular idea are explicitly mentioned. Galaxy Formation by its very nature has considerable overlap with other areas of cosmology such as the anisotropy of the cosmic background radiation, the question of the mass density of the universe, the nature of the large scale clustering, and detailed observations of galaxies. These are all topics covered by other reports to Commission 47 and the reader will therefore find only scant attention paid here to these important subjects.

1. GENERAL TRENDS DURING THE PERIOD

These past years have been dominated by two major developments. Firstly there has been the input to Cosmology from high energy physics which has provided us with the "inflationary" models for the early stages of the cosmic expansion and a plethora of exotic particles that might constitute the substantial amounts of "unseen mass" that are thought to pervade the universe. Secondly gravitational N-body simulations of cosmic evolution have provided a workbench for testing hypotheses for the formation of galaxies and the clustering process in general. The possible role of galaxy mergers has attracted a lot of attention. Observations of the large scale galaxy distribution have revealed a filamentary structure that is an important target for clustering theories. There has also been substantial input from observations of galaxies that have served to delineate specific chemical and dynamical processes taking place during the birth of a galaxy.

There have been a number of conference proceedings published during this period, these are cited in a separate bibliography. Review articles by Efstathiou and Silk (1983), Matzner (1984), and Peebles (1984c) have appeared. IAU Symposium N° 104 (see ref [E]) contains a remarkable variety of papers on the subject.

The material for this review covers theory and observation as relates to galaxy formation. The evolution that ensues once the final system is assembled and has reached a quasi stationary state will only be touched on here : thus no reference is made to the important issues of the chemical evolution of the disk of our Galaxy, dynamical friction in clusters of the relaxation processes that take place in globular clusters once they have formed. This is inevitably subjective and not without ambiguity, but it is essential in the interests of limiting the length and scope of the article !

2. EXOTIC PARTICLES AND DARK MATTER

There is growing (indirect) evidence for the existence of substantial amounts of dark invisible matter in the universe. Indeed, the inflationary model universes demand that the present universe be extremely close to the Einstein de Sitter case and so demand considerable quantities of dark nonbaryonic matter. (See Bean et al., 1983 for the current observational situation). Suggestions as to the nature of this material range from light .001 eV axions at the low mass end of the scale, through 100 eV neutrinos to primordial black holes at the other. The invention by high energy physicists of numerous exotic particles that could be of relevance to the earliest moments of the cosmic expansion leads naturally to reconsidering how galaxies might form it such particles were the major contributors to the present mass density.

The general features of galaxy formation in a universe dominated by massive (10 - 100 eV) neutrinos were outlined by Chernin (1981), Klinkhammer and Norman (1981), Doroshkevich and Khlopov (1981), Davis et al. (1981), Wasserman (1981, 1982). Problems of how the neutrinos got into galactic halos were examined by Melotte (1982, 1983) and by Bond et al. (1983).

Galaxy formation in a universe dominated by 1kev gravitinos was considered by Blumenthal et al. (1982), and by Bond et al. (1982b) who showed that the spectrum of fluctuations is flat between galactic and cluster scales in that theory.

Cosmological models that are presently dominated by axions are discussed for example by Axenides et al. (1983), Fukugita and Yoshimura (1983), Ipser and Sikivie (1983), Stecker ans Shafi (1983) and Turner et al. (1983). Models dominated by primordial black holes are considered by Freese et al. (1983). Turok (1983) considers fluctuations arising out of strings and Axenides and Brandenburger (1984) contemplate the consequences of para-photons.

3. N-BODY SIMULATIONS OF COSMIC EVOLUTION

Numerical N-body models have played an important role in seeing how nonlinear structures may have developed in the universe. Frenk et al. (1983) compared 1000 body models having different initial fluctuation spectra and concluded that the pancake-like theories gave a better visual impression for the appearance of the final state than did the hierarchical clustering models. White et al. (1983) also simulated neutrino dominated universes and concluded that such models appeared to be ruled out because they gave characteristic length scales that were unacceptably large. Dekel (1983) has modelled superclusters as large scale non dissipative pancakes. The filamentary structure is most clearly shown up in the very large simulations of Centrella and Melotte (1983) (the largest simulation to date) and of Miller (1983).

The origin of galaxy angular momentum has most recently been studied using N-body cosmological models by Efstathiou and Barnes (1984).

4. THE CLUSTERING AND SUPERCLUSTERING OF GALAXIES

While the N-Body simulations provide a graphic image of the clustering process, they do not actually explain why it develops the way it does. There have been a number of attempts to describe the clustering process by simple analytic models, the most recent of which is that of Hamilton and Saslaw (1984). With a simple thermodynamic ansatz they are able to reproduce the galaxy distribution functions of all orders which are in substantial agreement with observations. Schaeffer (1984) has also shown that these distribution functions can be obtained from the hypothesis that the N-point correlation functions are simply related to lower order correlation functions. Fry (1984) has discussed the evolution of the galaxy correlation hierarchy in perturbation theory.

The role of dynamical friction and mergers between galaxies in the evolution of the clusters themselves has been simulated by Roos (1981a,b), Roos and Aarseth (1982), Miller (1983), Farouki et al. (1983), Duncan et al. (1983), Cavaliere et al. (1984), Giuricin et al. (1984), Allen and Yabushita (1984) and Barnes (1984). Mass segregation arising during cluster evolution has been simulated by Barnes (1983) and by Farouki et al. (1983); the competition from tidal stripping has been considered by Merritt (1983). Analytic approximations for mass segregation based on simple N-body models were given by Farouki and Salpeter (1982). It has been suggested by Shaya and Tully (1984) that it is the supercluster tidal field that is responsible for the origin of galactic spin, and ultimately for the distinction between spirals and ellipticals.

A model for the evolution of a spherical supercluster has been given by Rivolo and Yahil (1983).

5. FORMATION OF GALAXIES AND GLOBULAR CLUSTERS

Peebles (1984a, 1984b) shows how naturally the characteristic scales of galaxies and globular clusters can emerge in a cosmology dominated by weakly interacting matter with negligible primeval pressure. A number of papers have appeared trying to follow the formation of individual galaxies having dynamically dominant dark massive halos. Jones and Wyse (1983a), Wyse (1983), treat the problem of the formation of the thick and thin disk components of galaxies, and Kashlinsky (1982, 1984a,b) looks at rotational properties. Fall (1982) has reviewed the significance of galactic rotation for theories of galaxy formation.

Silk (1983c) comments that the surface density versus velocity dispersion data for galaxies is telling us that dissipation is the key factor controlling galaxy evolution. (See also articles of Struck-Marcell, 1981; Faber, 1982; and Gunn, 1982). Wang and Scheuerle (1984) present a model showing how dissipation may work to govern the eventual morphological and dynamical properties of galaxies. The collapse of a protocloud, and its fragmentation through the generation of turbulence and shock waves is shown by Sabano and Tosa (1984) to lead to globular cluster sized fragments. (See also McCrea, 1982). The evolution of collapsing protogalactic clouds was earlier considered by Struck-Marcell (1982a,b) and by Miller and Smith (1981).

Epstein (1983, 1984) has discussed the origin of the galactic mass function in terms of the gravitational amplification of initially Poisson irregularities. Palmer (1983) has investigated the tidal interactions between protostructures; the effect may be in part responsible for the observed shapes of galaxies and clusters of galaxies.

There have been a number of relevant N-body simulations of galaxy formation. Dissipationless collapses have been modelled by Noguchi (1984). McGlynn (1984), van Albada (1982). Wilkinson and James (1982) looked at the dissipationless collapse of a rotating triaxial galaxy.

The question of the hierarchical merging of stellar systems to form ever larger ones is of considerable interest. Tremaine (1981) has given a clear account of this and the role of dynamical friction. Kashlinsky (1984a,b) discusses the problem of merging in rotating haloes, and in particular the relationship between the final spin and core radius of the merged system. The sinking of a satellite into a parent galaxy has been discussed in detail by White (1983) in response to a paper by Lin and Tremaine (1983). Numerical simulations of the merger process have been attempted by Villumsen (1982, 1983) using a collisionless N-body code. The merging and stripping of halos in binary galaxies was studied by Carlberg (1982). Gerhard (1981); and Farouki and Shapiro (1982), used N-body simulations to study the merging of two disk galaxies to assess the hypothesis that ellipticals could formed in that way. Negroponte and White (1983) simulated mergers between disk-halo like systems, including a set of particles that collide inelastically to simulate a gaseous component. The role of merging in the centers of galaxy clusters is discussed by Merritt (1984a,b) and by Malumuth and Richstone (1984), with differing conclusions.

5. THE ROLE OF OBSERVATION, ELLIPTICAL AND SPIRAL GALAXIES

The relationship between the absolute magnitude, central velocity dispersion and spectral line strength among elliptical galaxies is an important datum for theories of galaxy formation. It is generally thought that the galaxies form a two parameter system (Terlevich et al. 1981), the "second" parameter has been identified as surface brightness by Wyse and Jones (1984) or mass to light radio by Efstathiou and Fall (1984). Dressler (1984) has presented data on galaxies in the Coma and Virgo clusters that suggests that the second parameter effect may be environment dependent.

In spiral galaxies an important datum is the bulge to disk ratio (Whitmore, 1984). This is probably related to the so-called "thick disk" component that is thought to play an important role in the formation of the familiar "thin" disk. The model of Jones and Wyse (1983a) for the formation of a disk galaxy makes us of this and provides an explanation for the important Fisher-Tully relationship (see also Burstein and Sarazin, 1983). There is a very well defined infrared colour magnitude relation for spirals, but not for ellipticals (Wyse, 1982; Tully et al.,

1982); this may in part reflect a variation of bulge to disk ratio with absolute magnitude.

The relationship between environment and galaxy morphology is certainly an important clue to way in which galaxies have been formed. Postman and Geller (1984) have displayed a correlation of morphology with local galaxy density that extends over six orders of magnitude in density.

Davies et al. (1983) have studied the rotational properties of elliptical galaxies and suggested that rotation is dynamically more important in the intrinsically fainter ellipticals than in the brighter ones studied previously. Wyse and Jones (1984), however, have suggested that these observations are best explained in terms of rotation being more important in galaxies of higher surface brightness. It is interesting that the bulge components of disk galaxies are rotationally supported (Kormendy and Illingworth, 1982). (See Kormendy and Illingworth (1983) for a comparison of the velocity dispersion - absolute magnitude relationships for ellipticals and bulges).

Whether globular clusters are precursors of galaxies, or a consequence of the halo formation mechanism is quite unclear. The observational situation is reviewed by Freeman (1981), van den Bergh (1983, 1984), De Young et al. (1983) and by Pilachowski et al. (1983). It is possible that the globular clusters we see are the few survivors of an ongoing attrition by the tidal field of the Galaxy (Caputo and Castellani), 1984).

The situation as regards the origin of dwarf galaxies of various kinds is no less uncertain. Some way be the tidal debris of galactic encounters (Lynden-Bell, 1983; Gerola et al. 1983) or the dwarf spheroidals could be dwarf irregulars that have lost their gas by ram pressure sweeping (Lin and Faber, 1983).

6. PANCAKE THEORIES

The general features of pancake theories have been looked at in a number of papers (Arnold et al. 1982), and N-body experiments have proven particularly useful (Doroshkevich et al. 1983, Klypin and Shandarin, 1983).

The stability of shocked pancakes was studied by Jones et al. (1981) and by Palmer (1981a). Doroshkevich (1983) considered a gaseous pancake in the presence of massive neutrinos and Bond et al. (1984) did some numerical gasdynamic simulations of collapsing pancakes in a universe dominated "warm" or "hot" dark matter. Szalay et al. (1984) considered the effect of such collapse on the spectrum of the microwave background radiation. Chernin and Ushakov (1983) and Ushakov and Chernin (1983, 1984) have looked at the generation of angular momentum in the shock front and find values consistent with the angular momenta attributed to spiral galaxies. Gurevich and Verner (1983) looked at the origin of the angular momentam of ellipticals. Shapiro et al. (1983) did extensive numerical simulation of pancake hydrodynamics, including collisionless neutrinos.

Doroshkevich (1984) has calculated the expected separation between pancakes in neutrino universes and argued that this is consistent with the distribution of Ly-α absorption systems along the line of sight. Dekel (1982) on the other hand seeks to explain the observations in terms superclusters that collapsed recently and nondissipatively.

Peebles has offered some criticism of the pancake theory and suggested that the primeval departures from homogeneity may not have been Gaussian (Peebles 1982b, 1983). Schaeffer and Silk (1984a,b) attempt to counter some criticisms, especially those of White et al. (1983) concerning lengthscales.

The evidence for pancake theories mostly concerns the statistical nature of the large structure of the observed universe, and the impression of "filamentary" structures (Einasto et al. 1984). Interesting support for the idea has come from Djorgovski (1983) who claims that there is a genuine alignment of the galaxies in the Coma cluster.

7. THE FIRST STARS

What were the first gravitationally bound objects to form in our Universe, and when did they form ? The question is considered in general cosmological scenarios by Carr and Rees (1984a, 1984b), and with reference to objects having masses between 10 and 100 $M_{\odot}$ by Silk (1983a, 1984). The role of molecular cooling ad redshifts 30-1300 has been considered in detail by Lepp and Shull (1984) and by Shchekinov and Ehnteh1' (1983b). Izotov and Kolesnik (1984) study molecular hydrogen and suggest that the rotational J = 2 -> 0 transition should be looked for.

The contribution of pregalactic object to nucleosynthesis of light elements D and Li has been looked at by Rees (1984) and by Audouze and Silk (1983,1984). Bond et al. (1983a) and Carr et al. (1984) have studied the effect of hypothetical very massive primordial stars on He generation and the microwave background radiation. Tarbet and Rowan-Robinson (1982) imposed constraints on hypothetical populations of pregalactic stars.

A detailed hierarchical scenario for galactic evolution starting from primordial stars has been giben by J. Jones (1983). The collapse of a uniform metal poor protogalactic cloud can lead to a first generation of massive stars (Zinnecker and Drapatz, 1984). Silk (1983b) and Shchekinov and Ehnteh1' (1983a) looked at the general problem of thermal instability in collapsing primordial clouds. Kashlinsky and Rees (1983) describe a specific scenario for the formation of the first stars. Chiosi and Matteucci (1983) consider the chemical enrichment of the galactic disk on the basis of models of primordial (zero metal) stars.

8. OTHER THEORIES OF GALAXY FORMATION

There has been copnsiderable interest in the idea that the large scale structure was generated subsequent to the formation of galaxy sized systems by explosive events that were the consequence of galaxy formation or quasar like activity (Ostriker and Cowie, 1981; Ikeuchi, 1981). Hogan (1983, 1984), Hogan and Kaiser (1983) and Carr and Rees (1984c) have looked at various aspects of this and other nonstandard models. Time dependent numerical solutions of the effects of cosmic shocks that might lead to galaxy and cluster formation have been presented by Ikeuchi et al. (1983).

Lyubarsky and Sunyaev (1983) calculate the expected distortion of the microwave background radiation spectrum and the generation of spectral features when energy is injected into the universe at an early time. Freese et al. (1983) explore the consequences of assuming that the dark matter consists of planetary sized black holes. Clusters of these will form objects that may subsequently explode, leading to galaxy formation. See also Bertschinger (1983).

9. COSMIC VOIDS AND FILAMENTS

Enormous voids in the universe have attracted considerable attention (Silk, 1982; Silk et al. 1983). The existence of voids is an attractive aspect of the pancake theories (Doroshkevich et al. 1982; Zel'dovich et al. 1982). The large scale topology of the universe in the Pancake theory has been examined from a very general point of view by Arnold et al. (1982) and by Shandarin and Zel'dovich (1983). Analytic models for spherieal voids are discussed by Filmore and Goldreich (1984), Occhionero et al. (1983, 1984) and Sato (1982). Icke (1984) has commented that voids are likely to be roughly spherical, whereas condensations are likely to be filamentary. The numerical simulations of Melotte (1983b) show this. The evolution of voids and condensations of ellipsoidal shape were studied by numerical solution of the equations by Fujimoto (1983) and by Santangelo et al. (1983) respectively.

The shell that forms around the void is of interest (Hausman et al. (1983). The density profile in voids has been discussed by Palmer and Voglis (1983). More complex multi-void models ("honeycombs") have been presented by Hoffman et al. (1983) and there have been some N-body studies (Aarseth and Saslaw, 1982; Saslaw and Aarseth, 1982).

10. THE ORIGIN AND EVOLUTION OF INHOMOGENEITIES

The various "inflationary universe" scenarios have provided a strong stimulus for studying the origin of perturbations in terms of quantum fluctuations prior to and at the time of vacuum phase transitions (Lindley, 1984; Bardeen et al. 1983;

Hawking and Moss, 1983; Lukash and Novikov, 1983; Turner, 1983; Guth and Pi, 1982; Kompaneets et al. 1982; Chibisov and Mukhanov, 1982).

Khlopov and Polnarev (1983) show that when the universe is dominated by superheavy metastable particles, the evolution of small inhomogeneities leads to the formation of primordial black holes. The role of primordial black holes in the context of Grand Unified Theories has been discussed by Lindley (1982). Fluctuations arising out of quantum gravitational vacuum polarization have been evaluated by Starobinskij (1983). The generation of isotherm perturbations was considered by Bond et al. (1982a).

Collisionless damping of density perturbations in the early universe was discussed by Doroshkevich and Khlopov (1981), Davis et al. (1981), Wasserman (1981, 1982) and by Bond and Szalay (1983), and showed that massive neutrino and gravitino dominated cosmologies gave respectively supercluster and galaxy-like characteristic scales. Weakly nonlinear adiabatic and isothermal perturbations were considered by Vishniac 1982a,b, 1983), who also treated linear adiabatic perturbations as they cross the horizon of a universe with massive collisionless neutrinos (Vishniac 1982c). Cosmologies that deviate substantially from Friedmann models generally require numerical solution of the Einstein equations (Centrella 1983). The techniques for numerical modelling of planar inhomogeneous cosmologies have been presented by Centrella and Wilson (1983, 1984).

Brandenburger et al. (1983) have discussed fully the description of the evolution of fluctuations within general relativity. The evolution of adiabatic density perturbations through the recombination period was discussed in terms of an analytic model by Jones and Wyse (1983b). Kaiser (1983), Szymanski and Jaroszynski (1983), Zabotin and Nasel'skij (1982a,b, 1983) and Peebles (1981, 1982c,d) have calculated the expected anisotropy and polarization of the microwave background radiation arising from this epoch, using diverse assumptions about the cosmology. The evolution of perturbations subsequent to recombination in the weakly nonlinear regime has been calculated by Juszkizwick (1981) and in the nonlinear regime on the assumption that they have triaxial shape by Palmer (1981b).

REFERENCES

Aarseth, S.J. and Saslaw, W.C.: 1982, Astrophys. J. Lett., 258, L7.
Allen, A.J. and Yabushita S. : 1984, Astrophys. J. 278, 468.
Arnold, V.I., Shandarin, S.F. and Zel'dovich Ya. B. : 1982, Geophys. Astrophys. Fluid Dyn., 20, 111.
Audouze, J. and Silk, J. : 1983 see [D] below.
Audouze, J. and Silk, J. : 1984 see [C] below.
Axenides, M. and Brandenburger R. : 1984, Phys. Lett., 134B, 405.
Axenides, M., Brandenburger, R. and Turner, M.S. : 1983, Phys. Lett. 126B, 178.
Bardeen, J.M., Steinhardt, P.J. and Turner, M.S. : 1983, Phys. Rev. D28, 679.
Barnes, J. : 1983, Mon. Not. Astron. Soc. 203, 223.
Barnes, J. : 1984, Mon. Not. Astron. Soc. 208, 873.
Bean, A.J., Efstathiou, G., Ellis, R.S., Peterson, B.A. and Shanks, T. : 1983, Mon. Not. R. Astron. Soc. 205, 605.
Bertschinger, E : 1983, Astrophys. J. 268, 17.
Blumenthal, G.R., Pagels, H. and Primack, J.R.: 1982, Nature, 299, 37.
Bond, J.R., Kolb, E.W. and Silk, J. : 1982a, Astrophys. J., 255, 341.
Bond, J.R., Szalay, A.S. and Turner, M.S. : 1982b, Phys. Rev. Lett., 48, 1636.
Bond, J.R., Carr, B.J. and Arnett, W.D. : 1983a, Nature, 304, 514.
Bond, J.R., Szalay, A.S. and White, S.D.M. : 1983b, Nature, 301, 584.
Bond, J.R. and Szalay, A.S. : 1983, Astrophys. J., 274, 443.
Bond, J.R., Centrella, J.M., Szalay, A.S. and Wilson, J.R. : 1984 see [C] below.
Brandenburger, R., Kahn, R. and Press, W.H. : 1984, Phys. Rev., D28, 1809.
Burstein, D. and Sarazin, C.L. : 1983, Astrophys. J., 264, 427.
Caputo, F. and Castellani, V. : 1984, Mon. Not. R. Astron. Soc. , 207, 185.
Carlberg, R.G. : 1982, Mon. Not. R. Astron. Soc. , 199, 1159.
Carr, B.J., Bond, J.R. and Arnett, W.D. : 1984, Astrophys. J., 277, 445.
Carr, B.J. and Rees, M.J. : 1984a, Mon. Not. R. Astron. Soc., 206, 315.

Carr, B.J. and Rees, M.J. : 1984b, Mon. Not. R. Astron. Soc., 206, 767.
Carr, B.J. and Rees, M.J. : 1984c, Mon. Not. R. Astron. Soc., 206, 801.
Cavaliere, A., Santangelo, P., Tarquini, G. and Vittorio, N. : 1984 see [G] below.
Centrella, J.M. : 1983 see [L] below.
Centrella, J.M. and Melotte, A.L. : 1983, Nature, 305, 196.
Centrella, J.M. and Wilson, J. : 1983, Astrophys. J., 273, 428.
Centrella, J.M. and Wilson, J. : 1984, Astrophys. J. Suppl., 54, 229.
Chernin, A.D. : 1981, Astron. Zh., 58, 25 (Sov. A.J., 25).
Chernin, A.D. and Ushakov, A. Yu. : 1983, Astrophys. Space Sci., 96, 217.
Chibisov, G.V. and Mukhanov, V.F. : 1982, Mon. Not. R. Astron. Soc. , 200, 535.
Chiosi, C. and Matteucci, F. : 1983 see [D] below.
Davis, M., Lecar, M., Pryor, C. and Witten, E. : 1981, Astrophys. J., 250, 423.
Dekel, A.: 1982, Astrophys. J. Lett., 261, L13.
Dekel, A.: 1983, Astrophys. J., 264, 373.
De Young, D., Lind, K. and Strom, S.E. : 1983, Publ. Astron. Soc. Pacific, 95,401.
Djorgovski, S. : 1983, Astrophys. J. Lett., 274, L7.
Doroshkevich, A.G. : 1983, Pis'ma Astron. Zh., 9, 451.
Doroshkevich, A.G. : 1984, Astron. Zh., 61, 434.(Sov. A.J., 28).
Doroshkevich, A.G. and Khlopov, M. Yu. : 1981, Astron. Zh. 58, 913. (Sov.A.J.,25).
Doroshkevich, A.G. and Klypin, A.A. : 1981, Astron. Zh., 58, 225. (Sov. A.J., 25).
Doroshkevich, A.G., Kotok, E.V., Shandarin, S.F. and Sigov, Yu. S. : 1983, Mon. Not. R. Astron. Soc., 202, 537.
Dressler, A. : 1984, Astrophys. J., 281, 512.
Duncan, M.J., Farouki, R.T. and Shapiro, S.L. : 1983, Astrophys. J., 271, 22.
Efstathiou, G. and Silk, J. : 1983, Fundam. Cosmic Phys., 9, 1.
Efstathiou, G. and Barnes, J. : 1984, see (C] below.
Efstathiou, G. and Fall, S.M. : 1984, Mon. Not. R. Astron. Soc., 206, 453.
Einasto, J., Klypin, A.A., Saar, E. and Shandarin, S.F. : 1984, Mon. Not. R. Astron. Soc., 206, 529.
Epstein, R.I. : 1983, Mon. Not. R. Astron. Soc, 205, 207.
Epstein, R.I. : 1984, Astrophys. J., 281, 545.
Faber, S.M. : 1982 see [K].
Fall, S.M. : 1982 see [J].
Farouki, R.T. and Salpeter, E.E. : 1982, Astrophys. J., 253, 512.
Farouki, R.T. and Shapiro, S.L. : 1982, Astrophys. J., 259, 103.
Farouki, R.T., Shapiro, S.L. and Duncan, M.J. : 1983, Astrophys. J., 265, 597.
Farouki, R.T., Hoffman, G.L. and Salpeter, E.E. : 1983, Astrophys. J., 271, 11.
Filmore, J.A. and Goldreich, P. : 1984, Astrophys. J., 281, 9.
Freeman, K.C. : 1981 see [M].
Freese, K, Price, R. and Schramm, D.N. : 1983, Astrophys. J., 275, 405.
Frenck, C.S., White, S.D.M. and Davis, M. : 1983, Astrophys. J., 271, 417.
Fry, J. : 1984, Astrophys. J., 279, 499.
Fujimoto, M. : 1983, Publ. Astron. Soc. Japan : 1983, 35, 159.
Fukugita, M. and Yoshimura, M. : 1983, Phys. Lett. 127B, 181.
Gerhard, O.E. : 1981, Mon. Not. R. Astron. Soc., 197, 167.
Gerola, H., Carnevali, P. and Salpeter, E.E. : 1983, Astrophys. J. Lett., 268,L75.
Giuricin, G., Mardirossian, F., Mezzetti, M. and Santangelo, P. : 1984, Astrophys. J., 277, 38.
Gunn, J.E. : 1982 see [K].
Gurevich, L. Eh. and Verner, D.A. : 1983, Astron. Zh., 60, 851.
Guth, A.H. and Pi, S.-Y. : 1982, Phys. Rev. Lett., 49, 1110.
Hausman, M.A., Olson D.W. and Roth, B.D. : 1983, Astrophys. J., 270, 351.
Hawking, S.W. and Moss, I.G. : 1983, Nucl. Phys., B224, 180.
Hogan, C. : 1983, Mon. Not. R. Astron. Soc., 202, 1101.
Hogan, C. : 1984, see [C] below.
Hogan, C. and Kaiser, N. : 1983, Astrophys. J., 274, 7.
Hoffman,; G.L., Salpeter, E.E. and Wasserman, I. : 1983, Astrophys. J., 268, 527.
Icke, V. : 1984, Mon. Not. R. Astron. Soc., 206, 1P.
Ikeuchi, S. : Publ. Astron. Soc. Japan, 33, 211.

Ikeuchi, S., Tomisaka, K. and Ostriker, J.P. : 1983, Astrophys. J., 265, 583.
Ipser, J. and Sikivie, P. : 1983, Phys. Lett., B50, 925.
Izotov, Yu.I. and Kolesnik, I.G. : 1984, Astron. Zh., 61, 24.
Jones, B.J.T., Palmer, P.L. and Wyse, R.F.G. : 1981, Mon. Not. R. Astron. Soc., 197, 931.
Jones, B.J.T. and Wyse, R.F.G. : 1983a, Astron. Astrophys., 120, 165.
Jones, B.J.T. and Wyse, R.F.G. : 1983b, Mon. Not. R. Astron. Soc., 205, 983.
Jones, J.E. : 1983 see [I] below.
Juszkiewicz, R. : 1981, Mon. Not. R. Astron. Soc., 197, 931.
Kaiser, N. : 1983, Mon. Not. R. Astron. Soc., 202, 1169.
Kashlinsky, A. : 1982, Mon. Not. R. Astron. Soc., 200, 585.
Kashlinsky, A. : 1984a, Mon. Not. R. Astron. Soc., 208, 623.
Kashlinsky, A. : 1984b see [G] below.
Kashlinsky, A. and Rees, M.J. : 1983, Mon. Not. R. Astron. Soc., 205, 955.
Khlopov, M. Yu. and Polnarev, A.G. : 1983 see [H] below.
Klinkhammer, F.R. and Norman, C.A. : 1981, Astrophys. J. Lett., 245, L97.
Klypin, A.A. and Shandarin, S.F. : 1983, Mon. Not. R. Astron. Soc., 204, 891.
Kompaneets, D.A., Lukash, V.N. and Novikov, I.D. : 1982, Astron. Zh., 59, 424. (Sov. A.J., 26).
Kormendy, J. and Illingworth, G. : 1982, Astrophys. J., 256, 460.
Kormendy, J. and Illingworth, G. : 1983, Astrophys. J., 265, 632.
Lepp. S. and Shull, J.M. : 1984, Astrophys. J., 280, 465.
Lindley, D. : 1982, Mon. Not. R. Astron. Soc., 199, 775.
Lindley, D. : 1984, Nucl. Phys., B236, 522.
Lin, D.C.N. and Faber, S.M. : 1983, Astrophys. J. Lett., 266, L21.
Lin, D.C.N. and Tremaine, S. : 1983, Astrophys. J., 264, 364.
Lukash, V.N. and Novikov, I.D. : 1983, see [H] below.
Lybarsky, Y.E. and Sunyaev, R.A. : 1983, Astron. Astrophys., 123, 171.
Lynden-Bell, D. : 1983 see [F].
Malumuth, E.M. and Richstone, D.O. : 1984, Astrophys. J., 276, 413.
Matzner, R.A. : 1984, Publ. Astron. Soc. Pacific, 96, 189.
McCrea, W.H. : 1982 see [J].
McGlynn, T.A. : 1984, Astrophys. J., 281, 13.
Melotte, A.L. : 1982, Nature, 296, 721.
Melotte, A.L. : 1983a, Astrophys. J., 264, 59.
Melotte, A.L. : 1983b, Mon. Not. R. Astron. Soc., 204, 7P.
Merritt, D. : 1983, Astrophys. J., 264, 24.
Merritt, D. : 1984a, Astrophys. J., 276, 26.
Merritt, D. : 1984b, Astrophys. J. Lett., 280, L5.
Miller, G.E. : 1983, Astrophys. J., 268, 495.
Miller, R.H. : 1983, Astrophys. J., 270, 390.
Miller, R.H. and Smith, B.F. : 1981, Astrophys. J. : 244, 467.
Negroponte, J. and White, S.D.M. : 1983, Mont. Not. R. Astron. Soc., 205, 1009.
Noguchi, M. : 1984, Publ. Astron. Soc. Japan, 36, 1.
Occhionero, F., Santangelo, P. and Vittorio, N. : 1983, Astron. Astrophys.117,365.
Occhionero, F., Santangelo, P. and Vittorio, N. : 1984 see [G] below.
Ostriker, J.P. and Cowie, L.L. : 1981, Astrophys. J. Lett., 243, L131.
Palmer, P.L. : 1981a, Mon. Not. R. Astron. Soc., 197, 167.
Palmer, P.L. : 1981b, Mon. Not. R. Astron. Soc., 197, 721.
Palmer, P.L. : 1983, Mon. Not. R. Astron. Soc., 202, 561.
Palmer, P.L. and Voglis, N. : 1983, Mon. Not. R. Astron. Soc., 205, 543.
Peebles, P.J.E. : 1981, Astrophys. J., 248, 885.
Peebles, P.J.E. : 1982b, Astrophys. J., 258, 415.
Peebles, P.J.E. : 1982c, Astrophys. J., 259, 442.
Peebles, P.J.E. : 1982d, Astrophys. J. Lett., 263, L1.
Peebles, P.J.E. : 1983, Astrophys. J., 274, 1.
Peebles, P.J.E. : 1984a see [C] below.
Peebles, P.J.E. : 1984b, Astrophys. J., 277, 470.
Peebles, P.J.E. : 1984c, Science, 224, 1385.

Pilachowski, C.A., Sneden, C. and Wallerstein, G. : 1983, Astrophys. J. Suppl., 52, 241.
Postman, M. and Geller, M.J. : 1984, Astrophys. J., 281, 95.
Primack, J.R. and Blumenthal, G.G. : 1984 see [C] below.
Rees, M.J. : 1984 see [C] below.
Rivolo, A.R. and Yahil, A. : 1983, Astrophys. J., 274, 474.
Roos, N. : 1981a, Astron. Astrophys., 95, 349.
Roos, N. : 1981b, Astron. Astrophys., 104, 218.
Roos, N. and Aarseth, S.J. : 1982, Astron. Astrophys., 114, 41.
Sabano, Y. and Tosa, M. : 1984 see [A] below.
Santangelo, P., Occhionero, F. and Carnevali, P.: 1983, Astron. Astrophys. 126,403
Saslaw, W.C. and Aarseth, S.J. : 1982, Astrophys. J., 253, 470.
Saslaw, W.C. and Hamilton, A.J.S. : 1984, Astrophys. J., 276, 13.
Sato, H. : 1982, Prog. Theor. Phys. 68, 236.
Schaeffer, R. : 1984, Astron. Astrophys., 134, L15.
Schaeffer, R. and Silk, J. : 1984a, Astron. Astrophys., 130, 131.
Schaeffer, R. and Silk, J. : 1984b, Astrophys. J. Lett., 281, L13.
Shandarin, S.F. and Zel'dovich, Ya.B. : 1983, Comments Astrophys., 10, 33.
Shaya, E.J. and Tully, R.B. : 1984, Astrophys. J., 281, 56.
Shapiro, P.R., Struck-Marcell, C. and Melotte, A.L. : 1983, Astrophys.J., 275, 413
Shchekinov, Yu.A. and Ehntehl', M.B. : 1983a, Astron. Zh., 60, 640. (Sov. A.J. 27)
Shchekinov, Yu.A. and Ehntehl', M.B. : 1983b, Astron. Zh., 60, 1073.(Sov. A.J. 27)
Silk, J. : 1982, Nature, 295, 95.
Silk, J. : 1983a see [D] below.
Silk, J. : 1983b, Mon. Not. R. Astron. Soc., 205, 705.
Silk, J. : 1983c, Nature, 301, 574.
Silk, J. : 1984 see [C] below.
Silk, J., Szalay, A.S. and Zel'dovich, Ya.B. : 1983, Scientific American, 249, 56.
Starobinskij, A.A. : 1983, Pis'ma Astron. Zh., 10, 579. (Sov. A.J., 9).
Stecker, F.W. and Shafi, Q. : 1983, Phys. Rev. Lett., 50, 928.
Struck-Marcell, C. : 1981, Mon. Not. R. Astron. Soc., 197, 487.
Struck-Marcell, C. : 1982a, Astrophys. J., 259, 116.
Struck-Marcell, C. : 1982b, Astrophys. J., 259, 127.
Szalay, A.S., Bond, J.R. and Silk, J. : 1984 see [C] below.
Szymanski, M. and Jaroszynski, M. : 1983, Mon. Not. R. Astron. Soc., 205, 91.
Tarbet, P.W. and Rowan-Robinson : 1982, Nature, 298, 711.
Terlevich, R. : 1981, Mon. Not. R. Astron. Soc., 196, 381.
Tremaine, S. : 1981 see [M].
Tully, R.B., Mould, J.R. and Aaronson, M. : 1982, Astrophys. J., 257, 527.
Turner, M.S. : 1984 see [H] below.
Turner, M.S., Wilczek, F. and Zee, A. : 1983, Phys. Lett., 125B, 35.
Turok, N. : 1983, Phys. Lett. B126, 437.
Ushakov, A.Yu. and Chernin, A.D. : 1983, Astron. Zh., 60, 625 (Sov. A.J., 27).
Ushakov, A.Yu. and Chernin, A.D. : 1984, Astron. Zh., 61, 18. (Sov. A.J., 28).
van Albada, T.S. : 1982, Mon. Not. R. Astron. Soc., 201, 939.
van den Bergh, S. : 1983, Comments Astrophys., 10, 27.
van den Bergh, S. : 1984, Publ. Astron. Soc. Pacific, 96, 329.
Villumsen, J.V. : 1982, Mon. Not. R. Astron. Soc., 199, 493.
Villumsen, J.V. : 1983, Mon. Not. R. Astron. Soc., 240, 219.
Vishniac, E.T. : 1982a, Astrophys. J., 253, 457.
Vishniac, E.T. : 1982b, Astrophys. J., 253, 470.
Vishniac, E.T. : 1982c, Astrophys. J., 257, 456.
Vishniac, E.T. : 1983, Mon. Not. R. Astron. Soc., 203, 345.
Wasserman, I, : 1981, Astrophys. J., 248, 1.
Wasserman, I, : 1982, Astrophys. J., 248, 1.
White, S.D.M. : 1983, Astrophys. J., 274, 53.
White, S.D.M., Frenck C.S. and Davis, M. : 1983, Astrophys. J. Lett., 274, L1.
Whitmore, B.C. : 1984, Astrophys., 278, 61.
Wilkinson, A. and James, R.A. : Mon. Not. R. Astron. Soc., 199, 171.

Wyse, R.F.G. : 1982, Mon. Not. R. Astron. Soc., 199, 1P.
Wyse, R.F.G. : 1983, see [C].
Wyse, R.F.G. and Jones, B.J.T. : 1984 see [C] below.
Wyse, R.F.G. and Jones, B.J.T. : 1984, Astrophys. J., 286, 88.
Zabotin, N.A. and Nasel'skij, P.D. : 1982a, Astron. Zh., 59, 434.(Sov. A.J., 26).
Zabotin, N.A. and Nasel'skij, P.D. : 1982b, Astron. Zh., 59, 447.(Sov. A.J., 26).
Zabotin, N.A. and Nasel'skij, P.D. : 1983, Astron. Zh., 60, 430. (Sov. A.J., 27).
Zel'dovich, Ya.B., Einasto, J. and Shandarin, S.F. : 1982, Nature, 300, 407.
Zinnecker, H. and Drapatz, S. : 1984 see [C] below.

PROCEEDINGS OF MEETINGS AND CONFERENCES

[A] S. Aoki and Y. Yoshii
"Theoretical aspects on structure, activity, and evolution of galaxies, II"
Japanese symposium held Tokyo Astronomical Observatory, Tokyo, Japan
14-16 February, 1984.

[B] C. Chiosi and A. Renzini
"Stellar Nucleosynthesis"
Proceedings of the 3rd workshop of the Advanced School of Astronomy of the Ette Majorana Center for Scientific Culture.
Erice, Italy - May 11-21 1983
(Astrophys. and Space Sci. Library, 109; D. Reidel 1984).

[C] J. Audouze and T. Tran Thanh Van
"Formation and Evolution of Galaxies and Large Structures in the Universe"
3rd Moriond Astrophysics Meeting (NATO Advanced Study Institute)
La Plagne, France. March 1983
(D. Reidel, 1984).

[D] P.A. Shaver, D. Kunth and K. Kjar
Primordial Helium
Proceedings of an ESO Workshop held at Garching, F.R. Germany
2-3 February 1983.

[E] G.O. Abell and G. Chincarini
"The Early Evolution of the Universe and its Present Structure"
IAU Symposium N° 104
Kokymbari, Crete, Greece. 30 August-2 September 1982
(D. Reidel, 1983).

[F] E. Athanassoula
"Internal Kinematics and Dynamics of Galaxies"
IAU Symposium N° 100
Besançon, France. 9-13 August 1982
(D. Reidel, 1983).

[G] D. Gerbal and A. Masure
"Clustering in the Universe (L'Univers et ses Structures)"
Proceedings of a European Colloquium held at the Observatoire de Meudon
Paris, France, 1982
(Editions Frontières, 1983).

[H] G.W. Bibbons, S.W. Hawking and S. Siklos
"The very Early Universe"
Proceedings of the Nuffield Workshop held at Cambridge, UK
21 June-9 July 1982 (Cambridge University Press).

[I] J. Audouze and J. Tran Thanh Van
"The Birth of the Universe"
Proceedings of the 17th Rencontre de Moriond - Astrophysics Meeting
Les Arcs, Savoie, France. 14-26 March 1982
(Editions Frontières, 1982).

[J] A.W. Wolfendale
"Progress in Cosmology"
Proceedings of the Oxford International Symposium
Christ Church, Oxford, UK. 14-18 September 1981
(Astrophys. and Space Sci. Lib., 99; D. Reidel, 1982).

[K] H.A. Bruck, G.V. Coyne, M.S. Longair
"Astrophysical Cosmology"
Proceedings of the Study Week on Cosmology and Fundamental Physics
Citta del Vaticano, 28 September-2 October, 1981
(Pontificiae Academiae Scientiarum Scripta Varia, Nº 48, 1982).

[L] B.T.J. Jones and J.E. Jones
"The Origin and Evolution of Galaxies"
Proceedings of the NATO Advanced Study Institute
VIIth Course of the International School of Cosmology and Gravitation
Erice, Italy, 11-23 May 1981
(D. Reidel, 1983).

[M] S.M. Fall and D. Lynden-Bell
"The Structure and Evolution of Normal Galaxies"
NATO Advanced Study Institute held at the Institute of Astronomy.
Cambridge, UK, 3-15 August 1980
(Cambridge University Press, 1981).

5. Quasars and their evolution
(Maarten Schmidt)

OPTICAL SURVEYS

In the past three years several surveys of optically selected quasars have been published. The Palomar Bright Quasar Survey (Schmift & Green 1983) covers 10,714 deg^2 to a limiting magnitude varying with position in the sky, with an effective limiting magnitude B = 16.16. The survey lists 114 objects of which 92 are quasars with $M_B < -23$. All objects have redshifts less than 2.15 as they were selected for ultraviolet excess. Marshall et al. (1983) have extended the spectroscopic coverage of objects in the Braccesi field (AB) to B = 18.25. They also report first results in the deep Braccesi field (BF) to B = 19.2. Koo (1983) reports on a photometric and spectroscopic survey of quasars to B = 22.5 (see also Koo & Kron [1982]). The present state of quasar surveys is critically discussed by Smith (1983) in a very useful review.

EVOLUTION

If optical quasar surveys covered the entire optical Hubble diagram, then the luminosity function at given redshift could be derived from the apparent magnitude distribution at that redshift. The evolution of the luminosity function of quasars with redshift could then be studied directly. Only at a redshift around 2 is it possible at present to derive the luminosity function (Gaston 1983). At other redshifts the Hubble diagram is poorly covered by surveys as yet.

As a consequence, in discussing quasar evolution, assumptions have to be made about the nature of the redshift dependence. Some authors consider the variation of co-moving space density with redshift, while others use variation of luminosity and redshift to explain the observations (Cheney & Rowan-Robinson 1981a, Mathez & Nottale 1982, Woltjer & Setti 1982, Braccesi 1983, Marshall et al 1983, Schmidt & Green 1983, Veron 1983). It is generally agreed that pure density evolution is excluded by the relatively low surface density of quasars fainter than magnitude 20. Whether it is more useful to consider luminosity-dependent density evolution or luminosity evolution is best resolved by further surveys aimed to improve coverage of the Hubble diagram.

THE REDSHIFT LIMIT

Osmer (1982, 1983) reports on a survey for quasars in the redshift range 3.7-4.7 over an area of 5 deg^2. None of the emission-line objects found exhibited Lyman-alpha emission, and no quasars were detected. By extrapolation from surveys at redshifts below 3.5, Osmer expected to find between 9 and 22 quasars and he concludes that the space density must decrease beyond redshift 3.5. The evolution models of Schmidt and Green (1983) essentially agree with Osmer's expected number, based on smooth extrapolation from lower redshifts. Mathez and Nottale (1982) predict on the basis of their luminosity evolution model that Osmer should see few if any quasars with $z > 3.5$. However, their model does not correctly represent the observed numbers of quasars in grism surveys.

The cause of the apparent cutoff in quasar redshifts is as yet uncertain. It may represent the turn-on epoch of quasars, or a change in the line-emission envelope of quasars, or as noted by Ostriker and Heisler (1984) it may be caused by intervening dust.

X-RAY COUNTS AND BACKGROUND

Two X-ray source surveys were published that will greatly aid statistical studies of quasars and AGNs. Piccinotti et al. (1982) reports on the HEAO1-A2 survey covering 8.2 steradians above galactic latitude 20° to a limit of 3.1×10^{-11} ergs cm^{-2} s^{-1} in the 2-10 KeV band. The sample contains 61 extragalactic objects with redshifts. Maccacaro et al. (1982) and Stocke et al. (1983) published the first installment of the Medium Sensitivity Survey, based on 63 serendipitous sources detected on Einstein IPC exposures. Above declination -25° optical identifications are complete, showing that 56 per cent are AGNs.

The contribution of quasars and AGNs to the diffuse X-ray background has been discussed by Cheney and Rowan-Robinson (1981b), Kembhavi and Fabian (1982) and Setti and Woltjer (1982). Estimated contributions range from 40% to 100%. A remarkable problem exists, though, in the form of the spectra which are systematically different for the background and the discrete sources. As a consequence, constraints can be placed on the evolution and luminosity function of the discrete sources (Cavaliere et al. 1981, De Zotti et al. 1982, Worrall & Marshall 1983). Boldt and Leiter (1984) have derived the optical spectral properties of objects whose X-ray spectrum resembles that of the diffuse X-ray background.

ASSOCIATIONS

The association of quasars with superclusters has been discussed by Oort (1983) and Osmer (1983). None of the optically found superclusters contains a quasar. At large distances, the association of quasars and superclusters can only be detected from clustering of quasars among themselves. Webster (1982), Osmer (1983) and Oort (1983) note a concentration of 4 quasars near redshift 0.37 that could be part of a superstructure of diameter 140 Mpc. Shanks et al. (1983) and Boyle et al. (1983) find that faint ultraviolet-excess stars, which are frequently quasars, show evidence of clustering among themselves, but anti-clustering with respect to faint galaxies.

The association of quasars and galaxies continues to be controversial. Yee and Green (1984) find that galaxies tend to show clustering around quasars whose redshifts are less than 0.5. They find that the apparent magnitude distribution is consistent with a cosmological redshift of the quasars. On the other hand, Arp (1981, 1983) and Burbidge (1983) conclude from the association of quasars with bright galaxies that quasars are relatively nearby.

ABSORPTION SYSTEMS

Absorption lines observed in the spectra of quasars allow study of the absorbing clouds over a large range of redshift. Tytler (1982) finds that the redshift distribution of quasar absorption-line systems that are optically thick in the Lyman continuum matches that of a non-evolving population of absorption over the redshift range 0.4 - 3.5. Absorption systems showing metal lines such as C IV do not show evolution either, over the redshift range 1.2 - 2.3 (Young et al. 1982b, Boksenberg & Sargent 1983). In contrast, the Lyman-alpha absorption lines observed blueward of quasar Lyman-alpha emission exhibit evolution with redshift (Young et al. 1982a, Peterson 1983, Sargent & Boksenberg 1983). Sargent and Boksenberg consider various possible origins for the Lyman-alpha lines and conclude that they are primordial, intergalactic clouds of size between 1 and 100 kpc, not associated with galaxies.

REFERENCES

Arp, H.: 1981, Astrophys. J. 250, 31.
Arp, H.: 1983 in Quasars and Gravitational Lenses. Institut d'Astrophysique, Liège p. 307.
Avni, Y., Tananbaum, H. : 1982, Astrophys. J. Letters, 262, L17.
Boksenberg, A., Sargent, W.L.W.: 1983 in Quasars and Gravitational Lenses. Institut d'Astrophysique, Liège p. 500.
Bold, E., Leiter, D. : 1984, Astrophys. J. 276, 427.
Boyle, B.J., Fong, R., Shanks, T. : 1983 in Quasars and Gravitatioinal Lenses. Institut d'Astrophysique, Liège, p. 368.
Braccesi, A. : 1983 in IAU Symposium N° 104. Reidel, Dordrecht, p. 23.
Burbidge, G. : 1983 in IAU Symposium N° 104. Reidel, Dordrecht, p. 65.
Canizares, C.R. : 1983 in Quasarsand Gravitational Lenses. Institut d'Astrophysique, Liège, p. 126.
Cavaliere, A., Danese, L., De Zotti, G., Franceschini, A. : 1981, Astron. Astrophys. 97, 269.
Cheney, J.E., Rowan-Robinson, M. : 1981a, MNRAS 195, 497.
Cheney, J.E., Rowan-Robinson, M. : 1981b, MNRAS 197, 313.
De Zotti, G., Boldt, E.A., Cavaliere, A., Danese, L., Franchescini, AZ., Marshall, F.E., Swank, J.H., Szymkowiak, A.E. : 1982, Astrophys. J. 253, 47.
Gaston, B. : 1983, Astrophys. J. 272, 411.
Kembhavi, A.K., Fabian, A.C. : 1982, MNRAS 198, 921.
Koo, D.C. : 1983 in Quasars and Gravitational Lenses. Institut d'Astrophysique, Liège, p. 240.
Koo, D., Kron, R.G. : 1982, Astron. Astrophys. 105, 107.
Maccacaro, T., Feigelson, E.D., Fener, M., Giacconi, R., Gioia I.M., Griffiths R.E., Murray, S.S., Zamorani, G., Stocke, J., Liebert, J. : 1982, Astrophys. J. 253, 504.
Marshall, H.L., Avni, Y., Tananbaum, H., Zamorani, G. : 1983, Astrophys. J. 269, 35.
Marshall, H.L., Tananbaum, H., Zamorani, G., Huchra, J.P., Braccesi, A., Zitelli, V. : 1983, Astrophys. J. 269, 42.
Mathez, G., Nottale, L. : 1982, Astron. Astrophys. 113, 336.
Oort, J.H. : 1983 in Quasars and Gravitational Lenses. Institut d'Astrophysique, Liège, p. 301.
Osmer, P.S. : 1981, Astrophys. J. 247, 762.
Osmer, P.S. : 1982, Astrophys. J. 253, 28.
Osmer, P.S. : 1983 in Quasars and Gravitational Lenses. Institut d'Astrophysique, Liège, P. 51.
Osmer, P.S. : 1983 in IAU Symposium N° 104. Reidel, Dordrecht, p. 189.
Ostriker, J.P., Heisler, J. : 1984, Astrophys. J. 278, 1.
Peterson, B.A. : 1983, in Quasars and Gravitational Lenses. Institut d'Astrophysique, Liège, p. 563.
Piccinotti, G., Mushotzky, R.F., Boldt, E.A., Holt, S.S., Marshall, F.E., Serlemitos, P.J., Shafer, R.A. : 1982, Astrophys. J. 253, 485.
Reichert, G.A., Mason, K.D., Thorstensen, J.R., Bowyer, S. : 1982, Astrophys. J. 260, 437.
Sargent, W.L.W., Boksenberg, A. : 1983 un Quasars and Gravitational Lenses. Institut d'Astrophysique, Liège, p. 518.
Setti, G., Woltjer, L. : 1982 in Astrophysical Cosmology. Pontifica Academia Scientiarum, Vatican City, p. 315.
Shanks, T., Fong, R., Green, M.R., Clowes, R.G. : 1982, MNRAS 203, 181.
Smith, M.G. : 1983 in Quasars and Gravitational Lenses. Institut d'Astrophysique, Liège, p. 1.
Stocke, J., Liebert, J., Gioia, I.M., Griffiths, R.E., Maccacaro, T, Danziger I.J. Kunth, D., Lub, J. : 1983, Astrophys. J. 273, 458.
Tytler, D. : 1982, Nature 248, 427.
Veron, P. : 1983 in Quasars and Gravitational Lenses. Institut d'Astrophysique, Liège, p. 210.

Walsh, D. : 1983 in Quasars and Gravitational Lenses. Institut d'Astrophysique, Liège, p. 106.
Webster, A. : 1982 : MNRAS 199, 683.
Weymann, R., Foltz, C. : 1983 in Quasars and Gravitational Lenses. Institut d'Astrophysique, Liège, p. 538.
Woltjer, L., Setti, G. : 1982 in Astrophysical Cosmology. Pontifica Academia Scientiarum, Vatican City, p. 293.
Worrall, D.M., Marshall, F.K. : 1982 in Quasars and Gravitational Lenses. Institut d'Astrophysique, Liège, p. 430.
Yee, H.K.C., Green, R.F. : 1984, Astrophys. J. 280, 79.
Young, P., Sargent, W.L.W., Boksenberg, A. : 1982a, Astrophys. J. 252, 10.
Young, P., Sargent, W.L.W., Boksenberg, A. : 1982b, Astrophys. J. Suppl. 48, 455.

6. The cosmological evolution of radio sources

(R.A. Windhorst)

INTRODUCTION

Considerable progress has been made over the last few years in the study of the cosmological evolution of radio sources, both regarding the collection of deeper and higher quality observations and towards a better understanding of the physical processes that may cause the epoch dependent properties of the radio source populations on cosmological timescales. In this review it is attempted to give a fair coverage to the work of **all** groups that made relevant contributions to this field since 1981. A few very important papers from before 1980, that were not dealt with by the previous IAU commission, are covered as well. For more detailed reviews we refer to Katgert (1980), Kellerman (1980), van der Laan and Windhorst (1982), van der Laan et al. (1983), Longair (1978), Wall and Benn (1982), Wall (1983) and Windhorst (1984). Le latter review is a more detailed version of the current report to IAU commission 47 on Cosmology.

1. RADIO SOURCE SURVEYS

In this section we describe all new systematic radio surveys that have been made and published since 1980.

1a. Strong and Intermediate Strength Radio Surveys

1a1. Low frequencies : An update of the revised 3CR cataloque has been given by Laing, Riley and Longair (1983), who summarize all available morphological and radio spectral data on 3CR sources. The complete 3CR sample now consists of 173 radio sources with $S_{178} > 10$ Jy and angular size < 10'. High resolution mapping has been made for a complete sample of 59 intermediate strength Bologna sources with $1 < S_{408} < 2$ Jy in 0.11 sr by Allington-Smith (1982). A part of the 0.3 GHz Texas Survey, the +18° strip, has been published by Douglas et al. (1980).

1a2. High frequencies : The 2.7 GHz all sky equivalent of the 3CR cataloque is given by Peacock and Wall (1981) and Wall and Peacock (1984) and contains 233 sources with $S_{2.7} > 2.0$ Jy in 9.81 sr. An all sky sample at 5 GHz is given by Kühr et al. (1981a), which contains 518 radio sources with $S_{5.0} > 1.0$ Jy in 9.81 sr. The high frequency normalized differential source counts do not yet show any convergence at the Jansky level.

At somewhat lower flux levels, 5 GHz source samples have been collected in considerable patches of the sky. The Bonn 100 m sample of Kühr et al. (1981b) has 476 radio sources with $S_{5.0} > 250$ mJy in 0.401sr. The MIT-Green Bank survey con-contains about 6000 radio sources with $S_{5.0} > 50$ mJy in 1.87 sr (Bennett, 1983; Bennett et al., 1984). Other 5 GHz samples have been constructed by Owen et al. (1983) containing 480 radio sources with $S_{1.4} > 35$ mJy in 0.069 sr, and by Ledden et al. (1980), containing 237 sources with $S_{5.0} > 15$ mJy in 0.0096 sr. A southernhemisphere 5 GHz survey has been made with the Parkes 64 m telescope by Wall et al. (1982), who list 75 radio sources with $S_{5.0} > 32$ mJy in 0.0082 sr.

Only at the 50 mJy level do the normalized differential 5 GHz source counts show convincing evidence for convergence.

1b. Deep Radio Surveys

1b1. Low frequencies : A deep 408 and 1407 MHz survey of a north Galactic pole area has been made with the Cambridge one mile telescope by Benn et al. (1982), yielding 299 radio sources with $S_{0.4} > 9$ mJy in 12.2 deg^2 and 65 sources with $S_{1.4} > 1.4$ mJy in 1.2 deg^2. A 1.4 GHz VLA snapshot survey has been made by Condon et al. (1982), covering 11.0 deg^2 down to $S_{1.4}$ = 6mJy. A deep 1.4 GHz survey was made with the Westerbork 3 km array by Windhorst, van Heerde and Katgert (1984) in 9 selected areas with available deep multicolor Mayall 4 m plates. This Leiden-Berkeley Deep Survey has 306 radio sources in 5.52 deg^2 with $S_{1.4} > 0.6$ mJy.

The total normalized differential 1.4 GHz source counts, obtained by the various groups, are consistent to within 10% down to $S_{1.4} \sim 1$ mJy and in the range $100 > S_{1.4} > 5$ mJy they show the well known convergence.

Surveys at 610 MHz have also been made with the 1.5 km Westerbork array by Katgert-Merkelijn (1978, 1979), yielding 232 radio sources in 25 deg^2 down to $S_{0.6} > 22$ mJy. A survey of the Einstein X-ray deep survey areas has been made by Katgert-Merkelijn et al. (1984), yielding 451 radio sources with $S_{0.2} > 10$ mJy in 29.2 deg^2 and 369 sources at 21 cm with $S_{1.4} > 1.6$ mJk in 8.71 deg. A 610 MHz WSRT survey of M31 is described by Bystedt et al. (1984) and of the Cancer cluster by Valentijn (1980). A deep 610 MHz survey has been made of the Leiden-Berkeley Deep Survey areas with the WSRT 3 km array by Windhorst and Oppe (1984), yielding 520 radio sources in 12.3 deg^2 with $S_{0.6} > 2.4$ mJy.

If one takes account of population and resolution bias, all 610 MHz synthesis counts agree to within 20% down to $S_{0.6} \sim 2$ mJy and show the well known convergence down to $S_{0.6} \sim 10$ mJy.

1b2. High frequencies : Deeper 5 GHz surveys have been made with the Bonn 100m dish by Pauliny-Toth et al. (1980) and Maslowski et al. (1981, 1984), yielding in total 320 radio sources in 67 deg^2 with $S_{5.0} > 10$ mJy. For the north Galactic pole survey 5 GHz data have been obtained with the Bonn 100 m telescope by Benn et al. (1984). A 5 GHz P(D) analysis of NRAO 90 m scans has been made by Ledden et al. (1980) and of Parkes 64 m scans by Wall et al. (1982).

A systematic 5 GHz synthesis survey of background radio sources in Westerbork maps has been made by Willis and Miley (1979), yieldind 43 radio sources in 2.8 deg^2 with $S_{5.0} > 4.5$ mJy. A deep 5 GHz VLA survey has been made in 13 small areas by Bennett et al. (1983), who found 16 radio sources in 0.52 deg^2 with $S_{5.0} > 0.5$ mJy.

The 5 GHz source counts of these surveys (for a review see Maslowski et al., 1984) confirm the convergence below $S_{5.0}$ = 50 mJy, which however is not as strong as that seen in the low frequency surveys down to $S_{1.4}$ = 5 mJy.

1c. Ultradeep radio surveys and background fluctuations

1c1. Low frequencies : Ultradeep VLA counts were made in the Leiden-Berkeley area Lynx.2 by Windhorst et al. (1984c), yielding 94 radio sources in 0.58 deg^2 complete to $S_{1.4}$ = 225 µJy. Ultradeep Westerbork counts down to $S_{1.4}$ = 300 µJy of the same field were given by Oort and Windhorst (1984). Comparison with 1.4 GHz counts at brighter levels (section 1b1) show that below 5 mJy the 1.4 GHz normalized differential source counts show a remarkable resteepening, in the sense that an increasingly larger number of fainter radio sources is found at sub-mJy levels.

Similar ultradeep VLA 1.4 GHz counts have been made of another field by Condon and Mitchell (1984), who found 116 radio sources down to $S_{1.4}$ = 84 µJy. The upturn in the source counts from their data has the same slope as that of Windhorst et al. (1984c) and Oort and Windhorst (1984) in the Lynx.2 area and differs in amplitude by about 20%. Hence, the resteepening of the sub-mJy source counts is a property of the universe, its slope being well established with an uncertainty in amplitude of the order of 20%. A summary of the total 1.4 GHz source count, amalgamated from 10 different Westerbork surveys and updated to 1984, has been given by Katgert (1984).

1c2. High frequencies : Ultradeep 5 GHz sources counts have been made in one very small area with the VLA by Fomalont et al. (1984b), who detected 9 radio sources in 0.0051 deg^2 down to $S_{5.0}$ = 60 μJy. In addition, 10 surrounding areas were covered less deeply, yielding 25 radio sources in 0.051 deg^2 down to $S_{5.0}$ = 350 μJy. The 5 GHz source counts at the $S_{5.0} \geqslant 1$ mJy level are consistent with those of section 1b2, but at the $S_{5.0} \sim 100$ μJy level the slope of the normalized differential 5 GHz counts seems to be flatter than in the 1 - 50 mJy range, consistent with the upturn of source counts discovered at 1.4 GHz (see section 1c1).

From the same 5 GHz data Fomalont et al. (1984a) set an upper limit of 1×10^{-3} to the fluctuations in the background radiation at 5 GHz on scales of 18". A similar, but less strong limit of 5×10^{-3} has been set by Condon and Mitchell (1984) at 1.4 GHz on scales of 19". Several observed upper limits to fluctuations in the radio background are compared to model predictions by Danese et al. (1983).

2. SPECTRAL INDEX DISTRIBUTIONS

Spectral index distributions for various new source samples at different flux levels have been obtained over the last few years by the following groups. For the Green Bank 1.4 GHz survey two point spectral index distributions have been measured by Machalski and Maslowski (1982), for the MIT-Green Bank survey by Lawrence et al. (1983), for the NRAO 5 GHz survey by Condon and Ledden (1981), for the north Galactic pole survey by Benn et al. (1982, 1984), for the 1st and 2nd Westerbork survey by Katgert-Merkelijn (1979), for the Westerbork-Einstein deep survey by Katgert-Merkelijn et al. (1984), for the WSRT 5 GHz background survey by Willis and Miley (1979), for the Leiden-Berkeley Deep Survey by Windhorst and Oppe (1984), and for the ultradeep 5 GHz VLA survey by Fomalont et al. (1984b).

The general trend is that the spectral index distribution of radio sources selected at low frequencies ($\nu \leqslant 1.4$ GHz) is to first order independent of flux density. However, at the intermediate fluxes in **low** frequency surveys a larger fraction of **steep** (defined as having $\alpha \geqslant 0.5$) spectrum sources is seen, while at comparable flux levels in **high** frequency surveys a much larger fraction of **flat** spectrum sources is seen. The former relation has been reviewed using data from a variety of low frequency surveys by Gopal-Krishna and Steppe (1981, 1982, 1984). They notice the similar flux density dependence of the source surface densities and their median spectral index and explain this as a property of the radio sourde redshift distribution, in the sense that the most distant - and hence the most powerful - radio sources have the steepest radio spectra.

In principle it is difficult to decide whether the primary relation is between spectral index and radio power or between spectral index and redshift. Laing and Peacock (1980) argue that the primary relation is between spectral index and radio power. Katgert-Merkelijn, Lari and Padrielli (1980) argue that it may instead be between spectral index and redshift. A similar suggestion is obtained from the data of Tielens, Miley and Willis (1979), who found that the optical identification fraction of a complete sample of 4C sources on the Palomar Sky Survey decreases drastically for the sources with the steepest radio spectrum. This result has been modelled in terms of a spectral index dependent epoch dependent radio luminosity function by Blumenthal and Miley (1979). Similar spectral index dependent optical identification fraction have been found for samples at various flux levels, such as those of Lawrence et al. (1984), Kron, Koo and Windhorst (1984) and Machalski and Condon (1984). Other attempts have been made to model the spectral index dependent epoch dependent radio luminosity function by Peacock and Gull (1981) and by Peacock (1984) and Condon (1984b), who include more recent data (see section 7a).

3. ANGULAR SIZES

Systematic high resolution mapping has been carried out for the following radio source samples. The 1 Jansky sample of Allington-Smith (1982) has been mapped with the Cambridge 5 km telescope. A 4C sample has been mapped with the WSRT at 5 GHz by Kapahi (1981). An 5C subsample has been mapped by Downes et al. (1981) with the VLA and the Cambridge 5 km. Additionnal 5C samples have been similarly mapped by Fielden et al. (1983). Systematic high resolution VLA mapping of the 5 GHz MIT-Green Bank survey sources was carried out by Lawrence (1983). A detailed morphological classification of these 5 GHz sources is given by Lawrence et al. (1984). High resolution VLA mapping of the 1.4 GHz Green Bank sources is obtained by Machalski and Condon (1983a, 1983b) and by Machalski et al. 1982). Angular size distributions of 1.4 GHz sources ar the milliJansky level were also presented by Windhorst et al. (1984a) and by Oort and Windhorst (1984).

The main result from these systematic high resolution mapping programs is that the median radio source angular size decreases from 100" at $S_{1.4} \sim 10$ Jy to 20" at $S_{1.4} \sim 1$ Jy and to 10" at $S_{1.4} \sim 100$ mJy, while it stays around 10" down to $S_{1.4} \geqslant 10$ mJy. Below $S_{1.4} = 10$ mJy the median angular size seems to decrease more strongly again, reaching at most 5 - 7" around 1 mJy and is possibly even smaller.

These results have been interpreted by various groups in terms of the linear size and redshift distributions of radio sources, with or without linear size evolution. Downes et al. (1981) claimed that there is no convincing evidence for linear size evolution on the basis of 3CR, 4C, 5C and Ooty data. On the basis of the same data Kapahi and Subrahmanya (1982) argue that linear size evolution needs to be invoked à la $L(z) \propto L(z=0).(1+z)^n$, where $L(z)$ is the epoch dependent and $L(z=0)$ the local linear size and $n = 1.0 - 1.5$. On the basis of more VLA data, Fielden et al. (1983) conclude that size evolution may be requires for some, but not necessarily for all assumed redshift distributions. Allington-Smith (1984) gives a discussion of selection effects operating on the Θ_{med} - S relation. The 2.7 GHz sample of Peacock and Wall (1982) contains a rather large fraction of compact steep spectrum sources, which have to be taken into account when modelling the radio source angular size distributien and their linear size evolution. Okoye and Onuora (1982) and Onuora and Okoye (1983) analyze similar data and conclude similarly to Kapahi and Subrahmanya (1982) that some linear size evolution is required.

Despite the tremendous amount of high quality high resolution angular size data on radio sources in a wide range of flux densities the question of linear size evolution has not been resolved yet. The reason for this is that the local linear size distribution function is poorly known at **all** flux densities; it has only been measured at the brighter levels by Gavazzi and Perola (1978) and Katgert-Merkelijn et al. (1980).

4. OPTICAL IDENTIFICATIONS, PHOTOMETRY AND CLASSIFICATION

The advent of panoramic solid state detectors on large optical reflectors has considerably increased the success of identifying radio sources optically. Measuring accurate optical colors resulted in a more reliable classification.

Deep optical identifications of unidentified 3CR sources were done with a CCD detector in the 200" prime focus down to $r \sim 23.0$ by Gunn et al. (1981). Only 3% of the 3CR sources remained unidentified. A summary of all available 3CR identifications, photometry ans spectroscopic redshifts is given by Laing et al. (1983). Available optical identifications and spectroscopic redshifts of two 4C subsamples are summarized by Katgert-Merkelijn et al. (1980). Deep 200" CCD identifications of the 1 Jansky sample have been made by Allington-Smith et al. (1982), leading to 80% identifications fown to $r \sim 23.0$. Similar optical identifications of the Parkes 2.7 GHz bright source sample ($2_{2.7} \geqslant 1.5$ Jy) have been made by Peacock et al. (1981), who found 96% optical identifications down to $r \sim 23.0$. For the Parkes 5 GHz sample ($S_{5.0} \geqslant 32$ mJy) 40% optical identifications were found by Savage et al. (1982) on deep UK Schmidt plates down to $r \sim 22$. Optical identifications for various subsamples of the 1.4 GHz Green Bank surveys have been made on the Palomar Sky Survey by Machalski and Condon (1983a, 1983b), Machalski and Maslowski (1982) and Machalski et al. (1982), who find about 50% identifications down to r ~ 20. Identifications on the Palomar Sky Survey for the 5 GHz NRAO survey have been made by Condon and Ledden (1982), yielding 35% identifications for $S_{5.0} \geqslant 15$ mJy and for the 1.4 GHz Green Bank survey by Condon et al. (1982). Optical identifications have also been done on the Palomar Sky Survey for the MIT-Green Bank 5 GHz sources by Lawrence (1983) and Lawrence et al. (1984), who found 45% identifications for $S_{5.0} \geqslant 50$ mJy. For some 5C6 and 5C7 sources deep 200" prime focus CCD identifications have been obtained by Perryman et al. (1982), who found 40% optical identifications down to r ~ 23.0. Deep multicolor identifications have been made for the 1.4 GHz Leiden Berkeley Deep Survey sources on Mayall 4 m plates in the photographic passbands U, J, F and N by Windhorst, Kron and Koo (1984b), who found 53% optical identifications down to r ~ 22.5. A complete review of all available optical identifications of radio sources in literature, updated to 1983, is given by Veron and Veron (1983).

Swarup et al. (1982) summarized available data in literature on the fraction of optical identifications visible on the Palomar Sky Survey and showed that this fraction decreases from 65% at the 3CR level to 40% at the Bologna level and to 10-15% at the milliJansky level, a result that allowed these authors to rule out various canonical models for the cosmological evolution of radio sources.

Windhorst et al. (1984b) showed that identified radio sources with $1 \leqslant S_{1.4} \leqslant 100$ mJy consist for 20% of quasars, 78% galaxies and 2% of galactic radio stars. In contrast, radio sources at brighter levels ($S_{1.4} \geqslant 0.5$ Jy) consist for about 30% for quasars (Allington-Smith et al., 1982), while at the 3CR level 25% quasars are seen (Laing et al., 1983) and no radio stars.

Properly calibrated, homogeneous (photographic, photoelectric or CCD) photometry has been performed for only a few samples of radio source identifications : for some subsamples of the 3CR on deep CCD frames in r and i (Laing et al., 1983, and references therein), for the 1 Jansky sample on CCD frames in r (Allington Smith et al., 1982), for objects surrounding the 5C12 sources on multicolor 48" Schmidt plates (Grueff et al., 1984), for the 171 identifications in the Leiden Berkeley sample on Mayall 4 m plates in the passbands U, J, F and N (Kron, Koo and Windhorst, 1984) and for the ultradeep 1.4 GHz VLA sample on the same 4 m plates (Windhorst et al., 1984). We will follow the color classification of Kron et al. (1984), which is based upon 60 spectroscopic redshifts (section 5), to describe the nature of various radio source populations.

In survey areas of the order of several square degree, radio sources with $S_{1.4} \geqslant 10$ mJy are either quasars or luminous giant ellipticals, both often of the classical double radio source type. The giant elliptical radio galaxies have a small dispersion in absolute magnitude. The colors of the giant elliptical radio galaxies are without exception red at the brighter optical levels and become quickly redder at fainter magnitudes due to the K-correction. Below $S_{1.4} \leqslant 10$ mJy (Kron et al.,1984) find that a blue radio galaxy population becomes increasingly more important. These blue radio galaxies are mostly unresolved radio sources at arcsecond resolution and have often peculiar optical morphology, either indicative for interacting or merging galaxies or very compact galaxies. Their optical luminosities turn out to be 2 - 3 magnitudes fainter than those of the giant elliptical radio galaxies and have also a larger dispersion. Windhorst et al. (1984c) show that it is this blue radio population that dominates the radio source population at the sub-mJy level and that causes the upturn in the 1.4 GHz source counts below a few mJy.

Systematic near-infrared photometry of radio sources has been done by Lebofsky (1981) and Lilly and Longair (1982, 1984). The J-H and H-K colors of 3CR radio galaxies are consistent with those of passively or mildly evolving luminous giant elliptical galaxies out to redshifts of unity. The optical-infrared V-K and R-K colors show in some cases a blue excess at high redshifts, which may be due to some star formation at a later epoch, or due to the presence of emission lines or non-thermal contributions in these extremely powerful radio galaxies. Near-infrared photometry has also been obtained by Lilly et al. (1984) for the 1 Jansky sample, yielding several new radio source identifications not found previously on deep CCD frames.

Near-infrared photometry has been carried out for a sample of Jodrell Bank radio sources by Puschell et al. (1982) and for several dozen Leiden Berkeley Deep Survey radio sources by Thuan et al. (1984) and Katgert, Windhorst and Isaacman (1984). The near-infrared and optical-infrared colors of the milliJansky radio galaxies are similar to those of the 3CR sample confirming that the stellar population is the main component of the near-infrared spectrum of radio galaxies.

5. REDSHIFTS

In the last few years CCD spectrographs on large optical telescopes have made crucial contributions to the redshifts distribution od radio sources.

For the 3CR sample the spectroscopic redshift measurements have proceeded furthest, because most 3CR radio galaxies are relatively bright and the faintest ones have in general emission lines. Important contributions have been made by Spinrad et al. (1981), who found the first radio galaxy absorption redshift in excess of unity and by Spinrad (1982). Recently emission line redshifts have been measured for some 3CR radio galaxies as high as 1.82 (Spinrad and Djorgovski, 1984b). Another 6 redshifts were measured for faint 3CR radio galaxies by Perryman et al. (1984). The available 3CR spectroscopic redshifts are summarized by Ling et al. (1983) and including the latest measurements the complete sample of 173 3CR sources has now 151 spectroscopic redshifts measured (or 87%).

Noticeable is that the few radio galaxies known today with spectroscopic redshifts in excess of unity have very often narrow emission lines and sometimes spatially extended structure in their emission lines, with velocity dispersions up to 500 km sec^{-1} (Spinrad and Djorgovski, 1984a, 1984b). A similar result was found for one of the 12 radio galaxies from the 1 Jansky sample by Allington-Smith et al.(1984).

A systematic redshift survey has been undertaken during the last years in a long term project by Koo, Kron and Windhorst with the Kitt Peak 4m Cryogenic Camera. Using multiaperture techniques low resolution spectra are simultaneously obtained for field galaxies, radio galaxies and QSO's down to $r \sim 22.0$ in the selected areas where also the Leiden Berkeley Deep Survey was done. In total for a representative sample of more than 60 milliJansky radio sources spectroscopic redshifts are now available (Kron, Koo and Windhorst, 1984). Radio galaxies with $S_{1.4} \geqslant 10$ mJy are spectroscopically like giant elliptical galaxies, with a small dispersion in absolute magnitude $\langle M_F \rangle = -23.2 \pm 0.5$ for $H_0 = 50$), comparable to that measured at the 3CR level. The rather red colors of the giant elliptical radio galaxies are consistent with those of a non-evolving or at most a mildly evolving model spectral energy distribution out to redshifts of at least 0.6, such as those computed by Bruzual (1981, 1983). There is no evidence for drastic evolution of the spectral energy distribution of elliptical radio galaxies out to these redshifts, such as initially suggested by Katgert et al. (1979) and Windhorst et al. (1982) to explain the faint blue radio galaxies. Kron et al. (1984) also show that amongst this blue radio galaxy population, which dominates the source counts for $S_{1.4} \leqslant 10$ mJy, **no** very high redshifts ($z \geqslant 1$) are found, so that these objects are **not** the high redshift progenitors of elliptical galaxies in an early blue phase.

6. THE LOCAL RADIO LUMINOSITY FUNCTION OF RADIO GALAXIES

Since studies of the cosmological evolution of radio sources compare the radio luminosity function (RLF) at high redshift with the local RLF, **any** study of the cosmological evolution **must** properly consider the local RLF of all possiblerelevant radio source populations. Because the nature of radio sources is a priori inknown at **all** flux levels, only completely identified **radio** selected samples will provide an unbiassed view of the local RLF.

The local RLF of quasars and of elliptical, S0, spiral and irregular galaxies is reviewed by Fanti and Perola (1977). A more detailed determination of the local the bivariate (radio-optical) luminosity function of elliptical and S0 galaxies is made by Auriemma et al. (1977). Similar results have been found by Meier et al. (1979), who also noted that the RLF of elliptical galaxies apparently does not undergo significant evolution out to $z \sim 0.3$. This was also found by Katgert et al. (1979) from deep optical identifications of Westerbork sources. A further determination of the local RLF of elliptical and S0 galaxies has been given by Dressel (1981).

The local fractional RLF from "normal" spiral galaxies has been determined by Hummel (1980, 1981a) and by Gavazzi and Trinchieri (1981) from large samples of optically selected nearby galaxies. For interactring (spiral) galaxies a similar determination of the RLF has been made by Hummel (1981b). He and also Dressel (1981) and Heckman (1983) noted that interacting galaxies or galaxies in groups have statistically higher radio powers. The local 1.4 GHz RLF has been determined for Seyfert galaxies (both type I and II) by Meurs (1982), who shows that these objects have radio powers typically a factor 10 higher, but space densities at least a factor of 10 lower than those of normal spiral galaxies.

The local RLF for elliptical galaxies has also been determined from the ultradeep radio selected Leiden Berkeley Deep Survey sources by Windhorst (1984). A determination of the local RLF of the blue radio galaxy population that causes the upturn in the source counts below $S_{1.4} \leqslant 5$ mJy was also made. These objects have radio luminosities at least a factor of 5 higher than those of normal spiral galaxies, more like those of Seyferts, but space densities at least 20 x higher than those of Seyferts, in fact comparable th those of elliptical radio galaxies with $P < P^*$. This is consistent with the findings of Hummel (1981b) and Heckman (1983). The blue sub-milliJansky radio galaxy population has for $P < P^*$ the largest space density of all radio galaxy classes, which was not realized before from the RLF determinations from optically selected samples. It is therefore crucial to consider the blue radio galaxy population in the studies of cosmological evolution.

An important contribution to the physical undestanding of the local RLF of giant elliptical radio galaxies has been made by Valentijn and Bijleveld (1983), who determined the trivariate (radio-optical-X-ray) luminosity function using a homogeneous sample of cD galaxies in clusters and groups (Bijleveld and Valentijn, 1984). The cD galaxy mass is determined by the local galaxy density, while the optical luminosity of the cD galaxy is also proportional to its mass. The mass of the cD galaxy in turn determines the accretion rate of intrergalactic gas, which emits the hot X-rays, while a stronger accretion flow in the galaxy nucleus will indirectly also fuel a more powerful radio source.

7. STUDIES OF THE EPOCH DEPENDENT RADIO LUMINOSITY FUNCTION

In this section we discuss the various methods that have been employed over the last few years to describe and understand the cosmological evolution of radio sources. The central theme of these studies is the epoch dependent radio luminosity function, $\rho(\log P, z)$, which is the space density of various classes of radio sources as a function of radio power.

7a. Parametric and non-parametric methods

The epoch dependent radio luminosity function has been described by several parametrizations, such as the Schmidt type power laws ($\rho \propto (1+z)^n$) or exponential laws ($\rho \propto e^{n\tau(z)}$), where the exponents are constant with radio power for $P > P^c$ and zero for $P < P^c$ (Katgert, 1977) or some smooth function of radio power for $P > P^c$ (Wall, Pearson anf Longair, 1980, 1981), which allows the most powerful radio sources to evolve more strongly (differential evolution). The local quasar RLF had to be bootstrapped in these models for $P < 10^{27.5}$ W Hz^{-1}, because their local space density is so low, that cosmological volumes - where evolutionary effects are already important - are required for adequate statistical sampling. Another parametrization that explains multifrequency source counts and luminosity distributions has been given by Subrahmanya and Kapahi (1983).

In the power law models a redshift cut-off (at $z_{max} \sim 3.5$ or 5) had to be applied, in the other models integration was usually performed to $z = \infty$. These models thus imply the presence of a substantial number of primeval galaxies amongst the radio source population. In most parametric evolution is usually only assumed for $P > P^c$, where the latter is taken to be $\geq 10^{26.5}$ W Hz^{-1}.

Non-parametric models were first introduced by Robertson (1978, 1980), who adopted a purely free-form array of adjustable delta functions for the redshift dimension of $\rho(\log P,z)$, while the dependence on radio was only assumed for $P > P^c \sim 10^{26.5}$ W Hz^{-1}. Significant evolution was only required for $z \geq 0.3$. An two dimensional extension of this method has been given by Peacock and Gull (1981), who used limited series of expansion to fit $\rho(\log P,z)$ to multifrequency source counts, luminosity distributions, V/V_{max} data and optical identification statistics. The flat and steep spectrum populations are dealt with simultaneously and Peacock and Gull conclude that also the flat spectrum source population undergoes some evolution, although not as strong as the deep spectrum sources. Peacock (1984) includes new 2.7 GHz data and optical identification data at low flux levels. Especially the latter considerably improve the constraints to $\rho(\log P, z)$, requiring the presence of a redshift cut-off at $z_{max} \geq 2.5$, as described in more detail by the direct method in section 7b.

Another parametrization of $\rho(\log P,z)$ has been given by Condon (1984a, 1984b), who also uses multifrequency source counts, spectral index distributions, the local RLF and the 3CR redshift distribution. The method is similar to that of Blumenthal and Miley (1979). In Condon's model normal spiral galaxies must evolve in the same way as ellipticals in order to explain the upturn in the source counts. It was demonstrated by Windhorst et al. (1984) that instead the blue, interacting and merging population - with much higher radio powers than those of spirals already at local redshifts - are responsible for the upturn in the source counts at milliJansky levels. It should be kept in mind that some of the parametric models require to extrapolate the 3CR redshift distribution over more than 4 decades in flux density, which must be treated with great care.

7b. The direct method

The basic idea behind the direct method is to use all available optical identification information, and to use a statistical reliable distance indicator, such as apparent magnitude for giant elliptical radio galaxies, in those flux regimes where redshifts are mostly unknown. The method has been first employed by de Ruiter (1978) and Katgert et al. (1979), using the magnitude redshif relation as measured at the 3CR level for several faint Westerbork samples. As found by Kron et al. (1984), radio sources with $S_{1.4} \leq 10$ mJy are progressively more of the blue radio galaxy class, with optical luminosities typically 2-3 magnitudes below those of giant elliptical radio galaxies. As a consequence, the previous direct determinations of $\rho(\log P,z)$ from data below $S_{1.4} = 10$ mJy needed to be revised.

A more detailed direct determination of the epoch dependent radio luminosity function has been made by Windhorst (1984), who made a distinction between the RLF of red giant elliptical radio galaxies and the blue (interacting, merging) radio galaxies, according to the classification of radio galaxies of Kron et al. (1984). In addition, 9 other radio surveys with available redshifts or good optical identification data are used to sample the epoch dependent RLF in a wide range of radio power and redshifts. Spectroscopic redshifts were known and incorporated for 300 out of 685 radio galaxies.

The findings from previous Westerbork surveys are confirmed that there is no evidence for evolution in the RLF of giant elliptical radio galaxies out to $z \leq 0.3$ for a wide range of radio powers ($25.0 \leq P \leq 27.5$). For $z \geq 0.3$ evolution in the RLF of giant elliptical galaxies occurs for **all** radio powers $P > P^* = 10^{25}$ W Hz^{-1} somewhat differential with radio power in the sense that the evolution is stronger for the more powerful galaxies ($\log P \geq 26.0$). The RLF evolution for giant elliptical radio galaxies is mild, not more than a factor of 10 at $z \sim 0.8$. Most of the apparently stronger evolution at the highest radio powers ($\log P \geq 27.0$) at high redshifts, as derived from the bright source surveys alone, is due to the most powerful radio quasars.

For the blue radio galaxy population only a reliable determination of their **local** RLF is possible due to the lack of spectroscopy at faint magnitudes. Because the space density of the blue radio galaxy population is already very high at local redshifts, little evolution of their RLF is needed to explain the upturn in the sub-mJy source counts, although Windhorst (1984) argues that some mild RLF evolution may required to explain the slope of th upturn in the counts **and** the large surface densities of faint blue radio galaxies.

Little can be said about the RLF evolution of quasars by the direct method, since they are not standard candles. Since at no flux levels fainter than $S_{1.4} \leq$ 1 Jy complete spectroscopy exists for **radio selected** quasars as yet, their epoch dependent RLF remains largely unknown. In the direct method applied by Allington-Smith (1984) to the 1 Jansky sample, a Hubble relation is adopted for the quasars as derived from spectroscopy at the 3CR level ($S_{1.4} \geq 3$ Jy). In the direct method of Windhorst (1984) the contribution to the source counts is determined for quasars with $B < 24.0$, at which level mosts of the optical quasar luminosity function is seen out to $z \leq 3.5$. This quasar contribution is then subtracted from the total source counts.

Windhorst (1984) presents models that fill in the remaining (radio galaxy) source counts by extrapolating the epoch dependent RLF of the giant elliptical and blue radio galaxies to redshifts beyond those of the faintest observed optical identifications. The conclusion is that extension of the milliJansky radio galaxy population to redshifts well in excess of unity is **not required** by the existing data. More powerful radio galaxies can be extended to gradually higher redshifts, ranging from ~ 1.5 for $\log P_{1.4} \sim 26.5$ and ~ 2.0 for $\log P_{1.4} \geq 27.0$. In contrast, in several of the parametric and non-parametric models discussed in section 7a, the faint radio source redshift distribution needed to be extended to redshifts in excess of 5.0 mainly because RLF evolution was not considered for log $P \leq 26.5$.

The direct method has been applied by Jaffe (1982) to 13 VLA detections of the optically selected distant clusters of Gunn and Oke. He concludes, that although there is evidence for RLF evolution at the brighter radio powers, for $P \leq P^*$ there may well be negative evolution, this in contrast with the findings from the deep radio samples in the field.

7c. Physical methods

One elegant way to approach the physics behind the cosmological evolution of radio sources is an **ab initio** theoretical treatment of the epoch dependence of the physical processes that are potentially important in the behaviour of radio sources and the triggering of their nuclear engines.

In the model of Bailey and McDonald (1981) the epoch dependent stellar mass loss rate in galaxies powers the central engine, such that the critical galaxy mass, above which activity is possible, decreases at earlier epochs. Stocke and Perrenod (1981) argue that quasars can only originate in galactic nuclei if the surrounding intergalactic density is smaller that some critical value that decreases with epoch.

In the model of McMillan, Lightman and Cohn (1981) stellar collisions and tidal disruptions in a dense star cluster trigger the central engine in an epoch dependent way that depends on the mass and the dynamical evolution of the system. In the model of Cavaliere et al. (1982) the central engine accretes collisional debris from a dense star cluster until its exhaustion. In the models of Roos (1981a, 1981b) the effect of merging is considered on the tidal disruption rate of stars near the central engine. The intruding (smaller) galaxy perturbs the gravitational field of the other galaxy such that stars are scattered into loss cone orbits around the central engine of the latter. The local behaviour of this mechanism is considered by Roos (1984a), who also predicts a bivariate luminosity function consistent with the data. The epoch dependent effect of mergers on the luminosity function is modeled by Roos (1984b), who reproduces the total 1.4 GHz source counts and the optical QSO' counts.

7d. Relation to the spectral evolution of radio galaxies

The study of the evolution of the optical spectrum of radio galaxies may provide constraints to the epoch(s) of galaxy formation, hence confining the earliest possible epoch of nuclear activity. Inversely, constraints to the earliest possible epoch of radio source triggering may, when obtained from independent data, tell us something about the epoch of (radio) galaxy formation.

The spectral evolution history of radio galaxies was first studied by Kristian, Sandage and Westphal (1978), who provided a homogeneous set of BVR photometry for a sample of brightest cluster galaxies and 3CR radio galaxies and demonstrated that the powerful radio galaxies do not show evidence for spectral evolution for $z \leq 0.3$. Similar studies have been performed by Djorgovski, Spinrad and Marr (1984) and Djorgovski and Spinrad (1984), who combine recent samples of first ranked cluster and 3CR radio galaxies, some of the latter extending to redshifts well in excess of unity. In their Hubble diagram the 3CR radio galaxies at $z \geq 1$ appear to be 2-4 magnitude brighter than the non-evolving prediction, while their V-R color-redshift relation shows colors ≥ 1 magnitude bluer than those predicted by a non-evolving giant elliptical spectrum or Bruzual's (1983) passively evolving C-model. Since these 3CR sources are at $z \geq 1$, they are the most powerful radio galaxies in the universe. Consequently, there may also be major non-thermal contributions to their optical light, moreover because all their redshifts could only be measured thanks to the presence of emission lines.

Using also near-infrared photometry similar studies have been performed of 3CR radio galaxies by Lilly and Longair (1984) and of radio galaxies from the 1 Jansky sample by Lilly, Allington-Smith and Longair (1984). These authors show that the position of the horizontal branch is unchanged at a redshift of unity and suggest that the stellar population of these luminous radio galaxies has undergone about one magnitude luminosity evolution since $z \sim 1$ and that at least some fraction of them has optical-infrared colors blue enough to be consistent with star formation ongoing at significant later times than the epoch of formation. They argue that not all of the blue optical-infrared colors can be caused by non-thermal contributions or redshifted emission lines, since in some cases the blue excess seems to be spatially extended.

The spectral evolution of the faint radio galaxies in the Leiden Berkeley Deep Survey has been studied by Kron, Koo and Windhorst (1984) with photographic UJFN photometry and 60 spectroscopic redshifts. As discussed in section 4 and 5, radio sources with $S_{1.4} \geq 10$ mJy are mostly giant elliptical galaxies with colors that are out to $z \sim 0.6$ sufficiently red to be consistent with Bruzual's passively evolving C-model or at most his mildly evolving $\mu = 0.7$ model, that processes 70% of the galaxy mass into stars during the first Gyr after formation. Hence for at least a considerable fraction of the less powerful elliptical radio galaxies in the milliJansky sample the star formation occured very early on, presumably of the order of 15-16 Gyr ago or at $z \geq 4.0$ (for $H_0 = 50$ and $q_0 = 0$).

Summarizing, there seems to exist a population of very old elliptical radio galaxies in all samples that must have had most of their star formation very early on, presumably at $z \geq 4$. In addition it is possible, but not not yet unequivocally proven, that some considerable fraction of them had later ongoing star formation, perhaps extending to epochs as late as $z \sim 1$. Such objects have not yet been found in the milliJansky samples as yet. If blue milliJansky radio galaxies with $z \geq 1$ were to be found in significant numbers, they would provide crucial constraints to the epoch of galaxy formation and first triggering of the central engines.

REFERENCES

Allington-Smith, J.R. : 1982, Monthly Notices Roy. Astron. Soc. 199, 611.

Allington-Smith, J.R., Perryman, M.A.C., Longair, M.S., Gunn, J.E., Westphal, J.A.1982, Monthly Notices Roy. Astron. Soc. 201, 331.

Allington-Smith, J.R. : 1984a, Monthly Notices Roy. Astron. Soc. 209, 665.
Allington-Smith, J.R. : 1984b, Monthly Notices Roy. Astron. Soc. 210, 611.
Allington-Smith, J.R., Lilly, S.J., Longair, M.S. : 1984, Monthly Notices Roy. Astron. Soc., submitted.
Auriemma, C., Perola, G.C., Ekers, R.D., Fanti, R., Lari, C., Jaffe, W.J. : 1977, Astron. Astrophys. 57, 41.
Bailey, M.E., McDonald, J. : 1981, Nature 280, 659.
Benn, C.R., Grueff, G., Vigotti, M., Wall, J.V., : 1982, Monthly Notices Roy. Astron. Soc. 200, 747.
Benn, C.R., Wall, J.V., Grueff, G., Vigotti, M. : 1984, Monthly Notices Roy. Astron. Soc. 209, 683.
Bennet, C.L. : 1983, Ph.D. Thesis, Massachusets Institute of Technology.
Bennet, C.L., Lawrence, C.R., Garcia-Barreto, J.A., Hewitt, J.N., Burke, B.F. : 1983, Nature 301, 686.
Bennet, C.L., Lawrence, C.R., Hewitt, J.N., Mahoney, J., Burke, B.F. : 1984, preprint.
Bijleveld, W., Valentijn, E.A. : 1984, Astron. Astrophys. 125, 217.
Blumenthal, G., Miley, G. : 1979, Astron. Astrophys. 80, 13.
Bruzual, A., G.: 1981, Ph.D. Thesis, University of California, Berkeley.
Bruzual, A., G.: 1983, Astrophys. J., 273, 105.
Bystedt, J.E.V., Brinks, E., de Bruyn, A.G., Israel, F.P., Schwering, P.B.W., Shane, W.W., Walterbos, R.A.M. : 1984, Astron. Astrophys. Suppl. 56, 245.
Cavaliere, A., Giallongo, E., Messina, A., Vagnetti, F. : 1982, Astron. Astrophys. Letters 114, L1.
Condon, J.J., Condon, M.A., Hazard, C. : 1982, Astron. J., 87, 739.
Condon, J.J., Condon, M.A., Broderick, J.J., Davis, M.M. : 1982, Astron. J., 88, 20.
Condon, J.J., Ledden, J.E. : 1981, Astron. J. 86, 643.
Condon, J.J., Ledden, J.E. : 1982, Astron. J. 87, 219.
Condon, J.J., Mitchell, K.J. : 1984, Astron. J. 89, 5.
Condon, J.J. : 1984, Astrophys. J. 284, 44.
Condon, J.J. : 1984, preprint.
Danese, L., De Zotti, G., Mandolesi, N. : 1983, Astron. Astrophys. 121, 114.
Djorgovski, S., Spinrad, H., Marr, J., : 1984, in proceedings of the special IAU Colloquium on "New Aspects on Galaxy Photometry", ed. J.-L. Nieto, Lectures in Physics Series, Springer Verlag, in press.
Djorgovski, S., Spinrad, H. : 1984, preprint.
Douglas, J.N., Bash, F.N., Torrence, G.W., Wolfe, C. : 1980, "The Texas Survey : Preliminary + 18° - Strip", University of Texas Publications in Astronomy N° 17
Downes, A.J.B., Longair, M.S., Perryman, M.A.C. : 1981, Monthly Notices Roy. Astron. Soc.197, 593.
Downes, A.J.B., Peacock, J.A., Savage, A., Carrie, D.R. : 1984, Monthly Notices Roy. Astron. Soc., submitted.
Dressel L.L. : 1981, Astrophys. J. 245, 25.
Fanti, R., Perola, G.C. : 1977, in "Radio Astronomy and Cosmology", IAU Symposium N° 74, ed. D.L. Jauncey (Dordrecht, Reidel), p. 171.
Fielden, J., Downes, A.J.B., Allington-Smith, J.R., Benn, C.R., Longair, M.S., Perryman, M.A.C. : 1983, Monthly Notices Roy. Astron. Soc. 204, 289.
Folamont, E.B., Kellerman, K.I., Wall, J.V. : 1984a, Astrophys. J. Letters 277, L23.
Folamont, E.B., Kellerman, K.I., Wall, J.V., Weistrop, D. : 1984b, Science 225,23.
Gavazzi, G., Perola, G.C. : 1978, Astron. Astrophys. 66, 407.
Gavazzi, G., Trinchieri, G. : 1981, Astron. Astrophys. 97, 128.
Gopal-Krishna, Steppe, H. : 1981, Astron. Astrophys. 101, 315.
Gopal-Krishna, Steppe, H. : 1982, Astron. Astrophys. 113, 150.
Gopal-Krishna, Steppe, H. : 1984, Astron. Astrophys. submitted.
Grueff, G., Vigotti, M., Wall, J.V., Benn, C.R. : 1984, Monthly Notices Roy. Astron. Soc. 206, 475.
Gunn, J.E., Hoessel, J.G., Westphal, J.A., Perryman, M.A.C., Longair, M.S. : 1981,

Monthly Notices Roy. Astron. Soc. 194, 111.
Heckman, T.M. : 1983, Astrophys. J., 268, 628.
Hummel, E. : 1980, Ph.D. Thesis, University of Groningen.
Hummel, E. : 1981a, Astron. Astrophys., 93, 93.
Hummel, E. : 1981b, Astron. Astrophys., 96, 111.
Jaffe, W. : 1982, Astrophys. J. 262, 15.
Kapahi, V.K. : 1981, Astron. Astrophys. Suppl. 43, 381.
Kapahi, V.K. : Subrahmanya, C.R. : 1982, in "Extragalactic Radio Sources", IAU Symposium Nº 97, eds. D.S. Heeschen and C.M. Wade (Dordrecht, Reidel), p.401.
Katgert-Merkelijn, J.K. : 1978, Astron. Astrophys. Suppl., 31, 409.
Katgert-Merkelijn, J.K. : 1979, Astron. Astrophys., 73, 107.
Katgert-Merkelijn, J., Lari, C., Padrielli, L. : 1980, Astron. Astrophys. Suppl., 40, 91.
Katgert-Merkelijn, J., Robertson, J.G., Windhorst, R.A., Katgert, P.K. : 1984, Astron. Astrophys. Suppl., submitted.
Katgert, P. : 1977, Ph.D. Thesis, University of Leiden.
Katgert, P., de Ruiter, H.R., ven der Laan, H. : 1979, Nature 280, 20.
Katgert, P. : 1980, in "X-Ray Astronomy", eds. R. Giacconi and G. Setti (Dordrecht, Reidel), p. 253.
Katgert, P. Windhorst, R.A., Isaacman, R.B. : 1984, Monthly Notices Roy. Astron. Soc., in press.
Katgert, P. : 1984, Astron. Astrophys., submitted.
Kellerman, K.I. : 1980, Physica Scripta, 21, 664.
Kristian, J. Sandage, A., Westphal, J.A. : 1978, Astron. J. 221, 383.
Kron, R.G., Koo, D.C., Windhorst, R.A. : 1984, Astron. Astrophys., in press.
Kühr, H., Pauliny-Toth, I.I.K., Witzel, A., Schmidt, J. : 1981a, Astron. J. 86, 854.
Kühr, H., Witzel, A., Pauliny-Toth, I.I.K., Nauber, U. : 1981b, Astron. Astrophys. Suppl., 45, 367.
van der Laan, H., Windhorst, R.A. : 1982, in Proc. of the Vatican Study Week on Cosmology and Fundamental Physics, Astrophysical Cosmology, eds. H.A. Brück, G.V. Coyne, M.S. Longair (Vaticano : Pontificiae Academiae Scientiarum), p. 349.
van der Laan, H., Katgert, P. Windhorst, R.A., Oort, M.J.A. : 1983, in "The Early Evolution of the Universe and its Present Structure", IAU symposium Nº 104, eds. G.O. Abell, G. Chincarini (Dordrecht, Reidel), p. 73.
Laing, R.A., Peacock, J.A., : 1980, Monthly Notices Roy. Astron. Soc. 190, 203.
Laing, R.A., Riley, J.M., Longair, M.S. : 1983, Monthly Notices Roy. Astron. Soc. 204, 151.
Lawrence, C.R., Bennett, C.L., Garcia-Barreto, J.A., Greenfield, P.E., Burke B.F.: 1983, Astrophys. J. Suppl. 51, 67.
Lawrence, C.R. : 1983, Ph.D. Thesis, Massachusets Institute of Technology.
Lawrence, C.R., Bennett, C.L., Hewitt, J.N., Burke, B.F. : 1984, Astrophys. J. Letters 278, L95.
Lebofsky, M.J. : 1981, Astrophys. J. Letters, 245, L59.
Ledden, J.E., Broderick, J.J., Condon, J.J., Brown, R.L. : 1980, Astron. J. 85, 780.
Lilly, S.J., Longair, M.S. : 1982, Monthly Notices Roy. Astron. Soc. 199, 1053.
Lilly, S.J., Longair, M.S. : 1984, Monthly Notices Roy. Astron. Soc. 211, 833.
Lilly, S.J., Longair, M.S., Allington-Smith, J.R. : 1984, Monthly Notices Roy. Astron. Soc., submitted.
Longair, M.S. : 1978, in "Observational Cosmology", 8th Advanced Course of the Swiss Society of Astronomy and Astrophysics, eds. A. Maeder, L. Martinet and G. Tammann (Geneva Observatory), p. 127.
Machalski, J., Condon, J.J. : 1983a, Astron. J. 88, 143.
Machalski, J., Condon, J.J. : 1983b, Astron. J. 88, 1591.
Machalski, J., Condon, J.J. : 1984, Astron. J., submitted.
Machalski, J., Maslowski, J. : 1982, Astron. J. 87, 1132.
Machalski, J., Maslowski, J., Condon, J.J., Condon, M.A. : 1982, Astron. J. 87,

1150.
Maslowski, J., Pauliny-Toth, I.I.K., Witzel, A., Kühr, H. : 1981, Astron. Astrophys. 95, 285.
Maslowski, J., Pauliny-Toth, I.I.K., Witzel, A., Kühr, H. : 1984, Astron. Astrophys. 139, 85.
Meier, D.L., Ulrich, M.-H., Fanti, R., Gioia, I., Lari, C. : 1979, Astrophys. J. 229, 225.
Meurs, E.J.A. : 1982, Ph. D. Thesis, University of Leiden.
McMillan, S.L.W., Lightman, A.P., Cohn, H. : 1981, Astrophys. J. 251, 436.
Okoye, S.E., Onuora, L.I. : 1982, Astrophys. J. 260, 37.
Onuora, L.I., Okoye, S.E. : 1983, Astrophys. J. 270, 360.
Oort, M.J.A., Windhorst, R.A. : 1984, Astron. Astrophys. Suppl., in press.
Oort, M.J.A., Katgert, P., Windhorst, R.A. : 1984, Astron. Astrophys., submitted.
Owen, F.N., Condon, J.J., Ledden, J.E. : 1983, Astron. J. 88, 1.
Pauliny-Toth, I.I.K., Steppe, H. , Witzel, A. : 1980, Astron. Astrophys. 85, 329.
Peacock, J.A., Gull, S.F. : 1981, Monthly Notices Roy. Astron. Soc. 196, 611.
Peacock, J.A., Wall, J.V. : 1981, Monthly Notices Roy. Astron. Soc. 194, 331.
Peacock, J.A., Perryman, M.A.C., Longair, M.S., Gunn, J.E., Westphal, J.A. : 1981, Monthly Notices Roy. Astron. Soc. 194, 601.
Peacock, J.A., Wall, J.V. : 1982, Monthly Notices Roy. Astron. Soc. 198, 843.
Peacock, J.A. : 1984, Monthly Notices Roy. Astron. Soc. submitted.
Perryman, M.A.C., Longair, M.S., Allington-Smith, J.R., Fielden, J. : 1982, Monthly Notices Roy. Astron. Soc. 210, 957.
Perryman, M.A.C., Lilly, S.J., Longair, M.S., Downes, A.J.B. : 1984, Monthly Notices Roy. Astron. Soc. 209, 159.
Puschell, J.J., Owen, F.N., Laing, R.A. : 1982, Astrophys. J. Letters 257, L57.
Robertson, J.G. : 1978, Monthly Notices Roy. Astron. Soc. 182, 617.
Robertson, J.G. : 1980, Monthly Notices Roy. Astron. Soc. 190, 143.
Roos, N. : 1981, Ph.D. Thesis, University of Leiden.
Roos, N. : 1981, Astron. Astrophys. 104, 218.
Roos, N. : 1984a, Astron. Astrophys. in press.
Roos, N. : 1984b, Astron. Astrophys. in press.
de Ruiter, H.R. : 1978, Ph.D. Thesis, University of Leiden.
Savage, A., Bolton, J.G., Wall, J.V. : 1982, Monthly Notices Roy. Astron. Soc. 200, 1135.
Spinrad, H., Stauffer, J., Butcher, H.R. : 1981, Astrophys. J. 244, 382.
Spinrad, H. : 1982, Publ. Astron. Soc. Pacific. 94, 397.
Spinrad, H., Djorgovski, S. : 1984a, Astrophys. J. Letters 280, L9.
Spinrad, H., Djorgovski, S. : 1984b, Astrophys. J. Letters 285, L49.
Stocke, J.T., Perrenod, S.C. : 1981, Astrophys. J. 245, 375.
Subrahmanya, C.R., Kapahi, V.K. : 1983, in "The Early Evolution of the Universe and its Present Structure", IAU Symposium Nº 104, eds. G.O. Abell and G. Chincarini (Dordrecht, Reidel), p. 47.
Swarup, G., Subrahmanya, C.R., Venkatakrishna, K.L. : 1982, Astron. Astrophys. 107, 109.
Tielens, A.G.G.M., Miley, G.K., Willis, A.G. : 1979, Astron. Astrophys. Suppl., 35, 153.
Thuan, T.X., Windhorst, R.A., Puschell, J.J., Isaacman, R.B., Owen, F.N. : 1984, Astrophys. J. 285, 515.
Valentijn, E.A. : 1980, Astron. Astrophys. 89, 234.
Valentijn, E.A., Bijleveld, W. : 1983, Astron. Astrophys. 125, 223.
Veron-Cetty, M.P.; Veron, P. : 1983, Astron. Astrophys. Suppl. 53, 219.
Wall, J.V., Pearson, T.J., Longair, M.S. : 1980, Monthly Notices Roy. Astron. Soc. 193, 683.
Wall, J.V., Pearson, T.J., Longair, M.S. : 1981, Monthly Notices Roy. Astron. Soc. 196, 597.
Wall, J.V., Savage, A. Wright, A.E., Bolton, J.G. : 1982, Monthly Notices Roy. Astron. Soc. 200, 1123.
Wall, J.V., Scheuer, P.A.G., Pauliny-Toth, I.I.K., Witzel, A. : 1982, Monthly

Notices Roy. Astron. Soc. 198, 221.
Wall, J.V., Benn, C.R. : 1982, in "Extragalactic Radio Sources", IAU Symposium N° 97, eds. D.S. Heeschen and C.M. Wade (Reisel, Dordrecht) p. 441.
Wall, J.V. : 1983, in "The Origin and Evolution of Galaxies", VIIth Course of the International School of Cosmology and Gravitation, eds. B.J.T. Jones and J.E. Jones (Reidel, Dordrecht), p. 295.
Wall, J.V., Peacock, J.A. : 1984, Monthly Notices, Roy. Astron. Soc., submitted.
Willis, A.G., Miley, G.K. : 1979, Astron. Astrophys. 76, 65.
Windhorst, R.A., Kron, R.G., Koo, D.C., Katgert, P : 1982, in "Extragalactic Radio Sources", IAU Symposium N° 97, eds. D.S. Heeschen and C.M. Wade (Reidel, Dordrecht), p. 427.
Windhorst, R.A., van Heerde, G.M., Katgert, P. : 1984a, Astron. Astrophys. Suppl., 58, 1.
Windhorst, R.A., Kron, R.G., Loo, D.C. : 1984b, Astron. Astrophys. Suppl., 58, 39.
Windhorst, R.A., Milez, G.K., Owen, F.N., Kron, R.G., Koo, D.C. : 1984c, Astrophys. J., in press.
Windhorst, R.A., Oppe, J. : 1984, Astron. Astrophys. Suppl., submitted.
Windhorst, R.A. : 1984, Ph.D. Thesis, University of Leiden.
Windhorst, R.A. : 1984, "The cosmological evolution of radio sources", preprint.
Windhorst, R.A., Oort, M.J.A., Katgert, P. : 1984, Astron. Astrophys., submitted.

48. HIGH ENERGY ASTROPHYSICS (ASTROPHYSIQUE DE GRANDE ENERGIE)

PRESIDENT: Riccardo Giacconi
VICE-PRESIDENT: Catherine Cesarsky
ORGANIZING COMMITTEE: G. Clark, F. Pacini, P.A.G. Scheuer, E.E. Salpeter, R.A. Sunzaev, J. Truemper, L. Woltjer, Q. Qu

In view of the broad range of topics covered by Commission 48 and the consequent inevitable overlap with other commissions, it is not feasible to produce a comprehensive self-contained report. The commission therefore restricts its report to a selected list of accessible recent review articles and conference reports, where up-to-date summaries of various topics can be found. Such a list is given below.

A. REVIEW ARTICLES

Hillas, A.M., 1984, Origin of Ultrahigh Energy Cosmic Rays, Ann. Rev. Astron. Astrophys. 22.

Joss, P.C. & Rappaport, S.A., 1984, Neutron Stars in Interacting Binary Systems, Ann. Rev. Astron. Astrophys. 22.

Bradt, H.V.D. & McClintock, J., 1983, Galactic Astrophysics & Gamma Ray Astronomy, Space Science Reviews 37.

____________________, 1983, Optical Counterparts of Galactic X-Ray Sources. Ann. Rev. Astron. Astrophys. 22.

Bignami, G.F. & Hermsen, 1983, Galactic Gamma Ray Sources, Ann. Rev. Astron. & Astrophys. 21.

Holt, S.S. & McCray, 1982, Spectra of Cosmic X-ray Sources, Ann. Rev. Astron. & Astrophys. 20.

Forman, W. & Jones, C., 1982, X-ray Imaging of Clusters of Galaxies, Ann. Rev. Astron. & Astrophys. 20.

B. CONFERENCE PROCEEDINGS

10th Texas Symposium of Relativistic Astrophysics, 1982, ed. R. Ramaty and F.C. Jones: New York Academy of Sciences.

Galactic X-ray Sources, 1982, eds. P.W. Sanford, P. Laskandes and J. Salton; J. Wiley and Sons: Chichester.

IAU Symposium 101 "Supernova Remnants and Their X-ray Emmission", 1983, eds. J. Danziger, P. Gorenstein; D. Reidel: Dordrecht.

Composition and Origin of Cosmic Rays, 1983, ed. M.M. Shapiro; D. Reidel: Dordrecht.

49 THE INTERPLANETARY PLASMA AND THE HELIOSPHERE
(Plasma Interplanetaire et l'Heliosphere)

PRESIDENT	:	I.W. ROXBURGH
VICE PRESIDENT	:	S. GRZEDZIELSKI
ORGANIZING COMMITTEE	:	S. CUPERMAN, A.Z. DOLGINOV, H.J. FAHR, H.U. KELLER, F. PARESCE, G. SETTI, S.T. SUESS, C.S. WELLER

INTRODUCTION

The Commission is intended to study the origin of the solar wind, the heliosphere, i.e., the region dominated by the supersonic solar wind, and the heliospheric interface with the interstellar medium. The activities of the Commission cover both theoretical and observational aspects of these three regions.

This report covers work on large scale phenomena in the Heliosphere, and my thanks are due to Dr. S.T. Suess for providing the material.

LARGE SCALE PHENOMENA IN THE HELIOSPHERE

Overview

This section deals with the detection, description and modelling of large-scale phenomena. A comprehensive understanding of these phenomena generally requires data from several souces in order to obtain a global view. Presently, continuing analysis of radio observations, of space-craft data from inside and outside the orbit of the earth and of lengthy data sets from earth-orbiting spacecraft are producing new discoveries. Recent discoveries include magnetic clouds, compositional changes associated with coronal streamers and the ability to track streamers to 1 AU using type III radio bursts. The interplanetary signature of coronal transients continues to be unclear, but some ideas have been advanced.

In the following three sections, spatial gradients, stream interaction and the propagation of transient disturbances are discussed in A, the heliospheric current sheet in B and the interplanetary signature of coronal expansion and transients in C. Many of the results and associated theoretical developments are in reviews or summaries of recent research by Burlaga (1983), Cuperman (1983), Schwartz (1981), Schwenn (1983), Smith (1983) and Wu (1983). This research summary is not intended to be comprehensive. Several specific topics are highlighted, with representative references supplied for each.

Spatial Gradients and Dynamic Interactions

Pioneer 10 is now beyond the orbit of Pluto. Data from this spacecraft, together with data from Voyagers 1 and 2, Pioneer 11, Helios 1 and 2, IMP 6, 7 and 8, Explorers 34 and 35, and ISEE 3 are being used to define radial gradients of solar wind properties. Smith and Barnes (1983) review the spatial dependences of the solar wind plasma and magnetic field between 1 and 20 AU using the Pioneer data and compare the measurements with both the Parker theory for radial, azimuthally symmetric flow and the Goldstein-Jokipii theory including stream interaction. The observed radial gradients are consistent with the Parker model, except for the non-adiabatic temperature gradient. The field strength was found to decrease with distance from the heliospheric current sheet. No effects of the interstellar gas, e.g., mass loading, were detected. Gazis and Lazarus (1982) reported on a similar analysis of Voyager observations of the solar wind proton temperature between 1 and 10 AU, in comparison to IMP 8 results in Earth orbit. Again, the temperature gradient was observed to be sub-adiabatic. A detailed analysis of solar wind stream structure showed that the implied heating was occurring at the interface between high and low speed streams. Again, no effect of interstellar gas was reported.

Solar wind speeds between 1 and 15 AU were analyzed in more detail by Collard, Mihalov and Wolfe (1982). Pioneer 10 and 11 measurements were compared with IMP 6, 7 and 8 measurements made near the earth for six radial alignment periods between 1973 and 1978. The mean solar wind speed appeared to vary little with radius, but there were significant long-term temporal changes and changes with heliographic latitude. Changes in speed across streams decreased with distance and were extrapolated to almost disappear by 30-40 AU. Helios 1 and 2 data between 1974 and 1977 and between 0.3 and 1.0 AU were analyzed by Geranios (1982b), to show that the slow solar wind first increases its speed with distance, and then becomes more or less constant. The variation of solar wind speed with solar activity was also analyzed. The detailed complexity of the solar wind velocity structure was reported by Rhodes and Smith (1981). Using data from Explorers 34 and 35 and Mariner 5 during 1967,the existence of large, temporary, local north-south velocity gradients ranging up to 60 km/s/deg were implied. The momentum flux in the solar wind is the subject of a study by Steinitz (1983). He reports on Helios 2 data to show that the momentum flux does not depend on flowspeed nor on distance verifying the previously established invariance of momentum flux density. It is furthermore suggested that the momentum flux density carried by the solar wind may be constant over the solar cycle.

Considerable attention is being paid to solar wind composition due to better data from the new instruments. Marsch, et al (1982) have surveyed the radial variations of solar wind helium properties between 0.3 and 1 AU, using Helios 1 and 2 data. They reported extensively on the helium ion velocity distribuitons and derived parameters and placed strong restrictions on models of the interaction between solar wind protons and helium.

Further general studies of helium abundance variations are reported by Borrini, et al, 1983, using data from IMP 6, 7 and 8, from 1971 through 1978. Low and average values of helium abundance are identified with different characteristic plasma 'states'. Helium enhancements are identified with high magnetic field intensities and low proton temperatures. In another paper, Borrini, et al, (1982) report on an analysis of shock wave disturbances from the same spacecraft and time period. Helium enrichments are observed in association with 46% of the shocks, and those with helium enhancements tend to be the srongest shocks observed.

Results on large-scale variations in the interplanetary magnetic field between 1 and 5 AU were reported by Burlaga, et al, (1982), using Voyager 1 and 2 observations, for the years 1977 to 1979. During this interval, there were notable deviations from the Parker theory configuration. These were attributed to temporal fluctuations and to field fluctuations in the radial direction. Transverse fluctuations in the IMF were reported to decrease in amplitude with distance and to be consistent with the presence of undamped Alfven waves. In a more detailed study, Slavin, Smith and Thomas (1984) report on large-scale temporal and radial gradients in the IMF using Helio 1, 2, ISEE 3 and Pioneer 10, 11 data. In addition to specific temporal variations, the authors report a more rapid decrease in the IMF intensity with radius than predicted by classical Parker theory. The cause of this result is suggested to be meridional transport due either to MHD processes or stream interaction.

Remote observations of solar wind velocity using interplanetary scintillation observations over extended periods were reported by Sime and Rickett (1981) and reviewed by Sime (1983). A global view of the solar wind and its evolution throughout the solar cycle has been derived. The potential for sensing the morphology of propagating disturbances associated with eruptive solar phenomena has been demonstrated. Recent developments using spacecraft transmitters now permit analysis of the solar wind inside 50 solar radii.

Perhaps the most important consequence of these cumulative data on spatial gradients is that they are now providing new conditons that must be satisfied by models of the solar wind acceleration region. These observational constraints on solar wind acceleration mechanisms are reviewed and summarized by Neugebauer (1983). The theory for ordinary, spherical, symmetric flow has developed along predictable lines in recent years, with few surprises (see reviews by Schwarz, 1981; Cuperman, 1983; Leer, Holzer and Fla, 1982). One new aspect is the suggestion that standing shock waves of various types may exist near the sun. Whang (1982) infers that MHD slow shocks can exist in the vicinity of the sun as a consequence of nonradial outflow from coronal holes. Habbal and Tsinganos (1983) look at the possibility of ordinary shock transitions in the presence of extended momentum deposition. Several experiments are now being planned to look for these shocks and their consequences on the state of the solar wind.

Several specific analyses have been made of corotating features in the solar wind. Urlaga, Schwenn and Rosenbauer (1983) analyzed the dynamical evolution of interplanetary magnetic fields and flows, between 0.3 AU and 8.5 AU using data from Helios 1 and Voyager 1 during 1980.

Voyager 1 observed two streams which had little fine structure. The corresponding flows observed by Helios 1 were much more complex. The suggestion is that most of the fine-scale features were swept up and 'entrained' by the large-scale flow. Burlaga (1983) further reported on the detection of corotating pressure waves without fast streams in the solar wind. Voyager 1 and 2 data observed this phenomenon between 2 and 4 AU and it is suggested that the development of large-scale, nonlinear pressure waves at the expense of part of the kinetic energy of streams produces a qualitative change in the solar wind in the outer heliosphere. A very interesting new technique for mapping corotating events using radio data from ISEE 3 was described by Fainberg, Bougeret and Stone (1983). Type III radio storms are observed out to 0.5 -0.8 AU, at a rate of 2 to 3 storms per solar rotation near solar maximum. These storms correlate with type I and III radio storms close to the sun and are associated with an almost continuous injection of suprathermal electrons into the interplanetary medium. The storms are most likely the extension of streamers into the interplanetary medium, thereby permitting the identification of stream structure with the corresponding features on the sun.

Another application of ISEE 3 radio data has related type II solar radio events observed in the interplanetary medium to propagating interplanetary shock waves (Cane, et al., 1982). Type II events are thought to be produced in the vicinity of shock waves, and have long been used to estimate shock speed in the corona. With the extended frequency range available in space, type II bursts have been followed almost to 1 AU. The general topic of observation of interplanetary shocks is reviewed by Smith (1983).

Several authors reported on developments in the theory of stream interaction (see review by Wu, 1983). A kinematic model of stream interaction, together with superimposed kinetic distortions simulating transient disturbances, was developed by Hakamada and Akasofu (1982) and Akasofu and Hakamada (1983). Sakurai (1983) reported an application of an analytical method of characteristics to the azimuthally dependent solar wind. Wu, Dryer and Han (1983) describe a numerical model that, although intended for simulating propagating disturbances, can also be applied to corotating structures. This is an MHD model that incorporates a meridonal component of the magnetic field and flow vector. Pizzo (1982) also describes an MHD model of corotating streams - this being a completely three dimensional model.

In an application of MHD modelling, Pizzo (1981) demonstrated the use of numerical models in inverse mapping of solar wind flow structures from point of observation back to the sun. This exercise illustrated the strengths and limitations of such mapping in terms of the amount of steeping the stream front has undergone when observed and whether or not a shockhas formed at the interface.

In two separate articles (Sarris and Krimigis, 1982; Geranios 1982a) evidence is presented for the presence of magnetically closed regions in the solar wind. Geranios has analyzed Helios 1, 2 and IMP 8 data taken in 1974 and 1976 to identify 85 cases in which the proton temperature was very low.

In 50 of these cases, the IMF showed characteristics favourable for closed structures. Sarris and Krimigis present data on energetic particles injected by solar flares and infer that these particles are bouncing between two magnetic mirrors. They obtain estimates of the extent of these loops to distances of 3.5 AU from the sun. A related, and perhaps identical structure is described as 'magnetic clouds' by Klein and Burlaga (1982). A magnetic cloud is a structure with a radial dimension of about 1/4 AU in which the magnetic field strength is high and the magnetic field direction changes appreciably by rotation of one component parallel to a plane. Forty-five clouds were observed near the earth between 1967 and 1978, with at least one passing the earth every three months. Clouds have also been identified at Voyagers 1 and 2, between 2 and 4 AU, (Burlaga and Behannon, 1982). The relationship of these clouds to solar phenomena is discussed in section C.

Heliospheric Current Sheet

A great deal of effort has been expended on interplanetary observations of the current sheet and on comparing the observed sector structure and geometry of the current sheet with that which is expected from solar observations. Data on solar coronal structure exists in the form of coronagraph observations (Roselot and Fulconis, 1983; Wilcox and Hundhausen 1983), green line coronal intensity (Tyagun, 1983) and potential magnetic field models using photospheric data (Wilcox and Hundhausen, 1983), Burlaga Hundhausen and Zhao (1981), using Helios 1 and 2 data from 1976, identify the sector pattern with the maximum brightness curves in K coronameter data at 1.5 solar radii. Bruno, Burlaga and Hundhausen (1982) continue this approach into 1977, adding IMP 8 data and claim that the latitudinal extent of the current sheet is consistent with that suggested in the coronal brightness data. Villante and Bruno examine the Helio 2 data from 1976 to learn the local orientation of the current sheet. They find the boundaries to be at a large angle with respect to the ecliptic plane. Villante, Mariani and Francia (1982) examine the entire period from 1974 to 1978 to deduce how the sector pattern seems to evolve with time and heliographic latitude.

It is shown that the structure of the current sheet as predicted from the maximum coronal brightness data is quite similar to that predicted from potential magnetic field models of the corona (Wilcox and Hundhausen, 1983). The potential field models are now extended to a major portion of the solar cycle by Hoeksema, et al, 1982, 1983), and the structure of the heliospheric current sheet from 1974 to 1982 is thereby reconstructed. The structure and dynamics of the current sheet in the interplanetary medium is analysed in two very different papers by Thomas and Smith (1981) and Niedner (1982). Thomas and Smith show the interplanetary deformation of the current sheet as it is entrained by overtaking high speed streams. Niedner uses 'disconnection events' in the comet tails to directly map the three-dimensional properties of the current sheet.

Behannon, Burlaga and Hundhausen (1983) apply a minimum variance analysis to magnetic field data during current sheet passages to compute a local normal to the current sheet and compare the result to the inclination measure contrasts with the global measure determined with multiple spacecraft comparisons or with comet tails utilized in most of the above analyses.

Interplanetary Signature of Coronal Processes

A generalization of analyzing the heliospheric current sheet in comparison to what it is believed to be like near the sun is to study the total dynamics of coronal expansion in the context of interplanetary observations. Because the current sheet is a direct reflection of the largest scale coronal processes, an obvious first approach is to sort solar wind data in reference to the location of the current sheet. This approach has been used by Borrini, et al., (1981) to carefully document a solar velocity minimum, high plasma density, nearly identical proton and helium bulk velocities and low proton and helium kinetic temperatures at the current sheet. Several relatedand preceding studies have closely related the current sheet to coronal streamers (Gosling, et al., 1981; Sime and Rickett, 1981; Feldman, et al., 1981; and other, earlier studies). Therefore the pattern demonstrated by Borrini et al., is indicative of the 'signal' of a coronal streamer at 1 AU. Their results provide important constraints on models of helium dynamics in coronal streamers and the low speed solar wind.

Models of coronal streamers presently fall into two classes. The first, and most closely related to observations, is the use of potential field models described above in association with predictions of the topology of the heliospheric current sheet. The second class is true MHD models. Two advances in this type of model have been reported by Robertson (1983) and Steinolfson, Suess and Wu, (1982). Robertson incorporated the effects of thermal conduction into the approach used several years ago by Pneuman and Kopp (1971). Steinolfson, et al., described treating the solution for the steady state as an initial-boundary value problem and quoted results for a wide variety of magnetic field strengths in an axisymmetric model.

Acceptance of the relationship between coronal streamers, potential field models of the corona and the character of the interplanetary medium has led to more extensive efforts to sort interplanetary data using these relationships. Zhao and Hundhausen (1983) used interplanetary scintillation data from 1976 to show the minimum in solar wind speed at the current sheet and also an increase in the average solar wind speed with angular displacement from the current sheet. However, angular displacement from the current sheet is not a unique parameter for ordering the solar wind, in that the solar wind speed can be just as low far from the current sheet as at the current sheet. Suess, et al., (1984) added information on the strength of the magnetic field at the source surface in potential field models of the corona to improve the correlation between coronal structure and the character of the solar wind. There it was shown that strong field regions on the source surface were the sources of high speed solar wind.

A much more elusive question has been the interplanetary signature of coronal transients. Coronal transients have been studied for more than a decade, with only moderate success in relating them to interplanetary phenomena. A related question has been the origin at the sun of magnetic clouds, which were first reported by Klein and Burlaga (1982). One of the most striking results has been the observation of a head-on, or halo coronal transient (Howard, et al., (1982). This observation was made with an orbiting coronagraph and showed that at least some coronal transients were more bubble-shaped than loop-shaped. Schwenn (1983) has reviewed an analysis of Helios 1 and 2 data in comparison with data from the same orbiting coronagraph to show that interplanetary shock waves can often be associated with coronal transients. In contrast, Burlaga and Klein (1982) and Wilson and Hildner (1984) ask the question of whether magnetic clouds are manifestations of coronal transients. The answer seems to be yes, but better statistics are needed for a positive identification.

A different aspect of propagating interplanetary disturbances is composition anomolies associated with coronal processes. The work reported earlier by Borrini, et al., (1982), showing an association between strong shock waves and helium enrichments, was used to infer that coronal ejacta origina in solar regions where the field strength is strongest. The helium enrichment is often thought to be material from prominences or low levels in the corona in which helium has been enriched due to gravitation separation.

References

Akasofu, S.-I., and Hakamada, K.: 1983, *Geophys. Res. Lett., 10, 577*

Behannon, K.W., Burlaga, L.F., and Hundhausen, A.J.: 1983, *J. Geophys. Res., 88, 7837*

Borrini, G., Gosling, J.T., Bame, S.J., Feldman, W.C., and Wilcox J.M.: 1981, *J. Geophys. Res., 86, 4565*

Borrini, G., Gosling, J.T., Bame, S.J., and Feldman, W.C.: 1982, *J. Geophys. Res., 87, 4365*

Borrini, G., Gosling, J.T., Bame, S.J., and Feldman, W.C.: 1983, *Sol. Phys., 83, 367*

Bruno, R., Burlaga, L.F., and Hundhausen, A.J.:1982, *J. Geophys. Res. 87, 10339*

Burlaga, L.F., and Behannon, K.W.: 1982, *Sol. Phys., 81, 181*

Burlaga, L.F., and Klein, L: 1982, *Geophys. Res. Lett., 9, 1317*

Burlaga, L.F.: 1983, *Rev. Geophys and Space Phys., 21, 363*

Burlaga, L.F., Schwenn, R., and Rosenbauer, H.: 1983, *Geophys. Res. Lett., 10, 413*

Burlaga, L.F.: 1983, *J. Geophys. Res., 88, 6085*

Burlaga, L.F., Hundhausen,A.J., and Zhao,X. -P.: 1981, *J. Geophys. Res., 86, 8893*

Burlaga, L.F., Lepping, R.P., Behannon, K.W., Klein, L.W., and Neubauer, F.M.: 1982, *J. Geophys. Res., 87, 4345*

Cane, H.V., Stone, R.G., Fainberg, J., Steinberg, J.L., and Hoang, S.: 1982, *Sol. Phys., 78, 187*

Collard, H.R., Mihalov, J.D., and Wolfe, J.H.: 1982, *J. Geophys. Res., 87, 2203*
Cuperman, S.: 1983, *Space Sci Rev., 34, 63*
Feldman, W.C., Asbridge, J.R., Bame, S.J., Fenimore, E.E., and Gosling, J.T.: 1981, *J. Geophys. Res., 86, 5408*
Fainberg, J., Bougeret,J. -L., and Stone, R.G.: 1983, in *Solar Wind Five,* ed by M. Neugebauer, NASA Conference Publication 2280, pg 469
Gazis, P.R., and Lazarus, A.J.: 1982, *Geophys. Res. Lett. 9, 431*
Geranios, A.: 1982, *Astrophys Space Sci., 81, 103*
Geranios, A.: 1982, *Astrophys Space Sci., 81, 333*
Gosling, J.T., Borrini, G., Asbridge, J.R., Bame, S.J., Feldman, W.C., and Hansen, R.T.: 1981, *J.Geophys. Res., 86, 5438*
Habbal, S.R., and Tsinganos, K.: 1983, *J. Geophys Res, 88, 1965*
Hakamada, K, and Akasofu, S. -I.: 1982, *Space Sci Rev., 31, 3.*
Hoeksema, J.T., Wilcox, J.M., and Scherrer, P.H.: 1982, *J. Geophys. Res, 87, 10331*
Hoeksema, J.T., Wilcox, J.M., and Scherrer, P.H.: 1983, *J. Geophys. Res. 88, 9910*
Howard, R.A., Michels, D.J., Sheeley, N.R. Jr., and Koomen, M.J.: 1982, *Astrophys. J. Lett., 263, L 101*
Klein, L.W., and Burlaga, L.F.: 1982, *J. Geophys. Res., 87, 613*
Leer, E., Holzer, T.E., and Fla, T.: 1982, *Space Sci Rev., 33, 161*
Marsch, E., Muhlhauser, K. -H, Rosenbauer, J., Schwenn, R., and Neubauer, F.M.: 1982, *J. Geophys. Res., 87, 35*
Neugebauer, M.: 1983, in *Solar Wind Five,* ed by M. Neugebauer, NASA Conference Publication 2280, pg 135
Niedner, M. Jr.: 1982, *Astrophys. J. Supp. Ser., 48. 1*
Pizzo, V.J.: 1981, *J. Geophys. Res., 86, 6685*
Pizzo, V.J.: 1982, *J. Geophys. Res., 87, 4374*
Pneuman, G.W., and Kopp, R.A.: 1971, *Sol. Phys., 18, 258*
Rhodes, E.J. Jr.,and Smith, E.J.: 1981, *J.Geophys. Res., 86, 8877*
Robertson, B.J.: 1983, *Sol. Phys., 83, 64*
Rozelot, J.P., and Fulconis, M.: 1983, *Sol. Phys., 84, 77*
Sakurai, T.: 1983, *Mon. Not. R. Astron. Soc., 203, 1187*
Sarris, E.T., and Krimigis, S.M.: 1982, *Geophys. Res. Let., 9, 167*
Schwarz, S.J.: 1981, in *Solar Phenomena in Stars and Stellar Systems,* Proc. of Advanced Study Institute, Bonas, France, August 1980, Reidel, Pg 311
Schwenn, R.: 1983, *Space Sci Rev., 34, 85*
Sime, D.G., and Rickett, B.J.: 1981, *J. Geophys. Res., 86, 8869*
Sime, D.G.: 1983, in *Solar Wind Five,* ed by M. Neugebauer, NASA Conference Publication 2280, pg 453.
Slavin, J.A., Smith, E.J., and Thomas, B.T.: 1984, *Geophys Res Lett, 11 279*
Smith, E.J., and Barnes, A.: 1983, in *Solar Wind Five,* ed by M. Neugebauer NASA Conference Publication 2280, pg 251.
Smith, E.J.: 1983, *Space, Sci Rev, 34, 101*
Steinitz, R.: 1983, *Sol, Phys., 83, 379*
Steinolfson, R.S., Suess, S.T., and Wu, S.T. 1982, *Astrophys J, 255, 730*
Suess, S.T., Wilcox, J.M., Hoeksema, J.T., Henning, H., and Dryer, M.: 1984, *J. Geophys. Res., 89, 3957*

Thomas, B.T., and Smith, E.J.: 1981, *J. Geophys. Res. 86, 11105*
Tyagun, N.F.: 1983, *Soln. Dannye, Byull., 2, 78*
Villante, U., and Bruno, R.: 1982, *J. Geophys. Res., 87, 607*
Villante, U., Mariani, F., and Francia, P.: 1982, *J. Geophys Res. 87 249*
Whang, Y.C.: 1982, *Geophys Res. Lett., 9, 1081*
Wilcox, J.M., and Hundhausen, A.J.: 1983, *J. Geophys. Res., 88, 8095*
Wilson, R.M., and Hildner, E.: 1984, *Sol. Phys., 93, ****
Wu, S.T., Dryer, M., and Han, S.M.: 1983, *Sol. Pys., 84, 395*
Wu, S.T.: 1983, *Space Sci Rev., 34, 73*
Zhao, X. -P., and Hundhausen, A.J.: 1983, *J. Geophys. Res. 88 451*

50. IDENTIFICATION AND PROTECTION OF EXISTING AND POTENTIAL OBSERVATORY SITES

PRESIDENT: A. Hoag
VICE-PRESIDENT: S. van den Bergh
ORGANIZING COMMITTEE: C. A. Anguita, A. L. Ardeberg, F. Bertola, V. M. Blanco, R. Cayrel, D. L. Crawford, P. J. Edwards, W. Mattig, F. G. Smith, M. F. Walker

1. Identification of Observatory Sites

While a number of site testing campaigns are in progress throughout the world, the dominant efforts seem to be associated with plans for very large telescopes. These plans, improved telescope technology, and the realization that astronomical observations from space will put increasing demands on ground-based observing facilities have given new impetus to site identification and to the optimum utilization of existing sites. A most excellent review of the factors involved in site identification is now available in the **Proceedings of the ESO Workshop on Site Testing for Future Large Telescopes,** edited by A. Ardeberg and L. Woltjer, 1984. These **Proceedings** and Woolf's review (1982, **Ann. Rev. Astron. Astrophys.** 20, 367) provide a solid background of current site evaluation factors.

In the last triennium, an improved theoretical and experimental knowledge of turbulence, a widespread evaluation of Fried's parameter, r_0 (characteristic scale of wavefront coherence), and a growing awareness of image degradation caused by properties of the facilities themselves have engendered new interest in site evaluation studies.

A few summaries of site identification studies provided for this report follow. Other surveys are noted in the literature summary with **Astronomy and Astrophysics Abstracts** notation at the end of this report.

Professor Eduardo Brieva reports a site testing campaign has been initiated for the establishment of an astrophysical observing station for the Columbian National Observatory. Visual observations at four sites in the Columbian Andes and independent studies based on satellite images indicate that up to 250 useful nights per year may be obtainable at a site named Otún in the central cordillera.

In connection with plans for developing observational astronomy in India, Professor Bhattacharyya writes that several projects for finding new observatory sites are in progress. The Uttar Pradesh Observatory at Nainital is engaged in the task of identifying a suitable site for a projected 4-meter telescope. Following a survey of 40 locations in the Kumaon Hills of the Himalayan range, four locations have been instrumented for recording meteorological data. In western India, a site at 1700-meter elevation on Mount Abu has been selected for an infrared telescope and a site in the Ladhak region of the Himalayas at 3400-meter elevation is being evaluated as a potential good infrared site.

Dr. Elmar Brosterhus writes concerning a major survey in Saudi Arabia that has been organized with assistance from the National Research Council of Canada. Following an extensive preliminary survey, four sites were selected for comprehensive study with photoelectric image monitors and meteorological instruments. Data were gathered from May 1982 through all of 1983, at which time observing was terminated at the two least desirable sites. The remaining two installations are still operating. The image monitors, used on 15-cm aperture reflecting telescopes, accurately record image motion, size, and brightness. Extinction measures are obtained by observations of suitable stars.

In connection with plans for a very large telescope, Crawford completed a "Dark Sky Survey" (1983, KPNO special report), and Lynds and Goad have made a cartographic exploration of sites in the southwestern United States (1984, **Publ. Astron. Soc. Pacific** 96, 750). Woolf and Merrill have directed attention to Mount Graham in southeastern Arizona as a possible NNTT site (1984, **NOAO Technology Development Program Report** No. 4).

Dr. Zeki Aslan describes a long-term survey being conducted in Turkey. Meteorological records delineate two promising regions south of latitude 38°. A number of candidate sites have been selected by reconnaissance, and a four-year program of evaluation with Walker-type Polaris trail telescopes was started in 1982 following consultation with B. McInnes of the Royal Observatory, Edinburgh.

Special attention is again directed to the proceedings of the ESO Workshop on Site Testing for Future Large Telescopes and particularly to Walker's article on "High Quality Astronomical Sites Around the World" in those proceedings.

2. Extinction

Special reports on atmospheric extinction have been received from Jozef Tremko and G. W. Lockwood. The first of these concerns measures of extinction at the Skalnaté Pleso and Brno Observatories. Papoušek and Vetešník of Brno and Tremko and Juza of the Astronomical Institute of the Slovak Academy of Sciences report analyses of large numbers of UBV extinction measures made at Brno, an urban location, and at Skalnaté Pleso, a "mountain observatory," for the interval 1963-1980. The mean visual extinction at Skalnaté Pleso, altitude 1783 meters, is characteristic of good sites, $\langle k_V \rangle$ being the order of $0^{m}.15$. The yearly means show a variation with period of about 22 years and an amplitude of about $0^{m}.02$, seemingly in phase with sunspot activity. The minimum values of extinction in U, B, and V occurred at about 1975 near the time of minimum sunspot numbers. Observations are being continued to see if this relation persists.

As expected from the urban location, the extinction coefficients measured at Brno are large and show large disperson. However, a systematic decrease in extinction values has been observed over the interval. For example, $\langle k_V \rangle$ during 1964-1968 was about $0^{m}.48$, while $\langle k_V \rangle$ during 1977-1981 was about $0^{m}.35$, the trend being approximately linear.

Information communicated by G. W. Lockwood concerns spectrally resolved measurements of the El Chichon cloud during May 1982-August 1983 at Flagstaff (elevation 2200 meters). Measures by Lockwood, White, and Tüg show that a maximum extinction of $0^{m}.33$ at λ350 nm by stratospheric aerosols alone occurred on 15 May 1982. At this time, the cloud was essentially gray. Continuing extinction determinations by Lockwood show that observable effects of the El Chichon cloud persisted for more than a year.

Lockwood has analyzed precision **b** and **y** extinction determinations for more than 300 nights during 1972-1982. He finds no secular change, in spite of rapid growth of Flagstaff, and no solar cycle effect. There is a seasonal effect with almost pure Rayleigh scattering in January to an aerosol component of about $0^{m}.08$ in June.

The accompanying publication list shows that extinction monitoring is being done at a large number of sites.

3. Sky Brightness and Urban Sky Glow

A few special communications have been received, the most complete being by Carlos Torres, who described the situation at the Observatório Astrofísico

Brasileiro (altitude 1865 meters), which is located near Brasópoles in Minas Gerais State. He details the direction, distance, population, lighting technology, and terrain for 13 municipalities and towns, the largest of which is São Paulo with 10^7 inhabitants at a distance of 160 km. An azimuthal scan at 45° elevation in August 1982 yielded values of B_{sky} ranging from 21.5 to 22.1 mag·arcsec^{-2}, with maximum brightness in the general direction of São Paulo. Zenith sky brightness measured on 14 nights from July 1981 to July 1984 ranged from 22.2 to 23.0 mag·arcsec^{-2} in B and from 21.0 to 21.9 in V on clear moonless nights away from the galactic plane. A zenith sky spectrophotometric recording, also included in his report, shows only very weak mercury and sodium emission lines from urban lighting.

Dr. Yoshihide Kozai, director of the Tokyo Observatory, has sent sky measures for their Kiso Observatory (altitude 1130 meters), where a 105-cm Schmidt camera is operated. Total sky brightness at λ530 nm averaged from 1980 through 1984 is 21.9 mag·arcsec^{-2}, with very little dispersion. The average urban contribution is estimated to be 24.0 mag·arcsec^{-2}. Only a few small towns are visible at a distance of 20 km.

These zenith dark-sky values for V_{sky} of about 21.5 mag·arcsec^{-2} for Brazil and 21.9 mag·arcsec^{-2} for Japan may be compared to Walker's (1970, **Publ. Astron. Soc. Pacific** 82, 672) 1966 value for three California dark-sky sites of 21.6 and the 1964 value of 21.6 cited by Allen (1973, **Astrophysical Quantities**, third edition, Athlone, London). However, Dawson (**I.A.P.P.P Light Pollution Newsletter** No. 2) gives a value of 21.1 for Kitt Peak in 1980; Smyth reports (1982, **Occasional Reports of the Royal Observatory, Edinburgh** No. 8) a value of 21.1 for Siding Springs, and the value for the Lowell, Anderson Mesa dark-sky site was 21.1 in 1981- 1982 (Hoag, unpublished). These interesting differences, and the possibility that these values might be affected by solar activity, indicate the need for more extensive monitoring and reporting.

Dr. Carlo Blanco, director, Osservatorio Astrofisico di Catania (1725 meters) writes that urban lighting from Catania, distant 20 km, has markedly increased during the last decade. Recently, outdoor lights have been installed at hotels only 0.5 km distant; but low-pressure sodium lamps have been used at his request.

Professor S. Marx, Karl-Schwarzschild Observatorium, Tautenburg, sent a 1979 report (**Astron. Nachr.** Bd. 300, 209) describing an experiment we would all like to repeat at will in our localities. A portion of the outdoor lighting in Jena (10-15 km distant, 10^5 inhabitants) was purposely turned off for half an hour while the sky brightness was monitored by a variety of instruments. The sky brightness at λ578 nm decreased by 20%.

Dr. Roy Garstang has recently derived an improved formula for calculations of urban sky light (1984, **Observatory** 104, 196).

4. Radio Frequency Interference

Two kinds of interference were specifically brought to the attention of the Commission in 1984. The first concerns spacecraft transmissions in radio-frequency bands reserved for radio astronomy, and the second relates to interference from transmitters close to observatories.

Because of planned use of a space probe transmitter emitting at 1667.75 MHz, within the 1660-1670 MHz band reserved internationally for observations of spectral lines of the OH molecule, the Committee on Radio Frequencies of the U.S. National Academy of Sciences has sent the following resolution to organizations as URSI, IAU, COSPAR, IUCAF, and the ITU: the Committee "URGES that in the future all parties planning other active radio science or telecommunications experiments fully respect the international table of frequency allocations, and most particularly those bands allocated exclusively to passive services."

Professor Osório of the Observatório Astronómico do Porto writes that three nearby transmitters, with a total input power of 30 Kw, have caused severe interference problems with all electronic systems at the 76-cm telescope. Also, we were again reminded of the situation at Mount Wilson, where the 38 transmitters are located less than 1 km distant. In both of these situations the radiated power is orders of magnitude in excess of the acceptable level recommended by the Commission (cf. 1976, **Trans. IAU,** XVIB, 319). Dr. van den Bergh reports that extensive shielding of electronic components of the radial velocity scanner at the Dominion Astrophysical Observatory 48-inch telescope has been required because of a nearby transmitter. This kind of problem has caused technical difficulties at many observatories.

5. Protection of Sites

About 4 kg of U.S. literature on light pollution has been accumulated in the Commission files during the past three years. It would be impossible, as well as parochial, to attempt to treat it in detail. The public information effort has been extensive and greatly aided by amateurs.

Much of the professional effort has been concentrated in Hawaii and the Southwest and has been guided by a Committee on Light Pollution and Radio Interference of the American Astronomical Society chaired by D. L. Crawford. Impetus to the program has also been provided by a growing public and municipal awareness of energy costs for urban lighting. Because urban lighting questions involve aesthetics, economics, politics, and technological questions, the best compromises can be reached only by involving the public, governing bodies, municipal managers, manufacturers, lighting engineers, and utility company representatives.

An extensive report of a working group of the French Academy of Sciences on **La Protection des Observatoires Astronomique et Geophysiques** has been received from J. Kovalevsky. This compendium has the virtue of covering a broad scope of problems, including light pollution, sources of heat, radio-frequency interference, magnetic perturbations, ground vibrations, and chemical pollution.

IAU/CIE Publication No. 1, **Guide Lines for Minimising Urban Sky Glow near Astronomical Observatories**, 1980, continues to be a most significant aid to those concerned with protection of observing sites.

Publication List

SITE STUDIES

McInnes (30.082.014) Hawaii, Madeira, and Canary Islands.
Hartley **et al.** (30.082.015) Roque de los Muchachos microthermals.
Brandt, Wöhl (30.082.037) Solar site, Canary Islands.
discussion (30.082.068) Binchuan, China.
Gur'yanov (30.082.077) Mount Maydanak, surface microthermals, USSR.
Gur'yanov (30.082.078) Mount Maydanak, surface layer and seeing.
Gur'yanov (30.082.079) morning minimum of seeing effects.
Scheglov (30.082.085) r_0 distribution.
Khan (30.082.086) Mount Dushak-Erekdagh PSM seeing data, USSR.
Jarrett (31.082.006) Weather at Boyden Observatory, South Africa.
Sanguin (31.082.042) El Leoncito, Argentina.
Brandt, Wöhl (31.082.045) Solar site, Canary Islands.
Teleki (31.082.078) Astrometric site selection.
Smith (31.082.087) IAU Commission 50 report.

Shevchenko (32.082.015) Mount Maydanak, USSR.
Woolf (32.082.046) High-resolution imaging from the ground.
Shcheglov (32.082.074) PSM seeing comparisons for several mountains, USSR.
Khan (32.082.078) Seeing and scintillation, Mount Dushak-Erekdagh, USSR.
Popov **et al.** (32.082.079) Daytime star visibility, Mount Maydanak, USSR.
Banin **et al.** (33.082.001) Astroclimate, Bajkal Observatory, USSR.
Fourikis **et al.** (33.082.028) mm-wave sites, Australia.
Troyan (33.082.030) Day astroclimate, Terskol peak, USSR.
Klepikov, Khodzhadurdyev (33.082.034) Microthermals, Mount Dushak-Erekdagh.
Khan, Khodzhadurdyev (33.082.040) PSM seeing, Mount Dushak-Erekdagh.
Kahn (33.082.048) PSM results, Mount Dushak-Erekdagh.
Liu, Wu (33.082.068) Site testing for observing solar eclipses.
Omarev **et al.** (33.003.091) Astroclimate, Assy-Turgen plateau, USSR.
Ataev **et al.** (34.082.020) Seeing, Mount Sanglok, USSR.
Darchiya, Stajkov (34.082.021) Mount Elbrus, USSR.
Gubkin **et al.** (34.082.023) Astroclimate, Mount Elbrus, USSR.
Troyan, Osipov (34.082.029) Seeing for solar observations, Terskol peak, USSR.
Garcia, Yasukawa (34.082.036) Mauna Loa sky conditions, USA.
Ardeberg (34.082.048) Site selection for very large telescope, Chile.
Weigett (34.082.050) Speckle site testing.
Erokhin, Plyaskin (34.082.063) Sky and image quality at 6-m telescope, USSR.
Shcheglov, Boboev (34.082.097) PSM image motion, Kulab region, Tadzhik SSR.
Dyck, Howell (34.082.100) Infrared speckle seeing, Mauna Kea, USA.
Walker (34.082.103) Comparison of sites on Mauna Kea, USA.
Ferraz-Mello (34.082.113) Observatório Astrofísico Brasileiro.

ATMOSPHERIC PARAMETERS AND MEASURING TECHNIQUES

Azouit, Vernin (30.031.535) Vertical profile of turbulence.
Lund **et al.** (30.031.538) Evolution time of turbulence.
Allen, Barton (30.031.540) Infrared sky noise.
Kallistratova (30.031.627) SODAR in site testing.
Sabet-Peyman (30.031.633) Entropy estimation of image spread function.
Hubbard, Reitsema (30.082.096) Scintillation at two wavelengths.
Najbauer, Fedorishina (30.082.110) Scintillation.
Borgnino **et al.** (31.082.011) r_0 and radiosond, solar seeing.
Roddier (31.082.046) Effects of turbulence.
Dainty **et al.** (31.082.068) Scintillation, Mauna Kea.
Tarrius **et al.** (31.031.644) Sky noise.
Kutyrev **et al.** (32.082.075) Interferometric r_0 measures.
Kutyrev (32.082.076) r_0 and image diameter.
Gur'yanov, Khan (32.082.077) Seeing, ground layer, whole atmosphere.
Tugzhsuren **et al.** (32.082.083) Frequency spectra of scintillation.
Brunner (32.082.084) Effects of turbulence.
Alekseeva, Kamionko (32.082.087) Scintillation and spectrophotometry.
Vithal, Vats (32.062.115) Generation of turbulence by solar radiation.
Vernin, Azouit (34.082.008) Speckle formation.
Druesne **et al.** (34.082.009) Point-spread function from solar limb observations.
Ovezgel'dyev **et al.** (34.082.033) Forecasting astroclimate.
Roddier (34.082.049) Testing seeing quality.
Weigelt (34.082.050) Speckle site testing.
Vernin, Azouit (34.082.053) Remote sounding of turbulence.
Mariotti (34.082.058) Infrared speckle turbulence study.
Woolf, Ulich (34.082.089) Seeing and wind effects.
Lazorenko **et al.** (34.082.104) Image motion.

EXTINCTION

Spencer Jones (30.082.016) at Sutherland SAAO, South Africa.

discussion (30.082.017) at Binchuan, China.
Terez, Terez (31.021.007) Calculating extinction.
Smolyaninova **et al.** (31.082.007) Components of.
Naku, Chernobaj (31.082.008) at Kishinev, USSR.
Fajnberg **et al.** (31.082.044) at Byurakan, USSR.
Sholomitskij **et al.** (31.082.051) Submillimeter transmission, Shorbulak, USSR.
Conference (32.012.096) Atmospheric transmission.
Brand (32.082.014) at La Silla, Chile.
Gutiérrez-Moreno **et al.** (32.082.057) at Cerro Tololo, Chile.
Alekseeva, Novikov (32.082.073) at Pulkova, USSR.
Kopeika (32.082.088) Effects of aerosols.
Gutman **et al.** (32.082.097) Slant path extinction, White Sands, USA.
(1983, **Geophys. Res. Lett.** 10, 233. Turbidity over Africa - El Chichon eruption?)
Edwards (33.082.045) Measures of turbidity.
Lockwood **et al.** (33.082.046) Effects of El Chichon ash cloud.
Zambakos **et al.** (33.082.060) Smoke at Athens Observatory, Greece.
Sloane (33.082.062) Properties of aerosols.
Jiang **et al.** (33.082.070) at Luanpin and Xinglong, China.
Cardelli, Ackerman (34.082.024) St. Helen's ash cloud.
Garcia, Yasukawa (34.082.36) at Mauna Loa, USA.
Anderson (34.082.052) Asymmetries.
Wang (34.082.055) Atmospheric transparency over China.
Tomasi, Vitale (34.082.057) Extinction models for sun photometer measures.
Randall (34.082.059) Atmospheric transmittance from radiance measures.
Schmidtke (34.082.084) Calculation of extinction.
Miles (34.082.085) Transparency and the El Chichon eruption.
Lahulla (34.082.086) at Calar Alto, Spain.
Keen (34.082.093) Volcanic aerosols and lunar eclipses.
Hasan, Dzubay (34.082.109) Aerosol composition and extinction.
Chobanov (34.082.112) Mean transparency function.
Neizvestnyj (34.082.114) at SAO Academy of Sciences, USSR.

INFRARED

Humphries, Purkins (30.082.108) UKIRT, seeing in infrared.
Chou, Arking (31.082.013) Computing absorption by water vapor.
Barton (31.082.014) Water vapor absorption, Nimbus.
Melik-Alaverdyan **et al.** (31.082.043) Precipitable water, Byurakan and Selim, USSR.
Borghesi **et al.** (31.082.060) Water vapor survey, Italy.
Petri **et al.** (31.082.069) LIDAR remote water vapor measures.
Maiya, Dierich (31.082.088) Water vapor at two sites, India.
Sherwood, Greve (32.082.082) Water vapor at La Silla, Chile.
Han **et al.** (33.034.136) Water vapor at Purple Mountain, China.
Greve (33.082.053) Correlation of surface humidity and precipitable water.
Chandrasekhar **et al.** (34.082.010) Water vapor at Ladakh, India.
(Wallace **et al.** 1984, **Publ. Astron. Soc. Pacific** 96, 836. Water vapor at Kitt Peak.)

SKY BRIGHTNESS

Cayrel **et al.** (30.082.125) IAU guidelines for minimizing urban sky glow.
Lyutyj,Sharov (31.082.005) Night sky in the Crimea, USSR.
Upgren **et al.** (31.082.052) Measures of light pollution.
Tanabe (31.082.086) IAU Commission 21 report, Light of night sky.
Neizvestnyj (33.082.002) Night sky at SAO Academy of Sciences, USSR.
Berry (34.082.083) Techniques for measuring.
(Garstang, 1984 **Observatory** 104, 196. Improved scattering formula.)

COMMISSION 51: SEARCH FOR EXTRATERRESTRIAL LIFE (RECHERCHE DE LA VIE DANS L'UNIVERS)

PRESIDENT: M.D. PAPAGIANNIS

VICE PRESIDENTS: F.D. DRAKE and N.S. KARDASHEV

ORGANIZING COMMITTEE: R.D. BROWN, P. CONNES, G.D. GATEWOOD, L. GOLDBERG, J. JUGAKU, G. MARX, F. PACINI, M.J. REES, V.S. TROITSKY

I. INTRODUCTION

The possibility that life, primitive or advanced, might exist in other places of the Universe has occupied the minds of scientists and lay-people for thousands of years. It is only in the last 25 years, however, that we have finally begun to search for answers to this profound question using experimental techniques. The goal of Astronomy is to understand the origin and evolution of planets, stars, galaxies and of the Universe as a whole. The appearance of life is an integral part of this whole process and our picture of the Universe will never be complete until we will comprehend also the significance of life in the process of Cosmic Evolution.

The modern era in the search for extraterrestrial life, both primitive and advanced, originated in 1959. This was the year that the USSR launched the first Sputnik (October 4, 1959) that opened the new frontier of outer space, and was also the year Giuseppe Cocconi and Philip Morrison [1] published in NATURE their pioneering paper "Searching for Interstellar Communications" (September 19, 1959).

The advent of the space era made possible the exploration of our solar system where we have already searched for primitive life on the Moon (Apollo Missions in 1969) and on Mars (Viking Landers in 1975), and have also landed repeatedly on the infernal surface of Venus (Venera Missions). We have obtained excellent photographs and scientific data on planets and moons from Mercury to Saturn, with Uranus and the comet of Halley coming up soon. Though no signs of primitive life have been found as yet in our solar system, and the prospects for the future are not very encouraging, still these explorations have yielded valuable data on the formation of organic compounds of importance to life throughout our solar system, and have given us importnt insights on the origin and evolution of planets and moons. Parallel studies of interstellar molecules and interstellar grains are showing us that these organic compounds are quite common in the Universe, while the search for extrasolar planetary systems is now entering its critical stage. All this information will help us understand better the conditions that made it possible for life to originate on Earth and slowly, over billions of years, to evolve into an advanced technological civilization.

The search for extraterrestrial intelligence, SETI as it is commonly called, turned to other solar systems after the ideas of Percival Lowell (1855-1917) about Martian water-canals, etc, which had excited the imagination of many people during the first half of this century, were finally abandonned. In their 1954 paper to NATURE, Cocconi and Morrison suggested to search for radio signals from other stellar civilizations concentrating on the 21 cm line of atomic Hydrogen, which as they said "must be known to every observer in the Universe." The Hydrogen line was discovered by E.M. Purcell (Nobel Laureate for the discovery of NMR) and H.I. Ewen (his thesis student at Harvard) in 1951, and in 1959 was the only radio spectral line known. Today we know close to 100 radio lines but the Hydrogen line continues to dominate our searches, possibly because of inertia but also because Hydrogen is the first and most abundant chemical element in the Universe.

The last sentence in the Cocconi and Morrison paper read: "The probability of success is difficult to estimate, but if we never search the chance of success is zero." This challenge was answered almost immediately by Frank Drake [2], the current Vice President of our Commission, who was then a young (29) radioastronomer with the U.S. National Radio Astronomy Observatory (NRAO) in Green Bank, West Virginia. In April 1960, in his celebrated Project Ozma, Frank Drake used the 85 ft radio telescope of NRAO to search for radio signals at the Hydrogen line from the two nearby Sun-like stars, Epsilon Eridani (10.7 l.y.) and Tau Ceti (11.9 l.y.).

Figure 1 is from our recent IAU Symposium which brought together several of the protagonists mentioned above. Shown together from the Symposium Banquet are from left to right: Edward M. Purcell, Philip Morrison, Carl Sagan, Michael D. Papagiannis and Frank Drake.

II. THE NEW IAU COMMISSION

The acceptance of the idea that the search for extraterrestrial life is an important scientific task, and the realization that, though this is an interdisciplinary subject, the main thrust in the actual searches must come from the astronomers, led finally to the establishment of IAU Commission 51 - Search for Extraterrestrial Life in the 1982 General Assembly at Patras, Greece. Michael D. Papagiannis of Boston University was elected the first President, and Nikolai S. Kardashev of the Institute for Space Research of the USSR Academy of Sciences and Frank D. Drake of Cornell University were elected Vice Presidents.

The first exposure of the IAU to this new subject was made at the 1979 General Assembly in Montreal when M.D. Papagiannis organized a one-day Session on "Strategies for the Search for Life in the Universe". This was a very successful meeting, and when in an open evening Session Drake and Papagiannis reported on this Session to the general membership of the IAU, the large Auditorium of the University of Montreal was packed to standing room only, with more than 1.000 astronomers present from all around the world. The Proceedings of this meeting were published by D. Reidel Publ. Co. (Papagiannis [3]). The success of this event, and the interest demonstrated by many IAU members, led to the establishment of IAU Commission 51 in 1982, which makes it the youngest of all IAU Commissions.

The new Commission grew rapidly and in only two years its membership reached 250, 210 of which are astronomers members of the IAU and about 40 are IAU Consultants, i.e., distinguished scientists from related fields with common interests with IAU Commission 51. We established also a Commission Newsletter called BIOASTRONOMY NEWS, and we set out to organize our first IAU Symposium, which was held in Boston, USA June 18-21, 1984, i.e. in less than two years from the establishment of the new Commission.

III. IAU SYMPOSIUM 112

IAU Symposium 112 - The Search for Extraterrestrial Life: Recent Developments was hosted by Boston University, the home institution of the President of Commission 51, who will also be the Editor of the Proceedings of the Symposium (Papagiannis [4]). It was financially and morally supported by the IAU, NASA and Boston University, to all three of which we express our deepest appreciation. It was cosponosored by four other IAU Commissions: 15. Physical Study of Comets, Minor Planets, and Meteorites; 16. Physical Study of Planets; 24. Photographic Astrometry; 40. Radio Astronomy. It was also cosponsored by four other major international organizations with which IAU Commission 51 has established close working relationships. These four International Organizations were:
COSPAR (ISCU Committee on Space Research)
IAF/IAA (International Astronautical Federation/Intern.Acad. of Astronautics)
ISSOL (International Society for the Study of the Origin of Life)

IUBS (International Union of Biological Sciences)

The Scientific Organizing Committee, which included representatives of the above four organizations plus NASA, consisted of the following: M.D. Papagiannis (Chairman), J. Billingham (NASA), D. DeVincenzi (COSPAR), F.D. Drake, J. Jugaku, N.S. Kardashev, G. Marx, R. Pesek (IAF/IAA), C. Ponnamperuma (ISSOL), C. Sagan, O. Solbrig (IUBS), and V.S. Troitsky.

The Local Organizing Committee, on the other hand, was made up of faculty members from Harvard University, MIT, and Boston University. It consisted of the following: Philip Morrison (MIT) and Edward M. Purcell (HU) Co-Chairmen, Tom Bania (BU), Stephen Jay Gould (HU), Paul Horowitz (HU), Edward Lilley (HU), Michael D. Papagiannis (BU), and David Staelin (MIT). The Symposium was attended by about 150 participants from 18 different countries. Actually the Soviet participants, though they had preregistered, in the last minute they were not able to attend. They have sent in, however, their papers for the volume of the Proceedings, which will be published by D. Reidel Publ. Co. in their IAU Symposia Series, and will include more than 65 contributions. The scientific program of the Symposium consisted of seven half-day Sessions, the titles and Chairmen of which were:

I. THE SEARCH FOR OTHER PLANETARY SYSTEMS
Co-chairmen: Carl Sagan and Jun Jugaku

II. PLANETARY, INTERPLANETARY AND INTERSTELLAR ORGANIC MATTER
Co-Chairmen: William M. Irvine and Donald L. DeVincenzi

III. UNIVERSAL ASPECTS OF BIOLOGICAL EVOLUTION
Co-Chairmen: Lynn Margulis and John Billingham

IV. RADIO SEARCHES - RECENT OBSERVATIONS
Co-Chairmen: Edward M. Purcell and Edward Lilley

V. TECHNOLOGICAL PROGRESS IN RADIO SEARCHES
Co-Chairmen: Frank D. Drake and George Marx

VI. THE FERMI PARADOX AND ALTERNATIVE SEARCH STRATEGIES
Chairman: Philip Morrison

VII. SUMMARY AND CONCLUSIONS
Chairman: Harlan J. Smith

In addition there were several special events during the Symposium including: a welcoming reception at the Castle of Boston University hosted by Dr. G. Bannister the Dean of Arts and Sciences at Boston University, the Symposium Banquet in the new Hall of Flags of Boston University, with Carl Sagan as the Banquet Speaker, a special event at Boston's Museum of Science in which we honored Philip Morrison, and a visit by many of the participants to the Harvard-Smithsonian Oak-Ridge Observatory, about 40 miles (65 km) from Boston, where Prof. Paul Horowitz of Harvard is using its 84 ft radiotelescope as a SETI dedicated facility running a continuous, 24 hours per day search called "Project Sentinel".

The event at the Museum of Science was organized to honor Prof. Morrison for the 25th anniversary since the publication of the historic Cocconi and Morrison paper in NATURE. After a nice reception, Philip Morrison gave a thought provoking talk in which he reminisced about the difficulties int the beginning, expressed gratification on how far we have advanced in these 25 years, and shared with us some thoughts and ideas about the future. After his talk, IAU Commission 51 honored him with a commemorative plaque on which it was inscribed:

ON THE 25TH ANNIVERSARY OF THE HISTORIC PAPER
BY G. COCCONI AND P. MORRISON
"SEARCHING FOR INTERSTELLAR COMMUNICATIONS"
IAU COMMISSION 51 - SEARCH FOR EXTRATERRESTRIAL LIFE
AWARDS THIS PLAQUE TO PROFESSOR PHILIP MORRISON
IN RECOGNITION
OF INSPIRING LEADERSHIP AND OUTSTANDING CONTRIBUTIONS
IAU SYMPOSIUM 112 - BOSTON USA - JUNE 1984

Figure 2 shows Philip Morrison, who is holding the plaque in his hands, with the President, Michael D. Papagiannis, and the Vice President, Frank D. Drake, of IAU Commission 51, immediately after the presentation.

IV. HIGHLIGHTS OF RECENT DEVELOPMENTS

It is generally agreed that planets represent the cosmic wombs where life may start and develop, sheltered in the hospitable, rich in nutrients, protective environments of certain planets. It is also agreed that carbon is the most suitable chemical element to produce the complex molecular structures required by life, and water is the best agent to act as the medium for the chemistry of life. Though one might be accused of chauvinism for extrapolating into the whole Universe our own experience, and though several alternatives have been proposed in both the scientific and in the the science fiction literature, the fact remains that on many objective criteria a carbon based life on a planet with liquid water seems to be by far the most likely combination to occur in the Universe. As a result our search for extrasolar planetary systems and our work on the formation of organic compounds not only on Earth but also in other parts of our Solar System as well as in the interstellar space, are critical steps in our efforts to understand how common life might be.

In recent years there has been a lot of excitement in the search for other planetary systems. The Infra-Red Astronomy Satellite (IRAS), which was launched on January 25, 1983, was able to detect particle envelopes around a considerable number of young stars, possibly indicating planetary formation from preplanetary disks. The first such star found by the IRAS (Aumann, et al [5]) was Vega (alpha Lyrae, 25 l.y.), which was soon followed by Fomalhaut, epsilon Eridani and beta Pictoris. For this last one, Bradford Smith and Richard Tenile were able to obtain some high resolution pictures using a coronogrpah and a CCD camera with the 100-inch telescope of the Las Companas Observatory in Chile. When these pictures were computer enhanced they showed that the particle envelope around beta Pictoris was actually a tilted preplanetary disk. Preplanetary disks have also been detected using speckle interferometry and image processing techniques around several T-Tauri stars, such as HL Tauri (Beckwith, et al [6]). A disk was also discovered around the infrared source IRS 5 L1551 with the 45m Nobeyama radiotelescope of Japan. Another important discovery was made by Donald McCarthy, Frank Low and Ronald Probst, who using infrared speckle interferometry were able to observe a massive object called a brown dwarf, 20-50 times the mass of Jupiter, around the star Van Biersbroeck 8.

Meanwhile the Allegheny Observatory of the University of Pittsburgh is completing the installation of a new objective lens and several other improvements to their astrometric telescope. Combined with their multichannel Astrometric Photometer (MAP) it is expected to reach an accuracy of 0.002 arcsec and start in 1985 a systematic search for Jupiter-like planets in 100-200 stars in the neighborhood of our Sun. Several other groups are perfecting instruments and techniques to detect planets from the motions they induce to their central star. Promising is also the possible use of the Space Telescope in the search for extrasolar planets. The Faint Object camera of the ST, which will have a coronographic finger of 0.8 arcsec and an apodizing mask, might be used for the direct detection of planets. The fine Guidance System and the Wide Field/Planetary Camera might be used for astrometric planetary detection studies. The main problem with the Space Telescope is likely to be the lack of adequate time, because the requests already exceed the available time by a factor of 15. Planetary searches, however, are finally entering a new era and this field is likely to become one of the most exciting astronomical topics in the coming 10-20 years.

The study of organic matter in our Solar System and in the interplanetary space is beginning to confirm that the formation of complex organic molecules of importance to life is a rather common phenomenon in the Universe. R.D. Brown and his colleagues at Monach University in Australia have developed computer programs that can predict quite accurately the abundances of different organic compounds in interstellar space. They have also searched for glycine (Brown, et al [7]), the simplest amino acid, but without success. They feel, however, that they might have a better chance looking for glycinonitrile ($H_2N.CH_2.CN$). The study of interstellar grains and the accretion of complex organic compounds is actively pursued by J. May Greenberg [8] at the University of Leiden in Holland. When dark molecular clouds condense to form solar systems it seems that a significant fraction of these organic compounds can survive the process becoming incorporated in comets and in carbonaceous chondrite asteroids. C. Ponnamperuma [9] and his colleagues at the University of Maryland were able to detect the presence of all five Nitrogen bases, the key building blocks of DNA and RNA, in the Merchison meteorite, which is a carbonaceous chondrite meteorite that fell in Victoria, Australia on September 28, 1969.

Titan, the giant moon of Saturn, seems to be a very interesting example for the synthesis of complex organic compounds in our Solar System (Sagan and Thompson, [10]). It has a thick nitrogen atmosphere with a considerable admixture of methane, and probably oceans of liquid hydrocarbons on its surface. Irradiation of a simulated Titanian atmosphere produces complex organic solids, called tholins, which make the atmosphere of Titan opaque and give it its orange-reddish color. Tholins, which laboratory simulations show that they contain many of the organic building blocks central to life on Earth, precipitate and probably form thick deposits in the bottom of the hydrocarbon oceans of Titan. It is possible that Titan is preserving in deep freeze the products of the chemical evolution that led to the origin of life on Earth, which some day we might be able to study by landing on Titan.

We also learn a lot about the origin and evolution of life from our own planet. We now have clear evidence from sites in Western Australia and South Africa (Knoll and Barghoorn, [11]) that life was present on our planet 3.5 billion years ago and probably started a few hundred million years earlier. Since it took several hundred million years for our planet to undergo chemical differentiation, for its surface to cool off, and for the oceans to form, it seems almost surprising that life appeared so soon after that. As Stephen Jay Gould said at the Symposium, "we find life in the first rocks capable of holding it". Extremely interesting is also the work on molecular replication of Leslie Orgel of the Salk Institute. He is actually focussing on template-directed, non-enzymatic synthesis of oligonucleotides that are complementary to the template. He has succeeded, e.g., in using CCGCC to direct the synthesis of GGCGG with a 20% yield.

We still have a long way to go before we have adequate statistics on extrasolar planets, especially on small, solid planets like the Earth which are much harder to detect, and thus to be able to say how common planets with liquid water might be. The indications, however, are that life will rapidly originate on such planets because the processes of chemical evolution leading to prebiotic organic compounds such as sugars, aminoacids, etc., are of universal applicability. We also do not know yet what fraction of planats where life might originate are able to maintain this hospitable environment for billions of years (Papagiannis [12]) so that life may be able to follow its painfully slow process of biological evolution to an advanced civilization. Run-away glaciations or run-away green house effects might terminate prematurely the presence of life in a considerable number of such planets. If conditions hold, however, biological evolution is likely to produce forms that fit all available niches and then forms that are not restricted to any particular niche.

It is interesting that the biological evolution might be assisted by major astronomical catastrophies, such as major impacts by comets or asteroids (Alvarez, et al [13]), which produce the extinction of a large number of species thus making room for the appearance of new ones. In the case of the Earth there is evidence of a series of such extinctions (Raup and Sepkoski [14]) with a periodicity of about 26 million years. The most impressive of them was the one that occurred about 65 million years ago at the Cretaceous-Tertiary border which wiped out the dinosaurs and opened the road for the expansion and rapid evolution of mammals. It has been proposed that such periodic extinctions might be produced by a small companion star to our Sun for which the name Nemesis, an ancient Greek goddess of divine vengence and destruction, has been proposed (Davis, Hut and Muller [15]).

V. TECHNOLOGICAL PROGRESS IN SETI WORK

Assuming that sites for the origin and evolution of life are relatively common in the Galaxy, it is reasonable to expect that there might be a significant number of advanced technological civilizations interested to communicate with other fellow stellar civilizations. The number N of such advanced civilizations in our Galaxy has been the subject of extensive debates (see e.g. articles in [3]) and estimates vary from N =1, i.e. only ourselves, to many billions if the colonization of the Galaxy has taken place. The initial estimates, which were obtained with the so called Drake Equation, were of intermediate values placing N in the range of 10^5-10^6, i.e., approximately one civilization per million stars. On this basis many searches have been undertaken, primarily in the radio domain, though there have also been some searches in the optical and in the infrared. In the radio domain most searches have been at the 21 cm line of Hydrogen, though several other "magic frequencies" (radio spectral lines or their arithmatic combinations) have also been used. According to Jill Tarter [16], who maintains a log-book of all the SETI efforts around the world, from the first radio search by Frank Drake in 1960 to 1984, there have been 46 different searches, half of them in the last six years. The number, the sophistication, and the international participation in this effort have been increasing steadily. Radio searches have now been undertaken in the USA, USSR, Australia, Canada, France, Germany, and Holland, with Japan getting ready to join the group with its new Nobeyama millimeter radio telescope. It is estimated that more than 100,000 hours have been spent in these 46 search projects.

Radio searches today can be divided into three general categories: Directed, Shared and Dedicated. Directed searches usually employ one of the larger radio telescopes of the world for a specific SETI project of relatively short duration and most often are carried out in one or more of the "magic frequencies". Shared searches involve either the re-analysis of old data for SETI signals using data tapes that had been obtained for other purposes, or the parasitic on line analysis for SETI signals of the data obtained while the radio telescope is conducting an unrelated astronomical program. In this category belongs the SERENDIP II system developed by S. Bowyer and D. Werthimer of U.C.- Berkeley (Werthimer, et al [17]) which consists of an automated device that performs a real time search for narrow-band SETI signals, and can be attached in a parasitic mode to any radio telescope. Finally in the Dedicated category belong searches which are carried out by observatories fully committed to SETI work. These facilities are usually automated and are thus able to collect data continuously and on a 24 hour basis. There are two such professional observatories now in operation. The oldest is the Ohio State University Radio Observatory which has been a SETI dedicated facility since 1973 and is now under the direction of Robert Dixon [18]. The other is the Harvard-Smithsonian Oak Ridge Radio Observatory, near Boston, where Project Sentinel of Paul Horowitz [19] has been in operation since March 1983. In the last two years amateur groups have also built amateur-SETI dedicated facilities, two in the United States and one in Canada. Let us hope that they will be as successful with SETI as they have been in their searches for comets.

Important progress is also being made in the technological and planning aspects of SETI. In the hardware area two different 8 million channel spectrum analyzers are now being built. The one being constructed at Harvard by Paul Horowitz and his colleagues [20], for their Project Sentinel at the Oak Ridge Radio Observatory, is an 8.4 million channel ultra-narrowband spectrum analyzer. It will cover a range of 420 kHz with a resolution of 0.05 kz per channel, and will be using a swept receiver to cancel the effects of the Earth's rotation. It will extend the present frequency coverage by a factor of 200, thus alleviating the present strict requirements that the transmitting civilizations corrected their signals. for all possible Doppler shifts. It is expected to become operational within 1985.

The other megachannel spectrum analyzer is being built at Stanford for NASA by Allen Peterson, Kok Chen and Ivan Linscott [21]. It will have 8.25 million channels and a spectral bandwidth of about 8 MHz. The output bandwidths will be 1 Hz, 32 Hz, 1024 Hz and 74 KHz, and the output will consist of either complex samples or of power (sqaure-law-detected) samples. It will also have an accumulater to integrate the power of the output bands for periods up to 1000 seconds. A prototype unit with 74,000 channels has now been completed and will be undergoing tests in 1985. The entire system is expected to be completed by the end of this decade. This system will improve by a factor of 100 what has been available for SETI up to now, and will also be of significant value to radioastronomical work in general.

There are also active research and development programs for SETI related software. Signal recognition algorithms, to be used with the megachannel spectrum analyzer discussed above, are now being developed at the NASA-Ames Research Center by Berny Oliver, Kent Cullers, John Wolfe, and others. To detect an intelligent signal in the data coming out from the MCSA, the data must be searched by different algorithms particularly sensitive to signals concentrated in frequency and/or time, and also algorithms capable of detecting frequency drifts due to Doppler effects. Due to the extremely large quantities of data that will be pouring out from the MCSA, it is important to do most of the analysis in real time. The high sensitivity signal recognition techniques now under development at NASA-Ames (Cullers [22]) are also likely to find applications in many other areas, including some that are remote to Astronomy such as brain research.

NASA now has a very active SETI group concentrated mostly at its Ames Research Center, Moffett Field, California. SETI work started at NASA-Ames in 1970 and has continued ever since with lecture series, work shops, project studies such as Project Cyclops (Oliver and Billingham [23]), workshops, meetings of advisory groups, publications of volumes and reports, as well as experimental SETI work through observational programs and equipment development programs. John Billingham, who was the head of this group, and his colleagues deserve a lot of credit for keeping the flame for SETI burning on practically a shoestring during the difficult years of the 1970's and the early 1980's when the US Congress was adamantly opposed to any funding for SETI.

They did persevere, however, and finally won, because Congress at last withdrew in 1982 its opposition to SETI and starting with the fiscal year 1983 NASA is spending approximately 1.5-2.0 million dollars per year for SETI research and development. This supports the work on the MCSA at Stanford, an equally fine SETI group at the Jet Propulsion Laboratory in Pasadena, California, internal work at Ames, and finally NASA has started funding a few external projects on SETI, one of which was our IAU Symposium 112. SETI at NASA-Ames, which is still under the Extraterrestrial Research Division headed by John Billingham, has now a new Director, Dr. Bernard Oliver [24], a highly respected pioneer in this field with great technological and administrative experience (a past Vice President for Research of the Hewlett-Packard Co). In summary NASA-Ames has unquestionably become the world's center for SETI research.

Besides the development of hardware and software, NASA through its Ames and JPL groups has developed a comprehensive search program which will go into full gear with the completion of the MCSA now being built for NASA and Stanford. The search for intelligent radio signals is a multidimensional problem which is affected by several parameters (sky coverage, frequency coverage, sensitivity, etc.). To optimize the chances of success, NASA has developed a two component program. In the first one, called the Targeted Search, the emphasis is on sensitivity, while in the second called the All-Sky survey, the emphasis is on a much broader sky coverage at a much lower sensitivity. In both cases the idea of "magic frequencies" is given up in favor of a more comprehensive frequency coverage in the 1-10 GHz range. This, of course, will become possible thanks to the MCSA now under developments.

The Targeted Search (Seeger and Wolfe [25]) will examine about 800 targets (the 773 F, G, and K main sequence stars contained in the RGO catalogue, some additional stars with peculiar spectra, and possibly some galaxies), using radiotelescopes all around the world since these targets are scattered over the whole sky. The frequency range to be covered will be approximately from 1.2 GHz to 2.0 GHz, which is centered around the so-called "water hole" between the hydrogen (H)-line at 1.4 GHz and the hydroxyl (OH)-lines around 1.65 GHz. This region, on objective criteria, seems to be the most attractive for interstellar communications and might also have some additional significance since OH and H form water which probably is a basic ingredient for life throughout the Universe. This frequency range will be analyzed with a resolution of 1 Hz and an observing period of about 1,000 seconds per target. With only one MCSA with a bandwidth of 8 MHz, this project will take 5-10 years. The time, of course, can be shortened substantially if more than one MCSA were available, which seems quite possible because duplicating a successful instrument is much faster and much cheaper than developing the first one.

The All-Sky Survey (Klein and Gulkis [26]), will cover the entire sky but with a much lower sensitivity. It will also have a broader frequency coverage, ranging from about 1 GHz to approximately 10 GHz. The frequency resolution will probably be 32 Hz. This survey will probably use primarily smaller telescopes, which have broader beams. This speeds up the survey, at the expense of course of sensitivity. It is estimated that the All-Sky Survey will also take approximately 5 years. In both cases of course the biggest problem will be availability of telescope time and availability of additional units of the MCSA. An estimate of what time fractions of existing radiotelescopes might be available, including stations of the Deep Space Network in the United States, Australia and possibly Spain, indicates that with the availability of 2-3 MCSA's the entire project could be completed in less than 10 years and possibly in only 5 years.

VI. DISCUSSION ON STRATEGIES

It is fair to say that during the late 1970's there was a sort of a revolt within the SETI community, sparked primarily by a paper by Michael Hart [27]. It was based on the realization by a significant number of scientists that interstellar voyages at $V = 0.01\text{-}0.05c$, with large, self-sustained space colonies, will probably be within the capabilities of advanced technological civilizations. As a result a colonization wave could sweep through the entire galaxy in only about 10 million years, which is a very short period compared to the billions of years that our Galaxy has been capable of harboring advanced civilizations. It was also difficult to see how the complete colonization of the Galaxy, including our own Solar System, could have been avoided, if indeed our Galaxy had harbored close to one billion civilizations during the past 5 billion years, each with an average life-time of approximately one million years, as estimated by the proponents of the Drake equation.

The galactic colonization concept, however, leads to the question: "But then, where are they?", which is often called "the Fermi Paradox" because supposedly this question was first posed by the great nuclear physicist Enrico Fermi. Some have extended this argument to conclude that since they don't seem to be here, the colonization of the Galaxy must have not taken place simply because there was no one there to initiate it. Hence we must be alone in the entire Galaxy. Others argued that the galactic colonization has either not occurred because it is too expensive to be of any use [28], or that it has not yet been completed because it is a very slow process [29]. Others, on the other hand, have argued that extraterrestrials might have colonized our own Solar System, living probably in far away space colonies, but for certain reasons [30],[31] they have chosen not to make their presence known.

The important outcome from our IAU Symposium 112 was the realization and the acceptance that none of us can be certain of how civilizations far more advanced than ours are likely to act and to behave. Hence, though these debates are interesting and in many ways valuable, we must not allow them to slow down the momentum we have gained with so much effort. Instead we must forge ahead having one or two key search projects, such as the one developed by NASA, as the cornerstones of our search effort, but encouraging also the expression of alternative points of view through parallel experimental searches. This multiple-path approach or mixed strategy, as people called it at the Symposium, is a very healthy development for SETI which can only get stronger with a broader participation and with the exploration of several different paths toward the common goal.

VII. PLANS FOR THE FUTURE OF COMMISSION 51

During our symposium, IAU Commission 51 formulated several plans for the future [32], which were communicated to the Executive Committee of the IAU with a very favorable overall response. They include the following:

1. The next IAU Symposium of Commission 51, at the invitation of the Hungarian Academy of Sciences, will be held in Hungary in June 1987.
2. The next President of Commission 51 for the period 1985-88 will be Frank D. Drake.
3. The next Vice President of Commission 51 will be George Marx of Hungary.
4. The new 10 member Organizing Committee of Commission 51 will consist of the following: R.D. Brown (AUSTRALIA), P. Connes (FRANCE), G.D. Gatewood (USA), J. Jugaku (JAPAN), P. Feldman (CANADA), J. Mayo Greenberg (HOLLAND), N.S. Kardashev (USSR), P. Morrison (USA), M.D. Papagiannis (USA), and V.S. Troitsky (USSR).
5. Eight prominent scientists from other fields who are now serving as Consultants to our Commission will be elected full members of the IAU. They are: John Billingham, Donald DeVincenzi, Paul Horowitz, Anthony R. Martin, Bernard M. Oliver, R. Pesek, Cyril Ponnamperuma, and Edward M. Purcell.
6 Incorporate in the title of our Commission the name BIOASTRONOMY, which we feel describes in a concise and scientific manner the goals of our Commission, namely the astronomical search for life (bios) in the Universe.

VIII. CONCLUSIONS

IAU Commission 51, the youngest of IAU commission, which was established only in 1982, has made great strides forward in these three years and has acted as a catalyst and as a coordinator for the more cohesive expression of the many exciting new developments that are happening in this new field. The volume of the Proceedings [4] of our recent IAU symposium 112 will be a compendium and an excellent reference source for all these new developments. We look forward into the future of our new IAU Commission and of our young discipline with enthusiasm and great anticipation.

REFERENCES

1. Cocconi, G., and Morrison, P.: 1959, Nature, 184, p. 884.
2. Drake, F.D.: 1979, Cosmic Search 1, p.11.
3. Papagiannis, M.D.: 1980, Strategies for the Search for Life in the Universe, D. Reidel Publ. Co., Dordrecht, Holland.
4. Papagiannis, M.D.: 1985, IAU Symposium 112 - The Search for Extraterrestrial Life - Recent Developments, D. Reidel Publ. Co., Dordrecht, Holland.
5. Aumann, H.H., et al.: 1984, Ap.J. (Letters) 278, p. L23.
6. Beckwith, S., et al.: 1984, Ap.J. 278, p. 215.
7. Brown, R.D., et al.: 1979, MNRAS 186, p.5.
8. Greenberg, J.M.: 1984, Scientific American 250, No. 6, p. 124.
9. Ponnamperuma, C.: 1983, Cosmochemistry and the Origin of Life, D. Reidel Publ. Co., Dordrecht, Holland.
10. Sagan, C. and Thompson, W.R.: 1984, Icarus 59, p. 133.

11. Knoll, A.H. and Barghoorn, E.S.: 1977, Science 198, p. 396.
12. Papagiannis, M.D.: 1981, Origin of Life, ed. by Y. Wolman, D. Reidel Publ. Co., Dordrecht, Holland.
13. Alvarez, L.W., et al : 1980, Science 208, p. 1095.
14. Raup, D. and Sepkoski, J.J.: 1984, Proc. Nat. Acad. Sci. 81, p. 801.
15. Davis, M., Hut, P. and Muller, R.A.: 1984, Nature 308, p. 715.
16. Tarter, J.: 1985, IAU Symposium 112, ed. by M.D. Papagiannis, D. Reidel Publ. Co., Dordrecht, Holland.
17. Werthimer, D., Tarter, J., and Bowyer, S.: IAU Symposium 112, ed. by M.D. Papagiannis, D. Reidel Publ. Co., Dordrecht, Holland.
18. Dixon, R.S.: 1985, IAU Symposium 112, ed. by M.D. Papagiannis, D. Reidel Publ. Co., Dordrecht, Holland.
19. Horowitz, P. and Forster J.: 1985, IAU Symposium 112, ed. by M.D. Papagiannis, D. Reidel Publ. Co., Dordrecht, Holland.
20. Horowitz, P., Forster, J. and Linscott I.: 1985: IAU Symposium 112, ed. by M.D. Papagiannis, D. Reidel Publ. Co., Dordrecht, Holland.
21. Peterson, A.M., Chen, K.S. and Linscott I.R.: 1985, IAU Symposium 112, ed. by M.D. Papagiannis, D. Reidel Publ. Co., Dordrecht, Holland.
22. Cullers, K.D.: 1985, IAU Symposium 112, ed. by M.D. Papagiannis, D. Reidel Publ. Co., Dordrecht, Holland.
23. Oliver, B. and Billingham J.: 1973, Project Cyclops, NASA CR 11445.
24. Oliver, B.: 1985, IAU Symposium 112, ed. by M.D. Papagiannis, D. Reidel Publ. Co., Dordrecht, Holland.
25. Seeger, C.L. and Wolfe, J.H.: 1985, IAU Symposium 112, ed. by M.D. Papagiannis, D. Reidel Publ. Co., Dordrecht, Holland.
26. Klein, M.J. and Gulkes, S.: 1985, IAU Symposium 112, ed. by M.D. Papagiannis, D. Reidel Publ. Co., Dordrecht, Holland.
27. Hart, M.H.: 1975, Q.JL.R.Astr.Soc. 16, p. 128.
28. Drake, F.D. 1980, Strategies for the Search for Life in the Universe, ed. by M.D. Papagiannis, D. Reidel Publ. Co., Dordrecht, Holland, p. 29.
29. Newman, W.I. and Sagan, C.: 1981, Icarus 64, p. 293.
30. Ball, J.A.: 1973, Icarus 19, p. 347.
31 Papagiannis, M.D.: 1978, Q.Jl.R.Astr.Soc. 19, p. 277.
32. Papagiannis, M.D.: 1985, JBIS (in press).

FIGURE 1. A group of key IAU Symposium 112 participants. Left to right, Edward M. Purcell and Philip Morrison co-Chairmen L.O.C., Carl Sagan the Banquet Speaker, Michael D. Papagiannis President and Frank Drake Vice-President IAU Commission 51.

FIGURE 2. Philip Morrison holding the 25-th Anniversary Commemorative Plaque presented to him during the IAU Symposium 112. Standing by are the President Michael D. Papagiannis and the Vice-President Frank Drake of IAU Commission 51.

WORKING GROUP FOR PLANETARY SYSTEM NOMENCLATURE

(Committee of the Executive)

CHAIRMAN: H. Masursky
MEMBERS: K. Aksnes, G.E. Hunt, M. Ya. Marov, P.M. Millman, D. Morrison, T.C. Owen, V.V. Shevchenko, B.A. Smith, V.G. Tejfel
CONSULTANTS: J.M. Boyce, G.E. Burba, A.M. Komkov, J.D. Rosenthal

Since the last General Assembly in Patras, Greece, we have held three meetings of the Working Group. The 10th Meeting was held in Mzkheta, the ancient capital of Georgia, USSR, hosted by their Academy of Sciences on April 3-7, 1984. All members except one, who was represented by a member of his Task Group, were present at the very productive meeting.

We decided at that meeting (1) that subsequent meetings would be held even if only part of the membership could attend. Actions will be reviewed by correspondence: Decisions of a controversial nature will be delayed until those episodic meetings, with larger attendance, are held. The 11th Meeting was held at the COSPAR Meeting at Graz, Austria, on July 30, 1984 and the 12th was held at Kailua-Kona, Hawaii in connection with the Division of Planetary Science Meeeting of the American Astronomical Association on October 8, 1984. At the Graz meeting, the decision (2) was made to form a Task Group to name features on primitive or small bodies including asteroids and comets. We will propose the membership and Chairman of this Group at the IAU General Assembly Meeting at New Delhi, India, in November, 1985. We decided (3) to publish annually a computer listing of planetary nomenclature with limited distribution. This annual gazetteer of about 400 pages will appear as a softback U.S. Geological Survey open file report. An illustrated gazetteer will appear in 1986 as a widely distributed hardback national aeronautic and space administration special report with many maps and photo mosaics. Revised versions of the illustrated atlas and gazetteer will appear about every six years. Next year's gazetteers will contain revised coordinates of all named features. The coordinates published in the Transactions of the IAU customarily are taken from preliminary version maps that are semi-controlled. Subsequent maps include geodetic control nets and more sophisticated cartographic techniques, resulting in substantial changes in latitude and longitude locations of named features. These revisions will be listed in the annual printouts.

A large number of new features (4) on Mars, the Moon, Mercury and the satellites of the outer planets were approved and will be presented to the General Assembly in New Delhi.

The coordination in naming new satellites (Commission 20 and WGPSN) is working well: The orbit of additional satellites have been established (5) and will be presented at the General Assembly. Two new feature types were approved (6) for use on Venus: Corona, pl. Coronae, meaning multiple ringed features of low relief and Tessera, pl. Tesserae-intersection features that have a mosaic tile-like pattern. Also, two new feature types were approved for use on high resolution Mars maps: Labes, pl. Labes, meaning landslide and Fluctus, pl. Fluctus, meaning lava flow.

Additional ringlets and gaps in Saturn rings (7) will be named for scientists who contributed to ring studies. Name banks have been enlarged for all bodies including Uranus, new satellites of Uranus will be named for characters in

Shakespearean plays and Pope's Rape of the Lick as has been done in the past. Features on Ariel will be named for good spirits, fairies and nymphs from many cultures. Features on Umbriel will be named for bad spirits, gnomes and dwarfs. Features on Miranda will be named for heroines from world culture. Future meetings of the Working Group will be held in France in the summer of 1985 and in New Delhi.

The current modified list of members of the Working Group and Task Groups are as follows:

Working Group for the International Astronomical Union Planetary System Nomenclature:

President: H. Masursky
Members: K. Aksnes (Norway); G.E. Hunt (UK); M. Ya. Marov (USSR); P.M. Millman (Canada); D. Morrison (USA); T.C. Owen (USA); V.V. Shevchenko (USSR); B.A. Smith (USA); V.G. Tejfel (USSR)

Consultants: J.M. Boyce (USA); W.E. Brunk (USA); A.M. Komkor (USSR)

Task Group for Lunar Nomenclature:

V.V. Shevchenko (Chair) (USSR)
A. Dollfus (France)
F. El-Baz (USA)
H. Masursky (USA)
P.M. Millman (Canada)
S.K. Runcorn (UK)
E.A. Whitaker (USA)

Task Group for Mercury Nomenclature:

D. Morrison (Chair) (USA)
D.P. Campbell (USA)
M.E. Davies (USA)
A. Dollfus (France)
N.P. Erpylev (USSR)
J.E. Guest (UK)

Task Group for Venus Nomenclature:

M. Ya. Marov (Chair) (USSR)
A.T. Basilevsky (USSR)
D.B. Campbell (USA)
R.M. Goldstein (USA)
R.F. Jurgens (USA)
H. Masursky (USA)
G.H. Pettengill (USA)
Y.F. Tjuflin (Consultant) (USA)

Task Group for Mars Nomenclature:

B.A. Smith (Chair) (USA)
A. Dollfus (France)
M. Ya. Marov (USSR)
D. Ya. Martynov (USSR)
H. Masursky (USA)
S. Miyamoto (Japan)
C. Sagan (USA)

Task Group for Outer Solar System Nomenclature

T.C. Owen (Chair) (USA)
K. Aksnes (Norway)
R. Beebe (USA)
M.S. Bobrov (USSR)
A. Brahic (France)
M.E. Davies (USA)
N.P. Erpylev (USSR)
H. Masursky (USA)
B.A. Smith (USA)
V.J. Tejfel (USSR)
G. Bubba (Consultant) (USSR)